ESSENTIALS OF
Genetics

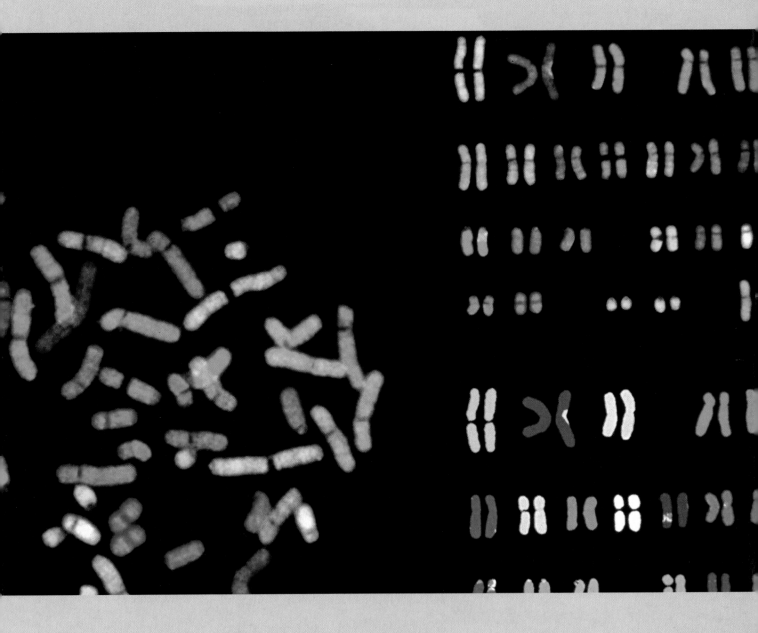

ESSENTIALS OF
Genetics

SIXTH EDITION

William S. Klug
The College of New Jersey

Michael R. Cummings
University of Illinois at Chicago

Charlotte A. Spencer
University of Alberta

with contributions by
Sarah M. Ward, Colorado State University

PEARSON

Prentice
Hall

Pearson Education International

Executive Editor: Gary Carlson
Manufacturing Manager: Alexis Heydt-Long
Manufacturing Buyer: Alan Fischer
Art Director: Maureen Eide
Creative Director: Juan Lopez
Director of Creative Service: Paul Belfanti
Senior Managing Editor, Art Production and Management: Patricia Burns
Manager, Production Technologies: Matthew Haas
Managing Editor, Art Management: Abigail Bass
Art Development Editor: Jay McElroy
Art Production Editor: Connie Long
Manager, Art Production: Sean Hogan
Assistant Manager, Art Production: Ronda Whitson
Illustrations: ESM Art Production, Lead Illustrators: Stacy Smith, Nathan Storck
Senior Media Editor: Patrick Shriner
Managing Editor, Media: Rich Barnes
Media Production Editor: John Cassar
Assistant Managing Editor, Science Supplements: Karen Bosch
Marketing Manager: Mandy Jellerichs
Project Manager: Crissy Dudonis
Cover and Interior Designer: Suzanne Behnke
Director, Image Reource Center: Melinda Reo
Manager, Rights and Permissions: Zina Arabia
Interior Image Specialist: Beth Boyd-Brenzel
Cover Image Specialist: Karen Sanatar
Cover Photos: NOBOU IIDA/Amana Images/Getty Images, S. Lowry/Univ Ulster/Stone/Getty Images,
    George Doyle/Stockbyte Platinum/Getty Images, MedioImages/Getty Images
Photo Research: Yvonne Gerin
Photo Editor: Debbie Hewitson
Production Services/Composition: Preparé Inc.

© 2007, 2005, 2002, 1999, 1996 by William S. Klug and Michael R. Cummings
Published by Pearson Education, Inc.
Pearson Prentice Hall
Pearson Education, Inc.
Upper Saddle River, NJ 07458

First edition © 1993 by Macmillan Publishing Company, a division of Macmillan, Inc.

Pearson Prentice Hall™ is a trademark of Pearson Education, Inc.

If you purchased this book within the United States or Canada you should be aware that it has been
wrongfully imported without the approval of the Publisher or the Author.

Printed in the United States of America

10 9 8 7 6 5 4 3 2 1

ISBN 0-13-241065-6

Pearson Education Ltd., *London*
Pearson Education Australia Pty., Limited, *Sydney*
Pearson Education Singapore, Pte. Ltd.
Pearson Education North Asia Ltd., *Hong Kong*
Pearson Education Canada, Ltd., *Toronto*
Pearson Educación de Mexico, S.A. de C.V.
Pearson Education—Japan, *Tokyo*
Pearson Education Malaysia, Pte. Ltd.
Pearson Education, Upper Saddle River, New Jersey

DEDICATION

In 1973, the pioneering evolutionary geneticist Theodosius Dobzhansky published an article, the title and thesis of which is now widely quoted: "Nothing in biology makes sense except in the light of evolution." In the middle of the first decade of the 21st century, as we continue to move further into the era of genomics, it is equally relevant to state that: "Our knowledge of biology will always remain incomplete until the genetic basis of all aspects of biological processes is made clear." We are rapidly moving closer to realizing this goal.

How lucky all of us who study genetics are as we become enlightened by the findings, past and present, that make up this discipline. It is one permeated with the sense of analytical thinking and discovery. Not a week passes without something of great interest and significance being reported in the scientific literature and by the media.

So, we dedicate this book to all those who have come to appreciate genetics as the core discipline in biology. And in particular, we single out the students just beginning their studies. We hope that this text provides valuable insights and inspiration as you expand your scientific horizons.

W. S. Klug

M. R. Cummings

C. A. Spencer

About the Authors

William S. Klug is currently a Professor of Biology at The College of New Jersey (formerly Trenton State College) in Ewing, New Jersey. He served as Chair of the Biology Department for 17 years, a position to which he was first elected in 1974. He received his B.A. degree in Biology from Wabash College in Crawfordsville, Indiana, and his Ph.D. from Northwestern University in Evanston, Illinois. Prior to coming to The College of New Jersey, he was on the faculty of Wabash College as an Assistant Professor, where he first taught genetics, as well as general biology and electron microscopy. His research interests have involved ultrastructural and molecular genetic studies of oogenesis in *Drosophila*. He has taught the genetics course as well as the senior capstone seminar course in human and molecular genetics to undergraduate biology majors for each of the last 37 years. He was the recent recipient of the first annual teaching award given at The College of New Jersey granted to the faculty member who most challenges students to achieve high standards. He also received the 2004 Outstanding Professor Award from Sigma Pi International, and in the same year, he was nominated as the Educator of the Year, an award given by the Research and Development Council of New Jersey.

Michael R. Cummings is currently Research Professor in the Department of Biological, Chemical, and Physical Sciences at Illinois Institute of Technology, Chicago, Illinois. For more than 25 years, he was a faculty member in the Department of Biological Sciences and in the Department of Molecular Genetics at the University of Illinois at Chicago. He has also served on the faculties of Northwestern University and Florida State University. He received his B.A. from St. Mary's College in Winona, Minnesota, and his M.S. and Ph.D. from Northwestern University in Evanston, Illinois. In addition to this text and its companion volumes, he has also written textbooks in human genetics and general biology for nonmajors. His research interests center on the molecular organization and physical mapping of the heterochromatic regions of human acrocentric chromosomes. At the undergraduate level, he teaches courses in Mendelian and molecular genetics, human genetics, and general biology, and has received numerous awards for teaching excellence given by university faculty, student organizations, and graduating seniors.

Charlotte A. Spencer is currently an Associate Professor in the Department of Oncology at the University of Alberta in Edmonton, Alberta, Canada. She has also served as a faculty member in the Department of Biochemistry at the University of Alberta. She received her B.Sc. in Microbiology from the University of British Columbia and her Ph.D. in Genetics from the University of Alberta, followed by postdoctoral training at the Fred Hutchinson Cancer Research Center in Seattle, Washington. Her research interests involve the regulation of RNA polymerase II transcription in cancer cells, cells infected with DNA viruses, and cells traversing the mitotic phase of the cell cycle. She has taught courses in Biochemistry, Genetics, Molecular Biology, and Oncology, at both undergraduate and graduate levels. She has contributed Genetics, Technology, and Society essays for several editions of *Concepts of Genetics* and *Essentials of Genetics*. In addition, she has written booklets in the Prentice Hall Exploring Biology series, which are aimed at the undergraduate nonmajor level.

Brief Contents

Contents

CHAPTER 8

Genetic Analysis and Mapping in Bacteria and Bacteriophages   164

GENETICS, TECHNOLOGY, AND SOCIETY

Bacterial Genes and Disease: From Gene Expression to Edible Vaccines   183

CHAPTER 9

DNA Structure and Analysis   188

GENETICS, TECHNOLOGY, AND SOCIETY

The Twists and Turns of the Helical Revolution   209

CHAPTER 10

DNA Replication and Synthesis   213

GENETICS, TECHNOLOGY, AND SOCIETY

Telomerase: The Key to Immortality?   231

CHAPTER 11

Chromosome Structure and DNA Sequence Organization   235

CHAPTER   12

The Genetic Code and Transcription   253

GENETICS, TECHNOLOGY, AND SOCIETY

Antisense Oligonucleotides: Attacking the Messenger   272

CHAPTER   13

Translation and Proteins   277

GENETICS, TECHNOLOGY, AND SOCIETY

Mad Cow Disease: The Prion Story   297

CHAPTER   14

Gene Mutation, DNA Repair, and Transposition   303

GENETICS, TECHNOLOGY, AND SOCIETY

In the Shadow of Chernobyl   324

C H A P T E R 1 9

Applications and Ethics of Genetic Engineering 425

C H A P T E R 2 0

Developmental Genetics 449

C H A P T E R 2 1

Quantitative Genetics 470

C H A P T E R 2 2

Population Genetics 489

Preface

Essentials of Genetics is written for courses requiring a text that is shorter and more basic than its more comprehensive companion, *Concepts of Genetics*. While coverage is thorough, current, and of high quality, *Essentials* is written to be more accessible to biology majors early in their undergraduate careers, as well as to students majoring in a number of other disciplines, including agriculture, chemistry, engineering, forestry, psychology, and wildlife management. Because the text is shorter than many other books, *Essentials of Genetics* is also more manageable in one-quarter and one-semester courses.

Goals

Although *Essentials of Genetics* is about 150 pages shorter than its companion volume, our goals during revision are the same for both books. Specifically, we seek to

- Emphasize concepts rather than excessive detail.

- Write clearly and directly to students in order to provide understandable explanations of complex analytical topics.

- Establish careful organization within and between chapters.

- Maintain constant emphasis on science as a way of illustrating how we know what we know.

- Propagate the rich history of genetics that so beautifully elucidates how information is acquired and extended within the discipline as it develops and grows.

- Emphasize problem solving, thereby guiding students to think analytically and to apply and extend their knowledge of genetics.

- Provide the most modern and up-to-date coverage of this exciting field.

- Create inviting, engaging, and pedagogically useful full-color figures enhanced by equally helpful photographs to support concept development.

- Provide outstanding Online Media Tutorials where students are guided in their understanding of important concepts by working through the best animations, tutorial exercises, and self-assessment tools available.

These goals serve as the cornerstones of *Essentials of Genetics*. This pedagogic foundation allows the book to accommodate courses with many different approaches and lecture formats. While the book presents a coherent table of contents that represents one approach to offering a course in genetics, chapters are nevertheless written to be independent of one another, allowing instructors to utilize them in various sequences. We believe that the varied approaches embodied in these goals together provide students with optimal support for their study of genetics.

Writing a textbook that achieves these goals and having the opportunity to continually improve on each new edition has been a labor of love for us. The creation of each of the six editions is a reflection of not only our passion for teaching genetics, but also the constructive feedback and encouragement provided by adopters, reviewers, and our students over the past decade and a half.

Features of the Sixth Edition

This edition has numerous features that continue to make *Essentials of Genetics* the most distinctive and pedagogically useful textbook available to students of genetics, including:

- **Now Solve This.** New to this edition is a feature entitled "Now Solve This." Interspersed throughout chapters, each entry functions to review immediately preceding concepts by focusing the student on solving a related problem at the end of the chapter. This not only serves as a review, but also reinforces the fact that genetics is about solving problems. Each entry includes a "Hint" that provides the student help in arriving at the solution, thus providing important pedagogic insight.

- **How Do We Know?** This feature is found at the end of the introduction in each chapter. Based on a modern approach to teaching biology referred to as "Science as a Way of Knowing," this section asks students to be aware of numerous important issues that have framed our thinking and to constantly ask themselves how we discovered this information. We believe that this approach, which has always been pursued in the text of each chapter but is now formalized, will enhance students' understanding of the topics covered in each chapter. The ideas behind this approach are explained in Chapter 1, and the execution begins in Chapter 2.

- **New Genetics, Technology, and Society Essays** As with every new edition, we have revised many of these essays and added new ones that reflect recent findings in genetics and their impact on society. Among the new essays are those that consider Tay-Sachs disease (Chapter 3), Fragile Chromosome Sites and Cancer (Chapter 6), Genetic Regulation and Human Disorders (Chapter 15), and Ownership of DNA (Chapter 18). These new additions supplement essays from past editions, including those that discuss stem cells, human sex selection, genetically modified foods, gene therapy, human cloning, mad cow disease, and endangered species, among many other topics.

- **Modernization of Topics** While we have updated each chapter to reflect the most current and significant findings in genetics, we have paid particular attention to the cutting-edge topics of genomics and proteomics (Chapter 18) as well as to the applications and ethics of genetic engineering

(Chapter 19). The impact of these topics on future genetic studies and on society will no doubt be enormous. In addition, we have further updated and extended our presentation of the rapidly advancing fields of gene regulation (Chapter 15) and cancer genetics (Chapter 16). An in-depth consideration of conservation genetics (Chapter 24) remains as another hallmark of our modern genetic coverage. This field, which attempts to assess and maintain genetic diversity in endangered species, remains at the forefront of genetic studies.

- **New Illustrations Throughout** The Sixth Edition features a whole new color palette as well as the redrafting of almost all figures to enhance their pedagogic value and artistic quality. Many figures feature "flow diagrams" that visually guide a student through experimental protocols and techniques.

- **New Photographs** An even greater number of photographs illustrate and enhance this edition.

- **Problem Solving** Students will find up to 30 Problems and Discussion Questions with varying degrees of difficulty at the end of the chapters. The most challenging problems are those that are based on research data derived from the literature of genetics. This is part of our continuing attempt to make the "connection" for students as to how the information constituting the discipline of genetics is acquired. As in past editions, brief solutions to even-numbered problems and questions appear in Appendix B, and more detailed solutions to all problems and questions appear in the *Student Handbook and Solutions Manual.*

- **Instructor and Student Media Address Real Needs** Support for lecture presentations and other teaching responsibilities has been increased, including electronic access to more text photos and tables and a greater variety of Power-Point offerings on the book's Instructor Resource Center on CD. Media found on the revamped Companion Website reflect the growing awareness that today's students must use their limited study time as wisely as possible.

Continued Emphasis on Concepts

As in its companion volume, *Essentials of Gentetics* continues to focus on the conceptual issues in genetics. Our experience with this book, reinforced by the many adopters with whom we have been in contact over the years, shows that students whose primary focus is on concepts more easily comprehend and take with them to succeeding courses the most important ideas in genetics as well as an analytic view of biological problem solving.

To aid students in identifying conceptual aspects of a major topic, each chapter begins with a new section called "Chapter Concepts," which outlines the most important ideas about to be presented. Within each chapter, as noted earlier, the "How Do We Know?" feature asks the student to connect concepts to experiments. Then, the "Now Solve This" feature asks students to link conceptual understanding in a more immediate

way to problem solving. In addition, each chapter ends with a "Chapter Summary" that enumerates the five to ten key points that have been covered. Collectively, these features help to ensure that students focus on and understand concepts of genetics as they confront the extensive vocabulary and the many important details of genetics. Carefully designed figures also support this approach throughout the book.

Problem Solving: Insights and Solutions

Genetics, more than any other discipline within biology, requires problem solving and analytical thinking. At the end of many chapters we include what has become an extremely popular and successful section called Insights and Solutions. In this section we stress:

Problem solving

Quantitative analysis

Analytical thinking

Experimental rationale

Problems or questions are posed, and detailed solutions or answers are provided. This feature primes students as they move on to the Problems and Discussion Questions that conclude each chapter.

The Genetics Web Invesigations section is available on the Companion Website. Each Web Investigation contains several Web-linked problems designed to enhance and extend the topics presented in the chapter. To complete these problems, students must actively participate in the exercises and virtual experiments. For reference, the estimated time required to solve the problem is noted at the beginning of the exercise.

Acknowledgments

All comprehensive texts are dependent on the valuable input provided by many colleagues. While we take full responsibility for any errors in this book, we gratefully acknowledge the help provided by those individuals who reviewed or otherwise contributed to the content and pedagogy of this and previous editions.

As these acknowledgments make clear, a text such as this is a collective enterprise. All of the above individuals deserve to share in any success this text enjoys. We want them to know that our gratitude is equaled only by the extreme dedication evident in their efforts. Many, many thanks to them all.

Contributors

We begin with special acknowledgments to those who have made direct contributions to this text. We particularly thank Sarah Ward at Colorado State University for creating Chapter 24 on Conservation Genetics, and also for providing specific input into the chapters involving quantitative and population genetics. In addition, Amanda Norvell revised several sections emphasizing eukaryotic molecular genetics, and Janet Morrison revised numerous aspects of our coverage of evolutionary genetics. Katherine Uyhazi wrote the essay on Tay-Sachs disease in humans (Chapter 3) and helped revise several other essays. Amanda and Janet are colleagues from The College of New Jersey, while Katherine, previously at The College of New Jersey, is now at Yale University. David Kass at Eastern Michigan University provided the original idea, drafted and helped construct the essay on Genetic Regulation found in Chapter 15. We also thank Mark Shotwell at Slippery Rock University for contributing several "Genetics, Technology, and Society" essays. As with previous editions, Elliott Goldstein from Arizona State University was always readily available to consult with us concerning the most modern findings in molecular genetics. We also express special thanks to Harry Nickla at Creighton University. In his role as author of the *Student Handbook* and the *Instructor's Manual*, he has written many new problems and authored the brief answers to Selected Answers that appear in Appendix B.

We are grateful to all of the above contributors not only for sharing their genetic expertise, but for their dedication to this project as well as the pleasant interactions they provided.

Proofreaders

Proofreading a manuscript of a 600-page text deserves more thanks than words can offer. Our utmost appreciation is extended to the three individuals who confronted this task with patience, diligence, and good humor:

Tamara Horton Mans, Princeton University

Arlene Larson, University of Colorado, Denver

Chaoyang Zeng, University of Wisconsin, Milwaukee

In addition, Betty Pessagno successfully confronted the task of copy-editing the entire manuscript. Her contribution to the overall quality of the text is very much appreciated.

Reviewers

Finally, we are pleased to have received input from many genetics colleagues during their reviews of both *Essentials of Genetics* and its companion volume *Concepts of Genetics*:

Robert W. Adkinson, Louisiana State University
Nancy Bachman, SUNY College at Oneonta
Hank W. Bass, Florida State University
Edward Berger, Dartmouth College
Andrew Bohonak, San Diego State University
Paul Bottino, University of Maryland
James Bricker, The College of New Jersey
Hugh Britten, University of South Dakota
Aaron Cassill, University of Texas, San Antonio
Joel Chandlee, University of Rhode Island
Michael John Christoffers, North Dakota State University
Mary Anne Clark, Texas Wesleyan University
Jimmy D. Clark, University of Kentucky
Gregory P. Copenhaver, University of North Carolina at Chapel Hill
Michael Crawford, University of Windsor
Jason Curtis, Purdue University North Central

Stephen J. D'Surney, University of Mississippi
Robert Duronio, University of North Carolina at Chapel Hill
Bert Ely, University of South Carolina
Scott Erdman, Syracuse University
Asim Esen, Virginia Tech
Nancy H. Ferguson, Clemson University
Linnea Fletcher, Austin Community College
Gail Fraizer, Kent State University
Kim Gaither, Oklahoma Christian University
David Galbreath, McMaster University
Alex Georgakilas, East Carolina University
Derek J. Girman, Sonoma State University
Elliott Goldstein, Arizona State University
Nels Granholm, South Dakota State University
Mark L. Hammond, Campbell University
Jon Herron, University of Washington
Karen Hicks, Kenyon College
Kent Holingser, University of Connecticut
David Hoppe, University of Minnesota, Morris
John A. Hunt, University of Hawaii, Honolulu
Cheryl Jorcyk, Boise State University
David Kass, Eastern Michigan University
Beth A. Krueger, Monroe Community College
M. A. Lachance, University of Western Ontario
Arlene Larson, University of Colorado at Denver
Hsiu-Ping Liu, Southwest Missouri State University
Paul F. Lurquin, Washington State University
Sally Mackenzie, University of Nebraska
Sandra Michael, SUNY Binghamton
Terry C. Matthews, Millikan University
Cynthia Moore, Washington University
Janet Morrison, The College of New Jersey
Michelle A. Murphy, University of Notre Dame
Marcia O'Connell, The College of New Jersey
Julie Olszewski, Shippensburg State University
Dan Panaccione, West Virginia University
Todd Rimkus, Marymount University
Laura Runyen-Janecky, University of Richmond
Margaret Saha, The College of William and Mary
Tom Savage, Oregon State University
Gerald Schlink, Missouri Southern State College
Randy Scholl, Ohio State University
Malcolm Schug, University of North Carolina, Greensboro
Ralph Seelke, University of Wisconsin, Superior
Gurel S. Sidhu, California State University
Theresa Spradling, University of Northern Iowa
Mark Sturtevant, Northern Arizona University
Christine Tachibana, University of Washington
Eileen Underwood, Bowling Green State University
Daniel Wang, University of Miami
Sarah Ward, Colorado State University
David Weaver, Auburn University
R. C. Woodruff, Bowling Green State University
Marie Wooten, Auburn University
Chaoyang Zeng, University of Wisconsin at Milwaukee

Media reviewers include:

Peggy Brickman, University of Georgia
Carol Chihara, University of San Francisco
Karen Hughes, University of Tennessee
Cheryl Ingram-Smith, Clemson University
David Kass, Eastern Michigan University
John Kemner, University of Washington
Arlene Larson, University of Colorado
John C. Osterman, University of Nebraska, Lincoln
Eric Stavney, DeVry University

Special thanks go to Mike Guidry of LightCone Interactive and Karen Hughes of the University of Tennessee for their original contributions to the media program.

Editorial and Production Input

At Prentice Hall, we express appreciation and high praise for the editorial guidance of Gary Carlson, whose ideas and efforts have helped to shape and refine the features of this and two previous editions of the text. He has worked tirelessly to provide us with reviews from leading specialists who are also dedicated teachers and to ensure that the pedagogy and design of the book are at the cutting edge of a rapidly changing discipline. We also appreciate the production efforts of those at Preparé Inc., whose quest for perfection is reflected throughout the text. In particular, Fran Daniele and Rosaria Cassinese provided an essential measure of sanity to the otherwise chaotic process of production. Without their work ethic and dedication, the text would never have come to fruition. Crissy Dudonis, biology project manager, has worked tirelessly to ensure that the book, supplements, and cutting-edge interactive media all work together seamlessly. Patrick Shriner managed the media to ensure a stronger emphasis on addressing the needs of today's instructors and students. Finally, the beauty and consistent presentation of the art work are the product of Artworks of York, PA. We particularly thank Patricia Burns and Jay McElroy for their efforts.

For the Student

Companion Website—www.prenhall.com/klug

Respect for the students' increasingly valuable study time is reflected in the features of the Companion Website, which have been designed to enable users of the Sixth Edition to focus on those chapter sections and topics where they need review or further explanation. The Online Study Guide provides students with a focused, section-by-section review of topic coverage that features concise summary points accompanied by key illustrations and probing review questions that offer hints and feedback. Interactive Chapter Quizzes give students a way to test their comprehension of chapter topics before exams. The grading function offers students precise feedback and text references that provide guidance for additional study.

The media's strict adherence to both the principles and specific lessons of the textbook means that students and instructors can be assured that study time isn't being squandered on media that confuses students and emphasizes extraneous topics.

Online Web Tutorials

The most sophisticated learning and tutorial package available for students of genetics, these online tutorials address the concepts that students find most difficult. Each of the tutorials is composed of animations and interactive exercises, along with a post-tutorial quiz of self-grading questions to reinforce the important concepts. Found on the student Companion Website, the tutorials offer timely and relevant support for students. Look for this icon in the margin of this page to identify media tutorials that support concepts presented in the textbook.

Student Handbook and Solutions Manual

Harry Nickla, Creighton University (0-13-224140-4)

This valuable handbook provides a detailed step-by-step solution or lengthy discussion for every problem in the text. The handbook also features additional study aids, including extra study problems, chapter outlines, vocabulary exercises, and an overview of how to study genetics.

For the Instructor

Instructor Resource Center on CD
(0-13-224139-0)

 The Instructor Resource Center on CD for the Sixth Edition offers adopters convenient access to a comprehensive and innovative set of lecture presentation and teaching tools. Developed to meet the needs of veteran and newer instructors alike, these resources include:

- The JPEG files of all text line drawings with labels enhanced for optimal projection results (as well as unlabeled versions)
- Pedagogically important textbook photos and all textbook tables as JPEG files
- The JPEG files of line drawings, photos, and tables pre-loaded into comprehensive PowerPoint® presentations for each chapter
- Over 100 Instructor Animations dynamically illustrating the most important concepts in the course pre-loaded into PowerPoint® presentations (PC and Macintosh versions)

Instructor's Resource Manual with Testbank
(0-13-224129-3)

This manual and testbank contains over 1000 questions and problems for use in preparing exams. The manual also provides optional course sequences, a guide to audiovisual supplements, and a section on searching the Web. The testbank portion of the manual is also available in electronic format.

TestGen EQ Computerized Testing Software
(0-13-224131-5)

In addition to the printed volume, the test questions are also available as part of the TestGen EQ Testing Software, a text-specific testing program that is networkable for administering tests. It also allows instructors to view and edit questions, export the questions as tests, and print them out in a variety of formats.

Transparencies
(0-13-224138-2)

The transparency package includes 200 figures from the text: 150 four-color transparencies from the text and 50 transparency masters. The font size of the labels has been increased and boldfaced for easy viewing from the back of the classroom.

Introduction to Genetics

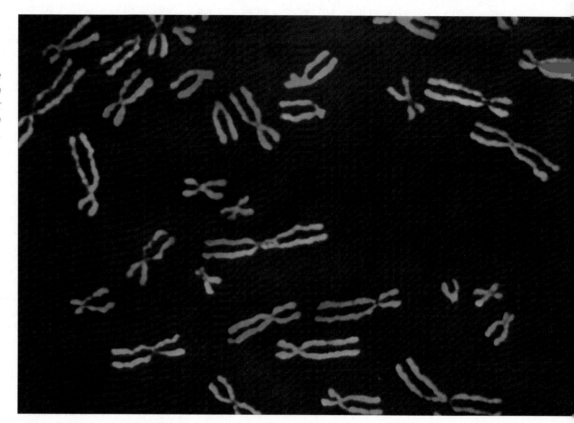

Human metaphase chromosomes, each composed of two sister chromatids joined at a common centromere.

CHAPTER CONCEPTS

- Using pea plants, Mendel discovered the fundamental principles of transmission genetics. Work by others showed that genes are on chromosomes and that mutant strains can be used to map genes on chromosomes.

- The discovery that DNA encodes genetic information, elucidation of the structure of DNA, and establishment of the mechanism of gene expression form the foundation of molecular genetics.

- The development of recombinant DNA technology revolutionized genetics and is the foundation for genome sequencing, including the Human Genome Project.

- Scientists use recombinant DNA technology to produce goods and services in a wide range of areas, including agriculture, medicine, and industry. The applications of genetic engineering has raised many legal and ethical issues involving the patenting of genetically modified organisms and relating to the use of gene therapy.

- Model organisms have been used in genetics since the early part of the twentieth century. The genetic knowledge gained from these organisms coupled with recombinant DNA technology and genomics makes these organisms useful as models to study human biology and diseases.

- Genetic technology is developing faster than policy, law, and convention in the use of this technology. Education and participation are key elements in the wise use of this technology.

From our perspective early in the twenty-first century, we can look back and ask when the interaction between genetic technology and society began to affect our lives. Did it begin when recombinant DNA technology was used to produce insulin, when the first food produced by genetic engineering reached the marketplace, or when gene therapy was first used to treat a genetic disorder? While each of these is an incremental step in using genetic knowledge in ways affecting society, we will focus on a case where the use of genetic technology directly affects the citizens of an entire country. This case captures the significant impact of genetics on society and provides a glimpse of future implications as more advances and applications are developed.

In December 1998, a controversy affecting the 270,000 residents of the island nation of Iceland was coming to a head. Following months of heated debate, the Icelandic Parliament passed a law granting deCODE, a biotechnology company with headquarters in Iceland, a license to create and operate a database called IHD (Icelandic Health Sector Database) containing detailed information drawn from medical records of all Iceland residents. The records in the database were to be encoded to ensure anonymity. The law also allows deCODE to cross-reference medical information from the IHD with a comprehensive genealogical database from the National Archives. In addition, deCODE can correlate information in these two databases with results of DNA profiles collected from Icelandic donors. The combination of medical history, genealogical, and genetic information is a powerful resource available exclusively to deCODE, which can market this information to researchers and companies for a period of 12 years.

This is not a scenario from a movie such as *Gattaca*, but a real example of the interaction between genetics and society as we begin a new century. The development and use of these databases in Iceland has generated similar projects in other countries as well. The largest is the "UK Biobank" effort launched in Great Britain in 2003. There, a huge database containing the genetic information of 500,000 Britons will be compiled from an initial group of 1.2 million residents. The database will be used to search for susceptibility genes that control complex traits. Similar projects to develop nationwide databases have been announced in Estonia, Latvia, Sweden, Singapore, and the Kingdom of Tonga, illustrating the global impact of genetic technology. In the United States, smaller-scale programs involving tens of thousands of individuals are underway at the Marshfield Clinic in Marshfield, Wisconsin; Northwestern University in Chicago, Illinois; and Howard University in Washington, D.C.

Why did deCODE select Iceland for such a project? For several reasons, the people of Iceland represent a unique case of genetic uniformity seldom seen or accessible to scientific investigation. This high degree of genetic relatedness derives from the founding of Iceland about 1000 years ago by a small population drawn mainly from Scandinavian and Celtic sources. Subsequent periodic population reductions by disease and natural disasters further reduced genetic diversity, and until the last few decades, there were few immigrants bringing new genes into the population. Thus, for geneticists trying to identify genes controlling complex disorders, the Icelandic population is a tremendous asset. Because the health-care system is state-

supported, medical records for all residents go back as far the early 1900s. Genealogical information is available in the National Archives and church records for almost every resident and for more than 500,000 of the estimated 750,000 individuals who have ever lived in Iceland. In spite of the associated controversies, the project already has a number of successes to its credit. Scientists at deCODE have identified genes associated with more than 25 of the most common diseases, including asthma, heart disease, stroke, and osteoporosis.

On the flip side of these successes are issues of privacy, consent, and commercialization—issues at the heart of many controversies arising from the applications of genetic technology. Scientists and nonscientists alike are debating the fate and control of genetic information and the role of law and society in decisions about how and when genetic technology is used. In less than a decade, technical advances are expected to reduce the cost of sequencing any person's genome to less than $1,000. Doctors may order sequencing of a patient's genome routinely, like asking for blood tests today. More than at any other time in the history of science, addressing the ethical questions surrounding an emerging technology is now as important as the information gained from that technology.

As you launch your study of genetics, remain sensitive to questions and issues like those just described. There has never been a more exciting time to be part of this science, but never has the need for caution and awareness of social issues been more apparent. Genetics has been misused before in the eugenics movement to justify social engineering through coercive and discriminative human rights violations. This text will enable you to achieve a thorough understanding of modern-day genetics and its underlying principles. With this understanding, you will be able to formulate your opinions based on facts rather than gut feelings and/or knee-jerk reactions. Along the way, enjoy your studies, but take your responsibilities as a novice geneticist very seriously.

1.1 From Mendel to DNA in Less Than a Century

Because genetic processes are unique to living beings and fundamental to the comprehension of life itself, genetics unifies biology and serves as its core. Thus, it is not surprising that genetics has a long, rich history. Our starting point for this history is a monastery garden in central Europe in the 1860s.

Mendel's Work on Transmission of Traits

In this garden (Figure 1–1) Gregor Mendel, an Augustinian monk, conducted a decade-long series of experiments using pea plants. In his work, Mendel showed that traits are passed from parents to offspring in quantitative and predictable ways. He concluded that traits in pea plants, such as height and flower color, are controlled by discrete units of inheritance we now call **genes**. He further concluded that genes controlling a trait exist in pairs and that members of a gene pair separate from each other during gamete formation. His work was published in 1866 but was largely unknown until it was partially duplicated and rediscovered by Carl Correns and others around 1900. Having been

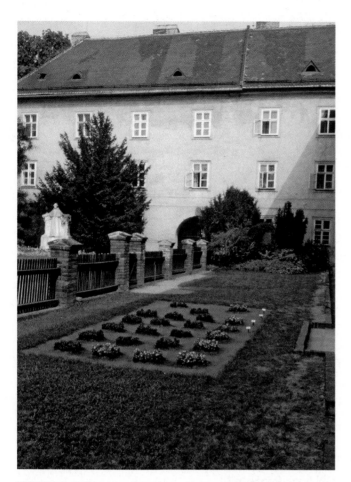

FIGURE 1–1 The monastery garden where Gregor Mendel conducted his experiments with garden peas. In 1866, Mendel put forward the major postulates of transmission genetics.

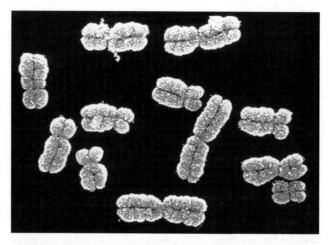

FIGURE 1–2 Colorized image of human mitotic chromosomes as visualized under the scanning electron microscope.

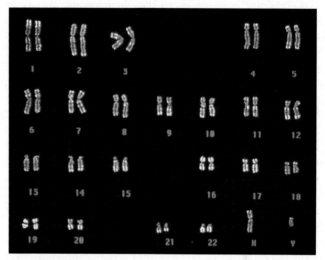

FIGURE 1–3 A colorized image of the human male chromosome set. Arranged in this way, the set is called a karyotype.

confirmed by others, Mendel's findings became recognized as the basis for the transmission of traits in pea plants and all other higher organisms including humans. His work forms the foundation for **genetics**, which is defined as the branch of biology concerned with the study of heredity and variation. The story of Gregor Mendel and the beginning of genetics is told in an engaging book, *The Monk in the Garden: The Lost and Found Genius of Gregor Mendel, the Father of Genetics*, by Robin M. Henig.

The Chromosome Theory of Inheritance: Uniting Mendel and Meiosis

Mendel did his work before the discovery of chromosomes. About 20 years after his work, advances in microscopy allowed researchers to identify chromosomes (Figure 1–2) and determine that each species has a characteristic number, called the **diploid number (2n)**. For example, humans have a diploid number of 46 (Figure 1–3). Chromosomes in diploid cells exist in pairs, called **homologous chromosomes**. Members of an autosomal pair are identical in size and location of the centromere, a structure to which spindle fibers attach during division.

Researchers in the last decades of the nineteenth century also described the behavior of chromosomes during two forms of cell division, **mitosis** and **meiosis**. In mitosis (Figure 1–4), chromosomes are copied and distributed so that each

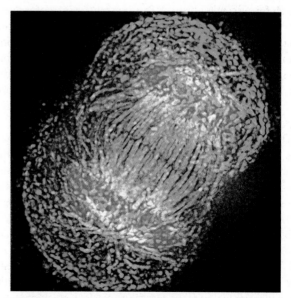

FIGURE 1–4 A stage in mitosis (anaphase) when the chromosomes (stained blue) move apart.

daughter cell receives a diploid set of chromosomes. Meiosis is associated with gamete formation in animals and spore formation in most plants. Cells produced by meiosis receive only one member of each homologous pair, constituting the **haploid (*n*) number** of chromosomes. This reduction in chromosome number is essential if the offspring arising from two gametes are to maintain a constant number of chromosomes characteristic of their parents and other members of their sexually reproducing species.

Early in the twentieth century, Walter Sutton and Theodore Boveri independently noted that genes and chromosomes have several properties in common and that the behavior of chromosomes during meiosis is identical to the behavior of genes during gamete formation. For example, genes and chromosomes exist in pairs, and members of a gene pair and members of a chromosome pair separate from each other during gamete formation. Based on these parallels, they each proposed that genes are carried on chromosomes (Figure 1–5). This proposal is the basis of the **chromosome theory of inheritance**, which states that inherited traits are controlled by genes residing on chromosomes faithfully transmitted through gametes, maintaining genetic continuity from generation to generation.

Genetic Variation

About the same time that the chromosome theory of inheritance was proposed, scientists began studying the inheritance of traits in the fruit fly, *Drosophila melanogaster*. A white-eyed fly (Figure 1–6) was discovered in a bottle containing normal (wild-type) red-eyed flies. This variation was produced by **mutation**, an inherited change in the gene controlling eye color. Mutations are defined as any heritable change and are the source of all genetic variation.

The variant eye color gene discovered in *Drosophila* represents an **allele** of a gene controlling eye color. Alleles are defined as alternative forms of a gene. Different alleles may produce differences in the observable features, or **phenotype** of an organism. The set of alleles carried by an organism is called the **genotype**. Using mutant genes as markers, geneticists were able to map the location of genes on chromosomes.

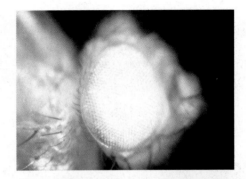

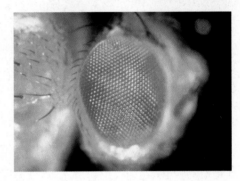

FIGURE 1–6 The normal red eye color in *Drosophila* (bottom) and the white-eyed mutant (top).

Chromosome I (the X chromosome):

- scute bristles, *sc*
- white eyes, *w*
- ruby eyes, *rb*
- crossveinless wings, *cv*
- singed bristles, *sn*
- lozenge eyes, *lz*
- vermilion eyes, *v*
- sable body, *s*
- scalloped wings, *sd*
- Bar eyes, *B*
- carnation eyes, *car*
- little fly, *lf*

FIGURE 1–5 Chromosome I (the X chromosome) of *D. melanogaster*, showing the location of many genes. Chromosomes can contain thousands of genes.

The Search for the Chemical Nature of Genes: DNA or Protein?

Work on white-eyed *Drosophila* showed that the mutant trait had a pattern of inheritance that could be traced to a single chromosome, confirming the idea that genes are carried on chromosomes. Once this was established, investigators turned their attention to identifying which chemical component of chromosomes carried genetic information. By the 1920s, it had been established that proteins and deoxyribonucleic acid (DNA) were the major chemical components of chromosomes. Proteins are the most abundant component in cells. There are a large number of different proteins, and because of their universal distribution in the nucleus and cytoplasm, many researchers thought proteins would be shown to be the carrier of genetic information.

In 1944, Avery, MacLeod, and McCarty, three researchers at the Rockefeller Institute in New York, showed that DNA was the carrier of genetic information in bacteria. This evidence, though clear-cut, failed to convince many influential scientists. Additional evidence for the role of DNA as a carrier of genetic information came from other researchers who worked with viruses called **bacteriophage** (or **phage**, for short), that infect and kill cells of the bacterium *Escherichia coli* (Figure 1–7). These viruses consist of a protein coat surrounding a DNA core. These experiments showed that the protein coat of the virus remains outside the cell, while the DNA enters the cell and directs the synthesis and assembly of more phage. This work was additional proof that DNA carries genetic information. Research over the next few years provided solid proof that DNA, not protein, is the genetic material, setting the stage for work to establish the structure of DNA.

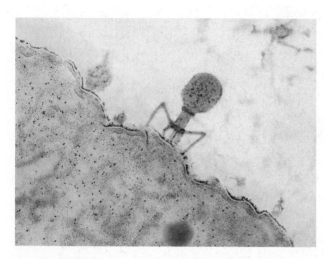

FIGURE 1–7 An electron micrograph showing a bacteriophage infecting a cell of the bacterium *E. coli.*

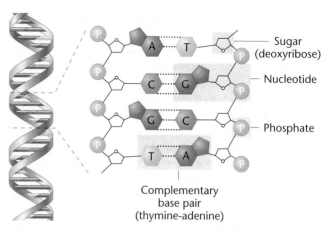

FIGURE 1–8 Summary of the structure of DNA, illustrating the nature of the double helix (on the left) and the chemical components making up each strand (on the right).

1.2 Discovery of the Double Helix Launched the Era of Molecular Genetics

Once it was accepted that nucleic acid in the form of DNA carries genetic information, efforts were focused on deciphering the structure of DNA and the mechanism by which information stored in this molecule is expressed to produce an observable phenotype giving rise to the science of molecular genetics. In the years after this was accomplished, researchers learned how to isolate and make copies of specific regions of DNA molecules, opening the way for the era of recombinant DNA technology.

The Structure of DNA and RNA

DNA is a long, ladderlike molecule that twists to form a double helix. Each strand of the helix is a linear molecule made up of subunits called **nucleotides**. In DNA, there are four different nucleotides. Each DNA nucleotide contains one of four nitrogenous bases—A (adenine), G (guanine), T (thymine), or C (cytosine). These four bases comprise the genetic alphabet, or **genetic code**, which in various combinations ultimately specifies the amino acid sequence of proteins. One of the great discoveries of the twentieth century was made in 1953 by James Watson and Francis Crick, who established that the two strands of DNA are exact complements of one another, such that the rungs of the ladder in the double helix always consist of either A═T or G≡C base pairs. As we shall see in a later chapter, this **complementary relationship** between adenine and thymine and between guanine and cytosine is critical to genetic function. This relationship serves as the basis for both the replication of DNA and for the basis of gene expression. During both processes, DNA strands serve as templates for the synthesis of complementary molecules. Two depictions of the structure and components of DNA are shown in Figure 1–8.

RNA, another nucleic acid, is chemically similar to DNA. RNA contains a different sugar (ribose vs. deoxyribose) in its nucleotides and contains the nitrogenous base uracil (U) in place of thymine. In addition, in contrast to the double helix of DNA, RNA is generally single stranded. Importantly, it can form complementary structures with a strand of DNA.

Gene Expression: From DNA to Phenotype

As noted earlier, nucleotide complementarity is one basis for gene expression. This process begins in the nucleus with the **transcription** of the nucleotide sequence of one strand of DNA into a complementary RNA sequence (Figure 1–9).

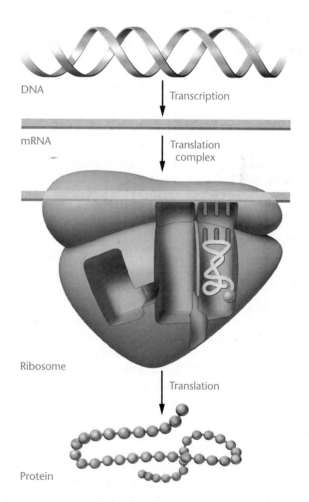

FIGURE 1–9 Gene expression involves transcription of DNA into mRNA (top) and the translation (center) of mRNA on a ribosome into a protein (bottom).

Once an RNA molecule is transcribed, it moves to the cytoplasm and directs the synthesis of proteins. This is accomplished when the RNA—called **messenger RNA**, or **mRNA**, for short—binds to a **ribosome**. The synthesis of proteins under the direction of mRNA is called **translation** (see bottom part of Figure 1–9). Proteins, as the end product of genes, are polymers made up of amino acid monomers. There are 20 different amino acids in proteins of all living organisms.

How can information contained in mRNA direct the insertion of specific amino acids into protein chains as they are synthesized? The information encoded in mRNA is called the genetic code, a linear series of triplet nucleotides. Each triplet, called a **codon**, is complementary to the information stored in DNA and specifies the insertion of a specific amino acid into a protein. This is accomplished by adapter molecules called **transfer RNAs (tRNAs)**. Within the ribosome, tRNAs recognize the information encoded in the mRNA codons and carry the proper amino acid for insertion into the protein during translation.

As the preceding discussion shows, DNA makes RNA. Most RNA proceeds to make protein, and allows the direction flow of information from DNA to RNA, and then to protein. These processes, known as the **central dogma** of genetics, occur with great specificity. Using an alphabet of only four letters (A, T, C, and G), genes direct the synthesis of highly specific proteins that collectively serve as the basis for all biological function.

Proteins and Biological Function

As we have mentioned, proteins are the end products of gene expression. These molecules are responsible for imparting the properties that we attribute to living systems. The diverse nature of biological function rests with the fact that proteins are made from 20 different amino acids. In a protein chain that is 100 amino acids long, and at each position there can be any of 20 amino acids, then the number of different protein molecules—each with an unique sequence is equal to

$$20^{100}$$

Because 20^{10} exceeds 5×10^{12}, or more than 5 trillion, imagine how large 20^{100} is! Obviously, evolution has seized on a class of molecules with the potential for enormous structural diversity to serve as the mainstay of biological systems. On the other hand, RNA or DNA consists of only 4 kinds of monomers. The simpler structure is beneficial for faithful storage and transmission of genetic codes.

The largest category of proteins includes **enzymes**. These molecules serve as biological catalysts, essentially allowing biochemical reactions to proceed at rates that sustain life under the conditions that exist on Earth. By lowering the energy of activation in reactions, metabolism proceeds under the direction of enzymes at body temperature.

There are countless proteins other than enzymes that are critical components of cells and organisms. These include **hemoglobin**, the oxygen-binding pigment in red blood cells; **insulin**, the pancreatic hormone; **collagen**, the connective tissue molecule; **keratin**, the structural molecule in hair; **histones**, proteins integral to chromosome structure in eukaryotes; **actin** and **myosin**,

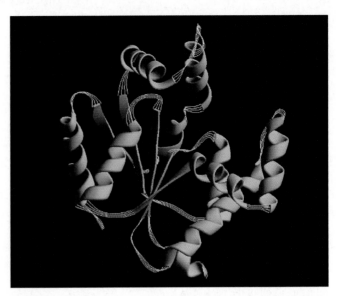

FIGURE 1–10 The three-dimensional conformation of a protein. The amino acid sequence of the protein is depicted as a ribbon.

the contractile muscle proteins; and **immunoglobulins**, the antibody molecules of the immune system. The potential for such diverse functions rests with the enormous variations in three-dimensional conformation of proteins (Figure 1–10). This conformation is determined by the linear sequence of amino acids constituting the molecule. To come full circle, this sequence is dictated by the stored information in the DNA of a gene that is transferred to RNA, which then directs the synthesis of a protein. DNA makes RNA that then makes protein.

Linking Genotype to Phenotype: Sickle-Cell Anemia

Once a protein is made, its action or location in a cell plays a role in producing a phenotype. When mutation alters a gene, it may abolish or modify the encoded protein's function and cause an altered phenotype. To trace the chain of events leading from the

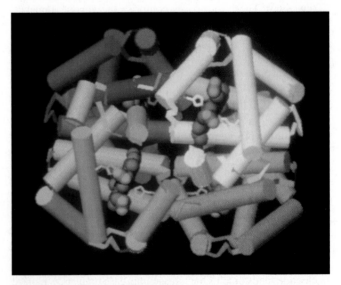

FIGURE 1–11 The hemoglobin molecule, showing the two alpha chains and the two beta chains. A mutation in the gene for the beta chain produces abnormal hemoglobin molecules and sickle-cell anemia.

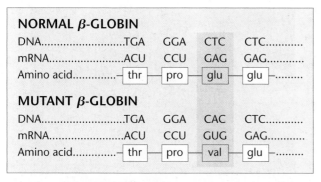

NORMAL β-GLOBIN				
DNA..........................TGA	GGA	CTC	CTC...........	
mRNA......................ACU	CCU	GAG	GAG...........	
Amino acid.............– thr –	pro –	glu –	glu –.........	

MUTANT β-GLOBIN				
DNA..........................TGA	GGA	CAC	CTC...........	
mRNA......................ACU	CCU	GUG	GAG...........	
Amino acid.............– thr –	pro –	val –	glu –.........	

FIGURE 1–12 A single nucleotide change in the DNA encoding the β-globin gene (CTC → CAC) leads to an altered mRNA codon (GAG → GUG) and the insertion of a different amino acid (glu → val), producing an altered version of the β-globin protein, causing sickle cell anemia.

synthesis of a protein to a phenotype, we will examine sickle-cell anemia, a human genetic disorder. Sickle-cell anemia is caused by a mutant form of hemoglobin, the protein that transports oxygen from the lungs to cells in the body (Figure 1–11). Hemoglobin is a composite molecule made up of two different proteins, α-globin and β-globin, each encoded by a different gene. Each functional hemoglobin molecule contains two α-globin and two β-globin proteins. In sickle-cell anemia, a mutation in the gene encoding β-globin causes an amino acid substitution in 1 of the 146 amino acids in the protein. Figure 1–12 shows part of the DNA sequence, mRNA codons, and amino acid sequence for the normal and mutant forms of β-globin. Notice that the mutation in sickle-cell anemia involves a change in one DNA nucleotide, leading to a change in codon 6 in the mRNA from GAG to GUG, which in turn changes amino acid number 6 in β-globin from glutamic acid to valine. The other 145 amino acids in the protein are not changed by this mutation.

Individuals with two mutant copies of the β-globin gene have sickle-cell anemia. Mutant β-globin proteins cause hemoglobin molecules in red blood cells to polymerize when oxygen concentration is low, forming long chains of hemoglobin that

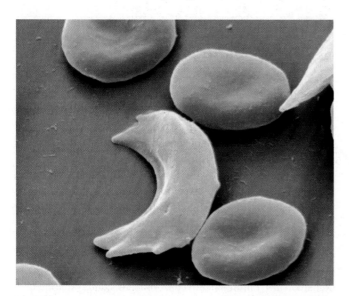

FIGURE 1–13 Normal red blood cells (round) and sickled red blood cells. The sickled cells block capillaries and small blood vessels.

distort the shape of red blood cells (Figure 1–13). When the blood cells are sickle shaped, they block blood flow in capillaries and small blood vessels, causing severe pain and damage to the heart, brain, muscles, and kidneys. Sickle-cell anemia can cause heart attacks and stroke, and can be fatal if left untreated. In addition, the deformed blood cells are fragile and break easily, causing anemia by reducing the number of red cells in circulation. Thus, all the symptoms of this disorder are caused by a change in a single nucleotide in a gene that changes one amino acid out of 146 in the β-globin molecule, emphasizing the close relationship between genotype and phenotype.

1.3 Genomics Grew Out of Recombinant DNA Technology

The era of recombinant DNA began in the early 1970s when researchers discovered that bacteria protect themselves from viral infection with enzymes that cut viral DNA at specific sites. When cut, the viral DNA cannot direct the synthesis of phage particles. Scientists quickly realized that such enzymes, called **restriction enzymes**, could be used to cut DNA from any organism at specific nucleotide sequences, producing a reproducible set of fragments. This set the stage for the development of DNA cloning, making large numbers of copies of these DNA fragments.

Making Recombinant DNA Molecules and Cloning DNA

Soon after it was discovered that restriction enzymes produce specific DNA fragments, methods were developed to insert these fragments into carrier DNA molecules called vectors to make **recombinant DNA** molecules and to transfer the recombinant DNA into bacterial cells to make hundreds or thousands of copies, or **clones**, of the combined vector and DNA fragments (Figure 1–14). These cloned copies can be recovered from the bacterial cells, and large amounts of the cloned DNA fragments can be isolated. Once large quantities of specific DNA fragments became available by cloning, they were used in many different ways: to isolate genes, to study their organization and expression, and to study their nucleotide sequence and evolution. In addition to preparing large amounts of specific DNA for research, recombinant DNA techniques were the foundation for the biotechnology industry (described in the next section of this chapter).

As techniques became more refined, it became possible to clone larger and larger DNA fragments, paving the way to clone an organism's **genome**, which is all the nuclear DNA carried by that organism. Collections of clones that contain an entire genome are called genomic libraries. Genomic libraries are now available for hundreds of organisms.

Sequencing Genomes: The Human Genome Project

Once genomic libraries became available, scientists began to consider ways to sequence all the clones in a genomic library in an organized way to spell out the nucleotide sequence of an organism's genome. The Human Genome Project began in 1990 as a federally sponsored international

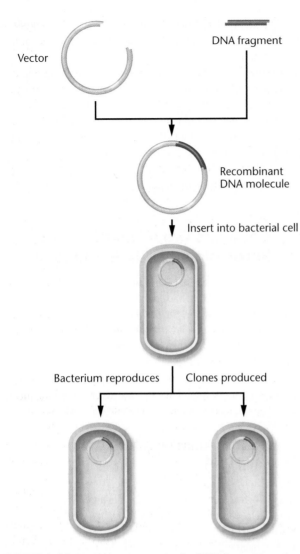

Vector

DNA fragment

Recombinant DNA molecule

Insert into bacterial cell

Bacterium reproduces   Clones produced

FIGURE 1–14 In cloning, a vector and a DNA fragment produced by cutting with a restriction enzyme are joined to produce a recombinant DNA molecule. This is transferred into a bacterial cell, where it is cloned as it replicates during the reproduction of the bacterial cell.

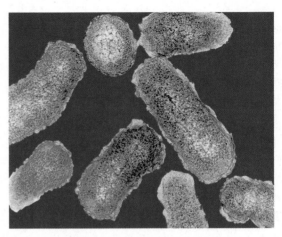

FIGURE 1–15 A colorized electron micrograph of *Haemophilus influenzae*, a bacterium that was the first free-living organism to have its genome sequenced. This bacterium causes respiratory infections and bacterial meningitis in humans.

As genome projects multiplied and more genome sequences were deposited in databases, a new discipline called **genomics** (the study of genomes) came into existence. Genomics uses nucleotide sequence information in databases to study the structure, function, and evolution of genes and genomes. Genomics is drastically changing biology from a laboratory-based science to one that combines lab experiments with information technology. Geneticists and other biologists can use information in databases containing nucleic acid sequences, protein sequences, and gene interaction networks to answer experimental questions in a matter of minutes instead of months and years.

Recombinant DNA technology has not only greatly accelerated the pace of research, generating new fields of study including genome projects and genomics, but has also given rise to many industrial and medical applications. These DNA-based technologies have grown to constitute a major component of the U.S. economy.

1.4   The Impact of Genetic Engineering Is Growing

Quietly and without much notice in the United States, biotechnology products and services have revolutionized many aspects of everyday life. Humans have used microorganisms, plants, and animals for thousands of years, but the development of recombinant DNA technology and genetic engineering allows us to genetically modify organisms in new ways and use them or their products to enhance our lives. Genetic engineering refers to the alteration of the genetic makeup of living organisms, using recombinant DNA technologies. Biotechnology is the commercial use of all types of living organisms or their products, including those that are genetically modified. Genetically engineered products are now found at the supermarket, in doctors' offices, drug stores, department stores, hospitals and clinics, on farms, in orchards, law enforcement, court-ordered child support, and even industrial chemicals. We will examine the impact of genetic engineering on a small crosssection of everyday life.

effort to sequence the human genome and the genomes of several model organisms used in genetics research. At about the same time, other genome projects, sponsored by industry, got underway. The first genome from a free-living organism, a bacterium (Figure 1–15), was sequenced and reported in 1995 by scientists at a biotechnology company.

In 2001, the publicly funded Human Genome Project and a private genome project sponsored by Celera Corporation reported the first draft of the human genome sequence, covering about 96 percent of the gene-containing portion of the genome. In 2003, the remaining portion of the gene-coding sequence was completed and the accuracy improved. Work is now focused on sequencing the remaining regions of the genome. Along the way, the genomes of five organisms used in genetic research, *Escherichia coli* (bacterium), *Saccharomyces cerevisiae* (yeast), *Caenorhabditis elegans* (a roundworm), the fruit fly (*Drosophila melanogaster*), and the mouse (*Mus musculus*) were also sequenced.

TABLE 1.1	Some Genetically Altered Traits in Crop Plants

Herbicide Resistance
Corn, Soybeans, Rice, Cotton, Sugarbeets, Canola

Insect Resistance
Corn, Cotton, Potato

Virus Resistance
Potato, Yellow Squash, Papaya

Altered Oil Content
Soybeans, Canola

Delayed Ripening
Tomato

FIGURE 1–16 Dolly, a Finn Dorset sheep cloned from the genetic material of an adult mammary cell, shown next to her first-born lamb, Bonnie.

Plants, Animals, and the Food Supply

The genetic modification of crop plants is one of the most rapidly expanding areas of genetic engineering. Attention has been focused on traits such as resistance to herbicides, insects, and viruses, enhancement of oil content, and others (Table 1.1). Currently, over a dozen genetically modified crop plants have been approved for commercial use in the United States, with dozens more in field trials. Herbicide-resistant corn and soybeans were first planted in the mid-1990s, and now about 40 percent of the corn crop and 80 percent of the soybean crop are genetically modified. In addition, more than 60 percent of the canola crop and 70 percent of the cotton crop are grown from genetically modified strains. It is estimated that more than 60 percent of the processed food in the United States contains ingredients from genetically modified crop plants.

This agricultural transformation is not without controversy. Critics are concerned that the use of herbicide-resistant crop plants will lead to dependence on chemical weed management and may eventually lead to herbicide-resistant weeds. Others are concerned that traits in genetically engineered crops could be transferred to wild plants in a way that leads to irreversible changes in the ecosystem. We will examine these concerns in Chapter 19.

Genetic engineering is also being used to nutritionally enhance crop plants. More than one-third of the world's population uses rice as a dietary staple, but most varieties of rice contain little or no vitamin A. Vitamin A deficiency causes more than 500,000 cases of blindness in children each year. A genetically engineered strain, called golden rice, has high levels of two compounds that the body converts to vitamin A. Golden rice is currently available for planting and more effective varieties of the strain are now in development. Other crops including wheat, corn, beans, and cassava are also being modified to enhance nutritional value by increasing their vitamin and mineral content.

Livestock such as sheep and cattle have been commercially cloned for more than 25 years, mainly by a method called embryo splitting. In 1996, Dolly the sheep (Figure 1–16) was cloned by a nuclear transfer, a method in which the nucleus of a differentiated adult cell is transferred into an egg that had its nucleus removed. This nuclear transfer method makes it pos-sible to produce dozens or hundreds of offspring with desir-able traits. Cloning by nuclear transfer has many applications in agriculture, sports, and medicine. Some desirable traits, such as high milk production or speed in race horses, do not appear until adulthood; animals with these traits can now be cloned using differentiated cells from adults with desirable traits. For medical applications, researchers have transferred human genes into animals so that as adults, they produce human proteins in their milk. By selecting and cloning ani-mals with high levels of human protein production, biophar-maceutical companies can produce a herd with uniformly high rates of protein production. Human proteins are used as drugs, and proteins from transgenic animals are now being tested as treatments for diseases such as emphysema. If suc-cessful, these proteins will soon be commercially available.

Who Owns Transgenic Organisms?

Once produced, can a transgenic plant or animal be patented? The answer is yes. The United States Supreme Court ruled in 1980 that living organisms can be patented, and the first organ-ism modified by recombinant DNA technology was patented in 1988 (Figure 1–17). Since then, dozens of plants and animals have been patented. The ethics of patenting living organ-isms is a contentious issue. Supporters of patenting argue that without the ability to patent the products of research to recover their costs, biotechnology companies will not invest in large-scale research and development. They further argue that patents represent an incentive to develop new products because companies will reap the benefits from taking risks to bring new products to market. Critics argue that patents for organisms such as crop plants will concentrate owner-ship of food production in the hands of a small number of biotechnology companies, making farmers economically dependent on seeds and pesticides produced by these com-panies, and reducing the genetic diversity of crop plants as

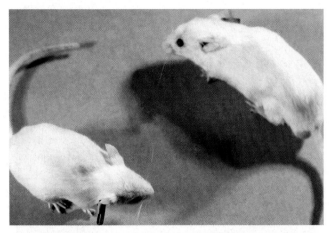

FIGURE 1–17 The first genetically altered organism to be patented, mice from the *onc* strain, genetically engineered to be susceptible to many forms of cancer. These mice were designed for studying cancer development and the design of new anticancer drugs.

farmers discard local crops that might harbor important genes for resistance to pests and disease. To resolve these and other issues about biotechnology and its uses, a combination of public awareness, education, enlightened social policy, and legislation are needed.

Genetic Engineering in Genetics and Medicine

Applications of genetic engineering, such as genetic testing and gene therapy are rapidly becoming an important part of medicine. These technologies promise to shape medical practice in the twenty-first century. The importance of developing tests and treatments for genetic diseases is underscored by the estimate that more than 10 million children or adults in the United States suffer from some form of genetic disorder and that every childbearing couple stands an approximately 3 percent risk of having a child with some form of genetic anomaly. The molecular basis for hundreds of genetic disorders is now known (Figure 1–18). Genes for sickle-cell anemia,

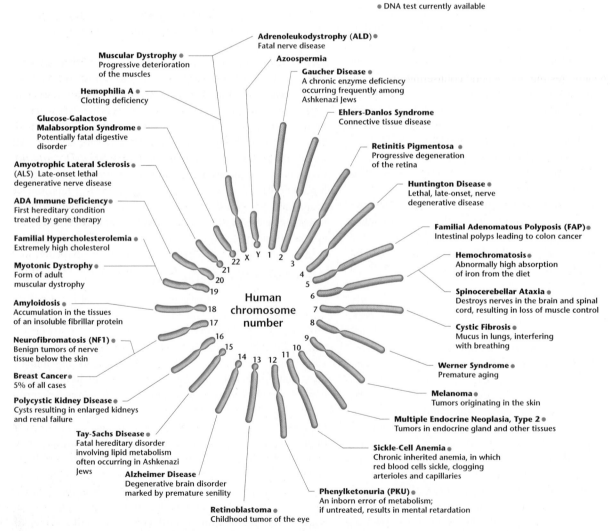

FIGURE 1–18 Diagram of the human chromosome set, showing the location of some genes whose mutant forms cause hereditary diseases. Conditions that can be diagnosed using DNA analysis are indicated by a red dot.

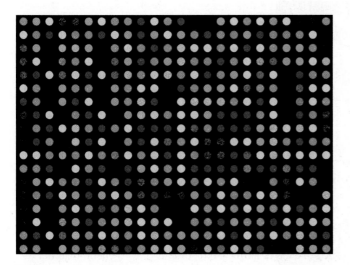

FIGURE 1–19 A DNA microarray. These arrays contains thousands of fields to which DNA molecules are attached. Using this microarray, we can test DNA from an individual to detect mutant copies of genes.

cystic fibrosis, hemophilia, muscular dystrophy, phenylke-tonuria, and many other disorders have been cloned. These cloned genes are used for the prenatal detection of affected fetuses. In addition, parents can also learn of their status as "carriers" of a large number of inherited disorders. The com-bination of genetic testing and genetic counseling gives cou-ples objective information on which they can base informed decisions about childbearing. At present, genetic testing is available for several hundred inherited disorders, and this number will grow as more genes are identified, isolated, and cloned. The use of genetic testing and other technologies, including gene therapy, has raised ethical concerns that have yet to be resolved.

Instead of testing one gene at a time to discover whether someone carries a mutant gene that can produce a disorder in his or her offspring, technology is being developed that will allow screening of an entire genome to determine an individ-ual's risk of developing a genetic disorder or of having a child with a genetic disorder. This technology uses devices called **DNA microarrays** or **DNA chips** (Figure 1–19). Each microarray can carry thousands of genes. In fact, microarrays carrying the human genome are now commercially available, making it possible to scan someone's entire genome to see which genetic diseases the individual carries or may develop. DNA microarrays have many other applications as well and are used to test for gene expression in cancer cells as a step in developing therapies tailored to specific forms of cancer.

In addition to testing for genetic disorders, clinicians can transfer normal genes into individuals affected with genetic disorders in a procedure known as **gene therapy**. Though ini-tially successful, therapeutic failures and patient deaths have slowed development of this technology. New methods of gene transfer may reduce the risks involved, and it seems certain that gene therapy will become an important tool in treating inherited disorders. As more is learned about the molecular basis of human diseases, more therapies can be developed. Much of this research on human genetic disorders now involves the use of model organisms.

1.5 Genetic Studies Rely on the Use of Model Organisms

After the rediscovery of Mendel's work in 1900, genetic research on a wide range of organisms confirmed that the principles of inheritance he described were of universal significance among plants and animals. Although work on the genetics of many different organisms continued, geneticists gradually centered their attention on a small number of organisms, including the fruit fly, *Drosophila*, and the mouse (*Mus musculus*) (Figure 1–20). Such organ-isms became popular for two main reasons. First, it was clear that genetic mechanisms were the same in most organisms, and second, these species had several advan-tages for genetic research. They were easy to grow, had relatively short life cycles, produced many offspring, and genetic analysis was fairly straightforward. Over time,

FIGURE 1–20 The first generation of model organisms in genetic analysis included (a) the mouse and (b) the fruit fly.

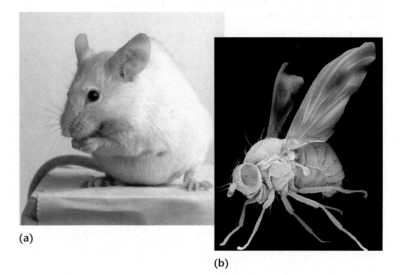

(a)

(b)

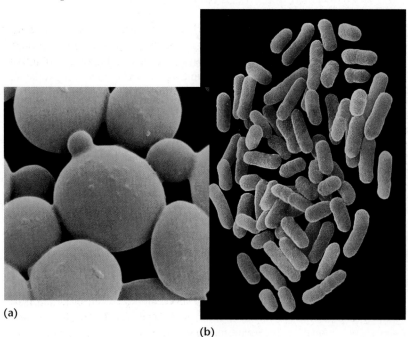

(a)

(b)

FIGURE 1–21 Microbes that have become model organisms for genetic studies include (a) the yeast *Saccharomyces* and (b) the bacterium *E. coli.*

researchers created a large catalog of mutant strains for each species. These mutations were carefully studied, characterized, and mapped. Serving as experimental models for other organisms, these species became **model organisms**, which we define as organisms used for the study of basic biological processes, including normal cellular events as well as genetic disorders and other diseases. Model organisms, as we will see in later chapters, are used to study many aspects of biology, including aging, cancer, the immune system, and behavior.

The Modern Set of Genetic Model Organisms

Gradually, geneticists added other species to their collection of model organisms, including microorganisms (the bacterium *Escherichia coli* and the yeast *Saccharomyces cerevisiae*) (Figure 1–21). Some of these were chosen for the reasons outlined earlier, while others were selected because they allowed certain aspects of genetics to be studied more easily.

More recently, three additional species have been developed as model organisms. Each was chosen to study some aspect of embryonic development. To study development and physiology, the nematode *Caenorhabditis elegans* [Figure 1–22(a)] was chosen as a model system. It is small, easy to grow, and has a nervous system with only a few hundred cells. *Arabidopsis thaliana* [Figure 1–22(b)] is a small plant with a short life cycle that can be grown in the laboratory. It was first used to study flower development but has become a model organism for the study of many other aspects of plant biology. The zebrafish, *Danio rerio* [Figure 1–22(c)], has several advantages for the study of vertebrate development; it is small, reproduces rapidly, and the egg, embryo, and larvae are all transparent. Although originally developed to study genetic mechanisms, modern model organisms are now also

being used to study the origin and mechanisms of many human diseases and to develop new and innovative therapies to treat these diseases.

Model Organisms and Human Diseases

The development of recombinant DNA technology and the results from genome sequencing projects have confirmed that all life has a common origin. As a result, genes with similar functions in different organisms are similar or identical in structure and DNA sequence. In addition, the ability to transfer genes across species has made it possible to develop models of human diseases in organisms ranging from bacteria, fungi, plants, and animals (Table 1.2). For these reasons, the Human Genome Project also sequenced the genomes of five model organ-

TABLE 1.2	Cellular Process and Human Diseases Studied Using Model Organisms	
Organism	**Cellular Processes**	**Human Diseases**
E. coli	DNA repair	colon cancer and other cancers
Yeast	Cell cycle	cancer, Werner syndrome
Drosophila	Cell signaling	cancer
C. elegans	Cell signaling	diabetes
Zebrafish	Developmental pathways	cardiovascular disease
Mouse	Gene expression	Lesch-Nyhan disease, cystic fibrosis, fragile-X syndrome, and many other diseases

(a)

(b)

(c)

FIGURE 1–22 Newer model organisms in genetics include (a) the roundworm *C. elegans*, (b) the plant *A. thaliana,* and (c) the zebrafish, *D. rerio*.

isms in addition to the human genome. Other genome projects have sequenced the genomes of other model organisms, as well as the genomes of hundreds of other organisms.

It may seem strange to study a human disease such as colon cancer by using *E. coli*, but the basic steps of DNA repair (the process of DNA repair is defective in some forms of colon cancer) is the same in both organisms, and the gene involved (*mutL* in *E. coli* and *MLH1* in humans) is the same in both organisms. More important, *E. coli* has the advantage of being easier to grow (the cells divide every 20 minutes), and it is easy to create and study new mutations in the *mutL* gene to help understand how it works. This knowledge may eventually lead to the development of drugs and other therapies to treat colon cancer in humans.

Other model organisms, including the fruit fly, *Drosophila melanogaster*, are being used to study specific human diseases. Mutant genes have been identified in *Drosophila* that produce phenotypes with abnormalities of the nervous system, including abnormalities of brain structure, adult-onset degeneration of the nervous system, and visual defects such as retinal degeneration. The information from genome sequencing projects indicates that almost all these genes have human counterparts. As an example, genes involved in a complex human disease of the retina called retinitis pigmentosa are identical to *Drosophila* genes involved in retinal degeneration. Study of these mutations in *Drosophila* is helping to dissect this complex disease and identify the function of the genes involved.

Using recombinant DNA technology, researchers are using *Drosophila* as a model to study diseases of the human nervous system by transferring human disease genes into flies. In addition to studying mutant human genes, the transgenic flies are used to study genes affecting the expression of the human disease genes and to test the effects of therapeutic drugs on the action of these genes, studies that are difficult or impossible to do in humans. This gene transfer approach is being used to study almost a dozen human neurodegenerative disorders, including Huntington disease, Machado-Joseph disease, myotonic dystrophy, and Alzheimer disease.

As you read through the text, you will encounter these model organisms again and again in the genetic analysis of basic biological processes. Remember, they not only have a rich history in genetics, but they are at the forefront in the study of human genetic disorders and infectious diseases. Keep in mind that understanding how a gene controls a process in yeast is relevant to understanding the same gene and the same process in normal human cells.

The development and use of model organisms is only one of the ways genetics and biotechnology are rapidly changing many aspects of everyday life. As discussed in the next section, we have yet to reach a consensus on how and when this technology is acceptable and useful.

1.6   We Live in the "Age of Genetics"

Genetics is no longer just a laboratory science in which researchers study fruit flies or yeast to learn about basic processes in cell function or development. As stated in the beginning of this chapter, genetics is the core of biology, and the method of choice in dissecting and understanding the functions and malfunctions of biological systems. As knowledge has increased, genetics has become intertwined with social issues. Genetics and its applications in biotech-nology are developing much faster than social convention, public policy, and law. Although other scientific disciplines are also expanding in knowledge, few have paralleled the growth of information occurring in genetics. While there has never been a more exciting time to study genetics, the impact of this discipline on society has never been more profound. We are confident that by the end of this course you will agree that the present represents the "Age of Genetics," and we urge you to think about and become a participant in the dialogue about genetics and its applica-tions in society.

GENETICS, TECHNOLOGY, AND SOCIETY

Genetics and Society: The Application and Impact of Science and Technology

As you embark on your study of genetics, we want you to be aware of a special feature of this text found at the conclusion of most chapters: **Genetics, Technology, and Society** essays. In these essays we provide coverage of a variety of topics derived from genetics that impact on the lives of each of us and thus on society in general. Genetics touches all aspects of modern life. Genetic technologies are rapidly chang-ing how we approach medicine, agricul-ture, law, the pharmaceutical industry, and biotechnology. We now use hun-dreds of genetic tests to diagnose and predict the course of disease and to detect genetic defects *in utero*. DNA-based methods allow scientists to trace the path of evolution taken by many species, including our own. We now design disease-resistant and drought-resistant crops, as well as more produc-tive farm animals, using gene transfer techniques. We apply DNA profiling methods to paternity testing and murder investigations. Biotechnologies based on information from genomics research have had dramatic effects on industry. The biotechnology industry doubles in size every decade, generating over 700,000 jobs and $50 billion in revenue each year.

Along with these rapidly changing gene-based technologies comes a chal-lenging array of ethical dilemmas. Who owns and controls genetic information? Are gene-enhanced agricultural plants and animals safe for humans and the envi-ronment? Do we have the right to patent organisms or the DNA derived from their genes and profit from their commercial-ization? How can we ensure that genomic technologies will be available to all and not just to the wealthy? What are the social issues that accompany the new reproductive technologies? Shall we allow medical insurance companies and employers to use genetic testing to screen applicants? We are now in a time when everyone needs to understand genetics in order to make complex personal and soci-etal choices.

The goal of the **Genetics, Technology, and Society** essays is to introduce topics that interface with society, including some of the new technologies based on genetics and genomics. As well, we will explore the relevant social and ethical issues. It is our hope that these essays will act as entry points for your exploration of the myriad of applications and societal implications of modern genetics. In the next column, we list the topics that serve as the basis of many of these essays (including the chapter in which each is found). Even should your genetics course not cover all chapters, we hope that you will find the essays in those chapters of interest. Good reading!

Tay-Sachs Disease (3)

Mitochondrial DNA and the Romanovs (4)

Human Sex Selection (5)

Fragile Chromosomes and Cancer (6)

Bacterial Genes and Disease (8)

The DNA Revolution (9)

Telomerase, Aging, and Cancer (10)

Antisense Technology (12)

Mad Cow Disease and Prions (13)

Chernobyl's Legacy (14)

Gene Regulation and Human Genetic Disease (15)

Breast Cancer (16)

Human Cloning (17)

DNA Patenting (18)

Gene Therapy (19)

The Stem Cell Debate (20)

The Green Revolution Revisited (21)

Tracking Humans out of Africa (22)

Eugenics Revisited (23)

Conserving the Florida Panther (24)

CHAPTER SUMMARY

1. Mendel's work on pea plants established the principles of the transmission of genes from parents to offspring and established the foundation for the science of genetics.
2. Genes and chromosomes are the fundamental units in the chromosomal theory of inheritance, which explains the transmission of genetic information controlling phenotypic traits.
3. Molecular genetics, based on the central dogma that DNA makes RNA—which makes protein—serves as the underpinnings of Mendelian genetics, referred to as transmission genetics.
4. Recombinant DNA technology allows genes from one organism to be spliced into vectors and cloned, serving as the basis for a far-reaching technology used in molecular genetics.
5. Genomics is one application of recombinant DNA technology in which the entire genetic makeup of an organism is sequenced and the structure and function of its genes are explored. The Human Genome Project is one example of genomics.

6. Genetic engineering has revolutionized agriculture, the pharmaceutical industry, and medicine. It has made possible the mass production of medically important gene products. Genetic testing allows detection of individuals with genetic disorders and those at risk of having affected children.
7. The use of model organisms in genetics has advanced our basic understanding of genetic mechanisms and, coupled with recombinant DNA technology, has been used to develop models of human genetic diseases.
8. Genetic technology is affecting many aspects of society. The development of policy and legislation is lagging behind the innovations and uses of biotechnology.

KEY TERMS

actin, 6
allele, 4
bacteriophage, 4
central dogma, 6
chromosome theory of inheritance, 4
clones, 7
codon, 6
collagen, 6
complementary relationship, 5
diploid number (2n), 3
DNA chips, 11
DNA microarrays, 11
enzymes, 6
gene therapy, 11

genes, 2
genetic code, 5
genetics, 3
genome, 7
genomics, 8
genotype, 4
haploid (n) number, 4
hemoglobin, 6
histones, 6
homologous chromosomes, 3
immunoglobulins, 6
insulin, 6
keratin, 6
meiosis, 3

messenger RNA (mRNA), 6
mitosis, 3
model organisms, 12
mutation, 4
myosin, 6
nucleotides, 5
phage, 4
phenotype, 4
recombinant DNA, 7
restriction enzymes, 7
ribosome, 6
transcription, 5
transfer RNA (tRNA), 6
translation, 6

PROBLEMS AND DISCUSSION QUESTIONS

1. Describe Mendel's conclusions about how traits are passed from generation to generation.
2. What is the chromosome theory of inheritance, and how is it related to Mendel's findings?
3. Define genotype and phenotype and describe how they are related.
4. What are alleles? If individuals carry genes in pairs, is it possible for more than two alleles of a gene to exist?
5. Given the state of knowledge at the time, why was it difficult for some scientists to accept that DNA is the carrier of genetic information?
6. Contrast chromosomes and genes.
7. How is genetic information encoded in a DNA molecule?
8. Describe the central dogma of molecular genetics and how it serves as the basis of modern genetics.
9. How many different proteins, each with a unique amino acid sequence, are possible in a protein that is five amino acids long?
10. Outline the roles played by restriction enzymes and vectors in cloning DNA.
11. What impact has biotechnology had on crop plants in the United States?
12. Summarize the arguments for and against patenting genetically modified organisms.
13. We all carry 20,000 to 25,000 genes in our genome. So far, patents have been issued for more than 6000 of these genes. Do you think that companies or individuals should be able to patent human genes? Why or why not?
14. How has the use of model organisms advanced our knowledge of the genes that control human diseases?
15. If you knew that a devastating late-onset inherited disease runs in your family and you could be tested for it at the age of 20, would you want to know if you are a carrier? Would your answers be likely to change when you reach age 40?

Mitosis and Meiosis

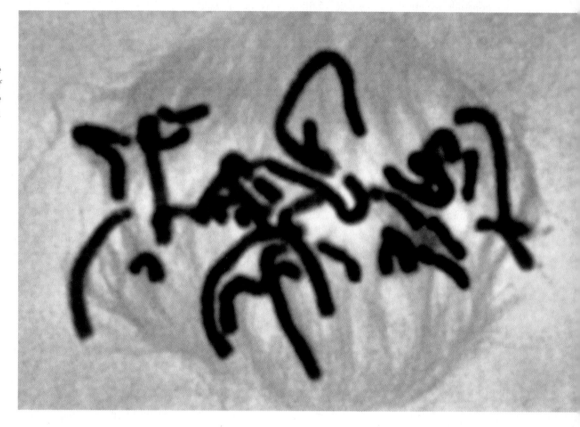

Chromosomes in the prometaphase stage of mitosis, from a cell in the flower of Haemanthus.

CHAPTER CONCEPTS

- Genetic continuity between cells and between organisms of sexually reproducing species is maintained through the processes of mitosis and meiosis, respectively.

- Diploid eukaryotic cells contain their genetic information in pairs of homologous chromosomes, with one member of each pair being derived from the maternal parent and one from the paternal parent.

- Mitosis provides a mechanism to distribute chromosomes that have duplicated into progeny cells during cell reproduction.

- Mitosis converts a diploid cell into two diploid daughter cells.

- The process of meiosis distributes one member of each homologous pair of chromosomes into each gamete or spore, thus reducing the diploid chromosome number to the haploid chromosome number.

- Meiosis generates genetic variability by distributing various combinations of maternal and paternal members of each homologous pair of chromosomes into gametes or spores.

- It is during the stages of mitosis and meiosis that the genetic material has been condensed into discrete structures called chromosomes.

In every living thing, there exists a substance referred to as the genetic material. Except in certain viruses, this material is composed of the nucleic acid, DNA. A molecule of DNA is organized into units called genes, the products of which direct the metabolic activities of cells. DNA, with its array of genes, is organized into structures called chromosomes, which serve as vehicles for transmitting genetic information. The manner in which chromosomes are transmitted from one generation of cells to the next and from organisms to their descendants must be exceedingly precise. In this chapter, we consider exactly how genetic continuity is maintained between cells and organisms.

Two major processes are involved in eukaryotes: **mitosis** and **meiosis**. Although the mechanisms of the two processes are similar in many ways, the outcomes are quite different. On one hand, mitosis leads to the production of two cells, each with the same number of chromosomes as the parent cell. Meiosis, on the other hand, reduces the genetic content and the number of chromosomes to precisely half. This reduction is essential if sexual reproduction is to occur without doubling the amount of genetic material at each generation. Strictly speaking, mitosis is that portion of the cell cycle during which the hereditary components are equally divided into daughter cells. Meiosis is part of a special type of cell division that leads to the production of sex cells: **gametes** or **spores**. This process is an essential step in the transmission of genetic information from an organism to its offspring.

Normally, chromosomes are visible only during mitosis and meiosis. When cells are not undergoing division, the genetic material making up chromosomes unfolds and uncoils into a diffuse network within the nucleus, generally referred to as chromatin. We will briefly review the structure of cells, emphasizing those components that are of particular significance to genetic function. We then devote the remainder of the chapter to the behavior of chromosomes during cell division.

How Do We Know?

In this chapter, we will focus on how chromosomes are distributed during cell division, both in dividing somatic cells (mitosis) and in gamete and spore-forming cells (meiosis). A key concept in genetics is the presence of chromosomes in cells as homologous pairs. As you study mitosis and meiosis, you should try to answer several fundamental questions:

1. How do we know that chromosomes exist in homologous pairs?

2. How did we know that DNA replication occurs during interphase, not early in mitosis?

3. How do we know about the events comprising the various stages of mitosis?

4. How do we know that mitotic chromosomes are derived from chromatin?

2.1 Cell Structure Is Closely Tied to Genetic Function

Before describing mitosis and meiosis, a brief review of the structure of cells will be helpful. Many components, such as the nucleolus, ribosome, and centriole, are involved directly or indirectly with genetic processes. Other components, like the mitochondria and chloroplasts, contain their own unique genetic information. It will also be useful to compare the structural differences between the prokaryotic bacterial cell and the eukaryotic cell. Variation in the structure and function of cells is dependent on specific genetic expression by each cell type.

Before 1940, our knowledge of cell structure was limited to what we could see with the light microscope. Around 1940, the transmission electron microscope was in its early stages of development, and by 1950, many details of cell ultrastructure had emerged. Under the electron microscope, cells were seen as highly organized, precise structures. A new world of whorled membranes, organelles, microtubules, granules, and filaments was revealed. These discoveries revolutionized thinking in the entire field of biology, but we will be concerned only with those aspects of cell structure that relate to genetic study. The typical animal cell shown in Figure 2–1 illustrates most of the structures we will discuss.

All cells are surrounded by a **plasma membrane**, an outer covering that defines the cell boundary and delimits the cell from its immediate external environment. This membrane is not passive but instead actively controls the movement of materials into and out of the cell. In addition to this membrane, plant cells have an outer covering called the **cell wall** whose major component is a polysaccharide called cellulose.

Many, if not most, animal cells have a covering over the plasma membrane, referred to as the **cell coat**. Consisting of glycoproteins and polysaccharides, the cell coat has a chemical composition that differs from comparable structures in either plants or bacteria. The cell coat, among other functions, provides biochemical identity at the surface of cells, and these forms of cellular identity are under genetic control. For example, various antigenic determinants such as the **AB** and **MN antigens** are found on the surface of red blood cells. In other cells, **histocompatibility antigens**, which elicit an immune response during tissue and organ transplants, are present. A variety of **receptor molecules** are also important components at the surface of cells. These are recognition sites that transfer specific chemical signals across the cell membrane into the cell.

The presence of a nucleus and other membranous organelles characterizes eukaryotic cells. The **nucleus** houses the genetic material, DNA, which is complexed with an array of acidic and basic proteins into thin fibers. During nondivisional phases of the cell cycle, these fibers are uncoiled and dispersed into **chromatin**. During mitosis and meiosis, chromatin fibers coil and condense into structures called **chromosomes**. Also present in the nucleus is the **nucleolus**, an amorphous component where ribosomal RNA (rRNA) is synthesized and where the initial stages of ribosomal assembly occur. The areas of

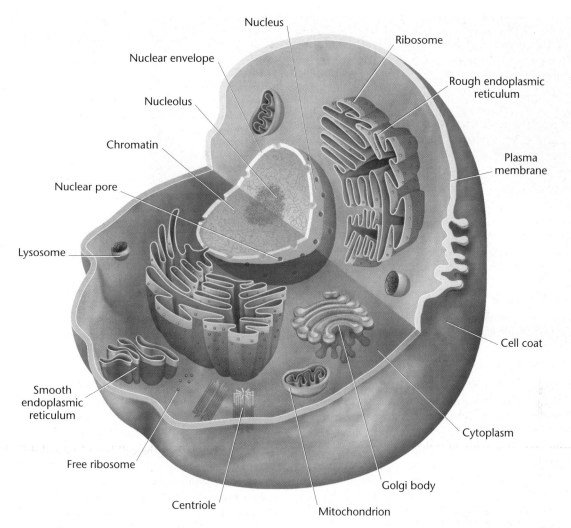

Nucleus

Ribosome

Nuclear envelope

Rough endoplasmic
reticulum

Nucleolus

Chromatin

Plasma
membrane

Nuclear pore

Lysosome

Cell coat

Smooth
endoplasmic
reticulum

Cytoplasm

Free ribosome

Golgi body

Centriole

Mitochondrion

FIGURE 2–1 A generalized animal cell. The cellular components discussed in the text are emphasized here.

DNA that encode rRNA are collectively referred to as the **nucleolus organizer region**, or the **NOR**.

Prokaryotes lack a nuclear envelope and membraneous organelles. In many bacteria such as *Escherichia coli*, the genetic material is present as a long, circular DNA molecule compacted into the **nucleoid** area. Part of the DNA may be attached to the cell membrane, but in general the nucleoid constitutes a large area throughout the cell. Although the DNA is compacted, it does not undergo the extensive coiling characteristic of the stages of mitosis where, in eukaryotes, chromosomes become visible. Nor is the DNA in these organisms associated as extensively with proteins as is eukaryotic DNA. Figure 2–2, in which two bacteria are forming during cell division, illustrates the nucleoid regions that house the bacterial chromosome. Prokaryotic cells do not have a distinct nucleolus, but they do contain genes that specify rRNA molecules.

The remainder of the eukaryotic cell enclosed by the plasma membrane, excluding the nucleus, is composed of **cytoplasm** and all associated cellular organelles. Cytoplasm is a nonparticulate, colloidal material referred to as the cytosol, which surrounds and encompasses the cellular organelles. Beyond these components, an extensive system of tubules and fila-ments comprising the cytoskeleton provides a lattice of support structures within the cytoplasm. Consisting primarily of tubulin-derived microtubules and actin-derived microfilaments, this structural framework maintains cell shape, facilitates cell mobility, and anchors the various organelles.

One such organelle, the membranous **endoplasmic reticulum (ER)**, compartmentalizes the cytoplasm, greatly increasing the surface area available for biochemical synthesis. The ER may appear smooth, in which case it serves as the site for synthesizing fatty acids and phospholipids, or it may appear rough because it is studded with ribosomes. Ribosomes serve as sites where genetic information contained in messenger RNA (mRNA) is translated into proteins.

Three other cytoplasmic structures are very important in the eukaryotic cell's activities: mitochondria, chloroplasts, and centrioles. **Mitochondria** are found in both animal and plant cells and are the sites of the oxidative phases of cell respiration. These chemical reactions generate large amounts of adenosine triphosphate (ATP), an energy-rich molecule. **Chloroplasts** are found in plants, algae, and some protozoans. This organelle is associated with photosynthesis, the major energy-trapping process on Earth. Both mitochondria and chloroplasts contain a type of DNA distinct from that

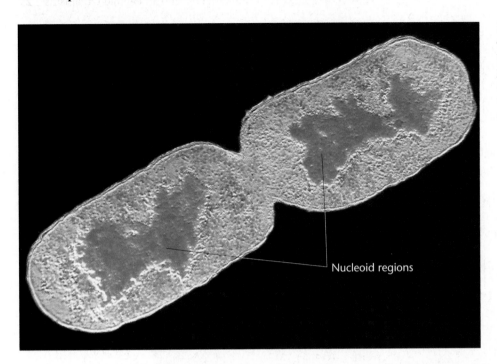

FIGURE 2–2 Color-enhanced electron micrograph of *E. coli* undergoing cell division. Particularly prominent are the two chromosomal areas (shown in red), called nucleoids, that have been partitioned into the daughter cells.

Nucleoid regions

found in the nucleus. Furthermore, these organelles can duplicate themselves and transcribe and translate their genetic information. It is interesting to note that the genetic machinery of mitochondria and chloroplasts closely resembles that of prokaryotic cells. This and other observations have led to the proposal that these organelles were once primitive free-living organisms that established a symbiotic relationship with a primitive eukaryotic cell. This theory, which describes the evolutionary origin of these organelles, is called the **endosymbiont hypothesis**.

Animal cells and some plant cells also contain a pair of complex structures called the **centrioles**. These cytoplasmic bodies, located in a specialized region called the centrosome, are associated with the organization of spindle fibers that function in mitosis and meiosis. In some organisms, the centriole is derived from another structure, the basal body, which is associated with the formation of cilia and flagella. Over the years, many reports have suggested that centrioles and basal bodies contain DNA, which could be involved in the replication of these structures. Currently, this is thought not to be the case.

The organization of **spindle fibers** by the centrioles occurs during the early phases of mitosis and meiosis. These fibers play an important role in the movement of chromosomes as they separate during cell division. They are composed of arrays of microtubules consisting of polymers of polypeptide subunits of the protein tubulin.

2.2 Chromosomes Exist in Homologous Pairs in Diploid Organisms

As we discuss the processes of mitosis and meiosis, it is important that you understand the concept of homologous chromosomes. Such an understanding will also be of criti-

cal importance in our future discussions of Mendelian genetics. Chromosomes are most easily visualized during mitosis. When they are examined carefully, distinctive lengths and shapes are apparent. Each chromosome contains a condensed or constricted region called the **centromere**, which establishes the general appearance of each chromosome. Figure 2–3 shows chromosomes with centromere placements at different points along their lengths. Extending from either side of the centromere are the arms of the chromosome. Depending on the position of the centromere, different arm ratios are produced. As Figure 2–3 illustrates, chromosomes are classified as **metacentric**, **submetacentric**, **acrocentric**, or **telocentric** on the basis of the centromere location. The shorter arm, by convention, is shown above the centromere and is called the **p arm** (p, for "petite"). The longer arm is shown below the centromere and is called the **q arm** (because q is the next letter in the alphabet).

When studying mitosis, several observations are of particular relevance. First, all somatic cells derived from members of the same species contain an identical number of chromosomes. In most cases, this represents the **diploid number (2*n*)**. When the lengths and centromere placements of the chromosomes are examined, a second general feature is apparent. Nearly all chromosomes exist in pairs with regard to these two criteria, and the members of each pair are called **homologous chromosomes**. So for each chromosome exhibiting a specific length and centromere placement, another exists with identical features.

There are exceptions to this rule. Most bacteria and viruses have only one chromosome, and organisms such as yeasts and molds and certain plants such as bryophytes (mosses) spend the predominant phase of the life cycle in the haploid stage. That is, they contain only one member of each homologous pair of chromosomes during most of their lives.

Centromere location	Designation	Metaphase shape	Anaphase shape
Middle	Metacentric	p arm — Centromere — q arm	← Migration →
Between middle and end	Submetacentric		
Close to end	Acrocentric		
At end	Telocentric		

FIGURE 2–3 Centromere locations and designations of chromosomes based on centromere location. Note that the shape of the chromosome during anaphase is determined by the position of the centromere.

Figure 2–4 illustrates the physical appearance of different pairs of homologous chromosomes. There, the human mitotic chromosomes have been photographed, cut out of the print, and matched up, creating a **karyotype**. As you can see, humans have a 2n number of 46, which on close examination exhibit a diversity of sizes and centromere placements. Note also that each of the 46 chromosomes is clearly a double structure consisting of two parallel **sister chromatids** connected by a common centromere. Had these chromosomes been allowed to continue dividing, the sister chromatids, which are replicas of one

another, would have separated into two new cells as division continued.

The haploid number (n) of chromosomes is equal to one-half the diploid number. Collectively, the genetic information contained in a haploid set of chromosomes constitutes the **genome** of the species. This, of course, includes copies of all genes as well as a large amount of noncoding DNA. The examples listed in Table 2.1 demonstrate the wide range of n values found in plants and animals.

Homologous pairs of chromosomes have important genetic similarities. They contain identical gene sites along their

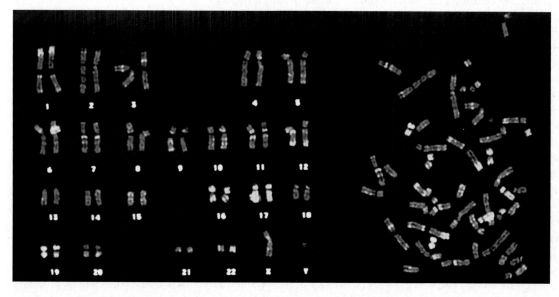

FIGURE 2–4 A metaphase preparation of chromosomes derived from a dividing cell of a human male (right), and the karyotype derived from the metaphase preparation (left). All but the X and Y chromosomes are present in homologous pairs. Each chromosome is clearly a double structure, constituting a pair of sister chromatids joined by a common centromere.

TABLE 2.1	The Haploid Number of Chromosomes for a Variety of Organisms	
Common Name	Scientific Name	Haploid Number
Black bread mold	*Aspergillus nidulans*	8
Broad bean	*Vicia faba*	6
Cat	*Felis domesticus*	19
Cattle	*Bos taurus*	30
Chicken	*Gallus domesticus*	39
Chimpanzee	*Pan troglodytes*	24
Corn	*Zea mays*	10
Cotton	*Gossypium hirsutum*	26
Dog	*Canis familiaris*	39
Evening primrose	*Oenothera biennis*	7
Frog	*Rana pipiens*	13
Fruit fly	*Drosophila melanogaster*	4
Garden onion	*Allium cepa*	8
Garden pea	*Pisum sativum*	7
Grasshopper	*Melanoplus differentialis*	12
Green alga	*Chlamydomonas reinhardi*	18
Horse	*Equus caballus*	32
House fly	*Musca domestica*	6
House mouse	*Mus musculus*	20
Human	*Homo sapiens*	23
Jimson weed	*Datura stramonium*	12
Mosquito	*Culex pipiens*	3
Mustard plant	*Arabidopsis thaliana*	5
Pink bread mold	*Neurospora crassa*	7
Potato	*Solanum tuberosum*	24
Rhesus monkey	*Macaca mulatta*	21
Roundworm	*Caenorhabditis elegans*	6
Silkworm	*Bombyx mori*	28
Slime mold	*Dictyostelium discoidium*	7
Snapdragon	*Antirrhinum majus*	8
Tobacco	*Nicotiana tabacum*	24
Tomato	*Lycopersicon esculentum*	12
Water fly	*Nymphaea alba*	80
Wheat	*Triticum aestivum*	21
Yeast	*Saccharomyces cerevisiae*	16
Zebrafish	*Danio rerio*	25

Web Tutorial 2.1
Mitosis and the Cell Cycle

lengths, each called a **locus** (pl., loci). Thus, they are identical in their genetic potential. In sexually reproducing organisms, one member of each pair is derived from the maternal parent (through the ovum), and one is derived from the paternal parent (through the sperm). Therefore, each diploid organism contains two copies of each gene as a consequence of **biparental inheritance**. As we shall see in the following chapters on transmission genetics, the members of each pair of genes, while influencing the same characteristic or trait, need not be identical. In a population of members of the same species, many different alternative forms of the same gene, called **alleles**, can exist.

The conceptual issues of haploid number, diploid number, and homologous chromosomes are important in understanding the process of meiosis. During the formation of gametes or spores, meiosis converts the diploid number of chromosomes to the haploid number. As a result, haploid gametes or

spores contain precisely one member of each homologous pair of chromosomes—that is, one complete haploid set. Following fusion of two gametes in fertilization, the diploid number is reestablished; that is, the zygote contains two complete sets of haploid chromosomes, one set from each parent. The constancy of genetic material is thus maintained from generation to generation.

There is one important exception to the concept of homologous pairs of chromosomes. In many species, one pair, the **sex-determining chromosomes**, is often not homologous in size, centromere placement, arm ratio, or genetic content. For example, in humans, while females carry two homologous X chromosomes, males carry one Y chromosome and one X chromosome (Figure 2–4). These X and Y chromosomes are not strictly homologous. The Y is considerably smaller and lacks most of the gene sites contained on the X. Nevertheless, they contain homologous regions and behave as homologs in meiosis so that gametes produced by males receive either one X or one Y chromosome.

2.3 Mitosis Partitions Chromosomes into Dividing Cells

The process of mitosis is critical to all eukaryotic organisms. In some single-celled organisms, such as protozoans and some fungi and algae, mitosis (as a part of cell division) provides the basis for asexual reproduction. Multicellular diploid organisms begin life as single-celled fertilized eggs called **zygotes**. The mitotic activity of the zygote and the subsequent daughter cells is the foundation for the development and growth of the organism. In adult organisms, mitotic activity is prominent in wound healing and other forms of cell replacement in certain tissues. For example, the epidermal skin cells of humans are continuously sloughed off and replaced. Cell division also results in the continuous production of reticulocytes (immature red blood cells) that eventually shed their nuclei and replenish the supply of red blood cells in vertebrates. In abnormal situations, somatic cells may lose control of cell division and form a tumor.

The genetic material is partitioned into daughter cells during nuclear division or **karyokinesis**. This process is quite complex and requires great precision. The chromosomes must first be exactly replicated and then accurately partitioned. The end result is the production of two daughter nuclei, each with a chromosome composition identical to that of the parent cell.

Karyokinesis is followed by cytoplasmic division, or **cytokinesis**. The less complex division of the cytoplasm requires a mechanism that partitions the volume into two parts, then encloses both new cells within a distinct plasma membrane. Cytoplasmic organelles either replicate themselves, arise from existing membrane structures, or are synthesized *de novo* (anew) in each cell. The subsequent proliferation of these structures is a reasonable and adequate mechanism for reconstituting the cytoplasm in daughter cells.

Following cell division, the initial size of each new daughter cell is approximately one-half the size of the parent cell. However, the nucleus of each new cell is not appreciably smaller than the nucleus of the original cell. Quantitative measurements of DNA confirm that there is an amount of genetic material in the daughter nuclei equivalent to that in the parent cell.

Interphase and the Cell Cycle

Many cells undergo a continuous alternation between division and nondivision. The events that occur from the completion of one division until the beginning of the next division constitute the **cell cycle** (Figure 2–5). We will consider the initial **interphase** stage of the cycle as the interval between divisions. It was once thought that the biochemical activity during interphase was devoted solely to the cell's growth and its normal function. However, we now know that another biochemical step critical to the ensuing mitosis occurs during interphase: *replication of the DNA of each chromosome*. This period during which DNA is synthesized occurs before the cell enters mitosis and is called the **S phase**. The initiation and completion of synthesis can be detected by monitoring the incorporation of radioactive precursors into DNA.

Investigations of this nature show two periods during interphase when no DNA synthesis occurs, one before and one after S phase. These are designated **G1 (gap I)** and **G2 (gap II)**, respectively. During both of these intervals, as well as during S, intensive metabolic activity, cell growth, and cell differentiation occur. By the end of G2, the volume of the cell has roughly doubled, DNA has been replicated, and mitosis (M) is initiated. Following mitosis, continuously dividing cells then repeat this cycle (G1, S, G2, M) over and over, as shown in Figure 2–5.

Much is known about the cell cycle based on *in vitro* (in glass) studies. When grown in culture, many cell types in different organisms traverse the complete cycle in about 16 hours. The actual process of mitosis occupies only a small part of the overall cycle, often less than an hour. The lengths of the S and G2 phases of interphase are fairly consistent among different cell types. Most variation is seen in the length

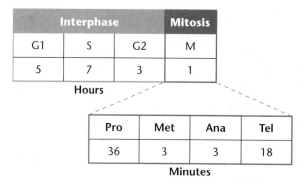

Interphase			Mitosis
G1	S	G2	M
5	7	3	1

Hours

Pro	Met	Ana	Tel
36	3	3	18

Minutes

FIGURE 2–6 The time spent in each phase of one complete cell cycle of a human cell in culture. Times vary according to cell types and conditions.

of time spent in the G1 stage. Figure 2–6 shows the length of these intervals in a typical cell.

G1 is of great interest in the study of cell proliferation and its control. At a point late in G1, all cells follow one of two paths. They either withdraw from the cycle, become quiescent, and enter the **G0 stage** (see Figure 2–5), or they become committed to initiating DNA synthesis and completing the cycle. Cells that enter G0 remain viable and metabolically active but are nonproliferative. Cancer cells apparently avoid entering G0 or pass through it very quickly. Other cells enter G0 and never reenter the cell cycle. Still others remain in G0, but they can be stimulated to return to G1 and thereby reenter the cycle.

Cytologically, interphase is characterized by the absence of visible chromosomes. Instead, the nucleus is filled with chromatin fibers that are formed as the chromosomes are uncoiled and dispersed after the previous mitosis [Figure 2–7(a)]. Once G1, S, and G2 are completed, mitosis is initiated. Mitosis is a dynamic period of vigorous and continual activity. For discussion purposes, the entire process is subdivided into discrete stages, and specific events are assigned to each one. These stages, in order of occurrence, are prophase, prometaphase, metaphase, anaphase, and telophase. They are diagrammed in Figure 2–7 along with photomicrographs of each stage.

Prophase

Often, over half of mitosis is spent in **prophase** [Figure 2–7(b)], a stage characterized by several significant activities. One of the early events in prophase of all animal cells involves the migration of two pairs of centrioles to opposite ends of the cell. These structures are found just outside the nuclear envelope in an area of differentiated cytoplasm called the **centrosome**. It is believed that each pair of centrioles consists of one mature unit and a smaller, newly formed centriole.

The direction of migration of the centrioles is such that two poles are established at opposite ends of the cell. Following their migration, the centrioles are responsible for organizing cytoplasmic microtubules into a series of spindle fibers that are formed and run between these poles to create an axis along which chromosomal separation occurs. Interestingly, the cells of most plants (there are a few exceptions), fungi, and certain algae seem to lack centrioles. Spindle fibers are

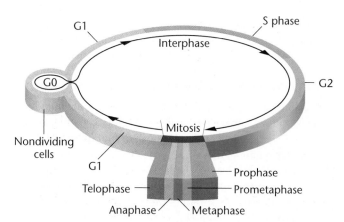

FIGURE 2–5 The intervals comprising an arbitrary cell cycle. Following mitosis, cells enter the G1 stage of interphase, initiating a new cycle. Cells may become nondividing (G0) or continue through G1, where they become committed to begin DNA synthesis (S) and complete the cycle (G2 and mitosis). Following mitosis, two daughter cells are produced and the cycle begins anew for both cells.

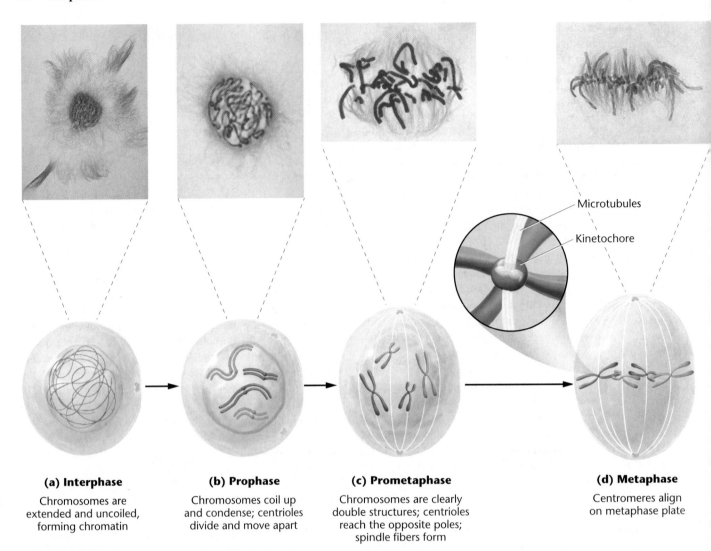

(a) Interphase

Chromosomes are
extended and uncoiled,
forming chromatin

(b) Prophase

Chromosomes coil up
and condense; centrioles
divide and move apart

(c) Prometaphase

Chromosomes are clearly
double structures; centrioles
reach the opposite poles;
spindle fibers form

(d) Metaphase

Centromeres align
on metaphase plate

FIGURE 2–7 Drawings depicting mitosis in an animal cell with a diploid number of 4. The events occurring in each stage are described in the text. Of the two homologous pairs of chromosomes, one contains longer, metacentric members and the other shorter, submetacentric members. The maternal chromosome and the paternal chromosome of each pair are shown in different colors. In (f), the late telophase stage in a plant cell illustrates the formation of the cell plate and lack of centrioles. The light micrographs illustrating each of the stages of mitosis are derived from the flower of *Haemanthus*, a plant that contains many more than 4 chromosomes.

nevertheless apparent during mitosis. Thus, centrioles are not universally responsible for the organization of spindle fibers.

As the centrioles migrate, the nuclear envelope begins to break down and gradually disappears. In a similar fashion, the nucleolus disintegrates within the nucleus. While these events are taking place, the diffuse chromatin fibers condense, continuing until distinct threadlike structures, the chromosomes, become visible. It becomes apparent near the end of prophase that each chromosome is actually a double structure split longitudinally except at the centromere (review Figure 2–4). The two parts of each chromosome are called **chromatids**. Because the DNA contained in each pair of chromatids represents the duplication of a single chromosome, these sister chromatids are genetically identical. Therefore, they are called sister chromatids. In humans, with a diploid number of 46, a cytological preparation of late prophase reveals 46 chromosomes

randomly distributed in the area formerly occupied by the nucleus.

Prometaphase and Metaphase

The distinguishing event of the ensuing stages is the migration of each chromosome, led by the centromeric region, to the equatorial plane. The equatorial plane, or **metaphase plate**, is the midline region of the cell, a plane that lies perpendicular to the axis established by the spindle fibers. In some descriptions, the term **prometaphase** refers to the period of chromosome movement [Figure 2–7(c)], and **metaphase** is applied strictly to the chromosome configuration after migration.

Migration is made possible by the binding of spindle fibers to a structure associated with the centromere of each chromosome called the **kinetochore**. This structure, consisting of multilayered plates of proteins, forms on opposite sides of each centromere, intimately associating with the two sister

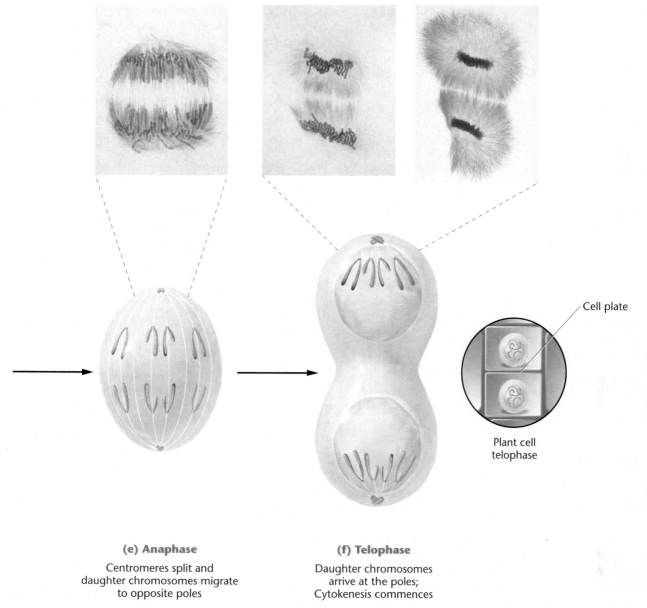

(e) Anaphase

Centromeres split and
daughter chromosomes migrate
to opposite poles

(f) Telophase

Daughter chromosomes
arrive at the poles;
Cytokenesis commences

FIGURE 2–7 (Continued)

chromatids of each chromosome. Once attached to micro-tubules making up the spindle fibers, the sister chromatids are now ready to be pulled to opposite poles during the ensuing anaphase stage.

At the completion of metaphase, each centromere is aligned at the metaphase plate, with the chromosome arms extending outward in a random array. This configuration is shown in Figure 2–7(d).

Anaphase

Events critical to chromosome distribution during mitosis occur during **anaphase**, the shortest stage of mitosis. During this phase, sister chromatids of each chromosome *disjoin* (separate) from each other and migrate to opposite ends of the cell. For complete disjunction to occur, each centromeric region must be split in two. This event signals the initiation of

anaphase. Once it occurs, each chromatid is referred to as a **daughter chromosome**.

Movement of daughter chromosomes to the opposite poles of the cell is dependent on the centromere–spindle fiber attachment. Recent investigations reveal that chromosome migration results from the activity of a series of specific pro-teins, generally called motor proteins. These proteins use the energy generated by the hydrolysis of ATP, and their activity is said to constitute **molecular motors** in the cell. These motors act at several positions within the dividing cell, but all of them are involved in the activity of microtubules and ultimately serve to propel the chromosomes to opposite ends of the cell. The centromeres of each chromosome *appear* to lead the way during migration, with the chromosome arms trailing behind. The location of the centromere determines the shape of the chromosome during separation, as you saw in Figure 2–3.

The steps that occur during anaphase are critical in providing each subsequent daughter cell with an identical set of chromosomes. In human cells, there would now be 46 chromosomes at each pole, one from each original sister pair. Figure 2–7(e) shows anaphase just prior to its completion.

Now Solve This

With the initial appearance of the feature we call "Now Solve This," a short introduction is in order. Occurring several times in this and all ensuing chapters, each entry identifies a problem from the Problems and Discussion Questions section at the end of the chapter. Each selection is related to the discussion just presented. A comment is made about the problem and then a Hint is offered. Each hint provides you with analytical insight that will be useful as you solve the problem. Here is the first one:

Problem 5 on page 36 involves an understanding of what happens to each pair of homologous chromosomes during mitosis.

Hint: The major issue in solving this problem is to understand that, throughout mitosis, members of each homologous pair do not pair up but instead behave individually.

Telophase

Telophase is the final stage of mitosis and is depicted in Figure 2–7(f). At its beginning, there are two complete sets of chromosomes, one set at each pole. The most significant event is **cytokinesis**, the division or partitioning of the cytoplasm. Cytokinesis is essential if two new cells are to be produced from one cell. The mechanism differs greatly in plant and animal cells. In plant cells, a **cell plate** is synthesized and laid down across the region of the metaphase plate. Animal cells, however, undergo a constriction of the cytoplasm in the same way a loop of string might be tightened around the middle of a balloon. The end result is the same: Two distinct cells are formed.

It is not surprising that the process of cytokinesis varies among cells of different organisms. Plant cells, which are more regularly shaped and are structurally rigid, require a mechanism for depositing new cell wall material around the plasma membrane. The cell plate, laid down during telophase, becomes the **middle lamella**. Subsequently, the primary and secondary layers of the cell wall are deposited between the cell membrane and middle lamella on both sides of the boundary between the two daughter cells. In animals, complete constriction of the cell membrane produces a **cell furrow** characteristic of newly divided cells.

Other events necessary for the transition from mitosis to interphase are initiated during late telophase. They are generally a reversal of the events that occurred during prophase. In each new cell, the chromosomes begin to uncoil and become diffuse chromatin once again while the nuclear envelope reforms around them. The spindle fibers disappear, and the nucleolus gradually reforms and becomes visible in the nucleus during early interphase. At the completion of telophase, the cell enters interphase.

2.4 Meiosis Reduces the Chromosome Number from Diploid to Haploid in Germ Cells and Spores

The process of meiosis, unlike mitosis, reduces the amount of genetic material to half. Whereas in diploids, mitosis produces daughter cells with a full diploid complement, meiosis produces gametes or spores with only one haploid set of chromosomes. During sexual reproduction, gametes then combine at fertilization to reconstitute the diploid complement found in parental cells. Figure 2–8 compares the two processes by following two pairs of homologous chromosomes.

Meiosis must be highly specific since, by definition, haploid gametes or spores contain precisely one member of each homologous pair of chromosomes. If it is successfully completed, meiosis ensures genetic continuity from generation to generation.

The process of sexual reproduction also ensures genetic variety among members of a species. As you study meiosis, you will see that this process results in gametes with many unique combinations of maternally and paternally derived chromosomes among the haploid complement. With such a tremendous genetic variation among the gametes, a large number of chromosome combinations are possible at fertilization. Furthermore, the meiotic event referred to as **crossing over** results in genetic exchange between members of each homologous pair of chromosomes. This creates intact chromosomes that are mosaics of the maternal and paternal homologs from which they arise, further enhancing the potential genetic variation in gametes and the offspring derived from them. Sexual reproduction therefore reshuffles the genetic material, producing offspring that often differ greatly from either parent. Thus, meiosis is the major form of genetic recombination within species.

An Overview of Meiosis

In the preceding discussion, we established what might be considered the goals of meiosis. Before we consider the phases of this process systematically, we will briefly examine how diploid cells give rise to haploid gametes or spores. You should refer to the meiosis I side of Figure 2–8 during the following discussion.

You have seen that in mitosis each paternally and maternally derived member of any homologous pair of chromosomes behaves autonomously during division. By contrast, early in meiosis, homologous chromosomes form pairs; that is, they synapse. Each synapsed structure is initially called a **bivalent**, which eventually gives rise to a unit, the **tetrad**, consisting of four chromatids. The presence of four chromatids demonstrates that *both* homologs (making up the bivalent) have, in fact, duplicated. Therefore, in order to achieve haploidy, two divisions are necessary.

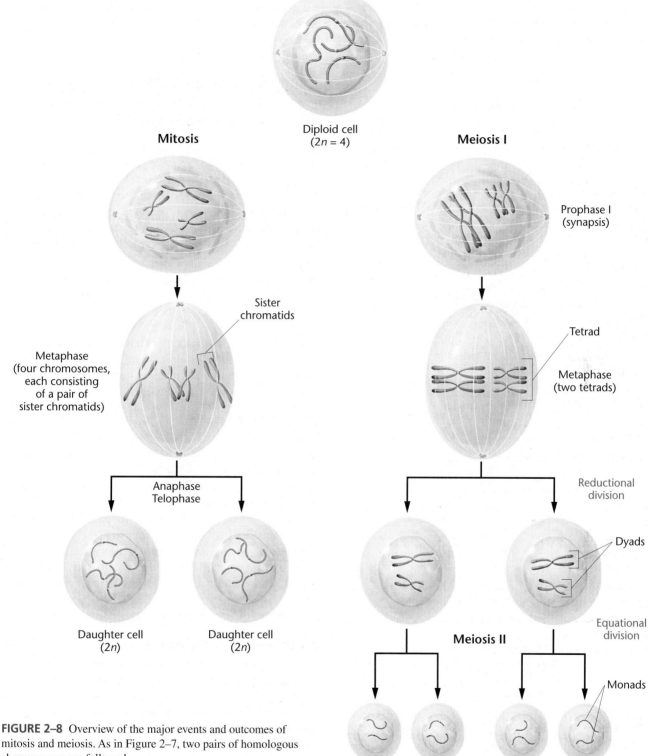

FIGURE 2–8 Overview of the major events and outcomes of mitosis and meiosis. As in Figure 2–7, two pairs of homologous chromosomes are followed.

The first division occurs in meiosis I and is described as a **reductional division** (because the number of centromeres, each representing one chromosome, is *reduced* by one-half). Components of each tetrad—representing the two homologs—separate, yielding two **dyads**. Each dyad is composed of two sister chromatids joined at a common centromere. The second division occurs during meiosis II and is described as an **equational division** (because the number of centromeres

remains *equal*). Here each dyad splits into two **monads** of one chromosome each. Thus, the two divisions potentially produce four haploid cells.

The First Meiotic Division: Prophase I

We turn now to a more detailed account of meiosis. As in mitosis, meiosis is a continuous process. We name the parts of each stage of division only to facilitate discussion. From a

Meiotic prophase I

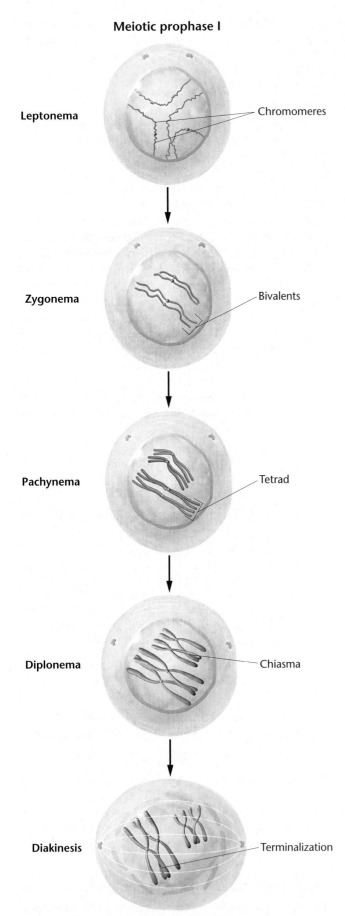

Leptonema ——————— Chromomeres

Zygonema ——————— Bivalents

Pachynema ——————— Tetrad

Diplonema ——————— Chiasma

Diakinesis ——————— Terminalization

FIGURE 2–9 The substages of meiotic prophase I for the chromosomes depicted in Figure 2–8.

genetic standpoint, three events characterize the initial stage, **prophase I** (Figure 2–9). First, as in mitosis, the chromatin present in interphase condenses and coils into visible chromosomes. Second, unlike mitosis, members of each homologous pair of chromosomes undergo **synapsis**. Third, crossing over occurs between synapsed homologs. Because of the complexity of these genetic events, prophase I is further divided into five sub-stages: leptonema, zygonema, pachynema, diplonema,* and diakinesis. As we discuss these substages, be aware that even though it is not immediately apparent in the earliest phases of meiosis, the DNA of chromosomes has already been replicated during the prior interphase.

Leptonema During the **leptotene stage**, the interphase chromatin material begins to condense, and the chromosomes, though still extended, become visible. Along each chromosome are **chromomeres**, localized condensations that resemble beads on a string. Recent evidence suggests that a process called homology search, which precedes and is essential to the initial pairing of homologs, begins during leptonema.

Zygonema The chromosomes continue to shorten and thicken during the **zygotene stage**. During the **homology search**, homologous chromosomes undergo initial alignment with one another. This so-called rough-pairing is complete by the end of zygonema. In yeast, homologs are separated by about 300 nm, and near the end of zygonema, structures called lateral elements are visible between paired homologs. As meiosis proceeds, the overall length of the lateral elements along the chromosome increases and a more extensive ultrastructural component, the **synaptonemal complex**, begins to form between the homologs.

At the completion of zygonema, the paired homologs are referred to as bivalents. Although both members of each bivalent have already replicated their DNA, it is not yet visually apparent that each member is a double structure. The number of bivalents in each species is equal to the haploid (*n*) number.

Pachynema In the transition from the zygotene to the **pachytene stage**, the chromosomes continue to coil and shorten, and further development of the synaptonemal complex occurs between the two members of each bivalent. This leads to synapsis, a more intimate pairing. Compared to the rough-pairing characteristic of yeast zygonema, the homologs are now separated by only 100 nm.

During pachynema, each homolog is first evident as a double structure, providing visual evidence of the earlier replication of the DNA of each chromosome. Thus, each bivalent contains four chromatids. As in mitosis, replicates are called sister chromatids, while chromatids from maternal and paternal members of a homologous pair are called nonsister chromatids. The four-membered structure is a tetrad, and each tetrad contains two pairs of sister chromatids.

*These are the noun forms of these substages. The adjective forms (leptotene, zygotene, pachytene, and diplotene) are also used.

Diplonema During the ensuing **diplotene stage**, it is even more apparent that each tetrad consists of two pairs of sister chromatids. Within each tetrad, each pair of sister chromatids begins to separate. However, one or more areas remain in contact where chromatids are intertwined. Each such area, called a **chiasma** (pl., chiasmata), is thought to represent a point where nonsister chromatids have undergone genetic exchange through the process of crossing over. Although the physical exchange between chromosome areas occurred during the previous pachytene stage, the result of crossing over is visible only when the duplicated chromosomes begin to separate. Crossing over is an important source of genetic variability, and new combinations of genetic material are formed during this process.

Diakinesis The final stage of prophase I is **diakinesis**. The chromosomes pull farther apart, but the nonsister chromatids remain loosely associated via the chiasmata. As separation proceeds, the chiasmata move toward the ends of the tetrad. This process of **terminalization** begins in late diplonema and is completed during diakinesis. During this final substage period, the nucleolus and nuclear envelope break down, and the two centromeres of each tetrad attach to the recently formed spindle fibers. By the completion of prophase I, the centromeres of each tetrad structure are present on the metaphase plate of the cell.

Metaphase, Anaphase, and Telophase I

The remainder of the meiotic process is depicted in Figure 2–10. After meiotic prophase I, steps similar to those of mitosis occur. In the first division, **metaphase I**, the chromosomes have maximally shortened and thickened. The terminal chiasmata of each tetrad are visible and appear to be the only factor holding the nonsister chromatids together. Each tetrad interacts with spindle fibers, facilitating movement to the metaphase plate. The alignment of each tetrad prior to the first anaphase is random. Half of each tetrad is pulled randomly to one or the other pole, and the other half then moves to the opposite pole.

During the stages of meiosis I, a single centromere holds each pair of sister chromatids together. It does *not* divide. At **anaphase I**, one-half of each tetrad (the dyad) is pulled toward each pole of the dividing cell. This separation process is the physical basis of **disjunction**, the separation of chromosomes from one another. Occasionally, errors in meiosis occur and separation is not achieved. The term **nondisjunction** describes such an error. At the completion of a normal anaphase I, a series of dyads equal to the haploid number is present at each pole.

If crossing over had not occurred in the first meiotic prophase, each dyad at each pole would consist solely of either paternal or maternal chromatids. However, the exchanges produced by crossing over create mosaic chromatids of paternal and maternal origin.

In many organisms, **telophase I** reveals a nuclear membrane forming around the dyads. Next, the nucleus enters into a short interphase period. If interphase occurs, the chromosomes do not replicate since they already consist of two chro-

matids. In other organisms, the cells go directly from anaphase I to meiosis II. In general, meiotic telophase is much shorter than the corresponding stage in mitosis.

The Second Meiotic Division

A second division, **meiosis II**, is essential if each gamete or spore is to receive only one chromatid from each original tetrad. The stages characterizing meiosis II are shown in the right half of Figure 2–10. During **prophase II**, each dyad is composed of one pair of sister chromatids attached by a common centromere. During **metaphase II**, the centromeres are positioned on the metaphase plate. When they divide, **anaphase II** is initiated, and the sister chromatids of each dyad are pulled to opposite poles. Because the number of dyads is equal to the haploid number, **telophase II** reveals one member of each pair of homologous chromosomes at each pole. Each chromosome is now a monad. Following cytokinesis in telophase II, four haploid gametes may result from a single meiotic event. At the conclusion of meiosis II, not only has the haploid state been achieved but if crossing over has occurred, each monad is a combination of maternal and paternal genetic information. As a result, the offspring produced by any gamete receives a mixture of genetic information originally present in his or her grandparents. Meiosis thus significantly increases the level of genetic variation in each ensuing generation.

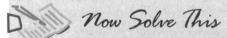

Now Solve This

Problem 14 on page 36 involves an understanding of what happens to the maternal and paternal members of each pair of homologous chromosomes during meiosis.

Hint: The major issue in solving this problem is to understand that maternal and paternal homologs synapse during meiosis. Once it is evident that each chromatid has duplicated, creating a tetrad in the early phases of meiosis, each original pair behaves as a unit and leads to two dyads during anaphase I.

2.5 The Development of Gametes Varies during Spermatogenesis and Oogenesis

Although events that occur during the meiotic divisions are similar in all cells that participate in gametogenesis in most animal species, there are certain differences between the production of a male gamete (spermatogenesis) and a female gamete (oogenesis). Figure 2–11 summarizes these processes.

Spermatogenesis takes place in the testes, the male reproductive organs. The process begins with the expanded growth of an undifferentiated diploid germ cell called a **spermatogonium**. This cell enlarges to become a **primary spermatocyte**, which undergoes the first meiotic division. The products of

FIGURE 2–10 The major events in meiosis in an animal with a diploid number of 4, beginning with metaphase I. Note that the combination of chromosomes in the cells produced following telophase II is dependent on the random alignment of each tetrad and dyad on the equatorial plate during metaphase I and metaphase II. Several other combinations, which are not shown, can also be formed. The events depicted here are described in the text.

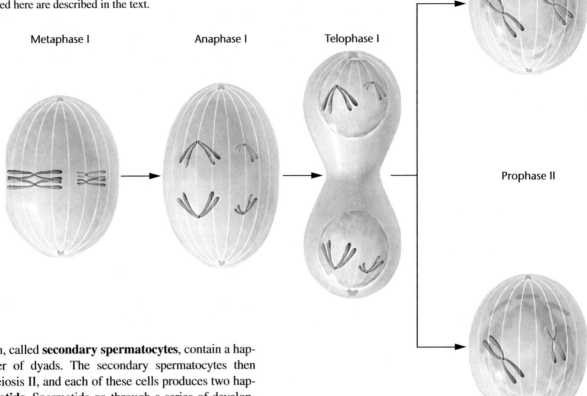

Metaphase I   Anaphase I   Telophase I   Prophase II

this division, called **secondary spermatocytes**, contain a haploid number of dyads. The secondary spermatocytes then undergo meiosis II, and each of these cells produces two haploid **spermatids**. Spermatids go through a series of developmental changes, **spermiogenesis**, and become highly specialized, motile **spermatozoa** or **sperm**. All sperm cells produced during spermatogenesis contain the haploid number of chromosomes and equal amounts of cytoplasm.

Spermatogenesis may be continuous or may occur periodically in mature male animals; its onset is determined by the species' reproductive cycle. Animals that reproduce year-round produce sperm continuously, whereas those whose breeding period is confined to a particular season produce sperm only during that time.

In animal **oogenesis**, the formation of **ova** (sing., ovum), or eggs, takes place in the ovaries, the female reproductive organs. The daughter cells resulting from the two meiotic divisions receive equal amounts of genetic material, but they do *not* receive equal amounts of cytoplasm. Instead, during each division almost all the cytoplasm of the **primary oocyte**, which is derived from the **oogonium**, is concentrated in one of the two daughter cells. This concentration of cytoplasm is necessary because a major function of the mature ovum is to nourish the developing embryo after fertilization.

During anaphase I in oogenesis, the tetrads of the primary oocyte separate, and the dyads move toward opposite poles. During telophase I, the dyads at one pole are pinched off with very little surrounding cytoplasm to form the **first polar body**. The first polar body may or may not divide again to produce two small haploid cells. The other daughter cell produced by this first meiotic division contains most of the cyto-

plasm and is called the **secondary oocyte**. The mature ovum will be produced from the secondary oocyte during the second meiotic division. During this division, the cytoplasm of the secondary oocyte again divides unequally, producing an **ootid** and a **second polar body**. The ootid then differentiates into the mature ovum.

Unlike the divisions of spermatogenesis, the two meiotic divisions of oogenesis may not be continuous. In some animal species, the two divisions may directly follow each other. In others, including humans, the first division of all oocytes begins in the embryonic ovary but arrests in prophase I. Many years later, meiosis resumes in each oocyte just prior to its ovulation. The second division is completed only after fertilization.

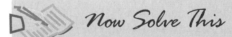

Now Solve This

Problem 9 on page 36 involves an understanding of meiosis during oogenesis.

Hint: To answer this question, you must take into account that crossing over occurred during meiosis I between each pair of homologs.

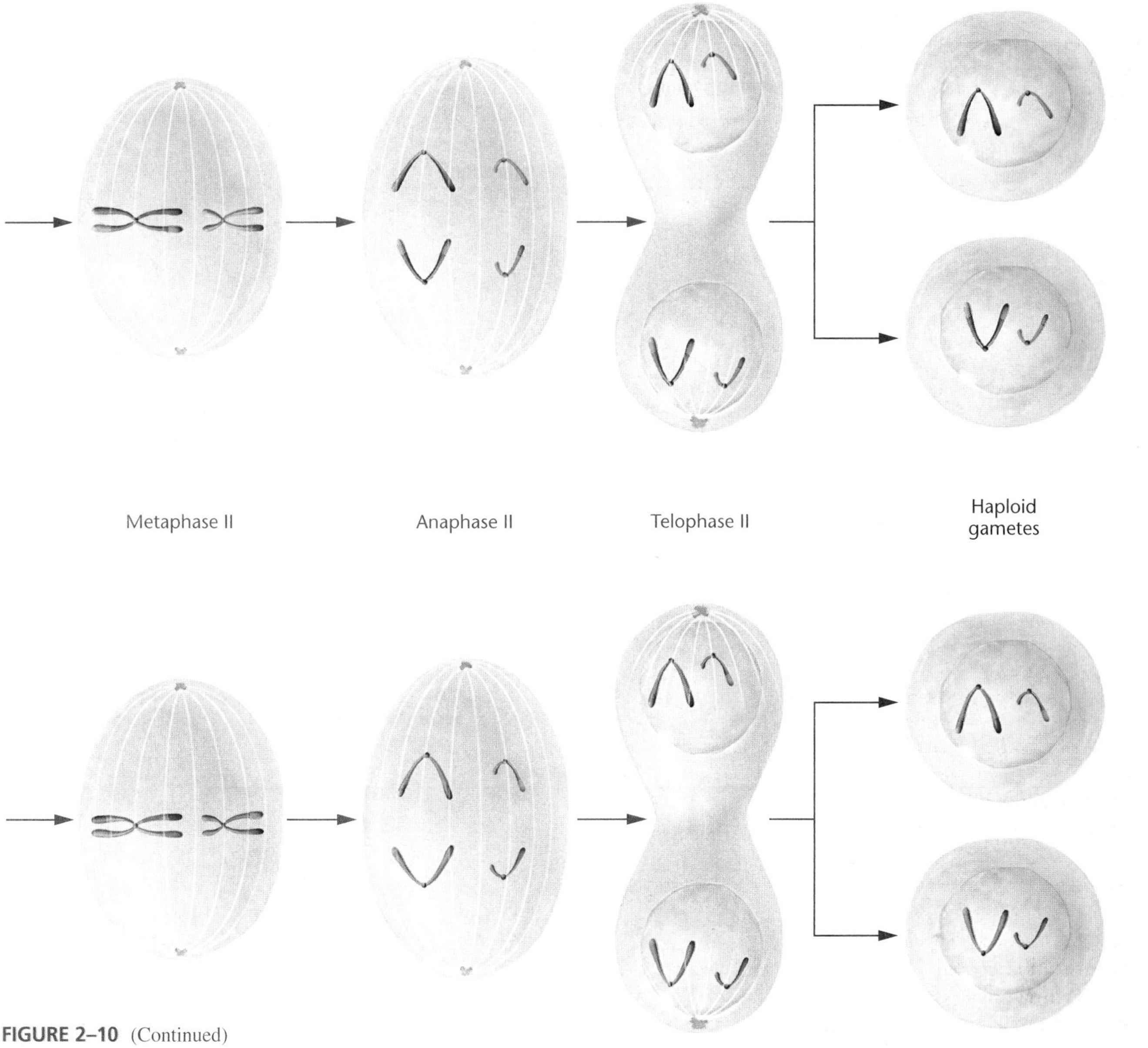

Metaphase II   Anaphase II   Telophase II   Haploid gametes

FIGURE 2–10 (Continued)

2.6 Meiosis Is Critical to the Successful Sexual Reproduction of All Diploid Organisms

The process of meiosis is critical to the successful sexual reproduction of all diploid organisms. It is the mechanism by which the diploid amount of genetic information is reduced to the haploid amount. In animals, meiosis leads to the formation of gametes, whereas in plants, haploid spores are produced, which in turn leads to the formation of haploid gametes.

Each diploid organism contains its genetic information in the form of homologous pairs of chromosomes. Each pair consists of one member derived from the maternal parent and one from the paternal parent. Following meiosis, haploid cells potentially contain either the paternal or maternal representative of each homologous pair of chromosomes. Howev-er, the process of crossing over, which occurs in the first meiotic prophase, reshuffles alleles between the maternal and paternal members of each homologous pair, which then segregate and assort independently into gametes. This results in the great amounts of genetic variation in gametes.

It is important to touch briefly on the significant role that meiosis plays in the life cycles of fungi and plants. In many fungi, the predominant stage of the life cycle consists of haploid vegetative cells. They arise through meiosis and proliferate by mitotic cell division. In multicellular plants, the life cycle alternates between the diploid **sporophyte stage** and the haploid **gametophyte stage** (Figure 2–12). While one or the other predominates in different plant groups during this "alternation of generations," the processes of meiosis and fertilization constitute the "bridges" between the sporophyte and gametophyte stages. Therefore, meiosis is an essential component of the life cycle of plants.

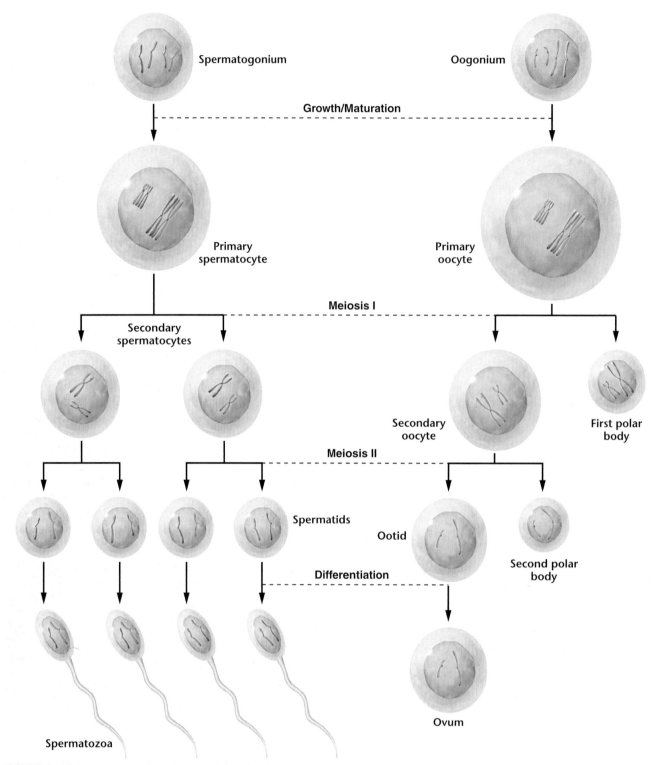

FIGURE 2–11 Spermatogenesis and oogenesis in animal cells.

2.7 Electron Microscopy Has Revealed the Cytological Nature of Mitotic and Meiotic Chromosomes

Thus far in this chapter, we have focused on mitotic and meiotic chromosomes, emphasizing their behavior during cell division and gamete formation. An interesting question is why chromosomes are invisible during interphase but visible during the various stages of mitosis and meiosis. Studies using electron microscopy clearly show why this is the case.

Recall that during interphase, only dispersed chromatin fibers are present in the nucleus [Figure 2–13(a)]. Once mitosis begins, however, the fibers coil and fold, condensing into typical metaphase chromosomes [Figure 2–13(b)]. If the fibers comprising the mitotic chromosome are loosened, the areas of greatest spreading reveal individual fibers similar to those seen in

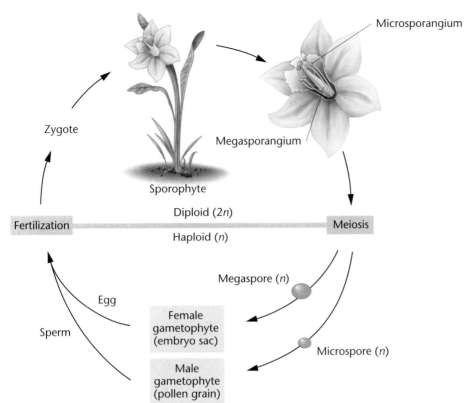

FIGURE 2–12 Alternation of generations between the diploid sporophyte ($2n$) and the haploid gametophyte (n) in a multicellular plant. The processes of meiosis and fertilization bridge the two phases of the life cycle. This is an angiosperm, where the sporophyte stage is the predominant phase.

interphase chromatin [Figure2–13(c)]. Very few fiber ends seem to be present, and in some cases, none can be seen. Instead, individual fibers always seem to loop back into the interior where they are twisted and coiled around one another, forming the regular pattern of the mitotic chromosome. Starting in late telophase of mitosis and continuing during G1 of interphase, chromosomes unwind to form the long fibers characteristic of chromatin, which consist of DNA and associated proteins, particularly proteins called histones. It is in this physical arrangement that DNA can most efficiently function during transcription and replication.

Electron microscopic observations of metaphase chromosomes in varying states of coiling led Ernest DuPraw to postulate the **folded-fiber model** shown in Figure 2–13(d). During metaphase, each chromosome consists of two sister chromatids joined at the centromeric region. Each arm of the chromatid appears to be a single fiber wound much like a skein of yarn. The fiber is composed of tightly coiled double-stranded DNA and protein. An orderly coiling-twisting-condensing process appears to be involved in the transition of the interphase chromatin to the more condensed, mitotic chromosomes. Geneticists believe that during the transition from interphase to prophase, a 5000-fold compaction occurs in the length of DNA within the chromatin fiber!

FIGURE 2–13
Comparison of (a) the chromatin fibers characteristic of the interphase nucleus with (b) and (c) metaphase chromosomes that are derived from chromatin during mitosis.
Part (d) diagrams the mitotic chromosome and its various components, showing how chromatin is condensed into it.
Parts (a) and (c) are transmission electron micrographs, while part (b) is a scanning electron micrograph.

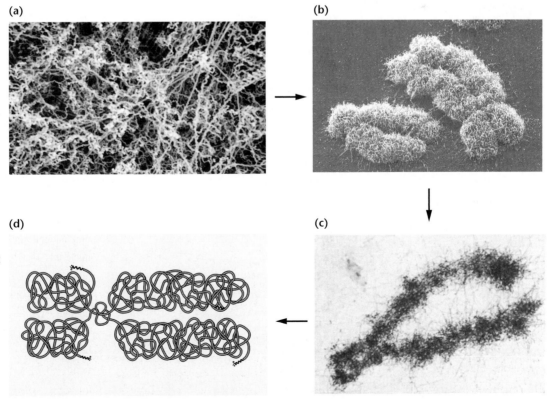

CHAPTER SUMMARY

1. The structure of cells is elaborate and complex. Many components of cells are involved directly or indirectly with genetic processes.

2. In diploid organisms, chromosomes exist in homologous pairs. Each pair shares the same size, centromere placement, and gene sites. One member of each pair is derived from the maternal parent and one is derived from the paternal parent.

3. Mitosis and meiosis are mechanisms by which cells distribute the genetic information contained in their chromosomes to progeny cells in a precise, orderly fashion.

4. Mitosis is but one part of the cell cycle, which is characteristic of all eukaryotes. The cell cycle consists of the stages G1, S, and G2, which precede mitosis.

5. Mitosis, or nuclear division, is the basis of cellular reproduction. Daughter cells are produced that are genetically identical to their progenitor cell.

6. Mitosis may be subdivided into discrete stages: prophase, prometaphase, metaphase, anaphase, and telophase. Condensation of chromatin into chromosome structures occurs during prophase. During prometaphase, chromosomes appear as double structures, each composed of a pair of sister chromatids. In metaphase, chromosomes line up on the metaphase plate of the cell. During anaphase, sister chromatids of each chromosome are pulled apart and directed toward opposite poles. Daughter-cell formation is completed at telophase and is characterized by cytokinesis, the division of the cytoplasm.

7. Meiosis, the underlying basis of sexual reproduction, results in the conversion of a diploid cell into a haploid gamete or spore. As a result of chromosome duplication and two subsequent divisions, each haploid cell receives one member of each homologous pair of chromosomes.

8. A major difference in meiosis exists between males and females. Spermatogenesis partitions the cytoplasmic volume equally and produces four haploid sperm cells. Oogenesis, on the other hand, accumulates the cytoplasm in one egg cell and reduces the other haploid sets of genetic material to polar bodies. The extra cytoplasm contributes to zygote development after fertilization.

9. Meiosis results in extensive genetic variation by virtue of the exchange during crossing over between maternal and paternal chromatids and their random segregation into gametes. In addition, meiosis plays an important role in the life cycles of fungi and plants, serving as the bridge between alternating generations.

10. Mitotic chromosomes are produced as a result of the coiling and condensation of the chromatin fibers that are characteristic of interphase.

KEY TERMS

AB antigens, 18

acrocentric, 20

allele, 22

anaphase, 25

anaphase I, 29

anaphase II, 29

biparental inheritance, 22

bivalent, 26

cell coat, 18

cell cycle, 23

cell furrow, 26

cell plate, 26

cell wall, 18

centriole, 20

centromere, 20

centrosome, 23

chiasma, 29

chloroplast, 19

chromatid, 24

chromatin, 18

chromomere, 28

chromosome, 18

crossing over, 26

cytokinesis, 22

cytoplasm, 19

daughter chromosome, 25

diakinesis, 29

diploid number (2n), 20

diplotene stage, 29

disjunction, 29

dyad, 27

endoplasmic reticulum (ER), 19

endosymbiont hypothesis, 20

equational division, 27

first polar body, 30

folded-fiber model, 33

gamete, 18

gametophyte stage, 31

genome, 21

G1 (gap I), 23

G2 (gap II), 23

G0 stage, 23

histocompatibility antigen, 18

homologous chromosome, 20

homology search, 28

interphase, 23

karyokinesis, 22

karyotype, 21

kinetochore, 24

leptotene stage, 28

locus, 22

meiosis, 18

meiosis II, 29

metacentric, 20

metaphase, 24

metaphase I, 29

metaphase II, 29

metaphase plate, 24

middle lamella, 26

mitochondria, 19

mitosis, 18

MN antigens, 18

molecular motor, 25

monad, 27

nondisjunction, 29

nucleoid, 19

nucleolus, 18

nucleolus organizer region (NOR), 19

nucleus, 18

oogenesis, 30

oogonium, 30

ootid, 30

INSIGHTS AND SOLUTIONS

With this initial appearance of "Insights and Solutions," it is appropriate to describe its value to you as a student. This section precedes the "Problems and Discussion Questions" in each chapter; it provides sample problems and solutions that demonstrate approaches useful in genetic analysis. The insights you gain will help you arrive at correct solutions to ensuing problems.

1. In an organism with a diploid number of $2n = 6$, how many individual chromosomal structures will align on the metaphase plate during (a) mitosis, (b) meiosis I, and (c) meiosis II? Describe each configuration.

Solution:

(a) In mitosis, where homologous chromosomes do not synapse, there will be six double structures, each consisting of a pair of sister chromatids. The number of structures is equivalent to the diploid number.

(b) In meiosis I, the homologs have synapsed, reducing the number of structures to three. Each is a tetrad and consists of two pairs of sister chromatids.

(c) In meiosis II, the same number of structures exist (three), but in this case they are dyads. Each dyad is a pair of sister chromatids. When crossing over has occurred, each chromatid may contain parts of one of its nonsister chromatids obtained during exchange in prophase I.

2. Disregarding crossing over, draw all possible alignment configurations that can occur during metaphase I for the chromosomes shown in Figure 2–10.

Solution: As shown in the diagram below, four cases are possible when $n = 2$.

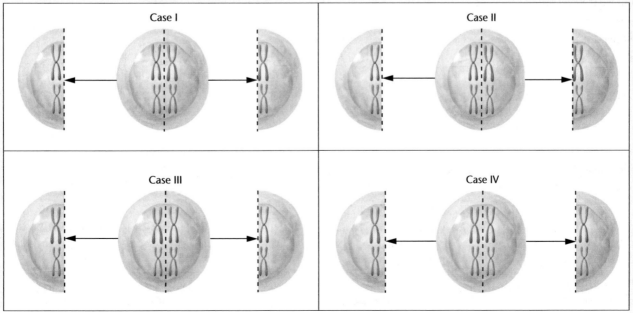

Case I   Case II

Case III   Case IV

Solution for #2

(Cont. on the next page)

3. For the chromosomes in the previous problem, assume one gene is present on both of the larger chromosomes with two alleles, *A* and *a*, as shown. Also assume a second gene with two alleles (*B*, *b*) is present on the smaller chromosomes. Calculate the probability of generating each gene combination (*AB, Ab, aB, ab*) following meiosis I.

Solution: As shown in the accompanying diagram,

Case I	*AB* and *ab*
Case II	*Ab* and *aB*
Case III	*aB* and *Ab*
Case IV	*ab* and *AB*

Total:		
AB = 2	(*p* = 1/4)
Ab = 2	(*p* = 1/4)
aB = 2	(*p* = 1/4)
ab = 2	(*p* = 1/4)

4. Describe the composition of a meiotic tetrad as it exists during prophase I, assuming no crossover event has occurred. What impact would a single crossover event have on this structure?

Solution: Such a tetrad contains four chromatids, existing as two pairs. Members of each pair are sister chromatids. They are held together by a common centromere. Members of one pair are maternally derived, whereas members of the other are paternally derived. Maternal and paternal members are nonsister chromatids. A single crossover event has the effect of exchanging a portion of a maternal *and* a paternal chromatid, leading to a chiasma, where the two chromatids overlap physically in the tetrad.

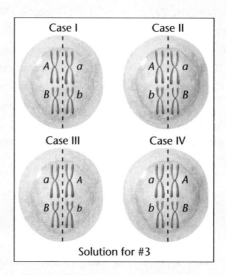

Solution for #3

PROBLEMS AND DISCUSSION QUESTIONS

1. What role do the following cellular components play in the storage, expression, or transmission of genetic information: (a) chromatin, (b) nucleolus, (c) ribosome, (d) mitochondrion, (e) centriole, (f) centromere?

2. Discuss the concepts of homologous chromosomes, diploidy, and haploidy. What characteristics are shared between two homologous chromosomes?

3. If two chromosomes of a species are the same length and have similar centromere placements, yet are not homologous, what is different about them?

4. Describe the events that characterize each stage of mitosis.

5. If an organism has a diploid number of 16, how many chromatids are visible at the end of mitotic prophase? How many chromosomes are moving to each pole during anaphase of mitosis?

6. How are chromosomes named on the basis of their centromere placement?

7. Contrast telophase in plant and animal mitosis.

8. Describe the phases of the cell cycle and the events that characterize each phase.

9. Examine Figure 2–11, which shows oogenesis in animal cells. Will the genotype of the second polar body (derived from meiosis II) always be identical to that of the ootid? Why or why not?

10. Contrast the end results of meiosis with those of mitosis.

11. Define and discuss these terms: (a) synapsis, (b) bivalent, (c) chiasmata, (d) crossing over, (e) chromomere, (f) sister chromatids, (g) tetrad, (h) dyad, (i) monad.

12. Contrast the genetic content and the origin of sister versus nonsister chromatids during their earliest appearance in prophase I of meiosis. How might the genetic content of these change by the time tetrads have aligned at the metaphase plate during metaphase I?

13. Given the end results of the two types of division, why is it necessary for homologs to pair during meiosis and not desirable for them to pair during mitosis?

14. An organism has a diploid number of 16 in a primary oocyte. (a) How many tetrads are present in prophase I? (b) How many dyads are present in prophase II? (c) How many monads migrate to each pole during anaphase II?

15. Contrast spermatogenesis and oogenesis. What is the significance of the formation of polar bodies?

16. Explain why meiosis leads to significant genetic variation while mitosis does not.

17. A diploid cell contains three pairs of homologous chromosomes designated C1 and C2, M1 and M2, and S1 and S2; no crossing over occurs. What possible combinations of chromosomes will be present in (a) daughter cells following mitosis, (b) the first meiotic metaphase, and (c) haploid cells following both divisions of meiosis?

18. Considering the preceding problem, predict the number of different haploid cells that will occur if a fourth chromosome pair (W1 and W2) is added.

19. During oogenesis in an animal species with a haploid number $n = 6$, one dyad undergoes nondisjunction during meiosis II. After the second meiotic division, the dyad ends up intact in the ovum. How many chromosomes are present in (a) the mature

ovum and (b) the second polar body? (c) Following fertilization by a normal sperm, what chromosome condition is created?

20. What is the probability that, in an organism with a haploid number of 10, a sperm will be formed which contains all 10 chromosomes whose centromeres were derived from maternal homologs?

21. During the first meiotic prophase, (a) when does crossing over occur; (b) when does synapsis occur; (c) during which stage are the chromosomes least condensed; and (d) when are chiasmata first visible?

22. Describe the role of meiosis in the life cycle of a plant.

23. Contrast the chromatin fiber with the mitotic chromosome. How are the two structures related?

24. Describe the folded-fiber model of the mitotic chromosome.

25. You are given a metaphase chromosome preparation (a slide) from an unknown organism that contains 12 chromosomes. Two are clearly smaller than the rest, appearing identical in length and centromere placement. Describe all that you can about these two chromosomes.

For Problems 26–31, consider a diploid cell that contains three pairs of chromosomes designated AA, BB, and CC. Each pair contains a maternal and a paternal member (e.g., A^m and A^p, etc.). Using these designations, demonstrate your understanding of mitosis and meiosis by drawing chromatid combinations as requested. Be sure to indicate when chromatids are paired as a result of replication and/or synapsis. You may wish to use a large piece of brown manila wrapping paper or a large cut-up paper bag and work with another student as you deal with these problems. Such cooperative learning may be a useful approach as you solve problems throughout the text.

26. In mitosis, what chromatid combination(s) will be present during metaphase? What combination(s) will be present at each pole at the completion of anaphase?

27. During meiosis I, assuming no crossing over, what chromatid combination(s) will be present at the completion of prophase? Draw all possible alignments of chromatids as migration begins during early anaphase.

28. Are there any possible combinations present during prophase of meiosis II other than those you drew in Problem 27? If so, draw them.

29. Draw all possible combinations of chromatids during anaphase in meiosis II.

30. Assume that during meiosis I, none of the *C* chromosomes disjoin at metaphase, but they separate into dyads (instead of monads) during meiosis II. How would this change the alignments that you constructed during the anaphase stages in meiosis I and II? Draw them.

31. Assume that each resultant gamete (Problem 30) participated in fertilization with a normal haploid gamete. What combinations will result? What percentage of zygotes will be diploid, containing one paternal and one maternal member of each chromosome pair?

3

Mendelian Genetics

Gregor Johann Mendel, who in 1866 put forward the major postulates of transmission genetics as a result of experiments with the garden pea.

CHAPTER CONCEPTS

- Inheritance is governed by information stored in discrete factors called genes.

- Genes are transmitted from generation to generation on vehicles called chromosomes.

- Chromosomes, which exist in pairs, provide the basis of biparental inheritance.

- During gamete formation, chromosomes are distributed according to postulates first described by Gregor Mendel, based on his nineteenth-century research with the garden pea.

- Mendelian postulates prescribe that homologous chromosomes segregate from one another and assort independently with other segregating homologs during gamete formation.

- Genetic ratios, expressed as probabilities, are subject to chance deviation and may be evaluated statistically.

- The analysis of pedigrees allows predictions involving the genetic nature of human traits.

Although inheritance of biological traits has been recognized for thousands of years, the first significant insights into the mechanisms involved occurred about 135 years ago. In 1866, Gregor Johann Mendel published the results of a series of experiments that would lay the foundation for the formal discipline of genetics. Mendel's work went largely unnoticed until the turn of the century, but in the ensuing years the concept of the gene as a distinct hereditary unit was established. The ways in which genes, as members of chromosomes, are transmitted to offspring and control traits were clarified. Research continued unabated throughout the twentieth century—indeed, studies in genetics, most recently at the molecular level, have remained continually at the forefront of biological research since the early 1900s.

When Mendel began his studies of inheritance using *Pisum sativum*, the garden pea, chromosomes and the role and mechanism of meiosis were totally unknown. Nevertheless, he determined that discrete **units of inheritance** exist and predicted their behavior during the formation of gametes. Subsequent investigators, with access to cytological data, were able to relate their observations of chromosome behavior during meiosis to Mendel's principles of inheritance. Once this correlation was made, Mendel's postulates were accepted as the basis for the study of what is known as **transmission genetics**.

How Do We Know?

In this chapter, we focus on how Mendel was able to derive the essential postulates that explain inheritance. As you study this topic, you should try to answer several fundamental questions:

1. How did Mendel know that unit factors existed as fundamental genetic components if he could not directly observe them?

2. How do we know that an organism expressing a dominant trait is homozygous or heterozygous?

3. In genetic data, how do we know that deviation from the expected ratio is due to chance rather than another independent factor?

4. How do we know how a trait is inherited in humans?

Mendel Used a Model Experimental Approach to Study Patterns of Inheritance

Johann Mendel was born in 1822 to a peasant family in the central European village of Heinzendorf. An excellent student in high school, he studied philosophy for several years afterward, and in 1843 he was admitted to the Augustinian Monastery of St. Thomas in Brno, now part of the Czech Republic, taking the name Gregor. In 1849, he was relieved of pastoral duties and accepted a teaching appointment that lasted several years. From 1851 to 1853, he attended the University of Vienna, where he studied physics and botany. He returned to Brno in 1854, where he taught physics and natural

science for the next 16 years. Mendel received support from the monastery for his studies and research throughout his life.

In 1856, Mendel performed his first set of hybridization experiments with the garden pea. The research phase of his career lasted until 1868, when he was elected abbot of the monastery. Although he retained his interest in genetics, his new responsibilities demanded most of his time. In 1884, Mendel died of a kidney disorder. The local newspaper paid him the following tribute:

> "His death deprives the poor of a benefactor, and mankind at large of a man of the noblest character, one who was a warm friend, a promoter of the natural sciences, and an exemplary priest."

Mendel first reported the results of some simple genetic crosses between certain strains of the garden pea in 1865. Although his was not the first attempt to provide experimental evidence pertaining to inheritance, Mendel's success where others failed can be attributed, at least in part, to his elegant model of experimental design and analysis.

Mendel showed remarkable insight into the methodology necessary for good experimental biology. He chose an organism that is easy to grow and hybridize artificially. The pea plant is self-fertilizing in nature but is easy to crossbreed experimentally. It reproduces well and grows to maturity in a single season. Mendel followed seven visible features (unit characters), each represented by two contrasting forms, or **traits** (Figure 3–1). For the character stem height, for example, he experimented with the traits *tall* and *dwarf*. He selected six other visibly contrasting pairs of traits involving seed shape and color, pod shape and color, and pod and flower arrangement. From local seed merchants, Mendel obtained true-breeding strains—those in which each trait appeared unchanged generation after generation in self-fertilizing plants.

There were several reasons for Mendel's success. In addition to his choice of a suitable organism, he restricted his examination to one or very few pairs of contrasting traits in each experiment. He also kept accurate quantitative records, a necessity in genetic experiments. From the analysis of his data, Mendel derived certain postulates that became principles of transmission genetics.

The results of Mendel's experiments were unappreciated until the turn of the century, well after his death. However, once Mendel's publications were rediscovered by geneticists investigating the function and behavior of chromosomes, the implications of his postulates were immediately apparent. He had discovered the basis for the transmission of hereditary traits!

The Monohybrid Cross Reveals How One Trait Is Transmitted from Generation to Generation

Mendel's simplest crosses involved only one pair of contrasting traits. Each such experiment is a **monohybrid cross**, which is made by mating true-breeding individuals from two parent strains, each exhibiting one of the two contrasting forms of the character under study. Initially, we examine the first

Character	Contrasting traits		F₁ results	F₂ results	F₂ ratio
Seeds	round/wrinkled		all round	5474 round 1850 wrinkled	2.96:1
	yellow/green		all yellow	6022 yellow 2001 green	3.01:1
Pods	full/constricted		all full	882 full 299 constricted	2.95:1
	green/yellow		all green	428 green 152 yellow	2.82:1
Flower color	violet/white	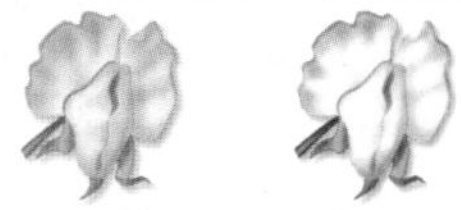	all violet	705 violet 224 white	3.15:1
Flower position	axial/terminal	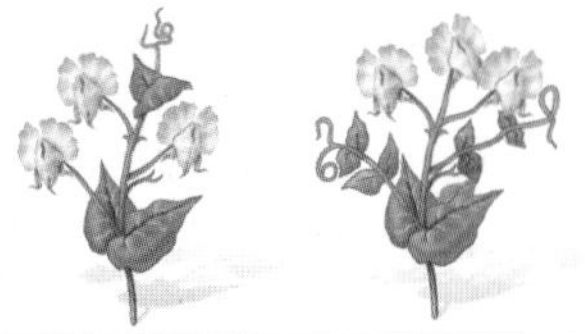	all axial	651 axial 207 terminal	3.14:1
Stem length	tall/dwarf		all tall	787 tall 277 dwarf	2.84:1

FIGURE 3-1 Seven pairs of contrasting traits and the results of Mendel's seven monohybrid crosses of the garden pea (*Pisum sativum*). In each case, pollen derived from plants exhibiting one trait was used to fertilize the ova of plants exhibiting the other trait. In the F₁ generation, one of the two traits was exhibited by all plants. The contrasting trait reappeared in approximately 1/4 of the F₂ plants.

generation of offspring of such a cross, and then we consider the results of **selfing**, the offspring of self-fertilizing individuals from this first generation. The original parents constitute the **P₁**, or **parental**, **generation**, their offspring are the **F₁**, or **first filial generation**, and the individuals resulting from the selfed F₁ generation are the **F₂**, or **second filial generation**. We can continue to follow subsequent generations.

The cross between true-breeding pea plants with tall stems and dwarf stems is representative of Mendel's monohybrid crosses. *Tall* and *dwarf* are contrasting traits of the character of stem height. Unless tall or dwarf plants are crossed together or with another strain, they will undergo self-fertilization and breed true, producing their respective traits generation after generation. However, when Mendel crossed tall plants with dwarf plants, the resulting F₁ generation consisted only of tall plants. When members of the F₁ generation were selfed, Mendel observed that 787 of 1064 F₂ plants were tall, while the remaining 277 were dwarf. Note that in this cross (Figure 3–1) the dwarf trait disappears in the *F₁*, only to reappear in the F₂ generation.

Genetic data are usually expressed and analyzed as ratios. In this particular example, many identical P₁ crosses were made, and many F₁ plants—all tall—were produced. Of the 1064 F₂ offspring, 787 were tall and 277 were dwarf—a ratio of 2.84:1.0, or about 3:1.

Mendel made similar crosses between pea plants exhibiting other pairs of contrasting traits; the results of these crosses are

shown in Figure 3–1. In every case, the outcome was similar to the tall/dwarf cross just described. All F₁ offspring were identical to one of the parents, but in the F₂ offspring, an approximate ratio of 3:1 was obtained. That is, three-fourths looked like the F₁ plants, while one-fourth exhibited the contrasting trait, which had disappeared in the F₁ generation.

We will point out one further aspect of Mendel's monohybrid crosses. In each cross, the F₁ and F₂ patterns of inheritance were similar regardless of which P₁ plant served as the source of pollen (sperm) and which served as the source of the ovum (egg). The crosses could be made either way—pollination of dwarf plants by tall plants or vice versa. These are called **reciprocal crosses**. Therefore, the results of Mendel's monohybrid crosses were not sex-dependent.

To explain these results, Mendel proposed the existence of particular unit factors for each trait. He suggested that these factors serve as the basic units of heredity and are passed unchanged from generation to generation, determining the various traits expressed by each individual plant. Using these general ideas, Mendel proceeded to hypothesize precisely how unit factors could account for the results of the monohybrid crosses.

Mendel's First Three Postulates

Using the consistent pattern of results in the monohybrid crosses, Mendel derived the following three postulates or principles of inheritance.

1. Unit Factors in Pairs

Genetic characters are controlled by unit factors that exist in pairs in individual organisms.

In the monohybrid cross involving tall and dwarf stems, a specific **unit factor** exists for each trait. Because the factors occur in pairs, three combinations are possible: two factors for tallness, two factors for dwarfness, or one factor for each trait. Every individual contains one of these three combinations, which determines stem height.

2. Dominance/Recessiveness

When two unlike unit factors responsible for a single character are present in a single individual, one unit factor is dominant to the other, which is said to be recessive.

In each monohybrid cross, the trait expressed in the F_1 generation is controlled by the **dominant** unit factor. The trait not expressed is controlled by the **recessive** unit factor. Note that this dominance/recessiveness relationship pertains only when unlike unit factors are present in pairs. The terms *dominant* and *recessive* are also used to designate traits. In this case, tall stems are said to be dominant over the recessive dwarf stems.

3. Segregation

During the formation of gametes, the paired unit factors separate or segregate randomly so that each gamete receives one or the other with equal likelihood.

If an individual contains a pair of like unit factors (e.g., both specific for tall), then all gametes receive one tall unit factor. If an individual contains unlike unit factors (e.g., one for tall and one for dwarf), then each gamete has a 50 percent probability of receiving either the tall or the dwarf unit factor.

These postulates provide a suitable explanation for the results of the monohybrid crosses. Let's use the tall/dwarf cross to illustrate. Mendel reasoned that P_1 tall plants contain identical paired unit factors, as do the P_1 dwarf plants. The gametes of tall plants all receive one tall unit factor as a result of **segregation**. Similarly, the gametes of dwarf plants all receive one dwarf unit factor. Following fertilization, all F_1 plants receive one unit factor from each parent: a tall factor from one and a dwarf factor from the other, reestablishing the paired relationship—but because tall is dominant to dwarf, all F_1 plants are tall.

When F_1 plants form gametes, the postulate of segregation demands that each gamete randomly receives either the tall or the dwarf unit factor. Following random fertilization events during F_1 selfing, four F_2 combinations result in equal frequency:

1. tall/tall
2. tall/dwarf
3. dwarf/tall
4. dwarf/dwarf

Combinations (1) and (4) result in tall and dwarf plants, respectively. According to the postulate of dominance/ recessiveness, combinations (2) and (3) both yield tall plants. Therefore, the F_2 is predicted to consist of 3/4 tall and 1/4 dwarf, or a ratio of 3:1. This is approximately what Mendel observed in the cross between tall and dwarf plants. A similar pattern was observed in each of the other monohybrid crosses (see Figure 3–1).

Modern Genetic Terminology

To illustrate the monohybrid cross and Mendel's first three postulates, we must first introduce several new terms as well as a symbol convention for the unit factors.

Traits such as tall or dwarf are visible expressions of the information contained in unit factors. The physical appearance of a trait is the **phenotype** of the individual. Mendel's unit factors represent units of inheritance called **genes** by modern geneticists. For any given character, such as plant height, the phenotype is determined by alternative forms of a single gene called **alleles**. For example, the unit factors representing tall and dwarf are alleles determining the height of the pea plant.

The convention we will use is to choose the first letter of the recessive trait to symbolize the character in question— the lowercase italic letter designates the allele for the recessive trait, and the uppercase italic letter designates the allele for the dominant trait. Thus, for Mendel's pea plants, we use *d* for the dwarf allele and *D* for the tall allele. When alleles are written in pairs to represent the two unit factors present in any individual (*DD*, *Dd*, or *dd*), these symbols are called the **genotype**. This term reflects the genetic makeup of an individual, whether it is haploid or diploid. By reading the genotype, we know the phenotype of the individual: *DD* and *Dd* are tall, and *dd* is dwarf. When both alleles are the same (*DD* or *dd*), the individual is **homozygous** or a **homozygote**; when the alleles are different (*Dd*), we use the term **heterozygous** or a **heterozygote**. These symbols and terms are used in Figure 3–2 to illustrate the monohybrid cross.

Because he operated without the hindsight that modern geneticists enjoy, Mendel's analytical reasoning must be considered a truly outstanding scientific achievement. On the basis of rather simple but precisely executed breeding experiments, he not only proposed that discrete particulate units of heredity exist, he also explained how they are transmitted from one generation to the next.

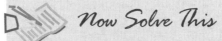

Now Solve This

Problem 5 on page 58 involves a Mendelian cross where you must determine the mode of inheritance and the genotypes of the parents in a number of instances.

Hint: The first step is to determine how many genes are involved. To do so, convert the data to ratios that are characteristic of Mendelian crosses. In the case of this problem, ask first whether any of the F_2 ratios match Mendel's 3:1 monohybrid ratio.

Punnett Squares

The genotypes and phenotypes resulting from the recombination of gametes during fertilization can be easily visualized by constructing a **Punnett square**, named after the person who first devised this approach, Reginald C. Punnett. Figure 3–3 demonstrates this method of analysis for our $F_1 \times F_1$ monohybrid cross. Each of the possible gametes is assigned to a column or a row; the vertical column represents those of the female parent, and the horizontal row represents those of the male parent. After putting the gametes into the rows and columns, the new generation is predicted by combining the male and female gametic information for each combination by entering the resulting genotypes in the boxes. This process thus lists all possible

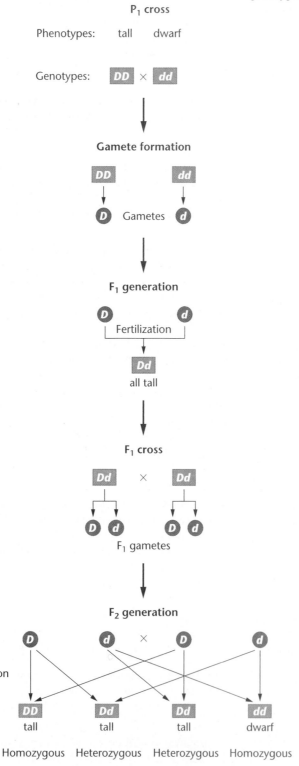

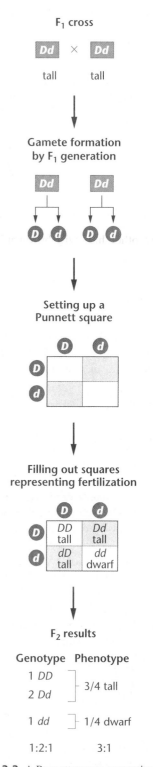

FIGURE 3-2 The monohybrid cross between tall (D) and dwarf (d) pea plants. Individuals are shown in rectangles, and gametes in circles.

FIGURE 3-3 A Punnett square generating the F_2 ratio of the $F_1 \times F_1$ cross shown in Figure 3–2.

random fertilization events. The genotypes and phenotypes of all potential offspring are ascertained by reading the entries in the boxes.

The Punnett square method is particularly useful when you are first learning about genetics and how to solve problems. Note the ease with which the 3:1 phenotypic ratio and the 1:2:1 genotypic ratio is derived in the F$_2$ generation in Figure 3–3.

The Testcross: One Character

Tall plants produced in the F$_2$ generation are predicted to be either the *DD* or *Dd* genotype. You might ask if there is a way to distinguish the genotype. Mendel devised a rather simple method that is still used today in breeding plants and animals: the **testcross**. The organism expressing the dominant phenotype, but of unknown genotype, is crossed to a known homozygous recessive individual. For example, as shown in Figure 3–4(a), if a tall plant of genotype *DD* is testcrossed to a dwarf plant, which must have the *dd* genotype, all offspring will be tall phenotypically and *Dd* genotypically. However, as shown in Figure 3–4(b), if a tall plant is *Dd* and it is crossed to a dwarf plant (*dd*), then one-half of the offspring will be tall (*Dd*) and the other half will be dwarf (*dd*). Therefore, a 1:1 tall/dwarf ratio demonstrates the heterozygous nature of the tall plant of unknown genotype. The test cross reinforced Mendel's conclusion that separate unit factors control traits.

Testcross results

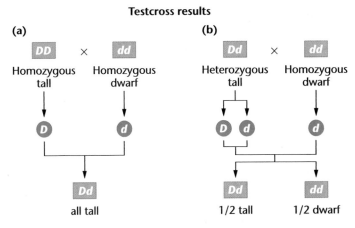

FIGURE 3-4 Test cross of a single character. In (a), the tall parent is homozygous, but in (b), the tall parent is heterozygous. The genotype of each tall P$_1$ plant can be determined by examining the offspring when each is crossed to a homozygous recessive dwarf plant.

3.3 Mendel's Dihybrid Cross Generated a Unique F$_2$ Ratio

As a natural extension of the monohybrid cross, Mendel also designed experiments in which he examined two characters simultaneously. Such a cross, involving two pairs of contrasting traits, is a **dihybrid cross**, or *two-factor cross*. For example, if pea plants having yellow seeds that are round are bred with those having green seeds that are wrinkled, the results shown in

How Mendel's Peas Become Wrinkled: A Molecular Explanation

Only recently, well over a hundred years after Mendel used wrinkled peas in his groundbreaking hybridization experiments, have we come to find out how the *wrinkled* gene makes peas wrinkled. The wild-type allele of the gene encodes a protein called **starch-branching enzyme (SBEI)**. This enzyme catalyzes the formation of highly branched starch molecules as the seed matures.

Wrinkled peas, which result from the homozygous presence of the mutant form of the gene, lack the activity of this enzyme. The production of branch points is inhibited during the synthesis of starch within the seed, which in turn leads to the accumulation of more sucrose and a higher water content while the seed develops. Osmotic pressure inside rises, causing the seed to lose water internally, and ultimately results in the wrinkled appearance of the seed at maturation. In contrast, developing seeds that bear at least one copy of the normal gene (being either homozygous or heterozygous for the dominant allele) synthesize starch and reach an

osmotic balance that minimizes the loss of water. The end result is a smooth-textured outer coat.

The *SBEI* gene has been cloned and analyzed, providing greater insight into the relationship between genotypes and phenotypes. Interestingly, the mutant gene contains a foreign sequence of some 800 base pairs that disrupts the normal

coding sequence. This foreign segment closely resembles other such sequences, called **transposable elements**. These sequences have the ability to move from place to place in the genome of organisms. Transposable elements have been found in maize (corn), parsley, and snapdragons, fruit flies, and humans, among many other organisms.

Wrinkled and round garden peas, the phenotypic traits in one of Mendel's monohybrid crosses.

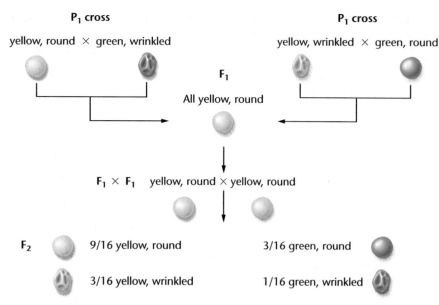

FIGURE 3-5 F_1 and F_2 results of Mendel's dihybrid crosses, where the plants on the top left with yellow, round seeds are crossed with plants having green, wrinkled seeds, and the plants on the top right with yellow, wrinkled seeds are crossed with plants having green, round seeds.

Figure 3–5 will occur: the F_1 offspring will all be yellow and round. It is therefore apparent that yellow is dominant to green and that round is dominant to wrinkled. When the F_1 individuals are selfed, approximately 9/16 of the F_2 plants express yellow and round, 3/16 express yellow and wrinkled, 3/16 express green and round, and 1/16 express green and wrinkled.

A variation of this cross is also shown in Figure 3–5. Instead of crossing one P_1 parent with both dominant traits (yellow, round) and one with both recessive traits (green, wrinkled), plants with yellow, wrinkled seeds are crossed with those with green, round seeds. In spite of the change in the P_1 phenotypes, both the F_1 and F_2 results remain unchanged. It will become clear in the next section why this is so.

Mendel's Fourth Postulate: Independent Assortment

We can most easily understand the results of a dihybrid cross if we consider it theoretically as consisting of two monohybrid crosses conducted separately. Think of the two sets of traits as

inherited independently of each other; that is, the chance of any plant having yellow or green seeds is not at all influenced by the chance that this plant will have round or wrinkled seeds. Thus, because yellow is dominant to green, all F_1 plants in the first theoretical cross would have yellow seeds. In the second theoretical cross, all F_1 plants would have round seeds because round is dominant to wrinkled. When Mendel examined the F_1 plants of the dihybrid cross, all were yellow and round, as we just predicted.

The predicted F_2 results of the first cross are 3/4 yellow and 1/4 green. Similarly, the second cross would yield 3/4 round and 1/4 wrinkled. Figure 3–5 shows that in the dihybrid cross, 12/16 F_2 plants are yellow while 4/16 are green, exhibiting the expected 3:1 (3/4:1/4) ratio. Similarly, 12/16 F_2 plants have round seeds while 4/16 have wrinkled seeds, again revealing the 3:1 ratio.

It is evident that the two pairs of contrasting traits are inherited independently, so we can predict the frequencies of all possible F_2 phenotypes by applying the **product law** of probabilities: *When two independent events occur simultaneously, the combined probability of the two outcomes is equal to the product of their individual probabilities of occurrence.* For example, the probability of an F_2 plant having yellow *and* round seeds is (3/4)(3/4), or 9/16, because 3/4 of all F_2 plants should be yellow and (3/4) of all F_2 plants should be round. In a like manner, the probabilities of the other three F_2 phenotypes can be calculated: yellow (3/4) and wrinkled (1/4) are predicted to be present together 3/16 of the time; green (1/4) and round (3/4) are predicted 3/16 of the time; and green (1/4) and wrinkled (1/4) are predicted 1/16 of the time. These calculations are shown in Figure 3–6.

It is now apparent why the F_1 and F_2 results are identical whether the initial cross is yellow, round plants bred with green, wrinkled plants, or if yellow, wrinkled plants are bred with green, round plants. In both crosses, the F_1 genotype of all plants is identical. Each plant is heterozygous for both gene pairs. As a result, the F_2 generation is also identical in both crosses.

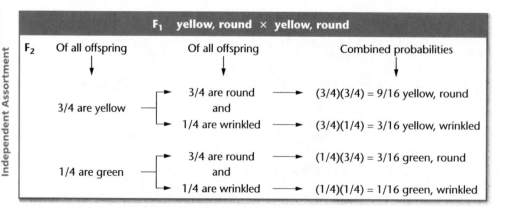

FIGURE 3-6 Computation of the combined probabilities of each F_2 phenotype for two independently inherited characters. The probability of each plant's being yellow or green is independent of the probability of its bearing round or wrinkled seeds.

On the basis of similar results in numerous dihybrid crosses, Mendel proposed a fourth postulate called **independent assortment**: *During gamete formation, segregating pairs of unit factors assort independently of each other.*

This postulate stipulates that segregation of any pair of unit factors occurs independently of all others. As a result of random segregation, each gamete receives one member of every pair of unit factors. For one pair, whichever unit factor is received does not influence the outcome of segregation of any other pair. Thus, according to the postulate of independent assortment, all possible combinations of gametes are formed in equal frequency.

The Punnett square in Figure 3–7 shows how independent assortment works in the formation of the F$_2$ generation.

FIGURE 3-7 Analysis of the dihybrid crosses shown in Figure 3–5. The F$_1$ heterozygous plants are self-fertilized to produce an F$_2$ generation, which is computed using a Punnett square. Both the phenotypic and genotypic F$_2$ ratios are shown.

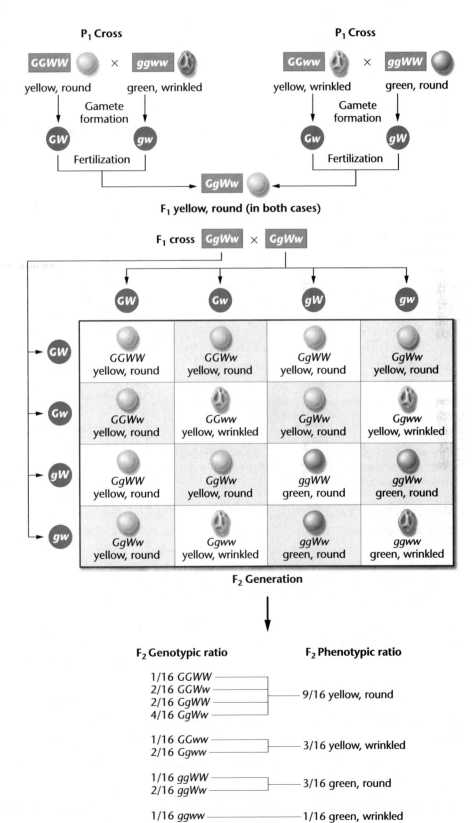

Examine the formation of gametes by the F_1 plants; segregation prescribes that every gamete receives either a G or g allele and a W or w allele. Independent assortment stipulates that all four combinations (GW, Gw, gW, and gw) will be formed with equal probabilities.

In every $F_1 \times F_1$ fertilization event, each zygote has an equal probability of receiving one of the four combinations from each parent. If many offspring are produced, 9/16 have yellow, round seeds, 3/16 have yellow, wrinkled seeds, 3/16 have green, round seeds, and 1/16 have green, wrinkled seeds, yielding what is designated as **Mendel's 9:3:3:1 dihybrid ratio**. This is an ideal ratio based on probability events involving segregation, independent assortment, and random fertilization. Because of deviation due strictly to chance, particularly if small numbers of offspring are produced, actual results are highly unlikely to match the ideal ratio.

The Testcross: Two Characters

The testcross can also be applied to individuals that express two dominant traits but whose genotypes are unknown. For example, the expression of the yellow, round seed phenotype in the F_2 generation just described may result from the $GGWW$, $GGWw$, $GgWW$, and $GgWw$ genotypes. If an F_2 yellow, round plant is crossed with a homozygous recessive green, wrinkled plant ($ggww$), analysis of the offspring will indicate the actual genotype of that yellow, round plant. Each of the above genotypes results in a different set of gametes, and in a testcross, a different set of phenotypes in the resulting offspring. You should work out the results of each of these four crosses to be sure you understand this concept.

Now Solve This

Problem 7 on page 59 involves a series of Mendelian dihybrid crosses where you must determine the genotypes of the parents in a number of instances.

Hint: In each case, write down everything that you know for certain. This reduces the problem to its bare essentials, clarifying what you need to determine. For example, the wrinkled, yellow plant in case (b) must be homozygous for the recessive wrinkled alleles and bear at least one dominant allele for the yellow trait. Having established this, you need only determine the remaining allele for cotyledon color.

3.4 The Trihybrid Cross Demonstrates That Mendel's Principles Apply to Inheritance of Multiple Traits

Thus far, we have considered inheritance by individuals of up to two pairs of contrasting traits. Mendel demonstrated that the identical processes of segregation and independent assortment apply to three pairs of contrasting traits in what is called a **trihybrid cross**, or *three-factor cross*.

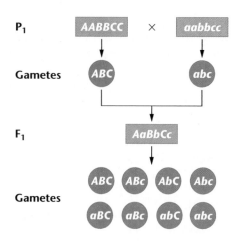

FIGURE 3-8 Formation of P_1 and F_1 gametes in a trihybrid cross.

Although a trihybrid cross is somewhat more complex than a dihybrid cross, its results are easily calculated if the principles of segregation and independent assortment are followed. For example, consider the cross shown in Figure 3–8, where the gene pairs of theoretical contrasting traits are represented by the symbols A, a, B, b, C, and c. In the cross between $AABBCC$ and $aabbcc$ individuals, all F_1 individuals are heterozygous for all three gene pairs. Their genotype, $AaBbCc$, results in the phenotypic expression of the dominant A, B, and C traits. When F_1 individuals serve as parents, each produces eight different gametes in equal frequencies. At this point, we could construct a Punnett square with 64 separate boxes and read out the phenotypes—but such a method is cumbersome in a cross involving so many factors. Therefore another method has been devised to calculate the predicted ratio.

The Forked-Line Method

It is much less difficult to consider each contrasting pair of traits separately and then to combine these results by using the **forked-line method**, first shown in Figure 3–6. This method, also called a **branch diagram**, relies on the simple application of the laws of probability established for the dihybrid cross. Each gene pair is assumed to behave independently during gamete formation.

When the monohybrid cross $AA \times aa$ is made, we know that:

1. All F_1 individuals have the genotype Aa and express the phenotype represented by the A allele, which is called the A phenotype in the following discussion.

2. The F_2 generation consists of individuals with either the A phenotype or the a phenotype in the ratio of 3:1.

The same generalizations can be made for the $BB \times bb$ and $CC \times cc$ crosses. Thus, in the F_2 generation, 3/4 of all organisms express phenotype A, 3/4 express B, and 3/4 express C. Similarly, 1/4 of all organisms express phenotype a, 1/4 express b, and 1/4 express c. The proportions of organisms that express each phenotypic combination can be predicted by assuming that fertilization, following the independent assortment of these three gene pairs during

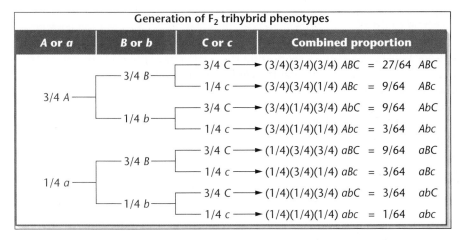

FIGURE 3-9 Generation of the F₂ trihybrid phenotypic ratio, using the forked-line method. This method is based on the expected probability of occurrence of each phenotype.

gamete formation, is a random process—we simply apply the product law of probabilities once again. Figure 3–9 uses the forked-line method to calculate the phenotypic proportions of the F_2 generation. They fall into the trihybrid ratio of 27:9:9:9:3:3:3:1. The same method can be used to solve crosses involving any number of gene pairs, *provided that all gene pairs assort independently of each other*. We shall see later that this is not always the case. However, it appeared to be true for all of Mendel's characters.

Now Solve This

Problem 15 on page 59 asks you to use the forked-line method to determine the outcome of a number of tri-hybrid crosses.

Hint: In using the forked-line method, consider each gene pair separately. For example, in this problem, first predict the outcome of each cross for *A/a* genes, then for the *B/b* genes, and finally, for the *C/c* genes. Then you are prepared to pursue the outcome of each cross using the forked-line method.

3.5 Mendel's Work Was Rediscovered in the Early Twentieth Century

Mendel's work, initiated in 1856, was presented to the Brünn Society of Natural Science in 1865 and published the following year. While his findings were often cited and discussed, their significance went unappreciated for about 35 years. Many reasons have been suggested to explain why the significance of his research was not immediately recognized.

Mendel's adherence to mathematical analysis of probability events was an unusual approach in those days for biological studies. Perhaps his approach seemed foreign to his contemporaries. More important, his conclusions did not fit well with existing theories on the cause of variation among organisms.

The source of natural variation intrigued students of evolutionary theory. These individuals, stimulated by the proposal developed by Charles Darwin and Alfred Russel Wallace, believed in **continuous variation**, whereby offspring were a *blend* of their parents' phenotypes. As we mentioned earlier, Mendel theorized that variation was due to discrete or particulate units, resulting in **discontinuous variation**. For example, Mendel proposed that the F_2 offspring of a dihybrid cross are expressing traits produced by new combinations of previously existing unit factors. As a result, Mendel's theories did not fit well with the evolutionists' preconceptions about causes of variation.

In the latter part of the nineteenth century, a remarkable observation set the scene for the rebirth of Mendel's work: Walter Flemming's discovery of chromosomes in the nuclei of salamander cells. In 1879, Flemming described the behavior of these threadlike structures during cell division. As a result of his findings and the work of many other cytologists, the presence of discrete units within the nucleus soon became an integral part of ideas about inheritance. It was this mind-set that prompted scientists to reexamine Mendel's findings.

In the early twentieth century, research led to renewed interest in Mendel's work. Hybridization experiments similar to Mendel's were performed independently by three botanists: Hugo de Vries, Karl Correns, and Erich Tschermak. De Vries's work demonstrated the principle of segregation in his experiments with several plant species. Apparently, he searched the existing literature and found that Mendel's work anticipated his own conclusions! Correns and Tschermak also reached conclusions similar to those of Mendel.

In 1902, two cytologists, Walter Sutton and Theodor Boveri, independently published papers linking their discoveries of the behavior of chromosomes during meiosis to the Mendelian principles of segregation and independent assortment. They pointed out that the separation of chromosomes during meiosis could serve as the cytological basis of these two postulates. Although they thought Mendel's unit factors were probably chromosomes rather than genes on chromosomes, their findings also reestablished the importance of Mendel's work, which became the basis of ensuing genetic investigations. Sutton and Boveri are credited with initiating the **chromosomal theory of inheritance**, which was developed during the next two decades.

Unit Factors, Genes, and Homologous Chromosomes

Because the correlation between Sutton's and Boveri's observations and Mendelian principles serves as the foundation for the modern interpretation of transmission genetics, we will examine this correlation in some depth before moving on to other topics.

As we know, each species possesses a specific number of chromosomes in each somatic cell nucleus (except in gametes). For diploid organisms, this number is called the **diploid number (2n)** and is characteristic of that species. During the formation of gametes, this number is precisely halved (n), and when two gametes combine during fertilization, the diploid number is reestablished. During meiosis, however, the chromosome number is not reduced in a random manner. It was apparent to early cytologists that the diploid number of chromosomes is composed of homologous pairs identifiable by their morphological appearance and behavior. The gametes contain one member of each pair—thus the chromosome complement of a gamete is quite specific, and the number of chromosomes in each gamete is equal to the haploid number.

With this basic information, we can see the correlation between the behavior of unit factors and chromosomes and genes. Figure 3–10 shows three of Mendel's postulates and the chromosomal explanation of each. Unit factors are really genes located on homologous pairs of chromosomes [Figure 3–10(a)]. Members of each pair of homologs separate, or segregate, during gamete formation [Figure 3–10(b)]. Two different alignments are possible, both of which are shown.

To illustrate the principle of independent assortment, we must distinguish between members of any given homologous pair of chromosomes. One member of each pair comes from

(a) Unit factors in pairs (first meiotic prophase)

Homologous chromosomes in pairs

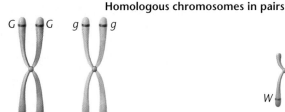

Genes are part of chromosomes

(b) Segregation of unit factors during gamete formation (first meiotic anaphase)

Homologs segregate during meiosis

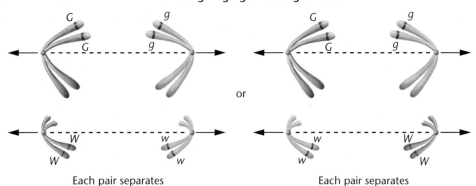

or

Each pair separates   Each pair separates

(c) Independent assortment of segregating unit factors (following many meiotic events)

Nonhomologous chromosomes assort independently

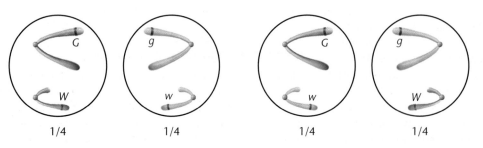

1/4   1/4   1/4   1/4

All possible gametic combinations are formed with equal probablility

FIGURE 3-10 Illustrated correlation between the Mendelian postulates of (a) unit factors in pairs, (b) segregation, and (c) independent assortment, showing the presence of genes located on homologous chromosomes and their behavior during meiosis.

the **maternal parent**, while the other member comes from the **paternal parent**. (We represent the different parental origins with different colors.) As shown in Figure 3–10(c), following independent segregation of each pair of homologs, each gamete receives one member from each pair of chromosomes. All possible combinations are formed with equal probability. If we add the symbols used in Mendel's dihybrid cross (G, g and W, w) to the diagram, we can see why equal numbers of the four types of gametes are formed. The independent behavior of Mendel's pairs of unit factors (G and W in this example) is due to their presence on separate pairs of homologous chromosomes.

Observations of the phenotypic diversity of living organisms make it logical to assume that there are many more genes than chromosomes. Therefore, each homolog must carry genetic information for more than one trait. The currently accepted concept is that a chromosome is composed of a large number of linearly ordered, information-containing *genes*. Mendel's unit factors (which determine tall or dwarf stems, for example) actually constitute a pair of genes located on one pair of homologous chromosomes. The location on a given chromosome where any particular gene occurs is called its **locus** (pl., loci). The different forms taken by a given gene, called *alleles* (G or g), contain slightly different genetic information that determines the same character (seed color in this case). Although we have examined only genes with two alternative alleles, most genes have more than two allelic forms. We conclude this section by reviewing the criteria necessary to classify two chromosomes as a homologous pair:

1. During mitosis and meiosis, when chromosomes are visible as distinct figures, both members of a homologous pair are the same size and exhibit identical centromere locations. The sex chromosomes are an exception.

2. During early stages of meiosis, homologous chromosomes form pairs, or synapse.

3. Although not generally microscopically visible, homologs contain identical, linearly ordered gene loci.

3.6 Independent Assortment Leads to Extensive Genetic Variation

One major consequence of independent assortment is the production by an individual of genetically dissimilar gametes. Genetic variation results because the two members of any homologous pair of chromosomes are rarely, if ever, genetically identical. Therefore, because independent assortment leads to the production of all possible chromosome combinations, extensive genetic diversity results.

We have seen that the number of possible gametes, each with different chromosome compositions, is 2^n, where n equals the haploid number. Thus, if a species has haploid number of $n = 4$, then $2^4 = 16$ different gamete combinations can be formed as a result of independent assortment. Although this number is not high, consider the human

species, where $n = 23$. If 2^{23} is calculated, we find that in excess of 8×10^6, or over 8 million, different types of gametes are represented. Because fertilization represents an event involving only one of approximately 8×10^6 possible gametes from each of two parents, each offspring represents only one of $(8 \times 10^6)^2$, or one of only 64×10^{12} potential genetic combinations! No wonder that, except for identical twins, each member of the human species demonstrates a distinctive appearance and individuality—this number of combinations is far greater than the number of humans who have ever lived on Earth! Genetic variation resulting from independent assortment has been extremely important to the process of evolution in all sexually reproducing organisms.

3.7 Laws of Probability Help to Explain Genetic Events

Recall that genetic ratios are expressed as probabilities—for example, 3/4 tall:1/4 dwarf. These values predict the outcome of each fertilization event, such that the probability of each zygote having the genetic potential for becoming tall is 3/4, while the potential for becoming dwarf is 1/4. Probabilities range from 0.0, when an event is *certain not to occur*, to 1.0, when an event is *certain to occur*. When two or more events occur independently but at the same time, we can calculate the probability of possible outcomes when they occur together. This is accomplished by applying the *product law*—the probability of two or more events occurring simultaneously is equal to the product of their individual probabilities. Two or more events are independent of one another if the outcome of each one does not affect the outcome of any of the others under consideration.

To illustrate the product law, consider the possible results if you toss a penny (P) and a nickel (N) at the same time and examine all combinations of heads (H) and tails (T) that can occur. There are four possible outcomes:

$$(P_H:N_H) = (1/2)(1/2) = 1/4$$
$$(P_T:N_H) = (1/2)(1/2) = 1/4$$
$$(P_H:N_T) = (1/2)(1/2) = 1/4$$
$$(P_T:N_T) = (1/2)(1/2) = 1/4$$

The probability of obtaining a head or a tail in the toss of either coin is 1/2 and is unrelated to the outcome of the toss of the other coin. Thus, all four possible combinations are predicted to occur with equal probability.

If we want to calculate the probability where the possible outcomes of two events are independent of one another but can be accomplished in more than one way, we apply the **sum law**. For example, what is the probability of tossing our penny and nickel and obtaining one head and one tail? In such a case, we do not care whether it is the penny or the nickel that comes up heads, provided the other coin has the alternative outcome. As we saw above, there are two ways in which the desired outcome can be accomplished, each with a probability of 1/4.

Thus, according to the sum law, the overall probability is equal to

$$(1/4) + (1/4) = 1/2$$

One-half of all coin tosses are predicted to yield the desired outcome.

These simple probability laws will be useful throughout our discussions of transmission genetics and for solving genetics problems. In fact, we already applied the product law when we used the forked-line method to calculate the phenotypic results of Mendel's dihybrid and trihybrid crosses. When we wish to know the results of a cross, we need only calculate the probability of each possible outcome. The results of this calculation then allow us to predict the proportion of offspring expressing each phenotype or each genotype.

An important point to remember when you deal with probability is that predictions of possible outcomes are based on large sample sizes. If we predict that 9/16 of the offspring of a dihybrid cross will express both dominant traits, it is very unlikely that, in a small sample, exactly 9 of every 16 offspring will express this phenotype. Instead, our prediction is that, of a large number of offspring, approximately 9/16 of them will do so. The deviation from the predicted ratio in smaller sample sizes is attributed to chance, a subject we examine in our discussion of statistics in the next section. As you shall see, the impact of deviation due strictly to chance diminishes as the sample size increases.

3.8   Chi-Square Analysis Evaluates the Influence of Chance on Genetic Data

Mendel's 3:1 monohybrid and 9:3:3:1 dihybrid ratios are hypothetical predictions based on the following assumptions: (1) Each allele is dominant or recessive; (2) segregation is operative; (3) independent assortment occurs; and (4) fertilization is random. The final two assumptions are influenced by chance events and are therefore subject to random fluctuation. This concept of **chance deviation** is most easily illustrated by tossing a single coin numerous times and recording the number of heads and tails observed. In each toss, there is a probability of 1/2 that a head will occur and a probability of 1/2 that a tail will occur. Therefore, the expected ratio of many tosses is 1:1. If a coin is tossed 1000 times, usually *about* 500 heads and 500 tails will be observed. Any reasonable fluctuation from this hypothetical ratio (e.g., 486 heads and 514 tails) is attributed to chance.

As the total number of tosses is reduced, the impact of chance deviation increases. For example, if a coin is tossed only four times, you would not be too surprised if all four tosses result in only heads or only tails. For 1000 tosses, however, 1000 heads or 1000 tails would be most unexpected. In fact, you might believe that such a result would be impossible. Actually, all heads or all tails in 1000 tosses can be predicted to occur with a probability of $(1/2)^{1000}$. Since $(1/2)^{20}$ is equivalent to less than one in a million times, an event occurring with a probability of $(1/2)^{1000}$ is virtually

impossible. Two major points are significant before we consider *chi-square analysis*:

1. The outcomes of independent assortment and fertilization, like coin tossing, are subject to random fluctuations from their predicted occurrences as a result of chance deviation.

2. As the sample size increases, the average deviation from the expected results decreases. Therefore, a larger sample size diminishes the impact of chance deviation on the final outcome.

Chi-Square Calculations and the Null Hypothesis

In genetics, the ability to evaluate observed deviation is a crucial skill. When we assume that data will fit a given ratio such as 1:1, 3:1, or 9:3:3:1, we establish what is called the **null hypothesis** (H_0). It is so named because the hypothesis assumes that *no real difference* exists between *measured values* (or ratio) and *predicted values* (or ratio). The apparent difference can be attributed purely to chance. The null hypothesis is evaluated using statistical analysis. On this basis, the null hypothesis may either (1) be rejected or (2) fail to be rejected. If it is rejected, the observed deviation from the expected result is not attributed to chance alone. The null hypothesis and the underlying assumptions leading to it must be reexamined. If the null hypothesis fails to be rejected, any observed deviations are attributed to chance.

One of the simplest statistical tests devised to assess the null hypothesis is **chi-square (χ^2) analysis**. This test takes into account the observed deviation in each component of an expected ratio as well as the sample size and reduces them to a single numerical value. The value for χ^2 is then used to estimate how frequently the observed deviation can be expected to occur strictly as a result of chance. The formula for chi-square analysis is

$$\chi^2 = \Sigma \frac{(o - e)^2}{e}$$

where o is the observed value for a given category, e is the expected value for that category, and Σ (the Greek letter sigma) represents the sum of the calculated values for each category of the ratio. Because $(o - e)$ is the deviation (d) in each case, the equation reduces to

$$\chi^2 = \Sigma \frac{d^2}{e}$$

Table 3.1(a) shows a χ^2 calculation for the F_2 results of a hypothetical monohybrid cross. To analyze these data, you work from left to right, calculating and entering the appropriate numbers in each column. Regardless of whether the deviation d is positive or negative, d^2 always becomes positive after the number is squared. In Table 3.1(b) the F_2 results of a hypothetical dihybrid cross are analyzed. Be sure that you understand how each number was calculated in the dihybrid example.

TABLE 3.1	Chi-Square Analysis

(a) Monohybrid Cross

Expected Ratio	Observed (o)	Expected (e)	Deviation (o − e)	Deviation^2 (d^2)	d^2/e
3/4	740	3/4(1000) = 750	740 − 750 = −10	$(-10)^2 = 100$	100/750 = 0.13
1/4	260	1/4(1000) = 250	260 − 250 = +10	$(+10)^2 = 100$	100/250 = 0.40
	Total = 1000				$\chi^2 = 0.53$
					$p = 0.48$

(b) Dihybrid Cross

Expected Ratio	(o)	(e)	(o − e)	(d^2)	d^2/e
9/16	587	567	+20	400	0.71
3/16	197	189	+8	64	0.34
3/16	168	189	−21	441	2.33
1/16	56	63	−7	49	0.78
	Total = 1008				$\chi^2 = 4.16$
					$p = 0.26$

The final step in chi-square analysis is to interpret the χ^2 value. To do so, you must initially determine the value of the **degrees of freedom (df)**, which is equal to $n - 1$, where n is the number of different categories into which each datum point may fall. For the 3:1 ratio, $n = 2$, so $df = 2 - 1 = 1$. For the 9:3:3:1 ratio, $n = 4$ and $df = 3$. Degrees of freedom must be taken into account because the greater the number of categories, the more deviation is expected as a result of chance.

Once you have determined the degrees of freedom, we can interpret the χ^2 value in terms of a corresponding **probability value (p)**. Since this calculation is complex, we usually take the p value from a standard table or graph. Figure 3–11 shows a wide range of χ^2 and p values for various degrees of freedom in both a graph and a table. Let's use the graph to determine the p value. The caption for Figure 3–11(b) explains how to use the table.

To determine p, execute the following steps:

1. Locate the χ^2 value on the abscissa (the horizontal or x-axis).

2. Draw a vertical line from this point up to the angled line on the graph representing the appropriate df.

3. Extend a horizontal line from this point to the left until it intersects the ordinate (the vertical or y-axis).

4. Estimate, by interpolation, the corresponding p value.

We used these steps for the monohybrid cross in Table 3.1(a) to estimate the p value of 0.48 shown in Figure 3–11(a). For the dihybrid cross, try this method to see if you can determine the p value. Since the χ^2 value is 4.16 and $df = 3$, an approximate p value is 0.26. Checking this result in the table

confirms that p values for both the monohybrid and dihybrid crosses are between 0.20 and 0.50.

Interpreting Probability Values

So far, we have been concerned with calculating χ^2 values and determining the corresponding p values. The most important aspect of chi-square analysis is understanding the meaning of the p value. Let's use the example of the dihybrid cross in Table 3.1(b) ($p = 0.26$). In these discussions, it is simplest to think of the p value as a percentage (e.g., $0.26 = 26\%$). In our example, the p value indicates that if we repeat the same experiment many times, 26 percent of the trials would be expected to exhibit chance deviation as great or greater than that seen in the initial trial. Conversely, 74 percent of the trials would show less deviation than initially observed as a result of chance. Thus, the p value reveals that a hypothesis (the 9:3:3:1 ratio in this case) is never proved or disproved absolutely. Instead, a relative standard is set that enables us to either *reject* or *fail to reject* the null hypothesis—this standard is most often a p value of 0.05. When applied to chi-square analysis, a p value less than 0.05 means that the observed deviation in the set of results will be obtained by chance alone less than 5 percent of the time. Such a p value indicates that the difference between the observed and predicted results is substantial and thus enables us to reject the null hypothesis.

On the other hand, p values of 0.05 or greater (0.05 to 1.0) indicate that the observed deviation will be obtained by chance alone 5 percent or more of the time. The conclusion is not to reject the null hypothesis. Thus, for the p value of 0.26, assessing the hypothesis that independent assortment accounts for the results fails to be rejected. Therefore, the observed deviation can be reasonably attributed to chance.

(a)

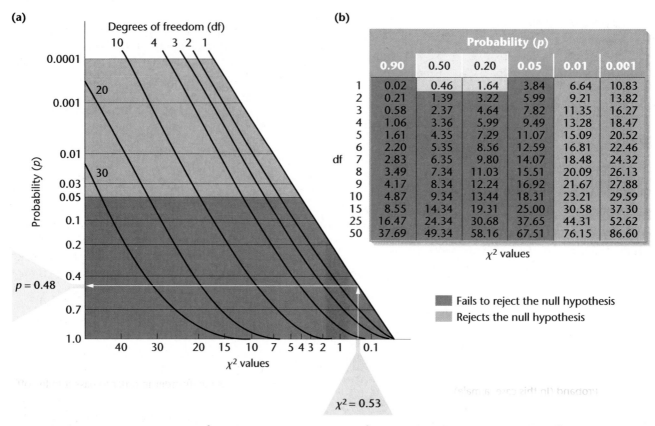

(b)

	Probability (p)					
	0.90	**0.50**	**0.20**	**0.05**	**0.01**	**0.001**
1	0.02	0.46	1.64	3.84	6.64	10.83
2	0.21	1.39	3.22	5.99	9.21	13.82
3	0.58	2.37	4.64	7.82	11.35	16.27
4	1.06	3.36	5.99	9.49	13.28	18.47
5	1.61	4.35	7.29	11.07	15.09	20.52
6	2.20	5.35	8.56	12.59	16.81	22.46
df 7	2.83	6.35	9.80	14.07	18.48	24.32
8	3.49	7.34	11.03	15.51	20.09	26.13
9	4.17	8.34	12.24	16.92	21.67	27.88
10	4.87	9.34	13.44	18.31	23.21	29.59
15	8.55	14.34	19.31	25.00	30.58	37.30
25	16.47	24.34	30.68	37.65	44.31	52.62
50	37.69	49.34	58.16	67.51	76.15	86.60

χ^2 values

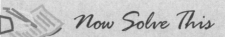 Fails to reject the null hypothesis

Rejects the null hypothesis

FIGURE 3-11 (a) Graph for converting χ^2 values to p values. (b) Table of χ^2 values for selected values of df and p. χ^2 values that lead to a p value of 0.05 or greater (darker blue areas) justify failure to reject the null hypothesis. Values leading to a p value of less than 0.05 (lighter blue areas) justify rejecting the null hypothesis. For example, using the table in part (b), where $\chi^2 = 0.53$ for 1 degree of freedom, the corresponding p value is between 0.20 and 0.50. The graph in (a) gives a more precise p value of 0.48 by interpolation. Thus, we fail to reject the null hypothesis.

A final note is relevant here for the case where the null hypothesis is rejected, that is, where $p \leq 0.05$. Suppose we are testing the null hypothesis that the data represented a 9:3:3:1 ratio, indicative of independent assortment. If the null hypothesis is rejected, what are alternative interpretations of the data? Researchers will reassess the assumptions that underlie the null hypothesis. In our example, we assumed that segregation operates faithfully for both gene pairs. We also assumed that fertilization is random and that the viability of all gametes is equal regardless of genotype—that is, that all gametes are equally likely to participate in fertilization. Finally, following fertilization, we assumed that all preadult stages and adult offspring are equally viable regardless of their genotype. If any of these assumptions is incorrect, the original hypothesis is not necessarily invalid.

An example will clarify this. Suppose our null hypothesis is that a dihybrid cross between fruit flies will result in 3/16 mutant wingless fly zygotes. However, not as many of the mutant embryos may survive their preadult development or as young adults, compared to flies whose genotype gives rise to wings. As a result, when the data are gathered, there are fewer than 3/16 wingless flies. Rejection of the null hypothesis alone is not cause for us to disregard the validity of the postulates of segregation and independent assortment, because other factors are operative.

> ### Now Solve This
>
> Problem 18 on page 59 asks you to apply χ^2 analysis to a set of data and determine whether the data fit several ratios.
>
> **Hint:** In calculating χ^2, first determine the expected outcomes using the predicted ratios. Then follow a stepwise approach, determining the deviation in each case, and calculating d^2/e for each category.

3.9 ## Pedigrees Reveal Patterns of Inheritance of Human Traits

We now explore how to determine the mode of inheritance of phenotypes in humans, where designed crosses are not possible and where relatively few offspring are available for study. The traditional way to study inheritance has been to construct a family tree, indicating the presence or absence of the trait in question for each member of each generation. Such a family tree is called a **pedigree**. By analyzing a pedigree, we may be able to predict how the trait under study is inherited—for example, is it due to a dominant or recessive allele? When many pedigrees for the same trait are studied, we can often ascertain the mode of inheritance.

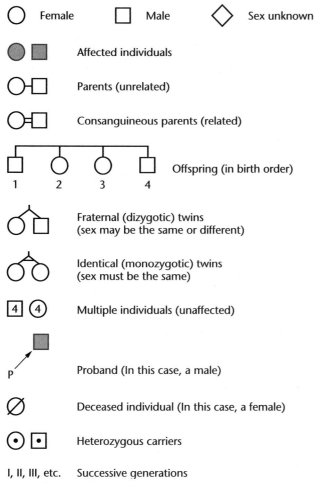

FIGURE 3-12 Conventions commonly encountered in human pedigrees.

Pedigree Conventions

A variety of conventions are commonly used in pedigree construction. Figure 3–12 illustrates a number of such conventions: circles represent females and squares designate males. Parents are connected by a single horizontal line, and vertical lines lead to their offspring. If the parents are related (**consanguineous**), such as first cousins, they are connected by a double line. Offspring are called **sibs** (short for **siblings**) and are connected by a horizontal **sibship line**. Sibs are placed from left to right according to birth order, and are labeled with Arabic numerals. Each generation is indicated by a Roman numeral. If the sex of an individual is unknown, a diamond is used. When a pedigree traces only a single trait, the circles, squares, and diamonds are shaded if the phenotype being considered is expressed and unshaded if not. In some pedigrees, those individuals that fail to express a recessive trait, but are known with certainty to be a heterozygous carrier, have a shaded dot within their unshaded circle or square. If an individual is deceased and the phenotype is unknown, a diagonal line is placed over the circle or square.

Twins are indicated by diagonal lines stemming from a vertical line connected to the sibship line. For **identical** (or **monozygotic**) **twins**, the diagonal lines are linked by a horizontal line. **Fraternal** (or **dizygotic**) **twins** lack this connecting line. A number within one of the symbols represents numerous sibs of the same or unknown phenotypes. The individual whose phenotype first brought attention to the investigation and construction of the pedigree is called the **proband** and is indicated by an arrow connected to the designation **p**. This term applies to either a male or a female.

Pedigree Analysis

In Figure 3–13, two pedigrees are shown. The first illustrates a representative pedigree for a trait that demonstrates autosomal recessive inheritance, such as **albinism**. The male parent of the first generation (I-1) is affected. Characteristic of a rare recessive trait with an affected parent, the trait "disappears" in the offspring of the next generation. Assuming recessiveness, we might predict that the unaffected female parent (I-2) is a homozygous normal individual because none of the offspring show the disorder. Had she been heterozygous, one half of the offspring would be expected to exhibit albinism, but none do. However, such a small sample (three offspring) prevents us from knowing for certain.

Further evidence supports the prediction of a recessive trait. If albinism were inherited as a dominant trait, individual II-3 would have to express the disorder in order to pass it to his offspring (III-3 and III-4), but he does not. Inspection of the offspring constituting the third generation (row III) provides still further support for the hypothesis that albinism is a recessive trait. If it is, parents II-3 and II-4 are both heterozygous, and approximately one-fourth of their offspring should be affected. Two of the six offspring do show albinism. This deviation from the expected ratio is not unexpected in crosses with few offspring. Once we are confident that albinism is inherited as an autosomal recessive trait, we could portray the II-3 and II-4 individuals with a shaded dot within their larger square and circle. Finally, we can note that, characteristic of pedigrees for autosomal traits, both males and females are affected with equal probability. In Chapter 4, we will examine a pedigree representing a gene located on the sex-determining X chromosome. We will see certain limitations imposed on the transmission of X-linked traits, such as that these traits are more prevalent in male offspring and are never passed from affected fathers to their sons.

The second pedigree illustrates the pattern of inheritance for a trait such as Huntington disease, which is caused by an autosomal dominant allele. The key to identifying such a pedigree that reflects a dominant trait is that all affected offspring will have a parent that also expresses the trait. It is also possible, by chance, that none of the offspring will inherit the dominant allele. If so, the trait will cease to exist in future generations. Like recessive traits, provided that the gene is autosomal, both males and females are equally affected.

When autosomal dominant diseases are rare within the population, and most are, then it is highly unlikely that affected individuals will inherit a copy of the mutant gene from both parents. Therefore, in most cases, affected individuals are heterozygous for the dominant allele. As a result, approximately one-half of the offspring inherit it. This is borne out in the second pedigree in Figure 3–13. Furthermore, if a mutation is dominant, and a single copy is sufficient to produce a mutant phenotype, homozygotes are

(a) Autosomal Recessive Trait

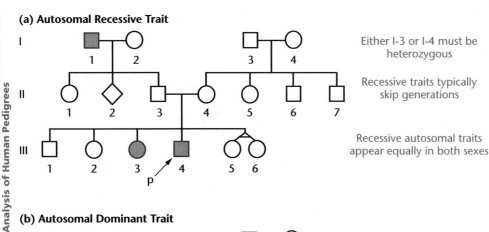

Either I-3 or I-4 must be heterozygous

Recessive traits typically skip generations

Recessive autosomal traits appear equally in both sexes

FIGURE 3-13 Representative pedigrees for two characteristics, each followed through three generations.

(b) Autosomal Dominant Trait

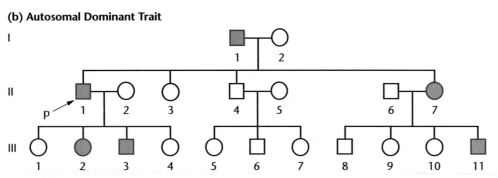

I-1 is heterozygous for a dominant allele

Dominant traits almost always appear in each generation.

Affected individuals all have an affected parent. Dominant autosomal traits appear equally in both sexes.

likely to be even more severely affected, perhaps even failing to survive. An illustration of this is the dominant gene for **familial hypercholesterolemia**. Heterozygotes display a defect in their receptors for low density lipoproteins, the so-called LDLs. As a result, too little cholesterol is taken up by cells from the blood, and elevated plasma levels of LDLs result. Such heterozygous individuals have heart attacks during the fourth decade of their life, or before. While heterozygotes have LDL levels about double that of a normal individual, rare homozygotes have been detected. They lack LDL receptors altogether, and have LDL levels nearly 10 times above the normal range. They are likely to have a heart attack very early in life, even before age five, and almost inevitably before they reach the age of 20.

Pedigree analysis of many traits has historically been an extremely valuable research technique in human genetic studies. However, the approach does not usually provide the certainty in drawing conclusions afforded by designed crosses yielding large numbers of offspring. Nevertheless, when many independent pedigrees of the same trait or disorder are analyzed, consistent conclusions can often be drawn. Table 3.2 lists numerous human traits and classifies them according to their recessive or dominant expression. The genes controlling some of these traits are located on the sex-determining chromosomes. We will discuss pedigrees for X-linked traits in Chapter 4.

TABLE 3.2	Representative Recessive and Dominant Human Traits
Recessive Traits	**Dominant Traits**
Albinism	Achondroplasia
Alkaptonuria	Brachydactyly
Ataxia telangiectasia	Congenital stationary night blindness
Color blindness	Ehler-Danlos syndrome
Cystic fibrosis	Hypotrichosis
Duchenne muscular dystrophy	Huntington disease
Galactosemia	Hypercholesterolemia
Hemophilia	Marfan syndrome
Lesch–Nyhan syndrome	Neurofibromatosis
Phenylketonuria	Phenylthiocarbamide (PTC) tasting
Sickle–cell anemia	Porphyria
Tay–Sachs disease	Widow's peak

Now Solve This

Problem 26 on page 59 asks you to examine a pedigree for myopia and predict whether the trait is dominant or recessive.

Hint: One of the first things to look for are individuals who express the trait, but neither of whose parents also expresses the trait. Such an observation makes it highly unlikely that the trait is dominant.

GENETICS, TECHNOLOGY, AND SOCIETY

Tay-Sachs Disease: The Molecular Basis of a Recessive Disorder in Humans

Tay-Sachs disease (TSD) is an inherited disorder that causes unalterable destruction of the central nervous system. This condition is particularly tragic because infants with TSD are unaffected at birth and appear to develop normally until they are about six months old. Parents, believing that they have a normal child, then must witness the progressive loss of mental and physical abilities. The disorder is severe, as afflicted infants eventually become blind, deaf, mentally retarded, and paralyzed, often within only a year or two. Most do not live beyond age five. Named for Warren Tay and Bernard Sachs, who first described the symptoms and associated them with the disorder in the late 1800s, Tay-Sachs disease clearly demonstrates a pattern of autosomal recessive inheritance. Two unaffected heterozygous parents, who most often have no immediate family history of the disorder, have a probability of one in four of having a Tay-Sachs child. While inheritance clearly displays a classic Mendelian pattern, the protein product of the affected gene has been identified, and we now have a clear understanding of the underlying molecular basis of the disorder. TSD results from the loss of activity of a single enzyme hexosaminidase A (Hex-A). This enzyme is normally found in lysosomes, organelles that break down large molecules for recycling by the cell. Hex-A is needed to break down the ganglioside GM2, a lipid component of nerve cell membranes. Without functional Hex-A, gangliosides accumulate within neurons in the brain and cause deterioration of the nervous system. Heterozygous carriers of TSD with one normal copy of the gene produce only about 50 percent of the normal amount of Hex-A, but they show no symptoms of the disorder. The observation that the activity of only one gene (one wild-type allele) is sufficient for the normal development and function of the nervous system explains and illustrates the molecular basis of recessive mutations. Only when both genes are disrupted by mutation is the mutant phenotype evident.

The gene responsible for Tay-Sachs disease is located on chromosome 15 and codes for the alpha subunit of the Hex-A enzyme. Hex-A, like many enzymes, displays quaternary protein structure, consisting of multiple polypeptide subunits. Since the gene was isolated in 1985, more than 50 different mutations have been identified that lead to TSD. Although the most common form of the disease is the infantile form, where no functional Hex-A is produced, there is also a rare late-onset form that occurs in patients with greatly reduced Hex-A activity. Late-onset TSD is not detectable until patients are in their twenties or thirties, and it is generally much less severe than the infantile form. Symptoms include hand tremors, speech impediments, muscle weakness, and loss of balance.

Tay-Sachs disease is almost a hundred times more common in Ashkenazi Jews—Jews of central or eastern European descent—than in the general population, and also has a higher incidence in French Canadians and in members of the Cajun population in Louisiana. In the United States, approximately one in every 27 Ashkenazi Jews is a heterozygous carrier of TSD. By contrast, the carrier rate in the general population and in Jews of Sephardic (Spanish or Portuguese) origin is approximately one in every 250.

Although there are currently no effective treatments for TSD, recent advances in carrier screening have helped to reduce the prevalence of the disorder in high-risk populations. Carriers can be identified by tests that measure Hex-A activity or by DNA-based tests that detect specific gene mutations. When both parents are carriers, prenatal diagnosis can be utilized with each pregnancy in order to detect affected fetuses. In addition, ganglioside synthesis inhibitors and Hex-A enzyme replacement therapy are currently being investigated as potential treatments for TSD newborns.

Reference

Fernandes, F., and Shapiro, B. 2004. Tay-Sachs Disease. *Arch. Neurol.* 61: 1466–1468.

CHAPTER SUMMARY

1. Over a century ago, Mendel studied inheritance patterns in the garden pea and established the principles of transmission genetics.

2. Mendel's postulates help describe the basis for the inheritance of phenotypic expression. He showed that unit factors, later called alleles, exist in pairs and exhibit a dominant/recessive relationship in determining the expression of traits.

3. Mendel postulated that unit factors must segregate during gamete formation, such that each gamete receives only one of the two factors with equal probability.

4. Mendel's postulate of independent assortment states that each pair of unit factors segregates independently of other such pairs. As a result, all possible combinations of gametes are formed with equal probability.

5. The discovery of chromosomes in the late 1800s, along with subsequent studies of their behavior during meiosis, led to the rebirth of Mendel's work, linking the behavior of his unit factors to that of chromosomes during meiosis.

6. The Punnett square and the forked-line methods are used to predict the probabilities of phenotypes (and genotypes) from crosses involving two or more gene pairs.

7. Genetic ratios are expressed as probabilities. Thus, deriving outcomes of genetic crosses requires an understanding of the laws of probability.

8. Statistical analysis is used to test the validity of experimental outcomes. In genetics, variations from the expected ratios due to chance deviations can be anticipated.

9. Chi-square analysis allows us to assess the null hypothesis, which states that there is no real difference between the expected and observed values. As such, it tests the probability of whether observed variations can be attributed to chance deviation.

10. Pedigree analysis is a method for studying the inheritance pattern of human traits over several generations. It frequently provides the basis for determining the mode of inheritance of human characteristics and disorders.

KEY TERMS

INSIGHTS AND SOLUTIONS

As a student, you will be asked to demonstrate your knowledge of transmission genetics by solving genetics problems. Success at this task represents not only comprehension of theory but its application to more practical genetic situations. Most students find problem solving in genetics to be challenging and rewarding. This section will provide you with basic insights into the reasoning essential to this process.

Genetics problems are in many ways similar to word problems in algebra. The approach to solving them is identical: (1) Analyze the problem carefully; (2) translate words into symbols, first defining each one; and (3) choose and apply a specific technique to solve the problem. The first two steps are critical. The third step is largely mechanical.

The simplest problems state all necessary information about the P_1 generation and ask you to find the expected ratios of the F_1 and F_2 genotypes and/or phenotypes. Always follow these steps when you encounter this type of problem:

1. Determine insofar as possible the genotypes of the individuals in the P_1 generation.

2. Determine what gametes may be formed by the P_1 parents.

3. Recombine gametes by the Punnett square or the forked-line methods, or if the situation is very simple, by inspection. Read the F_1 phenotypes.

4. Repeat the process to obtain information about the F_2 generation.

Determining the genotypes from the given information requires that you understand the basic theory of transmission genetics.

Consider this problem: *A recessive mutant allele, black, causes a very dark body in Drosophila (a fruit fly) when homozygous. The wild-type (normal) color is gray. What F_1 phenotypic ratio is predicted when a black female is crossed with a gray male whose father was black?*

To work out this problem, you must understand dominance and recessiveness, as well as the principle of segregation. Furthermore, you must use the information about the male parent's father. Here is one way to solve this problem:

1. The female parent is black, so she must be homozygous for the mutant allele (bb).

2. The male parent is gray; therefore, he must have at least one dominant allele (B). His father was black (bb), and he received one of the chromosomes bearing these alleles, so the male parent must be heterozygous (Bb).

With this information, the problem is simple:

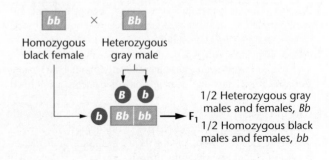

Apply this approach to the following problems.

1. Mendel found that full pods are dominant over constricted pods while round seeds are dominant over wrinkled seeds. One of his crosses was between full, round plants and constricted, wrinkled plants. From this cross, he obtained an F_1 generation that was all full and round. In the F_2 generation, Mendel obtained his classic 9:3:3:1 ratio. Using this information, determine the expected F_1 and F_2 results of a cross between homozygous constricted, round plants and full, wrinkled plants.

Solution: Define gene symbols for each pair of contrasting traits. Use the lowercase first letter of the recessive traits to designate those phenotypes and the uppercase first letter to designate the dominant traits. Thus, C and c indicate full and constricted, and W and w indicate round and wrinkled phenotypes, respectively.

Determine the genotypes of the P_1 generation, form the gametes, reconstitute the F_1 generation, and read off the phenotype(s):

$$P_1: \quad \underset{\text{constricted, round}}{ccWW} \quad \times \quad \underset{\text{full, wrinkled}}{CCww}$$
$$\downarrow \qquad\qquad\qquad \downarrow$$
$$\text{Gametes:} \quad cW \qquad\qquad\quad Cw$$
$$F_1: \quad\qquad \underset{\text{full, round}}{CcWw}$$

You can immediately see that the F_1 generation expresses both dominant phenotypes and is heterozygous for both gene pairs. Thus, you expect that the F_2 generation will yield the classic Mendelian ratio of 9:3:3:1. Let's work it out anyway just to confirm this, using the forked-line method. Both gene pairs are heterozygous and can be expected to assort independently, so we can predict the F_2 outcomes from each gene pair separately and then proceed with the forked-line method.

Every F_2 offspring is subject to the following probabilities:

$$Cc \times Cc \qquad\qquad Ww \times Ww$$
$$\downarrow \qquad\qquad\qquad \downarrow$$

$$\left.\begin{array}{c}CC\\Cc\\cC\end{array}\right\}\text{full} \qquad \left.\begin{array}{c}WW\\Ww\\wW\end{array}\right\}\text{round}$$

$$cc \;\;\text{constricted} \qquad\quad ww \;\;\text{wrinkled}$$

The forked-line method then confirms the 9:3:3:1 phenotypic ratio. Remember that this represents proportions of 9/16:3/16:3/16:1/16. Note that we are applying the product law as we compute the final probabilities:

$$3/4 \text{ full} \begin{cases} \text{---} 3/4 \text{ round} \xrightarrow{(3/4)(3/4)} 9/16 \text{ full, round} \\ \text{---} 1/4 \text{ wrinkled} \xrightarrow{(3/4)(1/4)} 3/16 \text{ full, wrinkled} \end{cases}$$

$$1/4 \text{ constricted} \begin{cases} \text{---} 3/4 \text{ round} \xrightarrow{(1/4)(3/4)} 3/16 \text{ constricted, round} \\ \text{---} 1/4 \text{ wrinkled} \xrightarrow{(1/4)(1/4)} 1/16 \text{ constricted, wrinkled} \end{cases}$$

2. In another cross involving parent plants of unknown genotype and phenotype, the following offspring were obtained.

F_1:   3/8 full, round

3/8 full, wrinkled

1/8 constricted, round

1/8 constricted, wrinkled

Determine the genotypes and phenotypes of the parents.

Solution: This problem is more difficult and requires keener insight because you must work backward. The best approach is to consider the outcomes of pod shape separately from those of seed texture.

Of all the plants, $3/8 + 3/8 = 3/4$ are full and $1/8 + 1/8 = 1/4$ are constricted. Of the various genotypic combinations that can serve as parents, which combination will give rise to a ratio of 3/4:1/4? This ratio is identical to Mendel's monohybrid F_2 results, and we can propose that both unknown parents share the same genetic characteristic as the monohybrid F_1 parents; they must both be heterozygous for the genes controlling pod shape and thus are Cc.

Before we accept this hypothesis, let's consider the possible genotypic combinations that control seed texture. If we consider this characteristic alone, we see that the traits are expressed in a ratio of $3/8 + 1/8 = 1/2$ round: $3/8 + 1/8 = 1/2$ wrinkled. To generate such a ratio, the parents cannot both be heterozygous, or their offspring would yield a 3/4 : 1/4 phenotypic ratio. They cannot both be homozygous, or all of their offspring would express a single phenotype. Thus, we are left with testing the hypothesis that one parent is homozygous and one is heterozygous for the alleles controlling texture. The potential case of $WW \times Ww$ does not work, since it yields only a single phenotype. This leaves us with the potential case of $Ww \times ww$. Offspring in such a mating will yield 1/2 Ww (round):1/2 ww (wrinkled), exactly the outcome we are seeking.

Now, let's combine the hypotheses and predict the outcome of the cross. In our solution, we use a dash (–) to indicate that the second allele may be either dominant or recessive, since we are only predicting phenotypes.

$$3/4 \; C\text{--} \begin{cases} \text{---} 1/2 \; Ww \rightarrow 3/8 \; C\text{--}Ww \text{ full, round} \\ \text{---} 1/2 \; ww \rightarrow 3/8 \; C\text{--}ww \text{ full, wrinkled} \end{cases}$$

$$1/4 \; cc \begin{cases} \text{---} 1/2 \; Ww \rightarrow 1/8 \; ccWw \text{ constricted, round} \\ \text{---} 1/2 \; ww \rightarrow 1/8 \; ccww \text{ constricted, wrinkled} \end{cases}$$

As you can see, this cross produces offspring according to our initial information, and we have solved the problem. Note that in this solution, we used genotypes in the forked-line method, in contrast to the use of phenotypes in the earlier solution.

3. Determine the probability that a plant of genotype $CcWw$ will be produced from parental plants with the genotypes $CcWw$ and $Ccww$.

Solution: The two gene pairs demonstrate straightforward dominance and recessiveness and assort independently during gamete formation. We need only calculate the individual probabilities of obtaining the two separate outcomes (Cc and Ww) and apply the product law to calculate the final probability:

$$Cc \times Cc \longrightarrow 1/4 \; CC : 1/2 \; Cc : 1/4 \; cc$$
$$Ww \times ww \longrightarrow 1/2 \; Ww : 1/2 \; ww$$
$$p = (1/2 \; Cc)(1/2 \; Ww) = 1/4 \; CcWw$$

4. In the laboratory, a genetics student crossed flies that had normal, long wings with flies expressing the *dumpy* mutation (truncated wings), which she believed was a recessive trait. In the F_1 generation, all flies had long wings. The following results were obtained in the F_2 generation:

792 long-winged flies

208 dumpy-winged flies

(Cont. on the next page)

The student tested the hypothesis that the dumpy wing is inherited as a recessive trait, using chi-square analysis of the F_2 data.

(a) What ratio was hypothesized?

(b) Did the analysis support the hypothesis?

(c) What do the data suggest about the *dumpy* mutation?

Solution:

(a) The student hypothesized that the F_2 data (792:208) fit Mendel's 3:1 monohybrid ratio for recessive genes.

(b) The initial step in χ^2 analysis is to calculate the expected results (e) if the ratio is 3:1. Then we can compute deviation $o - e$ (d) and the remaining numbers.

Ratio	o	e	d	d^2	d^2/e
3/4	792	750	42	1764	2.35
1/4	208	250	−42	1764	7.06

Total = 1000

$$\chi^2 = \sum \frac{d^2}{e}$$

$$= 2.35 + 7.06$$

$$= 9.41$$

We consult Figure 3–11 to determine the probability (p) and determine whether the deviations can be attributed to chance. There are two possible outcomes (n), so the degrees of freedom (df) = $n - 1$ or 1. The table in Figure 3–11(b) shows that p is a value between 0.01 and 0.001; the graph in Figure 3–11(a) gives an estimate of about 0.001. Since $p < 0.05$, we reject the null hypothesis. The data do not fit a 3:1 ratio.

(c) When we accepted Mendel's 3:1 ratio as a valid expression of the monohybrid cross, numerous assumptions were made. Examining our underlying assumptions may explain why the null hypothesis was rejected. We assumed that all genotypes are equally viable—that genotypes yielding long wings are equally likely to survive from fertilization through adulthood as the genotype yielding dumpy wings. Further study may reveal that dumpy-winged flies are somewhat less viable than normal flies. As a result, we would expect less than 1/4 of the total offspring to express dumpy wings. This observation is borne out in the data, although we have not proven that this is true.

PROBLEMS AND DISCUSSION QUESTIONS

When working out genetics problems in this and succeeding chapters, always assume that members of the P_1 generation are homozygous, unless the information given, or the data, indicates otherwise.

1. In a cross between a black and a white guinea pig, all members of the F_1 generation are black. The F_2 generation is made up of approximately 3/4 black and 1/4 white guinea pigs. Diagram this cross, and show the genotypes and phenotypes.

2. Albinism in humans is inherited as a simple recessive trait. Determine the genotypes of the parents and offspring for the following families. When two alternative genotypes are possible, list both. (a) Two nonalbino (normal) parents have five children, four normal and one albino. (b) A normal male and an albino female have six children, all normal.

3. In a problem involving albinism (see Problem 2), which of Mendel's postulates are demonstrated?

4. Why was the garden pea a good choice as an experimental organism in Mendel's work?

5. Pigeons exhibit a checkered or plain feather pattern. In a series of controlled matings, the following data were obtained:

			F₁ Progeny	
	P₁ Cross		Checkered	Plain
(a) checkered	×	checkered	36	0
(b) checkered	×	plain	38	0
(c) plain	×	plain	0	35

Then F_1 offspring were selectively mated with the following results. (The P_1 cross giving rise to each F_1 pigeon is indicated in parentheses.)

			F₂ Progeny	
F₁ × F₁ Crosses			Checkered	Plain
(d) checkered (a)	×	plain (c)	34	0
(e) checkered (b)	×	plain (c)	17	14
(f) checkered (b)	×	checkered (b)	28	9
(g) checkered (a)	×	checkered (b)	39	0

How are the checkered and plain patterns inherited? Predict the results of the $F_1 \times F_1$ mating from cross (b).

6. Mendel crossed peas having round seeds and yellow cotyledons with peas having wrinkled seeds and green cotyledons. All the F_1 plants had round seeds with yellow cotyledons. Diagram this cross through the F_2 generation, using both the Punnett square and forked-line methods.

7. Determine the genotypes of the parental plants by analyzing the phenotypes of the offspring from these crosses:

Parental Plants	Offspring
(a) round, yellow × round, yellow	3/4 round, yellow
	1/4 wrinkled, yellow
(b) round, yellow × wrinkled, yellow	6/16 wrinkled, yellow
	2/16 wrinkled, green
	6/16 round, yellow
	2/16 round, green
(c) round, yellow × wrinkled, green	1/4 round, yellow
	1/4 round, green
	1/4 wrinkled, yellow
	1/4 wrinkled, green

8. Are any of the crosses in Problem 7 testcrosses? If so, which one(s)?

9. Which of Mendel's postulates can be demonstrated in the crosses of Problem 7 but not in those in Problems 1 and 5? State this postulate.

10. Correlate Mendel's four postulates with what is now known about homologous chromosomes, genes, alleles, and the process of meiosis.

11. What is the basis for homology among chromosomes?

12. Distinguish between homozygosity and heterozygosity.

13. In *Drosophila*, gray body color is dominant over ebony body color, while long wings are dominant over vestigial wings. Work the following crosses through the F_2 generation, and determine the genotypic and phenotypic ratios for each generation. Assume that the P_1 individuals are homozygous:
    (a) gray, long × ebony, vestigial
    (b) gray, vestigial × ebony, long
    (c) gray, long × gray, vestigial

14. How many different types of gametes can be formed by individuals of the following genotypes? What are they in each case? (a) *AaBb*, (b) *AaBB*, (c) *AaBbCc*, (d) *AaBBcc*, (e) *AaBbcc*, and (f) *AaBbCcDdEe*?

15. Using the forked-line method, determine the genotypic and phenotypic ratios of these trihybrid crosses:
    (a) *AaBbCc* × *AaBBCC*, (b) *AaBBCc* × *aaBBCc*, and (c) *AaBbCc* × *AaBbCc*.

16. Mendel crossed peas with round, green seeds with peas having wrinkled, yellow seeds. All F_1 plants had seeds that were round and yellow. Predict the results of testcrossing these F_1 plants.

17. Shown are F_2 results of two of Mendel's monohybrid crosses. State a null hypothesis that you will test using chi-square analysis. Calculate the χ^2 value and determine the p value for both crosses, then interpret the p values. Which cross shows a greater amount of deviation?

(a) Full pods	882
Constricted pods	299
(b) Violet flowers	705
White flowers	224

18. In one of Mendel's dihybrid crosses, he observed 315 round, yellow; 108 round, green; 101 wrinkled, yellow; and 32 wrinkled, green F_2 plants. Analyze these data using chi-square analysis to see whether (a) they fit a 9:3:3:1 ratio; (b) the round, wrinkled traits fit a 3:1 ratio; or (c) the yellow, green traits fit a 3:1 ratio.

19. A geneticist, in assessing data that fell into two phenotypic classes, observed values of 250:150. He decided to perform chi-square analysis using two different null hypotheses: (a) The data fit a 3:1 ratio; and (b) the data fit a 1:1 ratio. Calculate the χ^2 values for each hypothesis. What can you conclude about each hypothesis?

20. The basis for rejecting any null hypothesis is arbitrary. The researcher can set more or less stringent standards by deciding to raise or lower the critical p value. Would the use of a standard of $p = 0.10$ be more or less stringent in failing to reject the null hypothesis? Explain.

21. Consider three independently assorting gene pairs, *A/a*, *B/b*, and *C/c*, where each demonstrates typical dominance (*A–*, *B–*, *C–*), and recessiveness (*aa*, *bb*, *cc*). What is the probability of obtaining an offspring that is *AABbCc* from parents that are *AaBbCC* and *AABbCc*?

22. What is the probability of obtaining a triply recessive individual from the parents shown in Problem 21?

23. Of all offspring of the parents in Problem 21, what proportion will express all three dominant traits?

24. For the following pedigree, predict the mode of inheritance and the resulting genotypes of each individual. Assume that the alleles *A* and *a* control the expression of the trait.

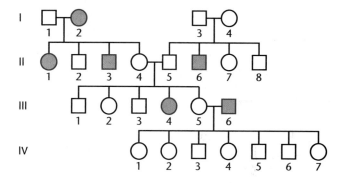

25. Which of Mendel's postulates are demonstrated by the pedigree in Problem 24? List and define these postulates.

26. The following pedigree follows the inheritance of myopia (nearsightedness) in humans. Predict whether the disorder is inherited as a dominant or a recessive trait. Based on your prediction, indicate the most probable genotype for each individual.

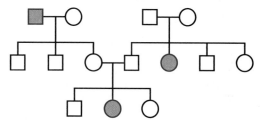

27. Draw all possible conclusions concerning the mode of inheritance of the trait expressed in each of the following limited pedigrees. (Each case is based on a different trait.)

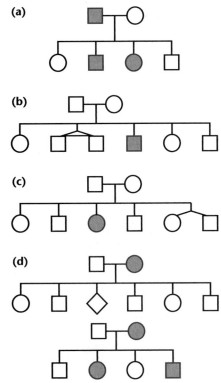

(a)

(b)

(c)

(d)

28. Two true-breeding pea plants are crossed. One parent is round, terminal, violet, constricted, while the other expresses the contrasting phenotypes of wrinkled, axial, white, full. The four pairs of contrasting traits are controlled by four genes, each located on a separate chromosome. In the F_1 generation, only round, axial, violet, and full are expressed. In the F_2 generation, all possible combinations of these traits are expressed in ratios consistent with Mendelian inheritance.

    (a) What conclusion can you draw about the inheritance of these traits based on the F_1 results?

    (b) Which phenotype appears most frequently in the F_2 results? Write a mathematical expression that predicts the frequency of occurrence of this phenotype.

    (c) Which F_2 phenotype is expected to occur least frequently? Write a mathematical expression that predicts this frequency.

    (d) How often is either P_1 phenotype likely to occur in the F_2 generation?

    (e) If the F_1 plant is testcrossed, how many different phenotypes will be produced, and how does this number compare to the number of different phenotypes in the F_2 generation discussed in part (b)?

29. Tay-Sachs disease (TSD) is an inborn error of metabolism that results in death, usually before the age of five. You are a genetic counselor, and you interview a phenotypically normal couple who consult you because the man had a female first cousin (on his father's side) who died from TSD, and the woman had a maternal uncle with TSD. There are no other known cases in either family, and none of the matings were/are between related individuals. Assume that this trait is rare in this population.

    (a) Using standard pedigree symbols, draw a pedigree of these individuals' families, showing the relevant individuals.

    (b) The couple asks you to calculate the probability that they both are heterozygous for the TSD allele.

    (c) They also want to know the probability that neither of them is heterozygous.

    (d) They also ask you for the probability that one of them is heterozygous but the other is not.

    [*Hint:* The answers to (b), (c), and (d) should add up to 1.0.]

30. The wild-type (normal) fruit fly, *Drosophila melanogaster*, has straight wings and long bristles. Mutant strains have been isolated with either curled wings or short bristles. The genes representing these two mutant traits are located on separate chromosomes. Carefully examine the data from the five crosses below. (a) For each mutation, determine whether it is dominant or recessive. In each case, identify which crosses support your answer; and (b) define gene symbols, and determine the genotypes of the parents for each cross.

		Number of Progeny			
Cross		straight wings, long bristles	straight wings, short bristles	curled wings, long bristles	curled wings, short bristles
1	straight, short × straight, short	30	90	10	30
2	straight, long × straight, long	120	0	40	0
3	curled, long × straight, short	40	40	40	40
4	straight, short × straight, short	40	120	0	0
5	curled, short × straight, short	20	60	20	60

31. To assess Mendel's law of segregation using tomatoes, a true-breeding tall variety (SS) is crossed with a true-breeding short variety (ss). The heterozygous tall plants (Ss) were crossed to produce the two sets of F_2 data shown below:

Set I	Set II
30 tall	300 tall
5 short	50 short

    (a) Using chi-square analysis, analyze the results for both data sets. Calculate χ^2 values, and estimate the p values in both cases.

    (b) From the analysis in part (a), what can you conclude about the importance of generating large data sets in experimental settings?

Modification of Mendelian Ratios

4

Pea

Inherited variation in comb shape of chickens, controlled by two pairs of genes.

Walnut

Rose

Single

■ CHAPTER CONCEPTS

- While alleles are transmitted from parent to offspring according to Mendelian principles, they often do not display the clear-cut dominant/recessive relationship observed by Mendel.

- In many cases, in contrast to Mendelian genetics, two or more genes are known to influence the phenotype of a single characteristic.

- Still another exception to Mendelian inheritance is the presence of genes on sex chromosomes, whereby one of the sexes contains only a single member of that chromosome.

- The result of the various exceptions to Mendelian principles is the occurrence of phenotypic ratios that differ from those resulting from standard monohybrid, dihybrid, and trihybrid crosses.

- Phenotypes are often the result of both genetics and the environment within which genes are expressed.

- Extranuclear inheritance, resulting from the expression of genes present in the DNA found in mitochondria and chloroplasts, modifies Mendelian inheritance patterns. Such genes are most often transmitted through the female gamete.

In Chapter 3, we discussed the simplest principles of transmission genetics. We saw that genes are present on homologous chromosomes and that these chromosomes segregate from each other and assort independently with other segregating chromosomes during gamete formation. These two postulates are the fundamental principles of gene transmission from parent to offspring. However, when gene expression does not adhere to a simple dominant/recessive mode or when more than one pair of genes influences the expression of a single character, the classic 3:1 and 9:3:3:1 ratios are usually modified. Although more complex modes of inheritance result, the fundamental principles set down by Mendel still hold true in these situations.

In this chapter, we will initially restrict our discussion to the inheritance of traits that are under the control of only one set of genes. In diploid organisms, which have homologous pairs of chromosomes, two copies of each gene influence such traits. The copies need not be identical because alternative forms of genes (alleles) occur within populations. How alleles influence phenotypes is our primary focus. We will then consider how a single phenotype can be controlled by more than one set of genes, a situation sometimes described as gene interaction.

Thus far, we have restricted our discussion to chromosomes other than the X and Y pair. By examining cases where genes are present on the X chromosome, illustrating X-linkage, we will see yet another modification of Mendelian ratios. Our discussion of modified ratios also includes the consideration of sex-limited and sex-influenced inheritance, cases where the sex of the individual, but not necessarily genes on the X chromosome, influences the phenotype. We will also consider how a given phenotype may vary depending on the overall environment in which a cell or an organism finds itself. This discussion points out that phenotypic expression depends on more than just the genotype of an organism. Finally, we conclude with a discussion of extranuclear inheritance, cases where DNA within organelles influences an organism's phenotype.

How Do We Know?

In this chapter, we will focus on how genes control phenotypes in ways that modify simple Mendelian inheritance patterns. As you study this topic, you should try to answer several fundamental questions:

1. What experimental approach did early geneticists use to explain inheritance patterns that did not fit typical Mendelian ratios?
2. How did geneticists establish that inheritance of some phenotypic characteristics involved the interaction of two or more gene pairs?
3. How do we know how many genes are involved in the inheritance of a trait?
4. What observations enabled us to distinguish between various modes of qualitative inheritance?
5. How do we know that specific genes are located on the sex-determining chromosomes (i. e., that X-linked inheritance exists)?
6. How was extranuclear inheritance discovered?

4.1 Alleles Alter Phenotypes in Different Ways

After Mendel's work was rediscovered in the early 1900s, researchers focused on the many ways in which genes influence an individual's phenotype. Each type of inheritance was more thoroughly investigated when observations of genetic data did not conform precisely to the expected Mendelian ratios, and hypotheses that modified and extended the Mendelian principles were proposed and tested with specifically designed crosses. The explanations were in accord with the principle that a phenotype is under the control of one or more genes located at specific loci on one or more pairs of homologous chromosomes.

To understand the various modes of inheritance, we must first examine the potential function of alleles. Alleles are alternative forms of the same gene. The allele that occurs most frequently in a population, the one that we arbitrarily designate as normal, is called the **wild-type allele** and is usually dominant (the allele for tall plants in the garden pea, for example). Its product is therefore functional in the cell. Wild-type alleles are responsible for the corresponding wild-type phenotype and are the standards against which all mutations at a particular locus are compared.

A mutant allele contains modified genetic information and often specifies an altered gene product. For example, in human populations, there are many known alleles of the gene that encode the β chain of human hemoglobin. All such alleles store information necessary for the synthesis of the β-chain polypeptide, but each allele specifies a slightly different form of the same molecule. Once the allele's product has been manufactured, the function of the product may or may not be altered.

The process of mutation is the source of alleles. For a new allele to be recognized when observing an organism, it must cause a change in the phenotype. A new phenotype results from a change in functional activity of the cellular product specified by that gene. Often, the mutation causes the diminution or the loss of the specific wild-type function. For example, if a gene is responsible for the synthesis of a specific enzyme, a mutation in that gene may ultimately change the conformation of this enzyme and reduce or eliminate its affinity for the substrate. Such a case is designated as a **loss of function mutation**. If the loss is complete, the mutation has resulted in what is called a **null allele**.

Conversely, other mutations may enhance the function of the wild-type product. Most often when this occurs, it is the result of increasing the quantity of the gene product. In such cases, the mutation may be affecting the regulation of transcription of the gene under consideration. Such cases are designated **gain of function mutations**, which generally result in dominant alleles since one copy in a diploid organism is sufficient to alter the normal phenotype. Examples of gain in function mutations include the genetic conversion of **proto-oncogenes**, which regulate the cell cycle, to **oncogenes**, where regulation is overridden by excess gene product. The result is the creation of a cancerous cell.

Having introduced the concept of gain or loss of function mutations, it is important to note that the possibility exists that a mutation will create an allele where no change in function can be detected. In this case, the mutation would not be immediately apparent since no phenotypic variation would be evident. However, such a mutation could be detected if the DNA sequence of the gene was examined directly.

Finally, we note here that while a phenotypic trait may be affected by a single mutation in one gene, traits are often influenced by more than one gene. For example, enzymatic reactions are most often part of complex metabolic pathways leading to the synthesis of an end product, such as an amino acid. Mutations in any of the various reactions have a common effect—the failure to synthesize the end product. Therefore, phenotypic traits related to the end product are often influenced by more than one gene.

In each of the many crosses discussed in the next few chapters, only one or a few gene pairs are involved. Keep in mind that in each cross discussed, all genes that are not under consideration are assumed to have no effect on the inheritance patterns described.

4.2 Geneticists Use a Variety of Symbols for Alleles

We have previously symbolized alleles for very simple Mendelian traits where the initial letter of the name of a recessive trait, lowercased and italic, denotes the recessive allele. The same letter in uppercase refers to the dominant allele. Thus, for tall and dwarf, where dwarf is recessive, D and d represent the alleles responsible for these respective traits. Mendel used upper- and lowercase letters such as these to symbolize his unit factors.

Another useful system was developed in genetic studies of the fruit fly *Drosophila melanogaster* to discriminate between wild-type and mutant traits. This system uses the initial letter, or a combination of two or three letters, of the name of the mutant trait. If the trait is recessive, lowercase is used; if it is dominant, uppercase is used. The contrasting wild-type trait is denoted by the same letter, but with a superscript $+$. For example, *ebony* is a recessive body color mutation in *Drosophila*. The normal wild-type body color is gray. Using this system, we denote *ebony* by the symbol e, while we denote gray by e^+. The responsible locus may be occupied by either the wild-type allele (e^+) or the mutant allele (e). A diploid fly may thus exhibit one of three possible genotypes:

e^+/e^+	gray homozygote (wild type)
e^+/e	gray heterozygote (wild type)
e/e	ebony homozygote (mutant)

The slash between the letters indicates that the two allele designations represent the same locus on two homologous chromosomes. If we instead consider a dominant wing mutation such as *Wrinkled* (*Wr*) wing in *Drosophila*, the three possible designations are Wr^+/Wr^+, Wr^+/Wr, and Wr/Wr. The latter two genotypes express the wrinkled-wing phenotype.

One advantage of this system is that further abbreviation can be used when convenient: The wild-type allele may simply be denoted by the $+$ symbol. With *ebony* as an example, the designations of the three possible genotypes become

$+/+$	gray homozygote (wild type)
$+/e$	gray heterozygote (wild type)
e/e	ebony homozygote (mutant)

Another variation is utilized when no dominance exists between alleles. We simply use uppercase italic letters and superscripts to denote alternative alleles (e.g., R^1 and R^2, L^M and L^N, I^A and I^B). Their use will become apparent later in this chapter.

Although we have adopted a standard convention for assigning genetic symbols, many diverse systems of genetic nomenclature are used to identify genes in various organisms. Usually, the symbol selected reflects the function of the gene or even a disorder caused by a mutant gene. For example, the yeast *cdk* is the abbreviation for the *cyclin dependent kinase* gene, whose product is involved in cell-cycle regulation. In bacteria, *leu*⁻ refers to a mutation that interrupts the biosynthesis of the amino acid leucine, where the wild-type gene is designated *leu*⁺. The symbol *dnaA* represents a bacterial gene involved in DNA replication (and DnaA is the protein made by that gene). In humans, capital letters are used to name genes: *BRCA1* represents the first gene associated with susceptibility to *br*east *ca*ncer. Although these different systems may seem complex, they are useful ways to symbolize genes.

4.3 Neither Allele Is Dominant in Incomplete, or Partial, Dominance

A cross between parents with contrasting traits may generate offspring with an intermediate phenotype. For example, if plants such as four-o'clocks or snapdragons with red flowers are crossed with white-flowered plants, the offspring have pink flowers. Some red pigment is produced in the F_1 intermediate pink-colored flowers. Therefore, neither red nor white flower color is dominant. This situation is known as **incomplete**, or **partial, dominance**.

If this phenotype is under the control of a single gene and two alleles where neither is dominant, the results of the F_1(pink) × F_1(pink) cross can be predicted. The resulting F_2 generation shown in Figure 4–1 confirms the hypothesis that only one pair of alleles determines these phenotypes. The genotypic ratio (1:2:1) of the F_2 generation is identical to that of Mendel's monohybrid cross. However, because neither allele is dominant, the phenotypic ratio is identical to the genotypic ratio. Note that because neither allele is recessive, we have chosen not to use upper- and lowercase letters as symbols. Instead, we denoted the red and white alleles as R^1 and R^2. We could have used W^1 and W^2 or still other designations such as C^W and C^R, where C indicates "color" and the W and R superscripts indicate white and red.

Clear-cut cases of incomplete dominance, which result in intermediate expression of the overt phenotype, are relatively rare. However, even when complete dominance seems apparent, careful examination of the gene product, rather than the phenotype, often reveals an intermediate level of gene expression. An

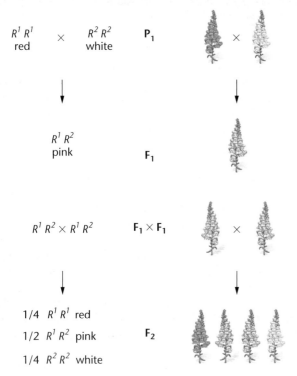

$R^1 R^1$ × $R^2 R^2$ **P₁**
red   white

↓   ↓

$R^1 R^2$
pink   **F₁**

$R^1 R^2$ × $R^1 R^2$ **F₁ × F₁**   ×

↓   ↓

1/4 $R^1 R^1$ red
1/2 $R^1 R^2$ pink   **F₂**
1/4 $R^2 R^2$ white

FIGURE 4–1 Incomplete dominance shown in the flower color of snapdragons.

example is the human biochemical disorder **Tay-Sachs disease**, in which homozygous recessive individuals are severely affected with a fatal lipid storage disorder, and neonates die during their first one to three years of life (see the GTS Essay in the previous chapter on page 55). There is almost no activity of the enzyme **hexosaminidase** in afflicted individuals, an enzyme normally involved in lipid metabolism. Heterozygotes, with only a single copy of the mutant gene, are phenotypically normal but express only about 50 percent of the enzyme activity found in homozygous normal individuals. Fortunately, this level of enzyme activity is adequate to achieve normal biochemical function—a situation not uncommon in enzyme disorders.

4.4 In Codominance, the Influence of Both Alleles in a Heterozygote Is Clearly Evident

If two alleles of a single gene are responsible for producing two distinct, detectable gene products, a situation different from incomplete dominance or dominance/recessiveness

arises. In this case, the joint expression of both alleles in a heterozygote is called **codominance**. The **MN blood group** in humans illustrates this phenomenon and is characterized by an antigen called a glycoprotein, found on the surface of red blood cells. In the human population, two forms of this glycoprotein exist, designated M and N; an individual may exhibit either one or both of them.

The MN system is under the control of an autosomal locus found on chromosome 4 and two alleles designated L^M and L^N. Humans are diploid, so three combinations are possible, each resulting in a distinct blood type:

Genotype	Phenotype
$L^M L^M$	M
$L^M L^N$	MN
$L^N L^N$	N

As predicted, a mating between two heterozygous MN parents may produce children of all three blood types, as follows:

$$L^M L^N \times L^M L^N$$

↓

$$1/4 \, L^M L^M$$
$$1/2 \, L^M L^N$$
$$1/4 \, L^N L^N$$

Once again the genotypic ratio, 1:2:1, is upheld.

Codominant inheritance is characterized by *distinct expression of the gene products of both alleles*. This characteristic distinguishes it from incomplete dominance, where heterozygotes express an intermediate, blended phenotype.

4.5 Multiple Alleles of a Gene May Exist in a Population

The information stored in any gene is extensive, and mutations can modify this information in many ways. Each change produces a different allele. Therefore, for any specific gene, the number of alleles within members of a population need not be restricted to two. When three or more alleles of the same gene are found, **multiple alleles** are present that create a unique mode of inheritance. It is important to realize that *multiple alleles can be studied only in populations*. An individual diploid organism has, at most, two homologous gene loci that may be occupied by different alleles of the same gene. However, among many members of a species, numerous alternative forms of the same gene can exist.

The ABO Blood Group

The simplest case of multiple alleles is that in which three alternative alleles of one gene exist. This situation is illustrated by the **ABO blood group** in humans, discovered by Karl Landsteiner in the early 1900s. The ABO system, like the MN blood group, is characterized by the presence of antigens on the surface of red blood cells. The A and B antigens are distinct from MN antigens and are under the control of a different gene, located on chromosome 9. As in the MN system, one combination of alleles in the ABO system exhibits a codominant mode of inheritance.

When individuals are tested using antisera that contain antibodies against the A or B antigen, four phenotypes are revealed. Each individual has either the A antigen (A phenotype), the B antigen (B phenotype), the A and B antigens (AB phenotype), or neither antigen (O phenotype). In 1924, it was hypothesized that these phenotypes were inherited as the result of three alleles of a single gene. This hypothesis was based on studies of the blood types of many different families.

Although different designations can be used, we use the symbols I^A, I^B, and I^O to distinguish these three alleles; the I designation stands for *isoagglutinogen*, another term for antigen. If we assume that the I^A and I^B alleles are responsible for the production of their respective A and B antigens and that I^O is an allele that does not produce any detectable A or B antigens, we can list the various genotypic possibilities and assign the appropriate phenotype to each:

Genotype	Antigen	Phenotype
$I^A I^A$	A ⎫	
$I^A I^O$	A ⎬	A
$I^B I^B$	B ⎫	
$I^B I^O$	B ⎬	B
$I^A I^B$	A, B	AB
$I^O I^O$	Neither	O

In these assignments the I^A and I^B alleles are dominant to the I^O allele, but are codominant to each other. Our knowledge of human blood types has several practical applications, the most important of which are compatible blood transfusions and organ transplantations.

The Bombay Phenotype

The biochemical basis of the ABO blood-type system has been carefully worked out. The A and B antigens are actually carbohydrate groups (sugars) that are bound to lipid molecules (fatty acids) protruding from the membrane of the red blood cell. The specificity of the A and B antigens is based on the terminal sugar of the carbohydrate group. Both the A and B antigens are derived from a precursor molecule called the **H substance**, to which one or two terminal sugars are added.

In extremely rare instances, first recognized in a woman in Bombay in 1952, the H substance is incompletely formed. As a result, it is an inadequate substrate for the enzyme that normally adds the terminal sugar. This condition results in the expression of blood type O and is called the **Bombay phenotype**. Research has revealed that this condition is due to a rare recessive mutation at a locus separate from that controlling the A and B antigens. The gene is now designated *FUT1* (encoding an enzyme, fucosyl transferase), and individuals that are homozygous for the mutation cannot synthesize the complete H substance. Thus, even though they may have the I^A and/or I^B alleles, neither the A nor B antigen can be added to the cell surface. This information explains why the woman in Bombay

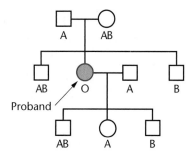

FIGURE 4–2 A partial pedigree of a woman with the Bombay phenotype. Functionally, her ABO blood group behaves as type O. Genetically, she is type B.

expressed blood type O even though one of her parents was type AB (thus she should not have been type O), and why she was able to pass the I^B allele to her children (Figure 4–2).

The *white* Locus in *Drosophila*

Many other phenotypes in plants and animals are known to be controlled by multiple allelic inheritance. In *Drosophila*, many alleles are known at practically every locus. The recessive mutation that causes white eyes, discovered by Thomas H. Morgan and Calvin Bridges in 1912, is one of over 100 alleles that can occupy this locus. In this allelic series, eye colors range from complete absence of pigment in the *white* allele, to deep ruby in the *white-satsuma* allele, to orange in the *white-apricot* allele, to a buff color in the *white-buff* allele. These alleles are designated w, w^{sat}, w^a, and w^{bf}, respectively. In each case, the total amount of pigment in these mutant eyes is reduced to less than 20 percent of that found in the brick-red, wild-type eye. Table 4.1 lists these and other *white* alleles and their color phenotypes.

Now Solve This

Problem 27 on page 92 involves a series of multiple alleles controlling coat color in rabbits.

Hint: Note particularly the hierarchy of dominance of the various alleles. Remember also that even though there can be more than two alleles in a population, an individual can have at most two of these. Thus, the allelic distribution into gametes adheres to the principle of segregation.

4.6 Lethal Alleles Represent Essential Genes

Many gene products are essential to an organism's survival. Mutations resulting in the synthesis of a gene product that is nonfunctional can often be tolerated in the heterozygous state; that is, one wild-type allele may be sufficient to produce enough of the essential product to allow survival. However,

TABLE 4.1	Some of the Alleles Present at the *white* Locus of *Drosophila melanogaster* and Their Eye-Color Phenotype	

Allele	Name	Eye Color
w	*white*	pure white
w^a	*white-apricot*	yellowish orange
w^{bf}	*white-buff*	light buff
w^{bl}	*white-blood*	yellowish ruby
w^{cf}	*white-coffee*	deep ruby
w^e	*white-eosin*	yellowish pink
w^{mo}	*white-mottled orange*	light mottled orange
w^{sat}	*white-satsuma*	deep ruby
w^{sp}	*white-spotted*	fine grain, yellow mottling
w^t	*white-tinged*	light pink

such a mutation behaves as a recessive **lethal allele**, and homozygous recessive individuals will not survive. The time of death will depend on when the product is essential. In mammals, for example, this might occur during development, early childhood, or even adulthood.

In some cases, the allele responsible for a lethal effect when homozygous may also result in a distinctive mutant phenotype when present heterozygously. It is behaving as a recessive lethal allele but is dominant with respect to the phenotype. For example, a mutation that causes a yellow coat in mice was discovered in the early part of this century. The yellow coat varies from the normal agouti (wild-type) coat phenotype, as shown in Figure 4–3. Crosses between the various combinations of the two strains yield unusual results:

Crosses				
(A) agouti	×	agouti	⟶	all agouti
(B) yellow	×	yellow	⟶	2/3 yellow: 1/3 agouti
(C) agouti	×	yellow	⟶	1/2 yellow: 1/2 agouti

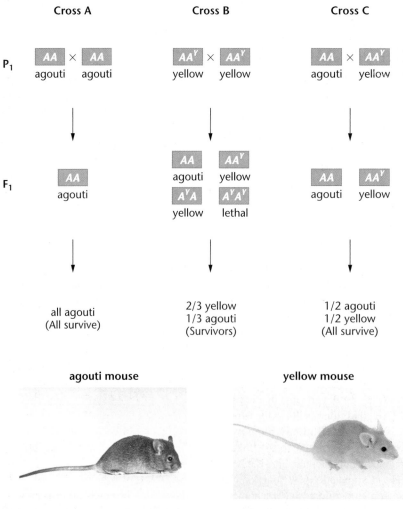

FIGURE 4–3 Inheritance patterns in three crosses involving the normal wild-type *agouti* allele (*A*) and the mutant *yellow* allele (A^Y) in the mouse. Note that the mutant allele behaves dominantly to the normal allele in controlling coat color, but it also behaves as a homozygous recessive lethal allele. The genotype $A^Y A^Y$ does not survive.

These results are explained on the basis of a single pair of alleles. With regard to coat color, the mutant *yellow* allele (A^Y) is dominant to the wild-type *agouti* allele (*A*), so heterozygous mice will have yellow coats. However, the *yellow* allele is also a homozygous recessive lethal. When present in two copies, the mice die before birth. Thus, there are no homozygous yellow mice. The genetic basis for these three crosses is shown in Figure 4–3.

In other cases, lethal alleles behave dominantly to their wild-type counterpart. In such dominant lethal alleles, the presence of just one copy of the allele results in the death of the individual. In humans, a disorder called **Huntington disease** (previously referred to as Huntington's chorea) is due to a dominant autosomal allele *H*, where the onset of the disease in heterozygotes (*Hh*) is delayed, usually well into adulthood. Affected individuals then undergo gradual nervous and motor degeneration until they die. This lethal disorder is particularly tragic because it has such a late onset, typically at about age 40. By that time, the affected individual may have produced a family, and each of their children has a 50 percent probability of inheriting the lethal allele, transmitting the allele to his or her offspring, and eventually developing the disorder. The American folk singer and composer Woody Guthrie (father of modern-day folk singer Arlo Guthrie) died from this disease at age 39.

Dominant lethal alleles are rarely observed. For these alleles to exist in a population, the affected individuals must reproduce before the lethal allele is expressed, as can occur in Huntington disease. If all affected individuals die before reaching reproductive age, the mutant gene will not be passed to future generations, and the mutation will disappear from the population unless it arises again as a result of a new mutation.

4.7 Combinations of Two Gene Pairs with Two Modes of Inheritance Modify the 9:3:3:1 Ratio

Each example discussed so far modifies Mendel's 3:1 F_2 monohybrid ratio. Therefore, combining any two of these modes of inheritance in a dihybrid cross will likewise modify the classical 9:3:3:1 ratio. Having established the foundation for the modes of inheritance of incomplete dominance, codominance, multiple alleles, and lethal alleles, we can now deal with the situation of two modes of inheritance occurring simultaneously. Mendel's principle of independent assortment applies to these situations, provided that the genes controlling each character are not linked on the same chromosome.

Consider, for example, a mating that occurs between two humans who are both heterozygous for the autosomal recessive gene that causes albinism and who are both of blood type AB. What is the probability of a particular phenotypic combination occurring in each of their children? Albinism is inherited in the simple Mendelian fashion, and the blood types are determined by the series of three multiple alleles, I^A, I^B, and I^O. The solution to this problem is

diagrammed in Figure 4–4, using the forked-line method. This dihybrid cross does not yield the classical four phenotypes in a 9:3:3:1 ratio. Instead, six phenotypes occur in a 3:6:3:1:2:1 ratio, establishing the expected probability for each phenotype. This is just one of the many variants of modified ratios that are possible when different modes of inheritance are combined.

4.8 Phenotypes Are Often Affected by More Than One Gene

Soon after Mendel's work was rediscovered, experimentation revealed that individual characteristics displaying discrete phenotypes are often under the control of more than one gene. This was a significant discovery because it revealed that genetic influence on the phenotype is often much more complex than envisioned by Mendel. Instead of single genes controlling the development of individual parts of the plant or animal body, it soon became clear that phenotypic characters can be influenced by the interactions of many different genes and their products.

The term **gene interaction** is often used to describe the idea that several genes influence a particular characteristic. This does not mean, however, that two or more genes, or their products, necessarily interact directly with one another to influence a particular phenotype. Rather, the cellular function of numerous gene products contributes to the development of a common phenotype. For example, the development of an organ such as the compound eye of an insect is exceedingly complex and leads to a structure with multiple phenotypic manifestations—such as specific size, shape, texture, and color. The development of the eye is a complex cascade of developmental events leading to its formation. This process exemplifies the developmental concept of **epigenesis**, whereby each step of development increases the complexity of this sensory organ and is under the control and influence of one or more genes.

Epistasis

Some of the best examples of gene interaction are those that reveal the phenomenon of **epistasis** (Greek for "stoppage"). Epistasis occurs when the expression of one gene or gene pair masks or modifies the expression of another gene or gene pair. Sometimes the genes involved control the expression of the same general phenotypic characteristic in an antagonistic manner, as when masking occurs. In other cases, however, the genes involved exert their influence on one another in a complementary, or cooperative, fashion.

For example, the homozygous presence of a recessive allele prevents or overrides the expression of other alleles at a second locus (or several other loci). In this case, the alleles at the first locus are said to be *epistatic* to those at the second locus, and the alleles at the second locus are *hypostatic* to those at the first locus. In another example, a single dominant allele at the first locus influences the expression of the alleles at a second gene locus. In a third example, two gene pairs complement one another such that

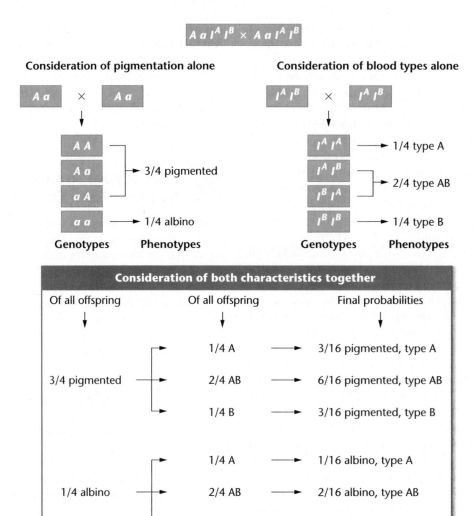

FIGURE 4–4 Calculation of the mating probabilities involving the ABO blood type and albinism in humans, using the forked-line method.

at least one dominant allele at each locus is required to express a particular phenotype.

The Bombay phenotype discussed earlier is an example of the homozygous recessive condition at one locus masking the expression of a second locus. There, we established that the homozygous presence of the mutant form of the *FUT1* gene masks the expression of the I^A and I^B alleles. Only individuals containing at least one wild-type *FUT1* allele can form the A or B antigen. As a result, individuals whose genotypes include the I^A or I^B allele and who lack a wild-type allele are of the type O phenotype, regardless of their potential to make either antigen. An example of the outcome of matings between individuals heterozygous at both loci is illustrated in Figure 4–5. If many such individuals have children, the phenotypic ratio of 3 A: 6 AB: 3 B: 4 O is expected in their offspring.

It is important to note the following points when examining this cross and the predicted phenotypic ratio:

1. A key distinction exists in this cross compared to the modified dihybrid cross shown in Figure 4–4: *only one*

characteristic—blood type—is being followed. In the modified dihybrid cross of Figure 4–4, blood type *and* skin pigmentation are followed as separate phenotypic characteristics.

2. Even though only a single character was followed, the phenotypic ratio is expressed in sixteenths. If we knew nothing about the H substance and the genes controlling it, we could still be confident that a second gene pair, other than that controlling the A and B antigens, is involved in the phenotypic expression. *When studying a single character, a ratio that is expressed in 16 parts (e.g., 3:6:3:4) suggests that two gene pairs are "interacting" during the expression of the phenotype under consideration.*

The study of gene interaction reveals inheritance patterns that modify the classical Mendelian dihybrid F_2 ratio (9:3:3:1) in other ways as well. In these examples, epistasis combines one or more of the four phenotypic categories in various ways. The generation of these four groups is reviewed in Figure 4–6, along with several modified ratios.

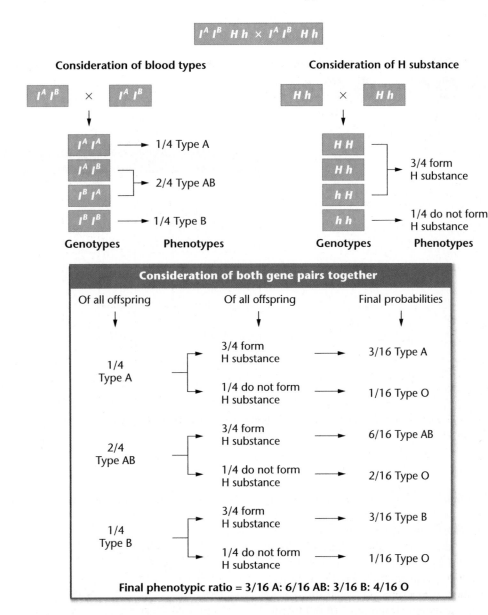

FIGURE 4–5 The outcome of a mating between individuals who are heterozygous at two genes determining their ABO blood type. Final phenotypes are calculated by considering both genes separately and then combining the results using the forked-line method.

As we discuss these and other examples, we will make several assumptions and adopt certain conventions:

1. In each case, distinct phenotypic classes are produced, each clearly discernible from all others. Such traits illustrate discontinuous variation, where phenotypic categories are discrete and qualitatively different from one another.

2. The genes considered in each cross are not linked and therefore assort independently of one another during gamete formation. To allow you to easily compare the results of different crosses, we designated alleles as *A*, *a* and *B*, *b* in each case.

3. When we assume that complete dominance exists between the alleles of any gene pair, such that *AA* and *Aa*, or *BB*

and *Bb* are equivalent in their genetic effects, we use the designations *A–* or *B–* for both combinations, where the dash (–) indicates that either allele may be present, without consequence to the phenotype.

4. All P₁ crosses involve homozygous individuals (e.g., *AABB* × *aabb*, *AAbb* × *aaBB*, or *aaBB* × *AAbb*). Therefore, each F₁ generation consists of only heterozygotes of genotype *AaBb*.

5. In each example, the F₂ generation produced from these heterozygous parents is our main focus of analysis. When two genes are involved (as in Figure 4–6), the F₂ genotypes fall into four categories: 9/16 *A–B–*, 3/16 *A–bb*, 3/16 *aaB–*, and 1/16 *aabb*. Because of dominance, all genotypes in each category have an equivalent effect on the phenotype.

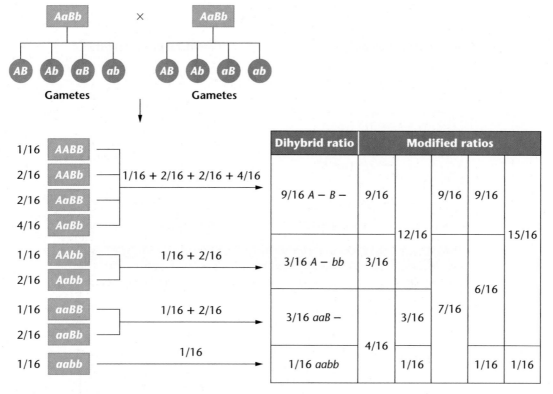

FIGURE 4–6 Generation of the various modified dihybrid ratios from the nine unique genotypes produced in a cross between individuals who are heterozygous at two genes.

Case 1 is the inheritance of coat color in mice (Figure 4–7). Normal wild-type coat color is agouti, a grayish pattern formed by alternating bands of pigment on each hair. Agouti is dominant to black (non-agouti) hair, which is caused by a recessive mutation, *a*. Thus, *A–* results in agouti, while *aa* yields black coat color. When it is homozygous, a recessive mutation, *b*, at a separate locus, eliminates pigmentation altogether, yielding albino mice (*bb*), regardless of the genotype at the other locus. The presence of at least one *B* allele allows pigmentation to occur in much the same way that the *H* allele in humans allows the expression of the ABO blood types. In a cross between agouti (*AABB*) and albino (*aabb*), members of the F₁ are all *AaBb* and have agouti coat color. In the F₂ progeny of a cross between two F₁ heterozygotes, the following genotypes and phenotypes are observed:

$$F_1: AaBb \times AaBb$$

$$\downarrow$$

F₂ Ratio	Genotype	Phenotype	Final Phenotypic Ratio
9/16	*A–B–*	agouti	
3/16	*A–bb*	albino	9/16 agouti
3/16	*aaB–*	black	4/16 albino
1/16	*aabb*	albino	3/16 black

We can envision gene interaction yielding the observed 9:3:4 F₂ ratio as a two-step process:

	Gene B		**Gene A**	
Precursor Molecule (colorless)	$\downarrow$ $\longrightarrow$ $B–$	Black Pigment	$\downarrow$ $\longrightarrow$ $A–$	Agouti Pattern

In the presence of a *B* allele, black pigment can be made from a colorless substance. In the presence of an *A* allele, the black pigment is deposited during the development of hair in a pattern that produces the agouti phenotype. If the *aa* genotype occurs, all of the hair remains black. If the *bb* genotype occurs, no black pigment is produced, regardless of the presence of the *A* or *a* alleles, and the mouse is albino. Therefore, the *bb* genotype masks or suppresses the expression of the *A* gene, thus demonstrating epistasis.

A second type of epistasis occurs when a dominant allele at one genetic locus masks the expression of the alleles at a second locus. For instance, Case 2 of Figure 4–7 deals with the inheritance of fruit color in summer squash. Here, the dominant allele *A* results in white fruit color regardless of the genotype at a second locus, *B*. In the absence of the dominant *A* allele (the *aa* genotype), *BB* or *Bb* results in yellow color, while *bb* results in green color. Therefore, if two white-colored double heterozygotes (*AaBb*) are crossed, this type of epistasis generates an interesting phenotypic ratio:

			F₂ Phenotypes				Modified ratio
Case	Organism	Character	9/16	3/16	3/16	1/16	
1	Mouse	Coat color	agouti	albino	black	albino	9:3:4
2	Squash	Color	white		yellow	green	12:3:1
3	Pea	Flower color	purple	white			9:7
4	Squash	Fruit shape	disc	sphere		long	9:6:1
5	Chicken	Color	white		colored	white	13:3
6	Mouse	Color	white-spotted	white	colored	white-spotted	10:3:3
7	Shepherd's purse	Seed capsule	triangular			ovoid	15:1
8	Flour beetle	Color	6/16 sooty : 3/16 red	black	jet	black	6:3:3:4

FIGURE 4–7 The basis of modified dihybrid F₂ phenotypic ratios, resulting from crosses between doubly heterozygous F₁ individuals. The four groupings of the F₂ genotypes shown in Figure 4–6 and across the top of this figure are combined in various ways to produce these ratios.

F₁: $AaBb \times AaBb$

F₂ Ratio	Genotype	Phenotype	Final Phenotypic Ratio
9/16	A–B–	white	12/16 white
3/16	A–bb–	white	
3/16	aaB–	yellow	3/16 yellow
1/16	aabb	green	1/16 green

Of the offspring, 9/16 are $A - B-$ and are thus white. The 3/16 bearing the genotypes $A - bb$ are also white. Finally, 3/16 are yellow ($aaB-$), while 1/16 are green ($aabb$); and we obtain the modified ratio of 12:3:1.

Our third type of gene interaction (Case 3 of Figure 4–7) was first discovered by William Bateson and Reginald Punnett (of Punnett square fame). It is demonstrated in a cross between two true-breeding strains of white-flowered sweet peas. Unexpectedly, the results of this cross yield all purple F₁ plants, and the F₂ plants occur in a ratio of 9/16 purple to 7/16 white. The proposed explanation suggests that the presence of at least one dominant allele of each of two gene pairs is essential for flowers to be purple. All other genotype combinations yield white flowers because the homozygous condition of *either* recessive allele masks the expression of the dominant allele at the other locus. The cross is shown as follows:

P₁: $AAbb \times aaBB$
white   white

F₁: All $AaBb$ (purple)

F₂ Ratio	Genotype	Phenotype	Final Phenotypic Ratio
9/16	A–B–	purple	9/16 purple
3/16	A–bb–	white	
3/16	aaB–	white	7/16 white
1/16	aabb	white	

We can now see how two gene pairs might yield such results:

Gene A   **Gene B**

Precursor Substance (colorless) →$_{A-}$ Intermediate Product (colorless) →$_{B-}$ Final Product (purple)

At least one dominant allele from each pair of genes is necessary to ensure both biochemical conversions to the final product, yielding purple flowers. In our cross, this will occur in 9/16 of the F₂ offspring. All other plants (7/16) have flowers that remain white.

The preceding examples illustrate how the products of two genes "interact" to influence the development of a common phenotype. In other instances, more than two genes and their products are involved in controlling phenotypic expression.

Novel Phenotypes

Other cases of gene interaction yield novel, or new, phenotypes in the F_2 generation, in addition to producing modified dihybrid ratios. Case 4 in Figure 4–7 depicts the inheritance of fruit shape in the summer squash *Cucurbita pepo*. When plants with disc-shaped fruit (*AABB*) are crossed to plants with long fruit (*aabb*), the F_1 generation all have disc fruit. However, in the F_2 progeny, fruit with a novel shape—sphere—appear, along with fruit exhibiting the parental phenotypes. A variety of fruit shapes are shown in Figure 4–8.

The F_2 generation, with a modified 9:6:1 ratio, is generated as follows:

$$F_1: \quad AaBb \quad \times \quad AaBb$$
$$\text{disc} \quad \downarrow \quad \text{disc}$$

F_2 Ratio	Genotype	Phenotype	Final Phenotypic Ratio
9/16	A–B–	disc	
3/16	A–bb–	sphere	9/16 disc
3/16	aaB–	sphere	6/16 sphere
1/16	aabb	long	1/16 long

In this example of gene interaction, both gene pairs influence fruit shape equally. A dominant allele at either locus ensures a sphere-shaped fruit. In the absence of dominant alleles, the fruit is long. However, if both dominant alleles (*A* and *B*) are present, the fruit displays a flattened, disc shape.

Now Solve This

Problem 7 on page 90 involves a single characteristic, flower color, that can take on one of three variations. You are asked to determine how many genes are involved in the inheritance of flower color and what genotypes are responsible for what phenotypes.

Hint: The most important information is the data provided. You must analyze the raw data and convert the numbers to a meaningful ratio. This will guide you in determining how many gene pairs are involved. Then you can categorize the genotypic ratio in a way to match the phenotypic ratio.

Other Modified Dihybrid Ratios

The remaining cases (5–8) in Figure 4–7 show additional modifications of the dihybrid ratio and provide still other examples of gene interactions. However, all eight cases have two things in common. First, we have not violated the principles of segregation and independent assortment to explain the inheritance pattern of each case. Therefore, the added complexity of inheritance in these exam-

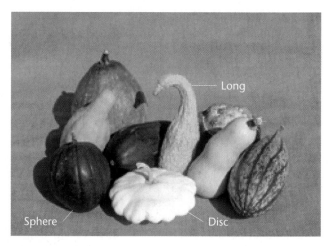

FIGURE 4–8 Summer squash exhibiting the fruit-shape phenotypes disc (white), long (orange gooseneck), and sphere (bottom left).

ples does not detract from the validity of Mendel's conclusions. Second, the F_2 phenotypic ratio in each example has been expressed in sixteenths. When similar observations are made in crosses where the inheritance pattern is unknown, it suggests to geneticists that two gene pairs are controlling the observed phenotypes. You should make the same inference in your analysis of genetics problems.

4.9 Complementation Analysis Can Determine If Two Mutations Causing a Similar Phenotype Are Alleles of the Same Gene

An interesting situation arises when two mutations, both of which produce a similar phenotype, are isolated independently. Suppose that two investigators independently isolate and establish a true-breeding strain of wingless *Drosophila* and demonstrate that each mutant phenotype is due to a recessive mutation. We might assume that both strains contain mutations in the same gene. However, since we know that many genes are involved in the formation of wings, mutations in any one of them might inhibit wing formation during development. The experimental approach called **complementation analysis** allows us to determine whether two such mutations are in the same gene—that is, whether they are alleles of the same gene or whether they represent mutations in separate genes.

Our analysis seeks to answer this simple question: *Are two mutations that yield similar phenotypes present in the same gene or in two different genes?* To find the answer, we cross the two mutant strains and analyze the F_1 generation. Two alternative outcomes and interpretations of this cross are shown in Figure 4–9. We discuss both cases, using the designations m^a for one of the mutations and m^b for the other one. Now we will determine experimentally whether or not m^a and m^b are alleles of the same gene.

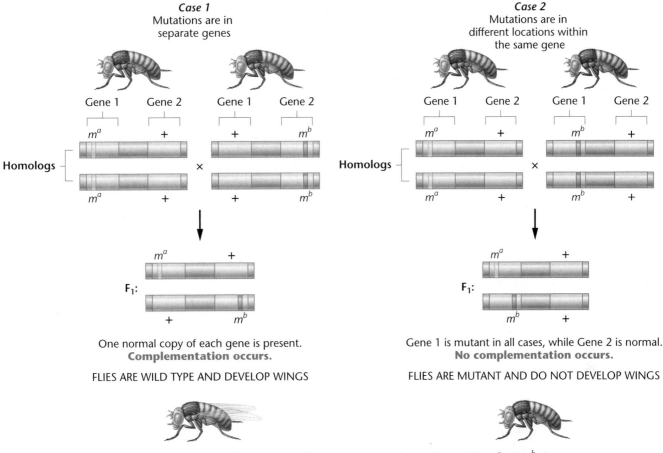

FIGURE 4–9   Complementation analysis of alternative outcomes of two wingless mutations in *Drosophila* (m^a and m^b). In Case 1, the mutations are not alleles of the same gene, while in Case 2, the mutations are alleles of the same gene.

Case 1.   *All offspring develop normal wings.*

Interpretation: The two recessive mutations are in separate genes and are not alleles of one another. Following the cross, all F_1 flies are heterozygous for both genes. *Complementation* is said to occur. Since each mutation is in a separate gene and each F_1 fly is heterozygous at both loci, the normal products of both genes are produced (by the one normal copy of each gene), and wings develop.

Case 2.   *All offspring fail to develop wings.*

Interpretation: The two mutations affect the same gene and are alleles of one another. Complementation does *not* occur. Since the two mutations affect the same gene, the F_1 flies are homozygous for the two mutant alleles (the m^a allele and the m^b allele). No normal product of the gene is produced, and in the absence of this essential product, wings do not form.

Complementation analysis, as originally devised by the *Drosophila* geneticist Edward B. Lewis, may be used to screen any number of individual mutations that result in the same phenotype. Such an analysis may reveal that only a single gene is involved or that two or more genes are involved. All mutations determined to be present in any single gene are said to fall into the same **complementation group**, and they will complement mutations in all other groups. When large numbers of mutations affecting the same trait are available

and studied using complementation analysis, it is possible to predict the total number of genes involved in the determination of that trait.

4.10   X-Linkage Describes Genes on the X Chromosome

In many animal and some plant species, one of the sexes contains a pair of unlike chromosomes that are involved in sex determination. In many cases, these are designated as the X and Y. For example, in both *Drosophila* and humans, males contain an X and a Y chromosome, whereas females contain two X chromosomes. While the Y chromosome must contain a region of pairing homology with the X chromosome if the two are to synapse and segregate during meiosis, much of the remainder of the Y chromosome in humans and other species is considered to be relatively inert genetically. Thus, it lacks most genes that are present on the X chromosome. As a result, genes present on the X chromosome exhibit unique patterns of inheritance in comparison with autosomal genes. The term **X-linkage** is used to describe these situations.

In the following discussion, we will focus on inheritance patterns resulting from genes present on the X but absent from the Y chromosome. This situation results in a modification of Mendelian ratios, the central theme of this chapter.

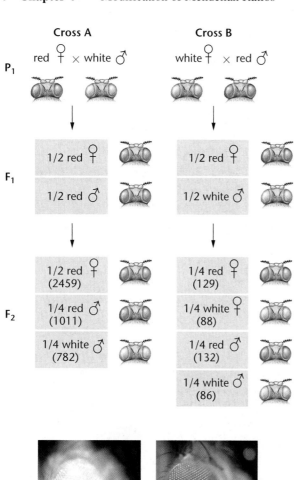

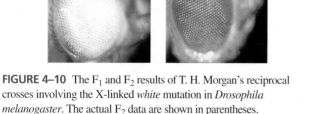

FIGURE 4–10 The F₁ and F₂ results of T. H. Morgan's reciprocal crosses involving the X-linked *white* mutation in *Drosophila melanogaster*. The actual F₂ data are shown in parentheses. The photographs show white eyes and the brick-red wild-type eye color.

X-Linkage in *Drosophila*

One of the first cases of X-linkage was documented by Thomas H. Morgan around 1920 during his studies of the *white* mutation in the eyes of *Drosophila* (Figure 4–10). The normal wild-type red eye color is dominant to white. We will use this case to illustrate **X-linkage**.

Morgan's work established that the inheritance pattern of the white-eye trait is clearly related to the sex of the parent carrying the mutant allele. Unlike the outcome of the typical monohybrid cross, reciprocal crosses between white- and red-eyed flies did not yield identical results. In contrast, in all of Mendel's monohybrid crosses, F₁ and F₂ data were similar regardless of which P₁ parent exhibited the recessive mutant trait. Morgan's analysis led to the conclusion that the *white* locus is present on the X chromosome rather than on one of the autosomes. As such, both the gene and the trait are said to be X-linked.

Results of reciprocal crosses between white-eyed and red-eyed flies are shown in Figure 4–10. The obvious differences in phe-

notypic ratios in both the F₁ and F₂ generations are dependent on whether or not the P₁ white-eyed parent was male or female.

Morgan was able to correlate these observations with the difference found in the sex chromosome composition between male and female *Drosophila*. He hypothesized that the recessive allele for white eyes is found on the X chromosome, but its corresponding locus is absent from the Y chromosome. Females thus have two available gene sites, one on each X chromosome, while males have only one available gene site on their single X chromosome.

Morgan's interpretation of X-linked inheritance, shown in Figure 4–11, provides a suitable theoretical explanation for his results. Since the Y chromosome lacks homology with most genes on the X chromosome, whatever alleles are present on the X chromosome of the males will be expressed directly in their phenotype. Males cannot be homozygous or heterozygous for X-linked genes, and this condition is referred to as being **hemizygous**.

One result of X-linkage is the **crisscross pattern of inheritance**, whereby phenotypic traits controlled by recessive X-linked genes are passed from homozygous mothers to all sons. This pattern occurs because females exhibiting a recessive trait carry the mutant allele on both X chromosomes. Because male offspring receive one of their mother's two X chromosomes and are hemizygous for all alleles present on that X, all sons will express the same recessive X-linked traits as their mother.

Morgan's work has taken on great historical significance. By 1910, the correlation between Mendel's work and the behavior of chromosomes during meiosis had provided the basis for the **chromosome theory of inheritance**. Work involving the X chromosome around 1920 is considered to be the first solid experimental evidence in support of this theory. In the ensuing two decades, these findings inspired further research, which provided indisputable evidence in support of this theory.

X-Linkage in Humans

In humans, many genes and the respective traits controlled by them are recognized as being linked to the X chromosome (see Table 4.2). These X-linked traits can be easily identified in a pedigree because of the crisscross pattern of inheritance. A pedigree for one form of human **color blindness** is shown in Figure 4–12. The mother in generation I passes the trait to all her sons but to none of her daughters. If the offspring in generation II marry normal individuals, the color-blind sons will produce all normal male and female offspring (III-1, 2, and 3); the normal-visioned daughters will produce normal-visioned female offspring (III-4, 6, and 7), as well as color-blind (III-8) and normal-visioned (III-5) male offspring.

The way in which X-linked genes are transmitted causes unusual circumstances associated with recessive X-linked disorders, in comparison to recessive autosomal disorders. For example, if an X-linked disorder debilitates or is lethal to the affected individual prior to reproductive maturation, the disorder occurs exclusively in males. This is so because the only sources of the lethal allele in the population are in heterozygous females who are "carriers" and do not express the disorder. They pass the allele to one-half of their sons, who develop the disorder because they are hemizygous but rarely, if ever, reproduce. Heterozygous females also pass the allele

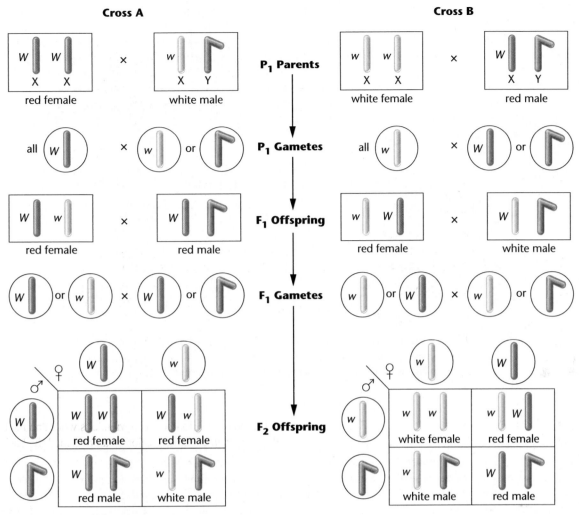

FIGURE 4–11 The chromosomal explanation of the results of the X-linked crosses shown in Figure 4–10.

TABLE 4.2	Human X-Linked Traits
Condition	**Characteristics**
Color blindness, deutan type	Insensitivity to green light.
Color blindness, protan type	Insensitivity to red light.
Fabry disease	Deficiency of galactosidase A; heart and kidney defects, early death.
G-6-PD deficiency	Deficiency of glucose-6-phosphate dehydrogenase, severe anemic reaction following intake of primaquines in drugs and certain foods, including fava beans.
Hemophilia A	Classical form of clotting deficiency; absence of clotting factor VIII.
Hemophilia B	Christmas disease; absence of clotting factor IX.
Hunter syndrome	Mucopolysaccharide storage disease resulting from iduronate sulfatase enzyme deficiency; short stature, clawlike fingers, coarse facial features, slow mental deterioration, and deafness.
Ichthyosis	Deficiency of steroid sulfatase enzyme; scaly dry skin, particularly on extremities.
Lesch-Nyhan syndrome	Deficiency of hypoxanthine-guanine phosphoribosyl transferase enzyme (HGPRT) leading to motor and mental retardation, self-mutilation, and early death.
Duchenne muscular dystrophy	Progressive, life-shortening disorder characterized by muscle degeneration and weakness; sometimes associated with mental retardation; absence of the protein dystrophin.

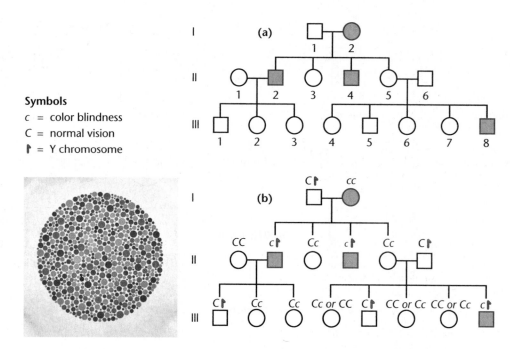

Symbols
c = color blindness
C = normal vision
Y = Y chromosome

FIGURE 4–12 (a) A human pedigree of the X-linked color-blindness trait. (b) The most probable genotype of each individual in the pedigree. The photograph is of an Ishihara color-blindness chart. Red-green color-blind individuals see a 3 rather than the figure 8 visualized by those with normal color vision.

to one-half of their daughters, who become carriers but do not develop the disorder. An example of such an X-linked disorder is Duchenne muscular dystrophy. The disease has an onset prior to age 6 and is often lethal around age 20. It normally occurs only in males.

Now Solve This

Problem 23 on page 91 asks you to determine if each of three pedigrees is consistent with X-linkage.

Hint: In X-linkage, because of hemizygosity, the genotype of males is immediately evident. Therefore, the key to solving this type of problem is to consider the possible genotypes of females that do not express the trait.

Problem 23 on page 91

4.11 In Sex-Limited and Sex-Influenced Inheritance, an Individual's Sex Influences the Phenotype

In other instances, inheritance patterns may be affected by the sex of an individual, although not necessarily by genes on the X chromosome. There are numerous examples in different organisms where the sex of the individual plays a determining role in the expression of certain phenotypes. In some cases, the expression of a specific phenotype is absolutely limited to one sex; in others, the sex of an individual influences the expression of a phenotype that is not limited to one sex or the other. This distinction differentiates **sex-limited inheritance** from **sex-influenced inheritance**.

In domestic fowl, tail and neck plumage is often distinctly different in males and females (Figure 4–13), demonstrating sex-limited inheritance. Cock feathering is longer, more curved, and pointed, while hen feathering is shorter and less curved. Inheritance of these feather phenotypes is controlled by a single pair of autosomal alleles whose expression is modified by the individual's sex hormones.

As shown in the following chart, hen feathering is due to a dominant allele, H, but regardless of the homozygous presence of the recessive h allele, all females remain hen-feathered. Only in males does the hh genotype result in cock feathering.

Genotype	Phenotype	
	Females	**Males**
HH	Hen-feathered	Hen-feathered
Hh	Hen-feathered	Hen-feathered
hh	Hen-feathered	Cock-feathered

In certain breeds of fowl, the hen-feathering or cock-feathering allele has become fixed in the population. In the Leghorn breed, all individuals are of the hh genotype; as a result, males always differ from females in their plumage. Sebright bantams are all HH, resulting in no sexual distinction in feathering phenotypes.

Another example of sex-limited inheritance involves the autosomal genes responsible for milk yield in dairy cattle. Regardless of the overall genotype that influences the quantity of milk production, those genes are obviously expressed only in females.

FIGURE 4–13 Hen feathering (left) and cock feathering (right) in domestic fowl. Note that the hen's feathers are shorter and less curved.

Cases of sex-influenced inheritance include pattern baldness in humans, horn formation in certain breeds of sheep (e.g., Dorset Horn sheep), and certain coat-color patterns in cattle. In such cases, autosomal genes are responsible for the contrasting phenotypes displayed by both males and females, but the expression of these genes is dependent on the hormonal constitution of the individual. Thus, the heterozygous genotype exhibits one phenotype in one sex and the contrasting one in the other. For example, **pattern baldness** in humans, where the hair is very thin on the top of the head (Figure 4–14), is inherited in this way:

Genotype	Phenotype	
	Females	**Males**
BB	Bald	Bald
Bb	Not bald	Bald
bb	Not bald	Not bald

Females can display pattern baldness, but this phenotype is much more prevalent in males. When females do inherit the *BB* genotype, the phenotype is less pronounced than in males and is expressed later in life.

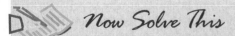

Now Solve This

Problem 12 on page 90 involves the inheritance of colored spots in cattle, and you are asked to analyze the F₁ and F₂ ratios to determine the mode of inheritance.

Hint: Note particularly that the data are differentiated into male and female offspring and that the ratios in the F₂ vary according to sex (i.e., 3/8 of the males are mahogany while only 1/8 of the females are mahogany, etc.). This should immediately alert you to consider the possible influences that sex differences impart on the outcome of crosses. In this case, you should consider whether X-linkage, sex-limited inheritance, or sex-influenced inheritance might be involved.

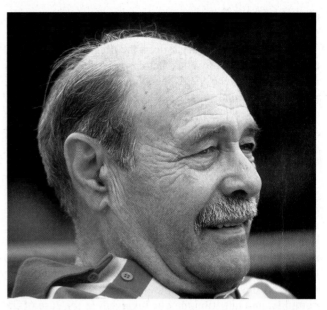

FIGURE 4–14 Pattern baldness, a sex-influenced autosomal trait in humans.

4.12 Genetic Background and the Environment Affect Phenotypic Expression

We now focus on **phenotypic expression**. In previous discussions, we assumed that the genotype of an organism is always directly expressed in its phenotype. For example, pea plants homozygous for the recessive *d* allele (*dd*) will always be dwarf. We discussed gene expression as though the genes operate in a closed system in which the presence or absence of functional products directly determines the collective phenotype of an individual. The situation is actually much more complex. Most gene products function within the internal milieu of the cell, and cells interact with one another in various ways. Furthermore, the organism exists under diverse environmental influences. Thus, gene expression and the resultant phenotype are often modified through the interaction between an individual's particular genotype and the external environment. Here, we deal with several important variables that are known to modify gene expression.

Penetrance and Expressivity

Some mutant genotypes are always expressed as a distinct phenotype, whereas others produce a proportion of individuals whose phenotypes cannot be distinguished from normal (wild type). The degree of expression of a particular trait can be studied quantitatively by determining the *penetrance* and *expressivity* of the genotype under investigation. The percentage of individuals that show at least some degree of expression of a mutant genotype defines the **penetrance** of the mutation. For example, the phenotypic expression of many mutant alleles in *Drosophila* can overlap with wild type. If 15 percent of mutant flies show the wild-type appearance, the mutant gene is said to have a penetrance of 85 percent.

By contrast, **expressivity** reflects the *range of expression* of the mutant genotype. Flies homozygous for the recessive mutant *eyeless* gene yield phenotypes that range from the presence of normal eyes to a partial reduction in size to the complete absence of one or both eyes (Figure 4–15). Although the average reduction of eye size is one-fourth to one-half, expressivity ranges from complete loss of both eyes to completely normal eyes.

Examples such as the expression of the *eyeless* gene provide the basis for experiments to determine the causes of phenotypic variation. If, on one hand, a laboratory environment is held constant and extensive phenotypic variation is still observed, other genes may be influencing or modifying the *eyeless* phenotype. On the other hand, if the genetic background is not the cause of the phenotypic variation, environmental factors such as temperature, humidity, and nutrition may be involved. In the case of the *eyeless* phenotype, experiments have shown that both genetic background and environmental factors influence its expression.

Genetic Background: Suppression and Position Effects

Although it is difficult to assess the specific effect of the **genetic background** and the expression of a gene responsible for determining a potential phenotype, two effects of genetic background have been well characterized.

First, the expression of other genes throughout the genome may have an effect on the phenotype produced by the gene in

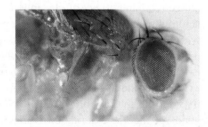
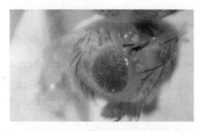

FIGURE 4–15 Variable expressivity, as shown in flies homozygous for the *eyeless* mutation in *Drosophila*. Gradations in phenotype range from wild type to partial reduction to eyeless.

question. The phenomenon of **genetic suppression** is an example. Mutant genes such as *suppressor of sable* (*su-s*), *suppressor of forked* (*su-f*), and *suppressor of Hairy-wing* (*su-Hw*) in *Drosophila* completely or partially restore the normal phenotype in an organism that is homozygous (or hemizygous) for the *sable, forked,* and *Hairy-wing* mutations, respectively. For example, flies hemizygous for both *forked* (a bristle mutation) and *su-f* have normal bristles. In each case, the suppressor gene causes the complete reversal of the expected phenotypic expression of the original mutation. Suppressor genes are excellent examples of the genetic background modifying primary gene effects.

Second, the physical location of a gene in relation to other genetic material may influence its expression. Such a situation is called a **position effect**. For example, if a region of a chromosome is relocated or rearranged (called a translocation or inversion event), normal expression of genes in that chromosomal region may be modified. This is particularly true if the gene is relocated to or near certain areas of the chromosome that are prematurely condensed and genetically inert, referred to as **heterochromatin**.

Temperature Effects

Chemical activity depends on the kinetic energy of the reacting substances, which in turn depends on the surrounding temperature. We can thus expect temperature to influence phenotypes. One example is the evening primrose, which produces red flowers when grown at 23°C and white flowers when grown at 18°C. An even more striking example is seen in Siamese cats and Himalayan rabbits, which exhibit dark fur in certain body regions where the body temperature is slightly cooler, particularly the nose, ears, and paws (Figure 4–16). In these animals, it appears that the enzyme responsible for pigment production is functional at the lower temperatures present in the extremities, but it loses its catalytic function at the slightly higher temperatures found throughout the rest of the body.

Mutations whose expression is affected by temperature are examples of **conditional mutations**. They are called **temperature-sensitive mutations**. Examples are known in viruses and a variety of organisms, including bacteria, fungi, and *Drosophila*. In extreme cases, an organism carrying a mutant allele may express a mutant phenotype when grown at one temperature but express the wild-type phenotype when reared at another temperature. This type of temperature effect is useful in studying mutations that interrupt essential processes during development and are thus normally detrimental to the organism. For example, if bacterial viruses are cultured under *permissive conditions* of 25°C, the gene product is functional, infection proceeds normally, and new viruses are produced; but if bacterial viruses carrying temperature-sensitive mutations infect bacteria cultured at 42°C (the *restrictive condition*), infection progresses up to the point where the essential gene product is required (e.g., for viral assembly) and then arrests. The use of temperature-sensitive mutations, which can be induced and isolated, has added immensely to the study of viral genetics.

(a)

(b)

FIGURE 4–16 (a) A Himalayan rabbit. (b) A Siamese cat. Both species show dark fur color on the snout, ears, and paws. The patches are due to the temperature-sensitive allele responsible for pigment production.

Onset of Genetic Expression

The age at which an organism expresses a gene corresponds to the normal sequence of growth and development. In humans, the prenatal, infant, preadult, and adult phases require different genetic information. As a result, many severe inherited disorders are not manifested until after birth. For example, **Tay-Sachs disease**, inherited as an autosomal recessive, is a lethal lipid metabolism disease involving an abnormal enzyme, hexosaminidase A. Newborns appear to be phenotypically normal for the first few months. Then developmental retardation, paralysis, and blindness ensue, and most affected children die around the age of three.

The **Lesch-Nyhan syndrome**, inherited as an X-linked recessive disease, is characterized by abnormal nucleic acid metabolism (biochemical salvage of nitrogenous purine bases), leading to the accumulation of uric acid in blood and tissues, mental retardation, palsy, and self-mutilation of the lips and fingers. The disorder is due to a mutation in the gene encoding hypoxanthine-guanine phosphoribosyl transferase (HGPRT). Newborns are normal for six to eight months prior to the onset of the first symptoms.

Still another example involves **Duchenne muscular dystrophy (DMD)**, an X-linked recessive disorder associated with progressive muscular wasting. It is not usually diagnosed until the child is three to five years old. Even with modern medical intervention, the disease is often fatal in the early 20s.

Perhaps the most age-variable of all inherited human disorders is **Huntington disease**. Inherited as an autosomal dominant, Huntington disease affects the frontal lobes of the cerebral cortex, where progressive cell death occurs over a period of more than a decade. Brain deterioration is accompanied by spastic uncontrolled movements, intellectual and emotional deterioration, and ultimately death. Onset of this disease has been reported at all ages, but it most frequently occurs between ages 30 and 50, with a mean onset age of 38 years.

These conditions support the concept that the critical expression of genes varies throughout the life cycle of all organisms, including humans. Gene products may play more essential roles at certain life stages, and it is likely that the internal physiological environment of an organism changes with age.

Genetic Anticipation

Interest in studying the genetic onset of phenotypic expression has intensified with the discovery of heritable disorders that exhibit a progressively earlier age of onset and an increased severity of the disorder in each successive generation. This phenomenon is called **genetic anticipation**.

Myotonic dystrophy (DM), the most common type of adult muscular dystrophy, clearly illustrates genetic anticipation. Individuals afflicted with this autosomal dominant disorder exhibit extreme variation in the severity of symptoms. Mildly affected individuals develop cataracts as adults but have little or no muscular weakness. Severely affected individuals demonstrate more extensive myopathy and may be mentally retarded. In its most extreme form, the disease is fatal just after birth. In 1989, C. J. Howeler and colleagues confirmed the correlation of increased severity and earlier onset with successive generations. They studied 61 parent-child pairs, and in 60 cases, age of onset was earlier and more severe in the child than in his or her affected parent.

In 1992, an explanation was put forward for the molecular cause of the mutation responsible for DM, as well as the basis of genetic anticipation. Interestingly, a particular region of the DM gene is repeated a variable number of times and is unstable. Normal individuals average about five copies of this region; minimally affected individuals have about 50 copies; and severely affected individuals have over 1000 copies. The most remarkable observation was that in successive generations, the size of the repeated segment increases. Although it is not yet clear how this expansion in size affects onset and phenotypic expression, the correlation is extremely strong. Several other inherited human disorders, including the fragile-X syndrome, Kennedy disease, and Huntington disease, also reveal an association between the size of specific regions of the responsible gene and disease severity.

Genomic Imprinting

There are some cases involving phenotypic expression where the phenotype depends on silencing of one or the other member of a gene pair following fertilization, depending on the parental origin of the chromosome on which a particular allele is located. This phenomenon is called **genomic (or parental) imprinting**. In some species, certain chromosomal regions and the genes contained within them are somehow "imprinted" depending on their parental origin, influencing whether specific genes are expressed or remain genetically silent. This leads to the direct phenotypic expression of the allele that is not silenced.

The imprinting step, the critical issue in understanding this phenomenon, is thought to occur before or during gamete formation, leading to differentially marked genes (or chromosome regions) in sperm-forming versus egg-forming tissues. The process differs from mutation because the imprint is eventually erased and can be reversed in succeeding generations as genes pass from mother to son to granddaughter, and so on.

The first example of genomic imprinting was discovered in 1991, when three specific mouse genes were shown to undergo imprinting. One is the gene-encoding insulinlike growth factor II (*Igf2*). A mouse that carries two normal wild-type alleles of this gene is normal in size, whereas a mouse that carries two mutant alleles lacks a growth factor and is dwarf. The size of a heterozygous mouse (one allele normal and one mutant; Figure 4–17) depends on the parental origin of the wild-type allele. The mouse will be normal in size if the wild-type allele comes from the father, but it will be dwarf if the wild-type allele comes from the mother. From this, we can deduce that the normal *Igf2* gene is imprinted during egg production but functions normally when it has passed through sperm-producing tissue in males.

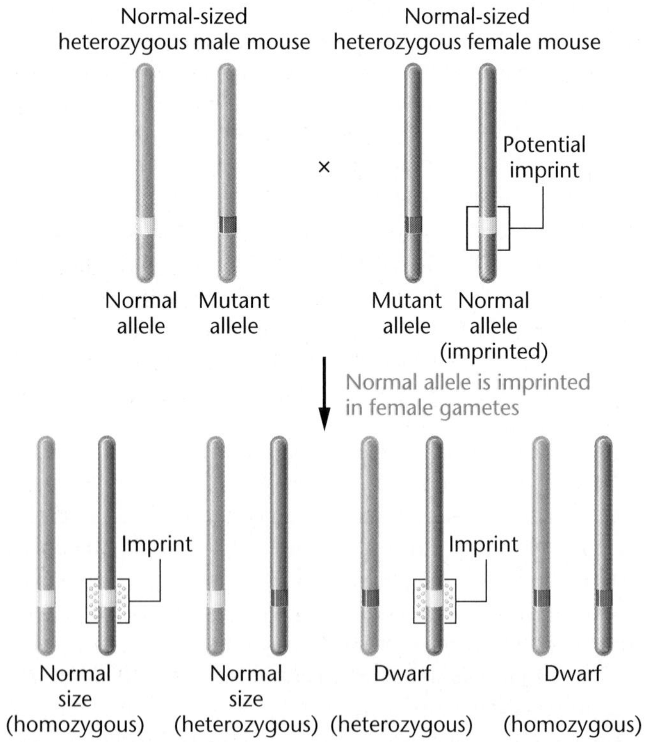

In humans, two distinct genetic disorders are thought to be caused by differential imprinting of the same region of chromosome 15. In both cases, the disorders appear to be the result of a deletion of an identical region (15q11–15) in one member of the chromosome-15 pair. The first disorder, **Prader-Willi syndrome** (**PWS**), results when the chromosome bearing the deletion is inherited through the sperm. If this chromosome is inherited through the ovum, a different disorder, **Angelman syndrome** (**AS**), results.

These two conditions exhibit different phenotypes. PWS entails mental retardation, a severe eating disorder marked by an uncontrollable appetite, obesity, and diabetes. AS also exhibits mental retardation, but involuntary muscle contractions (chorea) and seizures accompany the disorder. We can conclude that the involved region of chromosome 15 is imprinted differently in male and female gametes and that an undeleted maternal and paternal region are required for normal development.

Researchers are investigating how a region of a chromosome may be imprinted. While the initial signal in not yet clear, current thinking is that chemical methylation of certain nitrogenous bases found in DNA is involved in imprinting and gene silencing. Such an explanation is in keeping with the knowledge that methylation of cytosine in DNA inhibits gene activity.

4.13   Extranuclear Inheritance Modifies Mendelian Patterns

Throughout the history of genetics, occasional reports have challenged the basic tenet of Mendelian transmission genetics—that the phenotype is solely determined by nuclear genes located on the chromosomes of both parents. In this final section of the chapter, we consider several examples of inheritance patterns that vary from those predicted by the traditional biparental inheritance of nuclear genes, upon which Mendelian genetics is based. In the following cases, we will focus on two situations where transmission of genetic information is extranuclear. In the first category, an organism's phenotype is affected by the expression of genes contained in the DNA of mitochondria or chloroplasts rather than the nucleus, generally referred to as **organelle heredity**. In the second category, an organism's phenotype is determined by genetic information expressed in the gamete of one parent—usually the mother—such that, following fertilization, the zygote is influenced by gene products from only one of the parents during development. As a result, the phenotype is influenced not by the individual's genotype, but by the genotype of the mother. This general phenomenon is referred to as a **maternal effect**.

Initially, such observations met with skepticism. However, increasing knowledge of molecular genetics and the discovery of DNA in mitochondria and chloroplasts caused the phenomenon of **extranuclear inheritance** to be recognized as an important aspect of genetics.

FIGURE 4–17 The effect of imprinting on the mouse *Igf2* gene, which produces dwarf mice in the homozygous condition. Heterozygotes that receive an imprinted normal allele from their mother are dwarf.

FIGURE 4–18 Offspring from crosses between flowers from various branches of four-o'clock plants. The photograph illustrates the variation in flower color displayed by four-o'clocks as well as variegation, seen in the pink flowers.

	Location of Ovule		
Source of Pollen	White branch	Green branch	Variegated branch
White branch	White	Green	White, green, or variegated
Green branch	White	Green	White, green, or variegated
Variegated branch	White	Green	White, green, or variegated

Organelle Heredity: DNA in Chloroplasts and Mitochondria

Let us examine examples of inheritance patterns related to chloroplast and mitochondrial function. Before DNA was discovered in these organelles, the exact mechanism of transmission of the traits was not clear, except that their inheritance appeared to be linked to something in the cytoplasm rather than to genes in the nucleus. Furthermore, transmission was most often from the maternal parent through the ooplasm, causing the results of reciprocal crosses to vary. Such an extranuclear pattern of inheritance is now appropriately called **organelle heredity**.

Analysis of the inheritance patterns resulting from mutant alleles in chloroplasts and mitochondria has been difficult for two reasons. First, the function of these organelles is dependent on gene products from both nuclear and organelle DNA, making the discovery of the genetic origin of mutations affecting organelle function difficult. Second, many mitochondria and chloroplasts are contributed to each progeny. Thus, if only one or a few of the organelles contain a mutant gene in a cell with a population of mostly normal mitochondria, the corresponding mutant phenotype may not be revealed. This condition, referred to as **heteroplasmy**, may lead to normal cells since the organelles lacking the mutation provide the basis of wild-type function. Analysis is therefore much more complex than for Mendelian characters.

Chloroplasts: Variegation in Four-o'clock Plants

In 1908, Carl Correns (one of the rediscoverers of Mendel's work) provided the earliest example of inheritance linked to chloroplast transmission. Correns discovered a variant of the four-o'clock plant, *Mirabilis jalapa*, that had branches with either white, green, or variegated white-and-green leaves. The white areas in variegated leaves and the completely white leaves lack chlorophyll that provides the green color to normal leaves. Chlorophyll is the light-absorbing pigment made within chloroplasts.

Correns was curious about how inheritance of this phenotypic trait occurred. As shown in Figure 4–18, inheritance in all possible combinations of crosses is strictly determined by the phenotype of the ovule source. For example, if the seeds (representing the progeny) were derived from ovules on branches with green leaves, all progeny plants bore only green leaves, regardless of the phenotype of the source of pollen. Correns concluded that inheritance was transmitted through the cytoplasm of the maternal parent because the pollen, which contributes little or no cytoplasm to the zygote, had no apparent influence on the progeny phenotypes.

Since leaf coloration is related to the chloroplast, genetic information contained either in that organelle or somehow present in the cytoplasm and influencing the chloroplast must be responsible for the inheritance pattern. It now seems certain that the genetic "defect" that eliminates the green chlorophyll in the white patches on leaves is a mutation in the DNA housed in the chloroplast.

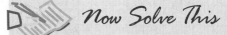

Now Solve This

In Problem 37 on page 93, you are asked to consider the genetic outcome of crosses involving a mutation in *Chlamydomonas* and to propose how inheritance of the trait occurs.

Hint: Consider the two possibilities that inheritance of the trait is uniparental vs. biparental, as related to the outcomes in this problem.

Mitochondrial Mutations: *poky* in *Neurospora* and *petite* in *Saccharomyces*

Mutations affecting mitochondrial function have been discovered and studied, revealing that they too contain a distinctive genetic system. As with chloroplasts, mitochondrial mutations are transmitted through the cytoplasm. In our current discussion, we will emphasize the link between mitochondrial mutations and the resultant extranuclear inheritance patterns.

In 1952, Mary B. Mitchell and Hershel K. Mitchell studied the bread mold *Neurospora crassa*. They discovered a slow-growing mutant strain and named it *poky*. Slow growth is associated with impaired mitochondrial function, specifically in relation to certain cytochromes essential for electron transport. Results of genetic crosses between wild-type and *poky* strains suggest that *poky* is an extranuclear trait inherited through the cytoplasm. If one mating type is *poky* and the other is wild type, all progeny colonies are *poky*. The reciprocal cross, where *poky* is transmitted by the other mating type, produces normal wild-type colonies.

Another extensive study of mitochondrial mutations has been performed with the yeast *Saccharomyces cerevisiae*. The first such mutation, described by Boris Ephrussi and his coworkers in 1956, was named *petite* because of the small size of the yeast colonies (Figure 4–19). Many independent *petite* mutations have since been discovered and studied, and all have a common characteristic—a deficiency in cellular respiration involving abnormal electron transport. This organism is a facultative anaerobe and can grow by fermenting glucose through glycolysis; thus, it may survive the loss of mitochondrial function by generating energy anaerobically.

The complex genetics of *petite* mutations has revealed that a small proportion are the result of nuclear mutations. They exhibit Mendelian inheritance and illustrate that mitochondria function depends on both nuclear and organellar gene products. The majority of them demonstrate cytoplasmic transmission, indicating mutations in the DNA of the mitochondria.

Mitochondrial Mutations: Human Genetic Disorders

Our knowledge of genetics of mitochondria has now greatly expanded. The DNA found in human mitochondria has been completely sequenced and contains 16,569 base pairs. Mitochondrial gene products have now been identified and include the following:

13 proteins, required for aerobic cellular respiration

22 transfer RNAs (tRNAs), required for translation

2 ribosomal RNAs (rRNAs), required for translation

Because a cell's energy supply is largely dependent on aerobic cellular respiration, disruption of any mitochondrial gene by mutation may potentially have a severe impact on that organism. We have seen this in our previous discussion of *petite* mutants in yeast, which would be a lethal mutation were it not for this organism's ability to respire anaerobically. In fact, mtDNA is particularly vulnerable to mutations, for two possible reasons. First, the ability to repair mtDNA damage appears not to be equivalent to that of nuclear DNA. Second, the concentration of highly mutagenic free radicals generated by cell respiration that accumulate in such a confined space very likely raises the mutation rate in mtDNA.

Fortunately, a zygote receives a large number of organelles through the egg, so if only one organelle or a few of them contains a mutation, its impact is greatly diluted by the many mitochondria that lack the mutation and function normally. If a deleterious mutation arises or is present in the initial population of organelles, adults will have cells with a variable mixture of both normal and abnormal organelles. From a genetic standpoint, this condition of heteroplasmy makes analysis quite difficult.

Several criteria must be met in order for a human disorder to be attributable to genetically altered mitochondria:

1. Inheritance must exhibit a maternal rather than Mendelian pattern.

2. The disorder must reflect a deficiency in the bioenergetic function of the organelle.

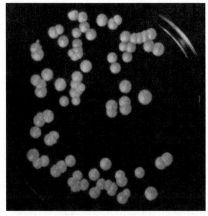

Normal colonies

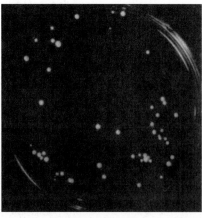

Petite colonies

FIGURE 4–19 Photos comparing normal versus *petite* colonies of the yeast *Saccharomyces cerevisiae*.

3. There must be a specific genetic mutation in one or more of the mitochondrial genes.

Several disorders in humans are known to demonstrate these characteristics. For example, **myoclonic epilepsy and ragged-red fiber disease (MERRF)** demonstrates a pattern of inheritance consistent with maternal transmission. Only the offspring of affected mothers inherit this disorder, while the offspring of affected fathers are normal. Individuals with this rare disorder express ataxia (lack of muscular coordination), deafness, dementia, and epileptic seizures. The disease is named for the presence of "ragged-red" skeletal-muscle fibers that exhibit blotchy red patches resulting from the proliferation of aberrant mitochondria (Figure 4–20). Brain function, which has a high energy demand, is also affected in this disorder, leading to the neurological symptoms described above.

The mutation that causes MERRF has now been identified and is in a mitochondrial gene whose altered product interferes with the capacity for translation of proteins within the organelle. This, in turn, leads to the various manifestations of the disorder.

The cells of MERRF individuals exhibit heteroplasmy, containing a mixture of normal and abnormal mitochondria. Different patients display different proportions of the two, and even different cells from the same patient exhibit various levels of abnormal mitochondria. Were it not for heteroplasmy, the mutation would very likely be lethal, testifying to the essential nature of mitochondrial function and its reliance on the genes encoded by DNA within the organelle.

A second human disorder, **Leber's hereditary optic neuropathy (LHON)**, also exhibits maternal inheritance as well as mitochondrial DNA lesions. The disease is characterized by sudden bilateral blindness. The average age of vision loss is 27, but onset is quite variable. Four mutations have been identified, all of which disrupt normal oxidative phosphorylation. Over 50 percent of cases are due to a mutation at a specific position in the mitochondrial gene encoding a subunit of NADH dehydrogenase, an enzyme essential to cell respiration. This mutation is transmitted maternally through the mitochondria to all offspring. It is interesting to note that in many instances of LHON, there is no family history; a significant number of cases are "sporadic," resulting from newly arisen mutations.

Maternal Effect: *Limnaea* Coiling

Another broad category of extranuclear control of gene expression, **maternal effect** (also referred to as maternal influence) is a situation where an offspring's phenotype for a particular trait is under the control of gene products present in the egg prior to fertilization. The nuclear genes of the female gamete are transcribed, and the gene products (either proteins or yet untranslated mRNAs) accumulate in the egg cytoplasm. After fertilization, these products are distributed among newly formed cells and influence the patterns or traits established during early development. The following case of snails will illustrate the influence of the maternal genome on particular traits.

One example of maternal effect involves the snail *Limnaea peregra*, where some strains have left-handed, or sinistrally, coiled shells (*dd*), while others have right-handed, or dextrally, coiled shells (*DD* or *Dd*). These snails are hermaphroditic and may undergo either cross- or self-fertilization, providing a variety of types of matings.

Figure 4–21 illustrates the results of reciprocal crosses between true-breeding snails. These crosses yield different outcomes, even though both are between sinistral and dextral organisms and produce all heterozygous offspring. Examination of the progeny reveals that their phenotypes depend on the genotypes of the female parents. If we adopt that conclusion as a working hypothesis, we can test it by examining the offspring in subsequent generations of self-fertilization events. In each case, the hypothesis is upheld. Maternal parents that are *DD* or *Dd* produce only dextrally coiled progeny. Maternal parents that are *dd* produce only sinistrally coiled progeny. The coiling pattern of the progeny is determined by the genotype of the parent producing the egg, *regardless of the phenotype of that parent*.

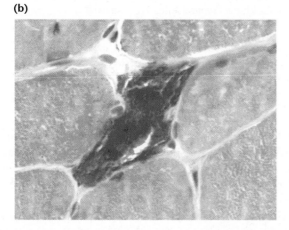

(a)

(b)

FIGURE 4–20 Ragged-red fibers in skeletal muscle cells from patients with MERRF. (a) The muscle fiber has mild proliferation (see red rim and speckled cytoplasm). (b) Marked proliferation where mitochondria have replaced most cellular structures.

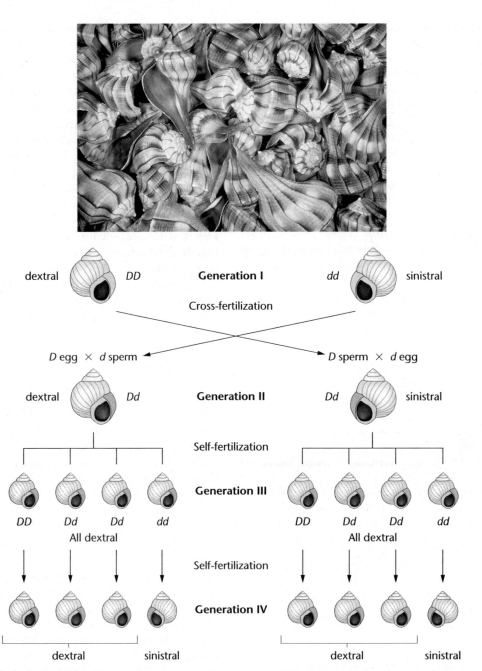

FIGURE 4–21 Inheritance of coiling in the snail *Limnaea peregra*. Coiling is either dextral (right-handed) or sinistral (left-handed). A maternal effect is evident in generations II and III, where the genotype of the maternal parent, rather than the offspring's own genotype, controls the phenotype of the offspring. The photograph shows a mixture of dextral and sinistral coiled snails.

Investigation of the developmental events in *Limnaea* reveals that the orientation of the spindle in the first cleavage division after fertilization determines the direction of coiling. Spindle orientation appears to be controlled by maternal genes acting on the developing eggs in the ovary. The orientation of the spindle, in turn, influences cell divisions following fertilization and establishes the permanent adult coiling pattern. The dextral allele (*D*) produces an active gene product that causes right-handed coiling. If ooplasm from dextral eggs is injected into uncleaved sinis-tral eggs, they cleave in a dextral pattern. However, in the converse experiment, sinistral ooplasm has no effect when injected into dextral eggs. Apparently, the sinistral allele is the result of a classic recessive mutation that encodes an inactive gene product.

We can conclude, therefore, that females that are either *DD* or *Dd* produce oocytes that synthesize the *D* gene product, which is stored in the ooplasm. Even if the oocyte contains only the *d* allele following meiosis and is fertilized by a *d*-bearing sperm, the resulting *dd* snail will be dextrally coiled.

Embryonic Development in *Drosophila*

A more recently documented example of maternal effect involves various genes that control embryonic development in *Drosophila melanogaster*. The genetic control of embryonic development in *Drosophila*, discussed in greater detail in Chapter 20, is a fascinating story. The protein products of the maternal-effect genes function to activate other genes, which may in turn activate still other genes. This cascade of gene activity leads to a normal embryo whose subsequent development yields a normal adult fly. The extensive work by Edward B. Lewis, Christiane Nüsslein-Volhard, and Eric Wieschaus (who shared the 1995 Nobel Prize for Physiology or Medicine for their findings) has clarified how these and other genes function. Genes that illustrate maternal effect have products that are synthesized by the developing egg and stored in the oocyte prior to fertilization. Following fertilization, these products specify molecular gradients that determine spatial organization as development proceeds.

For example, the gene *bicoid* (*bcd*) plays an important role in specifying the development of the anterior portion of the fly. Embryos derived from mothers who are homozygous for this mutation (bcd^-/bcd^-) fail to develop anterior areas that normally give rise to the head and thorax of the adult fly. Embryos whose mothers contain at least one wild-type allele (bcd^+) develop normally, even if the genotype of the embryo is homozygous for the mutation. Consistent with the concept of *maternal effect*, the *genotype of the female parent*, not the *genotype of the embryo*, determines the phenotype of the offspring.

When we return to our discussion of this general topic in Chapter 20, we will see examples of other genes illustrating maternal effect, as well as many "zygotic" genes whose expression occurs during early development and that behave genetically in the conventional Mendelian fashion.

GENETICS, TECHNOLOGY, AND SOCIETY

Mitochondrial DNA and the Mystery of the Romanovs

By most accounts, Nicholas II, the last Tsar of Russia, was a substandard monarch. He was accused of bungling during the Russo-Japanese War of 1904–1905, and his regime was plagued by corruption and incompetence. Even so, he probably didn't deserve the fate that befell him and his family one summer night in 1918. As we shall see, a full understanding of that event has relied on, of all things, mitochondrial DNA.

After being forced to abdicate in 1917, ending 300 years of Romanov rule, Tsar Nicholas and the imperial family were banished to Ekaterinburg in western Siberia. There, it was believed, they would be out of reach of the fiercely anti-imperialist Bolsheviks, who were then fighting to gain control of the country. But the Bolsheviks eventually caught up with the Romanovs. During a July night in 1918, Tsar Nicholas, Tsarina Alexandra (granddaughter of Queen Victoria), their five children (Olga, 22; Tatiana, 21; Marie, 19; Anastasia, 17; and Alexei, 13), their family doctor, and three of their servants were awakened and brought to a downstairs room of the house where they were held prisoner. There they were made to form a double row against the wall, presumably so that a photograph could be taken. Instead, 11 men with revolvers burst into the room and opened fire. After exhausting their ammunition, they proceeded to bayonet the bodies and smash their faces in with rifle butts. The corpses were then hauled away and flung down a mineshaft, only to be pulled out two days later and dumped into a shallow grave, doused with sulfuric acid, and covered over. There they rested for more than 60 years.

An air of mystery soon developed around the demise of the Romanovs. Did all of the children of Nicholas and Alexandra die with their parents that bloody night, or did the youngest daughter, Anastasia, get away? Over the years, the possibility that Anastasia miraculously escaped execution has inspired countless books, a Hollywood movie, a ballet, a Broadway play, and, most recently, an animated movie. In all of these retellings, Anastasia reemerges to claim her birthright as the only surviving member of the Romanovs. Adding to the puzzle was one Anna Anderson. Two years after being dragged from a Berlin canal after a suicide attempt in 1920, she began to claim to be the Grand Duchess Anastasia. Despite a history of mental instability and a curious inability to speak Russian, she managed to convince many people.

The unraveling of the mystery began in 1979, when a Siberian geologist and a Moscow filmmaker discovered four skulls they believed to belong to the Tsar's family. It wasn't until the summer of 1991, after the establishment of *glasnost*, that exhumation began. Altogether, almost 1000 bone fragments were recovered, which were reassembled into nine skeletons—five females and four males. Based on measurements of the bones and computer-assisted superimposition of the skulls onto photographs, the remains were tentatively identified as belonging to the murdered Romanovs. But there were still two missing bodies, one of the daughters (believed to be Anastasia) and the boy Alexei.

The next step in authenticating the remains involved studies of DNA that were conducted by Pavel Ivanov, the leading Russian forensic DNA analyst, in collaboration with Peter Gill of the British Forensic Science Service. Their goals were to establish family relationships among the remains, and then to determine, by comparisons with living relatives, whether the family group was in fact the Romanovs. They froze bone fragments from the nine skeletons in liquid nitrogen, ground them into a fine powder, and extracted small amounts of DNA,

(Cont. on the next page)

which comprised both nuclear DNA and mitochondrial DNA (mtDNA). Genomic DNA typing of each skeleton confirmed the familial relationships and showed that, indeed, one of the princesses and Alexei were missing. Now final proof that the bones belonged to the Romanovs awaited the analysis of mtDNA.

Mitochondrial DNA is ideal for forensic studies for several reasons. First, since all the mitochondria in a human cell are descended from the mitochondria present in the egg, mtDNA is transmitted strictly from mother to offspring, never from father to offspring. Therefore, mtDNA sequences can be used to trace maternal lineages without the complicating effects of meiotic crossing over, which recombines maternal and paternal nuclear genes every generation. In addition, mtDNA is small (~16,600 base pairs) and present in 500 to 1000 copies per cell, so it is much easier to recover intact than nuclear DNA.

Ivanov's group amplified two highly variable regions from the mtDNA isolated from all nine bone samples and determined the nucleotide sequences of these regions. By comparing these sequences with living relatives of the Romanovs, they hoped to establish the identity of the Ekaterinburg remains once and for all. The sequences from Tsarina Alexandra were an exact match with those from Prince Philip of England, who is her grandnephew, verifying her identity. However, authentication was more complicated for the Tsar.

The Tsar's sequences were compared with those from the only two living relatives who could be persuaded to participate in the study, Countess Xenia Cheremeteff-Sfiri (his great-grandniece) and James George Alexander Bannerman Carnegie, third Duke of Fife (a more distant relative, descended from a line of women stretching back to the Tsar's grandmother). These comparisons produced a surprise. At position 16,169 of the mtDNA, the Tsar seemed to have either one or another base, a C or a T. The Countess and the Duke, in contrast, both had only T. The conclusion was that Tsar Nicholas had two different populations of mitochondria in his cells, each with a different base at position 16,169 of its DNA. This condition, called heteroplasmy, is now believed to occur in 10 to 20 percent of humans.

This ambiguity between the presumed Tsar and his two living maternal relatives cast doubt on the identification of the remains. Fortunately, the Russian government granted a request to analyze the remains of the Tsar's younger brother, Grand Duke Georgij Romanov, who died in 1899 of tuberculosis. The Grand Duke's mtDNA was found to have the same heteroplasmic variant at position 16,169, either a C or a T. It was concluded that the Ekaterinburg bones were the doomed imperial family. With years of controversy finally resolved, the now-authenticated remains of Tsar Nicholas II, Tsarina Alexandra, and three of their daughters were buried in the St. Peter and Paul Cathedral in St. Petersburg on July 17, 1998, 80 years to the day after they were murdered.

This DNA analysis did not solve the mystery of the fate of Anastasia, however. Did she die with her parents, sisters, and brother in 1918? Or is it possible that Anna Anderson was telling the truth, that she was the escaped duchess? In a separate study, Anna Anderson's nuclear and mtDNA was recovered from intestinal tissue preserved from an operation performed five years before her death in 1984. Analysis of this DNA proved that she was not Anastasia, but rather a Polish peasant named Franziska Schanzkowska.

If Anastasia's remains were not among those of her parents and sisters and if Anna Anderson was an imposter, what did happen to Anastasia? Most evidence suggests that Anastasia and her brother Alexei were not found with the others in the mass grave because their bodies were burned over the grave site two days after the killings and the ashes scattered, never to be found again. Not exactly a Hollywood ending.

References

Gibbons, A. 1998. Calibrating the mitochondrial clock. *Science* 279: 28–29.

Ivanov, P., et al. 1996. Mitochondrial DNA sequence heteroplasmy in the Grand Duke of Russia Georgij Romanov establishes the authenticity of the remains of Tsar Nicholas II. *Nat. Genet.* 12: 417–420.

Masse, R. 1996. *The Romanovs: The final chapter.* New York: Ballantine Books.

Stoneking, M. et al. 1995. Establishing the identity of Anna Anderson Manahan. *Nat. Genet.* 9: 9–10.

CHAPTER SUMMARY

1. Since Mendel's work was rediscovered, the study of transmission genetics has expanded to include many alternative modes of inheritance. In many cases, phenotypes can be influenced by two or more genes.

2. Incomplete, or partial, dominance is exhibited when an intermediate phenotypic expression of a trait occurs in an organism that is heterozygous for two alleles.

3. Codominance is exhibited when distinctive expression of two alleles occurs in a heterozygous organism.

4. The concept of multiple alleles applies to populations, since a diploid organism may host only two alleles at any given locus. However, within a population many alternative alleles of the same gene can occur.

5. Lethal mutations usually result in the inactivation or lack of synthesis of gene products that are essential during an organism's development—these mutations can be recessive or dominant.

Some lethal genes, such as the one that causes Huntington disease, are not expressed until adulthood.

6. Mendel's classic F_2 ratio is often modified in instances where gene interaction controls phenotypic variation.

7. Epistasis may occur when two or more genes influence a single characteristic. Usually, the expression of one of the genes masks the expression of the other gene or genes.

8. Complementation analysis determines whether independently isolated mutations producing similar phenotypes are alleles of one another or whether they represent separate genes.

9. Genes located on the X chromosome display a unique mode of inheritance called X-linkage.

10. Sex-limited and sex-influenced inheritance occur when the sex of the organism affects the phenotype controlled by a gene located on an autosome.

11. Phenotypic expression is not always the direct reflection of the genotype. Penetrance measures the percentage of organisms in a given population that exhibit evidence of the corresponding mutant phenotype. Expressivity, on the other hand, measures the range of phenotypic expression of a given genotype.

12. The time of onset of gene expression in organisms varies when the need for certain gene products occurs at different periods during development, growth, and aging.

13. Genetic anticipation is a phenomenon where the onset of phenotypic expression occurs earlier and becomes more severe in each ensuing generation.

14. Genomic imprinting is a process whereby a region of either the paternal or maternal chromosome is modified (marked or imprinted), thereby affecting phenotypic expression. Expression therefore depends on which parent contributes a mutant allele.

15. Patterns of inheritance sometimes vary from that expected during the biparental transmission of nuclear genes. In such instances, phenotypes most often appear to result from extranuclear genetic information transmitted through the egg.

16. Organelle heredity is based on the genotypes of chloroplast and mitochondrial DNA as these organelles are transmitted to offspring. Chloroplast mutations affect the photosynthetic capabilities of plants, whereas mitochondrial mutations affect cells highly dependent on energy generated through cellular respiration. The resulting mutants display phenotypes related to the loss of function of these organelles.

17. Patterns of maternal effect result when nuclear gene products controlled by the maternal genotype of the egg influence early development.

KEY TERMS

ABO blood group, 64

Angelman syndrome (AS), 80

Bombay phenotype, 65

chromosome theory of inheritance, 74

codominance, 64

color blindness, 74

complementation analysis, 72

complementation group, 73

conditional mutation, 78

crisscross pattern of inheritance, 74

Duchenne muscular dystrophy (DMD), 79

epigenesis, 67

epistasis, 67

expressivity, 78

extranuclear inheritance, 80

gain of function mutations, 62

gene interaction, 67

genomic (or parental) imprinting, 80

genetic anticipation, 79

genetic background, 78

genetic suppression, 78

H substance, 65

hemizygous, 74

heterochromatin, 78

heteroplasmy, 81, 86

hexosaminidase, 64

Huntington disease, 67, 79

incomplete dominance, 63

Leber's hereditary optic neuropathy (LHON), 83

Lesch-Nyhan syndrome, 79

lethal allele, 66

loss of function mutation, 62

maternal effect, 80, 83

MN blood group, 64

multiple alleles, 64

myoclonic epilepsy and ragged-red fiber disease (MERRF), 83

myotonic dystrophy (DM), 79

null allele, 62

oncogenes, 62

organelle heredity, 80, 81

partial dominance, 63

pattern baldness, 77

penetrance, 77

phenotypic expression, 77

position effect, 78

Prader-Willi syndrome (PWS), 80

proto-oncogenes, 62

sex-influenced inheritance, 76

sex-limited inheritance, 76

Tay–Sachs disease, 64, 79

temperature-sensitive mutation, 78

wild-type allele, 62

X-linkage, 73, 74

INSIGHTS AND SOLUTIONS

Genetic problems take on added complexity if they involve two independent characters and multiple alleles, incomplete dominance, or epistasis. The most difficult types of problems are those that pioneering geneticists faced during laboratory or field studies. They had to determine the mode of inheritance by working backward from the observations of offspring to parents of unknown genotype.

1. Consider the problem of comb-shape inheritance in chickens, where walnut, pea, rose, and single are the observed distinct phenotypes. (See the opening photograph in this chapter on page 61.) How is comb shape inherited, and what are the genotypes of the P_1 generation of each cross? Use the following data:

Cross 1:	single	× single	⟶ all single
Cross 2:	walnut	× walnut	⟶ all walnut
Cross 3:	rose	× pea	⟶ all walnut
Cross 4:	F_1 × F_1 of Cross 3		
	walnut	× walnut	⟶ 93 walnut
			28 rose
			32 pea
			10 single

(*Cont. on the next page*)

Solution: At first glance, this problem appears quite difficult. However, applying a systematic approach and breaking the analysis into steps usually simplifies it. Our approach involves two steps. Analyze the data carefully for any useful information. Once you identify something that is clearly helpful, follow an empirical approach—that is, formulate a hypothesis and, in a sense, test it against the given data. Look for a pattern of inheritance that is consistent with all cases.

This problem gives two immediately useful facts. First, in cross 1, P_1 singles breed true. Second, while P_1 walnut breeds true in cross 2, a walnut phenotype is also produced in cross 3 between rose and pea. When these F_1 walnuts are crossed in cross 4, all four comb shapes are produced in a ratio that approximates 9:3:3:1. This observation immediately suggests a cross involving two gene pairs, because the resulting data display the same ratio as in Mendel's dihybrid crosses. Since only one trait is involved (comb shape), epistasis may be occurring. This could serve as your working hypothesis, and you must now propose how the two gene pairs "interact" to produce each phenotype.

If you call the allele pairs A, a and B, b, you can predict that because walnut represents 9/16 in cross 4, $A-B-$ will produce walnut. You might also hypothesize that in cross 2, the genotypes are $AABB \times AABB$, where walnut bred true. (Recall that $A-$ and $B-$ mean AA or Aa and BB or Bb, respectively.)

The phenotype representing 1/16 of the offspring of cross 4 is single; therefore; you could predict that this phenotype is the result of the $aabb$ genotype. This is consistent with cross 1.

Now you have only to determine the genotypes for rose and pea. The most logical prediction is that at least one dominant A or B allele combined with the double recessive condition of the other allele pair accounts for these phenotypes. For example,

$$A–bb \longrightarrow \text{rose}$$
$$aaB– \longrightarrow \text{pea}$$

If $AAbb$ (rose) is crossed with $aaBB$ (pea) in cross 3, all offspring will be $AaBb$ (walnut). This is consistent with the data, and you must now look at only cross 4. We predict these walnut genotypes to be $AaBb$ (as above), and from the cross $AaBb$ (walnut) $\times$ $AaBb$ (walnut) we expect

9/16	$A–B–$	(walnut)
3/16	$A–bb$	(rose)
3/16	$aaB–$	(pea)
1/16	$aabb$	(single)

Our prediction is consistent with the information given. The initial hypothesis of the epistatic interaction of two gene pairs proves consistent throughout, and the problem is solved.

This problem demonstrates the need for a basic theoretical knowledge of transmission genetics. Then, you can search for appropriate clues that will enable you to proceed in a stepwise fashion toward a solution. Mastering problem solving requires practice, but can give you a great deal of satisfaction. Apply this general approach to the following problems.

2. In radishes, flower color may be red, purple, or white. The edible portion of the radish may be long or oval. When only flower color is studied, no dominance is evident, and red $\times$ white crosses yield all purple. If these F_1 purples are interbred, the F_2 generation consists of 1/4 red: 1/2 purple: 1/4 white. Regarding radish shape, long is dominant to oval in a normal Mendelian fashion.

(a) Determine the F_1 and F_2 phenotypes from a cross between a true-breeding red, long radish and one that is white and oval. Be sure to define all gene symbols initially.

Solution: This is a modified dihybrid cross where the gene pair controlling color exhibits incomplete dominance. Shape is controlled conventionally. First, establish gene symbols:

RR = red	$O-$ = long
Rr = purple	oo = oval
rr = white	

$$P_1: \quad RROO \quad \times \quad rroo$$
$$\text{(red long)} \qquad \text{(white oval)}$$
$$F_1: \text{all } RrOo \text{ (purple long)}$$
$$F_1 \times F_1: RrOo \times RrOo$$

$$F_2: \begin{cases} 1/4\, RR \begin{cases} 3/4\, O– & 3/16\, RRO– & \text{red long} \\ 1/4\, oo & 1/16\, RRoo & \text{red oval} \end{cases} \\ 2/4\, Rr \begin{cases} 3/4\, O– & 6/16\, RrO– & \text{purple long} \\ 1/4\, oo & 2/16\, Rroo & \text{purple oval} \end{cases} \\ 1/4\, rr \begin{cases} 3/4\, O– & 3/16\, rrO– & \text{white long} \\ 1/4\, oo & 1/16\, rroo & \text{white oval} \end{cases} \end{cases}$$

Note that to generate the F_2 results, we have used the forked-line method. First, we consider the outcome of crossing F(Continued) the color genes ($Rr \times Rr$). Then the outcome of shape is considered ($Oo \times Oo$).

(b) A red oval plant was crossed with a plant of unknown genotype and phenotype, yielding the following offspring:

103 red long: 101 red oval

98 purple long: 100 purple oval

Determine the genotype and phenotype of the unknown plant.

Solution: The two characters appear to be inherited independently, so consider them separately. The data indicate a 1/4: 1/4: 1/4: 1/4 proportion. First, consider color:

P_1:	red $\times$??? (unknown)
F_1:	204 red (1/2)
	198 purple (1/2)

Because the red parent must be RR, the unknown must have a genotype of Rr to produce these results. Thus it is purple. Now, consider shape:

P_1:	oval $\times$??? (unknown)
F_1:	201 long (1/2)
	201 oval (1/2)

Since the oval plant must be oo, the unknown plant must have a genotype of Oo to produce these results. Thus it is long. The unknown plant is

$RrOo$ purple long

3. In humans, red-green color blindness is inherited as an X-linked recessive trait. A woman with normal vision whose father is color blind marries a male who has normal vision. Predict the color vision of their male and female offspring.

Solution: The female is heterozygous since she inherited an X chromosome with the mutant allele from her father. Her husband is normal. Therefore, the parental genotypes are

$Cc \times C\upharpoonright$ ($\upharpoonright$ represents the Y chromosome)

All female offspring are normal (CC or Cc). One-half of the male children will be color blind ($c\upharpoonright$), and the other half will have normal vision ($C\upharpoonright$).

4. Analyze the following hypothetical pedigree and determine the most consistent interpretation of how the trait is inherited and any inconsistencies:

Solution: The trait is passed from all affected male parents to all except one offspring, but it

is *never* passed maternally. Individual IV-7 (a female) is the only exception.

5. Can the explanation in Problem 1 be attributed to a gene on the Y chromosome? Defend your answer.

Solution: No, because male parents pass the trait to their daughters as well as to their sons.

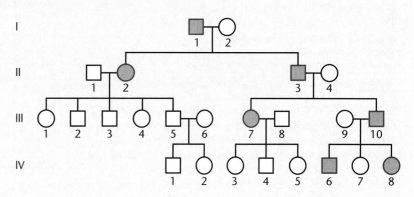

PROBLEMS AND DISCUSSION QUESTIONS

1. In Shorthorn cattle, coat color may be red, white, or roan. Roan is an intermediate phenotype expressed as a mixture of red and white hairs. The following data are obtained from various crosses:

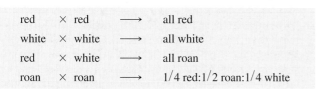

red	× red	⟶	all red
white	× white	⟶	all white
red	× white	⟶	all roan
roan	× roan	⟶	1/4 red:1/2 roan:1/4 white

How is coat color inherited? What are the genotypes of parents and offspring for each cross?

2. Contrast incomplete dominance and codominance.

3. With regard to the ABO blood types in humans, determine the genotypes of the male parent and female parent:

   Male parent: blood type B whose mother was type O
   Female parent: blood type A whose father was type B

   Predict the blood types of the offspring that this couple may have and the expected ratio of each.

4. In foxes, two alleles of a single gene, *P* and *p*, may result in lethality *(PP)*, platinum coat *(Pp)*, or silver coat *(pp)*. What ratio is obtained when platinum foxes are interbred? Is the *P* allele behaving dominantly or recessively in causing (a) lethality; (b) platinum coat color?

5. Three gene pairs located on separate autosomes determine flower color and shape as well as plant height. The first pair exhibits incomplete dominance, where color can be red, pink (the heterozygote), or white. The second pair leads to the dominant personate or recessive peloric flower shape, while the third gene pair produces either the dominant tall trait or the recessive dwarf trait. Homozygous plants that are red, personate, and tall are crossed with those that are white, peloric, and dwarf. Determine the F$_1$ genotype(s) and phenotype(s). If the F$_1$ plants are interbred, what proportion of the offspring will exhibit the same phenotype as the F$_1$ plants?

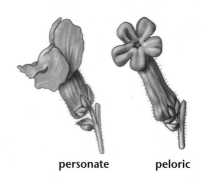

personate   **peloric**

6. As in the plants of Problem 5, color may be red, white, or pink; and flower shape may be personate or peloric. Determine the P$_1$ and F$_1$ genotypes for the following crosses:

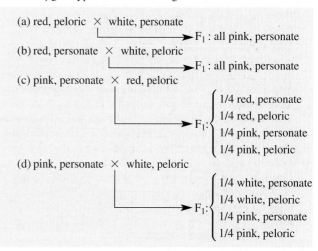

(a) red, peloric × white, personate
   ⟶ F$_1$: all pink, personate

(b) red, personate × white, peloric
   ⟶ F$_1$: all pink, personate

(c) pink, personate × red, peloric
   F$_1$: { 1/4 red, personate
            1/4 red, peloric
            1/4 pink, personate
            1/4 pink, peloric }

(d) pink, personate × white, peloric
   F$_1$: { 1/4 white, personate
            1/4 white, peloric
            1/4 pink, personate
            1/4 pink, peloric }

What phenotype ratios would result from crossing the F$_1$ of (a) with the F$_1$ of (b)?

7. In some plants, a red pigment, cyanidin, is synthesized from a colorless precursor. The addition of a hydroxyl group (—OH) to the cyanidin molecule causes it to become purple. In a cross between two randomly selected purple plants, the following results are obtained:

94 purple:31 red:43 colorless

How many genes are involved in determining these flower colors? Which genotypic combinations produce which phenotypes? Diagram the purple x purple cross.

8. The following genotypes of two independently assorting autosomal genes determine coat color in rats:

A–B–(gray); *A–bb* (yellow); *aaB–*(black); *aabb* (cream)

A third gene pair on a separate autosome determines whether any color will be produced. The *CC* and *Cc* genotypes allow color according to the expression of the *A* and *B* alleles. However, the *cc* genotype results in albino rats regardless of the *A* and *B* alleles present. Determine the F_1 phenotypic ratio of the following crosses: (a) *AAbbCC* × *aaBBcc*; (b) *AaBBCC* × *AABbcc*; (c) *AaBbCc* × *AaBbcc*.

9. Given the inheritance pattern of coat color in rats described in Problem 8, predict the genotype and phenotype of the parents that produced the following F_1 offspring: (a) 9/16 gray: 3/16 yellow: 3/16 black: 1/16 cream; (b) 9/16 gray: 3/16 yellow: 4/16 albino; (c) 27/64 gray:16/64 albino: 9/64 yellow: 9/64 black: 3/64 cream.

10. A husband and wife have normal vision, although both of their fathers are red-green color-blind, inherited as an X-linked recessive condition. What is the probability that their first child will be (a) a normal son, (b) a normal daughter, (c) a color-blind son, (d) a color-blind daughter?

11. In humans, the ABO blood type is under the control of autosomal multiple alleles. Red-green color blindness is a recessive X-linked trait. If two parents who are both type A and have normal vision produce a son who is color-blind and type O, what is the probability that their next child will be a female who has normal vision and is type O?

12. In spotted cattle, the colored regions may be mahogany or red. If a red female and a mahogany male, both derived from separate true-breeding lines, are mated and the cross is carried to an F_2 generation, the following results are obtained:

> F_1: 1/2 mahogany males: 1/2 red females
> F_2: 3/8 mahogany males: 1/8 red males:
> 1/8 mahogany females: 3/8 red females

When the reciprocal of the initial cross is performed (mahogany female and red male), identical results are obtained. Explain these results by postulating how the color is genetically determined, and diagram the crosses.

13. In cats, yellow coat color is determined by the *b* allele, and black coat color is determined by the *B* allele. The heterozygous condition results in a coat pattern known as tortoiseshell. These genes are X-linked. What kinds of offspring would be expected from a cross of a black male and a tortoiseshell female? What are the chances of getting a tortoiseshell male?

14. In *Drosophila*, an X-linked recessive mutation, *scalloped* (*sd*), causes irregular wing margins. Diagram the F_1 and F_2 results if
    (a) A scalloped female is crossed with a normal male.
    (b) A scalloped male is crossed with a normal female.
    Compare these results to those that would be obtained if the *scalloped* gene were autosomal.

15. Another recessive mutation in *Drosophila*, *ebony* (*e*), is on an autosome (chromosome 3) and causes darkening of the body compared with wild-type flies. What phenotypic F_1 and F_2 male and female ratios will result if a scalloped-winged female with normal body color is crossed with a normal-winged ebony male? Work this problem by both the Punnett square method and the forked-line method.

16. While *vermilion* is X-linked in *Drosophila* and causes eye color to be bright red, *brown* is an autosomal recessive mutation that causes the eye to be brown. Flies carrying both mutations lose all pigmentation and are white-eyed. Predict the F_1 and F_2 results of the following crosses:
    (a) vermilon females × brown males
    (b) brown females × vermilon males
    (c) white females × wild males

17. In pigs, coat color may be sandy, red, or white. A geneticist spent several years mating true-breeding pigs of all different color combinations, even obtaining true-breeding lines from different parts of the country. For crosses 1 and 4 below, she encountered a major problem: her computer crashed and she lost the F_2 data. She nevertheless persevered and, using the limited data shown here, was able to predict the mode of inheritance and the number of genes involved, and to assign genotypes to each coat color. Attempt to duplicate her analysis, based on the available data generated from the crosses shown.

Cross	P_1	F_1	F_2
1	sandy × sandy	All red	Data lost
2	red × sandy	All red	3/4 red:1/4 sandy
3	sandy × white	All sandy	3/4 sandy:1/4 white
4	white × red	All red	Data lost

When you have formulated a hypothesis to explain the mode of inheritance and assigned genotypes to the respective coat colors, predict the outcomes of the F_2 generations where the data were lost.

18. A geneticist from an alien planet that prohibits genetic research brought with him two true-breeding lines of frogs. One frog line croaks by *uttering* "rib-it rib-it" and has purple eyes. The other frog line croaks by *muttering* "knee-deep knee-deep" and has green eyes. He mated the two frog lines, producing F_1 frogs that were all utterers with blue eyes. A large F_2 generation then yielded the following ratios:

> 27/64 blue, utterer
> 12/64 green, utterer
> 9/64 blue, mutterer
> 9/64 purple, utterer
> 4/64 green, mutterer
> 3/64 purple, mutterer

(a) How many total gene pairs are involved in the inheritance of both eye color and croaking?
(b) Of these, how many control eye color, and how many control croaking?
(c) Assign gene symbols for all phenotypes, and indicate the genotypes of the P_1, F_1, and F_2 frogs.
(d) After many years, the frog geneticist isolated true-breeding lines of all six F_2 phenotypes. Indicate the F_1 and F_2 phenotypic ratios of a cross between a blue, mutterer and a purple, utterer.

19. In another cross, the frog geneticist from Problem 18 mated two purple, utterers, with the results shown here. What were the genotypes of the parents?

> 9/16 purple, utterers
> 3/16 purple, mutterers
> 3/16 green, utterers
> 1/16 green, mutterers

20. In cattle, coats may be solid white, solid black, or black-and-white spotted. When true-breeding solid whites are mated with true-breeding solid blacks, the F_1 generation consists of all solid white individuals. After many $F_1 \times F_1$ matings, the following ratio was observed in the F_2 generation:

> 12/16 solid white
> 3/16 black-and-white spotted
> 1/16 solid black

Explain the mode of inheritance governing coat color by determining how many gene pairs are involved and which genotypes yield which phenotypes. Is it possible to isolate a true-breeding strain of black-and-white spotted cattle? If so, what genotype would they have? If not, explain why not.

21. In the guinea pig, a locus controlling coat color may be occupied by any of four alleles: C (full color), c^k (sepia), c^d (cream), or c^a (albino). A progressive order of dominance exists among these alleles when they are present heterozygously: $C > c^k > c^d > c^a$. Determine the genotype of each individual, and predict the phenotypic ratios of the offspring in the following crosses:

(a) sepia $\times$ cream, where both had an albino parent
(b) sepia $\times$ cream, where the sepia individual had an albino parent and the cream individual had two sepia parents
(c) sepia $\times$ cream, where the sepia individual had two full-color parents and the cream individual had two sepia parents
(d) sepia $\times$ cream, where the sepia individual had two full-color parents and the cream individual had two full-color parents

22. In parakeets, two autosomal genes (located on different chromosomes) control the production of feather pigment. Gene B controls the production of blue pigment, and gene Y controls the production of yellow pigment. Known recessive mutations in each gene result in the loss of pigment synthesis. Two green parakeets are mated and produce green, blue, yellow, and albino progeny.

(a) Based on this information, explain the pattern of inheritance. Be sure to include the genotypes of the green parents and all four phenotypic classes, as well as the fraction of total progeny that each phenotypic class represents in your answer.

(b) The parental (green) parakeets are the progeny of a cross between two true-breeding strains. What two types of crosses between true-breeding strains could have produced the green parents? Indicate the genotypes and phenotypes for each cross.

23. Below are three pedigrees. For each, consider whether it could or could not be consistent with an X-linked recessive trait. In a sentence or two, indicate why or why not.

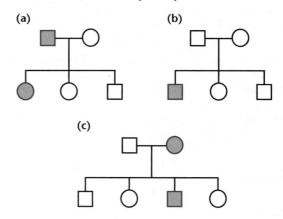

24. Consider the following three pedigrees, all involving the same human trait:

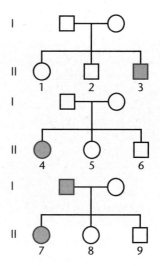

(a) Which sets of conditions, if any, can be excluded?
dominant and X-linked
dominant and autosomal
recessive and X-linked
recessive and autosomal

(b) For any set of conditions that you excluded, indicate the *single individual* in generation II (1–9) that was instrumental in your decision to exclude that condition. If none were excluded, answer "none apply."

(c) Given your conclusions in parts (a) and (b), indicate the *genotype* of individuals II-1, II-6, and II-9. If more than one possibility applies, list all possibilities. Use the symbols A and a for the genotypes.

25. Three autosomal recessive mutations in *Drosophila*, all with tan eye color ($r1$, $r2$, and $r3$), are independently isolated and subjected to complementation analysis. Of the results shown here, which, if any, are alleles of one another? Predict the results of the cross that is not shown—that is, $r2 \times r3$.

> Cross 1: $r1 \times r2 \longrightarrow F_1$: all wild-type eyes
> Cross 2: $r1 \times r3 \longrightarrow F_1$: all tan eyes

26. Labrador retrievers may be black, brown, or golden in color. While each color may breed true, many different outcomes occur if numerous litters are examined from a variety of matings,

where the parents are not necessarily true-breeding. The results below show some of the possibilities. Propose a mode of inheritance that is consistent with these data, and indicate the corresponding genotypes of the parents in each mating. Indicate as well the genotypes of dogs that breed true for each color.

(a) black	×	brown	⟶	all black
(b) black	×	brown	⟶	1/2 black
				1/2 brown
(c) black	×	brown	⟶	3/4 black
				1/4 golden
(d) black	×	golden	⟶	all black
(e) black	×	golden	⟶	4/8 golden
				3/8 black
				1/8 brown
(f) black	×	golden	⟶	2/4 golden
				1/4 black
				1/4 brown
(g) brown	×	brown	⟶	3/4 brown
				1/4 golden
(h) black	×	black	⟶	9/16 black
				4/16 golden
				3/16 brown

27. In rabbits, a series of multiple alleles controls coat color in the following way: C is dominant to all other alleles and causes full color. The chinchilla phenotype is due to the c^{ch} allele, which is dominant to all alleles other than C. The c^h allele, dominant only to c^a (albino), results in the Himalayan coat color. Thus, the order of dominance is $C > c^{ch} > c^h > c^a$. For each of the following three cases, the phenotypes of the P_1 generations of two crosses are shown, as well as the phenotype of one member of the F_1 generation.

P_1 **Phenotypes**		F_1 **Phenotypes**
Himalayan × Himalayan	⟶	albino
(a)	× ⟶	??
full color × albino	⟶	chinchilla
albino × chinchilla	⟶	albino
(b)	× ⟶	??
full color × albino	⟶	full color
chinchilla × albino	⟶	Himalayan
(c)	× ⟶	??
full color × albino	⟶	Himalayan

Determine the genotypes of the P_1 generation and the F_1 offspring for each case, and predict the results of making each cross between F_1 individuals as shown.

Full color   Chinchilla

Himalayan   Albino

28. Horses can be cremello (a light cream color), chestnut (a reddish brown color), or palomino (a golden color with white in the horse's tail and mane).

Chestnut   Palomino

Cremello

Of these phenotypes, only palominos never breed true. The following results have been observed:

cremello × palomino	⟶	1/2 cremello
		1/2 palomino
chestnut × palomino	⟶	1/2 chestnut
		1/2 palomino
palomino × palomino	⟶	1/4 chestnut
		1/2 palomino
		1/4 cremello

(a) From these results, determine the mode of inheritance by assigning gene symbols and indicating which genotypes yield which phenotypes.

(b) Predict the F_1 and F_2 results of many initial matings between cremello and chestnut horses.

29. Pigment in the mouse is produced only when the C allele is present. Individuals of the cc genotype have no color. If color is present, it may be determined by the A and a alleles. AA or Aa results in agouti color, while aa results in black coats.

(a) What F_1 and F_2 genotypic and phenotypic ratios are obtained from a cross between $AACC$ and $aacc$ mice?

(b) In three crosses shown below between agouti females whose genotypes were unknown and males of the $aacc$ genotype, what are the genotypes of the female parents for each of the following phenotypic ratios?

(1) 8 agouti	(2) 9 agouti	(3) 4 agouti
8 colorless	10 black	5 black
		10 colorless

30. Five human matings numbered 1–5 are shown in the following table. Included are both maternal and paternal phenotypes for ABO and MN blood-group antigen status.

Parental Phenotypes					Offspring	
(1)	A,	M	×	A, N	(a)	A, N
(2)	B,	M	×	B, M	(b)	O, N
(3)	O,	N	×	B, N	(c)	O, MN
(4)	AB,	M	×	O, N	(d)	B, M
(5)	AB,	MN	×	AB, MN	(e)	B, MN

Each mating resulted in one of the five offspring shown to the right (a–e). Match each offspring with one correct set of parents, using each parental set only once. Is there more than one set of correct answers?

31. In Dexter and Kerry cattle, animals may be polled (hornless) or horned. The Dexter animals have short legs, whereas the Kerry animals have long legs. When many offspring were obtained from matings between polled Kerrys and horned Dexters, half were found to be polled Dexters and half polled Kerrys. When these two types of F_1 cattle were mated to one another, the following F_2 data were obtained:

3/8 polled Dexters	3/8 polled Kerrys
1/8 horned Dexters	1/8 horned Kerrys

A geneticist was puzzled by these data and interviewed farmers who had bred these cattle for decades. She learned that Kerrys

Kerry cow

Dexter bull

were true breeding. Dexters, on the other hand, were not true-breeding and never produced as many offspring as Kerrys. Provide a genetic explanation for these observations.

32. What genetic criteria distinguish a case of extranuclear inheritance from (a) a case of Mendelian autosomal inheritance; (b) a case of X-linked inheritance?

33. In *Limnaea* (see Section 4.13), what results would you expect in a cross between a Dd dextrally coiled and a Dd sinistrally coiled snail, assuming cross-fertilization occurs as shown in Figure 4–21? What results would occur if the Dd dextral produced only eggs and the Dd sinistral produced only sperm?

34. In a cross of *Limnaea*, the snail contributing the eggs was dextral, but of unknown genotype. Both the genotype and the phenotype of the other snail are unknown. All F_1 offspring exhibited dextral coiling. Ten of the F_1 snails were allowed to undergo self-fertilization. One-half produced only dextrally coiled offspring, whereas the other half produced only sinistrally coiled offspring. What were the genotypes of the original parents?

35. The specification of the anterior–posterior axis in *Drosophila* embryos is initially controlled by various gene products that are synthesized and stored in the mature egg following oogenesis. Mutations in these genes result in abnormalities of the axis during embryogenesis, illustrating maternal effect. How do such mutations vary from those involved in organelle heredity that illustrate extranuclear inheritance? Devise a set of parallel crosses and expected outcomes involving mutant genes that contrast maternal effect and organelle heredity.

36. The maternal-effect mutation *bicoid* (*bcd*) is recessive. In the absence of the bicoid protein product, embryogenesis is not completed. Consider a cross between a female heterozygous for the bicoid (bcd^+/bcd^-) and a male homozygous for the mutation (bcd^-/bcd^-).

(a) How is it possible for a male homozygous for the mutation to exist?

(b) Predict the outcome (normal vs. failed embryogenesis) in the F_1 and F_2 generations of the cross described.

37. *Chlamydomonas*, a eukaryotic green alga, is sensitive to the antibiotic erythromycin, which inhibits protein synthesis in prokaryotes. There are two mating types in this alga, mt^+ and mt^-. If an mt^+ cell sensitive to the antibiotic is crossed with an mt^- cell that is resistant, all progeny cells are sensitive. The reciprocal cross (mt^+ resistant and mt^- sensitive) yields all resistant progeny cells. Assuming that the mutation for resistance is in the chloroplast DNA, what can you conclude from the results from these crosses?

CHAPTER

5

Sex Determination and Sex Chromosomes

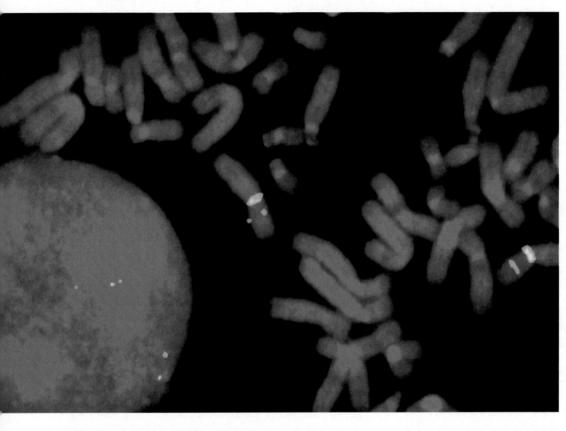

Human X chromosomes highlighted using fluorescence in situ hybridization (FISH), where specific probes bind to specific sequences of DNA. The green fluorescence is specific to the DNA of the X chromosome centromeres. The red fluorescence is specific to the DNA sequence of the Duchenne muscular dystrophy (DMD) gene, an X-linked gene.

CHAPTER CONCEPTS

■ Sexual reproduction, which greatly enhances genetic variation within species, requires mechanisms that result in sexual differentiation.

■ A wide variety of genetic mechanisms have evolved in organisms leading to sexual dimorphism.

■ Often, specific genes, usually on a single chromosome, cause maleness or femaleness during development.

■ In humans, the presence of extra X or Y chromosomes beyond the diploid number may be tolerated, but often leads to syndromes demonstrating distinctive phenotypes.

■ While segregation of sex-determining chromosomes should theoretically lead to a one-to-one sex ratio of males to females, in humans, this ratio greatly favors males at conception.

■ In mammals, females contain two X chromosomes compared to one in males, but the extra genetic information in females is compensated for by random inactivation of one of the X chromosomes early in development.

■ In some reptilian species, temperature during incubation of eggs determines the sex of offspring.

In the biological world, a wide range of reproductive modes and life cycles are recognized. Asexual organisms exist where no evidence of sexual reproduction is evident, while other species alternate between short periods of sexual reproduction and prolonged periods of asexual reproduction. In most diploid eukaryotes, however, sexual reproduction is the only natural mechanism that results in new members of a species. Orderly transmission of genetic units from parents to offspring, and the resultant phenotypic variability, relies on the processes of segregation and independent assortment that occur during meiosis. Meiosis produces haploid gametes so that, following fertilization, the resulting offspring maintain the diploid number of chromosomes characteristic of their species. Thus, meiosis ensures genetic constancy within members of the same species.

These events, which are involved in the perpetuation of all sexually reproducing organisms, ultimately depend on an efficient union of gametes during fertilization. In turn, successful mating between organisms, the basis for fertilization, relies on some form of sexual differentiation in organisms. Although not overtly apparent, this differentiation occurs as low on the evolutionary scale as bacteria and single-celled eukaryotic algae. In evolutionarily higher forms of life, the differentiation of the sexes is more evident as phenotypic dimorphism in the males and females of each species. The shield and spear (♂), the ancient symbols of iron and Mars, and the mirror (♀), the symbol of copper and Venus, symbolize the maleness and femaleness acquired by individuals.

Dissimilar or **heteromorphic chromosomes**, such as the XY pair, often characterize one sex or the other, resulting in their label as **sex chromosomes**. Nevertheless, genes, rather than chromosomes, ultimately serve as the underlying basis of sex determination. As we will see, some of these genes are present on sex chromosomes, but others are autosomal. Extensive investigation has revealed a wide variation in sex-chromosome systems—even in closely related organisms—suggesting that mechanisms controlling sex determination have undergone rapid evolution many times.

In this chapter, we review several representative modes of sexual differentiation by examining the life cycles of three organisms often studied in genetics: the green alga, *Chlamydomonas*; the maize plant, *Zea mays*; and the nematode or roundworm, *Caenorhabditis elegans* (most often referred to as *C. elegans*). These organisms contrast the different roles that sexual differentiation plays in the lives of diverse organisms. Then, we delve more deeply into what is known about the genetic basis for determining sexual differences, with a particular emphasis on two organisms: our own species, representing mammals, and *Drosophila*, on which pioneering sex-determining studies were performed.

How Do We Know?

In this chapter, we will focus on how sex is determined in a variety of organisms and several other sex-related phenomena. As you study this topic, you should try to answer several fundamental questions:

1. How do we know whether or not a heteromorphic chromosome such as the Y chromosome plays a crucial role in the determination of sex?

2. How do we know that *Drosophila* utilizes a different sex-determination mechanism from mammals, even though it has the same sex-chromosome compositions in males and females?

3. How do we know how mammals, including humans, solve the "dosage problem" caused by the presence of an X and Y chromosome in one sex and two X chromosomes in the other sex?

4. How did we learn that, although the sex ratio at birth in humans favors males slightly, the sex ratio at conception vastly favors them?

5. When the sex of an organism (such as certain reptiles) is determined environmentally, how do we know which environmental factors dictate the sex of these organisms?

5.1 Life Cycles Depend on Sexual Differentiation

In multicellular animals, it is important to distinguish between primary sexual differentiation, which involves only the gonads where gametes are produced, and secondary sexual differentiation, which involves the overall appearance of the organism, including clear differences in such organs as mammary glands and external genitalia. In plants and animals, the terms **unisexual**, **dioecious**, and **gonochoric** are equivalent; they refer to an individual containing only male *or* only female reproductive organs. Conversely, the terms **bisexual**, **monoecious**, and **hermaphroditic** refer to individuals containing both male *and* female reproductive organs, a common occurrence in both the plant and animal kingdoms. These organisms can produce both eggs and sperm. The term **intersex** is usually reserved for individuals of intermediate sexual differentiation, who are most often sterile.

Chlamydomonas

The life cycle of the green alga *Chlamydomonas* (Figure 5–1) is representative of organisms that exhibit only infrequent periods of sexual reproduction. Such organisms spend most of their life cycle in the haploid phase, asexually producing daughter cells by mitotic divisions. However, under unfavorable

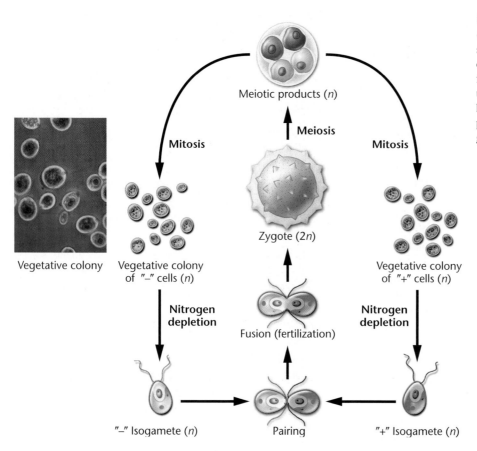

Vegetative colony

Meiotic products (*n*)

Mitosis   **Meiosis**   **Mitosis**

Vegetative colony
of "–" cells (*n*)

Zygote (2*n*)

Vegetative colony
of "+" cells (*n*)

**Nitrogen
depletion**

**Nitrogen
depletion**

Fusion (fertilization)

"–" Isogamete (*n*)   Pairing   "+" Isogamete (*n*)

FIGURE 5–1 The life cycle of *Chlamydomonas*. Unfavorable conditions stimulate the formation of isogametes of opposite mating types that may fuse in fertilization. The resulting zygote undergoes meiosis, producing two haploid cells of each mating type. The photograph shows vegetative cells of this green alga.

nutrient conditions, such as nitrogen depletion, certain daughter cells function as gametes. Following fertilization, a diploid zygote, which can withstand the unfavorable environment, is formed. When conditions become more suitable, meiosis ensues and haploid vegetative cells are again produced.

In such species, there is little visible difference between the haploid vegetative cells that reproduce asexually and the haploid gametes that are involved in sexual reproduction. The two gametes that fuse during mating are morphologically indistinguishable and are called **isogametes**. Species producing them are said to be **isogamous**.

In 1954, Ruth Sager and Sam Granik demonstrated that gametes in *Chlamydomonas* can be subdivided into two mating types. Working with clones derived from single haploid cells, they showed that cells from a given clone mate with cells from some but not all other clones. When they tested the mating abilities of large numbers of clones, all could be placed into one of two mating categories, either mt^+ or mt^-. "Plus" cells mate only with "minus" cells, and vice versa. Following fertilization and meiosis, the four haploid cells (zoospores) produced were found to consist of two plus types and two minus types.

Further experimentation established that plus and minus cells differ chemically. When extracts are prepared from cloned *Chlamydomonas* cells (or their flagella), and then added to cells of the opposite mating type, clumping or agglutination occurs. No such agglutination occurs if the extracts are added to cells of the mating type from which it was derived. These observations demonstrate that, despite the morphological similarities between isogametes, they are differentiated chemically. Therefore, in this alga, a primitive means of sex differentiation exists, even though there is no morphological indication that such differentiation has occurred.

Maize (*Zea mays*)

The life cycles of many plants alternate between the haploid gametophyte stage and the diploid sporophyte stage (see Figure 2–12). The processes of meiosis and fertilization link the two phases during the life cycle. *Zea mays*, or maize, familiar to you as corn, exemplifies a monoecious seed plant where the sporophyte phase and the morphological structures representing this stage predominate during the life cycle. Both male and female structures are present on the adult plant. Thus, sex determination occurs differently in different tissues of the same organism, as shown in the life cycle of this plant (Figure 5–2). The *stamens* (which collectively constitute the tassel) produce diploid microspore mother cells, each of which undergoes meiosis and gives rise to four haploid microspores. Each haploid microspore in turn develops into a mature male microgametophyte—the pollen grain—which contains two sperm nuclei.

Equivalent female diploid cells, known as megaspore mother cells, exist in the *pistil* of the sporophyte. Following meiosis, only one of the four haploid megaspores survives. It usually divides mitotically three times, producing a total of eight haploid nuclei enclosed in the embryo sac. Two of these nuclei unite near the center of the embryo sac, becoming the endosperm nuclei. At the micropyle end of the sac where the sperm enters, three nuclei remain: the oocyte nucleus and two

FIGURE 5–2 The life cycle of maize (*Zea mays*). The diploid sporophyte bears stamens and pistils that give rise to haploid microspores and megaspores, which develop into the pollen grain and the embryo sac that ultimately house the sperm and oocyte, respectively. Following fertilization, the embryo develops within the kernel and is nourished by the endosperm. Germination of the kernel gives rise to a new sporophyte (the mature corn plant), and the cycle repeats itself.

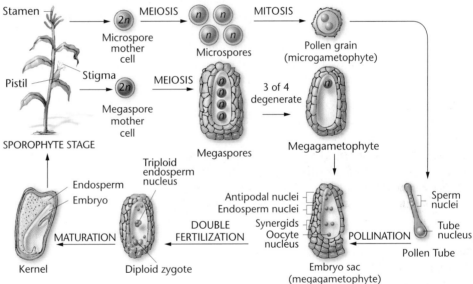

synergids. The other three antipodal nuclei cluster at the opposite end of the embryo sac.

Pollination occurs when pollen grains make contact with the silks (or stigma) of the pistil and develop extensive pollen tubes that grow toward the embryo sac. When contact is made at the micropyle, the two sperm nuclei enter the embryo sac. One sperm nucleus unites with the haploid oocyte nucleus, and the other sperm nucleus unites with two endosperm nuclei. This process, known as double fertilization, results in the diploid zygote nucleus and the triploid endosperm nucleus, respectively. Each ear of corn can contain as many as 1000 of these structures, each of which develops into a single kernel. Each kernel, if allowed to germinate, gives rise to a new plant, the sporophyte.

The mechanism of sex determination and differentiation in a monoecious plant like *Zea mays*, where the tissues that form both male and female gametes are of the same genetic constitution, was difficult to comprehend at first. However, the discovery of a large number of mutant genes that disrupt normal tassel and pistil formation supports the concept that normal products of these genes play an important role in sex determination by affecting the differentiation of male or female tissue in several ways.

For example, mutant genes that cause sex reversal provide valuable information. When homozygous, all mutations classified as *tassel seed* (*ts*) interfere with tassel production and induce the formation of female structures. Thus, a single gene can cause a normally monoecious plant to become functionally female. On the other hand, the recessive mutations *silkless* (*sk*) and *barren stalk* (*ba*) interfere with the development of the pistil, resulting in plants with only functional male reproductive organs.

Data gathered from studies of these and other mutants suggest that the products of many wild-type alleles of these genes interact in controlling sex determination. During development, certain cells are "determined" to become male or female structures. Following sexual differentiation into either male or female structures, male or female gametes are produced.

Caenorhabditis elegans

The roundworm *Caenorhabditis elegans* has become a popular organism in genetic studies [Figure 5–3(a)], particularly during the investigation of the genetic control of development. Its usefulness is based on the fact that the adult consists of exactly 959 cells, the precise lineage of which can be traced back to specific embryonic origins. Among many interesting mutant phenotypes that have been studied, behavioral modifications are a favorite topic of inquiry.

There are two sexual phenotypes in these worms: males, which have only testes, and hermaphrodites, which contain both testes and ovaries. During larval development of hermaphrodites, testes form that produce sperm, which is stored. Ovaries are also produced, but oogenesis does not occur until the adult stage is reached several days later. The eggs that are then produced are fertilized by the stored sperm in the process of self-fertilization.

The outcome of this process is quite interesting [Figure 5–3(b)]. The vast majority of organisms that result, like the parental worm, are hermaphrodites; less than 1 percent of the

offspring are males. As adults, males can mate with hermaphrodites, producing about one-half male and one-half hermaphrodite offspring.

The genetic signal that determines maleness rather than hermaphroditic development is provided by genes located on both the X chromosome and autosomes. *C. elegans* lacks a Y chromosome altogether—hermaphrodites have two X chromosomes, while males have only one X chromosome. It is believed that the ratio of X chromosomes to the number of sets of autosomes ultimately determines the sex of these worms. A ratio of 1.0 results in hermaphrodites and a ratio of 0.5 results in males. The absence of a heteromorphic Y chromosome is not uncommon in organisms.

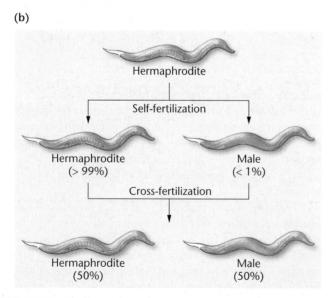

Now Solve This

Problem 22 on page 112 asks you to devise an experimental approach to elucidate the original findings regarding sex determination in a marine worm.

Hint: An obvious approach would be to attempt to isolate and devise experiments with the unknown factor affecting sex determination. An alternative approach, more in tune with genetic analysis, could involve the study of mutations that alter the normal outcomes.

5.2 X and Y Chromosomes Were First Linked to Sex Determination Early in the Twentieth Century

How sex is determined has long intrigued geneticists. In 1891, H. Henking identified a nuclear structure in the sperm of certain insects, which he labeled the X-body. Several years later, Clarance McClung showed that some grasshopper sperm contain an unusual genetic structure, which he called a heterochromosome, but the remainder of the sperm lack such a structure. He mistakenly associated the presence of the heterochromosome with the production of male progeny. In 1906, Edmund B. Wilson clarified the findings of Henking and McClung when he demonstrated that female somatic cells in the insect *Protenor* contain 14 chromosomes, including two X chromosomes. During oogenesis, an even reduction occurs, producing gametes with seven chromosomes, including one X chromosome. Male somatic cells, on the other hand, contain only 13 chromosomes, including one X chromosome. During spermatogenesis, gametes are produced containing either six chromosomes, without an X, or seven chromosomes, one of which is an X. Fertilization by X-bearing sperm results in female offspring, and fertilization by X-deficient sperm results in male offspring [Figure 5–4(a)].

The presence or absence of the X chromosome in male gametes provides an efficient mechanism for sex determination in this species and also produces a 1:1 sex ratio in the resulting offspring. This mechanism, now called the **XX/XO** or *Protenor* **mode of sex determination**, depends on the ran-

FIGURE 5–3 (a) Photomicrograph of an hermaphroditic nematode, *C. elegans*. (b) The outcomes of self-fertilization in a hermaphrodite, and a mating of a hermaphrodite and a male worm.

dom distribution of the X chromosome into one-half of the male gametes during segregation. As we saw earlier, *C. elegans* exhibits this system of sex determination.

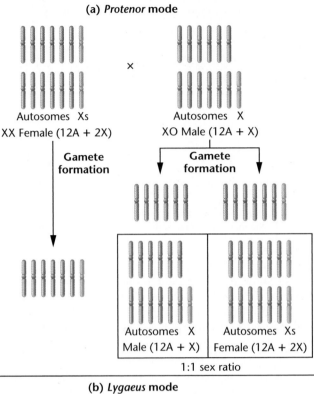

(a) *Protenor* mode

Autosomes Xs
XX Female (12A + 2X)

Autosomes X
XO Male (12A + X)

Gamete formation

Gamete formation

Autosomes X
Male (12A + X)

Autosomes Xs
Female (12A + 2X)

1:1 sex ratio

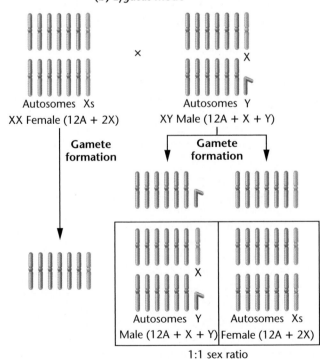

(b) *Lygaeus* mode

Autosomes Xs
XX Female (12A + 2X)

Autosomes Y
XY Male (12A + X + Y)

Gamete formation

Gamete formation

X

Autosomes Y
Male (12A + X + Y)

Autosomes Xs
Female (12A + 2X)

1:1 sex ratio

FIGURE 5–4 (a) The *Protenor* mode of sex determination, where the heterogametic sex (the male in this example) is XO and produces gametes with or without the X chromosome. (b) The *Lygaeus* mode of sex determination, where the heterogametic sex (again, the male in this example) is XY and produces gametes with either an X or a Y chromosome. In both cases, the chromosome composition of the offspring determines its sex.

Wilson also experimented with the hemipteran insect *Lygaeus turicus*, in which both sexes have 14 chromosomes. Twelve of these are autosomes (A). In addition, the females have two X chromosomes, while the males have only a single X and a smaller heterochromosome labeled the **Y chromosome**. Females in this species produce only gametes of the (6A + X) constitution, but males produce two types of gametes in equal proportions, (6A + X) and (6A + Y). Therefore, following random fertilization, equal numbers of male and female progeny are produced with distinct chromosome complements. This is called the *Lygaeus* or **XX/XY mode of sex determination** [Figure 5–4(b)].

In *Protenor* and *Lygaeus* insects, males produce unlike gametes. As a result, they are described as the **heterogametic sex**, and in effect, their gametes ultimately determine the sex of the progeny in those species. In such cases, the female, who has like sex chromosomes, is the **homogametic sex**, producing uniform gametes with regard to chromosome numbers and types.

The male is not always the heterogametic sex. In other organisms, the female produces unlike gametes, exhibiting either the *Protenor* (XX/XO) or *Lygaeus* (XX/XY) mode of sex determination. Examples include moths and butterflies, most birds, some fish, reptiles, amphibians, and at least one species of plants (*Fragaria orientalis*). To immediately distinguish situations in which the female is the heterogametic sex, some workers use the **ZZ/ZW** notation, where ZW is the heterogamous female, instead of the XX/XY notation.

The situation with fowl (chickens) demonstrates the difficulty in establishing which sex is heterogametic and whether the *Protenor* or *Lygaeus* mode is operative. While genetic evidence supported the hypothesis that the female is the heterogametic sex, the cytological identification of the sex chromosome was not accomplished until 1961 because of the large number of chromosomes (78) characteristic of chickens. When the sex chromosomes were finally identified, the female was shown to contain an unlike chromosome pair, including a heteromorphic chromosome (the W chromosome). Thus, in fowl, the female is indeed heterogametic and is characterized by the *Lygaeus* type of sex determination.

5.3 The Y Chromosome Determines Maleness in Humans

The first attempt to understand sex determination in our own species occurred almost 100 years ago and involved the examination of chromosomes present in dividing cells. Efforts were made to accurately determine the diploid chromosome number of humans, but because of the relatively large number of chromosomes, this proved to be quite difficult. In 1912, H. von Winiwarter counted 47 chromosomes in a spermatogonial metaphase preparation. It was believed that the sex-determining mechanism in humans was based on the presence of an extra chromosome in females, who were thought to have 48 chromosomes. However, in the 1920s, Theophilus Painter observed between 45 and 48 chromosomes in cells of testicular tissue and also discovered the small Y chromosome, which we now know occurs only in males. In his

(a)

(b)

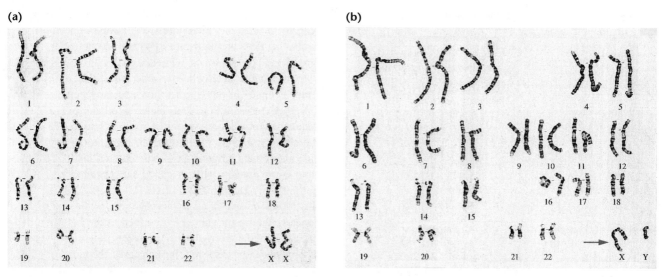

FIGURE 5–5 The traditional human karyotypes derived from a normal female and a normal male. Each contains 22 pairs of autosomes and two sex chromosomes. The female (a) contains two identical X chromosomes, while the male (b) contains one X and one Y chromosome.

original paper, Painter favored 46 as the diploid number in humans, but he later concluded incorrectly that 48 was the chromosome number in both males and females.

For 30 years, this number was accepted. Then, in 1956, Joe Hin Tjio and Albert Levan discovered a better way to prepare chromosomes. The improved technique led to a strikingly clear demonstration of metaphase stages showing that 46 is indeed the human diploid number. Later that same year, C. E. Ford and John L. Hamerton, also working with testicular tissue, confirmed this finding. The familiar karyotype of humans (Figure 5–5) is based on Tjio and Levan's technique.

Within the normal 23 pairs of human chromosomes, one pair was shown to vary in configuration in males and females. These two chromosomes were designated the X and Y sex chromosomes. The human female has two X chromosomes, and the human male has one X and one Y chromosome.

We might believe that this observation is sufficient to conclude that the Y chromosome determines maleness. However, several other interpretations are possible. The Y could play no role in sex determination; the presence of two X chromosomes could cause femaleness; or maleness could result from the lack of a second X chromosome. The evidence that clarified which explanation was correct emerged in the study of variations in the human sex chromosome composition. As such investigations reveal, the Y chromosome does indeed determine maleness in humans.

Klinefelter and Turner Syndromes

About 1940, scientists identified two human abnormalities characterized by aberrant sexual development, **Klinefelter syndrome (47,XXY)** and **Turner syndrome (45,X)**.* Indi-

viduals with Klinefelter syndrome have genitalia and internal ducts that are usually male, but their testes are rudimentary and fail to produce sperm. They are generally tall and have long arms and legs and large hands and feet. Although masculine development does occur, feminine sexual development is not entirely suppressed. Slight enlargement of the breasts (gynecomastia) is common, and the hips are often rounded. This ambiguous sexual development can lead to abnormal social development. Intelligence is often below the normal range as well.

In Turner syndrome, the affected individual has female external genitalia and internal ducts, but the ovaries are rudimentary. Other characteristic abnormalities include short stature (usually under 5 feet), skin flaps on the back of the neck, and underdeveloped breasts. A broad, shieldlike chest is sometimes noted. Intelligence is usually normal.

In 1959, the karyotypes of individuals with these syndromes were determined to be abnormal with respect to the sex chromosomes. Individuals with Klinefelter syndrome have more than one X chromosome. Most often they have an XXY complement in addition to 44 autosomes [Figure 5–6(a)], and people with this karyotype are designated 47,XXY. Individuals with Turner syndrome most often have only 45 chromosomes, including just a single X chromosome; thus, they are designated 45,X [Figure 5–6(b)]. Note the convention used in designating these chromosome compositions. The number states the total number of chromosomes present, and the information after the comma indicates the deviation from the normal diploid content. Both conditions result from **nondisjunction**, the failure of the chromosomes to segregate properly during meiosis (see Figure 6–1).

The Klinefelter and Turner karyotypes and their corresponding sexual phenotypes allow us to conclude that the Y chromosome determines maleness in humans. In its absence, the sex of the individual is female, even if only a single X chromosome is present. The presence of the Y chromosome in the individual with Klinefelter syndrome is sufficient to deter-

---

*Although the possessive form of the names of most syndromes (eponyms) is sometimes used (e.g., Klinefelter's), the current preference is to use the nonpossessive form, which we have adopted for all human syndromes and diseases.

(a)

(b)

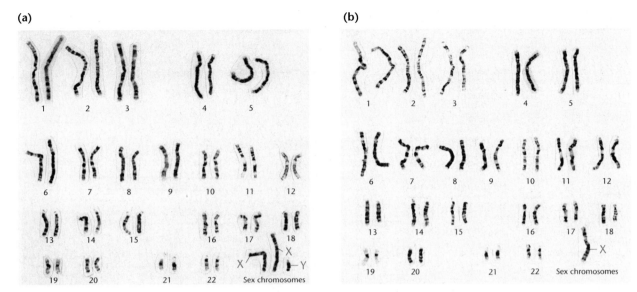

FIGURE 5–6 The karyotypes of individuals with (a) Klinefelter syndrome (47,XXY) and (b) Turner syndrome (45,X).

mine maleness, even though its expression is not complete. Similarly, in the absence of a Y chromosome, as in the case of individuals with Turner syndrome, no masculinization occurs.

Klinefelter syndrome occurs in about 2 of every 1000 male births. The karyotypes 48,XXXY, 48,XXYY, 49,XXXXY, and 49,XXXYY are similar phenotypically to 47,XXY, but manifestations are often more severe in individuals with a greater number of X chromosomes.

Turner syndrome can also result from karyotypes other than 45,X, including individuals called *mosaics* whose somatic cells display two different genetic cell lines, each exhibiting a different karyotype. Such cell lines result from a mitotic error during early development, the most common chromosome combinations being 45,X/46,XY and 45,X/46,XX. Thus, an embryo that began life with a normal karyotype can give rise to an individual whose cells show a mixture of karyotypes and who exhibits varying aspects of this syndrome.

Turner syndrome is observed in about 1 in 2000 female births, a frequency much lower than that for Klinefelter syndrome. One explanation for this difference is the observation that a substantial majority of 45,X fetuses die *in utero* and are aborted spontaneously. Thus, a similar frequency of the two syndromes may occur at conception.

47,XXX Syndrome

The presence of three X chromosomes along with a normal set of autosomes (**47,XXX**) results in female differentiation. This syndrome, which occurs in about 1 of 1200 female births, is highly variable in expression. Frequently, 47,XXX women are perfectly normal. In other cases, underdeveloped secondary sex characteristics, sterility, and mental retardation can occur. In rare instances, 48,XXXX and 49,XXXXX karyotypes have been reported. The syndromes associated with these karyotypes are similar to, but more pronounced than, the 47,XXX. Thus, in many cases, the presence of additional X chromosomes appears to disrupt the delicate balance of genetic information essential to normal female development.

47,XYY Condition

Another human condition involving the sex chromosomes, **47,XYY**, has also been intensively investigated. Studies of this condition, where the only deviation from diploidy is the presence of an additional Y chromosome in an otherwise normal male karyotype, were initiated in 1965 by Patricia Jacobs. She discovered that 9 of 315 males in a Scottish maximum security prison had the 47,XYY karyotype. These males were significantly above average in height and had been incarcerated as a result of antisocial (nonviolent) criminal acts. Of the nine males studied, seven were of subnormal intelligence, and all suffered personality disorders. Several other studies produced similar findings.

The possible correlation between this chromosome composition and criminal behavior piqued considerable interest and extensive investigations of the phenotype and frequency of the 47,XYY condition in both criminal and noncriminal populations ensued. Above-average height (usually over 6 feet) and subnormal intelligence have been generally substantiated, and the frequency of males displaying this karyotype is indeed higher in penal and mental institutions compared with unincarcerated males (see Table 5.1). A particularly relevant question involves the characteristics displayed by XYY males who are not incarcerated. The only nearly constant association is that such individuals are over 6 feet tall.

A study that addressed this issue was initiated to identify 47,XYY individuals at birth and to follow their behavioral patterns during preadult and adult development. By 1974, the two investigators, Stanley Walzer and Park Gerald, had identified about 20 XYY newborns in 15,000 births at Boston Hospital for Women. However, they soon came under great pressure to abandon their research. Those opposed to the study argued that the investigation could not be justified and might cause great harm to those individuals who displayed this karyotype. The opponents argued that (1) no association between the additional Y chromosome and abnormal behavior had been previously established in the population at large, and

TABLE 5.1	Frequency of XYY Individuals in Various Settings			XYY	
Setting	Restriction	Number Studied		Number	Frequency (%)
Control population	Newborns	28,366		29	0.10
Mental-penal	No height restriction	4239		82	1.93
Penal	No height restriction	5805		26	0.44
Mental	No height restriction	2562		8	0.31
Mental-penal	Height restriction	1048		48	4.61
Penal	Height restriction	1683		31	1.84
Mental	Height restriction	649		9	1.38

Source: Compiled from data presented in Hook, 1973, Tables 1–8. © 1973 by the American Association for the Advancement of Science.

(2) "labeling" these individuals in the study might become a self-fulfilling prophecy. That is, as a result of participation in the study, parents, relatives, and friends might treat individuals identified as 47,XYY differently, ultimately producing the expected antisocial behavior. Despite the support of a government funding agency and the faculty at Harvard Medical School, Walzer and Gerald abandoned the investigation in 1975.

More recently, it has become clear that many XYY males do not exhibit any form of antisocial behavior and lead normal lives. Therefore, we must conclude that there is a high, but not constant, correlation between the extra Y chromosome and the predisposition of these males to behavioral problems.

Sexual Differentiation in Humans

Once researchers established that, in humans, it is the Y chromosome that houses genetic information necessary for maleness, they attempted to pinpoint a specific gene or genes capable of providing the "signal" responsible for sex determination. Before we delve into this topic, it is useful to consider how sexual differentiation occurs in order to better comprehend how humans develop into sexually dimorphic males and females. During early development, every human embryo undergoes a period when it is potentially hermaphroditic. By the fifth week of gestation, gonadal primordia arise as a pair of ridges associated with each embryonic kidney. Primordial germ cells migrate to these ridges, where an outer cortex and inner medulla form. The cortex is capable of developing into an ovary, while the inner medulla may develop into a testis. In addition, two sets of undifferentiated male (Wolffian) and female (Mullerian) ducts exist in each embryo.

If the cells of the genital ridge have the XY constitution, development of the medullary region into a testis is initiated around the seventh week. However, in the absence of the Y chromosome, no male development occurs, and the cortex of the genital ridge subsequently forms ovarian tissue. Parallel development of the appropriate male or female duct system then occurs, and the other duct system degenerates. Substantial evidence indicates that in males, once testes differentiation is initiated, the embryonic testicular tissue secretes two hormones that are essential for continued male sexual differentiation.

In the absence of male development, as the twelfth week of fetal development approaches, the oogonia within the ovaries begin meiosis and primary oocytes can be detected. By the 25th week of gestation, all oocytes become arrested in meiosis and remain dormant until puberty is reached some 10 to 15 years later. In males, on the other hand, primary spermatocytes are not produced until puberty is reached.

The Y Chromosome and Male Development

The human Y chromosome, unlike the X, has long been thought to be mostly blank genetically. It is now known that this is not true, even though the Y chromosome contains far fewer genes than does the X. Current analysis has revealed numerous genes and regions with potential genetic function, some with and some without homologous counterparts on the X chromosome. For example, present on both ends of the Y chromosome are the so-called **pseudoautosomal regions (PARs)** that share homology with regions on the X chromosome and that synapse and recombine with it during meiosis. The presence of such a pairing region is critical to segregation of the X and Y chromosomes during male gametogenesis. The remainder of the chromosome, about 95 percent of it, does not synapse or recombine with the X chromosome. As a result, it was originally referred to as the *nonrecombining region of the Y (NRY)*. More recently, researchers have designated this region as the **male-specific region of the Y (MSY)**. As you will see, some portions of the MSY share homology with genes on the X chromosome, and some do not.

The human Y chromosome is diagrammed in Figure 5–7. The MSY is divided about equally between *euchromatic* regions that contain functional genes and *heterochromatic* regions that lack genes. Within euchromatin, adjacent to the PAR of the short arm of the Y chromosome, is a critical gene that controls male sexual development, called the *sex-determining region Y (SRY)*. In humans, the absence of a Y chromosome almost always leads to female development;

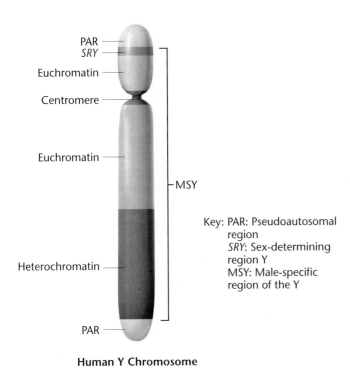

PAR

SRY

Euchromatin

Centromere

Euchromatin

MSY

Heterochromatin

Key: PAR: Pseudoautosomal region
SRY: Sex-determining region Y
MSY: Male-specific region of the Y

PAR

Human Y Chromosome

FIGURE 5–7 The regions of the human Y chromosome.

thus, this gene is absent from the X chromosome. *SRY* encodes a gene product that somehow triggers the undifferentiated gonadal tissue of the embryo to form testes. This product is called the **testis-determining factor** (**TDF**). *SRY* (or a closely related version) is present in all mammals thus far examined, indicative of its essential function throughout this diverse group of animals.

Our ability to identify the presence or absence of DNA sequences in rare individuals whose expected sex-chromosome composition does not correspond to their sexual phenotype has provided evidence that *SRY* is the gene responsible for male sex determination. For example, there are human males who have two X and no Y chromosomes. Often, attached to one of their X chromosomes is the region of the Y that contains *SRY*. There are also females who have one X and one Y chromosome. Their Y is almost always missing the *SRY* gene. These observations argue strongly in favor of the role of *SRY* in providing the primary signal for male development.

Further support of this conclusion involves an experiment using **transgenic mice**. These animals are produced from fertilized eggs injected with foreign DNA that is subsequently incorporated into the genetic composition of the developing embryo. In normal mice, a chromosome region designated *Sry* has been identified that is comparable to *SRY* in humans. When DNA containing only mouse *Sry* is injected into normal XX mouse eggs, most of the offspring develop into males.

The question of how the product of this gene triggers the embryonic gonadal tissue to develop into testes rather than ovaries is under extensive investigation. In humans, other autosomal genes are believed to be part of a cascade of genetic expression initiated by *SRY*. Examples include the *SOX9* gene and the *WT1* gene (on chromosome 11), originally identified

as an oncogene associated with Wilms tumor, which affects the kidney and gonads. Another, *SF1*, is involved in the regulation of enzymes affecting steroid metabolism. In mice, this gene is initially active in both the male and female bisexual genital ridge, persisting until the point in development when testis formation is apparent. At that time, its expression persists in males but is extinguished in females. The link between these various genes and sex determination brings us closer to a complete understanding of how males and females arise in humans.

Some very recent findings by David Page and his many colleagues have now provided a reasonably complete picture of the MSY region of the human Y chromosome. This work is based on information gained through the Human Genome Project, where the DNA of all chromosomes has now been sequenced. Page has spearheaded the detailed study of the Y chromosome for the past several decades.

The MSY consists of about 23 million base pairs (23 Mb) and can be divided into three regions. The first region is the *X-transposed region*. It comprises about 15 percent of the MSY and was originally derived from the X chromosome during human evolution (about 3 million to 4 million years ago). The X-transposed region is 99 percent identical to region Xq21 of the modern human X chromosome. Two genes, both with X-chromosome homologs, are present in this region.

The second area is designated the *X-degenerative region*. Comprising about 20 percent of the MSY, this region contains DNA sequences that are even more distantly related to those present on the X chromosome. The X-degenerative region contains 27 single-copy genes and a number of **pseudogenes** (genes whose sequence has degenerated sufficiently during evolution to render them nonfunctional). As with the genes present in the X-transposed region, all share some homology with counterparts on the X chromosome. These 27 genetic units contain 14 DNA regions that are capable of being transcribed, and each is present as a single copy. One of these is the *SRY* gene discussed above. Other X-degenerative genes that encode protein products are expressed ubiquitously in all tissues in the body, but *SRY* is expressed only in the testes.

The third area, the *ampliconic region*, contains about 30 percent of the MSY, including most of the genes closely associated with testes development. These genes lack counterparts on the X chromosome, and their expression is limited to the testes. There are 60 transcription units divided among 9 gene families in this region, most represented by multiple copies. Members of each family have nearly identical (>98%) DNA sequences. Each repeat unit is an **amplicon** and is contained within seven segments scattered across the euchromatic regions present on both the short and long arms of the Y chromosome. Genes in the ampliconic region encode proteins specific to the development and function of the testes, and the products of many of these genes are directly related to fertility in males. It is currently believed that a great deal of male sterility in our population can be linked to mutations in these genes.

This recent work has provided a comprehensive picture of the genetic information present on this unique chromosome. This information clearly refutes the so-called wasteland theory, prevalent only 20 years ago, that depicted the human Y chromosome as almost devoid of genetic information other than a gene or two that caused maleness. The knowledge we have gained provides the basis for a much clearer picture of how maleness is determined. In addition, this information provides important clues as to origin of the Y chromosome during human evolution.

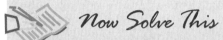

Now Solve This

Problem 28 on page 113 concerns itself with *SOX9*, a gene on an autosome, that when mutated appears to inhibit normal human male development. You are asked to draw conclusions based on numerous observations.

Hint: Some genes are activated and produce their normal product as a result of the expression of products of other genes found on different chromosomes—in this case, perhaps one that is on the Y chromosome.

5.4 The Ratio of Males to Females in Humans Is Not 1.0

The presence of heteromorphic sex chromosomes in one sex of a species but not the other provides a potential mechanism for producing equal proportions of male and female offspring. This potential is premised on the segregation of the X and Y (or Z and W) chromosomes during meiosis, such that one-half of the gametes of the heterogametic sex receive one of the chromosomes and one-half receive the other one. As we just learned in Section 5.3, small pseudoautosomal regions of pairing homology do exist at both ends of the X and the Y chromosomes in humans. Provided that both types of gametes are equally successful in fertilization and that the two sexes are equally viable during development, a 1:1 ratio of male and female offspring should result.

Given the potential for producing equal numbers of both sexes, the actual proportion of male to female offspring has been investigated and is referred to as the **sex ratio**. We can assess it in two ways. The **primary sex ratio** reflects the proportion of males to females conceived in a population. The **secondary sex ratio** reflects the proportion of each sex that is born. The secondary sex ratio is much easier to determine but has the disadvantage of not accounting for any disproportionate embryonic or fetal mortality.

When the secondary sex ratio in the human population was determined in 1969 using worldwide census data, it did not equal 1.0. For example, in the Caucasian population in the United States, the secondary ratio was a little less than 1.06, indicating that about 106 males were born for each 100 females. In 1995, this ratio dropped to slightly less than 1.05. In the African-American population in the United States, the ratio was 1.025. In other countries, the excess of male births was even greater than is reflected in these values. For example, in Korea, the secondary sex ratio was 1.15.

Despite these ratios, it is possible that the *primary sex ratio* is 1.0 and that it is altered between conception and birth. For the

secondary ratio to exceed 1.0, prenatal female mortality would have to be greater than prenatal male mortality. However, this hypothesis has been examined and shown to be false. In fact, just the opposite occurs. In a Carnegie Institute study, reported in 1948, the sex of approximately 6000 embryos and fetuses recovered from miscarriages and abortions was determined, and fetal mortality was actually higher in males. On the basis of the data derived from that study, the primary sex ratio in U.S. Caucasians was estimated to be 1.079. More recent data have estimated that this figure is much higher—between 1.20 and 1.60, suggesting that many more males than females are conceived in the human population.

It is not clear why such a radical departure from the expected primary sex ratio of 1.0 occurs. To come up with a suitable explanation, we must examine the assumptions on which the theoretical ratio is based:

1. Because of segregation, males produce equal numbers of X- and Y-bearing sperm.

2. Each type of sperm has equivalent viability and motility in the female reproductive tract.

3. The egg surface is equally receptive to both X- and Y-bearing sperm.

Although no direct experimental evidence contradicts any of these assumptions, the human Y chromosome is smaller than the X chromosome and therefore has less mass. Thus, it has been speculated that Y-bearing sperm are more motile than X-bearing sperm. If this is true, then the probability of a fertilization event leading to a male zygote is increased, providing one possible explanation for the observed primary ratio.

5.5 Dosage Compensation Prevents Excessive Expression of X-Linked Genes in Humans and Other Mammals

The presence of two X chromosomes in normal human females and only one X chromosome in normal human males is unique compared with the equal numbers of autosomes present in the cells of both sexes. On theoretical grounds alone, it is possible to speculate that this disparity should create a "genetic dosage" problem between males and females for all X-linked genes. There is the potential for females to produce twice as much of each gene product for all X-linked genes. The additional X chromosomes in both males and females exhibiting the various syndromes discussed earlier should compound this dosage problem even more. In this section, we describe research findings on X-linked gene expression that demonstrate a genetic mechanism allowing for **dosage compensation**.

Barr Bodies

Murray L. Barr and Ewart G. Bertram's experiments with female cats, and Keith Moore and Barr's subsequent study with humans, demonstrate a genetic mechanism in mammals that compensates for X-chromosome dosage disparities. Barr

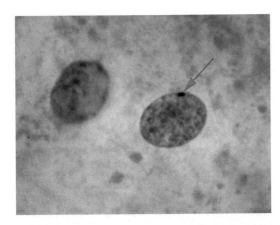

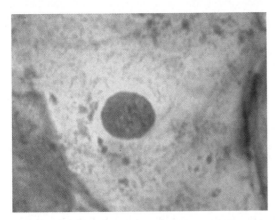

FIGURE 5–8 Photomicrographs comparing cheek epithelial cell nuclei from a male that fails to reveal Barr bodies (bottom) with a female that demonstrates a Barr body (indicated by the arrow in the top image). This structure, also called a sex chromatin body, represents an inactivated X chromosome.

(Figure 5–9). Therefore, the number of Barr bodies follows an [N − 1] rule, where N is the total number of X chromosomes present.

This mechanism of inactivating all but one X chromosome increases our understanding of dosage compensation, but it raises further questions concerning other matters. For example, because one of the two X chromosomes is inactivated in normal human females, why then is the Turner 45,X individual not entirely normal? Why aren't females with the triplo-X and tetra-X karyotypes (47,XXX and 48,XXXX, respectively) normal? Furthermore, in Klinefelter syndrome (47,XXY), X-chromosome inactivation effectively renders such individuals 46,XY. Why aren't these males unaffected by the additional X chromosome in their nuclei?

One possible explanation is that chromosome inactivation does not normally occur in the very early developmental stages of those cells destined to form gonadal tissues. Another possible explanation is that not all of each X chromosome forming a Barr body is inactivated. Recent studies have indeed demonstrated that as much as 15 percent of the human X-chromosomal genes actually escape inactivation. In either case, excessive expression of certain X-linked genes might still occur despite apparent inactivation of additional X chromosomes.

The Lyon Hypothesis

In mammalian females, one X chromosome is of maternal origin and the other is of paternal origin. Which one is inactivated? Is the inactivation random? Is the same chromosome inactive in all somatic cells? In 1961, Mary Lyon and Liane Russell independently proposed a hypothesis that answers these questions. They postulated that the inactivation of X chromosomes occurs randomly in somatic cells at a point early in embryonic development and that once inactivation

and Bertram observed a darkly staining body in interphase nerve cells of female cats that was absent in similar cells of males. In humans, this body can be easily demonstrated in female cells derived from the buccal mucosa or in fibroblasts, but not in similar male cells (Figure 5–8). This highly condensed structure, about 1 μm in diameter, lies against the nuclear envelope of interphase cells. It stains positively in the Feulgen reaction for DNA.

Current experimental evidence demonstrates that this structure, called a **sex chromatin body** or simply a **Barr body**, is an inactivated X chromosome. Susumo Ohno was the first to suggest that the Barr body arises from one of the two X chromosomes. This hypothesis is attractive because it provides a mechanism for dosage compensation. If one of the two X chromosomes is inactive in the cells of females, the dosage of genetic information that can be expressed in males and females is equivalent. Convincing but indirect evidence for this hypothesis comes from the study of the sex-chromosome syndromes described earlier in this chapter. Regardless of how many X chromosomes exist, all but one of them appear to be inactivated and can be seen as Barr bodies. For example, no Barr body is seen in Turner 45,X females; one is seen in Klinefelter 47,XXY males; two in 47,XXX females; three in 48,XXXX females; and so on

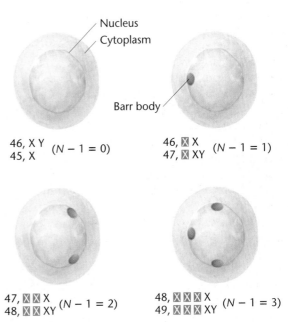

FIGURE 5–9 Occurrence of Barr bodies in various human karyotypes, where all X chromosomes except one (N − 1) are inactivated.

has occurred, all progeny cells have the same X chromosome inactivated.

This explanation, which has come to be called the **Lyon hypothesis**, was initially based on observations of female mice heterozygous for X-linked coat color genes. The pigmentation of these heterozygous females was mottled, with large patches expressing the color allele on one X chromosome and other patches expressing the allele on the other X chromosome. Indeed, such a phenotypic pattern would result if different X chromosomes were inactive in adjacent patches of cells. Similar mosaic patterns occur in the black and yellow-orange patches of female tortoiseshell and calico cats (Figure 5–10). Such X-linked coat color patterns do not occur in male cats because all their cells contain the single maternal X chromosome and are therefore hemizygous for only one X-linked coat color allele.

The most direct evidence in support of the Lyon hypothesis comes from studies of gene expression in clones of human fibroblast cells. Individual cells are isolated following biopsy and cultured *in vitro*. If each culture is derived from a single cell, it is called a **clone**. The synthesis of the enzyme *glucose-6-phosphate dehydrogenase* (G6PD) is controlled by an X-linked gene. Numerous mutant alleles of this gene have been detected, and their gene products can be differentiated from the wild-type enzyme by their migration pattern in an electrophoretic field.

Fibroblasts have been taken from females heterozygous for different allelic forms of *G6PD* and studied. The Lyon hypothesis predicts that if inactivation of an X chromosome occurs randomly early in development and is permanent in all progeny cells, such a female should show two types of clones, each showing only one electrophoretic form of G6PD, in approximately equal proportions.

In 1963, Ronald Davidson and his colleagues performed an experiment involving 14 clones from a single heterozygous female. Seven showed only one form of the enzyme, and seven showed only the other form. The most important finding was that none of the 14 clones showed both forms of the enzyme. Studies of G6PD in humans thus provide strong support for the random inactivation of either the maternal or paternal X chromosome.

The Lyon hypothesis is generally accepted as valid; in fact, the inactivation of an X chromosome into a Barr body is sometimes referred to as **lyonization**. One extension of the hypothesis is that mammalian females are mosaics for all heterozygous X-linked alleles—some areas of the body express only the maternally derived alleles, and others express only the paternally derived alleles. Two especially interesting examples involve **red-green color blindness** and **anhidrotic ectodermal dysplasia**, both X-linked recessive disorders. In the former case, hemizygous males are fully color-blind in all retinal cells. However, heterozygous females display mosaic retinas with patches of defective color perception and surrounding areas with normal color perception. Males hemizygous for anhidrotic ectodermal dysplasia show an absence of teeth, sparse hair growth, and lack of sweat glands. The skin of females heterozygous for this disorder reveals random patterns of tissue with and without sweat glands. In both examples, random inactivation of one or the other X chromosome early in the development of heterozygous females leads to these occurrences.

Now Solve This

Problem 32 on page 113 is concerned with Carbon Copy (CC), the first cloned cat, who was derived from a somatic nucleus of a calico cat. You are asked to comment on the likelihood that CC will appear identical to her genetic donor.

Hint: The donor nucleus was from a differentiated ovarian cell of an adult female cat, which itself had inactivated one of its X chromosomes.

(a)

(b)

FIGURE 5–10 (a) A calico cat, where the random distribution of orange and black patches demonstrates the Lyon hypothesis. The white patches are due to another gene (*S*), which distinguish calico cats from tortoiseshell cats (b), which lack the white patches.

The Mechanism of Inactivation: Imprinting

The least understood aspect of the Lyon hypothesis is the mechanism of chromosome inactivation. Somehow, either the DNA and/or the attached histone proteins of one (or more) of the mammalian X chromosomes of females is modified such that most genes that are part of that chromosome are silenced. Whatever the modification, a memory is created such that following chromosome replications and cell divisions the same chromosome remains inactivated. Such a step, whereby genetic expression of one homolog, but not the other, is affected, is referred to as **imprinting**. This term also applies to a number of similar instances whereby genetic expression is modified epigenetically.

Recent investigations are beginning to clarify this issue. In humans, a single region of the human X chromosome, called the **X-inactivation center** (**Xic**), has been discovered to be the major control unit. Genetic expression of this region, located on the proximal end of the p arm in humans, occurs only on the X chromosome that is inactivated. The constant association of expression of Xic and X-chromosome inactivation supports the conclusion that this region is an important genetic component in the inactivation process.

The Xic is about 1 Mb (10^6 base pairs) in length and contains four genes. One of these, ***X-inactive specific transcript*** (***XIST***), is now believed to represent the critical gene. Several interesting observations have been made regarding the RNA that is transcribed from it, with much of the underlying work having been done by using the equivalent gene in the mouse (*Xist*). First, the RNA product is quite large and lacks what is called an extended **open reading frame** (**ORF**). An ORF is comprised of information necessary for the translation of an RNA sequence into a protein. Thus, in this case, the RNA is transcribed but is not translated. It appears to serve a structural role in the nucleus, presumably in the mechanism of chromosome inactivation. This finding has led to the belief that the RNA products of *XIST* and *Xist* spread over and coat the X chromosome bearing the gene that produced it, creating some sort of molecular "cage" that entraps it, leading to its inactivation. Inactivation is therefore said to be cis-acting.

Second, transcription of *Xist* occurs initially at low levels on both the paternal and maternal X chromosomes. As the inactivation process begins, however, transcription continues and is enhanced only on the X chromosome(s) that becomes inactivated. In 1996, a research group led by Graeme Penny provided convincing evidence that transcription of *Xist* is the critical event in chromosome inactivation. These researchers were able to introduce a targeted deletion (7 kb) into this gene. As a result, the chromosome bearing this mutation lost its ability to become inactivated.

Several intriguing questions remain regarding imprinting leading to inactivation. In cells with more than two chromosomes, what sort of "counting" mechanism designates all but one X chromosome to be inactivated? What "blocks" the Xic of the active chromosome, preventing further transcription of *Xist*? How does imprinting impart a memory such that inactivation of the same X chromosome or chromosomes is subsequently maintained in progeny cells, as the Lyon hypothesis calls for? The inactivation signal must somehow remain stable as cells proceed through chromosome replication. Whatever the answers to these questions, we have taken an exciting step toward understanding how dosage compensation is accomplished in mammals.

5.6 The Ratio of X Chromosomes to Sets of Autosomes Determines Sex in *Drosophila*

Because males and females in *Drosophila melanogaster* (and other *Drosophila* species) have the same general sex chromosome composition as humans (males are XY and females are XX), we might assume that the Y chromosome also causes maleness in these flies. However, the elegant work of Calvin Bridges in 1916 showed this is not true. He studied flies with quite varied chromosome compositions, leading him to conclude that the Y chromosome is not involved in sex determination in this organism. Instead, Bridges proposed that both the X chromosomes and autosomes together play a critical role in sex determination. Recall that in the nematode, *C. elegans*, which lacks a Y chromosome, the sex chromosomes and autosomes are also both critical to sex determination.

The most telling observation that differentiates the mechanism operating in *Drosophila* from that in humans involves the XXY and XO sex chromosome compositions. Contrary to what was later discovered in humans, Bridges found that the XXY flies are normal females and the XO flies are sterile males. The presence of the Y chromosome in the XXY flies did not cause maleness, and its absence in the XO flies did not produce femaleness. From these data, he concluded that the Y chromosome in *Drosophila* lacks male-determining factors, but since the XO males are sterile, it does contain genetic information essential to male fertility.

Bridges was able to clarify the mode of sex determination in *Drosophila* by studying the progeny of triploid females ($3n$). These females have three copies each of the haploid complement of chromosomes. *Drosophila* has a haploid number of four, thereby displaying three pairs of autosomes in addition to its pair of sex chromosomes. Triploid females apparently originate from rare diploid eggs fertilized by normal haploid sperm. Triploid females have heavy-set bodies, coarse bristles, and coarse eyes, and they can be fertile. Because of the odd number of each chromosome (3), during meiosis a wide range of chromosome complements is distributed into gametes that give rise to offspring with a variety of abnormal chromosome constitutions. A correlation among the sexual morphology, chromosome composition, and Bridges' interpretation is shown in Figure 5–11.

Bridges realized that the critical factor in determining sex is the ratio of X chromosomes to the number of haploid sets of autosomes (A) present. Normal (2X:2A) and triploid (3X:3A) females each have a ratio equal to 1.0, and both are fertile. As the ratio exceeds unity (3X:2A, or 1.5, for example), what was once called a *superfemale* is produced. Because such females are most often inviable, they are now more appropriately called **metafemales**.

FIGURE 5–11 Chromosome compositions, the ratios of X chromosomes to sets of autosomes, and the resultant sexual morphology in *Drosophila melanogaster*. The normal diploid male chromosome composition is shown as a reference on the left (XY/2A).

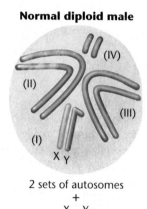

Normal diploid male

(IV)

(II)

(III)

(I)

X Y

2 sets of autosomes
+
X   Y

Chromosome composition	Chromosome formulation	Ratio of X chromosomes to autosome sets	Sexual morphology
	$3X/2A$	1.5	Metafemale
	$3X/3A$	1.0	Female
	$2X/2A$	1.0	Female
	$3X/4A$	0.75	Intersex
	$2X/3A$	0.67	Intersex
	$X/2A$	0.50	Male
	$XY/2A$	0.50	Male
	$XY/3A$	0.33	Metamale

Normal (XY:2A) and sterile (XO:2A) males each have a ratio of 1:2 or 0.5. When the ratio decreases to 1:3, or 0.33, as in the case of an XY:3A male, infertile **metamales** result. Other flies recovered by Bridges in these studies contained an X:A ratio intermediate between 0.5 and 1.0. These flies were generally larger, and they exhibited a variety of morphological abnormalities and rudimentary bisexual gonads and genitalia. They were invariably sterile and expressed both male and female morphology, thus being designated as **intersexes**.

Bridges' results indicate that in *Drosophila*, factors that cause a fly to develop into a male are not localized on the sex chromosomes but are instead found on the autosomes. Some female-determining factors, however, are localized on the X chromosomes. Thus, with respect to primary sex determination, male gametes containing one of each autosome plus a Y chromosome result in male offspring not because of the presence of the Y but because they fail to contribute an X chromosome. This mode of sex determination is explained by the **genic balance theory**. Bridges proposed that a threshold for maleness is reached when the X:A ratio is 1:2 (X:2A), but that the presence of an additional X chromosome (XX:2A) alters this balance and results in female differentiation.

Numerous mutant genes have been identified that are involved in sex determination in *Drosophila*. The recessive autosomal gene *transformer* (*tra*), discovered over 50 years ago by Alfred H. Sturtevant, clearly demonstrates that a single autosomal gene can have a profound impact on sex determination. Females homozygous for *tra* are transformed into sterile males, but homozygous males are unaffected.

More recently, another gene, *Sex-lethal* (*Sxl*), has been shown to play a critical role and serves as a "master switch" in sex determination. Activation of the X-linked *Sxl* gene, which relies on a ratio of X chromosomes to sets of autosomes that equals 1.0, is essential to female development. In the absence of activation, resulting, for example, from an X:A ratio of 0.5, male development occurs. It is interesting to note that mutations that inactivate the *Sxl* gene kill female embryos but have no effect on male embryos, consistent with the role of the gene. Although it is not yet exactly clear how this ratio influences the *Sxl* locus, we do have some insights into the question. The *Sxl* locus is part of a hierarchy of gene expression and exerts control over still other genes, including *tra* and *dsx* (*doublesex*), as well as others. The wild-type allele of *tra* is activated by the product of *Sxl* only in females, which in turn influences the expression of *dsx*. Depending on how the initial RNA transcript of *dsx* is processed (spliced), the resultant dsx protein activates either male-specific or female-specific genes required for sexual differentiation. Each step in this regulatory cascade requires a form of processing called **RNA splicing**, in which portions of the RNA are removed and the remaining fragments "spliced" back together prior to translation into a protein. In the case of the *Sxl* gene, its transcript can be spliced in several different ways, a phenomenon called **alternative splicing**. Two different RNA transcripts are produced in females and males, respectively. In potential females, the transcript is active and initiates a cascade of regulatory gene expression, ultimately leading to female differentiation. In potential males, the transcript is inactive, leading to different gene activity, whereby male differentiation occurs.

5.7 Temperature Variation Controls Sex Determination in Reptiles

We conclude this chapter by discussing several cases involving reptiles where the environment—specifically temperature—has a profound influence on sex determination. The investigations leading to this information may well come closer to revealing the true nature of the primary basis of sex determination than any finding previously discussed.

In many species of reptiles, sex is predetermined at conception by sex-chromosome composition, as is the case in many of the organisms already considered in this chapter. For example, in many snakes, including vipers, a ZZ/ZW mode is in effect, where the female is the heterogamous sex (ZW). However, in boas and pythons, it is impossible to distinguish one sex chromosome from the other in either sex. In lizards, both the XX/XY and the ZZ/ZW systems are found, depending on the species. However, in other reptilian species, including all crocodiles, most turtles, and some lizards, sex determination is achieved according to the incubation temperature of eggs during a critical period of embryonic development.

Three distinct patterns of temperature-dependent sex determination emerge (cases I–III in Figure 5–12). In case I, low temperatures yield 100 percent females and high temperatures yield 100 percent males; just the opposite occurs in case II. In case III, low *and* high temperatures yield 100 percent females, while intermediate temperatures yield various proportions of males. The third pattern is seen in various species of crocodiles, turtles, and lizards, although some members of these groups are known to exhibit the first two patterns.

Two observations are noteworthy. First, under certain temperatures in all three patterns, both male and female offspring result; second, the pivotal temperature (T_p) is a fairly narrow range, usually less than 5°C and sometimes only 1°C. The central question raised by these observations is: What metabolic or physiological parameters are being affected by temperature that lead to the differentiation of one sex or the other?

The answer is thought to involve steroids (mainly estrogens) and the enzymes involved in their synthesis. Studies clearly demonstrate that the effects of temperature on estrogens, androgens, and inhibitors of the enzymes controlling their synthesis are involved in the sexual differentiation of ovaries and testes. One enzyme in particular, *aromatase*, converts androgens (male hormones such as testosterone) to estrogens (female hormones such as estradiol). The activity of this enzyme is correlated with the pathway followed during gonadal differentiation activity and is high in developing ovaries and low in developing testes. Researchers in this field, including Claude Pieau and colleagues, have proposed that a thermosensitive factor mediates the transcription of the reptilian aromatase gene that leads to temperature-dependent sex determination. Several other genes are likely to be involved in this mediation.

The involvement of sex steroids in gonadal differentiation has also been documented in birds, fish, and amphibians. Thus, sex-determining mechanisms involving estrogens seem to be characteristic of nonmammalian vertebrates. The regulation of such a system, through temperature dependent in many reptiles, appears to be controlled by sex chromosomes (XX/XY or ZZ/ZW) in many of these other organisms.

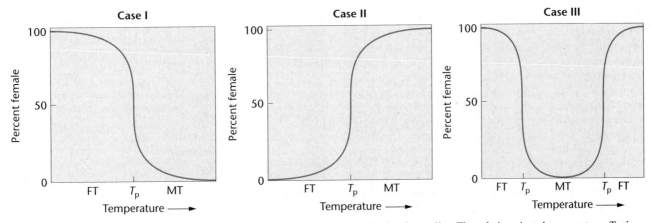

FIGURE 5–12 Three different patterns of temperature-dependent sex determination in reptiles. The relative pivotal temperature, T_p, is crucial to sex determination during a critical point in embryonic development; FT = female-determining temperature and MT = male-determining temperature.

GENETICS, TECHNOLOGY, AND SOCIETY

A Question of Gender: Sex Selection in Humans

The desire to choose a baby's gender is as pervasive as human nature itself. Throughout time, people have resorted to varied and sometimes bizarre methods to achieve the preferred gender of their offspring. In medieval Europe, prospective parents would place a hammer under the bed to help them conceive a boy, or a pair of scissors to conceive a girl. Other practices were based on the ancient belief that semen from the right testicle created male offspring and that from the left testicle created females. Men in ancient Greece would lie on their right side during intercourse in order to conceive a boy. Up until the eighteenth century, European men would tie off (or remove) their left testicle to increase the chances of getting a male heir.

In some cultures, efforts to control the sex of offspring has a darker side—female infanticide. In ancient Greece, the murder of female infants was so common that the male:female ratio in some areas approached 4:1. Some societies, even in present times, still practice female infanticide. In some parts of rural India, hundreds of families admit to this practice, even as late as the 1990s. In 1997, the World Health Organization reported population data showing that about 50 million women were "missing" in China, likely because of selective abortion of female fetuses and institutionalized neglect of female children. The practice of female infanticide arises from poverty and age-old traditions. In some Indian cultures, sons work and provide income and security, whereas daughters not only contribute no income but require large dowries when they marry. Under these conditions, it is easy to see why females are held in low esteem.

In recent times, sex-specific abortion has replaced much of the traditional female infanticide. Amniocentesis and ultrasound techniques have become lucrative businesses that provide prenatal sex determination. Studies in India estimate that hundreds of thousands of fetuses are aborted each year because they are female. As a result of sex-selective abortion, the male:female ratio in India was 1000:927 in 1991. In some regions, the ratio is as high as 1000:600. Although sex determination and selective abortion of female fetuses was outlawed in India and China in the mid-1990s, the practice is thought to continue.

In Western industrial countries, advances in genetics and reproductive technology offer parents ways to select their children's gender prior to implantation–methods called preimplantation gender selection (PGS). Following *in vitro* fertilization, embryos can be biopsied and assessed for gender. Only sex-selected embryos are then implanted. The simplest method involves separating X- and Y-chromosome bearing spermatozoa. Sperm are sorted based on their DNA content. Because of the different size of the X and Y chromosomes, X-bearing sperm contain 2.8 to 3.0 percent more DNA than Y-bearing sperm. Sperm samples are treated with a fluorescent DNA stain and then passed single file through a laser beam in a fluorescence-activated cell sorter (FACS) machine. The machine separates the sperm into two fractions based on the intensity of their DNA fluorescence. Through this method, human sperm can be separated into X-and Y-chromosome fractions, with enrichments of about 85 percent and 75 percent, respectively. The sorted sperm are then used for standard intrauterine insemination. The Genetics & IVF Institute (Fairfax, Virginia) is presently using this PGS technique in an FDA-approved clinical trial. As of January 2002, a total of 419 human pregnancies have resulted from the method. The company has reported an approximate 80 percent success rate in producing the desired gender.

The emerging PGS methods raise a number of legal and ethical issues. Some feel that prospective parents have the legal right to use sex-selection techniques as part of their fundamental procreative liberty. Others believe that this liberty does not extend to custom-designing a child to the parents' specifications. Proponents state that the benefits far outweigh any dangers to offspring or society. The medical use of PGS is a clear case of benefit. People at risk for trans- mitting X-linked diseases such as hemophilia or Duchenne muscular dystrophy can now enhance their chance of conceiving a female child who will not express the disease. As there are over 500 known X-linked diseases and they are expressed in about 1 in 1000 live births, PGS could greatly reduce suffering for many families.

The greatest number of people undertake PGS for nonmedical reasons—to "balance" their families. It is possible that the ability to intentionally select the desired sex of an offspring may reduce overpopulation and economic burdens for families who would repeatedly reproduce to get the desired gender. In some cases, PGS may reduce the number of abortions. It is also possible that PGS may increase the happiness of both parents and children, as the children would be more "wanted."

On the other hand, some argue that PGS serves neither the individual nor the common good. It is argued that PGS is inherently sexist, based on the concept of the superiority of one sex over another, and leads to an increase in linking a child's worth to gender. Some fear that large-scale PGS will reinforce sex discrimination and lead to sex-ratio imbalances. Others feel that sexism and discrimination are not caused by sex ratios and would be better addressed through education and economic equality measures for men and women. The experience so far in Western countries suggests that sex-ratio imbalances would not result from PGS. Over half of U.S. couples who use PGS request female offspring. However, the consequences of widespread PGS in some Asian countries may be more problematic. Both India and China already have sex-ratio imbalances, which contribute to some socially undesirable side effects, such as the presence of millions of adult men who are unable to marry.

Some critics of PGS argue that this technology may contribute to social and economic inequality if it is made available only to those who can afford it. Other critics fear that our approval of

PGS will open the door to accepting other genetic manipulations of children for socially acceptable characteristics. It is difficult to predict the full effects that PGS will bring to the world. But the gender-selection genie is now out of the bottle and will be unwilling to step back in time.

References

Sills, E. S., Kirman, I., Thatcher, S. S. III, and Palermo, G. D. 1998. Sex-selection of human spermatozoa: Evolution of current techniques and applications. *Arch. Gynecol. Obstet.* 261: 109–115.

Robertson, J. A. 2001. Preconception gender selection. *Am. J. Bioethics* 1: 2–9.

Web Site

Microsort technique, Genetics & IVF Institute, Fairfax, Virginia. **http://www.microsort.net**

Female Infanticide, Gendercide Watch. **http://www.gendercide.org/case_infanticide.html**

CHAPTER SUMMARY

1. Sexual reproduction ultimately relies on some form of sexual differentiation, which is achieved by a variety of sex-determining mechanisms.

2. The genetic basis of sexual differentiation is often related to different chromosome compositions in the two sexes. The heterogametic sex either lacks one chromosome or contains a unique heteromorphic chromosome, usually referred to as the Y or W chromosome.

3. In humans, the study of individuals with altered sex chromosome compositions has established that the Y chromosome is responsible for male differentiation. The absence of the Y chromosome leads to female differentiation. Similar studies in *Drosophila* have excluded the Y chromosome in such a role, instead demonstrating that a balance between the number of X chromosomes and sets of autosomes is the critical factor.

4. The primary sex ratio in humans substantially favors males at conception. During embryonic and fetal development, male mortality is higher than that of females. The secondary sex ratio at birth still favors males by a small margin.

5. Dosage compensation mechanisms limit the expression of X-linked genes in females, who have two X chromosomes, as compared to males who have only one X. In mammals, compensation is achieved by the inactivation of either the maternal or paternal X chromosome early in development. This process results in the formation of Barr bodies in female somatic cells.

6. The Lyon hypothesis states that early in development, inactivation is random between the maternal and paternal X chromosomes. All subsequent progeny cells inactivate the same X as their progenitor cell. Mammalian females thus develop as mosaics with respect to their expression of heterozygous X-linked alleles.

7. In many reptiles, the incubation temperature at a critical time during embryogenesis is responsible for sex determination. Temperature influences the activity of enzymes involved in the metabolism of steroids related to sexual differentiation.

KEY TERMS

alternative splicing, 109
amplicon, 103
anhidrotic ectodermal dysplasia, 106
Barr body, 105
bisexual, 95
clone, 106
dioecious, 95
dosage compensation, 104
47,XXX, 101
47,XYY, 101
genic balance theory, 108
gonochoric, 95
hermaphroditic, 95
heterogametic sex, 99
heteromorphic chromosomes, 95
homogametic sex, 99
imprinting, 107

intersex, 95, 108
isogametes, 96
isogamous, 96
Klinefelter syndrome (47,XXY), 100
Lygaeus mode of sex determination (XX/XY), 99
Lyon hypothesis, 106
lyonization, 106
male-specific region of the Y (MSY), 102
metafemale, 107
metamale, 108
monoecious, 95
nondisjunction, 100
open reading frame (ORF), 107
preimplantation gender selection (PGS), 110
primary sex ratio, 104
Protenor mode of sex determination (XX/XO), 98

pseudoautosomal regions (PARs), 102
pseudogenes, 103
red-green color blindness, 106
RNA splicing, 109
secondary sex ratio, 104
sex chromatin body, 105
sex chromosomes, 95
sex ratio, 104
sex-determining region Y (SRY), 102
testis-determining factor (TDF), 103
transgenic mice, 103
Turner syndrome (45,X), 100
unisexual, 95
X-inactivation center (Xic), 107
X-inactive specific transcript (XIST), 107
Y chromosome, 99
ZZ/ZW, 99

INSIGHTS AND SOLUTIONS

1. In *Drosophila*, the X chromosomes can attach to one another ($\widehat{XX}$) such that they always segregate together. Some flies contain both an attached X chromosome and a Y chromosome. (a) What sex would such a fly be? Explain why this is so. (b) Given the answer to part (a), predict the sex of the offspring in a cross between this fly and a normal one of the opposite sex. (c) If the offspring of part (b) are allowed to interbreed, what will be the outcome?

Solution:

(a) The fly will be a female. The ratio of X chromosomes to sets of autosomes will be 1.0, leading to normal female development. The Y chromosome has no influence on sex determination in *Drosophila*.

(b) All flies will have two sets of autosomes, but each offspring will have one of the following sex chromosome compositions:

(1) $\widehat{XX}$X → a metafemale with 3 X's (a trisomic)

(2) $\widehat{XX}$Y → a female like her mother

(3) XY → a normal male

(4) YY → no development occurs

(c) A true-breeding stock will be created that maintains the attached-X females generation after generation.

2. The Xg cell-surface antigen is coded for by a gene located on the X chromosome. No equivalent gene exists on the Y chromosome. Two codominant alleles of this gene have been identified: *Xg1* and *Xg2*. A woman of genotype *Xg2/Xg2* marries a man of genotype *Xg1/Y*, and they produce a son with Klinefelter syndrome of genotype *Xg1/Xg2/Y*. Using proper genetic terminology, briefly explain how this individual was generated. In which parent and in which meiotic division did the mistake occur?

Solution: Because the son with Klinefelter syndrome is *Xg1/Xg2/Y*, he must have received both the *Xg1* allele and the Y chromosome from his father. Therefore, nondisjunction must have occurred during meiosis I in the father.

PROBLEMS AND DISCUSSION QUESTIONS

1. As related to sex determination, what is meant by (a) homomorphic and heteromorphic chromosomes; (b) isogamous and heterogamous organisms?
2. Contrast the life cycle of a plant such as *Zea mays* with an animal such as *C. elegans.*
3. Discuss the role of sexual differentiation in the life cycles of *Chlamydomonas*, *Zea mays*, and *C. elegans.*
4. Distinguish between the concepts of sexual differentiation and sex determination.
5. Contrast the *Protenor* and *Lygaeus* modes of sex determination.
6. Describe the major difference between sex determination in *Drosophila* and in humans.
7. What specific observations (evidence) support the conclusions you have drawn about sex determination in *Drosophila* and humans?
8. Describe how nondisjunction in human female gametes can give rise to Klinefelter and Turner syndrome offspring following fertilization by a normal male gamete.
9. An insect species is discovered in which the heterogametic sex is unknown. An X-linked recessive mutation for *reduced wing* (*rw*) is discovered. Contrast the F_1 and F_2 generations from a cross between a female with reduced wings and a male with normal-sized wings when (a) the female is the heterogametic sex; (b) the male is the heterogametic sex.
10. Based on your answers in Problem 9, is it possible to distinguish between the *Protenor* and *Lygaeus* mode of sex determination based on the outcome of these crosses?
11. When cows have twin calves of unlike sex (fraternal twins), the female twin is usually sterile and has masculinized reproductive organs. This calf is referred to as a freemartin. In cows, twins may share a common placenta and thus fetal circulation. Predict why a freemartin develops.
12. An attached-X female fly, $\widehat{XX}$Y (see the "Insights and Solutions" box), expresses the recessive X-linked *white* eye phenotype. It is crossed to a male fly that expresses the X-linked recessive miniature wing phenotype. Determine the outcome of this cross regarding the sex, eye color, and wing size of the offspring.
13. Assume that rarely, the attached X chromosomes in female gametes become unattached. Based on the parental phenotypes in Problem 12, what outcomes in the F_1 generation would indicate that this has occurred during female meiosis?
14. It has been suggested that any male-determining genes contained on the Y chromosome in humans cannot be located in the limited region that synapses with the X chromosome during meiosis. What might be the outcome if such genes were located in this region?
15. What is a Barr body, and where is it found in a cell?
16. Indicate the expected number of Barr bodies in interphase cells of the following individuals: Klinefelter syndrome; Turner syndrome; and karyotypes 47,XYY, 47,XXX, and 48,XXXX.
17. Define the Lyon hypothesis.
18. Can the Lyon hypothesis be tested in a human female who is homozygous for one allele of the X-linked *G6PD* gene? Why, or why not?
19. Predict the potential effect of the Lyon hypothesis on the retina of a human female heterozygous for the X-linked red-green color-blindness trait.
20. Cat breeders are aware that kittens expressing the X-linked calico coat pattern and tortoiseshell pattern are almost invariably females. Why?
21. What does the apparent need for dosage compensation mechanisms suggest about the expression of genetic information in normal diploid individuals?
22. The marine echiurid worm *Bonellia viridis* is an extreme example of the environment's influence on sex determination. Undifferentiated larvae either remain free-swimming and differentiate into females or they settle on the proboscis of an adult female and become males. If larvae that have been on a female proboscis for a short period are removed and placed in seawater, they develop as intersexes. If larvae are forced to develop in an aquarium where pieces of proboscises have been placed, they develop into males. Contrast this mode of sexual differentiation with that of mammals. Suggest further experimentation to elucidate the mechanism of sex determination in *B. viridis.*

23. How do we know that the primary sex ratio in humans is as high as 1.20 to 1.60?

24. Devise as many hypotheses as you can that might explain why so many more human male conceptions than human female conceptions occur.

25. In mice, the *Sry* gene (see Section 5.3) is located on the Y chromosome very close to one of the pseudoautosomal regions that pairs with the X chromosome during male meiosis. Given this information, propose a model to explain the generation of unusual males who have two X chromosomes (with an *Sry*-containing piece of the Y chromosome attached to one X chromosome).

26. The genes encoding the red and green color-detecting proteins of the human eye are located next to one another on the X chromosome and probably arose during evolution from a common ancestral pigment gene. The two proteins demonstrate 76 percent homology in their amino acid sequences. A normal-visioned woman with one copy of each gene on each of her two X chromosomes has a red color-blind son who was shown to contain one copy of the green-detecting gene and no copies of the red-detecting gene. Devise an explanation at the chromosomal level (during meiosis) that explains these observations.

27. The X-linked dominant mutation in the mouse, *Testicular feminization* (*Tfm*), eliminates the normal response to the testicular hormone testosterone during sexual differentiation. An XY mouse bearing the *Tfm* allele on the X chromosome develops testes, but no further male differentiation occurs—the external genitalia of such an animal are female. From this information, what might you conclude about the role of the *Tfm* gene product and the X and Y chromosomes in sex determination and sexual differentiation in mammals? Can you devise an experiment, assuming you can "genetically engineer" the chromosomes of mice, to test and confirm your explanation?

28. Campomelic dysplasia (CMD1) is a congenital human syndrome, featuring malformation of bone and cartilage. It is caused by an autosomal dominant mutation of a gene located on chromosome 17. Consider the following observations in sequence, and in each case, draw whatever appropriate conclusions are warranted.
    (a) Of those with the syndrome who are karyotypically 46,XY, approximately 75 percent are sex reversed, exhibiting a wide range of female characteristics.
    (b) The nonmutant form of the gene, called *SOX9*, is expressed in the developing gonad of the XY male but not the XX female.
    (c) The *SOX9* gene shares 71 percent amino acid coding sequence homology with the Y-linked *SRY* gene.
    (d) CMD1 patients who exhibit a 46,XX karyotype develop as females, with no gonadal abnormalities.

29. In the wasp, *Bracon hebetor*, a form of parthenogenesis (where unfertilized eggs initiate development) resulting in haploid organisms is not uncommon. All haploids are males. When offspring arise from fertilization, females almost invariably result. P. W. Whiting has shown that an X-linked gene with nine multiple alleles (X_a, X_b, etc.) controls sex determination. Any homozygous or hemizygous condition results in males, and any heterozygous condition results in females. If an X_a/X_b female mates with an X_a male and lays 50 percent fertilized and 50 percent unfertilized eggs, what proportion of male and female offspring will result?

30. Shown at the top of the next column are two graphs that plot the percentage of males occurring against the atmospheric temperature during the early development of fertilized eggs in (a) snapping turtles and (b) most lizards. Interpret these data as they relate to the effect of temperature on sex determination.

(a) Snapping turtles

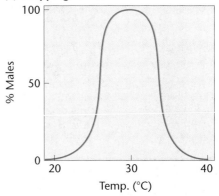

(b) Most lizards

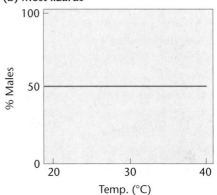

31. CC (Carbon Copy), the first cloned cat, was created from an ovarian cell taken from her genetic donor, Rainbow. The diploid nucleus from the cell was extracted and then injected into an enucleated egg. The resulting zygote was then allowed to divide in a petri dish and the cloned embryo was implanted in the uterus of a surrogate mother cat, who gave birth to CC. Rainbow is a calico cat. CC's surrogate mother is a tabby. Geneticists were very interested in the outcome of cloning a calico cat, because they were not certain if the cat would have patches of orange and black, just orange, or just black. Taking into account the Lyon hypothesis, explain the basis of the uncertainty.

CC (Carbon Copy), the first cloned cat, shown as a kitten along with her surrogate mother.

32. Let's assume hypothetically that Carbon Copy from Problem 31 is indeed a calico with black and orange patches, along with the patches of white characterizing a calico cat. Would you expect CC to appear identical to Rainbow? Explain why or why not.

33. When Carbon Copy was born (see Problem 31), she had black patches and white patches, but completely lacked any orange patches. Knowledgeable students of genetics were not surprised at this outcome. Starting with the somatic ovarian cell used as the source of the nucleus in the cloning process, explain how this outcome occurred.

6

Chromosome Mutations: Variation in Number and Arrangement

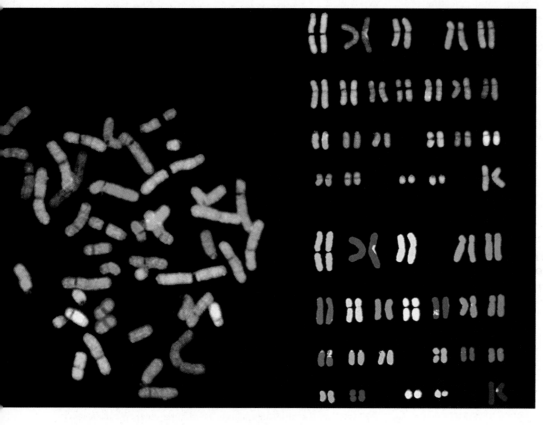

Spectral karyotyping of human chromosomes utilizing differentially labeled "painting" probes.

CHAPTER CONCEPTS

- During meiosis, the failure of chromosomes to properly separate results in variation in the chromosome composition of gametes and subsequently in offspring arising from such gametes.

- Plants often tolerate the abnormal content of genetic information but manifest unique phenotypes. Such genetic variation has been an important factor in the evolution of plants.

- In animals, genetic information exists in a delicate equilibrium, whereby the gain or loss of a chromosome, or part of a chromosome, in an otherwise diploid organism often leads to lethality or to an abnormal phenotype.

- The rearrangement of genetic information within the genome of a diploid organism may be tolerated by that organism but may affect the viability of gametes and the phenotypes of organisms arising from those gametes.

- Chromosomes in humans contain regions susceptible to breakage known as fragile sites, which lead to abnormal phenotypes.

In previous chapters, we have emphasized how mutations and the resulting alleles affect an organism's phenotype and how traits are passed from parents to offspring according to Mendelian principles. In this chapter, we look at phenotypic variation that results from more substantial changes than alterations of individual genes—modifications at the level of the chromosome.

Although most members of diploid species normally contain precisely two haploid chromosome sets, many known cases vary from this pattern. Modifications include a change in the total number of chromosomes, the deletion or duplication of genes or segments of a chromosome, and rearrangements of the genetic material either within or among chromosomes. Taken together, such changes are called **chromosome mutations** or **chromosome aberrations**, to distinguish them from gene mutations. Because the chromosome is the unit of genetic transmission, according to Mendelian laws, chromosome aberrations are passed to offspring in a predictable manner, resulting in many unique genetic outcomes.

Because the genetic component of an organism is delicately balanced, even minor alterations of either content or location of genetic information within the genome can result in some form of phenotypic variation. More substantial changes may be lethal, particularly in animals. Throughout this chapter, we consider many types of chromosomal aberrations, the phenotypic consequences for the organism that harbors an aberration, and the impact of the aberration on the offspring of an affected individual. We will also discuss the role of chromosome aberrations in the evolutionary process.

6.1 Specific Terminology Describes Variations in Chromosome Number

Variation in chromosome number ranges from the addition or loss of one or more chromosomes to the addition of one or more haploid sets of chromosomes. Before we embark on our discussion, it is useful to clarify the terminology that describes such changes. In the general condition known as **aneuploidy**, an organism gains or loses one or more chromosomes but not a complete set. The loss of a single chromosome from an otherwise diploid genome is called *monosomy*. The gain of one chromosome results in *trisomy*. These changes are contrasted with the condition of **euploidy**, where complete haploid sets of chromosomes are present. If more than two sets are present, the term **polyploidy** applies. Organisms with three sets are specifically *triploid*, those with four sets are *tetraploid*, and so on. Table 6.1 provides an organizational framework for you to follow as we discuss each of these categories of aneuploid and euploid variation and the subsets within them.

6.2 Variation in the Number of Chromosomes Results from Nondisjunction

As we consider cases that include the gain or loss of chromosomes, it is useful to examine how such aberrations originate. For instance, how do the syndromes arise where the number of sex-determining chromosomes in humans is altered, as described in

? How Do We Know?

In this chapter, we will focus on chromosomal mutations resulting from a change in number or arrangement of chromosomes. As you study this topic, you should try to answer several fundamental questions:

1. How do we know that changes in chromosome number or structure result in specific mutant phenotypes?

2. How do we determine the chromosome content in somatic nuclei of individuals?

3. How do we know that the extra chromosome causing Down syndrome is usually maternal in origin?

4. In *Drosophila*, how do we know that the mutant *Bar* eye phenotype is due to a duplication?

5. How do we know that gene duplication is a source of new genes during evolution?

TABLE 6.1	Terminology for Variation in Chromosome Numbers
Term	**Explanation**
Aneuploidy	$2n \pm x$ chromosomes
Monosomy	$2n - 1$
Disomy	$2n$
Trisomy	$2n + 1$
Tetrasomy, pentasomy, etc.	$2n + 2, 2n + 3$, etc.
Euploidy	Multiples of n
Diploidy	$2n$
Polyploidy	$3n, 4n, 5n, \ldots$
Triploidy	$3n$
Tetraploidy, pentaploidy, etc.	$4n, 5n$, etc.
Autopolyploidy	Multiples of the same genome

Chapter 5? As you may recall, the gain (47,XXY) or loss (45,X) of an X chromosome from an otherwise diploid genome affects the phenotype, resulting in **Klinefelter syndrome** or **Turner syndrome**, respectively (see Figure 5–6). Human females may contain extra X chromosomes (e.g., 47,XXX, 48,XXXX), and some males contain an extra Y chromosome (47,XYY).

Chromosomal variation originates as the result of an error during meiosis, a phenomenon referred to as **nondisjunction**, whereby paired homologs fail to disjoin during segregation. This process disrupts the normal distribution of chromosomes into gametes. The results of nondisjunction during meiosis I and meiosis II for a single chromosome of a diploid organism are shown in Figure 6–1. As you can see, for the affected chromosome, abnormal gametes can form that contain either two members or none at all. Fertilizing these with a normal haploid gamete produces a zygote with either three members (trisomy) or only one member (monosomy) of this chromosome. As we shall see, nondisjunction leads to a variety of aneuploid conditons in humans and other organisms.

Now Solve This

Problem 14 on page 134 considers a female with Turner syndrome who expresses hemophilia, as did her father. You are asked which of her parents was responsible for the nondisjunction event leading to her syndrome.

Hint: The parent who contributed a gamete with an X chromosome underwent normal meiosis.

6.3 ## Monosomy, the Loss of a Single Chromosome, May Have Severe Phenotypic Effects

We turn now to a consideration of variations in the number of autosomes and the genetic consequence of such changes. The most common examples of aneuploidy, where an organism has a chromosome number other than an exact multiple of the haploid set, are cases in which a single chromosome is either added to, or lost from, a normal diploid set. The loss of one chromosome produces a $2n - 1$ complement called **monosomy**.

Although monosomy for the X chromosome occurs in humans, as we have seen in 45,X Turner syndrome, monosomy for any of the autosomes is not usually tolerated in humans or other animals. In *Drosophila*, flies that are monosomic for the very small chromosome IV develop more slowly, exhibit reduced body size, and have impaired viability. Monosomy for the larger chromosomes II and III is apparently lethal because such flies have never been recovered.

The failure of monosomic individuals to survive is at first quite puzzling, since at least a single copy of every gene is present in the remaining homolog. However, if just one of those genes is represented by a lethal allele, the unpaired chromosome condition leads to the death of the organism. This occurs because monosomy unmasks recessive lethals that are tolerated in heterozygotes carrying the corresponding wild-type alleles.

Aneuploidy is better tolerated in the plant kingdom. Monosomy for autosomal chromosomes has been observed in maize, tobacco, the evening primrose (*Oenothera*), and the jimson weed (*Datura*) among many other plants. Neverthe-

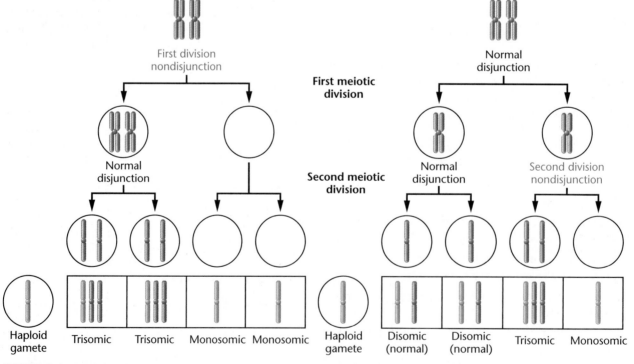

FIGURE 6–1 Nondisjunction during the first and second meiotic divisions. In both cases, some of the gametes that are formed either contain two members of a specific chromosome or lack that chromosome. After fertilization by a gamete with a normal haploid content, monosomic, disomic (normal), or trisomic zygotes are produced.

less, such monosomic plants are usually less viable than their diploid derivatives. Haploid pollen grains, which undergo extensive development before participating in fertilization, are particularly sensitive to the lack of one chromosome and are seldom viable.

Cri-du-chat Syndrome

In humans, autosomal monosomy has not been reported beyond birth. Individuals with such chromosome complements are undoubtedly conceived, but none apparently survive embryonic and fetal development. There are, however, examples of survivors with partial monosomy, where only a small part of one chromosome is lost. Such a condition is also sometimes referred to as a **segmental deletion**. One case was first reported by Jérôme LeJeune in 1963 when he described the clinical symptoms of the **cri-du-chat** (cry of the cat) **syndrome**. This syndrome is associated with the loss of part of the short arm of chromosome 5 (Figure 6–2). Thus, the genetic constitution may be designated as **46,5p−**, meaning that the individual has all 46 chromosomes but that some of the *p* arm (the petite arm) of one member of the chromosome 5 pair is missing.

Infants with this syndrome exhibit anatomic malformations, including gastrointestinal and cardiac complications, and they are mentally retarded. Abnormal development of the glottis and larynx is also characteristic of individuals with this syndrome. As a result, the infant has a distinctive, unusual cry, one that is similar to the meowing of a kitten, giving the syndrome its name.

Since 1963, hundreds of cases of cri-du-chat syndrome have been reported worldwide. An incidence of 1 in 50,000 live births has been estimated. Most often, the loss of chromosomal material is a sporadic event occurring in gametes. The length of the short arm that is deleted varies somewhat and may be as extensive as 60 percent of the petite arm. Longer deletions appear to have a greater impact on the physical, psychomotor, and mental skill levels of those children who survive. Although the effects of the syndrome are severe, most individuals achieve motor and language skills and may be home-cared. In 2004, it was reported that the portion of the chromosome that is missing contains the *TERT* gene, which encodes telomerase reverse transcriptase, an enzyme essential for the maintenance of telomeres during DNA replication. Whether the absence of this gene on one homolog is related to the multiple phenotypes of cri-du-chat infants is still unknown.

6.4 Trisomy Involves the Addition of a Chromosome to a Diploid Genome

In general, the effects of **trisomy** $(2n + 1)$ parallel those of monosomy. However, the addition of an extra chromosome produces somewhat more viable individuals in both animal and plant species than does the loss of a chromosome. In animals, this is often true, provided that the chromosome involved is relatively small. However, the addition of a large autosome to the diploid complement in both *Drosophila* and humans has severe effects and is usually lethal during development.

In plants, trisomic individuals are viable, but their phenotype may be altered. A classical example involves the jimson weed, *Datura*, whose diploid number is 24. Twelve primary trisomic conditions are possible, and examples of each one have been recovered. Each trisomy alters the phenotype of the plant's capsule sufficiently to produce a unique phenotype. These capsule phenotypes were first thought to be caused by mutations in one or more genes.

Still another example is seen in the rice plant (*Oryza sativa*), which has a haploid number of 12. Trisomic strains for each chromosome have been isolated and studied—the plants of 11 strains can be distinguished from one another and from wild-type plants. Trisomics for the longer chromosomes are the most distinctive, and the plants grow more slowly. This is in keeping with the belief that larger chromosomes cause greater genetic imbalance than smaller ones. Leaf structure, foliage, stems, grain morphology, and plant height also vary among the various trisomies.

Down Syndrome

The only human autosomal trisomy in which a significant number of individuals survive longer than a year past birth was discovered in 1866 by Langdon Down. The condition is now known to result from trisomy of chromosome 21, one of

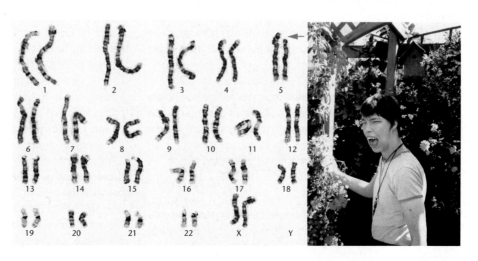

FIGURE 6–2 A representative karyotype and a photograph of a child exhibiting cri-du-chat syndrome (46,5p–). In the karyotype, the arrow identifies the absence of a small piece of the short arm of one member of the chromosome 5 homologs.

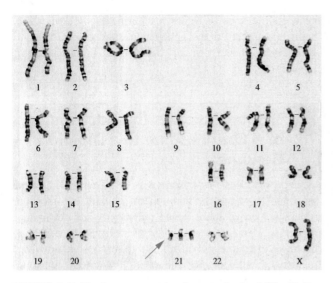

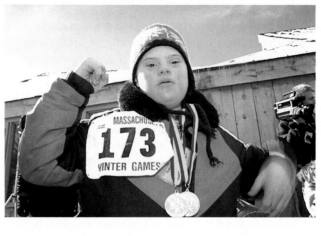

FIGURE 6–3 The karyotype and a photograph of a child with Down syndrome. In the karyotype, three members of the G-group chromosome 21 are present, creating the 47,21 + condition.

the G group* (Figure 6–3), and is called **Down syndrome** or simply **trisomy 21** (designated **47,21 +**). This trisomy is found in approximately 1 infant in every 800 live births.

The overt phenotype of these individuals is so similar that they bear a striking resemblance to one another. They display a prominent epicanthic fold in the corner of the upper eyelid and are characteristically short. They may have round heads with flat faces, a protruding furrowed tongue, which causes the mouth to remain partially open, and short, broad hands with fingers showing characteristic palm and fingerprint patterns. Physical, psychomotor, and mental development is retarded, and poor muscle tone is characteristic. Children afflicted with Down syndrome are prone to respiratory disease and heart malformations, and they show an incidence of leukemia approximately 20 times higher than that of the normal population. Their life expectancy is shortened, although individuals are known to survive into their fifties.

Careful medical scrutiny and treatment throughout their lives have extended their survival significantly. A striking observation is that the death of older Down syndrome adults is frequently due to Alzheimer's disease, although the onset of this disease occurs at a much earlier age than it does in the normal population.

Typical of other conditions referred to as a syndrome, there are many phenotypic characteristics that may be present, but any single affected individual usually expresses only a subset of these. In the case of Down syndrome, there are 12 to 14 such characteristics, but each individual, on average, expresses only 6 to 8 of them. Because this condition is common in our population, achieving a comprehensive understanding of the underlying genetic basis has long been a research goal. Investigations have led to the concept of a "critical region" of

chromosome 21, which may contain the genes that are dosage sensitive in trisomy and that are responsible for the many phenotypes associated with the syndrome. This hypothetical portion of the chromosome has been called the **Down syndrome critical region (DSCR)**. A mouse model was created in 2004 that is trisomic for the DSCR, but such mice did not exhibit the characteristics of the syndrome. Nevertheless, this remains an important investigative approach.

The most frequent origin of this trisomic condition is through nondisjunction of chromosome 21 during meiosis. Failure of paired homologs to disjoin during anaphase I or failure of chromatids to disjoin during anaphase II can result in male or female gametes with the $n + 1$ chromosome composition. Following fertilization with a normal gamete, the trisomic condition is created. Chromosome analysis has shown that while the additional chromosome can be derived from either the mother or father, the ovum is the source in 95 percent of the cases.

Before techniques able to distinguish paternal from maternal homologs were developed, this conclusion was supported by other indirect evidence derived from studies of the age of mothers giving birth to Down syndrome infants. Figure 6–4 shows the relationship between maternal age and the incidence of Down syndrome newborns. While the frequency is about 1 in 1000 at maternal age 30, a 10-fold increase to a frequency of 1 in 100 is noted at age 40. The frequency increases still further to about 1 in 50 at age 45. In spite of these statistics, it is important to point out that, overall, more than half of affected births occur to women who are under 35 years of age, primarily because there are many more pregnancies in this group of women.

Although the nondisjunctional event that produces Down syndrome seems more likely to occur during oogenesis in women between the ages of 35 and 45, we do not know with certainty why this is so. However, one observation may be relevant. In human females, all primary oocytes have been formed by birth. Therefore, once ovulation begins, each succeeding ovum has

---

* On the basis of size and centromere placement, human autosomal chromosomes are divided into seven groups: A (1–3), B (4–5), C (6–12), D (13–15), E (16–18), F (19–20), and G (21–22).

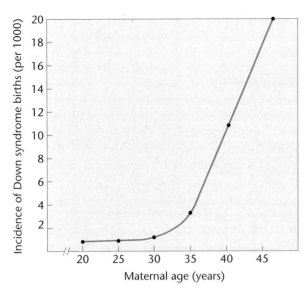

FIGURE 6–4 Incidence of Down syndrome births related to maternal age.

been arrested in meiosis for about a month longer than the one preceding it. Women 30 or 40 years old produce ova that are significantly older and arrested longer than those they ovulated 10 or 20 years previously. However, it is not yet known whether ovum age is the cause of the increased incidence of nondisjunction leading to Down syndrome.

These statistics pose a serious problem for the woman who becomes pregnant late in her reproductive years. Genetic counseling early in the pregnancy serves two purposes. First, it informs the parents about the probability that their child will be affected and educates them about Down syndrome. Although some individuals with Down syndrome must be institutionalized, others benefit greatly from special education programs and can be cared for at home. Furthermore, these children are noted for their affectionate, loving nature. Second, a genetic counselor may recommend a prenatal diagnostic technique such as **amniocentesis** or **chorionic villus sampling (CVS)**. These techniques require the removal and culture of fetal cells. The karyotype of the fetus is then determined by cytogenetic analysis. If the fetus is diagnosed as having Down syndrome, a therapeutic abortion is one option the parents may consider.

Because Down syndrome appears to be caused by a random error—nondisjunction of chromosome 21 during maternal or paternal meiosis—the disorder is not expected to be inherited. Nevertheless, Down syndrome occasionally runs in families. This condition, **familial Down syndrome**, involves a *translocation* of chromosome 21, another type of chromosomal aberration, which we will discuss in Section 6.10.

Viability in Human Aneuploid Conditions

The reduced viability of individuals with recognized monosomic and trisomic conditions is evident. Only two other trisomies in humans survive to term. Both **Patau and Edwards syndromes** (**47,13** + and **47,18** +, respectively) result in severe malformations and early lethality. Figure 6–5 illustrates the abnormal karyotype and the many defects characterizing Patau infants.

Such observations lead us to believe that many other aneuploid conditions arise but that the affected fetuses do not survive to term. This observation has been confirmed by karyotypic analysis of spontaneously aborted fetuses. These studies reveal some rather striking statistics. At least 15 to 20 percent of all conceptions terminate in spontaneous abortion (some estimates are considerably higher). About 30 percent of all spontaneously aborted fetuses demonstrate some form of chromosomal anomaly, and approximately 90 percent of all chromosomal anomalies are terminated prior to birth as a result of spontaneous abortion.

A large percentage of fetuses demonstrating chromosomal abnormalities are aneuploids. The aneuploid with highest incidence among abortuses is the 45,X condition, which produces an infant with Turner syndrome if the fetus survives to term.

An extensive review of this subject by David H. Carr also reveals that a significant percentage of aborted fetuses are trisomic for one of the chromosome groups. Trisomies for every human chromosome have been recovered. Monosomies are seldom found, however, even though nondisjunction should produce $n - 1$ gametes with a frequency equal to $n + 1$ gametes. This finding suggests that gametes lacking a single chromosome are functionally impaired to a serious degree or that the embryo dies so early in its development that recovery occurs infrequently. Various forms of polyploidy and other miscellaneous chromosomal anomalies were also found in Carr's study.

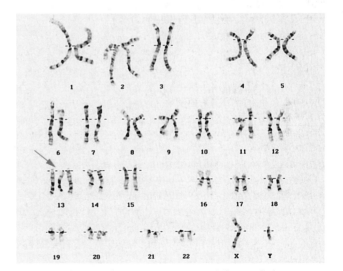

Mental retardation	Microcephaly
Growth failure	Cleft lip and palate
Low-set deformed ears	Polydactyly
	Deformed finger nails
Deafness	Kidney cysts
Atrial septal defect	Double ureter
Ventricular septal defect	Umbilical hernia
	Developmental uterine abnormalities
Abnormal polymorphonuclear granulocytes	Cryptorchidism

FIGURE 6–5 The karyotype and phenotypic description of an infant with Patau syndrome, where three members of the D-group chromosome 13 are present, creating the 47,13+ condition.

These observations support the hypothesis that normal embryonic development requires a precise diploid complement of chromosomes to maintain the delicate equilibrium in the expression of genetic information. The prenatal mortality of most aneuploids provides a barrier against the introduction of these genetic anomalies into the human population.

6.5 Polyploidy, in Which More Than Two Haploid Sets of Chromosomes Are Present, Is Prevalent in Plants

The term *polyploidy* describes instances in which more than two multiples of the haploid chromosome set are found. The naming of polyploids is based on the number of sets of chromosomes found: A triploid has $3n$ chromosomes; a tetraploid has $4n$; a pentaploid, $5n$; and so forth. Several general statements can be made about polyploidy. This condition is relatively infrequent in many animal species but is well known in lizards, amphibians, and fish, and is much more common in plant species. Usually, odd numbers of chromosome sets are not reliably maintained from generation to generation because a polyploid organism with an uneven number of homologs often does not produce genetically balanced gametes. For this reason, triploids, pentaploids, and so on, are not usually found in plant species that depend solely on sexual reproduction for propagation.

Polyploidy originates in two ways: (1) The addition of one or more extra sets of chromosomes, identical to the normal haploid complement of the same species, resulting in **autopolyploidy**; or (2) the combination of chromosome sets from different species occurring as a consequence of hybridization, resulting in **allopolyploidy** (from the Greek word *allo*, meaning "other" or "different"). The distinction between auto- and allopolyploidy is based on the genetic origin of the extra chromosome sets, as shown in Figure 6–6.

In our discussion of polyploidy, we use the following symbols to clarify the origin of additional chromosome sets. For example, if A represents the haploid set of chromosomes of any organism, then

$$A = a_1 + a_2 + a_3 + a_4 + \cdots + a_n$$

where a_1, a_2, and so on, are individual chromosomes and n is the haploid number. A normal diploid organism is represented simply as AA.

Autopolyploidy

In autopolyploidy, each additional set of chromosomes is identical to the parent species. Therefore, triploids are represented as AAA, tetraploids are $AAAA$, and so forth.

Autotriploids arise in several ways. A failure of all chromosomes to segregate during meiotic divisions can produce a diploid gamete. If such a gamete is fertilized by a haploid gamete, a zygote with three sets of chromosomes is produced. Or, rarely, two sperm may fertilize an ovum, resulting in a triploid zygote. Triploids are also produced under experimental conditions by crossing diploids with tetraploids. Diploid organisms produce gametes with n chromosomes, while tetraploids produce $2n$ gametes. Upon fertilization, the desired triploid is produced.

Because they have an even number of chromosomes, **autotetraploids** ($4n$) are theoretically more likely to be found in nature than are autotriploids. Unlike triploids, which often produce genetically unbalanced gametes with odd numbers of chromosomes, tetraploids are more likely to produce balanced gametes when involved in sexual reproduction.

How polyploidy arises naturally is of great interest to geneticists. In theory, if chromosomes have replicated, but the parent cell never divides and instead reenters interphase, the chromosome number will be doubled. That this very likely occurs is supported by the observation that tetraploid cells can be produced experimentally from diploid cells. This is accomplished by applying cold or heat shock to meiotic cells or by applying colchicine to somatic cells undergoing mitosis. **Colchicine**, an alkaloid derived from the autumn crocus, interferes with spindle formation, and thus replicated chromosomes cannot separate at anaphase and do not migrate to the poles. When colchicine is removed, the cell can reenter interphase. When the paired sister chromatids separate and uncoil, the nucleus contains twice the diploid number of chromosomes and is therefore $4n$. This process is shown in Figure 6–7.

In general, autopolyploids are larger than their diploid relatives. This increase seems to be due to larger cell size rather than greater cell number. Although autopolyploids do not contain new or unique information compared with their diploid relatives, the flower and fruit of plants are often increased in size, making such varieties of greater horticultural or commercial value. Economically important triploid plants include several potato species of the genus *Solanum*, Winesap apples, commercial bananas, seedless watermelons, and the cultivated tiger lily *Lilium tigrinum*. These plants are propagated asexually. Diploid bananas contain hard seeds, but the commercial, triploid, "seedless" variety has edible seeds. Tetraploid alfalfa, coffee, peanuts, and McIntosh apples are also of economic value because they are

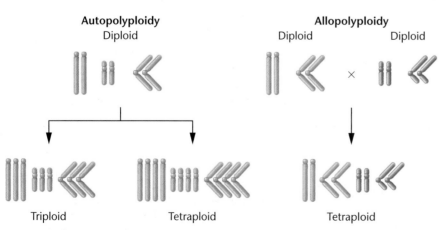

FIGURE 6–6 Contrasting chromosome origins of an autopolyploid versus an allopolyploid karyotype.

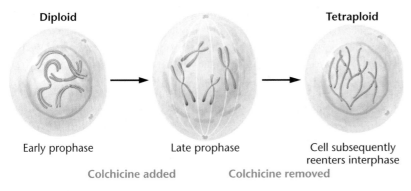

Diploid → **Tetraploid**

Early prophase | Late prophase | Cell subsequently reenters interphase

Colchicine added | Colchicine removed

FIGURE 6–7 The potential involvement of colchicine in doubling the chromosome number. Two pairs of homologous chromosomes are shown. While each chromosome had replicated its DNA earlier during interphase, the chromosomes do not appear as double structures until late prophase. When anaphase fails to occur normally, the chromosome number doubles if the cell reenters interphase.

either larger or grow more vigorously than do their diploid or triploid counterparts. Many varieties of the most popular varieties of hosta plant are tetraploid. In each case, leaves are thicker and larger, the foliage is more vivid, and the plant grows more vigorously. The commercial strawberry is an octoploid.

We have long been curious about how cells with increased ploidy values, where no new genes are present, express different phenotypes from their diploid counterparts. Our current ability to examine gene expression using modern biotechnology has provided some interesting insights. For example, Gerald Fink and his colleagues have been able to create strains of the yeast *Saccaromyces cerevisiae* with one, two, three, or four copies of the genome. Thus, each strain contains identical genes (they are said to be isogenic) but different ploidy values. These scientists then examined the expression levels of all genes during the entire cell cycle of the organism. Using the rather stringent standards of a 10-fold increase or decrease of gene expression, Fink and coworkers proceeded to identify 10 cases where, as ploidy increased, gene expression was increased at least 10-fold and 7 cases where it was reduced by a similar level.

One of these genes provides insights into how polyploid cells are larger than their haploid or diploid counterparts. In polyploid yeast, two **G1 cyclins**, Cln1, and Pcl1 are repressed as ploidy increases, while the size of the yeast cells increases. This is explained based on the observation that G1 cyclins facilitate the cell's movement through G1, which is delayed when expression of these genes is repressed. The polyploid cell stays in the G1 phase longer and, on average, grows to a larger size before it moves beyond the G1 stage of the cell cycle. Yeast cells also show different morphology as ploidy increases. Several of the other genes, repressed as ploidy increases, have been linked to cytoskeletal dynamics that account for the morphological changes.

Allopolyploidy

Polyploidy can also result from hybridizing two closely related species. If a haploid ovum from a species with chromosome sets AA is fertilized by sperm from a species with sets BB, the resulting hybrid is AB, where $A = a_1, a_2, a_3, \ldots a_n$ and $B = b_1, b_2, b_3, \ldots b_n$. The hybrid plant may be sterile because of its inability to produce viable gametes. Most often, this occurs when some or all of the a and b chromosomes are not homologous and therefore cannot synapse in meiosis. As a result, unbalanced genetic conditions result. If, however, the new AB genetic combination undergoes a natural or induced chromosomal doubling, two copies of all a chromosomes and two copies of all b chromosomes will be present, and they will pair during meiosis. As a result, a fertile $AABB$ tetraploid is produced. These events are shown in Figure 6–8. Since this polyploid contains the equivalent of four haploid genomes derived from separate species, such an organism is called an **allotetraploid**. When both original species are known, an equivalent term, **amphidiploid**, is preferred in describing the allotetraploid.

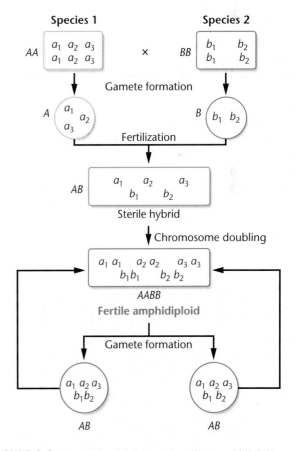

Species 1 × **Species 2**

AA | $a_1\ a_2\ a_3$ / $a_1\ a_2\ a_3$ | BB | $b_1\ b_2$ / $b_1\ b_2$

Gamete formation

A | $a_1\ a_2\ a_3$ | B | $b_1\ b_2$

Fertilization

AB | $a_1\ a_2\ a_3$ / $b_1\ b_2$

Sterile hybrid

Chromosome doubling

$a_1 a_1\ a_2 a_2\ a_3 a_3$ / $b_1 b_1\ b_2 b_2$
$AABB$
Fertile amphidiploid

Gamete formation

$a_1 a_2 a_3$ / $b_1 b_2$ — AB | $a_1 a_2 a_3$ / $b_1 b_2$ — AB

FIGURE 6–8 The origin and propagation of an amphidiploid. Species 1 contains genome A consisting of three distinct chromosomes, a_1, a_2, and a_3. Species 2 contains genome B consisting of two distinct chromosomes, b_1 and b_2. Following fertilization between members of the two species and chromosome doubling, a fertile amphidiploid containing two complete diploid genomes ($AABB$) is formed.

Amphidiploid plants are often found in nature. Their reproductive success is based on their potential for forming balanced gametes. Since two homologs of each specific chromosome are present, meiosis occurs normally (Figure 6–8) and fertilization successfully propagates the plant sexually. This discussion assumes the simplest situation, where none of the chromosomes in set *A* are homologous to those in set *B*. In amphidiploids, formed from closely related species, some homology between *a* and *b* chromosomes is likely. Allopolyploids are rare in most animals because mating behavior is most often species-specific, and thus the initial step in hybridization is unlikely to occur.

A classical example of amphidiploidy in plants is the cultivated species of American cotton, *Gossypium* (Figure 6–9). This species has 26 pairs of chromosomes: 13 are large and 13 are much smaller. When it was discovered that Old World cotton had only 13 pairs of large chromosomes, allopolyploidy was suspected. After an examination of wild American cotton revealed 13 pairs of small chromosomes, this speculation was strengthened. J. O. Beasley reconstructed the origin of cultivated cotton experimentally by crossing the Old World strain with the wild American strain, then treating the hybrid with colchicine to double the chromosome number. The result of these treatments was a fertile amphidiploid variety of cotton. It contained 26 pairs of chromosomes as well as characteristics similar to the cultivated variety.

Amphidiploids often exhibit traits of both parental species. An interesting example, but one with no practical economic importance, is that of the hybrid formed between the radish *Raphanus sativus* and the cabbage *Brassica oleracea*. Both species have a haploid number $n = 9$. The initial hybrid consists of nine *Raphanus* and nine *Brassica* chromosomes $(9R + 9B)$. Although hybrids are almost always sterile, some fertile amphidiploids $(18R + 18B)$ have been produced. Unfortunately, the root of this plant is more like the cabbage and its shoot more like the radish; had the converse occurred, the hybrid might have been of economic importance.

A much more successful commercial hybridization uses the grasses wheat and rye. Wheat (genus *Triticum*) has a basic haploid genome of seven chromosomes. In addition to normal diploids $(2n = 14)$, cultivated autopolyploids exist, including tetraploid $(4n = 28)$ and hexaploid $(6n = 42)$ species. Rye (genus *Secale*) also has a genome consisting of seven chromosomes. The only cultivated species is the diploid plant $(2n = 14)$.

Using the technique outlined in Figure 6–8, geneticists have produced various hybrids. When tetraploid wheat is crossed with diploid rye and the F_1 plants are treated with colchicine, a hexaploid variety $(6n = 42)$ is obtained; the hybrid, designated *Triticale* represents a new genus. Fertile hybrid varieties derived from various wheat and rye species can be crossed or backcrossed. These crosses have created many variations of the genus *Triticale*. The hybrid plants demonstrate characteristics of both wheat and rye. For example, certain hybrids combine the high-protein content of wheat with rye's high content of the amino acid lysine. (The lysine content is low in wheat and thus is a limiting nutritional factor.) Wheat is considered to be a high-yielding grain, whereas rye is noted for its versatility of growth in unfavorable environments. *Triticale* species, which combine both traits, have the potential of significantly increasing grain production. Programs designed to improve crops through hybridization have long been under way in several developing countries.

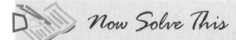

Now Solve This

In Problem 7 on page 133 you are asked to consider the sterility of a hybrid plant derived from two different species that is more ornate than either of its parents.

Hint: Allopolyploid plants are often sterile when they contain an odd number of each chromosome, resulting in unbalanced gametes during meiosis.

6.6 Variation Occurs in the Composition and Arrangement of Chromosomes

The second general class of chromosome aberrations includes structural changes that delete, add, or rearrange substantial portions of one or more chromosomes. Included in this broad category are deletions and duplications of genes or part of a chromosome and rearrangements of genetic material in which a chromosome segment is inverted, exchanged with a segment of a nonhomologous chromosome, or merely transferred to another chromosome. Exchanges and transfers are called translocations, in which the location of a gene is altered within the genome. These types of chromosome alterations are illustrated in Figure 6–10.

In most instances, these structural changes are due to one or more breaks along the axis of a chromosome, followed by either the loss or rearrangement of genetic material. Chromosomes can break spontaneously, but the rate of breakage may increase in cells exposed to chemicals or radiation. Although

FIGURE 6–9 The pods of the amphidiploid form of *Gossypium*, the cultivated American cotton plant.

FIGURE 6–10 An overview of the five different types of gain, loss, or rearrangement of chromosome segments.

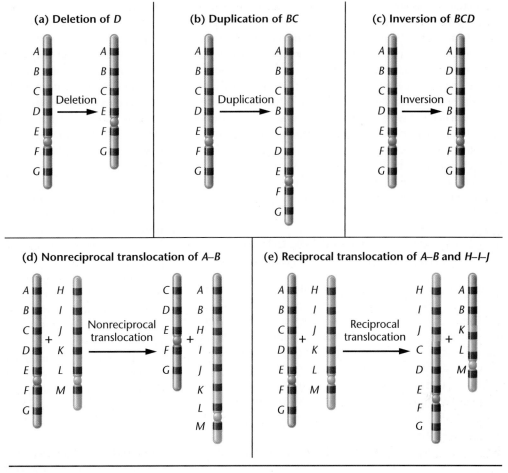

(a) Deletion of *D*

(b) Duplication of *BC*

(c) Inversion of *BCD*

(d) Nonreciprocal translocation of *A–B*

(e) Reciprocal translocation of *A–B* and *H–I–J*

the actual ends of chromosomes, known as telomeres, do not readily fuse with newly created ends of "broken" chromosomes or with other telomeres, the ends produced at points of breakage are "sticky" and can rejoin other broken ends. If breakage and rejoining do not reestablish the original relationship and if the alteration occurs in germ plasm, the gametes will contain the structural rearrangement, which is heritable.

If the aberration is found in one homolog, but not the other, the individual is said to be *heterozygous for the aberration*. In such cases, unusual but characteristic pairing configurations are formed during meiotic synapsis. These patterns are useful in identifying the type of change that has occurred. If no loss or gain of genetic material occurs, individuals bearing the aberration "heterozygously" are likely to be unaffected phenotypically. However, the unusual pairing arrangements often lead to gametes that are duplicated or deficient for some chromosomal regions. When this occurs, the offspring of "carriers" of certain aberrations have an increased probability of demonstrating phenotypic changes.

6.7 A Deletion Is a Missing Region of a Chromosome

When a chromosome breaks in one or more places and a portion of it is lost, the missing piece is called a **deletion** (or a **deficiency**). The deletion can occur either near one end or within the interior of the chromosome. These are **terminal** and **intercalary deletions**, respectively [Figure 6–11(a) and (b)].

(a) Origin of terminal deletion

(b) Origin of intercalary deletion

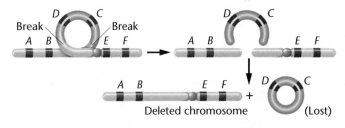

(c) Formation of deficiency loop

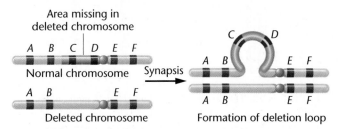

FIGURE 6–11 Origins of (a) a terminal and (b) an intercalary deletion. In (c), pairing occurs between a normal chromosome and one with an intercalary deletion by looping out the undeleted portion to form a deletion (or compensation) loop.

The portion of the chromosome that retains the centromere region is usually maintained when the cell divides, whereas the segment without the centromere is eventually lost in progeny cells following mitosis or meiosis. For synapsis to occur between a chromosome with a large intercalary deletion and a normal homolog, the unpaired region of the normal homolog must "buckle out" into a **deletion**, or **compensation, loop** [Figure 6–11(c)].

As noted in our discussion of the cri-du-chat syndrome, where only part of the short arm of chromosome 5 is lost, deletion of a portion of a chromosome need not be very great before the effects become severe. If even more genetic information is lost as a result of a deletion, the aberration is often lethal, and these chromosome mutations never become available for study.

6.8   A Duplication Is a Repeated Segment of a Chromosome

When any part of the genetic material—a single locus or a large piece of a chromosome—is present more than once in the genome, it is called a duplication. As in deletions, pairing in heterozygotes can produce a compensation loop. Duplications may arise as the result of unequal crossing over between synapsed chromosomes during meiosis (Figure 6–12) or through a replication error prior to meiosis. In the former case, both a duplication and a deletion are produced.

We consider three interesting aspects of duplications. First, they may result in gene redundancy. Second, as with deletions, duplications may produce phenotypic variation. Third, according to one convincing theory, duplications have also been an important source of genetic variability during evolution.

Gene Redundancy and Amplification— Ribosomal RNA Genes

Although many gene products are not needed in every cell of an organism, other gene products are known to be essential components of all cells. For example, ribosomal RNA must be

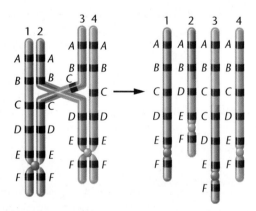

FIGURE 6–12 The origin of duplicated and deficient regions of chromosomes as a result of unequal crossing over. The tetrad on the left is mispaired during synapsis. A single crossover between chromatids 2 and 3 results in the deficient (chromosome 2) and duplicated (chromosome 3) chromosomal regions shown on the right. The two chromosomes uninvolved in the crossover event remain normal in gene sequence and content.

present in abundance to support protein synthesis. The more metabolically active a cell is, the higher the demand for this molecule. We might hypothesize that a single copy of the gene encoding rRNA is inadequate in many cells. Studies using the technique of molecular hybridization, which enables us to determine the percentage of the genome that codes for specific RNA sequences, show that our hypothesis is correct. Indeed, multiple copies of genes code for rRNA. Such DNA is called **rDNA**, and the general phenomenon is called **gene redundancy**. For example, in the common intestinal bacterium *Escherichia coli* (*E. coli*), about 0.7 percent of the haploid genome consists of rDNA—this is the equivalent of seven copies of the gene. In *Drosophila melanogaster*, 0.3 percent of the haploid genome, equivalent to 130 gene copies, consists of rDNA. Although the presence of multiple copies of the same gene is not restricted to those coding for rRNA, we will focus on them in this section.

In some cells, particularly oocytes, even the normal redundancy of rDNA is insufficient to provide adequate amounts of rRNA needed to construct ribosomes. Oocytes store abundant nutrients, including huge quantities of ribosomes, for use by the embryo during early development. More ribosomes are included in oocytes than in any other cell type. By considering how the amphibian *Xenopus laevis* acquires this abundance of ribosomes, we shall see a second way in which the amount of rRNA is increased. This phenomenon is called **gene amplification**.

The genes that code for rRNA are located in an area of the chromosome known as the **nucleolar organizer region** (**NOR**). The NOR is intimately associated with the nucleolus, which is a processing center for ribosome production. Molecular hybridization analysis has shown that each NOR in the frog *Xenopus* contains the equivalent of 400 redundant gene copies coding for rRNA. Even this number of genes is apparently inadequate to synthesize the vast amount of ribosomes that must accumulate in the amphibian oocyte to support development following fertilization.

To further amplify the number of rRNA genes, the rDNA is selectively replicated, and each new set of genes is released from its template. Because each new copy is equivalent to an NOR, multiple small nucleoli are formed around each NOR in the oocyte. As many as 1500 of these "micronucleoli" have been observed in a single oocyte. If we multiply the number of micronucleoli (1500) by the number of gene copies in each NOR (400), we see that amplification in *Xenopus* oocytes can result in over half a million gene copies! If each copy is transcribed only 20 times during the maturation of the oocyte, in theory, sufficient copies of rRNA are produced to result in well over 12 million ribosomes.

The *Bar* Mutation in *Drosophila*

Duplications can cause phenotypic variation that might at first appear to be caused by a simple gene mutation. The *Bar* eye phenotype in *Drosophila* (Figure 6–13) is classic example. Instead of the normal oval eye shape, *Bar*-eyed flies have narrow, slitlike eyes. This phenotype is inherited in the same way as a dominant X-linked mutation.

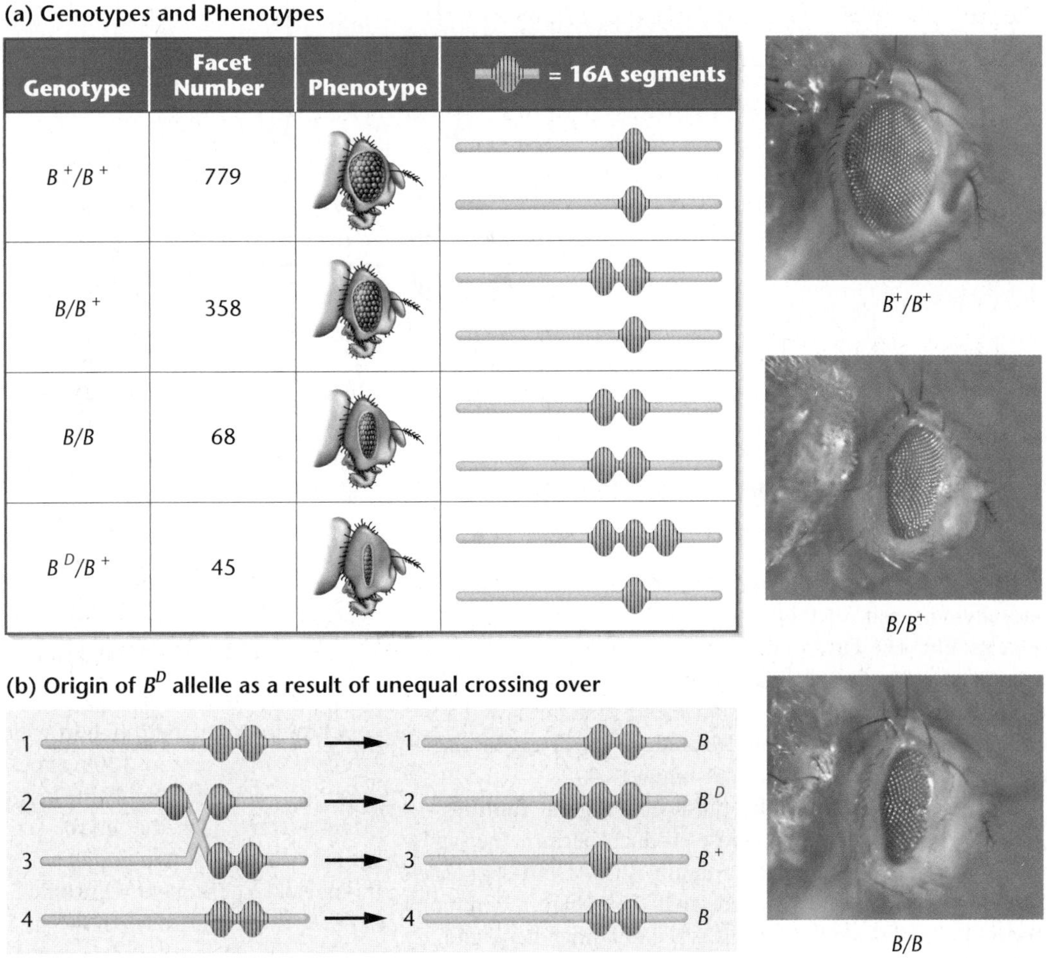

FIGURE 6–13 The duplication genotypes and resultant *Bar* eye phenotypes in *Drosophila*. Photographs show two *Bar* eye phenotypes and the wild type (B^+/B^+).

(a) Genotypes and Phenotypes

Genotype	Facet Number	Phenotype	▦ = 16A segments
B^+/B^+	779		
B/B^+	358		
B/B	68		
B^D/B^+	45		

B^+/B^+

B/B^+

(b) Origin of B^D allele as a result of unequal crossing over

B/B

In the early 1920s, Alfred H. Sturtevant and Thomas H. Morgan discovered and investigated this "mutation." Normal wild-type females (B^+/B^+) have about 800 facets in each eye. Heterozygous females (B/B^+) have about 350 facets, while homozygous females (B/B) average only about 70 facets. Females were occasionally recovered with even fewer facets and were designated as *double Bar* (B^D/B^+).

About 10 years later, Calvin Bridges and Herman J. Muller compared the polytene X chromosome banding pattern of the *Bar* fly with that of the wild-type fly. These chromosomes contain specific banding patterns that have been well categorized into regions. As shown in Figure 6–13, their studies revealed that one copy of region 16A is present on both X-chromosomes of wild-type flies but that this region was duplicated in *Bar* flies and triplicated in *double Bar* flies. These observations provided evidence that the *Bar* phenotype is not the result of a simple chemical change in the gene but is instead a duplication.

The Role of Gene Duplication in Evolution

During the study of evolution, it is intriguing to speculate on the possible mechanisms of genetic variation. The origin of unique gene products present in more recently evolved organisms but absent in ancestral forms is a topic of particular interest. In other words, how do "new" genes arise?

In 1970, Susumo Ohno published a provocative monograph, *Evolution by Gene Duplication*, in which he suggested that gene duplication is essential to the origin of new genes during evolution. Ohno's thesis is based on the supposition that the gene products of many genes, present as only a single copy in the genome, are indispensable to the survival of members of any species during evolution. Therefore, unique genes are not free to accumulate mutations sufficient to alter their primary function and give rise to new genes.

However, if an essential gene is duplicated in the germ line, major mutational changes in this extra copy will be tolerated in future generations because the original gene provides the genetic information for its essential function. The duplicated copy will be free to acquire many mutational changes over extended periods of time. Over short intervals, the new genetic information may be of no practical advantage. However, over long evolutionary periods, the duplicated gene may change sufficiently so that its product assumes a divergent role in the cell. The new function may impart an "adaptive" advantage to organisms, enhancing their fitness. Ohno has outlined a mechanism through which sustained genetic variability may have originated.

Ohno's thesis is supported by the discovery of genes that have a substantial amount of their DNA sequence in common,

FIGURE 6–14 One possible origin of a pericentric inversion.

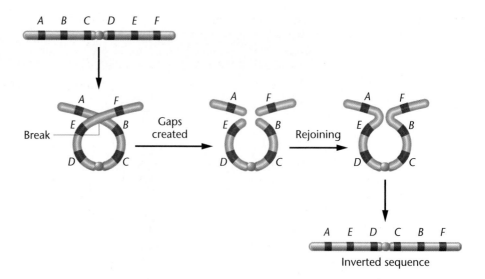

Inverted sequence

but whose gene products are distinct. For example, trypsin and chymotrypsin fit this description, as do myoglobin and hemoglobin. The DNA sequence is so similar (homologous) in each case that we may conclude that members of each pair of genes arose from a common ancestral gene through duplication. During evolution, the related genes diverged sufficiently that their products became unique.

Other support includes the presence of **gene families**—groups of contiguous genes whose products perform the same function. Again, members of a family show DNA-sequence homology sufficient to conclude that they share a common origin. Examples are the various types of human hemoglobin polypeptide chains, as well as the immunologically important T-cell receptors and antigens encoded by the major histocompatibility complex.

Many recent findings derived from our ability to sequence entire genomes lend support to the idea that gene duplication has been a common feature of evolutionary progression. For example, Jurg Spring has compared a large number of genes in *Drosophila* and their counterparts in humans. In 50 genes studied, the fruit fly has only one copy of each, while there are multiple copies present in the human genome. Many other investigations have provided similar findings.

A new debate has begun concerning a second aspect of Ohno's thesis—that major evolutionary jumps, such as the transition from invertebrates to vertebrates, may have involved the duplication of entire genomes. Ohno has suggested that this might have occurred several times during the course of evolution. Although it seems clear that genes, and even segments of chromosomes, have been duplicated, there is not yet any compelling evidence to convince evolutionary biologists that this was clearly the case. However, from the standpoint of gene expression, duplicating all genes proportionately seems more likely to be tolerated than duplicating just a portion of them. Whatever may be the case, it is interesting to examine evidence based on new technology that tests a hypothesis that was put forward over 30 years ago, long before that technology was available.

6.9   Inversions Rearrange the Linear Gene Sequence

The **inversion**, another class of structural variation, is a type of chromosomal aberration in which a segment of a chromosome is turned around 180 degrees within a chromosome. An inversion does not involve a loss of genetic information but simply rearranges the linear gene sequence. An inversion requires breaks at two points along the length of the chromosome and subsequent reinsertion of the inverted segment. Figure 6–14 illustrates how an inversion might arise. By forming a chromosomal loop prior to breakage, the newly created "sticky ends" are brought close together and rejoined.

The inverted segment may be short or quite long and may or may not include the centromere. If the centromere is not part of the rearranged chromosome segment, it is a **paracentric inversion**. If the centromere is part of the inverted segment, it is described as a **pericentric inversion**, which is the type shown in Figure 6–14.

Although inversions appear to have a minimal impact on the individuals bearing them, their consequences are of great interest to geneticists. Organisms that are heterozygous for inversions may produce aberrant gametes that have a major impact on their offspring.

Consequences of Inversions During Gamete Formation

If only one member of a homologous pair of chromosomes has an inverted segment, normal linear synapsis during meiosis is not possible. Organisms with one inverted chromosome and one noninverted homolog are called **inversion heterozygotes**. Pairing between two such chromosomes in meiosis is accomplished only if they form an **inversion loop** (Figure 6–15).

If crossing over does not occur within the inverted segment of the inversion loop, the homologs will segregate, which results in two normal and two inverted chromatids that are distributed into gametes. However, if crossing over does occur within the inversion loop, abnormal chromatids are produced.

The effect of a single exchange (SCO) event within a paracentric inversion is diagrammed in Figure 6–15(a).

As in any meiotic tetrad, a single crossover between nonsister chromatids produces two parental chromatids and two recombinant chromatids. When the crossover occurs within a paracentric inversion, however, one recombinant **dicentric chromatid** (two centromeres), and one recombinant **acentric chromatid** (lacking a centromere) are produced. Both contain duplications and deletions of chromosome segments as well. During anaphase, an acentric chromatid moves randomly to one pole or the other or may be lost, while a dicentric chromatid is pulled in two directions. This polarized movement produces dicentric bridges that are cytologically recognizable. A dicentric chromatid usually breaks at some point so that part of the chromatid goes into one gamete and part into another gamete during the reduction divisions. Therefore, gametes containing either recombinant chromatid are deficient in genetic material. When such a gamete participates in fertilization, the zygote most often develops abnormally, if at all.

A similar chromosomal imbalance is produced as a result of a crossover event between a chromatid bearing a pericentric inversion and its non-inverted homolog, as shown in Figure 6–15(b). The recombinant chromatids that are directly involved in the exchange have duplications and deletions. In plants, gametes receiving such aberrant chromatids fail to develop normally, leading to aborted pollen or ovules. Thus, lethality occurs prior to fertilization, and inviable seeds result. In animals, the gametes have developed prior to the meiotic error, so fertilization is more likely to occur in spite of the chromosome error. However, the end result is the production of inviable embryos following fertilization. In both cases, viability is reduced.

Since viable offspring do not result in either plants or animals, it *appears* as if the inversion suppresses crossing over, because offspring bearing crossover gametes are not recovered. Actually, in inversion heterozygotes, the inversion has the effect of *suppressing the recovery of crossover products* when chromosome exchange occurs within the inverted region. If crossing over always occurred within a paracentric or pericentric inversion, 50 percent of the gametes would be ineffective. The viability of the resulting zygotes is therefore greatly diminished. Furthermore, up to one-half of the viable gametes have the inverted chromosome, and the inversion will be perpetuated within the species. The cycle will be repeated continuously during meiosis in future generations.

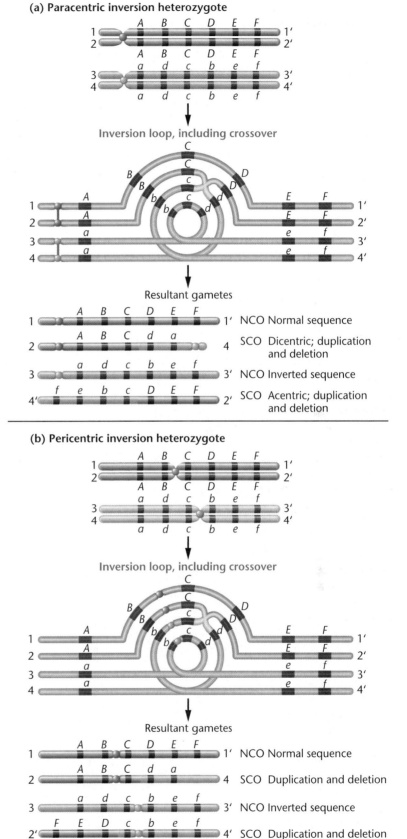

FIGURE 6–15 (a) The effects of a single crossover (SCO) within an inversion loop in a paracentric inversion heterozygote, where two altered chromosomes are produced, one acentric and one dicentric. Both chromosomes also contain duplicated and deficient regions. (b) The effects of a crossover in a pericentric inversion heterozygote, where two altered chromosomes are produced, both with duplicated and deficient regions.

Now Solve This

Problem 23 on page 134 considers a prospective male parent with a family history of stillbirths and malformed babies. He was shown to contain an inversion covering 70 percent of one member of chromosome 1. You are asked to explain the cause of the stillbirths.

Hint: Human chromosome 1 is metacentric; therefore, an inversion covering 70 percent of the chromosome is no doubt a pericentric inversion.

6.10 Translocations Alter the Location of Chromosomal Segments in the Genome

Translocation, as the name implies, is the movement of a chromosomal segment to a new location in the genome. Reciprocal translocation, for example, involves the exchange of segments between two nonhomologous chromosomes. The least complex way for this event to occur is for two nonhomologous chromosome arms to come close to each other so that an exchange is facilitated. Figure 6–16(a) shows a simple reciprocal translocation in which only two breaks are required. If the exchange includes internal chromosome segments, four breaks are required, two on each chromosome.

The genetic consequences of reciprocal translocations are, in several instances, similar to those of inversions. For example, genetic information is not lost or gained. Rather, there is only a rearrangement of genetic material. The presence of a translocation does not, therefore, directly alter the viability of individuals bearing it.

Homologs that are heterozygous for a reciprocal translocation undergo unorthodox synapsis during meiosis. As shown in Figure 6–16(b), pairing results in a crosslike configuration. As with inversions, genetically unbalanced gametes are also produced as a result of this unusual alignment during meiosis. In the case of translocations, however, aberrant gametes are not necessarily the result of crossing over. To see how unbalanced gametes are produced, focus on the homologous centromeres in Figure 6–16(b) and Figure 6–16(c). According to the principle of independent assortment, the chromosome containing centromere 1 migrates randomly toward one pole of the spindle during the first meiotic anaphase; it travels along with *either* the chromosome having centromere 3 *or* the chromosome having centromere 4. The chromosome with centromere 2 moves to the other pole along with *either* the chromosome containing centromere 3 *or* centromere 4. This results in four potential meiotic products. The 1,4 combination contains chromosomes that are not involved in the translocation. The 2,3 combination, however, contains translocated chromosomes. These contain a complete complement of genetic information and are balanced. The other two potential products, the 1,3 and 2,4 combinations,

contain chromosomes displaying duplicated and deleted segments. To simplify matters, crossover exchanges are ignored here.

When incorporated into gametes, the resultant meiotic products are genetically unbalanced. If they participate in fertilization, lethality often results. As few as 50 percent of the progeny of parents that are heterozygous for a reciprocal translocation survive. This condition, called *semisterility*, has an impact on the reproductive fitness of organisms, thus playing a role in evolution. Furthermore, in humans, such an unbalanced condition results in partial monosomy or trisomy, leading to a variety of birth defects.

(a) Possible origin of a reciprocal translocation

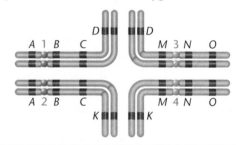

(b) Synapsis of translocation heterozygote

(c) Two possible segregation patterns leading to gamete formation

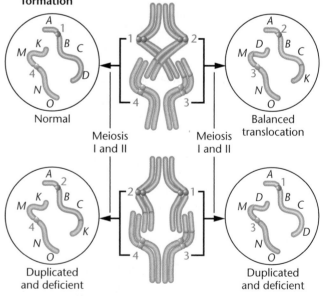

FIGURE 6–16 (a) Possible origin of a reciprocal translocation. (b) Synaptic configuration formed during meiosis in an individual that is heterozygous for the translocation. (c) Two possible segregation patterns, one of which leads to a normal and a balanced gamete (called alternate segregation) and one that leads to gametes containing duplications and deficiencies (called adjacent segregation).

Translocations in Humans: Familial Down Syndrome

Research conducted since 1959 has revealed numerous translocations in members of the human population. One common type of translocation involves breaks at the extreme ends of the short arms of two nonhomologous acrocentric chromosomes. These small segments are lost, and the larger segments fuse at their centromeric region. This type of translocation produces a new, large submetacentric or metacentric chromosome, often called a **Robertsonian translocation**.

One such translocation accounts for cases in which Down syndrome is familial (inherited). Earlier in this chapter, we pointed out that most instances of Down syndrome are due to trisomy 21. This chromosome composition results from nondisjunction during meiosis in one parent. Trisomy accounts for over 95 percent of all cases of Down syndrome. In such instances, the chance of the same parents producing a second affected child is extremely low. However, in the remaining families with a Down child, the syndrome occurs in a much higher frequency over several generations.

Cytogenetic studies of the parents and their offspring from these unusual cases explain the cause of **familial Down syndrome**. Analysis reveals that one of the parents contains a 14/21, D/G translocation (Figure 6–17). That is, one parent has the majority of the G-group chromosome 21 translocated to one end of the D-group chromosome 14. This individual is phenotypically normal, even though he or she has only 45 chromosomes. During meiosis, one-fourth of the individual's gametes have two copies of chromosome 21: a normal chromosome and a second copy translocated to chromosome 14. When such a gamete is fertilized by a standard haploid gamete, the resulting zygote has 46 chromosomes but three copies of chromosome 21. These individuals exhibit Down syndrome. Other potential surviving offspring contain either the standard diploid genome (without a translocation) or the balanced translocation like the parent. Both cases result in normal individuals. Knowledge of translocations has allowed geneticists to resolve the seeming paradox of an inherited trisomic phenotype in an individual with an apparent diploid number of chromosomes.

It is interesting to note that the "carrier," who has 45 chromosomes and exhibits a normal phenotype, does not contain the *complete* diploid amount of genetic material. A small region is lost from both chromosomes 14 and 21 during the translocation event. This occurs because the ends of both chromosomes have broken off prior to their fusion. These specific regions are known to be two of many chromosomal locations housing multiple copies of the genes encoding rRNA, the major component of ribosomes. Despite the loss of up to 20 percent of these genes, the carrier is unaffected.

6.11 Fragile Sites in Humans Are Susceptible to Chromosome Breakage

We conclude this chapter with a brief discussion of the results of an intriguing discovery made around 1970 during observations of metaphase chromosomes prepared following human cell culture. In cells derived from certain individuals, a specific area along one of the chromosomes failed to stain, giving the appearance of a gap. In other individuals whose chromosomes displayed such morphology, the gaps appeared at other positions within the set of chromosomes. Such areas eventually became known as **fragile sites**, since they appeared to be susceptible to chromosome breakage when cultured in the absence of certain chemicals such as folic acid, which is normally present in the culture medium. Fragile sites were at first considered curiosities, until a strong association was subsequently shown to exist between one of the sites and a form of mental retardation.

The cause of the fragility at these sites is unknown. Because they represent points along the chromosome that are susceptible to breakage, these sites may indicate regions where the

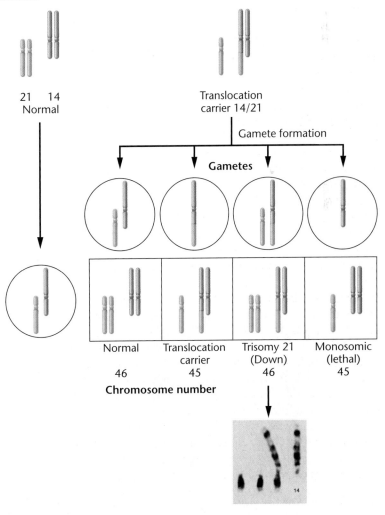

FIGURE 6-17 Chromosomal involvement and translocation in familial Down syndrome. The photograph shows the relevant chromosomes from a trisomy 21 offspring produced by a translocation carrier parent.

chromatin is not tightly coiled. Note that even though almost all studies of fragile sites have been carried out *in vitro* using mitotically dividing cells, clear associations have been established between several of these sites and the corresponding altered phenotype, including mental retardation and cancer.

Fragile X Syndrome (Martin-Bell Syndrome)

Most fragile sites do not appear to be associated with any clinical syndrome. However, individuals bearing a folate-sensitive site on the X chromosome (Figure 6–18) exhibit the **fragile X syndrome** (or **Martin-Bell syndrome**), the most common form of inherited mental retardation. This syndrome affects about 1 in 4000 males and 1 in 8000 females. Females carrying only one fragile X chromosome can be mentally retarded because it is a dominant trait. Fortunately, the trait is not fully expressed, as only about 30 percent of fragile X females are retarded, whereas about 80 percent of fragile X males are mentally retarded. In addition to mental retardation, affected males have characteristic long, narrow faces with protruding chins, enlarged ears, and increased testicular size.

A gene that spans the fragile site may be responsible for this syndrome. This gene, known as *FMR-1*, is one of a growing number of genes that have been discovered in which a sequence of three nucleotides is repeated many times, expanding the size of the gene. This phenomenon, called **trinucleotide repeats**, is also recognized in other human disorders, including Huntington disease. In *FMR-1*, the trinucleotide sequence CGG is repeated in an untranslated area adjacent to the coding sequence of the gene (called the "upstream" region). The number of repeats varies immensely within the human population, and a high number correlates directly with expression of fragile X syndrome. Normal individuals have between 6 and 54 repeats, whereas those with 55 to 230 repeats are considered "carriers" of the disorder. More than 230 repeats leads to expression of the syndrome.

It is thought that when the number of repeats reaches this level, the CGG regions of the gene become chemically modified so that the bases within and around the repeat are methylated, causing inactivation of the gene. The normal product of the gene is an RNA-binding protein, FMRP, known to be expressed in the brain. Evidence is now accumulating that directly links the absence of the protein with the cognitive defects associated with the syndrome.

The protein is prominent in cells of the developing brain and normally shuttles between the nucleus and cytoplasm, transporting mRNAs to ribosomal complexes. Its absence in dendritic cells seems to prevent the translation of other critical proteins during the development of the brain, presumably those produced from mRNAs that are transported by FMRP. Ultimately, the result is impairment of synaptic activity related to learning and memory. Molecular analysis has shown that FMRP is very selective, binding specifically to mRNAs that have a high guanine content. These RNAs form unique secondary structures called **G-quartets**, for which the protein has a strong binding affinity. The fact that possession of such quartets is important to recognition by FMRP becomes evident by examining cases where the formation of these stacked quartets is inhibited (using Li^+). In such cases, FMRP binding to these RNAs is abolished.

The *Drosophila* homolog (*dfrx*) of the human *FRM-1* gene has now been identified and used as a model for experimental investigation. The normal protein product of the fruit fly displays chemical properties that parallel its human counterpart. Mutations have been generated and studied that either overexpress the gene or eliminate its expression altogether. The latter case is designated a null mutation, where the gene has been "knocked out" using a molecular procedure now common in DNA biotechnology. The results of these studies show that the normal gene functions during the formation of synapses and that abnormal synaptic function accompanies the loss of *dfrx* expression. Knockout mice lacking this gene also display a similar phenotype. Ultimately, then, a closely related version of this gene is conserved in humans, *Drosophila*, and mice. In all three species, it has been shown to be essential to synaptic maturation and the subsequent establishment of proper cell communication in the brain. In humans, the neurological features of the syndrome are linked to the absence of its genetic expression.

From a genetic standpoint, perhaps the most interesting aspect of fragile X syndrome is the instability of the CGG repeats. An individual with 6 to 54 repeats transmits a gene containing the same number to his or her offspring. However, those with 55 to 230 repeats, while not at risk to develop the syndrome, may transmit to their offspring a gene with an increased number of repeats. The number of repeats continues to increase in future generations, demonstrating the phenomenon known as **genetic anticipation**. Once the threshold of 230 repeats is exceeded, expression of the malady becomes more severe in each successive generation as the number of trinucleotide repeats increases.

While the mechanism that leads to the trinucleotide expansion has not yet been established, several factors are known that influence the instability. Most significant is the observation that expansion from the carrier status (55 to 230 repeats) to the syndrome status (over 230 repeats) occurs during the transmission of the gene by the maternal parent but not by the paternal parent. Furthermore, several reports suggest that male offspring are more likely to receive the increased repeat size leading to the syndrome than are female offspring. Obviously, we have much to learn about the genetic basis of instability and expansion of DNA sequences.

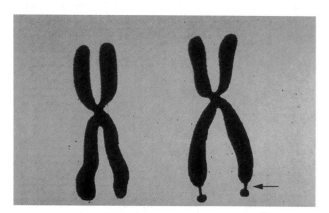

FIGURE 6–18 A normal human X chromosome (left) contrasted with a fragile X chromosome (right). The "gap" region (near the bottom of the chromosome) is associated with the fragile X syndrome.

GENETICS, TECHNOLOGY, AND SOCIETY

The Link between Fragile Sites and Cancer

While the study of the fragile X syndrome first brought unstable chromosome regions to the attention of geneticists, a second link between a fragile site and a human disorder was reported in 1996 by Carlo Croce, Kay Huebner, and their colleagues. They demonstrated an association between an autosomal fragile site and lung cancer. Many years of investigation led this research group to postulate that the defect associated with this fragile site contributes to the formation of a variety of different tumor types. Croce and Huebner first showed that the *FHIT* gene (standing for *f*ragile *h*istidine *t*riad), located within the well-defined fragile site designated as *FRA3B* (on the p arm of chromosome 3), is often altered or missing in cells taken from tumors of individuals with lung cancer. Molecular analysis of numerous mutations showed that the DNA had been broken and incorrectly fused back together, resulting in deletions within the *FHIT* gene. In most cases, these mutations resulted in the loss of the normal protein product of this gene. More extensive studies have now revealed that this protein is absent in cells of many other cancers, including those of the esophagus, breast, cervix, liver, kidney, pancreas, colon, and stomach. Genes such as *FHIT* that are located within fragile regions undoubtedly exhibit an increased susceptibility to mutations and deletions.

We now have a better understanding of fragile sites, with more than 80 such regions identified in the Human Genome Data Base. Fragile sites are said to be recombinogenic, based on the observations that within these chromosomal regions, alterations such as translocations, sister-chromatid exchanges (SCEs), and other rearrangements between chromosomes frequently occur. It is now believed that breaks and gaps associated with fragile sites are induced by agents that inhibit DNA replication. Such gaps and breaks are therefore the result of DNA that has been incompletely replicated. Synthesis of DNA within these regions during the S phase of the cell cycle is delayed as compared to nonfragile chromosomal regions. This hypothesis has been strengthened in the case of the *FRA3B* region, now shown to be characterized by delayed DNA repli-

cation. As a result, the DNA in this region is not replicated by the time the cell enters the G_2 phase and subsequently initiates mitosis. The resulting metaphase chromosomes reveal distinct cytogenetic gaps and breaks within the fragile region. If these breaks are incorrectly repaired during cell cycle checkpoints, cancer-specific mutations may occur if the region contains genes involved in cell-cycle control.

FHIT is a good example of a mutant gene within a fragile site and has now been designated as a tumor-suppressor gene. The product of the normal gene is believed to recognize genetic damage in cells and to induce apoptosis, whereby a potentially malignant cell is targeted to undergo programmed cell death and is effectively eliminated. The failure to remove such cells as a result of mutations in the *FHIT* gene causes individuals to be particularly sensitive to carcinogen-induced damage, creating a susceptibility to cancer. For example, cigarette smokers who develop lung cancer demonstrate an increased expression of the FRA3B fragile region compared to the normal population. Still in question, however, is whether the alteration in the fragile site, leading to inactivation of the gene, causes cancer, *or* whether the behavior of malignant cells somehow induces breaks at fragile sites, subsequently inactivating or deleting this gene. Another important question under investigation is the extent of molecular polymorphism at the fragile site within the human population, causing some individuals to be more susceptible to the effects of carcinogens than others.

More recently, Muller Fabrii and Kay Huebner, working with others in Croce's lab, have identified and studied another fragile site, with most interesting results. Found within the FRA16D site on chromosome 16 is the *WWOX* gene. Like the *FHIT* gene, it has been implicated in a diverse number of human cancers. In particular, like *FHIT*, it has been found to be either lost or genetically silenced in the large majority of lung tumors, as well as in cancer tissue of the breast, ovary, prostate, bladder, esophagus, and pancreas. When the gene is present, but silent, its DNA is thought to be heavily methylated, rendering it inactive. Furthermore, the gene is also thought to

behave as a tumor suppressor, providing a surveillance function by recognizing cancer cells and inducing apoptosis, effectively eliminating them before malignant tumors can be initiated.

Fabrii, Huebner, and colleagues have attempted to demonstrate the tumor suppressor role of the normal gene by using recombinant DNA technology and successfully delivering it into cancer cells and monitoring for apoptotic action. Using a number of cancer cell lines in culture, they were able to successfully introduce the gene into a viral vector and to infect these cells. The majority of those with working copies of the gene underwent cell suicide, while those that lacked working copies of the gene did not. Encouraged by these results, they then turned to *in vivo* experiments. "Nude" mice, which lack a functional immune system and are favorites in this type of experiment, had treated and untreated cancer cells transplanted under their skin. The results were striking. Control mice that were infected with cells containing the virus that lacked the *WWOX* gene developed skin tumors, while about 70 percent of the mice who had received cells with an active *WWOX* gene remained tumor-free. These findings reveal that restoration of gene function in cells that have previously lost it will not only inhibit further reproduction of cancer cells but, in fact, destroy them.

This is an important example of pioneering work in gene therapy, representing significant and exciting studies that may prove valuable in cancer therapy. While applications of the research will not prevent cancer initially, approaches using these genes found in fragile sites may someday play an important role in combination with other cancer therapies utilized to treat a variety of malignancies.

References

Fabbri, M. et. al. 2005. *WWOX* gene restoration prevents lung cancer growth *in vitro* and *in vivo*. *Proc. Nat. Acad. Sci.* 102: 15611–15616.

Huebner, K., and Croce, C. (2003). Cancer and the *FRA3B/FHIT* fragile locus: it's a HIT. *British J. Cancer* 88: 1501–1506.

CHAPTER SUMMARY

1. Investigations into the uniqueness of each organism's chromosomal constitution is enhancing our understanding of genetic variation. Alterations of the precise diploid content of chromosomes are called chromosomal aberrations or chromosomal mutations.

2. Deviations from the expected chromosomal number, or mutations in the structure of the chromosome, are inherited in predictable Mendelian fashion; they often result in inviable organisms or substantial changes in the phenotype.

3. Aneuploidy is the gain or loss of one or more chromosomes from the diploid complement, resulting in conditions of monosomy, trisomy, tetrasomy, and so on. Studies of monosomic and trisomic disorders are increasing our understanding of the delicate genetic balance that enables normal development.

4. When complete sets of chromosomes are added to the diploid genome, polyploidy occurs. These sets can have identical or diverse genetic origin, creating either autopolyploidy or allopolyploidy, respectively.

5. Large segments of the chromosome can be modified by deletions or duplications. Deletions can produce serious conditions such as the cri-du-chat syndrome in humans, whereas duplications can be particularly important as a source of redundant or new genes.

6. Inversions and translocations, while altering the gene order along chromosomes, initially cause little or no loss of genetic information or deleterious effects. However, heterozygous combinations may result in genetically abnormal gametes following meiosis, with lethality often ensuing.

7. Fragile sites in human mitotic chromosomes have sparked research interest because one such site on the X chromosome is associated with the most common form of inherited mental retardation. Another fragile site, located on chromosome 3, has been linked to lung cancer.

KEY TERMS

acentric chromatid, 127
allopolyploidy, 120
allotetraploid, 121
amniocentesis, 119
amphidiploid, 121
aneuploidy, 115
autopolyploidy, 120
autotetraploid, 120
autotriploid, 120
chorionic villus sampling (CVS), 119
chromosome aberration, 115
chromosome mutation, 115
colchicine, 120
compensation loop, 124
cri-du-chat syndrome (46,5p−), 117
deficiency, 123
deletion, 123
deletion loop, 124
dicentric chromatid, 127
Down syndrome (47,21+), 118

Down syndrome critical region (DSCR), 118
Edwards syndrome (47,18+), 119
euploidy, 115
familial Down syndrome, 119, 129
FHIT gene, 131
FMR-1 gene, 130
FRA3B fragile site, 131
fragile site, 129
fragile X syndrome, 130
G1 cyclins, 121
gene amplification, 124
gene family, 126
gene redundancy, 124
gene therapy, 131
genetic anticipation, 130
G-quartet, 130
intercalary deletion, 123
inversion, 126
inversion heterozygote, 126

inversion loop, 126
Klinefelter syndrome, 116
Martin-Bell syndrome, 130
monosomy, 116
nondisjunction, 116
nucleolar organizer region (NOR), 124
paracentric inversion, 126
pericentric inversion, 126
Patau syndrome (47,13+), 119
polyploidy, 115
rDNA, 124
Robertsonian translocation, 129
segmental deletion, 117
terminal deletion, 123
translocation, 128
trinucleotide repeat, 130
trisomy, 117
trisomy 21, 118
Turner syndrome, 116

INSIGHTS AND SOLUTIONS

1. In a cross using maize that involves three genes, *a*, *b*, and *c*, a heterozygote (*abc*/+++) is testcrossed to *abc*/*abc*. Even though the three genes are separated along the chromosome, thus predicting that crossover gametes and the resultant phenotypes should be observed, only two phenotypes are recovered: *abc* and +++. In addition, the cross produced significantly fewer viable plants than expected. Can you propose why no other phenotypes were recovered and why the viability was reduced?

Solution: One of the two chromosomes may contain an inversion that overlaps all three genes, effectively precluding the recovery of any "crossover" offspring. If this is a paracentric inversion and the genes are clearly separated (assuring that a significant number of crossovers occurs between them), then numerous acentric and dicentric chromosomes will form, resulting in the observed reduction in viability.

2. A male *Drosophila* from a wild-type stock is discovered to have only seven chromosomes, whereas normally $2n = 8$. Close examination reveals that one member of chromosome IV (the smallest chromosome) is attached to (translocated to) the distal end of chromosome II and is missing its centromere, thus accounting for the reduction in chromosome number. (a) Diagram all members of chromosomes II and IV during synapsis in meiosis I.

Solution:

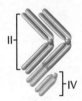

II-

IV

(b) If this male mates with a female with a normal chromosome composition who is homozygous for the recessive chromosome IV mutation *eyeless* (*ey*), what chromosome compositions will occur in the offspring regarding chromosomes II and IV?

Solution:

Normal female   Translocation male

(1) (2) (3) (4)

(c) Referring to the diagram in the solution to part (b), what phenotypic ratio will result regarding the presence of eyes, assuming all abnormal chromosome compositions survive?

Solution:

(1) normal (heterozygous)

(2) eyeless (monosomic, contains chromosome IV from mother)

(3) normal (heterozygous; trisomic and may die)

(4) normal (heterozygous; balanced translocation)

The final ratio is 3/4 normal: 1/4 eyeless.

PROBLEMS AND DISCUSSION QUESTIONS

1. For a species with a diploid number of 18, indicate how many chromosomes will be present in the somatic nuclei of individuals that are haploid, triploid, tetraploid, trisomic, and monosomic.

2. Define these pairs of terms, and distinguish between them.

   aneuploidy/euploidy
   monosomy/trisomy
   Patau syndrome/Edwards syndrome
   autopolyploidy/allopolyploidy
   autotetraploid/amphidiploid
   paracentric inversion/pericentric inversion

3. Contrast the relative survival times of individuals with Down, Patau, and Edwards syndromes. Speculate as to why such differences exist.

4. What evidence suggests that Down syndrome is more often the result of nondisjunction during oogenesis rather than during spermatogenesis?

5. What evidence indicates that humans with aneuploid karyotypes occur at conception but are usually inviable?

6. Contrast the fertility of an allotetraploid with an autotriploid and an autotetraploid.

7. When two plants belonging to the same genus but different species are crossed, the F_1 hybrid is more viable and has more ornate flowers. Unfortunately, this hybrid is sterile and can only be propagated by vegetative cuttings. Explain the sterility of the hybrid. How might a horticulturist attempt to reverse its sterility?

8. Describe the origin of cultivated American cotton.

9. Predict how the synaptic configurations of homologous pairs of chromosomes might appear when one member is normal and the other member has sustained a deletion or duplication.

10. Inversions are said to "suppress crossing over." Is this terminology technically correct? If not, restate the description accurately.

11. Contrast the genetic composition of gametes derived from tetrads of inversion heterozygotes where crossing over occurs within a paracentric versus a pericentric inversion.

12. Discuss Ohno's hypothesis on the role of gene duplication in the process of evolution.

13. What roles have inversions and translocations played in the evolutionary process?

14. A human female with Turner syndrome also expresses the X-linked trait hemophilia, as did her father. Which of her parents underwent nondisjunction during meiosis, giving rise to the gamete responsible for the syndrome?

15. The primrose, *Primula kewensis*, has 36 chromosomes that are similar in appearance to the chromosomes in two related species, *P. floribunda* ($2n = 18$) and *P. verticillata* ($2n = 18$). How could *P. kewensis* arise from these species? How would you describe *P. kewensis* in genetic terms?

16. Certain varieties of chrysanthemums contain 18, 36, 54, 72, and 90 chromosomes; all are multiples of a basic set of nine chromosomes. How would you describe these varieties genetically? What feature is shared by the karyotypes of each variety? A variety with 27 chromosomes has been discovered, but it is sterile. Why?

17. *Drosophila* may be monosomic for chromosome 4, yet remain fertile. Contrast the F_1 and F_2 results of the following crosses involving the recessive chromosome 4 trait, *bent* bristles:
   (a) monosomic IV, bent bristles × diploid, normal bristles
   (b) monosomic IV, normal bristles × diploid, bent bristles.

18. Mendelian ratios are modified in crosses involving autotetraploids. Assume that one plant expresses the dominant trait green seeds and is homozygous (*WWWW*). This plant is crossed to one with white seeds that is also homozygous (*wwww*). If only one dominant allele is sufficient to produce green seeds, predict the F_1 and F_2 results of such a cross. Assume that synapsis between chromosome pairs is random during meiosis.

19. Having correctly established the F_2 ratio in Problem 18, predict the F_2 ratio of a "dihybrid" cross involving two independently assorting characteristics (e.g., $P_1 = WWWWAAAA \times wwwwaaaa$).

20. In a cross between two varieties of corn, $gl_1gl_1 Ws_3Ws_3$ (egg parent) × $Gl_1Gl_1 ws_3ws_3$ (pollen parent), a triploid offspring was produced with the genetic constitution $Gl_1Gl_1gl_1Ws_3ws_3ws_3$. From which parent, egg or pollen, did the $2n$ gamete originate? Is another explanation possible? Explain.

21. What is the effect of a rare double crossover (a) within a pericentric inversion present heterozygously, (b) within a paracentric inversion present heterozygously?

22. The outcome of a single crossover between nonsister chromatids in the inversion loop of an inversion heterozygote varies depending on whether the inversion is of the paracentric or pericentric type. What differences are expected?

23. A couple planning their family are aware that through the past three generations on the husband's side a substantial number of stillbirths have occurred and several malformed babies were born who died early in childhood. The wife has studied genetics and urges her husband to visit a genetic counseling clinic, where a complete karyotype-banding analysis is performed. Although the tests show that he has a normal complement of 46 chromosomes, banding analysis reveals that one member of the chromosome 1 pair (in group A) contains an inversion covering 70 percent of its length. The homolog of chromosome 1 and all other chromosomes show the normal banding sequence. (a) How would you explain the high incidence of past stillbirths? (b) What can you predict about the probability of abnormality/normality of their future children? (c) Would you advise the woman that she will have to bring each pregnancy to term to determine whether the fetus is normal? If not, what else can you suggest?

24. Briefly compare the fate of gametes in plants and animals carrying aberrant chromatids or unbalanced chromosomal complements.

25. A woman who sought genetic counseling is found to be heterozygous for a chromosomal rearrangement between the second and third chromosomes. Her chromosomes, compared to those in a normal karyotype, are diagrammed here:
   (a) What kind of chromosomal aberration is shown?
   (b) Using a drawing, demonstrate how these chromosomes would pair during meiosis. Be sure to label the different segments of the chromosomes.
   (c) This woman is phenotypically normal. Does this surprise you? Why or why not? Under what circumstances might you expect a phenotypic effect of such a rearrangement?

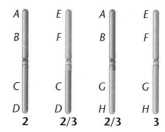

26. The woman in Problem 25 has had two miscarriages. She has come to you, an established genetic counselor, with these questions: (a) Is there a genetic explanation of her frequent miscarriages? (b) Should she abandon her attempts to have a child of her own? (c) If not, what is the chance that she could have a normal child? Provide an informed response to her concerns.

27. In a recent cytogenetic study on 1021 cases of Down syndrome, 46 were the result of translocations, the most frequent of which was symbolized as t(14;21). What does this symbol represent, and how many chromosomes would you expect to be present in t(14;21) Down syndrome individuals?

28. A boy with Klinefelter syndrome is born to a mother who is phenotypically normal and a father who has the X-linked skin condition called anhidrotic ectodermal dysplasia. The mother's skin is completely normal with no signs of the skin abnormality. In contrast, her son has patches of normal skin and patches of abnormal skin. (a) Which parent contributed the abnormal gamete? (b) Using the appropriate genetic terminology, describe the meiotic mistake that occurred. Be sure to indicate in which division the mistake occurred. (c) Using the appropriate genetic terminology, explain the son's skin phenotype.

29. To investigate the origin of nondisjunction, 200 human oocytes that had failed to be fertilized during *in vitro* fertilization procedures were subsequently examined (Angel, R. 1997. *Am. J. Hum. Genet.* 61:23–32). These oocytes had completed meiosis I and were arrested in metaphase II (MII). The majority (67%) had a normal MII-metaphase complement, showing 23 chromosomes, each consisting of two sister chromatids joined at a common centromere. The remaining oocytes all had abnormal chromosome compositions. Surprisingly, when trisomy was considered, none of the abnormal oocytes had 24 chromosomes. (a) Interpret these results in regard to the origin of trisomy, as it relates to nondisjunction, and when it occurs. Why are the results surprising? (b) A large number of the abnormal oocytes contained 22 1/2 chromosomes; that is, 22 chromosomes plus a single chromatid representing the 1/2 chromosome. What chromosome compositions will result in the zygote if such oocytes proceed through meiosis and are fertilized by normal sperm? (c) How could the complement of 22 1/2 chromosomes arise? Provide a drawing that includes several pairs of MII chromosomes. (d) Do your answers support or dispute the generally accepted theory regarding nondisjunction and trisomy, as outlined in Figure 6–1?

30. In a human genetic study, a family with five phenotypically normal children was investigated. Two were "homozygous" for a Robertsonian translocation between chromosomes 19 and 20 (they contained two identical copies of the fused chromosome). These children have only 44 chromosomes but a complete genetic complement. Three of the children were "heterozygous" for the translocation and contained 45 chromosomes, with one translocated chromosome plus a normal copy of both chromosomes 19 and 20. Two other pregnancies resulted in stillbirths. It was later discovered that the parents were first cousins. Based on this information, determine the chromosome compositions of the parents. What led to the stillbirths? Why was the discovery that the parents were first cousins a key piece of information in understanding the genetics of this family?

7 Linkage and Chromosome Mapping in Eukaryotes

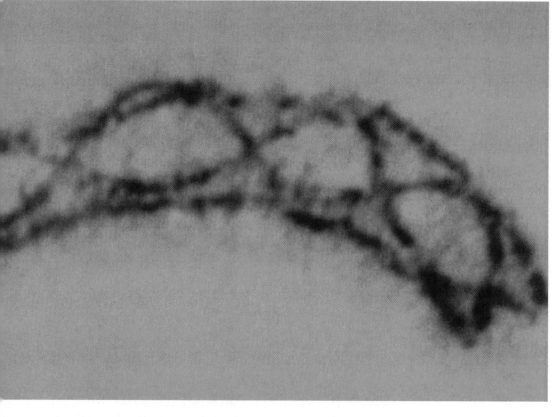

Chiasmata between synapsed homologs during the first meiotic prophase of meiosis.

CHAPTER CONCEPTS

- Chromosomes in eukaryotes contain many genes whose locations are fixed along the length of the chromosomes.

- Unless separated by crossing over, alleles present on a chromosome segregate as a unit during gamete formation.

- Crossing over between homologs during meiosis creates recombinant gametes with different combinations of alleles that enhance genetic variation.

- Crossing over between homologs serves as the basis for the construction of chromosome maps.

- Genetic maps depict relative locations of genes on chromosomes in a species.

W alter Sutton, along with Theodor Boveri, was instrumental in uniting the fields of cytology and genetics. As early as 1903, Sutton pointed out the likelihood that there must be many more "unit factors" than chromosomes in most organisms. Soon thereafter, genetics investigations revealed that certain genes segregate as if they were somehow joined or linked together. Further investigations showed that such genes are part of the same chromosome and may indeed be transmitted as a single unit. We now know that most chromosomes contain a very large number of genes. Those that are part of the same chromosome are said to be *linked* and to demonstrate **linkage** in genetic crosses.

Because the chromosome, not the gene, is the unit of transmission during meiosis, linked genes are not free to undergo independent assortment. Instead, the alleles at all loci of one chromosome should, in theory, be transmitted as a unit during gamete formation. However, in many instances this does not occur. During the first meiotic prophase, when homologs are paired or synapsed, a reciprocal exchange of chromosome segments can take place. This **crossing over** results in the reshuffling, or recombination, of the alleles between homologs and always occurs during the tetrad stage.

The degree of crossing over between any two loci on a single chromosome is proportional to the distance between them. Therefore, depending on which loci are being studied, the percentage of recombinant gametes varies. This correlation allows us to construct chromosome maps, which give the relative locations of genes on chromosomes.

In this chapter, we will discuss linkage, crossing over, and chromosome mapping in more detail. We will conclude by entertaining the rather intriguing question of why Mendel, who studied seven genes in an organism with seven chromosomes, did not encounter linkage. Or did he?

How Do We Know?

In this chapter, we will focus on genes located on the same eukaryotic chromosome, a concept called linkage. We are particularly interested in how such linked genes are transmitted to offspring and how the results of such transmission enable us to map them. As you study this topic, you should try to answer several fundamental questions:

1. How do we know that specific genes are linked on a single chromosome, in contrast to being located on separate chromosomes?

2. How was it determined that linked genes on homologous chromosomes are recombined during meiosis?

3. How do we know the order of, and intergenic distance between, genes found on the same chromosome?

4. In organisms where designed matings do not occur (such as humans), how do we know that genes are linked, and how do we map them?

5. How do we know that crossing over results from a physical exchange between chromatids?

7.1 Genes Linked on the Same Chromosome Segregate Together

A simplified overview of the major theme of this chapter is given in Figure 7–1, which contrasts the meiotic consequences of (a) independent assortment, (b) linkage *without* crossing over, and (c) linkage *with* crossing over. In Figure 7–1(a), we see the results of independent assortment of two pairs of chromosomes, each containing one heterozygous gene pair. No linkage is exhibited. When a large number of meiotic events are observed, four genetically different gametes are formed in equal proportions, and each contains a different combination of alleles of the two genes.

Now let's compare these results with what occurs if the same genes are linked on the same chromosome. If no crossing over occurs between the two genes [Figure 7–1(b)] only two genetically different gametes are formed. Each gamete receives the alleles present on one homolog or the other, which is transmitted intact as the result of segregation. This case demonstrates *complete linkage*, which produces only **parental** or **noncrossover gametes**. The two parental gametes are formed in equal proportions. Though complete linkage between two genes seldom occurs, it is useful to consider the theoretical consequences of this concept.

Figure 7–1(c) shows the results of crossing over between two linked genes. As you can see, this crossover involves only two nonsister chromatids of the four chromatids present in the tetrad. This exchange generates two new allele combinations, called **recombinant** or **crossover gametes**. The two chromatids not involved in the exchange result in noncrossover gametes, like those in Figure 7–1(b). The frequency with which crossing over occurs between any two linked genes is generally proportional to the distance separating the respective loci along the chromosome. In theory, two randomly selected genes can be so close to each other that crossover events are too infrequent to easily detect. As shown in Figure 7–1(b), this complete linkage produces only parental gametes. On the other hand, if a small but distinct distance separates two genes, few recombinant and many parental gametes will be formed. As the distance between the two genes increases, the proportion of recombinant gametes increases and that of the parental gametes decreases.

As we will discuss later in this chapter, when the loci of two linked genes are far apart, the number of recombinant gametes approaches, but does not exceed, 50 percent. If 50 percent recombinants occur, the result is a 1:1:1:1 ratio of

Now Solve This

Problem 9 on page 160 asks you to contrast the results of a testcross when two genes are unlinked versus linked, and when they are linked, if they are very far apart or relatively close together.

Hint: The results are indistinguishable when two genes are unlinked compared to the case where they are linked but so far apart that crossing over always intervenes between them during meiosis.

**(a)   Independent assortment: Two genes on
two different homologous pairs of chromosomes**

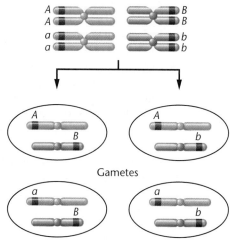

Gametes

**(b)   Linkage: Two genes on a single pair
of homologs; no exchange occurs**

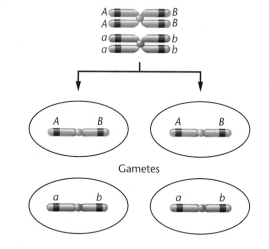

Gametes

**(c)   Linkage: Two genes on a single pair
of homologs; exchange occurs between
two nonsister chromatids**

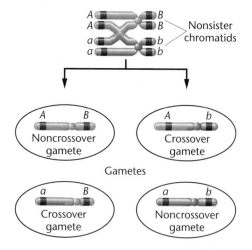

Gametes

FIGURE 7–1 Results of gamete formation where two heterozygous genes are (a) on two different pairs of chromosomes; (b) on the same pair of homologs, but where no exchange occurs between them; and (c) on the same pair of homologs, where an exchange occurs between two nonsister chromatids.

the four types (two parental and two recombinant gametes). In this case, transmission of two linked genes is indistinguishable from that of two unlinked, independently assorting genes. That is, the proportion of the four possible genotypes is identical, as shown in Figures 7–1(a) and (c).

The Linkage Ratio

If complete linkage exists between two genes because of their close proximity, and organisms heterozygous at both loci are mated, a unique F_2 phenotypic ratio results, which we designate the **linkage ratio**. To illustrate this ratio, let's consider a cross involving the closely linked, recessive, mutant genes *brown* eye (*bw*) and *heavy* wing vein (*hv*) in *Drosophila melanogaster* (Figure 7–2). The normal, wild-type alleles bw^+ and hv^+ are both dominant and result in red eyes and thin wing veins, respectively.

In this cross, flies with mutant brown eyes and normal thin wing veins are mated to flies with normal red eyes and mutant heavy wing veins. In more concise terms, brown-eyed flies are crossed with heavy-veined flies. If we extend the system of genetic symbols established in Chapter 4, linked genes are represented by placing their allele designations above and below a single or double horizontal line. Those above the line are located at loci on one homolog, and those below are located at the homologous loci on the other homolog. Thus, we represent the P_1 generation as follows:

$$P_1: \frac{bw\ hv^+}{bw\ hv^+} \times \frac{bw^+\ hv}{bw^+\ hv}$$

brown, thin   red, heavy

These genes are located on an autosome, so no distinction between males and females is necessary.

In the F_1 generation, each fly receives one chromosome of each pair from each parent. All flies are heterozygous for both gene pairs and exhibit the dominant traits of red eyes and thin wing veins:

$$F_1: \frac{bw\ \ hv^+}{bw^+\ hv}$$

red, thin

As shown in Figure 7–2(a), when the F_1 generation is interbred, each F_1 individual forms only parental gametes because of complete linkage. After fertilization, the F_2 generation is produced in a 1:2:1 phenotypic and genotypic ratio. One-fourth of this generation shows brown eyes and thin wing veins; one-half shows both wild-type traits, namely, red eyes and thin wing veins; and one-fourth shows red eyes and heavy wing veins. In more concise terms, the ratio is 1 brown: 2 wild:1 heavy. Such a 1:2:1 ratio is characteristic of complete linkage. Complete linkage is usually observed only when genes are very close together and the number of progeny is relatively small.

Figure 7–2(b) also gives the results of a testcross with the F_1 flies. Such a cross produces a 1:1 ratio of brown, thin and red, heavy flies. Had the genes controlling these traits been incompletely linked or located on separate autosomes, the testcross would have produced four phenotypes rather than two.

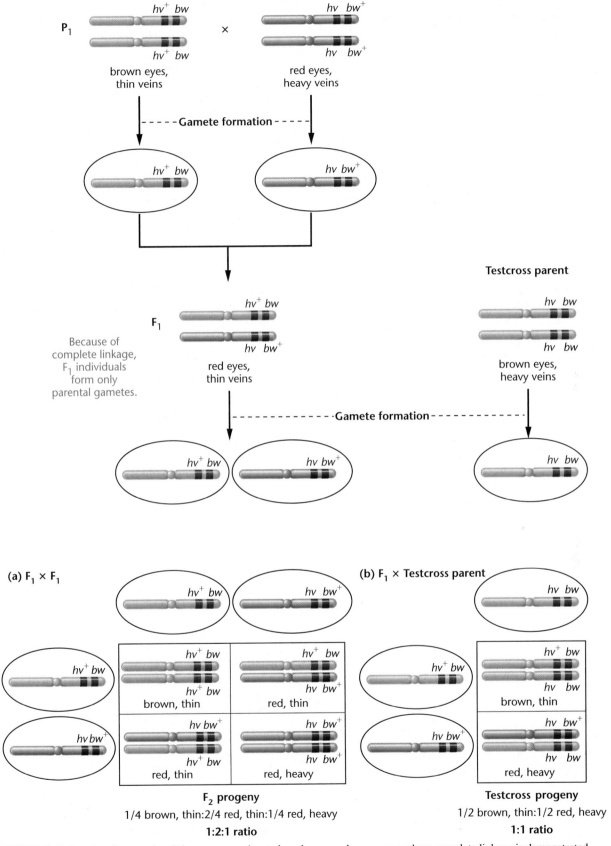

FIGURE 7–2 Results of a cross involving two genes located on the same chromosome where complete linkage is demonstrated. (a) The F$_2$ results of the cross. (b) The results of a testcross involving the F$_1$ progeny.

When large numbers of mutant genes present in any given species are investigated, genes located on the same chromosome show evidence of linkage to one another. As a result, **linkage groups** can be established, one for each chromosome. In theory, the number of linkage groups should correspond to the haploid number of chromosomes. In diploid organisms in which large numbers of mutant genes are available for genetic study, this correlation has been confirmed.

7.2 Crossing Over Serves as the Basis of Determining the Distance Between Genes During Mapping

It is highly improbable that two randomly selected genes linked on the same chromosome will be so close to one another along the chromosome that they demonstrate complete linkage. Instead, crosses involving two such genes almost always produce a percentage of offspring resulting from recombinant gametes. This percentage is variable and depends on the distance between the two genes along the chromosome. This phenomenon was first explained around 1910 by two *Drosophila* geneticists, Thomas H. Morgan and his undergraduate student, Alfred H. Sturtevant.

Morgan, Sturtevant, and Crossing Over

In his studies, Morgan investigated numerous *Drosophila* mutations located on the X chromosome. When he analyzed crosses involving only one trait, he deduced the mode of X-linked inheritance. However, when he made crosses involving two X-linked genes, his results were initially puzzling. For example, female flies expressing the mutant *yellow* body (*y*) and *white* eyes (*w*) alleles were crossed with wild-type males (gray bodies and red eyes). The F_1 females were wild type, while the F_1 males expressed both mutant traits. In the F_2, the vast majority of the offspring showed the expected parental phenotypes—either yellow-bodied, white-eyed flies or wild-type flies (gray-bodied, red-eyed). However, the remaining flies, less than 1.0 percent, were either yellow-bodied with red eyes or gray-bodied with white eyes. It was as if the two mutant alleles had somehow separated from each other on the homolog during gamete formation in the F_1 female flies. This cross is illustrated in cross A of Figure 7–3, using data later compiled by Sturtevant.

When Morgan studied other X-linked genes, the same basic pattern was observed, but the proportion of the unexpected F_2 phenotypes differed. For example, in a cross involving the mutant *white* eye, *miniature* wing (*m*) alleles, the majority of the F_2 again showed the parental phenotypes, but a much higher proportion of the offspring appeared as if the mutant genes had separated during gamete formation. This is illustrated in cross B of Figure 7–3, again using data subsequently compiled by Sturtevant.

Morgan was faced with two questions: (1) What was the source of gene separation, and (2) why did the frequency of the apparent separation vary depending on the genes being studied? The answer he proposed for the first question was based on his knowledge of earlier cytological observations made by F. A. Janssens and others. Janssens had observed that synapsed homologous chromosomes in meiosis wrapped around each other, creating **chiasmata** (sing., *chiasma*) where points of overlap are evident (see the photo on p. 136). Morgan proposed that chiasmata could represent points of genetic exchange.

In the crosses shown in Figure 7–3, Morgan postulated that if an exchange occurs during gamete formation between the mutant genes on the two X chromosomes of the F_1 females, the unique phenotypes will occur. He suggested that such exchanges led to **recombinant gametes** in both the *yellow–white* cross and the *white–miniature* cross, in contrast to the **parental gametes** that have undergone no exchange. On the basis of this and other experiments, Morgan concluded that linked genes exist in a linear order along the chromosome and that a variable amount of exchange occurs between any two genes during gamete formation.

In answer to the second question, Morgan proposed that two genes located relatively close to each other along a chromosome are less likely to have a chiasma form between them than if the two genes are farther apart on the chromosome. Therefore, the closer two genes are, the less likely a genetic exchange will occur between them. Morgan was the first to propose the term *crossing over* to describe the physical exchange leading to recombination.

Sturtevant and Mapping

Morgan's student, Alfred H. Sturtevant, was the first to realize that his mentor's proposal could be used to map the sequence of linked genes. According to Sturtevant,

> In a conversation with Morgan … I suddenly realized that the variations in strength of linkage, already attributed by Morgan to differences in the spatial separation of the genes, offered the possibility of determining sequences in the linear dimension of a chromosome. I went home and spent most of the night (to the neglect of my undergraduate homework) in producing the first chromosomal map.

Sturtevant compiled data from numerous crosses made by Morgan and other geneticists involving recombination between the genes represented by the *yellow*, *white*, and *miniature* mutants. These data are shown in Figure 7–3. The following recombination between each pair of these three genes, published in Sturtevant's paper in 1913, is as follows:

(1) *yellow–white*	0.5%
(2) *white–miniature*	34.5%
(3) *yellow–miniature*	35.4%

Because the sum of (1) and (2) approximately equals (3), Sturtevant suggested that the recombination frequencies between linked genes are additive. On this basis, he predicted that the order of the genes on the X chromosome is *yellow–white–miniature*. In arriving at this conclusion, he reasoned as follows: The *yellow* and *white* genes are apparently close to each other because the recombination frequency is

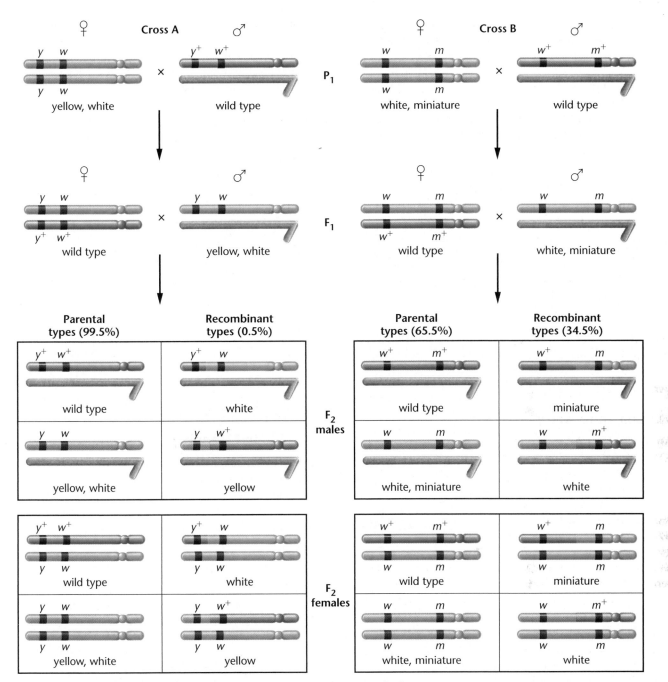

FIGURE 7–3 Data representing the F$_1$ and F$_2$ results of crosses involving the *yellow* body (*y*), *white* eye (*w*) mutations (cross A), and the *white* eye, *miniature* wing (*m*) mutations (cross B), as compiled by Sturtevant. In cross A, 0.5 percent of the F$_2$ flies (males and females) demonstrate recombinant phenotypes, which express either *white* or *yellow*. In cross B, 34.5 percent of the F$_2$ flies (males and females) demonstrate recombinant phenotypes, which express either *miniature* or *white*.

low. However, both of these genes are much farther apart from *miniature* gene because the *white–miniature* and *yellow–miniature* combinations show larger recombination frequencies. Because *miniature* shows more recombination with *yellow* than with *white* (35.4% versus 34.5%), it follows that *white* is located between the other two genes, not outside of them.

Sturtevant knew from Morgan's work that the frequency of exchange could be used as an estimate of the distance between two genes or loci along the chromosome. He constructed a **chromosome map** of the three genes on the X

chromosome, setting 1 map unit (mu) equal to 1 percent recombination between two genes.* In the preceding example, the distance between *yellow* and *white* is thus 0.5 mu, and the distance between *yellow* and *miniature* is 35.4 mu. It follows that the distance between *white* and *miniature* should be $35.4 - 0.5 = 34.9$ mu. This estimate is close to the actual frequency of recombination between *white* and *miniature* (34.5%). The map for these three genes is shown

---

*In honor of Morgan's work, 1 map unit is now referred to as a centimorgan (*cM*).

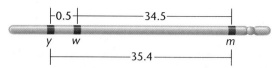

FIGURE 7–4 A map of the *yellow* body (*y*), *white* eye (*w*), and *miniature* wing (*m*) genes on the X chromosome of *Drosophila melanogaster*. Each number represents the percentage of recombinant offspring produced in one of three crosses, each involving two different genes.

in Figure 7–4. The fact that these do not add up perfectly is due to the imprecision of independently-conducted mapping experiments.

In addition to these three genes, Sturtevant considered two other genes on the X chromosome and produced a more extensive map that included all five genes. He and a colleague, Calvin Bridges, soon began a search for autosomal linkage in *Drosophila*. By 1923, they had clearly shown that linkage and crossing over are not restricted to X-linked genes but can also be demonstrated with autosomes. During this work, they made another interesting observation. Crossing over in *Drosophila* was shown to occur only in females. The fact that no crossing over occurs in males made genetic mapping much less complex to analyze in *Drosophila*. However, crossing over does occur in both sexes in most other organisms.

Although many refinements in chromosome mapping have been developed since Sturtevant's initial work, his basic principles are considered to be correct. These principles are used to produce detailed chromosome maps of organisms for which large numbers of linked mutant genes are known. Sturtevant's findings are also historically significant to the broader field of genetics. In 1910, the **chromosomal theory**

of inheritance was still widely disputed—even Morgan was skeptical of this theory before he conducted his experiments. Research has now firmly established that chromosomes contain genes in a linear order and that these genes are the equivalent of Mendel's unit factors.

Single Crossovers

Why should the relative distance between two loci influence the amount of recombination and crossing over observed between them? During meiosis, a limited number of crossover events occur in each tetrad. These recombinant events occur randomly along the length of the tetrad. Therefore, the closer two loci reside along the axis of the chromosome, the less likely any single crossover event will occur between them. The same reasoning suggests that the farther apart two linked loci, the more likely a random crossover event will occur between them.

In Figure 7–5(a), a single crossover occurs between two nonsister chromatids but not between the two loci; therefore, the crossover is not detected because no recombinant gametes are produced. In Figure 7–5(b), where two loci are quite far apart, crossover does occur between them, yielding recombinant gametes.

When a single crossover occurs between two nonsister chromatids, the other two chromatids of the tetrad are not involved in this exchange and enter the gamete unchanged. Even if a single crossover occurs 100 percent of the time between two linked genes, recombination is subsequently observed in only 50 percent of the potential gametes formed. This concept is diagrammed in Figure 7–6. Theoretically, if we consider only single exchanges and observe 20 percent recombinant gametes, crossing over actually occurred in 40 percent of the

(a)

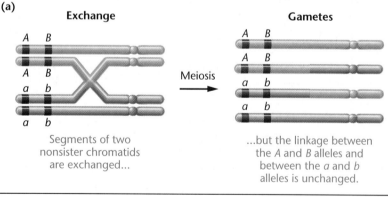

Segments of two nonsister chromatids are exchanged...

...but the linkage between the *A* and *B* alleles and between the *a* and *b* alleles is unchanged.

(b)

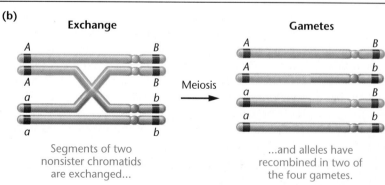

Segments of two nonsister chromatids are exchanged...

...and alleles have recombined in two of the four gametes.

FIGURE 7–5 Two examples of a single crossover between two nonsister chromatids and the gametes subsequently produced. In (a), the exchange does not alter the linkage arrangement between the alleles of the two genes, only parental gametes form, and the exchange goes undetected. In (b), the exchange separates the alleles and results in recombinant gametes, which are detectable.

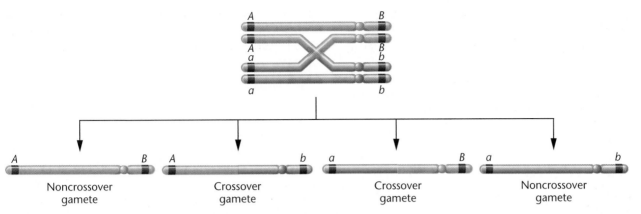

FIGURE 7–6 The consequences of a single exchange between two nonsister chromatids occurring in the tetrad stage. Two noncrossover (parental) and two crossover (recombinant) gametes are produced.

tetrads. Under these conditions, the general rule is that the percentage of tetrads involved in an exchange between two genes is twice the percentage of recombinant gametes produced. Therefore, the theoretical limit of observed recombination due to crossing over is 50 percent.

When two linked genes are more than 50 mu apart, a crossover can theoretically be expected to occur between them in 100 percent of the tetrads. If this prediction were achieved, each tetrad would yield equal proportions of the four gametes shown in Figure 7–6, just as if the genes were on different chromosomes and assorting independently. However, this theoretical limit is seldom achieved.

7.3 Determining the Gene Sequence During Mapping Relies on the Analysis of Multiple Crossovers

The study of single crossovers between two linked genes provides the basis of determining the *distance* between them. However, when many linked genes are studied, their *sequence* along the chromosome is more difficult to determine. Fortunately, the discovery that multiple exchanges occur between

the chromatids of a tetrad has facilitated the process of producing more extensive chromosome maps. As we shall see next, when three or more linked genes are investigated simultaneously, it is possible to determine first the sequence of the genes and then the distances between them.

Multiple Crossovers

It is possible that in a single tetrad, two, three, or more exchanges will occur between nonsister chromatids as a result of several crossover events. Double exchanges of genetic material result from **double crossovers (DCOs)**, as shown in Figure 7–7. For a double exchange to be studied, three gene pairs must be investigated, each heterozygous for two alleles. Before we determine the frequency of recombination among all three loci, let's review some simple probability calculations.

As we have seen, the probability of a single exchange occurring between the *A* and *B* or the *B* and *C* genes relates directly to the distance between the respective loci. The closer *A* is to *B* and *B* is to *C*, the less likely a single exchange will occur between either of the two sets of loci. In the case of a double crossover, two separate and independent events or exchanges must occur simultaneously. The mathematical probability of

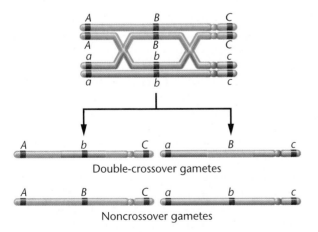

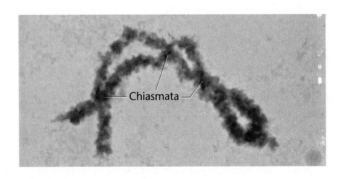

FIGURE 7–7 Consequences of a double exchange between two nonsister chromatids. Because the exchanges involve only two chromatids, two noncrossover gametes and two double-crossover gametes are produced. The photograph shows several chiasmata found in a tetrad isolated during the first meiotic prophase. See also the chapter opening photograph on p.136.

two independent events occurring simultaneously is equal to the product of the individual probabilities (the product law).

Suppose that crossover gametes resulting from single exchanges are recovered 20 percent of the time ($p = 0.20$) between *A* and *B*, and 30 percent of the time ($p = 0.30$) between *B* and *C*. The probability of recovering a double-crossover gamete arising from two exchanges (between *A* and *B*, and between *B* and *C*) is predicted to be $(0.20)(0.30) = 0.06$, or 6 percent. It is apparent from this calculation that the frequency of double-crossover gametes is always expected to be much lower than that of either single-crossover class of gametes.

If three genes are relatively close together along one chromosome, the expected frequency of double-crossover gametes is extremely low. For example, suppose the *A–B* distance in Figure 7–7 is 3 mu and the *B–C* distance is 2 mu. The expected double-crossover frequency is $(0.03)(0.02) = 0.0006$, or 0.06 percent. This translates to only 6 events in 10,000. Thus, in a mapping experiment where closely linked genes are involved, very large numbers of offspring are required to detect double-crossover events. In this example, it is unlikely that a double crossover will be observed even if 1000 offspring are examined. Thus, it is evident that if four or five genes are being mapped, even fewer triple and quadruple crossovers can be expected to occur.

Now Solve This

Problem 14 on page 161 asks you to contrast the results of crossing over when the arrangement of alleles along the homologs differs in an organism heterozygous for three genes.

Hint: Homologs enter noncrossover gametes unchanged, and all crossover results must be derived from the noncrossover sequence of alleles.

Three-Point Mapping in *Drosophila*

The information in the preceding section enables us to map three or more linked genes in a single cross. To illustrate the mapping process in its entirety, we examine two situations involving three linked genes in two quite different organisms.

To execute a successful mapping cross, three criteria must be met:

1. The genotype of the organism producing the crossover gametes must be heterozygous at all loci under consideration.

2. The cross must be constructed so that genotypes of all gametes can be determined accurately by observing the phenotypes of the resulting offspring. This is necessary because the gametes and their genotypes can never be observed directly. To overcome this problem, each phenotypic class must reflect the genotype of the gametes of the parents producing it.

3. A sufficient number of offspring must be produced in the mapping experiment to recover a representative sample of all crossover classes.

These criteria are met in the three-point mapping cross from *D. melanogaster* shown in Figure 7–8. In this cross, three X-linked recessive mutant genes—*yellow* body color (*y*), *white* eye color (*w*), and *echinus* eye shape (*ec*)—are considered. To diagram the cross, *we must assume some theoretical sequence, even though we do not yet know if it is correct.* In Figure 7–8, we initially assume the sequence of the three genes to be *y–w–ec*. If this is incorrect, our analysis will demonstrate this and reveal the correct sequence.

In the P$_1$ generation, males hemizygous for all three wild-type alleles are crossed to females that are homozygous for all three recessive mutant alleles. Therefore, the P$_1$ males are wild type with respect to body color, eye color, and eye shape. They are said to have a *wild-type phenotype*. The females, on the other hand, exhibit the three mutant traits—yellow body color, white eyes, and echinus eye shape.

This cross produces an F$_1$ generation consisting of females that are heterozygous at all three loci and males that, because of the Y chromosome, are hemizygous for the three mutant alleles. Phenotypically, all F$_1$ females are wild type, while all F$_1$ males are yellow, white, and echinus. The genotype of the F$_1$ females fulfills the first criterion for mapping the three linked genes; that is, it is heterozygous at the three loci and can serve as the source of recombinant gametes generated by crossing over. Note that because of the genotypes of the P$_1$ parents, all three mutant alleles are on one homolog and all three wild-type alleles are on the other homolog. *Other arrangements are possible.* For example, the heterozygous F$_1$ female might have the *y* and *ec* mutant alleles on one homolog and the *w* allele on the other. This would occur if, in the P$_1$ cross, one parent was yellow, echinus and the other parent was white.

In our cross, the second criterion is met by virtue of the gametes formed by the F$_1$ males. Every gamete contains either an X chromosome bearing the three mutant alleles or a Y chromosome, which is genetically inert for the three loci being considered. Whichever type participates in fertilization, the genotype of the gamete produced by the F$_1$ female will be expressed phenotypically in the F$_2$ male and female offspring derived from it. Thus, all F$_1$ noncrossover and crossover gametes can be detected by observing the F$_2$ phenotypes.

With these two criteria met, we can now construct a chromosome map from the crosses shown in Figure 7–8. First, we determine which F$_2$ phenotypes correspond to the various noncrossover and crossover categories. To determine the noncrossover F$_2$ phenotypes, we must identify individuals derived from the parental gametes formed by the F$_1$ female. Each such gamete contains an X chromosome *unaffected by crossing over*. As a result of segregation, approximately equal proportions of the two types of gametes and, subsequently, the F$_2$ phenotypes, are produced. Because they derive from a heterozygote, the genotypes of the two parental gametes and the resultant F$_2$ phenotypes complement one another. For

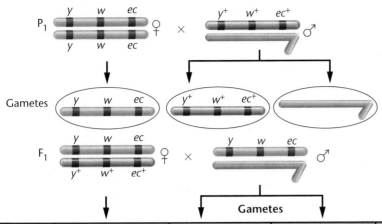

Origin of female gametes	Gametes			F₂ phenotype	Observed Number	Category, total, and percentage
NCO	① y w ec	y w ec / y w ec	y w ec	y w ec	4685	Non-crossover
	② y⁺ w⁺ ec⁺	y⁺ w⁺ ec⁺ / y w ec	y⁺ w⁺ ec⁺	y⁺ w⁺ ec⁺	4759	9444 94.44%
SCO	③ y w⁺ ec⁺	y w⁺ ec⁺ / y w ec	y w⁺ ec⁺	y w⁺ ec⁺	80	Single crossover between y and w
	④ y⁺ w ec	y⁺ w ec / y w ec	y⁺ w ec	y⁺ w ec	70	150 1.50%
SCO	⑤ y w ec⁺	y w ec⁺ / y w ec	y w ec⁺	y w ec⁺	193	Single crossover between w and ec
	⑥ y⁺ w⁺ ec	y⁺ w⁺ ec / y w ec	y⁺ w⁺ ec	y⁺ w⁺ ec	207	400 4.00%
DCO	⑦ y w⁺ ec	y w⁺ ec / y w ec	y w⁺ ec	y w⁺ ec	3	Double crossover between y and w and between w and ec
	⑧ y⁺ w ec⁺	y⁺ w ec⁺ / y w ec	y⁺ w ec⁺	y⁺ w ec⁺	3	6 0.06%

Map of y, w, and ec loci
⊢1.56⊣ 4.06 ⊣

FIGURE 7–8 A three-point mapping cross involving the *yellow* (*y* or *y⁺*), *white* (*w* or *w⁺*), and *echinus* (*ec* or *ec⁺*) genes in *D. melanogaster*. NCO, SCO, and DCO refer to noncrossover, single-crossover, and double-crossover groups, respectively. Centromeres are not included on the chromosomes, and only two nonsister chromatids are shown initially.

example, if one is wild type, the other is completely mutant. This is the case in the cross being considered. In other situations, if one chromosome shows one mutant allele, the second chromosome shows the other two mutant alleles, and so on. They are therefore called **reciprocal classes** of gametes and phenotypes.

The two noncrossover phenotypes are most easily recognized because *they exist in the greatest proportion*. Figure 7–8 shows that gametes 1 and 2 are present in the greatest numbers. Therefore, flies that express yellow, white, and echinus phenotypes and flies that are normal (or wild type) for all three characters constitute the noncrossover category and represent 94.44 percent of the F_2 offspring.

The second category that can be easily detected is represented by the double-crossover phenotypes. Because of their low probability of occurrence, *they must be present in the least numbers*. Remember that this group represents two independent but simultaneous single-crossover events. Two reciprocal phenotypes can be identified: gamete 7, which shows the mutant traits yellow, echinus but normal eye color; and gamete 8, which shows the mutant trait white but normal body color and eye shape. Together these double-crossover phenotypes constitute only 0.06 percent of the F_2 offspring.

The remaining four phenotypic classes represent two categories resulting from single crossovers. Gametes 3 and 4, reciprocal phenotypes produced by single-crossover events occurring between the *yellow* and *white* loci, are equal to 1.50 percent of the F_2 offspring; gametes 5 and 6, constituting 4.00 percent of the F_2 offspring, represent the reciprocal phenotypes resulting from single-crossover events occurring between the *white* and *echinus* loci.

The map distances separating the three loci can now be calculated. The distance between *y* and *w* or between *w* and *ec* is equal to the percentage of all detectable exchanges occurring between them. For any two genes under consideration, this includes all appropriate single crossovers as well as all double crossovers. *The latter are included because they represent two simultaneous single crossovers.* For the *y* and *w* genes, this includes gametes 3, 4, 7, and 8, totaling 1.50% + 0.06%, or 1.56 mu. Similarly, the distance between *w* and *ec* is equal to the percentage of offspring resulting from an exchange between these two loci: gametes 5, 6, 7, and 8, totaling 4.00% + 0.06%, or 4.06 mu. The map of these three loci on the X chromosome is shown at the bottom of Figure 7–8.

Determining the Gene Sequence

In the preceding example, the sequence (or order) of the three genes along the chromosome was assumed to be *y–w–ec*. Our analysis shows this sequence to be consistent with the data. However, in most mapping experiments the gene sequence is not known, and this constitutes another variable in the analysis. In our example, had the gene sequence been unknown, it could have been determined using a straightforward method.

This method is based on the fact that there are only three possible arrangements, each containing one of the three genes between the other two:

(I)	*w–y–ec*	(*y* in the middle)
(II)	*y–ec–w*	(*ec* in the middle)
(III)	*y–w–ec*	(*w* in the middle)

Use the following steps during your analysis to determine the gene order:

1. Assuming any one of the three orders, first determine the *arrangement of alleles* along each homolog of the heterozygous parent giving rise to noncrossover and crossover gametes (the F_1 female in our example).

2. Determine whether a double-crossover event occurring within that arrangement will produce the *observed double-crossover phenotypes*. Remember that these phenotypes occur least frequently and are easily identified.

3. If this order does not produce the predicted phenotypes, try each of the other two orders. One must work!

These steps are shown in Figure 7–9, using our *y–w–ec* cross. The three possible arrangements are labeled I, II, and III, as shown above.

1. Assuming that *y* is between *w* and *ec*, arrangement I of alleles along the homologs of the F_1 heterozygote is

$$\frac{w \quad y \quad ec}{w^+ \quad y^+ \quad ec^+}$$

We know this because of the way in which the P_1 generation was crossed: The P_1 female contributes an X chromosome bearing the *w*, *y*, and *ec* alleles, while the P_1 male contributes an X chromosome bearing the w^+, y^+, and ec^+ alleles.

2. A double crossover within that arrangement yields the following gametes:

$$\frac{w \quad y^+ \quad ec}{} \quad \text{and} \quad \frac{w^+ \quad y \quad ec^+}{}$$

Following fertilization, if *y* is in the middle, the F_2 double-crossover phenotypes will correspond to these gametic genotypes, yielding offspring that express the white, echinus phenotype and offspring that express the yellow phenotype. Instead, however, determination of the actual double-crossover phenotypes reveals them to be yellow, echinus flies and white flies. *Therefore, our assumed order is incorrect.*

3. If we consider arrangement II with the ec/ec^+ alleles in the middle or arrangement III with the w/w^+ alleles in the middle:

$$\text{(II)} \frac{y \quad ec \quad w}{y^+ \quad ec^+ \quad w^+} \quad \text{or} \quad \text{(III)} \frac{y \quad w \quad ec}{y^+ \quad w^+ \quad ec^+}$$

we see that arrangement II again provides *predicted* double-crossover phenotypes that *do not* correspond to the *actual* (observed) double-crossover phenotypes. The pre-

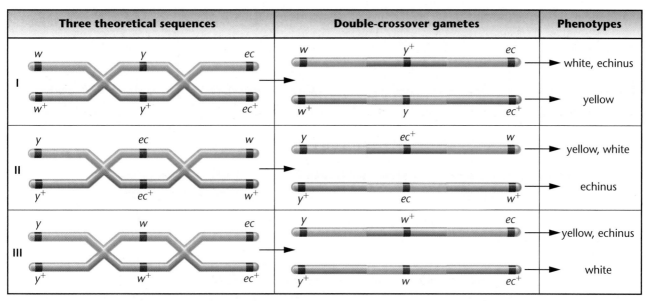

FIGURE 7–9 The three possible sequences of the *white*, *yellow*, and *echinus* genes, the results of a double crossover in each case, and the resulting phenotypes produced in a testcross. For simplicity, the two noncrossover chromatids of each tetrad are omitted.

dicted phenotypes are yellow, white flies and echinus flies in the F$_2$ generation. *Therefore, this order is also incorrect.* However, arrangement III produces the observed phenotypes—yellow, echinus flies and white flies. *Therefore, this arrangement, with the w gene in the middle, is correct.*

To summarize, this method is rather straightforward: First determine the arrangement of alleles on the homologs of the heterozygote yielding the crossover gametes by locating the reciprocal noncrossover phenotypes. Then, test each of three possible orders to determine which yields the observed double-crossover phenotypes—*the one that does so represents the correct order.*

A Mapping Problem in Maize

Having established the basic principles of chromosome mapping, we now consider a related problem in maize (corn), in which the gene sequence and interlocus distances are unknown.

This analysis differs from the preceding example in two ways. First, the previous mapping cross involved X-linked genes. Here, we consider autosomal genes. Second, in the discussion of this cross we have changed our use of symbols, as first suggested in Chapter 4. Instead of using the gene symbols and superscripts (e.g., bm^+, v^+, and pr^+), we simply use + to denote each wild-type allele. This system is easier to manipulate but requires a better understanding of mapping procedures.

When we look at three autosomally linked genes in maize, the experimental cross must still meet the same three criteria we established for the X-linked genes in *Drosophila*: (1) One parent must be heterozygous for all traits under consideration; (2) the gametic genotypes produced by the heterozygote must be apparent from observing the phenotypes of the offspring; and (3) a sufficient sample size must be available for complete analysis.

In maize, the recessive mutant genes *brown* midrib (*bm*), *virescent* seedling (*v*), and *purple* aleurone (*pr*) are linked on

chromosome 5. Assume that a female plant is known to be heterozygous for all three traits, but we do not know (1) the arrangement of the mutant alleles on the maternal and paternal homologs of this heterozygote, (2) the sequence of genes, or (3) the map distances between the genes. What genotype must the male plant have to allow successful mapping? To meet the second criterion, the male must be homozygous for all three recessive mutant alleles. Otherwise, offspring of this cross showing a given phenotype might represent more than one genotype, making accurate mapping impossible.

Figure 7–10 diagrams this cross. As shown, we know neither the arrangement of alleles nor the sequence of loci in the heterozygous female. Several possibilities are shown, but we have yet to determine which is correct. We don't know the sequence in the testcross male parent either, and so we must designate it randomly. Note that we have initially placed *v* in the middle. *This may or may not be correct.*

The offspring are arranged in groups of two for each pair of reciprocal phenotypic classes. The two members of each reciprocal class are derived from no crossing over (NCO), one of two possible single-crossover events (SCO), or a double crossover (DCO).

To solve this problem, refer to Figures 7–10 and 7–11 as you consider the following questions.

1. *What is the correct heterozygous arrangement of alleles in the female parent?* Determine the two noncrossover classes, those that occur with the highest frequency. In this case, they are $+ v\ bm$ and $pr+ +$. Therefore, the alleles on the homologs of the female parent must be arranged as shown in Figure 7–11(a). These homologs segregate into gametes, unaffected by any recombination event. Any other arrangement of alleles will not yield the observed noncrossover classes. (Remember that $+ v\ bm$ is equivalent to $\underline{pr^+\ v\ bm}$, and that $pr+ +$ is equivalent to $\underline{pr\ v^+\ bm^+}$).

(a) Some possible allele arrangements and gene sequences in a heterozygous female

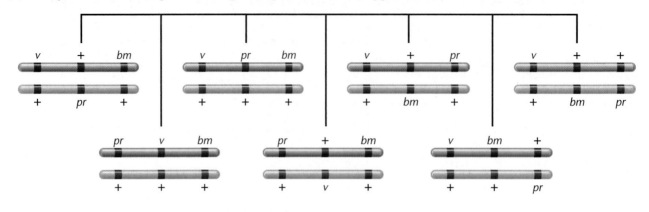

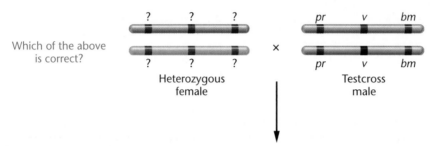

Which of the above is correct?

Heterozygous female × Testcross male

(b) Actual results of mapping cross*

Phenotypes of offspring	Number	Total and percentage	Exchange classification
+ v bm	230	467 42.1%	Noncrossover (NCO)
pr + +	237		
+ + bm	82	161 14.5%	Single crossover (SCO)
pr v +	79		
+ v +	200	395 35.6%	Single crossover (SCO)
pr + bm	195		
pr v bm	44	86 7.8%	Double crossover (DCO)
+ + +	42		

* The sequence *pr – v – bm* may or may not be correct.

FIGURE 7–10 (a) Some possible allele arrangements and gene sequences in a heterozygous female. The data from a three-point mapping cross depicted in (b), where the female is testcrossed, determine which combination of arrangement and sequence is correct [see Figure 7–11(d)].

2. *What is the correct sequence of genes?* We know that the arrangement of alleles is

$$\frac{+\quad v\quad bm}{pr\quad +\quad +}$$

But is the gene sequence correct? That is, will a double-crossover event yield the observed double-crossover phenotypes after fertilization? *Observation shows that it will not* [Figure 7–11(b)]. Now try the other two orders [Figures 7–11(c) and (d)] *maintaining the same arrangement*:

$$\frac{+\quad bm\quad v}{pr\quad +\quad +}\quad \text{or} \quad \frac{v\quad +\quad bm}{+\quad pr\quad +}$$

Only the order on the right yields the observed double-crossover gametes [Figure 7–11(d)]. Therefore, the *pr* gene is in the middle. From this point on, work the problem using this arrangement and sequence, with the *pr* locus in the middle.

3. *What is the distance between each pair of genes?* Having established the sequence of loci as *v–pr–bm*, we can determine the distance between *v* and *pr* and between *pr*

Allele arrangement and sequence	Testcross phenotypes	Explanation
(a) + v bm pr + +	+ v bm and pr + +	Noncrossover phenotypes provide the basis of determining the correct arrangement of alleles on homologs
(b) + v bm pr + +	+ + bm and pr v +	Expected double-crossover phenotypes if *v* is in the middle
(c) + bm v pr + +	+ + v and pr bm +	Expected double-crossover phenotypes if *bm* is in the middle
(d) v + bm + pr +	v pr bm and + + +	Expected double-crossover phenotypes if *pr* is in the middle (This is the *actual situation*.)
(e) v + bm + pr +	v pr + and + + bm	Given that (a) and (d) are correct, single-crossover phenotypes when exchange occurs between *v* and *pr*
(f) v + bm + pr +	v + + and + pr bm	Given that (a) and (d) are correct, single-crossover phenotypes when exchange occurs between *pr* and *bm*

(g) Final map

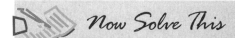

FIGURE 7–11 Producing a map of the three genes in the testcross of Figure 7–10, where neither the arrangement of alleles nor the sequence of genes in the heterozygous female parent is known.

and *bm*. Remember that the map distance between two genes is calculated on the basis of all detectable recombination events occurring between them. This includes both single- and double-crossover events.

Figure 7–11(e) shows that the phenotypes *v pr +* and *+ +bm* result from single crossovers between the *v* and *pr* loci, accounting for 14.5 percent of the offspring [according to data in Figure 10(b)]. By adding the percentage of double crossovers (7.8%) to the number obtained for single crossovers, the total distance between the *v* and *pr* loci is calculated to be 22.3 mu.

Figure 7–11(f) shows that the phenotypes *v + +* and *+ pr bm* result from single crossovers between the *pr* and *bm* loci, totaling 35.6 percent. Added to the double crossovers (7.8%), the distance between *pr* and *bm* is calculated to be 43.4 mu. The final map for all three genes in this example is shown in Figure 7–11(g).

Now Solve This

Problem 20 on page 162 asks you to solve a three-point mapping problem where only six phenotypic categories are observed, when eight categories are typical of such a cross.

Hint: If the distances between each pair of genes is relatively small, reciprocal pairs of single crossovers will be evident, but the sample size may be too small to recover the predicted number of double crossovers, excluding their appearance. You should write the missing gametes down, assign them as double crossovers, and record zeroes for their frequency of appearance.

7.4   As the Distance Between Two Genes Increases, Mapping Estimates Become More Inaccurate

So far, we have assumed that crossover frequencies are directly proportional to the distance between any two loci along the chromosome. However, it is not always possible to detect all crossover events. A case in point is a double exchange that occurs between the two loci in question. As shown in Figure 7–12(a), if a double exchange occurs, the original arrangement of alleles on each nonsister homolog is recovered. Therefore, even though crossing over has occurred, it is impossible to detect. This phenomenon is true for all even-numbered exchanges between two loci.

Furthermore, as a result of complications posed by **multiple-strand exchanges**, mapping determinations usually underestimate the actual distance between two genes. The farther apart two genes are, the greater the probability that undetected crossovers will occur. While the discrepancy is minimal for two genes relatively close together, the degree of inaccuracy increases as the distance increases, as shown in the graph of recombination frequency versus map distance in Figure 7–12(b). There, the theoretical frequency where a direct correlation between recombination and map distance exists is contrasted with the actual frequency observed as the distance between two genes increases. The most accurate maps are constructed from experiments where genes are relatively close together.

Interference and the Coefficient of Coincidence

As shown in our maize example, we can predict the expected frequency of multiple exchanges, such as double crossovers, once the distance between genes is established. For example,

in the maize cross, the distance between v and pr is 22.3 mu, and the distance between pr and bm is 43.4 mu. If the two single crossovers that make up a double crossover occur independently of one another, we can calculate the expected frequency of double crossovers (DCO_{exp}):

$$DCO_{exp} = (0.223) \times (0.434) = 0.097 = 9.7\%$$

Often in mapping experiments, the observed DCO frequency is less than the expected number of DCOs. In the maize cross, for example, only 7.8 percent DCOs are observed when 9.7 percent are expected. **Interference** (**I**), the phenomenon where a crossover event in one region of the chromosome inhibits a second event in nearby regions, causes this reduction.

To quantify the disparities that result from interference, we calculate the **coefficient of coincidence (C)**:

$$C = \frac{\text{Observed DCO}}{\text{Expected DCO}}$$

In the maize cross, we have

$$C = \frac{0.078}{0.097} = 0.804$$

Once we have found C, we can quantify interference using the simple equation

$$I = 1 - C$$

In the maize cross, we have

$$I = 1.000 - 0.804 = 0.196$$

If interference is complete and no double crossovers occur, then $I = 1.0$. If fewer DCOs than expected occur, I is a posi-

(a) Two-strand double exchange

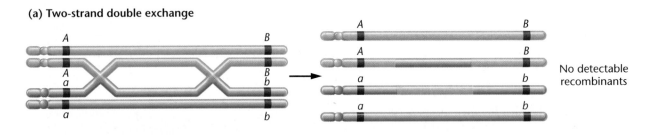

No detectable recombinants

(b)

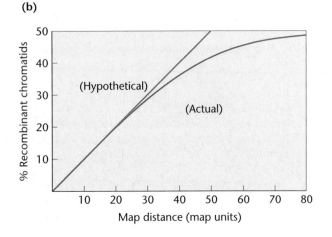

% Recombinant chromatids

(Hypothetical)

(Actual)

Map distance (map units)

FIGURE 7–12 (a) A double crossover is undetected because no rearrangement of alleles occurs. (b) The theoretical and actual percentage of recombinant chromatids versus map distance. The straight line shows the theoretical relationship if a direct correlation between recombination and map distance exists. The curved line is the actual relationship derived from studies of *Drosophila*, *Neurospora*, and *Zea mays*.

tive number and positive interference has occurred. If more DCOs than expected occur, *I* is a negative number and negative interference has occurred. In the maize example, *I* is a positive number (0.196), indicating that 19.6 percent fewer double crossovers occurred than expected.

Positive interference is most often observed in eukaryotic systems. In general, the closer genes are to one another along the chromosome, the more positive interference occurs. In fact, interference in *Drosophila* is often complete within a distance of 10 mu, and no multiple crossovers are recovered. This observation suggests that physical constraints preventing the formation of closely aligned chiasmata contribute to interference. This interpretation is consistent with the finding that interference decreases as the genes in question are located far-

ther apart. In the maize cross in Figures 7–10 and 7–11, the three genes are relatively far apart and 80 percent of the expected double crossovers are observed.

7.5 *Drosophila* Genes Have Been Extensively Mapped

In organisms such as *Drosophila*, maize, and the mouse, where large numbers of mutations have been discovered and where mapping crosses are possible, extensive chromosome maps have been constructed. Figure 7–13 shows partial maps for the four chromosomes (I, II, III, and IV) of *Drosophila*.

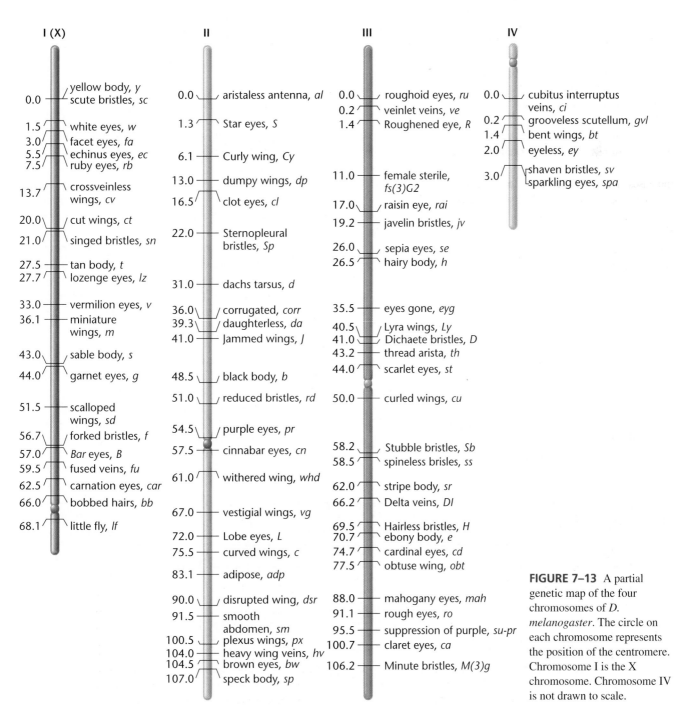

FIGURE 7–13 A partial genetic map of the four chromosomes of *D. melanogaster*. The circle on each chromosome represents the position of the centromere. Chromosome I is the X chromosome. Chromosome IV is not drawn to scale.

Virtually every morphological feature of the fruit fly has been subjected to mutations. The locus of each mutant gene is first localized to one of the four chromosomes (or linkage groups) and then mapped in relation to all other genes present on that chromosome. Based on cytological evidence, the relative lengths of these genetic maps correlate roughly with the relative physical lengths of these chromosomes.

7.6 Lod Score Analysis and Somatic Cell Hybridization Were Historically Important in Creating Human Chromosome Maps

In humans, where neither designed matings nor large numbers of offspring are available, the earliest linkage studies were based on pedigree analysis. Attempts were made to establish whether a trait was X-linked or autosomal. For autosomal traits, geneticists tried to distinguish whether pairs of traits demonstrate linkage or independent assortment. In this way, it was hoped that a human gene map could be created.

The difficulty arises, however, when two genes of interest are separated on a chromosome such that recombinant gametes are formed, obscuring linkage in a pedigree. In such cases, the demonstration of linkage is enhanced by an approach that relies on probability, called the **lod score method**. First devised by J. B. S. Haldane and C. A. Smith in 1947, and refined by Newton Morton in 1955, the lod score (*log* of the *od*ds favoring linkage) assesses the probability that a particular pedigree involving two traits reflects linkage or not. First, the probability is calculated that family data (pedigrees) concerning two or more traits conform to the transmission of traits without linkage. Then the probability is calculated that the identical family data following these same traits result from linkage with a specified recombination frequency. The ratio of these probabilities expresses the "odds" for and against linkage.

Accuracy using the lod score method is limited by the extent of the family data, but nevertheless represents an important advance in assigning human genes to specific chromosomes and constructing preliminary human chromosome maps. The initial results were discouraging because of the method's inherent limitations and the relatively high haploid number of human chromosomes (23), and by 1960, almost no linkage or mapping information had become available.

In the 1960s, a new technique, **somatic cell hybridization**, proved to be an immense aid in assigning human genes to their respective chromosomes. This technique, first discovered by Georges Barsky, relies on the fact that two cells in culture can be induced to fuse into a single hybrid cell. Barsky used two mouse-cell lines, but it soon became evident that cells from different organisms could also be fused. When fusion occurs, an initial cell type called a **heterokaryon** is produced. The hybrid cell contains two nuclei in a common cytoplasm. By using the proper techniques, it is possible to fuse human and mouse cells, for example, and isolate the hybrids from the parental cells.

As the heterokaryons are cultured *in vitro*, two interesting changes occur. The nuclei eventually fuse, creating a **synkaryon**. Then, as culturing is continued for many generations, chromosomes from one of the two parental species are gradually lost. In the case of the human–mouse hybrid, human chromosomes are lost randomly until eventually the synkaryon has a full complement of mouse chromosomes and only a few human chromosomes. As we shall see, it is the preferential loss of human chromosomes (rather than mouse chromosomes) that makes possible the assignment of human genes to the chromosomes on which they reside.

The experimental rationale is straightforward. If a specific human gene product is synthesized in a synkaryon containing one to three human chromosomes, then the gene responsible for that product must reside on one of the human chromosomes remaining in the hybrid cell. On the other hand, if the human gene product is not synthesized in a synkaryon, the gene responsible is not present on those human chromosomes that remain in this hybrid cell. Ideally, a panel of 23 hybrid-cell lines, each with just one unique human chromosome, would allow the immediate assignment of any human gene for which the product could be characterized.

In practice, a panel of cell lines, each with several remaining human chromosomes, is most often used. The correlation of the presence or absence of each chromosome with the presence or absence of each gene product is called **synteny testing**. Consider, for example, the hypothetical data provided in Figure 7–14, where four gene products (A, B, C, and D) are tested in relationship to eight human chromosomes. Let's analyze the gene that produces product A.

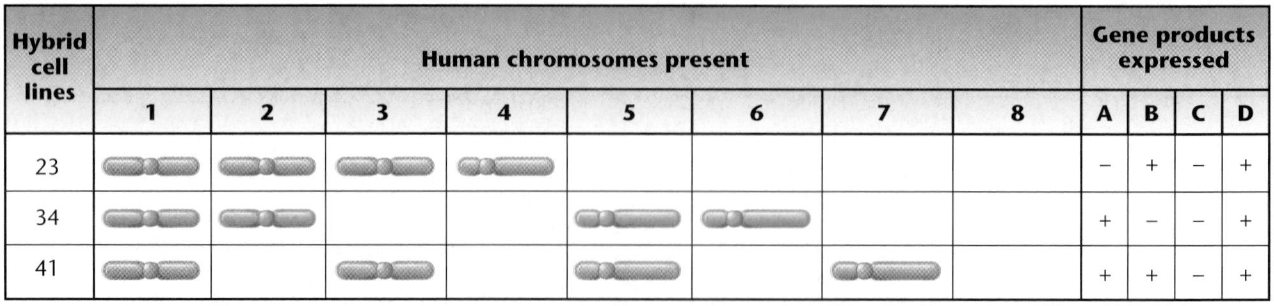

Hybrid cell lines	Human chromosomes present								Gene products expressed			
	1	2	3	4	5	6	7	8	A	B	C	D
23	▭	▭	▭	▭					−	+	−	+
34	▭	▭			▭	▭			+	−	−	+
41	▭		▭		▭		▭		+	+	−	+

FIGURE 7–14 A hypothetical grid of data used in synteny testing to assign genes to their appropriate human chromosomes. Three somatic hybrid-cell lines, designated 23, 34, and 41, have each been scored for the presence or absence of human chromosomes 1–8, as well as for their ability to produce the hypothetical human gene products A, B, C, and D.

1. Product A is not produced by cell line 23, but chromosomes 1, 2, 3, and 4 are present in cell line 23. Therefore, we rule out the presence of gene *A* on those four chromosomes and conclude that it must be on chromosome 5, 6, 7, or 8.

2. Product A is produced by cell line 34, which contains chromosomes 5 and 6 but not 7 and 8. Therefore, gene *A* is on chromosome 5 or 6, but cannot be on 7 or 8 because they are absent even though product A is produced.

3. Product A is also produced by cell line 41, which contains chromosome 5 but not chromosome 6. Using similar reasoning we see that gene *A* must be on chromosome 5.

Using a similar approach, we can assign gene *B* to chromosome 3. Perform this analysis for yourself to demonstrate that this is correct.

Gene *C* presents a unique situation. The data indicate that it is not present on chromosomes 1–7. While it might be on chromosome 8, no direct evidence supports this conclusion, and other panels are needed. We leave gene *D* for you to analyze—on what chromosome does it reside?

Using this technique, researchers have assigned literally hundreds of human genes to one chromosome or another. Some of the assignments shown in Figure 7–15 were either derived or confirmed with the use of somatic cell hybridization techniques. To map genes for which the products have yet

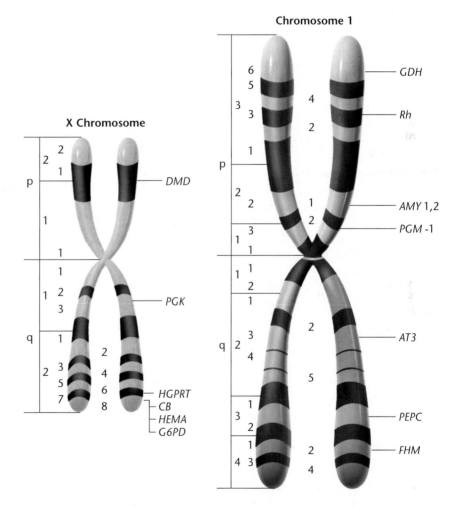

Key

AMY	Amylase (salivary and pancreatic)
AT3	Antithrombin (clotting factor IV)
CB	Color Blindness
DMD	Duchenne Muscular Dystrophy
FHM	Fumarate Hydratase (mitochondrial)
GDH	Glucose Dehydrogenase
G6PD	Glucose-6-Phosphate Dehydrogenase
HEMA	Hemophilia A (classic)
HGPRT	Hypoxanthine-Guanine-Phosphoribosyl Transferase (Lesch–Nyhan syndrome)
PEPC	Peptidase C
PGK	Phosphoglycerate Kinase
PGM	Phosphoglucomutase
Rh	Rhesus Blood Group (erythroblastosis fetalis)

FIGURE 7–15 Representative regional gene assignments for human chromosome 1 and the X chromosome. Many assignments were initially derived using somatic cell hybridization techniques.

to be discovered, researchers have had to rely on other approaches. For example, by using recombinant DNA technology in conjunction with pedigree analysis, it has been possible to assign the genes responsible for **Huntington disease**, **cystic fibrosis**, and **neurofibromatosis** to their respective chromosomes, 4, 7, and 17.

7.7   Linkage and Mapping Studies Can Be Performed in Haploid Organisms

Many of the single-celled eukaryotes are haploid during the vegetative stages of their life cycle. The alga *Chlamydomonas* and the mold *Neurospora* demonstrate this genetic condition. But these organisms do form reproductive cells that fuse during fertilization, producing a diploid zygote. However, this structure soon undergoes meiosis, resulting in haploid vegetative cells that then propagate by mitotic divisions. In genetic studies, small haploid organisms have several important advantages compared with diploid eukaryotes. They can be cultured and manipulated in genetic crosses much more easily. In addition, a haploid organism contains only a single allele of each gene, which is expressed directly in the phenotype. This greatly simplifies genetic analysis. As a result, organisms such as *Chlamydomonas* and *Neurospora* serve as the subjects of research investigations in many areas of genetics, including linkage and mapping studies.

In order to perform genetics experiments with such organisms, crosses are made, and following fertilization, the meiotic structures may be isolated. Because all four meiotic products give rise to spores, the structures bearing these products (asci) are called **tetrads**. *The term* tetrad *has a different meaning here than earlier when it was used to describe a precise chromatid configuration in meiosis.* Individual tetrads are isolated, and the resultant cells are grown and analyzed separately from those of other tetrads. In the results we are about to describe, the data reflect the proportion of tetrads that show one combination of genotypes, the proportion that show another combination, and so on. Such experimentation is called **tetrad analysis**.

Gene-to-Centromere Mapping

When a single gene in *Neurospora* is analyzed (Figure 7–16), the data can be used to calculate the map distance between the gene and the centromere. This process is sometimes referred to as **mapping the centromere**. It is accomplished by experimentally determining the frequency of recombination using tetrad data. Note that once the four meiotic products of the tetrad are formed, a mitotic division occurs, resulting in eight ordered products (ascospores). If no crossover event occurs between the gene under study and the centromere, the pattern of ascospores contained within an ascus (pl., asci) appears as shown in Figure 7–16(a), *aaaa++++*.*

The pattern (++++aaaa*) can also be formed. However, it is indistinguishable from (*aaaa++++*).

This pattern represents **first-division segregation** because the two alleles are separated during the first meiotic division. However, a crossover event will alter this pattern, as shown in Figure 7–16(b), *aa++aa++*, and Figure 7–16(c), *++aaaa++*. Two other recombinant patterns occur but are not shown: *++aa++aa* and *aa++++aa*. These depend on the chromatid orientation during the second meiotic division. These four patterns reflect **second-division segregation** because the two alleles are not separated until the second meiotic division. Usually, the ordered tetrad data are condensed to reflect the genotypes of the identical ascospore pairs, and six combinations are possible.

First-Division Segregation	Second-Division Segregation	
aa++	*a+a+*	*+aa+*
++aa	*+a+a*	*a++a*

To calculate the distance between the gene and the centromere, a large number of asci resulting from a controlled cross are counted. We then use these data to calculate the distance (*d*):

$$d = \frac{1/2 \ (\text{second-division segregant asci})}{\text{total asci scored}}$$

The distance (*d*) reflects the percentage of recombination and is only half the number of second-division segregant asci. This is because crossing over in each occurs in only two of the four chromatids during meiosis.

To illustrate, we use *a* for albino and + for wild type in *Neurospora*. In crosses between the two genetic types, suppose we observe 65 first-division segregants, and 70 second-division segregants. The distance between *a* and the centromere is

$$d = \frac{(1/2)(70)}{135} = 0.259$$

or about 26 mu.

As the distance increases to 50 mu, all asci should reflect second-division segregation. However, numerous factors prevent this. As in diploid organisms, mapping accuracy based on crossover events is greatest when the gene and centromere are relatively close together.

We can also analyze haploid organisms to distinguish between linkage and independent assortment of two genes— mapping distances between gene loci are calculated once linkage is established. As a result, detailed maps of organisms such as *Neurospora* and *Chlamydomonas* are now available.

7.8   Other Aspects of Genetic Exchange

Careful analysis of crossing over during gamete formation allows us to construct chromosome maps in both diploid and haploid organisms. However, we should not lose sight of the real biological significance of crossing over, which is to generate genetic variation in gametes and, subsequently, in the

FIGURE 7–16 Three ways in which different ascospore patterns can be generated in *Neurospora*. Analysis of these patterns is the basis of gene-to-centromere mapping. The photograph shows a variety of ascospore arrangements within *Neurospora* asci.

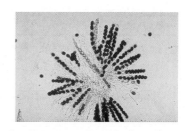

Condition	Four-strand stage	Chromosomes following meiosis	Chromosomes following mitotic division	Ascospores in ascus
(a) No crossover	*a* *a* / + +	*a* / *a* / + / +	*a* *a* *a* *a* + + + +	*a* *a* *a* *a* + + + +
			First-division segregation	
(b) One form of crossover in four-strand stage	*a* *a* + +	*a* / + / *a* / +	*a* *a* + + *a* *a* + +	*a* *a* + + *a* *a* + +
			Second-division segregation	
(c) An alternate crossover in four-strand stage	*a* *a* + +	+ / *a* / *a* / +	+ + *a* *a* *a* *a* + +	+ + *a* *a* *a* *a* + +
			Second-division segregation	

offspring derived from the resultant eggs and sperm. Because of the critical role of crossing over in generating variation, the study of genetic exchange is a key topic for study in genetics. But many important questions remain. For example, does crossing over involve an actual exchange of chromosome arms? Does exchange occur between paired sister chromatids during mitosis? We shall briefly consider possible answers to these questions.

Cytological Evidence for Crossing Over

Once genetic mapping was understood, it was of great interest to investigate the relationship between chiasmata observed in meiotic prophase I and crossing over. For example, are chiasmata visible manifestations of crossover events? If so, then crossing over in higher organisms appears to result from an actual physical exchange between homologous chromosomes. That this is the case was demonstrated independently in the 1930s by Harriet Creighton and Barbara McClintock in *Zea mays*, and by Curt Stern in *Drosophila*.

Since the experiments are similar, we will consider only the work with maize. Creighton and McClintock studied two linked genes on chromosome 9. At one locus, the alleles *colorless* (*c*) and *colored* (*C*) control endosperm coloration. At the other locus, the alleles *starchy* (*Wx*) and *waxy* (*wx*) control the carbohydrate characteristics of the endosperm. The maize plant studied is heterozygous at both loci. The key to this experiment is that one of the homologs contains two unique cytological markers. The markers consist of a densely stained knob at one end of the chromosome and a translocated piece of another chromosome (8) at the other end. The arrangements of alleles and cytological markers can be detected cytologically and are shown in Figure 7–17.

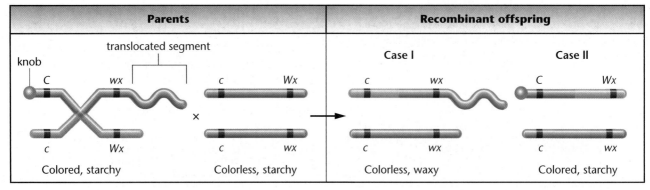

Parents		Recombinant offspring	
		Case I	**Case II**
Colored, starchy	Colorless, starchy	Colorless, waxy	Colored, starchy

FIGURE 7–17 The phenotypes and chromosome compositions of parents and recombinant offspring in Creighton and McClintock's experiment in maize. The knob and translocated segment are the cytological markers that established that crossing over involves an actual exchange of chromosome arms.

Creighton and McClintock crossed this plant to a plant homozygous for the *colored* allele (*c*) and heterozygous for the endosperm alleles. They obtained a variety of different phenotypes in the offspring, but they were most interested in a crossover result involving the chromosome with the unique cytological markers. They examined the chromosomes of this plant with the colorless, waxy phenotype (case I in Figure 7–17) for the presence of the cytological markers. If physical exchange between homologs accompanies genetic crossing over, the translocated chromosome will still be present, but the knob will not—this is exactly what happened. In a second plant (case II), the phenotype colored, starchy should result from either nonrecombinant gametes or from crossing over. Some of the plants then ought to contain chromosomes with the dense knob but not the translocated chromosome. This condition was also found, and the conclusion that a physical exchange takes place was again supported. Along with Stern's findings with *Drosophila*, this work clearly established that crossing over has a cytological basis.

Sister Chromatid Exchanges

Knowing that crossing over occurs between synapsed homologs in meiosis, we might ask whether a similar physical exchange occurs between homologs during mitosis. While homologous chromosomes do not usually pair up or synapse in somatic cells (*Drosophila* is an exception), each individual chromosome in prophase and metaphase of mitosis consists of two identical sister chromatids, joined at a common centromere. Surprisingly, several experimental approaches have demonstrated that reciprocal exchanges similar to crossing over occur between sister chromatids. These **sister chromatid exchanges (SCEs)** do not produce new allelic combinations, but evidence is accumulating that attaches significance to these events.

Identification and study of SCEs are facilitated by several modern staining techniques. In one technique, cells replicate for two generations in the presence of the thymidine analog **bromodeoxyuridine (BUdR)**.* Following two rounds of

replication, each pair of sister chromatids has one member with one strand of DNA "labeled" with BUdR and the other member with both strands labeled with BUdR. Using a differential stain, chromatids with the analog in both strands stain less brightly than chromatids with BUdR in only one strand. As a result, SCEs are readily detectable if they occur. In Figure 7–18, numerous instances of SCE events are clearly evident. These sister chromatids are sometimes referred to as **harlequin chromosomes** because of their patchlike appearance.

While the significance of SCEs is still uncertain, several observations have led to great interest in this phenomenon. We know, for example, that agents that induce chromosome damage (viruses, X-rays, ultraviolet light, and certain chem-

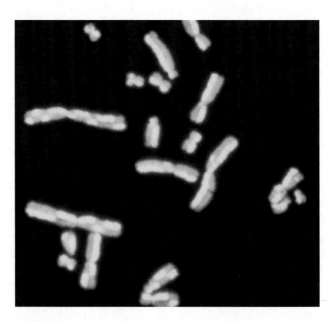

FIGURE 7–18 Light micrograph of sister chromatid exchanges (SCEs) in mitotic chromosomes. Sometimes called harlequin chromosomes because of their patchlike appearance, chromatids with the thymidine analog BUdR in both DNA strands fluoresce *less* brightly than do those with the analog in only one strand. These chromosomes were stained with 33258-Hoechst reagent and acridine orange and then viewed using fluorescence microscopy.

*The abbreviation BrdU is also used to denote bromodeoxyuridine.

ical mutagens) increase the frequency of SCEs. The frequency of SCEs is also elevated in **Bloom syndrome**, a human disorder caused by a mutation in the *BLM* gene on chromosome 15. This rare, recessively inherited disease is characterized by prenatal and postnatal retardation of growth, a great sensitivity of the facial skin to the sun, immune deficiency, a predisposition to malignant and benign tumors, and abnormal behavior patterns. The chromosomes from cultured leukocytes, bone marrow cells, and fibroblasts derived from homozygotes are very fragile and unstable compared to those of homozygous and heterozygous normal individuals. Increased breaks and rearrangements between nonhomologous chromosomes are observed in addition to excessive amounts of sister chromatid exchanges. Work by James German and colleagues suggests that the *BLM* gene encodes an enzyme called DNA helicase, which is best known for its role in DNA replication.

7.9 Did Mendel Encounter Linkage?

We conclude this chapter by examining a modern-day interpretation of the experiments that form the cornerstone of transmission genetics—Mendel's crosses with garden peas.

Some observers believe that Mendel had extremely good fortune in his classic experiments. He did not encounter apparent linkage relationships between the seven mutant characters in any of his crosses. Had Mendel obtained highly variable data characteristic of linkage and crossing over, these unorthodox ratios might have hindered his successful analysis and interpretation.

The article by Stig Blixt, reprinted in its entirety in the following box, demonstrates the inadequacy of this hypothesis. As we shall see, some of Mendel's genes were indeed linked. We leave it to Stig Blixt to enlighten you as to why Mendel did not detect linkage.

Why Didn't Gregor Mendel Find Linkage?

It is quite often said that Mendel was very fortunate not to run into the complication of linkage during his experiments. He used seven genes, and the pea has only seven chromosomes. Some have said that had he taken just one more, he would have had problems. This, however, is a gross oversimplification. The actual situation, most probably, is shown in Table 7.1. This shows that Mendel worked with three genes in chromosome 4, two genes in chromosome 1, and one gene in each of chromosomes 5 and 7. It seems at first glance that, out of the 21 dihybrid combinations Mendel theoretically could have studied, no fewer than four (that is, *a–i*, *v–fa*, *v–le*, *fa–le*) ought to have resulted in linkages. However, as found in hundreds of crosses and shown by the genetic map of the pea, *a* and *i* in chromosome 1 are so distantly located on the chromosome that no linkage is normally detected. The same is true for *v* and *le* on the one hand, and *fa* on the other, in chromosome 4. This leaves *v–le*, which ought to have shown linkage.

Mendel, however, seems not to have published this particular combination and thus, presumably, never made the appropriate cross to obtain both genes segregating simultaneously. It is therefore not so astonishing that Mendel did not run into the complication of linkage, although he did not avoid it by choosing one gene from each chromosome.

STIG BLIXT

Weibullsholm Plant Breeding Institute, Landskrona, Sweden, and Centro Energia Nucleate na Agricultura, Piracicaba, SP, Brazil.

Source: Reprinted by permission from *Nature*, Vol. 256, p. 206. © 1975 Macmillan Magazines Limited.

TABLE 7.1 Relationship between modern genetic terminology and character pairs used by Mendel

Character Pair Used by Mendel	Alleles in Modern Terminology	Located in Chromosome
Seed color, yellow–green	*I–i*	1
Seed coat and flowers, colored–white	*A–a*	1
Mature pods, smooth expanded–wrinkled indented	*V–v*	4
Inflorescences, from leaf axis–umbellate in top of plant	*Fa–fa*	4
Plant height, 0.5–1 m	*Le–le*	4
Unripe pods, green–yellow	*Gp–gp*	5
Mature seeds, smooth–wrinkled	*R–r*	7

CHAPTER SUMMARY

1. Genes located on the same chromosome are said to be linked. Alleles located on the same homolog, therefore, may be transmitted together during gamete formation. However, crossing over between homologs during meiosis results in the reshuffling of alleles and thereby contributes to genetic variability within gametes.

2. Early in the twentieth century, geneticists realized that crossing over provides an experimental basis for mapping the location of linked genes relative to one another along the chromosome.

3. Somatic cell hybridization techniques have made possible linkage and mapping analysis of human genes.

4. Mapping studies may also be performed with haploid organisms such as *Chlamydomonas* and *Neurospora*.

5. Cytological investigations of both maize and *Drosophila* reveal that crossing over involves a physical exchange of segments between nonsister chromatids.

6. An exchange of genetic material between sister chromatids can occur during mitosis as well. These events are referred to as sister chromatid exchanges (SCEs). An elevated frequency of such events is seen in the human disorder Bloom syndrome.

7. Evidence now suggests that several of the genes studied by Mendel are, in fact, linked. However, in such cases, the genes are sufficiently far apart to prevent the detection of linkage.

KEY TERMS

Bloom syndrome, 157
bromodeoxyuridine (BUdR), 156
chiasmata, 140
chromosomal theory of inheritance, 142
chromosome map, 141
coefficient of coincidence (*C*), 150
complete linkage, 137
crossing over, 137
crossover gametes, 137
cystic fibrosis, 154
double crossover (DCO), 143
first-division segregation, 154

harlequin chromosomes, 156
heterokaryon, 152
Huntington disease, 154
interference (*I*), 150
linkage, 137
linkage group, 140
linkage ratio, 138
lod score method, 152
mapping the centromere, 154
multiple-strand exchange, 150
neurofibromatosis, 154

noncrossover gametes, 137, 140
parental gametes, 137
reciprocal classes, 146
recombinant gametes, 137, 140
second-division segregation, 154
sister chromatid exchanges (SCEs), 156
somatic cell hybridization, 152
synkaryon, 152
synteny testing, 152
tetrad, 154
tetrad analysis, 154

INSIGHTS AND SOLUTIONS

1. In rabbits, black color (*B*) is dominant to brown (*b*), while full color (*C*) is dominant to *chinchilla* (*c^ch*). The genes controlling these traits are linked. Rabbits that are heterozygous for both traits and express black, full color are crossed to rabbits that express brown, chinchilla with the following results:

31	brown, chinchilla
34	black, full
16	brown, full
19	black, chinchilla

Determine the arrangement of alleles in the heterozygous parents and the map distance between the two genes.

Solution: This is a two-point map problem, where the two most prevalent reciprocal phenotypes are the noncrossovers. The less frequent reciprocal phenotypes arise from a single crossover. The arrangement of alleles is derived from the noncrossover phenotypes because they enter gametes intact.

The single crossovers give rise to 35/100 offspring (35%). Therefore, the distance between the two genes is 35 mu.

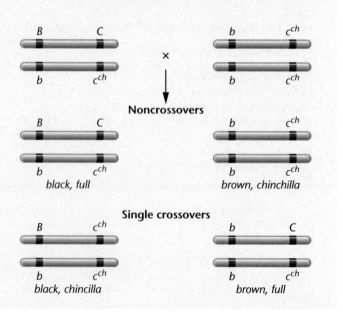

2. In *Drosophila*, *Lyra* (*Ly*) and *Stubble* (*Sb*) are dominant mutations located at locus 40 and 58, respectively, on chromosome III. A recessive mutation with bright red eyes is discovered and shown also to be located on chromosome III. A map is obtained by crossing a female who is heterozygous for all three mutations to a male that is homozygous for the *bright-red* mutation (which we will call *br*), and the data in the table are generated. Determine the location of the *br* mutation on chromosome III.

Phenotype			Number
(1) *Ly*	*Sb*	*br*	404
(2) +	+	+	422
(3) *Ly*	+	+	18
(4) +	*Sb*	*br*	16
(5) *Ly*	+	*br*	75
(6) +	*Sb*	+	59
(7) *Ly*	*Sb*	+	4
(8) +	+	*br*	2
		Total =	1000

Solution: First, determine the arrangement of the alleles on the homologs of the heterozygous crossover parent (the female in this case). To do this, locate the most frequent reciprocal phenotypes, which arise from the noncrossover gametes—these are phenotypes (1) and (2). Each phenotype represents the arrangement of alleles on one of the homologs. Therefore, the arrangement is

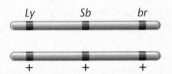

Second, find the correct sequence of the three loci along the chromosome. This is done by determining which sequence yields the observed double-crossover phenotypes, which are the least frequent reciprocal phenotypes (7 and 8). If the sequence is correct as written, then the double crossover depicted here,

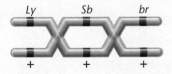

will yield *Ly* + *br* and + *Sb* + as phenotypes. Inspection shows that these categories (5 and 6) are actually single crossovers, not double crossovers. Therefore, the sequence is incorrect, as written. Only two other sequences are possible: The *br* gene is either to the left of *Ly* (case A), or it is between *Ly* and *Sb* (case B).

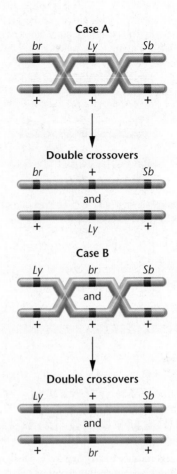

Comparison with the actual data shows that case B is correct. The double-crossover gametes yield flies that express *Ly* and *Sb* but not *br*, or express *br* but not *Ly* and *Sb*. Therefore, the correct arrangement and sequence are shown below.

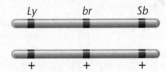

Once this sequence is found, determine the location of *br* relative to *Ly* and *Sb*. A single crossover between *Ly* and *br*, as shown here,

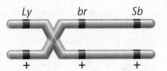

yields flies that are *Ly* + + and + *br Sb* (phenotypes 3 and 4). Therefore, the distance between the *Ly* and *br* loci is equal to

$$(18 + 16 + 4 + 2)/1000 = 40/1000 = 0.04 = 4 \text{ mu}$$

(*Cont. on the next page*)

Remember to add the double crossovers because they represent two single crossovers occurring simultaneously. You need to know the frequency of all crossovers between *Ly* and *br*, so they must be included.

Similarly, the distance between the *br* and *Sb* loci is derived mainly from single crossovers between them.

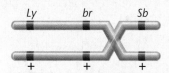

This event yields *Ly br +* and *+ + Sb* phenotypes (phenotypes 5 and 6). Therefore, the distance equals

$$(75 + 59 + 4 + 2)/1000 = 140/1000 = 0.14 = 14 \text{ mu}$$

The final map shows that *br* is located at locus 44, since *Lyra* and *Stubble* are known.

3. Refer to Figure 7–13, and predict what gene (which we called *br*) was discovered on chromosome III in Problem 2. Suggest an experimental cross that could confirm your prediction.

Solution: Inspection of Figure 7–13 reveals that the mutation *scarlet* (*st*) is present at locus 44.0, so it is reasonable to hypothesize that the *bright-red* eye mutation is an allele at the *scarlet* locus.

To test this hypothesis, you could perform complementation analysis (see Chapter 4) by crossing females expressing the *bright-red* mutation with known *scarlet* males. If the two mutations are alleles, no complementation will occur and all progeny will reveal a *bright-red* mutant eye phenotype. If complementation occurs, all progeny will express normal brick-red (wild-type) eyes, since the bright-red mutation and *scarlet* are at different loci (they are probably very close together). In such a case, all progeny will be heterozygous at both the *bright-red* eye and the *scarlet* loci and will not express either mutation because they are both recessive. This type of complementation analysis is called an **allelism test**.

PROBLEMS AND DISCUSSION QUESTIONS

1. What is the significance of crossing over (which leads to genetic recombination) to the process of evolution?
2. Describe the cytological observation that suggests that crossing over occurs during the first meiotic prophase.
3. Why does more crossing over occur between two distantly linked genes than between two genes that are very close together on the same chromosome?
4. Why is a 50 percent recovery of single-crossover products the upper limit, even when crossing over *always* occurs between two linked genes?
5. Why are double-crossover events expected less frequently than single-crossover events?
6. What is the proposed basis for positive interference?
7. What three essential criteria must be met in order to execute a successful mapping cross?
8. The genes *dumpy* wings (*dp*), *clot eyes* (*cl*), and *apterous* wings (*ap*) are linked on chromosome II of *Drosophila*. In a series of two-point mapping crosses, the genetic distances shown below were determined. What is the sequence of the three genes?

dp–ap	42
dp–cl	3
ap–cl	39

9. Consider two hypothetical recessive autosomal genes *a* and *b*, where a heterozygote is testcrossed to a double-homozygous mutant. Predict the phenotypic ratios under the following conditions:
   (a) *a* and *b* are located on separate autosomes.
   (b) *a* and *b* are linked on the same autosome but are so far apart that a crossover always occurs between them.
   (c) *a* and *b* are linked on the same autosome but are so close together that a crossover almost never occurs.
   (d) *a* and *b* are linked on the same autosome about 10 mu apart.
10. Colored aleurone in the kernels of corn is due to the dominant allele *R*. The recessive allele *r*, when homozygous, produces colorless aleurone. The plant color (not kernel color) is controlled by another gene with two alleles, *Y* and *y*. The dominant *Y* allele results in green color, whereas the homozygous presence of the recessive *y* allele causes the plant to appear yellow. In a testcross between a plant of unknown genotype and phenotype and a plant that is homozygous recessive for both traits, the following progeny were obtained:

colored, green	88
colored, yellow	12
colorless, green	8
colorless, yellow	92

Explain how these results were obtained by determining the exact genotype and phenotype of the unknown plant, including the precise association of the two genes on the homologs (i. e., the arrangement).
11. In the cross shown here, involving two linked genes, *ebony* (*e*) and *claret* (*ca*), in *Drosophila*, where crossing over does not occur in males, offspring were produced in a (2 + :1 *ca*:1 *e*) phenotypic ratio:

$$\frac{e \quad ca^+}{e^+ \quad ca} \times \frac{e \quad ca^+}{e^+ \quad ca}$$

These genes are 30 mu apart on chromosome III. What did crossing over in the female contribute to these phenotypes?

12. With two pairs of genes involved (*P, p* and *Z, z*), a testcross (to *ppzz*) with an organism of unknown genotype indicated that the gametes were produced in these proportions: $PZ = 42.4\%$; $Pz = 6.9\%$; $pZ = 7.1\%$; and $pz = 43.6\%$. Draw all possible conclusions from these data.

13. In a series of two-point map crosses involving five genes located on chromosome II in *Drosophila*, the following recombinant (single-crossover) frequencies were observed:

pr–adp	29
pr–vg	13
pr–c	21
pr–b	6
adp–b	35
adp–c	8
adp–vg	16
vg–b	19
vg–c	8
c–b	27

(a) If the *adp* gene is present near the end of chromosome II (locus 83), construct a map of these genes.

(b) In another set of experiments, a sixth gene (*d*) was tested against *b* and *pr*, and the results were $d-b = 17\%$ and $d-pr = 23\%$. Predict the results of two-point maps between *d* and *c*, *d* and *vg*, and *d* and *adp*.

14. Two different female *Drosophila* were isolated, each heterozygous for the autosomally linked genes *black* body (*b*), *dachs* tarsus (*d*), and *curved* wings (*c*). These genes are in the order *d–b–c*, with *b* closer to *d* than to *c*. Shown below is the genotypic arrangement for each female, along with the various gametes formed by both. Identify which categories are noncrossovers (NCO), single crossovers (SCO), and double crossovers (DCO) in each case. Then, indicate the relative frequency in which each will be produced.

Female A	Female B
d b +	d + +
——————	——————
+ + c	+ b c
↓ Gamete formation	↓

(1) *d b c*	(5) *d + +*	(1) *d b +*	(5) *d b c*
(2) *+ + +*	(6) *+ b c*	(2) *+ + c*	(6) *+ + +*
(3) *+ + c*	(7) *d + c*	(3) *d + c*	(7) *d + +*
(4) *d b +*	(8) *+ b +*	(4) *+ b +*	(8) *+ b c*

15. In *Drosophila*, a cross was made between females expressing the three X-linked recessive traits, *scute* bristles (*sc*), *sable* body (*s*), and *vermilion* eyes (*v*) and wild-type males. All females were wild type in the F$_1$, while all males expressed all three mutant traits. The cross was carried to the F$_2$ generation, and 1000 offspring were counted, with the results shown in the table. No de-

termination of sex was made in the F$_2$ data. (a) Using proper nomenclature, determine the genotypes of the P$_1$ and F$_1$ parents. (b) Determine the sequence of the three genes and the map distance between them. (c) Are there more or fewer double crossovers than expected? (d) Calculate the coefficient of coincidence; does this represent positive or negative interference?

Phenotype			Offspring
sc	*s*	*v*	314
+	+	+	280
+	*s*	*v*	150
sc	+	+	156
sc	+	*v*	46
+	*s*	+	30
sc	*s*	+	10
+	+	*v*	14

16. A cross in *Drosophila* involved the recessive, X-linked genes *yellow* body (*y*), *white* eyes (*w*), and *cut* wings (*ct*). A yellow-bodied, white-eyed female with normal wings was crossed to a male whose eyes and body were normal, but whose wings were cut. The F$_1$ females were wild type for all three traits, while the F$_1$ males expressed the yellow-body, white-eye traits. The cross was carried to F$_2$ progeny, and only male offspring were tallied. On the basis of the data shown here, a genetic map was constructed. (a) Diagram the genotypes of the F$_1$ parents. (b) Construct a map, assuming that *w* is at locus 1.5 on the X chromosome. (c) Were any double-crossover offspring expected? (d) Could the F$_2$ female offspring be used to construct the map? Why or why not?

Phenotype			Male Offspring
y	+	*ct*	9
+	*w*	+	6
y	*w*	*ct*	90
+	+	+	95
+	+	*ct*	424
y	*w*	+	376
y	+	+	0
+	*w*	*ct*	0

17. In *Drosophila*, *Dichaete* (*D*) is a mutation on chromosome III with a dominant effect on wing shape. It is lethal when homozygous. The genes *ebony* body (*e*) and *pink* eye (*p*) are recessive mutations on chromosome III. Flies from a Dichaete stock were crossed to homozygous ebony, pink flies, and the F$_1$ progeny with a Dichaete phenotype were backcrossed to the ebony, pink homozygotes. Using the results of this backcross shown in the table, (a) diagram the cross, showing the genotypes of the parents and offspring of both crosses. (b) What is the sequence and interlocus distance between these three genes?

Phenotype	Number
Dichaete	401
ebony, pink	389
Dichaete, ebony	84
pink	96
Dichaete, pink	2
ebony	3
Dichaete, ebony, pink	12
wild type	13

18. *Drosophila* females homozygous for the third chromosomal genes *pink* eye (*p*) and *ebony* body (*e*) were crossed with males homozygous for the second chromosomal gene *dumpy* wings (*dp*). Because these genes are recessive, all offspring were wild type (normal). F_1 females were testcrossed to triply recessive males. If we assume that the two linked genes (*p* and *e*) are 20 mu apart, predict the results of this cross. If the reciprocal cross were made (F_1 males—where no crossing over occurs—with triply recessive females), how would the results vary, if at all?

19. In *Drosophila*, the two mutations *Stubble* bristles (*Sb*) and *curled* wings (*cu*) are linked on chromosome III. *Sb* is a dominant gene that is lethal in a homozygous state, and *cu* is a recessive gene. If a female of the genotype

$$\frac{Sb \quad cu}{+ \quad +}$$

is to be mated to detect recombinants among her offspring, what male genotype would you choose as her mate?

20. In *Drosophila*, a heterozygous female for the X-linked recessive traits *a*, *b*, and *c* was crossed to a male that phenotypically expressed *a*, *b*, and *c*. The offspring occurred in the phenotypic ratios in the following table, and no other phenotypes were observed. (a) What is the genotypic arrangement of the alleles of these genes on the X chromosome of the female? (b) Determine the correct sequence, and construct a map of these genes on the X chromosome. (c) What progeny phenotypes are missing, and why?

+	*b*	*c*	460
a	+	+	450
a	*b*	*c*	32
+	+	+	38
a	+	*c*	11
+	*b*	+	9

21. Are sister chromatid exchanges effective in producing genetic variability in an individual? in the offspring of individuals?

22. What conclusion can be drawn from the observations that in male *Drosophila*, no crossing over occurs and that during meiosis, synaptonemal complexes (ultrastructural components found between synapsed homologs in meiosis) are not seen in males but are observed in females where crossing over occurs?

23. An organism of the genotype *AaBbCc* was testcrossed to a triply recessive organism (*aabbcc*). The genotypes of the progeny are in the following table.

AaBbCc	20	AaBbcc	20
aabbCc	20	aabbcc	20
AabbCc	5	Aabbcc	5
aaBbCc	5	aaBbcc	5

(a) Assuming simple dominance and recessiveness in each gene pair, if these three genes were all assorting independently, how many genotypic and phenotypic classes would result in the offspring, and in what proportion?

(b) Answer part (a) again, assuming the three genes are so tightly linked on a single chromosome that no crossover gametes were recovered in the sample of offspring.

(c) What can you conclude from the *actual* data about the location of the three genes in relation to one another?

24. Based on our discussion of the potential inaccuracy of mapping (see Figure 7–12), would you revise your answer to Problem 23(c)? If so, how?

25. In a plant, fruit color is either red or yellow, and fruit shape is either oval or long. Red and oval are the dominant traits. Two plants, both heterozygous for these traits, were testcrossed, with the results shown below. Determine the location of the genes relative to one another and the genotypes of the two parental plants.

	Progeny	
Phenotype	Plant A	Plant B
red, long	46	4
yellow, oval	44	6
red, oval	5	43
yellow, long	5	47
Total	100	100

26. In a plant heterozygous for two gene pairs (*Ab/aB*), where the two loci are linked and 25 mu apart, two such individuals were crossed together. Assuming that crossing over occurs during the formation of both male and female gametes and that the *A* and *B* alleles are dominant, determine the phenotypic ratio of the offspring.

27. In a cross in *Neurospora* involving two alleles, *B* and *b*, the tetrad patterns in the following table were observed. Calculate the distance between the gene and the centromere.

Tetrad Pattern	Number
BBbb	36
bbBB	44
BbBb	4
bBbB	6
BbbB	3
bBBb	7

28. In Creighton and McClintock's experiment demonstrating that crossing over involves physical exchange between chromosomes (see Section 7.8), explain the importance of the cytological markers (the translocated segment and the chromosome knob) in the experimental rationale.

29. A number of human–mouse somatic cell hybrid clones were examined for the expression of specific human genes and the presence of human chromosomes—the results are summarized in this table. Assign each gene to the chromosome upon which it is located.

| | **Hybrid-Cell Clone** | | | | | |
	A	B	C	D	E	F
Genes (expressed or not)						
ENO1 (enolase-1)	−	+	−	+	+	−
MDH1 (malate dehydrogenase-1)	+	+	−	+	−	+
PEPS (peptidase S)	+	−	+	−	−	−
PGM1 (phosphoglucomutase-1)	−	+	−	+	+	−
Chromosomes (present or absent)						
1	−	+	−	+	+	−
2	+	+	−	+	−	+
3	+	+	−	−	+	−
4	+	−	+	−	−	−
5	−	+	+	+	+	+

30. A female of genotype

$$\frac{a \quad b \quad c}{+ \quad + \quad +}$$

produces 100 meiotic tetrads. Of these, 68 show no crossover events. Of the remaining 32, 20 show a crossover between a and b, 10 show a crossover between b and c, and 2 show a double crossover between a and b and between b and c. Of the 400 gametes produced, how many of each of the 8 different genotypes will be produced? Assuming the order a–b–c and the allele arrangement shown above, what is the map distance between these loci?

31. *D. melanogaster* has one pair of sex chromosomes (XX or XY) and three autosomes (chromosomes II, III, and IV). A genetics student discovered a male fly with very short (*sh*) legs. Using this male, the student was able to establish a pure-breeding stock of this mutant and found that it was recessive. She then incorpo-

rated the mutant into a stock containing the recessive gene *black* (*b*, body color, located on chromosome II) and the recessive gene *pink* (*p*, eye color, located on chromosome III). A female from the homozygous black, pink, short stock was then mated to a wild-type male. The F_1 males of this cross were all wild type and were then backcrossed to the homozygous *b, p, sh* females. The F_2 results appeared as shown in the table that follows, and no other phenotypes were observed. (a) Based on these results, the student was able to assign *sh* to a linkage group (a chromosome). Determine which chromosome, and include step-by-step reasoning. (b) The student repeated the experiment, making the reciprocal cross: F_1 females backcrossed to homozygous *b, p, sh* males. She observed that 85 percent of the offspring fell into the given classes, but that 15 percent of the offspring were equally divided among $b+p$, $b++$, $+shp$, and $+sh+$ phenotypic males and females. How can these results be explained, and what information can be derived from these data?

Phenotype	Female	Male
wild	63	59
pink*	58	65
black, short	55	51
black, pink, short	69	60

*Pink indicates that the other two traits are wild type (normal). Similarly, black, short offspring are wild type for eye color.

32. In *Drosophila*, a female fly is heterozygous for three mutations, *Bar* eyes (*B*), *miniature* wings (*m*), and *ebony* body (*e*). (Note that *Bar* is a dominant mutation.) The fly is crossed to a male with normal eyes, miniature wings, and ebony body. The results of the cross are shown below. Interpret the results of this cross. If you conclude that linkage is involved between any of the genes, determine the map distance(s) between them.

miniature	111	Bar	117
wild	29	Bar, miniature	26
Bar, ebony	101	ebony	35
Bar, miniature, ebony	31	miniature, ebony	115

Genetic Analysis and Mapping in Bacteria and Bacteriophages

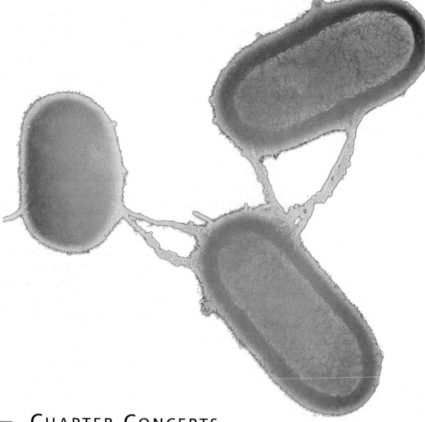

Transmission electron micrograph of conjugating E. coli.

CHAPTER CONCEPTS

- Bacterial genomes are most often contained in a single circular chromosome.

- Bacteria have developed numerous ways in which they can exchange and recombine genetic information between individual cells, including conjugation, transformation, and transduction.

- The ability to undergo conjugation and to transfer the bacterial chromosome from one cell to another is governed by the presence of genetic information contained in the DNA of a "fertility," or F factor.

- The F factor can exist autonomously in the bacterial cytoplasm as a plasmid, or it can integrate into the bacterial chromosome, where it facilitates the transfer of the host chromosome to the recipient cell, leading to genetic recombination.

- During conjugation, genetic recombination provides the basis for mapping bacterial genes.

- Bacteriophages are viruses that have bacteria as their hosts.

- During infection of the bacterial host, bacteriophage DNA is injected into the host cell, where it is replicated and directs the the reproduction of the bacteriophage.

- During bacteriophage infection, following replication, the phage DNA may undergo recombination, which may serve as the basis for intergenic mapping.

In this chapter, we shift from the consideration of mapping genetic information in eukaryotes to a discussion of the analysis and mapping of genes in **bacteria** (prokaryotes) and **bacteriophages**, viruses that use bacteria as their hosts. The study of bacteria and bacteriophages has been essential to the accumulation of knowledge in many areas of genetic study. For example, much of what we know about molecular genetics, recombinational phenomena, and gene structure was initially derived from experimental work with them. Furthermore, our extensive knowledge of bacteria and their resident plasmids has led to their widespread use in DNA cloning and other recombinant DNA studies.

Bacteria and their viruses are especially useful research organisms in genetics for several reasons. They have extremely short reproductive cycles—literally hundreds of generations, giving rise to billions of genetically identical bacteria or phages, can be produced in short periods of time. Furthermore, they can be studied in pure cultures. That is, a single species or mutant strain of bacteria or one type of virus can be isolated and investigated independently of other similar organisms.

In this chapter, we focus on genetic recombination and chromosome mapping. Complex processes have evolved in bacteria and bacteriophages that facilitate the transfer of genetic information between individual cells within populations. As we shall see, these processes are the basis for the chromosome mapping analysis that forms the cornerstone of molecular genetic investigations of bacteria and the viruses that invade them.

How Do We Know?

In this chapter, we will focus on genetic systems present in bacteria and the viruses that use bacteria as hosts (bacteriophages). In particular, we will discuss mechanisms by which bacteria and their phages undergo genetic recombination, the basis of chromosome mapping. As you study these topics, you should try to answer several fundamental questions:

1. How do we know that genes exist in bacteria and bacteriophages?

2. How do we know that bacteria undergo genetic recombination, allowing the transfer of genes from one organism to another?

3. How do we know that genetic recombination between bacteria involves cell contact and that cell contact precedes the transfer of genes from one bacterium to another?

4. How did we learn that the mechanism of genetic recombination differs between bacteria and bacteriophages?

5. How do we know that bacteriophages recombine genetic material through transduction and that cell contact is not essential for transduction to occur?

8.1 Bacteria Mutate Spontaneously and Grow at an Exponential Rate

It has long been known that pure cultures of bacteria give rise to cells that exhibit heritable variation, particularly with respect to growth under unique environmental conditions. Prior to 1943, the source of this variation was hotly debated. The majority of bacteriologists believed that environmental factors induced changes in certain bacteria that led to their adaptation to the new conditions. For example, strains of *Escherichia coli* are known to be sensitive to infection by the bacteriophage T1. Infection by this bacteriophage leads to the virus reproducing at the expense of the bacterial cell, from which new phages are released as the host cell is disrupted, or lysed. If a plate of *E. coli* is uniformly sprayed with T1, almost all cells are lysed. Rare *E. coli* cells, however, survive infection and are not lysed. If these cells are isolated and established in pure culture, all of the descendants are resistant to T1 infection. The **adaptation hypothesis**, put forth to explain this type of observation, implies that the interaction of the phage and bacterium is essential to the acquisition of immunity. In other words, the phage "induces" resistance in the bacteria.

On the other hand, the existence of **spontaneous mutations**, which occur in the presence or the absence of bacteriophage T1, suggested an alternative model to explain the origin of resistance in *E. coli*. In 1943, Salvador Luria and Max Delbruck presented the first convincing evidence that bacteria, like eukaryotic organisms, are capable of spontaneous mutation. This experiment, referred to as the **fluctuation test**, marks the initiation of modern bacterial genetic study. Spontaneous mutation is now considered the primary source of genetic variation in bacteria.

Mutant cells that arise spontaneously in otherwise pure cultures can be isolated and established independently from the parent strain by using established selection techniques. As a result, mutations for almost any desired characteristic can now be induced and isolated. Because bacteria and viruses usually contain only a single chromosome and are therefore haploid, all mutations are expressed directly in the descendants of mutant cells, adding to the ease with which these microorganisms can be studied.

Bacteria are grown in a liquid culture medium or in a petri dish on a semisolid agar surface. If the nutrient components of the growth medium are simple and consist only of an organic carbon source (such as a glucose or lactose) and a variety of ions, including Na^+, K^+, Mg^{2+}, Ca^{2+}, and NH^{4+}, present as inorganic salts, it is called **minimal medium**. To grow on such a medium, a bacterium must be able to synthesize all essential organic compounds (e.g., amino acids, purines, pyrimidines, sugars, vitamins, and fatty acids). A bacterium that can accomplish this remarkable biosynthetic feat—one that we ourselves cannot duplicate—is a **prototroph**. It is said to be wild type for all growth requirements. On the other hand, if a bacterium loses, through mutation, the ability to synthesize one or more organic components, it is an **auxotroph**. For example, if it

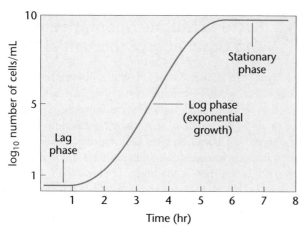

FIGURE 8–1 Typical bacterial population growth curve showing the initial lag phase, the subsequent log phase where exponential growth occurs, and the stationary phase that occurs when nutrients are exhausted.

loses the ability to make histidine, then this amino acid must be added as a supplement to the minimal medium for growth to occur. The resulting bacterium is designated as a *his⁻* auxotroph, in contrast to its prototrophic *his⁺* counterpart.

To study mutant bacteria quantitatively, an inoculum of bacteria is placed in liquid culture medium. A graph of the characteristic growth pattern is shown in Figure 8–1. Initially, during the **lag phase**, growth is slow. Then, a period of rapid growth, called the **logarithmic (log) phase**, ensues. During this phase, cells divide continually with a fixed time interval between cell divisions, resulting in exponential growth. When a cell density of about 10^9 cells/mL is reached, nutrients and oxygen become limiting and the cells cease dividing; at this point, the cells enter the **stationary phase**. The doubling time during the log phase can be as short as 20 minutes. Thus, an initial inoculum of a few thousand cells easily achieves maximum cell density during an overnight incubation.

Cells grown in liquid medium can be quantified by transferring them to semisolid medium in a petri dish. Following incubation and many divisions, each cell gives rise to a visible colony on the surface of the medium. If the number of colonies is too great to count, then a series of successive dilutions (a technique called **serial dilution**) of the original liquid culture is made and plated, until the colony number is reduced to the point where it can be counted (Figure 8–2). This technique allows the number of bacteria present in the original culture to be calculated.

As an example, let's assume that the three dishes in Figure 8–2 represent serial dilutions of 10^{-3}, 10^{-4}, and 10^{-5} (from left to right). We need only select the dish in which the number of colonies can be counted accurately. Because each colony arose from a single bacterium, the number of colonies multiplied by the dilution factor represents the number of bacteria in each milliliter of the initial inoculum used to start the serial dilutions. In Figure 8–2, the rightmost dish has 15 colonies. The dilution factor for a 10^{-5} dilution is 10^5. Therefore, the initial number of bacteria was 15×10^5 per mL.

8.2 Conjugation Is One Means of Genetic Recombination in Bacteria

Development of techniques that allowed the identification and study of bacterial mutations led to detailed investigations of the arrangement of genes on the bacterial chromosome. Such studies began in 1946 when Joshua Lederberg and Edward Tatum showed that bacteria undergo **conjugation**, a parasexual process in which the genetic information from one bacterium is transferred to and recombined with that of another bacterium (see the chapter opening photograph). Like meiotic crossing over in eukaryotes, genetic recombination in bacteria enabled the development of methodology for chromosome mapping. Note that the term **genetic recombination**, as applied to bacteria and bacteriophages, leads to the *replacement* of one or more genes present in one strain with those from a genetically distinct strain. While this is somewhat different from our use of genetic recombination in eukaryotes, where the term describes crossing over that results in *reciprocal exchange events*, the overall effect is the same: Genetic information is transferred from one chromosome to another, resulting in an altered genotype. Two other phenomena that result in the transfer of genetic information from one bacterium to another, *transformation* and *transduction*, have helped us determine the arrangement of genes on the bacterial chromosome. We will discuss these processes in later sections of this chapter.

Lederberg and Tatum's initial experiments were performed with two multiple-auxotroph strains of *E. coli* K12. Strain A

FIGURE 8–2 Results of the serial dilution technique and subsequent culture of bacteria. Each dilution varies by a factor of 10. Each colony is derived from a single bacterial cell.

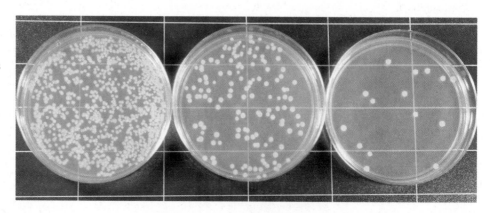

FIGURE 8–3 Genetic recombination of two auxotrophic strains producing prototrophs. Neither auxotroph grows on minimal medium, but prototrophs do, suggesting that genetic recombination has occurred.

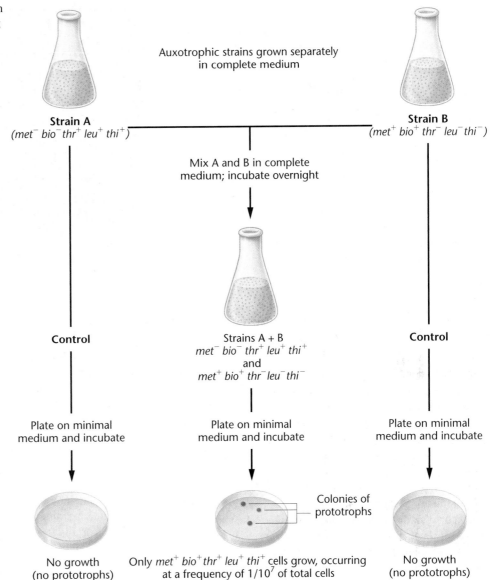

Auxotrophic strains grown separately in complete medium

Strain A
(*met⁻ bio⁻ thr⁺ leu⁺ thi⁺*)

Strain B
(*met⁺ bio⁺ thr⁻ leu⁻ thi⁻*)

Mix A and B in complete medium; incubate overnight

Control

Strains A + B
met⁻ bio⁻ thr⁺ leu⁺ thi⁺
and
met⁺ bio⁺ thr⁻ leu⁻ thi⁻

Control

Plate on minimal medium and incubate

Plate on minimal medium and incubate

Plate on minimal medium and incubate

Colonies of prototrophs

No growth
(no prototrophs)

Only *met⁺ bio⁺ thr⁺ leu⁺ thi⁺* cells grow, occurring at a frequency of $1/10^7$ of total cells

No growth
(no prototrophs)

required methionine and biotin in order to grow, while strain B required threonine, leucine, and thiamine (Figure 8–3). Therefore, neither strain would grow on minimal medium. The two strains were first grown separately in supplemented media, and cells from both were mixed and grown together for several more generations and then plated on minimal medium. Any bacterial cells that grew on minimal medium were prototrophs. It is highly improbable that any of the cells that contained two or three mutant genes underwent spontaneous mutation simultaneously at two or three independent locations, leading to wild-type cells. Therefore, the researchers assumed that any prototrophs recovered arose as a result of some form of genetic exchange and recombination between the two mutant strains.

In this experiment, prototrophs were recovered at a rate of $1/10^7$ (or 10^{-7}) cells plated. The controls for this experiment involved separate plating of cells from strains A and B on minimal medium. No prototrophs were recovered. Based on these observations, Lederberg and Tatum proposed that genetic exchange had occurred.

F⁺ and F⁻ Bacteria

The findings of Lederberg and Tatum were soon followed by numerous experiments that elucidated the genetic basis of conjugation. It quickly became evident that different strains of bacteria are involved in a unidirectional transfer of genetic material. When cells serve as donors of parts of their chromosomes, they are designated as **F⁺ cells** (F for "fertility"). Recipient bacteria receive the donor chromosome material (now known to be DNA), and recombine it with part of their own chromosome. They are designated as **F⁻ cells**.

Experimentation subsequently established that cell contact is essential for chromosome transfer to occur. Support for this concept was provided by Bernard Davis, who designed the Davis U-tube for growing F⁺ and F⁻ cells shown in Figure 8–4. At the base of the tube is a sintered glass filter with a pore size that allows passage of the liquid medium but that is too small to allow the passage of bacteria. The F⁺ cells are placed on one side of the filter, and F⁻ cells on the other side. The medium is moved back and forth across the filter so that the cells share a common medium

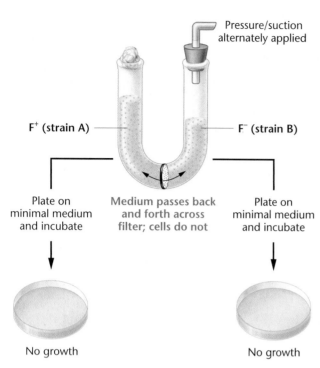

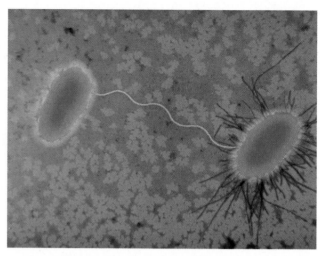

FIGURE 8–5 An electron micrograph of conjugation between an F$^+$ *E. coli* cell and an F$^-$ cell. The sex pilus linking them is clearly visible.

FIGURE 8–4 When strain A and B auxotrophs are grown in a common medium but separated by a filter, as in this Davis U-tube apparatus, no genetic recombination occurs and no prototrophs are produced.

during bacterial incubation. When Davis plated samples from both sides of the tube on minimal medium, no prototrophs were found, and he logically concluded that *physical contact between cells of the two strains is essential to genetic recombination.* We now know that this physical interaction is the initial step in the process of conjugation established by a structure called the **F pilus** (or **sex pilus**; pl. pili). Bacteria often have many pili, which are tubular extensions of the cell. After contact is initiated between mating pairs (Figure 8–5), chromosome transfer is now possible.

Later evidence established that F$^+$ cells contain a **fertility factor** (**F factor**) that confers the ability to donate part of their chromosome during conjugation. Experiments by Joshua and Esther Lederberg and by William Hayes and Luca Cavalli-Sforza show that certain conditions eliminate the F factor in otherwise fertile cells. However, if these "infertile" cells are then grown with fertile donor cells, the F factor is regained.

The conclusion that the F factor is a mobile element is further supported by the observation that, following conjugation and genetic recombination, recipient cells always become F$^+$. Thus, in addition to the *rare* cases of gene transfer from the bacterial chromosome (genetic recombination), the F factor itself is passed to *all* recipient cells. On this basis, the initial cross of Lederberg and Tatum (see Figure 8–3) is as follows:

Strain A   **Strain B**
F$^+$ ×   **F$^-$**
Donor   **Recipient**

Characterization of the F factor confirmed these conclusions. Like the bacterial chromosome, though distinct from it,

the F factor has been shown to consist of a circular, double-stranded DNA molecule, equivalent to about 2 percent of the bacterial chromosome (about 100,000 nucleotide pairs). There are 19 genes contained within the F factor, whose products are involved in the transfer of genetic information, including those essential to the formation of the sex pilus.

Geneticists believe that transfer of the F factor during conjugation involves separation of the two strands of its double helix and movement of one of the two strands into the recipient cell. Both strands, one moving across the conjugation tube and one remaining in the donor cell, are replicated. The result is that both the donor *and* the recipient cells become F$^+$. This process is diagrammed in Figure 8–6.

To summarize, an *E. coli* cell may or may not contain the F factor. When this factor is present, the cell is able to form a sex pilus and potentially serve as a donor of genetic information. During conjugation, a copy of the F factor is almost always transferred from the F$^+$ cell to the F$^-$ recipient, converting the recipient to the F$^+$ state. The question remained as to exactly why such a low proportion of cells involved in these matings (10^{-7}) also results in genetic recombination. The answer awaited further experimentation.

As you soon shall see, the F factor is in reality an autonomous genetic unit called a *plasmid*. However, in our historical coverage of its discovery, we will continue to refer to it as a factor.

Hfr Bacteria and Chromosome Mapping

Subsequent discoveries not only clarified how genetic recombination occurs but also defined a mechanism by which the *E. coli* chromosome could be mapped. Let's address chromosome mapping first.

In 1950, Cavalli-Sforza treated an F$^+$ strain of *E. coli* K12 with nitrogen mustard, a chemical known to induce mutations. From these treated cells, he recovered a genetically altered strain of donor bacteria that underwent recombination at a rate of $1/10^4$ (or 10^{-4})—1000 times more frequently than the original F$^+$ strains. In 1953, Hayes isolated another strain that demonstrated an elevated frequency. Both strains were designated **Hfr**, for **high-frequency recombination**.

FIGURE 8–6 An F⁺ × F⁻ mating demonstrating how the recipient F⁻ cell converts to F⁺. During conjugation, the DNA of the F factor replicates with one new copy entering the recipient cell, converting it to F⁺. The bars added to the F factors follow their clockwise rotation during replication. Newly replicated DNA is depicted by a lighter shade of blue as the F factor is transferred.

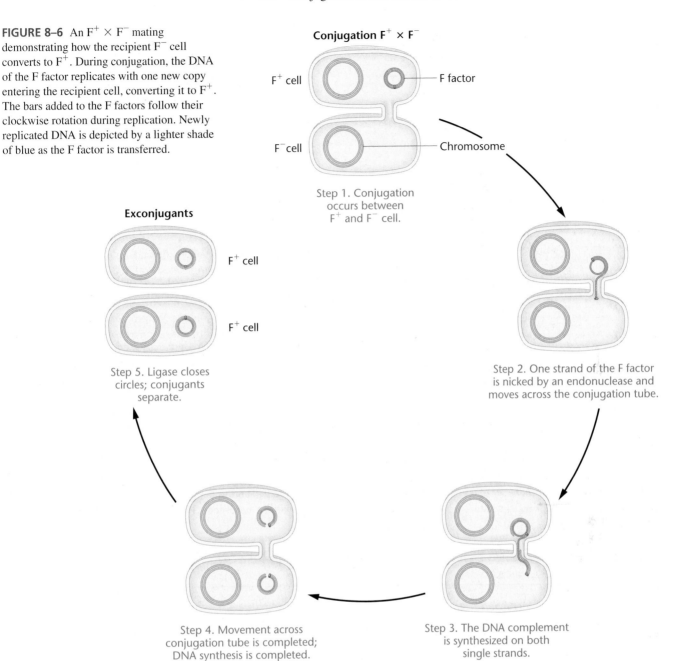

Conjugation F⁺ × F⁻

F⁺ cell — F factor

F⁻ cell — Chromosome

Step 1. Conjugation occurs between F⁺ and F⁻ cell.

Step 2. One strand of the F factor is nicked by an endonuclease and moves across the conjugation tube.

Step 3. The DNA complement is synthesized on both single strands.

Step 4. Movement across conjugation tube is completed; DNA synthesis is completed.

Exconjugants

F⁺ cell

F⁺ cell

Step 5. Ligase closes circles; conjugants separate.

Because Hfr cells behave as donors, they are a special class of F⁺ cells.

Another important difference was noted between Hfr strains and the original F⁺ strains. If the donor is from an Hfr strain, recipient cells, though sometimes displaying genetic recombination, never become Hfr; that is, they remain F⁻. In comparison, then,

$$F^+ \times F^- \longrightarrow F^+ \quad \text{(low rate of recombination)}$$
$$Hfr \times F^- \longrightarrow F^- \quad \text{(higher rate of recombination)}$$

Perhaps the most significant characteristic of Hfr strains is the *nature of recombination*. In any given strain, certain genes are more frequently recombined than others, and some not at all. This *nonrandom* pattern was shown to vary between Hfr strains. Although these results were puzzling, Hayes interpreted them to mean that some physiological alteration of the F factor had occurred, resulting in the production of Hfr strains of *E. coli*.

In the mid-1950s, experimentation by Ellie Wollman and François Jacob elucidated the difference between Hfr and F⁺ and showed how Hfr strains allow genetic mapping of the *E. coli* chromosome. In their experiments, Hfr and F⁻ strains with suitable marker genes were mixed, and recombination of specific genes was assayed at different times. To accomplish this, a culture containing a mixture of an Hfr and an F⁻ strain was first incubated, and samples were removed at various intervals and placed in a blender. The shear forces in the blender separated conjugating bacteria so that the transfer of the chromosome was terminated. The cells were then assayed for genetic recombination.

This process, called the **interrupted mating technique**, demonstrated that specific genes of a given Hfr strain were transferred and recombined sooner than others. The graph in Figure 8–7 illustrates this point. During the first eight minutes after the two strains were mixed, no genetic recombination was detected. At about 10 minutes, recombination of the *azi*

Hfr H (*thr⁺ leu⁺ aziᴿ tonˢ lac⁺ gal⁺*)
×
F⁻ (*thr⁻ leu⁻ aziˢ tonᴿ lac⁻ gal⁻*)

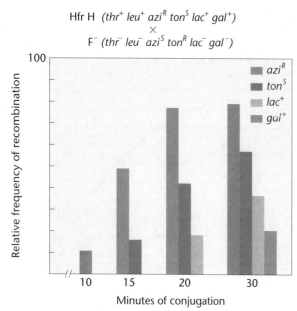

FIGURE 8–7 The progressive transfer during conjugation of various genes from a specific Hfr strain of *E. coli* to an F⁻ strain. Certain genes (*azi* and *ton*) transfer more quickly than others and recombine more frequently. Others (*lac* and *gal*) take longer to transfer and recombinants are found at a lower frequency.

gene was detected, but no transfer of the *tonˢ*, *lac⁺*, or *gal⁺*, genes was noted. By 15 minutes, 50 percent of the recombinants were *aziᴿ*, and 15 percent were *tonˢ*; but none was *lac⁺* or *gal⁺*. Within 20 minutes, the *lac⁺* was found among the recombinants; and within 25 minutes, *gal⁺* was also being

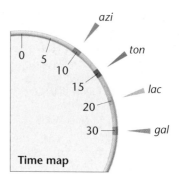

FIGURE 8–8 A time map of the genes studied in the experiment depicted in Figure 8–7.

transferred. Wollman and Jacob had demonstrated *an ordered transfer of genes* that correlated with the length of time conjugation proceeded.

It appeared that the chromosome of the Hfr bacterium was transferred linearly and that the gene order and distance between genes, as measured in minutes, could be predicted from such experiments (Figure 8–8). This information served as the basis for the first genetic map of the *E. coli* chromosome. Minutes in bacterial mapping are equivalent to map units in eukaryotes.

Wollman and Jacob then repeated the same type of experiment with other Hfr strains, obtaining similar results with one important difference. While genes were always transferred linearly with time, as in their original experiment, the order in which genes entered seemed to vary from Hfr strain to Hfr strain [see Figure 8–9(a)]. When they reexamined the entry rate of genes, and thus the genetic maps for each strain, a definite pattern emerged. The major difference between each

(a)

Hfr strain	Order of transfer (earliest) → (latest)							
H	thr –	leu –	azi –	ton –	pro –	lac –	gal –	thi
1	leu –	thr –	thi –	gal –	lac –	pro –	ton –	azi
2	pro –	ton –	azi –	leu –	thr –	thi –	gal –	lac
7	ton –	azi –	leu –	thr –	thi –	gal –	lac –	pro

(b)

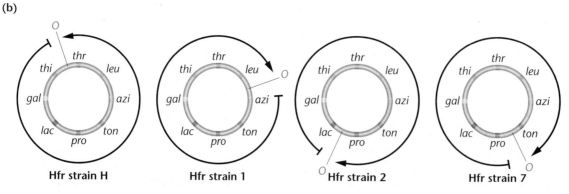

FIGURE 8–9 (a) The order of gene transfer in four Hfr strains, suggesting that the *E. coli* chromosome is circular. (b) The point where transfer originates (*O*) is identified in each strain. The origin is determined by the point of integration into the chromosome of the F factor, and the direction of transfer is determined by the orientation of the F factor as it integrates. The arrowheads indicate the points of initial transfer.

strain was simply the point of origin (*O*) and the direction in which entry proceeded from that point [Figure 8–9(b)].

To explain these results, Wollman and Jacob postulated that the *E. coli* chromosome is circular (a closed circle, with no free ends). If the point of origin (*O*) varies from strain to strain, a different sequence of genes will be transferred in each case. But what determines *O*? They proposed that in various Hfr strains, the F factor integrates into the chromosome at different points and that its position determines the site of *O*. A case of integration is shown in step 1 of Figure 8–10. During conjugation between an Hfr and an F⁻ cell, the position of the F factor determines the initial point of transfer

FIGURE 8–10 Conversion of F⁺ to an Hfr state occurs by integrating the F factor into the bacterial chromosome. The point of integration determines the origin (*O*) of transfer. During conjugation, an enzyme nicks the F factor, now integrated into the host chromosome, initiating transfer of the chromosome at that point. Conjugation is usually interrupted prior to complete transfer. Only the *A* and *B* genes are transferred to the F⁻ cell, which may recombine with the host chromosome. Newly replicated DNA of the chromosome is depicted by a lighter shade of orange.

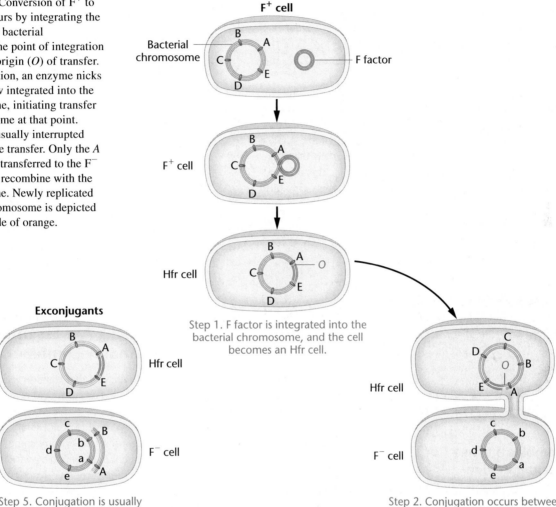

Step 1. F factor is integrated into the bacterial chromosome, and the cell becomes an Hfr cell.

Step 2. Conjugation occurs between an Hfr and F⁻ cell. The F factor is nicked by an enzyme, creating the origin of transfer of the chromosome (*O*).

Step 5. Conjugation is usually interrupted before the chromosome transfer is complete. Here, only the *A* and *B* genes have been transferred.

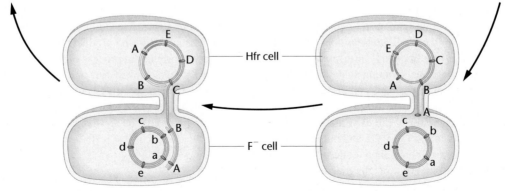

Step 4. Replication begins on both strands as chromosome transfer continues. The F factor is now on the end of the chromosome adjacent to the origin.

Step 3. Chromosome transfer across the conjugation tube begins. The Hfr chromosome rotates clockwise.

(steps 2 and 3). Those genes adjacent to O are transferred first, and the F factor becomes the last part that can be transferred (step 4). However, conjugation rarely, if ever, lasts long enough to allow the entire chromosome to pass across the conjugation tube (step 5). *This proposal explains why recipient cells, when mated with Hfr cells, remain F⁻.*

Figure 8–10 also depicts the way in which the two strands making up a DNA molecule behave during transfer, allowing for the entry of one strand of DNA into the recipient (see step 3). Following replication, the entering DNA now has the potential to recombine with its homologous region of the host chromosome. The DNA strand that remains in the donor also undergoes replication.

Use of the interrupted mating technique with different Hfr strains allowed researchers to map the entire *E. coli* chromosome. Mapped in time units, strain K12 (or *E. coli* K12) was shown to be 100 minutes long. While modern genome analysis of the *E. coli* chromosome has now established the presence of just over 4000 protein-coding sequences, this original mapping procedure established the location of approximately 1000 genes.

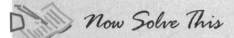
Now Solve This

Problem 25 on page 186 involves an understanding of how the bacterial chromosome is transferred during conjugation, leading to recombination and mapping. You are asked to draw a map of the bacterial chromosome.

Hint: Chromosome transfer is strain-specific and depends on the position within the chromosome where the F factor is integrated.

Recombination in F⁺ × F⁻ Matings: A Reexamination

The preceding experiment helped geneticists better understand how genetic recombination occurs during F⁺ × F⁻ matings. Recall that recombination occurs much less frequently than in Hfr × F⁻ matings and that random gene transfer is involved. The current belief is that when F⁺ and F⁻ cells are mixed, conjugation occurs readily and each F⁻ cell involved in conjugation with an F⁺ cell receives a copy of the F factor, *but no genetic recombination occurs.* However, at an extremely low frequency in a population of F⁺ cells, the F factor integrates spontaneously from the cytoplasm to a random point in the bacterial chromosome, converting the F⁺ cell to the Hfr state as we saw in Figure 8–10. Therefore, in F⁺ × F⁻ matings, the extremely low frequency of genetic recombination (10^{-7}) is attributed to the rare, newly formed Hfr cells, which then undergo conjugation with F⁻ cells. Because the point of integration of the F factor is random, the gene or genes that are transferred by any newly formed Hfr donor *will also appear to be random within the larger F⁺/F⁻ population.* The recipient bacterium

will appear as a recombinant but will remain F⁻. If it subsequently undergoes conjugation with an F⁺ cell, it will then be converted to F⁺.

The F′ State and Merozygotes

In 1959, during experiments with Hfr strains of *E. coli*, Edward Adelberg discovered that the F factor could lose its integrated status, causing the cell to revert to the F⁺ state (Figure 8–11, step 1). When this occurs, the F factor frequently carries several adjacent bacterial genes along with it (step 2). Adelberg labeled this condition **F′** to distinguish it from F⁺ and Hfr. F′, like Hfr, is thus another special case of F⁺, but this conversion is from Hfr to F′.

The presence of bacterial genes within a cytoplasmic F factor creates an interesting situation. An F′ bacterium behaves like an F⁺ cell by initiating conjugation with F⁻ cells (Figure 8–11, step 3). When this occurs, the F factor, containing chromosomal genes, is transferred to the F⁻ cell (step 4). As a result, whatever chromosomal genes are part of the F factor are now present as duplicates in the recipient cell (step 5) because the recipient still has a complete chromosome. This creates a partially diploid cell called a **merozygote**. Pure cultures of F′ merozygotes can be established. They have been extremely useful in the study of bacterial genetics, particularly in genetic regulation.

8.3 Rec Proteins Are Essential to Bacterial Recombination

Once researchers established that a unidirectional transfer of DNA occurs between bacteria, they became interested in determining how the actual recombination event occurs in the recipient cell. Just how does the donor DNA replace the comparable region in the recipient chromosome? As with many systems, the biochemical mechanism by which recombination occurs was deciphered through genetic studies. Major insights were gained as a result of isolating a group of mutations representing **rec genes**.

The first relevant observation involved a series of mutant genes labeled *recA*, *recB*, *recC*, and *recD*. The first mutant gene, *recA*, diminished genetic recombination in bacteria 1000-fold, nearly eliminating it altogether; the other *rec* mutations reduced recombination by about 100 times. Clearly, the normal wild-type products of these genes must play some essential role in the process of recombination.

Researchers looked for, and subsequently isolated, several functional gene products present in normal cells but missing in mutant cells and showed that they play a role in genetic recombination. The first is called the **RecA protein**.* The

---

*Note that the names of bacterial genes use lowercase letters and are italicized, while the corresponding gene products begin with capital letters and are not italicized, as illustrated by the *recA* gene and RecA protein.

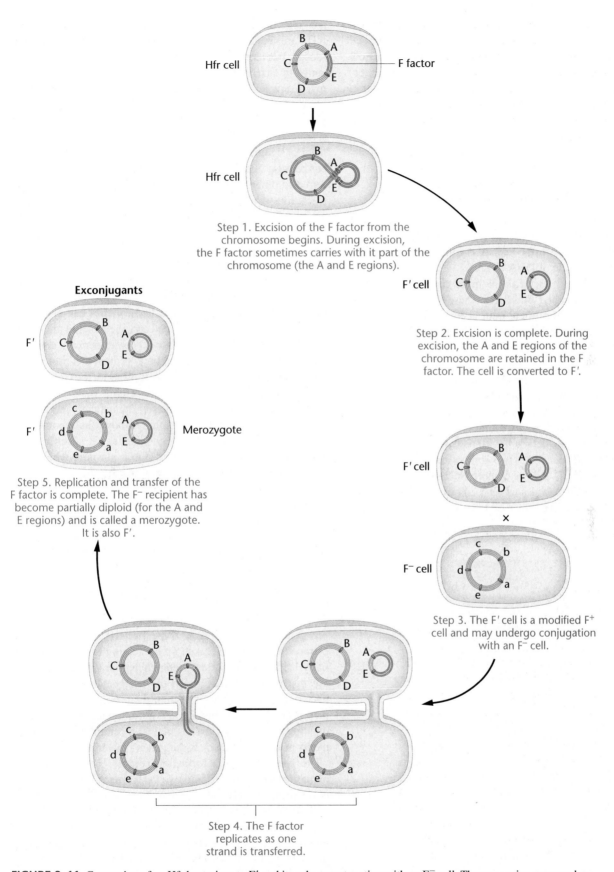

Step 1. Excision of the F factor from the chromosome begins. During excision, the F factor sometimes carries with it part of the chromosome (the A and E regions).

Step 2. Excision is complete. During excision, the A and E regions of the chromosome are retained in the F factor. The cell is converted to F'.

Step 3. The F' cell is a modified F⁺ cell and may undergo conjugation with an F⁻ cell.

Step 4. The F factor replicates as one strand is transferred.

Step 5. Replication and transfer of the F factor is complete. The F⁻ recipient has become partially diploid (for the A and E regions) and is called a merozygote. It is also F'.

Exconjugants

Merozygote

FIGURE 8–11 Conversion of an Hfr bacterium to F' and its subsequent mating with an F⁻ cell. The conversion occurs when the F factor loses its integrated status. During excision from the chromosome, it carries with it one or more chromosomal genes (*A* and *E*). Following conjugation with an F⁻ cell, the recipient cell becomes partially diploid and is called a merozygote; it also behaves as an F⁺ donor cell.

second is a more complex protein called the **RecBCD protein**, an enzyme consisting of polypeptide subunits encoded by three other *rec* genes. The roles of these proteins have now been elucidated *in vitro*. This genetic research has considerably extended our knowledge of the process of recombination and underscores the value of isolating mutations, establishing their phenotypes, and determining the biological role of the normal, wild-type gene.

8.4   F Factors Are Plasmids

In the preceding sections, we examined the extrachromosomal heredity unit called the F factor. When it exists autonomously in the bacterial cytoplasm, the F factor is composed of a double-stranded closed circle of DNA [Figure 8–12(a)]. These characteristics place the F factor in the more general category of genetic structures called **plasmids**. These structures contain one or more genes and often, quite a few. Their replication depends on the same enzymes that replicate the chromosome of the host cell, and they are distributed to daughter cells along with the host chromosome during cell division.

Plasmids are generally classified according to the genetic information specified by their DNA. The F factor plasmid confers fertility and contains the genes essential for sex pilus formation, on which genetic recombination depends. Other examples of plasmids include the R and Col plasmids.

Most **R plasmids** consist of two components: the **resistance transfer factor (RTF)** and one or more **r-determinants** [Figure 8–12(b)]. The RTF encodes genetic information essential to transferring the plasmid between bacteria, and the r-determinants are genes that confer resistance to antibiotics or mercury. While RTFs are similar in a variety of plasmids from different bacterial species, r-determinants are specific for resistance to one class of antibiotic and vary widely. Resistance to tetracycline, streptomycin, ampicillin, sulfonamide, kanamycin, or chloramphenicol is most frequently encountered. Sometimes these occur in a single plasmid, conferring multiple resistance to several antibiotics [Figure 8–12(b)]. Bacteria bearing these plasmids are of great medical significance not only because of their multiple resistance but because of the ease with which the plasmids can be transferred to other bacteria. Sometimes, a bacterial cell contains r-determinant plasmids but no RTF—the cell is resistant but cannot transfer the genetic information for resistance to recipient cells. The most commonly studied plasmids, however, contain the RTF as well as one or more r-determinants.

The **Col plasmid**, ColE1, (derived from *E. coli*), is clearly distinct from R plasmids. It encodes one or more proteins that are highly toxic to bacterial strains that do not harbor the same plasmid. These proteins, called **colicins**, can kill neighboring bacteria, and bacteria that carry the plasmid are said to be colicinogenic. Present in 10 to 20 copies per cell, a gene in the Col plasmid encodes an immunity protein that protects the host cell from the toxin. Unlike an R plasmid, the Col plasmid is not usually transmissible to other cells.

Interest in plasmids has increased dramatically because of their role in the genetic technology known as recombinant DNA research. As we will see in Chapter 17, specific genes from any source can be inserted into a plasmid, which may then be inserted into a bacterial cell. As the altered cell replicates its DNA and undergoes division, the foreign gene is also replicated, thus cloning the foreign genes.

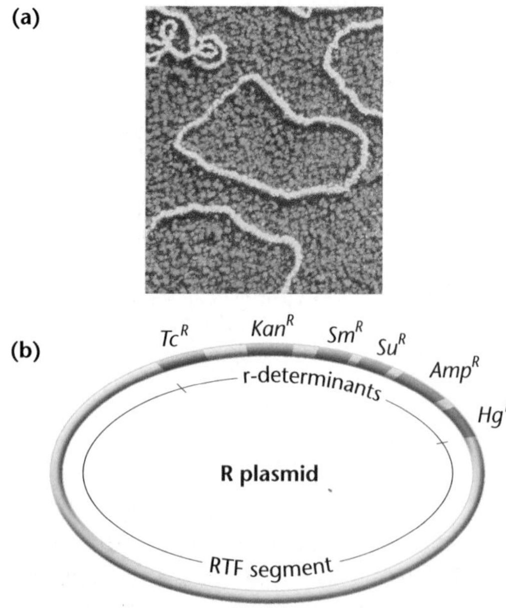

8.5   Transformation Is Another Process Leading to Genetic Recombination in Bacteria

Transformation provides another mechanism for recombining genetic information in some bacteria. Small pieces of extracellular DNA are taken up by a living bacterium, potentially leading to a stable genetic change in the recipient cell. We discuss transformation in this chapter because in those bacterial species where it occurs, the process can be used to map bacterial genes, though in a more limited way than conjugation. Transformation has played a central role in experiments proving that DNA is the genetic material.

The process of transformation (Figure 8–13) consists of numerous steps divided into two categories: (1) entry of DNA into a recipient cell and (2) recombination of the donor DNA with its homologous region in the recipient chromosome. In a population of bacterial cells, only those in the particular physiological state of **competence** take up DNA. Entry is thought to occur at a limited number of receptor sites on the surface of the bacterial cell (Figure 8–13, step 1). Passage into the cell is an active process that requires energy and specific transport molecules. This model is supported by the fact that substances that inhibit energy production or protein synthesis in the recipient cell also inhibit the transformation process.

(a)

(b)

FIGURE 8–12 (a) Electron micrograph of a plasmid isolated from *E. coli*. (b) An R plasmid containing resistance transfer factors (RTFs) and multiple r-determinants (Tc, tetracycline; Kan, kanamycin; Sm, streptomycin; Su, sulfonamide; Amp, ampicillin; and Hg, mercury).

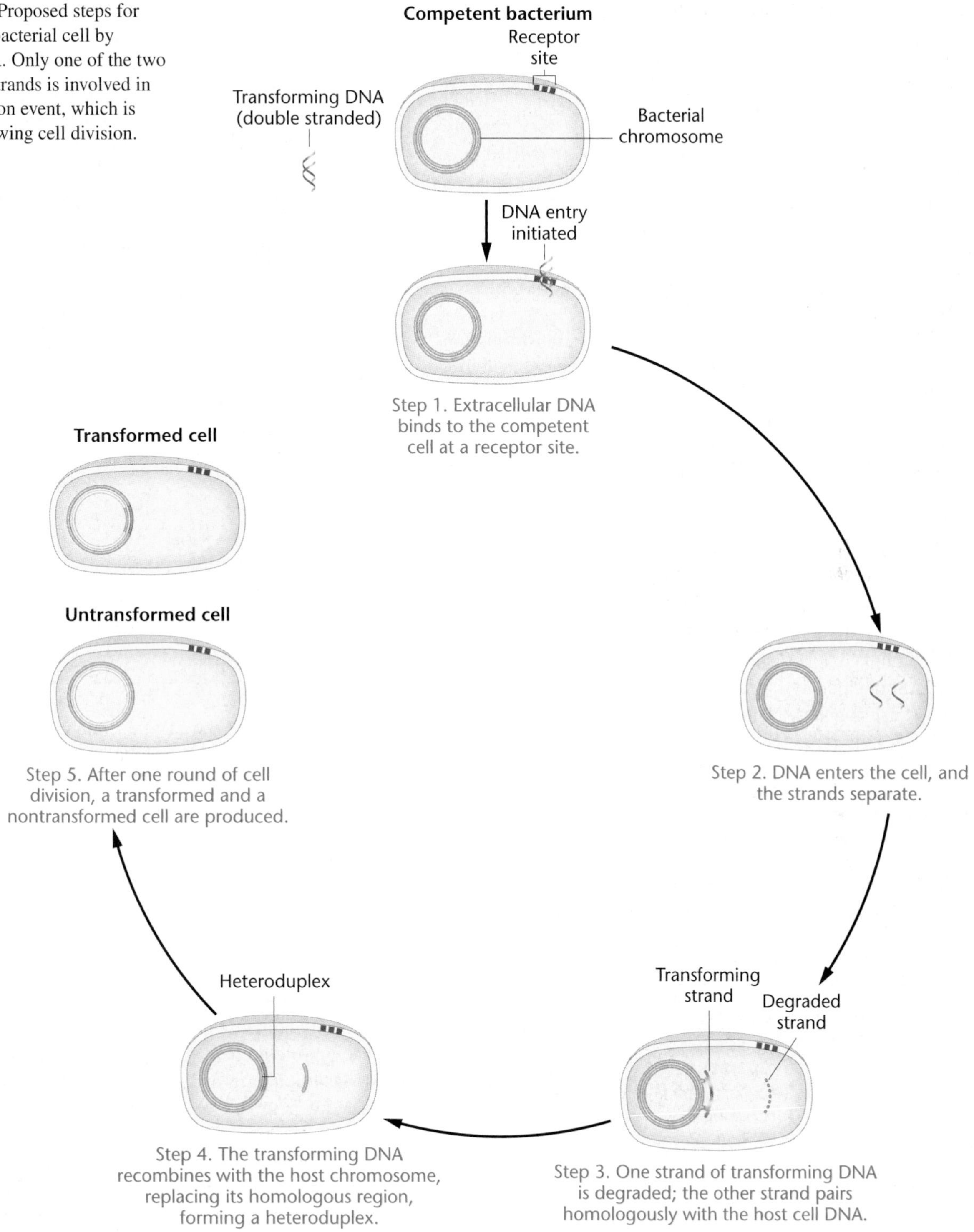

FIGURE 8–13 Proposed steps for transforming a bacterial cell by exogenous DNA. Only one of the two entering DNA strands is involved in the transformation event, which is completed following cell division.

Competent bacterium

Receptor site

Transforming DNA (double stranded)

Bacterial chromosome

DNA entry initiated

Step 1. Extracellular DNA binds to the competent cell at a receptor site.

Transformed cell

Untransformed cell

Step 5. After one round of cell division, a transformed and a nontransformed cell are produced.

Step 2. DNA enters the cell, and the strands separate.

Heteroduplex

Transforming strand   Degraded strand

Step 4. The transforming DNA recombines with the host chromosome, replacing its homologous region, forming a heteroduplex.

Step 3. One strand of transforming DNA is degraded; the other strand pairs homologously with the host cell DNA.

Soon after entry, one of the two strands of the double helix is digested by nucleases, leaving only a single strand to participate in transformation (Figure 8–13, steps 2 and 3). The surviving strand of DNA then aligns with its complementary region of the bacterial chromosome. In a process involving several enzymes, the segment replaces its counterpart in the chromosome, which is excised and degraded (step 4).

For recombination to be detected, the transforming DNA must be derived from a different strain of bacteria that bears some genetic variation, such as a mutation. Once it is integrated into the chromosome, the recombinant region contains one host strand (present originally) and one mutant strand. Because these strands are from different sources, this helical region is referred to as a **heteroduplex**. Following one round of DNA replication, one chromosome is restored to its original configuration, and the other contains the mutant gene. Following cell division, one untransformed cell (nonmutant) and one transformed cell (mutant) are produced (step 5).

Transformation and Linked Genes

For DNA to be effective in transformation, it must include between 10,000 and 20,000 nucleotide pairs, a length equal to about 1/200 of the *E. coli* chromosome—this size is sufficient to encode several genes. Genes adjacent to or very close to one another on the bacterial chromosome can be carried on a single segment of this size. Consequently, a single event can result in the **cotransformation** of several genes simultaneously. Genes that are close enough to each other to be cotransformed are *linked*. In contrast to *linkage groups* in eukaryotes, which indicate all genes on a single chromosome, linkage here refers to the close proximity of genes.

If two genes are not linked, simultaneous transformation occurs only as a result of two independent events involving two distinct segments of DNA. As in double crossovers in eukaryotes, the probability of two independent events occurring simultaneously is equal to the product of the individual probabilities. Thus, the frequency of two unlinked genes being transformed simultaneously is much lower than if they are linked.

Studies have shown that a variety of bacteria readily undergo transformation (e.g., *Hemophilus influenzae, Bacillus subtilis, Shigella paradysenteriae, Diplococcus pneumoniae,* and *E. coli*). Under certain conditions, the relative distances between linked genes can be determined from transformation data. Although analysis is more complex, such data are interpreted in a manner analogous to chromosome mapping in eukaryotes.

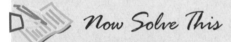

Now Solve This

Problem 9 on page 186 involves an understanding of how transformation can be used to determine if bacterial genes are "linked" in close proximity to one another. You are asked to predict the location of two genes, relative to one another.

Hint: Cotransformation occurs according to the laws of probability. Two "unlinked" genes are transformed as a result of two separate events. In such a case, the probability of that occurrence is equal to the product of the individual probabilities.

8.6 Bacteriophages Are Bacterial Viruses

Bacteriophages, or **phages** as they are commonly known, are viruses that have bacteria as their hosts. During their reproduction, phages can be involved in still another mode of bacterial genetic recombination called **transduction**. To understand this process, we must consider the genetics of bacteriophages, which themselves undergo recombination.

A great deal of genetic research has been done using bacteriophages as a model system, making them a worthy subject of discussion. In this section, we will first examine the structure and life cycle of one type of bacteriophage. We then dis-

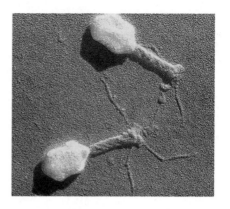

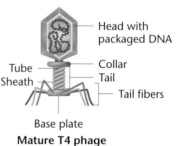

Mature T4 phage

FIGURE 8–14 The structure of bacteriophage T4 includes an icosahedral head filled with DNA, a tail consisting of a collar, tube, sheath, base plate, and tail fibers. During assembly, the tail components are added to the head, and then tail fibers are added.

cuss how these phages are studied during their infection of bacteria. Finally, we contrast two possible modes of behavior once the initial phage infection occurs. This information is background for our discussion of *transduction* and *bacteriophage recombination*.

Phage T4: Structure and Life Cycle

Bacteriophage T4 is one of a group of related bacterial viruses referred to as T-even phages. It exhibits the intricate structure shown in Figure 8–14. The phage T4's genetic material (DNA) is contained within an icosahedral (a polyhedron with 20 faces) protein coat, making up the head of the virus. The DNA is sufficient in quantity to encode more than 150 average-sized genes. The head is connected to a tail that contains a collar and a contractile sheath surrounding a central core. Tail fibers, which protrude from the tail, contain binding sites in their tips that specifically recognize unique areas of the outer surface of the cell wall of the bacterial host, *E. coli*.

The life cycle of phage T4 (Figure 8–15) is initiated when the virus binds by adsorption to the bacterial host cell. Then, an ATP-driven contraction of the tail sheath causes the central core to penetrate the cell wall. The DNA in the head is extruded, and it moves across the cell membrane into the bacterial cytoplasm. Within minutes, all bacterial DNA, RNA, and protein synthesis in the host cell is inhibited, and synthesis of viral molecules begins. At the same time, degradation of the host DNA is initiated.

A period of intensive viral gene activity characterizes infection. Initially, phage DNA replication occurs, leading to a pool of viral DNA molecules. Then, the components of the

FIGURE 8–15 Life cycle of bacteriophage T4.

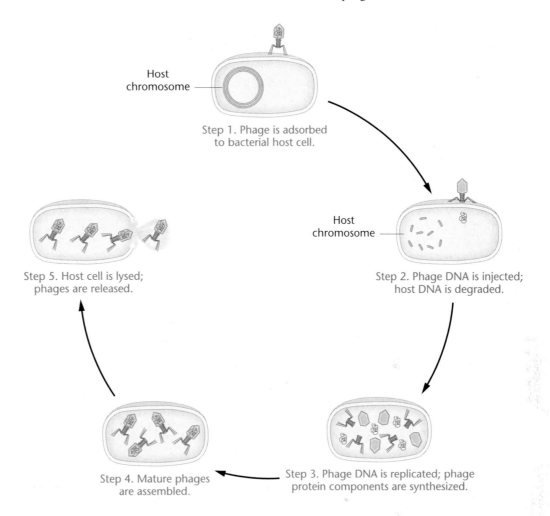

Step 1. Phage is adsorbed to bacterial host cell.

Host chromosome

Host chromosome

Step 2. Phage DNA is injected; host DNA is degraded.

Step 5. Host cell is lysed; phages are released.

Step 4. Mature phages are assembled.

Step 3. Phage DNA is replicated; phage protein components are synthesized.

head, tail, and tail fibers are synthesized. The assembly of mature viruses is a complex process that has been well studied by William Wood, Robert Edgar, and others. Three sequential pathways occur: (1) DNA packaging as the viral heads are assembled, (2) tail assembly, and (3) tail fiber asssembly. Once DNA is packaged into the head, it combines with the tail components, to which tail fibers are added. Total construction is a combination of self-assembly and enzyme-directed processes.

When approximately 200 new viruses have been constructed, the bacterial cell is ruptured by the action of the enzyme lysozyme (a phage gene product), and the mature phages are released from the host cell. The new phages infect other available bacterial cells, and the process repeats itself over and over again.

The Plaque Assay

Bacteriophages and other viruses have played a critical role in our understanding of molecular genetics. During infection of bacteria, enormous quantities of bacteriophages can be obtained for investigation. Often, over 10^{10} viruses are produced per milliliter of culture medium. Many genetic studies rely on the ability to quantify the number of phages produced following infection under specific culture conditions. The **plaque assay** is a routinely used technique, which is invaluable in mutational and recombinational studies of bacteriophages.

This assay is shown in Figure 8–16, where actual plaque morphology is also shown. A serial dilution of the original virally-infected bacterial culture is performed first. Then, a 0.1-mL sample (an *aliquot*) from a dilution is added to melted nutrient agar (about 3 mL) into which a few drops of a healthy bacterial culture have been added. The solution is then poured evenly over a base of solid nutrient agar in a petri dish and allowed to solidify before incubation. A clear area called a **plaque** occurs wherever a single virus initially infected one bacterium in the culture (the lawn) that has grown up during incubation. The plaque represents clones of the single infecting bacteriophage, created as reproduction cycles are repeated. If the dilution factor is too low, the plaques are plentiful, and they will fuse, lysing the entire lawn—which has occurred in the 10^{-3} dilution of Figure 8–16. On the other hand, if the dilution factor is increased, plaques can be counted and the density of viruses in the initial culture can be estimated,

initial phage density = (plaque number/mL) × (dilution factor)

Using the results shown in Figure 8–16, 23 phage plaques are derived from the 0.1-mL aliquot of the 10^{-5} dilution. Therefore, we estimate that there are 230 phages/mL *at this dilution* (since the initial aliquot was 0.1 mL). The initial phage density in the undiluted sample, where 23 plaques are observed from 0.1 mL, is then calculated as

initial phage density = $(230/\text{mL}) \times (10^5) = 230 \times 10^5/\text{mL}$

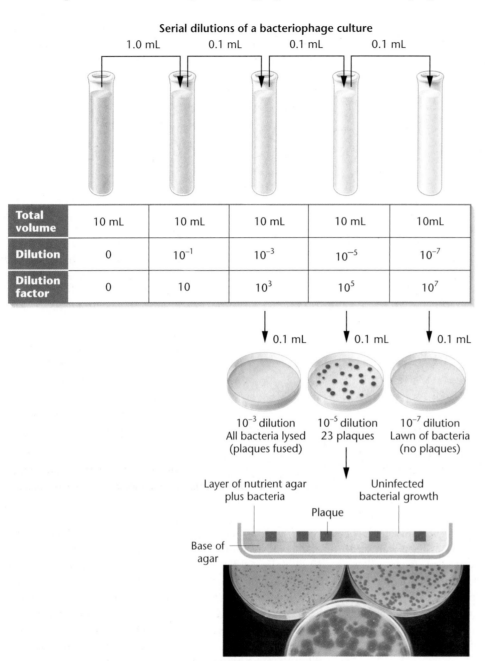

Serial dilutions of a bacteriophage culture

| | 1.0 mL | 0.1 mL | 0.1 mL | 0.1 mL |

Total volume	10 mL	10 mL	10 mL	10 mL	10mL
Dilution	0	10^{-1}	10^{-3}	10^{-5}	10^{-7}
Dilution factor	0	10	10^3	10^5	10^7

10^{-3} dilution
All bacteria lysed
(plaques fused)

10^{-5} dilution
23 plaques

10^{-7} dilution
Lawn of bacteria
(no plaques)

Layer of nutrient agar plus bacteria

Uninfected bacterial growth

Plaque

Base of agar

FIGURE 8–16 A plaque assay for bacteriophage analysis. Serial dilutions of a bacterial culture infected with bacteriophages are first made. Then three of the dilutions (10^{-3}, 10^{-5}, and 10^{-7}) are analyzed using the plaque assay technique. Each plaque represents the initial infection of one bacterial cell by one bacteriophage. In the 10^{-3} dilution, so many phages are present that all bacteria are lysed. In the 10^{-5} dilution, 23 plaques are produced. In the 10^{-7} dilution, the dilution factor is so great that no phages are present in the 0.1-mL sample, and thus no plaques form. From the 0.1-mL sample of the 10^{-5} dilution, the original bacteriophage density is calculated to be 23 × 10 × 10^5 phages/mL (23 × 10^6, or 2.3 × 10^7). The photograph shows phage T2 plaques on lawns of *E. coli*.

Because this figure is derived from the 10^{-5} dilution, we can also estimate that there will be only 0.23 phage/0.1 mL in the 10^{-7} dilution. Thus, when 0.1 mL from this tube is assayed, it is predicted that no phage particles will be present. This possibility is borne out in Figure 8–16, where an intact lawn of bacteria lacking any plaques is depicted. The dilution factor is simply too great.

Lysogeny

The relationship between a virus and a bacterium does not always result in viral reproduction and lysis. As early as the 1920s, it was known that a virus can enter a bacterial cell and coexist with it. The precise molecular basis of this relationship is now well understood. Upon entry, the viral DNA is integrated into the bacterial chromosome instead of replicating in the bacterial cytoplasm, a step that characterizes the developmental stage referred to as **lysogeny**. Subsequently,

each time the bacterial chromosome is replicated, the viral DNA is also replicated and passed to daughter bacterial cells following division. No new viruses are produced, and no lysis of the bacterial cell occurs. However, under certain stimuli, such as chemical or ultraviolet light treatment, the viral DNA loses its integrated status and initiates replication, phage reproduction, and lysis of the bacterium.

Several terms are used to describe this relationship. The viral DNA that integrates into the bacterial chromosome is a **prophage**. Viruses that either lyse the cell or behave as a prophage are **temperate phages**. Those that only lyse the cell are referred to as **virulent phages**. A bacterium harboring a prophage is **lysogenic**; that is, it is capable of being lysed as a result of induced viral reproduction. The viral DNA, which can either replicate in the bacterial cytoplasm or become integrated into the bacterial chromosome, is thus classified as an **episome**.

Transduction Is Virus-Mediated Bacterial DNA Transfer

In 1952, Norton Zinder and Joshua Lederberg were investigating possible recombination in the bacterium *Salmonella typhimurium*. Although they recovered prototrophs from mixed cultures of two different auxotrophic strains, investigation revealed that recombination was occurring in a manner different from that attributable to the presence of an F factor, as in *E. coli*. What they had discovered was a process of bacterial recombination mediated by bacteriophages and now called **transduction**.

The Lederberg-Zinder Experiment

Lederberg and Zinder mixed the *Salmonella* auxotrophic strains LA-22 and LA-2 together, and when the mixture was plated on minimal medium, they recovered prototrophic cells. The LA-22 strain was unable to synthesize the amino acids phenylalanine and tryptophan (phe^-, trp^-), and LA-2 could not synthesize the amino acids methionine and histidine (met^-, his^-). Prototrophs (phe^+, trp^+, met^+, his^+) were recovered at a rate of about $1/10^5$ (10^{-5}) cells.

Although these observations at first suggested that the recombination involved was the type observed earlier in conjugative strains of *E. coli*, experiments using the Davis U-tube soon showed otherwise (Figure 8–17). The two auxotrophic strains were separated by a sintered glass filter, thus preventing cell contact but allowing growth to occur in a common medi-

um. Surprisingly, when samples were removed from both sides of the filter and plated independently on minimal medium, prototrophs *were* recovered, but only from the side of the tube containing LA-22 bacteria. Recall that if conjugation were responsible, the conditions in the Davis U-tube would be expected to *prevent* recombination altogether (see Figure 8–4).

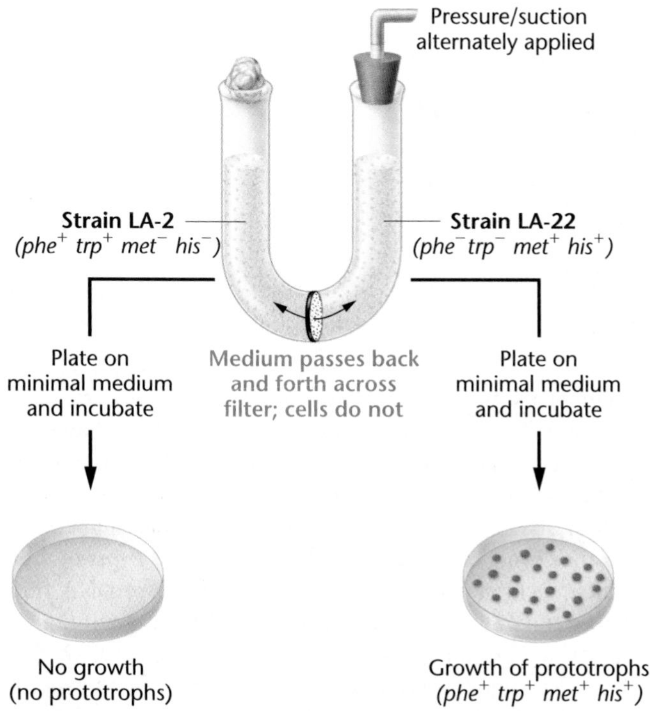

Since LA-2 cells appeared to be the source of the new genetic information (phe^+ and trp^+), how that information crossed the filter from the LA-2 cells to the LA-22 cells, allowing recombination to occur, was a mystery. The unknown source was designated simply as a **filterable agent** (**FA**).

Three observations were used to identify the FA:

1. The FA was produced by the LA-2 cells only when they were grown in association with LA-22 cells. If LA-2 cells were grown independently and that culture medium was then added to LA-22 cells, recombination did not occur. Therefore, LA-22 cells play some role in the production of FA by LA-2 cells and do so only when they share a common growth medium.

2. The addition of DNase, which enzymatically digests DNA, did not render the FA ineffective. Therefore, the FA is not naked DNA, ruling out transformation.

3. The FA could not pass across the filter of the Davis U-tube when the pore size was reduced below the size of bacteriophages.

Aided by these observations and aware that temperate phages can lysogenize *Salmonella*, researchers proposed that the genetic recombination event is mediated by bacteriophage P22, present initially as a prophage in the chromosome of the LA-22 *Salmonella* cells. They hypothesized that P22 prophages rarely enter the vegetative or lytic phase, reproduce, and are released by the LA-22 cells. Such P22 phages, being much smaller than a bacterium, then cross the filter of the U-tube and subsequently infect and lyse some of the LA-2 cells. In the process of lysis of LA-2, these P22 phages occasionally package a region of the LA-2 chromosome in their heads. If this region contains the phe^+ and trp^+ genes and the phages subsequently pass back across the filter and infect LA-22 cells, these newly lysogenized cells will behave as prototrophs. This process of transduction, whereby bacterial recombination is mediated by bacteriophage P22, is diagrammed in Figure 8–18.

The Nature of Transduction

Further studies have revealed the existence of transducing phages in other species of bacteria. For example, *E. coli* can be transduced by phages P1 and λ, and *B. subtilis* and *Pseudomonas aeruginosa* can be transduced by the phages SPO1 and F116, respectively. The details of several different modes of transduction have also been established. Even though the initial discovery of transduction involved a temperate phage and a lysogenized bacterium, the same process can occur during the normal lytic cycle. Sometimes a small piece of bacterial DNA is packaged *along with* the viral chromosome so that the transducing phage contains both viral and bacterial DNA. In such cases, only a few bacterial genes are present in the transducing phage. However, when *only*

Pressure/suction
alternately applied

Strain LA-2
(phe^+ trp^+ met^- his^-)

Strain LA-22
(phe^- trp^- met^+ his^+)

Plate on
minimal medium
and incubate

Medium passes back
and forth across
filter; cells do not

Plate on
minimal medium
and incubate

No growth
(no prototrophs)

Growth of prototrophs
(phe^+ trp^+ met^+ his^+)

FIGURE 8–17 The Lederberg-Zinder experiment using *Salmonella*. After placing two auxotrophic strains on opposite sides of a Davis U-tube, Lederberg and Zinder recovered prototrophs from the side with the LA-22 strain, but not from the side containing the LA-2 strain.

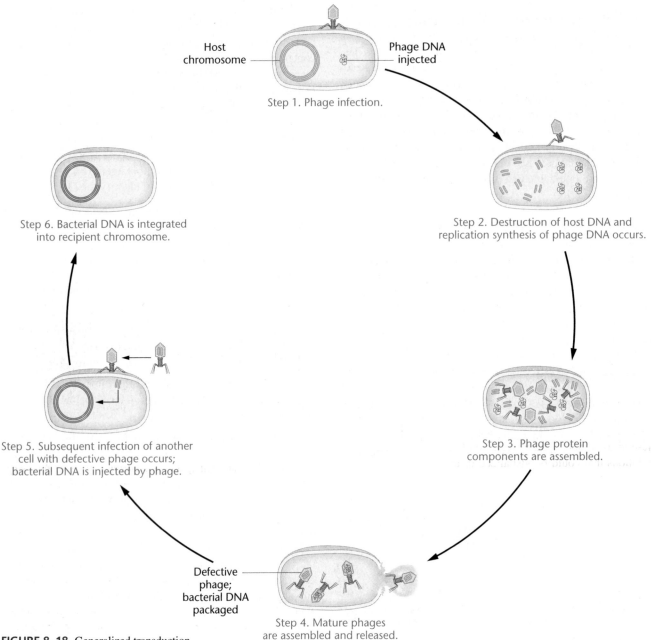

Host chromosome — Phage DNA injected

Step 1. Phage infection.

Step 2. Destruction of host DNA and replication synthesis of phage DNA occurs.

Step 3. Phage protein components are assembled.

Defective phage; bacterial DNA packaged

Step 4. Mature phages are assembled and released.

Step 5. Subsequent infection of another cell with defective phage occurs; bacterial DNA is injected by phage.

Step 6. Bacterial DNA is integrated into recipient chromosome.

FIGURE 8–18 Generalized transduction.

bacterial DNA is packaged, regions as large as 1 percent of the bacterial chromosome can become enclosed in the viral head. In either case, the ability to infect a host cell is unrelated to the type of DNA in the phage head, making transduction possible.

When bacterial rather than viral DNA is injected into the bacterium, it either remains in the cytoplasm or recombines with the homologous region of the bacterial chromosome. If the bacterial DNA remains in the cytoplasm, it does not replicate but is transmitted to one progeny cell following each division. When this happens, only a single cell, partially diploid for the transduced genes, is produced—a phenomenon called *abortive transduction*. If the bacterial DNA recombines with its homologous region of the bacterial chromosome, *complete transduction* occurs, where the transduced genes are replicated as part of the chromosome and passed to all daughter cells.

Both abortive and complete transduction are subclasses of the broader category of *generalized transduction*, which is characterized by the random nature of DNA fragments and genes that are transduced. Each fragment of the bacterial chromosome has a finite but small chance of being packaged in the phage head. Most cases of generalized transduction are of the abortive type; some data suggest that complete transduction occurs 10 to 20 times less frequently.

Transduction and Mapping

Like transformation, generalized transduction was used in linkage and mapping studies of the bacterial chromosome. The fragment of bacterial DNA involved in a transduction event is large enough to include numerous genes. As a result, two genes that closely align (are linked) on the bacterial chromosome can be simultaneously transduced, a process called **cotransduction**. Two genes that are not close enough to one another along the chromosome to be included on a single DNA fragment require two independent events to be transduced into a single cell. Since this occurs with a much lower probability than cotransduction, linkage can be determined.

By concentrating on two or three linked genes, transduction studies can also determine the precise order of these genes. The closer linked genes are to each other, the greater the frequency of cotransduction. Mapping studies involving three closely aligned genes can thus be executed, and the analysis of such an experiment is predicated on the same rationale underlying other mapping techniques.

8.8 Bacteriophages Undergo Intergenic Recombination

Around 1947, several research teams demonstrated that genetic recombination can be detected in bacteriophages. This led to the discovery that gene mapping can be performed in these viruses. Such studies relied on finding numerous phage mutations that could be visualized or assayed. As in bacteria and eukaryotes, these mutations allow genes to be identified and followed in mapping experiments. Thus, before considering recombination and mapping in these bacterial viruses, we briefly introduce several of the mutations that were studied.

Bacteriophage Mutations

Phage mutations often affect the morphology of the plaques formed following lysis of bacterial cells. For example, in 1946, Alfred Hershey observed unusual T2 plaques on plates of *E. coli* strain B. Normal T2 plaques are small and have a clear center surrounded by a diffuse (nearly invisible) halo, but the unusual plaques were larger and possessed a distinctive outer perimeter (Figure 8–19). When the viruses were isolated from these plaques and replated on *E. coli* B cells, the resulting plaque appearance was identical. Thus, the plaque phenotype was an inherited trait resulting from the reproduction of mutant phages. Hershey named the mutant *rapid lysis* (*r*) because the plaques were larger, apparently resulting from a more rapid or more efficient life cycle of the phage. We now know that in wild-type phages, reproduction is inhibited once a particular-sized plaque has been formed. The *r* mutant T2 phages overcome this inhibition, producing larger plaques.

Salvador Luria discovered another bacteriophage mutation, *host range* (*h*). This mutation extends the range of bacterial hosts that the phage can infect. Although wild-type T2 phages can infect *E. coli* B (a unique strain), they normally

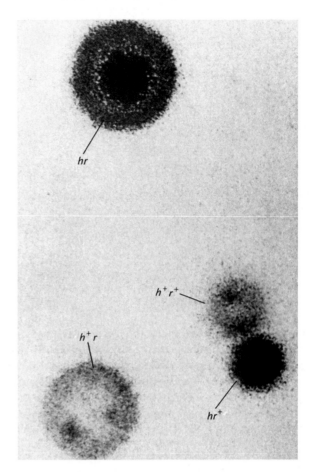

FIGURE 8–19 Plaque morphology phenotypes observed following simultaneous infection of *E. coli* by two strains of phage T2, $h^+ r$ and $h r^+$. In addition to the parental genotypes, recombinant plaques $h r$ and $h^+ r^+$ are shown.

cannot attach or be adsorbed to the surface of *E. coli* B-2 (a different strain). The *h* mutation, however, provides the basis for adsorption and subsequent infection of *E. coli* B-2. When grown on a mixture of *E. coli* B and B-2, the center of the *h* plaque appears much darker than the h^+ plaque (Figure 8–19).

Table 8.1 lists other types of mutations that have been isolated and studied in the T-even series of bacteriophages (e.g., T2, T4, T6). These mutations are important to the study of genetic phenomena in bacteriophages.

Mapping in Bacteriophages

Genetic recombination in bacteriophages was discovered during **mixed infection experiments** in which two distinct mutant strains were allowed to *simultaneously* infect the same bacterial culture. These studies were designed so that the number of viral particles sufficiently exceeded the number of bacterial cells to ensure simultaneous infection of most cells by both viral strains. If two loci are involved, recombination is referred to as intergenic.

For example, in one study using the T2/*E. coli* system, the parental viruses were of either the $h^+ r$ (wild-type host range,

TABLE 8.1	Some Mutant Types of T-even Phages
Name	**Description**
minute	Small plaques
turbid	Turbid plaques on *E. coli* B
star	Irregular plaques
UV-sensitive	Alters UV sensitivity
acriflavin-resistant	Forms plaques on acriflavin agar
osmotic shock	Withstands rapid dilution into distilled water
lysozyme	Does not produce lysozyme
amber	Grows in *E. coli* K12, but not B
temperature-sensitive	Grows at 25°C, but not at 42°C

TABLE 8.2	Results of a Cross Involving the *h* and *r* Genes in Phage T2 $(hr^+ \times h^+r)$	
Genotype	**Plaques**	**Designation**
$h\ r^+$	42 ⎱	
$h^+\ r$	34 ⎰	Parental progeny 76%
$h^+\ r^+$	12 ⎱	
$h\ r$	12 ⎰	Recombinants 24%

Source: Data derived from Hershey and Rotman, 1949.

rapid lysis) or $h\ r^+$ (extended host range, normal lysis) genotype. If no recombination occurred, these two parental genotypes would be the only expected phage progeny. However, the recombinants $h^+\ r^+$ and $h\ r$ were detected in addition to the parental genotypes (see Figure 8–19). As with eukaryotes, the percentage of recombinant plaques divided by the total number of plaques reflects the relative distance between the genes.

recombinational frequency = $(h^+r^+ + h\ r)$/total plaques $\times$ 100

Sample data for the *h* and *r* loci are shown in Table 8.2.

Similar recombinational studies have been conducted with numerous mutant genes in a variety of bacteriophages. Data are analyzed in much the same way as in eukaryotic mapping experiments. Two- and three-point mapping crosses are possible, and the percentage of recombinants in the total number of phage progeny is calculated. This value is proportional to the relative distance between two genes along the DNA molecule constituting the chromosome.

Investigations into phage recombination support a model similar to that of eukaryotic crossing over—a breakage and reunion process between the viral chromosomes. A fairly clear picture of the dynamics of viral recombination has emerged. After the early phase of infection, the chromosomes of the phages begin replication. As this stage progresses, a pool of phage chromosomes accumulates in the bacterial

cytoplasm. If double infection by phages of two genotypes has occurred, then the pool of chromosomes initially consists of the two parental types. Genetic exchange between these two types will occur before, during, and after replication, producing recombinant chromosomes.

In the case of the $h^+\ r$ and $h\ r^+$ example discussed here, recombinant $h^+\ r^+$ and $h\ r$ chromosomes are produced. Each of these chromosomes can undergo replication, with new replicates exchanging with each other and with parental chromosomes. Furthermore, recombination is not restricted to exchanges between two chromosomes—three or more can be involved simultaneously. As phage development progresses, chromosomes are randomly removed from the pool and packed into the phage head, forming mature phage particles. Thus, a variety of parental and recombinant genotypes are represented in progeny phages.

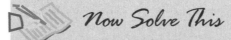

Now Solve This

Problem 21 on page 186 involves an understanding of intergenic mapping as performed in bacteriophages. You are asked to construct a map from recombination data.

Hint: Mapping is based on recombination occurring during simultaneous infection of a bacterium by two or more strain-specific bacteriophages. Mapping in phages is similar to three-point mapping in eukaryotes. The probability of recombination varies directly with the distance between two genes along the bacteriophage chromosome.

GENETICS, TECHNOLOGY, AND SOCIETY

Bacterial Genes and Disease: From Gene Expression to Edible Vaccines

Using an expanding toolbox of molecular genetic tools, scientists are tackling some of the most serious bacterial diseases affecting our own species. As we will see, our newly acquired understanding of bacterial genes is leading directly to exciting new treatments based on edible vaccines. The story of vaccines against cholera and hepatitis B are models for this research. We will focus on cholera.

The causative agent of cholera is *Vibrio cholerae*, a curved, rod-shaped bacterium found mostly in rivers and oceans. Most genetic strains of *V. cholerae* are harmless; only a few are pathogenic. Infection occurs when a person drinks water or eats food contaminated with pathogenic *V. cholerae*. Once in the digestive system, these bacteria colonize the small intestine and produce proteins called enterotoxins that invade the mucosal cells lining the intestine. This triggers a massive secretion of water and dissolved salts resulting in violent diarrhea, severe dehydration, muscle cramps, lethargy, and often death. The enterotoxin contains two polypeptides, called the A and B subunits, encoded by two separate genes.

Cholera remains a leading cause of death of infants and children throughout the Third World, where basic sanitation is lacking and water supplies are often contaminated. In July of 1994, 70,000 cases of cholera, leading to 12,000 fatalities, were reported among the Rwandans crowded into refugee camps in Goma, Zaire. And after an absence of over 100 years, cholera reappeared in Latin America in 1991, spreading from Peru to Mexico and claiming more than 10,000 lives. In 2001, more than 40 cholera outbreaks in 28 countries were reported to the World Health Organization.

A new genetic technology is emerging to attack cholera. This technology focuses on genetically engineered plants that act as vaccines. When a foreign gene is inserted into the plant genome, transcribed, and translated, a transgenic plant is produced that contains the foreign gene product. Immunization then requires eating these plants as the antigen stimulates the production of antibodies to protect against bacterial infection. Since it is the B subunit of the cholera enterotoxin that binds to intestinal cells, attention has focused on this polypeptide as the antigen, reasoning that antibodies against it will prevent toxin binding and render the bacteria harmless.

Leading the efforts to develop an edible vaccine has been Charles Arntzen and associates at Cornell University. To test the system, Arntzen is using the B subunit of an *E. coli* enterotoxin, which is similar in structure and immunological properties to the cholera protein. The first step was to obtain the DNA clone of the gene encoding the B subunit and to attach it to a promoter that would induce transcription in all tissues of the plant. Second, the hybrid gene was introduced into potato plants by means of *Agrobacterium*-mediated transformation. Arntzen and his colleagues chose the potato so they could assay the effectiveness of the antigen in the edible part of the plant, the tuber. Analysis showed that the engineered plants expressed their new gene and produced the enterotoxin B subunit. The third step was to feed mice a few grams of these genetically engineered tubers. Arntzen found that the mice produced specific antibodies against the B subunit and secreted these into the small intestine. And, most critically, mice later fed purified enterotoxin were protected from its effects and did not develop the symptoms of cholera. In clinical trials conducted using humans in 1998, almost all of the volunteers developed an immune response, and none experienced adverse side effects.

The Arntzen group is also producing edible vaccines in bananas and tomatoes. Bananas have several advantages over potatoes. They can be grown almost anywhere throughout the tropical or subtropical developing countries of the world, and unlike potatoes, bananas are usually eaten raw, avoiding the potential inactivation of the antigenic proteins by cooking. Finally, bananas are well liked by infants and children, making this approach to immunization a more feasible one. If all goes as planned, it may someday be possible to immunize all Third World children against cholera.

Arntzen's experimentation has served as a model for other research efforts involving a variety of human diseases. A group from the John P. Robarts Institute in Ontario, Canada, has shown that potato-produced vaccines can prevent juvenile diabetes in mice. Meristem Therapeutics, based in France, is in clinical trials using corn geared toward alleviating the effects of cystic fibrosis. An Australian research team has successfully produced tobacco plants that contain a protein found in the measles virus. After demonstrating the induction of immunity by feeding mice extracts of the tobacco leaves, testing of the measles vaccine has begun on primates. Tobacco plants are also being used to produce the antiviral protein interleukin 10 to treat Crohn's disease. Similar efforts are underway to create vaccines against rabies, anthrax, tetanus, and AIDS.

Although edible vaccines appear to have a promising future, many issues remain to be addressed. One problem has been dosage. Plants may vary in the concentration of the antigen. The amount of the potato or the banana ingested might also vary, altering the dose consumed. One approach is to shift to capsules containing plant extracts in order to standardize doses of the antigen. There are also concerns over potential environmental hazards associated with growing transgenic crops. In particular, measures must be taken to prevent transgenes from being introduced into native plant populations. In addition, citizens of some countries are morally opposed to genetically engineered foods. If these obstacles can be overcome, edible vaccines hold great promise for the amelioration of many human diseases.

References

Ariza, L. M. 2005. Defensive eating. *Scien. Amer.* May 2005, p. 25.

Haq, T. A., Mason, H. S., Clements, J. D., and Arntzen, C. J. 1995. Oral immunization with a recombinant bacterial antigen produced in transgenic plants. *Science* 268: 714–776.

Mason, H. S., Warzecha, H., Mor, T., and Arntzen, C. J. 2002. Edible plant vaccines: Applications for prophylactic and therapeutic molecular medicine. *Trends Mol. Med.* 8: 24–329.

CHAPTER SUMMARY

1. Inherited phenotypic variation in bacteria results from spontaneous mutation.

2. Genetic recombination in bacteria can result from three different modes: conjugation, transformation, and transduction.

3. Conjugation is initiated by a bacterium housing a plasmid called the F factor. If the F factor is in the cytoplasm of a donor cell (F^+), the recipient F^- cell receives a copy of the F factor, converting it to the F^+ status.

4. If the F factor is integrated into the donor cell chromosome (Hfr), recombination is initiated with the recipient F^- cell and genetic information flows unidirectionally to it. Time mapping of the bacterial chromosome is based on the orientation and position of the F factor in the donor chromosome.

5. The products of a group of genes designated as *rec* are directly involved in recombination between the invading DNA and the recipient bacterial chromosome.

6. Plasmids, such as the F factor, are autonomously replicating DNA molecules found in the bacterial cytoplasm. Some plasmids contain unique genes conferring antibiotic resistance, as well as those necessary for their transfer during conjugation.

7. In the phenomenon of transformation, which does not require cell contact, exogenous DNA enters the cell and recombines with the host chromosome of a recipient bacterium. Linkage mapping of closely aligned genes may be performed using this process.

8. Bacteriophages (viruses that infect bacteria) demonstrate a defined life cycle during which they reproduce within the host cell. They are studied using the plaque assay.

9. Bacteriophages can be lytic, where they infect the host cell and reproduce, then lyse the host cell; or they can lysogenize the host cell, where they infect it and integrate their DNA into the host chromosome, but do not reproduce.

10. Transduction is virus-mediated bacterial recombination. When a lysogenized bacterium subsequently reenters the lytic cycle, the new bacteriophages serve as the vehicles for the transfer of host (bacterial) DNA. In the process of generalized transduction, a random part of the bacterial chromosome is transferred.

11. Transduction is also used for bacterial linkage and mapping studies.

12. Various mutant phenotypes, including plaque morphology and host range, have been studied in bacteriophages. These have served as the basis for investigating genetic exchange and mapping in these viruses.

KEY TERMS

adaptation hypothesis, 165
auxotroph, 165
bacteria, 165
bacteriophage, 165, 176
cholera, 183
Col plasmid, 174
colicin, 174
competence, 174
conjugation, 166
cotransduction, 181
cotransformation, 176
edible vaccines, 183
episome, 178
F^- cell, 167
F^+ cell, 167
F' cell, 172
F factor, 168
F pilus, 168

fertility factor, 168
filterable agent (FA), 179
fluctuation test, 165
genetic recombination, 166
heteroduplex, 175
high-frequency recombination (Hfr), 168
interrupted mating technique, 169
lag phase, 166
logarithmic (log) phase, 166
lysogenic, 178
lysogeny, 178
merozygote, 172
minimal medium, 165
mixed infection experiments, 181
phage, 176
plaque, 177
plaque assay, 177
plasmid, 174

prophage, 178
prototroph, 165
r-determinants, 174
R plasmids, 174
RecA protein, 172
RecBCD protein, 174
rec genes, 172
resistance transfer factor (RTF), 174
serial dilution, 166
sex pilus, 168
spontaneous mutation, 165
stationary phase, 166
temperate phage, 178
transduction, 176, 179
transformation, 174
virulent phage, 178

INSIGHTS AND SOLUTIONS

1. Time mapping is performed in a cross involving the genes *his*, *leu*, *mal*, and *xyl*. The recipient cells are auxotrophic for all four genes. After 25 minutes, mating is interrupted with the results in recipient cells shown below. Diagram the positions of these genes relative to the origin (*O*) of the F factor and to one another.

90% are *xyl*⁺

80% are *mal*⁺

20% are *his*⁺

none are *leu*⁺

Solution: The *xyl* gene is transferred most frequently, so it is closest to *O* (very close). The *mal* gene is next and reasonably close to *xyl*, followed by the more distant *his* gene. The *leu* gene is far beyond these three, since no recovered recombinants include it. The diagram shows these relative locations along a piece of the circular chromosome.

2. In four Hfr strains of bacteria, all derived from an original F⁺ culture grown over several months, a group of hypothetical genes is studied and shown to transfer in the orders shown in the following table. (a) Assuming *B* is the first gene along the chromosome, determine the sequence of all genes shown. (b) One strain creates an apparent dilemma. Which one is it? Explain why the dilemma is only apparent, not real.

Hfr Strain	Order of Transfer					
1	E	R	I	U	M	B
2	U	M	B	A	C	T
3	C	T	E	R	I	U
4	R	E	T	C	A	B

Solution:

(a) The sequence is found by overlapping the genes in each strain.

Strain 2	U	M	B	A	C	T		
Strain 3				C	T	E	R	I	U
Strain 1					E	R	I	U	M	B

Starting with *B* in strain 2, the gene sequence is *BACTERIUM*.

(b) Strain 4 creates a dilemma, which is resolved when we realize that the F factor is integrated in the opposite orientation. Thus, the genes enter in the opposite sequence, starting with gene *R*.

$$\overrightarrow{RETCAB}$$

3. Three strains of bacteria, each bearing a separate mutation, a^-, b^-, or c^-, are the sources of donor DNA in a transformation experiment. Recipient cells are wild type for those genes but express the mutation d^-. (a) Based on the data below and assuming that the location of the *d* gene precedes the *a*, *b*, and *c* genes, propose a linkage map for these four genes. (b) If the donor DNA is wild type and the recipient cells are either $a^- b^-$, $a^- c^-$, or $b^- c^-$, in which case would wild-type transformants be expected most frequently?

DNA Donor	Recipient	Transformants	Frequency of Transformants
$a^- d^+$	$a^+ d^-$	$a^+ d^+$	0.21
$b^- d^+$	$b^+ d^-$	$b^+ d^+$	0.18
$c^- d^+$	$c^+ d^-$	$c^+ d^+$	0.63

Solution:

(a) These data reflect the relative distances between each of the *a*, *b*, and *c* genes and the *d* gene. The *a* and *b* genes are about the same distance from the *d* gene and are thus tightly linked to one another. The *c* gene is more distant. Assuming that the *d* gene precedes the others, the map looks like this:

(b) Because the *a* and *b* genes are closely linked, they most likely cotransform in a single event. Thus, recipient cells of $a^- b^-$ are most likely to convert to wild type.

PROBLEMS AND DISCUSSION QUESTIONS

1. Distinguish among the three modes of recombination in bacteria.
2. With respect to F^+ and F^- bacterial matings, (a) How was it established that physical contact was necessary? (b) How was it established that chromosome transfer was unidirectional? (c) What is the genetic basis of a bacterium being F^+?
3. List all of the differences between $F^+ \times F^-$ and Hfr $\times F^-$ bacterial crosses.
4. List all of the differences between F^+, F^-, Hfr, and F' bacteria.
5. Describe the basis for chromosome mapping in the Hfr $\times F^-$ crosses.
6. Why are the recombinants produced from an Hfr $\times F^-$ cross rarely if ever F^+?
7. Describe the origin of F' bacteria and merozygotes.
8. Describe the mechanism of transformation.
9. In a transformation experiment involving a recipient bacterial strain of genotype $a^- b^-$ the results below were obtained. What can you conclude about the location of the a and b genes relative to each other?

	Transformants (%)		
Transforming DNA	a^+b^-	a^-b^+	a^+b^+
$a^+ b^+$	3.1	1.2	0.04
$a^+ b^-$ and $a^- b^+$	2.4	1.4	0.03

10. In a transformation experiment, donor DNA was obtained from a prototroph bacterial strain $(a^+ b^+ c^+)$, and the recipient was a triple auxotroph $(a^- b^- c^-)$. What general conclusions can you draw about the linkage relationships among the three genes from the following transformant classes that were recovered?

$a^+ b^- c^-$	180
$a^- b^+ c^-$	150
$a^+ b^+ c^-$	210
$a^- b^- c^+$	179
$a^+ b^- c^+$	2
$a^- b^+ c^+$	1
$a^+ b^+ c^+$	3

11. The bacteriophage genome consists primarily of genes encoding proteins that make up the head, collar and tail, and tail fibers. When these genes are transcribed following phage infection, how are these proteins synthesized, since the phage genome lacks genes essential to ribosome structure?
12. Describe the temporal sequence of the bacteriophage life cycle.
13. In the plaque assay, what is the precise origin of a single plaque?
14. In the plaque assay, exactly what makes up a single plaque?
15. A plaque assay is perfomed beginning with 1.0 mL of a solution containing bacteriophages. This solution is serially diluted three times by taking 0.1 mL and adding it to 9.9 mL of liquid medium. The final dilution is plated and yields 17 plaques. What is the initial density of bacteriophages in the original 1.0 mL?
16. Describe the difference between the lytic cycle and lysogeny when bacteriophage infection occurs.
17. Define prophage.

18. Explain the observations that led Zinder and Lederberg to conclude that the prototrophs recovered in their transduction experiments were not the result of Hfr-mediated conjugation.
19. Describe generalized transduction and distinguish between abortive and complete transduction.
20. Describe how generalized transduction can be used to map the bacterial chromosome. How does cotransduction play a role in mapping?
21. Two theoretical genetic strains of a virus $(a^- b^- c^-$ and $a^+ b^+ c^+)$ are used to simultaneously infect a culture of host bacteria. Of 10,000 plaques scored, the genotypes in the following table were observed. Determine the genetic map of these three genes on the viral chromosome.

$a^+ b^+ c^+$	4100	$a^- b^+ c^-$	160
$a^- b^- c^-$	3990	$a^+ b^- c^+$	140
$a^+ b^- c^-$	740	$a^- b^- c^+$	90
$a^- b^+ c^+$	670	$a^+ b^+ c^-$	110

22. Describe the conditions under which genetic recombination may occur in bacteriophages.
23. If a single bacteriophage infects one *E. coli* cell present in a culture of bacteria and, upon lysis, yields 200 viable viruses, how many phages will exist in a single plaque if three more lytic cycles occur?
24. A phage-infected bacterial culture was subjected to a series of dilutions, and a plaque assay was performed in each case, with the following results. What conclusion can be drawn in the case of each dilution?

Dilution Factor	Assay Results
10^4	All bacteria lysed
10^5	14 plaques
10^6	0 plaques

25. When the interrupted mating technique was used with five different strains of Hfr bacteria, the order of gene entry during recombination shown in the following table was observed. On the basis of these data, draw a map of the bacterial chromosome. Do the data support the concept of circularity?

Hfr Strain	Order				
1	T	C	H	R	O
2	H	R	O	M	B
3	M	O	R	H	C
4	M	B	A	K	T
5	C	T	K	A	B

26. In *B. subtilis*, linkage analysis of two mutant genes affecting the synthesis of the two amino acids, tryptophan (typ_2^-), and tyrosine (tyr_1^-), was performed using transformation. (E. Nester, M. Schafer, and J. Lederberg (1963) *Genetics* 48: 529–551.) Examine the data in the table that follows, and draw all possible conclusions regarding linkage.

	Donor DNA	Recipient Cell	Transfor-mants	Number of Transformants
			$trp^+ tyr^-$	196
Part A	$trp_2^+ tyr_1^+$	$trp_2^- tyr_1^-$	$trp^- tyr^+$	328
			$trp^+ tyr^+$	367
	$trp_2^+ tyr_1^-$		$trp^+ tyr^-$	190
Part B	and	$trp_2^- tyr_1^-$	$trp^- tyr^+$	256
	$trp_2^- tyr_1^+$		$trp^+ tyr^+$	2

27. What is the role of part B in the experiment shown in Problem 26?

28. An Hfr strain is used to map three genes in an interrupted mating experiment. The cross is $Hfr/a^+b^+c^+rif^s \times F^-/a^-b^-c^-rif^r$ (no map order is implied in the listing of the alleles; rif = the antibiotic rifampicin). The a^+ gene is required for biosynthesis of nutrient A, the b^+ gene for nutrient B, and c^+ for nutrient C. The minus alleles are auxotrophs for these nutrients. The cross is initiated at time = 0, and, at various intervals, the mating mixture is plated on three types of medium. Each plate contains minimal medium (MM) *plus* rifampicin *plus* the specific supplements indicated in the following table (results for each time point are shown as number of colonies growing on each plate). (a) What is the purpose of rifampicin in the experiment? (b) Based on these data, determine the approximate location on the chromosome of the a, b, and c genes relative to one another and to the F factor. (c) Can the location of the rif gene be determined in this experiment? If not, design an experiment to determine the location of rif relative to the F factor and to gene b.

	Time of Interruption (min)			
Supplements Added to MM	5	10	15	20
Nutrients A and B	0	0	4	21
Nutrients B and C	0	5	23	40
Nutrients A and C	4	25	60	82

29. In a cotransformation experiment using various combinations of genes, two at a time, the following data were produced. Determine which genes are linked and to which others.

Successful Cotransformation	Unsuccessful Cotransformation
a and d; b and c; b and f	a and b; a and c; a and f
	d and b; d and c; d and f
	a and e; b and e; c and e
	d and e; f and e

30. In Problem 29, another gene, g, was also studied. It demonstrated positive cotransformation when tested with gene f. Predict the results of experiments when it was tested with genes a, b, c, d, and e.

31. Bacterial conjugation, mediated mainly by conjugative plasmids like F, represents a potential health threat through the sharing of genes for pathogenicity or antibiotic resistance. Given that more than 400 different species of bacteria co-inhabit a healthy human gut and more than 200 co-inhabit human skin, Dionisio and colleagues. investigated the ability of plasmids to undergo between-species conjugal transfer. Data (Dionisio et al. 2002. *Genetics* 162: 1525–1532) involving various species of the enterobacterial genus, *Escherichia*, are presented in the following table. The data are presented as "log base 10" values, where for example, -2.0 would be equivalent to 10^{-2} as a rate of transfer. Assume that all differences between values are statistically significant. (a) What general conclusion(s) can be drawn from these data? (b) In what species is within-species transfer most likely? In what species pair is between-species transfer most likely? (c) What is the significance of these findings in terms of human health?

	Donor			
Recipient	*E. chrysanthemi*	*E. blattae*	*E. fergusonii*	*E. coli*
E. chrysanthemi	-2.4	-4.7	-5.8	-3.7
E. blattae	-2.0	-3.4	-5.2	-3.4
E. fergusonii	-3.4	-5.0	-5.8	-4.2
E. coli	-1.7	-3.7	-5.3	-3.5

32. A study was conducted in an attempt to determine which functional regions of a particular conjugative transfer gene (*tra*I) are involved in the transfer of plasmid R27 in *Salmonella enterica*. The R27 plasmid is of significant clinical interest because it is capable of encoding multiple-antibiotic resistance to typhoid fever. To identify functional regions responsible for conjugal transfer, an analysis (modified from Lawley et al. 2002. *J. Bacteriol.* 184: 2173–2180) was conducted whereby particular regions of the *tra*I gene were mutated and tested for their impact on conjugation. Shown at the bottom of the page is a map of the regions tested and believed to be involved in conjugative transfer of the R27 plasmid. Similar shading indicates related function, and the numbers correspond to each functional region subjected to mutation analysis. Above the map is a table showing the effects of these mutations on R27 conjugation. (a) Given the data, do all functional regions appear to influence conjugative transfer similarly? (b) Which regions appear to have the most impact on conjugation? (c) Which regions appear to have a limited impact on conjugation? (d) What general conclusions might one draw from these data?

Effects of Mutations in Functional Regions of Transfer Region 1 (*tra*I) on R27 Conjugation

R27 Mutation in Region	Conjugative Transfer	Relative Conjugation Frequency (%)
1	+	100
2	+	100
3	−	0
4	+	100
5	−	0
6	−	0
7	+	12
8	−	0
9	−	0
10	−	0
11	+	13
12	−	0
13	−	0
14	−	0

1 2 3 4   5   6   7 8 9 10| 12 13   14
11

DNA Structure and Analysis

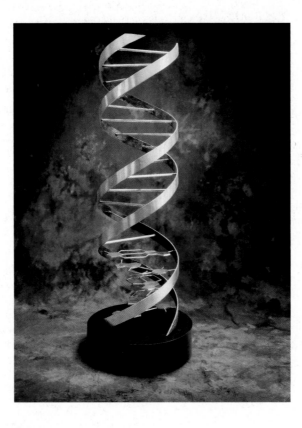

Bronze sculpture of the Watson-Crick double-helical DNA.

■ CHAPTER CONCEPTS

- ■ With the exception of some viruses, DNA serves as the genetic material in all living organisms on earth.

- ■ According to the Watson–Crick model, DNA exists in the form of the right-handed double helix.

- ■ The strands of the double helix are antiparallel and held together by hydrogen bonding between complementary nitrogenous bases.

- ■ The structure of DNA provides the basis for storing and expressing genetic information.

- ■ RNA has many similarities to DNA but exists mostly as a single-stranded molecule.

- ■ In some viruses, RNA serves as the genetic material.

Earlier in the text, we discussed the existence of genes on chromosomes that control phenotypic traits and the way in which the chromosomes are transmitted through gametes to future offspring. Logically, genes must contain some sort of information, which, when passed to a new generation, influences the form and characteristics of the offspring; we refer to this as the **genetic material**. We might also conclude that this same information in some way directs the many complex processes that sustain the adult form.

Until 1944, it was not clear what chemical component of the chromosome makes up genes and constitutes the genetic material. Because chromosomes were known to have both a nucleic acid and a protein component, both were candidates. In 1944, however, there emerged direct experimental evidence that the nucleic acid DNA serves as the informational basis for heredity.

Once the importance of DNA in genetic processes was realized, work intensified with the hope of discerning not only the structural basis of this molecule but also the relationship of its structure to its function. Between 1944 and 1953, many scientists sought information that might answer the most significant and intriguing question in the history of biology: How does DNA serve as the genetic basis for the living process? Researchers believed the answer depended strongly on the chemical structure of the DNA molecule, given the complex but orderly functions ascribed to it.

These efforts were rewarded in 1953 when James Watson and Francis Crick set forth their hypothesis for the double-helical nature of DNA. The assumption that the molecule's functions would be clarified more easily once its general structure was determined proved to be correct. In this chapter, we initially review the evidence that DNA is the genetic material and then discuss the elucidation of its structure.

9.1 The Genetic Material Must Exhibit Four Characteristics

For a molecule to serve as the genetic material, it must possess four major characteristics: **replication, storage of information, expression of information,** and **variation by mutation.** Replication of the genetic material is one facet of the cell cycle, a fundamental property of all living organisms. Once the genetic material of cells replicates and is doubled in amount, it must then be partitioned equally into daughter cells. During the formation of gametes, the genetic material is also replicated but is partitioned so that each cell gets only one-half of the original amount of genetic material—the process of meiosis. Although the products of mitosis and meiosis differ, these processes are both part of the more general phenomenon of cellular reproduction.

The characteristic of storage can be viewed as a repository of genetic information that may or may not be expressed. It is clear that while most cells contain a complete complement of DNA, at any point in time they express only part of its genetic potential. For example, bacteria "turn on" many genes in response to specific environmental conditions, and turn them off when conditions change. In vertebrates, skin cells may display active melanin genes but never activate their hemoglobin genes; digestive cells activate many genes specific to their function but do not activate their melanin genes.

Expression of the stored genetic information is the complex process of **information flow** within the cell (Figure 9–1). The initial event is the **transcription** of DNA, in which three main types of RNA molecules are synthesized: messenger RNA (mRNA), ribosomal RNA (rRNA), and transfer RNA (tRNA). Of these, mRNAs are translated into proteins. Each mRNA is the product of a specific gene and directs the synthesis of a different protein. **Translation** occurs in conjunc-

---

? How Do We Know?

In this chapter, we will focus on DNA, the molecule that stores genetic information in all living things. We shall define its structure and delve into how we analyze this molecule. As you study this topic, you should try to answer several fundamental questions:

1. How were we able to determine that DNA, and not some other molecule, serves as the genetic material in bacteria and bacteriophages?

2. How do we know that DNA also serves as the genetic material in eukaryotes such as humans?

3. How do we know that the structure of DNA is in the form of a right-handed double-helical molecule?

4. How do we know that complementary base pairs in DNA really exist?

5. How do we know that G pairs with C and that A pairs with T as complementary base pairs are formed?

6. How do we know that repetitive DNA sequences exist in eukaryotes?

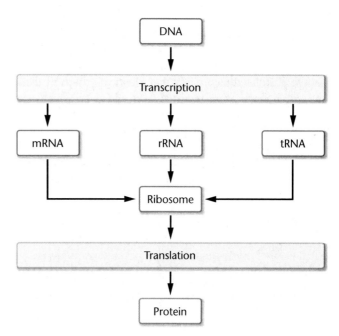

FIGURE 9–1 Simplified view of information flow (the central dogma) involving DNA, RNA, and proteins within cells.

tion with ribosomes and involves tRNA, which adapts the chemical information in mRNA to the amino acids that make up proteins. Collectively, these processes form the **central dogma of molecular genetics**: "DNA makes RNA, which makes proteins."

The genetic material is also the source of variability among organisms through the process of mutation. If a mutation—a change in the chemical composition of DNA—occurs, the alteration is reflected during transcription and translation, affecting the specific protein. If a mutation is present in gametes, it may be passed to future generations and, with time, become distributed throughout the population. Genetic variation, which also includes alterations of chromosome number and rearrangements within and between chromosomes, provides the raw material for the process of evolution.

9.2   Until 1944, Observations Favored Protein as the Genetic Material

The idea that genetic material is physically transmitted from parent to offspring has been accepted for as long as the concept of inheritance has existed. Beginning in the late nineteenth century, research into the structure of biomolecules progressed considerably, setting the stage for describing the genetic material in chemical terms. Although both proteins and nucleic acid were major candidates for the role of the genetic material, until the 1940s many geneticists favored proteins. This is not surprising, since a diversity of proteins was known to be abundant in cells, and much more was known about protein chemistry.

DNA was first studied in 1868 by a Swiss chemist, Friedrick Miescher. He isolated cell nuclei and derived an acid substance containing DNA that he called **nuclein**. As investigations progressed, however, DNA, which was shown to be present in chromosomes, seemed to lack the chemical diversity necessary to store extensive genetic information. This conclusion was based largely on Phoebus A. Levene's observations in 1910 that DNA contained approximately equal amounts of four similar molecules called *nucleotides*. Levene postulated incorrectly that identical groups of these four components were repeated over and over, which was the basis of his **tetranucleotide hypothesis** for DNA structure. Attention was thus directed away from DNA, favoring proteins. However, in the 1940s, Erwin Chargaff showed that Levene's proposal was incorrect when he demonstrated that most organisms do not contain precisely equal propor-

tions of the four nucleotides. We shall see later that the structure of DNA accounts for Chargaff's observations.

9.3   Evidence Favoring DNA as the Genetic Material Was First Obtained During the Study of Bacteria and Bacteriophages

Oswald Avery, Colin MacLeod, and Maclyn McCarty's 1944 publication on the chemical nature of a "transforming principle" in bacteria was the initial event that led to the acceptance of DNA as the genetic material. Their work, along with subsequent findings of other research teams, constituted the first direct experimental proof that DNA, and not protein, is the biomolecule responsible for heredity. It marked the beginning of the era of molecular genetics, a period of discovery in biology that made biotechnology feasible and has moved us closer to understanding the basis of life. The impact of the initial findings on future research and thinking paralleled that of the publication of Darwin's theory of evolution and the subsequent rediscovery of Mendel's postulates of transmission genetics. Together, these events constitute the three great revolutions in biology.

Transformation Studies

The research that provided the foundation for Avery, MacLeod, and McCarty's work was initiated in 1927 by Frederick Griffith, a medical officer in the British Ministry of Health. He experimented with several different strains of the bacterium *Diplococcus pneumoniae.** Some were **virulent strains**, which cause pneumonia in certain vertebrates (notably humans and mice), while others were **avirulent strains**, which do not cause illness.

The difference in virulence depends on the existence of a polysaccharide capsule; virulent strains have this capsule, whereas avirulent strains do not. The nonencapsulated bacteria are readily engulfed and destroyed by phagocytic cells in the animal's circulatory system. Virulent bacteria, which possess the polysaccharide coat, are not easily engulfed; they multiply and cause pneumonia.

The presence or absence of the capsule causes a visible difference between colonies of virulent and avirulent strains. Encapsulated bacteria form **smooth colonies** (*S*) with a shiny surface when grown on an agar culture plate; nonencapsulated strains produce **rough colonies** (*R*) (Figure 9–2). Thus, virulent and avirulent strains are easily distinguished by standard microbiological culture techniques.

Each strain of *Diplococcus* may be one of dozens of different types called **serotypes**. The specificity of the serotype is due to the detailed chemical structure of the polysaccharide constituent of the thick, slimy capsule. Serotypes are identified by immunological techniques and are usually designated by Roman numerals. Griffith used the avirulent type II*R* and the virulent type III*S* in his critical experiments. Table 9.1 summarizes the characteristics of these strains.

TABLE 9.1	Strains of *Diplococcus pneumoniae* Used by Frederick Griffith in His Original Transformation Experiments		
Serotype	Colony Morphology	Capsule	Virulence
II*R*	Rough	Absent	Avirulent
III*S*	Smooth	Present	Virulent

*This organism is now designated *Streptococcus pneumonia.*

FIGURE 9–2 Griffith's transformation experiment. The photographs show bacterial cells that exhibit capsules (type III*S*) or not (type II*R*).

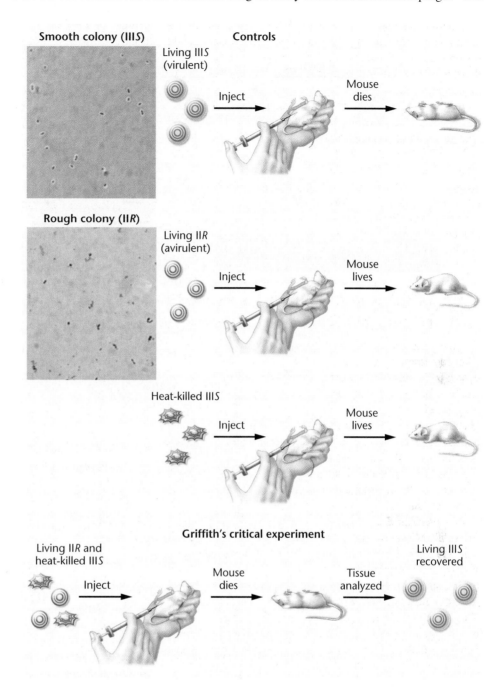

Griffith knew from the work of others that only living virulent cells produced pneumonia in mice. If heat-killed virulent bacteria were injected into mice, no pneumonia resulted, just as living avirulent bacteria failed to produce the disease. Griffith's critical experiment (Figure 9–2) involved injecting mice with living II*R* (avirulent) cells combined with heat-killed III*S* (virulent) cells. Since neither cell type caused death in mice when injected alone, Griffith expected that the double injection would not kill the mice. But, after five days, all of the mice that had received both types of cells were dead. Paradoxically, analysis of their blood revealed large numbers of living type III*S* bacteria.

As far as could be determined, these III*S* bacteria were identical to the III*S* strain from which the heat-killed cell preparation had been made. Control mice, injected only with living avirulent II*R* bacteria, did not develop pneumonia and remained healthy.

This ruled out the possibility that the avirulent II*R* cells simply changed (or mutated) to virulent III*S* cells in the absence of the heat-killed III*S* bacteria. Instead, some type of interaction had taken place between living II*R* and heat-killed III*S* cells.

Griffith concluded that the heat-killed III*S* bacteria somehow converted live avirulent II*R* cells into virulent III*S* cells. Calling the phenomenon **transformation**, he suggested that the **transforming principle** might be some part of the polysaccharide capsule or a compound required for capsule synthesis, although the capsule alone did not cause pneumonia. To use Griffith's term, the transforming principle from the dead III*S* cells served as a "pabulum" for the II*R* cells.

Griffith's work led bacteriologists and other physicians to explore the phenomenon of transformation. By 1931, Henry Dawson and his coworkers showed that transformation could

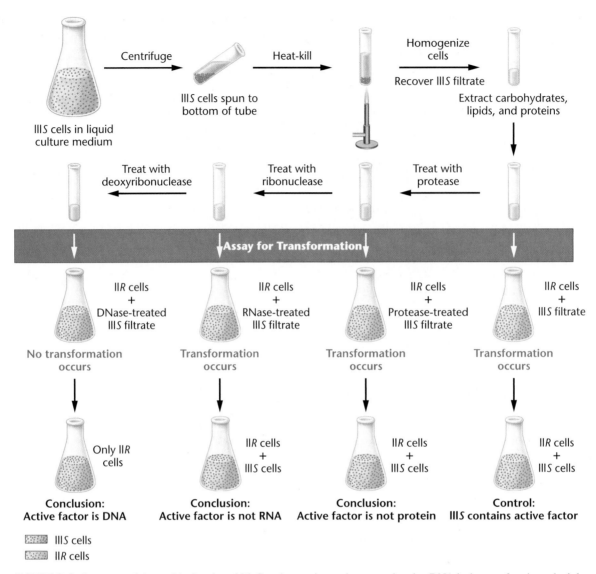

FIGURE 9–3 Summary of Avery, MacLeod, and McCarty's experiment demonstrating that DNA is the transforming principle.

occur *in vitro* (in a test tube containing only bacterial cells). That is, injection into mice was not necessary for transformation to occur. By 1933, Lionel J. Alloway had refined the *in vitro* experiments using extracts from *S* cells added to living *R* cells. The soluble filtrate from the heat-killed III*S* cells was as effective in inducing transformation as were the intact cells. Alloway and others did not view transformation as a genetic event, but rather as a physiological modification of some sort. Nevertheless, the experimental evidence that a chemical substance was responsible for transformation was quite convincing.

Then, in 1944, after 10 years of work, Avery, MacLeod, and McCarty published their results in what is now regarded as a classic paper in the field of molecular genetics. They reported that they had obtained the transforming principle in a highly purified state and that beyond reasonable doubt it was DNA.

The details of their work are illustrated in Figure 9–3. The researchers began their isolation procedure with large quantities (50–75 L) of liquid cultures of type III*S* virulent cells. The cells were centrifuged, collected, and heat-killed. Following various chemical treatments, a soluble filtrate was derived from these cells, which retained the ability to induce transformation

of type II*R* avirulent cells. The soluble filtrate was treated with a protein-digesting enzyme, called a protease, and an RNA-digesting enzyme, called ribonuclease. Such treatment destroyed the activity of any remaining protein and RNA. Nevertheless, transforming activity still remained. They concluded that neither protein nor RNA was responsible for transformation. The final confirmation came with experiments using crude samples of the DNA-digesting enzyme **deoxyribonuclease**, isolated from dog and rabbit sera. Digestion with this enzyme destroyed transforming activity present in the filtrate—thus, Avery and his coworkers were certain that the active transforming principle in these experiments was DNA.

The great amount of work, the confirmation and reconfirmation of the conclusions, and the logic of the experimental design involved in the research of these three scientists are truly impressive. Their conclusion in the 1944 publication, however, was stated very simply: "The evidence presented supports the belief that a nucleic acid of the desoxyribose* type is the fundamental unit of the transforming principle of *Pneumococcus* type III."

_____

*Desoxyribose is now spelled deoxyribose.

They also immediately recognized the genetic and biochemical implications of their work. They suggested that the transforming principle interacts with the II*R* cell and gives rise to a coordinated series of enzymatic reactions that culminates in the synthesis of the type III*S* capsular polysaccharide. They emphasized that, once transformation occurs, the capsular polysaccharide is produced in successive generations. Transformation is therefore heritable, and the process affects the genetic material.

Transformation, originally introduced in chapter 8, has now been shown to occur in *Hemophilus influenzae*, *Bacillus subtilis*, *Shigella paradysenteriae*, and *Escherichia coli*, among many other microorganisms. Transformation of numerous genetic traits other than colony morphology has been demonstrated, including those that resist antibiotics. These observations further strengthened the belief that transformation by DNA is primarily a genetic event rather than simply a physiological change. We will pursue this idea in the "Insights and Solutions" section at the end of this chapter.

The Hershey-Chase Experiment

The second major piece of evidence supporting DNA as the genetic material was provided during the study of the bacterium *E. coli* and one of its infecting viruses, bacteriophage T2. Often referred to simply as a *phage*, the virus consists of a protein coat surrounding a core of DNA. Electron micrographs reveal that the phage's external structure is composed of a hexagonal head plus a tail. Figure 9–4 shows the life cycle of a T-even bacteriophage such as T2, as it was known in 1952. Recall that the phage adsorbs to the bacterial cell and that some component of the phage enters the bacterial cell. Following infection, the viral information "commandeers" the cellular machinery of the host and undergoes viral reproduction. In a reasonably short time, many new phages are constructed and the bacterial cell is lysed, releasing the progeny viruses.

In 1952, Alfred Hershey and Martha Chase published the results of experiments designed to clarify the events leading to phage reproduction. Several of the experiments clearly established the independent functions of phage protein and

nucleic acid in the reproduction process of the bacterial cell. Hershey and Chase knew from this existing data that:

1. T2 phages consist of approximately 50 percent protein and 50 percent DNA.

2. Infection is initiated by adsorption of the phage by its tail fibers to the bacterial cell.

3. The production of new viruses occurs within the bacterial cell.

It appeared that some molecular component of the phage, DNA and/or protein, entered the bacterial cell and directed viral reproduction. Which was it?

Hershey and Chase used radioisotopes to follow the molecular components of phages during infection. Both ^{32}P and ^{35}S, radioactive forms of phosphorus and sulfur, respectively, were used. DNA contains phosphorus but not sulfur, so ^{32}P

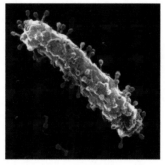

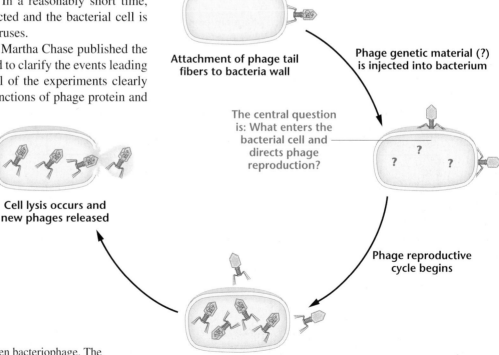

FIGURE 9–4 Life cycle of a T-even bacteriophage. The electron micrograph shows an *E. coli* cell during infection by numerous T2 phages (shown in blue).

effectively labels DNA. Because proteins contain sulfur but not phosphorus, ^{35}S labels protein. *This is a key point in the experiment.* If *E. coli* cells are first grown in the presence of *either* ^{32}P *or* ^{35}S and then infected with T2 viruses, the progeny phage will have *either* a labeled DNA core *or* a labeled protein coat, respectively. These radioactive phages can be isolated and used to infect unlabeled bacteria (Figure 9–5).

When labeled phage and unlabeled bacteria were mixed, an *adsorption complex* was formed as the phages attached their tail fibers to the bacterial wall. These complexes were isolated and

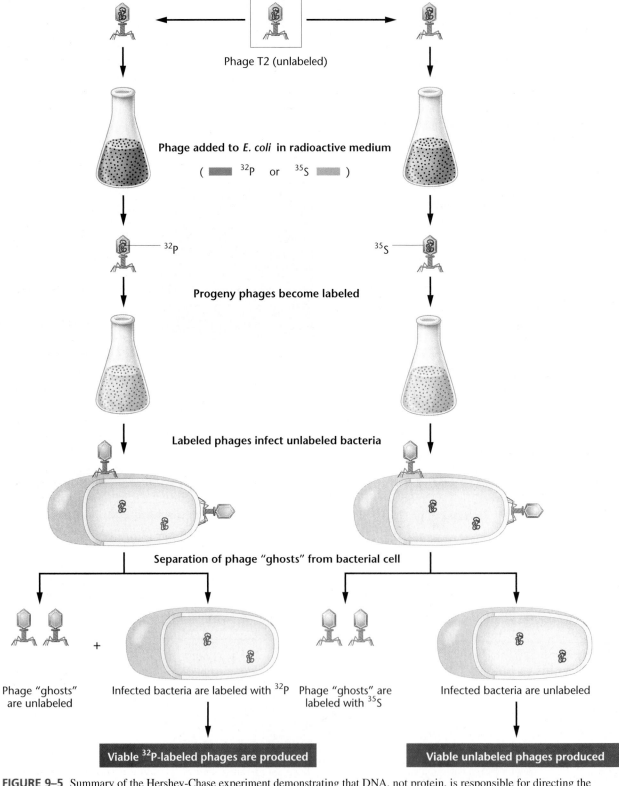

FIGURE 9–5 Summary of the Hershey-Chase experiment demonstrating that DNA, not protein, is responsible for directing the reproduction of phage T2 during the infection of *E. coli*.

subjected to a high shear force by placing them in a blender. This force strips off the attached phages, which can then be analyzed separately (Figure 9–5). By tracing the radioisotopes, Hershey and Chase were able to demonstrate that most of the ^{32}P-labeled DNA had transferred into the bacterial cell following adsorption; on the other hand, almost all of the ^{35}S-labeled protein remained outside the bacterial cell and was recovered in the phage "ghosts" (empty phage coats) after the blender treatment. After separation, the bacterial cells, which now contained viral DNA, were eventually lysed as new phages were produced. These progeny contained ^{32}P but not ^{35}S.

Hershey and Chase interpreted these results as indicating that the protein of the phage coat remains outside the host cell and is not involved in the production of new phages. On the other hand, and most important, phage DNA enters the host cell and directs phage reproduction. Hershey and Chase had demonstrated that the genetic material in phage T2 is DNA, not protein.

These experiments, along with those of Avery and his colleagues, provided convincing evidence that DNA is the molecule responsible for heredity. This conclusion has since served as the cornerstone of the field of molecular genetics.

Now Solve This

Problem 7 on page 211 asks about applying the protocol of the Hershey-Chase experiment to the investigation of transformation. You are asked about the possible success of attempting to do so.

Hint: As you attempt to apply the Hershey-Chase protocol, remember that in transformation, exogenous DNA enters the soon-to-be transformed cell and no cell-to-cell contact is involved in the process.

Transfection Experiments

During the eight years following the publication of the Hershey-Chase experiment, additional research with bacterial viruses provided even more solid proof that DNA is the genetic material. In 1957, several reports demonstrated that if *E. coli* is treated with the enzyme lysozyme, the outer wall of the cell can be removed without destroying the bacterium. Enzymatically treated cells are naked, so to speak, and contain only the cell membrane as the outer boundary of the cell—these structures are called **protoplasts** (or **spheroplasts**). John Spizizen and Dean Fraser independently reported that by using protoplasts, they were able to initiate phage multiplication with disrupted T2 particles. That is, provided protoplasts were used, a virus did not have to be intact for infection to occur.

Similar but more refined experiments were reported in 1960 using only DNA purified from bacteriophages. This process of infection by only the viral nucleic acid, called **transfection**, proves conclusively that phage DNA alone contains all the necessary information for producing mature viruses. Thus, the evidence that DNA serves as the genetic material in all organisms was further strengthened, even though all direct evidence had been obtained from bacterial and viral studies.

9.4   Indirect and Direct Evidence Supports the Concept that DNA Is the Genetic Material in Eukaryotes

In 1950, eukaryotic organisms were not amenable to the types of experiments that used bacteria and viruses to demonstrate that DNA is the genetic material. Nevertheless, it was generally assumed that the genetic material would be a universal substance and also serve this role in eukaryotes. Initially, support for this assumption relied on several circumstantial observations that, taken together, indicated that DNA is also the genetic material in eukaryotes. Subsequently, direct evidence established unequivocally the central role of DNA in genetic processes.

Indirect Evidence: Distribution of DNA

The genetic material should be found where it functions—in the nucleus as part of chromosomes. Both DNA and protein fit this criterion. However, protein is also abundant in the cytoplasm, while DNA is not. Both mitochondria and chloroplasts are known to perform genetic functions, and DNA is also present in these organelles. Thus, DNA is found only where primary genetic function is known to occur. Protein, however, is found everywhere in the cell. These observations are consistent with the interpretation favoring DNA over protein as the genetic material.

Because it had been established earlier that chromosomes within the nucleus contain the genetic material, a correlation was expected between the ploidy (*n*, 2*n*, etc.) of cells and the quantity of the molecule that functions as the genetic material. Meaningful comparisons can be made between gametes (sperm and eggs) and somatic or body cells. The latter are recognized as being diploid (2*n*) and containing twice the number of chromosomes as gametes, which are haploid (*n*).

Table 9.2 compares the amount of DNA found in haploid sperm and the diploid nucleated precursors of red blood cells from a variety of organisms. The amount of DNA and the number of sets of chromosomes are closely correlated. No consistent correlation can be observed between gametes and diploid cells for proteins, thus again favoring DNA over proteins as the genetic material of eukaryotes.

TABLE 9.2	DNA Content of Haploid Versus Diploid Cells of Various Species (in picograms)	
Organism*	***n***	***2n***
Human	3.25	7.30
Chicken	1.26	2.49
Trout	2.67	5.79
Carp	1.65	3.49
Shad	0.91	1.97

*Sperm (*n*) and nucleated precursors to red blood cells (2*n*) were used to contrast ploidy levels.

Indirect Evidence: Mutagenesis

Ultraviolet (UV) light is one of many agents capable of inducing mutations in the genetic material. Simple organisms such as yeast and other fungi can be irradiated with various wavelengths of UV light, and the effectiveness of each wavelength can then be measured by the number of mutations it induces. When the data are plotted, an **action spectrum** of UV light as a mutagenic agent is obtained. This action spectrum can then be compared with the **absorption spectrum** of any molecule suspected to be genetic material (Figure 9–6). *The molecule serving as the genetic material is expected to absorb at the wavelengths shown to be mutagenic.*

UV light is most mutagenic at the wavelength (λ) of about 260 nanometers (nm), and both DNA and RNA strongly absorb UV light at 260 nm. On the other hand, protein absorbs most strongly around 280 nm, yet no significant mutagenic effects are observed at this wavelength. This indirect evidence also supports the idea that a nucleic acid is the genetic material and tends to exclude protein.

Direct Evidence: Recombinant DNA Studies

Although the circumstantial evidence described above does not constitute direct proof that DNA is the genetic material in eukaryotes, these observations spurred researchers to forge ahead, basing their work on this hypothesis. Today, there is no doubt of the validity of this conclusion. DNA *is* the genetic material in eukaryotes. The strongest evidence is provided by molecular analysis utilizing **recombinant DNA technology**. In this procedure, segments of eukaryotic DNA corresponding to specific genes are isolated and spliced into bacterial DNA. This complex can then be inserted into a bacterial cell, and its

genetic expression is monitored. If a eukaryotic gene is introduced, the presence of the corresponding eukaryotic protein product demonstrates directly that this DNA is present and functional in the bacterial cell. This has been shown to be the case in countless instances. For example, the products of the human genes specifying insulin and interferon are produced by bacteria after the human genes that encode these proteins are inserted. As the bacterium divides, the eukaryotic DNA replicates along with the host DNA and is distributed to the daughter cells, which also express the human genes and synthesize the corresponding proteins. The availability of vast amounts of DNA coding for specific genes, derived from recombinant DNA research, has led to other direct evidence that DNA serves as the genetic material. Work done in the laboratory of Beatrice Mintz demonstrated that DNA encoding the human β-globin gene, when microinjected into a fertilized mouse egg, is later found to be present and expressed in adult mouse tissue, and it is transmitted to and expressed in that mouse's progeny. These mice are examples of **transgenic animals**. Other work introduced rat DNA encoding a growth hormone into fertilized mouse eggs. About one-third of the resultant mice grew to twice their normal size, indicating that foreign DNA was present and functional. Subsequent generations of mice inherited this genetic information and also grew to a large size. This clearly demonstrates that DNA meets the requirement of expression of genetic information in eukaryotes. Later we shall discuss exactly how DNA is stored, replicated, expressed, and mutated.

9.5 RNA Serves as the Genetic Material in Some Viruses

Some viruses contain an RNA core rather than a DNA core. In these viruses, it appears that RNA serves as the genetic material—an exception to the general rule that DNA performs this function. In 1956, it was demonstrated that when purified RNA from **tobacco mosaic virus** (**TMV**) was spread on tobacco leaves, the characteristic lesions caused by viral infection subsequently appeared on the leaves. Thus, it was concluded that RNA is the genetic material of this virus.

In 1965 and 1966, Norman Pace and Sol Spiegelman demonstrated further that RNA from the phage Qβ can be isolated and replicated *in vitro*. Replication depends on an enzyme, **RNA replicase**, which is isolated from host *E. coli* cells following normal infection. When the RNA replicated *in vitro* is added to *E. coli* protoplasts, infection and viral multiplication (transfection) occur. Thus, RNA synthesized in a test tube serves as the genetic material in these phages by directing the production of all the components necessary for viral reproduction.

One other group of RNA-containing viruses bears mention. These are the **retroviruses**, which replicate in an unusual way. Their RNA serves as a template for the synthesis of the complementary DNA molecule. The process, **reverse transcription**, occurs under the direction of an RNA-dependent DNA polymerase enzyme called **reverse transcriptase**. This DNA intermediate can be incorporated into the genome of the host cell, and when the host DNA is transcribed, copies of the original retroviral RNA chromosomes are produced.

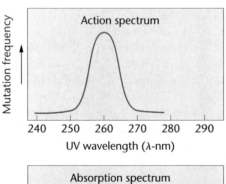

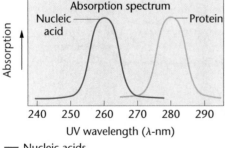

— Nucleic acids
— Proteins

FIGURE 9–6 Comparison of the action spectrum, which determines the most effective mutagenic UV wavelength, and the absorption spectrum, which shows the range of wavelengths where nucleic acids and proteins absorb UV light.

Retroviruses include the human immunodeficiency virus (HIV), which causes AIDS, as well as RNA tumor viruses.

9.6   The Structure of DNA Holds the Key to Understanding Its Function

Having established that DNA is the genetic material in all living organisms (except certain viruses), we turn now to the structure of this nucleic acid. In 1953, James Watson and Francis Crick proposed that the structure of DNA is in the form of a double helix. Their proposal was published in a short paper in the journal *Nature*, reprinted in its entirety (see p. 202). In a sense, this publication was the finish of a highly competitive scientific race to obtain what some consider to be the most significant finding in the history of biology. This race, as recounted in Watson's book *The Double Helix*, demonstrates the human interaction, genius, frailty, and intensity involved in the scientific effort that eventually led to the elucidation of DNA structure.

The data available to Watson and Crick, crucial to the development of their proposal, came primarily from two sources: (1) base composition analysis of hydrolyzed samples of DNA and (2) X-ray diffraction studies of DNA. Watson and Crick's analytical success can be attributed to model building that conformed to the existing data. If the correct solution to the structure of DNA is viewed as a puzzle, Watson and Crick, working in the Cavendish Laboratory in Cambridge, England, were the first to fit the pieces together successfully.

Nucleic Acid Chemistry

Before turning to this work, a brief introduction to nucleic acid chemistry is in order. This chemical information was well known to Watson and Crick during their investigation and served as the basis of their model building.

DNA is a nucleic acid, and nucleotides are the building blocks of all nucleic acid molecules. Sometimes called mononucleotides, these structural units have three essential components: a **nitrogenous base**, a **pentose sugar** (a five-carbon sugar), and a **phosphate group**. There are two kinds of nitrogenous bases: the nine-member double-ring **purines** and the six-member single-ring **pyrimidines**.

Two types of purines and three types of pyrimidines are found in nucleic acids. The two purines are **adenine** and **guanine**, abbreviated A and G respectively. The three pyrimidines are **cytosine, thymine**, and **uracil** (respectively, C, T, and U). The chemical structures of the five bases are shown in Figure 9–7(a). Both DNA and RNA contain A, G, and C, but

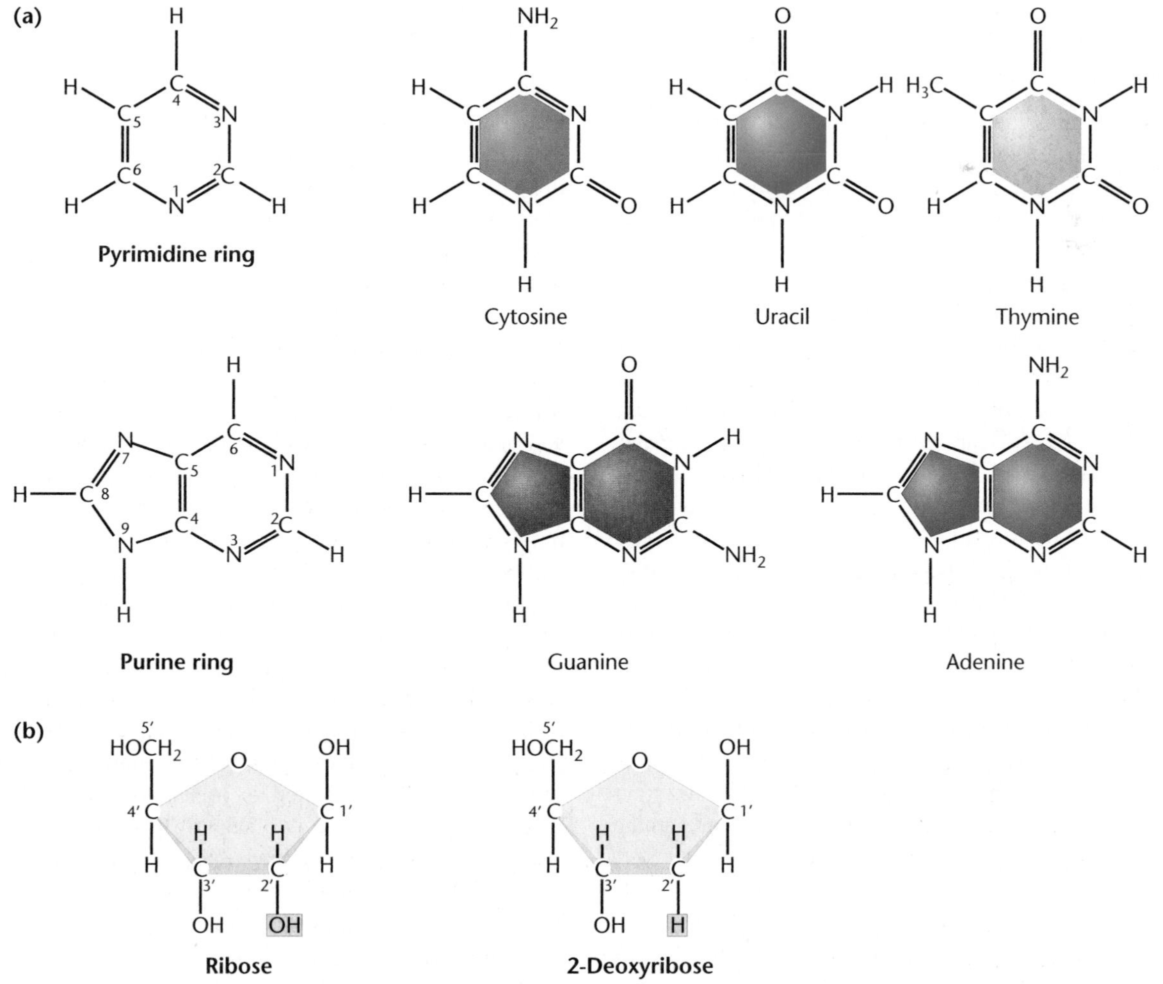

FIGURE 9–7   (a) Chemical structures of the pyrimidines and purines that serve as the nitrogenous bases in RNA and DNA. (b) Chemical ring structures of ribose and 2-deoxyribose, which serve as the pentose sugars in RNA and DNA, respectively.

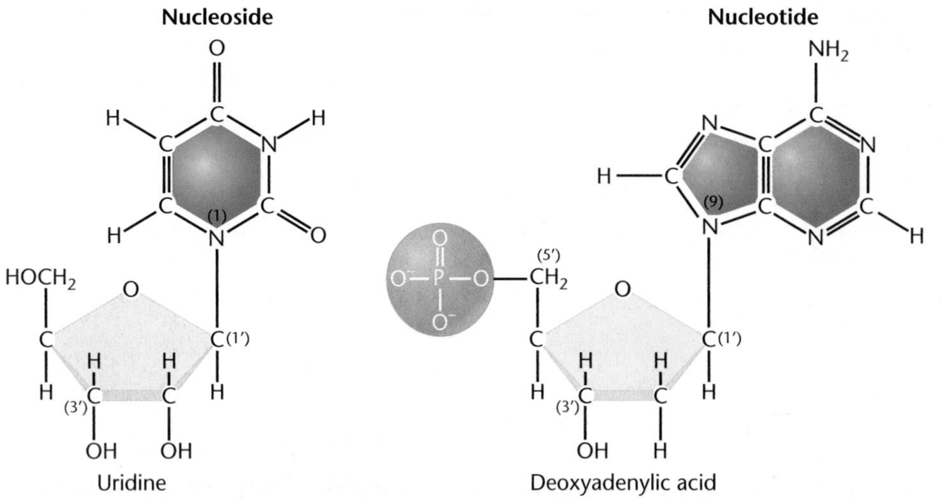

Nucleoside

Nucleotide

Uridine

Deoxyadenylic acid

FIGURE 9–8 Structures and names of the nucleosides and nucleotides of RNA and DNA.

Ribonucleosides	Ribonucleotides
Adenosine	Adenylic acid
Cytidine	Cytidylic acid
Guanosine	Guanylic acid
Uridine	Uridylic acid
Deoxyribonucleosides	**Deoxyribonucleotides**
Deoxyadenosine	Deoxyadenylic acid
Deoxycytidine	Deoxycytidylic acid
Deoxyguanosine	Deoxyguanylic acid
Deoxythymidine	Deoxythymidylic acid

only DNA contains the base T and only RNA contains the base U. Each nitrogen or carbon atom of the ring structures of purines and pyrimidines is designated by a number. Note that corresponding atoms in the purine and pyrimidine rings are numbered differently.

The pentose sugars found in nucleic acids give them their names. **Ribonucleic acids (RNA)** contain **ribose**, while **deoxyribonucleic acids (DNA)** contain **deoxyribose**. Figure 9–7(b) shows the chemical structures for these two pentose sugars. Each carbon atom is distinguished by a number with a prime sign ($'$). As you can see in Figure 9–7(b), compared with ribose, deoxyribose has a hydrogen atom rather than a hydroxyl group at the C-$2'$ position. The absence of a hydroxyl group at the C-$2'$ position thus distinguishes DNA from RNA. In the absence of the C-$2'$ hydroxyl group, the sugar is more specifically named **2-deoxyribose**.

If a molecule is composed of a purine or pyrimidine base and a ribose or deoxyribose sugar, the chemical unit is called a **nucleoside**. If a phosphate group is added to the nucleoside, the molecule is now called a **nucleotide**. Nucleosides and nucleotides are named according to the specific nitrogenous base (A, G, C, T, or U) that is part of the molecule. The structure of a nucleotide and the nomenclature used in naming DNA nucleotides and nucleosides are shown in Figure 9–8.

The bonding between components of a nucleotide is highly specific. The C-$1'$ atom of the sugar is involved in the chemical linkage to the nitrogenous base. If the base is a purine, the N-9 atom is covalently bonded to the sugar; if the base is a pyrimidine, the N-1 atom bonds to the sugar. In deoxyribonucleotides,

the phosphate group may be bonded to the C-$2'$, C-$3'$, or C-$5'$ atom of the sugar. The C-$5'$-phosphate configuration is shown in Figure 9–8. It is by far the most prevalent one in biological systems and the one found in DNA and RNA.

Nucleotides are also described by the term **nucleoside monophosphate (NMP)**. The addition of one or two phosphate groups results in **nucleoside diphosphates (NDP)** and **triphosphates (NTP)**, respectively, as shown in Figure 9–9. The triphosphate form is significant because it is the precursor molecule during nucleic acid synthesis within the cell. In addition, adenosine triphosphate (ATP) and guanosine triphosphate (GTP) are important in cell bioenergetics because of the large amount of energy involved in adding or removing the terminal phosphate group. The hydrolysis of ATP or GTP to ADP or GDP and inorganic phosphate (P_i) is accompanied by the release of a large amount of energy in the cell. When these chemical conversions are coupled to other reactions, the energy produced is used to drive them. As a result, ATP and GTP are involved in many cellular activities.

The linkage between two mononucleotides involves a phosphate group linked to two sugars. A **phosphodiester bond** is formed as phosphoric acid is joined to two alcohols (the hydroxyl groups on the two sugars) by an ester linkage on both sides. Figure 9–10 shows the resultant phosphodiester bond in DNA. Each structure has a C-$3'$ end and a C-$5'$ end. Two joined nucleotides form a **dinucleotide**; three nucleotides, a **trinucleotide**; and so forth. Short chains consisting of up to 20 nucleotides or so are called **oligonucleotides**; longer chains are **polynucleotides**.

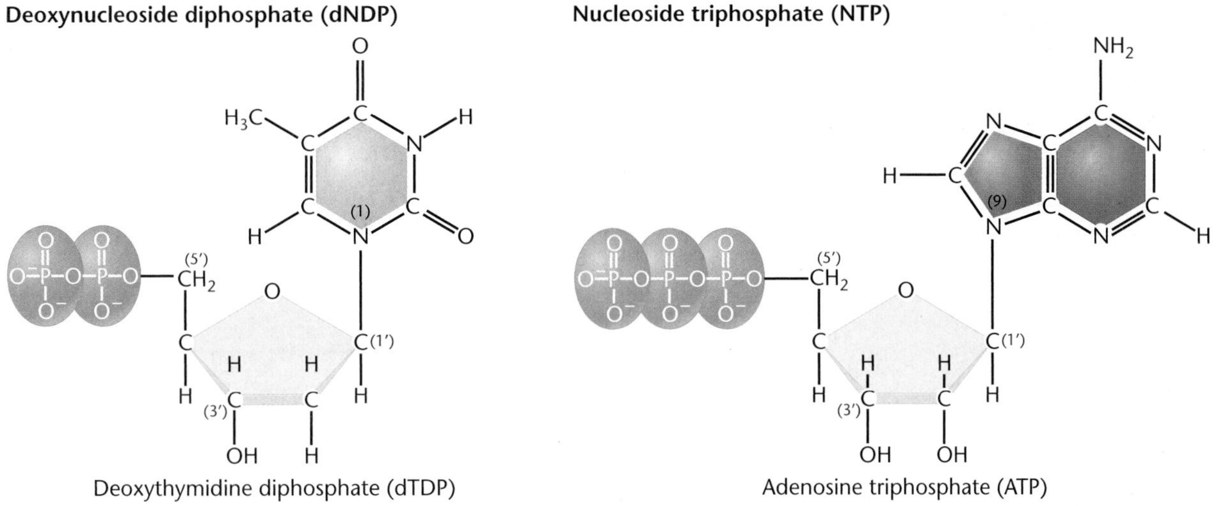

Deoxynucleoside diphosphate (dNDP)   **Nucleoside triphosphate (NTP)**

Deoxythymidine diphosphate (dTDP)   Adenosine triphosphate (ATP)

FIGURE 9–9 Basic structures of nucleoside diphosphates and triphosphates. Deoxythymidine diphosphate and adenosine triphosphate are diagrammed here.

Long polynucleotide chains account for the large molecular weight of DNA and explain its most important property—storage of vast quantities of genetic information. If each nucleotide position in this long chain can be occupied by any one of four nucleotides, extraordinary variation is possible. For example, a polynucleotide only 1000 nucleotides in length can be arranged 4^{1000} different ways, each one different from all other possible sequences. This potential variation in molecular structure is essential if DNA is to store the vast

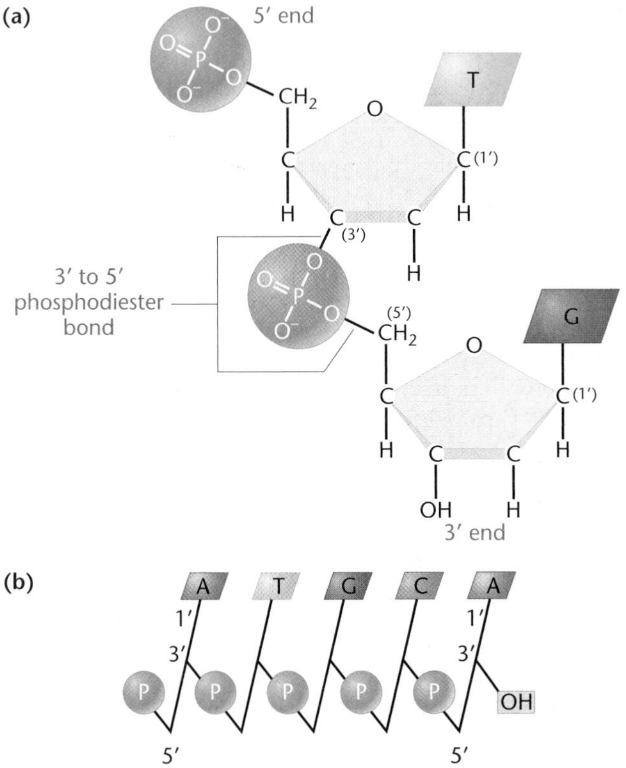

FIGURE 9–10 (a) Linkage of two nucleotides by the formation of a C-3′–C-5′ (3′–5′) phosphodiester bond, producing a dinucleotide. (b) Shorthand notation for a polynucleotide chain.

amounts of chemical information necessary to direct cellular activities.

Base Composition Studies

Between 1949 and 1953, Erwin Chargaff and his colleagues used chromatographic methods to separate the four bases in DNA samples from various organisms. Quantitative methods were then used to determine the amounts of the four nitrogenous bases from each source. Table 9.3(a) lists some of Chargaff's original data. Parts (b) and (c) show more recently derived information that reinforces Chargaff's findings. As we shall see, Chargaff's data were critical to the successful model of DNA put forward by Watson and Crick. On the basis of these data, the following conclusions may be drawn.

1. The amount of adenine residues is proportional to the amount of thymine residues in DNA (columns 1, 2, and 5). Also, the amount of guanine residues is proportional to the amount of cytosine residues (columns 3, 4, and 6).

2. Based on this proportionality, the sum of the purines (A + G) equals the sum of the pyrimidines (C + T), as shown in column 7.

3. The percentage of (G + C) does not necessarily equal the percentage of (A + T). As you can see, this ratio varies greatly between different organisms, as shown in column 8, and in Part (c).

These conclusions indicate a definite pattern of base composition of DNA molecules. The data provided the initial clue to the DNA puzzle. They also directly refuted Levene's tetranucleotide hypothesis, which stated that all four bases are present in equal amounts.

X-Ray Diffraction Analysis

When fibers of a DNA molecule are subjected to X-ray bombardment, the X-rays scatter according to the molecule's atomic structure. The pattern of scatter can be captured as spots on photographic film and analyzed, particularly for the

| **TABLE 9.3** | DNA Base Composition Data |

(a) Chargaff's Data*

	Molar Proportions[†]			
	1	2	3	4
Source	A	T	G	C
Ox thymus	26	25	21	16
Ox spleen	25	24	20	15
Yeast	24	25	14	13
Avian tubercle bacilli	12	11	28	26
Human sperm	29	31	18	18

(c) G + C Content in Several Organisms

Organism	G + C (%)
Phage T2	36.0
Drosophila	45.0
Maize	49.1
Euglena	53.5
Neurospora	53.7

(b) Base Compositions of DNAs from Various Sources

	Base Composition				Base Ratio		Combined Base Ratios	
	1	2	3	4	5	6	7	8
Source	A	T	G	C	A:T	G:C	(A + G):(C + T)	(A + T):(C + G)
Human	30.9	29.4	19.9	19.8	1.05	1.00	1.04	1.52
Sea urchin	32.8	32.1	17.7	17.3	1.02	1.02	1.02	1.58
E. coli	24.7	23.6	26.0	25.7	1.04	1.01	1.03	0.93
Sarcina lutea	13.4	12.4	37.1	37.1	1.08	1.00	1.04	0.35
T7 bacteriophage	26.0	26.0	24.0	24.0	1.00	1.00	1.00	1.08

Source: Chargaff, 1950.
[†]Moles of nitrogenous constituent per mole of P (often, the recovery was less than 100%).

overall shape of and regularities within the molecule. This process, **X-ray diffraction analysis**, was applied successfully to the study of protein structure by Linus Pauling and other chemists. The technique had been attempted on DNA as early as 1938 by William Astbury. By 1947, he had detected a periodicity of 3.4 angstroms (Å)* within the structure of the molecule, which suggested to him that the bases were stacked like coins on top of one another.

Between 1950 and 1953, Rosalind Franklin, working in the laboratory of Maurice Wilkins, obtained improved X-ray data from more purified samples of DNA (Figure 9–11). Her work confirmed the 3.4 Å periodicity seen by Astbury and suggested that the structure of DNA was some sort of helix. However, she did not propose a definitive model. Pauling had analyzed the work of Astbury and others and proposed incorrectly that DNA is a triple helix.

The Watson-Crick Model

Watson and Crick published their analysis of DNA structure in 1953 (see p. 202). By building models under the constraints of the information just discussed, they proposed the double-helical form of DNA shown in Figure 9–12(a). This model has these major features:

1. Two long polynucleotide chains are coiled around a central axis, forming a right-handed double helix.

2. The two chains are antiparallel; that is, their C-5′ to C-3′ orientations run in opposite directions.

3. The bases of both chains are flat structures, lying perpendicular to the axis; they are "stacked" on one another, 3.4 Å (0.34 nm) apart, and located on the inside of the structure.

4. The nitrogenous bases of opposite chains are *paired* as the result of hydrogen bonds; in DNA, only A $=$ T and G $\equiv$ C pairs occur.

5. Each complete turn of the helix is 34 Å (3.4 nm) long; thus, 10 bases exist per turn in each chain.

6. In any segment of the molecule, alternating larger **major grooves** and smaller **minor grooves** are apparent along the axis.

7. The double helix measures 20 Å (2.0 nm) in diameter.

The nature of *base pairing* (point 4 above) is the most genetically significant feature of the model. Before we discuss it in detail, several other important features warrant emphasis. First, the antiparallel nature of the two chains is a key part of the double-helix model. While one chain runs in the 5′ to 3′ orientation (what seems right side

*Today, measurement in nanometers (nm) is favored (1 nm = 10 Å).

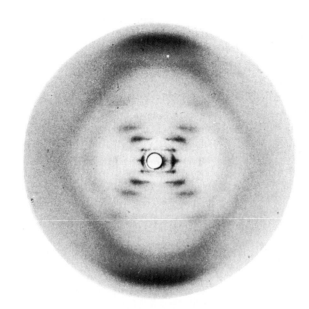

FIGURE 9–11 X-ray diffraction photograph of purified DNA fibers. The strong arcs on the periphery show closely spaced aspects of the molecule, providing an estimate of the periodicity of nitrogenous bases, which are 3.4 Å apart. The inner cross pattern of spots shows the grosser aspect of the molecule, indicating its helical nature.

up to us), the other chain is in the 3′ to 5′ orientation (and thus appears upside down). This is illustrated in Figure 9–12(b) and (c). Given the constraints of the bond angles of the various nucleotide components, the double helix could not be constructed easily if both chains ran parallel to one another.

The key to the model proposed by Watson and Crick is the specificity of base pairing. Chargaff's data suggested that the amounts of A equaled T and that the amounts of G equaled C. Watson and Crick realized that if A pairs with T and C pairs with G, this would account for these proportions and that such pairing could occur as a result of hydrogen bonding between base pairs [Figure 9–12(c)], providing the chemical stability necessary to hold the two chains together. Arranged in this way, both major and minor grooves become apparent along the axis. Further, a purine (A or G) opposite a pyrimidine (T or C) on each "rung of the spiral staircase" of the proposed double helix accounts for the 20 Å (2 nm) diameter suggested by X-ray diffraction studies.

The specific A = T and G ≡ C base pairing is the basis for **complementarity**. This term describes the chemical affinity provided by hydrogen bonding between the bases. As we shall see, complementarity is very important in DNA replication and gene expression.

FIGURE 9–12 (a) The DNA double helix as proposed by Watson and Crick. The ribbonlike strands constitute the sugar–phosphate backbones, and the horizontal rungs constitute the nitrogenous base pairs, of which there are 10 per complete turn. The major and minor grooves are shown. The solid vertical bar represents the central axis. (b) A detailed view labeled with the bases, sugars, phosphates, and hydrogen bonds of the helix. (c) A demonstration of the antiparallel nature of the helix and the horizontal stacking of the bases.

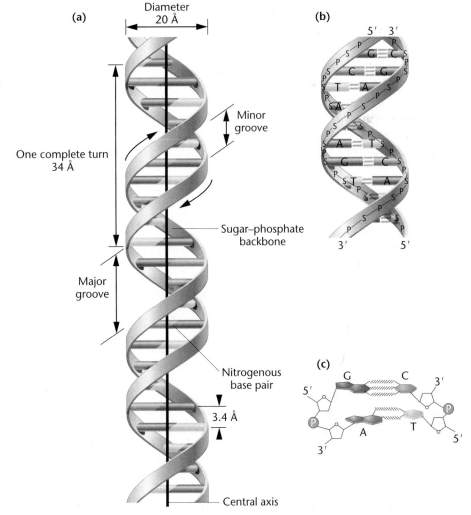

Molecular Structure of Nucleic Acids: A Structure for Deoxyribose Nucleic Acid

We wish to suggest a structure for the salt of deoxyribose nucleic acid (DNA). This structure has novel features which are of considerable biological interest. A structure for nucleic acid has already been proposed by Pauling and Corey.[1] They kindly made their manuscript available to us in advance of publication.

Their model consists of three intertwined chains, with the phosphates near the fibre axis, and the bases on the outside. In our opinion, this structure is unsatisfactory for two reasons: (1) We believe that the material which gives the X-ray diagrams is the salt, not the free acid. Without the acidic hydrogen atoms it is not clear what forces would hold the structure together, especially as the negatively charged phosphates near the axis will repel each other. (2) Some of the van der Waals distances appear to be too small.

Another three-chain structure has also been suggested by Fraser (in the press). In his model the phosphates are on the outside and the bases on the inside, linked together by hydrogen bonds. This structure as described is rather ill-defined, and for this reason we shall not comment on it.

We wish to put forward a radically different structure for the salt of deoxyribose nucleic acid. This structure has two helical chains each coiled round the same axis. We have made the usual chemical assumptions, namely, that each chain consists of phosphate diester groups joining β-D-deoxyribofuranose residues with $3'$, $5'$ linkages. The two chains (but not their bases) are related by a dyad perpendicular to the fibre axis. Both chains follow right-handed helices, but owing to the dyad the sequences of the atoms in the two chains run in opposite directions. Each chain loosely resembles Furberg's[2] model No. 1; that is, the bases are on the inside of the helix and the phosphates on the outside. The configuration of the sugar and the atoms near it is close to Furberg's "standard configuration," the sugar being roughly perpendicular to the attached base. There is a residue on each chain every 3.4 Å in the z-direction. We have assumed an angle of 36° between adjacent residues

in the same chain, so that the structure repeats after 10 residues on each chain, that is, after 34 Å. The distance of a phosphate atom from the fibre axis is 10 Å. As the phosphates are on the outside, cations have easy access to them.

The structure is an open one, and its water content is rather high. At lower water content we would expect the bases to tilt so that the structure could become more compact.

The novel feature of the structure is the manner in which the two chains are held together by the purine and pyrimidine bases. The planes of the bases are perpendicular to the fibre axis. They are joined together in pairs, a single base from one chain being hydrogen-bonded to a single base from the other chain, so that the two lie side by side with identical z-coordinates. One of the pair must be a purine and the other a pyrimidine for bonding to occur. The hydrogen bonds are made as follows: purine position 1 to pyrimidine position 1; purine position 6 to pyrimidine position 6.

If it is assumed that the bases only occur in the structure in the most plausible tautomeric forms (that is, with the keto rather than the enol configuration), it is found that only specific pairs of bases can bond together. These pairs are: adenine (purine) with thymine (pyrimidine), and guanine (purine) with cytosine (pyrimidine).

In other words, if an adenine forms one member of a pair, on either chain, then on these assumptions the other member must be thymine; similarly for guanine and cytosine. The sequence of bases on a single chain does not appear to be restricted in any way. However, if only specific pairs of bases can be formed, it follows that if the sequence of bases on one chain is given, then the sequence on the other chain is automatically determined.

It has been found experimentally[3,4] that the ratio of the amounts of adenine to thymine, and the ratio of guanine to cytosine, are always very close to unity for deoxyribose nucleic acid.

It is probably impossible to build this structure with a ribose sugar in place of deoxyribose, as the extra oxygen atom would make too close a van der Waals contact.

The previously published X-ray data[5,6] on deoxyribose nucleic acid are insufficient for a rigorous test of our structure. So far as we can tell, it is roughly compatible with the experimental data, but it must be regarded as unproved until it has been checked against more exact results. Some of these are given in the following communications. We were not aware of the details of the results presented there when we devised our structure, which rests mainly though not entirely on published experimental data and stereochemical arguments.

It has not escaped our notice that the specific pairing we have postulated immediately suggests a possible copying mechanism for the genetic material. Full details of the structure, including the conditions assumed in building it, together with a set of co-ordinates for the atoms, will be published elsewhere.

We are much indebted to Dr. Jerry Donohue for constant advice and criticism, especially on interatomic distances. We have also been stimulated by a knowledge of the general nature of the unpublished experimental results and ideas of Dr. M. H. F. Wilkins, Dr. R. E. Franklin and their co-workers at King's College, London. One of us (J. D. W.) has been aided by a fellowship from the National Foundation for Infantile Paralysis.

J. D. WATSON

F. H. C. CRICK
Medical Research Council Unit for the Study of the Molecular Structure of Biological Systems, Cavendish Laboratory, Cambridge, England

[1]Pauling L., and Corey, R. B., *Nature*, 171, 346 (1953); *Proc. U.S. Nat. Acad. Sci.*, 39, 84 (1953).
[2]Furberg, S., *Acta Chem. Scand.*, 6, 634 (1952).
[3]Chargaff, E., for references see Zamenhof, S., Brawerman, G., and Chargaff, E., *Biochim. et Biophys. Acta*, 9, 402 (1952).
[4]Wyatt, G. R., *J. Gen. Physiol.*, 36, 201 (1952).
[5]Astbury, W. T., *Symp. Soc. Exp. Biol. 1, Nucleic Acid*, 66 (Camb. Univ. Press, 1947).
[6]Wilkins, M. H. F., and Randall, J. T., *Biochim. et Biophys. Acta*, 10, 192 (1953).

It is appropriate to inquire into the nature of a hydrogen bond and to ask whether it is strong enough to stabilize the helix. A **hydrogen bond** is a very weak electrostatic attraction between a covalently bonded hydrogen atom and an atom with an unshared electron pair. The hydrogen atom assumes a partial positive charge, while the unshared electron pair—characteristic of covalently bonded oxygen and nitrogen atoms—assumes a partial negative charge. These opposite charges are responsible for the weak chemical attractions. As oriented in the double helix, adenine forms two hydrogen bonds with thymine, and guanine forms three hydrogen bonds with cytosine. Although two or three individual hydrogen bonds are energetically very weak, 2000 to 3000 bonds in tandem (typical of two long polynucleotide chains) provide great stability to the helix.

Another stabilizing factor is the arrangement of sugars and bases along the axis. In the Watson-Crick model, the *hydrophobic* ("water-fearing") nitrogenous bases are stacked almost horizontally on the interior of the axis and are thus shielded from water. The *hydrophilic* ("water-loving") sugar–phosphate backbone is on the outside of the axis, where both components can interact with water. These molecular arrangements provide significant chemical stabilization to the helix.

A more recent and accurate analysis of the form of DNA that served as the basis for the Watson-Crick model reveals a minor structural difference. A precise measurement of the number of base pairs (bp) per turn has demonstrated a value of 10.4 bp rather than the 10.0 bp predicted by Watson and Crick. In the classical model, each base pair is rotated 36° around the helical axis relative to the adjacent base pair, whereas the new finding requires a rotation of 34.6°. Thus, there are slightly more than 10 bp per turn.

The Watson-Crick model had an immediate effect on the emerging discipline of molecular biology. Even in their initial 1953 article, the authors noted, "It has not escaped our notice that the specific pairing we have postulated immediately suggests a possible copying mechanism for the genetic material." Two months later, Watson and Crick pursued this idea in a second article in *Nature*, suggesting a specific mechanism of replication of DNA—the **semiconservative mode of replication**. The second article alluded to two new concepts: (1) the storage of genetic information in the sequence of the bases, and (2) the mutations or genetic changes that would result from an alteration of the bases. These ideas have received vast amounts of experimental support since 1953 and are now universally accepted.

Watson and Crick's synthesis of ideas was highly significant with regard to subsequent studies of genetics and biology. The nature of the gene and its role in genetic mechanisms could now be viewed and studied in biochemical terms. Recognition of their work, along with that of Wilkins, led to their receipt of the Nobel Prize in Physiology or Medicine in 1962. Unfortunately, Rosalind Franklin had died in 1958 at the age of 37, making her contributions ineligible for consideration since the award is not given posthumously. The Nobel Prize was to be one of many such awards bestowed for work in the field of molecular genetics.

9.7 Alternative Forms of DNA Exist

Under different conditions of isolation, several conformational forms of DNA have been recognized. At the time Watson and Crick performed their analysis, two forms—**A-DNA** and **B-DNA**—were known. Watson and Crick's analysis was based on X-ray studies of B-DNA performed by Franklin, which is present under aqueous, low-salt conditions and is believed to be the biologically significant conformation.

While DNA studies around 1950 relied on the use of X-ray diffraction, more recent investigations have been performed using **single-crystal X-ray analysis**. The earlier studies achieved limited resolution of about 5 Å, but single crystals diffract X-rays at about 1 Å, near atomic resolution. As a result, every atom is "visible" and much greater structural detail is available during analysis.

With this modern technique, A-DNA, which is prevalent under high-salt or dehydration conditions, has now been scrutinized. In comparison to B-DNA (Figure 9–13), A-DNA is slightly more compact, with 11 bp in each complete turn of the helix, which is 23 Å (2.3 nm) in diameter. It is also a right-handed helix, but the orientation of the bases is somewhat different—they are tilted and displaced laterally in relation to the axis of the helix. As a result, the appearance of the major and minor grooves is modified. It seems doubtful that A-DNA occurs *in vivo* (under physiological conditions).

Other forms of DNA (e.g., C-, D-, E-, and most recently, P-DNA) are now known, but it is **Z-DNA** that has drawn the most attention. Discovered by Andrew Wang, Alexander Rich, and their colleagues in 1979 when they examined a small synthetic DNA fragment containing only G $\equiv$ C base pairs, Z-DNA takes on the rather remarkable configuration of a *left-handed double helix* (Figure 9–13). Like A- and B-DNA, Z-DNA consists of two antiparallel chains held together by Watson-Crick base pairs. Beyond these characteristics, Z-DNA is quite different. The left-handed helix is 18 Å (1.8 nm) in diameter, contains 12 bp per turn, and assumes a zigzag conformation (hence its name). The major groove present in B-DNA is nearly eliminated in Z-DNA.

Speculation abounds over the possibility that regions of Z-DNA exist in the chromosomes of living organisms. The unique helical arrangement could provide an important recognition point for the interaction with other molecules. However, it is still not clear whether Z-DNA occurs *in vivo*.

Still other forms of DNA have been studied, including P-DNA, named after Linus Pauling. It is produced by artificial "stretching" of DNA, creating a longer, more narrow version with the phosphate groups on the interior.

Now Solve This

Problem 34 on page 212 asks you to construct a protocol in order to analyze DNA from a newly discovered source.

Hint: Knowing the nature and relative compositions of the nitrogenous bases of the unknown DNA will provide particularly important experimental information at the outset.

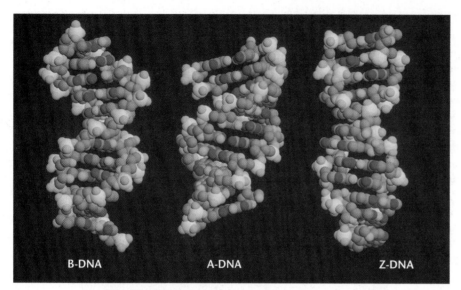

FIGURE 9–13 The top half of the figure shows computer-generated space-filling models of B-DNA, A-DNA, and Z-DNA. Below the photograph is an artist's rendering showing the orientation of the base pairs of B-DNA and A-DNA. Note that in B-DNA the base pairs are perpendicular to the helix, while they are tilted and pulled away from the helix in A-DNA.

B-DNA   A-DNA   Z-DNA

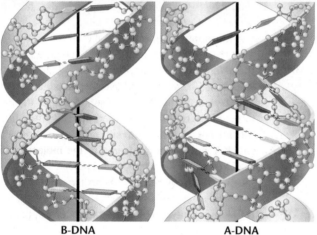

B-DNA   A-DNA

9.8 The Structure of RNA Is Chemically Similar to DNA, but Single-Stranded

The structure of RNA molecules resembles DNA, with several important exceptions. Although RNA also has nucleotides linked with polynucleotide chains, the sugar ribose replaces deoxyribose, and the nitrogenous base uracil replaces thymine. Another important difference is that most RNA is single-stranded, although there are two important exceptions. First, RNA molecules sometimes fold back on themselves to form double-stranded regions of complementary base pairs. Second, some animal viruses that have RNA as their genetic material contain double-stranded helices.

As established earlier (see Figure 9–1), three major classes of cellular RNA molecules function during the expression of genetic information: **ribosomal RNA (rRNA)**, **messenger RNA (mRNA)**, and **transfer RNA (tRNA)**. These molecules all originate as complementary copies of one of the two strands of DNA segments during the process of transcription. That is, their nucleotide sequence is complementary to the deoxyribonucleotide sequence of DNA, which served as the template for their synthesis. Because uracil replaces thymine in RNA, uracil is complementary to adenine during transcription and RNA base pairing.

Table 9.4 characterizes these major forms of RNA, as found in prokaryotic and eukaryotic cells. Different RNAs are distinguished according to their sedimentation behavior in a centrifugal field and their size, as measured by the number of nucleotides each contains. Sedimentation behavior depends on a molecule's density, mass, and shape, and its measure is called the **Svedberg coefficient** (S). While higher S values almost always designate molecules of greater molecular weight, the correlation is not direct; that is, a twofold increase in molecular weight does not lead to a twofold increase in S. This is because, in addition to a molecule's mass, the size and shape of the molecule also affect its rate of sedimentation (S). As you can see in Table 9.4, a wide variation exists in the size of the three classes of RNA.

Ribosomal RNA is generally the largest of these molecules (reflected in its S values) and usually constitutes about 80 percent of all RNA in the cell. Ribosomal RNAs are important structural components of **ribosomes**, which function as nonspecific workbenches where proteins are synthesized during translation. The various forms of rRNA found in prokaryotes and eukaryotes differ distinctly in size.

Messenger RNA molecules carry genetic information from the DNA of the gene to the ribosome. The mRNA molecules vary considerably in size, which reflects the variation in the

TABLE 9.4	RNA Characterization				
RNA Class	**Total RNA* (%)**	**Components (Svedberg Coefficient)**	**Eukaryotic (E) or Prokaryotic (P)**	**Number of Nucleotides**
Ribosomal (rRNA)	80	5S	P and E	120
		5.8S	E	160
		16S	P	1542
		18S	E	1874
		23S	P	2904
		28S	E	4718
Transfer (tRNA)	15	4S	P and E	75–90
Messenger (mRNA)	5	varies	P and E	100–10,000

*In *E. coli.*

size of the protein encoded by the mRNA as well as the gene serving as the template for transcription of mRNA.

Transfer RNA, the smallest class of these RNA molecules, carries amino acids to the ribosome during translation. Since more than one tRNA molecule interacts simultaneously with the ribosome, the molecule's smaller size facilitates these interactions.

These RNAs represent the major forms of the molecule involved in genetic expression, but other unique RNAs exist that perform various roles, expecially in eukaryotes. For example, **small nuclear RNA (snRNA)** participates in processing mRNAs. **Telomerase RNA** is involved in DNA replication at the ends of chromosomes (the telomeres), and **antisense RNA, microRNA (miRNA)**, and **short interfering RNA (siRNA)** are involved in gene regulation. DNA stores genetic information, while RNA most often functions in the expression of that information.

9.9 Many Analytical Techniques Have Been Useful During the Investigation of DNA and RNA

Since 1953, the role of DNA as the genetic material and the role of RNA in transcription and translation have been clarified through detailed analysis of nucleic acids. Let's consider several methods of analysis of these molecules that have been particularly important. Many of them are based on the unique nature of the hydrogen bond that is so integral to the structure of nucleic acids.

Hydrogen bonds impart an interesting and important set of qualities to the chemical behavior of nucleic acids under both laboratory and physiological conditions. For example, if DNA is isolated and subjected to heating, the double helix is denatured and unwinds. If a mixture of single strands that are complementary to each other are slowly cooled, they reassociate and re-form the helix. In the laboratory, these transformations can be "tracked" by using a spectrophotometer and monitoring the absorption of UV light (or optical density, OD) at 260 nm (OD_{260}).

During unwinding, the viscosity of DNA decreases and UV absorption increases (called the **hyperchromic shift**). A melting profile, in which OD_{260} is plotted against temperature, is shown for two DNA molecules in Figure 9–14. The midpoint of each curve is called the **melting temperature (T_m)**, where 50 percent of the strands have unwound. The molecule with a higher T_m has a higher percentage of G $\equiv$ C base pairs than A $=$ T base pairs compared to the molecule with the lower T_m, since G $\equiv$ C pairs share three hydrogen bonds compared to the two bonds between A $=$ T pairs.

Molecular Hybridization Techniques

The property of denaturation/renaturation of nucleic acids is the basis for one of the most powerful and useful techniques in molecular genetics—**molecular hybridization**. Provided that a reasonable degree of base complementarity exists, under the proper temperature conditions, two nucleic acid

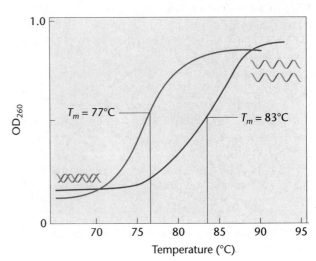

FIGURE 9–14 A graph of increase in UV absorption versus temperature (the hyperchromic effect) for two DNA molecules with different G $\equiv$ C contents. The molecule with a melting point (T_m) of 83°C has a greater G $\equiv$ C content than the molecule with T_m = 77°C.

strands from different sources will join. As a result, molecular hybridization is possible between DNA strands from different species, and between DNA and RNA strands. For example, an RNA molecule will hybridize with the segment of DNA from which it was transcribed or with a DNA molecule from a different species, as long as its nucleotide sequence is nearly complementary.

The technique can even be performed using the DNA in cytological preparations as the "target" for hybrid formation. This process is called *in situ* **molecular hybridization**. Mitotic or interphase cells are first fixed to slides and then subjected to hybridization conditions. Single-stranded DNA or RNA is added, and hybridization is monitored. The nucleic acid that is added may be either radioactive or contain a fluorescent label to allow its detection. In the former case, autoradiography is used.

Figure 9–15 illustrates the use of a fluorescent label. A "probe," a short fragment of DNA that is complementary to DNA in the chromosomes' centromere regions, has been hybridized. Fluorescence occurs only in the centromere regions and thus identifies each one along its chromosome. Because fluorescence is used, the technique is known by the acronym **FISH** (**fluorescent** *in situ* **hybridization**). The use of this technique to identify chromosomal locations housing specific genetic information has been a valuable addition to geneticists' repertoire of experimental techniques.

Reassociation Kinetics and Repetitive DNA

In one extension of molecular hybridization procedures, the *rate of reassociation* of complementary single DNA strands is analyzed. This technique, **reassociation kinetics**, was first refined and studied by Roy Britten and David Kohne.

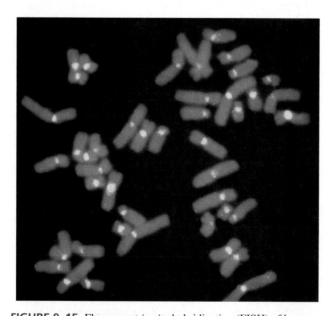

FIGURE 9–15 Fluorescent *in situ* hybridization (FISH) of human metaphase chromosomes. The probe, specific to centromeric DNA, produces a yellow fluorescence signal indicating hybridization. The red fluorescence is produced by propidium iodide counterstaining of chromosomal DNA.

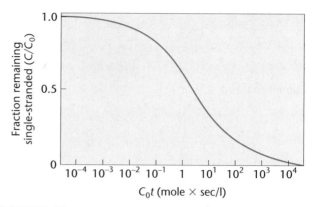

FIGURE 9–16 The ideal time course for reassociation of DNA (C/C_0) when, at time zero, all of the DNA consists of unique fragments of single-stranded complements. Note that the x-axis ($C_0 t$) is scaled logarithmically.

The DNA used in such studies is first fragmented into small pieces by shearing forces introduced during its isolation. The resultant DNA fragments have an average size of several hundred base pairs. The fragments are then dissociated into single strands by heating (denatured), and when the temperature is lowered, reassociation is monitored. During reassociation, pieces of single-stranded DNA randomly collide. If they are complementary, a stable double strand is formed; if not, they separate and are free to encounter other DNA fragments. The process continues until all possible matches are made.

The results of one such experiment are graphed in Figure 9–16. The percentage of reassociation of DNA fragments is plotted against a logarithmic scale of normalized time, a function referred to as $C_0 t$, where C_0 is the initial concentration of DNA single strands, and t is the time.

A great deal of information can be obtained from studies that compare the reassociation of DNA of different organisms. For example, we can compare the point in the reaction when one-half of the DNA is present as double-stranded fragments. This point is called the **half reaction time ($C_0 t_{1/2}$)**. Provided that all of the DNA fragments contain unique nucleotide sequences and all are about the same size, $C_0 t_{1/2}$ varies directly in relation to the total length of the DNA.

Figure 9–17 compares DNA from three sources, each with a different genome size. As genome size increases, the curves obtained have a similar shape but are shifted farther and farther to the right, indicative of an extended reassociation time. Reassociation occurs more slowly in larger genomes because it takes longer for initial matches if there are greater numbers of unique DNA fragments. The collisions are random, therefore the more sequences that are present, the greater the number of mismatches before all correct matches are made.

When reassociation kinetics in eukaryotic organisms with much larger genome sizes was first studied, a surprising observation was made. Rather than exhibiting a reduced rate of reassociation, the data revealed that *some*

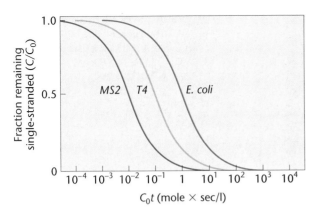

FIGURE 9–17 The reassociation rates (C/C_0) of DNA derived from phage MS2, phage T4, and *E. coli*. The genome of T4 is larger than MS2, and that of *E. coli* is larger than T4.

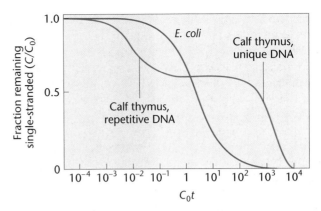

FIGURE 9–18 The $C_0 t$ curve of calf thymus DNA compared with *E. coli*. The repetitive fraction of calf DNA reassociates more quickly than that of *E. coli*, while the more complex, unique calf DNA takes longer to reassociate than that of *E. coli*.

DNA segments reassociated even more rapidly than those derived from *E. coli*. The remaining DNA, as expected because of its greater size and complexity, took longer to reassociate.

For example, Britten and Kohne examined DNA from calf thymus tissue. Based on their observations (Figure 9–18), they correctly hypothesized that the rapidly reassociating fraction might represent *repetitive sequences* in the calf genome. This interpretation would explain why these segments reassociate so rapidly—multiple copies of the same sequence are much more likely to make matches, thus reassociating more quickly than single copies. On the other hand, the remaining DNA segments consist of unique nucleotide sequences (present only once) in the genome. Because calf thymus DNA has many more unique sequences than *E. coli*, their reassociation takes longer.

Multiple copies of a sequence in the genome are collectively known as **repetitive DNA**. Repetitive DNA is prevalent in eukaryotic genomes and is key to our understanding of how genetic information is organized in chromosomes. Careful study has shown that various levels of repetition exist. In some cases, short DNA sequences are repeated over a million times. In other cases, longer sequences are repeated only a few times, or intermediate levels of sequence redundancy are present. The discovery of repetitive DNA was one of the first clues that much of the DNA in eukaryotes is not contained in the genes that encode proteins.

9.10 Nucleic Acids Can Be Separated Using Electrophoresis

The final, essential, technique in the analysis of nucleic acids that we will discuss is **electrophoresis**. This technique separates different-sized fragments of DNA and RNA chains and is invaluable in current research investigations in molecular genetics.

In general, electrophoresis separates the molecules in a mixture by causing them to migrate under the influence of an electric field. A sample is placed on a porous substance (a piece of filter paper or a semisolid gel), which is then placed in a solution that conducts electricity. If two molecules have approximately the same shape and mass, the one with the greater net charge will migrate more rapidly toward the electrode of opposite polarity.

As electrophoretic technology developed from its initial application to protein separation, researchers discovered that using gels of varying pore sizes significantly improved the resolution of this research technique. This advance is particularly useful for mixtures of molecules with a similar charge–mass ratio but of different sizes. For example, two polynucleotide chains of different lengths (e.g., 10 versus 20 nucleotides) are both negatively charged based on the phosphate groups of the nucleotides. Both chains move to the positively charged pole (the anode), but the charge–mass ratio is the same for each chain, and separation due to the electric field is minimal. However, using porous medium such as a **polyacrylamide gel** or an **agarose gel**, which can be prepared with various pore sizes, enables us to separate the two molecules. *Smaller molecules migrate at a faster rate through the gel than larger molecules* (Figure 9–19). The key to separation is based on the matrix (pores) of the gel, which restricts migration of larger molecules more than it restricts smaller molecules. The resolving power is so great that polynucleotides that vary by just one nucleotide in length may be separated. Once electrophoresis is complete, bands representing the variously sized molecules are identified either by autoradiography (if a component of the

---

Now Solve This

Problem 28 on page 212 asks you to extrapolate information about $C_0 t$ analysis of DNA to the overall size of the DNA molecule.

Hint: Absolute $C_0 t$ values are directly proportional to the number of base pairs making up a DNA molecule.

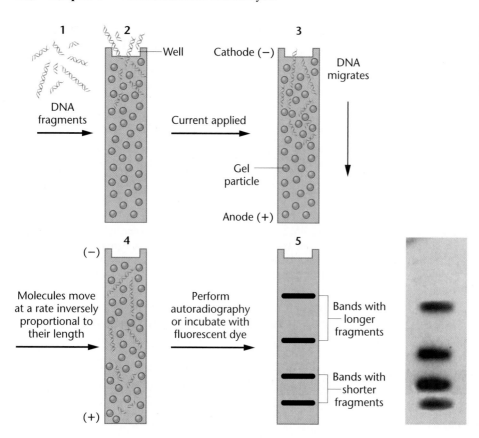

FIGURE 9–19 Electrophoretic separation of a mixture of DNA fragments that vary in length. The photograph at the bottom right shows an agarose gel with DNA bands.

molecule is radioactive) or by the use of a fluorescent dye that binds to nucleic acids.

Electrophoretic separation of nucleic acids is at the heart of a variety of other commonly used research techniques. Of particular note are the various "blotting" techniques (e.g., Southern blots and Northern blots), as well as DNA sequencing methods, which we will discuss in detail later in the text (see chapter 17).

CHAPTER SUMMARY

1. The existence of a genetic material capable of replication, storage, expression, and mutation is deducible from observed patterns of inheritance in organisms. Both proteins and nucleic acids were considered as possible candidates for the genetic material. Proteins were initially favored.

2. Transformation studies, as well as experiments using bacteria infected with bacteriophages, strongly suggested that DNA is the genetic material for bacteria and most viruses.

3. Initially, only indirect observations supported the hypothesis that DNA controls inheritance in eukaryotes. These included DNA distribution in the cell, quantitative analysis of DNA, and UV-induced mutagenesis. More recent recombinant DNA techniques, as well as experiments with transgenic mice, have provided direct experimental evidence that the eukaryote genetic material is DNA.

4. RNA serves as the genetic material in some viruses, including bacteriophages as well as some plant and animal viruses.

5. By the 1950s, many scientists sought to determine the structure of DNA. These efforts culminated in 1953 with Watson and Crick's proposal. Based on base composition and X-ray diffraction data, they constructed a model. The key features of their model include two antiparallel polynucleotide chains held together in a right-handed double helix by the hydrogen bonds formed between complementary bases. To date, the basic tenets of the double-helix model have held true.

6. RNA varies from DNA by virtue of almost always being single-stranded, containing uracil rather than thymine, and having ribose rather than deoxyribose as its constituent sugar.

7. The structure of DNA lends itself to various forms of analysis, which have in turn led to studies of the functional aspects of the genetic machinery. Absorption of UV light, denaturation–reassociation, and electrophoresis procedures are important tools in the study of nucleic acids. Reassociation kinetics analysis enabled geneticists to postulate the existence of repetitive DNA in eukaryotes, where certain nucleotide sequences are present many times in the genome.

GENETICS, TECHNOLOGY, AND SOCIETY

The Twists and Turns of the Helical Revolution

Western civilization is frequently transformed by new scientific ideas that overturn our self-concepts and permanently alter our relationships with each other and the rest of the animate world. For 50 years, we have been in the midst of such a revolution—one as significant as those triggered by Darwin's theory of evolution or the Copernican rejection of Ptolemy's earth-centered universe.

The revolution began in April 1953 with Watson and Crick's discovery of the molecular structure of DNA. Their discovery that the DNA molecule consists of a twisted double helix, held together by weak bonds between specific pairs of bases, suddenly provided elegant solutions to age-old questions about the mechanisms of heredity, mutation, and evolution. Some of the greatest mysteries of life could be explained by the beauty and simplicity of a helix that replicates and shuffles the code of life.

After 1953, the double helix rapidly became the focus of modern science. Aware of DNA's helical structure, molecular biologists quickly devised methods to purify, mutate, cut, and paste DNA in the test tube. They spliced DNA molecules from one organism into those of another and then introduced these chimeric molecules into bacteria or cells in culture. They read the nucleotide sequences of genes and modified the traits of bacteria, fungi, fruit flies, and mice by removing and mutating their genes, or by introducing genes from other organisms. On the fiftieth anniversary of Watson and Crick's double-helical DNA model, the Human Genome Project announced the completion of the largest DNA project so far—sequencing the entire human genome.

In a mere 50 years, the helical revolution has touched the lives of millions of people. We can now test for simple genetic diseases, such as Tay-Sachs, cystic fibrosis, and sickle-cell anemia. We can manufacture large quantities of medically important proteins, such as insulin and growth hormone, using DNA technologies. DNA forensic tests help convict criminals, exonerate the innocent, and establish paternity. By following a trail of DNA sequences, anthropologists can now trace human origins back in time and place.

The helical revolution has profoundly altered our views of ourselves and our world. Although scientists dismiss the idea that humans are simply the products of their genes, popular culture endows DNA with almost magical powers. Genes are said to explain personality, career choice, criminality, intelligence—even fashion preferences and political attitudes. Advertisements hijack the language of genetics in order to grant inanimate objects a "genealogy" or "genetic advantage." Popular culture speaks of DNA as an immortal force, with the ability to affect morality and fate. The double helix is proclaimed as the essence of life, with the power to shape our future. Simple genetic explanations for our behavior appear to have more resonance for us than explanations involving social influences, economic factors, or free will. The beauty, symmetry, and biological significance of the double helix have insinuated themselves into art, movies, advertising, and music. Paintings, sculpture, films such as *Jurassic Park* and *Boys from Brazil*—even video games and perfumes—use the language and imagery of genetics to confer upon the DNA molecule all the power and fears of modern technology.

But what of the future? Can we predict how the double helix and genetics will shape our world over the next 50 years? Although prophecy is certainly a risky business, some scientific developments seem assured. With the completion of the Human Genome Project, we will undoubtedly identify more and more of the genes that control normal and abnormal processes. In turn, this will enhance our ability to diagnose and predict genetic diseases. Over the next 50 years, we can look forward to biotechnologies as complex as gene therapies, prenatal diagnoses, and screening programs for susceptibilities to diseases as complicated as cancer and heart disease. We will continue to expand the applications of genetic engineering to agriculture as we manipulate plant and animal genes for enhanced productivity, disease resistance, and flavor.

The helical revolution will also continue to transform our concepts of ourselves and other creatures. As the human genome is compared to the genomes of other animals, it will become increasingly evident that we are closely related genetically to the rest of the animate world. The nucleotide sequence of the human genome differs only about 1 percent from that of chimpanzees, and some of our genes are virtually identical to homologous genes in plants, animals, and bacteria. As we realize the extent of our genetic kinship, extending over billions of years in a chain from the first life on earth, it is possible that this knowledge will alter our relationships with animals and with each other. When more genes are identified that contribute to phenotypic traits as simple as eye color and as complicated as intelligence or sexual orientation, it is possible that we will define ourselves even more as genetic beings and even less as creatures of free will or as the products of our environment.

During the first half of the current century, we will inevitably be faced with the practical and philosophical consequences of the DNA revolution. Will society harness DNA for everyone's benefit, or will this new genetic knowledge be used as a vehicle for discrimination? At the same time that modern genetics grants us more dominion over life, will it paradoxically increase our feelings of powerlessness? Will our new DNA-centered self-concepts increase our compassion for all life forms, or will it increase our perceived separation from the natural world? We will make our choices, and human history will proceed.

Reference

Dennis, C., and Campbell, P. 2003. The eternal molecule. (Introduction to a series of feature articles commemorating the 50th anniversary of the discovery of DNA structure). *Nature* 421: 396.

Web Site

A Revolution at 50. [*The New York Times* articles, on the 50th anniversary of the discovery of DNA structure.] **http://www.nytimes.com/indexes/2003/02/25/health/genetics/index.html**

KEY TERMS

INSIGHTS AND SOLUTIONS

In contrast to preceding chapters, this chapter does not emphasize genetic problem solving. Instead, it recounts some of the initial experimental analyses that launched the era of molecular genetics. Quite fittingly, then, our Insights and Solutions section shifts its emphasis to experimental rationale and analytical thinking, an approach that will continue throughout the remainder of the text, whenever appropriate.

1. Based strictly on the transformation analysis of Avery, MacLeod, and McCarty, what objection might be made to the conclusion that DNA is the genetic material? What other conclusion might be considered?

Solution: Based solely on their results, we could conclude that DNA is essential for transformation. However, DNA might have been a substance that caused capsular formation by converting nonencapsulated cells *directly* to cells with a capsule. That is, DNA may simply have played a catalytic role in capsular synthesis, leading to cells that display smooth, type III colonies.

2. What observations argue against this objection?

Solution: First, transformed cells pass the trait on to their progeny cells, thus supporting the conclusion that DNA is responsible for heredity, not for the direct production of poly-saccharide coats. Second, subsequent transformation studies over the next five years showed that other traits, such as antibiotic resistance, could be transformed. Therefore, the transforming factor has a broad general effect, not one specific to polysaccharide synthesis.

3. If RNA were the universal genetic material, how would this have affected the Avery experiment and the Hershey-Chase experiment?

Solution: In the Avery experiment, deoxyribonuclease (DNase), rather than ribonuclease (RNase), would have eliminated transformation. Had this occurred, Avery and his colleagues would have concluded that RNA was the transforming factor. Hershey and Chase would have obtained identical results, since ^{32}P would also label RNA but not protein.

4. Sea urchin DNA, which is double-stranded, contains 17.5 percent of its bases in the form of cytosine (C). What percentages of the other three bases are expected to be present in this DNA?

Solution: The amount of C equals G, so guanine is also present at 17.5 percent. The remaining bases, A and T, are present in equal amounts, and together they represent the remaining bases (100–35). Therefore, A = T = 65/2 = 32.5 percent.

PROBLEMS AND DISCUSSION QUESTIONS

1. The functions ascribed to the genetic material are replication, expression, storage, and mutation. What does each of these terms mean?

2. Discuss the reasons why proteins were generally favored over DNA as the genetic material before 1940. What was the role of the tetranucleotide hypothesis in this controversy?

3. Contrast the various contributions made to our understanding of transformation by Griffith, Alloway, and Avery.

4. When Avery and his colleagues had obtained what was concluded to be the transforming factor from the III*S* virulent cells, they treated the fraction with proteases, ribonuclease, and deoxyribonuclease, followed by the assay for retention or loss of transforming ability. What were the purpose and results of these experiments? What conclusions were drawn?

5. Why were ^{32}P and ^{35}S chosen in the Hershey-Chase experiment? Discuss the rationale and conclusions of this experiment.

6. Does the design of the Hershey-Chase experiment distinguish between DNA and RNA as the molecule serving as the genetic material? Why or why not?

7. Would an experiment similar to that performed by Hershey and Chase work if the basic design were applied to the phenomenon of transformation? Explain why or why not.

8. What observations are consistent with the conclusion that DNA serves as the genetic material in eukaryotes? List and discuss them.

9. What are the exceptions to the general rule that DNA is the genetic material in all organisms? What evidence supports these exceptions?

10. Draw the chemical structure of the three components of a nucleotide, and then link them together. What atoms are removed from the structures when the linkages are formed?

11. How are the carbon and nitrogen atoms of the sugars, purines, and pyrimidines numbered?

12. Adenine may also be named 6-amino purine. How would you name the other four nitrogenous bases, using this alternative system? (O is oxy, and CH_3 is methyl.)

13. Draw the chemical structure of a dinucleotide composed of A and G. Opposite this structure, draw the dinucleotide composed of T and C in an antiparallel (or upside-down) fashion. Form the possible hydrogen bonds.

14. Describe the various characteristics of the Watson-Crick double-helix model for DNA.

15. What evidence did Watson and Crick have at their disposal in 1953? What was their approach in arriving at the structure of DNA?

16. What might Watson and Crick have concluded, had Chargaff's data from a single source indicated the following base composition?

	A	T	G	C
%	29	19	21	31

Why would this conclusion be contradictory to Wilkins and Franklin's data?

17. How do covalent bonds differ from hydrogen bonds? Define base complementarity.

18. List three main differences between DNA and RNA.

19. What are the three major types of RNA molecules? How is each related to the concept of information flow?

20. What component of the nucleotide is responsible for the absorption of ultraviolet light? How is this technique important in the analysis of nucleic acids?

21. What is the physical state of DNA after being denatured by heat?

22. What is the hyperchromic effect? How is it measured? What does T_m imply?

23. Why is T_m related to base composition?

24. Compare the curves below, representing reassociation kinetics. What can be said about the DNA represented by each set of data compared with *E. coli*?

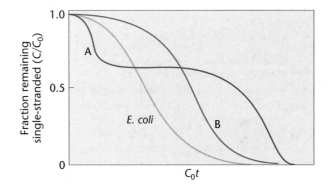

25. What is the chemical basis of molecular hybridization?

26. What did the Watson-Crick model suggest about the replication of DNA?

27. A genetics student was asked to draw the chemical structure of an adenine- and thymine-containing dinucleotide derived from DNA. His answer is shown below. The student made more than six major errors. One of them is circled, numbered 1, and explained. Find five others. Circle them, number them 2–6, and briefly explain each by following the example given.

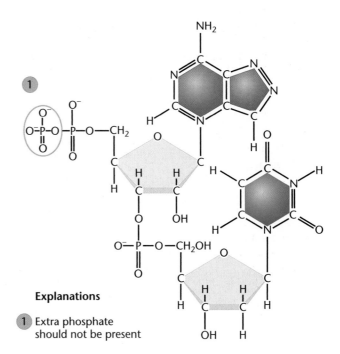

Explanations

1 Extra phosphate should not be present

28. The DNA of the bacterial virus T4 produces a $C_0 t_{1/2}$ of about 0.5 and contains 10^5 nucleotide pairs in its genome. How many nucleotide pairs are present in the genome of the virus MS2 and the bacterium *E. coli*, whose respective DNAs produce $C_0 t_{1/2}$ values of 0.001 and 10.0?

29. A primitive eukaryote was discovered that displayed a unique nucleic acid as its genetic material. Analysis revealed the following observations:

    (a) X-ray diffraction studies display a general pattern similar to DNA, but with somewhat different dimensions and more irregularity.

    (b) A major hyperchromic shift is evident upon heating and monitoring UV absorption at 260 nm.

    (c) Base composition analysis reveals four bases in the following proportions:

Adenine	= 8%	Hypoxanthine	= 18%
Guanine	= 37%	Xanthine	= 37%

    (d) About 75 percent of the sugars are deoxyribose, while 25 percent are ribose.

    Attempt to solve the structure of this molecule by postulating a model that is consistent with the foregoing observations.

30. With the information given in this chapter on B- and Z-DNA and the nature of helices, carefully analyze the structures shown here, and draw conclusions about the helical nature of areas (a) and (b). Which is right-handed and which is left-handed?

(a)    (b)

31. One of the most common spontaneous lesions that occurs in DNA under physiological conditions is the hydrolysis of the amino group of cytosine, converting it to uracil. What would be the effect on DNA structure if a uracil group replaced cytosine?

32. In some organisms, cytosine is methylated at carbon 5 of the pyrimidine ring after it is incorporated into DNA. If a 5-methyl cytosine is then hydrolyzed, as described in Problem 31, what base will be generated?

33. *Newsdate: March 1, 2015.* A unique creature has been discovered during exploration of outer space. Recently, its genetic material has been isolated and analyzed, and is similar in some ways to DNA in chemical makeup. It contains in abundance the 4-carbon sugar erythrose and a molar equivalent of phosphate groups. In addition, it contains six nitrogenous bases: adenine (A), guanine (G), thymine (T), cytosine (C), hypoxanthine (H), and xanthine (X). These bases exist in the following relative proportions:

$$A = T = H \quad \text{and} \quad C = G = X$$

X-ray diffraction studies have established a regularity in the molecule and a constant diameter of about 30 Å.

Together, these data have suggested a model for the structure of this molecule. (a) Propose a general model of this molecule, and briefly describe it. (b) What base-pairing properties must exist for H and for X in the model? (c) Given the constant diameter of 30 Å, do you think *either* (i) both H and X are purines or both pyrimidines, *or* (ii) one is a purine and one is a pyrimidine?

34. You are provided with DNA samples from two newly discovered bacterial viruses. Based on the various analytical techniques discussed in this chapter, construct a research protocol that would be useful in characterizing and contrasting the DNA of both viruses. Indicate the type of information you hope to obtain for each technique included in the protocol.

35. During electrophoresis, DNA molecules can easily be separated according to size because all DNA molecules have the same charge–mass ratio and the same shape (long rod). Would you expect RNA molecules to behave in the same manner as DNA during electrophoresis? Why or why not?

DNA Replication and Synthesis

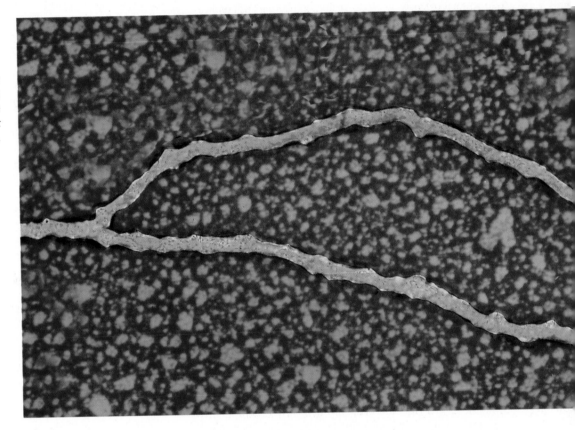

Transmission electron micrograph of human DNA from a HeLa cell, illustrating a replication fork characteristic of active DNA replication.

CHAPTER CONCEPTS

- Genetic continuity between parental and progeny cells is maintained by semiconservative replication of DNA, as predicted by the Watson-Crick model.

- Semiconservative replication uses each strand of the parent double helix as a template, and each newly replicated double helix includes one "old" and one "new" strand of DNA.

- DNA synthesis is a complex but orderly process, occurring under the direction of a myriad of enzymes and other proteins.

- DNA synthesis involves the polymerization of nucleotides into polynucleotide chains.

- DNA synthesis is similar in prokaryotes and eukaryotes, but more complex in eukaryotes.

- In eukaryotes, DNA synthesis at the ends of chromosomes (telomeres) poses a special problem, overcome by a unique RNA-containing enzyme, telomerase.

- Genetic recombination, an important process leading to the exchange of segments between DNA molecules, occurs under the direction of a group of enzymes.

Following Watson and Crick's proposal for the structure of DNA, scientists focused their attention on how this molecule is replicated. Replication is an essential function of the genetic material and must be executed precisely if genetic continuity between cells is to be maintained following cell division. It is an enormous, complex task. Consider for a moment that more than 3×10^9 (3 billion) base pairs exist within the 23 chromosomes of the human genome. To duplicate faithfully the DNA of just one of these chromosomes requires a mechanism of extreme precision. Even an error rate of only 10^{-6} (one in a million) will still create 3000 errors (obviously an excessive number) during each replication cycle of the genome. While it is not error free, and much of evolution would not have occurred if it were, an extremely accurate system of DNA replication has evolved in all organisms.

As Watson and Crick noted in the concluding paragraph of their 1953 paper (reprinted on page 202), their proposed model of the double helix provided the initial insight into how replication occurs. Called *semiconservative replication*, this mode of DNA duplication was soon to receive strong support from numerous studies of viruses, prokaryotes, and eukaryotes. Once the general *mode* of replication was clarified, research to determine the precise details of *DNA synthesis* intensified. What has since been discovered is that numerous enzymes and other proteins are needed to copy a DNA helix. Because of the complexity of the chemical events during synthesis, this subject remains an extremely active area of research.

In this chapter, we will discuss the general mode of replication, as well as the specific details of DNA synthesis. The research leading to such knowledge is another link in our understanding of life processes at the molecular level.

? How Do We Know?

In this chapter, we will focus on how DNA is replicated and synthesized. We shall elucidate the general mechanism of replication and describe how DNA is synthesized when it is copied. As you study this topic, you should try to answer several fundamental questions:

1. What is the experimental basis for concluding that DNA replicates semiconservatively in both prokaryotes and eukaryotes?

2. How was it demonstrated that DNA synthesis occurs under the direction of DNA polymerase?

3. How do we know the requirements of DNA polymerase in directing DNA synthesis?

4. How do we know that *in vivo* DNA synthesis occurs in the 5′ to 3′ direction?

5. What discoveries served as the basis for concluding that DNA polymerase I (Kornberg's enzyme) is not the universal *in vivo* replicating enzyme in bacteria such as *E. coli*?

6. How do we know that DNA synthesis is discontinuous on one of the two template strands?

7. What observations reveal that a "telomere problem" exists during eukaryotic DNA replication, and how did we learn of the solution to this problem?

10.1 DNA Is Reproduced by Semiconservative Replication

It was apparent to Watson and Crick that, because of the arrangement and nature of the nitrogenous bases, each strand of a DNA double helix could serve as a template for the synthesis of its complement (Figure 10–1). They proposed that, if the helix were unwound, each nucleotide along the two parent strands would have an affinity for its complementary nucleotide. As we learned in Chapter 9, the complementarity is due to the potential hydrogen bonds that can be formed. If thymidylic acid (T) were present, it would "attract" adenylic acid (A); if guanidylic acid (G) were present, it would attract cytidylic acid (C); likewise, A would attract T, and C would attract G. If these nucleotides were then covalently linked into polynucleotide chains along both templates, the result would be the production of two identical double strands of DNA. Each replicated DNA molecule would consist of one "old" and one "new" strand, hence the reason for the name **semiconservative replication**.

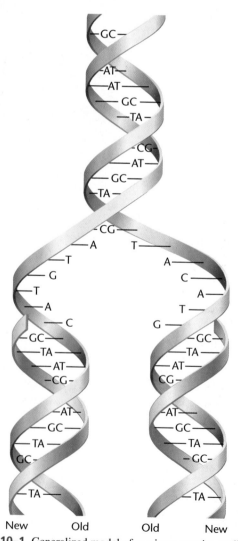

New   Old   Old   New

FIGURE 10–1 Generalized model of semiconservative replication of DNA. New synthesis is shown in purple.

Two other theoretical modes of replication are possible that also rely on the parental strands as a template (Figure 10–2). In **conservative replication**, complementary polynucleotide chains are synthesized as described earlier. Following synthesis, however, the two newly created strands then come together and the parental strands reassociate. The original helix is thus "conserved."

In the second alternative mode, called **dispersive replication**, the parental strands are dispersed into two new double helices following replication. Hence, each strand consists of both old and new DNA. This mode would involve cleavage of the parental strands during replication. It is the most complex of the three possibilities and is therefore considered to be least likely to occur. It could not, however, be ruled out as an experimental model. Figure 10–2 shows the theoretical results of a single round of replication by each of the three different modes.

The Meselson-Stahl Experiment

In 1958, Matthew Meselson and Franklin Stahl published the results of an experiment providing strong evidence that semiconservative replication is the mode used by bacterial cells to produce new DNA molecules. They grew *E. coli* cells for many generations in a medium that had $^{15}NH_4Cl$ (ammonium chloride) as the only nitrogen source. A "heavy" isotope of nitrogen, ^{15}N contains one one more neutron than the naturally occurring ^{14}N isotope; thus, molecules containing ^{15}N are more dense than those containing ^{14}N. Unlike radioactive iso-

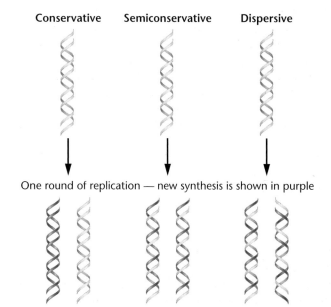

FIGURE 10–2 Results of one round of replication of DNA for each of the three possible modes by which replication could be accomplished.

topes, ^{15}N is stable. After many generations, almost all nitrogen-containing molecules, including the nitrogenous bases of DNA, contained the heavier isotope in the *E. coli* cells.

Critical to the success of this experiment, DNA containing ^{15}N can be distinguished from DNA containing ^{14}N. The experimental procedure involves the use of a technique referred to as **sedimentation equilibrium centrifugation**, or as it is also called, *density gradient centrifugation*. Samples are forced by centrifugation through a density gradient of a heavy metal salt, such as cesium chloride. Molecules of DNA will reach equilibrium when their density equals the density of the gradient medium. In this case, ^{15}N-DNA will reach this point at a position closer to the bottom of the tube than will ^{14}N-DNA.

In this experiment (Figure 10–3), uniformly labeled ^{15}N cells were transferred to a medium containing only $^{14}NH_4Cl$.

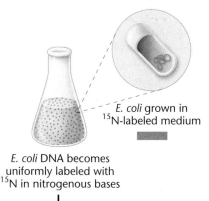

E. coli grown in ^{15}N-labeled medium

E. coli DNA becomes uniformly labeled with ^{15}N in nitrogenous bases

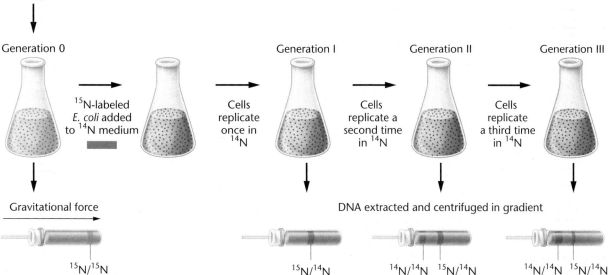

Generation 0

^{15}N-labeled *E. coli* added to ^{14}N medium

Generation I
Cells replicate once in ^{14}N

Generation II
Cells replicate a second time in ^{14}N

Generation III
Cells replicate a third time in ^{14}N

Gravitational force

DNA extracted and centrifuged in gradient

$^{15}N/^{15}N$ $^{15}N/^{14}N$ $^{14}N/^{14}N$ $^{15}N/^{14}N$ $^{14}N/^{14}N$ $^{15}N/^{14}N$

FIGURE 10–3 The Meselson–Stahl experiment.

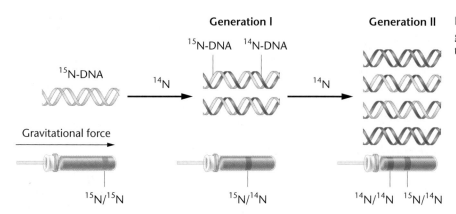

FIGURE 10–4 The expected results of two generations of semiconservative replication in the Meselson–Stahl experiment.

Thus, all "new" synthesis of DNA during replication contained only the "lighter" isotope of nitrogen. The time of transfer to the new medium was taken as time zero ($t = 0$). The *E. coli* cells were allowed to replicate over several generations, with cell samples removed after each replication cycle. DNA was isolated from each sample and subjected to sedimentation equilibrium centrifugation.

After one generation, the isolated DNA was present in only a single band of intermediate density—the expected result for semiconservative replication in which each replicated molecule was composed of one new ^{14}N-strand and one old ^{15}N-strand (Figure 10–4). This result was not consistent with the prediction of conservative replication, in which two distinct bands would occur, and thus this mode may be rejected.

After two cell divisions, DNA samples again showed two density bands—one intermediate band and one lighter band corresponding to the ^{14}N position in the gradient. Similar results occurred after a third generation, except that the proportion of the band increased. This was again consistent with the interpretation that replication is semiconservative.

You may have realized that a molecule exhibiting intermediate density is also consistent with dispersive replication. However, Meselson and Stahl ruled out this mode of replication on the basis of two observations. First, after the first generation of replication in an ^{14}N-containing medium, they isolated the hybrid molecule and heat denatured it. Recall from Chapter 9 that heating will separate a duplex into single strands. When the densities of the single strands of the hybrid were determined, they exhibited *either* an ^{15}N profile *or* an ^{14}N profile, but *not* an intermediate density. This observation is consistent with the semiconservative mode but inconsistent with the dispersive mode.

Furthermore, if replication were dispersive, *all* generations after $t = 0$ would demonstrate DNA of an intermediate density. In each generation after the first, the ratio of ^{15}N/^{14}N would decrease, and the hybrid band would become lighter and lighter, eventually approaching the ^{14}N band. This result was not observed. The Meselson-Stahl experiment provided conclusive support for semiconservative replication in bacteria and tended to rule out both the conservative and dispersive modes.

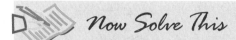

Now Solve This

Problem 2 on page 233 asks you to describe the role of ^{15}N during the Meselson-Stahl experiment.

Hint: Remember that ^{15}N is a stable, heavy isotope of nitrogen that can be distinguished from ^{14}N using centrifugation techniques.

Semiconservative Replication in Eukaryotes

In 1957, the year before the work of Meselson and his colleagues was published, J. Herbert Taylor, Philip Woods, and Walter Hughes presented evidence that semiconservative replication also occurs in eukaryotic organisms. They experimented with root tips of the broad bean *Vicia faba*, which are an excellent source of dividing cells. These researchers were able to monitor the process of replication by labeling DNA with ^{3}H-thymidine, a radioactive precursor of DNA, and performing autoradiography.

Autoradiography is a common technique that, when applied cytologically, pinpoints the location of a radioisotope in a cell. In this procedure, a photographic emulsion is placed over a histological preparation containing cellular material (root tips, in this experiment), and the preparation is stored in the dark. The slide is then developed, much as photographic film is processed. Because the radioisotope emits energy, following development the emulsion turns black at the approximate point of emission. The end result is the presence of dark spots or "grains" on the surface of the section, identifying the location of newly synthesized DNA within the cell.

Taylor and his colleagues grew root tips for approximately one generation in the presence of the radioisotope and then placed them in unlabeled medium in which cell division continued. At the conclusion of each generation, they arrested the cultures at metaphase by adding colchicine (a chemical derived from the crocus plant that poisons the spindle fibers) and then examined the chromosomes by autoradiography. They found radioactive thymidine only in association with chromatids that contained newly synthesized DNA. Figure 10–5 illustrates the replication of a single chromosome over two division cycles, including the distribution of grains.

FIGURE 10–5 The Taylor–Woods–Hughes experiment, demonstrating the semiconservative mode of replication of DNA in root tips of *Vicia faba*. A portion of the plant is shown in the top photograph. (a) An unlabeled chromosome proceeds through the cell cycle in the presence of ^{3}H-thymidine. As it enters mitosis, both sister chromatids of the chromosome are labeled, as shown, by autoradiography. After a second round of replication (b), this time in the absence of ^{3}H-thymidine, only one chromatid of each chromosome is expected to be surrounded by grains. Except where a reciprocal exchange has occurred between sister chromatids (c), the expectation was upheld. The micrographs are of the actual autoradiograms obtained in the experiment.

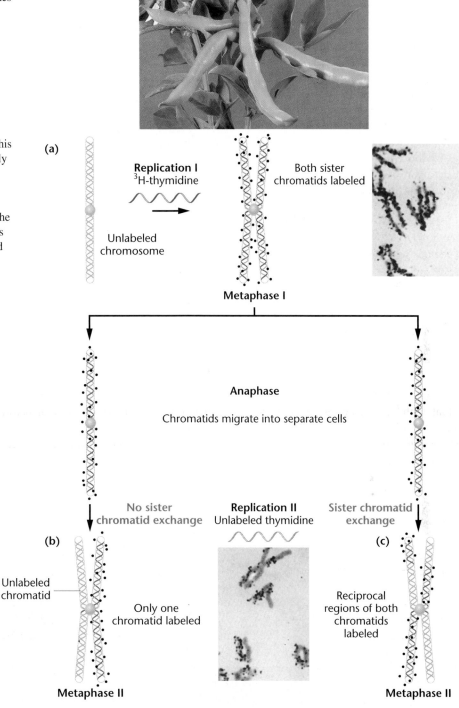

These results are compatible with the semiconservative mode of replication. After the first replication cycle in the presence of the isotope, both sister chromatids show radioactivity, indicating that each chromatid contains one new radioactive DNA strand and one old unlabeled strand. After the second replication cycle, *which takes place in unlabeled medium*, only one of the two sister chromatids of each chromosome should be radioactive, because half of the parent strands are unlabeled. With only the minor exceptions of *sister chromatid exchanges* (discussed in Chapter 7), this result was observed.

Together, the Meselson-Stahl experiment and the experiment by Taylor, Woods, and Hughes soon led to the general acceptance of the semiconservative mode of replication. Later studies with other organisms reached the same conclusion and also strongly supported Watson and Crick's proposal for the double-helix model of DNA.

Origins, Forks, and Units of Replication

To enhance our understanding of semiconservative replication, let's briefly consider a number of relevant issues. The first concerns the **origin of replication**. Where along the chromosome is DNA replication initiated? Is there only a single origin, or does DNA synthesis begin at more than one point? Is any given point of origin random, or is it located at a specific region along the chromosome? Second, once replication begins, does it proceed in a single direction or in both directions away from the origin? In other words, is replication **unidirectional** or **bidirectional**?

To address these issues, we need to introduce two terms. First, at each point along the chromosome where replication is occurring, the strands of the helix are unwound, creating what is called a **replication fork**. Such a fork will initially appear at the point of origin of synthesis and then move along the DNA duplex as replication proceeds. If replication is bidirectional, two such forks will be present, migrating in opposite directions away from the origin. The second term refers to the length of DNA that is replicated following one initiation event at a single origin. This is a unit referred to as the **replicon**. In *E. coli*, the replicon consists of the entire genome of 4.2 Mb (4.2 million base pairs).

The evidence is clear regarding the origin and direction of replication. John Cairns tracked replication in *E. coli*, using radioactive precursors of DNA synthesis and autora-

diography. He was able to demonstrate that in *E. coli* there is only a single region, called *oriC*, where replication is initiated. The presence of only a single origin is characteristic of bacteria, which have only one circular chromosome. Since DNA synthesis in bacteriophages and bacteria originates at a single point, the entire chromosome constitutes one replicon.

Still other results, again relying on autoradiography, demonstrated that replication is bidirectional, moving away from *oriC* in both directions (Figure 10–6). This creates two replication forks that migrate farther and farther apart as replication proceeds. These forks eventually merge as semiconservative replication of the entire chromosome is completed at a termination region, called *ter*.

Later in this chapter we will see that in eukaryotes, each chromosome contains multiple points of origin.

10.2 DNA Synthesis in Bacteria Involves Five Polymerases, as well as Other Enzymes

The determination that replication is semiconservative and bidirectional indicates only the *pattern* of DNA duplication and the association of finished strands with one another once synthesis is completed. A more complex issue is how the actual *synthesis* of long complementary polynucleotide chains occurs on a DNA template. As in most molecular biological studies, this question was first approached by using microorganisms. Research began about the same time as the Meselson-Stahl work, and the topic is still an active area of investigation. What is most apparent in this research is the tremendous complexity of the biological synthesis of DNA.

DNA Polymerase I

Studies of the enzymology of DNA replication were first reported by Arthur Kornberg and colleagues in 1957. They isolated an enzyme from *E. coli* that was able to direct DNA synthesis in a cell-free (*in vitro*) system. The enzyme is called **DNA polymerase I**, as it was the first of several similar enzymes to be isolated.

Kornberg determined that there were two major requirements for *in vitro* DNA synthesis under the direction of DNA polymerase I: (1) all four deoxyribonucleoside triphosphates (dNTPs) and (2) template DNA. If any one of the four deoxyribonucleoside triphosphates was omitted from the reaction, no measurable synthesis occurred. If derivatives of these precursor molecules other than the nucleoside triphosphate were used (nucleotides or nucleoside diphosphates), synthesis also did not occur. If no template DNA was added, synthesis of DNA occurred but was reduced greatly.

Most of the synthesis directed by Kornberg's enzyme appeared to be exactly the type required for semiconservative replication. The reaction is summarized in Figure 10–7, which depicts the addition of a single nucleotide. The enzyme has since been shown to consist of a single polypeptide containing 928 amino acids.

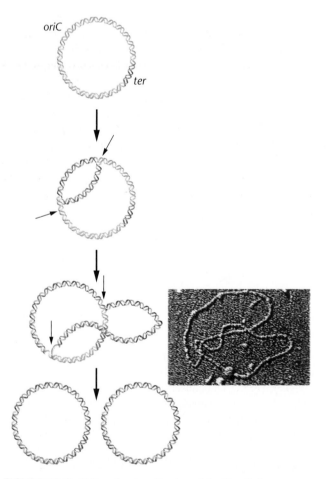

FIGURE 10–6 Bidirectional replication of the *E. coli* chromosome. The thin black arrows identify the advancing replication forks. The micrograph is of a bacterial chromosome in the process of replication, comparable to the figure next to it.

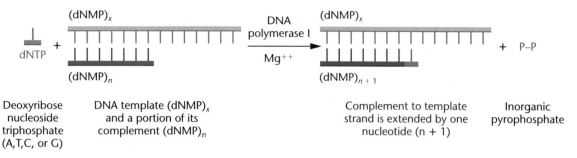

FIGURE 10–7 The chemical reaction catalyzed by DNA polymerase I. During each step, a single nucleotide is added to the growing complement of the DNA template, using a nucleoside triphosphate as the substrate. The release of inorganic pyrophosphate drives the reaction energetically.

The way in which each nucleotide is added to the growing chain is a function of the specificity of DNA polymerase I. As shown in Figure 10–8, the precursor dNTP contains the three phosphate groups attached to the 5′-carbon of deoxyribose. As the two terminal phosphates are cleaved during synthesis, the remaining phosphate attached to the 5′-carbon is covalently linked to the 3′-OH group of the deoxyribose to which it is added. Thus, **chain elongation** occurs in the **5′ to 3′ direction** by the addition of one nucleotide at a time to the growing 3′ end. Each step provides a newly exposed 3′-OH group that can participate in the next addition of a nucleotide as DNA synthesis proceeds.

Having shown how DNA was synthesized, Kornberg sought to demonstrate the accuracy, or fidelity, with which the enzyme replicated the DNA template. Because the technology to determine the nucleotide sequences of the template and the product was not yet available in 1957, he initially had to rely on several indirect methods.

One of Kornberg's approaches was to compare the nitrogenous base compositions of the DNA template with those of the recovered DNA product. Table 10.1 shows Kornberg's base composition analysis of three DNA templates. Within experimental error, the base composition of each product agreed with the template DNAs used. These data, along with

other types of comparisons of template and product, suggested that the templates were replicated faithfully.

Synthesis of Biologically Active DNA

Despite Kornberg's extensive work, not all researchers were convinced that DNA polymerase I was the enzyme that replicates DNA within cells (*in vivo*). Their reservations were that *in vitro* synthesis was much slower than the *in vivo* rate, that the enzyme was much more effective replicating single-stranded DNA than double-stranded DNA, and that the enzyme appeared to be able to *degrade* DNA as well as to *synthesize* it—that is, the enzyme exhibits **exonuclease activity**.

Uncertain of the true cellular function of DNA polymerase I, Kornberg pursued another approach. He reasoned that if the enzyme could be used to synthesize **biologically active DNA** *in vitro*, then DNA polymerase I must be the major catalyzing force for DNA synthesis within the cell. The term *biological activity* means that the DNA synthesized is capable of supporting metabolic activities and directs the reproduction of the organism from which it was originally duplicated.

In 1967, Kornberg, Mehran Goulian, and Robert Sinsheimer showed that the DNA of the small **bacteriophage φX174** could be completely copied by DNA polymerase I *in vitro* and that the

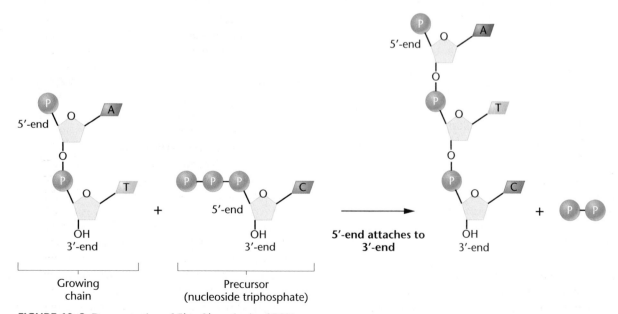

FIGURE 10–8 Demonstration of 5′ to 3′ synthesis of DNA.

TABLE 10.1	Base Composition of the DNA Template and the Product of Replication in Kornberg's Early Work				
Organism	Template or Product	%A	%T	%G	%C
T2	Template	32.7	33.0	16.8	17.5
	Product	33.2	32.1	17.2	17.5
E. coli	Template	25.0	24.3	24.5	26.2
	Product	26.1	25.1	24.3	24.5
Calf	Template	28.9	26.7	22.8	21.6
	Product	28.7	27.7	21.8	21.8

Source: Kornberg (1960).

TABLE 10.2	Properties of Bacterial DNA Polymerases I, II, and III		
Properties	I	II	III
Initiation of chain synthesis	−	−	−
5′–3′ polymerization	+	+	+
3′–5′ exonuclease activity	+	+	+
5′–3′ exonuclease activity	+	−	−
Molecules of polymerase/cell	400	?	15

new product could be isolated and used to transfect *E. coli*. This resulted in the production of mature phages under the direction of the synthetic DNA, thus demonstrating biological activity.

This demonstration of biological activity was viewed as a precise assessment of faithful copying. If even a single error had occurred to alter the base sequence of any of the 5386 nucleotides constituting the ϕX174 chromosome, the change might easily have caused a mutation that would prohibit the production of viable phages.

DNA Polymerase II, III, IV, and V

Although DNA synthesized under the direction of polymerase I demonstrated biological activity, a more serious reservation about the enzyme's true biological role was raised in 1969. Paula DeLucia and John Cairns discovered a mutant strain of *E. coli* that was deficient in polymerase I activity. The mutation was designated *polA1*. In the absence of the functional enzyme, this mutant strain of *E. coli* still duplicated its DNA and successfully reproduced. However, the cells were deficient in their ability to repair DNA. For example, the mutant strain is highly sensitive to ultraviolet light (UV) and radiation, both of which damage DNA and are mutagenic. Nonmutant bacteria are able to repair a great deal of UV-induced damage.

These observations led to two conclusions:

1. At least one other enzyme that is responsible for replicating DNA *in vivo* is present in *E. coli* cells.

2. DNA polymerase I may serve a secondary function *in vivo*. This function is now believed by Kornberg and others to be critical to the *fidelity* of DNA synthesis, but the enzyme does not actually synthesize the entire complementary strand during replication.

To date, four other unique DNA polymerases have been isolated from cells lacking polymerase I activity and from normal cells that contain polymerase I. Table 10.2 contrasts several characteristics of DNA polymerase I with **DNA polymerase II** and **III**. While none of the three can *initiate* DNA synthesis on a template, all three can *elongate* an existing DNA strand, called a **primer**.

The DNA polymerase enzymes are all large proteins exhibiting a molecular weight in excess of 100,000 Daltons

(Da). All three possess 3′ to 5′ exonuclease activity, which means that they have the potential to polymerize in one direction and then pause, reverse their direction, and excise nucleotides just added. As we will discuss later in the chapter, this activity provides a capacity to proofread newly synthesized DNA and to remove and replace incorrect nucleotides.

DNA polymerase I also demonstrates 5′ to 3′ exonuclease activity. This activity allows the enzyme to excise nucleotides, starting at the end at which synthesis begins and proceeding in the same direction of synthesis. Two final observations probably explain why Kornberg isolated polymerase I and not polymerase III: polymerase I is present in greater amounts than is polymerase III, and it is also much more stable.

What then are the roles of the polymerases *in vivo*? Polymerase III is the enzyme responsible for the 5′ to 3′ polymerization essential to *in vivo* replication. Its 3′ to 5′ exonuclease activity also provides a proofreading function that is activated when it inserts an incorrect nucleotide. When this occurs, synthesis stalls and the polymerase "reverses course," excising the incorrect nucleotide. Then, it proceeds back in the 5′ to 3′ direction, synthesizing the complement of the template strand. Polymerase I is believed to be responsible for removing the primer, as well as for the synthesis that fills gaps produced after this removal. Its exonuclease activities also allow for its participation in DNA repair. Polymerase II, as well as **polymerase IV and V**, are involved in various aspects of repair of DNA that has been damaged by external forces, such as ultraviolet light. Polymerase II is encoded by a gene activated by disruption of DNA synthesis at the replication fork.

We end this section by emphasizing the complexity of the DNA polymerase III molecule. Its active form, called a **holoenzyme**, consists of a dimer containing 10 different polypeptide subunits (Table 10.3) and has a molecular weight of 900,000 Da. The largest subunit, α, has a molecular weight of 140,000 Da and, along with subunits ϵ and θ, constitutes the core enzyme responsible for the polymerization activity. The α subunit is responsible for nucleotide polymerization on the template strands, whereas the ϵ subunit of the core enzyme possesses the 3′ to 5′ exonuclease activity.

A second group of five subunits (γ, δ, δ', χ, and ψ) forms what is called the γ complex, which is involved in "loading" the enzyme onto the template at the replication fork. This enzymatic function requires energy and is dependent on the hydrolysis of ATP. The β subunit serves as a "clamp" and prevents the core enzyme from falling off the template during polymerization. Finally, the τ subunit functions to dimerize two core polymerases facilitating simultaneous synthesis of both strands of the helix at

TABLE 10.3	Subunits of the DNA Polymerase III Holoenzyme	
Subunit	**Function**	**Groupings**
α	5′–3′ polymerization	Core enzyme: Elongates
ϵ	3′–5′ exonuclease	polynucleotide chain
θ	core assembly	and proofreads
γ		
δ		
δ'	Loads enzyme on template (Serves as clamp loader)	γ complex
χ		
ψ		
β	Sliding clamp structure (processivity factor)	
τ	Dimerizes core complex	

the replication fork. The holoenzyme and several other proteins at the replication fork together form a huge complex (nearly as large as a ribosome) known as the **replisome**. We consider the function of DNA polymerase III in more detail later in this chapter.

10.3 Many Complex Issues Must Be Resolved During DNA Replication

We have thus far established that in bacteria and viruses replication is semiconservative and bidirectional along a single replicon. We also know that synthesis is catalyzed by DNA polymerase III and occurs in the 5′ to 3′ direction. Bidirectional synthesis creates two replication forks that move in opposite directions away from the origin of synthesis. As we can see in the following points, many issues must still be resolved in order to provide a comprehensive understanding of DNA replication:

1. A mechanism must exist by which the helix undergoes localized unwinding and is stabilized in this "open" configuration so that synthesis may proceed along both strands.

2. As unwinding and subsequent DNA synthesis proceed, increased coiling creates tension further down the helix, which must be reduced.

3. A primer of some sort must be synthesized so that polymerization can commence under the direction of DNA polymerase III. Surprisingly, RNA, not DNA, serves as the primer.

4. Once the RNA primers have been synthesized, DNA polymerase III begins to synthesize the DNA complement of both strands of the parent molecule. Because the two strands are antiparallel to one another, continuous synthesis in the direction that the replication fork moves is possible along only one of the two strands. On the other strand, synthesis is discontinuous in the opposite direction.

5. The RNA primers must be removed prior to completion of replication. The gaps that are temporarily created must be filled with DNA complementary to the template at each location.

6. The newly synthesized DNA strand that fills each temporary gap must be joined to the adjacent strand of DNA.

7. While DNA polymerases accurately insert complementary bases during replication, they are not perfect and, occasionally, incorrect bases are added to the growing strand. A proofreading mechanism that also corrects errors is an integral process during DNA synthesis.

As we consider these points, examine Figures 10–9, 10–10, 10–11, and 10–12 to see how each issue is resolved. Figure 10–13 summarizes the model of DNA synthesis.

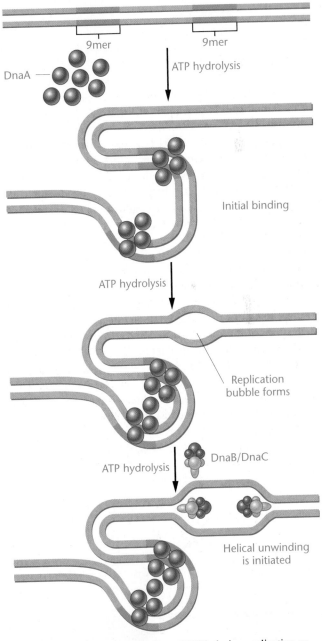

FIGURE 10–9 Helical unwinding of DNA during replication as accomplished by DnaA, DnaB, and DnaC proteins. Initial binding of many monomers of DnaA occurs at DNA sites containing repeating sequences of 9 nucleotides, called 9mers. Not illustrated are 13mers, which are also involved.

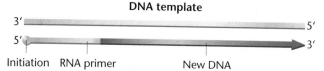

DNA template

Initiation   RNA primer   New DNA

FIGURE 10–10 The initiation of DNA synthesis. A complementary RNA primer is first synthesized, to which DNA is added. All synthesis is in the 5′ to 3′ direction. Eventually, the RNA primer is replaced with DNA under the direction of DNA polymerase I.

Unwinding the DNA Helix

As discussed earlier, there is a single point of origin along the circular chromosome of most bacteria and viruses at which DNA synthesis is initiated. This region of the *E. coli* chromosome has been particularly well studied. Called *oriC*, it consists of 245 base pairs characterized by repeating sequences of 9 and 13 bases (called **9mers** and **13mers**). As shown in Figure 10–9, one particular protein, called **DnaA** (because it is encoded by the gene called *dnaA*), is responsible for the initial step in unwinding the helix. A number of subunits of the DnaA protein bind to each of several 9mers. This step facilitates the subsequent binding of **DnaB** and **DnaC** proteins that further open and destabilize the helix. Proteins such as these, which require the energy supplied by the hydrolysis of ATP in order to break hydrogen bonds and denature the double helix, are called **helicases**. Other proteins, called **single-stranded binding proteins (SSBPs)**, stabilize this open conformation.

As unwinding proceeds, a coiling tension is created ahead of the replication fork, often producing **supercoiling**. In circular molecules, supercoiling may take the form of added twists and turns of the DNA, much like the coiling you can create in a rubber band by stretching it out and then twisting one end. Such supercoiling can be relaxed by **DNA gyrase**, a member of a larger group of enzymes referred to as **DNA topoisomerases**. The gyrase makes either single- or double-stranded "cuts" and also catalyzes localized movements that have the effect of "undoing" the twists and knots created during supercoiling. The strands are then resealed. These various reactions are driven by the energy released during ATP hydrolysis.

Together, the DNA, the polymerase complex, and associated enzymes make up an array of molecules that participate in DNA synthesis and are part of what we have previously called the *replisome*.

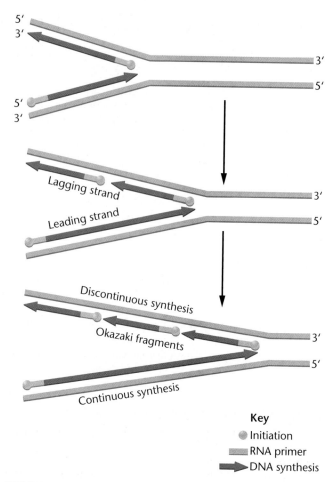

Key
- Initiation
- RNA primer
- DNA synthesis

FIGURE 10–11 Opposite polarity of DNA synthesis along the two strands, necessary because the two strands of DNA run antiparallel to one another and DNA polymerase III synthesizes in only one direction (5′ to 3′). On the lagging strand, synthesis must be discontinuous, resulting in the production of Okazaki fragments. On the leading strand, synthesis is continuous. RNA primers are used to initiate synthesis on both strands.

Initiation of DNA Synthesis Using an RNA Primer

Once a small portion of the helix is unwound, the initiation of synthesis may occur. As we have seen, DNA polymerase III requires a primer with a free 3′-hydroxyl group in order to elongate a polynucleotide chain. Since none is available in a circular chromosome, this prompted researchers to

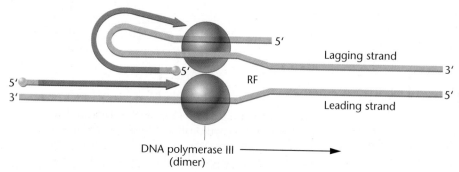

FIGURE 10–12 Illustration of how concurrent DNA synthesis may be achieved on both the leading and lagging strands at a single replication fork (RF). The lagging template strand is "looped" in order to invert the physical direction of synthesis, but not the biochemical direction. The enzyme functions as a dimer, with each core enzyme achieving synthesis on one or the other strand.

FIGURE 10–13
Summary of DNA synthesis at a single replication fork. Various enzymes and proteins essential to the process are shown.

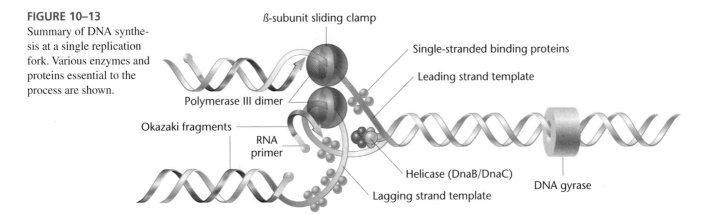

ß-subunit sliding clamp

Polymerase III dimer

Okazaki fragments

RNA primer

Single-stranded binding proteins

Leading strand template

Helicase (DnaB/DnaC)

DNA gyrase

Lagging strand template

investigate how the first nucleotide could be added. It is now clear that RNA serves as the primer that initiates DNA synthesis.

A short segment of RNA (about 10 to 12 nucleotides long), complementary to DNA, is first synthesized on the DNA template. Synthesis of the RNA is directed by a form of RNA polymerase called **primase**, which does not require a free 3′ end to initiate synthesis. It is to this short segment of RNA that DNA polymerase III begins to add deoxyribonucleotides, initiating DNA synthesis. A conceptual diagram of initiation on a DNA template is shown in Figure 10–10. At a later point, the RNA primer must be clipped out and replaced with DNA. This is thought to occur under the direction of DNA polymerase I. Recognized in viruses, bacteria, and several eukaryotic organisms, RNA priming is a universal phenomenon during the initiation of DNA synthesis.

Continuous and Discontinuous DNA Synthesis

We must now reconsider the fact that the two strands of a double helix are **antiparallel** to each other—that is, one runs in the 5′–3′ direction, while the other has the opposite 3′–5′ polarity. Because DNA polymerase III synthesizes DNA in only the 5′–3′ direction, synthesis along an advancing replication fork occurs in one direction on one strand and in the opposite direction on the other.

As a result, as the strands unwind and the replication fork progresses down the helix (Figure 10–11), only one strand can serve as a template for **continuous DNA synthesis**. This newly synthesized DNA is called the **leading strand**. As the fork progresses, many points of initiation are necessary on the opposite DNA template, resulting in **discontinuous DNA synthesis** of the **lagging strand**.* Evidence supporting the occurrence of discontinuous DNA synthesis was first provided by Reiji and Tuneko Okazaki. They discovered that when bacteriophage DNA is replicated in *E. coli*, some of the newly formed DNA that is hydrogen bonded to the template strand is present as small fragments containing 1000 to 2000 nucleotides. RNA primers are part of each such fragment.

These pieces, now called **Okazaki fragments**, are converted into longer and longer DNA strands of higher molecular weight as synthesis proceeds.

Discontinuous synthesis of DNA requires enzymes that both remove the RNA primers and unite the Okazaki fragments into the lagging strand. As we have noted, DNA polymerase I removes the primers and replaces the missing nucleotides. Joining the fragments is the work of **DNA ligase**, which is capable of catalyzing the formation of the phosphodiester bond that seals the nick between the discontinuously synthesized strands. The evidence that DNA ligase performs this function during DNA synthesis is strengthened by the observation of a ligase-deficient mutant strain (*lig*) of *E. coli*, in which a large number of unjoined Okazaki fragments accumulate.

Concurrent Synthesis Occurs on the Leading and Lagging Strands

Given the model just discussed, we might ask how DNA polymerase III synthesizes DNA on both the leading and lagging strands. Can both strands be replicated simultaneously at the same replication fork, or are the events distinct, involving two separate copies of the enzyme? Evidence suggests that both strands can be replicated simultaneously. As Figure 10–12 illustrates, if the lagging strand forms a loop, nucleotide polymerization can occur on both template strands under the direction of a dimer of the enzyme. After the synthesis of 1000 to 2000 nucleotides, the monomer of the enzyme on the lagging strand will encounter a completed Okazaki fragment, at which point it releases the lagging strand. A new loop is then formed with the lagging template strand, and the process is repeated. Looping inverts the orientation of the template but not the direction of actual synthesis on the lagging strand, which is always in the 5′ to 3′ direction.

Another important feature of the holoenzyme that facilitates synthesis at the replication fork is a dimer of the β subunit that forms a clamplike structure around the newly formed DNA duplex. This β-subunit clamp prevents the **core enzyme** (the α, ε, and θ subunits that are responsible for catalysis of nucleotide addition) from falling off the template as polymerization proceeds. Because the entire holoenzyme moves along the parent duplex, advancing the replication fork, the β-subunit dimer is often referred to as a *sliding clamp*.

---

*Because DNA synthesis is continuous on one strand and discontinuous on the other, the term **semidiscontinuous synthesis** is sometimes used to describe the overall process.

Now Solve This

In Problem 28 on page 234 DNA synthesis in a hypothetical organism is observed in which no Okazaki fragments are produced and a telomere problem exists. You are asked to suggest a DNA model consistent with these two features.

Hint: This observation suggests that DNA synthesis is continuous on both strands.

Proofreading and Error Correction During DNA Replication

The underpinning of DNA replication is the synthesis of a new strand that is precisely complementary to the template strand at each nucleotide position. Although the action of DNA polymerases is very accurate, synthesis is not perfect and a noncomplementary nucleotide is occasionally inserted erroneously. To compensate for such inaccuracies, the DNA polymerases all possess $3'$ to $5'$ exonuclease activity. This property imparts the potential for them to detect and excise a mismatched nucleotide (in the $3'$–$5'$ direction). Once the mismatched nucleotide is removed, $5'$ to $3'$ synthesis can again proceed. This process, called **proofreading**, increases the fidelity of synthesis by a factor of about 100. In the case of the holoenzyme form of DNA polymerase III, the epsilon (ϵ) subunit is directly involved in the proofreading step. In strains of *E. coli* with a mutation that has rendered the ϵ subunit nonfunctional, the error rate (the mutation rate) during DNA synthesis is increased substantially.

10.4   A Coherent Model Summarizes DNA Replication

We can now combine the various aspects of DNA replication occurring at a single replication fork into a coherent model, as shown in Figure 10–13. At the advancing fork, a helicase is unwinding the double helix. Once unwound, single-stranded binding proteins associate with the strands, preventing the reformation of the helix. In advance of the replication fork, DNA gyrase functions to diminish the tension created as the helix supercoils. Each half of the dimeric polymerase is a core enzyme bound to one of the template strands by a β-subunit sliding clamp. Continuous synthesis occurs on the leading strand, while the lagging strand must loop around in order for simultaneous (concurrent) synthesis to occur on both strands. Not shown in the figure, but essential to replication on the lagging strand, is the action of DNA polymerase I and DNA ligase, which together replace the RNA primers with DNA and join the Okazaki fragments, respectively.

Because the investigation of DNA synthesis is still an extremely active area of research, this model will no doubt be extended in the future. In the meantime, it provides a summary of DNA synthesis against which genetic phenomena can be interpreted.

10.5   Replication Is Controlled by a Variety of Genes

Much of what we know about DNA replication in viruses and bacteria is based on genetic analysis of the process. For example, we have already discussed studies involving the *polA1* mutation, which revealed that DNA polymerase I is not the major enzyme responsible for replication. Many other mutations interrupt or seriously impair some aspect of replication, such as the ligase-deficient and the proofreading-deficient mutations mentioned previously. Because such mutations are lethal ones, genetic analysis frequently uses **conditional mutations**, which are expressed under one condition but not under a different condition. For example, a **temperature-sensitive mutation** may not be expressed at a particular *permissive* temperature. When mutant cells are grown at a *restrictive* temperature, the mutant phenotype is expressed and can be studied. By examining the effect of the loss of function associated with the mutation, the investigation of such temperature-sensitive mutants can provide insight into the product and the associated function of the normal, nonmutant gene.

As shown in Table 10.4, a variety of genes in *E. coli* specify the subunits of the DNA polymerases and encode products involved in specification of the origin of synthesis, helix unwinding and stabilization, initiation and priming, relaxation of supercoiling, repair, and ligation. The discovery of such a large group of genes attests to the complexity of the process of replication, even in the relatively simple prokaryote. Given the enormous quantity of DNA that must be unerringly replicated in a very brief time, this level of complexity is not unexpected. As we will see in the next section, the process is even more involved and therefore more difficult to investigate in eukaryotes.

TABLE 10.4	Some of the Various *E. coli* Genes and Their Products or Role in Replication
Gene	**Product or Role**
polA	DNA polymerase I
polB	DNA polymerase II
dnaE, N, Q, X, Z	DNA polymerase III subunits
dnaG	Primase
dnaA, I, P	Initiation
dnaB, C	Helicase at *oriC*
gyrA, B	Gyrase subunits
lig	DNA ligase
rep	DNA helicase
ssb	Single-stranded binding proteins
rpoB	RNA polymerase subunit

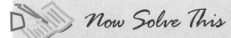

Now Solve This

Problem 21 on page 233 involves several temperature-sensitive mutations in *E. coli*, and you are asked to interpret the action of the gene that has mutated based on the phenotype that results.

Hint: Each mutation has disrupted one of the many steps essential to DNA synthesis. In each case, the mutant phenotype provides the clue as to which enzyme or function is affected.

10.6 Eukaryotic DNA Synthesis Is Similar to Synthesis in Prokaryotes but More Complex

Research has shown that eukaryotic DNA is replicated in a manner similar to that of bacteria. In both systems, double-stranded DNA is unwound at replication origins, replication forks are formed, and bidirectional DNA synthesis creates leading and lagging strands from templates under the direction of DNA polymerase. Eukaryotic polymerases have the same fundamental requirements for DNA synthesis as do bacterial systems: four deoxyribonucleoside triphosphates, a template, and a primer. However, because eukaryotic cells contain much more DNA per cell, because eukaryotic chromosomes are linear rather than circular, and because this DNA is complexed with proteins, eukaryotes face many complexities not encountered by bacteria. Thus, the process of DNA synthesis is more complicated in eukaryotes and more difficult to study. However, a great deal is now known about the process.

Multiple Replication Origins

The most obvious difference between eukaryotic and prokaryotic DNA replication is that eukaryotic chromosomes contain multiple replication origins, in contrast to the single site that is part of the *E. coli* chromosome. Multiple origins, visible under the electron microscope as "replication bubbles" that form as the helix opens up (Figure 10–14), each provide two potential replication forks. Multiple origins are essential if replication of the entire genome of a typical eukaryote is to be completed in a reasonable time. Recall that (1) eukaryotes have much greater amounts of DNA than bacteria—for example, yeast has three times as much DNA, and *Drosophila* has 40 times as much as *E. coli*—and (2) the rate of synthesis by eukaryotic DNA polymerase is much slower—only about 2000 nucleotides per minute, a rate 25 times less than the comparable bacterial enzyme. Under these conditions, replication from a single origin of a typical eukaryotic chromosome might take days to complete! However, replication of entire eukaryotic genomes is usually accomplished in a matter of hours.

Insights are now available concerning the molecular nature of the multiple origins and the initiation of DNA synthesis at these sites. Most information has originally been derived from

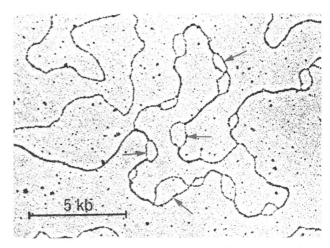

FIGURE 10–14 A demonstration of the multiple origins of replication along a eukaryotic chromosome. Each origin is apparent as a replication bubble along the axis of the chromosome. Arrows identify some of these replication bubbles.

the study of yeast (e.g., *Saccharomyces cerevisiae*), which has between 250 and 400 replicons per genome; subsequent studies have used mammalian cells, which have as many as 25,000 replicons. The origins in yeast have been isolated and are called **autonomously replicating sequences (ARSs)**. They consist of a unit containing a consensus sequence of 11 base pairs, flanked by other short sequences involved in efficient initiation. As we know from Chapter 2, DNA synthesis is restricted to the S phase of the eukaryotic cell cycle. Research has shown that the many origins are not all activated at once; instead, clusters of 20 to 80 adjacent replicons are activated sequentially throughout the S phase until all DNA is replicated.

How the polymerase finds the ARSs among so much DNA is an obvious recognition problem. The solution involves a mechanism that is initiated prior to the S phase. During the G1 phase of the cell cycle, all ARSs are initially bound by a group of specific proteins (six in yeast), forming what is called the **origin recognition complex (ORC)**. Mutations in either the ARSs or in any of the genes encoding these proteins of the ORC abolish or reduce initiation of DNA synthesis. Since these recognition complexes are formed in G1, but synthesis is not initiated at these sites until S, there must be still other proteins involved in the actual initiation signal. The most important of these proteins are specific kinases, key enzymes involved in phosphorylation, which are integral parts of cell-cycle control. When bound along with ORC, a pre-replication complex is formed that is accessible to DNA polymerase. After these kinases are activated, they serve to complete the initiation complex, directing localized unwinding and triggering DNA synthesis. Activation also inhibits reformation of the prereplication complexes once DNA synthesis has been completed at each replicon. This mechanism is important because it distinguishes segments of DNA that have completed replication from segments of unreplicated DNA, thus maintaining orderly and efficient replication. This ensures that replication only occurs once along each stretch of DNA during each cell cycle.

Eukaryotic DNA Polymerases

The most complex aspect of eukaryotic replication is the array of polymerases involved in directing DNA synthesis. As we will see, many different forms of the enzyme have been isolated and studied. However, only four are actually involved in replication of DNA, while the remainder are involved in repair processes. In order for the polymerases to have access to DNA, the topology of the helix must first be modified. As synthesis is triggered at each origin site, the double strands are opened up within an AT-rich region, allowing the entry of a helicase enzyme that proceeds to further unwind the double-stranded DNA. Before polymerases can begin synthesis, histone proteins complexed to the DNA (which form the characteristic nucleosomes of chromatin, as discussed in Chapter 11) also must be stripped away or otherwise modified. As DNA synthesis then proceeds, histones reassociate with the newly formed duplexes, reestablishing the characteristic nucleosome pattern (Figure 10–15). In eukaryotes, the synthesis of new histone proteins is tightly coupled to DNA synthesis during the S phase of the cell cycle.

The nomenclature and characteristics of six different polymerases are summarized in Table 10.5. Of these, three (Pol α, δ, and ϵ) are essential to nuclear DNA replication in eukaryotic cells. Two (Pol β and ζ) are thought to be involved in DNA repair (and still other repair forms have been discovered). The sixth form (Pol γ) is involved in the synthesis of mitochondrial DNA. Presumably, its replication function is limited to that organelle even though it is encoded by a nuclear gene. All but one of the six forms of the enzyme (the β form) consist of multiple subunits. Different subunits perform different functions during replication.

Pol α and δ are the major forms of the enzyme involved in initiation and elongation during nuclear DNA synthesis, so we concentrate our discussion on these. Two of the four subunits of the Pol α enzyme function in the synthesis of the RNA primers during the initiation of synthesis of both the leading and lagging strands. After the addition of about 10 ribonucleotides, another subunit functions to further elongate the RNA primer by adding 20 to 30 complementary deoxyribonucleotides. Pol α is said to possess low **processivity**, a term that essentially reflects the length of DNA that is synthesized by an enzyme before it

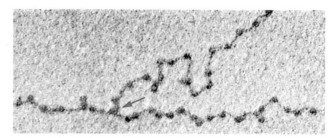

FIGURE 10–15 An electron micrograph of a eukaryotic replicating fork demonstrating the presence of histone-protein-containing nucleosomes on both branches.

dissociates from the template. Once the primer is in place, an event known as **polymerase switching** occurs, whereby Pol α dissociates from the template and is replaced by Pol δ. This form of the enzyme possesses high processivity and elongates the leading and lagging strands. It also possesses 3′ to 5′ exonuclease activity, which provides it with the potential to proofread. Pol ϵ, the third essential form, possesses the same general characteristics as Pol δ, but it is believed to operate under different cellular conditions, or to be restricted to synthesis of the lagging strand. In yeast, mutations that render Pol ϵ inactive are lethal, attesting to its essential function during replication.

The process described applies to both leading and lagging strand synthesis. On both strands, RNA primers must be replaced with DNA. On the lagging strand, the Okazaki fragments, which are about 10 times smaller (100 to 150 nucleotides) in eukaryotes than in prokaryotes, must be ligated.

To accommodate the increased number of replicons, eukaryotic cells contain many more DNA polymerase molecules than do bacteria. While *E. coli* has about 15 copies of DNA polymerase III per cell, there may be up to 50,000 copies of the α form of DNA polymerase in animal cells. As has been pointed out, the presence of greater numbers of smaller replicons in eukaryotes compared with bacteria compensates for the slower rate of DNA synthesis in eukaryotes. *E. coli* requires 20 to 40 minutes to replicate its chromosome, while *Drosophila*, with 40 times more DNA, is known during embryonic cell divisions to accomplish the same task in only 3 minutes.

TABLE 10.5	Properties of Eukaryotic DNA Polymerases					
	Polymerase α	Polymerase β	Polymerase δ	Polymerase ϵ	Polymerase γ	Polymerase ζ
Location	Nucleus	Nucleus	Nucleus	Nucleus	Mitochondria	Nucleus
3′–5′ Exonuclease activity	No	No	Yes	Yes	Yes	No
Essential to nuclear replication?	Yes	No	Yes	Yes	No	No

The Ends of Linear Chromosomes Are Problematic During Replication

A final difference that exists between prokaryotic and eukaryotic DNA synthesis involves the nature of the chromosomes. Unlike the closed, circular DNA of bacteria and most bacteriophages, eukaryotic chromosomes are linear. During replication, they face a special problem at the "ends" of these linear molecules, called the **telomeres**.

Before addressing this problem, we need to establish several things about telomeres. These structures consist of long stretches of a short repeating sequence of DNA bound to specific **telomere-associated proteins**. The unique quality of telomeres serves to preserve the integrity and stability of chromosomes. Telomeres are essential because the double-stranded "ends" of DNA molecules at the termini of chromosomes potentially resemble what are called **double-stranded breaks (DSBs)** that can occur if the chromosome should be fragmented internally. In such a case, the resulting double-stranded ends can fuse to other such ends and, if not, are vulnerable to degradation. Thus, telomeres are believed to protect the ends of intact eukaryotic chromosomes from such a fate.

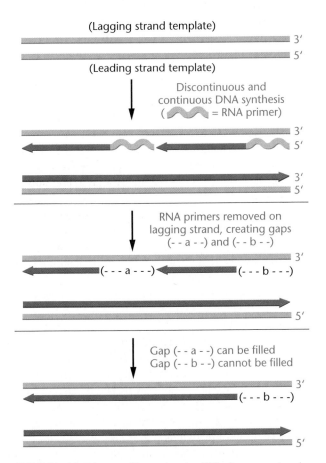

FIGURE 10–16 Diagram illustrating the difficulty encountered during the replication of the ends of linear chromosomes. A gap (- -b- -) is left following synthesis on the lagging strand.

Now, consider semiconservative replication near the end of a double-stranded DNA molecule. While synthesis of the leading strand proceeds to the end, a difficulty arises on the lagging strand once the RNA primer is removed (Figure 10–16). Normally, the newly created gap would be filled starting with the addition of a nucleotide to the existing 3'-OH group provided during discontinuous synthesis (to the right of gap (b) in Figure 10–16). However, since it is the end of the DNA molecule, there is no preceding Okazaki fragment present to provide the 3'-OH group. Thus, a gap remains on the lagging strand during each successive round of synthesis that will shorten the end of the chromosome by the length of the RNA primer. Because of this significant problem, we can suppose that a molecular solution would have been developed early in evolution and subsequently be conserved and shared by almost all eukaryotes, which is indeed the case.

A unique eukaryotic enzyme called **telomerase**, first discovered in the ciliated protozoan *Tetrahymena*, has helped us understand the solution. Before we look at how the enzyme actually works, let's look at what is accomplished. As pointed out above, telomeric DNA in eukaryotes is always found to consist of many short, repeated nucleotide sequences. For example, the template producing the lagging strand at the end of the *Tetrahymena* chromosome contains many repeats of the sequence 5'-TTGGGG-3'.

As illustrated in Figure 10–17, telomerase is capable of adding several repeats of this six-nucleotide sequence to the 3' end of the lagging strand template (using 5'–3' synthesis), thus preventing chromosome shortening. These repeats fold back on themselves forming a "hairpin loop" that is stabilized by unorthodox hydrogen bonding between opposite guanine residues (G=G), creating a free 3'-OH group on the end of the hairpin that can serve as a substrate for DNA polymerase. This makes it possible to subsequently fill the gap that would otherwise shorten the chromosome. The hairpin loop can then be cleaved off at its terminus, and the potential loss of DNA in each subsequent replication cycle is averted.

Detailed investigation by Elizabeth Blackburn and Carol Greider of how the *Tetrahymena* telomerase enzyme accomplishes the synthesis yielded an extraordinary finding. The enzyme is highly unusual in that it is a *ribonucleoprotein*, containing within its molecular structure a short piece of RNA that is essential to its catalytic activity. The RNA component serves as both a "guide" and a template for the synthesis of its DNA complement, a process referred to as **reverse transcription**. In *Tetrahymena*, the RNA contains several repeats of the sequence 5'-CCCCAA-3', the complement of the repeating telomeric DNA sequence that must be synthesized (5'-TTGGGG-3').

We can envision that part of the RNA sequence of the enzyme base pairs with the telomeric DNA, with the remainder of the RNA overlapping the end of the lagging strand template. Reverse transcription of this overlapping sequence, which synthesizes DNA on an RNA template, will lead to the extension of the lagging strand template, as shown (Figure 10–17). It is believed that the enzyme is then translocated toward the end of the lagging strand template, as shown and the sequence of events is repeated. In yeast, whose synthesis

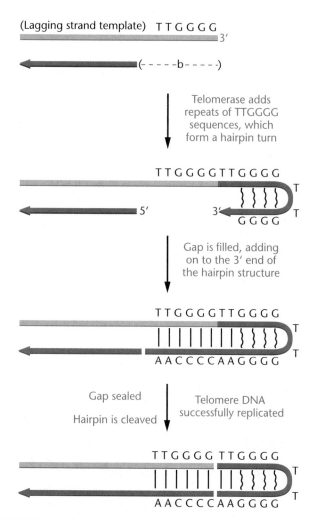

FIGURE 10–17 The predicted solution to the problem posed in Figure 10–16. The enzyme telomerase directs synthesis of the TTGGGG sequences, resulting in the formation of a hairpin structure. The gap can now be filled, and, following cleavage of the hairpin structure, the process averts the creation of a gap during replication of the ends of linear chromosomes.

has been investigated in detail, specific proteins are responsible for terminating synthesis, thus determining the number of repeating units added to the end of the lagging strand template. Similar proteins no doubt function in the same way in other eukaryotes.

Analogous enzyme functions have now been found in almost all eukaryotes studied. In humans, the telomeric DNA sequence that creates the lagging strand and is repeated is 5′-TTAGGG-3′, differing from *Tetrahymena* by only one nucleotide.

As we shall see in Chapter 11, telomeric DNA sequences have been highly conserved throughout evolution, reflecting the critical function of telomeres. In the essay at the end of this chapter, we will see that telomere shortening has been linked to a molecular mechanism involved in cellular aging. In fact, in most eukaryotic somatic cells, telomerase is not active and thus, with each cell division, the telomeres of each chromosome shorten. After many divisions, the telomere is seriously eroded and the cell

loses the capacity for further division. Malignant cells, on the other hand, maintain telomerase activity and are immortalized.

10.8   DNA Recombination, Like DNA Replication, Is Directed by Specific Enzymes

We now turn to a topic previously discussed in Chapter 7—genetic recombination. There, we pointed out that the process of crossing over between homologs depends on breakage and rejoining of the DNA strands. Now that we have discussed the chemistry and replication of DNA, we can consider how recombination occurs at the molecular level. In general, the information that follows pertains to genetic exchange between any two homologous double-stranded DNA molecules, whether they be viral or bacterial chromosomes or eukaryotic homologs during meiosis. Genetic exchange at equivalent positions along two chromosomes with substantial DNA sequence homology is referred to as **general** or **homologous recombination**.

Several models attempt to explain homologous recombination, and they all share certain common features. First, all are based on proposals put forth independently by Robin Holliday and Harold L. K. Whitehouse in 1964. They also depend on the complementarity between DNA strands for the precision of exchange. Finally, each model relies on a series of enzymatic processes in order to accomplish genetic recombination.

One such model is shown in Figure 10–18. It begins with two paired DNA duplexes or homologs [Step (a)] in each of which an endonuclease introduces a single-stranded nick at an identical position [Step (b)]. The ends of the strands produced by these cuts are then displaced and subsequently pair with their complements on the opposite duplex [Step (c)]. A ligase then seals the loose ends [Step (d)], creating hybrid duplexes called **heteroduplex DNA molecules**. The exchange creates a cross-bridged structure. The position of this cross bridge can then move down the chromosome by a process referred to as branch migration [Step (e)], which occurs as a result of a zipperlike action as hydrogen bonds are broken and then reformed between complementary bases of the displaced strands of each duplex. This migration yields an increased length of heteroduplex DNA on both homologs.

If the duplexes now separate [Step (f)] and the bottom portions rotate 180° [Step (g)], an intermediate planar structure called a χ form—the characteristic **Holliday structure**—is created. If the two strands on opposite homologs previously uninvolved in the exchange are now nicked by an endonuclease [Step (h)] and ligation occurs [Step (i)], recombinant duplexes are created. Note that the arrangement of alleles is altered as a result of recombination.

Evidence supporting this model includes the electron microscopic visualization of χ-form planar molecules from bacteria in which four duplex arms are joined at a single point of exchange [Figure 10–18(Step g)]. Further important

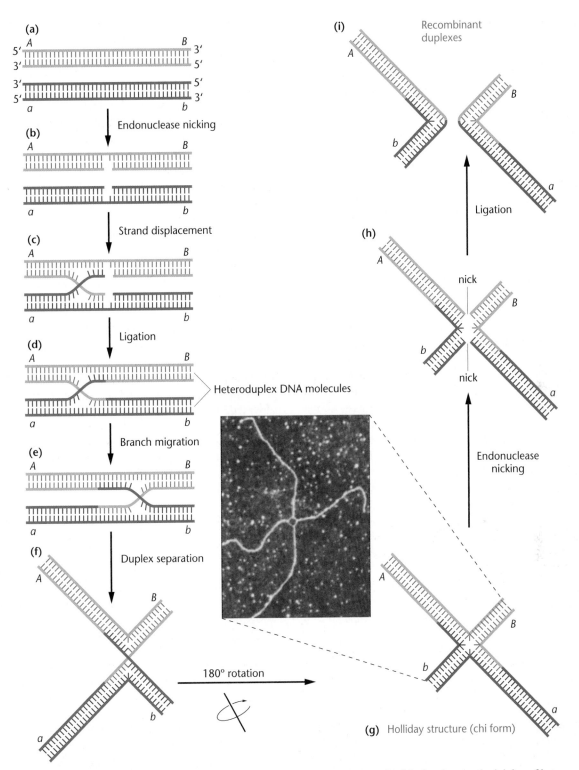

FIGURE 10–18 Model depicting how genetic recombination can occur as a result of the breakage and rejoining of heterologous DNA strands. Each stage is described in the text. The electron micrograph shows DNA in a χ-form structure similar to the diagram in (g); the DNA is an extended Holliday structure, derived from the *Col*E1 plasmid of *E. coli*. *David Dressler, Oxford University, England.*

evidence comes from the discovery of the **RecA protein** in *E. coli*. The molecule promotes the exchange of reciprocal single-stranded DNA molecules as occurs in Step (c) of the model. RecA also enhances the hydrogen-bond formation during strand displacement, thus initiating heteroduplex

formation. Finally, many other enzymes essential to the nicking and ligation process have also been discovered and investigated. The products of the *recB*, *recC*, and *recD* genes are thought to be involved in the nicking and unwinding of DNA. Numerous mutations that prevent genetic recombination have been found in viruses and bacteria. These mutations represent genes whose products play an essential role in this process.

Gene Conversion, a Consequence of DNA Recombination

A modification of the preceding model has helped us to better understand a unique genetic phenomenon known as **gene conversion**. Initially found in yeast by Carl Lindegren and in *Neurospora* by Mary Mitchell, gene conversion is characterized by a *nonreciprocal* genetic exchange between two closely linked genes. For example, if we were to cross two *Neurospora* strains, each bearing a separate mutation ($a + \times + b$) a *reciprocal* recombination between the genes would yield spore pairs of the $++$ and the ab genotypes. However, a nonreciprocal exhange yields one pair without the other. Working with pyridoxine mutants, Mitchell observed several asci-containing spore pairs with the $++$ genotype, but not the reciprocal product (ab). Because the frequency of these events was higher than the predicted mutation rate and consequently could not be accounted for by mutation, they were called "gene conversions." They were so named because it appeared that one allele had somehow been "converted" to another in which genetic exchange had also occurred. Similar findings come from studies of other fungi as well.

Gene conversion is now considered to be a consequence of the process of DNA recombination. One possible explanation interprets conversion as a mismatch of base pairs during heteroduplex formation, as shown in Figure 10–19. Mismatched regions of hybrid strands can be repaired by the excision of one of the strands and the synthesis of the complement by using the remaining strand as a template. Excision may occur in either one of the strands, yielding two possible "corrections." One repairs the mismatched base pair and "converts" it to restore the original sequence. The other also corrects the mismatch but does so by copying the altered strand, creating a base-pair substitution. Conversion may have the effect of creating identical alleles on the two homologs that were different initially.

In our example in Figure 10–19, suppose that the ($G\equiv C$) pair on one of the two homologs was responsible for the mutant allele, while the ($A=T$) pair was part of the wild-type gene sequence on the other homolog. Conversion of the ($G\equiv C$) pair to ($A=T$) would have the effect of changing the mutant allele to wild type, just as Mitchell originally observed.

Gene-conversion events have helped to explain other puzzling genetic phenomena in fungi. For example, when mutant and wild-type alleles of a single gene are crossed, asci should yield equal numbers of mutant and wild-type spores. However, exceptional asci with 3:1 or 1:3 ratios are sometimes observed. These ratios can be understood in terms of gene conversion. The phenomenon also has been detected during mitotic events in fungi, as well as during the study of unique compound chromosomes in *Drosophila*.

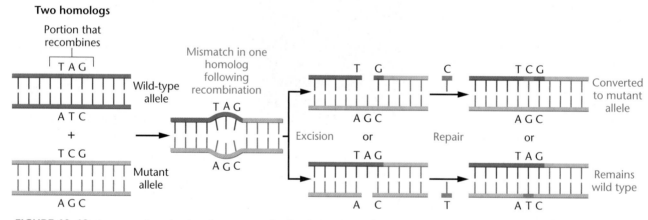

FIGURE 10–19 A proposed mechanism that accounts for the phenomenon of gene conversion. A base-pair mismatch occurs in one of the two homologs (bearing the mutant allele) during heteroduplex formation, which accompanies recombination in meiosis. During excision repair, one of the two mismatched strands is removed and the complement is synthesized. In one case, the mutant base pair is preserved. When it is subsequently included in a recombinant spore, the mutant genotype will be maintained. In the other case, the mutant base pair is converted to the wild-type sequence. When included in a recombinant spore, the wild-type genotype will be expressed, leading to a nonreciprocal exchange ratio.

GENETICS, TECHNOLOGY, AND SOCIETY

Telomerase: The Key to Immortality?

Humans, as all multicellular organisms, grow old and die. As we age, our immune systems become less efficient, wound healing is impaired, and tissues and organs lose resilience. It has always been a mystery why we go through these age-related declines, and why each species has a characteristic finite life span. Why do we grow old? Can we reverse this march to mortality? Some recent discoveries suggest that the answers to these questions may lie at the ends of our chromosomes.

The study of human aging begins with a study of human cells growing in culture dishes. Like the organisms from which the cells are taken, cells in culture have a finite life span. This "replicative senescence" was noted more than 30 years ago by Leonard Hayflick. He reported that average human fibroblasts lose their ability to grow and divide after about 50 cell divisions. These senescent cells remain metabolically active but can no longer proliferate. Eventually, they die. Although we don't know whether cellular senescence directly causes organismal aging, the evidence is suggestive. For example, cells from young people go through more divisions in culture than do cells from older people; human fetal cells divide 60 to 80 times before undergoing senescence, whereas cells from older adults divide only 10 to 20 times. In addition, cells from species with short life spans stop growing after fewer divisions than do cells from species with longer life spans; mouse cells divide 10 to 15 times in culture, but tortoise cells undergo over 100 divisions. Moreover, cells from patients with genetic premature aging syndromes (such as Werner syndrome) undergo fewer divisions in culture than do cells from normal patients.

Another characteristic of aging cells is that their telomeres become shorter. Telomeres are the tips of linear chromosomes and consist of several thousand repeats of a short DNA sequence (TTAGGG in humans). Telomeres help preserve the structural integrity of chromosomes by protecting their ends from degrading or from fusing to other chromosomes. Telomeres are created and maintained by telomerase—a remarkable RNA-containing enzyme that adds telomeric DNA sequences onto the ends of linear chromosomes. Telomerase also solves the "end-replication" problem—which asserts that

linear DNA molecules become shorter at each replication, because DNA polymerase cannot synthesize new DNA at the 3′ ends of each parent strand. By adding numerous telomeric repeat sequences onto the 3′ ends of chromosomes, telomerase prevents the chromosomes from shrinking into oblivion. Unfortunately, most somatic cells contain little, if any, telomerase. As a result, telomere length decreases every time a normal cell divides. Telomere shortening may act as a clock that counts cell divisions and instructs the cell to stop dividing.

Could we gain perpetual youth and vitality by increasing our telomere lengths? A recent study suggests that it may be possible to reverse senescence by artificially increasing the amount of telomerase in our cells. When the investigators introduced cloned telomerase genes into normal human cells in culture, telomeres lengthened by thousands of base pairs and the cells continued to grow long past their senescence point. These observations confirm that telomere length acts as a cellular clock. In addition, they suggest that some of the atrophy of tissues that accompanies old age may someday be reversed by activating telomerase genes. However, before we rush out to buy telomerase pills, we must consider a possible consequence of cellular immortality: cancer.

Although normal cells undergo senescence after a specific number of cell divisions, cancer cells do not. It is thought that cancers arise after several genetic mutations accumulate in a cell. These mutations disrupt the normal checks and balances that control cell growth and division. It therefore seems logical that cancer cells would also stop the normal aging clock. If their telomeres became shorter after each cell division, tumor cells would eventually succumb to aging and cease growth. However, if they synthesized telomerase, they would arrest the ticking of the senescence clock and become immortal. In keeping with this idea, more than 90 percent of human tumor cells contain telomerase activity and have stable telomeres. The correlation between uncontrolled tumor cell growth and the presence of telomerase activity is so strong that telomerase assays are being developed as diagnostic markers for cancer. Although there is currently some debate about whether the presence of telomerase is a prerequisite for, or simply a consequence of, cell transformation, it is possible that acquiring telomerase activity may be an

important step in the development of a cancer cell. In support of this hypothesis, recent studies show that the induction of telomerase activity, when combined with the inactivation of a tumor suppressor gene (p16^{INK4a}), results in cell immortalization—an essential step toward tumor development. Therefore, any attempt to increase telomerase activity in normal cells carries the risk of enhancing the development of tumors.

An attractive possibility is that telomerase may be an ideal target for anticancer drugs. Drugs that inhibit telomerase might destroy cancer cells by allowing their telomeres to shorten, thereby forcing the cells into senescence. Because most normal human cells do not express telomerase, such a therapy might be specific for tumor cells and hence less toxic than most current anticancer drugs. Such antitelomerase, anticancer drugs are currently under development by Geron Corporation. Although we don't yet know whether this approach will work in animals, it appears to work in cultured tumor cells. Tumor cells that are treated with an antitelomerase agent lose telomeric sequences and die after about 25 cell divisions.

Before antitelomerase drugs can be developed and used on humans, several questions must be answered. Is telomerase required by some normal human cells (such as lymphocytes and germ cells)? If so, antitelomerase drugs may be unacceptably toxic. Could some cancer cells compensate for the loss of telomerase by using other telomere-lengthening mechanisms (such as recombination)? If so, antitelomerase drugs may be doomed to failure. Even if we inhibit telomerase activity in tumor cells, could they undergo multiple divisions before reaching senescence and still damage the host?

Will telomerase allow us to both arrest cancers *and* reverse the descent into old age? Time will tell.

References

Bodnar, A. G., et al. 1998. Extension of life-span by introduction of telomerase into normal human cells. *Science* 279: 349–352.

deLange, T. 1998. Telomeres and senescence: Ending the debate. *Science* 279: 334–335.

Kiyono, T., et al. 1998. Both RB/p16^{INK4a} inactivation and telomerase activity are required to immortalize human epithelial cells. *Nature* 396: 84–88.

CHAPTER SUMMARY

1. In theory, three modes of DNA replication are possible: semi-conservative, conservative, and dispersive. Although all three rely on base complementarity, semiconservative replication is the most straightforward and was predicted by Watson and Crick.

2. In 1958, Meselson and Stahl resolved this question in favor of semiconservative replication in *E. coli*, showing that newly synthesized DNA consists of one old strand and one new strand. Taylor, Woods, and Hughes used root tips of the broad bean to demonstrate semiconservative replication in eukaryotes.

3. During the same period, Kornberg isolated the enzyme DNA polymerase I from *E. coli* and showed that it is capable of directing *in vitro* DNA synthesis, provided that a template and precursor nucleoside triphosphates are supplied.

4. The subsequent discovery of the *polA1* mutant strain of *E. coli*, capable of DNA replication despite its lack of polymerase I activity, cast doubt on the enzyme's *in vivo* replicative function. DNA polymerases II and III were then isolated. Polymerase III has been identified as the enzyme responsible for DNA replication *in vivo*.

5. During the process of DNA synthesis, the double helix unwinds, forming a replication fork at which synthesis begins. Proteins stabilize the unwound helix and assist in relaxing the coiling tension created ahead of the replication.

6. Synthesis is initiated at specific sites along each template strand by the enzyme primase, which results in short segments of RNA that provide suitable 3′ ends upon which DNA polymerase III can begin polymerization.

7. Because of the antiparallel nature of the double helix, polymerase III synthesizes DNA continuously on the leading strand in a 5′–3′ direction. On the opposite strand, called the lagging strand, synthesis results in short Okazaki fragments that are later joined by DNA ligase.

8. DNA polymerase I removes and replaces the RNA primer with DNA, which is joined to the adjacent polynucleotide by DNA ligase.

9. The isolation of numerous phage and bacterial mutant genes affecting many of the molecules involved in the replication of DNA has helped to define the complex genetic control of the entire process.

10. DNA replication in eukaryotes is similar to, but more complex than, replication in prokaryotes. Multiple replication origins exist, and multiple forms of DNA polymerase direct DNA synthesis.

11. Replication at the ends (telomeres) of linear molecules poses a special problem in eukaryotes that can be solved by a unique RNA-containing enzyme called telomerase.

12. Homologous recombination between DNA molecules relies on a series of enzymes that can cut, realign, and reseal DNA strands. The phenomenon of gene conversion may be best explained in terms of mismatch repair synthesis following exchanges.

KEY TERMS

antiparallel strands, 223

autonomously replicating sequences (ARSs), 225

autoradiography, 216

bateriophage φX174, 219

bidirectional replication, 218

biologically active DNA, 219

chain elongation, 219

chi form (χ), 228

conditional mutation, 224

conservative replication, 215

continuous DNA synthesis, 223

core enzyme, 223

discontinuous DNA synthesis, 223

dispersive replication, 215

DnaA protein, 222

DnaB protein, 222

DnaC protein, 222

DNA gyrase, 222

DNA ligase, 223

DNA polymerase I, 218

DNA polymerase II and III, 220

DNA polymerase IV and V, 220

DNA topoisomerase, 222

exonuclease activity, 219

5′ to 3′ direction, 219

gene conversion, 230

genetic recombination, 228

Holliday structure, 228

helicase, 222

heteroduplex DNA molecules, 228

holoenzyme, 220

homologous recombination, 228

lagging DNA strand, 223

leading DNA strand, 223

9mer, 222

nucleosome, 226

Okazaki fragment, 223

origin of replication, 217

origin recognition complex (ORC), 225

polymerase switching, 226

primase, 223

primer, 220

processivity, 226

proofreading, 224

RecA protein, 229

replication fork, 218

replicative senescence, 231

replicon, 218

replisome, 221

reverse transcription, 227

sedimentation equilibrium centrifugation, 215

semiconservative replication, 214

semidiscontinuous synthesis, 223

single-stranded binding proteins (SSBPs), 222

telomerase, 227

telomere, 227

temperature-sensitive mutation, 224

3′ to 5′ exonuclease activity, 224

unidirectional replication, 218

INSIGHTS AND SOLUTIONS

1. Predict the theoretical results of conservative and dispersive replication of DNA under the conditions of the Meselson-Stahl experiment. Follow the results through two generations of replication after cells have been shifted to an ^{14}N-containing medium, using the following sedimentation pattern.

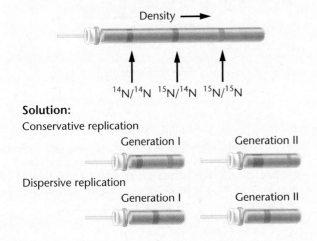

Density ⟶

$^{14}N/^{14}N$ $^{15}N/^{14}N$ $^{15}N/^{15}N$

Solution:
Conservative replication

Generation I   Generation II

Dispersive replication

Generation I   Generation II

2. Mutations in the *dnaA* gene of *E. coli* are lethal and can only be studied following the isolation of conditional, temperature-sensitive mutations. Such mutant strains grow nicely and replicate their DNA at the permissive temperature of 18°C, but they do not grow or replicate their DNA at the restrictive temperature of 37°C. Two observations were useful in determining the function of the DnaA protein product. First, *in vitro* studies using DNA templates that have unwound do not require the DnaA protein. Second, if intact cells are grown at 18°C and are then shifted to 37°C, DNA synthesis continues at this temperature until one round of replication is completed and then stops. What do these observations suggest about the role of the *dnaA* gene product?

Solution: At 18°C (the permissive temperature), the mutation is not expressed and DNA synthesis begins. Following the shift to the restrictive temperature, the already-initiated DNA synthesis continues, but no new synthesis can begin. Because the DnaA protein is not required for synthesis of unwound DNA, these observations suggest that, *in vivo*, the DnaA protein plays an essential role in DNA synthesis by interacting with the intact helix and somehow facilitating the localized denaturation necessary for synthesis to proceed.

PROBLEMS AND DISCUSSION QUESTIONS

1. Compare conservative, semiconservative, and dispersive modes of DNA replication.
2. Describe the role of ^{15}N in the Meselson-Stahl experiment.
3. In the Meselson-Stahl experiment, which of the three modes of replication could be ruled out after one round of replication? After two rounds?
4. Predict the results of the experiment by Taylor, Woods, and Hughes if replication were (a) conservative and (b) dispersive.
5. Reconsider Problem 33 in Chapter 9. In the model you proposed, could the molecule be replicated semiconservatively? Why? Would other modes of replication work?
6. What are the requirements for *in vitro* synthesis of DNA under the direction of DNA polymerase I?
7. In Kornberg's initial experiments, it was rumored that he grew *E. coli* in Anheuser-Busch beer vats. (He was working at Washington University in St. Louis.) Why do you think this might have been helpful to the experiment?
8. How did Kornberg assess the fidelity of DNA by polymerase I in copying DNA template?
9. Which characteristics of DNA polymerase I raised doubts that its *in vivo* function is the synthesis of DNA leading to complete replication?
10. One of Kornberg's tests demonstrating that DNA polymerase I is the enzyme used *in vivo* involved the replication of ϕX174 DNA. What is meant by "biologically active" DNA?
11. What was the significance of the *polA1* mutation?
12. Summarize and compare the properties of DNA polymerase I, II, and III.
13. List and describe the function of the 10 subunits constituting DNA polymerase III. Distinguish between the holoenzyme and the core enzyme.
14. Distinguish between (a) unidirectional and bidirectional synthesis, and (b) continuous and discontinuous synthesis of DNA.
15. List the proteins that unwind DNA during *in vivo* DNA synthesis. How do they function?
16. Define and indicate the significance of (a) Okazaki fragments, (b) DNA ligase, and (c) primer RNA during DNA replication.
17. Outline the current model for DNA synthesis.
18. Why is DNA synthesis expected to be more complex in eukaryotes than in bacteria? How is DNA synthesis similar in the two types of organisms?
19. If the analysis of DNA from two different microorganisms demonstrated very similar base compositions, are the DNA sequences of the two organisms also nearly identical?
20. Suppose that *E. coli* synthesizes DNA at a rate of 100,000 nucleotides per minute and takes 40 minutes to replicate its chromosome. (a) How many base pairs are present in the entire *E. coli* chromosome? (b) What is the physical length of the chromosome in its helical configuration—that is, what is the circumference of the chromosome if it were opened into a circle?
21. Several temperature-sensitive mutant strains of *E. coli* display the following characteristics. Predict what enzyme or function is being affected by each mutation.
    (a) Newly synthesized DNA contains many mismatched base pairs.
    (b) Okazaki fragments accumulate, and DNA synthesis is never completed.
    (c) No initiation occurs.
    (d) Synthesis is very slow.
    (e) Supercoiled strands remain after replication, which is never completed.
22. Define gene conversion, and describe how this phenomenon is related to genetic recombination.

23. Many of the gene products involved in DNA synthesis were initially defined by studying mutant *E. coli* strains that could not synthesize DNA. (a) The *dnaE* gene encodes the α subunit of DNA polymerase III. What effect is expected from a mutation in this gene? How could the mutant strain be maintained? (b) The *dnaQ* gene encodes the ε subunit of DNA polymerase. What effect is expected from a mutation in this gene?

24. In 1994, telomerase activity was discovered in human cancer cell lines. Although telomerase is not active in human somatic tissue, this discovery indicated that humans do contain the genes for telomerase proteins and telomerase RNA. Since inappropriate activation of telomerase can cause cancer, why do you think the genes coding for this enzyme have been maintained in the human genome throughout evolution? Are there any types of human body cells where telomerase activation would be advantageous or even necessary? Explain.

25. The genome of *D. melanogaster* consists of approximately 1.7×10^8 base pairs. DNA synthesis occurs at a rate of 30 base pairs per second. In the early embryo, the entire genome is replicated in five minutes. How many bidirectional origins of synthesis are required to accomplish this feat?

26. Assume a hypothetical organism in which DNA replication is conservative. Design an experiment similar to that of Taylor, Woods, and Hughes that will unequivocally establish this fact. Using the format established in Figure 10–5, draw sister chromatids and illustrate the expected results establishing this mode of replication.

27. DNA polymerases in all organisms add only 5′ nucleotides to the 3′ end of a growing DNA strand, never to the 5′ end. One possible reason for this is the fact that most DNA polymerases have a proofreading function that would not be *energetically* possible if DNA synthesis occurred in the 3′ to 5′ direction. (a) Sketch the reaction that DNA polymerase would have to catalyze if DNA synthesis occurred in the 3′ to 5′ direction. (b) Consider the information in your sketch and speculate as to why proofreading would be problematic.

28. An alien organism was investigated. It displayed characteristics of eukaryotes. When DNA replication was examined, two unique features were apparent: (1) no Okazaki fragments were observed; and (2) there was a telomere problem (i.e., telomeres shortened) but only on one end of the chromosome. Create a model of DNA that is consistent with both of these observations.

29. Assume that the sequence of bases given below is present on one nucleotide chain of a DNA duplex and that the chain has opened up at a replication fork. Synthesis of an RNA primer occurs on this template starting at the base that is underlined. (a) If the RNA primer consists of eight nucleotides, what is its base sequence? (b) In the intact RNA primer, which nucleotide has a free 3′-OH terminus?

3′.....GGCTACC<u>T</u>GGATTCA.....5′

30. Given the diagram below, assume that the phase G1 chromosome on the left underwent one round of replication in ^{3}H-thymidine and the metaphase chromosome on the right had both chromatids labeled. Which of the replicative models (conservative, dispersive, semiconservative) could be eliminated by this observation?

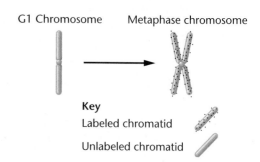

31. Consider the figure of a dinucleotide below. (a) Is it DNA or RNA? (b) Is the arrow closest to the 5′ or the 3′ end? (c) Suppose that the molecule was cleaved with the enzyme spleen diesterase, which breaks the covalent bond connecting the phosphate to C-5′. After cleavage, to which nucleoside is the phosphate now attached, (A or T)?

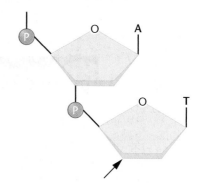

32. DNA is allowed to replicate in moderately radioactive ^{3}H-thymidine for several minutes and is then switched to a highly radioactive medium for several more minutes. Synthesis is stopped, and the DNA is subjected to autoradiography and electron microscopy. Interpret as much as you can regarding DNA replication from the drawing of the electron micrograph presented here.

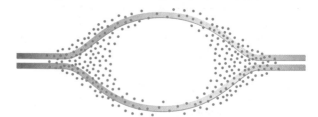

Chromosome Structure and DNA Sequence Organization

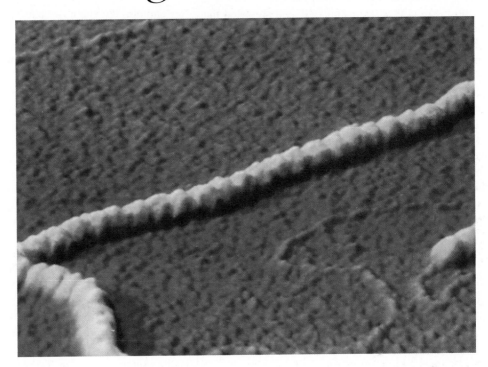

A chromatin fiber viewed using a scanning transmission electron microscope (STEM).

CHAPTER CONCEPTS

- Genetic information in viruses, bacteria, mitochondria, and chloroplasts is most often contained in a short, circular DNA molecule, relatively free of associated proteins.

- Eukaryotic cells, in contrast to viruses and bacteria, contain relatively large amounts of DNA organized into nucleosomes and present during most of the cell cycle as chromatin fibers.

- During division stages, uncoiled chromatin fibers characteristic of interphase coil up and condense into chromosomes.

- Eukaryotic genomes are characterized by both unique and repetitive DNA sequences.

- The vast majority of eukaryotic genomes consists of noncoding DNA sequences.

Once geneticists understood that DNA houses genetic information, it became very important to determine how DNA is organized into genes and how these basic units of genetic function are organized into chromosomes. In short, the major question had to do with how the genetic material was organized as it makes up the genome of organisms. There has been much interest in this question because knowledge of the organization of the genetic material and associated molecules is important to understanding many other areas of genetics. For example, the way in which the genetic information is stored, expressed, and regulated must be related to the molecular organization of the genetic molecule, DNA. How genomic organization varies in different organisms—from viruses to bacteria to eukaryotes—will undoubtedly provide a better understanding of the evolution of organisms on Earth.

In this chapter, we focus on the various ways DNA is organized into chromosomes. We first survey what we know about chromosomes in viruses and bacteria and in two cellular organelles: mitochondria and chloroplasts. Then, we will examine the large specialized structures called polytene and lampbrush chromosomes. In the second half of the chapter, we discuss how eukaryotic chromosomes are organized. For example, how is DNA complexed with proteins to form chromatin, and how are the chromatin fibers, characteristic of interphase, condensed into chromosome structures visible during mitosis and meiosis? We conclude the chapter by examining aspects of DNA sequence organization characteristic of eukaryotic genomes.

? How Do We Know?

In this chapter, we will focus on chromosome structure and the way DNA is organized within chromosomes. As you study this topic, you should try to answer several fundamental questions:

1. How have we come to know so much about chromosomes?

2. How do we know that mitochondria and chloroplasts contain DNA?

3. How do we know that puffs are areas of active transcription in polytene chromosomes?

4. How do we know about the organization of DNA within chromosomes, given that DNA has a diameter of only 2 nm?

5. How do we know that satellite DNA in eukaryotes consists of repetitive sequences and has been derived from regions of the centromere?

11.1 Viral and Bacterial Chromosomes Are Relatively Simple DNA Molecules

The chromosomes of viruses and bacteria are much less complicated than those in eukaryotes. They usually consist of a single nucleic acid molecule, unlike the multiple chromosomes comprising the genome of higher forms. The chromosomes are largely devoid of associated proteins and contain relatively little genetic information. These characteristics have greatly simplified analysis, and we now have a fairly comprehensive view of the structure of viral and bacterial chromosomes.

The chromosomes of viruses consist of a nucleic acid molecule—either DNA or RNA—that can be either single- or double-stranded. They can exist as circular structures (closed loops), or they can take the form of linear molecules. For example, the single-stranded DNA of the **φX174 bacteriophage** and the double-stranded DNA of the **polyoma virus** are closed loops housed within the protein coat of the mature viruses. The **bacteriophage lambda (λ)**, on the other hand, possesses a linear double-stranded DNA molecule prior to infection, which closes to form a ring upon its infection of the host cell. Still other viruses, such as the T-even series of bacteriophages, have linear double-stranded chromosomes of DNA that do not form circles inside the bacterial host. Thus, circularity is not an absolute requirement for replication in viruses.

Viral nucleic acid molecules have been seen with the electron microscope. Figure 11–1 shows a mature bacteriophage λ and its double-stranded DNA molecule in the circular configuration. One constant feature shared by viruses, bacteria, and eukaryotic cells is the ability to package an exceedingly long DNA molecule into a relatively small volume. In λ, the DNA is 17 μm long and must fit into the phage head, which is less than 0.1 μm on any side. Table 11.1 compares the length of the chromosomes of several viruses to the size of their head structure. In each case, a similar packaging feat must be accomplished. Compare the dimensions given for phage T2 with the micrograph of both the DNA and the viral particle shown in Figure 11–2. Seldom does the space available in the head of a virus exceed the chromosome volume by more than a factor of two. In many cases, almost all of the space is filled, indicating nearly perfect packing. Once packed within the head, the genetic material is functionally inert until it is released into a host cell.

Bacterial chromosomes are also relatively simple in form. They always consist of a double-stranded DNA molecule, compacted into a structure sometimes referred to as the **nucleoid**. *Escherichia coli*, the most extensively studied bacterium, has a large circular chromosome measuring approximately 1200 μm (1.2 mm) in length that may occupy up to one-third of the volume of the cell. When the cell is gently lysed and the chromosome is released, it can be visualized under the electron microscope (Figure 11–3).

DNA in bacterial chromosomes is associated with several types of **DNA-binding proteins**. Two, called **HU and H1 proteins**, are small but abundant in the cell and contain a high percentage of positively charged amino acids that can bond ionically to the negative charges of the phosphate groups in DNA. Although these proteins are structurally similar to molecules called histones that are associated with eukaryotic DNA, they clearly do not play a similar

(a)

(b)

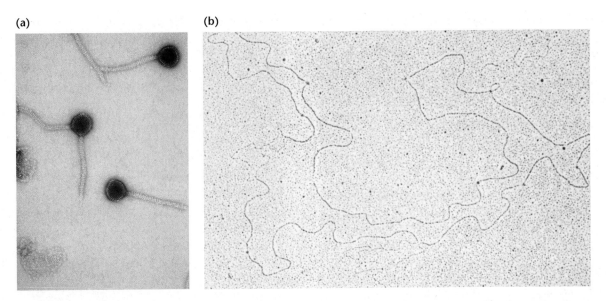

FIGURE 11–1 Electron micrographs of phage λ (left) and the DNA that was isolated from it (right). The chromosome is 17 μm long. Note that the phages are magnified about five times more than the DNA.

TABLE 11.1	The Genetic Material of Representative Viruses and Bacteria				
			Nucleic Acid		**Overall Size of Viral Head or Bacteria (μm)**
	Organism	**Type**	**SS or DS***	**Length (μm)**	
Viruses	φX174	DNA	SS	2.0	0.025 × 0.025
	Tobacco mosaic virus	RNA	SS	3.3	0.30 × 0.02
	Phage λ	DNA	DS	17.0	0.07 × 0.07
	T2 phage	DNA	DS	52.0	0.07 × 0.10
Bacteria	*Haemophilus influenzae*	DNA	DS	832.0	1.00 × 0.30
	Escherichia coli	DNA	DS	1200.0	2.00 × 0.50

*SS = single-stranded, DS = double-stranded.

role in compacting DNA. Unlike the tightly packed chromosome present in the head of a virus, the bacterial chromosome is *not* functionally inert and can be readily replicated and transcribed.

Now Solve This

Problem 2 on page 251 involves the consideration of viral chromosomes that are linear in the bacteriophage, but circularize after they enter the bacterial host cell. You are asked to consider the advantages of circular DNA molecules vs. linear molecules during replication.

Hint: Recall from Chapter 10 that the enzyme telomerase is essential to replication in eukaryotes.

11.2 Mitochondria and Chloroplasts Contain DNA Similar to Bacteria and Viruses

Numerous studies demonstrate that both **mitochondria** and **chloroplasts** contain their own DNA and genetic system for expressing this information. This was first suggested by the discovery of mutations in yeast, other fungi, and plants that produce altered phenotypes that can be linked to these organelles (see Chapter 4). Transmission of such traits was found to occur through the cytoplasm rather than through chromosomes in the nucleus. Because both mitochondria and chloroplasts are inherited through the maternal cytoplasm in most organisms, these observations suggested that the organelles house their own DNA, which, when mutated, may be responsible for the altered phenotypes.

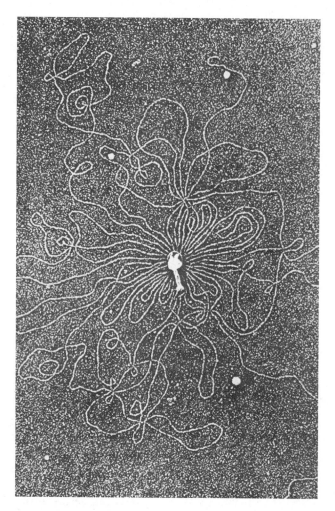

FIGURE 11–2 Electron micrograph of bacteriophage T2, which has had its DNA released by osmotic shock. The chromosome is 52 μm long.

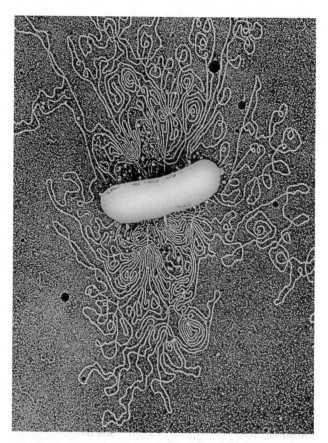

FIGURE 11–3 Electron micrograph of the bacterium *E. coli*, which has had its DNA released by osmotic shock. The chromosome is 1200 μm long.

Thus, geneticists set out to look for more direct evidence of DNA in these organelles. Electron microscopists not only documented the presence of DNA in both organelles, they also saw DNA in a form quite unlike that seen in the nucleus of the eukaryotic cells that house these organelles. This DNA looked remarkably similar to that seen in viruses and bacteria. This similarity, along with other observations, led to the idea that mitochondria and chloroplasts arose independently more than a billion years ago from free-living, prokaryote-like organisms that possessed the ability to undergo aerobic respiration (mitochondria) or photosynthesis (choroplasts). This theory, called the **endosymbiotic hypothesis**, was championed by Lynn Margulis and others. They proposed that the prokaryotes were engulfed by larger primitive eukaryotic cells, which lacked these bioenergetic functions. A symbiotic relationship developed whereby the prokaryotic organisms eventually lost their ability to function independently, while the eukaryotic host cells gained the ability to either respire aerobically or undergo photosynthesis. Although many questions remain unanswered, the basic tenets of this theory are widely accepted.

In the following sections, we shall explore what is known about the DNA found in these cellular organelles and examine what is known about the genes present on these DNA molecules.

Molecular Organization and Gene Products of Mitochondrial DNA

Extensive information is now available on the molecular aspects and gene products of **mitochondrial DNA (mtDNA)**. In most eukaryotes, mtDNA is a double-stranded closed circle (Figure 11–4) that replicates semiconservatively and is free of the chromosomal proteins characteristic of eukaryotic DNA. In size, mtDNA differs greatly among organisms, as demonstrated in Table 11.2. In a variety of animals, including humans, mtDNA consists of about 16,000 to 18,000 bp (16–18 kb). However, yeast (*Saccharomyces*) contains 75 kb, while up to 367 kb may be present in plant mitochondria such as in *Arabidopsis*. Vertebrates have 5 to 10 such DNA molecules per organelle, while plants have 20 to 40 copies per organelle.

We can say several things about mtDNA. With only rare exceptions, introns, the noncoding regions of genes, do not appear to be present in mitochondrial genes, and there are few or no gene repetitions. Expression of mitochondrial genes uses several modifications of the otherwise standard genetic code. Replication of mtDNA is dependent on enzymes encoded by nuclear DNA. In humans, mtDNA encodes 2 ribosomal

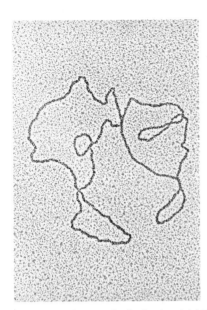

FIGURE 11–4 Electron micrograph of mitochondrial DNA (mtDNA) derived from *Xenopus laevis*.

TABLE 11.3	Sedimentation Coefficients of Mitochondrial Ribosomes	
Kingdom	**Examples**	**Sedimentation Coefficient (S)**
Animalia	Vertebrates	55–60
	Insects	60–71
Protista	*Euglena*	71
	Tetrahymena	80
Fungi	*Neurospora*	73–80
	Saccharomyces	72–80
Plantae	Maize	77

TABLE 11.2	The Size of mtDNA in Different Organisms
Organism	**Size (kb)**
Human	16.6
Mouse	16.2
Xenopus (frog)	18.4
Drosophila (fruit fly)	18.4
Saccharomyces (yeast)	75.0
Pisum sativum (pea)	110.0
Arabidopsis (mustard plant)	367.0

RNAs (rRNAs), 22 transfer RNAs (tRNAs), as well as numerous polypeptides essential to the cellular respiratory functions of the organelles. In almost every case, the polypeptides are part of multichain proteins, and the other polypeptides of each protein are encoded in the nucleus, synthesized in the cytoplasm, and then transported into the organelle. Thus, the protein-synthesizing apparatus and the molecular components for cellular respiration are jointly derived from nuclear and mitochondrial genes.

As expected then, ribosomes found in the organelle are different from those present in the neighboring cytoplasm. Table 11.3 shows that mitochondrial ribosomes of different species vary considerably in their sedimentation coefficients, ranging from 55*S* to 80*S*. Since many are closer in their coefficient to bacterial counterparts than eukaryotes, this also supports the endosymbiont hypothesis.

The nuclear-coded gene products essential to biological activity in mitochondria include DNA and RNA polymerases, initiation and elongation factors essential for translation, ribo-

somal proteins, aminoacyl tRNA synthetases, and several tRNA species. These imported components are distinct from their cytoplasmic counterparts, even though both sets are coded by nuclear genes. For example, the synthetase enzymes essential for charging mitochondrial tRNA molecules (a process essential to translation) show a distinct affinity for the mitochondrial tRNA species as compared to the cytoplasmic tRNAs. Similar affinity has been shown for the initiation and elongation factors. Furthermore, while bacterial and nuclear RNA polymerases are known to be composed of numerous subunits, the mitochondrial variety consists of only one polypeptide chain. This polymerase is generally susceptible to antibiotics that inhibit bacterial RNA synthesis but not to eukaryotic inhibitors. The contributions of nuclear and mitochondrial gene products are shown in Figure 11–5.

Molecular Organization and Gene Products of Chloroplast DNA

Chloroplasts, like mitochondria, contain an autonomous genetic system distinct from that found in the nucleus and cytoplasm. This system includes DNA as a source of genetic information and a complete protein-synthesizing apparatus. Also similar to mitochondria, the molecular components of the chloroplast translation apparatus are jointly derived from both nuclear and organelle genetic information. **Chloroplast DNA (cpDNA)**, shown in Figure 11–6, is much larger than mitochondrial DNA—usually 100 to 225 kb in length. Nevertheless, it also shares similarities to DNA found in prokaryotic cells. It is circular, double-stranded, replicated semiconservatively, and free of the associated proteins characteristic of eukaryotic DNA. Compared with nuclear DNA from the same organism, it invariably shows a different buoyant density and base composition.

In the green alga *Chlamydomonas*, there are about 75 copies of the chloroplast DNA molecule per organelle, and each copy of DNA is 195,000 bp (195 kb) in length. In higher plants such as the sweet pea, multiple copies of the DNA molecule are present in each organelle, but the molecule is considerably smaller than that in *Chlamydomonas*, at 134 kb. Genetic recombination between the multiple copies of DNA within chloroplasts has been documented in *Chlamydomonas*.

FIGURE 11–5 Gene products that are essential to mitochondrial function. Those shown entering the organelle are derived from the cytoplasm and encoded by the nucleus.

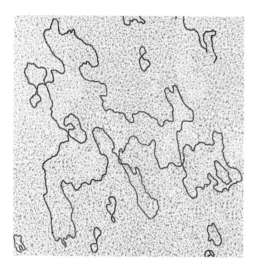

FIGURE 11–6 Electron micrograph of chloroplast DNA obtained from lettuce.

Some chloroplast gene products function during translation. In a variety of higher plants (beans, lettuce, spinach, maize, and oats), two sets of the genes coding for the ribosomal RNAs—5S, 16S, and 23S rRNA—are present. In addition, chloroplast DNA codes for at least 25 tRNA species and a number of ribosomal proteins specific to the chloroplast ribosomes. These ribosomes have a sedimentation coefficient slightly less than 70S, similar to that of bacteria. Even though chloroplast ribosomal proteins are encoded by both nuclear and chloroplast DNA, most, if not all, such proteins are distinct from their counterparts in cytoplasmic ribosomes.

Still other chloroplast genes specific to the photosynthetic function have been identified. Mutations in these genes may inactivate photosynthesis in chloroplasts with this mutation. One of the major photosynthetic enzymes is ribulose-1-5-bisphosphate carboxylase (RuBP). Interestingly, the small subunit of this enzyme is encoded by a nuclear gene, whereas the large subunit is encoded by cpDNA.

The great difference in size in cpDNA compared to mtDNA can be partly explained by an increased number of genes. However, the biggest difference appears to be due to the presence of long noncoding sequences of DNA as well as duplications of many DNA sequences. This observation is indicative of the independent evolution that occurred in chloroplasts and mitochondria following their initial invasion of a primitive eukaryote-like cell.

11.3 Specialized Chromosomes Reveal Variations in the Organization of DNA

We now consider two cases of genetic organization that demonstrate the specialized forms that eukaryotic chromosomes can take. Both types—*polytene chromosomes* and *lampbrush chromosomes*—are so large that their organization was discerned using light microscopy long before we understood how mitotic chromosomes form from interphase chromatin. The study of these chromosomes provided many of our initial insights into the arrangement and function of the genetic information.

Polytene Chromosomes

Giant **polytene chromosomes** are found in various tissues (salivary, midgut, rectal, and malpighian excretory tubules) in the larvae of some flies and in several species of protozoans and plants. Such structures were first observed by E. G. Balbiani in 1881. The vast amount of information obtained from studies of these genetic structures provided a model system for subsequent investigations of chromosomes. What is particularly intriguing about polytene chromosomes is that they can be seen in the nuclei of interphase cells.

Each polytene chromosome is 200 to 600 μm long, and when they are observed under the light microscope, they reveal a linear series of alternating bands and interbands (Figure 11–7). The banding pattern is distinctive for each chromosome in any given species. Individual bands are sometimes called **chromomeres**, a generalized term describing lateral condensations of material along the axis of a chromosome.

Extensive study using electron microscopy and radioactive tracers led to an explanation for the unusual appearance of these chromosomes. First, polytene chromosomes represent paired homologs. This is highly unusual since they are present in somatic cells, where in most organisms, chromosomal material is normally dispersed as chromatin and homologs are not paired. Second, their large size and distinctiveness result from the many DNA strands that compose them. The DNA of these paired

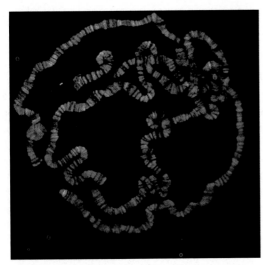

FIGURE 11–7 Polytene chromosomes derived from larval salivary gland cells of *Drosophila*.

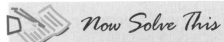

Problem 8 on page 251 involves polytene chromosomes that are cultured in ³H-thymidine and subjected to autoradiography. You are asked to predict the pattern of grains that will result.

Hint: ³H-thymidine will only be incorporated during the synthesis of DNA.

homologs undergoes many rounds of replication, *but without strand separation or cytoplasmic division*. As replication proceeds, chromosomes contain 1000 to 5000 DNA strands that remain in precise parallel alignment with one another. Apparently, the parallel register of so many DNA strands gives rise to the distinctive band pattern along the axis of the chromosome.

The presence of bands on polytene chromosomes was initially interpreted as the visible manifestation of individual genes. The discovery that the strands present in bands undergo localized uncoiling during genetic activity further strengthened this view. Each such uncoiling event results in what is called a **puff** because of its appearance (Figure 11–8). That puffs are visible manifestations of gene activity (transcription that produces RNA) is evidenced by their high rate of incorporation of radioactively labeled RNA precursors, as assayed by autoradiography. Bands that are not extended into puffs incorporate fewer radioactive precursors or none at all.

The study of bands during development in insects, such as *Drosophila* and the midge fly *Chironomus*, reveals differential gene activity. A characteristic pattern of band formation, which is equated with gene activation, is observed as development proceeds. Despite attempts to resolve the issue, it is not yet clear how many genes are contained in each band; however, we do know that a band can contain up to 10^7 bp of DNA, certainly enough DNA to encode 50 to 100 average-sized genes.

Lampbrush Chromosomes

Another specialized chromosome that has given us insight into chromosomal structure is the **lampbrush chromosome**, so named because it resembles the brushes used to clean kerosene-lamp chimneys in the nineteenth century. Lampbrush chromosomes were first discovered in 1892 in the oocytes of sharks and are now known to be characteristic of most vertebrate oocytes as well as the spermatocytes of some insects. Therefore, they are meiotic chromosomes. Most experimental work has been done with material taken from amphibian oocytes.

These chromosomes are easily isolated from oocytes in the diplotene stage of the first prophase of meiosis, where they are active in directing the metabolic activities of the developing cell. The homologs are seen as synapsed pairs held together by chiasmata. However, instead of condensing, as most meiotic chromosomes do, lampbrush chromosomes often extend to lengths of 500 to 800 μm. Later in meiosis, they revert to their normal length of 15 to 20 μm. Based on these observations, lampbrush chromosomes are interpreted as extended, uncoiled versions of the normal meiotic chromosomes.

The two views of lampbrush chromosomes in Figure 11–9 provide significant insights into their morphology. Part (a) shows the meiotic configuration under the light microscope. The linear axis of each structure contains a large number of condensed areas, and as with polytene chromosomes, these are referred to as *chromomeres*. Emanating from each chromomere is a pair of **lateral loops**, which give the chromosome its distinctive appearance. In part (b), the scanning electron micrograph (SEM) reveals adjacent loops present along one of the two axes of the chromosome. As with bands

FIGURE 11–8 Photograph of a puff within a polytene chromosome. The diagram depicts the uncoiling of strands within a band (B) region to produce a puff (P) in polytene chromosomes. Interband regions (IB) are also labeled.

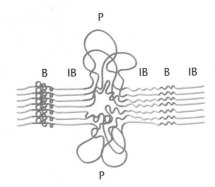

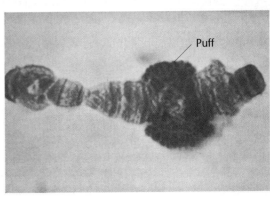

(a)

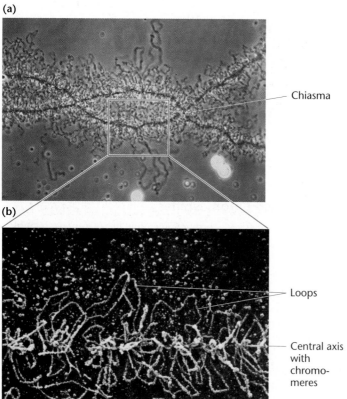

Chiasma

(b)

Loops

Central axis with chromomeres

FIGURE 11–9 Lampbrush chromosomes derived from amphibian oocytes. Part (a) is a photomicrograph; part (b) is a scanning electron micrograph.

in polytene chromosomes, much more DNA is present in each loop than is needed to encode a single gene. This SEM provides a clear view of the chromomeres and the chromosomal fibers emanating from them. Each chromosomal loop is thought to be composed of one DNA double helix, while the central axis is composed of two DNA helices. This hypothesis is consistent with the belief that each meiotic chromosome is composed of a pair of sister chromatids. Studies using radioactive RNA precursors have revealed that the loops are active in the synthesis of RNA. The lampbrush loops, in a manner similar to puffs in polytene chromosomes, represent DNA that has been uncoiled from the central chromomere axis during transcription.

11.4 DNA Is Organized into Chromatin in Eukaryotes

We now turn our attention to the way DNA is organized in eukaryotic chromosomes. Our focus will be on conventional eukaryotic cells, in which chromosomes are visible only during mitosis. After chromosome separation and cell division, cells enter the interphase stage of the cell cycle, during which time the components of the chromosome uncoil and are present in the form referred to as **chromatin**. While in interphase, the chromatin is dispersed in the nucleus, and the DNA of each chromosome is replicated. As the cell cycle progresses, most cells reenter mitosis, whereupon chromatin coils into

visible chromosomes once again. This condensation represents a length contraction of some 10,000 times for each chromatin fiber.

The organization of DNA during the transitions just described is much more intricate and complex than in viruses or bacteria, which never exhibit a process similar to mitosis. This is due to the greater amount of DNA per chromosome, as well as the presence of a large number of proteins associated with eukaryotic DNA. For example, while DNA in the *E. coli* chromosome is 1200 μm long, the DNA in each human chromosome ranges from 19,000 to 73,000 μm in length. In a single human nucleus, all 46 chromosomes contain sufficient DNA to extend almost 2 meters. This genetic material, along with its associated proteins, is contained within a nucleus that usually measures about 5 to 10 μm in diameter.

Such intricacy parallels the structural and biochemical diversity of the many types of cells in a multicellular eukaryotic organism. Different cells assume specific functions based on highly specific biochemical activity. Although all cells carry a full genetic complement, different cells activate different sets of genes, so a highly ordered regulatory system governing the readout of the information must exist. Such a system must in some way be imposed on or related to the molecular structure of the genetic material.

Because of the limitation of light microscopy, early studies of the structure of eukaryotic genetic material concentrated on intact chromosomes, preferably large ones, such as the polytene and lampbrush chromosomes. Subsequently, new techniques for biochemical analysis, as well as the examination of relatively intact eukaryotic chromatin and mitotic chromosomes under the electron microscope, have greatly enhanced our understanding of chromosome structure.

Chromatin Structure and Nucleosomes

As we have seen, the genetic material of viruses and bacteria consists of strands of DNA or RNA that are nearly devoid of proteins. In eukaryotic chromatin, a substantial amount of protein is associated with the chromosomal DNA in all phases of the eukaryotic cell cycle. The associated proteins are divided into basic, positively charged **histones** and less positively charged nonhistones. The histones clearly play the most essential structural role of all the proteins associated with DNA. Histones contain large amounts of the positively charged amino acids lysine and arginine, making it possible for them to bond electrostatically to the negatively charged phosphate groups of nucleotides. Recall that a similar interaction has been proposed for several bacterial proteins. The five main types of histones are shown in Table 11.4.

The general model for chromatin structure is based on the assumption that chromatin fibers, composed of DNA and protein, undergo extensive coiling and folding as they are condensed within the cell nucleus. X-ray diffraction studies confirm that histones play an important role in chromatin structure. Chromatin produces regularly spaced diffraction rings, suggesting that repeating structural units occur along the chromatin axis. If the histone molecules are chemically removed from chromatin, the regularity of this diffraction pattern is disrupted.

TABLE 11.4	Categories and Properties of Histone Proteins	
Histone Type	**Lysine-Arginine Content**	**Molecular Weight (Da)**
H1	Lysine-rich	23,000
H2A	Slightly lysine-rich	14,000
H2B	Slightly lysine-rich	13,800
H3	Arginine-rich	15,300
H4	Arginine-rich	11,300

(a)

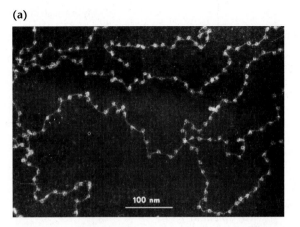

(b)

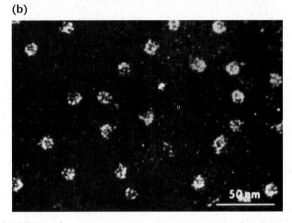

FIGURE 11–10 (a) Dark-field electron micrograph of nucleosomes present in chromatin derived from a chicken erythrocyte nucleus. (b) Dark-field electron micrograph of nucleosomes produced by micrococcal nuclease digestion.

A basic model for chromatin structure was worked out in the mid-1970s. Several observations were particularly relevant to the development of this model:

1. Digestion of chromatin by certain endonucleases, such as micrococcal nuclease, yields DNA fragments that are approximately 200 bp in length or multiples thereof. This demonstrates that enzymatic digestion is not random, for if it were, we would expect a wide range of fragment sizes. Thus, chromatin consists of some type of repeating unit, each of which is protected from enzymatic cleavage, except where any two units are joined. It is the area between units that is attacked and cleaved by the endonuclease.

2. Electron microscopic observations of chromatin reveal that chromatin fibers are composed of linear arrays of spherical particles (Figure 11–10). Discovered by Ada and Donald Olins, the particles occur regularly along the axis of a chromatin strand and resemble beads on a string. These particles, initially referred to as v-bodies (v is the Greek letter nu), are now called **nucleosomes**. This conforms nicely to the earlier observation, which suggests the existence of repeating units.

3. Studies of precise interactions of histone molecules and DNA in the nucleosomes constituting chromatin show that histones H2A, H2B, H3, and H4 occur as two types of tetramers, $(H2A)_2 \cdot (H2B)_2$ and $(H3)_2 \cdot (H4)_2$. Roger Kornberg predicted that each repeating nucleosome unit consists of one of each tetramer (creating an octamer) in association with about 200 bp of DNA. Such a structure is consistent with previous observations and provides the basis for a model that explains the interaction of histones and DNA in chromatin.

4. When nuclease digestion time is extended, some of the 200 bp of DNA are removed from the nucleosome, creating a **nucleosome core particle** consisting of 147 bp. The DNA lost in this prolonged digestion is responsible for linking nucleosomes together. This linker DNA is associated with the fifth histone, H1.

5. On the basis of this information, as well as on X-ray and neutron-scattering analyses of crystallized core particles by John T. Finch, Aaron Klug, and others, a detailed

model of the nucleosome was put forward in 1984. In this model, the 147-bp DNA core is coiled around an octamer of histones in a left-handed superhelix, which completes about 1.7 turns per nucleosome.

The extensive investigation of nucleosomes provides the basis for predicting how the chromatin fiber within the nucleus is formed and how it coils up into a mitotic chromosome. The 2-nm DNA molecule is initially coiled into a nucleosome about 11 nm in diameter [Figure 11–11(a)], consistent with the longer dimension of the ellipsoidal nucleosome. Significantly, the formation of the nucleosome represents the first level of packing, whereby the DNA helix is reduced to about one-third its original length.

In the nucleus, the chromatin fiber seldom, if ever, exists in the extended form described here. Instead, the 11-nm chromatin fiber is further packed into a thicker 30-nm fiber, initially called a **solenoid** [Figure 11–11(b)]. This larger fiber consists of numerous closely coiled nucleosomes, creating the second level of packing. Solenoids condense the eukaryotic fiber by a factor of five. The exact details of this structure are not completely clear, but 30-nm chromatin fibers are characteristically seen under the electron microscope.

In the transition to the mitotic chromosome, still another level of packing occurs. The 30-nm fiber forms a series of

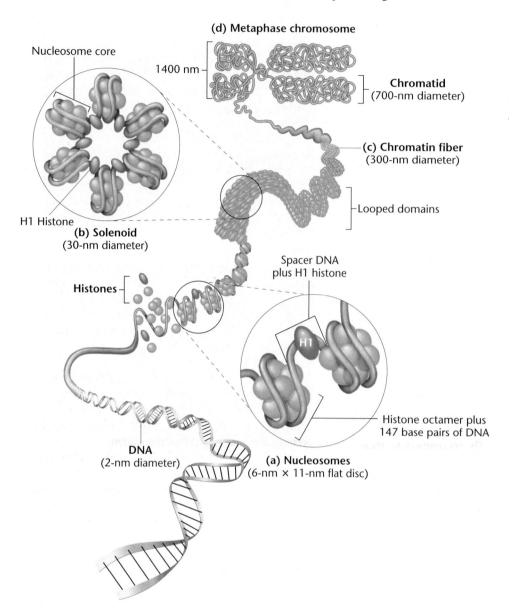

(d) Metaphase chromosome

1400 nm

Chromatid
(700-nm diameter)

(c) Chromatin fiber
(300-nm diameter)

Looped domains

Nucleosome core

H1 Histone

(b) Solenoid
(30-nm diameter)

Spacer DNA
plus H1 histone

Histones

H1

Histone octamer plus
147 base pairs of DNA

DNA
(2-nm diameter)

(a) Nucleosomes
(6-nm × 11-nm flat disc)

FIGURE 11–11 General model of the association of histones and DNA in the nucleosome, showing how the chromatin fiber can coil into a more condensed structure, ultimately producing a metaphase chromosome.

looped domains that condense the structure into the chromatin fiber, which is 300 nm in diameter [Figure 11–11(c)]. The fibers are then coiled into the chromosome arms that constitute a chromatid, which is part of the metaphase chromosome [Figure 11–11(d)]. While this figure shows the chromatid arms to be 700 nm in diameter, this value undoubtedly varies among different organisms. At a value of 700 nm, a pair of sister chromatids comprising a chromosome measures about 1400 nm.

The importance of the organization of DNA into chromatin and chromatin into mitotic chromosomes can be illustrated by considering a human cell that stores its genetic material in a nucleus that is about 5 to 10 μm in diameter. The haploid genome contains 3.2×10^9 base pairs of DNA distributed among 23 chromosomes. The diploid cell contains twice that amount. At 0.34 nm per base pair, this amounts to an enormous length of DNA (as stated earlier, almost 2 m). One estimate is that about 25×10^6 nucleosomes per nucleus are complexed with the DNA present in a typical human nucleus.

In the overall transition from a fully extended DNA helix to the extremely condensed status of the mitotic chromosome, a packing ratio (the ratio of DNA length to the length of the structure containing it) of about 500:1 must be achieved. In fact, our model accounts for a ratio only one-tenth that. Obviously, the larger fiber can be further bent, coiled, and packed, as even greater condensation occurs during the formation of a mitotic chromosome.

Chromatin Remodeling

As with many significant findings in genetics, the study of nucleosomes has answered some important questions, but at the same time has also led us to new ones. For example, in the preceding discussion, we established that histone proteins play an important structural role in packaging DNA into the nucleosomes that make up chromatin. While solving the structural problem of how to organize a huge amount of DNA within the eukaryotic nucleus, a new problem was apparent: *the chromatin fiber, when complexed with histones and folded into various levels of compaction, makes the DNA inacces-*

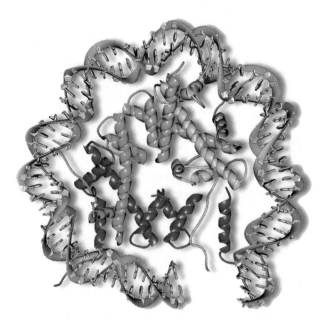

FIGURE 11–12 The nucleosome core particle derived from X-ray crystal analysis at 2.8 Å resolution. The double-helical DNA surrounds four pairs of histones.

sible to interaction with important nonhistone proteins. The variety of proteins that function in enzymatic and regulatory roles during the processes of replication and gene expression must interact directly with DNA. To accommodate these protein–DNA interactions, chromatin must be induced to change its structure, a process called **chromatin remodeling**. In the case of replication and gene expression, chromatin must relax its compact structure but be able to reverse the process during periods of inactivity.

Insights into how different states of chromatin structure may be achieved were forthcoming in 1997, when Timothy Richmond and members of his research team were able to significantly improve the level of resolution in X-ray diffraction studies of nucleosome crystals (from 7 Å in the 1984 studies to 2.8 Å in the 1997 studies). One model based on their work is shown in Figure 11–12. At this resolution, most atoms are visible, thus revealing the subtle twists and turns of the superhelix of DNA that encircles the histones. Recall that the double-helical ribbon represents 147 bp of DNA surrounding four pairs of histone proteins. This configuration is repeated over and over in the chromatin fiber and is the principal packaging unit of DNA in the eukaryotic nucleus.

The work of Richmond and colleagues, extended to a resolution of 1.9 Å in 2003, has revealed the details of the location of each histone entity within the nucleosome. Of particular interest to chromatin remodeling is that unstructured **histone tails** are not packed into the folded histone domains within the core of the nucleosome. For example, tails devoid of any secondary structure extending from histones H3 and H2B protrude through the minor groove channels of the DNA helix. The tails of histone H4 appear to make a connection with adjacent nucleosomes. Histone tails also provide potential targets for a variety of chemical modifications that may be linked to genetic functions along the chromatin fiber, including the regulation of gene expression.

Several of these potential chemical modifications are now recognized as important to genetic function. One of the best-studied histone modifications involves **acetylation**, resulting from the action of *histone acetyltransferase (HAT)*. This enzyme adds an acetyl group to the positively charged amino group present on the side chain of the amino acid lysine, effectively changing the net charge of the protein by neutralizing the positive charge. Lysine is in abundance in histones, and it has been known for some time that acetylation is linked to gene activation. It appears that high levels of acetylation open the chromatin fiber, an effect that occurs in regions of active genes and decreases in inactive regions. A well-known example involves the inactive X chromosome forming a Barr body in mammals, in which histone H4 is known to be greatly underacetylated.

The two other important chemical modifications include the **methylation** and **phosphorylation** of amino acids that are part of histones. These chemical processes result from the action of enzymes called *methyltransferases* and *kinases*, respectively. Methyl groups can be added to both arginine and lysine of histones, and this change has been correlated with the activation of genes. Phosphate groups can be added to the hydroxyl groups of the amino acids serine and histidine, introducing a negative charge on the protein. During the cell cycle, increased phosphorylation, particularly of histone H3, is known to occur at characteristic times. Such chemical modification is believed to be related to the cycle of chromatin unfolding and condensation that occurs during and after DNA replication.

Interestingly, while methylation of histones in nucleosomes is positively correlated with gene activity in eukaryotes, methylation of the nitrogenous base cytosine within polynucleotide chains of DNA, forming **5-methyl cytosine**, is negatively correlated with gene activity. Methylation occurs most often when the nucleotide cytidylic acid is next to the nucleotide guanylic acid (forming what is called a **CpG island**).

This discussion extends our knowledge of nucleosomes and chromatin organization and may serve as a general introduction to the concept of chromatin remodeling. A great deal more work must be done to elucidate the specific involvement of chromatin remodeling during genetic processes. In particular, the way in which the modifications are influenced by regulatory molecules within cells will provide important insights into our understanding of gene expression. What is clear is that the dynamic forms in which chromatin exists are vitally important to the way that all genetic processes directly involving DNA are executed. We will return to a more detailed discussion of the role of chromatin remodeling when we consider the regulation of eukaryotic gene expression in Chapter 15.

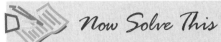

Now Solve This

In Problem 20 on page 251 you are asked to consider the extent to which the entire diploid content of human DNA can fit into the nucleus.

Hint: Assuming the nucleus is a perfect sphere, start with the formula $V = (4/3) \pi r^3$.

Heterochromatin

Evidence that the DNA of each eukaryotic chromosome consists of one continuous double-helical fiber along its entire length might suggest that the whole chromosome is structurally uniform. However, in the early part of the twentieth century, it was observed that some parts of the chromosome remain condensed and stain deeply during interphase, but most parts are uncoiled and do not stain. In 1928, the terms **heterochromatin** and **euchromatin** were coined to describe the parts of chromosomes that remain condensed and those that are uncoiled, respectively.

Subsequent investigation revealed a number of characteristics that distinguish heterochromatin from euchromatin. Heterochromatic areas are genetically inactive because they either lack genes or contain genes that are repressed. Also, heterochromatin replicates later during the S phase of the cell cycle than euchromatin does. The discovery of heterochromatin provided the first clues that parts of eukaryotic chromosomes do not always encode proteins. Instead, some chromosome regions are thought to be involved in maintenance of the chromosome's structural integrity and in other functions, such as chromosome movement during cell division.

Heterochromatin is characteristic of the genetic material of eukaryotes. Early cytological studies showed that areas of the centromeres are composed of heterochromatin. The ends of chromosomes, called *telomeres*, are also heterochromatic. In some cases, whole chromosomes are heterochromatic. Such is the case with the mammalian Y chromosome, much of which is genetically inert. And, as we discussed in Chapter 5, the inactivated X chromosome in mammalian females is condensed into an inert heterochromatic Barr body. In some species, such as mealy bugs, all of the chromosomes in one entire haploid set are heterochromatic.

When certain heterochromatic areas from one chromosome are translocated to a new site on the same or another nonhomologous chromosome, genetically active areas sometimes become genetically inert if they lie adjacent to the translocated heterochromatin. This influence on existing euchromatin is one example of what is more generally referred to as a **position effect**. That is, the position of a gene or group of genes relative to all other genetic material may affect their expression.

11.5 Eukaryotic Genomes Demonstrate Complex Sequence Organization Characterized by Repetitive DNA

Thus far, we have looked at how DNA is organized into chromosomes in bacteriophages, bacteria, and eukaryotes. We now begin an examination of what we know about the organization of DNA sequences within the chromosomes making up an organism's genome, placing our emphasis on eukaryotes. Once we establish the pattern of genome organization, we will focus on how the genes themselves are organized within chromosomes in a later chapter (see chapter 18).

We have already established that, in addition to single copies of unique DNA sequences that comprise genes, a great deal of the DNA sequences within chromosomes is repetitive in nature and that various levels of repetition occur within the genome of organisms. Many studies have now provided insights into **repetitive DNA**, demonstrating various classes of these sequences and their organization within the genome. Figure 11–13 outlines the various categories of repetitive DNA. Some functional genes are present in more than one copy and are therefore repetitive in nature. However, the majority of repetitive sequences are nongenic, and in fact, most serve no known function. We explore three main categories: (1) heterochromatin found associated with centromeres and making up telomeres, (2) tandem repeats of both short and long DNA sequences, and (3) transposable sequences that are interspersed throughout the genome of eukaryotes.

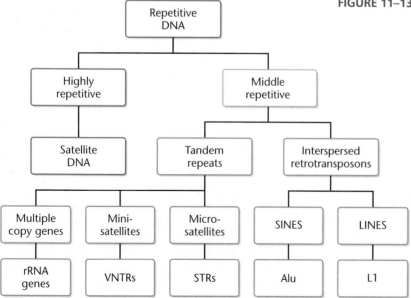

FIGURE 11–13 An overview of the categories of repetitive DNA.

Repetitive DNA and Satellite DNA

The nucleotide composition of the DNA (e.g., the percentage of $G \equiv C$ versus $A = T$ pairs) of a particular species is reflected in its density, which can be measured with sedimentation equilibrium centrifugation. When eukaryotic DNA is analyzed in this way, the majority is present as a single main band, or peak, of fairly uniform density. However, one or more additional peaks represent DNA that differs slightly in density. This component, called **satellite DNA**, represents a variable proportion of the total DNA, depending on the species. A profile of main-band and satellite DNA from the mouse is shown in Figure 11–14. By contrast, prokaryotes contain only main-band DNA.

The significance of satellite DNA remained an enigma until the mid-1960s, when Roy Britten and David Kohne developed the technique for measuring the reassociation kinetics of DNA that had previously been dissociated into single strands. They demonstrated that certain portions of DNA reassociated more rapidly than others. They concluded that rapid reassociation was characteristic of multiple DNA fragments composed of identical or nearly identical nucleotide sequences—the basis for the descriptive term **repetitive DNA** (see Chapter 9).

When satellite DNA is subjected to analysis by reassociation kinetics, it falls into the category of **highly repetitive DNA**, which is known to consist of relatively short sequences repeated a large number of times. Further evidence suggests that these sequences are present as tandem repeats clustered in very specific chromosomal areas known to be heterochromatic—the regions flanking centromeres. This was discovered in 1969 when several researchers, including Mary Lou Pardue and Joe Gall, applied *in situ* **molecular hybridization** to the study of satellite DNA. This technique involves the molecular hybridization between an isolated fraction of radioactively labeled DNA or RNA probes and the DNA contained in the chromosomes of a cytological preparation. Following the hybridization procedure, autoradiography is performed to locate the chromosome areas complementary to the fraction of DNA or RNA.

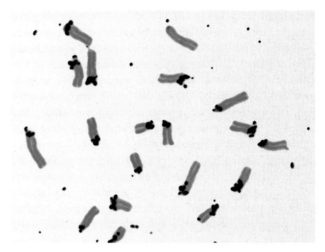

FIGURE 11–15 *In situ* molecular hybridization between RNA transcribed from mouse satellite DNA and mitotic chromosomes. The grains in the autoradiograph localize the chromosome regions (the centromeres) containing satellite DNA sequences.

Pardue and Gall demonstrated that radioactive probes made from mouse satellite DNA hybridize with the DNA of centromeric regions of mouse mitotic chromosomes (Figure 11–15). Several conclusions were drawn: Satellite DNA differs from main-band DNA in its molecular composition, as established by buoyant density studies. It is composed of repetitive sequences. Finally, satellite DNA is found in the heterochromatic centromeric regions of chromosomes.

Centromeric and Telomeric DNA Sequences

Separation of chromatids is essential to the fidelity of chromosome distribution during mitosis and meiosis. Most estimates of infidelity during mitosis are exceedingly low: 1×10^{-5} to 1×10^{-6} or 1 error per 100,000 to 1 million cell divisions. As a result, it has been generally assumed that analysis of the DNA sequence of centromeric regions would provide insights into the rather remarkable features of this chromosomal region. The minimal region of the centromere that supports the function of chromosomal segregation is designated the **CEN region**, and this function is now reasonably clear. Within this heterochromatic region of the chromosome, the DNA binds a platform of proteins forming the centromere, which in multicellular organisms includes the **kinetochore** that binds to the spindle fiber during division.

The CEN regions of the yeast *Saccharomyces cerevisiae* were the first to be studied. Each centromere serves an identical function, so it is not surprising that CENs from different chromosomes were found to be remarkably similar in their organization. The CEN region of yeast chromosomes consists of about 120 bp, which can be divided into three sequential regions. The first and third regions (I and III) are relatively short and highly conserved, consisting of only 9 bp and 11 bp, respectively. Region II, which is larger (~100 bp) and extremely rich in A and T (up to 95%), varies in sequence among different chromosomes.

Mutational analysis suggests that regions I and II are less critical to centromere function than is region III. Mutations in the former regions are often tolerated without loss of function, but certain

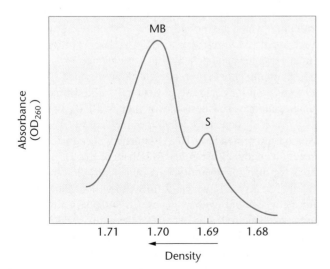

FIGURE 11–14 Separation of main-band (MB) and satellite (S) DNA from the mouse, using ultracentrifugation in a CsCl gradient.

mutations in region III can disrupt centromere function altogether. Thus, the DNA of this region appears to be essential to the eventual binding to the spindle fiber (yeast do not have kinetechores).

The amount of DNA associated with the centromeres of multicellular eukaryotes is much more extensive than in yeast. Recall from our prior discussion that highly repetitive satellite DNA is localized in the centromere regions of mice. Such sequences, absent from yeast but characteristic of most multicellular organisms, vary considerably in size. For example, in *Drosophila* the CEN region is found within some 200 to 600 kb of DNA, much of which is highly repetitive. In humans, one of the most recognized satellite DNA sequences is the **alphoid family**, found mainly in the centromere regions. Alphoid sequences, each about 170 bp in length, are present in tandem arrays of up to 1 million base pairs. Embedded somewhere in this repetitive DNA are more specific sequences that are critical to centromere function. The role of highly repetitive DNA present in centromere regions remains unclear.

The other prominent structural component of chromosomes is the **telomere**, found at the ends of linear chromosomes. Telomeres provide stability to the chromosome by rendering chromosome ends generally inert in interactions with other linear chromosome ends, and with enzymes that use double-stranded DNA ends as substrates (such as repair enzymes). It is thought that some aspect of the molecular structure of telomeres must be unique compared with most other chromosome regions. As with centromeres, the analysis of **telomeric DNA sequences** was first approached by investigating the smaller chromosomes of simple eukaryotes, such as protozoans and yeast. The idea that all telomeres of all chromosomes in a given species might share a related nucleotide sequence has now been borne out.

As we discussed in the previous chapter, telomeric DNA sequences are repetitive in nature, consisting of short tandem repeats. In the ciliate *Tetrahymena*, over 50 tandem repeats of the hexanucleotide sequence GGGGTT occur. In humans, the sequence GGGATT is repeated many times. The number varies in different organisms, and there may be as many as 1000 repeats in some species.

The analysis of telomeric DNA sequences has shown this pattern to be highly conserved throughout evolution, reflecting the critical role they play in maintaining the integrity of chromosomes. How exactly this occurs is still under investigation. One theory, based on *in vitro* studies, is that the repetitive GT-rich sequence characteristic of telomeres may be extended on one of the two strands, forming a single-stranded tail at the end of the telomere. Since guanine has a tendency to base pair with other guanine residues, it is further believed that base pairing occurs within this tail, creating a closed "loop" that effectively seals off the ends of chromosomes. If this model is correct, the absence of any free ends would contribute to the stability and integrity of the chromosome.

Middle Repetitive Sequences: VNTRs and STRs

A brief review of still another prominent category of repetitive DNA sheds more light on our understanding of the organization of the eukaryotic genome. In addition to highly repetitive DNA, which constitutes about 5 percent of the human genome (and 10% of the mouse genome), a second category, **middle** (or **moderately**) **repetitive DNA**, recognized by C_0t analysis (Chapter 9), is fairly well characterized. Because we are learning a great deal about the human genome, we will use our own species to illustrate this category of DNA in genome organization.

Although middle repetitive DNA does include some duplicated genes (such as those encoding ribosomal RNA), most prominent in this category are either noncoding tandemly repeated or interspersed sequences. No function has been ascribed to these components of the genome. An example includes those called **variable number tandem repeats (VNTRs)**. The repeating DNA sequence of VNTRs may be 15 to 100 bp long and is found within and between genes. Many such clusters are dispersed throughout the genome, and they are often referred to as **minisatellites**.

The number of tandem copies of each specific sequence at each location varies in individuals, creating localized regions of 1000 to 20,000 bp (1–20 kb) in length. The variation in size (length) of these regions between individuals in humans was originally the basis for the forensic technique referred to as **DNA fingerprinting**.

Another group of tandemly repeated sequences consists of di-, tri-, tetra-, and pentanucleotides, also referred to as **microsatellites** or **short tandem repeats (STRs)**. Like VNTRs, they are dispersed throughout the genome and vary among individuals in the number of repeats present at any site. For example, in humans, the most common microsatellite is the dinucleotide $(CA)_n$, where n equals the number of repeats. Most commonly, n is between 5 and 50. These clusters have served as useful molecular markers during genome analysis and are now the sequence of choice in DNA fingerprinting.

Repetitive Transposed Sequences: SINEs and LINEs

Still another category of repetitive DNA consists of sequences that are interspersed throughout the genome, rather than being tandemly repeated. They can be either short or long, and many have the added distinction of being **transposable sequences**, which are mobile and can potentially move to different locations within the genome. A large portion of the human genome is composed of these sequences.

For example, **short interspersed elements**, or **SINEs**, are 100 to 500 bp long and are present 1.5 million times in the human genome. The best-known human SINE is a set of closely related sequences called the ***Alu* family** (the name is based on the presence of DNA sequences recognized by the restriction endonuclease *Alu*I). Members of this DNA family, which are also found in other mammals, are 200 to 300 bp long and dispersed rather uniformly throughout the genome, both between and within genes. In humans, this family encompasses more than 5 percent of the entire genome.

Members of the *Alu* family are sometimes transcribed. The role of this RNA is not certain, but it may relate to their mobility in the genome. In fact, *Alu* sequences are thought to have arisen from an RNA element whose DNA complement was dispersed throughout the genome as a result of the activity of reverse transcriptase (an enzyme that synthesizes DNA on an RNA template).

The group of **long interspersed elements (LINEs)** represents still another category of repetitive transposable DNA sequences. LINEs are usually about 6 kb in length and in the human genome are present 850,000 times. The most prominent example in humans is the **L1 family**. Members of this sequence family are about 6400 bp long and are present up to 100,000 times.

The basis for transposition of L1 elements is now clear. The L1 DNA sequence is first transcribed into an RNA molecule. The RNA then serves as the template for the synthesis of the DNA complement via the enzyme reverse transcriptase. This enzyme is encoded by a portion of the L1 sequence. The new L1 copy then integrates into the DNA of the chromosome at a new site. Because this mechanism of transposition resembles that used by retroviruses, LINEs are referred to as **retrotransposons**.

These repetitive elements represent a significant portion of human DNA. SINEs constitute about 13 percent of the human genome, while LINEs constitute up to 21 percent. Both types of elements share the organizational feature of the repeating sequences being combined with unique sequences within the DNA of each entity.

Middle Repetitive Multiple Copy Genes

In some cases, middle repetitive DNA includes functional genes tandemly present in multiple copies. For example, many copies exist of the genes encoding ribosomal RNA. *Drosophila* has 120 copies per haploid genome. Single genetic units encode a large precursor molecule that is processed into the 5.8S, 18S, and 28S rRNA components. In humans, multiple copies of this gene are clustered on the p arm of the acrocentric chromosomes 13, 14, 15, 21, and 22. Multiple copies of the genes encoding 5S rRNA are transcribed separately from multiple clusters found together on the terminal portion of the p arm of chromosome 1.

11.6 The Vast Majority of a Eukaryotic Genome Does Not Encode Functional Genes

Given the preceding information involving various forms of repetitive DNA in eukaryotes, we can pose an intriguing question: *What proportion of the eukaryotic genome actually encodes functional genes?*

As we have seen, taken together, the various forms of highly repetitive and moderately repetitive DNA comprise a substantial portion of the human genome. In addition to repetitive DNA, many of the single-copy DNA sequences appear to be noncoding. For example, there are many instances of what we call **pseudogenes**. These are DNA sequences that represent evolutionary vestiges of duplicated copies of genes that have undergone significant mutational alteration. As a result, although they show some homology to their parent gene, they are usually not transcribed because of insertions and deletions throughout their structure.

The proportion of the genome consisting of repetitive DNA varies among organisms, but one feature seems to be shared: Only a very small part of the genome actually codes for proteins. For example, the genes encoding proteins in sea urchin occupy less than 10 percent of the genome. In *Drosophila*, only 5 to 10 percent of the genome is occupied by genes coding for proteins. In humans, it appears that the estimated 20,000 to 25,000 functional genes occupy only about 1 to 3 percent of the total DNA sequence making up the genome. We will pursue this topic in much more detail in Chapter 18—Genomics and Proteomics.

CHAPTER SUMMARY

1. The organization of the molecular components that form chromosomes is essential to understanding the function of the genetic material. Largely devoid of associated proteins, bacteriophage and bacterial chromosomes contain DNA molecules in a form equivalent to the Watson-Crick model.

2. Mitochondria and chloroplasts contain DNA that encodes products essential to their biological function. This DNA is remarkably similar in form and appearance to some bacterial and bacteriophage DNA, lending support to the endosymbiotic hypothesis, which suggests that these organelles were once free-living prokaryote-like organisms.

3. Polytene and lampbrush chromosomes are examples of specialized structures that have extended our knowledge of genetic organization and function.

4. The eukaryotic chromatin fiber is a nucleoprotein organized into repeating units called nucleosomes. Composed of 147 bp of DNA and an octamer of four types of histones, the nucleosome facilitates the conversion of the extended chromatin fiber characteristic of interphase into the highly condensed chromosome seen in mitosis.

5. The structural heterogeneity of the chromosome axis has been established as a result of both biochemical and cytological investigation. Heterochromatin, prematurely condensed in interphase, is genetically inert. The centromeric and telomeric regions, parts of the Y chromosome, and the Barr body are examples.

6. Eukaryotic genomes demonstrate complex sequence organization characterized by numerous categories of repetitive DNA.

7. DNA analysis reveals unique nucleotide sequences in both the centromere and telomere regions of eukaryotic chromosomes, which no doubt impart the heterochromatic characteristic that they share and play a role in their respective functions.

8. Repetitive DNA consists of either tandem repeats clustered in various regions of the genome or single sequences interspersed randomly throughout the genome. In the former group, the size of each cluster varies among individuals, providing a form of genetic identity. The latter group of sequences may be short like *Alu* or long such as in L1, and they are transposable elements.

9. The vast majority of most eukaryotic genomes does not encode functional genes.

KEY TERMS

INSIGHTS AND SOLUTIONS

A previously undiscovered single-cell organism was found living at a great depth on the ocean floor. Its nucleus contained only a single linear chromosome with 7×10^6 nucleotide pairs of DNA coalesced with three types of histonelike proteins. Consider the following questions:

1. A short micrococcal nuclease digestion yielded DNA fractions of 700, 1400, and 2100 bp. Predict what these fractions represent. What conclusions can be drawn?

Solution: The chromatin fiber may consist of a variation of nucleosomes containing 700 bp of DNA. The 1400- and 2100-bp fractions, respectively, represent 2 and 3 linked nucleosomes. Enzymatic digestion may have been incomplete, leading to the latter two fractions.

2. The analysis of individual nucleosomes reveals that each unit contained one copy of each protein and that the short linker DNA contained no protein bound to it. If the entire chromosome consists of nucleosomes (discounting any linker DNA), how many are there, and how many total proteins are needed to form them?

Solution: Since the chromosome contains 7×10^6 bp of DNA, the number of nucleosomes, each containing 7×10^2 bp, is equal to

$$7 \times 10^6 / 7 \times 10^2 = 10^4 \text{ nucleosomes}$$

The chromosome contains 10^4 copies of each of the 3 proteins, for a total of 3×10^4 proteins.

3. Further analysis revealed the organism's DNA to be a double helix similar to the Watson-Crick model but containing 20 bp per complete turn of the right-handed helix. The physical size of the nucleosome was exactly double the volume occupied by that found in all other known eukaryotes, by virtue of increasing the distance along the fiber axis by a factor of two. Compare the degree of compaction of this organism's nucleosome to that found in other eukaryotes.

Solution: The unique organism compacts a length of DNA consisting of 35 complete turns of the helix (700 bp per nucleosome/20 bp per turn) into each nucleosome. The normal eukaryote compacts a length of DNA consisting of 20 complete turns of the helix (200 bp per nucleosome/10 bp per turn) into a nucleosome one-half the volume of that in the unique organism. The degree of compaction is therefore less in the unique organism.

4. No further coiling or compaction of this unique chromosome occurs in the unique organism. Compare this to a eukaryotic chromosome. Do you think an interphase human chromosome 7×10^6 bp in length would be a shorter or longer chromatin fiber?

Solution: The length of the unique chromosome is compacted into 10^4 nucleosomes, each with an axis length twice that of the eukaryotic fiber. The eukaryotic fiber consists of

$$7 \times 10^6 / 2 \times 10^2 = 3.5 \times 10^4 \text{ nucleosomes}$$

which is 3.5 times more than the unique organism. However, they are compacted by a factor of five in each solenoid. Therefore, the chromosome of the unique organism is a longer chromatin fiber.

PROBLEMS AND DISCUSSION QUESTIONS

1. Contrast the sizes of the chromosomes of bacteriophage λ and T2 with that of *E. coli*. How does this relate to the relative size and complexity of the phages and bacteria?

2. Bacteriophages and bacteria almost always contain their DNA as circular (closed loops) chromosomes. Phage λ is an exception, maintaining its DNA in a linear chromosome within the viral particle. However, as soon as it is injected into a host cell, it circularizes before replication begins. Taking into account the information in Chapter 10, what advantage exists in replicating circular DNA molecules compared to linear molecules?

3. Contrast the appearance of the DNA associated with mitochondria and chloroplasts.

4. Compare the size of DNA and the encoded gene products in mitochondria and chloroplasts.

5. While protein synthesis occurs within mitochondria and chloroplasts, not all necessary gene products are encoded by the organellar DNA. Explain the origin of the genetic machinery within the organelles.

6. Mitochondria and chloroplasts contain ribosomal RNA molecules that are unlike those found in the adjoining cytoplasm. How does this observation relate to the endosymbiotic hypothesis?

7. Describe how giant polytene chromosomes are formed.

8. Salivary gland cells from *Drosophila* are isolated and placed in the presence of radioactive thymidylic acid. Autoradiography is performed, revealing polytene chromosomes. Predict the distribution of the grains along the chromosomes.

9. What genetic process is occurring in a puff of a polytene chromosome? How do we know this experimentally?

10. Describe the structure of LINE sequences. Why are LINEs referred to as retrotransposons?

11. During what genetic process are lampbrush chromosomes present in vertebrates?

12. Why might we predict that the organization of eukaryotic genetic material will be more complex than that of viruses or bacteria?

13. Describe the sequence of research findings that led to the development of the model of chromatin structure.

14. What is the molecular composition and arrangement of the components in the nucleosome?

15. Describe the transitions that occur as nucleosomes are coiled and folded, ultimately forming a chromatid.

16. Provide a comprehensive definition of heterochromatin, and list as many examples as you can.

17. Mammals contain a diploid genome consisting of at least 10^9 bp. If this amount of DNA is present as chromatin fibers, where each group of 200 bp of DNA is combined with 9 histones into a nucleosome and each group of 6 nucleosomes is combined into a solenoid, achieving a final packing ratio of 50, determine (a) the total number of nucleosomes in all fibers, (b) the total number of histone molecules combined with DNA in the diploid genome, and (c) the combined length of all fibers.

18. Assume that a viral DNA molecule is a 50-μm-long circular strand of a uniform 20 Å diameter. If this molecule is contained in a viral head that is a 0.08-μm-diameter sphere, will the DNA molecule fit into the viral head, assuming complete flexibility of the molecule? Justify your answer mathematically.

19. How many base pairs are in a molecule of phage T2 DNA 52 μm long?

20. If a human nucleus is 10 μm in diameter, and it must hold as much as 2 m of DNA, which is complexed into nucleosomes that are 11 nm in diameter during full extension, what percentage of the volume of the nucleus is occupied by the genetic material?

21. Sun and others (2002. *Proc. Natl. Acad. Sci. (USA)* 99: 8695–8700) studied *Drosophila* in which two normally active genes, w^+ (wild-type allele of the *white*-eye gene) and hsp26 (a heat-shock gene), were introduced (using a plasmid vector) into euchromatic and heterochromatic chromosomal regions. The relative activity of each gene was assessed, and an approximation of the data obtained is shown below. Considering three characteristics of heterochromatin, which is/are supported by the experimental data?

Activity (relative percentage)		
Gene	Euchromatin	Heterochromatin
*hsp*26	100%	31%
w^+	100%	8%

22. Using molecular methods to "label" chromosomes with fluorescent dyes, Nagele and colleagues (1995. *Science* 270: 1831–1835) observed a precise nuclear positioning of chromosomes during an early stage of mitosis (prometaphase) in human fibroblast cells. Below is a sketch modified from this research that describes the relative positions of chromosomes 7, 8, 16, and X. Homologous chromosomes share the same color. Assuming that this pattern is consistent among other cells in humans, what conclusions can be drawn regarding the nuclear positions of the chromosomes during interphase and during the initial phases of mitosis? How could chromosomal localization influence gene function during interphase?

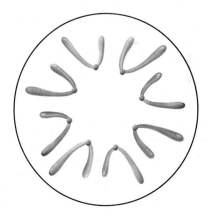

23. While much remains to be learned about the role of nucleosomes and chromatin structure and function, recent research indicates that *in vivo* chemical modification of histones is associated with changes in gene activity. For example, Bernstein and others (2000. *Proc. Natl. Acad. Sci. (USA)* 97: 5340–5345) determined that acetylation of H3 and H4 is associated with 21.1 percent and 13.8 percent increase in yeast gene activity, respectively, and that yeast heterochromatin is hypomethylated relative to the genome average. Speculate on the significance of these findings in terms of nucleosome–DNA interactions and gene activity.

24. In an article entitled *Nucleosome positioning at the replication fork*, Lucchini and others (2002. *EMBO* 20: 7294–7302) state, "both the 'old' randomly segregated nucleosomes as well as the 'new' assembled histone octamers rapidly position themselves (within seconds) on the newly replicated DNA strands." Given this statement, how would one compare the distribution of nucleosomes and DNA in newly replicated chromatin? How could

one experimentally test the distribution of nucleosomes on newly replicated chromosomes?

25. The human genome contains approximately 10^6 copies of an *Alu* sequence, one of the best-studied classes of short interspersed elements (SINEs), per haploid genome. Individual *Alu*s share a 282-nucleotide consensus sequence followed by a 3'-adenine-rich tail region (Schmid. 1998. *Nuc. Acids Res.* 26: 4541–4550). Given that there are approximately 3×10^9 bp per human haploid genome, about how many base pairs are spaced between each *Alu* sequence?

26. Below is a diagram of the general structure of the bacteriophage λ chromosome. Speculate on the mechanism by which it forms a closed ring upon infection of the host cell.

5'GGGCGGCGACCT—double-stranded region——3'

3'—double-stranded region——CCCGCCGCTGGA5'

27. Tandemly repeated DNA sequences with a repeat sequence of one to six base pairs [e.g., $(GACA)_n$] are called microsatellites and are common in eukaryotes. A particular subset, the trinucleotide repeat, is of great interest because of the role it plays in human neurodegenerative disorders (Huntington disease, myotonic dystrophy, spinal-bulbar muscular atrophy, spinocerebellar ataxia, and fragile X syndrome). Below are data modified from Toth and colleagues (2000. *Gen. Res.* 10: 967–981) regarding the location of microsatellites within and between genes. What general conclusions can be drawn from these data?

Percentage of Microsatellite DNA Sequences Within Genes and Between Genes

Taxonomic Group	Within Genes	Between Genes
Primates	7.4	92.6
Rodents	33.7	66.3
Arthropods	46.7	53.3
Yeasts	77.0	23.0
Other fungi	66.7	33.3

28. More information from the research effort in Problem 27 produced data regarding the pattern of the length of such repeats within genes. Each value in the following table represents the number of times a microsatellite of a particular sequence length, one to six bases long, is found within genes. For instance, in primates, a dinucleotide sequence (GC, for example) is found 10 times, while a trinucleotide is found 1126 times. In fungi, a repeat motif composed of 6 nucleotides (GACACC, for example) is found 219 times whereas a tetranucleotide repeat (GACA, for example) is found only 2 times. Analyze and interpret these data by indicating what general pattern is apparent regarding the distribution of various microsatellite lengths within genes. Of what significance might this general pattern be?

Distribution of Microsatellites by Unit Length Within Genes

Taxonomic Group	Length of Repeated Motif (bp)					
	1	2	3	4	5	6
Primates	49	10	1126	29	57	244
Rodents	62	70	1557	63	116	620
Arthropods	12	34	1566	0	21	591
Yeasts	36	19	706	7	52	330
Other fungi	9	4	381	2	35	219

29. In spite of the considerable medical and biological significance of repetitive DNA sequences, the factors that determine their genesis and genomic distribution remain uncertain. Misalignment of repetitive DNA strands and DNA polymerase slippage have been described as mechanisms causing variation in repeat number of *existing* repeats. Until recently, there has been little information relating to the *initial* creation of a microsatellite genomic region. Wilder and Hollocher (2001. *Mol. Biol. Evol.* 18: 384–392) sequenced DNA surrounding numerous tetranucleotide microsatellite regions in several strains within two species of *Drosophila* and observed the sequences shown below. (a) Identify the microsatellite tetranucleotide motif. Is it a perfect motif? Using Pu to represent a purine and Py to represent a pyrimidine, symbolize the tetranucleotide repeat in a form similar to $(CPuGPy)_n$. (b) What is the sequence of the nonmicrosatellite region? Is it a perfectly conserved region among all the species and strains listed?

Species (strain)	Base Sequence
D. nigrodunni-1	5'-TCGATATAGCCATGTCCGTCTGTCCGTCTGT
D. nigrodunni-2	5'-TCGATATAGCCATGTCCGTCTGTCCGTCTGT
D. nigrodunni-3	5'-TCGATATAGCAATGTCCGTCTGTCCGTCTGT
D. dunni-1	5'-TCGATATAGCAATGTCCGTCTGTCCGTCTGT
D. dunni-2	5'-TCGATATAGCCATGTCCGTCTGTCCGTCTGT

30. Regarding the findings and your analysis of data in Problem 29, what significance might there be to having a highly conserved nonmicrosatellite region flanking a specific microsatellite type?

The Genetic Code and Transcription

Electron micrograph visualizing the process of transcription.

■ CHAPTER CONCEPTS

- Genetic information is stored in DNA using a triplet code that is nearly universal to all living things on Earth.

- The genetic code is initially transferred from DNA to RNA during the process of transcription.

- Once transferred to RNA, the genetic code exists as triplet codons, using the four ribonucleotides in RNA as the letters composing it.

- Using four different letters taken three at a time, 64 triplet sequences are possible. Most all encode one of the 20 amino acids present in proteins, which serve as the end products of most genes.

- Several codons provide signals that initiate and terminate protein synthesis.

- The process of transcription is similar but more complex in eukaryotes compared to prokaryotes and bacteriophages that infect them.

The linear sequence of deoxyribonucleotides making up DNA ultimately dictates the components constituting proteins, the end product of most genes. The central question is how such information stored as a nucleic acid is decoded into a protein. Figure 12–1 gives a simplified overview of how this transfer of information occurs. In the first step in gene expression, information on one of the two strands of DNA (the template strand) is transferred into an RNA complement through transcription. Once synthesized, this RNA acts as a "messenger" molecule bearing the coded information—hence its name, messenger RNA (mRNA). The mRNAs then associate with ribosomes, where decoding into proteins takes place.

In this chapter, we focus on the initial phases of gene expression by addressing two major questions. First, how is genetic information encoded? Second, how does the transfer from DNA to RNA occur, thus defining the process of transcription? As you shall see, ingenious analytical research established that the genetic code is written in units of three letters—ribonucleotides present in mRNA that reflect the stored information in genes. Most all triplet code words direct the incorporation of a specific amino acid into a protein as it is synthesized. As we can predict based on our prior discussion of the replication of DNA, transcription is also a complex process dependent on a major polymerase enzyme and a cast of supporting proteins. We will explore what is known about transcription in bacteria and then contrast this prokaryotic model with the differences found in eukaryotes. Together, the

information in this and the next chapter provides a comprehensive picture of molecular genetics, which serves as the most basic foundation for understanding living organisms. In Chapter 13, we will address how translation occurs and discuss the structure and function of proteins.

? How Do We Know?

In this chapter, we will focus on how the genetic information stored in DNA encodes proteins, the end products of most genes. We shall also elucidate the general mechanism by which genetic information in DNA is transfered to RNA, the process of transcription. As you study this topic, you should try to answer several fundamental questions:

1. Why did geneticists believe, even before experimental evidence was obtained, that the genetic code is a triplet?

2. Once the triplet nature of the code seemed very likely, how were the 64 possible triplet codes deciphered; that is, how was it predicted which three-letter codes specify each amino acid?

3. How do we know that expression of the information encoded in DNA involves an RNA intermediate?

4. After the genetic code was predicted experimentally, how did we confirm that these predictions were correct?

5. How do we know that the initial transcript of a eukaryotic gene contains noncoding sequences that must be removed before accurate translation into proteins can occur?

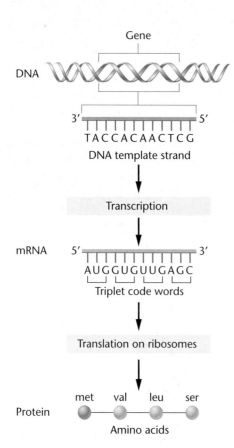

FIGURE 12–1 Flowchart illustrating how genetic information encoded in DNA produces protein.

12.1 The Genetic Code Exhibits a Number of Characteristics

Before we consider the various analytical approaches that led to our current understanding of the genetic code, let's summarize the general features that characterize it.

1. The genetic code is written in linear form, using the ribonucleotide bases that compose mRNA molecules as "letters." The ribonucleotide sequence is derived from the complementary nucleotide bases in DNA.

2. Each "word" within the mRNA contains three ribonucleotide letters. With only three exceptions, each group of three ribonucleotides, called a **codon**, specifies one amino acid; the code is thus a **triplet codon**.

3. The code is **unambiguous**—each triplet specifies only a single amino acid.

4. The code is **degenerate**; that is, a given amino acid can be specified by more than one triplet codon. This is the case for 18 of the 20 amino acids.

5. The code contains one "start" and three "stop" signals, triplets that **initiate** and **terminate** translation.

6. No internal punctuation (such as a comma) is used in the code. Thus, the code is said to be **commaless**. Once translation of mRNA begins, the codons are read one after the other, with no breaks between them.

7. The code is **nonoverlapping**. Once translation commences, any single ribonucleotide at a specific location within the mRNA is part of only one triplet.

8. The code is nearly **universal**. With only minor exceptions, almost all viruses, prokaryotes, archaea, and eukaryotes use a single coding dictionary.

12.2 Early Studies Established the Basic Operational Patterns of the Code

In the late 1950s, before it became clear that mRNA is the intermediate that transfers genetic information from DNA to proteins, researchers thought that DNA itself might directly encode proteins during their synthesis. Because ribosomes had already been identified, the initial thinking was that information in DNA was transferred in the nucleus to the RNA of the ribosome, which served as the template for protein synthesis in the cytoplasm. This concept soon became untenable as accumulating evidence indicated that there was an unstable intermediate template. The RNA of ribosomes, on the other hand, was extremely stable. As a result, in 1961 researchers postulated the existence of **messenger RNA(mRNA)**. Once mRNA was discovered, it was clear that even though genetic information is stored in DNA, the code that is translated into proteins resides in RNA. The central question then was how only four letters—the four nucleotides—could specify 20 words—the amino acids.

The Triplet Nature of the Code

In the early 1960s, Sidney Brenner argued on theoretical grounds that the code had to be a triplet since three-letter words represent the minimal use of four letters to specify 20 amino acids. A code of four nucleotides, taken two at a time, for example, provides only 16 unique code words (4^2). A triplet code yields 64 words (4^3)—clearly more than the 20 needed—and is much simpler than a four-letter code which specifies 256 words (4^4).

Experimental evidence supporting the triplet nature of the code was subsequently derived from research by Francis Crick and his colleagues. Using phage T4, they studied **frameshift mutations**, which result from the addition or deletion of one or more nucleotides within a gene and subsequently the mRNA transcribed by it. The gain or loss of letters shifts the *frame of reading* during translation. Crick and his colleagues found that the gain or loss of one or two nucleotides caused a frameshift mutation, but when three nucleotides were involved, the frame of reading was reestablished (Figure 12–2). This would not occur if the code was anything other than a triplet. This work also suggested that most triplet codes are not blank, but rather encode amino acids, supporting the concept of a degenerate code.

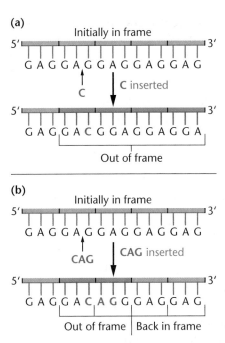

FIGURE 12–2 The effect of frameshift mutations on a DNA sequence with the repeating triplet sequence GAG. (a) The insertion of a single nucleotide shifts all subsequent triplet reading frames. (b) The insertion of three nucleotides changes only two triplets, but the frame of reading is then reestablished to the original sequence.

12.3 Studies by Nirenberg, Matthaei, and Others Deciphered the Code

In 1961, Marshall Nirenberg and J. Heinrich Matthaei deciphered the first specific coding sequences, which served as a cornerstone for the complete analysis of the genetic code. Their success, as well as that of others who made important contributions to breaking the code, was dependent on the use of two experimental tools—an *in vitro* (**cell-free**) **protein-synthesizing system** and an enzyme, **polynucleotide phosphorylase**, which enabled the production of synthetic mRNAs. These mRNAs are templates for polypeptide synthesis in the cell-free system.

Cell-Free Polypeptide Synthesis

In the cell-free system, amino acids are incorporated into polypeptide chains. This *in vitro* mixture must contain the essential factors for protein synthesis in the cell: ribosomes, tRNAs, amino acids, and other molecules essential to translation (see Chapter 13). In order to follow (or trace) protein synthesis, one or more of the amino acids must be radioactive. Finally, an mRNA must be added, which serves as the template that will be translated.

In 1961, mRNA had yet to be isolated. However, use of the enzyme polynucleotide phosphorylase allowed the artificial synthesis of RNA templates, which could be added to the cell-free system. This enzyme, isolated from bacteria, catalyzes the reaction shown in Figure 12–3. Discovered in 1955 by Marianne Grunberg-Manago and Severo Ochoa, the enzyme functions metabolically in bacterial cells to degrade RNA.

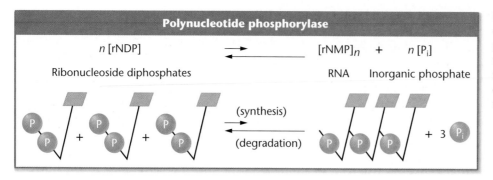

FIGURE 12–3 The reaction catalyzed by the enzyme polynucleotide phosphorylase. Note that the equilibrium of the reaction favors the degradation of RNA but can be "forced" in the direction favoring synthesis.

However, *in vitro*, with high concentrations of ribonucleoside diphosphates, the reaction can be "forced" in the opposite direction to synthesize RNA, as shown.

In contrast to RNA polymerase, polynucleotide phosphorylase does not require a DNA template. As a result, each addition of a ribonucleotide is random, based on the relative concentration of the four ribonucleoside diphosphates added to the reaction mixtures. The probability of the insertion of a specific ribonucleotide is proportional to the availability of that molecule, relative to other available ribonucleotides. *This point is absolutely critical to understanding the work of Nirenberg and others in the ensuing discussion.*

The cell-free system for protein synthesis and the availability of synthetic mRNAs provided a means of deciphering the ribonucleotide composition of various triplets encoding specific amino acids.

The Use of Homopolymers

In their initial experiments, Nirenberg and Matthaei synthesized **RNA homopolymers**, each with only one type of ribonucleotide. Therefore, the mRNA added to the *in vitro* system was either UUUUUU..., AAAAAA..., CCCCCC..., or GGGGGG.... They tested each mRNA and were able to determine which, if any, amino acids were incorporated into newly synthesized proteins. To do this, the researchers labeled 1 of the 20 amino acids added to the *in vitro* system and conducted a series of experiments, each with a different radioactively labeled amino acid.

For example, in experiments using ^{14}C-phenylalanine (Table 12.1), Nirenberg and Matthaei concluded that the message poly U (polyuridylic acid) directed the incorporation of only phenylalanine into the homopolymer polyphenylalanine. Assuming the validity of a triplet code, they determined the first specific codon assignment—UUU codes for phenylalanine. Using similar experiments, they quickly found that AAA codes for lysine and CCC codes for proline. Poly G was not an adequate template, probably because the molecule folds back upon itself. Thus, the assignment for GGG had to await other approaches.

Note that the specific triplet codon assignments were possible only because homopolymers were used. This method yields only the composition of triplets, but since three identical letters can have only one possible sequence (e.g., UUU), the actual codons were identified.

Mixed Copolymers

With these techniques in hand, Nirenberg and Matthaei, and Ochoa and coworkers turned to the use of **RNA heteropolymers**. In this type of experiment, two or more different ribonucleoside diphosphates are added in combination to form the artificial message. The researchers reasoned that if they knew the relative proportion of each type of ribonucleoside diphosphate, they could predict the frequency of any particular triplet codon occurring in the synthetic mRNA. If they then added the mRNA to the cell-free system and ascertained the percentage of any particular amino acid present in the new protein, they could analyze the results and predict the *composition* of triplets specifying particular amino acids.

This approach is shown in Figure 12–4. Suppose that A and C are added in a ratio of 1A:5C. The insertion of a ribonucleotide at any position along the RNA molecule during its synthesis is determined by the ratio of A:C. Therefore, there is a 1/6 chance for an A and a 5/6 chance for a C to occupy each position. On this basis, we can calculate the frequency of any given triplet appearing in the message.

For AAA, the frequency is $(1/6)^3$, or about 0.4 percent. For AAC, ACA, and CAA, the frequencies are identical—that is, $(1/6)^2(5/6)$, or about 2.3 percent for each triplet. Together, all three 2A:1C triplets account for 6.9 percent of the total three-letter sequences. In the same way, each of three 1A:2C triplets accounts for $(1/6)(5/6)^2$, or 11.6 percent (or a total of 34.8%); CCC is represented by $(5/6)^3$, or 57.9 percent of the triplets.

By examining the percentages of any given amino acid incorporated into the protein synthesized under the direction of this message, we can propose probable base compositions for each amino acid (Figure 12–4). Since proline appears 69 percent of the time, we could propose that proline is encoded by CCC (57.9%) and one triplet of 2C:1A (11.6%). Histidine, at 14 percent, is probably

TABLE 12.1	Incorporation of ^{14}C-Phenylalanine into Protein
Artificial mRNA	**Radioactivity (counts/min)**
None	44
Poly U	39,800
Poly A	50
Poly C	38

Source: After Nirenberg and Matthaei (1961).

FIGURE 12–4 Results and interpretation of a mixed copolymer experiment where a ratio of 1A:5C is used (1/6A:5/6C).

Possible compositions	Possible triplets	Probability of occurrence of any triplet	Final %
3A	AAA	$(1/6)^3 = 1/216 = 0.4\%$	0.4
1C:2A	AAC ACA CAA	$(5/6)(1/6)^2 = 5/216 = 2.3\%$	$3 \times 2.3 = 6.9$
2C:1A	ACC CAC CCA	$(5/6)^2(1/6) = 25/216 = 11.6\%$	$3 \times 11.6 = 34.8$
3C	CCC	$(5/6)^3 = 125/216 = 57.9\%$	57.9
			100.0

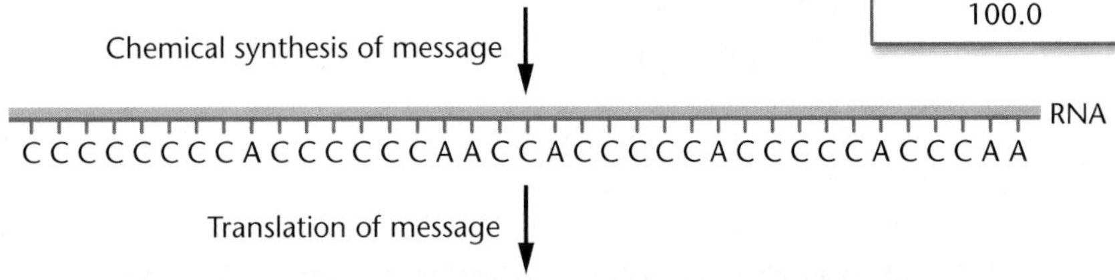

Chemical synthesis of message ↓

————————————————————————————————— RNA
CCCCCCCCACCCCCCAACCACCCCCACCCCCACCCAA

Translation of message ↓

Percentage of amino acids in protein		Probable base-composition assignments
Lysine	<1	AAA
Glutamine	2	1C:2A
Asparagine	2	1C:2A
Threonine	12	2C:1A
Histidine	14	2C:1A, 1C:2A
Proline	69	CCC, 2C:1A

coded by one 2C:1A (11.6%) and one 1C:2A (2.3%). Threonine, at 12 percent, is likely coded by only one 2C:1A. Asparagine and glutamine each appear to be coded by one of the 1C:2A triplets, and lysine appears to be coded by AAA.

Using as many as all four ribonucleotides to construct the mRNA, the researchers conducted many similar experiments. Although determining the *composition* of the triplet code words for all 20 amino acids represented a significant breakthrough, the *specific sequences* of triplets were still unknown—other approaches were needed.

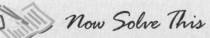

Now Solve This

Problem 24 on page 275 asks you to analyze a reciprocal pair of mixed copolymer experiments and to predict codon compositions to the amino acids they encode.

Hint: The data set generated in the reciprocal experiment is essential to solving the problem because analysis of the initial data set will provide you with more than one possible answer. However, there is only one answer consistent with both sets of data.

The Triplet Binding Assay

It was not long before more advanced techniques were developed. In 1964, Nirenberg and Philip Leder developed the **triplet binding assay**, which led to specific assignments of triplets. The technique took advantage of the observation that

ribosomes, when presented *in vitro* with an RNA sequence as short as three ribonucleotides, will bind to it and form a complex similar to that found *in vivo*. The triplet acts like a codon in mRNA, attracting the complementary sequence within tRNA (Figure 12–5). The triplet sequence in tRNA that is complementary to a codon of mRNA is an **anticodon**.

Although it was not yet feasible to chemically synthesize long stretches of RNA, triplets of known sequence could be synthesized in the laboratory to serve as templates. All that was needed was a method to determine which tRNA–amino acid was bound to the triplet RNA–ribosome complex. The test system Nirenberg and Leder devised was quite simple. The amino acid to be tested was made radioactive, and a charged tRNA was produced. Because codon compositions were known, researchers could narrow the range of amino acids that should be tested for each specific triplet.

The radioactively charged tRNA, the RNA triplet, and ribosomes were incubated together and then passed through a nitrocellulose filter, which retains the larger ribosomes but not the other smaller components, such as unbound charged tRNA. If radioactivity is not retained on the filter, an incorrect amino acid has been tested. But if radioactivity remains on the filter, it is retained because the charged tRNA has bound to the triplet associated with the ribosome. When this occurs, a specific codon assignment can be made.

Work proceeded in several laboratories, and in many cases clear-cut, unambiguous results were obtained. Table 12.2, for example, shows 26 triplets assigned to 9 amino acids. However, in some cases, the degree of triplet binding was inefficient and assignments were not possible. Eventually, about 50 of the 64

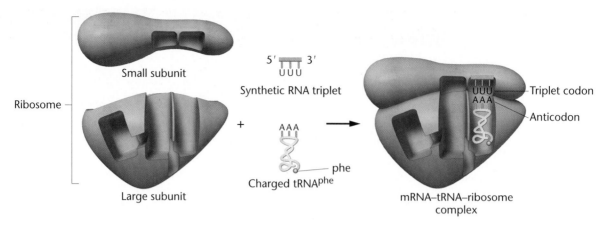

FIGURE 12–5 Illustration of the behavior of the components during the triplet binding assay. The UUU triplet RNA sequence acts as a codon, attracting the complementary AAA anticodon of the charged tRNA^phe, which together are bound by the subunits of the ribosome.

TABLE 12.2	Amino Acid Assignments to Specific Trinucleotides Derived from the Triplet Binding Assay
Trinucleotides	**Amino Acid**
AAA AAG	Lysine
AUG	Methionine
AUU AUG AUA	Isoleucine
CCG CCA CCU CCC	Proline
CUC CUA CUG CUU	Leucine
GAA GAG	Glutamic acid
UCA UCG UCU UCC	Serine
UGU UGC	Cysteine
UUA UUG	Leucine
UUU UUC	Phenylalanine

12–6, a dinucleotide made in this way is converted to an mRNA with two repeating triplets. A trinucleotide is converted to an mRNA with three potential triplets, depending on the point at which initiation occurs, and a tetranucleotide creates four repeating triplets.

When these synthetic messages were added to a cell-free system, the predicted number of amino acids incorporated was upheld. Several examples are shown in Table 12.3. When

TABLE 12.3	Amino Acids Incorporated Using Repeated Synthetic Copolymers of RNA	
Repeating Copolymer	**Codons Produced**	**Amino Acids in Polypeptides**
UG	UGU	Cysteine
	GUG	Valine
AC	ACA	Threonine
	CAC	Histidine
UUC	UUC	Phenylalanine
	UCU	Serine
	CUU	Leucine
AUC	AUC	Isoleucine
	UCA	Serine
	CAU	Histidine
UAUC	UAU	Tyrosine
	CUA	Leucine
	UCU	Serine
	AUC	Isoleucine
GAUA	GAU	None
	AGA	None
	UAG	None
	AUA	None

triplets were assigned. These specific assignments of triplets to amino acids led to two major conclusions. First, the genetic code is *degenerate*; that is, one amino acid can be specified by more than one triplet. Second, the code is *unambiguous*. That is, a single triplet specifies only one amino acid. As you shall see later in this chapter, these conclusions have been upheld with only minor exceptions. The triplet binding technique was a major innovation in deciphering the genetic code.

Repeating Copolymers

Yet another innovative technique used to decipher the genetic code was developed in the early 1960s by Gobind Khorana, who chemically synthesized long RNA molecules consisting of short sequences repeated many times. First, he created shorter sequences (e.g., di-, tri-, and tetranucleotides), which were then replicated many times and finally joined enzymatically to form the long polynucleotides. As shown in Figure

FIGURE 12–6 The conversion of di-, tri-, and tetranucleotides into repeating copolymers. The triplet codons that are produced in each case are shown.

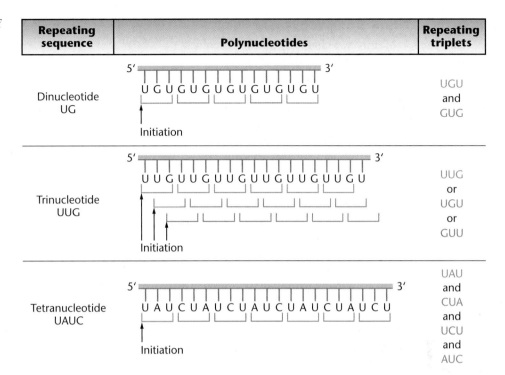

Repeating sequence	Polynucleotides	Repeating triplets
Dinucleotide UG	5′ UGUGUGUGUGUGUGU 3′ Initiation	UGU and GUG
Trinucleotide UUG	5′ UUGUUGUUGUUGUUGUUGU 3′ Initiation	UUG or UGU or GUU
Tetranucleotide UAUC	5′ UAUCUAUCUAUCUAUCUAUCU 3′ Initiation	UAU and CUA and UCU and AUC

the data were combined with those on composition assignment and triplet binding, specific assignments were possible.

One example of specific assignments made from this data demonstrates the value of Khorana's approach. Consider the following three experiments in concert with one another. The repeating trinucleotide sequence UUCUUCUUC ... produces three possible triplets: UUC, UCU, and CUU, depending on the initiation point. When placed in a cell-free translation system, the polypeptides containing phenylalanine (phe), serine (ser), and leucine (leu) are produced. On the other hand, the repeating dinucleotide sequence UCUCUCUC ... produces the triplets UCU and CUC with the incorporation of leucine and serine into the polypeptide. These results indicate that the triplets UCU and CUC specify leucine and serine, but they don't indicate which triplet specifies which amino acid. We can further conclude that *either* the CUU *or* the UUC triplet also encodes leucine or serine, while the other encodes phenylalanine.

To derive more specific information, let's examine the results of using the repeating tetranucleotide sequence UUAC, which produces the triplets UUA, UAC, ACU, and CUU. The CUU triplet is one of the two that interest us. Three amino acids are incorporated: leucine, threonine, and tyrosine. Because CUU must specify only serine or leucine and because, of these two, only leucine appears, we can conclude that CUU specifies leucine.

Once this is established, we can logically determine all other assignments. Of the two triplet pairs remaining (UUC and UCU from the first experiment and UCU and CUC from the second experiment), whichever triplet is common to both must encode serine. This is UCU. By elimination, we find that UUC encodes phenylalanine and CUC encodes leucine. While the logic must be followed carefully, four specific triplets encoding three different amino acids have been assigned from these experiments.

From these interpretations, Khorana reaffirmed triplets that were already deciphered and filled in gaps left from other approaches. For example, the use of two tetranucleotide sequences, GAUA and GUAA, suggested that at least two triplets were *termination codons*. He reached this conclusion because neither of these repeating sequences directed the incorporation of more than a few amino acids into a polypeptide. Since there are no triplets common to both messages, he predicted that each repeating sequence would contain at least one triplet that terminates protein synthesis. Table 12.3 lists the possible triplets for the poly-(GAUA) sequence, of which UAG is a termination codon.

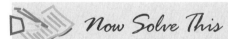

Now Solve This

Problem 4 on page 274 asks you to consider several outcomes of repeating copolymer experiments.

Hint: In a repeating copolymer of RNA, initiation of translation can occur at any of several ribonucleotides. Determining the number of triplet codons produced by each possible initiation point is the key to solving this problem.

12.4 The Coding Dictionary Reveals the Function of the 64 Triplets

The various techniques used to decipher the genetic code have yielded a dictionary of 61 triplet codon–amino acid assignments. The remaining three triplets are termination signals and do not specify any amino acid.

Degeneracy and the Wobble Hypothesis

A general pattern of triplet codon assignments becomes apparent when we look at the genetic coding dictionary. Figure 12–7 designates the assignments in a particularly illustrative form first suggested by Francis Crick.

Most evident is that the code is degenerate, as the early researchers predicted. That is, almost all amino acids are specified by two, three, or four different codons. Three amino acids (serine, arginine, and leucine) are each encoded by six different codons. Only tryptophan and methionine are encoded by single codons.

Also evident is the *pattern* of degeneracy. Most often, in a set of codons specifying the same amino acid, the first two letters are the same, with only the third differing. Crick discerned a pattern in the degeneracy at the third position, and in 1966, he postulated the **wobble hypothesis**.

Crick's hypothesis first predicted that the initial two ribonucleotides of triplet codes are more critical than the third in attracting the correct tRNA. He postulated that hydrogen bonding at the third position of the codon–anticodon interaction is less constrained and need not adhere as specifically to the established base-pairing rules. The wobble hypothesis thus proposes a more flexible set of base-pairing rules at the third position of the codon (Table 12.4).

This relaxed base-pairing requirement, or "wobble," allows the anticodon of a single form of tRNA to pair with more than one triplet in mRNA. Consistent with the wobble hypothesis and degeneracy, U at the first position (the 5′ end) of the

TABLE 12.4	Codon–Anticodon Base-Pairing Rules

Base at first position (5′ end) of tRNA	Base at third position (3′ end) of mRNA
A	U
C	G
G	C or U
U	A or G
I	A, U, or C

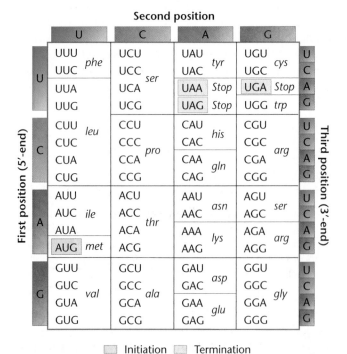

tRNA anticodon may pair with A or G at the third position (the 3′ end) of the mRNA codon, and G may likewise pair with U or C. Inosine (I), one of the modified bases found in tRNA, may pair with C, U, or A. Applying these wobble rules, a minimum of about 30 different tRNA species is necessary to accommodate the 61 triplets specifying an amino acid. If nothing more, wobble can be considered a potential economy measure, provided that the fidelity of translation is not compromised. Current estimates are that 30 to 40 tRNA species are present in bacteria and up to 50 tRNA species exist in animal and plant cells.

The Ordered Nature of the Code

Still another observation has become apparent in the pattern of codon sequences and their corresponding amino acids, leading to the description referred to as an **ordered genetic code**. Chemically similar amino acids often share one or two "middle" bases in the different triplets encoding them. For example, either U or C is often present in the second position of triplets that specify hydrophobic amino acids, including valine and alanine, among others. Two codons (AAA and AAG) specify the positively charged amino acid lysine. If only the middle letter of these codons is changed from A to G (AGA and AGG), the positively charged amino acid arginine is specified. Hydrophilic amino acids, such as serine and threonine, are specified by triplet codons, with G or C in the second position.

The chemical properties of amino acids will be discussed in more detail in Chapter 13. The end result of an "ordered" code is that it buffers the potential effect of mutation on protein function. While many mutations of the second base of triplet codons result in a change of one amino acid to another, the change is often to an amino acid with similar chemical properties. In such cases, protein function may not be noticeably altered.

FIGURE 12–7 The coding dictionary. AUG encodes methionine, which initiates most polypeptide chains. All other amino acids except tryptophan, which is encoded only by UGG, are represented by two to six triplets. The triplets UAA, UAG, and UGA are termination signals and do not encode any amino acids.

Initiation and Termination

Initiation of protein synthesis is a highly specific process. In bacteria (in contrast to the *in vitro* experiments discussed earlier), the initial amino acid inserted into all polypeptide chains is a modified form of methionine—*N*-**formylmethionine** (**fmet**). Only one codon, AUG, codes for methionine, and it is sometimes called the **initiator codon**. However, when AUG appears internally in mRNA rather than at an initiating position, unformylated methionine is inserted into the polypeptide chain. Rarely, another triplet, GUG, specifies methionine during initiation, although it is not clear why this happens, since GUG normally encodes valine.

In bacteria, either the formyl group is removed from the initial methionine upon the completion of protein synthesis or the entire formylmethionine residue is removed. In eukaryotes, methionine is also the initial amino acid during polypeptide synthesis. However, it is not formylated.

As mentioned in the preceding section, three other triplets (UAG, UAA, and UGA) serve as **termination codons**, punctuation signals that do not code for any amino acid. They are not recognized by a tRNA molecule, and translation terminates when they are encountered. Mutations that produce any of the three triplets internally in a gene will also result in termination. Consequently, only a partial polypeptide has been synthesized when it is prematurely released from the ribosome. When such a change occurs in the DNA, it is called a **nonsense mutation**.

12.5 The Genetic Code Has Been Confirmed in Studies of Bacteriophage MS2

The various aspects of the genetic code discussed thus far yield a fairly complete picture. The code is triplet in nature, degenerate, unambiguous, and commaless, but it contains punctuation with respect to start and stop signals. These individual principles have been confirmed by a detailed analysis of the RNA-containing **bacteriophage MS2** by Walter Fiers and his coworkers.

MS2 is a bacteriophage that infects *E. coli*. Its nucleic acid (RNA) contains only about 3500 ribonucleotides, making up only three genes. These genes specify a coat protein, an RNA-directed replicase, and a maturation protein (the A protein). This simple system of a small genome and a few gene products enabled Fiers and his colleagues to sequence the genes and their products. The amino acid sequence of the coat protein was completed in 1970, and the nucleotide sequence of the gene and several nucleotides on each of its ends was reported in 1972.

When the chemical constitution of genes and their encoded proteins are compared, they are found to exhibit **colinearity**. That is, based on the coding dictionary, the linear sequence of nucleotides, and thus the sequence of triplet codons, corresponds precisely with the linear sequence of amino acids in the protein. Furthermore, the codon for the first amino acid is AUG, the common initiator codon. The codon for the last amino acid is followed by two consecutive termination codons, UAA and UAG.

By 1976, the other two genes and their protein products were sequenced. The analysis clearly showed that the genetic code in this virus was identical to that established in bacterial systems. Other evidence suggested that the code was also identical in eukaryotes, thus providing confirmation of what seemed to be a universal genetic code.

12.6 The Genetic Code Is Nearly Universal

Between 1960 and 1978, it was generally assumed that the genetic code would be found to be universal, applying equally to viruses, bacteria, archaea, and eukaryotes. Certainly, the nature of mRNA and the translation machinery seemed to be very similar in these organisms. For example, cell-free systems derived from bacteria can translate eukaryotic mRNAs. Poly U stimulates synthesis of polyphenylalanine in cell-free systems when the components are derived from eukaryotes. Many recent studies involving recombinant DNA technology (see Chapter 17) reveal that eukaryotic genes can be inserted into bacterial cells, which are then transcribed and translated. Within eukaryotes, mRNAs from mice and rabbits have been injected into amphibian eggs and efficiently translated. For the many eukaryotic genes that have been sequenced, notably those for hemoglobin molecules, the amino acid sequence of the encoded proteins adheres to the coding dictionary established from bacterial studies.

However, several 1979 reports on the coding properties of DNA derived from mitochondria (**mtDNA**) of yeast and humans undermined the principle of the universality of the genetic language. Since then, mtDNA has been examined in many other organisms.

Cloned mtDNA fragments have been sequenced and compared with the amino acid sequences of various mitochondrial proteins, revealing several exceptions to the coding dictionary (Table 12.5). Most surprising is that the codon

TABLE 12.5		Exceptions to the Universal Code	
Triplet	Normal Code Word	Altered Code Word	Source
UGA	Termination	Tryptophan	Human and yeast mitochondria; *Mycoplasma*
CUA	Leucine	Threonine	Yeast mitochondria
AUA	Isoleucine	Methionine	Human mitochondria
AGA	Arginine	Termination	Human mitochondria
AGG	Arginine	Termination	Human mitochondria
UAA	Termination	Glutamine	*Paramecium; Tetrahymena; Stylonychia*
UAG	Termination	Glutamine	*Paramecium*

UGA, normally specifying termination, specifies the insertion of tryptophan during translation in yeast and human mitochondria. In yeast mitochondria, threonine is inserted instead of leucine when CUA is encountered in mRNA. In human mitochondria, AUA, which normally specifies isoleucine, directs the internal insertion of methionine.

In 1985, several other exceptions to the standard coding dictionary were discovered in the bacterium *Mycoplasma capricolum* and in the nuclear genes of the protozoan ciliates *Paramecium*, *Tetrahymena*, and *Stylonychia*. For example, as shown in Table 12.5, one alteration converts the termination codons (UGA) to tryptophan, yet several others convert the normal termination codon (UAA and UAG) to glutamine. These changes are significant because both a prokaryote and several eukaryotes are involved, representing distinct species that have evolved separately over a long period of time.

Note the apparent pattern in several of the altered codon assignments. The change in coding capacity involves only a shift in recognition of the third, or wobble, position. For example, AUA specifies isoleucine in the cytoplasm and methionine in the mitochondrion, but in the cytoplasm, methionine is specified by AUG. Similarly, UGA calls for termination in the cytoplasm, but it specifies tryptophan in the mitochondrion; in the cytoplasm, tryptophan is specified by UGG. It has been suggested that such changes in codon recognition may represent an evolutionary trend toward reducing the number of tRNAs needed in mitochondria; only 22 tRNA species are encoded in human mitochondria, for example. However, until more examples are found, the differences must be considered to be exceptions to the previously established general coding rules.

12.7 Different Initiation Points Create Overlapping Genes

Earlier we stated that the genetic code is nonoverlapping—each ribonucleotide in an mRNA is part of only one codon. However, this characteristic of the code does not rule out the possibility that a single mRNA may have multiple initiation points for translation. If so, these points could theoretically create several different reading frames within the same mRNA, thus specifying more than one polypeptide and leading to the concept of **overlapping genes**.

That this might actually occur in some viruses was suspected when phage φX174 was carefully investigated. The circular DNA chromosome consists of 5386 nucleotides, which should encode a maximum of 1795 amino acids, sufficient for five or six proteins. However, this small virus in fact synthesizes 11 proteins consisting of more than 2300 amino acids. A comparison of the nucleotide sequence of the DNA and the amino acid sequences of the polypeptides synthesized has clarified the apparent paradox. At least four cases of multiple initiation have been discovered, creating overlapping genes.

For example, there are polypeptides designated as K, B, C, D, and E. The sequences specifying the K and B polypeptides are initiated with separate reading frames within the sequence specifying the A polypeptide. The *K* gene sequence overlaps

into the adjacent gene sequence specifying the C polypeptide. The E sequence is out of frame with, but initiated in, that of the D polypeptide. As described, six different polypeptides may be created from a DNA sequence that might otherwise have specified only three polypeptides.

A similar situation has been observed in other viruses, including phage G4 and the animal virus SV40. Like φX174, phage G4 contains a circular single-stranded DNA molecule. The use of overlapping reading frames optimizes the use of a limited amount of DNA present in these small viruses. However, such an approach to storing information has a distinct disadvantage in that a single mutation may affect more than one protein and thus increase the chances that the change will be deleterious or lethal. In the case we just discussed, a single mutation near the junction of genes *A* and *C* could affect three proteins (the A, C, and K proteins). It may be for this reason that overlapping genes are not common in other organisms.

12.8 Transcription Synthesizes RNA on a DNA Template

Even while the genetic code was being studied, it was quite clear that proteins were the end products of many genes. Thus, while some geneticists attempted to elucidate the code, other research efforts focused on the nature of genetic expression. The central question was how DNA, a nucleic acid, could specify a protein composed of amino acids.

The complex multistep process begins with the transfer of genetic information stored in DNA to RNA. The process by which RNA molecules are synthesized on a DNA template is called **transcription**. It results in an mRNA molecule complementary to the gene sequence of one of the double helix's two strands. Each triplet codon in the mRNA is, in turn, complementary to the anticodon region of its corresponding tRNA as the amino acid is correctly inserted into the polypeptide chain during translation. The significance of transcription is enormous, for it is the initial step in the process of **information flow** within the cell. The idea that RNA is involved as an intermediate molecule in the process of information flow between DNA and protein was suggested by the following observations:

1. DNA is, for the most part, associated with chromosomes in the nucleus of the eukaryotic cell. However, protein synthesis occurs in association with ribosomes located outside the nucleus in the cytoplasm. Therefore, DNA does not appear to participate directly in protein synthesis.

2. RNA is synthesized in the nucleus of eukaryotic cells, where DNA is found, and is chemically similar to DNA.

3. Following its synthesis, most RNA migrates to the cytoplasm, where protein synthesis (translation) occurs.

4. The amount of RNA is generally proportional to the amount of protein in a cell.

Collectively, these observations suggested that genetic information, stored in DNA, is transferred to an RNA intermediate, which directs the synthesis of proteins. As with most new ideas in molecular genetics, the initial supporting experimental evidence was based on studies of bacteria and their phages. It was clearly established that during initial infection, RNA synthesis preceded phage protein synthesis and that the RNA is complementary to phage DNA.

The results of these experiments agree with the concept of a messenger RNA (mRNA) being made on a DNA template and then directing the synthesis of specific proteins in association with ribosomes. This concept was formally proposed by François Jacob and Jacques Monod in 1961 as part of a model for gene regulation in bacteria. Since then, mRNA has been isolated and studied thoroughly. There is no longer any question about its role in genetic processes.

12.9 RNA Polymerase Directs RNA Synthesis

To prove that RNA can be synthesized on a DNA template, it was necessary to demonstrate that there is an enzyme capable of directing this synthesis. By 1959, several investigators, including Samuel Weiss, had independently isolated such a molecule from rat liver. Called **RNA polymerase**, it has the same general substrate requirements as does DNA polymerase, the major exception being that the substrate nucleotides contain the ribose rather than the deoxyribose form of the sugar. Unlike DNA polymerase, no primer is required to initiate synthesis; the initial base remains as a nucleoside triphosphate (NTP). The overall reaction summarizing the synthesis of RNA on a DNA template can be expressed as

$$n(\text{NTP}) \xrightarrow[\text{enzyme}]{\text{DNA}} (\text{NMP})_n + n(\text{PP}_i)$$

As this equation reveals, nucleoside triphosphates (NTPs) are substrates for the enzyme, which catalyzes the polymerization of nucleoside monophosphates (NMPs), or nucleotides, into a polynucleotide chain $(\text{NMP})_n$. Nucleotides are linked during synthesis by $5'$-to-$3'$ phosphodiester bonds (see Figure 9–10). The energy created by cleaving the triphosphate precursor into the monophosphate form drives the reaction, and inorganic pyrophosphates (PP$_i$) are produced.

A second equation summarizes the sequential addition of each ribonucleotide as the process of transcription progresses:

$$(\text{NMP})_n + \text{NTP} \xrightarrow[\text{enzyme}]{\text{DNA}} (\text{NMP})_{n+1} + \text{PP}_i$$

Here each transcription step involves the addition of one ribonucleotide (NMP) to the growing polyribonucleotide chain $(\text{NMP})_n$, using a nucleoside triphosphate (NTP) as the precursor.

RNA polymerase from *E. coli* has been extensively characterized and shown to consist of subunits designated α, β, β', and σ. The active form of the enzyme, the **holoenzyme**, contains the subunits α_2, β, β', σ and has a molecular weight of almost 500,000 Da. Of these subunits, it is the β and β' polypeptides that provide the catalytic basis and active site for transcription. As we shall see, another subunit called the σ (**sigma**) **factor** plays a regulatory function in the initiation of RNA transcription.

While there is but a single form of the enzyme in *E. coli*, there are several different σ factors, creating variations of the polymerase holoenzyme. On the other hand, eukaryotes display three distinct forms of RNA polymerase, each consisting of a greater number of polypeptide subunits than in bacteria.

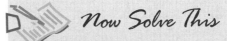

Now Solve This

Problem 23 on page 275 asks you to consider the outcome of the transfer of complementary information from DNA to RNA and the amino acids encoded by this information.

Hint: In RNA, uracil is complementary to adenine, and while DNA stores genetic information in the cell, the code that is translated is contained in the RNA complementary to the template strand of DNA making up a gene.

Promoters, Template Binding, and the σ Subunit

Transcription results in the synthesis of a single-stranded RNA molecule complementary to a region along only one strand of the DNA double helix. For simplicity, let's call the transcribed DNA strand the **template strand** and its complement the **partner strand**.

The initial step is **template binding** (Figure 12–8). In bacteria, the site of this initial binding is established when the RNA polymerase σ subunit recognizes specific DNA sequences called **promoters**. These regions are located in the $5'$ region upstream from the point of initial transcription of a gene. It is believed that the enzyme "explores" a length of DNA until it recognizes the promoter region and binds to about 60 nucleotide pairs of the helix, 40 of which are upstream from the point of initial transcription. Once this occurs, the helix is denatured or unwound locally, making the DNA template accessible to the action of the enzyme. The point at which transcription actually begins is called the **transcription start site**.

The importance of promoter sequences cannot be overemphasized. They govern the efficiency of the initiation of transcription. In bacteria, both strong promoters and weak promoters have been discovered. These promoters cause a variation in time of initiation from once every 1 to 2 seconds to only once every 10 to 20 minutes. Mutations in promoter sequences may severely reduce the initiation of gene expression. Because the interaction of promoters with RNA polymerase governs transcription, the nature of the binding between them is at the heart of discussions concerning genetic regulation, the subject of Chapter 15. While we will pursue more detailed information involving promoter–enzyme interactions, we must address three points here.

The first point is the concept of **consensus sequences** of DNA. These sequences are similar (homologous) in different genes of the same organism or in one or more genes of related organisms. Their conservation throughout evolution attests to the critical

(a) Transcription components

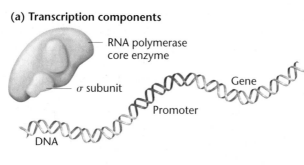

(b) Template binding and initiation of transcription

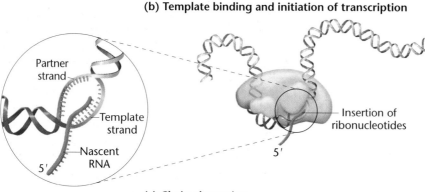

(c) Chain elongation

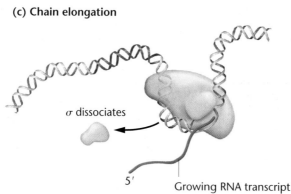

FIGURE 12–8 The early stages of transcription in prokaryotes, showing (a) the components of the process; (b) template binding at the −10 site involving the σ subunit of RNA polymerase and subsequent initiation of RNA synthesis; and (c) chain elongation, after the σ subunit has dissociated from the transcription complex and the enzyme moves along the DNA template.

nature of their role in biological processes. Two such sequences have been found in bacterial promoters. One, TATAAT, is located 10 nucleotides upstream from the site of initial transcription (the **−10 region,** or **Pribnow box**). The other, TTGACA, is located 35 nucleotides upstream (the **−35 region**). Mutations in either region diminish transcription, often severely.

Sequences such as these are said to be *cis*-**acting elements**. The use of the term *cis* is drawn from organic chemistry nomenclature, meaning "next to" or on the same side as, in contrast to being "across from," or *trans*, to other functional groups. In molecular genetics, then, *cis*-elements are adjacent parts of the same DNA molecule. This is in contrast to ***trans*-acting factors**, molecules that bind to these DNA elements. As we will soon see, in most eukaryotic genes studied, a consensus sequence comparable to that in the −10 region has been recognized. Because it is rich in adenine and thymine residues, it is called the **TATA box**.

The second point is that the degree of RNA polymerase binding to different promoters varies greatly, which causes the variable gene expression mentioned earlier. Currently, this is attributed to sequence variation in the promoters.

A final general point involves the σ subunit in bacteria. The major form is designated as σ^{70}, based on its molecular weight of 70 kilodaltons (kDa). The promoters of most bacterial genes recognize this form; however, several alternative forms of RNA polymerase in *E. coli* have unique σ subunits associated with them (e.g., σ^{32}, σ^{54}, σ^{S}, and σ^{E}). Each form recognizes different promoter sequences, which in turn provides specificity to the initiation of transcription.

Initiation, Elongation, and Termination of RNA Synthesis

Once it has recognized and bound to the promoter [Figure 12–8(b)], RNA polymerase catalyzes **initiation**, the insertion of the first 5′-ribonucleoside triphosphate, which is complementary to the first nucleotide at the start site of the DNA template strand. As we noted earlier, no primer is required. Subsequent ribonucleotide complements are inserted and linked by phosphodiester bonds as RNA polymerization proceeds. This process of **chain elongation** [Figure 12–8(c)], continues in a 5′ to 3′ extension, creating a temporary DNA/RNA duplex whose chains run antiparallel to one another.

After a few ribonucleotides have been added to the growing RNA chain, the σ subunit dissociates from the holoenzyme and elongation proceeds under the direction of the core enzyme. In *E. coli,* this process proceeds at the rate of about 50 nucleotides/second at 37°C.

Eventually, the enzyme traverses the entire gene until it encounters a specific nucleotide sequence that acts as a termination signal. The termination sequences, about 40 base pairs in length, are extremely important in prokaryotes because of the close proximity of the end of one gene and the upstream sequences of the adjacent gene. In some cases, the termination of synthesis is dependent on the **termination factor,** ρ **(rho)**—a large hexameric protein that physically interacts with the growing RNA transcript. At the point of termination, the transcribed RNA molecule is released from the DNA template and the core polymerase enzyme dissociates. The synthesized RNA molecule is precisely complementary to a DNA sequence representing the template strand of a gene. Wherever an A, T, C, or G residue existed, a corresponding U, A, G, or C residue, respectively, is incorporated into the RNA molecule. These RNA molecules ultimately provide the information leading to the synthesis of all proteins present in the cell.

In bacteria, groups of genes whose products are related are often clustered along the chromosome. In many such cases, they are contiguous, and all but the last gene lack the encoded signals for termination. The result is that during transcription, a large mRNA is produced that encodes more than one protein. Since genes in bacteria are sometimes called cistrons, the RNA is called a **polycistronic mRNA**. The products of genes transcribed in this fashion are usually all needed by the cell at the same time, so this is an efficient way to transcribe and subsequently translate the needed genetic information. In eukaryotes, **monocistronic mRNAs** are the rule.

12.10 Transcription in Eukaryotes Differs from Prokaryotic Transcription in Several Ways

Much of our knowledge of transcription has been derived from studies of prokaryotes. Most of the general aspects of the mechanics of these processes are similar in eukaryotes, but there are several notable differences:

1. Transcription in eukaryotes occurs within the nucleus under the direction of three separate forms of RNA polymerase. Unlike prokaryotes, in eukaryotes the RNA transcript is not free to associate with ribosomes prior to the completion of transcription. For the mRNA to be translated, it must move out of the nucleus into the cytoplasm.

2. Initiation of transcription of eukaryotic genes requires that the compact chromatin fiber, characterized by nucleosome coiling (Chapter 11), must be uncoiled and the DNA made accessible to RNA polymerase and other regulatory proteins. This transition is referred to as **chromatin remodelling**, reflecting the dynamics involved in the conformational change that occurs as the DNA helix is opened.

3. Initiation and regulation of transcription involve a more extensive interaction between *cis*-acting upstream DNA sequences and *trans*-acting protein factors involved in stimulating and initiating transcription. In addition to promoters, other control units, called **enhancers**, may be located in the 5′ regulatory region upstream from the initiation point, but they have also been found within the gene or even in the 3′ downstream region beyond the coding sequence.

4. Maturation of eukaryotic mRNA from the primary RNA transcript involves many complex stages referred to generally as "processing." An initial processing step involves the addition of a 5′ cap and a 3′ tail to most transcripts destined to become mRNAs. Other extensive modifications occur to the internal nucleotide sequence of eukaryotic RNA transcripts that eventually serve as mRNAs. The initial (or primary) transcripts are most often much larger than those that are eventually translated. Sometimes called **pre-mRNAs**, they are part of a group of molecules found only in the nucleus—a group referred to collectively as **heterogeneous nuclear RNA (hnRNA)**. Such RNA molecules are of variable but large size and are complexed with proteins, forming **heterogeneous nuclear ribonucleoprotein particles (hnRNPs)**. Only about 25 percent of hnRNA molecules are converted to mRNA. In those that are converted, substantial amounts of the ribonucleotide sequence are excised, and the remaining segments are spliced back together prior to nuclear export and translation. This phenomenon has given rise to the concepts of **split genes** and **splicing** in eukaryotes.

In the remainder of this chapter we will look at the basic details of transcription in eukaryotic cells. The process of transcription is highly regulated, determining which DNA sequences are copied into RNA and when and how frequently they are transcribed. We will return to topics directly related to regulation of eukaryotic transcription in Chapter 15.

Initiation of Transcription in Eukaryotes

The recognition of certain highly specific DNA regions by RNA polymerase is the basis of orderly genetic function in all cells. Eukaryotic RNA polymerase exists in three unique forms, each of which transcribes different types of genes, as indicated in Table 12.6. Each enzyme is larger and more complex than the single prokaryotic polymerase, consisting of 2 large subunits and 10 to 15 smaller subunits. In regard to the initial template binding step and promoter regions, most is known about polymerase II, which transcribes all mRNAs in eukaryotes.

TABLE 12.6	RNA Polymerases in Eukaryotes	
Form	**Product**	**Location**
I	rRNA	Nucleolus
II	mRNA, snRNA	Nucleoplasm
III	5S rRNA, tRNA	Nucleoplasm

Several *cis*-acting elements common to eukaryotic genes aid the efficient initiation of transcription by polymerase II. One such element found within the promoter region is called the **Goldberg-Hogness (TATA) box**, located about 35 nucleotide pairs upstream (−35) from the start point of transcription. The consensus sequence is a heptanucleotide consisting solely of A and T residues (TATAAAA). The sequence and function are analogous to that found in the −10 promoter region of prokaryotic genes. Because this region is common to most eukaryotic genes, the TATA box is thought to be nonspecific and is responsible only for fixing the site of transcription initiation by facilitating denaturation of the helix. Such a conclusion is supported by the fact that A═T base pairs are less stable than G≡C pairs.

A second *cis*-acting element is the **CAAT box**, located upstream within the promoter of many genes at about 80 nucleotides from the start of transcription (−80). It contains the consensus sequence GGCCAATCT. Still other upstream regulatory regions have been found, and most genes contain one or more of them. They influence the efficiency of the promoter, along with the TATA and CAAT box. The locations of these elements are based on studies of deletions of particular regions of promoters that reduce the efficiency of transcription.

DNA regions called **enhancers** represent yet another *cis*-acting element. Although their locations can vary, enhancers are often found farther upstream than the regions we have already mentioned or even downstream or within the gene. Thus, they can modulate transcription from a distance. Although they may not participate directly in template binding, they are essential to highly efficient initiation of transcription.

Complementing the *cis*-acting regulatory sequences are various ***trans*-acting factors** that facilitate template binding and, therefore, the initiation of transcription. These proteins are referred to as **transcription factors**. They are essential because RNA polymerase II cannot bind directly to eukaryotic promoter sites and initiate transcription without them. The transcription factors involved with human RNA polymerase II binding have been well characterized and are designated TFIIA, TFIIB, and so on. One of these, TFIID, binds directly to the TATA-box sequence and is sometimes called the **TATA-binding protein (TBP)**. TFIID consists of about 10 polypeptide subunits. Once initial binding to DNA occurs, at least seven other transcription factors bind sequentially to TFIID, forming an extensive pre-initiation complex, which is then bound by RNA polymerase II.

Transcription factors with similar activity have been discovered in a variety of eukaryotes, including *Drosophila* and yeast. These factors appear to supplant the role of the σ factor in the prokaryotic enzyme and play an important part in eukaryotic gene regulation.

Heterogeneous Nuclear RNA and Its Processing: Caps and Tails

The genetic code is written in the ribonucleotide sequence of mRNA, which originated in the template strand of DNA, where complementary sequences of deoxyribonucleotides exist. In bacteria, the relationship between DNA and RNA appears to be quite direct. The DNA base sequence is transcribed into an mRNA sequence, which is then translated into an amino acid sequence according to the genetic code. By contrast, complex processing of mRNA in eukaryotes occurs before it is transported to the cytoplasm to participate in translation.

By 1970, accumulating evidence showed that eukaryotic mRNA is transcribed initially as a precursor molecule much larger than that which is translated. This notion was based on the observation by James Darnell and his coworkers of heterogeneous nuclear RNA (hnRNA) in mammalian nuclei that contained nucleotide sequences common to the smaller mRNA molecules present in the cytoplasm. They proposed that the initial transcript of a gene results in a large RNA molecule that must first be processed in the nucleus before it appears in the cytoplasm as a mature mRNA molecule. The various processing steps are summarized in Figure 12–9.

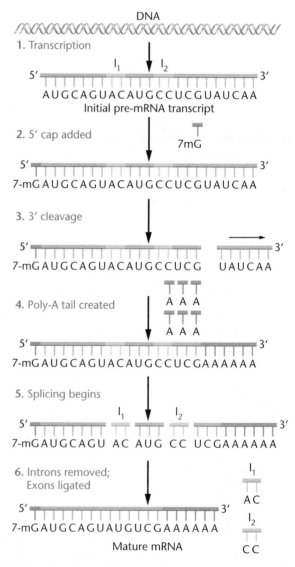

FIGURE 12–9 Posttranscriptional RNA processing in eukaryotes. Heterogeneous nuclear RNA (pre-mRNA) is converted to mRNA, which contains a 5′ cap and a 3′ poly-A tail. The introns are then spliced out.

The initial **posttranscriptional modification** of eukaryotic RNA transcripts destined to become mRNAs involves the 5'-end of these molecules, where a **7-methylguanosine (7mG) cap** is added (Figure 12–9, step 2). The cap, which is added even before the initial transcript is complete, appears to be important to subsequent processing within the nucleus, perhaps by protecting the 5'-end of the molecule from nuclease attack. Subsequently, this cap may be involved in the transport of mature mRNAs across the nuclear membrane into the cytoplasm. The cap is fairly complex and distinguished by a unique 5' to 5' bonding between the cap and the initial ribonucleotide of the RNA. Some eukaryotes also contain a methyl group (CH_3) added to the 2' carbon of the ribose sugars of the first two ribonucleotides of the RNA.

Further insights into the processing of RNA transcripts during the maturation of mRNA came from the discovery that both hnRNAs and mRNAs have a stretch of as many as 250 adenylic acid residues at their 3' end. Such **poly-A sequences** are added after the 7mG cap has been added. First, the 3' end of the initial transcript is cleaved enzymatically at a point 10 to 35 ribonucleotides from a highly conserved AAUAAA sequence (step 3). Then polyadenylation occurs by the sequential addition of a poly-A sequence (step 4). Poly A has been found at the 3' end of almost all mRNAs studied in a variety of eukaryotic organisms. The exceptions seem to be the products of histone genes.

While the AAUAAA sequence is not found on all eukaryotic transcripts, mutations in that sequence of those transcripts that do have it cannot add the poly-A tail. In the absence of this tail, the RNA transcripts are rapidly degraded. Therefore, both the 5' cap and the 3' poly-A tail are critical if an mRNA **RNA** transcript is to be further processed and transported to the cytoplasm.

12.11 The Coding Regions of Eukaryotic Genes Are Interrupted by Intervening Sequences Called Introns

One of the most exciting breakthroughs in the history of molecular genetics occurred in 1977 when Susan Berget, Philip Sharp, and Richard Roberts presented direct evidence that the genes of animal viruses contain *internal* nucleotide sequences that are not expressed in the amino acid sequence of the proteins they encode. These internal DNA sequences are present in initial RNA transcripts, but they are removed before the mature mRNA is translated (Figure 12–9, steps 5 and 6). These nucleotide segments are called **intervening sequences** (I_1 and I_2 in Figure 12–9), and the genes that contain them are **split genes**. DNA sequences that are not represented in the final mRNA product are also called **introns** (*int* for intervening), and those retained and expressed are called **exons** (*ex* for expressed). Splicing involves the removal of the ribonucleotide sequences present in introns as a result of an excision process and the rejoining of exons.

Similar discoveries were soon made in a variety of eukaryotic genes. Two approaches have been most fruitful. The first involves the molecular hybridization of purified, functionally mature mRNAs with DNA containing the genes specifying that mRNA. Hybridization between nucleic acids that are not perfectly complementary results in **heteroduplexes**, in which introns present in the DNA but absent in the mRNA loop out and remain unpaired. Such structures can be visualized with the electron microscope, as shown in Figure 12–10. The chicken ovalbumin shown in the figure is a heteroduplex with seven loops (A–G), representing seven introns whose sequences are present in DNA but not in the final mRNA.

The second approach provides more specific information. It involves a direct comparison of nucleotide sequences of DNA with those of mRNA and the correlation with amino acid sequences. Such an approach allows the precise identification of all intervening sequences.

Thus far, most eukaryotic genes have been shown to contain introns. One of the first to be identified was the **β-globin gene** in mice and rabbits, studied independently by Philip Leder and Richard Flavell. Both the mouse and rabbit genes contain two introns of approximately the same size and same location within the gene. The rabbit gene is diagrammed in

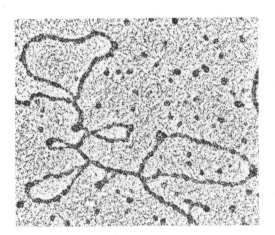

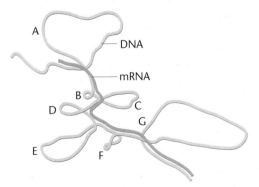

FIGURE 12–10 An electron micrograph and interpretive drawing of the hybrid molecule (heteroduplex) formed between the template DNA strand of the chicken ovalbumin gene and the mature ovalbumin mRNA. Seven DNA introns, labeled A–G, produce unpaired loops.

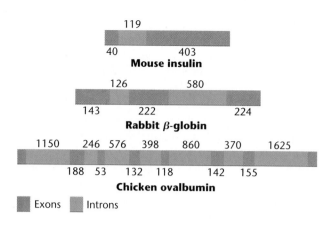

FIGURE 12–11 Intervening sequences in various eukaryotic genes. The numbers indicate the number of nucleotides present in various intron and exon regions.

Figure 12–11, revealing the intron locations. Similar introns have been found in the β-globin gene in all mammals examined.

The **ovalbumin gene** of chickens has been extensively characterized by Bert O'Malley in the United States and Pierre Chambon in France. As shown in Figure 12–11, the gene contains seven introns. In fact, the majority of the gene's DNA sequence is composed of introns and is thus "silent." The initial RNA transcript is nearly three times the length of the mature mRNA. Compare the ovalbumin gene in Figures 12–10 and 12–11. Can you match the unpaired loops in Figure 12–10 with the sequence of introns specified in Figure 12–11?

The list of genes containing intervening sequences is long. In fact, few eukaryotic genes seem to lack introns. An extreme example of the number of introns in a single gene is found in the gene coding for one subunit of collagen, the major connective tissue protein in vertebrates. The *pro-α-2(1) collagen* gene contains 50 introns. The precision of cutting and splicing that occurs must be extraordinary if errors are not to be introduced into the mature mRNA. What is equally noteworthy is a comparison of the size of genes with the size of the final mRNA once the introns are removed. As shown in Table 12.7, about 15 percent of the collagen gene consists of exons that finally appear in mRNA. For other proteins, an even more extreme picture emerges. Only about 8 percent of the albumin gene remains to be translated, and in the largest human gene known, dystrophin (the missing protein product in Duchenne muscular dystrophy), less than 1 percent of the gene sequence is retained in the mRNA. Two other human genes are also included in Table 12.7.

While the vast majority of eukaryotic genes examined thus far contain introns, there are several exceptions. Notably, the genes coding for histones and interferon do not appear to contain introns. It is not clear why or how the genes encoding these molecules have been maintained throughout evolution without acquiring the extraneous information characteristic of almost all other genes.

Splicing Mechanisms: Autocatalytic RNAs

The discovery of split genes led to intensive attempts to elucidate the mechanism by which introns of RNA are excised and exons are spliced back together, and a great deal of progress has been made. Interestingly, it appears that somewhat different mechanisms exist for different types of RNA, as well as for RNAs produced in mitochondria and chloroplasts.

Introns can be categorized into several groups based on their splicing mechanisms. Group I, represented by introns that are part of the primary transcript of rRNAs, requires no additional components for intron excision; the intron itself is the source of the enzymatic activity necessary for its own removal. This amazing discovery, which contradicted expectations, was made in 1982 by Thomas Cech and his colleagues during a study of the ciliate protozoan *Tetrahymena*. Reflecting their autocatalytic properties, RNAs that are capable of splicing themselves are sometimes called **ribozymes**.

The **self-excision process** is shown in Figure 12–12. Chemically, two nucleophilic reactions (called transesterification reactions) take place. The first is an interaction between guanosine, which acts as a cofactor in the reaction, and the primary transcript [Figure 12–12(a)]. The 3'-OH group of guanosine is transferred to the nucleotide adjacent to the 5'-end of the intron [Figure 12–12(b)]. The second reaction involves the interaction of the newly acquired 3'-OH group on the left-hand exon and the phosphate on the 3' end of the right intron [Figure 12–12(c)]. The intron is spliced out, and the two exon regions are ligated, leading to the mature RNA [Figure 12–12(d)].

Self-excision of group I introns, as described, is now known to apply to pre-rRNAs from other protozoans. Self-excision also seems to govern the removal of introns present in the primary mRNA and tRNA transcripts produced in the mitochondria and chloroplasts. These are referred to as group II introns. Like group I molecules, splicing involves two autocatalytic reactions leading to the excision of introns. However, guanosine is not involved as a cofactor in group II introns.

Splicing Mechanisms: The Spliceosome

Introns are a major component of nuclear-derived pre-mRNA transcripts. Compared with the other RNAs we have discussed, introns in nuclear-derived mRNA can be much larger—up to 20,000 nucleotides—and they are more plentiful. Their removal appears to require a much more complex mechanism, which has been more difficult to define.

TABLE 12.7	Comparing Human Gene Size, mRNA Size, and the Number of Introns		
Gene	**Gene Size (kb)**	**mRNA Size (kb)**	**Number of Introns**
Insulin	1.7	0.4	2
Collagen [*pro-α-2(1)*]	38.0	5.0	50
Albumin	25.0	2.1	14
Phenylalanine hydroxylase	90.0	2.4	12
Dystrophin	2000.0	17.0	50

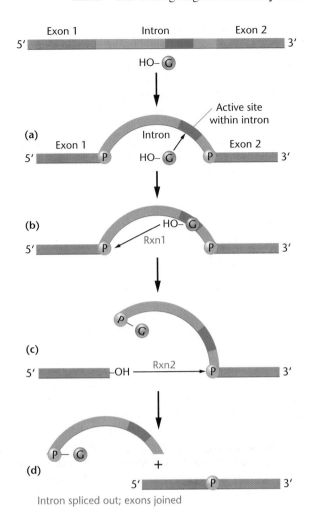

FIGURE 12–12 Splicing mechanism of pre-rRNA involving group I introns that are removed from the initial transcript. The process is one of self-excision involving two transesterification reactions.

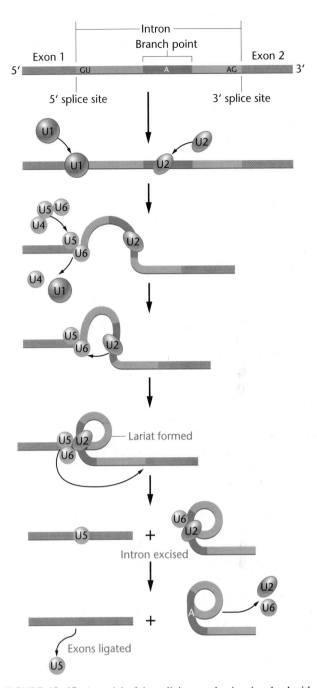

FIGURE 12–13 A model of the splicing mechanism involved with the removal of an intron from a pre-mRNA. Excision is dependent on various snRNAs (U1, U2, ... , U6) that combine with proteins to form snurps, which are part of the spliceosome. The lariat structure in the intermediate stage is characteristic of this mechanism.

Nevertheless, many clues are now emerging, and the model in Figure 12–13 illustrates the removal of one intron. The nucleotide sequences near the perimeters of this type of intron are often similar. Many begin at the 5′ end with a GU dinucleotide sequence and terminate at the 3′ end with an AG dinucleotide sequence. These, as well as other consensus sequences shared by introns, attract specific molecules that form a molecular complex, a **spliceosome**, essential to splicing. Spliceosomes have been identified in extracts of yeast as well as mammalian cells and are very large, being 40*S* in yeast and 60*S* in mammals. Perhaps the most essential component of spliceosomes is the unique set of **small nuclear RNAs** (**snRNAs**). These RNAs are usually 100 to 200 nucleotides or less and are often complexed with proteins to form **small nuclear ribonucleoproteins** (**snRNPs**, or **snurps**). These are found only in the nucleus. Because they are rich in uridine residues, the snRNAs have been arbitrarily designated U1, U2, ... , U6.

The snRNA of U1 bears a nucleotide sequence that is homologous to the 5′ end of the intron. Base pairing resulting from this homology promotes binding that represents the initial step in formation of the spliceosome. After the other snurps (U2, U4, U5, and U6) are added, splicing commences.

As with group I splicing, two transesterification reactions are involved. The first involves the interaction of the 3′-OH group from an adenine (A) residue present within the **branch point** region of the intron. The A residue attacks the 5′ splice site, cutting the RNA chain. In a subsequent step involving several other snurps, an intermediate structure is formed and the second reaction ensues, linking the cut 5′ end of the intron to the A. This results in the formation of the characteristic loop structure called a *lariat*, which contains the excised intron. The exons are then ligated, and the snurps are released.

The processing involved in splicing represents a potential regulatory step during gene expression. For example, several cases are known where introns present in pre-mRNAs *derived from the same gene* are spliced *in more than one way*, thereby yielding different collections of exons in the mature mRNA. This process of **alternative splicing** yields a group of mRNAs that, upon translation, results in a series of related proteins called **isoforms**. A growing number of examples have been found in organisms ranging from viruses to *Drosophila* to humans. Alternative splicing of pre-mRNAs provides the basis for producing related proteins from a single gene, thus increasing the number of gene products that can be derived from an organism's genome.

12.12 RNA Editing Modifies the Final Transcript

In the late 1980s, still another quite unexpected form of post-transcriptional RNA processing was discovered. In this form, called **RNA editing**, the nucleotide sequence of a pre-mRNA is actually changed prior to translation. As a result, the ribonucleotide sequence of the mature RNA differs from the sequence encoded in the exons of the DNA from which the RNA was transcribed.

Other variations of RNA editing exist, but there are two main types—**substitution editing**, in which the identities of individual nucleotide bases are altered; and **insertion/deletion editing**, in which nucleotides are added to or subtracted from the total number of bases. Substitution editing is used in some nuclear-derived eukaryotic RNAs and is prevalent in mitochondrial and chloroplast RNAs transcribed in plants. *Physarum polycephalum*, a slime mold, uses both substitution and insertion/deletion editing for its mitochondrial mRNAs.

Trypanosoma, a parasite that causes African sleeping sickness, and its relatives use extensive insertion/deletion editing in mitochondrial RNAs. The number of uridines added to an individual transcript can make up more than 60 percent of the coding sequence, usually forming the initiation codon and placing the rest of the sequence into the proper reading frame. Insertion/deletion editing in trypanosomes is directed by **gRNA (guide RNA) templates**, which are transcribed from the mitochondrial genome. These small RNAs are complementary to the edited region of the final, edited mRNAs. They base pair with the pre-edited mRNAs to direct the editing machinery to make the correct changes.

The most well-studied examples of substitutional editing occur in mammalian nuclear-encoded mRNA transcripts. Apolipoprotein B (apo B) exists in both a long and a short form, although a single gene encodes both proteins. In human intestinal cells, apo B mRNA is edited by a single C-to-U change, which converts a CAA glutamine codon into a UAA stop codon and terminates the polypeptide at approximately half its genomically encoded length. The editing is performed by a complex of proteins that bind to a "mooring sequence" on the mRNA transcript just downstream of the editing site. In a second system, the synthesis of subunits constituting the glutamate receptor channels (GluR) in mammalian brain tissue is also affected by RNA editing. In this case, adenosine (A) to inosine (I) editing occurs in pre-mRNAs prior to their translation, where I is read as guanosine (G) during translation. A family of three ADAR (adenosine deaminase acting on RNA) enzymes is believed to be responsible for editing various sites within the glutamate channel subunits. The double-stranded RNAs required for editing by ADARs are provided by intron/exon pairing of the GluR mRNA transcripts. The editing changes alter the physiological parameters (solute permeability and desensitization response time) of the receptors containing the subunits.

The importance of RNA editing resulting from the action of ADARs is most apparent when we examine situations in which these enzymes have lost their functional capacity as a result of mutation. In several investigations, the loss of function was shown to have a lethal impact in mice. In one study, embryos heterozygous for a defective *ADAR1* gene die during embryonic development as a result of a defective hematopoietic system. In another study, mice with two defective copies of *ADAR2* progress through development normally but are prone to epileptic seizures and die while still in the weaning stage. Their tissues contain the unedited version of one of the GluR products. The defect leading to death is believed to be in the brain. Heterozygotes for the mutation are normal.

Findings such as these in mammals have established that RNA editing provides still another important mechanism of posttranscriptional modification and that this process is not restricted to small or asexually reproducing genomes such as mitochondria. Several new examples of RNA editing have been found each year since its discovery, and this trend is likely to continue. Furthermore, the process has important implications for the regulation of genetic expression.

12.13 Transcription Has Been Visualized by Electron Microscopy

We conclude this chapter by presenting a striking visual demonstration of the transcription process based on electron microscope studies. Figure 12–14 shows a micrograph and interpretive drawing derived from studies of the bacterium *E. coli*. Multiple strands of RNA are seen to emanate from different points along a central DNA axis serving as the template. Many RNA strands result because numerous transcription events are occurring simultaneously along the gene. Progressively longer RNA strands are found farther downstream from the point of initiation of transcription (on the right in the micrograph), whereas the shortest strands are closest to the point of initiation (on the left in the micrograph).

Another interesting phenomenon is captured in this study. Because prokaryotes lack nuclei, cytoplasmic ribosomes are not separated physically from the chromosome. As a result, ribosomes are free to attach to *partially* transcribed mRNA molecules and initiate translation. The longer RNA strands demonstrate the greatest number of ribosomes.

E. coli

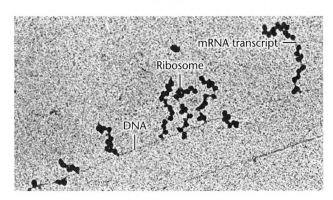

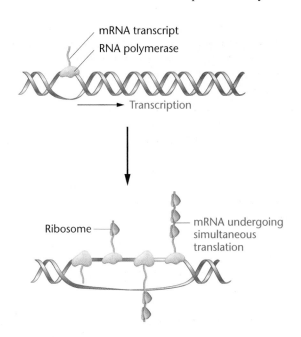

FIGURE 12–14 An electron micrograph and interpretive drawing of simultaneous transcription of a gene in *E. coli.* As each transcript is forming, ribosomes attach, initiating simultaneous translation along each strand. *O.L. Miller, Jr. Barbara A. Hamkalo, C.A. Thomas, Jr. Science 169: 392–395, 1970 by the American Association for the Advancement of Science. F:2*

Visualization of transcription confirms our expectations based on the biochemical analysis of this process. You might now wish to consult the chapter opening photograph on p. 253 to again visualize transcription. This photograph depicts a single gene that encodes rRNA in the newt *Notophthalmus.* Simultaneous transcription is occurring from bottom to top. As each rRNA molecule is created, it "flops" to one side or the other along the vertical DNA axis.

CHAPTER SUMMARY

1. The genetic code stored in DNA is copied to RNA, where it is used to direct the synthesis of polypeptide chains. The code is degenerate, unambiguous, nonoverlapping, and commaless.

2. The complete coding dictionary, determined by using various experimental approaches, reveals that of the 64 possible codons, 61 encode the 20 amino acids found in proteins, while 3 triplets terminate translation. One of the 61 codons is the initiation codon and specifies methionine.

3. The observed pattern of degeneracy often involves only the third letter of a triplet series. It led Francis Crick to propose the wobble hypothesis.

4. Confirmation for the coding dictionary, including codons for initiation and termination, was obtained by comparing the complete nucleotide sequence of phage MS2 with the amino acid sequences of the corresponding proteins. Other findings support the belief that, with minor exceptions, the code is universal for all organisms.

5. In some bacteriophages, multiple initiation points may occur during the transcription of RNA, resulting in multiple reading frames and overlapping genes.

6. Transcription—the initial step in gene expression—describes the synthesis, under the direction of RNA polymerase, of a strand of RNA complementary to a DNA template.

7. The processes of transcription, like DNA replication, can be subdivided into the stages of initiation, elongation, and termination. Also like DNA replication, the process relies on base-pairing affinities between complementary nucleotides.

8. Initiation of transcription is dependent on an upstream (5′) DNA region, called the promoter, that represents the initial binding site for RNA polymerase. Promoters contain specific DNA sequences, such as the TATA box, that are essential to polymerase binding.

9. Transcription is more complex in eukaryotes than in prokaryotes. The primary transcript is a pre-mRNA that must be modified in various ways before it can be efficiently translated. Processing, which produces a mature mRNA, includes the addition of a 7mG cap and a poly-A tail, and the removal, through splicing, of intervening sequences, or introns. RNA editing of pre-mRNA prior to its translation also occurs in some systems.

GENETICS, TECHNOLOGY, AND SOCIETY

Antisense Oligonucleotides: Attacking the Messenger

Standard chemotherapies for diseases such as cancer and AIDS are often accompanied by toxic side effects. Conventional therapeutic drugs target both normal and diseased cells, with diseased or infected cells being only slightly more susceptible than the patient's normal cells. Scientists have long wished for a magic bullet that could seek out and destroy the virus or cancer cells, leaving normal cells alive and healthy. Over the last decade, one particularly promising candidate for magic-bullet status has emerged—the antisense oligonucleotide.

Antisense therapies have arisen through an understanding of the molecular biology of gene expression. Gene expression is a two-step process. First, a single-stranded messenger RNA (mRNA) is copied from the template strand of the duplex DNA molecule. Second, the mRNA is complexed with ribosomes, and its coded information is translated into the amino acid sequence of a polypeptide. Normally, a gene is transcribed into RNA from only one strand of the DNA duplex. The resulting RNA is known as *sense RNA*. However, it is sometimes possible for the other DNA strand to be copied into RNA. The RNA produced by the transcription of the "wrong" strand of DNA is called *antisense RNA*. As with complementary strands of DNA molecules, complementary strands of RNA can form double-stranded molecules.

The formation of duplex structures between a sense and an antisense RNA may affect the sense RNA in at least two ways. Binding antisense RNA to sense RNA may physically block ribosome binding or elongation, hence inhibiting translation. Or binding antisense RNA to sense RNA may trigger the degradation of the sense RNA because double-stranded RNA molecules are attacked by intracellular ribonucleases. In both cases, gene expression is blocked.

What makes the antisense approach so exciting is its potential specificity. Scientists can design antisense RNA (or DNA) molecules of known nucleotide sequence and can then synthesize large amounts of these antisense nucleic acids *in vitro*. Usually, oligonucleotides of up to 20 nucleotides are used. It is theoretically possible to treat cells with these synthetic antisense oligonucleotides, have the oligonucleotides enter the cell and bind precise target mRNAs, and turn off the synthesis of specific proteins. If a specific protein is necessary for virus reproduction or cancer cell growth (but is not necessary in normal cells), the antisense oligonucleotide should have only therapeutic effects.

In the past few years, laboratory tests of antisense drugs have been so promising that many clinical trials are in progress. In addition, one antisense drug (Vitravene©, from ISIS Pharmaceuticals and CIBA Vision Corporation) has been approved for sale in the United States. Vitravene is an antisense oligonucleotide used to treat cytomegalovirus-induced retinitis in AIDS patients. Cytomegalovirus (CMV) is a common virus that infects most people. Although it causes few problems in people with normal immune systems, it can cause serious symptoms in people with impaired immune systems (such as AIDS). Up to 40 percent of AIDS patients develop retinitis or blindness as a result of CMV infections of the eye. Clinical trials indicate that the introduction of antisense CMV oligonucleotides into the eyes of AIDS patients with CMV-induced retinitis significantly delays the disease progression compared to untreated groups. Laboratory studies suggest that antisense oligonucleotides may trigger the destruction of CMV mRNA and interfere with the adsorption of the virus to the surface of host cells.

Antisense oligonucleotides may also act as antiinflammatory drugs. Clinical trials are in progress to test antisense compounds in the treatment of asthma, rheumatoid arthritis, psoriasis, ulcerative colitis, and transplant rejection. One interesting antisense oligonucleotide (now in clinical trials conducted by ISIS and Boehringer Ingelheim) binds to the mRNA that encodes ICAM-1, a cell-surface glycoprotein that helps activate immune system and inflammatory cells. It is often overexpressed in body tissues that exhibit extreme inflammatory responses. Researchers hope that antisense oligonucleotides might temper inflammatory responses by reducing the expression of ICAM-1. The results of a recent clinical trial suggest that antisense ICAM-1 may be an effective treatment for the debilitating, chronic Crohn's disease, a form of inflammatory bowel disease that affects about 200,000 people in the United States. In one clinical trial, almost half of the Crohn's disease patients went into remission after treatment with antisense ICAM-1, whereas none of the placebo group did so. In this trial, the antisense ICAM-1 drugs appeared to be well tolerated and safe.

Some interesting clinical trials are in progress to test antisense oligonucleotides as treatments for some types of cancer. These oligonucleotides are designed to reduce the synthesis of proteins that are either overexpressed in cancer cells or are present as mutant forms. These drugs will be evaluated as treatments for tumors of the ovary, prostate, breast, brain, colon, and lung. Other antisense drugs in the pipeline are designed to attack the hepatitis B, hepatitis C, human papilloma, and AIDS viruses. If antisense therapeutics pass the scrutiny of these clinical trials, we may indeed have acquired a molecular magic bullet to use in the battle against a variety of diseases.

More recently, another gene-silencing technology, analogous to the antisense principle, has emerged. This technology, known as RNA interference (RNAi), uses short double-stranded RNA molecules (~20 nucleotides long) called short interfering RNA (siRNA) to target specific mRNAs within cells. One strand of the short RNA recognizes a sequence within the mRNA, leading to the degradation or silencing of the mRNA. Although RNAi is highly efficient *in vitro* and is used in laboratory research, the application of RNAi to human disease is less advanced than the use of oligonucleotide-based antisense technologies. RNAi is discussed in more detail in Chapter 15.

References

Dykxhoorn, D. M., Novina, C. D., and Sharp, P. A. 2003. Killing the messenger: Short RNAs that silence gene expression. *Nature Revs. Molec. Cell Biol.* 4: 457–467.

Nyce, J. W., and Metzger, W. J. 1997. DNA antisense therapy for asthma in an animal model. *Nature* 385: 721–725.

Roush, W. 1997. Antisense aims for a renaissance. *Science* 276: 1192–1193.

KEY TERMS

INSIGHTS AND SOLUTIONS

1. Calculate how many triplet codons would be possible had evolution seized on six bases (three complementary base pairs) rather than four bases within the structure of DNA. Would six bases accommodate a two-letter code, assuming 20 amino acids and start and stop codons?

Solution: Six things taken three at a time will produce $(6)^3$ or 216 triplet codes. If the code was a doublet, there would be $(6)^2$ or 36 two-letter codes, more than enough to accommodate 20 amino acids and start and stop signals.

2. In a heteropolymer experiment using $1/2C : 1/4A : 1/4G$, how many different triplets will occur in the synthetic RNA molecule? How frequently will the most frequent triplet occur?

Solution: There will be $(3)^3$, or 27, triplets produced. The most frequent will be CCC, present $(1/2)^3$, or 1/8, of the time.

3. In a regular copolymer experiment, where UUAC is repeated over and over, how many different triplets will occur in the synthetic RNA, and how many amino acids will occur in the polypeptide when this RNA is translated? (Consult Figure 12–7.)

Solution: The synthetic RNA will repeat four triplets—UUA, CUU, ACU, UAC,—over and over. Because both UUA and CUU encode leucine, while ACU and UAC encode threonine and tyrosine, respectively, the polypeptides synthesized under the directions of this RNA would contain three amino acids in the repeating sequence leu-leu-thr-tyr.

4. Actinomycin D inhibits DNA-dependent RNA synthesis. This antibiotic is added to a bacterial culture where a specific protein is being monitored. Compared to a control culture, where no antibiotic is added, translation of the protein declines over a period of 20 minutes, until no further protein is made. Explain these results.

Solution: The mRNA, which is the basis for the translation of the protein, has a lifetime of about 20 minutes. When actinomycin D is added, transcription is inhibited and no new mRNAs are made. Those already present support the translation of the protein for up to 20 minutes.

PROBLEMS AND DISCUSSION QUESTIONS

1. Early proposals regarding the genetic code considered the possibility that DNA served directly as the template for polypeptide synthesis. In eukaryotes, what difficulties would such a system pose? What observations and theoretical considerations argue against such a proposal?

2. In studies of frameshift mutations, Crick, Barnett, Brenner, and Watts-Tobin found that either three nucleotide insertions or deletions restored the correct reading frame. (a) Assuming the code is a triplet, what effect would the addition or loss of six nucleotides have on the reading frame? (b) If the code were a sextuplet (consisting of six nucleotides), would the reading frame be restored by the addition or loss of three, six, or nine nucleotides?

3. In a mixed copolymer experiment using polynucleotide phosphorylase, 3/4G:1/4C was added to form the synthetic message. Using the resulting amino acid composition of the ensuing protein shown below, (a) indicate the percentage of time each possible triplet will occur in the message, and (b) determine one consistent base-composition assignment for the amino acids present. (c) Considering the wobble hypothesis, predict as many specific triplet assignments as possible.

Glycine	36/64	(56%)
Alanine	12/64	(19%)
Arginine	12/64	(19%)
Proline	4/64	(6%)

4. When repeating copolymers are used to form synthetic mRNAs, dinucleotides produce a single type of polypeptide that contains only two different amino acids. On the other hand, a trinucleotide sequence produces three different polypeptides, each consisting of only a single amino acid. Why? What will be produced when a repeating tetranucleotide is used?

5. The mRNA formed from the repeating tetranucleotide UUAC incorporates only three amino acids, but the use of UAUC incorporates four amino acids. Why?

6. In studies using repeating copolymers, AC ... incorporates threonine and histidine, and CAACAA ... incorporates glutamine, asparagine, and threonine. What triplet code can definitely be assigned to threonine?

7. In a coding experiment using repeating copolymers (as shown in Table 12.3), the data below were obtained. AGG is known to code for arginine. Taking into account the wobble hypothesis, assign each of the four remaining different triplet codes to its correct amino acid.

Copolymer	Codons Produced	Amino Acids in Polypeptide
AG	AGA, GAG	arg, glu
AAG	AGA, AAG, GAA	lys, arg, glu

8. In the triplet binding assay technique, radioactivity remains on the filter when the amino acid corresponding to the experimental triplet is labeled. Explain the basis of this technique.

9. When the amino acid sequences of insulin isolated from different organisms were determined, some differences were noted. For example, alanine was substituted for threonine, serine was substituted for glycine, and valine was substituted for isoleucine at corresponding positions in the protein. List the single-base changes that could occur in triplets to produce these amino acid changes.

10. In studies of the amino acid sequence of wild-type and mutant forms of tryptophan synthetase in *E. coli*, the following changes have been observed:

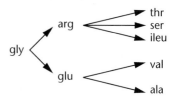

Determine a set of triplet codes in which only a single nucleotide change produces each amino acid change.

11. Why doesn't polynucleotide phosphorylase (Ochoa's enzyme) synthesize RNA *in vivo*?

12. Refer to Table 12.1. Can you hypothesize why a mixture of (Poly U) + (Poly A) would not stimulate incorporation of ^{14}C-phenylalanine into protein?

13. Predict the amino acid sequence produced during translation of the short theoretical mRNA sequences below. (Note that the second sequence was formed from the first by a deletion of only one nucleotide.) What type of mutation gave rise to sequence 2?

Sequence 1:   5'-AUGCCGGAUUAUAGUUGA-3'
Sequence 2:   5'-AUGCCGGAUUAAGUUGA-3'

14. A short RNA molecule was isolated that demonstrated a hyperchromic shift indicating secondary structure. Its sequence was determined to be

5'-AGGCGCCGACUCUACU-3'

(a) Propose a two-dimensional model for this molecule. (b) What DNA sequence would give rise to this RNA molecule through transcription? (c) If the molecule were a tRNA fragment containing a CGA anticodon, what would the corresponding codon be? (d) If the molecule were an internal part of a message, what amino acid sequence would result from it following translation? (Refer to the code chart in Figure 12–7.)

15. A glycine residue exists at position 210 of the tryptophan synthetase enzyme of wild-type *E. coli*. If the codon specifying glycine is GGA, how many single-base substitutions will result in an amino acid substitution at position 210, and what are they? How many will result if the wild-type codon is GGU?

16. Shown here is a theoretical viral mRNA sequence:

5'-AUGCAUACCUAUGAGACCCUUGGA-3'

(a) Assuming that it could arise from overlapping genes, how many different polypeptide sequences can be produced? Using the chart in Figure 12–7, what are the sequences? (b) A base substitution mutation that altered the sequence in part (a) eliminated the synthesis of all but one polypeptide. The altered sequence is shown below. Use Figure 12–7 to determine why it was altered.

5'-AUGCAUACCUAUGUGACCCUUGGA-3'

17. Most proteins have more leucine than histidine residues but more histidine than tryptophan residues. Correlate the number of codons for these three amino acids with this information.

18. Define the process of transcription. Where does this process fit into the central dogma of molecular genetics?

19. What was the initial evidence for the existence of mRNA?

20. Describe the structure of RNA polymerase in bacteria. What is the core enzyme? What is the role of the σ subunit?

21. In a written paragraph, describe the abbreviated chemical reactions that summarize RNA polymerase-directed transcription.

22. Messenger RNA molecules are very difficult to isolate from prokaryotes because they are quickly degraded. Can you suggest a reason why this occurs? Eukaryotic mRNAs are more stable and exist longer in the cell than do prokaryotic mRNAs. Is this an advantage or disadvantage for a pancreatic cell making large quantities of insulin?

23. Deoxyribonucleotide sequences derived from a template strand of DNA are shown below. (a) Determine the mRNA sequence that would be derived from transcription of each DNA sequence. (b) Using Figure 12–7, determine the amino acid sequence that is encoded by each mRNA. (c) For sequence 1, what is the sequence of the partner DNA strand?

Sequence 1:	5′-CTTTTTTGCCAT-3′
Sequence 2:	5′-ACATCAATAACT-3′
Sequence 3:	5′-TACAAGGGGTTCT-3′

24. In a mixed copolymer experiment, messages were created with either 4/5C:1/5A or 4/5A:1/5C. These messages yielded proteins with the following amino acid compositions shown in the table below. Using these data, predict the most specific coding composition for each amino acid.

4/5C:1/5A		4/5A:1/5C	
Proline	63.0%	Proline	3.5%
Histidine	13.0%	Histidine	3.0%
Threonine	16.0%	Threonine	16.6%
Glutamine	3.0%	Glutamine	13.0%
Asparagine	3.0%	Asparagine	13.0%
Lysine	0.5%	Lysine	50.0%
	98.5%		99.1%

25. Shown below are the amino acid sequences of the wild-type and three mutant forms of a short protein. (a) Using Figure 12–7, predict the type of mutation that created each altered protein. (b) Determine the specific ribonucleotide change that led to the synthesis of each mutant protein. (c) The wild-type RNA consists of nine triplets. What is the role of the ninth triplet? (d) For the first eight wild-type triplets, which, if any, can you determine specifically from an analysis of the mutant proteins? In each case, explain why or why not. (e) Another mutation (mutant 4) is isolated. Its amino acid sequence is unchanged, but mutant cells produce abnormally low amounts of the wild-type proteins. As specifically as you can, predict where this mutation exists in the gene.

Wild type:	met-trp-tyr-arg-gly-ser-pro-thr
Mutant 1:	met-trp
Mutant 2:	met-trp-his-arg-gly-ser-pro-thr
Mutant 3:	met-cys-ile-val-val-val-gln-his

26. The genetic code is degenerate. Amino acids are encoded by either 1, 2, 3, 4, or 6 triplet codons. (See Figure 12–7.) An interesting question is whether the frequency of triplet codes is in any way correlated with how frequently amino acids appear in proteins. That is, is the genetic code optimized for its intended use? Some approximations of the frequency of appearance of nine amino acids in proteins in *E. coli* are shown in the table below. (a) Determine how many triplets encode each amino acid in the table. (b) Devise a way to graphically express the data from part (a) and the table. (c) Analyze your data to determine what, if any, correlations can be drawn between the relative frequency of amino acids making up proteins with the number of triplets for each. Write a paragraph that states your specific and general conclusions. (d) How would you proceed with your analysis if you wanted to pursue this problem further?

Amino Acid	Percentage of Appearance in Proteins
met	2
cys	2
gln	5
pro	5
arg	5
ile	6
glu	7
ala	8
leu	10

27. As described in Chapter 11, Alu elements proliferate in the human genome by a process called retrotransposition, in which the *Alu* DNA sequence is transcribed into RNA, copied into double-stranded DNA, and then inserted back into the genome at a site distant from that of its "parent" *Alu* gene. This seems to have been an extremely efficient process, since *Alu* genes have proliferated to about 10^6 copies in the human genome. This efficiency is largely due to the fact that *Alu* elements, like many small structural RNAs, carry their promoter sequences *within the transcribed region of the gene*, rather than 5′ to the transcription start site. If *Alu* elements carried promoters upstream to the transcription site, similar to those of protein-coding genes, what would happen once they were retrotransposed? Would a retrotransposed *Alu* gene be able to proliferate? Explain.

28. DNA and RNA base compositions were analyzed from a hypothetical bacterial species with the results shown in the following table. (a) On the basis of these data, what can you conclude about the DNA and RNA of the organism? (b) Are the data consistent with the Watson-Crick model of DNA? (c) Is the RNA single-stranded or double-stranded? (d) If we assume that the entire length of DNA has been transcribed, do the data suggest that RNA has been derived from the transcription of one or both DNA strands. Can we determine this from these data?

Ratio	DNA	RNA
(A + G)/(T + C)	1.0	
(A + T)/(C + G)	1.2	
(A + G)/(U + C)		1.3
(A + U)/(C + G)		1.2

29. M. Klemke and others (2001. *EMBO J.* 20: 3849–3860.) discovered an interesting coding phenomenon in an exon within a neurologic hormone receptor gene in mammals, by which two protein entities (XLαs and ALEX) appear to be produced from the same exon. Below is the DNA sequence of the exon's 5′ end derived from a rat. The lowercase letters represent the initial coding portion for the XLαs protein, and the uppercase letters indicate the portion where the ALEX entity initiates. (For simplicity, and to correspond with the RNA coding dictionary, it is customary to represent the noncoding, partner strand of the DNA segment.)

> 5′-gtcccaaccatgcccaccgatcttccgcctgcttctgaagATGCGGGCCCAG

(a) Convert the noncoding DNA sequence to the coding RNA sequence. (b) Locate the initiator codon within the XLαs segment. (c) Locate the initiator codon within the ALEX segment. Are the two initiator codons in frame? (d) Provide the amino acid sequence for each coding sequence. In the region of overlap, are the two amino acid sequences the same? (e) Are there any evolutionary advantages to having the same DNA sequence code for two protein products? Are there any disadvantages?

30. The concept of consensus sequences of DNA was introduced in this chapter as those sequences that are similar (homologous) in different genes of the same organism or in genes of different organisms. Examples were the Pribnow box and the −35 region in prokaryotes and the TATA-box region in eukaryotes. Work by Novitsky and colleagues (2002. Virology. *J. Virol.* 76: 5435–5451) indicates that among 73 isolates of HIV-Type 1C (a major contributor to the AIDS epidemic), a GGGNNNNNCC consensus sequence exists (where N equals any nitrogenous base) in the promoter-enhancer region of the NF-κB transcription factor, a *cis*-acting motif that is critical in initiating HIV transcription in human macrophages. The authors contend that finding this and other conserved sequences may be of value in designing an AIDS vaccine. What advantages would finding these consensus sequences confer? What disadvantages?

31. Theoretically, antisense oligodeoxynucleotides (relatively short, 7 to 20 nucleotides, single-stranded DNAs) can selectively block disease-causing genes. They do so by base pairing with complementary regions of the RNA transcripts and inhibiting their function. The RNA/DNA hybrid is recognized by intracellular RNase H, and the RNA portion of the hybrid is degraded. Cancer genes have often been chosen as potential targets for antisense drugs. Data from Cho and coworkers (2001. *Proc. Natl. Acad. Sci.* (USA) 98: 9819–9823) (see the table below) established that in response to a single population of antisense oligodeoxynucleotides that target a specific kinase mRNA in a prostate cancer cell line, there was as much as a 20-fold (plus and minus) response differential of unrelated gene expression (as measured by mRNA production). (a) Diagram your concept of the intracellular action of an antisense oligodeoxynucleotide. (b) Diagram your concept of the action of RNase H. (c) What might cause nonrelated gene expression to be decreased, as in the case of collagen and catalase? (d) What might cause nonrelated gene expression to be increased, as in the case of myosin and G protein? (e) Given the data presented here, what drawbacks might you expect in the use of antisense therapy for genetic diseases?

Gene Assayed	Change in Expression
Myosin light chain	+14.4x
G protein receptor	+3.7x
Collagen	−4.0x
Catalase	−6.7x

32. Recent observations indicate that alternative splicing is a common mechanism for eukaryotes to expand their repertoire of gene functions. Studies by Xu and colleagues (2002. *Nuc. Acids Res.* 30: 3754–3766.) indicate that approximately 50 percent of human genes use alternative splicing, and approximately fifteen percent of disease-causing mutations involve aberrant alternative splicing. Different tissues show remarkably different frequencies of alternative splicing, with the brain accounting for approximately 18 percent of such events. (a) What does alternative splicing mean, and what is an isoform? (b) What evolutionary strategy does alternative splicing offer, and why might some tissues engage in more alternative splicing than others?

Translation and Proteins

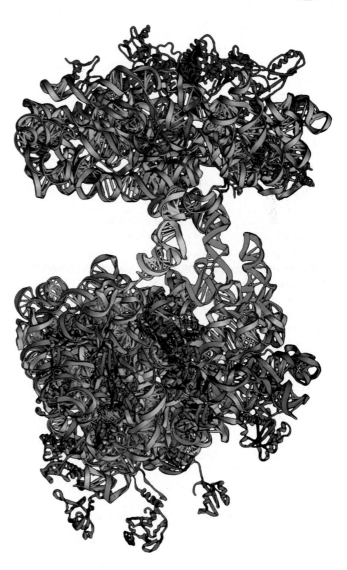

Crystal structure of a Thermus thermophilus 70S ribosome containing three bound transfer RNAs.

(Reprinted from the front cover of Science, Vol. 292, May 4, 2001 "Crystal structure of a Thermus thermophilus 70S ribosome containing three bound transfer RNAs (top) and exploded views showing its different molecular components (middle and bottom). Image provided by Dr. Albion Baucom (baucom@biology.ucsc.edu). Copyright American Association for the Advancement of Science.)

CHAPTER CONCEPTS

- The ribonucleotide sequence of messenger RNA (mRNA) reflects genetic information stored in DNA that makes up genes and corresponds to the amino acid sequences in proteins encoded by those genes.

- The process of translation decodes the information in mRNA, leading to the synthesis of polypeptide chains.

- Translation involves the interactions of mRNA, tRNA, ribosomes, and a variety of translation factors essential to the initiation, elongation, and termination of the polypeptide chain.

- Proteins, the final product of most genes, achieve a three-dimensional conformation that is based on the primary amino acid sequences of the polypeptide chains making up each protein.

- The function of any protein is closely tied to its three-dimensional structure, which can be disrupted by mutation.

We have already learned that a genetic code exists that stores information in the form of triplet nucleotides in DNA and that this information is initially expressed through the process of transcription into a messenger RNA that is complementary to one strand of the DNA helix. However, the final product of gene expression, in most instances, is a polypeptide chain consisting of a linear series of amino acids whose sequence has been prescribed by the genetic code. In this chapter, we will examine how the information present in mRNA is processed to create polypeptides, which then fold into protein molecules. We will also review the evidence that confirms that proteins are the end products of gene expression and briefly discuss the various levels of protein structure, diversity, and function. This information extends our understanding of gene expression and provides an important foundation for interpreting how the mutations that arise in DNA can result in the diverse phenotypic effects observed in organisms.

? How Do We Know?

In this chapter, we will focus on how genetic information, transferred from DNA to mRNA, is expressed through the production of proteins, the end product of most gene expression. As you study this topic, you should try to answer several fundamental questions:

1. How do we know that the process of translation occurs in association with ribosomes?

2. How have the mechanics of translation been deciphered?

3. How do we know that proteins are the end products of genetic expression?

4. How did we come to know that most genes ultimately encode a single polypeptide chain?

5. How do we know that the structure of a protein is intimately related to the function of that protein?

13.1 Translation of mRNA Depends on Ribosomes and Transfer RNAs

Translation of mRNA is the biological polymerization of amino acids into polypeptide chains. This process, alluded to in our discussion of the genetic code in Chapter 12, occurs only in association with ribosomes, which serve as nonspecific workbenches. The central question in translation is how triplet ribonucleotides of mRNA direct specific amino acids into their correct position in the polypeptide. This question was answered once **transfer RNA (tRNA)** was discovered. This class of molecules adapts specific triplet codons in mRNA to their correct amino acids. The *adaptor hypothesis* for the role of tRNA was postulated by Francis Crick in 1957.

In association with a ribosome, mRNA presents a triplet codon that calls for a specific amino acid. A specific tRNA molecule contains within its nucleotide sequence three con-

secutive ribonucleotides complementary to the codon, called the **anticodon**, which can base-pair with the codon. Another region of this tRNA is covalently bonded to its corresponding amino acid.

Inside the ribosome, hydrogen bonding of tRNAs to mRNA holds the amino acids in proximity so that a peptide bond can be formed. This process occurs over and over as mRNA runs through the ribosome and amino acids are polymerized into a polypeptide. Before we discuss the actual process of translation, let's first consider the structures of the ribosome and tRNA.

Ribosomal Structure

Because of its essential role in the expression of genetic information, the **ribosome** has been extensively analyzed. One bacterial cell contains about 10,000 ribosomes, and a eukaryotic cell contains many times more. Electron microscopy reveals that the bacterial ribosome is about 25 μm at its largest diameter and consists of two subunits, one large and one small. Both subunits consist of one or more molecules of rRNA and an array of **ribosomal proteins**. When the two subunits are associated with each other in a single ribosome, the structure is sometimes called a **monosome**.

The specific differences between prokaryotic and eukaryotic ribosomes are summarized in Figure 13–1. The subunit and rRNA components are most easily isolated and characterized on the basis of their sedimentation behavior in sucrose gradients (their rate of migration, or Svedberg coefficient S, which reflects their density, mass, and shape). In prokaryotes, the monosome is a 70S particle; in eukaryotes, it is approximately 80S. Sedimentation coefficients, which reflect the variable rate of migration of different-sized particles and molecules, are not additive. For example, the prokaryotic 70S monosome consists of a 50S and a 30S subunit, and the eukaryotic 80S monosome consists of a 60S and a 40S subunit.

The larger subunit in prokaryotes consists of a 23S rRNA molecule, a 5S rRNA molecule, and 31 ribosomal proteins. In the eukaryotic equivalent, a 28S rRNA molecule is accompanied by a 5.8S and 5S rRNA molecule and 49 proteins. The smaller prokaryotic subunits consist of a 16S rRNA component and 21 proteins. In the eukaryotic equivalent, an 18S rRNA component and 33 proteins are found. The approximate molecular weights (MW) and the numbers of nucleotides of these components are also shown in Figure 13–1.

Regarding the components of the ribosome, it is now clear that the RNA molecules perform the all-important catalytic functions associated with translation. The many proteins, whose functions were long a mystery, are thought to promote the binding of the various molecules involved in translation and in general, to fine-tune the process. This conclusion is based on the observation that some of the catalytic functions in ribosomes still occur in experiments involving "ribosomal protein-depleted" ribosomes.

Molecular hybridization studies have established the degree of redundancy of the genes coding for the rRNA components. The *E. coli* genome contains seven copies of a single sequence that encodes all three components—23S, 16S, and 5S. The initial transcript of each set of these genes produces a

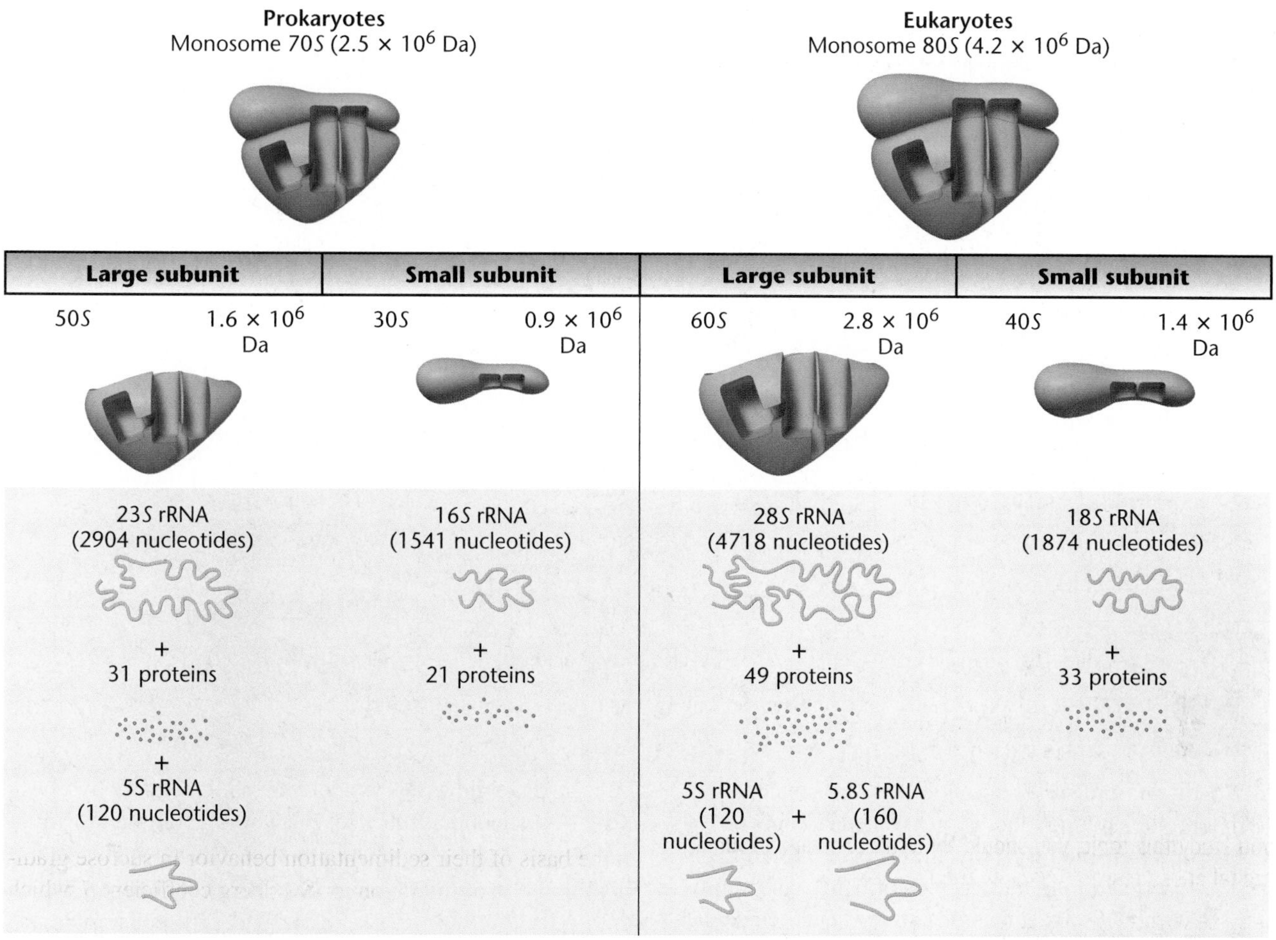

Prokaryotes
Monosome 70S (2.5 × 10^6 Da)

Eukaryotes
Monosome 80S (4.2 × 10^6 Da)

Large subunit		Small subunit		Large subunit		Small subunit	
50S	1.6 × 10^6 Da	30S	0.9 × 10^6 Da	60S	2.8 × 10^6 Da	40S	1.4 × 10^6 Da

23S rRNA (2904 nucleotides)	16S rRNA (1541 nucleotides)	28S rRNA (4718 nucleotides)	18S rRNA (1874 nucleotides)
+ 31 proteins	+ 21 proteins	+ 49 proteins	+ 33 proteins
+ 5S rRNA (120 nucleotides)		5S rRNA (120 nucleotides) + 5.8S rRNA (160 nucleotides)	

FIGURE 13–1 A comparison of the components in prokaryotic and eukaryotic ribosomes.

30S RNA molecule that is enzymatically cleaved into these smaller components. Coupling of the genetic information encoding these three rRNA components ensures that after multiple transcription events, equal quantities of all three will be present as ribosomes are assembled.

In eukaryotes, many more copies of a sequence encoding the 28S, 18S, and 5.8S components are present. In *Drosophila*, approximately 120 copies per haploid genome are each transcribed into a molecule of about 34S. This molecule is then processed into the 28S, 18S, and 5.8S rRNA species. These species are homologous to the three rRNA components of *E. coli*. In *Xenopus laevis*, over 500 copies of the 34S component are present per haploid genome. In mammalian cells, the initial transcript is even larger at 45S. The rRNA genes, called **rDNA**, are part of the moderately repetitive DNA fraction and are present in clusters at various chromosomal sites.

Each cluster in eukaryotes consists of tandem repeats, with each unit separated by a noncoding spacer DNA sequence. In humans, these gene clusters have been localized near the ends of chromosomes 13, 14, 15, 21, and 22. The unique 5S rRNA component of eukaryotes is not part of this larger transcript. Instead, genes coding for this ribosomal component are distinct and located separately. In humans, a gene cluster encoding them has been located on chromosome 1.

Despite the detailed knowledge available about the structure and genetic origin of the ribosomal components, a complete

understanding of the function of these components has thus far eluded geneticists. This is not surprising; the ribosome is the largest and perhaps the most intricate of all cellular structures. For example, the bacterial monosome has a combined molecular weight of 2.5 million Da.

tRNA Structure

Because of their small size and stability in the cell, transfer RNAs (tRNAs) have been investigated extensively and are the best-characterized RNA molecules. They are composed of only 75 to 90 nucleotides, displaying a nearly identical structure in bacteria and eukaryotes. In both types of organisms, tRNAs are transcribed as larger precursors, which are cleaved into mature 4S tRNA molecules. In *E. coli*, for example, tRNA^{tyr} (the superscript identifies the specific tRNA and the cognate amino acid that binds to it) is composed of 77 nucleotides, yet its precursor contains 126 nucleotides.

In 1965, Robert Holley and his colleagues reported the complete sequence of tRNA^{ala} isolated from yeast. They found that a number of nucleotides are unique to tRNA. As shown in Figure 13–2, each nucleotide is a modification of one of the four nitrogenous bases normally present in RNA (G, C, A, and U). These include inosinic acid (which contains the purine hypoxanthine), ribothymidylic acid, and pseudouridine, among others. These modified structures, variously

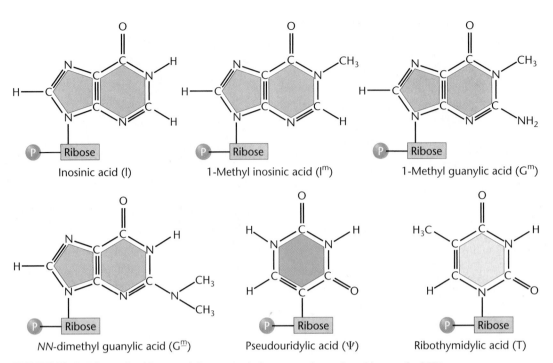

FIGURE 13–2 Ribonucleotides containing unusual nitrogenous bases found in transfer RNA.

referred to as *unusual*, *rare*, or *odd bases*, are created *after* transcription, illustrating the more general concept of **posttranscriptional modification**. In this case, the unmodified base is inserted during transcription of tRNA, and subsequently, enzymatic reactions catalyze the chemical modifications to the base.

Holley's sequence analysis led him to propose the two-dimensional **cloverleaf model of tRNA**. It was known that tRNA demonstrates a secondary structure due to base pairing. Holley discovered that he could arrange the linear model in such a way that several stretches of base pairing would result. This arrangement created a series of paired stems and unpaired loops resembling the shape of a cloverleaf. Loops consistently contained modified bases that did not generally form base pairs. Holley's model is shown in Figure 13–3.

The triplets GCU, GCC, and GCA specify alanine; therefore, Holley looked for an anticodon sequence complementary to one of these codons in his tRNA^{ala} molecule. He found it in the form of CGI (the 3' to 5'-direction) in one loop of the cloverleaf. The nitrogenous base I (inosinic acid) can form hydrogen bonds with U, C, or A, the third members of the alanine triplets. Thus, the **anticodon loop** was established.

Studies of other tRNA species reveal many constant features. At the 3'-end, all tRNAs contain the sequence (. . . pCpCpA-3'). This is the end of the molecule where the amino acid is covalently joined to the terminal adenosine residue. All tRNAs contain the nucleotide (5'-Gp . . .) at the other end of the molecule. In addition, the lengths of various stems and loops are very similar. Each tRNA that has been examined also contains an anticodon complementary to the known amino acid codon for which it is specific, and all anticodon loops are present in the same position of the cloverleaf.

Because the cloverleaf model was predicted strictly on the basis of nucleotide sequence, there was great interest in the X-ray crystallographic examination of tRNA, which reveals a three-dimensional structure. By 1974, Alexander Rich and his

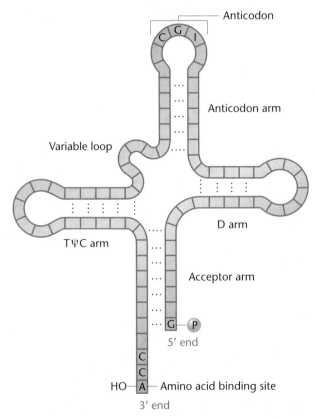

FIGURE 13–3 Holley's two-dimensional cloverleaf model of transfer RNA. Blocks represent nitrogenous bases.

colleagues in the United States, and J. Roberts, B. Clark, Aaron Klug, and their colleagues in England had succeeded in crystallizing tRNA and performing X-ray crystallography at a resolution of 3 Å. At this resolution, the pattern formed by individual nucleotides is discernible.

As a result of these studies, a complete three-dimensional model of tRNA is shown in Figure 13–4. At one end of the molecule is the anticodon loop and stem, and at the other end is the 3′-acceptor region where the amino acid is bound. Geneticists speculate that the shapes of the intervening loops may be recognized by the specific enzymes responsible for adding amino acids to tRNAs—a subject to which we now turn our attention.

Charging tRNA

Before translation can proceed, the tRNA molecules must be chemically linked to their respective amino acids. This activation process, called **charging**, occurs under the direction of enzymes called **aminoacyl tRNA synthetases**. There are 20 different amino acids, so there must be at least 20 different tRNA molecules and as many different enzymes. In theory, because there are 61 triplets that encode amino acids, there could be 61 specific tRNAs and enzymes. However, because of the ability of the third member of a triplet code to "wobble," it is now thought that there are but 31 different tRNAs. It is also believed that there are only 20 synthetases, one for each amino acid, regardless of the greater number of corresponding tRNAs.

The charging process is outlined in Figure 13–5. In the initial step, the amino acid is converted to an activated form, reacting with ATP to create an **aminoacyladenylic acid**. A covalent linkage is formed between the 5′-phosphate group of ATP and the carboxyl end of the amino acid. This molecule

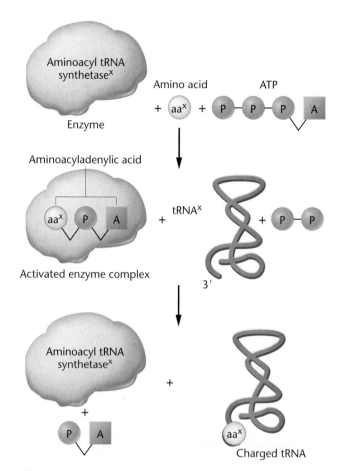

FIGURE 13–5 Steps involved in charging tRNA. The superscript x denotes that only the corresponding specific tRNA and specific aminoacyl tRNA synthetase enzyme are involved in the charging process for each amino acid.

remains associated with the synthetase enzyme, forming a complex that then reacts with a specific tRNA molecule. During this next step, the amino acid is transferred to the appropriate tRNA and bonded covalently to the adenine residue at the 3′ end. The charged tRNA may now participate directly in protein synthesis. Aminoacyl tRNA synthetases are highly specific enzymes because they recognize only one amino acid and the subset of corresponding tRNAs called **isoaccepting tRNAs**. This is crucial if fidelity of translation is to be maintained.

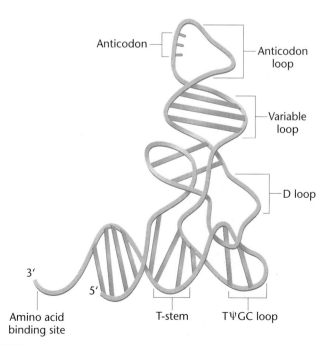

FIGURE 13–4 A three-dimensional model of transfer RNA.

Now Solve This

Problem 27 on page 301 asks you to consider experimental findings attempting to determine whether the anticodon of tRNA or the amino acid added to the tRNA during charging is more critical to the accuracy of translation.

Hint: In this experiment, when the triplet codon in mRNA calls for cysteine, alanine is inserted during translation, even though it is the "incorrect" amino acid.

TABLE 13.1	Various Protein Factors Involved During Translation in *E. coli*	
Process	**Factor**	**Role**
Initiation of translation	IF1	Stabilizes 30*S* subunit
	IF2	Binds fmet-tRNA to 30*S*-mRNA complex; binds to GTP and stimulates hydrolysis
	IF3	Binds 30*S* subunit to mRNA
Elongation of polypeptide	EF-Tu	Binds GTP; brings aminoacyl-tRNA to the A site of ribosome
	EF-Ts	Generates active EF-Tu
	EF-G	Stimulates translocation; GTP-dependent
Termination of translation and release of polypeptide	RF1	Catalyzes release of the polypeptide chain from tRNA and dissociation of the translocation complex; specific for UAA and UAG termination codons
	RF2	Behaves like RF1; specific for UGA and UAA codons
	RF3	Stimulates RF1 and RF2

13.2 Translation of mRNA Can Be Divided into Three Steps

In a way similar to transcription, the process of translation can be best described by breaking it into discrete phases. We will consider three phases, each with its own illustration, but keep in mind that translation is a dynamic, continuous process. You should correlate the following discussion with the step-by-step characterization in the figures. Many of the protein factors and their roles in translation are summarized in Table 13.1.

Initiation

Initiation of translation is depicted in Figure 13–6. Recall that the ribosome serves as a nonspecific workbench for the translation process. Most ribosomes, when they are not involved in translation, are dissociated into their large and small subunits. Initiation of translation in *E. coli* involves the small ribosome subunit, an mRNA molecule, a specific charged tRNA, GTP, Mg^{++}, and at least three proteinaceous

Initiation

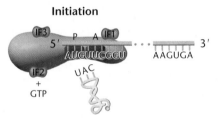

Step 1. mRNA binds to small subunit along with initiation factors (IF1, 2, 3)

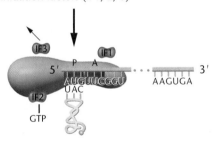

Initiation complex

Step 2. Initiator tRNA^{fmet} binds to mRNA codon in P site; IF3 released

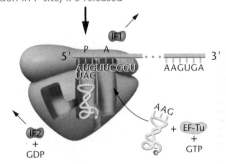

Step 3. Large subunit binds to complex; IF1 and IF2 released; EF-Tu binds to tRNA, facilitating entry into A site

Translation components

FIGURE 13–6 Initiation of translation. The components are depicted at the left of the figure.

initiation factors (IFs) that enhance the binding affinity of the various translational components. In prokaryotes, the initiation codon of mRNA (AUG) calls for the modified amino acid **formylmethionine (fmet)**.

The small ribosomal subunit binds to several initiation factors, and this complex then binds to mRNA (Step 1). In bacteria, this binding involves a sequence of up to six ribonucleotides (AGGAGG, not shown), which *precedes* the initial AUG start codon of mRNA. This sequence (containing only purines and called the **Shine-Dalgarno sequence**) base-pairs with a region of the 16S rRNA of the small ribosomal subunit, facilitating initiation.

Another initiation protein then enhances the binding of charged fmet-tRNA to the small subunit in response to the AUG triplet (Step 2). This step sets the reading frame so that all subsequent groups of three ribonucleotides are translated accurately. This aggregate represents the **initiation complex**, which then combines with the large ribosomal subunit. At this point, a molecule of GTP is hydrolyzed, providing the required energy, and the initiation factors are released (Step 3).

Elongation

The second phase of translation, elongation, is depicted in Figure 13–7. Once both subunits of the ribosome are assembled with the mRNA, binding sites for two charged tRNA molecules are formed. These are the **P (peptidyl) site** and the **A (aminoacyl) site**. The charged initiator tRNA binds to the P site, provided that the AUG triplet of mRNA is in the corresponding position of the small subunit.

Increasing the growing polypeptide chain by one amino acid is called **elongation**. The sequence of the second triplet in mRNA dictates which charged tRNA molecule will become positioned at the A site (Step 1). Once it is present, an enzyme within the large subunit of the ribosome, called **peptidyl transferase**, catalyzes the formation of the peptide bond, which links the two amino acids (Step 2). At the same time, the covalent bond between the amino acid and the tRNA occupying the P site is hydrolyzed (broken) producing a dipeptide, which is attached to the 3′ end of tRNA still residing in the A site.

Before elongation can be repeated, the tRNA attached to the P site, which is now uncharged, must be released from the large subunit. The uncharged tRNA moves transiently through a third site on the ribosome, called the **E (exit) site**. The entire *mRNA–tRNA–aa2–aa1 complex* then shifts in the direction of the P site by a distance of three nucleotides (Step 3). This event requires several protein **elongation factors (EFs)** as

Elongation

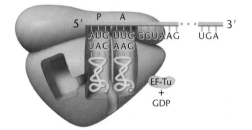

Step 1. Second charged tRNA has entered A site, facilitated by EF-Tu; first elongation step commences

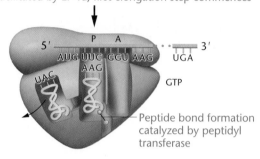

Step 2. Peptide bond forms; uncharged tRNA moves to the E site and subsequently out of the ribosome; the mRNA has been translocated three bases to the left, resulting in the tRNA bearing the dipeptide to shift into the P site.

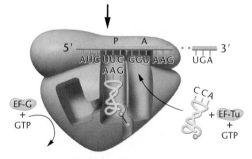

Step 3. The first elongation step is complete, facilitated by EF-G. The third charged tRNA is ready to enter the A site.

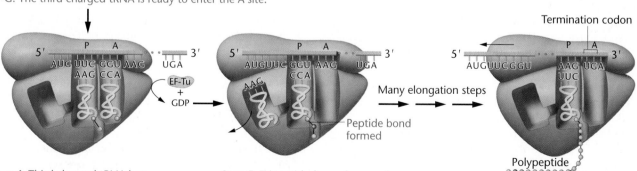

Step 4. Third charged tRNA has entered A site, facilitated by EF-Tu; second elongation step begins

Step 5. Tripeptide formed; second elongation step completed; uncharged tRNA moves to E site

Step 6. Polypeptide chain synthesized and exiting ribosome

FIGURE 13–7 Elongation of the growing polypeptide chain during translation.

well as the energy derived from hydrolysis of GTP. The result is that the third triplet of mRNA is now in a position to accept another specific charged tRNA into the A site (Step 4). One simple way to distinguish the two sites is to remember that, *following the shift*, the P site (P for peptide) contains a tRNA attached to a peptide chain, whereas the A site (A for amino acid) contains a tRNA with an amino acid attached.

The sequence of elongation is repeated over and over (Steps 4 and 5). An additional amino acid is added to the growing polypeptide chain each time the mRNA advances through the ribosome. Once a polypeptide chain of reasonable size is assembled (about 30 amino acids), it begins to emerge from the base of the large subunit, as illustrated in Step 6. A tunnel exists within the large subunit, from which the elongating polypeptide emerges.

As we have seen, the role of the small subunit during elongation is one of "decoding" the triplets present in mRNA, while that of the large subunit is peptide bond synthesis. The efficiency of the process is remarkably high. The observed error rate is only about 10^{-4}. At this rate, an incorrect amino acid will occur only once in every 20 polypeptides of an average length of 500 amino acids. In *E. coli*, elongation occurs at a rate of about 15 amino acids per second at 37°C.

Termination

Termination, the third phase of translation, is depicted in Figure 13–8. Termination of protein synthesis is signaled by one or more of three triplet codes in the A site: UAG, UAA, or UGA. These codons do not specify an amino acid, nor do they call for a tRNA in the A site. They are called **stop codons, termination codons**, or **nonsense codons**. The finished polypeptide is therefore still attached to the terminal tRNA at the P site, and the A site is empty. The termination codon signals the action of **GTP-dependent release factors**, which cleave the polypeptide chain from the terminal tRNA, releasing it from the translation complex (Step 1). Once cleavage occurs, the tRNA is released from the ribosome, which then dissociates into its subunits (Step 2). If a termination codon appears in the middle of an mRNA molecule as a result of mutation, cleavage occurs, and the polypeptide chain is prematurely terminated.

Polyribosomes

As elongation proceeds and the initial portion of mRNA has passed through the ribosome, this mRNA is free to associate with another small subunit to form a second initiation complex. This process can be repeated several times with a single mRNA and results in what are called **polyribosomes** or just **polysomes**.

Polyribosomes can be isolated and analyzed following a gentle lysis of cells. The photos in Figure 13–9 show these complexes as seen under an electron microscope. In Figure 13–9(a), you can see the thin lines of mRNA between the individual ribosomes. The micrograph in Figure 13–9(b) is even more remarkable, for it shows the polypeptide chains emerging from the ribosomes during translation. The formation of polysome complexes represents an efficient use of the

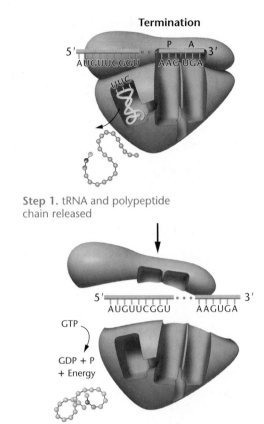

Termination

Step 1. tRNA and polypeptide chain released

Step 2. GTP-dependent termination factors activated; components separate; polypeptide folds into protein

FIGURE 13–8 Termination of the process of translation.

components available for protein synthesis during a particular unit of time. Using the analogy of a song recorded on a tape and a tape recorder, in polysome complexes one tape (mRNA) would be played simultaneously by several recorders (the ribosomes), but at any given moment, each song (the polypeptide being synthesized in each ribosome) would be at a different point in the lyrics.

13.3 Crystallographic Analysis Has Revealed Many Details about the Functional Prokaryotic Ribosome

Our knowledge of the process of translation and the structure of the ribosome, as described in the previous sections, is based primarily on biochemical and genetic observations, in addition to visualization of ribosomes under the electron microscope. Because of the tremendous size and complexity of the functional ribosome during active translation, obtaining the crystals needed to perform X-ray diffraction studies was extremely difficult. Nevertheless, great strides have been made in the past several years. First, the individual ribosomal subunits were crystallized and examined in several laboratories, most prominently that of V. Ramakrishnan. Then, in 2001, the crystal structure of the intact 70S ribosome, complete with associated mRNA and tRNAs, was examined by

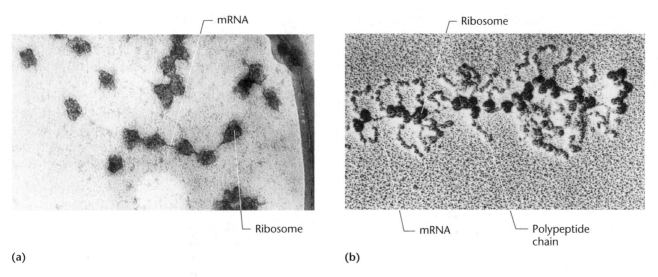

FIGURE 13–9 Polyribosomes as seen under the electron microscope. Those in (a) were derived from rabbit reticulocytes engaged in the translation of hemoglobin mRNA. The polyribosomes in (b) were taken from giant salivary gland cells of the midgefly, *Chironomus thummi*. Note that the nascent polypeptide chains are apparent as they emerge from each ribosome. Their length increases as translation proceeds from left (5′) to right (3′) along the mRNA.

Harry Noller and his colleagues—in essence, the entire translational complex was seen at the atomic level. Both Ramakrishnan and Noller derived the ribosomes from the bacterium *Thermus thermophilus*.

Many noteworthy observations have been made from these investigations. One of the models based on Noller's findings is shown as the opening photograph of this chapter (p. 277). For example, the sizes and shapes of the subunits, measured at atomic dimensions, are in agreement with earlier estimates based on high-resolution electron microscopy. Furthermore, the shape of the ribosome changes during different functional states, attesting to the dynamic nature of the process of translation. A great deal has also been learned about the prominence and location of the RNA components of the subunits. For example, about one-third of the 16S RNA is responsible for producing a flat projection within the smaller 30S subunit referred to as the "platform," which modulates movement of the mRNA–tRNA complex during translocation.

Information also supports the concept that RNA is the major "player" in the ribosome during translation. The interface between the two subunits, considered to be the location in the ribosome where polymerization of amino acids occurs, is composed almost exclusively of RNA. In contrast, the numerous ribosomal proteins are found mostly on the periphery of the ribosome. These observations confirm what has been predicted on genetic grounds—the catalytic steps that join amino acids during translation occur under the direction of RNA, not proteins.

Another interesting finding involves the actual location of the three sites predicted to house tRNAs during translation. All three sites (A, P, and E) have been identified, and in each case, the RNA of the ribosome makes direct contact with the various loops and domains of the tRNA molecule. This observation points to the importance of the different regions of tRNA and helps us understand why the specific three-dimensional conformation of all tRNA molecules has been preserved throughout evolution.

A final observation takes us back almost 50 years, to when Francis Crick proposed the wobble hypothesis. The Ramakrishnan research group has identified the precise location along the 16S RNA of the 30S subunit involved in the decoding step between mRNA and tRNA. At this location, two particular nucleotides of the 16S RNA actually flip out and probe the codon–anticodon region and are also believed to check the accuracy of base pairing during this interaction. According to the wobble hypothesis, the stringency of this step is high for the first two base pairs but less stringent for the third (or wobble) base pair.

These landmark studies provide us with a much better picture of the dynamic changes that must occur within the ribosome during translation. However, numerous questions still remain about ribosome structure and function. In particular, the role of the many ribosomal proteins has yet to be clarified. Nevertheless, the models that are emerging based on the work of Noller, Ramakrishnan, and their many colleagues provide us with a much better understanding of the mechanism of translation.

13.4 Translation Is More Complex in Eukaryotes

The general features of the model we just discussed were initially derived from investigations of the translation process in bacteria. As we have seen, one main difference between translation in prokaryotes and eukaryotes is that in the latter, translation occurs on larger ribosomes whose rRNA and protein components are more complex than those of prokaryotes (see Figure 13–1).

Several other differences are also important. Eukaryotic mRNAs are much longer-lived than their prokaryotic counterparts. Most exist for hours rather than minutes prior to degradation by nucleases in the cell; thus they remain available much longer to orchestrate protein synthesis.

Several aspects involving the initiation of translation are also different in eukaryotes. First, as we discussed in Chapter 12, the 5′ end of mRNA is capped with a 7-methylguanosine (7mG) residue at maturation. The presence of the 7mG cap, absent in prokaryotes, is essential to efficient translation, since RNAs that lack the cap are translated poorly. In addition, most eukaryotic mRNAs contain a short recognition sequence that surrounds the initiating AUG codon—5′-ACCAUGG. Named after Marilyn Kozak, who discovered it, this **Kozak sequence** appears to function during initiation in the same way that the Shine-Dalgarno sequence functions in prokaryotic mRNA. Both greatly facilitate the initial binding of mRNA to the small subunit of the ribosome.

Another difference is that the amino acid formylmethionine is not required to initiate eukaryotic translation. However, as in prokaryotes, the AUG triplet, which encodes methionine, is essential to the formation of the translational complex, and a unique transfer RNA ($tRNA_i^{met}$) is used during initiation.

Protein factors similar to those in prokaryotes guide the initiation, elongation, and termination of translation in eukaryotes. Many of these eukaryotic factors are clearly homologous to their counterparts in prokaryotes. However, a greater number of factors are usually required during each step, and some are more complex than in prokaryotes.

Finally, recall that in eukaryotes a large proportion of the ribosomes are found in association with the membranes that make up the endoplasmic reticulum (forming the rough ER). Such membranes are absent from the cytoplasm of prokaryotic cells. This association in eukaryotes facilitates the secretion of newly synthesized proteins from the ribosomes directly into the channels of the endoplasmic reticulum. Recent studies using cryo-electron microscopy have established how this occurs. A **tunnel**

in the large subunit of ribosomes begins near the point where the two subunits interface and exits near the back of the large subunit. The location of the tunnel within the large subunit is the basis for the belief that it provides the conduit for the movement of the newly synthesized polypeptide chain out of the ribosome. In studies in yeast, newly synthesized polypeptides enter the ER through a membrane channel formed by a specific protein, Sec61. This channel is perfectly aligned with the exit point of the ribosomal tunnel. In prokaryotes, the polypeptides are released by the ribosome directly into the cytoplasm.

13.5 The Initial Insight that Proteins Are Important in Heredity Was Provided by the Study of Inborn Errors of Metabolism

Let's consider how we know that proteins are the end products of genetic expression. The first insight into the role of proteins in genetic processes was provided by observations made by Sir Archibald Garrod and William Bateson early in the twentieth century. Garrod was born into an English family of medical scientists. His father was a physician with a strong interest in the chemical basis of rheumatoid arthritis, and his eldest brother was a leading zoologist in London. It is not surprising, then, that as a practicing physician, Garrod became interested in several human disorders that seemed to be inherited. Although he also studied albinism and cystinuria, we shall describe his investigation of the disorder **alkaptonuria**. Individuals afflicted with this disorder have an important metabolic pathway blocked (Figure 13–10). As a result, they cannot metabolize the alkapton 2,5-dihydroxyphenylacetic acid, also known as homogentisic acid. Homogentisic acid accumulates in cells and

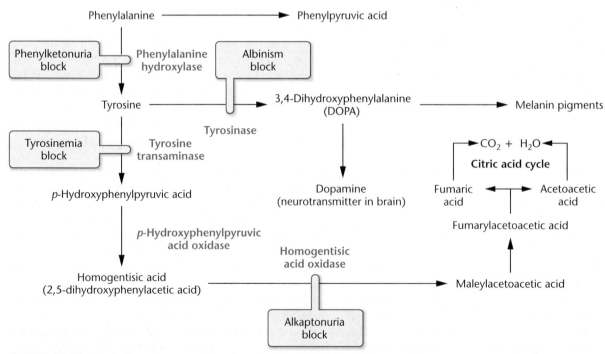

FIGURE 13–10 Metabolic pathway involving phenylalanine and tyrosine. Various metabolic blocks resulting from mutations lead to the disorders phenylketonuria, alkaptonuria, albinism, and tyrosinemia.

tissues and is excreted in the urine. The molecule's oxidation products are black and easily detectable in the diapers of newborns. The products tend to accumulate in cartilaginous areas, causing the ears and nose to darken. The deposition of homogentisic acid in joints leads to a benign arthritic condition. This rare disease is not serious, but it persists throughout an individual's life.

Garrod studied alkaptonuria by either increasing dietary protein or adding the amino acids phenylalanine or tyrosine to the diet; both of these amino acids are chemically related to homogentisic acid. Under these conditions, homogentisic acid levels increased in the urine of alkaptonurics but not in unaffected individuals. Garrod concluded that normal individuals are able to break down, or catabolize, this alkapton, but afflicted individuals are not. By studying the disorder's pattern of inheritance, Garrod further concluded that alkaptonuria is inherited as a simple recessive trait.

On the basis of these conclusions, Garrod hypothesized that hereditary information controls chemical reactions in the body and that the inherited disorders he studied are the result of alternative modes of metabolism. While the terms *genes* and *enzymes* were not familiar during Garrod's time, he used the corresponding concepts *unit factors* and *ferments*. Garrod published his initial observations in 1902.

Only a few geneticists, including Bateson, were familiar with or referred to Garrod's work. Garrod's ideas fit nicely with Bateson's belief that inherited conditions are caused by the lack of some critical substance. In 1909, Bateson published *Mendel's Principles of Heredity*, in which he linked Garrod's ferments with heredity. However, for almost 30 years, most geneticists failed to see the relationship between genes and enzymes. Garrod and Bateson, like Mendel, were ahead of their time.

Phenylketonuria

The inherited human metabolic disorder, **phenylketonuria (PKU)**, results when another reaction in the pathway shown in Figure 13–10 is blocked. Described first in 1934, this disorder can result in mental retardation and is transmitted as an autosomal recessive disease. Afflicted individuals are unable to convert the amino acid phenylalanine to the amino acid tyrosine. These molecules differ by only a single hydroxyl group (OH) that is present in tyrosine but absent in phenylalanine. The reaction is catalyzed by the enzyme **phenylalanine hydroxylase**, which is inactive in affected individuals and active at about a 30 percent level in heterozygotes. The enzyme functions in the liver. The normal blood level of phenylalanine is about 1 mg/100 mL; phenylketonurics show levels as high as 50 mg/100 mL.

As phenylalanine accumulates, it can be converted to phenylpyruvic acid and subsequently to other derivatives. These are less efficiently resorbed by the kidney and tend to spill into the urine more quickly than phenylalanine. Both phenylalanine and its derivatives subsequently enter the cerebrospinal fluid, resulting in elevated levels in the brain. The presence of these substances during early development is thought to cause mental retardation.

Screening newborns for PKU is routine throughout the United States, preventing retardation through early detection. Phenylketonuria occurs in approximately one in 11,000 births. When the condition is detected in the analysis of an infant's blood, a strict dietary regimen is instituted. A low-phenylalanine diet can reduce by-products such as phenylpyruvic acid, and the abnormalities characterizing the disease can be diminished.

Our knowledge of inherited metabolic disorders such as alkaptonuria and phenylketonuria has caused a revolution in medical thinking and practice. Human diseases, once believed to be solely attributable to the action of invading microorganisms, viruses, or parasites, clearly can have a genetic basis. We now know that hundreds of medical conditions are caused by errors in metabolism resulting from mutant genes. These human biochemical disorders include all classes of organic biomolecules.

13.6 Studies of *Neurospora* Led to the One-Gene:One-Enzyme Hypothesis

In two separate investigations beginning in 1933, George Beadle provided the first convincing experimental evidence that genes are directly responsible for the synthesis of enzymes. The first investigation, conducted in collaboration with Boris Ephrussi, involved *Drosophila* eye pigments. Together, they confirmed that mutant genes that alter the eye color of fruit flies could be linked to biochemical errors that, in all likelihood, involved the loss of enzyme function. Encouraged by these findings, Beadle then joined with Edward Tatum to investigate nutritional mutations in the pink bread mold *Neurospora crassa*. This investigation led to the **one-gene:one-enzyme hypothesis**.

Beadle and Tatum: *Neurospora* Mutants

In the early 1940s, Beadle and Tatum chose to work with *Neurospora* because much was known about its biochemistry and because mutations could be induced and isolated with relative ease. By inducing mutations, they produced strains that had genetic blocks of reactions essential to the growth of the organism.

Beadle and Tatum knew that this mold could manufacture nearly everything necessary for normal development. For example, using rudimentary carbon and nitrogen sources, this organism can synthesize 9 water-soluble vitamins, 20 amino acids, numerous carotenoid pigments, and all essential purines and pyrimidines. Beadle and Tatum irradiated asexual conidia (spores) with X-rays to increase the frequency of mutations and allowed them to be grown on "complete" medium containing all the necessary growth factors (e.g., vitamins and amino acids). Under such growth conditions, a mutant strain unable to grow on minimal medium was able to grow by virtue of supplements present in the enriched complete medium. All the cultures were then transferred to minimal medium. If growth occurred on the minimal medium, the organisms were able to synthesize

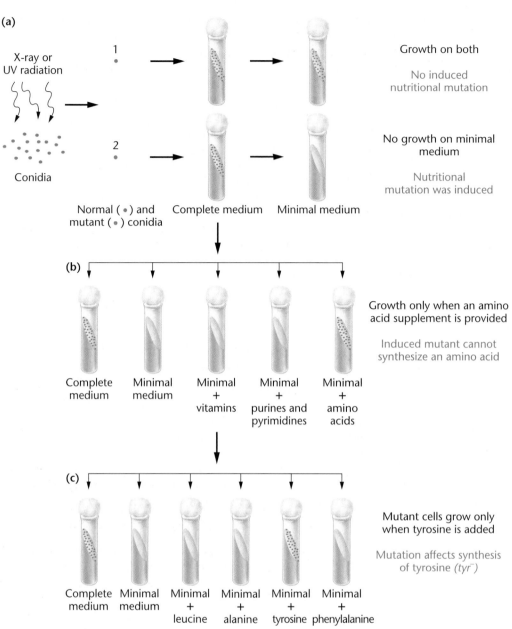

(a)

X-ray or
UV radiation

Conidia

Normal (•) and   Complete medium   Minimal medium
mutant (•) conidia

Growth on both

No induced
nutritional mutation

No growth on minimal
medium

Nutritional
mutation was induced

(b)

Complete   Minimal   Minimal   Minimal   Minimal
medium   medium   +   +   +
            vitamins   purines and   amino
                        pyrimidines   acids

Growth only when an amino
acid supplement is provided

Induced mutant cannot
synthesize an amino acid

(c)

Complete   Minimal   Minimal   Minimal   Minimal   Minimal
medium   medium   +   +   +   +
                    leucine   alanine   tyrosine   phenylalanine

Mutant cells grow only
when tyrosine is added

Mutation affects synthesis
of tyrosine *(tyr⁻)*

FIGURE 13–11
Induction, isolation, and characterization of a nutritional auxotrophic mutation in *Neurospora*. (a) Most conidia are not affected, but one conidium (shown in red) contains a mutation. In (b) and (c), the precise nature of the mutation is established and found to involve the biosynthesis of tyrosine.

all the necessary growth factors themselves, and the researchers concluded that the culture did not contain a nutritional mutation. If no growth occurred, then they concluded that the culture contained a nutritional mutation, and the only task remaining was to determine its type. These results are shown in Figure 13–11(a).

Many thousands of individual spores from this procedure were isolated and grown on complete medium. In subsequent tests on minimal medium, many cultures failed to grow, indicating that a nutritional mutation had been induced. To identify the mutant type, the mutant strains were then tested on a series of different minimal media [Figure 13–11(b)], each containing groups of supplements, and subsequently on media containing single vitamins, purines, pyrimidines, or amino acids [Figure 13–11(c)] until one specific supplement that permitted growth was found. Beadle and Tatum reasoned that the supplement that restored growth would be the molecule that the mutant strain could not synthesize.

The first mutant strain they isolated required vitamin B_6 (pyridoxine) in the medium, and the second required vitamin B_1 (thiamine). Using the same procedure, Beadle and Tatum eventually isolated and studied hundreds of mutants deficient in the ability to synthesize other vitamins, amino acids, or other substances.

The findings derived from testing over 80,000 spores convinced Beadle and Tatum that genetics and biochemistry have much in common. It seemed likely that each nutritional mutation caused the loss of the enzymatic activity that facilitated an essential reaction in wild-type organisms. It also appeared that a mutation could be found for nearly any enzymatically controlled reaction. Beadle and Tatum had thus provided sound experimental evidence for the hypothesis that *one gene specifies one enzyme*, an idea alluded to over 30 years earlier by Garrod and Bateson. With modifications, this concept was to become another major principle of genetics.

Genes and Enzymes: Analysis of Biochemical Pathways

The one-gene:one-enzyme concept and its attendant methods have been used over the years to work out many details of metabolism in *Neurospora*, *E. coli*, and a number of other microorganisms. One of the first metabolic pathways to be investigated in detail was that leading to the synthesis of the amino acid arginine in *Neurospora*. By studying seven mutant strains, each requiring arginine for growth (arg^-), Adrian Srb and Norman Horowitz ascertained a partial biochemical pathway that leads to the synthesis of this molecule. Their work demonstrates how genetic analysis can be used to establish biochemical information.

Srb and Horowitz tested each mutant strain's ability to grow if either citrulline or ornithine, two compounds with close chemical similarity to arginine, was used to supplement a minimal medium. If either compound was able to substitute for arginine, they reasoned that it must be involved in the biosynthetic pathway of arginine. They found that both molecules could be substituted in one or more strains.

Of the seven mutant strains, four of them (*arg4–7*) grew if supplied with either citrulline, ornithine, or arginine. Two of them (*arg2* and *arg3*) grew if supplied with citrulline or arginine. One strain (*arg1*) would grow only if arginine was supplied—neither citrulline nor ornithine could substitute for it. From these experimental observations, the following pathway and metabolic blocks for each mutation were deduced:

$$\text{Precursor} \xrightarrow[\text{Enzyme A}]{arg4–7} \text{Ornithine} \xrightarrow[\text{Enzyme B}]{arg2 \text{ and } arg3} \text{Citrulline} \xrightarrow[\text{Enzyme C}]{arg1} \text{Arginine}$$

The logic supporting these conclusions is as follows: If mutants *arg4* through *arg7* can grow regardless of which of the three molecules is supplied as a supplement to minimal medium, the mutations preventing growth must cause a metabolic block that occurs *prior* to the involvement of ornithine, citrulline, or arginine in the pathway. When any one of these three molecules is added, *its presence bypasses the block*. As a result, both citrulline and ornithine appear to be involved in the biosynthesis of arginine. However, the sequence of their participation in the pathway cannot be determined on the basis of these data.

On the other hand, the *arg2* and *arg3* mutations grow if supplied with citrulline, but not if they are supplied with only ornithine. Therefore, ornithine must be synthesized in the pathway *prior to the block*. Its presence will not overcome the block. Citrulline, however, *does overcome the block*, so it must be synthesized beyond the point of blockage. Therefore, the conversion of ornithine to citrulline represents the correct sequence in the pathway.

Finally, we can conclude that *arg1* represents a mutation preventing the conversion of citrulline to arginine. Neither ornithine nor citrulline can overcome the metabolic block because both participate earlier in the pathway.

Together, these reasons support the sequence of biosynthesis outlined here. Since Srb and Horowitz's work in 1944, the detailed pathway has been worked out and the enzymes controlling each step have been characterized; an abbreviated metabolic pathway is shown in Figure 13–12.

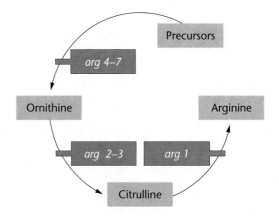

FIGURE 13–12 Abbreviated pathway resulting in the biosynthesis of arginine in *Neurospora*.

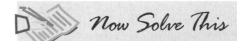

Now Solve This

Problem 12 on page 300 asks you to analyze data to establish a biochemical pathway in the bacterium *Salmonella*.

Hint: Apply the same principles and approach used to decipher biochemical pathways in *Neurospora*.

13.7 Studies of Human Hemoglobin Established that One Gene Encodes One Polypeptide

The concept of the one-gene:one-enzyme hypothesis that developed in the early 1940s was not immediately accepted by all geneticists. This is not surprising because it was not yet clear how mutant enzymes could cause variation in many phenotypic traits. For example, *Drosophila* mutants demonstrate altered eye size, wing shape, wing-vein pattern, and so on. Plants exhibit mutant varieties of seed texture, height, and fruit size. How an inactive mutant enzyme could result in such phenotypes puzzled many geneticists.

Two factors soon modified the one-gene:one-enzyme hypothesis. First, while *nearly all enzymes are proteins, not all proteins are enzymes*. As the study of biochemical genetics progressed, it became clear that all proteins are specified by the information stored in genes, leading to the more accurate phraseology, **one-gene:one-protein hypothesis**. Second, proteins often show a substructure consisting of two or more polypeptide chains. This is the basis of the quaternary protein structure, which we will discuss later in this chapter. Because each distinct polypeptide chain is encoded by a separate gene, a more accurate statement of Beadle and Tatum's basic tenet is **one-gene:one-polypeptide chain hypothesis**. These modifications of the original hypothesis became apparent during the analysis of hemoglobin structure in individuals afflicted with sickle-cell anemia.

(a)

(b)

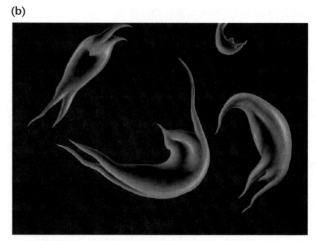

FIGURE 13–13 A comparison of erythrocytes from (a) healthy individuals, and (b) those with sickle-cell anemia.

Sickle-cell Anemia

The first direct evidence that genes specify proteins other than enzymes came from work on mutant hemoglobin molecules found in humans afflicted with the disorder **sickle-cell anemia**. Affected individuals have erythrocytes that, under low oxygen tension, become elongated and curved because of the polymerization of hemoglobin. The sickle shape of these erythrocytes is in contrast to the biconcave disc shape characteristic in unaffected individuals (Figure 13–13). Those with the disease suffer attacks when red blood cells aggregate in the venous side of capillary systems, where oxygen tension is very low. As a result, a variety of tissues are deprived of oxygen and suffer severe damage. When this occurs, an individual is said to experience a sickle-cell crisis. If left untreated, a crisis can be fatal. The kidneys, muscles, joints, brain, gastrointestinal tract, and lungs can be affected.

In addition to suffering crises, these individuals are anemic because their erythrocytes are destroyed more rapidly than are normal red blood cells. Compensatory physiological mechanisms include increased red blood cell production by bone marrow, along with accentuated heart action. These mechanisms lead to abnormal bone size and shape, as well as dilation of the heart.

In 1949, James Neel and E. A. Beet demonstrated that the disease is inherited as a Mendelian trait. Pedigree analysis revealed three genotypes and phenotypes controlled by a single pair of alleles, Hb^A and Hb^S. Unaffected and affected individuals result from the homozygous genotypes Hb^AHb^A and Hb^SHb^S, respectively. The red blood cells of heterozygotes, who exhibit the **sickle-cell trait** but not the disease, undergo much less sickling because over half of their hemoglobin is normal. Although they are largely unaffected, heterozygotes are "carriers" of the defective gene, which is transmitted on average to 50 percent of their offspring.

In that same year, Linus Pauling and his coworkers provided the first insight into the molecular basis of the disease. They showed that hemoglobins isolated from diseased and normal individuals differ in their rates of electrophoretic migration. In this technique, charged molecules migrate in an electric field. If the net charge of two molecules is different, their rates of migration will be different. On this basis, Pauling and his colleagues concluded that a chemical difference exists between normal **(HbA)** and sickle-cell **(HbS)** hemoglobin.

Figure 13–14(a) shows the migration pattern of hemoglobin derived from individuals of all three possible genotypes when it was subjected to **starch gel electrophoresis**. The gel provides the supporting medium for the molecules during migration. In this experiment, samples were placed at a point of origin between a cathode ($-$) and an anode ($+$), and an electric field was applied. The migration pattern revealed that all of the molecules moved toward the anode, indicating a net negative charge. However, HbA migrated farther than HbS, suggesting that its net negative charge was greater. The electrophoretic pattern of hemoglobin derived from carriers revealed the presence of both HbA and HbS and confirmed their heterozygous genotype.

Pauling's findings suggested two possibilities. It was known that hemoglobin consists of four nonproteinaceous, iron-containing *heme groups* and a *globin portion* that contains four polypeptide chains. The alteration in net charge in HbS had to be due, theoretically, to a chemical change in one of these components.

Work carried out between 1954 and 1957 by Vernon Ingram resolved this question. He demonstrated that the chemical change occurs in the primary structure of the globin portion of the hemoglobin molecule. Using the **protein fingerprinting technique** shown in Figure 13–14, Ingram showed that HbS differs in amino acid composition compared to HbA. Human adult hemoglobin contains two identical α chains of 141 amino acids and two identical β chains of 146 amino acids in its quaternary structure.

The fingerprinting technique involves enzymatic digestion of the protein into peptide fragments. The mixture is then placed on absorbent paper and exposed to an electric field, where migration occurs according to net charge. The paper is then turned at a right angle to its first exposure and placed in a solvent, where chromatographic action causes the migration of the peptides in the second direction. The end result is a two-dimensional separation of the peptide fragments into a distinctive pattern of spots or a "fingerprint." Ingram's work revealed that HbS and HbA differed by only a single peptide fragment [Figure 13–14(b)]. Further analysis revealed just a single amino acid change: Valine was substituted for glutamic acid at the sixth position of the β chain, thus accounting for the peptide difference [Figure 13–14(c)].

FIGURE 13–14 Investigation of hemoglobin derived from Hb^AHb^A, Hb^AHb^S, and Hb^SHb^S individuals by using electrophoresis, protein fingerprinting, and amino acid analysis. Hemoglobin from individuals with sickle-cell anemia (Hb^SHb^S) (a) migrates differently in an electrophoretic field, (b) shows an altered peptide in fingerprint analysis, and (c) shows an altered amino acid, valine, at the sixth position in the β chain. During electrophoresis, heterozygotes (Hb^AHb^S) reveal both forms of hemoglobin.

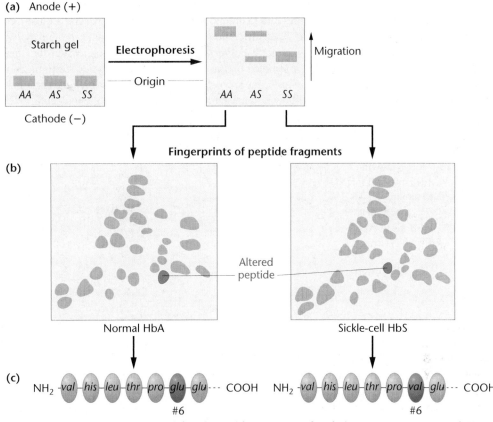

The significance of this discovery has been multifaceted. It clearly establishes that a single gene provides the genetic information for a single polypeptide chain. Studies of HbS also demonstrate that a mutation can affect the phenotype by directing a single amino acid substitution. Also, by providing the explanation for sickle-cell anemia, the concept of **inherited molecular disease** was firmly established. Finally, this work has led to a thorough study of human hemoglobins, which has provided valuable genetic insights.

In the United States, sickle-cell anemia is found almost exclusively in the African-American population. It affects about 1 in every 625 African-American infants. Currently, about 50,000 to 75,000 individuals are afflicted. In 1 of about every 145 African-American married couples, both partners are heterozygous carriers. In these cases, each of their children has a 25 percent chance of having the disease.

Human Hemoglobins

Having introduced human hemoglobins in an historical context, it may be useful to extend our discussion about these molecules in our species. Molecular analysis reveals that a variety of hemoglobin molecules are produced in humans at different stages of the life cycle. All are tetramers consisting of numerous combinations of seven distinct polypeptide chains, each encoded by a separate gene. The expression of these various genes is developmentally regulated.

Almost all adult hemoglobin consists of **HbA**, which contains two α and two β chains. Recall that the mutation in sickle-cell anemia involves the β chain. HbA represents about 98

percent of all hemoglobin found in an adult's erythrocytes after the age of six months. The remaining 2 percent consists of **HbA$_2$**, a minor adult component. This molecule contains two α chains and two **delta (δ) chains**. The δ chain is very similar to the β chain, consisting of 146 amino acids.

During embryonic and fetal development, a completely different set of polypeptides are found in hemoglobin. The earliest set detected is in embryos and is called **Gower 1**. It contains two **zeta (ζ) chains**, which are similar to α chains, and two **epsilon (ε) chains**, which are similar to β chains. By eight weeks gestation, the embryonic form is gradually replaced by another hemoglobin molecule with still different polypeptides. This molecule is called **HbF**, or **fetal hemoglobin**, and consists of two alpha chains and two **gamma (γ) chains**. There are two different types of γ chains designated $^G\gamma$ and $^A\gamma$. Both are similar to β chains and differ from each other by only a single amino acid. These persist until birth, and gradually, HbF is replaced with HbA and HbA$_2$.

13.8 Protein Structure Is the Basis of Biological Diversity

Having established that the genetic information is stored in DNA and influences cellular activities through the proteins it encodes, we turn now to a brief discussion of protein structure. How can these molecules play such a critical role in determining the complexity of cellular activities? As we shall see, the fundamental aspects of the structure of proteins provide the basis for incredible complexity and diversity. At the outset, we should differentiate between **polypeptides** and **proteins**. Both are molecules composed of amino acids. They differ, however, in their state of assembly and functional capacity. Polypeptides are the precursors of proteins. As it is assembled on the ribosome during translation, the molecule is called a *polypeptide*. When released from the ribosome following translation, a polypeptide folds up and assumes a higher order of structure. When this occurs, a three-dimensional conformation emerges. In many cases, several polypeptides interact to produce this conformation. When the final conformation is achieved, the molecule is now fully functional and is appropriately called a *protein*. Its three-dimensional conformation is essential to the function of the molecule.

FIGURE 13–15 Chemical structures and designations of the 20 amino acids found in living organisms, divided into four major categories. Each amino acid has two abbreviations; that is, alanine is designated either ala or A (a universal nomenclature).

1. Nonpolar: Hydrophobic

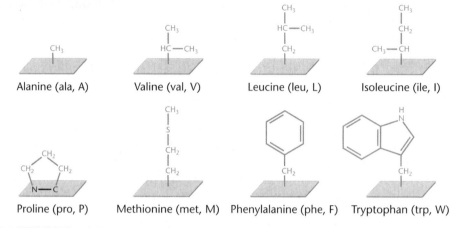

Alanine (ala, A) Valine (val, V) Leucine (leu, L) Isoleucine (ile, I)

Proline (pro, P) Methionine (met, M) Phenylalanine (phe, F) Tryptophan (trp, W)

2. Polar: Hydrophilic

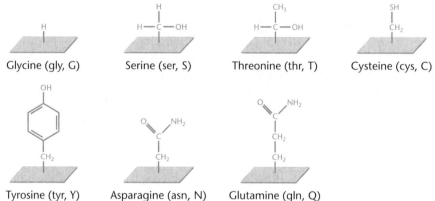

Glycine (gly, G) Serine (ser, S) Threonine (thr, T) Cysteine (cys, C)

Tyrosine (tyr, Y) Asparagine (asn, N) Glutamine (gln, Q)

Amino group

R

$H_3N^+ - C - C$ Carboxyl group

H

Amino acid structure

3. Polar: positively charged (basic)

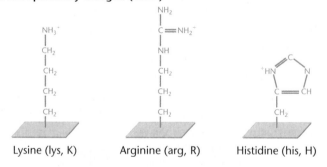

Lysine (lys, K) Arginine (arg, R) Histidine (his, H)

4. Polar: negatively charged (acidic)

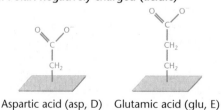

Aspartic acid (asp, D) Glutamic acid (glu, E)

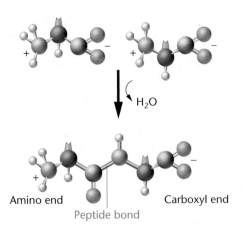

FIGURE 13–16 Peptide bond formation between two amino acids, resulting from a dehydration reaction.

The polypeptide chains of proteins, like nucleic acids, are linear nonbranched polymers. There are 20 amino acids that serve as the subunits (the building blocks) of proteins. Each amino acid has a **carboxyl group**, an **amino group**, and an **R (radical) group** (a side chain) bound covalently to a **central carbon (C) atom**. The R group gives each amino acid its chemical identity. Figure 13–15 shows the 20 R groups, which exhibit a variety of configurations and can be divided into four main classes: (1) *nonpolar* (hydrophobic), (2) *polar* (hydrophilic), (3) *positively charged*, and (4) *negatively charged*. Because polypeptides are often long polymers and because each position may be occupied by any 1 of the 20 amino acids with their unique chemical properties, enormous variation in chemical conformation and activity is possible. For example, if an average polypeptide is composed of 200 amino acids (molecular weight of about 20,000 Da), 20^{200} different molecules, each with a unique sequence, can be created using the 20 different building blocks.

Around 1900, German chemist Emil Fischer determined the manner in which the amino acids are bonded together. He showed that the amino group of one amino acid reacts with the carboxyl group of another amino acid during a dehydration reaction, releasing a molecule of H_2O. The resulting covalent bond is a **peptide bond** (Figure 13–16). Two amino acids linked together constitute a **dipeptide**, three a **tripeptide**, and so on. Once 10 or more amino acids are linked by peptide bonds, the chain is referred to as a polypeptide. Generally, no matter how long a polypeptide is, it will contain a free amino group at one end (the N-terminus) and a free carboxyl group at the other end (the C-terminus).

Four levels of protein structure are recognized: primary, secondary, tertiary, and quaternary. The sequence of amino acids in the linear backbone of the polypeptide constitutes its **primary structure**. It is specified by the sequence of deoxyribonucleotides in DNA via an mRNA intermediate. The primary structure of a polypeptide helps determine the specific characteristics of the higher orders of organization as a protein is formed.

The **secondary structure** refers to a regular or repeating configuration in space assumed by amino acids closely aligned in the polypeptide chain. In 1951, Linus Pauling and Robert Corey predicted, on theoretical grounds, an **α helix** as one type of secondary structure. The α-helix model [Figure 13–17(a)] has since been confirmed by X-ray crystallographic studies. The helix is composed of a spiral chain of amino acids stabilized by hydrogen bonds; it is rodlike and has the greatest possible theoretical stability.

The side chains (the R groups) of amino acids extend outward from the helix, and each amino acid residue occupies a vertical distance of 1.5 Å in the helix. There are 3.6 residues per turn. While left-handed helices are theoretically possible, all proteins seen with an α helix are right-handed.

FIGURE 13–17 (a) The right-handed α helix, which represents one form of secondary structure of a polypeptide chain. (b) The β-pleated sheet, an alternative form of secondary structure of polypeptide chains. To maintain clarity, not all atoms are shown.

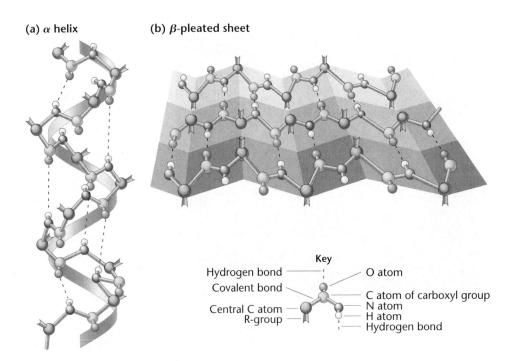

Also in 1951, Pauling and Corey proposed a second structure, the **β-pleated sheet**. In this model, a single-polypeptide chain folds back on itself or several chains run in either parallel or antiparallel fashion next to one another. Each such structure is stabilized by hydrogen bonds formed between atoms on adjacent chains [Figure 13–17(b)]. A zigzagging plane is formed in space with adjacent amino acids 3.5 Å apart.

As a general rule, most proteins demonstrate a mixture of α-helix and β-pleated-sheet structures. Globular proteins, most of which are round in shape and water soluble, usually contain a core of β-pleated-sheet structure as well as many areas with α-helical structures. The more structurally rigid proteins, many of which are water insoluble, rely on more extensive β-pleated-sheet regions for their rigidity. For example, **fibroin**, the protein made by the silk moth, depends extensively on this form of secondary structure.

The secondary structure describes the arrangement of amino acids within certain areas of a polypeptide chain, but the **tertiary structure** defines the three-dimensional conformation of the entire chain in space. Each protein twists and turns and loops around itself in a very particular fashion, characteristic of the specific protein. A model of the three-dimensional tertiary structure of the respiratory pigment myoglobin is shown in Figure 13–18. Three aspects of this level of structure are most important in stabilizing the molecule and in determining its conformation.

1. Covalent disulfide bonds form between closely aligned cysteine residues to make the unique amino acid cystine.

2. Usually, the polar hydrophilic R groups are located on the surface, where they can interact with water.

3. The nonpolar hydrophobic R groups are usually located on the inside of the molecule, where they interact with one another, avoiding interaction with water.

The three-dimensional conformation achieved by any protein is a product of the *primary structure* of the polypeptide. Thus, the genetic code need only specify the sequence of amino

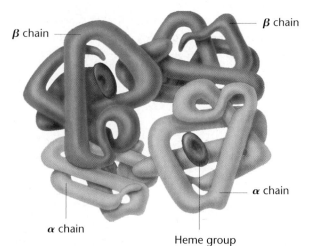

FIGURE 13–19 The quaternary level of protein structure as seen in hemoglobin. Four chains (two α and two β) interact with four heme groups to form the functional molecule.

acids to encode information that leads ultimately to the final conformation of proteins. The three stabilizing factors depend on the location of each amino acid relative to all others in the chain. As the polypeptide is folded, the most thermodynamically stable conformation possible results. This level of organization is essential because the specific function of any protein is directly related to its tertiary structure.

The **quaternary structure** of proteins applies only to those composed of more than one polypeptide chain and indicates the position of the various chains in relation to one another. This type of protein is *oligomeric*, and each chain is a *protomer* or, less formally, a *subunit*. Protomers have conformations that facilitate their fitting together in a specific complementary fashion. Hemoglobin, an oligomeric protein consisting of four polypeptide chains, has been studied in great detail—its quaternary protein structure is shown in Figure 13–19. Most enzymes, including DNA and RNA polymerase, demonstrate quaternary structure.

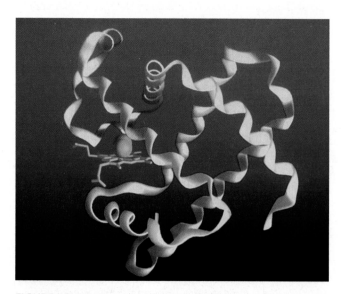

FIGURE 13–18 The tertiary level of protein structure in a respiratory pigment, myoglobin. The bound oxygen atom is shown in red.

Now Solve This

Problem 30 on page 301 asks you to consider the potential impact of several amino acid substitutions that result due to mutations in one of the genes encoding one of the chains making up human hemoglobin.

Hint: When considering the three amino acids (glutamic acid, lysine, and valine), consider the net charge of each R group.

13.9 Posttranslational Modification Alters the Final Protein Product

Polypeptide chains, like RNA transcripts, are often modified after they have been synthesized. This additional processing is broadly described as **posttranslational modification**. Although many of these alterations are detailed biochemical transformations beyond the scope of this discussion, you should be aware that they occur and that they are critical to the functional capability of the final protein product. Several examples of posttranslational modification are as follows.

1. *The N-terminus amino acid is usually removed or modified.* For example, either the formyl group or the entire formylmethionine residue in bacterial polypeptides is usually removed enzymatically. In eukaryotic polypeptide chains, the amino group of the initial methionine residue is often removed, and the amino group of the N-terminal residue may be modified (acetylated).

2. *Individual amino acid residues are sometimes modified.* For example, phosphates may be added to the hydroxyl groups of certain amino acids, such as tyrosine. Modifications such as these create negatively charged residues that may form an ionic bond with other molecules. The process of phosphorylation is extremely important in regulating many cellular activities and is a result of the action of enzymes called **kinases**. At other amino acid residues, methyl groups or acetyl groups may be added enzymatically, which can also affect the function of the modified polypeptide chain.

3. *Carbohydrate side chains are sometimes attached.* These are added covalently, producing **glycoproteins**, an important category of extracellular molecules, such as those specifying the antigens in the ABO blood-type system in humans.

4. *Polypeptide chains may be trimmed.* For example, insulin is first translated into a longer molecule that is enzymatically trimmed to its final 51-amino-acid form.

5. *Signal sequences are removed.* At the N-terminal end of some proteins, a sequence of up to 30 amino acids is found that plays an important role in directing the protein to the location in the cell in which it functions. This is called a **signal sequence**, and it determines the final destination of a protein in the cell. The process is called **protein targeting**. For example, proteins whose fate involves secretion or proteins that are to become part of the plasma membrane are dependent on specific sequences for their initial transport into the lumen of the endoplasmic reticulum. While the signal sequence of various proteins with a common destination might differ in their primary amino acid sequence, they share many chemical properties. For example, those destined for secretion all contain a string of up to 15 hydrophobic amino acids preceded by a positively charged amino acid at the N-terminus of the signal sequence. Once the polypeptides are transported, but prior to achieving their functional status as proteins, the signal sequence is enzymatically removed from these polypeptides.

6. *Polypeptide chains are often complexed with metals.* The tertiary and quaternary levels of protein structure often include and are dependent on metal atoms. The function of the protein is thus dependent on the molecular complex that includes both polypeptide chains and metal atoms. Hemoglobin, containing four iron atoms along with four polypeptide chains, is a good example.

These types of posttranslational modifications are obviously important in achieving the functional status specific to any given protein. Because the final three-dimensional structure of the molecule is intimately related to its specific function, how polypeptide chains ultimately fold into their final conformations is also an important topic. For many years, it was thought that protein folding was a spontaneous process whereby the molecule achieved maximum thermodynamic stability, based largely on the combined chemical properties inherent in the amino acid sequence of the polypeptide chain(s) composing the protein. However, numerous studies have shown that, for many proteins, folding is dependent upon members of a family of still other, ubiquitous proteins called **chaperones**. Chaperone proteins (sometimes called *molecular chaperones* or *chaperonins*) function to facilitate the folding of other proteins. While the mechanism by which chaperones function is not yet clear, like enzymes, they do not become part of the final product. Initially discovered in *Drosophila*, in which they are called **heat-shock proteins**, chaperones have been discovered in a variety of organisms, including bacteria, animals, and plants. Ultimately, protein folding is a critically important process, not only because misfolded proteins may be nonfunctional, but also because improperly folded proteins can be dangerous. It is becoming clear that some disorders, such as the **spongiform encephalopathies** (**mad cow disease** in cattle and **Creutzfeldt–Jakob disease** in humans) are caused by the presence of incorrectly folded neural proteins. Currently, many laboratories are focused on trying to understand how protein folding occurs normally, as well as how the presence of misfolded polypeptides "poisons" the folding of normal polypeptides and ultimately causes cell death and disease.

13.10   Protein Function Is Directly Related to the Structure of the Molecule

The essence of life on Earth rests at the level of diverse cellular function. One can argue that DNA and RNA simply serve as vehicles to store and express genetic information. However, proteins are at the heart of cellular function. And it is the capability of cells to assume diverse structures and functions that distinguishes most eukaryotes from less evolutionarily advanced organisms such as bacteria. Therefore, an introductory understanding of protein function is critical to a complete view of genetic processes.

Proteins are the most abundant macromolecules found in cells. As the end products of genes, they play many diverse roles. For example, the respiratory pigments **hemoglobin** and **myoglobin** transport oxygen, which is essential for cellular metabolism. **Collagen** and **keratin** are structural proteins associated with the skin, connective tissue, and hair of organisms. **Actin** and **myosin** are contractile proteins, found in abundance in muscle tissue. Still other examples are the **immunoglobulins**, which function in the immune system of vertebrates; **transport proteins**, involved in movement of molecules across membranes; some of the **hormones** and their **receptors**, which regulate various types of chemical activity; and **histones**, which bind to DNA in eukaryotic organisms.

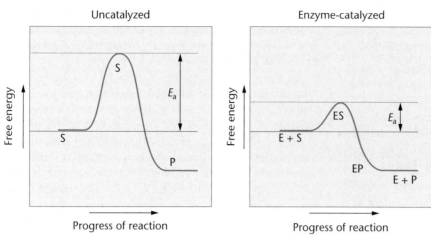

E = Enzyme
S = Substrate
P = Product

FIGURE 13–20 Energy requirements of an uncatalyzed versus an enzymatically catalyzed chemical reaction. The energy of activation (E_a) necessary to initiate the reaction is substantially lower as a result of catalysis.

The largest group of proteins with a related function are the **enzymes**. Since we have referred to these molecules throughout this chapter, it may be useful to extend our discussion and include a more detailed description of their biological role.

Enzymes specialize in catalyzing chemical reactions within living cells. They increase the rate at which a chemical reaction reaches equilibrium but do not alter the end point of the chemical equilibrium. Their remarkable, highly specific catalytic properties largely determine the metabolic capacity of any cell type. The specific functions of many enzymes involved in the genetic and cellular processes of cells are described throughout this text.

Biological catalysis is a process whereby the **energy of activation** (E_a) for a given reaction is lowered (Figure 13–20). The energy of activation is the increased kinetic energy state that molecules usually must reach before they react with one another. This state can be attained as a result of elevated temperatures, but enzymes allow biological reactions to occur at lower physiological temperatures. In this way, enzymes make life as we know it possible.

The catalytic properties and specificity of an enzyme are determined by the chemical configuration of the molecule's **active site**. This site is associated with a crevice, a cleft, or a pit on the surface of the enzyme, which binds the reactants, or substrates, facilitating their interaction. Enzymatically catalyzed reactions control metabolic activities in the cell. Each reaction is either catabolic or anabolic. **Catabolism** is the degradation of large molecules into smaller, simpler ones with the release of chemical energy. **Anabolism** is the synthetic phase of metabolism and yields the various components that make up nucleic acids, proteins, lipids, and carbohydrates.

13.11 Proteins Consist of Functional Domains

We conclude our discussion of proteins by briefly discussing the important finding that regions made up of specific amino acid sequences are associated with specific functions in protein molecules. Such sequences, usually between 50 and 300 amino acids, constitute **protein domains** and are represented by modular portions of the protein that fold into stable, unique conformations independently of the rest of the molecule. Different domains impart different functional capabilities. Some proteins contain only a single domain, while others contain two or more.

The significance of domains rests at the tertiary structure level of proteins. Each such modular unit can be a mixture of secondary structures, including both α helices and β-pleated sheets. The unique conformation that is assumed in a single domain imparts a specific function to the protein. For example, a domain may serve as the catalytic basis of an enzyme, or it may impart the capability to bind to a specific ligand as part of a membrane or another molecule. Thus, in the study of proteins, you will hear of *catalytic domains*, *DNA-binding domains*, and so on. A protein must be envisioned as being composed of a series of structural and functional modules. Obviously, the presence of multiple domains in a single protein increases the versatility of each molecule and adds to its functional complexity.

Exon Shuffling

An interesting proposal to explain the genetic origin of protein domains was put forward by Walter Gilbert in 1977. Gilbert suggested that the functional regions of genes in higher organisms consist of collections of exons originally present in ancestral genes that were brought together through recombination during the course of evolution. Referring to this process as **exon shuffling**, Gilbert proposed that exons, like protein domains, are also modular, and that during evolution, exons may have been reshuffled between genes in eukaryotes, with the result that different genes share similar domains.

Since 1977, a serious research effort has been directed toward the analysis of gene structure. In 1985, more direct evidence in favor of Gilbert's proposal of exon modules was presented. For example, the human gene encoding the membrane receptor for low-density lipoproteins (LDL) was isolated and sequenced. The **LDL receptor protein** is essential to the transport of plasma cholesterol into the cell. It mediates endocytosis and seems to have numerous functional domains. These include the capa-

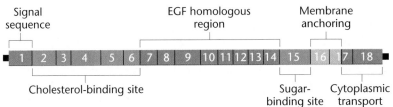

Signal sequence

EGF homologous region

Membrane anchoring

| 1 | 2 | 3 | 4 | 5 | 6 | 7 | 8 | 9 | 10 | 11 | 12 | 13 | 14 | 15 | 16 | 17 | 18 |

Cholesterol-binding site

Sugar-binding site

Cytoplasmic transport

FIGURE 13–21 The 18 exons making up the gene encoding the LDL receptor protein are organized into five functional domains and one signal sequence.

bility to bind specifically to the LDL substrates and to interact with other proteins at different levels of the membrane during transport across it. In addition, this receptor protein is modified posttranslationally by the addition of a carbohydrate; a domain must exist that links to this carbohydrate.

Detailed analysis of the gene encoding this protein supports the concept of exon modules and their shuffling during evolution. The gene is quite large—45,000 nucleotides—and contains 18 exons. These represent only slightly fewer than 2600 nucleotides. These exons are related to the functional domains of the protein *and* appear to have been recruited from other genes during evolution.

Figure 13–21 shows these relationships. The first exon encodes a signal sequence that is removed from the protein before the LDL receptor becomes part of the membrane. The next five exons represent the domain specifying the binding site for cholesterol. This domain is made up of a sequence of 40 amino acids repeated seven times. The next domain consists of a sequence of 400 amino acids bearing a striking homology to the peptide-hormone epidermal growth factor (EGF) in mice. This region is encoded by eight exons and contains three repetitive sequences of 40 amino acids. A similar sequence is also found in three blood-clotting proteins. The fifteenth exon specifies the domain for the posttranslational addition of the carbohydrate, while the next two specify regions of the protein that are part of the membrane, anchoring the receptor to specific sites on the cell surface.

These observations concerning the LDL exons constitute fairly compelling support for the theory of exon shuffling during evolution. Certainly, there is no disagreement that protein domains are responsible for specific molecular interactions.

GENETICS, TECHNOLOGY, AND SOCIETY

Mad Cow Disease: The Prion Story

In March 1996, the British government announced that a new brain disease had killed 10 young Britons, and that the victims might have acquired the disease by eating affected beef. Bovine spongiform encephalopathy (BSE), popularly known as mad cow disease, slowly destroys brain cells and is always fatal. Recent studies confirm that BSE and the human disease, new variant Creutzfeldt-Jakob disease (nvCJD), are so similar at the molecular and pathological levels that they are very likely to be the same disease. Mad cow disease triggered political turmoil in Europe, a worldwide ban on British beef, and the near-collapse of the $8.9 billion British beef industry. The European Union demanded the slaughter and incineration of 4.7 million British cattle, a campaign that cost the government more than $12 billion in compensation to farmers, milk import, and the purchase of stock for new herds. Most European countries, as well as the Middle East, Japan, and Canada, have discovered BSE in at least one cow. Although most nvCJD

cases have occurred in Britain, cases have also appeared in France, Italy, Ireland, South America, Canada, and the United States. More than 125 people have now died of nvCJD, and epidemiologists estimate that many more cases will appear in the next two decades.

BSE, nvCJD, and Creutzfeldt-Jakob disease (CJD) are all members of a group of neurological diseases known as transmissible spongiform encephalopathies (TSEs), which affect animals and humans. In this group of diseases, the affected brain tissue eventually resembles a sponge (hence, spongiform) and is riddled with proteinaceous deposits. Victims of the disease lose motor function, become demented, and ultimately die. CJD and nvCJD differ symptomatically in that victims of nvCJD display psychological symptoms such as depression or anxiety prior to developing neurological symptoms such as shaking or paralysis. Also, patients with nvCJD are usually young (16–40 years), whereas CJD normally affects people over 55. A number of CJD cases arise spontaneously and randomly,

at a rate of one per million per year worldwide, but CJD can also be inherited as an autosomal dominant condition.

CJD can be transmitted through corneal or nervous tissue grafts or by injection of a growth hormone derived from human pituitary glands. Kuru, a CJD-like disease of the Fore people of New Guinea, was transmitted from person to person through ritualistic cannibalism. Transmissible spongiform encephalopathies in animals include scrapie (sheep and goats), chronic wasting disease (deer and elk), and BSE. Like Kuru, BSE is passed from animal to animal by ingestion of diseased animal remains, particularly neural tissue. The epidemic of BSE in Britain occurred because diseased cows and sheep were processed and fed to cattle as a protein supplement. In 1998, the British government banned the use of cows and sheep as feed for other cows and sheep, and the epidemic has subsided. The European Union has now banned the use of feeds containing animal products for all livestock. In contrast, the United States and Canada still allow nonruminant animals to

(Cont. on the next page)

consume feed containing ruminants and ruminant animals to consume feed containing nonruminants, as well as certain ruminant by-products, including blood, gelatin, and fat. However, these regulations may change as a result of the discovery of a BSE case in Canada in 2003. In addition, recent studies suggest that BSE may be transmitted via blood or tallow.

For many years, the analysis of spongiform encephalopathies defied the best efforts of scientists. The diseases are difficult to study for a number of reasons. First, they require injection of affected brain material into the brains of experimental animals, and the diseases take months or years to develop. In addition, the infectious agent is apparently not a virus or bacterium, and affected animals do not develop antibodies against these mysterious agents.

There are no treatments for the diseases, and the only way to make a firm diagnosis is to examine brain tissue after death. The infectious material is unaffected by radiation or nucleases that damage nucleic acids; however, it is destroyed by some reagents that hydrolyze or modify proteins. In the early 1980s, American scientist Stanley Prusiner purified the infectious agent and concluded that it consists of only protein. He proposed that scrapie is spread by an infectious protein particle that he called a prion. His hypothesis was dismissed by most scientists, as the idea of an infectious agent with no DNA or RNA as genetic material was heretical. However, Prusiner and others presented evidence supporting the prion hypothesis, and the notion that the disease can be transmitted by an infectious particle that contains no genetic material has gained acceptance.

If prions are composed of protein only, how do they cause disease? The answer may be as strange as the disease itself. The protein that makes up a prion (PrP) is a version of a normal protein that is synthesized in neurons and found in the brains of all adult animals. The difference between normal PrP and prion PrP lies in their secondary protein structures. Normal, noninfectious PrP folds into α helices, whereas infectious prion PrP folds into β-pleated sheets. When a normal PrP molecule contacts a prion PrP molecule, the normal protein is somehow refolded into the abnormal PrP conformation. Once the normal PrP molecule has been transformed into an abnormal PrP molecule, it spreads its lethal conformation to neighboring normal PrP molecules, and the process takes off in a chain reaction. Normal PrP is a soluble protein that is easily destroyed by heat or enzymes that digest proteins. However, abnormal infectious PrP is insoluble in detergents, resists both heat and protease digestion, and is nearly indestructible. Hence, spongiform encephalopathies can be considered diseases of secondary protein structure.

Many urgent questions need to be addressed. How extensive is BSE contamination of the world's food supply? How many humans are affected with nvCJD but don't show symptoms? Can humans or animals act as asymptomatic carriers of prion diseases? Can prions exist in other parts of the body besides the brain and spinal cord, and if so, can nvCJD be spread through blood transfusions, from mother to fetus, or by sterilized surgical instruments? Can we develop diagnostic tests and therapies for BSE and nvCJD? Are we near the end of the BSE story, or is it just beginning?

References

Balter, M. 2000. Tracking the human fallout from "mad cow disease." *Science* 289: 1452–1454.

Belay, E.D. 1999. Transmissible spongiform encephalopathies in humans. *Annu. Rev. Microbiol.* 53: 283–314.

Spencer, C. A. 2004. *Mad Cows and Cannibals: A Guide to the Transmissible Spongiform Encephalopathies.* Upper Saddle River, NJ: Pearson/Prentice-Hall.

Web Sites

Online compilation of news reports about mad cow disease, **http:// organic-consumers.org/madcow.htm**

The Official Mad Cow Disease Home Page (a compilation of news reports), **http://www.mad-cow.org**.

CHAPTER SUMMARY

1. Translation describes the synthesis of polypeptide chains, under the direction of mRNA and in association with ribosomes. This process ultimately converts the information stored in the genetic code of the DNA that makes up a gene into the corresponding sequence of amino acids making up the polypeptide.

2. Translation is a complex energy-requiring process that also depends on charged tRNA molecules and numerous protein factors. Transfer RNA (tRNA) serves as the adaptor molecule between an mRNA triplet and the appropriate amino acid.

3. The processes of translation, like transcription, can be subdivided into the stages of initiation, elongation, and termination. Translation relies on base-pairing affinities between complementary nucleotides and is more complex in eukaryotes than in prokaryotes.

4. The first insight that proteins are the end products of gene expression was provided by the study of inherited metabolic disorders in humans early in the twentieth century by Garrod. Inborn errors of metabolism leading to cystinuria, albinism, and alkaptonuria were the basis of his studies.

5. The investigation of nutritional requirements in *Neurospora* by Beadle and colleagues made it clear that mutations cause the loss of enzyme activity. Their work led to the one-gene:one-enzyme hypothesis.

6. The one-gene:one-enzyme hypothesis was later revised as the one-gene:one-polypeptide chain hypothesis. Pauling and Ingram's investigations of hemoglobins from patients with sickle-cell anemia led to the discovery that one gene directs the synthesis of only one polypeptide chain.

7. Proteins, the end products of gene expression, demonstrate four levels of structural organization that together provide the chemical basis for their three-dimensional conformation, which is the basis of the molecule's function.

8. Of the myriad functions performed by proteins, the most influential role is assumed by enzymes. These highly specific, cellular catalysts play a central role in the production of all classes of molecules in living systems.

9. Proteins consist of one or more functional domains, which are often shared by different molecules. The origin of these domains may be the result of exon shuffling during evolution.

KEY TERMS

actin, 295

active site, 296

alkaptonuria, 286

α helix, 293

amino group, 293

A (aminoacyl) site, 283

aminoacyl tRNA synthetases, 281

aminoacyladenylic acid, 281

anabolism, 296

anticodon, 278

anticodon loop, 280

β-pleated sheet, 294

biological catalysis, 296

bovine spongiform encephalopathy (BSE), 297

carboxyl group, 293

catabolism, 296

central carbon (C) atom, 293

chaperone, 295

charging (of tRNA), 281

cloverleaf model of tRNA, 280

collagen, 295

Creutzfeldt–Jakob disease, 295

delta (δ) chains, 291

dipeptide, 293

E (exit) site, 283

elongation, 283

elongation factors (EFs), 283

energy of activation (E_a), 296

enzyme, 296

epsilon (ε) chains, 291

exon shuffling, 296

fetal hemoglobin, 291

fibroin, 294

formylmethionine (fmet), 283

gamma (γ) chains, 291

glycoprotein, 295

Gower 1, 291

GTP-dependent release factor, 284

HbA, 290, 291

HbA$_2$, 291

HbF 291

HbS, 290

heat-shock proteins (HSPs), 295

hemoglobin, 295

histone, 295

hormones, 295

immunoglobulin, 295

inherited molecular disease, 291

initiation, 282

initiation complex, 283

initiation factors (IFs), 283

isoaccepting tRNA, 281

keratin, 295

kinase, 295

Kozak sequence, 286

kuru, 297

LDL receptor protein, 296

mad cow disease, 295

monosome, 278

myoglobin, 295

myosin, 295

new variant Creutzfeldt-Jacob disease, 297

nonsense codon, 284

one-gene:one-enzyme hypothesis, 287

one-gene:one-polypeptide chain hypothesis, 289

one-gene:one-protein hypothesis, 289

peptide bond, 293

peptidyl transferase, 283

phenylalanine hydroxylase, 287

phenylketonuria (PKU), 287

polypeptide, 292

polyribosome, 284

polysome, 284

posttranscriptional modification, 280

posttranslational modification, 294

P (peptidyl) site, 283

primary structure of proteins, 293

prion, 298

protein, 292

protein domain, 296

protein fingerprinting technique, 290

protein targeting, 295

quaternary structure of proteins, 294

rDNA, 279

receptors, 295

ribosomal proteins, 278

ribosome, 278

R (radical) group, 293

scrapie, 297

secondary structure of proteins, 293

Shine-Dalgarno sequence, 283

sickle-cell anemia, 290

sickle-cell trait, 290

signal sequence, 295

spongiform encephalopathies, 295

starch gel electrophoresis, 290

stop codon, 284

structure of protein, 294

termination, 284

termination codon, 284

tertiary structure (of proteins), 294

transfer RNA (tRNA), 278

translation, 278

transport protein, 295

tripeptide, 293

tunnel, 286

zeta (ζ) chains, 291

INSIGHTS AND SOLUTIONS

1. The growth responses in the following chart were obtained using four mutant strains of *Neurospora* and the related compounds A, B, C, and D. None of the mutations grows on minimal medium. Draw all possible conclusions from this data.

	Growth Supplement			
Mutation	A	B	C	D
1	−	−	−	−
2	+	+	−	+
3	+	+	−	−
4	−	+	−	−

Solution: Nothing can be concluded about mutation *1* except that it lacks some essential growth factor, perhaps even unrelated to the biochemical pathway represented by mutations *2, 3,* and *4.* Nor can anything be concluded about compound C. If it is involved in the pathway, it is a product that was synthesized prior to compounds A, B, and D.

We now analyze these three compounds and the control of their synthesis by the enzymes encoded by mutations *2, 3,* and *4.* Because product B allows growth in all three cases, it may be considered the "end product"—it bypasses the block in all three instances. Using similar reasoning, product A precedes B in the pathway, since it bypasses the block in two of the three steps, and product D precedes B; yielding a partial solution:

$$C(?) \longrightarrow D \longrightarrow A \longrightarrow B$$

Now let's determine which mutations control which steps. Since mutation *2* can be alleviated by products D, B, and A, it must control a step prior to all three products, perhaps the direct conversion to D (although we cannot be certain). Mutation *3* is alleviated by B and A, so its effect must precede them in the pathway. Thus, we assign it as controlling the conversion of D to A. Likewise, we can assign mutation *4* to the conversion of A to B, leading to a more complete solution:

$$C(?) \xrightarrow{2(?)} D \xrightarrow{3} A \xrightarrow{4} B$$

PROBLEMS AND DISCUSSION QUESTIONS

1. List and describe the role of all of the molecular constituents present in a functional polyribosome.
2. Contrast the roles of tRNA and mRNA during translation, and list all enzymes that participate in the transcription and translation process.
3. Francis Crick proposed the adaptor hypothesis for the function of tRNA. Why did he choose that description?
4. During translation, what molecule bears the anticodon? the codon?
5. The α chain of eukaryotic hemoglobin is composed of 141 amino acids. What is the minimum number of nucleotides in an mRNA coding for this polypeptide chain?
6. Summarize the steps involved in charging tRNAs with their appropriate amino acids.
7. Each transfer RNA requires at least four specific recognition sites that must be inherent in its tertiary protein structure in order for it to carry out its role. What are these sites?
8. Discuss the potential difficulties involved in designing a diet to alleviate the symptoms of phenylketonuria.
9. Phenylketonurics cannot convert phenylalanine to tyrosine. Why don't these individuals exhibit a deficiency of tyrosine?
10. Phenylketonurics are often more lightly pigmented than are normal individuals. Can you suggest a reason why this is so?
11. The synthesis of flower pigments is known to be dependent on enzymatically controlled biosynthetic pathways. Postulate the role of mutant genes and their products in producing the observed phenotypes in the crosses shown.

(a) P_1:   white strain A $\times$ white strain B
     F_1:   all purple
     F_2:   9/16 purple: 7/16 white

(b) P_1:   white $\times$ pink
     F_1:   all purple
     F_2:   9/16 purple: 3/16 pink: 4/16 white

12. A series of mutations in the bacterium *Salmonella typhimurium* results in the requirement of either tryptophan or some related molecule in order for growth to occur. From the data shown here, suggest a biosynthetic pathway for tryptophan.

	Growth Supplement				
Mutation	Minimal Medium	Anthranilic Acid	Indole Glycerol Phosphate	Indole	Tryptophan
trp-8	−	+	+	+	+
trp-2	−	−	+	+	+
trp-3	−	−	−	+	+
trp-1	−	−	−	−	+

13. The study of biochemical mutants in organisms such as *Neurospora* has demonstrated that some pathways are branched. The following data illustrate the branched nature of the pathway

resulting in the synthesis of thiamine. Why don't the data support a linear pathway? Can you postulate a pathway for the synthesis of thiamine in *Neurospora*?

Growth Supplement

Mutation	Minimal Medium	Pyrimidine	Thiazole	Thiamine
thi-1	–	–	+	+
thi-2	–	+	–	+
thi-3	–	–	–	+

14. Explain why the one-gene:one-enzyme hypothesis is no longer considered to be totally accurate.

15. Why is an alteration of electrophoretic mobility interpreted as a change in the primary structure of the protein under study?

16. Hemoglobin is a tetramer consisting of two α and two β chains. How does this information relate to each of the four levels of protein structure?

17. Using sickle-cell anemia as a basis, describe what is meant by a genetic or inherited molecular disease. What are the similarities and dissimilarities between this type of a disorder and a disease caused by an invading microorganism?

18. Contrast the contributions Pauling and Ingram made to our understanding of the genetic basis for sickle-cell anemia.

19. Hemoglobins from two individuals are compared by starch gel electrophoresis and with protein fingerprinting. Electrophoresis reveals no difference in migration, but fingerprinting shows an amino acid difference. How is this possible?

20. Assuming that each nucleotide is 0.34 nm long in mRNA, how many triplet codes can simultaneously occupy space in a ribosome that is 20 nm in diameter?

21. Review the concept of colinearity in Section 12.5 (pp. 261) and consider the following question: Certain mutations called *amber* in bacteria and viruses result in premature termination of polypeptide chains during translation. Many *amber* mutations have been detected at different points along the gene that codes for a head protein in phage T4. How might this system be further investigated to demonstrate and support the concept of colinearity?

22. In your opinion, which of the four levels of protein organization is the most critical to a protein's function? Defend your choice.

23. List and describe the function of as many non-enzymatic proteins as you can that are unique to eukaryotes.

24. How does an enzyme function? Why are enzymes essential for living organisms?

25. Shown below are several amino acid substitutions in the α and β chains of human hemoglobin. Use the genetic code table in Figure 12–7 to determine how many of them can occur as a result of a single nucleotide change.

Hb Type	Normal Amino Acid	Substituted Amino Acid
HbJ Toronto	ala	asp (α-5)
HbJ Oxford	gly	asp (α-15)
Hb Mexico	gln	glu (α-54)
Hb Bethesda	tyr	his (β-145)
Hb Sydney	val	ala (β-67)
HbM Saskatoon	his	tyr (β-63)

26. Early detection and adherence to a strict dietary regime has relieved much of the mental retardation that once occurred in people afflicted with phenylketonuria (PKU). Now, affected individuals often lead normal lives and have families. For various reasons, some individuals adhere less rigorously to their diet as they get older. Predict the effect of such dietary neglect on the newborns of mothers with PKU.

27. In 1962, F. Chapeville and others (1962. *Proc. Natl. Acad. Sci. (USA)* 48: 1086–1093) reported an experiment in which they isolated radioactive ^{14}C-cysteinyl-tRNA^{cys} (charged tRNA^{cys} + cysteine). They then removed the sulfur group from the cysteine, creating alanyl-tRNA^{cys} (charged tRNA^{cys} + alanine). When alanyl-tRNA^{cys} was added to a synthetic mRNA calling for cysteine, but not alanine, a polypeptide chain was synthesized containing alanine. What can you conclude from this experiment?

28. Three independently assorting genes are known to control the biochemical pathway here that provides the basis for flower color in a hypothetical plant:

$$\text{colorless} \xrightarrow{A-} \text{yellow} \xrightarrow{B-} \text{green} \xrightarrow{C-} \text{speckled}$$

Homozygous recessive mutations, which interrupt each step, are known. Determine the phenotypic results in the F_1 and F_2 generations resulting from the P_1 crosses involving true-breeding plants given below.

(a) speckled	(*AABBCC*)	$\times$	yellow (*AAbbCC*)
(b) yellow	(*AAbbCC*)	$\times$	green (*AABBcc*)
(c) colorless	(*aaBBCC*)	$\times$	green (*AABBcc*)

29. How would the results in cross (a) of Problem 28 vary if genes *A* and *B* were linked with no crossing over between them? How would the results of cross (a) vary if genes *A* and *B* were linked and 20 map units apart?

30. HbS results from the amino acid change of glutamic acid to valine at the number 6 position in the β chain of human hemoglobin. HbC is the result of a change at the same position in the β chain, but lysine replaces glutamic acid. Return to the genetic code table in Figure 12–7, and determine whether single-nucleotide changes can account for these mutations. Then turn to Figure 13–15, and examine the R groups in the amino acids glutamic acid, valine, and lysine. Describe the chemical differences between the three amino acids and predict how the changes might alter the structure of the molecule and lead to altered hemoglobin function.

31. HbS results in anemia and resistance to malaria, whereas in those with HbA, the parasite *Plasmodium falciparum* invades red blood cells and causes the disease. Predict whether those with HbC are likely to be anemic and whether they would be resistant to malaria.

32. Deep in a previously unexplored South American rain forest, a species of plant was discovered with true-breeding varieties whose flowers were either pink, rose, orange, or purple. A very astute plant geneticist made a single cross, carried to the F_2 generation, as shown on the next page. Based solely on these data, he was able to propose both a mode of inheritance for flower pigmentation and a biochemical pathway for the synthesis of these pigments. Carefully study the data. Create your own hypothesis to explain the mode of inheritance, and then propose a biochemical pathway consistent with your hypothesis. What other crosses would enable you to test the hypothesis?

P$_1$:	purple × pink
F$_1$:	all purple
F$_2$:	27/64 purple
	16/64 pink
	12/64 rose
	9/64 orange

33. The emergence of antibiotic-resistant strains of *Enterococci* and transfer of resistant genes to other bacterial pathogens have highlighted the need for new generations of antibiotics to combat serious infections. To grasp the range of potential sites for the action of existing antibiotics, sketch the components of the translation machinery (e.g., see step 3 of Figure 13–6), and using a series of numbered pointers, indicate the specific location for the action of the antibiotics shown in the following table.

Antibiotic	Action
1. Streptomycin	Binds to 30*S* ribosomal subunit.
2. Chloramphenicol	Inhibits peptidyl transferase of 70*S* ribosome
3. Tetracycline	Inhibits binding of charged tRNA to ribosome
4. Erythromycin	Binds to free 50*S* particle and prevents formation of 70*S* ribosome
5. Kasugamycin	Inhibits binding of tRNA^{fmet}
6. Thiostrepton	Prevents translocation by inhibiting EF-G

34. Development of antibiotic resistance by pathogenic bacteria represents a major health concern. One potential new antibiotic is evernimicin, which was isolated from *Micromonospora carbonaceae*. Evernimicin is an oligosaccharide antibiotic with activity against a broad range of gram-positive pathogenic bacteria. To determine the mode of action of this drug, Adrian and others (2000. *Antimicrob. Ag. and Chemo.* 44: 3101–3106) ana-lyzed 23*S* ribosomal DNA mutants that showed reduced sensitivity to evernimicin. They and others discovered two classes of mutants that conferred resistance: 23*S* rRNA nucleotides 2475–2483 and ribosomal protein L16. This suggests that these two ribosomal components are structurally and functionally linked. It turns out that the tRNA anticodon stem-loop appears to bind to the A site of the ribosome at rRNA bases 2465–2485. This finding conforms to the proposed function of L16, which appears to be involved in attracting the aminoacyl stem of the tRNA to the ribosome at its A site. Using your sketch of the translation machinery from Problem 33 along with this information, designate where the proposed antibacterial action of evernimicin is likely to occur.

35. The flow of genetic information from DNA to protein is mediated by messenger RNA. If you introduce short DNA strands (called antisense oligonucleotides) that are complementary to mRNAs, hydrogen bonding may occur and "label" the DNA/RNA hybrid for ribonuclease-H degradation of the RNA. Lloyd and others (2001. *Nuc. Acids Res.* 29: 3664–3673) compared the effect of different-length antisense oligonucleotides upon ribonuclease-H–mediated degradation of tumor necrosis factor (*TNFα*) mRNA. TNFα exhibits antitumor and proinflammatory activities. The graph below indicates the efficacy of various-sized antisense oligonucleotides in causing ribonuclease-H cleavage. (a) Describe how antisense oligonucleotides interrupt the flow of genetic information in a cell. (b) What general conclusion is apparent in the graph? (c) What factors other than oligonucleotide length are likely to influence antisense efficacy *in vivo*?

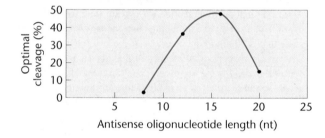

Gene Mutation, DNA Repair, and Transposition

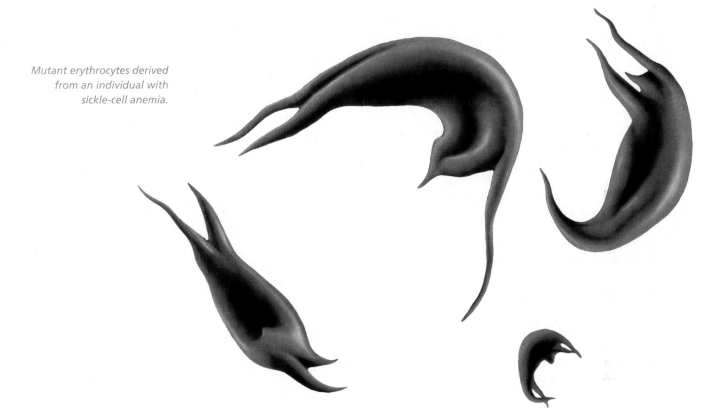

Mutant erythrocytes derived from an individual with sickle-cell anemia.

CHAPTER CONCEPTS

- Mutations are the source of genetic variation and the basis for natural selection. They are also the source of genetic damage that contributes to cell death, genetic diseases, and cancer.

- Mutations have a wide range of effects on organisms depending on the type and location of the nucleotide change within the gene and the genome.

- Mutations can occur spontaneously as a result of natural biological and chemical processes, or they can be induced by external factors, such as chemicals or radiation.

- The rates of spontaneous mutation vary between organisms and between genes within an organism.

- Organisms invoke a number of DNA repair mechanisms to counteract mutations. These mechanisms range from proofreading of replication errors to base excision and homologous recombination repair.

- Mutations in genes whose products control DNA repair lead to genome hypermutability and human DNA repair diseases, such as xeroderma pigmentosum.

- Transposable elements create mutations by moving into and out of chromosomes. They can induce mutations within coding regions and in gene regulatory regions, and cause chromosome breaks.

- Geneticists induce gene mutations as the first step in classical genetic analysis.

The ability of DNA molecules to store, replicate, transmit, and decode information is the basis of genetic function. But equally important is the capacity of DNA to make mistakes. Without the variation that arises from changes in DNA sequences, there would be no phenotypic variability, no adaptation to environmental changes, and no evolution. Gene mutations are the source of most new alleles and are the origin of genetic variation within populations. On the downside, they are also the source of genetic changes that can lead to cell death, genetic diseases, and cancer.

Mutations also provide the basis for genetic analysis. The phenotypic variability resulting from mutations allows geneticists to identify and study the genes responsible for the modified trait. In genetic investigations, mutations act as identifying "markers" for genes so that they can be followed during their transmission from parents to offspring. Without phenotypic variability, genetic analysis would be impossible. For example, if all pea plants displayed a uniform phenotype, Mendel would have had no foundation for his research.

In Chapter 6, we examined mutations in large regions of chromosomes—chromosomal mutations. In this chapter, we will explore mutations primarily at the level of changes in the base-pair sequence of DNA within individual genes—**gene mutations**. We will also describe how the cell defends itself from mutations using various mechanisms of DNA repair. The chapter also includes a discussion of transposable elements, sometimes called "jumping genes" or transposons, which can disrupt gene function and carry genetic information within and between DNA molecules. The chapter concludes with a description of how geneticists use mutations to identify genes and analyze gene functions.

How Do We Know?

In this chapter, we will focus on one of the major sources of genetic variation: gene mutations. We will discuss how mutations arise and how we study them. We will also focus on the repair mechanisms that have evolved to counteract the DNA damage that leads to altered phenotypes. As you study this topic, you should try to answer several fundamental questions:

- How do we know that changes in DNA sequence cause phenotypic variation?
- How do we know that mutations occur spontaneously?
- How do we know that certain chemicals and wavelengths of radiation induce mutations in DNA?
- How do we know that DNA repair mechanisms detect and correct the majority of spontaneous and induced mutations?
- How do we know that mobile elements exist within an organism's genome?
- How do we know that certain human genetic diseases are caused by expansions of trinucleotide repeat sequences?

Web Tutorial 14.1
Mutations at the DNA Level

14.1 Mutations Are Classified in Various Ways

A mutation can be defined as an alteration in DNA sequence. Any base-pair change in any part of a DNA molecule can be considered a mutation. A mutation may comprise a single base-pair substitution, a deletion or insertion of one or more base pairs, or a major alteration in the structure of a chromosome.

Mutations may occur within regions of a gene that code for protein or within noncoding regions of a gene such as introns and regulatory sequences. Mutations may or may not bring about a detectable change in phenotype. The extent to which a mutation changes the characteristics of an organism depends on where the mutation occurs and the degree to which the mutation alters the function of the gene product.

Mutations can occur within somatic cells or within germ cells. Those that occur in germ cells are heritable and are the basis for the transmission of genetic diversity and evolution, as well as genetic diseases. Those that occur in somatic cells are not transmitted to the next generation, but may lead to altered cellular function or tumors.

Because of the wide range of types and effects of mutations, geneticists classify mutations according to several different schemes. These organizational schemes are not mutually exclusive. In this section, we outline some of the ways in which gene mutations are classified.

Spontaneous and Induced Mutations

All mutations are either spontaneous or induced, although these two categories overlap to some degree. **Spontaneous mutations** are those that happen naturally. No specific agents are associated with their occurrence, and they are generally assumed to be random changes in the nucleotide sequences of genes. Many of these mutations arise as a result of normal biological or chemical processes in the organism that alter the structure of nitrogenous bases. Often, spontaneous mutations occur during the enzymatic process of DNA replication, as we discuss later in this chapter.

In contrast to spontaneous mutations, mutations that result from the influence of extraneous factors are considered to be **induced mutations**. Induced mutations may be the result of either natural or artificial agents. For example, radiation from cosmic and mineral sources and ultraviolet radiation from the sun are energy sources to which most organisms are exposed and, as such, may be factors that cause induced mutations. In addition to various forms of radiation, numerous natural and synthetic chemical agents are also mutagenic.

Classification Based on Location of Mutation

Mutations may be classified according to the cell type or chromosomal locations in which they occur. **Somatic mutations** may occur in any cell in the body except germ cells. **Germ-line mutations** occur in gametes. **Autosomal mutations** occur within genes located on the autosomes, whereas **X-linked mutations** occur within genes located on the X chromosome.

Mutations arising in somatic cells are not transmitted to future generations. When a recessive autosomal mutation occurs in a somatic cell of a diploid organism, it is unlikely to result in a detectable phenotype. The expression of most such mutations is likely to be masked by the wild-type allele within that cell. Somatic mutations will have a greater impact if they are dominant or, in males, if they are X-linked, since such mutations are most likely to be immediately expressed. Similarly, the impact of dominant or X-linked somatic mutations will be more noticeable if they occur early in development, when a small number of undifferentiated cells replicate to give rise to several differentiated tissues or organs. Dominant mutations that occur in cells of adult tissues are often masked by the thousands upon thousands of nonmutant cells in the same tissue that perform the normal function.

Mutations in gametes are of greater significance because they are transmitted to offspring as part of the germ line. They have the potential of being expressed in all cells of an offspring. Inherited dominant autosomal mutations will be expressed phenotypically in the first generation. X-linked recessive mutations arising in the gametes of a homogametic female may be expressed in hemizygous male offspring. This will occur provided that the male offspring receives the affected X chromosome. Because of heterozygosity, the occurrence of an autosomal recessive mutation in the gametes of either males or females (even one resulting in a lethal allele) may go unnoticed for many generations, until the resultant allele has become widespread in the population. Usually, the new allele will become evident only when a chance mating brings two copies of it together into the homozygous condition.

Classification Based on Type of Molecular Change

Geneticists often classify gene mutations in terms of the nucleotide changes that create the mutation. A change of one base pair to another in a DNA molecule is known as a **point mutation**, or **base substitution** (Figure 14–1). A change of one nucleotide of a triplet within a protein-coding portion of a gene may result in the creation of a new triplet that codes for a different amino acid in the protein product. If this occurs, the mutation is known as a **missense mutation**. A second possible outcome is that the triplet will be changed into a stop codon, resulting in the termination of translation of the protein. This is known as a **nonsense mutation**. If the point mutation alters a codon but does not result in a change in the amino acid at that position in the protein (due to degeneracy of the genetic code), it can be considered a **silent mutation**.

You will often see two other terms used to describe base substitutions. If a pyrimidine replaces a pyrimidine or a purine replaces a purine, a **transition** has occurred. If a purine and a pyrimidine are interchanged, a **transversion** has occurred.

Another type of change is the insertion or deletion of one or more nucleotides at any point within the gene. As illustrated in Figure 14–1, the loss or addition of a single letter causes all of the subsequent three-letter words to be changed. These are called **frameshift mutations** because the frame of triplet reading during translation is altered. A frameshift mutation will occur when any number of bases are added or deleted, except multiples of three, which would reestablish the initial frame of reading. The analogy in Figure 14–1 demonstrates that insertions and deletions have the potential to change all the subsequent triplets in a gene. It is possible that one of the many altered triplets will be UAA, UAG, or UGA, the translation termination codons. When one of these triplets is encountered during translation, polypeptide synthesis is terminated at that point. Obviously, the results of frameshift mutations can be very severe, especially if they occur early in the coding sequence.

Classification Based on Phenotypic Effects

Depending on their type and location, mutations can have a wide range of phenotypic effects, from silent mutations to dominant lethals.

As discussed in Chapter 4, a **loss-of-function mutation** is one that eliminates the function of the gene product. Any type of mutation, from a point mutation to deletion of the entire gene, may lead to a loss of function. These mutations are also known as **null mutations**. It is possible for a loss-of-function mutation to be either dominant or recessive. A dominant loss-of-function mutation may result from the presence of a defective protein product that binds to, or inhibits the action of, the normal gene product, which is also present in the same organism. A **gain-of-function** mutation results in a gene product with a new function. This may be due to a change in the amino acid sequence of the protein that confers a new activity, or it may result from a mutation in the regulatory region of the gene, leading to expression of the gene at an abnormal level, time, or place. Most gain-of-function mutations are dominant. Because eukaryotic genomes consist mainly of noncoding regions, the vast majority of mutations are likely to occur in the large portions of the genome that do not contain genes. These are considered to be **neutral mutations**, because they do not affect gene products.

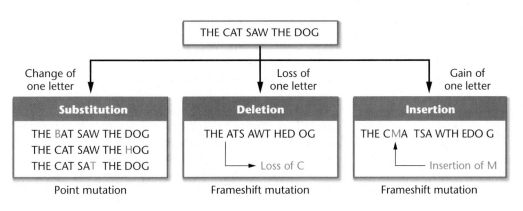

FIGURE 14–1 Analogy of the effects of substitution, deletion, and insertion of one letter in a sentence composed of three-letter words, demonstrating point and frameshift mutations.

The most easily observed mutations are those affecting a **morphological trait**. These mutations are also known as **visible mutations** and are recognized by their ability to alter a normal or wild-type phenotype. For example, all of Mendel's pea characters and many genetic variations encountered in *Drosophila* fit this designation, since they cause obvious changes to the morphology of the organism.

An additional category of mutations includes those that exhibit nutritional or biochemical effects. In bacteria and fungi, a typical **nutritional mutation** results in a loss of ability to synthesize an amino acid or vitamin. In humans, sickle-cell anemia and hemophilia are examples of diseases resulting from **biochemical mutations**. Although such mutations do not always affect morphological characters, they can have an effect on the well-being and survival of the affected individual.

Still another category consists of mutations that affect the behavior patterns of an organism. For example, the mating behavior or circadian rhythms of animals can be altered. The primary effect of **behavioral mutations** is often difficult to analyze. For example, the mating behavior of a fruit fly may be impaired if it cannot beat its wings. However, the defect may be in the flight muscles, the nerves leading to them, or the brain, where the nerve impulses that initiate wing movements originate.

Another group of mutations may affect the regulation of gene expression. For example, as we will see in the *lac* operon discussed in Chapter 15, a regulatory gene can produce a product that controls the transcription of other genes. A mutation in a regulatory gene or a gene control region can disrupt normal regulatory processes and inappropriately activate or inactivate a gene. Our knowledge of genetic regulation has been dependent on the study of such **regulatory mutations**.

It is also possible that a mutation may interrupt a process that is essential to the survival of the organism. In this case, it is referred to as a **lethal mutation**. For example, a mutant bacterium that has lost the ability to synthesize an essential amino acid will cease to grow and eventually will die when placed in a medium lacking that amino acid. Various inherited human biochemical disorders are also good examples of lethal mutations. For example, Tay-Sachs disease and Huntington disease are caused by mutations that result in lethality, but at different points in the life cycle of humans.

Another interesting aspect of mutations is when their expression depends on the environment in which the organism finds itself. Such mutations are called **conditional mutations**, because the mutation is present in the genome of an organism but can be detected only under certain conditions. Among the best examples of conditional mutations are **temperature-sensitive mutations**. At a "permissive" temperature, the mutant gene product functions normally, but it loses its function at a different, "restrictive" temperature. Therefore, when the organism is shifted from the permissive to the restrictive temperature, the impact of the mutation becomes apparent.

14.2   The Spontaneous Mutation Rate Varies Greatly among Organisms

Several general conclusions can be made regarding spontaneous mutation rates in a variety of organisms. First, the rate of spontaneous mutation is exceedingly low for all organisms.

Second, the rate varies considerably between different organisms. Third, even within the same species, the spontaneous mutation rate varies from gene to gene.

Viral and bacterial genes undergo spontaneous mutation at an average of about 1 in 100 million (10^{-8}) cell divisions. *Neurospora* exhibits a similar rate, but maize, *Drosophila*, and humans demonstrate a rate several orders of magnitude higher. The genes studied in these groups average between 1/1,000,000 and 1/100,000 (10^{-6} to 10^{-5}) mutations per gamete formed. Mouse genes are another order of magnitude higher in their spontaneous mutation rate, 1/100,000 to 1/10,000 (10^{-5} to 10^{-4}). It is not clear why such a large variation occurs in mutation rates. The variation between organisms may reflect the relative efficiencies of their DNA proofreading and repair systems. We will discuss these systems later in the chapter.

Now Solve This

Problem 17 on page 327 provides data involving a bacterial mutation that interrupts the biosynthesis of the amino acid leucine. You are asked to calculate the spontaneous mutation rate.

Hint: You must first determine the number of bacteria per milliliter in the original cultures grown under the two conditions by converting the "dilution" data back to the "undiluted" data. For example, if six colonies were observed in one ml, after a 1000-fold dilution, the original cell concentration was 6×10^3 bacteria/ml.

14.3   Spontaneous Mutations Arise from Replication Errors and Base Modifications

In this section, we will outline some of the processes that lead to spontaneous mutations. It is useful to keep in mind, however, that many of the base modifications that occur during spontaneous mutagenesis also occur, at a higher rate, during induced mutagenesis.

DNA Replication Errors

As we learned in Chapter 10, the process of DNA replication is imperfect. Occasionally, DNA polymerases insert incorrect nucleotides in a replicated strand of DNA. Although DNA polymerases can correct most of these replication errors using their inherent 3′ to 5′ exonuclease proofreading capacity, misincorporated nucleotides may persist after replication. If these errors are not detected and corrected by DNA repair mechanisms, they may lead to mutations. The fact that bases can take several forms, known as **tautomers**, also increases the chance of mispairing during DNA replication. Tautomeric shifts are discussed below. Replication errors due to mispairing predominantly lead to point mutations.

Replication Slippage

In addition to point mutations, DNA replication can lead to the introduction of small insertions or deletions. These mutations can occur when one strand of the DNA template loops out and becomes displaced during replication, or when DNA polymerase slips or stutters during replication. If a loop occurs in the template strand during replication, DNA polymerase may miss the looped-out nucleotides and a small deletion in the new strand will be introduced. If DNA polymerase repeatedly introduces nucleotides that are not present in the template strand, an insertion of one or more nucleotides will be introduced, creating an unpaired loop on the newly synthesized strand. Insertions and deletions may lead to frameshift mutations, amino acid additions, or deletions in the gene product.

Replication slippage can occur anywhere in the DNA but appears to have a distinct preference for regions containing repeated sequences. Repeat sequences are hot-spots for DNA mutation and in some cases contribute to hereditary diseases (discussed in Section 14.5). The hypermutability of repeat sequences in noncoding regions of the genome also forms the basis for current methods of forensic DNA analysis.

Tautomeric Shifts

Purines and pyrimidines can exist in tautomeric forms—that is, nitrogenous bases in alternate chemical forms, each differing by only a single proton shift in the molecule. The biologically important tautomers involve the keto–enol forms of thymine and guanine, and the amino–imino forms of cytosine and adenine. These shifts change the bonding structure of the molecule, allowing hydrogen bonding with noncomplementary bases. Hence, **tautomeric shifts** may lead to permanent base-pair changes and mutations. Figure 14–2 compares normal base-pairing relationships with rare unorthodox pairings. Anomalous $T \equiv G$ and $C = A$ pairs, among others, may be formed.

A mutation occurs during DNA replication when a transiently formed tautomer in the template strand pairs with a non-complementary base. In the next round of replication, the "mismatched" members of the base pair are separated, and each becomes the template for its normal complementary base. The end result is a point mutation (Figure 14–3).

Depurination and Deamination

Some of the most common causes of spontaneous mutations are due to DNA base damage. **Depurination** involves the loss of one of the nitrogenous bases in an intact double-helical DNA molecule. Most frequently, such an event involves purines— either guanine or adenine. These bases may be lost if the glycosidic bond linking the 1'-C of the deoxyribose and the number 9 position of the purine ring is broken. This leads to the creation of an **apurinic (AP) site** on one strand of the DNA. Geneticists estimate that thousands of such spontaneous lesions are formed daily in the DNA of mammalian cells in culture. If AP sites are not repaired, there will be no base at that position to act as a template during DNA replication. As a result, DNA polymerase may introduce a nucleotide at random at that site.

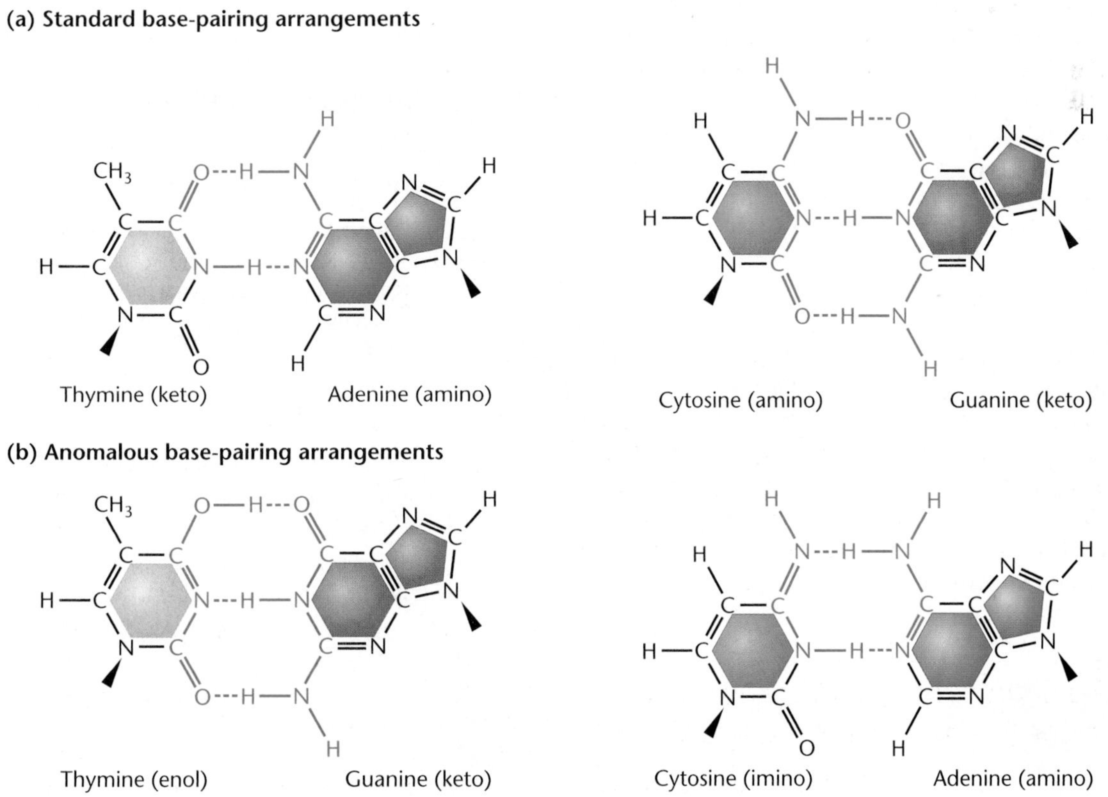

(a) Standard base-pairing arrangements

Thymine (keto)   Adenine (amino)   Cytosine (amino)   Guanine (keto)

(b) Anomalous base-pairing arrangements

Thymine (enol)   Guanine (keto)   Cytosine (imino)   Adenine (amino)

FIGURE 14–2 Standard base-pairing relationships (a), compared with anomalous base-pairing that occurs as a result of tautomeric shifts (b). The long triangle indicates the point at which the base bonds to the pentose sugar.

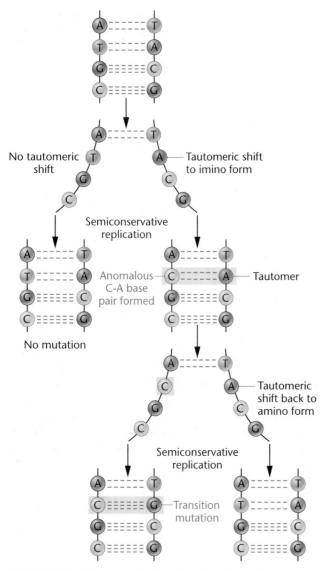

FIGURE 14–3 Formation of a A = T to G ≡ C transition mutation as a result of a tautomeric shift in adenine.

In the process of **deamination**, an amino group is converted to a keto group in cytosine or adenine (Figure 14–4). In these cases, cytosine is converted to uracil, and adenine is changed to hypoxanthine. The major effect of these changes is an alteration in the base-pairing specificities of these two bases during DNA replication. For example, cytosine normally pairs with guanine. Following its conversion to uracil, which pairs with adenine, the original G ≡ C pair is converted to an A = U pair and then, following an additional replication, is converted to an A = T pair. When adenine is deaminated, the original A = T pair is converted to a G ≡ C pair because hypoxanthine pairs naturally with cytosine. Deamination may occur spontaneously or as a result of treatment with chemical mutagens such as nitrous acid (HNO_2).

Oxidative Damage

DNA may also suffer damage from the by-products of normal cellular processes. These by-products include reactive oxygen species (electrophilic oxidants) that are generated during normal aerobic respiration. For example, superoxides ($O_2^{\cdot-}$) hydroxyl radicals (·OH), and hydrogen peroxide (H_2O_2) are created during cellular metabolism and are constant threats to the integrity of DNA. **Reactive oxidants**, also generated by exposure to high-energy radiation, can produce more than 100 different types of chemical modifications in DNA, including modifications to bases that lead to mispairing during replication.

Transposons

Transposable genetic elements (transposons) are often inserted into a gene, disrupting the normal coding sequence. Thus, these elements may be agents of spontaneous mutations in both eukaryotes and prokaryotes. Transposons will be discussed later in this chapter (Section 14.8).

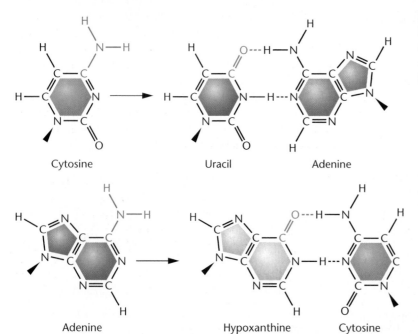

FIGURE 14–4 Deamination of cytosine and adenine, leading to new base pairing and mutation. Cytosine is converted to uracil, which base-pairs with adenine. Adenine is converted to hypoxanthine, which base-pairs with cytosine.

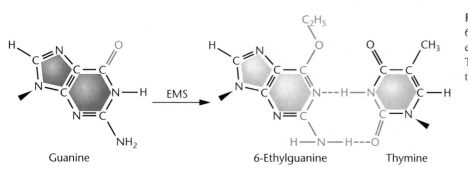

FIGURE 14–5 Similarity of 5-bromouracil (5-BU) structure to thymine structure. In the common keto form, 5-BU base-pairs normally with adenine, behaving as an analog. In the rare enol form, it pairs anomalously with guanine.

Thymine | 5-Bromouracil (keto form) | 5-Bromouracil (enol form)

5-BU (keto form) | Adenine

5-BU (enol form) | Guanine

Base Analogs

These mutagenic chemicals can substitute for purines or pyrimidines during nucleic acid biosynthesis. For example, **5-bromouracil (5-BU)**, a derivative of uracil, behaves as a thymine analog and is halogenated at the number 5 position of the pyrimidine ring. If 5-BU is chemically linked to deoxyribose, the nucleoside analog **bromodeoxyuridine (BrdU)** is formed. Figure 14–5 compares the structure of this analog with that of thymine. The presence of the bromine atom in place of the methyl group increases the probability that a tautomeric shift will occur. If 5-BU is incorporated into DNA in place of thymine and a tautomeric shift to the enol form occurs, 5-BU base-pairs with guanine. After one round of replication, an $A = T$ to $G \equiv C$ transition results. Furthermore, the presence of 5-BU within DNA increases the sensitivity of the molecule to ultraviolet (UV) light, which itself is mutagenic.

There are other **base analogs** that are mutagenic. For example, **2-amino purine (2-AP)** can act as an analog of adenine. In addition to its base-pairing affinity with thymine, 2-AP can also base-pair with cytosine, leading to possible transitions from $A = T$ to $G \equiv C$ following replication.

14.4 Induced Mutations Arise from DNA Damage Caused by Chemicals and Radiation

All cells on Earth are exposed to a plethora of agents called **mutagens**, that have the potential to damage DNA and cause mutations. Some of these agents, such as some fungal toxins, cosmic rays, and ultraviolet light, are natural components of our environment. Others, including some industrial pollutants, medical X-rays, and chemicals within tobacco smoke, can be considered as unnatural or human-made additions to our modern world. On the positive side, geneticists harness some mutagens in order to analyze genes and gene functions, as discussed in Section 14.9. The mechanisms by which some of these natural and unnatural agents lead to mutations are outlined in this section.

Alkylating Agents and Acridine Dyes

The sulfur-containing mustard gases, discovered during World War I, were some of the first groups of chemical mutagens identified in chemical warfare studies. Mustard gases are **alkylating agents**—that is, they donate an alkyl group, such as CH_3 or CH_3CH_2, to amino or keto groups in nucleotides. Ethylmethane sulfonate (EMS), for example, alkylates the keto groups in the number 6 position of guanine and in the number 4 position of thymine. As with base analogs, base-pairing affinities are altered and transition mutations result. For example, 6-ethylguanine acts as an analog of adenine and pairs with thymine (Figure 14–6).

FIGURE 14–6 Conversion of guanine to 6-ethylguanine by the alkylating agent ethylmethane sulfonate (EMS). The 6-ethylguanine base-pairs with thymine.

Guanine | 6-Ethylguanine | Thymine

Chemical mutagens called **acridine dyes** cause frameshift mutations. As illustrated in Figure 14–1, frameshift mutations result from the addition or removal of one or more base pairs in the polynucleotide sequence of the gene. Acridine dyes such as proflavin and acridine orange are about the same dimensions as nitrogenous base pairs and **intercalate**, or wedge, between the purines and pyrimidines of intact DNA. Intercalation introduces contortions in the DNA helix that may cause deletions and insertions during DNA replication or repair, leading to frameshift mutations.

Ultraviolet Light

All radiation on Earth consists of a series of electromagnetic components of varying wavelength (Figure 14–7). Referred to as the **electromagnetic spectrum**, the energy associated with the various components varies inversely with wavelength. Everything longer than, and including, visible light is benign when it interacts with most organic molecules. However, everything of shorter wavelength than visible light, being inherently more energetic, has the potential to disrupt organic molecules. As we know, purines and pyrimidines absorb **ultraviolet (UV) radiation** most intensely at a wavelength of about 260 nm. One major effect of UV radiation on DNA is the creation of **pyrimidine dimers**, particularly those involving two thymine residues (Figure 14–8). The dimers distort the DNA conformation and inhibit normal replication. As a result, errors can be introduced in the base sequence of DNA during replication. When UV-induced dimerization is extensive, it is responsible (at least in part) for the killing effects of UV radiation on cells.

Ionizing Radiation

Within the electromagnetic spectrum, energy varies inversely with wavelength. As shown in Figure 14–7, **X-rays, gamma rays**, and **cosmic rays** have even shorter wavelengths and are more energetic than UV radiation. As a result, they penetrate

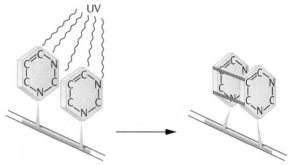

Dimer formed between adjacent thymidine residues along a DNA strand

FIGURE 14–8 Induction of a thymine dimer by UV radiation, leading to distortion of the DNA. The covalent crosslinks occur between the atoms of the pyrimidine ring.

deeply into tissues, causing ionization of the molecules encountered along the way.

As X-rays penetrate cells, electrons are ejected from the atoms of molecules encountered by the radiation. Thus, stable molecules and atoms are transformed into **free radicals** and reactive ions that can directly or indirectly affect the genetic material, altering the purines and pyrimidines in DNA and resulting in point mutations. **Ionizing radiation** is also capable of breaking phosphodiester bonds, disrupting the integrity of chromosomes, and producing a variety of chromosomal aberrations, such as deletions, translocations, and chromosomal fragmentation.

Figure 14–9 shows a graph of the percentage of induced X-linked recessive lethal mutations versus the dose of X-rays administered. There is a linear relationship between X-ray dose and the induction of mutation; for each doubling of the dose, twice as many mutations are induced. Because the line intersects near the zero axis, this graph suggests that even very small doses of radiation are mutagenic.

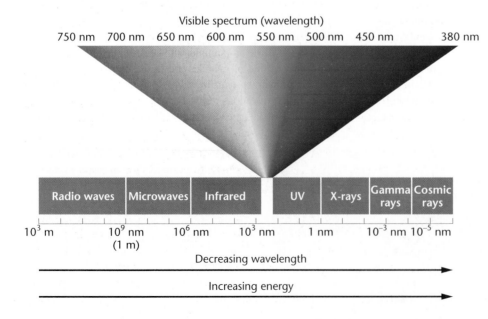

Visible spectrum (wavelength)

| 750 nm | 700 nm | 650 nm | 600 nm | 550 nm | 500 nm | 450 nm | | 380 nm |

| Radio waves | Microwaves | Infrared | | UV | X-rays | Gamma rays | Cosmic rays |

10^3 m 10^9 nm 10^6 nm 10^3 nm   1 nm 10^{-3} nm 10^{-5} nm
              (1 m)

Decreasing wavelength

Increasing energy

FIGURE 14–7 The components of the electromagnetic spectrum and their associated wavelengths.

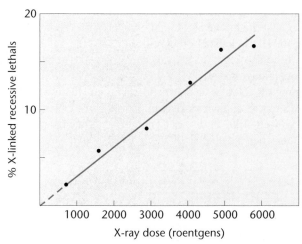

FIGURE 14–9 Plot of the percentage of X-linked recessive mutations induced in *Drosophila* by increasing doses of X-rays. If extrapolated, the graph intersects the zero axis as shown by the dashed line.

Now Solve This

Problem 23 on page 327 asks you to describe ways in which the chemotherapy drug melphalan kills cancer cells and how these cells can repair the DNA damage caused by this drug.

Hint: To answer this question, consider the effect of the alkylation of guanine on base pairing during DNA replication. In Section 14.7, you will learn about the ways in which cells repair the types of mutations introduced by alkylating agents.

14.5 Genomics and Gene Sequencing Have Enhanced Our Understanding of Mutations in Humans

So far, we have discussed the molecular basis of mutation largely in terms of nucleic acid chemistry. Historically, scientists learned about mutations by analyzing the amino acid sequences of proteins within populations that show substantial diversity. This diversity, which arises during evolution, is a reflection of changes in the triplet codons following substitution, insertion, or deletion of one or more nucleotides in the DNA sequences of protein-coding genes.

As our ability to analyze DNA more directly increases, we are able to examine the actual nucleotide sequence of genes and surrounding DNA regions, and to gain greater insights into the nature of mutations. Modern techniques for rapid sequencing of DNA have greatly extended our knowledge of mutations. In this section, we describe how several types of mutations directly affect human health.

ABO Blood Groups

The **ABO blood group system** is based on a series of antigenic determinants found on erythrocytes and other cells, particularly epithelial cells. As we discussed in Chapter 4, there are three alleles of the *I* gene that encodes the **glycosyltransferase**

enzyme. The H substance is modified to either the A or B antigen, as a result of the activity of the glycosyltransferase enzyme encoded by the I^A or I^B allele, respectively. Failure to modify the H substance results from the presence of the null I^O allele.

When the DNAs of the I^A and I^B alleles are compared, four consistent single-nucleotide substitutions are found. The resulting changes in the amino acid sequence of the glycosyltransferase gene product lead to altered gene product function and, consequently, the two different modifications of the H substance.

The I^O allele situation is unique and interesting. Individuals homozygous for this allele have type O blood, lack glycosyltransferase activity, and fail to modify the H substance. Analysis of the DNA of the I^O allele shows one consistent change that is unique compared with the sequences of the other alleles—the deletion of a single nucleotide early in the coding sequence, causing a frameshift mutation. A complete messenger RNA is transcribed, but at translation, the reading frame shifts at the point of the deletion and continues out of frame for about 100 nucleotides before a stop codon is encountered. At this point, the glycosyltransferase polypeptide chain terminates prematurely, resulting in a nonfunctional product.

These findings provide a direct molecular explanation of the ABO allele system and the basis for the biosynthesis of the corresponding antigens. The molecular basis for the antigenic phenotypes is clearly the result of mutations within the gene encoding the glycosyltransferase enzyme.

Muscular Dystrophy

Muscular dystrophies are genetic diseases characterized by progressive muscle weakness and degeneration. There are many types of muscular dystrophy, and they differ in their severities, onsets, and genetic causes. Two related forms of muscular dystrophy—**Duchenne muscular dystrophy (DMD)** and **Becker muscular dystrophy (BMD)**—are recessive, X-linked conditions. DMD is the more severe of the two diseases, with a rapid progression of muscle degeneration and involvement of the heart and lungs. Males with DMD usually lose the ability to walk by the age of 12 and may die in their early 20s. The incidence of 1 in 5000 live male births makes DMD one of the most common life-shortening hereditary diseases. In contrast, BMD does not involve the heart or lungs and progresses slowly, from adolescence to the age of 50 or more.

The gene responsible for DMD and BMD—the *dystrophin* gene—is unusually large, consisting of about 2.5 million base pairs. In normal (unaffected) individuals, transcription and subsequent processing of the *dystrophin* initial transcript results in a messenger RNA containing only about 14,000 bases (14 kb). It is translated into the protein **dystrophin**, which consists of 3685 amino acids.

In recent years, geneticists have determined the molecular basis of mutations leading to BMD and DMD. With few exceptions, DMD mutations change the reading frame of the dystrophin gene, whereas BMD mutations usually do not. Studies have shown that about two-thirds of mutations in the dystrophin gene that lead to DMD and BMD are deletions and duplications. Only one-third are point mutations. The majority of DMD mutations lead to premature termination of translation. This, in turn, leads to degradation of the improperly translated dystrophin transcript and the nearly complete absence of dystrophin protein. In contrast, the majority of BMD gene mutations alter the internal sequence of the dystrophin transcript and protein, but do not alter the translation reading frame. As a result, modified dystrophin protein is produced, preventing the severe consequences of DMD.

Fragile X Syndrome, Myotonic Dystrophy, and Huntington Disease

Beginning about 1990, molecular analysis of the genes responsible for a number of inherited human disorders provided a remarkable set of observations. Researchers discovered that some mutant genes contain expansions of **trinucleotide repeat sequences**, usually from fewer than 15 copies in normal individuals to a large number in affected individuals. About 20 human disorders, including a number of developmental and degenerative diseases, appear to be caused by the presence of abnormally large numbers of repeat sequences within specific genes.

For example, the genes responsible for fragile X syndrome, myotonic dystrophy, and Huntington disease contain a specific trinucleotide DNA sequence repeated many times. Although these trinucleotide repeat sequences are also present in the nonmutant (normal) allele of each gene, the mutant alleles contain significant increases in the number of times the trinucleotide is repeated.

We begin with a short discussion of **fragile X syndrome**, described in detail in Chapter 6. The responsible gene, *FMR-1*, may contain several to several thousand copies of the trinucleotide sequence CGG, located in the 5' untranslated region of the gene. Individuals with up to 55 copies are normal and do not display the mental retardation associated with the syndrome. Individuals with 55 to 230 copies are considered carriers. Although they are normal, their offspring may bear even more copies and express the syndrome. The large regions of CGG repeats in the gene's regulatory region result in loss of expression of the FMRP protein, thought to be an RNA-binding protein affecting brain cell function.

Myotonic dystrophy is the most common form of adult muscular dystrophy, affecting 1 in 8000 individuals. The dominantly inherited disorder is not as severe as DMD, and it is highly variable in both symptoms and age of onset. Mild myotonia (atrophy and weakness) of the musculature of the face and extremities is most common. Cataracts, reduced cognitive ability, and cutaneous and intestinal tumors are also part of the syndrome.

The affected gene, *MDPK*, encodes a serine–threonine protein kinase, MDPK. This protein is the product of 15 exons of the gene, the last of which encodes the 3'-untranslated sequence of the mRNA. It is this sequence that houses multiple copies of the trinucleotide CTG. Individuals with 5 to 35 copies of the CTG repeat are normal, and the number of copies is stable from generation to generation. Individuals with more than about 150 copies exhibit symptoms ranging from mild to severe, with onset occurring anywhere between birth and age 60. Both the severity and onset are directly related to the size of the repeated sequence. The mechanism by which these repeated regions cause myotonic dystrophy is still uncertain.

Huntington disease (HD) is a fatal neurodegenerative disease, inherited as an autosomal dominant. The gene responsible for HD is located on chromosome 4. The gene contains the trinucleotide CAG sequence, repeated 11 to 34 times in normal individuals. The CAG repeat is located within the coding region of the gene and encodes a polyglutamine tract. The CAG repeat sequence exists in significantly increased numbers (up to 120) in diseased individuals. Much earlier onset occurs when the number of copies is closer to the upper range.

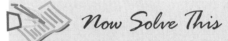

Now Solve This

Problem 24 on page 327 asks you to speculate on why the *dystrophin* gene appears to suffer a large number of mutations.

Hint: In answering this question, you may want to think about the size of the gene and its organization of introns and exons.

14.6 The Ames Test Is Used to Assess the Mutagenicity of Compounds

There is great concern about the possible mutagenic properties of any chemical that enters the human body, whether through the skin, the digestive system, or the respiratory tract. Examples are materials in air and water pollution, food preservatives and additives, artificial sweeteners, herbicides, pesticides, and pharmaceutical products. Mutagenicity may be tested in various organisms, including fungi, plants, and cultured mammalian cells; however, the most common test uses bacteria.

The **Ames test** (Figure 14–10) uses any of four strains of the bacterium *Salmonella typhimurium* that have been selected for their sensitivity to specific types of mutagenesis. For example, one strain is used to detect base-pair substitutions, and three strains detect various frameshift mutations. Each mutant strain is unable to synthesize histidine and therefore requires histidine for growth (*his*⁻ strains). The assay measures the frequency of reverse mutation, which yields wild-type bacteria (*his*⁺ revertants). These *Salmonella* strains also have a high sensitivity to mutagens due to the presence of mutations in genes involved in excision repair and synthesis of the lipopolysaccharide barrier that coats bacteria and protects them from external substances.

Many substances entering the human body are relatively innocuous until activated metabolically, usually in the liver, to a more chemically reactive product. Thus, the Ames test includes a step in which the test compound is incubated *in vitro* in the presence of a mammalian liver extract. Alternatively, test compounds may be injected into a mouse in which they are modified by liver enzymes and then recovered for use in the Ames test.

In the initial use of Ames testing in the 1970s, a large number of known **carcinogens** (cancer-causing agents) were examined, and more than 80 percent of these were shown to be strong mutagens. This is not surprising, as the transformation of cells to the malignant state occurs as a result of muta-

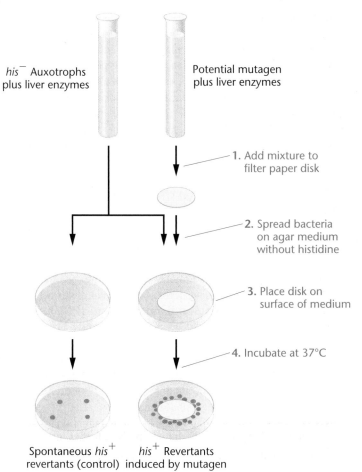

his⁻ Auxotrophs plus liver enzymes

Potential mutagen plus liver enzymes

1. Add mixture to filter paper disk

2. Spread bacteria on agar medium without histidine

3. Place disk on surface of medium

4. Incubate at 37°C

Spontaneous *his*⁺ revertants (control)

his⁺ Revertants induced by mutagen

FIGURE 14–10 The Ames test, which screens compounds for potential mutagenicity.

tions. Although a positive response in the Ames test does not prove that a compound is carcinogenic, the Ames test is useful as a preliminary screening device. The Ames test is used extensively during the development of industrial and pharmaceutical chemical compounds.

14.7 Organisms Use DNA Repair Systems to Counteract Mutations

Living systems have evolved a variety of elaborate repair systems that counteract both spontaneous and induced DNA damage. **DNA repair** systems are absolutely essential to the maintenance of the genetic integrity of organisms and, as such, to the survival of organisms on Earth. The balance between mutation and repair likely manifests itself in the observed mutation rates of individual genes and organisms. Of foremost concern in humans is the potential to counteract genetic damage that results in genetic diseases and cancer. The link between defective DNA repair and cancer susceptibility is described in Chapter 16.

We now embark on a review of some systems of DNA repair. Since the field is expanding rapidly, our goal here is to survey the major approaches that organisms use to counteract genetic damage.

Proofreading and Mismatch Repair

Some of the most common types of mutations arise during DNA replication when an incorrect nucleotide is inserted by DNA polymerase. The enzyme in bacteria (DNA polymerase III) makes an error approximately once every 100,000 insertions, leading to an error rate of 10^{-5}. Fortunately, the enzyme polices its own synthesis by **proofreading** each step, catching 99 percent of those errors. During polymerization, when an incorrect nucleotide is inserted, the enzyme has the potential to recognize the error and "reverse" its direction and behave as a 3′ to 5′ exonuclease, cutting out the incorrect nucleotide and then replacing it with the correct one. This improves the efficiency of replication one hundredfold, creating only $1/10^7$ mismatches immediately following DNA replication, for a final error rate of 10^{-7}.

To cope with those errors that remain after proofreading, another mechanism, called **mismatch repair**, may be activated. As in repair of other DNA lesions, the alterations or mismatches are detected, the incorrect nucleotides are removed, and the incorrect nucleotides replaced with the correct ones. But a special problem exists during the correction of a mismatch. How does the repair system recognize which strand is correct (the template) and which contains the incorrect base (the newly synthesized strand)? If the mismatch is recognized, but no discrimination occurs, the excision will be random, and the strand bearing the correct base will be clipped out 50 percent of the time. Hence, strand discrimination by a repair enzyme is a critical step.

The process of strand discrimination has been elucidated at least in some bacteria, including *E. coli*, and is based on **DNA methylation**. These bacteria contain an enzyme, adenine methylase, which recognizes the DNA sequence

$$5'\ldots GATC\ldots 3'$$
$$3'\ldots CTAG\ldots 5'$$

as a substrate during DNA replication, adding a methyl group to each of the adenine residues.

Following a round of replication, the newly synthesized strand remains temporarily unmethylated, as the methylase lags behind the DNA polymerase. At this point, the repair enzyme recognizes the mismatch and binds to the unmethylated (newly synthesized) DNA strand. A nick is made by an endonuclease enzyme, either 5′ or 3′ to the mismatch on the unmethylated strand. The nicked DNA strand is then unwound and degraded by an exonuclease, until the region of the mismatch is reached. Finally, DNA polymerase fills in the gap created by the exonuclease, using the correct DNA strand as a template. DNA ligase then seals the gap.

A series of *E. coli* gene products, Mut H, L, and S, are involved in mismatch repair. Mutations in any of these genes result in bacterial strains deficient in mismatch repair. While the preceding mechanism is based on studies of *E. coli*, similar mechanisms involving homologous proteins exist in yeast and in mammals.

Postreplication Repair and the SOS Repair System

Many other types of repair have been discovered, illustrating the diversity of mechanisms that have evolved to overcome DNA damage. One system, called **postreplication repair**

Postreplication repair

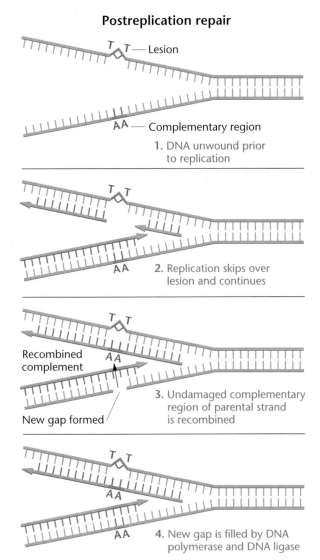

FIGURE 14–11 Postreplication repair occurs if DNA replication has skipped over a lesion such as a thymine dimer. Through the process of recombination, the correct complementary sequence is recruited from the parental strand and inserted into the gap opposite the lesion. The new gap is filled by DNA polymerase and DNA ligase.

responds *after* damaged DNA has escaped repair and failed to be completely replicated, thus its name. As illustrated in Figure 14–11, when DNA bearing a lesion of some sort (such as a pyrimidine dimer) is being replicated, DNA polymerase may stall at the lesion and then skip over it, leaving a gap on the newly synthesized strand. To correct the gap, the RecA protein directs a recombinational exchange with the corresponding region on the undamaged parental strand of the same polarity (the "donor" strand). When the undamaged segment of DNA replaces the gapped segment, this transfers the gap to the donor strand. This gap can be filled by repair synthesis as replication proceeds. Because a recombinational event is involved in this type of DNA repair, postreplication repair is also referred to as **homologous recombination repair**.

Still another repair pathway in *E. coli* is called the **SOS repair system**, which also responds to damaged DNA but in a different way. In the presence of a large number of DNA mis-

matches and gaps during replication, bacteria can induce the expression of about 20 genes (including *lexA, recA,* and *uvr*) whose products allow DNA replication to occur even in the presence of these types of lesions. As this type of repair is a last resort to DNA damage, it is known as SOS repair. During SOS repair, DNA synthesis becomes error-prone, inserting random and possibly incorrect nucleotides in places that would normally stall DNA replication. As a result, SOS repair itself becomes mutagenic—although it may allow the cell to survive DNA damage that might otherwise kill it.

Photoreactivation Repair: Reversal of UV Damage in Prokaryotes

As illustrated in Figure 14–8, UV light is mutagenic as a result of the creation of pyrimidine dimers. UV-induced damage to *E. coli* DNA can be partially reversed if, following irradiation, the cells are exposed briefly to light in the blue range of the visible spectrum. The process is dependent on the activity of a protein called **photoreactivation enzyme (PRE)**. The enzyme's mode of action is to cleave the bonds between thymine dimers, thus directly reversing the effect of UV radiation on DNA (Figure 14–12). Although the enzyme will associate with a dimer in the dark, it must absorb a photon of light to cleave the dimer. In spite of the potential for diminishing UV-induced mutations, photoreactivation repair is not absolutely essential in *E. coli*, since a mutation creating a null allele in the gene coding for PRE is not lethal. Nonetheless, the enzyme is detectable in many organisms including bacte-

Photoreactivation repair

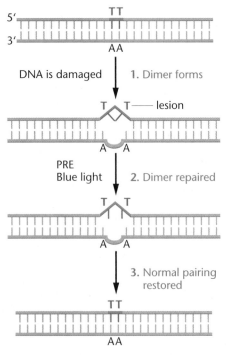

FIGURE 14–12 Damaged DNA repaired by photoreactivation repair. The bond creating the thymine dimer is cleaved by the photoreactivation enzyme (PRE), which must be activated by blue light in the visible spectrum.

Web Tutorial 14.4

DNA Repair: Proofreading, Mismatch, and SOS

ria, fungi, plants, and some vertebrates—but not humans. Hence, humans and other organisms that lack photoreactivation repair must rely on other repair mechanisms to reverse the effects of UV radiation.

Base and Nucleotide Excision Repair

In addition to PRE, light-independent repair systems exist in all prokaryotes and eukaryotes. The basic mechanisms involved in these types of repair—referred to as **excision repair**—consist of the following three steps and can be described generally as "cut-and-paste" systems.

1. The distortion or error present on one of the two strands of the DNA helix is recognized and enzymatically clipped out by an endonuclease. Excisions in the phosphodiester backbone usually include a number of nucleotides adjacent to the error as well, leaving a gap on one strand of the helix.

2. A DNA polymerase fills in the gap by inserting deoxyribonucleotides complementary to those on the intact strand, using it as a replicative template. The enzyme adds these bases to the free 3′-OH end of the clipped DNA. In *E. coli*, this is usually performed by DNA polymerase I.

3. DNA ligase seals the final "nick" that remains at the 3′-OH end of the last base inserted, closing the gap.

There are two types of excision repair: base excision repair and nucleotide excision repair. **Base excision repair (BER)** corrects damage to nitrogenous bases created by spontaneous hydrolysis or agents that chemically alter them. The first step in the BER pathway in *E. coli* involves the recognition of the chemically altered base by **DNA glycosylases**, which are specific to different types of DNA damage (Figure 14–13). For example, the enzyme uracil DNA glycosylase recognizes the presence of uracil in DNA. The enzyme first cuts the glycosidic bond between the base and the sugar, creating an apyrimidinic (AP) site. Such a sugar with a missing base is then recognized by an enzyme called **AP endonuclease**. The endonuclease makes a cut in the phosphodiester backbone at the AP site. This creates a distortion in the DNA helix that is recognized by the excision-repair system, which is then activated, leading to the correction of the error.

Although much has been learned about DNA glycosylases in *E. coli*, much less is known about the DNA glycosylases in eukaryotes. In addition, unlike nucleotide excision pathways, there is no human or animal disease model available that has a defective BER pathway, so it is difficult to assess the importance of BER in eukaryotes.

Nucleotide excision repair (NER) pathways repair "bulky" lesions in DNA that alter or distort the double helix, such as the UV-induced pyrimidine dimers discussed previously. The NER pathway (Figure 14–14) was first discovered in *E. coli*

Base excision repair

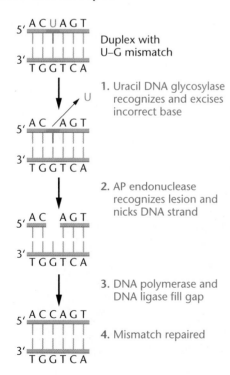

FIGURE 14–13 Base excision repair (BER) accomplished by uracil DNA glycosylase, AP endonuclease, DNA polymerase, and DNA ligase. Uracil is recognized as a noncomplementary base, excised, and replaced with the complementary base (C).

Nucleotide excision repair

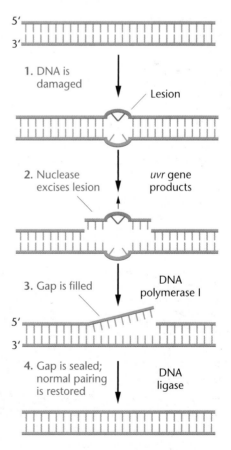

FIGURE 14–14 Nucleotide excision repair (NER) of a UV-induced thymine dimer. During repair, 13 bases are excised in prokaryotes and 28 bases are excised in eukaryotes.

by Paul Howard-Flanders and coworkers, who isolated several independent mutants that are sensitive to UV radiation. One group of genes was designated *uvr* (ultraviolet repair) and included the *uvrA, uvrB,* and *uvrC* mutations. In the NER pathway, the *uvr* gene products are involved in recognizing and clipping out lesions in the DNA. Usually, a very specific number of nucleotides is clipped out around both sides of the lesion. In *E. coli,* usually a total of 13 nucleotides are removed, which include the lesion. The repair is then completed by DNA polymerase I and DNA ligase, in a manner similar to that occurring in BER. The undamaged strand opposite the lesion is used as a template for the replication resulting in repair.

Nucleotide Excision Repair and Xeroderma Pigmentosum in Humans

The mechanism of nucleotide excision repair (NER) in eukaryotes is much more complicated than that in prokaryotes and involves many more proteins, encoded by about 30 genes. Much of what is known about the system in humans has come from detailed studies of individuals with **xeroderma pigmentosum (XP)**, a rare recessive genetic disorder that predisposes individuals to severe skin abnormalities. These individuals have lost their ability to undergo NER; as a result, individuals suffering from XP who are exposed to the UV radiation in sunlight exhibit reactions that range from initial freckling and skin ulceration to the development of skin cancer.

The condition is severe and may be lethal, although early detection and protection from sunlight can arrest it (Figure 14–15). The repair of UV-induced lesions was investigated *in vitro*, using human fibroblast cell cultures derived from XP and normal individuals. (Fibroblasts are undifferentiated connective tissue cells.) The results of these studies suggested that the XP phenotype could be caused by more than one mutant gene.

In 1968, James Cleaver showed that cells from XP patients were deficient in **unscheduled DNA synthesis** (DNA synthesis other than that occurring during chromosome replication), which is elicited in normal cells by UV radiation. Because this type of synthesis is thought to represent the activity of DNA polymerization during excision repair, the lack of unscheduled DNA synthesis in XP patients suggested that XP may be a deficiency in excision repair.

The link between XP and excision repair was further strengthened by studies using **somatic cell hybridization**. Fibroblast cells from any two unrelated XP patients, when grown together in tissue culture, can fuse together, forming a **heterokaryon**. A heterokaryon is a single cell with two nuclei but a common cytoplasm. After fusion of two XP cells, excision repair, as assayed by unscheduled DNA synthesis, can be measured. If the mutation in each of the two XP cells occurs in the same gene, the heterokaryon will still be unable to undergo excision repair. This is because there is no normal copy of the relevant gene present in the heterokaryon. However, if excision repair occurs in the heterokaryon, the mutations in the two XP cells must have been present in two different genes. Hence, the two mutants are said to demonstrate **complementation**. Complementation occurs because the heterokaryon has at least one normal copy of each gene in the fused cell. By fusing XP cells from a large number of XP patients, it was possible to determine how many genes contribute to the XP phenotype.

Based on many studies, XP patients have been divided into seven **complementation groups**, suggesting that at least seven different genes are involved in excision repair in humans. These seven human genes, *XPA* to *XPG* (*X*eroderma *P*igmentosum gene *A* to *G*), have now been identified, and a homologous gene for each has been identified in yeast. Approximately 20 percent of XP patients do not fall into any of the seven complementation groups. These patients manifest similar symptoms, but their fibroblasts do not demonstrate defective excision repair. There is some evidence that these XP cells are less efficient in normal DNA replication.

As a result of the study of defective genes in xeroderma pigmentosum, a great deal is now known about how NER counteracts DNA damage in normal cells. The first step in humans is the recognition of the damaged DNA by proteins encoded by the *XPC, XPE,* and *XPA* genes. These proteins then recruit the remainder of the repair proteins to the site of DNA damage. The *XPB* and *XPD* genes encode helicases and are also components of the basal transcription factor, TFIIH. The *XPF* and *XPG* genes encode nucleases. The excision repair complex containing these and other factors excises a 28-nucleotide-long fragment from the DNA strand that contains the lesion.

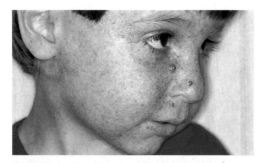
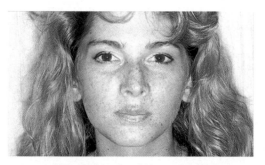

FIGURE 14–15 Two individuals with xeroderma pigmentosum (XP). The 4-year-old boy on the left shows marked skin lesions induced by sunlight. Mottled redness (erythema) and irregular pigment changes in response to cellular injury are apparent. Two nodular cancers are present on his nose. The 18-year-old girl on the right has been carefully protected from sunlight since her diagnosis of xeroderma pigmentosum in infancy. Several cancers have been removed, and she has worked as a successful model.

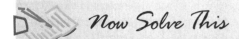

Now Solve This

Problem 27 on page 327 examines the results of heterokaryon analysis of seven cell lines derived from xeroderma pigmentosum patients. You are asked to determine the number of complementation groups represented by these data.

Hint: Complementation (+) results only when the two mutations being examined are in different complementation groups (genes).

Double-Strand Break Repair in Eukaryotes

Thus far, we have discussed repair pathways that deal with damage or error within one strand of DNA. We conclude our discussion of DNA repair by considering what happens when both strands of the DNA helix are cleaved—as a result of exposure to ionizing radiation, for example. While such damage occurs in bacteria, we will focus our discussion on eukaryotic cells.

In these cases, a specialized form of DNA repair, the **DNA double-strand break repair (DSB repair)** pathway, is activated and is responsible for reattaching the two DNA strands. Recently, interest has grown in DSB repair because defects in this pathway are associated with X-ray hypersensitivity and immune deficiency. Such defects may also underlie familial disposition to breast and ovarian cancer.

One pathway involved in double-strand break repair is homologous recombination repair. In a process similar to that of postreplication repair, damaged DNA is recombined and replaced with homologous undamaged DNA. In this case, the first step in the process involves the activity of an enzyme that recognizes the double-strand break, then digests back the 5′ ends of the broken DNA helix, leaving overhanging 3′ ends. These overhanging ends then interact with a region of an undamaged sister chromatid, aligning the homologous sequences and allowing exchange of the damaged DNA region with the undamaged DNA region. This interaction of two sister chromatids is necessary because, when both strands of one helix are broken, there is no undamaged parental DNA strand available to use as the source of the complementary template DNA sequence during repair. Therefore, the genetic information present in the homologous region of the sister chromatid is "recruited" to replace the damaged double-stranded break. Any gaps remaining after exchange can be filled by DNA polymerase and ligase, using correct and unbroken DNA strands as templates in those regions. The process usually occurs during the late S/G2 phase of the cell cycle, after DNA replication, a time when sister chromatids are available to be used as repair templates.

A second pathway, called **nonhomologous recombinational repair**, or **end joining**, also repairs double-strand breaks. However, as the name implies, the mechanism does not recruit a homologous region of DNA during repair. This system is activated in G1, prior to DNA replication. End joining involves a complex of three proteins including DNA-dependent protein kinase. These proteins bind to the free ends of the broken DNA, resulting in the ends being trimmed and ligated back together. Because some nucleotide sequence is lost in the process of end joining, it is an error-prone repair system. In addition, if more than one chromosome suffers a double-strand break, the wrong ends could be joined together, leading to abnormal chromosome structures, such as those discussed in Chapter 6.

14.8 Transposable Elements Move within the Genome and May Disrupt Genetic Function

Transposable elements, also known as **transposons** or "jumping genes," can move or transpose within and between chromosomes, inserting themselves into various locations within the genome.

Transposons are present in the genomes of all organisms from bacteria to humans. Not only are they ubiquitous, but they also comprise large portions of some eukaryotic genomes. For example, recent genomic sequencing has revealed that almost 50 percent of the human genome is derived from transposable elements. Some organisms with unusually large genomes, such as salamanders and barley, contain hundreds of thousands of copies of various types of transposable elements. Although the function of these elements is unknown, data from human genome sequencing suggest that some genes may have evolved from transposons and that the activity of transposons may help to modify and reshape the genome. In this sense, some transposable elements may confer benefits on their hosts. Transposable elements have also proved to be valuable tools in genetic research. Geneticists have harnessed transposons as mutagens, as cloning tags, and as vehicles for introducing foreign DNA into model organisms.

In this section, we discuss transposable elements as naturally occurring mutagens. The movement of transposons from one place in the genome to another has the capacity to disrupt genes and cause mutations, as well as to create chromosomal damage such as double-strand breaks.

Insertion Sequences

There are two types of transposable elements in bacteria: insertion sequences and bacterial transposons. **Insertion sequences (IS elements)** can move from one location to another and, if they insert into a gene or gene regulatory region, may cause mutations.

IS elements were first identified during analyses of mutations in the *gal* operon of *E. coli*. Researchers discovered that certain mutations in this operon were due to the presence of several hundred base pairs of extra DNA inserted into the beginning of the operon. Surprisingly, the segment of mutagenic DNA could spontaneously excise from this location, restoring wild-type function to the *gal* operon. Subsequent research revealed that several other DNA elements could behave in a similar fashion, inserting into bacterial chromosomes and affecting gene function.

IS elements are relatively short, not exceeding 2000 bp (2 kb). The first insertion sequence to be characterized in *E. coli*, IS1, is about 800 bp long. Other IS elements such as IS2, 3, 4, and 5 are about 1250 to 1400 bp in length. IS elements are present in multiple copies in bacterial genomes. For example, the

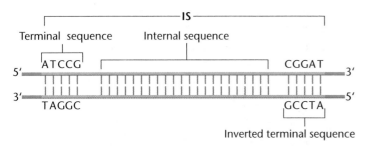

FIGURE 14–16 An insertion sequence (IS), shown in purple. The terminal sequences are perfect inverted repeats of one another.

E. coli chromosome contains five to eight copies of IS1, five copies each of IS2 and IS3, as well as copies of IS elements on plasmids such as F factors.

All IS elements contain two features that are essential for their movement. First, they contain a gene that encodes an enzyme called **transposase**. This enzyme is responsible for making staggered cuts in chromosomal DNA, into which the IS element can insert. Second, the ends of IS elements contain **inverted terminal repeats (ITRs)**. ITRs are short segments of DNA that have the same nucleotide sequence as each other but are oriented in the opposite direction (Figure 14–16). Although Figure 14–16 shows the ITRs to consist of only a few nucleotides, IS ITRs usually contain about 20 to 40 nucleotide pairs. ITRs are essential for transposition and act as recognition sites for the binding of the transposase enzyme.

Bacterial Transposons

Bacterial transposons (**Tn elements**) are larger than IS elements and contain protein-coding genes that are unrelated to their transposition. Some Tn elements, such as Tn10, are comprised of a drug-resistance gene flanked by two IS elements present in opposite orientations. The IS elements encode the transposase enzyme that is necessary for transposition of the Tn element. Other types of Tn elements, such as Tn3, have shorter inverted repeat sequences at their ends and encode their transposase enzyme from a transposase gene located in the middle of the Tn element. Like IS elements, Tn elements are mobile in both bacterial chromosomes and in plasmids, and can cause mutations if they insert into genes or gene regulatory regions.

Tn elements are currently of particular interest because they can introduce multiple drug resistance onto bacterial plasmids. These plasmids, called R factors, may contain many Tn elements conferring simultaneous resistance to heavy metals, antibiotics, and other drugs. These elements can move from plasmids onto bacterial chromosomes and can spread multiple drug resistance between different strains of bacteria.

The *Ac–Ds* System in Maize

About 20 years before the discovery of transposons in bacteria, Barbara McClintock discovered mobile genetic elements in corn plants (maize). She did this by analyzing the genetic behavior of two mutations, *dissociation* (*Ds*) and *activator* (*Ac*), expressed in either the endosperm or aleurone layers. She then correlated her genetic observations with cytological examinations of the maize chromosomes. Initially, McClintock determined that *Ds* was

located on chromosome 9. If *Ac* was also present in the genome, *Ds* induced breakage at a point on the chromosome adjacent to its own location. If chromosome breakage occurred in somatic cells during their development, progeny cells often lost part of the broken chromosome, causing a variety of phenotypic effects.

Subsequent analysis suggested to McClintock that both *Ds* and *Ac* elements sometimes moved to new chromosomal locations. While *Ds* moved only if *Ac* was also present, *Ac* was capable of autonomous movement. Where *Ds* came to reside determined its genetic effects—that is, it might cause chromosome breakage, or it might inhibit expression of a certain gene. In cells in which *Ds* caused a gene mutation, *Ds* might move again, restoring the gene mutation to wild type.

Figure 14–17 illustrates the types of movements and effects brought about by *Ds* and *Ac* elements. In McClintock's

(a) In absence of *Ac*, *Ds* is not transposable.

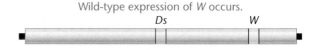

(b) When *Ac* is present, *Ds* may be transposed.

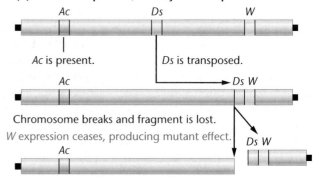

(c) *Ds* can move into and out of another gene

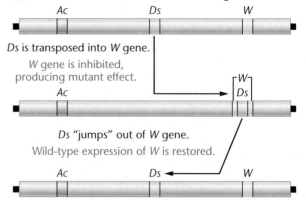

FIGURE 14–17 Effects of *Ac* and *Ds* elements on gene expression. (a) If *Ds* is present in the absence of *Ac*, there is normal expression of a distantly located hypothetical gene *W*. (b) In the presence of *Ac*, *Ds* transposes to a region adjacent to *W*. *Ds* can induce chromosome breakage, which may lead to loss of a chromosome fragment bearing the *W* gene. (c) In the presence of *Ac*, *Ds* may transpose into the *W* gene, disrupting *W* gene expression. If *Ds* subsequently transposes out of the *W* gene, *W* gene expression may return to normal.

FIGURE 14–18 A comparison of the structure of an *Ac* element and three *Ds* elements. The imperfect inverted repeats are at the ends of the *Ac* element, and are shown in pink. The transposase gene is in the open reading frame, ORF 1. No function has yet been assigned to ORF 2. Noncoding regions are designated Nc. As this illustration shows, *Ds-a* appears to be an *Ac* element containing a small deletion in the gene encoding the transposase enzyme.

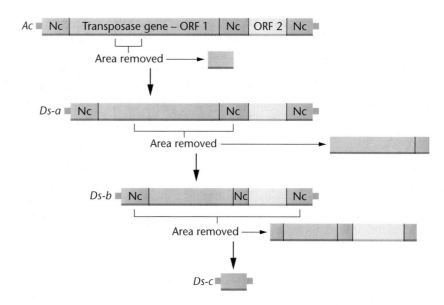

original observation, pigment synthesis was restored in cells in which the *Ds* element jumped out of chromosome 9. McClintock concluded that the *Ds* and *Ac* genes were **mobile controlling elements**. We now commonly refer to them as transposable elements, a term coined by another great maize geneticist, Alexander Brink.

Several *Ac* and *Ds* elements have now been analyzed, and the relationship between the two elements has been clarified (Figure 14–18). The *Ac* element is 4563 nucleotides long, and its structure is strikingly similar to that of bacterial transposons. The *Ac* sequence contains two 11-base-pair imperfect ITRs, two open reading frames (ORFs), and three noncoding regions. One of the two ORFs encodes the *Ac* transposase enzyme. The first *Ds* element studied (*Ds-a*) is nearly identical to *Ac* except for a 194-bp deletion within the transposase gene. The deletion of part of the transposase gene in the *Ds-a* element explains its dependence on the *Ac* element for transposition. Several other *Ds* elements have also been sequenced, and each contains an even larger deletion within the transposase gene. In each case, however, the ITRs are retained.

Although the significance of Barbara McClintock's mobile controlling elements was not fully appreciated following her initial observations, molecular analysis has since verified her conclusions. She was awarded the Nobel Prize in Physiology or Medicine in 1983.

Copia Elements in *Drosophila*

In 1975, David Hogness and his colleagues David Finnegan, Gerald Rubin, and Michael Young identified a class of DNA elements in *Drosophila melanogaster* that they designated as *copia*. These elements are transcribed into "copious" amounts of RNA (hence their name). *Copia* elements are present in 10 to 100 copies in the genomes of *Drosophila* cells. Mapping studies show that they are transposable to different chromosomal locations and are dispersed throughout the genome.

Each *copia* element consists of approximately 5000 to 8000 bp of DNA, including a long **direct terminal repeat (DTR)** sequence of 267 bp at each end. Within each DTR is an inverted terminal repeat (ITR) of 17 bp (Figure 14–19). The

short ITR sequences are characteristic of *copia* elements. The DTR sequences are found in other transposons in other organisms, but they are not universal.

Insertion of *copia* is dependent on the presence of the ITR sequences and seems to occur preferentially at specific target sites in the genome. The *copia*-like elements demonstrate regulatory effects at the point of their insertion in the chromosome. Certain mutations, including those affecting eye color and segment formation, are due to *copia* insertions within genes. For example, the eye-color mutation *white–apricot*, is caused by an allele of the *white* gene which contains a *copia* element within the gene. Transposition of the *copia* element out of the *white-apricot* allele can restore the allele to wild type.

Copia elements are only one of approximately 30 families of transposable elements in *Drosophila*, each of which is present in 20 to 50 copies in the genome. Together, these families constitute about 5 percent of the *Drosophila* genome and over half of the middle repetitive DNA of this organism. One study suggests that 50 percent of all visible mutations in *Drosophila* are the result of the insertion of transposons into otherwise wild-type genes.

P Element Transposons in *Drosophila*

Perhaps the most significant *Drosophila* transposable elements are the ***P* elements**. These were discovered while studying the phenomenon of **hybrid dysgenesis**, a condition characterized by sterility, elevated mutation rates, and chromosome rearrangements in the offspring of crosses between certain strains of fruit flies. Hybrid dysgenesis is caused by high rates of *P* element

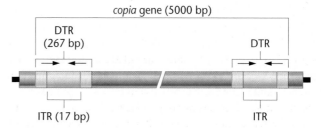

FIGURE 14–19 Structural organization of a *copia* transposable element in *Drosophila melanogaster*, showing the terminal repeats.

transposition in the germ line, in which transposons insert themselves into or near genes, thereby causing mutations. *P* elements range from 0.5 to 2.9 kb long, with 31-bp ITRs. Full-length *P* elements encode at least two proteins, one of which is the transposase enzyme that is required for transposition, and another is a repressor protein that inhibits transposition. The transposase gene is expressed only in the germ line, accounting for the tissue specificity of *P* element transposition. Strains of flies that contain full-length *P* elements inserted into their genomes are resistant to further transpositions due to the presence of the repressor protein encoded by the *P* elements.

Mutations can arise from several kinds of insertional events. If a *P* element inserts into the coding region of a gene, it can terminate transcription of the gene and destroy normal gene expression. If it inserts into the promoter region of a gene, it can affect the level of expression of the gene. Insertions into introns can affect splicing or cause the premature termination of transcription.

Geneticists have harnessed *P* elements as tools for genetic analysis. One of the most useful applications of *P* elements is as vectors to introduce transgenes into *Drosophila*—a technique known as **germ-line transformation**. *P* elements are also used to generate mutations and to clone mutant genes. In addition, researchers are perfecting methods to target *P* element insertions to precise single-chromosomal sites, which should increase the precision of germ-line transformation in the analysis of gene activity.

Transposable Elements in Humans

The human genome, like that of other eukaryotes, is riddled with DNA derived from transposons. Recent genomic sequencing data reveal that approximately half of the human genome is comprised of transposable element DNA.

As we saw in Chapter 11, the major families of human transposons are the long interspersed elements and short interspersed elements (**LINES** and **SINES**, respectively). LINES consist of DNA elements of about 6 kilobase pairs in length, present in up to 850,000 copies. In all, LINES account for 21 percent of human genomic DNA. SINES are about 100 to 500 bp long with about 1.5 million copies present in human cells. SINES comprise about 13 percent of human genomic DNA. Other families of transposable elements account for a further 11 percent of the human genome. As coding sequences comprise only about 1 percent of the human genome, there is about forty to fifty times more transposable element DNA in the human genome than DNA in functional genes.

Although most human transposons appear to be inactive, the potential mobility and mutagenic effects of transposable elements have far-reaching implications for human genetics, as can be seen in a recent example of a transposon "caught in the act." The case involves a male child with hemophilia. One cause of hemophilia is a defect in blood-clotting factor VIII, the product of an X-linked gene. Haig Kazazian and his colleagues found LINES inserted at two points within the gene. Researchers were interested in determining if one of the mother's X chromosomes also contained this specific LINE. If so, the unaffected mother would be heterozygous and pass

the LINE-containing chromosome to her son. The surprising finding was that the LINE sequence was *not* present on either of her X chromosomes but *was* detected on chromosome 22 of both parents. This suggests that this mobile element may have transposed from one chromosome to another in the gamete-forming cells of the mother, prior to being transmitted to the son.

LINE insertions into the human *dystrophin* gene have resulted in at least two separate cases of Duchenne muscular dystrophy. In one case, a transposon inserted into exon 48 and in another case, a transposon inserted into exon 44, both leading to frameshift mutations and premature termination of translation of the dystrophin protein. There are also reports that LINES have inserted into the *APC* and *c-myc* genes, leading to mutations that may have contributed to the development of some colon and breast cancers. In the latter cases, the transposition had occurred within one or a few somatic cells.

SINE insertions are also responsible for a number of cases of human disease. In one case, an **Alu element** integrated into the *BRCA2* gene, inactivating this tumor suppressor gene and leading to a familial case of breast cancer. Other genes that have been mutated by *Alu* integrations are the *factor IX* gene (leading to hemophilia B), the *ChE* gene (leading to acholinesterasemia), and the *NF1* gene (leading to neurofibromatosis).

14.9 Geneticists Use Mutations to Identify Genes and Study Gene Function

We conclude this chapter by considering one of the major goals of genetics: to understand what genes are and how they work. In classical genetic analysis, geneticists attempt to answer these questions by first collecting a number of individuals that display mutations affecting the phenotype of interest. From there, they determine whether the phenotype is controlled by one or more genes, which genes are responsible for which steps in the pathway leading to the phenotype, and how each gene product controls the phenotype at the biochemical level. Starting from a series of mutants, geneticists identify, clone, and characterize the function of genes.

In order to dissect the genes and processes that regulate biological functions, geneticists induce mutations in various model organisms. A good **model organism** must be easy to grow, have a short generation time, produce abundant progeny, and be readily mutagenized and crossed. The most extensively used model organisms are bacteria (*E. coli*), budding yeasts (*Saccharomyces cerevisiae*), fruit flies (*Drosophila melanogaster*), nematodes (*Caenorhabditis elegans*), mustard plants (*Arabidopsis thaliana*), and mice (*Mus musculus*).

In this section, we will focus on some of the methods that geneticists use to generate and detect gene mutations, as the first step toward a full genetic analysis.

Generating Mutants with Radiation, Chemicals, and Transposon Insertion

Geneticists sometimes examine spontaneous, naturally occurring mutations; however, as we saw in Section 14.2, mutations are generally rare in nature. Researchers would need to screen millions of individuals in order to detect one relevant spontaneous mutation. Mutant hunting is greatly enhanced by subjecting the model organism to mutagenesis before screening for mutations. The goal is to create one mutation at random in the genome of each individual in the population, so that one gene product is disrupted in each individual, leaving the rest of the genome wild-type.

As we saw earlier in this chapter, any substance that alters a DNA sequence is mutagenic. In genetic analysis, researchers select certain mutagens, depending on the type of mutation desired. For example, ionizing radiation causes chromosome breaks, deletions, translocations, and other major rearrangements. Mutations created by radiation are likely to be null muations and have severe effects on the phenotype. In contrast, ultraviolet light and certain chemicals such as ethyl methane sulfonate (EMS) and nitrosoguanidine cause single base-pair changes or small deletions and insertions. With these mutagens, a range of mild to severe mutant phenotypes may emerge, depending on where in the gene the lesion occurs. It is more likely that single base-pair mutations will result in the creation of conditional mutations, such as temperature-sensitive mutations, which are particularly useful for the study of essential gene functions.

Geneticists also use transposons to create mutations. The random insertion of a transposon, such as a *Drosophila P* element, into the genome can cause major disruptions of gene function. These large insertions usually create null mutations if they transpose into a gene's open reading frame. They may create a number of different effects on gene expression if they transpose into a gene's regulatory sequences, either in flanking regions around the gene or within the gene's introns.

Screening for Mutants

The next step in a genetic analysis involves detecting those individuals, within the mutagenized population, who display mutations in the gene or genes affecting the phenotype of interest. The most frequently used method to detect mutants is a **genetic screen**. A genetic screen may involve the visual or biochemical examination of large numbers of mutagenized organisms. For example, Mendel formed the scientific basis for transmission genetics by screening thousands of individual pea plants for visible phenotypic features. Similarly, modern biochemical analysis of the proteins in maize endosperm led to the detection of the *opaque-2* mutant strain, which contains significantly more lysine than that in nonmutant strains. Since maize protein is usually low in lysine, this mutation significantly improves the nutritional value of the plant.

It is easy to see how dominant mutations can be detected during a genetic screen, in either haploid or diploid organisms. As long as the dominant mutation is not lethal at an early stage of development, the phenotype will be visible immediately in any organism that bears a dominant mutation in the relevant gene. The mutant can then be crossed, and heterozygous or homozygous stocks can be maintained.

Most mutations, however, result in a loss of gene function, and most of these loss-of-function mutations are recessive. Organisms that have a haploid phase in their life cycle, such as yeast, have a significant advantage for recessive mutant detection, as the mutant phenotype will be immediately evident in the mutated haploid individual. The mutation can then be crossed and analyzed in diploid phases of the life cycle.

Diploid organisms that are heterozygous for a recessive allele will not show the mutant phenotype. Geneticists have devised some intricate strategies to detect and recover recessive mutations in diploid organisms. An example of one of these methods is the **attached-X** method for detecting recessive mutations on the X chromosome of *Drosophila*. This method was developed by Hermann Muller in the 1920s, in order to demonstrate that X-rays cause mutations in *Drosophila*.

Attached-X females have two X chromosomes attached to a single centromere and one Y chromosome, in addition to the normal diploid complement of autosomes. When attached-X females are mated to males with normal sex chromosomes (XY), four types of progeny result: triplo-X females that die, viable attached-X females, YY males that also die, and viable XY males. Figure 14–20 shows how P_1 males that have been treated with a mutagenic agent produce F_1 male offspring that express any recessive mutation induced on the X chromosome. In this screening procedure, the mutant phenotype is expressed in the first generation.

Geneticists have devised similar techniques to detect recessive mutations on autosomes of *Drosophila* and other diploid

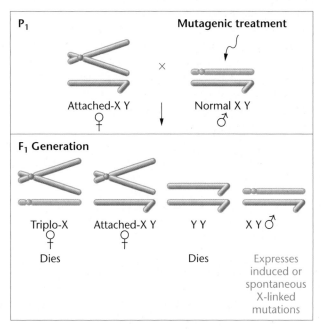

FIGURE 14–20 The attached-X method for detection of induced morphological mutations in *Drosophila*.

organisms. These methods involve crossing mutagenized individuals with special strains bearing a number of visible and lethal gene markers or chromosomes containing specific deletions. After three or more generations, the recessive lethal mutation may be revealed, but the mutant strain can be maintained in a heterozygous state. Although these techniques are cumbersome to perform, they are fairly efficient. Many of the developmental mutations in *Drosophila* that are discussed in Chapter 20 were identified using these techniques.

Selecting for Mutants

In a genetic screen, each organism must be individually examined for the phenotype of interest. Often, tens of thousands of individuals must be screened to find the required number of mutants. Though effective, this strategy is obviously slow and labor-intensive. In some cases, a different method, **selection**, can be employed to reduce the amount of time and labor. The goal of selection is to create conditions that remove the wild-type or irrelevant mutant organisms from the population, leaving only the mutants that are sought. Usually, this is accomplished by killing or inhibiting the growth of individuals that do not display the relevant phenotype.

Selection is simplest if the mutant phenotype enhances survival under certain conditions. For example, a mutant gene that confers resistance to a drug can be easily selected by growing the mutagenized population on a medium containing the drug. Also, a mutation that corrects a defect in a metabolic pathway, such as an inability to metabolize galactose, can be selected if the organism is grown in a medium that contains only galactose as a carbon source.

Detection in Humans

Because humans are obviously not suitable experimental organisms, techniques developed for the detection of mutations in organisms such as bacteria and plants are not available to human geneticists. To determine the mutational basis for any human characteristic or disorder, geneticists may first analyze a **pedigree** that traces the family history as far back as possible. If a trait is shown to be inherited, the pedigree may also be useful in determining whether the mutant allele is behaving as a dominant or a recessive and whether it is X-linked or autosomal.

Dominant mutations are the simplest to detect. If they are present on the X chromosome, affected fathers pass the trait to all of their daughters. If dominant mutations are autosomal, approximately 50 percent of the offspring of an affected heterozygous individual are expected to show the trait. Figure 14–21 shows a pedigree illustrating the initial occurrence of an autosomal dominant mutation for cataracts of the eye. The parents in Generation I were unaffected, but one of their three offspring (Generation II) developed cataracts, so the original mutation presumably occurred in a gamete of one of the parents. The affected female, the proband, produced two children, of which the male child was affected (Generation III). Of his six offspring (Generation IV), four were also affected. These observations are consistent with, but do not prove, an autosomal dominant mode of inheritance. However, the high percentage of affected offspring in Generation IV favors this conclusion. Also, the presence of an unaffected daughter in this generation (Generation IV) argues against X-linkage, because she received her X chromosome from her affected father. Hence, the conclusion is sound, provided that the mutant allele is completely penetrant.

X-linked recessive mutations can also be detected by pedigree analysis, as discussed in Chapter 4. The most famous case of an X-linked recessive mutation in humans is that of **hemophilia**, which was found in the descendants of Britain's Queen Victoria. Inspection of the pedigree (Figure 14–22) leaves little doubt that Victoria was heterozygous for the trait. Her father was not affected, and there is no evidence to show that her mother was a carrier, as was Victoria. The recessive mutation for hemophilia has occurred many times in human populations, but the political consequences of the mutation that occurred in the British royal family were sweeping. Two of Queen Victoria's daughters were carriers, and they passed hemophilia into the Russian and Spanish royal families. Many historians have posited that Tsarina Alexandra's attempts to deal with her son's hemophilia resulted in Rasputin's presence in the royal household and may have contributed to the start of the Russian Revolution.

It is also possible to determine the recessive nature of alleles by pedigree analysis. Because this type of mutation is "hidden" when heterozygous, it is not unusual for the trait to

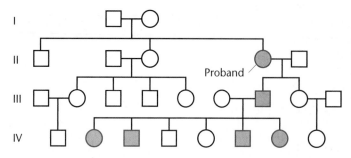

FIGURE 14–21 A hypothetical pedigree of inherited cataracts of the eye in humans.

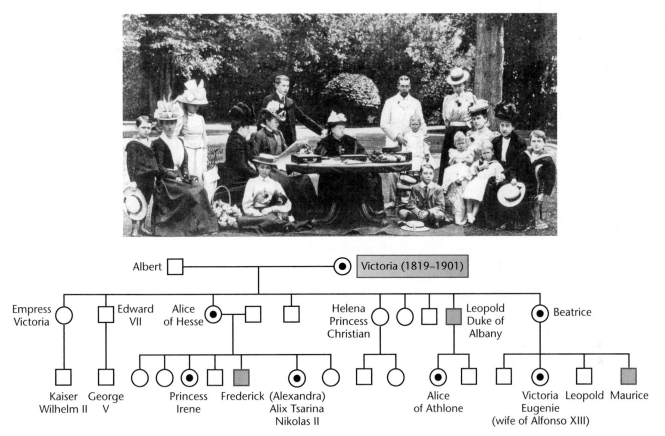

FIGURE 14–22 A partial pedigree of hemophilia in the British royal family descended from Queen Victoria. The pedigree is typical of the transmission of X-linked recessive traits. Circles with a dot in them indicate presumed female carriers heterozygous for the trait. The photograph shows Queen Victoria (seated at the table, center) and some of her immediate family.

appear only intermittently through a number of generations. A mating between an affected individual and a homozygous normal individual will produce unaffected heterozygous carrier children. Matings between two carriers will produce, on average, one-fourth affected offspring.

In addition to pedigree analysis, human cells may now be routinely cultured *in vitro*. This procedure has allowed the study of many more mutations than any other form of analysis. Analysis of enzyme activity, protein migration in electrophoretic fields, and direct sequencing of DNA and proteins are among the techniques that have identified mutations and demonstrated the wide genetic variation among individuals in human populations. As we saw earlier in the chapter, initial studies of the human disorder xeroderma pigmentosum relied on the ***in vitro* cell culture** technique. More recently, genomics and reverse genetic techniques have expanded the techniques available for geneticists to study human mutations.

Once a mutant gene has been identified and cloned, sequence analysis of mutant and normal genes from affected and unaffected individuals may provide the molecular basis of the disease or mutant phenotype. In addition, knowledge of the mutant gene sequence may open the way to developing specific genetic tests and gene-based therapeutics.

Now Solve This

Problem 26 on page 327 asks you to explain how X-linked hemophilia could have arisen in the British royal family.

Hint: To answer this question, you might consider the ways in which mutations occur, in what types of cells they can occur, and how they are inherited.

GENETICS, TECHNOLOGY, AND SOCIETY

In the Shadow of Chernobyl

On April 26, 1986, Reactor 4 of the Chernobyl Nuclear Power Station exploded and ejected 6.7 tons of radioactive material into the surrounding countryside and throughout the Northern Hemisphere. The accident immediately killed 30 emergency workers and caused acute radiation sickness in more than 200 others. In the nine days following the initial explosion, as the reactor's temperature approached meltdown, radioactive fission products including radioactive iodine, xenon, strontium, and cesium, were released into the atmosphere. Fallout traveled through central Europe, reaching Finland and Sweden three days after the initial explosion and the United Kingdom and North America within a week. Millions of people were exposed to measurable amounts of radioactivity. People living within 30 km (about 18 miles) of Chernobyl were exposed to high levels of radioactivity prior to their evacuation 36 hours after the accident. Equally high were the exposures of some of the 200,000 military and civilian workers who were sent to Chernobyl to decontaminate the area and encase the shattered reactor in a sarcophagus. A total of 62 deaths have so far been attributed to the Chernobyl accident.

The Chernobyl incident was the world's largest accidental release of radioactive material. The question that remains is whether Chernobyl's radioactive pollution directly threatens the long-term health of millions of people.

More is known about the effects of ionizing radiation on human health than any other toxic agent (except perhaps cigarette smoke). Ionizing radiation (such as X-rays and gamma rays) is a subset of radiation that possesses sufficient energy to eject electrons from atoms. Ionizing radiation can damage any cellular component, alter nucleotides, and induce double-strand breaks in DNA. These DNA lesions can lead to mutations or chromosomal translocations.

High doses of ionizing radiation increase the risk of developing certain cancers. The survivors of the atomic-bomb blasts of Hiroshima and Nagasaki showed an increased incidence of leukemia within two years of the bombing. Breast cancers increased 10 years after exposure, as did cancers of the lung, thyroid, colon, ovary, stomach, and nervous system. Since ionizing radiation induces DNA damage, the offspring of the survivors were expected to show increases in birth defects, yet these increases were not detected.

The problem in extrapolating from Hiroshima to Chernobyl is the difference in dose. In atomic-bomb survivors, the cancer rate increased among people exposed to at least 200 mSv (mSv = millisievert, a unit dose of absorbed radiation). It is estimated that people living in the most contaminated areas close to Chernobyl may have received about 50 mSv, and some cleanup workers were exposed to approximately 250 mSv. Outside the Chernobyl area, radiation doses were estimated at 0.4 to 0.9 mSv in Germany and Finland, 0.01 mSv in the United Kingdom, and 0.0006 mSv in the United States.

To put these numbers in perspective, the average dose for medical diagnostic procedures (such as chest and dental X-rays) is 0.39 mSv per year. A person's radiation exposure from natural background sources (cosmic rays, rocks, and radon gas) is about 2 to 3 mSv per year. Smokers expose themselves to an average dose of about 2.8 mSv per year from the intake of naturally occurring radioactive materials in tobacco smoke.

It appears that mutation rates among plants and animals exposed to Chernobyl's radioactive waste may be two- to tenfold higher than normal. Also, Chernobyl cleanup workers exhibit a 25 percent higher mutation frequency at the *HPRT* locus. Mutation rates at minisatellites among children born in polluted areas are two fold higher than those from controls in the United Kingdom. But do these increased mutation rates translate into health effects? So far, we cannot answer that question.

Estimates of 26 leukemia cases higher than the 25 to 30 that would occur spontaneously were projected in the 115,000 people evacuated from Chernobyl. It was also estimated that up to 17,000 additional cancers might occur among Europeans, above the 123 million that would occur normally. At present, however, there have been no detectable increases in leukemias or solid tumors in contaminated areas of the former Soviet Union, Finland, or Sweden, or in the 200,000 Chernobyl cleanup workers. In addition, there have been no detectable fertility problems or increases in birth defects. Due to uncertainties about the effects of low doses of radiation, estimates of future cancer deaths due to Chernobyl vary from 4000 to more than 50,000.

Despite the lack of detectable increases in leukemias and solid tumors, one type of cancer has increased in the Chernobyl area. The rate of childhood thyroid cancer has reached more than 100 cases per million children per year, whereas normal rates are expected to be between 0.5 and 3 cases per million children per year. About 4,000 cases of thyroid cancer have developed in people who were children at the time of the accident. These cases likely developed as a result of the children drinking milk contaminated with radioactive iodine. Luckily, the survival rate from thyroid cancer is about 99%, and only fifteen children have died from the disease.

The most profound effects of Chernobyl have been psychological. Chernobyl cleanup workers demonstrated a 50 percent increase in suicides and a detectable increase in smoking- and alcohol-related disease. This mirrors other studies showing that 45 percent of people living within 300 km of Chernobyl believe that they have a radiation-induced illness. Health effects such as depression, sleep disturbance, hypertension, and altered perception have been documented. Posttraumatic stress may be a greater threat to health than the actual radiation exposure from the accident. People feel that they live in constant danger and are simply awaiting the results of a cancer "lottery." Even if cancer rates and genetic defects do not increase dramatically, the indirect health effects from the Chernobyl Nuclear Power Station explosion have been and continue to be immense.

References

Peplow, M. 2006. Counting the dead. *Nature* 440: 982–983.

Williams, D. and Baverstock, K. 2006. Too soon for a final diagnosis. *Nature* 440: 993–994.

The Chernobyl Forum: 2003-2005. Chernobyl's Legacy: Health, Environmental, and Socio-economic Impacts. Report from eight UN Agencies including International Atomic Energy Agency, World Health Organization and the UN Development Program. September, 2005.

http://www.iaea.org/Publications/ Booklets/Chernobyl/chernobyl.pdf

CHAPTER SUMMARY

1. Mutations provide the foundation for genetic variation, evolution, and genetic analysis.

2. A mutation is defined as an alteration in DNA sequence. A mutation may or may not create a detectable phenotype.

3. Mutations may be either spontaneous or induced, somatic or germ-line, autosomal or sex-linked.

4. Mutations can have many different effects on gene function depending on the type of nucleotide changes that comprise the mutation. Point mutations within coding regions of a gene may lead to missense mutations or nonsense mutations. Insertions or deletions within coding regions may lead to frameshift mutations and premature termination of translation.

5. The phenotypic effects of mutations may be varied, from loss-of-function mutations and gene knockouts to gain-of-function mutations that confer a new activity to a gene product.

6. Spontaneous mutation rates vary considerably between organisms and between different genes in the same organism.

7. Spontaneous mutations arise from many mechanisms, from errors during DNA replication to damage caused to DNA bases as a result of normal cellular metabolism.

8. Mutations can be induced by many types of chemicals and radiation. These agents can damage both bases and the sugar-phosphate backbone of DNA molecules.

9. The human ABO blood groups owe their variation to point mutations or frameshift mutations within the gene encoding the enzyme that modifies the H substance.

10. The most severe forms of muscular dystrophy are caused by duplication or deletion mutations that create changes in the translation reading frame of the *dystrophin* gene.

11. Mutations in the copy numbers of trinucleotide repeats in some genes lead to human diseases such as the fragile X syndrome, myotonic dystrophy, and Huntington disease.

12. The Ames test allows scientists to estimate the mutagenicity and cancer-causing potential of chemical agents by following the rate of mutation in specific strains of bacteria.

13. Organisms counteract both spontaneous and induced mutations using DNA repair systems. Errors in DNA synthesis can be repaired by proofreading during replication, mismatch repair, and postreplication repair.

14. Other types of DNA damage can be repaired by photoreactivation repair, SOS repair, base excision repair, nucleotide excision repair, and double-strand break repair.

15. Mutations in genes controlling nucleotide excision repair in humans can lead to diseases such as xeroderma pigmentosum.

16. Transposable elements can move within a genome and create mutations if they insert within or near genes. They have also been harnessed as genetic dissection tools to create mutations, clone genes, and introduce foreign genes into model organisms.

17. A range of genetic and biochemical techniques are used to induce and analyze mutations in model organisms. In contrast, the study of human mutations often begins with pedigree analysis.

KEY TERMS

ABO blood group system, 311

acridine dyes, 310

activator (Ac), 318

alkylating agent, 309

Alu element, 320

Ames test, 312

2-amino-purine (2-AP), 309

AP endonuclease, 315

apurinic (AP) site, 307

attached-X, 321

autosomal mutations, 304

base analogs, 309

base excision repair (BER), 315

base substitution, 305

Becker muscular dystrophy (BMD), 311

behavioral mutations, 306

biochemical mutations, 306

bromodeoxyuridine (BrdU), 309

5-bromouracil (5-BU), 309

carcinogens, 312

complementation, 316

complementation groups, 316

conditional mutations, 306

copia, 319

cosmic rays, 310

deamination, 308

depurination, 307

direct terminal repeat (DTR), 319

dissociation (Ds), 318

DNA double-strand break repair (DSB repair), 317

DNA glycosylases, 315

DNA methylation, 313

DNA repair, 313

Duchenne muscular dystrophy (DMD), 311

dystrophin, 311

electromagnetic spectrum, 310

end joining, 317

excision repair, 315

fragile X syndrome, 312

frameshift mutations, 305

free radicals, 310

gain-of-function mutation, 305

gamma rays, 310

gene mutations, 304

genetic screen, 321

germ-line mutations, 304

germ-line transformation, 320

glycosyltransferase, 311

hemophilia, 322

heterokaryon, 316

homologous recombination repair, 314

Huntington disease (HD), 312

hybrid dysgenesis, 319

induced mutations, 304

insertion sequences (IS elements), 317

intercalate, 310

in vitro cell culture, 323

ionizing radiation, 310

lethal mutation, 306

LINES, 320

loss-of-function mutation, 305

mismatch repair, 313

missense mutation, 305

mobile controlling elements, 319

model organism, 320

morphological trait, 306

mutagens, 309

myotonic dystrophy, 312

neutral mutations, 305

nonhomologous recombination repair, 317

nonsense mutation, 305

nucleotide excision repair (NER), 315

null mutation, 305

nutritional mutation, 306

P element, 319

photoreactivation enzyme (PRE), 314

INSIGHTS AND SOLUTIONS

1. The base analog 2-amino purine (2-AP) substitutes for adenine during DNA replication, but it may base-pair with cytosine. The base analog 5-bromouracil (5-BU) substitutes for thymidine, but it may base-pair with guanine. Follow the double-stranded trinucleotide sequence shown here through three rounds of replication, assuming that, in the first round, both analogs are present and become incorporated wherever possible. In the second and third round of replication, they are removed. What final sequences occur?

Solution:

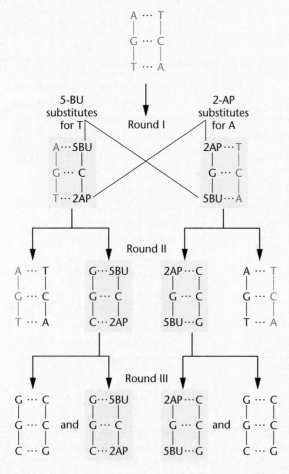

2. A rare dominant mutation expressed at birth was studied in humans. Records showed that six cases were discovered in 40,000 live births. Family histories revealed that in two cases, the mutation was already present in one of the parents. Calculate the spontaneous mutation rate for this mutation. What are some underlying assumptions that may affect our conclusions?

Solution: Only four cases represent a new mutation. Because each live birth represents two gametes, the sample size is from 80,000 meiotic events. The rate is equal to

$$\frac{4}{80,000} = \frac{1}{20,000} = 5 \times 10^{-5}$$

We have assumed that the mutant gene is fully penetrant and is expressed in each individual bearing it. If it is not fully penetrant, our calculation may be an underestimate because one or more mutations may have gone undetected. We have also assumed that the screening was 100 percent accurate. One or more mutant individuals may have been "missed," again leading to an underestimate. Finally, we assumed that the viability of the mutant and nonmutant individuals is equivalent and that they survive equally *in utero*. Therefore, our assumption is that the number of mutant individuals at birth is equal to the number at conception. If this were not true, our calculation would again be an underestimate.

3. Consider the following estimates:

(a) There are 5.5×10^9 humans living on this planet.

(b) Each individual has about 30,000 (0.3×10^5) genes.

(c) The average mutation rate at each locus is 10^{-5}.

How many spontaneous mutations are currently present in the human population? Assuming that these mutations are equally distributed among all genes, how many new mutations have arisen at each gene in the human population?

Solution: First, since each individual is diploid, there are two copies of each gene per person, each arising from a separate gamete. Therefore, the total number of spontaneous mutations is

$(2 \times 0.3 \times 10^5 \text{ genes/individual})$
$\qquad \times (5.5 \times 10^9 \text{ individuals}) \times (10^{-5} \text{ mutations/gene})$
$\qquad = (0.6 \times 10^5) \times (5.5 \times 10^9) \times (10^{-5}) \text{ mutations}$
$\qquad = 3.3 \times 10^9 \text{ mutations in the population}$
$\qquad 3.3 \times 10^9 \text{ mutations}/0.3 \times 10^5 \text{ genes}$
$\qquad = 11 \times 10^4 \text{ mutations per gene in the population}$

PROBLEMS AND DISCUSSION QUESTIONS

1. Discuss the importance of mutations in genetic studies.
2. Why would a mutation in a somatic cell of a multicellular organism escape detection?
3. Most mutations are thought to be deleterious. Why, then, is it reasonable to state that mutations are essential to the evolutionary process?
4. Why is a random mutation more likely to be deleterious than beneficial?
5. Most mutations in a diploid organism are recessive. Why?
6. What is meant by a conditional mutation?
7. Describe a tautomeric shift and how it may lead to a mutation.
8. Contrast and compare the mutagenic effects of deaminating agents, alkylating agents, and base analogs.
9. Acridine dyes induce frameshift mutations. Why are frameshift mutations likely to be more detrimental than point mutations, in which a single pyrimidine or purine has been substituted?
10. Why are X-rays more potent mutagens than UV radiation?
11. Compare the mechanisms by which UV radiation and X-rays cause mutations.
12. Contrast the various types of DNA repair mechanisms known to counteract the effects of UV radiation. What is the role of visible light in repairing UV-induced mutations?
13. Mammography is an accurate screening technique for the early detection of breast cancer in humans. Because this technique uses X-rays diagnostically, it has been highly controversial. Can you explain why? What reasons justify the use of X-rays for such a type of medical diagnosis?
14. Explain the molecular basis of fragile X syndrome, myotonic dystrophy, and Huntington disease and how the types of mutations involved in each relate to the severity of the disorders.
15. Describe how the Ames test screens for potential environmental mutagens. Why is it thought that a compound that tests positively in the Ames test may also be carcinogenic?
16. What genetic defects result in the disorder xeroderma pigmentosum (XP) in humans? How do these defects create the phenotypes associated with the disorder?
17. In a bacterial culture in which all cells are unable to synthesize leucine (*leu*⁻) a potent mutagen is added, and the cells are allowed to undergo one round of replication. At that point, samples are taken, a series of dilutions is made, and the cells are plated on either minimal medium or minimal medium containing leucine. The first culture condition (minimal medium) allows the growth of only *leu*⁺ cells, while the second culture condition (minimum medium with leucine added) allows the growth of all cells. The results of the experiment are as follows:

Culture Condition	Dilution	Colonies
minimal medium	10^{-1}	18
minimal + leucine	10^{-7}	6

What is the rate of mutation at the locus involved with leucine biosynthesis?

18. Compare the various transposable genetic elements in bacteria, maize, *Drosophila*, and humans. What properties do they share?
19. The initial discovery of IS elements in bacteria revealed the presence of an element upstream (5′) of three genes controlling galactose metabolism. All three genes were affected simultaneously, although there was only one IS insertion. Offer an explanation as to why this might occur.

20. *Ty*, a transposable element in yeast, contains an open reading frame (ORF) that encodes the enzyme reverse transcriptase. This enzyme synthesizes DNA from an RNA template. Speculate on the role of this gene product in the transposition of *Ty* within the yeast genome.
21. It has been noted that most transposons in humans and other organisms are located in noncoding regions of the genome— regions such as introns, pseudogenes, and stretches of particular types of repetitive DNA. There are several ways to interpret this observation. Describe two possible interpretations. Which interpretation do you favor? Why?
22. Speculate on how improved living conditions and medical care in the developed nations might affect human mutation rates, both neutral and deleterious.
23. The cancer drug melphalan is an alkylating agent of the mustard gas family. It acts in two ways: by causing alkylation of guanine bases and by crosslinking DNA strands together. Describe two ways in which melphalan might kill cancer cells. What are two ways in which cancer cells could repair the DNA damaging effects of melphalan?
24. Muscular dystrophies are some of the most common inherited diseases in humans, resulting from a large number of different mutations in the *dystrophin* gene. Speculate on why this gene appears to suffer so many mutations.
25. Describe how the process of DNA replication can lead to expansion of trinucleotide repeat regions in the gene responsible for Huntington disease.
26. The origin of the mutation that led to hemophilia in Queen Victoria's family is controversial. Her father did not have X-linked hemophilia, and there is no evidence that any members of her mother's family had the condition. What are some possibilities to explain how the mutation arose? What types of mutations could lead to the disease?
27. Presented here are hypothetical findings from studies of heterokaryons formed from seven human xeroderma pigmentosum cell strains:

	XP1	XP2	XP3	XP4	XP5	XP6	XP7
XP1	−						
XP2	−	−					
XP3	−	−	−				
XP4	+	+	+	−			
XP5	+	+	+	+	−		
XP6	+	+	+	+	−	−	
XP7	+	+	+	+	−	−	−

Note: + = complementing; − = noncomplementing

These data are measurements of the occurrence of unscheduled DNA synthesis in the fused heterokaryon. None of the strains alone shows any unscheduled DNA synthesis. What does unscheduled DNA synthesis represent? Which strains fall into the same complementation groups? How many different groups are revealed based on these data? What can we conclude about the genetic basis of XP from these data?

28. Imagine yourself as one of the team of geneticists who launched the study of the genetic effects of high-energy radiation on the surviving Japanese population immediately following the atomic-bomb attacks at Hiroshima and Nagasaki in 1945. Demonstrate your insights into both chromosomal and gene mutation by outlining a comprehensive short-term and long-term study that addresses the topic of this study. Be sure to include strategies for considering the effects on both somatic and germ-line tissues.

29. Cystic fibrosis (CF) is a severe autosomal recessive disorder in humans that results from a chloride ion–channel defect in epithelial cells. More than 500 mutations have been identified in the 24 exons of the responsible gene (*CFTR*, for cystic fibrosis transmembrane regulator), including dozens of different missense mutations and frameshift mutations, as well as numerous splice-site defects. Although all affected CF individuals demonstrate chronic obstructive lung disease, there is variation in pancreatic enzyme insufficiency (PI). Speculate which types of observed mutations are likely to give rise to less severe symptoms of CF, including only minor PI. Some of the 300 sequence alterations that have been detected within the exon regions of the *CFTR* gene do not give rise to cystic fibrosis. Taking into account your accumulated knowledge of the genetic code, gene expression, protein function, and mutation, describe why this might be so, as if you were explaining it to a freshman biology major.

30. Electrophilic oxidants are known to create modified bases in DNA named 7,8-dihydro-8-oxoguanine (oxoG). Whereas guanine base-pairs with cytosine, oxoG base-pairs with either cytosine or adenine.

    (a) What are the sources of reactive oxidants within cells that cause this type of base alteration?

    (b) Drawing on your knowledge of nucleotide chemistry, draw the structure of oxoG, and, below it, draw guanine. Opposite guanine, draw cytosine, including the hydrogen bonds that allow these two molecules to base-pair. Does the structure of oxoG, in contrast to guanine, provide any hint as to why it base-pairs with adenine?

    (c) Assume that an unrepaired oxoG lesion is present in the helix of DNA opposite cytosine. Following several rounds of replication, predict the type of mutation that will occur.

    (d) Which DNA repair mechanisms might work to counteract an oxoG lesion? Which of these is likely to be most effective? (*Reference*: Bruner, S. D., et al. 2000. *Nature* 403: 859–862.)

31. Among Betazoids in the world of *Star Trek*®, the ability to read minds is under the control of a gene called *mindreader* (abbreviated *mr*). Most Betazoids can read minds, but rare recessive mutations in the *mr* gene result in two alternative phenotypes: *delayed-receivers* and *insensitives*. Delayed-receivers have some mind-reading ability but perform the task much more slowly than normal Betazoids. Insensitives cannot read minds at all.

Betazoid genes do not have introns, so the gene only contains coding DNA. It is 3332 nucleotides in length and Betazoids use a four-letter genetic code.

The following table shows some data from five unrelated *mr* mutations.

Description of Mutation		Phenotype
mr-1	Nonsense mutation in codon 829	delayed-receiver
mr-2	Missense mutation in codon 52	delayed-receiver
mr-3	Deletion of nucleotides 83–150	delayed-receiver
mr-4	Missense mutation in codon 192	insensitive
mr-5	Deletion of nucleotides 83–93	insensitive

For each mutation, provide a plausible explanation for why it gives rise to its associated phenotype and not to the other phenotype. For example, hypothesize why the *mr-1* nonsense mutation in codon 829 gives rise to the milder delayed-receiver phenotype rather than the more severe insensitive phenotype. Then repeat this type of analysis for the other mutations. (More than one explanation is possible, so be creative within *plausible* bounds!)

32. Skin cancer carries a lifetime risk nearly equal to that of all other cancers combined. Following is a graph (modified from Kraemer. 1997. *Proc. Natl. Acad. Sci. (USA)* 94:11–14), depicting the age of onset of skin cancers in patients with or without XP, where cumulative percentage of skin cancer is plotted against age. The non-XP curve is based on 29,757 cancers surveyed by the National Cancer Institute, and the curve representing those with XP is based on 63 skin cancers from the Xeroderma Pigmentosum Registry. (a) Provide an overview of the information contained in the graph. (b) Explain why individuals with XP show such an early age of onset.

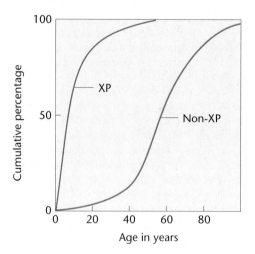

Regulation of Gene Expression

Model showing how the lac *repressor (red) and catabolite activating protein (dark blue in center of DNA loop) bind to the* lac *operon promoter, creating a 93-base-pair repression loop in the* lac *regulatory DNA.*

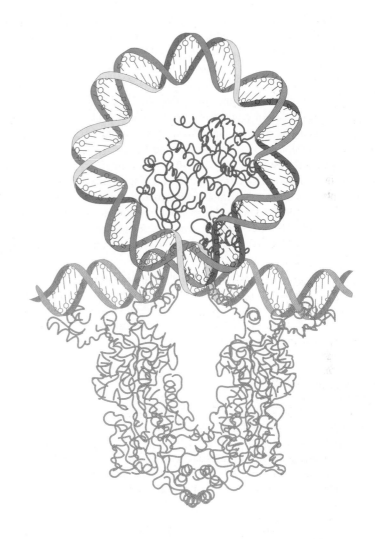

CHAPTER CONCEPTS

- Expression of genetic information is regulated by intricate regulatory mechanisms that exert control over transcription, mRNA stability, translation, and posttranslational modifications.

- In prokaryotes, genes that encode proteins with related functions tend to be organized in clusters, and are often under coordinated control. One such cluster, including its regulatory sequences, is called an operon.

- Transcription within operons is either inducible or repressible and is often regulated by the metabolic substrate or end product of the pathway.

- Eukaryotic gene regulation is more complex than prokaryotic gene regulation.

- The organization of eukaryotic chromatin in the nucleus plays a role in regulating gene expression. Chromatin must be remodeled to provide access to regulatory DNA sequences within it.

- Eukaryotic transcription initiation requires the assembly of transcription regulators at enhancer sites and the assembly of basal transcription complexes at promoter sites.

- Eukaryotic gene expression is also regulated at multiple posttranscriptional steps, including alternative splicing of pre-mRNA and control of mRNA stability.

We have previously established how DNA is organized into genes, how genes store genetic information, and how this information is expressed through the processes of transcription and translation. We now consider one of the most fundamental questions in molecular genetics: *How is gene expression regulated?* It is clear that not all genes are expressed at all times in all situations. For example, detailed analysis of proteins in *E. coli* shows that some proteins are present in as few as 5 to 10 molecules per cell, whereas others, such as ribosomal proteins and the many proteins involved in the glycolytic pathway, are present in as many as 100,000 copies per cell. Although many prokaryotic gene products are present continuously at a basal level (a few copies), the level of these products can increase dramatically when required. In multicellular eukaryotes, differential gene expression is also essential and is at the heart of embryonic development and maintenance of the adult state. Cells of the pancreas, for example, do not synthesize retinal pigment, and retinal cells do not make insulin. The activation and repression of gene expression is a delicate balancing act for an organism; expression of a gene at the wrong time, in the wrong cell type, or in abnormal amounts can lead to a deleterious phenotype, cancer, or cell death—even when the gene itself is normal. Clearly, fundamental regulatory mechanisms must exist to control the expression of genetic information.

In this chapter, we will explore regulation of gene expression in both prokaryotes and eukaryotes. We will outline some of the general features of gene regulation, including the *cis*-acting DNA elements and *trans*-acting factors that influence the regulation of transcription and how these components interact. We will also consider the role of posttranscriptional mechanisms of regulating gene expression in eukaryotes.

How Do We Know?

In this chapter, we will focus on how prokaryotic and eukaryotic organisms regulate the expression of genetic information stored in DNA. We will explore a number of mechanisms by which the process of transcription is either activated or repressed. In addition, we will examine some posttranscriptional mechanisms of gene regulation in eukaryotes. As you study this topic, you should try to answer several fundamental questions:

1. How do we know that promoter regions control the initiation of transcription in prokaryotes and eukaryotes?

2. How do we know that both prokaryotic and eukaryotic transcription regulatory proteins bind to DNA within promoters and enhancers of genes whose transcription they control?

3. How do we know that bacteria regulate the expression of certain genes in response to changes in their environment?

4. How do we know that bacterial genes are often arranged in clusters that are coordinately regulated?

5. How do we know that physical interactions occur between the *lac* repressor and inducer molecules that control the *lac* operon in *E. coli*?

6. How do we know that promoters and enhancers exist in eukaryotic genes?

15.1 Prokaryotes Regulate Gene Expression in Response to Environmental Conditions

Regulation of gene expression has been extensively studied in prokaryotes, particularly in *E. coli*. We have learned that highly efficient genetic mechanisms have evolved to turn transcription of specific genes on and off, depending on the cell's metabolic need for the respective gene products. Not only do bacteria respond to changes in their environment, but they also regulate gene transcription in order to synthesize products required for a variety of normal cellular activities (including the replication, recombination, and repair of their DNA) and for cell division. In this chapter, we will focus on prokaryotic gene regulation at the level of transcription, which is the predominant level of regulation in prokaryotes. Keep in mind, however, that posttranscriptional regulation also occurs in bacteria. We will defer discussion of posttranscriptional gene regulation to subsequent sections dealing with eukaryotic gene regulation.

The idea that microorganisms regulate the synthesis of gene products is not a new one. As early as 1900, it was shown that when lactose (a galactose and glucose-containing disaccharide) is present in the growth medium of yeast, the organisms synthesize enzymes required for lactose metabolism. When lactose is absent, the enzymes are not manufactured. Soon thereafter, investigators were able to generalize that bacteria also adapt to their environment, producing certain enzymes only when specific chemical substrates are present. Such enzymes are referred to as **inducible**, reflecting the role of the substrate, which serves as the **inducer** in enzyme production. In contrast, enzymes that are produced continuously, regardless of the chemical makeup of the environment, are called **constitutive**.

More recent investigation has revealed a contrasting system whereby the presence of a specific molecule inhibits gene expression. This is usually true for molecules that are end products of anabolic biosynthetic pathways. For example, the amino acid tryptophan can be synthesized by bacterial cells. If a sufficient supply of tryptophan is present in the environment or culture medium, it is energetically inefficient for the organism to synthesize the enzymes necessary for tryptophan production. A mechanism has evolved whereby tryptophan plays a role in repressing transcription of genes that encode the appropriate biosynthetic enzymes. In contrast to the inducible system controlling lactose metabolism, the system governing tryptophan expression is said to be **repressible**.

Regulation, whether it is inducible or repressible, may be under either **negative** or **positive control**. Under negative control, gene expression occurs *unless it is shut off by some form of a regulator molecule*. In contrast, under positive control, transcription occurs *only if a regulator molecule directly stimulates RNA production*. In theory, either type of control or a combination of the two can govern inducible or repressible systems.

15.2 Lactose Metabolism in *E. coli* Is Regulated by an Inducible System

Beginning in 1946 with the studies of Jacques Monod and continuing through the next decade with significant contributions by Joshua Lederberg, François Jacob, and Andre L'woff, genetic and biochemical evidence involving lactose metabolism was amassed. Insights were provided into the way in which gene activity is repressed when lactose is absent, but induced when it is available. In the presence of lactose, concentrations of the enzymes responsible for lactose metabolism increase rapidly from a few molecules to thousands per cell. The enzymes responsible for lactose metabolism are thus *inducible*, and lactose serves as the *inducer*.

In prokaryotes, genes that code for enzymes with related functions (e.g., genes involved with lactose metabolism) tend to be organized in clusters on the bacterial chromosome, and transcription of these genes is often under the coordinated control of a single transcription regulatory region. The location of this regulatory region is almost always upstream of the gene cluster it controls, and we refer to the regulatory region as a ***cis*-acting site**. *Cis*-acting regulatory regions bind molecules that control transcription of the gene cluster. Such molecules are called ***trans*-acting molecules**. Actions at the *cis*-acting regulatory site determine whether the genes are transcribed into RNA and thus whether the corresponding enzymes or other protein products are synthesized from the mRNA. Binding of a *trans*-acting molecule at a *cis*-acting site can regulate the gene cluster either negatively (by turning off transcription) or positively (by turning on transcription of genes in the cluster). In this section, we discuss how transcription of such bacterial gene clusters is coordinately regulated.

Structural Genes

As illustrated in Figure 15–1, three genes and an adjacent regulatory region constitute the **lactose**, or *lac*, **operon**. Together, the entire gene cluster functions in an integrated fashion to provide a rapid response to the presence or absence of lactose.

Genes coding for the primary structure of the enzymes are called **structural genes**. There are three structural genes in the *lac* operon. The *lacZ* gene encodes **β-galactosidase**, an enzyme whose primary role is to convert the disaccharide lactose to the monosaccharides glucose and galactose (Figure 15–2). This conversion is essential if lactose is to

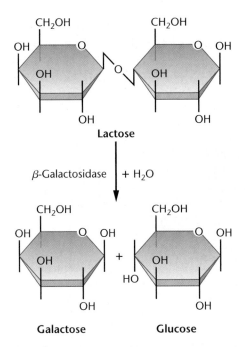

FIGURE 15–2 The catabolic conversion of the disaccharide lactose into its monosaccharide units, galactose and glucose.

serve as the primary energy source in glycolysis. The second gene, *lacY*, specifies the primary structure of **permease**, an enzyme that facilitates the entry of lactose into the bacterial cell. The third gene, *lacA*, codes for the enzyme **transacetylase**. Although its physiological role is still not completely clear, it may be involved in the removal of toxic by-products of lactose digestion from the cell.

To study the genes coding for these three enzymes, researchers isolated numerous mutations in order to eliminate the function of one or the other enzyme. Such *lac⁻* mutants were first isolated and studied by Joshua Lederberg. Mutant cells that fail to produce active β-galactosidase (*lacZ⁻*) or permease (*lacY⁻*) are unable to use lactose as an energy source. Mapping studies by Lederberg established that all three genes are closely linked or contiguous to one another on the bacterial chromosome, in the order *Z–Y–A*. (See Figure 15–1.) All three genes are transcribed as a single unit, resulting in a polycistronic mRNA (Figure 15–3). This results in the coordinate regulation of all three genes, since a single-message RNA is simultaneously translated into all three gene products.

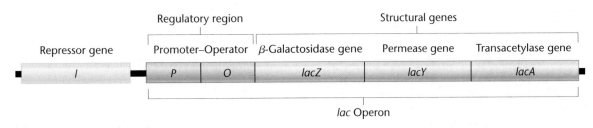

FIGURE 15–1 A simplified overview of the genes and regulatory units involved in the control of lactose metabolism. (This region of DNA is not drawn to scale.) A more detailed model will be developed later in this chapter.

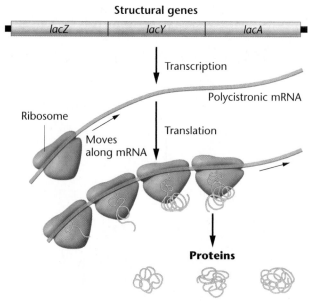

FIGURE 15–3 The structural genes of the *lac* operon are transcribed into a single polycistronic mRNA, which is translated simultaneously by several ribosomes into the three enzymes encoded by the operon.

The Discovery of Regulatory Mutations

How does lactose stimulate transcription of the *lac* operon and induce the synthesis of the related enzymes? A partial answer came from studies using **gratuitous inducers**, chemical analogs of lactose such as the sulfur analog **isopropylthiogalactoside (IPTG)**, shown in Figure 15–4. Gratuitous inducers behave like natural inducers, but they do not serve as substrates for the enzymes that are subsequently synthesized.

What, then, is the role of lactose in gene regulation? The answer to this question required the study of a class of mutations called **constitutive mutants**. In cells bearing these types of mutations, enzymes are produced regardless of the presence or absence of lactose. Studies of the constitutive mutation *lacI⁻* mapped the mutation to a site on the bacterial chromosome close to, but distinct from, the *lacZ, lacY,* and *lacA* genes. This mutation defined the *lacI* gene, which is appropriately called a **repressor gene**. Another set of constitutive mutations producing identical effects to those of *lacI⁻* is present in a region immediately adjacent to the structural genes. This class of mutations, designated *lacOᶜ*,

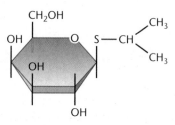

FIGURE 15–4 The gratuitous inducer isopropylthiogalactoside (IPTG).

is located in the **operator region** of the operon. In both types of constitutive mutants, enzymes are continually produced, inducibility is eliminated, and gene regulation is lost.

The Operon Model: Negative Control

Around 1960, Jacob and Monod proposed a scheme involving negative control called the **operon model**, whereby a group of genes is regulated and expressed together as a unit. As we saw in Figure 15–1, the *lac* operon they proposed consists of the *Z, Y,* and *A* structural genes, as well as the adjacent sequences of DNA referred to as the *operator region*. They argued that the *lacI* gene regulates the transcription of the structural genes by producing a **repressor molecule**, and that the repressor is **allosteric**, meaning that it reversibly interacts with another molecule, causing both a conformational change in the repressor's three-dimensional shape and a change in its chemical activity. Figure 15–5 illustrates the components of the *lac* operon as well as the action of the *lac* repressor in the presence and absence of lactose.

Jacob and Monod suggested that the repressor normally binds to the DNA sequence of the operator region. When it does so, it inhibits the action of RNA polymerase, effectively repressing the transcription of the structural genes [Figure 15–5(b)]. However, when lactose is present, this sugar binds to the repressor and causes an allosteric conformational change. This change renders the repressor incapable of interacting with operator DNA [Figure 15–5(c)]. In the absence of the repressor–operator interaction, RNA polymerase transcribes the structural genes, and the enzymes necessary for lactose metabolism are produced. Because transcription occurs only when the repressor *fails* to bind to the operator region, regulation is said to be under *negative control*.

The operon model invokes a series of molecular interactions between proteins, inducers, and DNA to explain the efficient regulation of structural gene expression. In the absence of lactose, the enzymes encoded by the genes are not needed, and expression of genes encoding these enzymes is repressed. When lactose is present, it indirectly induces the transcription of the structural genes by interacting with the repressor.* If all lactose is metabolized, none is available to bind to the repressor, which is again free to bind to operator DNA and repress transcription.

Both the *I⁻* and *Oᶜ* constitutive mutations interfere with these molecular interactions, allowing continuous transcription of the structural genes. In the case of the *I⁻* mutant, seen in Figure 15–6(a), the repressor protein is altered or absent and cannot bind to the operator region, so the structural genes are always transcribed. In the case of the *Oᶜ* mutant [Figure 15–6(b)], the nucleotide sequence of the operator DNA is altered and will not bind with a normal repressor molecule. The result is the same: the structural genes are always transcribed.

---

*Technically, allolactose, an isomer of lactose, is the inducer. When lactose enters the bacterial cell, some of it is converted to allolactose by the β-galactosidase enzyme.

FIGURE 15–5 The components of the wild-type *lac* operon (a) and the response of the *lac* operon to the absence (b) and presence (c) of lactose.

(a) Components

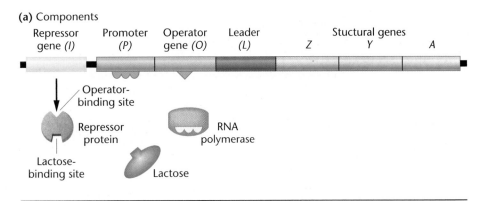

(b) $I^+\ O^+\ Z^+\ Y^+\ A^+$ (wild type) — no lactose present — repressed

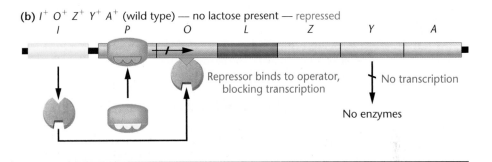

(c) $I^+\ O^+\ Z^+\ Y^+\ A^+$ (wild type) — lactose present — induced

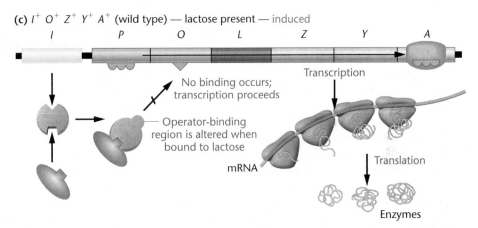

Genetic Proof of the Operon Model

The operon model leads to three major predictions that can be tested to determine its validity. The major predictions to be tested are that (1) the *I* gene produces a diffusible product; (2) the *O* region is involved in regulation but does not produce a product; and (3) the *O* region must be adjacent to the structural genes in order to regulate transcription.

The construction of partially diploid bacteria allows us to assess these assumptions, particularly those that predict *trans*-acting regulatory molecules. For example, as introduced in Chapter 8, the F plasmid may contain chromosomal genes, in which case it is designated F′. When an F⁻ cell acquires such a plasmid, it contains its own chromosome plus one or more additional genes present in the plasmid. This creates a host cell, called a **merozygote**, that is diploid for those genes. The use of such a plasmid makes it possible, for example, to introduce an I^+ gene into a host cell whose genotype is I^- or to introduce an O^+ region into a host cell of genotype O^C. The Jacob-Monod operon model predicts how regulation should be affected in such cells. Adding an I^+ gene to an I^- cell should restore

inducibility because the normal wild-type repressor, which is a *trans*-acting factor, would be produced by the inserted I^+ gene. Adding an O^+ region to an O^C cell should have no effect on constitutive enzyme production, since regulation depends on the presence of an O^+ region immediately adjacent to the structural genes—that is, O^+ is a *cis*-acting regulator.

Results of these experiments are shown in Table 15.1, where Z represents the structural genes. The inserted genes are listed after the designation F′. In both cases described here, the Jacob–Monod model is upheld (part B of Table 15.1). Part C shows the reverse experiments, where either an I^- gene or an O^C region is added to cells of normal inducible genotypes. As the model predicts, inducibility is maintained in these partial diploids.

Another prediction of the operon model is that certain mutations in the *I* gene should have the opposite effect of I^-. That is, instead of being constitutive because the repressor can't bind the operator, mutant repressor molecules should be produced that cannot interact with the inducer, lactose. As a result, the repressor would always bind to the operator sequence, and the structural genes

(a) $I^- O^+ Z^+ Y^+ A^+$ (mutant repressor gene) — no lactose present — constitutive

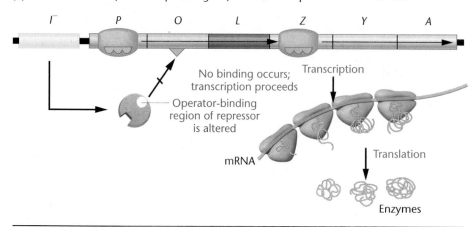

FIGURE 15–6 The response of the *lac* operon in the absence of lactose when a cell bears either the I^- (a) or the O^C (b) mutation.

(b) $I^+ O^c Z^+ Y^+ A^+$ (mutant operator gene) — no lactose present — constitutive

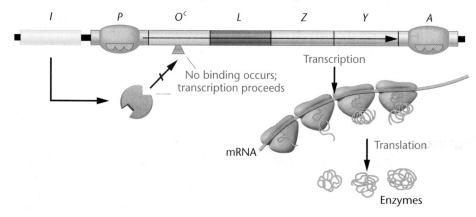

would be permanently repressed (Figure 15–7). If this were the case, the presence of an additional I^+ gene would have little or no effect on repression.

In fact, such a mutation, I^S, was discovered wherein the operon is "superrepressed," as shown in part D of Table 15.1. An additional I^+ gene does not effectively relieve repression of gene activity. These observations are consistent with the idea that the repressor contains separate DNA binding domains and inducer binding domains. The binding of lactose to the inducer binding domain causes an allosteric change in the DNA binding domain.

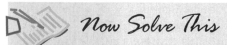

Now Solve This

Problem 6 on page 354 asks you to predict the outcome of gene expression under the conditions of varying genotypes and cellular conditions.

Hint: Determine initially whether the repressor is active or inactive based on whether the gene encoding it is wild type or mutant. Then consider the impact of the presence or absence of lactose.

Isolation of the Repressor

Although Jacob and Monod's operon theory succeeded in explaining many aspects of genetic regulation in prokaryotes, the nature of the repressor molecule was not known when their landmark paper was published in 1961. While they had assumed that the

allosteric repressor was a protein, RNA was also a candidate because activity of the molecule required the ability to bind to DNA. Despite many attempts to isolate and characterize the hypothetical repressor molecule, no direct chemical evidence was

TABLE 15.1	A Comparison of Gene Activity (+ or −) in the Presence or Absence of Lactose for Various *E. coli* Genotypes		
		Presence of β-Galactosidase Activity	
	Genotype	**Lactose Present**	**Lactose Absent**
	$I^+O^+Z^+$	+	−
A.	$I^+O^+Z^-$	−	−
	$I^-O^+Z^+$	+	+
	$I^+O^CZ^+$	+	+
B.	$I^-O^+Z^+/F'I^+$	+	−
	$I^+O^CZ^+/F'O^+$	+	+
C.	$I^+O^+Z^+/F'I^-$	+	−
	$I^+O^+Z^+/F'O^C$	+	−
D.	$I^SO^+Z^+$	−	−
	$I^SO^+Z^+/F'I^+$	−	−

Note: In parts B to D, most genotypes are partially diploid, containing an F factor plus attached genes (F').

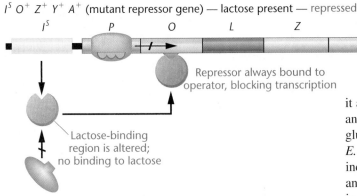

$I^S\ O^+\ Z^+\ Y^+\ A^+$ (mutant repressor gene) — lactose present — repressed

Repressor always bound to operator, blocking transcription

Lactose-binding region is altered; no binding to lactose

FIGURE 15–7 The response of the *lac* operon in the presence of lactose in a cell bearing the I^S mutation.

immediately forthcoming. A single *E. coli* cell contains no more than 10 or so molecules of the *lac* repressor; therefore, direct chemical identification of 10 molecules in a population of millions of proteins and RNAs in a single cell presented a tremendous challenge. Nevertheless, in 1966, Walter Gilbert and Benno Müller-Hill reported the isolation of the *lac* repressor. Once the repressor was purified, it was shown to have various characteristics of a protein. The isolation of the repressor thus confirmed the operon model, which had been put forward strictly on genetic grounds.

15.3 The Catabolite-Activating Protein (CAP) Exerts Positive Control over the *lac* Operon

As we discussed at the beginning of this section, the role of β-galactosidase is to cleave lactose into its components, glucose and galactose. Then, for galactose to be used by the cell,

it also is converted to glucose. What if the cell found itself in an environment that contained an ample amount of lactose *and* glucose? Given that glucose is the preferred carbon source for *E. coli*, it would not be energetically efficient for a cell to induce transcription of the *lac* operon, make β-galactosidase, and metabolize lactose, since what it really needs—glucose—is already present. As we shall see next, a molecule called the **catabolite-activating protein (CAP)** is involved in effectively repressing the expression of the *lac* operon when glucose is present. This inhibition is called **catabolite repression**.

To understand CAP and its role in regulation, let's backtrack for a moment. When the *lac* repressor is bound to the inducer, the *lac* operon is activated and RNA polymerase transcribes the structural genes. As stated in Chapter 12, transcription is initiated as a result of the binding that occurs between RNA polymerase and the nucleotide sequence of the promoter region, found upstream (5′) from the initial coding sequences. Within the *lac* operon, the promoter is found between the *I* gene and the operator region (*O*). (See Figure 15–1.) Careful examination has revealed that RNA polymerase binding is never very efficient unless CAP is also present to facilitate the process.

The mechanism is summarized in Figure 15–8. In the absence of glucose and under inducible conditions, CAP exerts positive control by binding to the CAP site, facilitating RNA polymerase binding at the promoter and thus transcription. Therefore, for maximal transcription of the structural genes, the repressor must

FIGURE 15–8 Catabolite repression. (a) In the absence of glucose, cAMP levels increase, resulting in the formation of a CAP–cAMP complex, which binds to the CAP site of the promoter, stimulating transcription. (b) In the presence of glucose, cAMP levels decrease, CAP–cAMP complexes are not formed, and transcription is not stimulated.

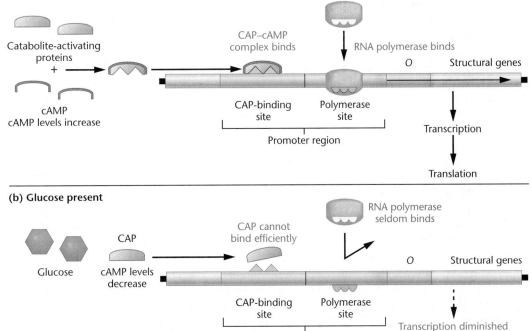

(a) Glucose absent

Catabolite-activating proteins + cAMP

cAMP levels increase

CAP-binding site

Promoter region

CAP–cAMP complex binds

RNA polymerase binds

Polymerase site

O   Structural genes

Transcription

Translation

(b) Glucose present

Glucose

CAP

cAMP levels decrease

CAP cannot bind efficiently

CAP-binding site

Promoter region

RNA polymerase seldom binds

Polymerase site

O   Structural genes

Transcription diminished

Translation diminished

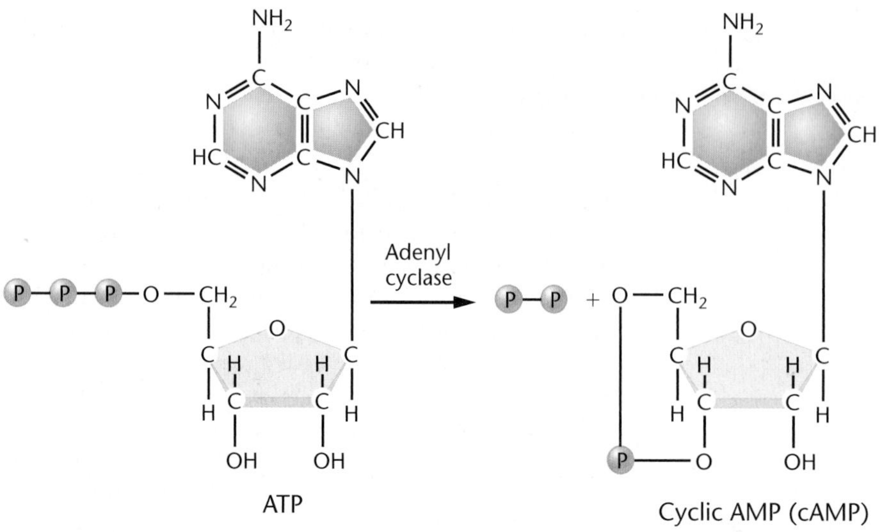

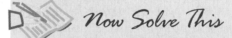

NH$_2$

Adenyl
cyclase

ATP

Cyclic AMP (cAMP)

FIGURE 15–9 The formation of cAMP from ATP, catalyzed by adenyl cyclase.

be bound by lactose (so as not to repress operon expression), *and* CAP must be bound to the CAP-binding site.

This leads to the central question about CAP. What role does glucose play in inhibiting CAP binding when it is present? The answer involves still another molecule, **cyclic adenosine monophosphate (cAMP)**, upon which CAP binding is dependent. In order to bind to the *lac* operon promoter, CAP must be bound to cAMP. The level of cAMP is itself dependent on an enzyme, **adenyl cyclase**, which catalyzes the conversion of ATP to cAMP. (See Figure 15–9.)

The role of glucose in catabolite repression is to inhibit the activity of adenyl cyclase, causing a decline in the level of cAMP in the cell. Under this condition, CAP cannot form the CAP–cAMP complex essential to the positive control of transcription of the *lac* operon.

The structures of CAP and cAMP–CAP have been examined by using X-ray crystallography. CAP is a dimer that inserts into adjacent regions of a specific nucleotide sequence of the DNA making up the *lac* promoter. The cAMP–CAP complex, when bound to DNA, bends it, causing it to assume a new conformation.

Binding studies in solution further clarify the mechanism of gene activation. Alone, neither cAMP–CAP nor RNA polymerase has a strong affinity to bind to *lac* promoter DNA, nor does either molecule have a strong affinity to bind to the other. However, when both are together in the presence of the *lac* promoter DNA, a tightly bound complex is formed, an example of what is called **cooperative binding**. In the case of cAMP–CAP and the *lac* operon, the phenomenon illustrates the high degree of specificity that is involved in the genetic regulation of just one small group of genes.

Regulation of the *lac* operon by catabolite repression results in efficient energy use because the presence of glucose will override the need for the metabolism of lactose, should it also be available to the cell. In contrast to the negative regulation conferred by the *lac* repressor, the action of cAMP-CAP constitutes positive regulation. Thus, a combination of positive and negative regulatory mechanisms determine transcription levels of the *lac* operon. Catabolite repression involving CAP has also been observed for other inducible operons, including those controlling the metabolism of galactose and arabinose.

Now Solve This

Problem 10 on page 354 asks you to assess the level of gene activity in the *lac* operon when considering various combinations of the presence or absence of both lactose and glucose.

Hint: You must keep in mind that regulation involving lactose is a negative control system, while regulation involving glucose is a positive control system.

15.4   The Tryptophan (*trp*) Operon in *E. coli* Is a Repressible Gene System

Although the process of induction had been known for some time, it was not until 1953 that a repressible system was discovered. Studies on the biosynthesis of the essential amino acid tryptophan revealed that, if tryptophan is present in sufficient quantity in the growth medium, the enzymes necessary for its synthesis (such as **tryptophan synthase**) are not produced. It is energetically advantageous for bacteria to repress expression of genes involved in tryptophan synthesis when ample tryptophan is present in the growth medium.

Further investigation showed that a series of enzymes encoded by five contiguous genes on the *E. coli* chromosome is involved in tryptophan synthesis. These genes are part of an operon, and in the presence of tryptophan, all are coordinately repressed, and none of the enzymes is produced. Because of the great similarity between this repression and the induction of enzymes for lactose metabolism, a repression model resembling that of the *lac* system was proposed (Figure 15–10).

The model suggests the presence of a *normally inactive repressor* that alone cannot interact with the operator region of the operon. However, the repressor is an allosteric molecule that can bind to tryptophan. When tryptophan is present, the resultant complex of repressor and tryptophan attains a new conformation that binds to the operator, repressing transcription. Thus, when tryptophan, the end product of this anabolic pathway, is present, the system is repressed and enzymes are not made.

FIGURE 15–10 (a) The components involved in the regulation of the tryptophan operon. (b) Regulatory conditions are depicted that involve either activation or (c) repression of the structural genes. In the absence of tryptophan, an inactive repressor is made that cannot bind to the operator (*O*), thus allowing transcription to proceed. In the presence of tryptophan, it binds to the repressor, causing an allosteric transition to occur. This complex binds to the operator region, leading to repression of the operon.

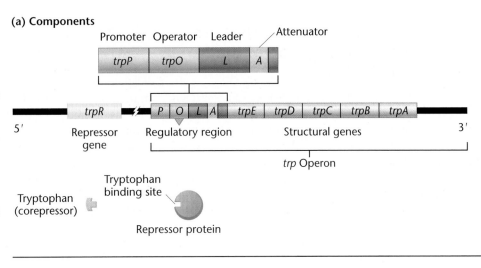

(a) Components

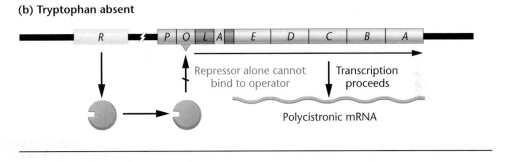

(b) Tryptophan absent

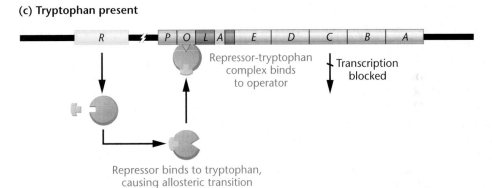

(c) Tryptophan present

Since the regulatory complex inhibits transcription of the operon, this repressible system is under negative control. And as tryptophan participates in repression, it is referred to as a **corepressor** in this regulatory scheme.

Evidence for the *trp* Operon

Support for the concept of a repressible operon is based primarily on the isolation of two distinct categories of constitutive mutations. The first class, *trpR⁻*, maps at a considerable distance from the structural genes. This locus represents the gene coding for the repressor. Presumably, the mutation either inhibits the interaction of the repressor with tryptophan or inhibits repressor formation entirely. Whichever the case, no repression is present in cells with the *trpR⁻* mutation. As expected, if the *trpR⁺* gene encodes a functional repressor molecule, the presence of a copy of this gene will restore repressibility.

The second constitutive mutant is analogous to the *O^C* mutant of the lactose operon, because it maps immediately adjacent to

the structural genes. Furthermore, the addition of a wild-type operator gene into mutant cells (as an external element) does not restore repression. This is predictable if the mutant operator no longer interacts with the repressor–tryptophan complex.

The entire *trp* operon has now been well defined, as shown in Figure 15–10. Five contiguous structural genes (*trp E, D, C, B,* and *A*) are transcribed as a polycistronic message directing translation of the enzymes that catalyze the biosynthesis of tryptophan. As in the *lac* operon, a promoter region (*trpP*) represents the binding site for RNA polymerase, and an operator region (*trpO*) binds the repressor. In the absence of binding, transcription is initiated within the overlapping *trpP–trpO* region and proceeds along a **leader sequence** 162 nucleotides prior to the first structural gene (*trpE*). Within that leader sequence, still another regulatory site has been demonstrated, called an *attenuator*—the subject of the next section of this chapter. As we shall see, this regulatory unit is an integral part of the control mechanism of the operon.

Attenuation Is a Critical Process during the Regulation of the *trp* Operon in *E. coli*

Charles Yanofsky, his coworker Kevin Bertrand, and their colleagues observed that, even when tryptophan is present and the *trp* operon is repressed, initiation of transcription still occurs, leading to synthesis of the initial portion of the mRNA (the 5′-leader sequence). Hence, while the activated repressor binds to the operator region, it does not strongly inhibit the *initial expression* of the operon; this suggests that there must be a subsequent mechanism by which tryptophan somehow inhibits transcription of the entire operon. Yanofsky discovered that, following initiation of transcription, *in the presence of high concentrations of tryptophan*, mRNA synthesis is usually terminated at a point about 140 nucleotides along the transcript. This process is called **attenuation**,* indicative of the effect of diminishing genetic expression of the operon.

When tryptophan is absent, or present in very low concentrations, transcription is initiated and *not* subsequently terminated, instead continuing beyond the leader sequence along the DNA encoding the structural genes, starting with the *trpE* gene. As a result, a polycistronic mRNA is produced, and the enzymes essential to the biosynthesis of tryptophan are subsequently translated. Identification of the site involved in attenuation was made possible by the isolation of various deletion mutations in the region 115 to 140 nucleotides into the leader sequence. Such mutations abolish attenuation. This site is referred to as the **attenuator**. An explanation of how attenuation occurs and how it is overcome, put forward by Yanofsky and colleagues, is summarized in Figure 15–11. The initial DNA sequence that is transcribed gives rise to an mRNA molecule that has the potential to fold into two mutually exclusive stem-loop structures referred to as "hairpins." In the presence of excess tryptophan, the hairpin that is formed behaves as a **terminator** structure, and transcription is almost always terminated prematurely. On the other hand, if tryptophan is scarce, the alternative structure referred to as the **antiterminator hairpin** is formed. Transcription is allowed to proceed past the antiterminator hairpin region, and the entire mRNA is subsequently produced. These hairpins are illustrated in Figure 15–11(c).

The question is how the absence (or a low concentration) of tryptophan allows attenuation to be bypassed. A key point in Yanofsky's model is that the leader transcript must be translated in order for the antiterminator hairpin to form. Yanofsky discovered that the leader transcript includes two triplets (UGG) that encode tryptophan preceded upstream by the initial AUG sequence that prompts the initiation of translation by ribosomes. When adequate tryptophan is present, charged tRNA^{trp} is present. As a result, translation proceeds past these

triplets, and the *terminator hairpin* is formed, as illustrated in Figure 15–11(d). If cells are starved of tryptophan, charged tRNA^{trp} is unavailable. As a result, the ribosome "stalls" during translation of the UGG triplets. This event induces the formation of the antiterminator hairpin within the leader transcript, as shown in Figure 15–11(d). As a result, attenuation is overcome and transcription proceeds, leading to expression of the entire set of structural genes.

Attenuation appears to be a mechanism common to other bacterial operons in *E. coli* that regulate the enzymes essential to the biosynthesis of amino acids, including those involved in threonine, histidine, leucine, and phenylalanine metabolism. As with the *trp* operon, attenuators in these operons contain multiple codons calling for the amino acid being regulated, which prompts the "stalling" of translation if the appropriate amino acid is missing. For example, the leader sequence in the histidine operon codes for seven histidine residues in a row.

Eukaryotic Gene Regulation Differs from That in Prokaryotes

In contrast to prokaryotic gene regulation, regulation of gene expression in eukaryotes can occur at many additional levels (Figure 15–12). These levels encompass the steps that regulate the initiation of gene transcription, and extend to the numerous mechanisms that convert nascent mRNAs into fully modified and functional gene products.

The complex nature of gene regulation in eukaryotes is essential for their existence as higher organisms. Many eukaryotes are multicellular organisms that function because they contain a wide range of specialized cell types. While virtually all cells in a eukaryotic organism contain a complete genome, only a subset of genes is expressed in any particular cell type. For example, some white blood cells express genes encoding certain immunoglobulins, allowing these cells to synthesize antibodies that defend the organism from infection and foreign agents. However, skin, kidney, and liver cells do not express immunoglobulins. Pancreatic islet cells synthesize and secrete insulin in response to the presence of blood sugars; however, these cells do not manufacture immunoglobulins; nor do they synthesize insulin when it is not required. Eukaryotic cells, as part of multicellular organisms, do not grow solely in response to the availability of nutrients. Instead, they express genes in the correct temporal and spatial manner in order to carefully regulate their cell division. Loss of regulation over cell growth and division may lead to developmental defects, cancer, or death.

There are several other reasons why gene regulation involves more steps in eukaryotes than in prokaryotes:

1. Eukaryotic cells contain a much greater amount of DNA than do prokaryotic cells, and this DNA is complexed with histones and other proteins to form highly compact chromatin structures. Eukaryotic cells modify this structural organization in order to influence gene expression.

---

*The word "attenuation" is derived from the verb *attenuate*, meaning "to reduce in strength, weaken, or impair."

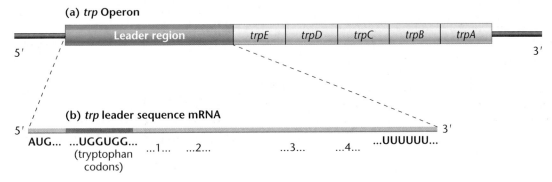

(a) *trp* Operon

(b) *trp* leader sequence mRNA

(c) Two potential stem-loop structures can form within the *trp* leader

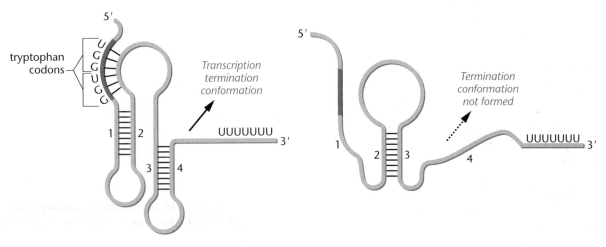

(d) Position of ribosome determines which stem-loop forms

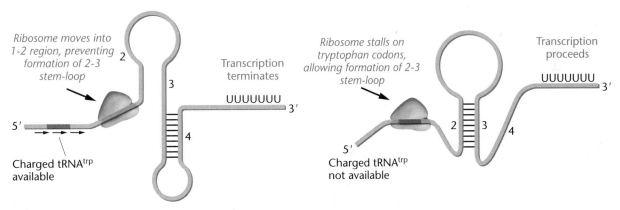

FIGURE 15–11 Attenuation in the *trp* operon. (a) The *trp* operon, showing the leader region, followed by the five *trp* genes. (b) The leader region in the mRNA is enlarged to show the translation start codon (AUG), the two tryptophan codons (UGGUGG), and the string of U residues at the end of the leader. The four regions marked 1, 2, 3, and 4 indicate the four sequence regions that have potential to form stems by base pairing. (c) The two possible stem-loop structures that can form within the leader sequence mRNA. The first conformation creates a stem-loop involving regions 3 and 4, followed by a string of U residues. This conformation triggers transcription termination by RNA polymerase as it passes this region. The second conformation creates a stem-loop involving regions 2 and 3 that does not allow formation of the terminator signal. (d) During translation, the ribosome either moves past the tryptophan codons or stalls, depending on the presence of tRNA^trp. The position of the ribosome determines whether the terminator structure forms, or whether transcription can proceed through the terminator region.

2. The mRNAs of most eukaryotic genes must be spliced, capped and polyadenylated prior to transport from the nucleus; hence, each of these processes can be regulated in order to influence the numbers and types of mRNAs available for translation.

3. Genetic information in eukaryotes is carried on many chromosomes (rather than just one), and these chromosomes are enclosed within a double-membrane-bound nucleus. Hence, transport of RNAs into the cytoplasm can be regulated in order to modulate the availability of mRNAs for translation.

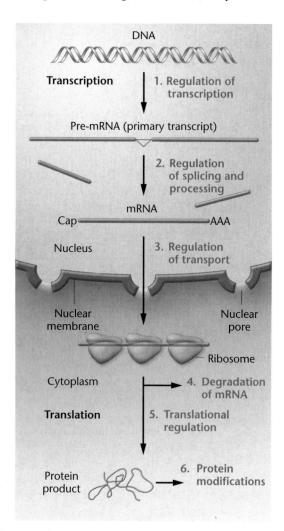

FIGURE 15–12 Various levels of regulation that are possible during the expression of genetic material in eukaryotes.

4. Eukaryotic mRNAs can have a wide range of half-lives ($t_{1/2}$). In contrast, the majority of prokaryotic mRNAs decay very rapidly. Rapid turnover of mRNAs allows prokaryotic cells to rapidly respond to environmental changes. In eukaryotes, the complement of mRNAs in each cell type can be more subtly manipulated by altering mRNA decay rates over a larger range.

5. In eukaryotes, translation rates can be modulated, as well as the way proteins are processed and modified.

In the following sections, we outline some of the major ways in which eukaryotic gene expression is regulated. As most eukaryotic genes are regulated, at least in part, at the transcriptional level, we will emphasize transcriptional control, although we will discuss other levels of control as well. In addition, we will limit our discussion to regulation of genes transcribed by RNA polymerase II. As we previously described in Chapter 12, eukaryotic genes use three RNA polymerases for transcription. (Prokaryotic genes are all transcribed by a single RNA polymerase.) RNA polymerase II transcribes all mRNAs and some small nuclear RNAs, whereas RNA polymerases I and III transcribe ribosomal

RNAs, some small nuclear RNAs, and transfer RNAs. The promoter for each type of polymerase has a different nucleotide sequence and binds different transcription factors.

15.7 **Eukaryotic Gene Expression Is Influenced by Chromosome Organization and Chromatin Modifications**

During interphase of the cell cycle, chromosomes are unwound and cannot be seen as intact structures by light microscopy. In the interphase nucleus, each chromosome occupies a discrete domain called a **chromosome territory** that separates it from other chromosomes. In general, gene-poor chromosomes are located at the nuclear periphery, and gene-dense chromosomes are located internally. Channels between chromosomes are called **interchromosomal compartments**. Each compartment is a space between chromosomes that contains little or no DNA.

Chromosome organization appears to be continuously rearranged so that transcriptionally active genes are cycled to the edge of chromosome territories at the border of the interchromosomal domain channels. Although evidence suggests that transcription of many, if not most, genes occurs when they are in direct contact with the border between the chromosome and the interchromosomal compartment, other evidence suggests that transcription can also take place within chromosomal territories. Once a gene has moved to the edge of a chromosome territory, chromatin structure must be remodeled, in order to make promoter sites accessible to the transcription machinery.

Chromatin Remodeling

The DNA in eukaryotic chromosomes is combined with histones and nonhistone proteins to form chromatin. As described in Chapter 11, chromatin is characterized by the presence of repeating structures called nucleosomes. Nucleosomal DNA is further compacted in 30 nm fibers and higher order structures. The presence of these compact chromatin structures is inhibitory to many processes, including transcription, replication, and DNA repair. The ability of the cell to alter the association of DNA with chromatin structures is essential to allow regulatory proteins to access DNA. Hence, chromatin modification, referred to as **chromatin remodeling**, is an important step in gene regulation. Chromatin remodeling appears to be a prerequisite for transcription of eukaryotic genes, although it can occur simultaneously with transcription initiation and elongation.

Chromatin remodeling is carried out by a diverse group of protein complexes that have ATP-ase activity. These remodeling complexes are recruited to specific genes that are tagged for transcription by the presence of transcription activators on their promoter regions, or by the presence of acetylated histones or methylated DNA regions.

Nucleosome remodeling complexes may alter nucleosome structure by several different mechanisms. As seen

(a) Alteration of DNA-protein contacts

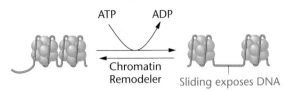

(b) Alteration of the DNA path

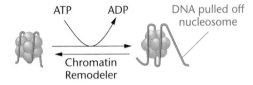

(c) Remodeling of nucleosome core particle

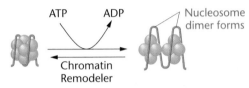

FIGURE 15–13 Three mechanisms that might be used to alter nucleosome structure by ATP hydrolysis-dependent remodeling complexes. (a) DNA–histone contacts may be loosened. (b) The path of the DNA around an unaltered nucleosome core particle may be altered. (c) The composition of the nucleosome core particle may be altered, with loss of specific histones such as H2A and H2B.

in Figure 15–13, this can entail altering the contacts between the DNA and the nucleosome histone proteins, thus causing the nucleosome to slide farther down the DNA molecule. Alternatively, the path of the DNA around the nucleosome core particle may be altered, pulling the DNA off the nucleosome. Another possibility is that the structure of the nucleosome core particle itself may be altered, producing a nucleosome dimer. Any of these modifications "open up" the DNA template, allowing transcription to be initiated and elongated.

Histone Modification

A second mechanism of chromatin alteration is histone modification. One such modification is acetylation, a chemical alteration of the histone component of nucleosomes that is catalyzed by **histone acetyltransferase enzymes (HATs)**. When an acetate group is added to specific basic amino acids on the histone tails, the attraction between the basic histone protein and acidic DNA is lessened. The loosening of histones with DNA facilitates chromatin remodeling catalyzed by ATP-dependent chromatin remodeling complexes. As we learned above, chromatin remodeling opens chromatin and makes promoter regions available for binding to transcription factors that initiate the chain of events leading to gene transcription. In addition to acetylation, histones can be modified by phosphorylation and methylation. These structural alterations occur at specific amino acid residues in histones. Some evidence suggests that specific histone modifications may

facilitate the binding of regulatory proteins involved in gene regulation.

DNA Methylation

Another type of change in chromatin that plays a role in gene regulation involves adding or removing methyl groups on bases in DNA. The DNA of most eukaryotic organisms can be modified after DNA replication by the enzyme-mediated addition of methyl groups to bases and sugars. **DNA methylation** most often involves cytosine. In the genome of any given eukaryotic species, approximately 5 percent of the cytosine residues are methylated. However, the extent of methylation can be tissue specific and can vary from less than 2 percent to more than 7 percent.

Evidence for the role of methylation in eukaryotic gene expression is somewhat indirect and is based on a number of observations. First, an inverse relationship exists between the degree of methylation and the degree of expression. That is, low amounts of methylation are associated with high levels of gene expression, and high levels of methylation are associated with low levels of gene expression. In mammalian females, the inactivated X chromosome, which is almost totally inactive, has a higher level of methylation than does the active X chromosome. Within the inactive X, regions that escape inactivation have much lower levels of methylation than those seen in adjacent inactive regions.

Second, methylation patterns are tissue specific and, once established, are heritable for all cells of that tissue. Perhaps the strongest evidence for the role of methylation in gene expression comes from studies using base analogs. The nucleotide 5-azacytidine is incorporated into DNA in place of cytidine and cannot be methylated, causing the undermethylation of the sites where it is incorporated. The incorporation of 5-azacytidine causes changes in the pattern of gene expression and can stimulate expression of alleles on inactivated X chromosomes.

Although the available evidence indicates that the absence of methyl groups in DNA is related to increases in gene expression, methylation cannot be regarded as a general mechanism for gene regulation because DNA methylation is not present in all eukaryotes.

How might methylation affect gene regulation? One possibility comes from the observation that certain proteins bind to 5-methyl cytosine without regard to the DNA sequence. These proteins could recruit transcription repressor proteins or repressive chromatin remodeling complexes. Methylation may also directly repress transcription by interfering with the binding of transcription activators.

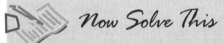

Now Solve This

Question 28 on page 356 asks you to interpret the effect of methylation of DNA on the expression of a gene.

Hint: Remember that the location of various regulatory sequences outside the gene will affect the results.

FIGURE 15–14 Expression of eukaryotic genes is controlled by regulatory elements directly adjacent to the gene, such as promoters, and by sequences that can be far from the transcriptional unit, such as enhancers and silencers.

15.8 Eukaryotic Transcription Is Regulated at Specific *Cis*-Acting Sites

Eukaryotic genes have several types of *cis*-regulatory sequences that control transcription, including promoters, silencers, and enhancers (Figure 15–14). Transcription of DNA into an mRNA molecule is a complex, highly regulated process involving several different types of DNA sequences, the interactions of many proteins, chromatin remodeling, and the bending and looping of DNA sequences. We will begin by discussing some of the DNA sequences involved in regulating eukaryotic gene transcription.

Promoters

Promoters are nucleotide sequences that serve as recognition sites for the transcription machinery. They are necessary in order for transcription to be initiated at a basal level. Promoters are located immediately adjacent to the genes they regulate. These regions are usually several hundred nucleotides in length and specify the site at which transcription begins and the direction of transcription along the DNA.

The promoters of most eukaryotic genes contain several elements, including TATA, CAAT, and GC boxes, as well as the transcription start site (Figure 15–15). The **TATA box**, along with the start site, is referred to as the **core promoter**. Located

about 25 to 30 bases upstream from the transcription start site (designated as −25 to −30), the TATA box consists of a 7- to 8-bp consensus sequence (a sequence conserved in most genes studied) composed of AT base pairs often flanked on either side by GC-rich regions. Genetic analysis of TATA sequences shows that mutations within TATA sequences reduce transcription and that deletions may alter the initiation point of transcription (Figure 15–16).

Many promoter regions also contain **CAAT boxes**. These elements have the consensus sequence CAAT or CCAAT. The CAAT box frequently appears 70 to 80 bp upstream from the start site. Mutational analysis suggests that CAAT boxes are critical to the promoter's ability to facilitate transcription. Mutations on either side of this element have no effect on transcription, whereas mutations within the CAAT sequence dramatically lower the rate of transcription (Figure 15–16). Another element often seen in promoter regions, called the **GC box**, has the consensus sequence GGGCGG and is often located at about position −110. The CAAT and GC elements bind transcription factors and function somewhat like enhancers, which we will cover in the next section.

To sum up, each eukaryotic gene contains a number of promoter elements that specify the basal level of transcription from that gene. Different genes may have different subsets of these promoter elements and they may vary in location and organization from gene to gene.

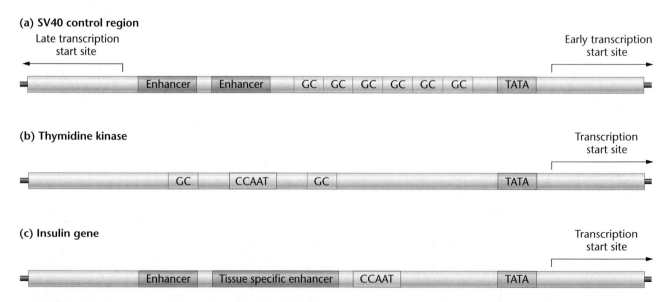

(a) SV40 control region

Late transcription start site

Early transcription start site

Enhancer | Enhancer | GC | GC | GC | GC | GC | GC | TATA

(b) Thymidine kinase

Transcription start site

GC | CCAAT | GC | TATA

(c) Insulin gene

Transcription start site

Enhancer | Tissue specific enhancer | CCAAT | TATA

FIGURE 15–15 Organization of transcription regulatory regions in several genes expressed in eukaryotic cells, illustrating the variable nature, number, and arrangement of controlling elements.

FIGURE 15–16 Summary of the effects of point mutations in the promoter region on transcription of the β-globin gene. Each line represents the level of transcription produced by a single-nucleotide mutation (relative to wild type) in a separate experiment. Dots represent nucleotides in which no mutation was obtained. Note that mutations within the specific elements of the promoter have the greatest effect on the level of transcription.

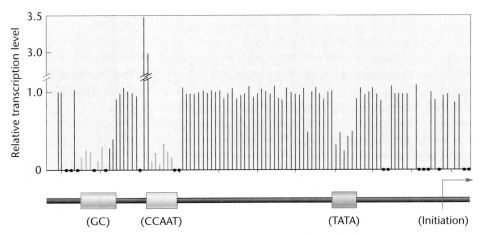

Enhancers and Silencers

Transcription of eukaryotic genes is regulated not only by promoters, but also by DNA sequences called **enhancers**. Enhancers can be located on either side of a gene, at some distance from the gene, or even within the gene. They are called *cis* regulators because they function when adjacent to the structural genes they regulate, as opposed to *trans* regulators (e.g., binding proteins), which can regulate a gene on any chromosome. While promoter sequences are essential for basal-level transcription, enhancers are necessary for the full level of transcription. In addition, enhancers are responsible for time- and tissue-specific gene expression.

Enhancers typically interact with multiple regulatory proteins and transcription factors and can increase the efficiency of transcription initiation from an associated promoter. Within enhancers, binding sites are often found for positive as well as negative gene regulators. Thus, there is some degree of analogy between enhancers and operator regions in prokaryotes. However, enhancers appear to be much more complex in both structure and function. Other features that distinguish enhancers from promoters include the following:

1. The position of an enhancer is not fixed; it can be upstream, downstream, or within the gene it regulates.

2. Its orientation can be inverted without significant effect on its action.

3. If an enhancer is experimentally moved to another location in the genome, or if an unrelated gene is placed near an enhancer, the transcription of the adjacent gene is enhanced.

An example of an enhancer located *within* the gene it regulates is found in the immunoglobulin heavy-chain gene, in which an enhancer is located in an intron between two coding regions. This enhancer is active only in cells expressing the immunoglobulin genes, indicating that tissue-specific gene expression can be modulated through enhancers. Internal enhancers have been discovered in other eukaryotic genes, including the immunoglobulin light-chain gene. An example of a downstream enhancer is the β-globin gene enhancer. In chickens, an enhancer located between the β-globin gene and the ε-globin gene works in one direction to control transcription of the ε-globin gene during embryonic development and in the opposite direction to regulate expression of the β-globin gene during adult life.

Like promoters, enhancers are modular and often contain several different short DNA sequences. For example, the enhancer of the SV40 virus (which is transcribed inside a eukaryotic cell) has a complex structure consisting of two adjacent 72-bp sequences located some 200 bp upstream from a transcriptional start point. One of these is shown in Figure 15–17. Each of the two 72-bp regions contains multiple sequence motifs that contribute to the maximum rates of transcription. If one or the other of these regions is deleted, there is no effect on transcription; but if both are deleted, *in vivo* transcription is greatly reduced.

Another type of transcription regulatory element, the **silencer**, acts upon eukaryotic genes to repress the level of transcription initiation. Like enhancers, silencers are short DNA sequence elements, located in regions surrounding a promoter, that affect the rate of transcription from that promoter. They often act in tissue- or temporal-specific ways to

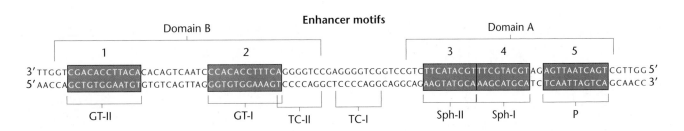

FIGURE 15–17 DNA sequence for the SV40 enhancer. The blue boxed sequences are those required for maximum enhancer effect. The brackets below the sequence name the various sequence motifs within this region. The two domains of the enhancer (A and B) are indicated.

control gene expression. An example of a silencer element is that found in the human thyrotropin-β gene. The thyrotropin-β gene encodes one subunit of the thyrotropin hormone, and this gene is expressed only in thyrotrophs of the pituitary gland. Transcription of this gene is restricted to thyrotrophs due to the actions of a silencer element located −140 bp upstream from the transcription start site. The silencer binds a cellular factor known as Oct-1, which in the context of the thyrotropin-β promoter represses transcription in all cell types except thyrotrophs. In thyrotrophs, the action of the silencer is overcome by an enhancer element located over 1.2 kb upstream of the promoter.

15.9 Eukaryotic Transcription Is Regulated by Transcription Factors that Bind to *Cis*-Acting Sites

It is generally accepted that *cis*-acting regulatory sites—including promoters, enhancers, and silencers—influence transcription initiation by acting as binding sites for specific transcription regulatory proteins. These transcription regulatory proteins, known as **transcription factors**, can have diverse and complicated effects on transcription. Some transcription factors are expressed in tissue-specific ways, thereby regulating their target genes for tissue-specific expression. Similarly, some transcription factors are expressed in cells at certain times during development or in response to external physiological signals, which regulates the temporal expression of their target genes. In some cases, a transcription factor that binds a *cis*-acting site and regulates a certain gene may be present in a cell and may even bind to its appropriate *cis*-acting site but will only become active when modified structurally (for example, by phosphorylation or by binding to a co-activator such as a hormone). These modifications to transcription factors can also be regulated in tissue- or temporal-specific ways. In addition, different transcription factors may compete for binding to a given DNA sequence or to two overlapping sequences. In these cases, transcription factor concentrations and the strength with which each factor binds to the DNA will dictate which factor binds. The same site may also bind different factors in two different tissues. Finally, multiple transcription factors that bind several enhancers and promoter elements within a gene regulatory region can interact with each other to fine-tune the levels and timing of transcription initiation.

The Human Metallothionein IIA Gene: Multiple *Cis*-Acting Elements and Transcription Factors

The human metallothionein IIA gene (hMTIIA) provides an example of how one gene can be transcriptionally regulated through the interplay of multiple promoter and enhancer elements, and the transcription factors that bind to them. The product of the hMTIIA gene is a protein that binds to heavy metals such as zinc and cadmium, thereby protecting cells from the toxic effects of high levels of these metals. The gene is expressed at low levels in all cells but is transcriptionally induced to high levels when cells are in the presence of heavy metals and steroid hormones such as glucocorticoids.

The cis-acting regulatory elements controlling the hMTIIA gene include promoter, enhancer and silencer elements (Figure 15–18). Each *cis*-acting element is a short DNA sequence that has specificity for binding to one or more transcription factors.

The hMTIIA gene contains the core promoter elements TATA and start site, which specify the start of transcription. The basal element, GC, binds the SP1 factor, which is present in most eukaryotic cells and hence stimulates transcription at low levels in most cells. Basal levels of expression are also regulated by the BLE (basal element) and ARE (AP factor response element) regions. These *cis*-elements bind to the activator proteins 1, 2, and 4 (AP1, AP2, and AP4) transcription factors, which are present in various levels in different cell types and can be activated in response to extracellular growth signals. The BLE element contains overlapping binding sites for the AP1 and AP4 factors, providing some degree of selectivity in how these factors stimulate transcription of hMTIIA when bound to the BLE in different cell types. High levels of transcription induction are conferred by the MRE (metal response element) and GRE (glucocorticoid response element). The metal-inducible transcription factor (MTF1) binds to the MRE element in response to the presence of heavy metals. The glucocorticoid receptor protein binds to the GRE, but only when the receptor is bound to the glucocorticoid steroid hormone. The glucocorticoid receptor is normally located in the cytoplasm of the cell; however, when glucocorticoid hormone enters the cytoplasm, it binds to the

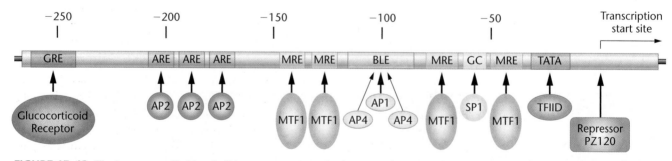

FIGURE 15–18 The human metallothionein IIA gene promoter and enhancer regions, containing multiple *cis*-acting regulatory sites. The transcription factors controlling both basal and induced levels of MTIIA transcription, and their binding sites, are indicated below the gene, and are described in the text.

receptor and causes a conformational change that allows the receptor to enter the nucleus, bind to the GRE, and stimulate hMTIIA transcription. In addition to induction, transcription of the hMTIIA gene can be repressed by the actions of the repressor protein PZ120, which binds over the transcription start region.

The presence of multiple regulatory elements and transcription factors that bind to them allows the hMTIIA gene to be transcriptionally induced or repressed in response to subtle changes in both extracellular and intracellular conditions.

Functional Domains of Eukaryotic Transcription Factors

Transcription factors are proteins that bind to DNA and activate (or repress) transcription initiation. To do this, they have two functional domains (clusters of amino acids that carry out a specific function). One domain binds to DNA sequences present in the regulatory site (**DNA-binding domain**), and the other activates or represses transcription through protein–protein interactions (***trans*-activating** or ***trans*-repression domain**). The *trans*-activating or *trans*-repression domains bind to RNA polymerase or to other transcription factors at the promoter, as we will discuss in the next section.

The domains of eukaryotic transcription factors take on several forms. The DNA-binding domains have characteristic three-dimensional structural patterns or motifs. There are several classes of these motifs, including the helix–turn–helix (HTH), zinc finger, and basic leucine zipper (bZIP) motifs.

The first DNA-binding domain to be discovered was the **helix–turn–helix (HTH)** motif. This motif, also present in prokaryotic transcription factors, is characterized by a geometric conformation rather than a distinctive amino acid sequence (Figure 15–19). The presence of two adjacent α-helices separated by a "turn" of several amino acids enables the protein to bind to DNA (hence the name of the motif). Amino acid residues outside the HTH motif are also important in regulating DNA recognition and binding. The HTH motif is an important feature of homeobox-containing transcription factors. The **homeobox** is a stretch of 180 bp that encodes a 60-amino-acid sequence known as a **homeodomain**. Transcription factors that contain homeodomains are conserved throughout eukaryotes and are important regulators of animal development.

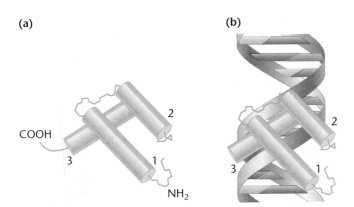

FIGURE 15–19 A *helix–turn–helix* motif in which
(a) three planes of the α-helix of the protein are established, and
(b) these domains bind in the grooves of the DNA molecule.

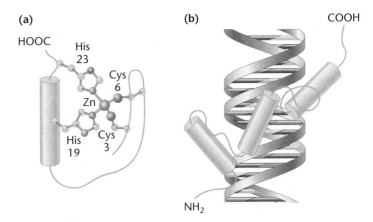

FIGURE 15–20 (a) A zinc finger in which cysteine and histidine residues bind to a Zn^{++} atom. (b) This loops the amino acid chain out into a fingerlike configuration that binds in the major groove of the DNA helix.

The **zinc-finger** motif is found in a wide range of transcription factors that regulate gene expression related to cell growth, development, and differentiation. For example, the MTF1 metal-inducible transcription factor discussed in the previous section contains six zinc fingers that comprise the DNA binding region of this factor. A typical zinc-finger protein contains clusters of two cysteines and two histidines at repeating intervals (Figure 15–20). The consensus amino acid repeat is $CysN_{2-4}CysN_{12-14}HisN_3His$ (where N is any amino acid). The interspersed cysteine and histidine residues covalently bind zinc atoms, folding the amino acids into loops known as zinc fingers. Each finger consists of approximately 23 amino acids (with a loop of 12 to 14 amino acids between the Cys and His residues) and a linker between loops consisting of 7 or 8 amino acids. The amino acids in the loop interact with, and bind to, specific DNA sequences.

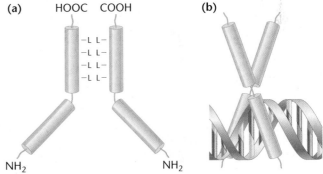

FIGURE 15–21 (a) A leucine zipper results from the interactions of leucine residues at every other turn of the α-helix in facing stretches of two polypeptide chains. (b) When the α-helical regions form a leucine zipper, the regions beyond the zipper form a Y-shaped region that grips the DNA in a scissorlike configuration.

A third type of DNA-binding domain is the **basic leucine zipper (bZIP)** motif. This DNA-binding domain contains the **leucine zipper**, a region that allows protein–protein dimerization. The leucine zipper contains four leucine residues flanked by basic amino acids. When two such molecules dimerize, the leucine residues "zip" together (Figure 15–21). The dimer contains two basic α-helical regions adjacent to the zipper that bind to phosphate residues and specific bases in DNA, making the dimer look like a pair of scissors. The transcription factor AP1 is an example of a bZIP transcription factor. AP1 is a heterodimer consisting of two different proteins of the c-fos and c-jun families. The activities of AP1 can be modulated by the mix and match of different family members of c-fos and c-jun that go into the composition of each AP1 molecule.

Transcription factors also contain domains that interact with proteins in the basal transcription complex and control the level of transcription initiation. These *trans*-activating domains, distinct from the DNA-binding domains, can occupy from 30 to 100 amino acids. In addition, many transcription factors also contain domains that bind **coactivators**, such as hormones or small metabolites that regulate their activity.

15.10 Eukaryotic Transcription Factors Regulate Transcription Through Interactions with Basal Transcription Factors

We have now reviewed the basic steps in eukaryotic transcription regulation. First, chromatin must be remodeled and modified in such a way that transcription factors can bind to their specific *cis*-acting sites. Second, transcription factors bind to *cis*-acting sites and bring about a wide range of positive and negative effects on the transcription rate—often in response to extracellular signals or in tissue- or time-specific ways. However, the question remains: how do these *cis*-acting regulatory elements and their DNA binding transcription factors act, in order to influence transcription initiation? To answer this question, we will first discuss how eukaryotic RNA polymerase II and its basal transcription factors assemble at promoters.

Formation of the Transcription Initiation Complex

A series of proteins called **basal** or **general transcription factors** are essential to initiate either basal-level or induced levels of transcription. These proteins are not part of the RNA polymerase II molecule but assemble at the promoter in a specific order, forming a transcriptional **pre-initiation complex** (**PIC**) which in turn provides a platform for the polymerase to recognize and bind to the promoter (Figure 15–22). To initiate formation of the PIC, a complex called **TFIID** binds to the TATA box through one of its subunits, called **TBP** (*TATA*

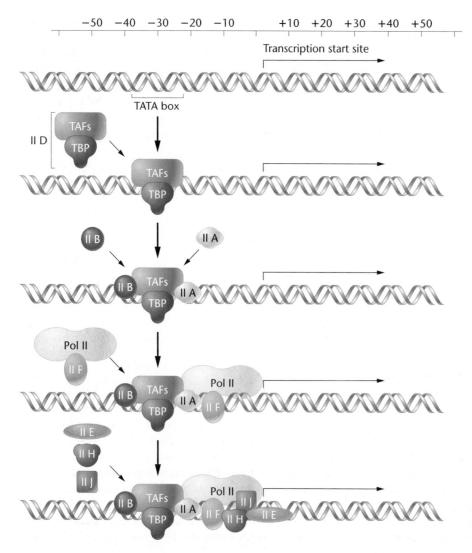

FIGURE 15–22 The assembly of basal transcription factors required for the initiation of transcription by RNA polymerase II.

FIGURE 15–23 Formation of DNA loops allows factors that bind to enhancers at a distance from the promoter to interact with regulatory proteins in the pre-initiation complex and to maximize transcription.

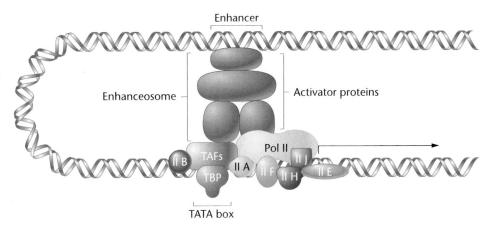

Binding Protein). The TFIID complex is composed of TBP plus approximately 13 proteins called *TAFs* (*TATA Associated Factors*). TFIID responds to contact with transcription activator proteins by binding additional factors, such as TFIIB, TFIIA, TFIIF, TFIIH, and TFIIJ. RNA polymerase II is recruited to the complex along with TFIIF. Once all of the basal transcription factors have assembled at a promoter, and RNA polymerase II has made stable contacts with the start site of transcription, the DNA double helix is melted at the start site and RNA polymerase begins transcription.

Interactions of the Basal Transcription Factors with Transcription Activators

Although the basal transcription factors form a platform for the binding of RNA polymerase II and the initiation of transcription at a basal level, transcription factors that bind to gene regulatory elements modulate the rate of transcription. There are several possible mechanisms by which transcription factors appear to alter the rate of transcription. In many cases, the first step involves the assembly of activator proteins, which are bound to the DNA of regulatory elements, with other proteins in a complex known as an **enhanceosome**. The second step, which appears to be essential for regulation of transcription rate, is the interaction of transcription factors bound at a promoter or enhancer with one or more members of the basal transcription factors in the PIC (Figure 15–23). In order for proteins in these regions to interact, DNA loops out or bends between the promoter and enhancer regions. By influencing the rate of PIC assembly, transcription factors stimulate (or repress) the rate of transcription initiation. In addition to PIC formation, transcription activators may also stimulate chromatin remodeling, stabilize the binding of basal factors to DNA, increase the rate of DNA unwinding within the gene, and accelerate the release of RNA polymerase from the promoter.

In summary, the picture of transcription regulation in eukaryotes is complex, but several important generalizations can be drawn. First, alterations in chromatin structure allow the binding of transcription factors. Second, the regulation of transcription by transcription factors is largely positive, although transcription repressors are also components of transcription regulation. In addition, different transcription factors may compete for binding to a given DNA sequence or to two over-

lapping sequences. The same sequence may bind different factors in different tissues or at different times or in response to different extracellular signals. Transcription factor concentrations or the efficiency with which each factor binds to the DNA may dictate which factor binds. Finally, the rate of transcription initiation can be affected simultaneously by multiple enhancers, silencers and the multiplicity of factors that may bind to these elements in different circumstances.

Now Solve This

Problem 25 on page 355 asks you to interpret results generated by experiments conducted in two types of *in vitro* transcription systems using a number of DNA fragments carrying various portions of a gene promoter.

Hint: Consider the types of promoter and enhancer elements that exist in eukaryotic gene regulatory regions and their locations. Also, think about what types of proteins bind to these elements and whether the *in vitro* transcription systems would contain these proteins.

15.11 Alternative Splicing and mRNA Stability Can Regulate Eukaryotic Gene Expression

Regulation of gene expression can occur at many points along the pathway from DNA to protein. Although transcriptional control is a major type of gene regulation in eukaryotes, **posttranscriptional regulation** is also widely used. Eukaryotic nuclear RNA transcripts are modified prior to translation, noncoding introns are removed, the remaining exons are precisely spliced together, and the mRNA is modified by the addition of a cap at the 5' end and a poly-A tail at the 3' end. The message RNA is then exported to the cytoplasm where it is translated and degraded. Each of these processing steps can be regulated to control the quantity of functional mRNA available for synthesis of a protein product. We will examine two mechanisms of gene regulation that are especially important in eukaryotes—alternative splicing of identical primary

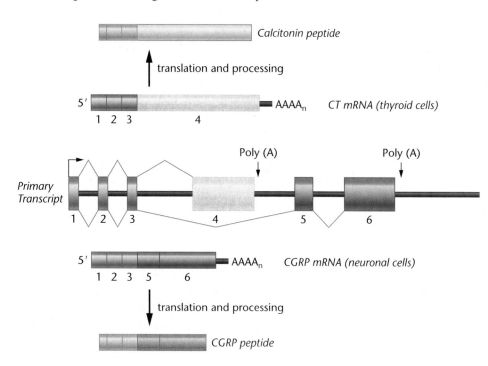

FIGURE 15–24 Alternative splicing of the *CT/CGRP* gene transcript. The primary transcript is shown in the middle of the diagram and contains six exons. The primary transcript can be spliced into two different mRNAs, both containing the first three exons. The *CT* mRNA contains exon 4 with polyadenylation occurring at the end of the fourth exon. The *CGRP* mRNA contains exons 5 and 6, and polyadenylation occurs at the end of exon 6. After mRNA processing in thyroid cells, the *CT* mRNA is translated, and the protein is processed into the calcitonin peptide. In neuronal cells, the *CGRP* mRNA is translated, and the protein product is processed into the CGRP peptide.

transcripts to give multiple mRNAs and regulation of the stability of mRNA itself. In a subsequent section, we will discuss a newly discovered method of posttranscriptional gene regulation—RNA silencing.

Alternative Splicing Pathways for mRNA

Alternative splicing can generate different forms of mRNA from identical pre-mRNA molecules, so that expression of one gene can give rise to a number of proteins, with similar or different functions. Changes in splicing patterns can have many different effects on the translated protein. Small changes can alter enzymatic activity, receptor binding capacity, or protein localization in the cell. Changes in splicing patterns are important events in development, apoptosis, axon-to-axon connection in the nervous system, and many other processes. Mutations that affect regulation of splicing are the basis of several genetic disorders.

Alternative splicing increases the number of proteins that can be made from each gene. As a result, the number of proteins that an organism can make (its **proteome**) is not the same as the number of genes in the genome, and protein diversity can exceed gene number by an order of magnitude. Alternative splicing is found in all metazoans but is especially com-

mon in vertebrates, including humans. It has been estimated that 30 to 60 percent of the genes in the human genome use alternative splicing. Thus, humans can produce several hundred thousand different proteins (or perhaps more) from the 25,000 to 30,000 or so genes in the haploid genome.

Figure 15–24 presents an example of alternative splicing of the pre-mRNA transcribed from the calcitonin/calcitonin gene-related peptide gene (*CT/CGRP* gene). In thyroid cells, the *CT/CGRP* primary transcript is spliced in such a way that the mature mRNA contains the first four exons only. In these cells, the exon 4 polyadenylation signal is used to process the mRNA and add the poly-A tail. The mRNA is translated into the calcitonin peptide, a 32-amino acid peptide hormone that is involved in regulating calcium. In the brain and peripheral nervous system, the *CT/CGRP* primary transcript is spliced to include exons 5 and 6, but not exon 4. In these cells, the exon 6 polyadenylation site is recognized. The *CGRP* mRNA encodes a 37-amino acid peptide with hormonal activities in a wide range of tissues. By using alternative splicing, two peptide hormones with different structures, locations, and functions are synthesized from the same gene. Even more complex alternative splicing patterns occur in some genes, such as the example in Figure 15–25.

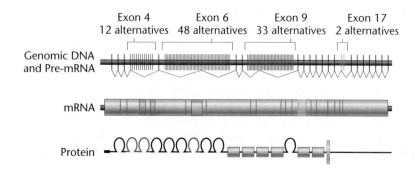

FIGURE 15–25 (Top) Organization of the *Dscam* gene in *Drosophila melanogaster* and the transcribed pre-mRNA. The *Dscam* gene encodes a protein that guides axon growth during development. Each mRNA will contain one of the 12 possible exons for exon 4 (red), one of the 48 possible exons for exon 6 (blue), one of 33 for exon 9 (green), and one of 2 for exon 17 (yellow). If all possible combinations of these exons are used, the *Dscam* gene can encode 38,016 different versions of the DSCAM protein.

Controlling mRNA Stability

After mRNA precursors are processed and transported, they enter the population of cytoplasmic mRNA molecules, from which mRNAs are recruited for translation. All mRNA molecules have a characteristic life span (called a half-life, or $t_{1/2}$); they are degraded some time after they are synthesized, usually in the cytoplasm. The lifetimes of different mRNA molecules vary widely. Some are degraded within minutes after synthesis, whereas others last hours, or even months and years (in the case of mRNAs stored in oocytes). RNA stability can be regulated to affect the levels of protein product.

One way in which mRNA stability is controlled is through translational stability—translation of the message affects its stability. One of the best studied examples of translational stability control is the synthesis of α- and β-tubulins, the subunit components of eukaryotic microtubules. In this system, the presence of free cytoplasmic tubulin subunits shuts off tubulin translation by causing degradation of tubulin mRNA that is in the act of being translated. Free tubulin subunits are present in the cytoplasm of cells if they are not assembling microtubules.

Work by Don Cleveland and his colleagues revealed how the stability of tubulin mRNA is regulated during its translation (Figure 15–26). They propose that the first four amino acids of the tubulin gene product constitute a recognition element to which regulatory factors—the α- and β-tubulins—bind. The interaction of tubulin subunits with the nascent protein chain activates an RNase that may be either a ribosomal component or a nonspecific cytoplasmic RNase. The RNase degrades the tubulin mRNA in the act of translation, shutting down tubulin biosynthesis.

Translation-coupled mRNA turnover has been proposed as a regulatory mechanism for other genes, including histones, some transcription factors, lymphokines, and cytokines.

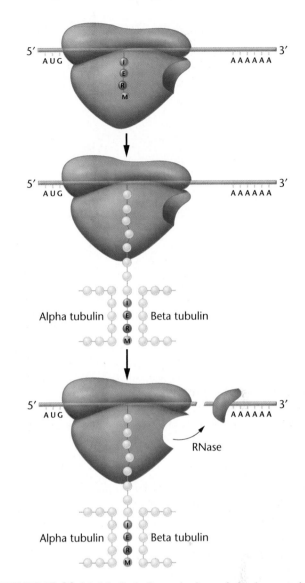

FIGURE 15–26 Model of tubulin synthesis regulation by control of tubulin mRNA stability.

Now Solve This

Problem 30 on page 356 asks you to interpret data showing that a deletion of one extra exon by alternative splicing could rescue protein production from the dystrophin gene.

Hint: To answer this question, consider how the deletion or addition of certain numbers of base pairs in DNA can affect reading frames in the resulting mRNA. You might also consider the various outcomes for both the mRNA and the encoded protein that could result from frameshift mutations.

15.12 RNA Silencing Controls Gene Expression in Several Ways

In the last several years, the discovery that regulatory RNA molecules play an important role in controlling gene expression has given rise to a new field of research. First discovered in plants, short RNA molecules, ~21 nucleotides long, are now known to regulate gene expression in the cytoplasm by repressing translation of mRNAs and degrading mRNAs. More recently, similar RNAs have been shown to act in the nucleus to alter chromatin structure and bring about **gene silencing**.

The best-studied form of RNA silencing is called **RNA interference (RNAi)** in animals and **posttranscriptional gene silencing (PTGS)** in plants. This process begins with a gene-specific double-stranded RNA (about 70 nucleotides long) that is processed by a protein (called Dicer) with double-stranded RNAse activity. Dicer has two catalytic domains and functions as a dimeric enzyme (Figure 15–27). One of the catalytic domains in each monomer is inactive, and alignment and cutting by the active domains result in cleavage of the RNA at about 21-nucleotide intervals. The product is a short (~21 nucleotide) RNA called **short interfering RNA (siRNA)**. The siRNA unwinds into sense and antisense single strands. The antisense strand combines with a protein complex, **RISC (*R*NA-*I*nduced *S*ilencing *C*omplex)** that recognizes, binds to, and cleaves mRNA containing sequences

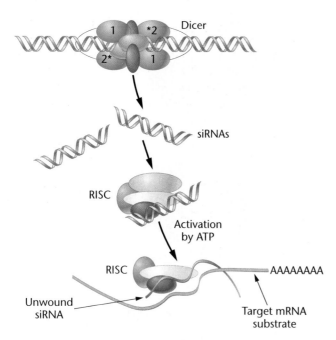

FIGURE 15–27 The action of Dicer and RISC (RNA-induced silencing complex). Dicer binds to double-stranded RNA molecules and cleaves them into ~21 nucleotide molecules called small interfering RNAs (siRNAs). These bind to a multiprotein RISC complex and are unwound to form single-stranded molecules that target mRNAs with complementary sequences, marking them for degradation.

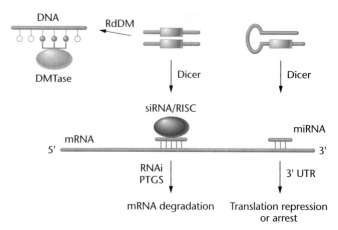

FIGURE 15–28 Mechanisms of gene regulation by RNA gene silencing. In the cytoplasm, two systems operate to silence genes. (Middle) In siRNA mediated silencing, a precursor RNA molecule is processed by Dicer, a protein with RNAse activity to form an antisense single-stranded RNA that combines with a protein complex with endonuclease activity. siRNA/RISC (RNA-induced silencing complex) binds to mRNAs with complementary sequences and cuts the mRNA into fragments that are degraded. This process is called RNAi in animal cells and posttranscriptional gene silencing (PTGS) in plants. (Right) A partially double-stranded precursor is processed by Dicer to yield microRNA (miRNA) that binds to complementary 3′-untranslated regions (UTRs) of mRNA, inhibiting translation. (Left) Small RNAs, processed by Dicer, play a role in RNA-directed DNA methylation (RdDM). These RNAs combine with DNA methyl transferases (DMTases) to methylate cytosine residues in promoter regions (purple circles), silencing genes.

complementary to the antisense strand of the siRNA. RISC cuts the mRNA at or near the middle of the region paired with the siRNA, and the mRNA fragments are degraded.

RNAi plays an important role in defending cells against invading viruses and in silencing transposons. Preliminary evidence indicates that RNAi also plays roles in the control of development.

A second form of RNA silencing is mediated by short RNA molecules called **microRNAs (miRNAs)**. miRNAs are short, single-stranded RNAs about 20 to 25 nucleotides long. They are derived from longer (about 100-nucleotide) precursor RNAs that form hairpin structures. These RNA precursors are cleaved by Dicer, and the resulting miRNAs pair with internal or 3′-untranslated regions (UTRs) of other mRNA molecules. This pairing results in either a block to translation or the targeting of the mRNA for degradation (Figure 15–28).

In *C. elegans* and in vertebrates, miRNA genes may constitute up to 0.2 to 0.5 percent of the genome and are transcribed from several hundred genes. The expression pattern for many miRNA genes is consistent with a role for these RNAs in controlling development. Most of the targets of *Arabidopsis* miRNAs are RNAs encoding transcription factors that are important in determining cell fate in the developing plant.

More recently, it has been discovered that short RNAs can target specific regions of the genome for chromatin modification. siRNAs and miRNAs depend on RNA–RNA sequence recognition and binding. However, RNA can also base-pair with DNA, and short RNAs are involved in several forms of genome modifications that regulate gene expression. These include **RNA-directed DNA methylation (RdDM)** of cyto-

sine (Figure 15–28). RdDM is a highly specific process and is limited to the region of RNA–DNA pairing. In RdDM, CG dinucleotides and other C residues in promoter regions are methylated, leading to gene silencing.

Several other types of small, endogenous RNAs have been discovered and have yet to be classified, making it seem likely that there are many more functionally distinct classes of small RNAs to be discovered that play important roles in regulating genome structure and function.

RNA Silencing in Biotechnology

Recently, geneticists have applied RNAi as a powerful research tool. RNAi technology allows investigators to specifically create single-gene defects without resorting to the creation of heritable mutations. In addition to their specificity, RNAi-mediated gene-silencing technologies are relatively inexpensive and allow rapid analysis of gene function. Researchers hope to use RNAi technologies to systematically knock out the function of each gene in certain model organisms. The technique may also be useful for gene therapy, in which specific disease genes can be silenced.

To use RNAi as a research tool, investigators introduce short synthetic double-stranded RNA molecules into cells. These molecules trigger the same RNA-degradation pathways that operate during natural RNA interference. The antisense strand from the double-stranded RNA hybridizes to the complementary mRNA in the cells, targeting the mRNA for degradation or translation inhibition.

RNAi methods have been used successfully to investigate gene functions in organisms that are usually difficult to manipulate genetically. These organisms include mosquitoes, trypanosomes, and mammalian cells. RNAi has also been used to silence the expression of genes that are aberrantly expressed in cancer cells. When these cancer-related genes are silenced, the cancer cells lose their ability to form tumors when introduced into mice. Although RNAi technologies, sometimes referred to as "genetics without mutations," are just beginning to be developed, they hold great promise for both investigative and therapeutic uses.

GENETICS, TECHNOLOGY, AND SOCIETY

Gene Regulation and Human Genetic Disorders

When we think of human genetic disorders, we usually think of mutations that alter the coding portion of a gene or changes in chromosome number or structure that disrupt the delicate balance of gene expression. However, gene expression is a complex process requiring a number of sequential steps. Both the DNA encoding genes and the mRNA products of genes undergo epigenetic modifications that alter gene function. These alterations in turn may result in a variety of genetic disorders.

For example, DNA methylation, one of the key examples of epigenetic processing, may lead to gene silencing, and aberrant methylation is often associated with cancer. As a result, assessment of DNA methylation is now a burgeoning field in cancer diagnosis and possible treatment. Recent advancements allow for the non-invasive diagnosis of methylation status from exfoliated cells in urine, sputum, and stool samples. From a urine analysis, aberrant methylation of *laminin-5*, a gene critical to transport of molecules out of the nucleus, appears to be a useful marker to distinguish invasive from non-invasive bladder cancers. Methylation of the *p16*, *MGMT*, and *RASSF1A* genes in DNA derived from sputum appear to be sensitive markers for lung cancer. Geneticists have also used stool samples to analyze point mutations in tumor suppressor genes, and have investigated whether they can be used for determining DNA methylation status for these genes as well.

The mechanisms leading to altered DNA methylation in cancer cells are unclear, possibly the result of somatic cell mutations involving the breakdown of methylation machinery. Widespread demethylation (hypomethylation) of the genome is a phenomenon found across cancer types and appears to be associated with an increase in genetic instability and activation of certain genes involved in tumor growth and metastatic potential. Hypermethylation of DNA at specific chromosomal regions has also been associated with various cancers, seemingly resulting in the silencing of tumor suppressor genes such as *pRB*, *BRCA1*, and *APC* genes involved in retinoblastoma, breast cancer, and colon cancer, respectively.

Molecular techniques for determining methylation changes in tumor cells may increase the accuracy of cancer diagnosis. For example, the childhood kidney cancer called Wilms tumor occurs when a gene that normally halts cell division in the developing kidney is altered. Pockets of kidney cells continue dividing as they would otherwise do in the fetus, causing kidney tumors. Normal kidney cells proliferate until the kidney reaches normal size and then they cease growing. Normal and tumor tissues can be distinguished based on the degree of DNA methylation. DNA in the cells of normal kidney tissue is highly methylated at specific chromosomal locations; however, in tumor cell DNA, methylation at these loci is severely reduced or absent.

Once aberrant methylation patterns are diagnosed, treatments could involve reversing the abnormal DNA methylation using drugs for anti-tumor therapy. Additionally, there is potential for cancer prevention, as linkage of tobacco-smoke exposure to the methylation of a specific gene has been implicated in non-small-cell lung cancer. In the future, DNA methylation-based diagnostics may potentially change the way physicians diagnose and treat cancer patients.

As we learn more about the process of gene regulation, it is apparent that mutations affecting all steps in the process can result in genetic disorders. For example, one of the first steps in eukaryotic gene expression involves remodeling chromatin to make it accessible for transcription. Remodeling involves the action of the ATP-dependent SWI/SNF protein complex. Mutations of genes encoding the enzymes in the SWI/SNF complex cause diseases with a wide range of phenotypes. These include Williams syndrome, an autosomal dominant disorder causing heart and lung defects as well as growth deficiencies. In this case, the affected gene encodes the *Daxx* cotranscription factor that targets chromatin remodeling to specific promoters, and may act by causing aberrant methylation of repetitive DNA sequences.

Histone acetylation and other epigenetic modifications are also important contributors to regulating gene expression. Histone acetyltransferases (HATs) and histone deacetylases (HDACs) cooperate with the SWI/SNF factors to remodel chromatin as a first step in transcription. Mutation in a histone acetylase gene *CREB* (*cAMP-responsive element binding protein*) is responsible for Rubinstein-Taybi syndrome, an autosomal dominant disorder exhibiting growth retardation, facial and hand abnormalities, and mental retardation. In general, inactivation of histone acetylases or deacetylases is associated with developmental disorders with a broad range of complex phenotypes. Abnormal activation of HATs or HDACs is also associated with several forms of cancer.

From what we are learning about the molecular mechanisms of human genetic disorders, it is becoming clear that mutations in genes associated with epigenetic processing, chromatin remodeling, and

(Cont. on the next page)

gene expression result in complex, aberrant phenotypes. Work with model organisms is revealing how genetic and epigenetic factors contribute to the phenotypes of complex diseases, and coupled with information from the Human Genome Project, we are entering a new and exciting era in human genetics research.

References

Cho, K.S., Elizondo, L.I., and Boerkoel, C.F. 2004. Advances in chromatin remodeling and human disease. *Curr. Opin. Genet. Develop.* 14: 308–315.

Gabellini, D., Green, M.R., and Tupler, R. 2004. When enough is enough: genetic diseases associated with transcription-al derepression. *Curr. Opin. Genet. Develop.* 14: 301–307.

Kisseljova, N.P., and Kisseljov F.L. 2005. DNA demethylation and carcinogenesis. *Biochemistry (Moscow)* 70: 743–752.

Laird, P.L. 2005. Cancer epigenetics. *Hum. Mol. Genet.* 14:R65–R76.

CHAPTER SUMMARY

1. Genetic regulation exists so that the complete genome is not continuously transcribed in every cell under all conditions.

2. The study of the *lac* operon in *E. coli* pioneered the study of gene regulation in bacteria. Genes involved in the metabolism of lactose are coordinately regulated by a negative control system that responds to the presence or absence of lactose.

3. The catabolite-activating protein (CAP) exerts positive control over *lac* gene expression. It does so by interacting with RNA polymerase at the *lac* promoter and by responding to the levels of cyclic AMP in the bacterial cell.

4. The biosynthesis of tryptophan in *E. coli* involves a number of enzymes that are encoded by genes that are present in an operon. In contrast to the inducible operon controlling lactose metabolism, the *trp* operon is repressible. In the presence of tryptophan, the repressor binds to the regulatory region of the *trp* operon and represses transcription initiation.

5. An additional regulatory step, referred to as attenuation, regulates the *trp* operon. The mechanism depends on alternative hairpin structures that form under different conditions in the leader sequence of the mRNA. Attenuation occurs when tryptophan is abundant and causes premature termination of transcription within a leader sequence that precedes the structural genes.

6. Gene regulation in eukaryotes can occur at any of the steps involved in gene expression, from chromatin modifications to transcription initiation and posttranscriptional processing.

7. Eukaryotic gene regulation at the level of chromatin may involve gene-specific chromatin remodeling, histone modifications, or DNA modifications. These modifications may either allow or inhibit access of promoters and enhancers to transcription factors, resulting in increased or decreased levels of transcription initiation.

8. Eukaryotic transcription is regulated at gene-specific promoter and enhancer elements. These *cis*-acting DNA sites may act constitutively or may be active in tissue- or temporal-specific ways.

9. Transcription factors influence transcription rates by binding to *cis*-acting regulatory sites within or adjacent to a gene promoter. They are thought to act by enhancing or repressing the association of basal transcription factors at the core promoter. They may also assist in chromatin remodeling.

10. Posttranscriptional gene regulation can involve alternative splicing of nascent RNA, RNA transport, or changes in mRNA stability. Alternative splicing increases the number of gene products encoded by a single gene. Changes in mRNA stability regulate the quantity of mRNA available for translation.

11. RNA silencing is a posttranscriptional mechanism of gene regulation that affects the translatability or stability of mRNA. It acts through the hybridization of small antisense RNAs to specific regions of an mRNA.

KEY TERMS

adenyl cyclase, 336
allosteric, 332
alternative splicing, 348
antiterminator hairpin, 338
attenuation, 338
attenuator, 338
basal transcription factors, 346
basic leucine zipper (bZIP), 345
β-galactosidase, 331
CAAT boxes, 342
catabolite-activating protein (CAP), 335
catabolite repression, 335
chromatin remodeling, 340
chromosome territory, 340
cis-acting site 331
coactivators 346
constitutive 330

constitutive mutants, 332
cooperative binding, 336
core promoter, 342
corepressor, 337
cyclic adenosine monophosphate (cAMP), 336
DNA-binding domain, 345
DNA methylation, 341
enhanceosome, 347
enhancers, 343
GC box, 342
gene silencing, 349
general transcription factors, 346
gratuitous inducers, 332
helix–turn–helix (HTH), 345
histone acetyltransferases, (HATs), 341
homeobox, 345

homeodomain, 345
inducer, 330
inducible, 330
interchromosomal compartments, 340
isopropylthiogalactoside (IPTG), 332
lac operon, 331
lactose, 331
leader sequence, 337
leucine zipper, 345
merozygote, 333
microRNAs (miRNAs), 350
negative control, 330
operator region, 332
operon model, 332
permease, 331
positive control, 330

INSIGHTS AND SOLUTIONS

1. A theoretical operon (*theo*) in *E. coli* contains several structural genes encoding enzymes that are involved sequentially in the biosynthesis of an amino acid. Unlike the *lac* operon, in which the repressor gene is separate from the operon, the gene encoding the regulator molecule is contained within the *theo* operon. When the end product (the amino acid) is present, it combines with the regulator molecule, and this complex binds to the operator, repressing the operon. In the absence of the amino acid, the regulatory molecule fails to bind to the operator, and transcription proceeds.

Characterize this operon, then consider the following mutations, as well as the situation in which the wild-type gene is present along with the mutant gene in partially diploid cells (F′):

(a) Mutation in the operator region.

(b) Mutation in the promoter region.

(c) Mutation in the regulator gene.

In each case, will the operon be active or inactive in transcription, assuming that the mutation affects the regulation of the *theo* operon? Compare each response with the equivalent situation of the *lac* operon.

Solution: The *theo* operon is repressible and under negative control. When there is no amino acid present in the medium (or the environment), the product of the regulatory gene cannot bind to the operator region, and transcription proceeds under the direction of RNA polymerase. The enzymes necessary for the synthesis of the amino acid are produced, as is the regulator molecule. If the amino acid *is* present, or is present after sufficient synthesis occurs, the amino acid binds to the regulator, forming a complex that interacts with the operator region, causing repression of transcription of the genes within the operon.

The *theo* operon is similar to the tryptophan system, except that the regulator gene is within the operon rather than separate from it. Therefore, in the *theo* operon, the regulator gene is itself regulated by the presence or absence of the amino acid.

(a) As in the *lac* operon, a mutation in the *theo* operator gene inhibits binding with the repressor complex, and transcription occurs constitutively. The presence of an F′ plasmid bearing the wild-type allele would have no effect, since it is not adjacent to the structural genes.

(b) A mutation in the *theo* promoter region would no doubt inhibit binding to RNA polymerase and therefore inhibit transcription. This would also happen in the *lac* operon. A wild-type allele present in an F′ plasmid would have no effect.

(c) A mutation in the *theo* regulator gene, as in the *lac* system, may inhibit either its binding to the repressor or its binding to the operator gene. In both cases, transcription will be constitutive, because the *theo* system is repressible. Both cases result in the failure of the regulator to bind to the operator, allowing transcription to proceed. In the *lac* system, failure to bind the corepressor lactose would permanently repress the system. The addition of a wild-type allele would restore repressibility, provided that this gene was transcribed constitutively.

2. Regulatory sites for eukaryotic genes are usually located within a few hundred nucleotides of the transcription start site, but they can be located up to several kilobases away. DNA sequence-specific binding assays have been used to detect and isolate protein factors present at low concentrations in nuclear extracts. In these experiments, short DNA molecules containing DNA-binding sequences are attached to material on a column, and nuclear extracts are passed over the column. The idea is that if proteins that specifically bind to the DNA sequence are present in the nuclear extract, they will bind to the DNA, and they can be recovered from the column after all other nonbinding material has been washed away. Once a DNA-binding protein has been isolated and identified, the problem is to devise a general method for screening cloned libraries for the genes encoding the DNA-binding factors. Determining the amino acid sequence of the protein and constructing synthetic oligonucleotide probes are time consuming and useful for only one factor at a time. Knowing the strong affinity for binding between the protein and its DNA-recognition sequence, how would you screen for genes encoding binding factors?

Solution: Several general strategies have been developed, and one of the most promising was devised by Steve McKnight's laboratory at the Fred Hutchinson Cancer Center. The cDNA isolated from cells expressing the binding factor is cloned into the lambda vector, gt11. Plaques of this library, containing proteins derived from expression of cDNA inserts, are adsorbed onto nitrocellulose filters and probed with double-stranded radioactive DNA corresponding to the binding site. If a fusion protein corresponding to the binding factor is present, it will bind to the DNA probe. After the unbound probe is washed off, the filter is subjected to autoradiography and the plaques corresponding to the DNA-binding proteins can be identified. An added advantage of this strategy is filter recycling by washing the bound DNA from the filters prior to their reuse. Such an ingenious procedure is similar to the colony-hybridization and plaque-hybridization procedures described for library screening in Chapter 17, and it provides a general method for isolating genes encoding DNA-binding factors.

■ PROBLEMS AND DISCUSSION QUESTIONS

1. Contrast the need for the enzymes involved in lactose and tryptophan metabolism in bacteria when lactose and tryptophan, respectively, are (a) present and (b) absent.
2. Contrast positive versus negative control of gene expression.
3. Contrast the role of the repressor in an inducible system and in a repressible system.
4. Even though the *lac* Z, Y, and A structural genes are transcribed as a single polycistronic mRNA, each gene contains the appropriate initiation and termination signals essential for translation. Predict what will happen when a cell growing in the presence of lactose contains a deletion of one nucleotide (a) early in the Z gene and (b) early in the A gene.
5. For the *lac* genotypes shown in the accompanying table, predict whether the structural gene (Z) is constitutive, permanently repressed, or inducible in the presence of lactose.

Genotype	Constitutive	Repressed	Inducible
$I^+O^+Z^+$			X
$I^-O^+Z^+$			
$I^+O^CZ^+$			
$I^-O^+Z^+/F'I^+$			
$I^+O^CZ^+/F'O^+$			
$I^SO^+Z^+$			
$I^SO^+Z^+/F'I^+$			

6. For the genotypes and conditions (lactose present or absent) shown in the accompanying table, predict whether functional enzymes, nonfunctional enzymes, or no enzymes are made.

Genotype	Condition	Functional Enzyme Made	Nonfunctional Enzyme Made	No Enzyme Made
$I^+O^+Z^+$	No lactose			X
$I^+O^CZ^+$	Lactose			
$I^-O^+Z^-$	No lactose			
$I^-O^+Z^-$	Lactose			
$I^-O^+Z^+/F'I^+$	No lactose			
$I^+O^CZ^+/F'O^+$	Lactose			
$I^+O^+Z^-/F'I^+O^+Z^+$	Lactose			
$I^-O^+Z^-/F'I^+O^+Z^+$	No lactose			
$I^SO^+Z^+/F'O^+$	No lactose			
$I^+O^CZ^+/F'O^+Z^+$	Lactose			

7. The locations of numerous *lacI⁻* and *lacIˢ* mutations have been determined within the DNA sequence of the *lacI* gene. Among these, *lacI⁻* mutations were found to occur in the 5'-upstream region of the gene, while *lacIˢ* mutations were found to occur farther downstream in the gene. Are the locations of the two types of mutations within the gene consistent with what is known about the function of the repressor that is the product of the *lacI* gene?

8. Explain why catabolite repression is used in regulating the *lac* operon and describe how it fine-tunes β-galactosidase synthesis.
9. Describe experiments that would confirm whether or not two transcription regulatory molecules act through the mechanism of cooperative binding.
10. Predict the level of genetic activity of the *lac* operon as well as the status of the *lac* repressor and the CAP protein under the cellular conditions listed in the accompanying table.

	Lactose	Glucose
(a)	−	−
(b)	+	−
(c)	−	+
(d)	+	+

11. Predict the effect on the inducibility of the *lac* operon of a mutation that disrupts the function of (a) the *crp* gene, which encodes the CAP protein, and (b) the CAP-binding site within the promoter.
12. Describe the role of attenuation in the regulation of tryptophan biosynthesis.
13. Attenuation of the *trp* operon was viewed as a relatively inefficient way to achieve genetic regulation when it was first discovered in the 1970s. Since then, however, attenuation has been found to be a relatively common regulatory strategy. Assuming that attenuation is a relatively inefficient way to achieve genetic regulation, what might explain its widespread use?
14. In a theoretical operon, genes A, B, C, and D represent the repressor gene, the promoter sequence, the operator gene, and the structural gene, *but not necessarily in that order*. This operon is concerned with the metabolism of a theoretic molecule (tm). From the data provided in the accompanying table, first decide whether the operon is inducible or repressible. Then, assign A, B, C, and D to the four parts of the operon. Explain your rationale. (AE = active enzyme; IE = inactive enzyme; NE = no enzyme)

Genotype	tm Present	tm Absent
$A^+B^+C^+D^+$	AE	NE
$A^-B^+C^+D^+$	AE	AE
$A^+B^-C^+D^+$	NE	NE
$A^+B^+C^-D^+$	IE	NE
$A^+B^+C^+D^-$	AE	AE
$A^-B^+C^+D^+/F'A^+B^+C^+D^+$	AE	AE
$A^+B^-C^+D^+/F'A^+B^+C^+D^+$	AE	NE
$A^+B^+C^-D^+/F'A^+B^+C^+D^+$	AE + IE	NE
$A^-B^+C^+D^-/F'A^+B^+C^+D^+$	AE	NE

15. A bacterial operon is responsible for the production of the biosynthetic enzymes needed to make the theoretical amino acid tisophane (tis). The operon is regulated by a separate gene, *R*, deletion of which causes the loss of enzyme synthesis. In the wild-type condition, when tis is present, no enzymes are made; in the absence of tis, the enzymes are made. Mutations in the operator gene (O^-) result in repression regardless of the presence of tis.

 Is the operon under positive or negative control? Propose a model for (a) repression of the genes in the presence of tis in wild-type cells and (b) the mutations.

16. A marine bacterium is isolated and is shown to contain an inducible operon whose genetic products metabolize oil when it is encountered in the environment. Investigation demonstrates that the operon is under positive control and that there is a *reg* gene whose product interacts with an operator region (*o*) to regulate the structural genes designated *sg*.

 In an attempt to understand how the operon functions, a constitutive mutant strain and several partial diploid strains were isolated and tested with the results shown here:

Host Chromosome	F′ Factor	Phenotype
wild type	none	inducible
wild type	*reg* gene from mutant strain	inducible
wild type	operon from mutant strain	constitutive
mutant strain	*reg* gene from wild type	constitutive

 Draw all possible conclusions about the mutation as well as the nature of regulation of the operon. Is the constitutive mutation in the *trans*-acting *reg* element or in the *cis*-acting *o* operator element?

17. The SOS repair genes in *E. coli* are negatively regulated by the *lexA* gene product, called the LexA repressor. When a cell sustains extensive damage to its DNA, the LexA repressor is inactivated by the *recA* gene product (RecA), and transcription of the SOS genes is increased dramatically.

 One of the SOS genes is the *uvrA* gene. You are studying the function of the *uvrA* gene product in DNA repair. You isolate a mutant strain that shows constitutive expression of the UvrA protein. You name this mutant strain *uvrA^C*. The following simple diagram shows the *lexA* and *uvrA* operons:

(a) Describe two different mutations that would result in a *uvrA* constitutive phenotype. Indicate the actual genotypes involved.

(b) Outline a series of genetic experiments, using partial diploid strains, that would allow you to determine which of the two possible mutations you have isolated.

18. A fellow student considers the issues in Problem 17 and argues that there is a more straightforward, nongenetic experiment that could differentiate between the two types of mutations. While the experiment requires no fancy genetics, you must be able to easily assay the products of the other SOS genes. Propose such an experiment.

19. Why is gene regulation assumed to be more complex in a multicellular eukaryote than in a prokaryote? Why is the study of this phenomenon in eukaryotes more difficult?

20. List and define the levels of gene regulation discussed in this chapter.

21. Distinguish between the regulatory elements referred to as promoters and enhancers.

22. Is the binding of a transcription factor to its DNA recognition sequence necessary and sufficient for an initiation of transcription at a regulated gene? What else plays a role in this process?

23. Write an essay comparing the control of gene regulation in eukaryotes and prokaryotes at the level of initiation of transcription. How do the regulatory mechanisms work? What are the similarities and differences in these two types of organisms regarding the specific components of the regulatory mechanisms? Address how the differences or similarities relate to the biological context of the control of gene expression.

24. In the autoregulation of tubulin synthesis, two models for the mechanism were proposed: (1) the tubulin subunits bind to the mRNA, and (2) the subunits interact with the nascent tubulin polypeptides. To distinguish between these two models, Cleveland and colleagues introduced mutations into the 13-base regulatory element of the β-tubulin gene. Some of the mutations resulted in amino acid substitution, while others did not. In addition, they shifted the reading frame of the intact 13-base sequence. Presented here are the results of a mutagenesis study of the mRNA:

Wild type	met AUG	arg AGG	glu GAA	lys ATC	*Autoregulation* +
Second	_____	UGG	_____		−
codon	_____	GGG	_____		−
mutations	_____	CGG	_____		+
	_____	AGA	_____		+
	_____	AGC	_____		−
Third	_____	_____	GAC	_____	+
codon	_____	_____	AAC	_____	−
mutations	_____	_____	UAU	_____	−
	_____	_____	UAC	_____	−

 Which of the two models is supported by the results? What experiments would you do to confirm this?

25. You are interested in studying transcription factors and have developed an *in vitro* transcription system by using a defined segment of DNA that is transcribed under the control of a eukaryotic promoter. The transcription of this DNA occurs when you add purified RNA polymerase II, TFIID (the TATA binding factor), and TFIIB and TFIIE (which bind to RNA polymerase). You perform a series of experiments that compare the efficiency of transcription in the "defined system" with the efficiency of transcription in a crude nuclear extract. You test the two systems with your template DNA and with

various deletion templates that you have generated. The following are the results of your study:

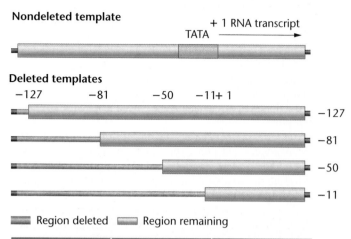

Nondeleted template

TATA + 1 RNA transcript

Deleted templates

−127 −81 −50 −11+ 1

−127

−81

−50

−11

■ Region deleted ▨ Region remaining

DNA added	Nuclear extract	Purified system
undeleted	++++	+
−127 deletion	++++	+
−81 deletion	++++	+
−50 deletion	+	+
−11 deletion	o	o

+ Low efficiency transcription
++++ High efficiency transcription
o No transcription

(a) Why is there no transcription from the −11 deletion template?

(b) How do the results for the nuclear extract and the defined system differ from the *undeleted* template? How would you interpret these results?

(c) For the various deleted templates, compare the results with the nuclear extract and the purified system. How would you interpret the results with the deleted templates? Be very specific about what you can conclude from these data.

26. While it is customary to consider transcriptional regulation in eukaryotes as resulting from the positive or negative influence of different factors binding to DNA, a more complex picture is emerging. For instance, Ducret and others (1999. *Mol. and Cell. Biol.* 19: 7076–7087) described the action of a transcriptional repressor (Net) that is regulated by nuclear export. Under neutral conditions, Net inhibits transcription of target genes; however, when phosphorylated, Net stimulates transcription of target genes. When stress conditions exist in a cell (ultraviolet light or heat shock), Net is excluded from the nucleus, and target genes are transcribed. Devise a model that includes diagrams that provide a consistent explanation for these three conditions.

27. DNA supercoiling occurs when coiling tension is generated ahead of the replication fork where it is relieved by DNA gyrase. Supercoiling may also be involved in transcription regulation. Liu and colleagues (2001. *Proc. Natl. Acad. Sci. [USA]* 98: 14,883–14,888) discovered that transcriptional enhancers operating over a long distance (2500 base pairs) are dependent on DNA supercoiling, while enhancers operating over shorter distances (110 base pairs) are not so dependent. Using a diagram, suggest a way in which supercoiling may positively influence enhancer activity over long distances.

28. DNA methylation is commonly associated with reduction of transcription. Irvine and colleagues (2002. *Mol. and Cell. Biol.* 22: 6689–6696) studied the impact of the location of DNA methylation relative to gene activity in human cells. The following data describe the relative expression of a reporter gene (luciferase) with variable DNA methylation outside and within the transcription region. What general conclusions can be drawn from these data?

DNA Segment	Patch Size of Methylation (kb)	Number of Methylated CpGs	Relative Luciferase Expression
Outside transcription unit (0–7.6 kb away)	0.0	0	490X
	2.0	100	290X
	3.1	102	250X
	12.1	593	2X
Inside transcription unit	0.0	0	490X
	1.9	108	80X
	2.4	134	5X
	12.1	593	2X

29. In some organisms, including mammals, there is an inverse relationship between the presence of 5-methylcytosine (m^5C) in CpG sequences and gene activity. In addition, m^5C may be involved in the recruitment of proteins that repress chromatin, thus segregating the genome into transcriptionally active and inactive regions. Overall, genomic DNA is relatively poor in CpG sequences due to the accumulation of m^5C compared to thymine transitions; however, unmethylated CpG islands are often associated with genes. Oakes and colleagues (2003. *Proc. Natl. Acad. Sci.* 100: 1775–1780) have determined that patterns of DNA methylation in spermatozoa vary with age in rats and suggest that such age-related alterations in DNA methylation may be one mechanism underlying age-related abnormalities in mammals. Consider the above information and provide an explanation that relates paternal age-related alterations in DNA methylation and birth abnormalities.

30. Philips and Cooper (2000. *Cell. Mol. Life Sci.* 57: 235–249) have estimated that approximately 15 percent of disease-causing mutations involve errors in alternative splicing. However, there is an interesting case where an exon deletion appears to be a phenomenon that enhances dystrophin production in muscle cells of DMD (Duchenne muscular dystrophy) patients. It turns out that a deletion of exon 45 is the most frequent DMD-causing mutation. But some individuals with Becker muscular dystrophy (BMD), a milder form of muscular dystrophy, have deletions in exons 45 *and* 46, a condition that was experimentally verified by van Deutekom and van Ommen (2003. *Nat. Rev. Genetics* 4: 774–783). Given that deletions often cause frameshift mutations, provide a possible explanation for the enhanced production of dystrophin in the presence of exon 45 *and* 46 deletions in BMD patients.

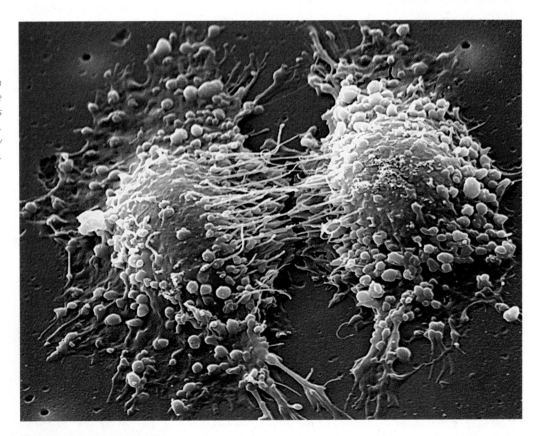

Cell-Cycle Regulation and Cancer

CHAPTER 16

Colored scanning electron micrograph of two prostate cancer cells in the final stages of cell division (cytokinesis). The cells are still joined by strands of cytoplasm.

CHAPTER CONCEPTS

- Cancer is a group of genetic diseases affecting fundamental aspects of cellular function, including DNA repair, the cell cycle, apoptosis, differentiation, cell migration and cell–cell contact.

- Most cancer-causing mutations occur in somatic cells; only about 1 percent of cancers have a hereditary component.

- Mutations in cancer-related genes lead to abnormal proliferation and loss of control over how cells spread and invade surrounding tissues.

- The development of cancer is a multistep process requiring mutations in genes controlling many aspects of cell proliferation and metastasis.

- Cancer cells show high levels of genomic instability, leading to the accumulation of multiple mutations in cancer-related genes.

- Mutations in proto-oncogenes and tumor suppressor genes contribute to the development of cancers.

- Oncogenic viruses introduce oncogenes into infected cells and stimulate cell proliferation.

- Environmental agents contribute to cancer by damaging DNA.

357

ancer is the second leading cause of death in Western countries, surpassed only by heart disease. It strikes people of all ages, and one out of three people will experience a cancer diagnosis sometime in his or her lifetime. Each year, more than 1 million cases of cancer are diagnosed in the United States and more than 500,000 people die from the disease (Table 16.1).

Over the last 30 years, scientists have discovered that cancer is a genetic disease, characterized by an interplay of mutant forms of oncogenes and tumor suppressor genes leading to the uncontrolled growth and spread of cancer cells. While some of these mutations may be inherited, as we will see, most mutations that lead to cancer occur in somatic cells that then divide and lead to tumors. Completion of the Human Genome Project is opening the door to a wealth of new information about the mutations that trigger a cell to become cancerous. This new understanding of cancer genetics is also leading to new gene-specific treatments, some of which are now entering clinical trials. Some scientists predict that gene therapies will replace chemotherapies within the next 25 years.

The goal of this chapter is to highlight our current understanding of the nature and causes of cancer. As we will see, cancer is a genetic disease that arises from mutations in genes controlling many basic aspects of cellular function. We will examine the relationship between genes and cancer, and consider how mutations, chromosomal changes, and environmental agents play roles in the development of cancer.

❓ How Do We Know?

In this chapter, we will focus on cancer as a genetic disease, with an emphasis on the relationship between cancer and DNA damage, as well as on the multiple genetic steps that lead to cancer. We also discuss how cancer cells show defects in cell-cycle regulation. We conclude with an investigation of the roles played by viruses and environmental agents in the development of cancer. As you study this topic, you should try to answer several fundamental questions:

1. How do we know that cancers arise from a single cell that contains mutations?

2. How do we know that cancer development requires more than one mutation?

3. How do we know that cancer cells contain defects in DNA repair?

4. How do we know that most cancers are not hereditary?

5. How do we know that some viruses contain genes that contribute to the development of cancer?

16.1 Cancer Is a Genetic Disease at the Level of Somatic Cells

Perhaps the most significant development in understanding the causes of cancer is the realization that cancer is a genetic disease. Genomic alterations that are associated with cancer range from single-nucleotide substitutions to large-scale chromosome rearrangements, amplifications, and deletions (Figure 16–1). However, unlike other genetic diseases, cancer is caused by mutations that occur predominantly in somatic cells. Only about 1 percent of cancers are associated with germ-line mutations that increase a person's susceptibility to certain types of cancer. Another important difference between cancer and other genetic diseases is that cancers rarely arise from a single mutation, but from the accumulation of many mutations—as many as six to ten. The mutations that lead to cancer affect multiple cellular functions, including repair of DNA damage, cell division, apoptosis, cellular differentiation, migratory behavior, and cell–cell contact.

What Is Cancer?

Clinically, cancer is defined as a large number of complex diseases, up to a hundred, that behave differently depending on the cell types from which they originate. Cancers vary in their ages of onset, growth rates, invasiveness, prognoses, and responsive-

(a)

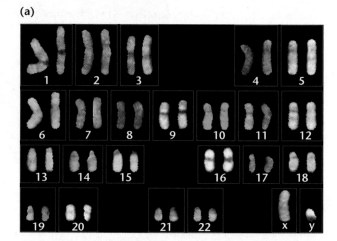

(b)

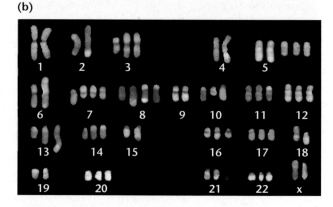

FIGURE 16–1 (a) Spectral karyotype of a normal cell. (b) Karyotype of a cancer cell showing translocations, deletions, and aneuploidy—characteristic features of cancer cells.

TABLE 16.1 Cancer Probabilities in the United States

Cancer Site	Gender	Birth to 39	Age 40–59	Age 60–79	Birth to Death
All sites	Male	1 in 62	1 in 12	1 in 3	1 in 2
	Female	1 in 52	1 in 11	1 in 4	1 in 3
Breast	Female	1 in 235	1 in 25	1 in 15	1 in 8
Prostate	Male	<1 in 10,000	1 in 53	1 in 7	1 in 6
Lung, bronchus	Male	1 in 3300	1 in 92	1 in 17	1 in 13
	Female	1 in 3180	1 in 120	1 in 25	1 in 17
Colon, rectum	Male	1 in 1500	1 in 124	1 in 29	1 in 18
	Female	1 in 1900	1 in 149	1 in 33	1 in 18

Source: American Cancer Society

ness to treatments. However, at the molecular level, all cancers exhibit common characteristics that unite them as a family.

All cancer cells share two fundamental properties: (1) abnormal cell growth and division (**cell proliferation**), and (2) abnormalities in the normal restraints that keep cells from spreading and invading other parts of the body (**metastasis**). In normal cells, these functions are tightly controlled by genes that are expressed appropriately in time and place. In cancer cells, these genes are either mutated or are expressed inappropriately.

It is this combination of uncontrolled cell proliferation and metastatic spread that makes cancer cells dangerous. When a cell simply loses genetic control over cell growth, it may grow into a multicellular mass, a **benign tumor**. Such a tumor can often be removed by surgery and may cause no serious harm. However, if cells in the tumor also acquire the ability to break loose, enter the bloodstream, invade other tissues, and form secondary tumors (**metastases**), they become malignant. **Malignant tumors** are difficult to treat and may become life threatening. As we will see later in the chapter, there are multiple steps and genetic mutations that convert a benign tumor into a dangerous malignant tumor.

The Clonal Origin of Cancer Cells

Although malignant tumors may contain billions of cells, and may invade and grow in numerous parts of the body, all cancer cells in the primary and secondary tumors are clonal, meaning that they originated from a common ancestral cell that accumulated numerous specific mutations. This is an important concept in understanding the molecular causes of cancer and has implications for its diagnosis.

Numerous data support the concept of cancer clonality. For example, reciprocal chromosomal translocations are characteristic of many cancers, including leukemias and lymphomas (two cancers involving white blood cells). Cancer cells from patients with **Burkitt's lymphoma** show reciprocal translocations between chromosome 8 (with translocation breakpoints at or near the *c-myc* gene) and chromosomes 2, 14, or 22 (with translocation breakpoints at or near one of the immunoglobulin genes). Each Burkitt's lymphoma patient exhibits unique break-

points in their *c-myc* and immunoglobulin gene DNA sequences; however, all lymphoma cells within that patient contain identical translocation breakpoints. This demonstrates that all cancer cells in each case of Burkitt's lymphoma arise from a single cell, and this cell passes on its genetic aberrations to its progeny.

Another demonstration that cancer cells are clonal is their pattern of X-chromosome inactivation. As explained in Chapter 5, female humans are mosaic, with some cells containing an inactivated paternal X chromosome and other cells containing an inactivated maternal X chromosome. X-chromosome inactivation occurs early in development and occurs at random. All cancer cells within a tumor, both primary and metastatic, within one female individual, contain the same inactivated X chromosome. This supports the concept that all the cancer cells in that patient arose from a common ancestral cell.

Cancer As a Multistep Process, Requiring Multiple Mutations

Although we know that cancer is a genetic disease initiated by mutations that lead to uncontrolled cell proliferation and metastasis, a single mutation is not sufficient to transform a normal cell into a tumor-forming (tumorigenic), malignant cell. If it were sufficient, then cancer would be far more prevalent than it is. In humans, mutations occur spontaneously at a rate of about 10^{-6} mutations per gene, per cell division, mainly due to the intrinsic error rates of DNA replication. As there are approximately 10^{16} cell divisions in a human body during a lifetime, a person might suffer up to 10^{10} mutations per gene somewhere in the body, during his or her lifetime. However, only about one person in three will suffer from cancer.

The phenomenon of age-related cancer is another indication that cancer develops from the accumulation of several mutagenic events in a single cell. The incidence of most cancers rises exponentially with age. If only a single mutation were sufficient to convert a normal cell to a malignant one, then cancer incidence would appear to be independent of age. The age-related incidence of cancer suggests that up to 10 independent mutations, occurring randomly and with a low probability, are necessary before a cell is transformed into a malignant cancer cell. Another indication that cancer is a multistep process is the delay that occurs between exposure to **carcinogens** (cancer-causing agents) and the appearance of the cancer. For example, an incubation period of five to eight years separated exposure of people to the radiation of the atomic explosions at Hiroshima and Nagasaki and the onset of leukemias.

The multistep nature of cancer development is supported by the observation that cancers often develop in progressive steps, from tumors containing mildly aberrant cells to those that are increasingly tumorigenic and malignant. This progressive nature of cancer is illustrated by the development of colon cancer, as discussed later in this chapter (see Section 16.6).

Each step in **tumorigenesis** (the development of a malignant tumor) appears to be the result of one or more genetic alterations that release the cells progressively from the controls that normally operate on proliferation and malignancy. As we will discover in subsequent sections of this chapter, the genes that undergo mutations leading to cancer (called oncogenes and tumor suppressor genes) are those that control DNA damage repair, the cell cycle, cell–cell contact, and programmed cell death.

We will now investigate each of these fundamental processes, the genes that control them and how mutations in these genes may lead to cancer.

16.2 Cancer Cells Contain Genetic Defects Affecting Genomic Stability and DNA Repair

Cancer cells show higher than normal rates of mutation, chromosomal abnormalities, and genomic instability. In fact, many researchers believe that the fundamental defect in cancer cells is a derangement of the cells' normal ability to repair DNA damage. This loss of genomic integrity leads to a general increase in the mutation rate in every gene, including specific genes that control aspects of cell proliferation, programmed cell death, and cell–cell contact. In turn, the accumulation of mutations in genes controlling these processes leads to cancer. The high level of genomic instability seen in cancer cells is known as the **mutator phenotype**.

Genomic instability in cancer cells manifests itself by the presence of gross defects such as translocations, aneuploidy, chromosome loss, DNA amplification, and chromosome deletions (Figures 16–1 and 16–2). Cancer cells that are grown in cultures in the lab also show a great deal of genomic instability—duplicating, losing, and translocating chromosomes or parts of chromosomes. Often cancer cells show specific chromosomal defects that are used to diagnose the type and stage of the cancer. For example, leukemic white blood cells from patients with **chronic myelogenous leukemia (CML)** bear a specific translocation, in which the *C-ABL* gene on chromosome 9 is translocated into the *BCR* gene on chromosome 22. This translocation creates a structure known as the **Philadelphia chromosome** (Figure 16–3). The *BCR-ABL* fusion gene codes for a chimeric BCR-ABL protein. The normal ABL protein is a **protein kinase** that acts within signal transduction pathways, transferring growth factor signals from the external environment to the nucleus. The BCR-ABL protein is an abnormal signal transduction molecule in

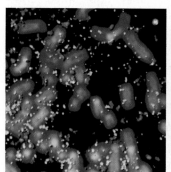
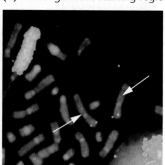

(a) Double minutes   **(b)** Heterogeneous staining region

FIGURE 16–2 DNA amplifications in neuroblastoma cells. (a) Two cancer genes (*MYCN* in red and *MDM2* in green) are amplified as small DNA fragments that remain separate from chromosomal DNA within the nucleus. These units of amplified DNA are known as double minute chromosomes. Normal chromosomes are stained blue. (b) Multiple copies of the *MYCN* gene are amplified within one large region called a heterogeneous staining region (green). Single copies of the *MYCN* gene are visible as green dots at the ends of the normal parental chromosomes (white arrows). Normal chromosomes are stained red.

CML cells, constantly stimulating these cells to proliferate even in the absence of external growth signals.

In keeping with the concept of the cancer mutator phenotype, a number of inherited cancers are caused by defects in genes that control DNA repair. For example, xeroderma pigmentosum (XP) is a rare hereditary disorder that is characterized by extreme sensitivity to ultraviolet light and other carcinogens. Patients with XP often develop skin cancer. Cells from patients with XP are defective in nucleotide excision repair, with mutations appearing in any one of seven genes whose products are necessary to carry out DNA repair. XP cells are impaired in their ability to repair DNA lesions such as thymine dimers induced by UV light. The

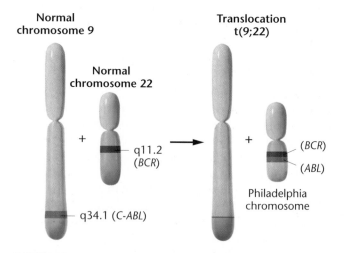

FIGURE 16–3 A reciprocal translocation involving the long arms of chromosomes 9 and 22 results in the formation of a characteristic chromosome, the Philadelphia chromosome, which is associated with chronic myelogenous leukemia (CML). The t(9;22) translocation results in the fusion of the *C-ABL* proto-oncogene on chromosome 9 with the *BCR* gene on chromosome 22. The fusion protein is a powerful hybrid molecule that allows cells to escape control of the cell cycle, contributing to the development of CML.

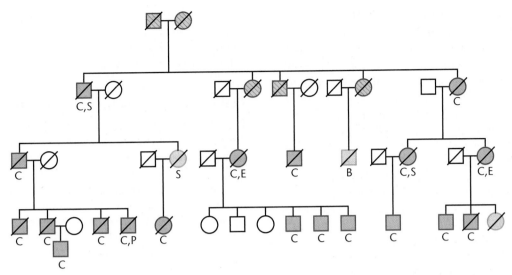

FIGURE 16–4 Pedigree of a family with HNPCC. Families with HNPCC are defined as those in which at least three relatives in two generations have been diagnosed with colon cancer, with one relative diagnosed at less than 50 years of age. Colon cancer: C; stomach cancer: S; endometrial cancer: E; pancreatic cancer: P; bladder/urinary cancer: B. Blue symbols indicate family members with colon cancer; diagonal stripes mean that diagnosis is uncertain; orange symbols indicate other tumors. Symbols with slashes indicate deceased individuals. Reprinted with permission from Aaltonen et al. Clues to the pathogenesis of familial colorectal cancer. *Science* 260: 812-816. Copyright 1993 AAAS.

relationship between XP and genes controlling nucleotide excision repair is also described in Chapter 14.

Another hereditary cancer, **hereditary nonpolyposis colorectal cancer (HNPCC)**, is also caused by mutations in genes controlling DNA repair. HNPCC is an autosomal dominant syndrome, affecting about one in every 200 people (Figure 16–4). Patients affected by HNPCC have an increased risk of developing colon, ovary, uterine, and kidney cancers. Cells from patients with HNPCC show higher than normal mutation rates and genomic instability. At least eight genes are associated with HNPCC, and four of these genes control aspects of DNA mismatch repair. Inactivation of any of these four genes—*MSH2*, *MSH6*, *MLH1*, and *MLH3*—causes a rapid accumulation of genome-wide mutations and the subsequent development of colorectal and other cancers.

The observation that hereditary defects in genes controlling nucleotide excision repair and DNA mismatch repair lead to high rates of cancer lends support to the idea that the mutator phenotype is a significant contributor to the development of cancer.

Now Solve This

Problem 17 on page 374 asks you to consider how the BCR-ABL hybrid fusion protein, found in CML leukemic white blood cells, could be used as a target for cancer therapy.

Hint: Most cancer therapies, including radiation and chemotherapies, aim to kill cells that are constantly dividing. However, many normal cells in the body also divide and are killed by cancer therapies, leading to side effects. The BCR-ABL fusion protein is found only in CML white blood cells. As you try to answer this problem, you may wish to learn more about the drug Gleevec (see http://www.nci.nih.gov/newscenter/qandagleevec).

16.3 Cancer Cells Contain Genetic Defects Affecting Cell-Cycle Regulation

One of the fundamental aberrations in all cancer cells is a loss of control over cell proliferation. Cell proliferation is the process of cell growth and division that is essential for all development and tissue repair in multicellular organisms. Although some cells, such as epidermal cells of the skin or blood-forming cells in the bone marrow, continue to grow and divide throughout the organism's lifetime, most cells in adult multicellular organisms are in a nondividing, quiescent, and differentiated state. Differentiated cells are those that are specialized for a specific function, such as photoreceptor cells of the retina or muscle cells of the heart. The most extreme examples of nonproliferating cells are nerve cells, which divide little, if at all, even to replace damaged tissue. In contrast, many differentiated cells, such as those in the liver and kidney, are able to grow and divide when stimulated by extracellular signals and growth factors. In this way, multicellular organisms are able to replace dead and damaged tissue. However, the growth and differentiation of cells must be strictly regulated; otherwise, the integrity of organs and tissues would be compromised by inappropriate types and quantities of cells. Normal regulation over cell proliferation involves a large number of gene products that control steps in the cell cycle, programmed cell death, and the response of cells to external growth signals. In cancer cells, many of the genes that control these functions are mutated or aberrantly expressed, leading to uncontrolled cell proliferation.

In this section, we will review steps in the cell cycle, some of the genes that control the cell cycle, and how these genes, when mutated, lead to cancer.

The Cell Cycle and Signal Transduction

The cellular events that occur from one cell division to the next comprise the cell cycle (Figure 16–5). The interphase stage of the cell cycle is the interval between mitotic divisions. During this time, the cell grows and replicates its DNA. During G1, the cell prepares for DNA synthesis by accumulating the enzymes and molecules required for DNA replication. G1 is followed by S phase, during which the cell's chromosomal DNA is replicated. During G2, the cell continues to grow and prepare for division. During M phase, the duplicated chromosomes condense, sister chromosomes separate to opposite poles, and the cell divides in two.

In early to mid-G1, the cell makes a decision either to enter the next cell cycle or to withdraw from the cell cycle into quiescence. Continuously dividing cells do not exit the cell cycle but proceed through G1, S, G2, and M phases until they receive signals to stop growing. If the cell stops proliferating, it enters the G0 phase of the cell cycle. During G0, the cell remains metabolically active but does not grow or divide. Most differentiated cells in multicellular organisms can remain in this G0 phase indefinitely. Some, such as neurons, never reenter the cell cycle. In contrast, cancer cells are unable to enter G0 and instead, they continuously cycle. Their rate of proliferation is not necessarily any greater than a normal proliferating cell; however, they are not able to become quiescent at the appropriate time or place.

Cells in G0 can often be stimulated to reenter the cell cycle by external growth signals. These signals are delivered to the cell by molecules such as growth factors and hormones that bind to cell-surface receptors, which then relay the signal from the plasma membrane to **signal transduction** molecules located in the cytoplasm. Ultimately, signal transduction initiates a program of gene expression that propels the cell out of G0 back into the cell cycle. Cancer cells often have defects in signal transduction pathways. Sometimes, abnormal signal transduction molecules send continuous growth signals to the nucleus even in the absence of external growth signals. In addition, malignant cells may not respond to external signals from surrounding cells—signals that would normally inhibit cell proliferation within a mature tissue.

Cell-Cycle Control and Checkpoints

In normal cells, progress through the cell cycle is tightly regulated, and the completion of each step is necessary prior to initiating the next step. There are at least three distinct points in the cell cycle at which the cell monitors external signals and internal equilibrium before proceeding to the next stage of the cell cycle. These are the **G1/S**, the **G2/M**, and **M checkpoints** (Figure 16–5). At the G1/S checkpoint, the cell monitors its size and determines if its DNA has been damaged. If the cell has not achieved an adequate size, or if the DNA has been damaged, further progress through the cell cycle is halted until these conditions are corrected. If cell size and DNA integrity are normal, the G1/S checkpoint is traversed, and the cell proceeds to S phase. The second important checkpoint is the G2/M checkpoint, where physiological conditions in the cell are monitored prior to entering mitosis. If DNA replication or repair of any DNA damage has not been completed, the cell cycle arrests until these processes are complete. The third major checkpoint occurs during mitosis and is called the M checkpoint. At this checkpoint, both the successful formation of the spindle-fiber system and the attachment of spindle fibers to the kinetochores associated with the centromeres are monitored. If spindle fibers are not properly formed or attachment is inadequate, mitosis is arrested.

In addition to regulating the cell cycle at checkpoints, the cell controls progress through the cell cycle through two classes of proteins: **cyclins** and **cyclin-dependent kinases (CDKs)**. The cell synthesizes and destroys cyclin proteins in a precise pattern during the cell cycle (Figure 16–6). When a cyclin is present, it binds to a specific CDK, triggering activity of the CDK/cyclin complex. The CDK/cyclin complexes then selectively phosphorylate and activate other proteins that in turn bring about the changes necessary to advance the cell through the cell cycle. For example, in G1 phase, CDK4/cyclin D complexes activate proteins that stimulate transcription of genes whose products (such as DNA polymerase δ and DNA ligase) are required for DNA replication during S phase. Another CDK/cyclin complex, CDK1/cyclin B, phosphorylates a number of proteins that bring about the events of early mitosis, such as nuclear membrane breakdown, chromosome condensation and cytoskeletal reorganization. Mitosis can only be completed, however, when cyclin B is degraded and the protein phosphorylations characteristic of M phase are reversed. Although a large number of different protein kinases exist in cells, only a few are involved in cell-cycle regulation.

Both cell-cycle checkpoints and cell-cycle control molecules are genetically regulated. In general, the cell cycle is regulated by an interplay of genes whose products either promote or suppress cell division. Mutation or mis-expression of any of the genes controlling the cell cycle contributes to the

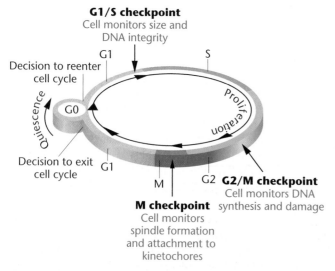

G1/S checkpoint
Cell monitors size and
DNA integrity

Decision to reenter
cell cycle

Quiescence

G0

Decision to exit
cell cycle

G1

S

Proliferation

M

G2 **G2/M checkpoint**
Cell monitors DNA
synthesis and damage

M checkpoint
Cell monitors
spindle formation
and attachment to
kinetochores

FIGURE 16–5 Checkpoints and proliferation decision points monitor the progress of the cell through the cell cycle.

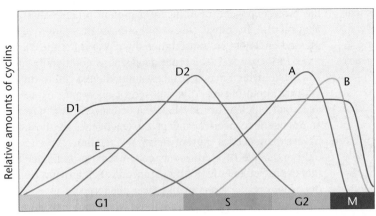

FIGURE 16–6 Relative expression times and amounts of cyclins during the cell cycle. Cyclin D1 accumulates early in G1 and is expressed at a constant level through most of the cycle. Cyclin E accumulates in G1, reaches a peak, and declines by mid-S phase. Cyclin D2 begins accumulating in the last half of G1, reaches a peak just after the beginning of S, and then declines by early G2. Cyclin A appears in late G1, accumulates through S phase, peaks at the G2/M transition, and is rapidly degraded. Cyclin B peaks at the G2/M transition and declines rapidly in M phase.

development of cancer in several ways. For example, if genes that control the G1/S or G2/M checkpoints are mutated, the cell may continue to cycle before repairing DNA damage. This may lead to the accumulation of mutations in genes whose products control cell proliferation or metastasis. Similarly, if genes that control progress through the cell cycle, such as those that encode the cyclins, are expressed inappropriately, the cell may cycle continuously and may be unable to exit the cell cycle into G0.

As already described, if DNA replication, repair, or chromosome assembly is aberrant, normal cells halt their progress through the cell cycle until the condition is corrected. This reduces the number of mutations and chromosomal abnormalities that accumulate in normal proliferating cells. However, if DNA or chromosomal damage is so severe that repair is impossible, the cell may initiate a second line of defense—a

process called **apoptosis**, or **programmed cell death**. Apoptosis is a genetically controlled process whereby the cell commits suicide. Apoptosis is also initiated during normal multicellular development in order to eliminate certain cells that do not contribute to the final adult organism. The steps in apoptosis are the same for damaged cells and for cells eliminated during development: Nuclear DNA becomes fragmented, internal cellular structures are disrupted, and the cell dissolves into small spherical structures known as apoptotic bodies (Figure 16–7). In the final step, the apoptotic bodies are engulfed by the immune system's phagocytic cells. A series of proteases called **caspases** are responsible for initiating apoptosis and for digesting intracellular components. Apoptosis is genetically controlled in that regulation of specific gene products such as the Bcl2 and BAX proteins can trigger or prevent apoptosis. By removing damaged cells,

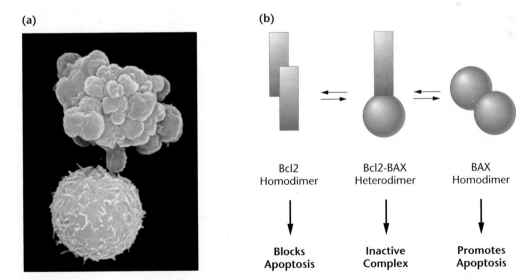

FIGURE 16–7 (a) A normal white blood cell (bottom) and a white blood cell undergoing apoptosis (top). Apoptotic bodies appear as grape-like clusters on the cell surface. (b) The relative concentrations of the Bcl2 and BAX proteins regulate apoptosis. A normal cell contains a balance of Bcl2 and BAX, which form inactive heterodimers. A relative excess of Bcl2 results in Bcl2 homodimers, which prevent apoptosis. Cancer cells with Bcl2 overexpression are resistant to chemotherapies and radiation therapies. A relative excess of BAX results in BAX homodimers, which induce apoptosis. In normal cells, activated p53 protein induces transcription of BAX and inhibits transcription of Bcl2, leading to cell death. In many cancer cells, p53 is defective, preventing the apoptotic pathway from removing the cancer cells.

programmed cell death reduces the number of mutations that are propagated to the next generation, including those in cancer-causing genes. The same genes that regulate cell-cycle checkpoints can trigger apoptosis. As we will see, these genes are mutated in a many cancers.

16.4 Many Cancer-Causing Genes Disrupt Control of the Cell Cycle

Two general categories of genes are mutated or mis-expressed in cancer cells—the proto-oncogenes and the tumor suppressor genes (Table 16.2). **Proto-oncogenes** are genes whose products are important for normal cell functions, especially cell growth and division. They do this by encoding transcription factors that stimulate expression of other genes, signal tranduction molecules that stimulate cell division, or cell-cycle regulators that move the cell through the cell cycle. When normal cells become quiescent and cease division, they repress the expression or activity of most proto-oncogene products. In cancer cells, one or more proto-oncogenes are altered in such a way that their activities cannot be controlled in a normal fashion. This is sometimes due to a mutation in

the proto-oncogene resulting in a protein product that acts abnormally. In other cases, proto-oncogenes are overexpressed or cannot be transcriptionally repressed at the correct time. In these cases, the proto-oncogene is continually in an "on" state, which may constantly stimulate the cell to divide. When a proto-oncogene is mutated or aberrantly expressed and contributes to the development of cancer, it is known as an **oncogene** (cancer-causing gene). Oncogenes are those that have experienced a gain-of-function alteration. As a result, only one allele of a proto-oncogene needs to be mutated or mis-expressed in order to trigger uncontrolled growth. Hence, oncogenes confer a dominant cancer phenotype.

Tumor suppressor genes are those whose products normally regulate cell-cycle checkpoints and initiate the process of apoptosis. In normal cells, proteins encoded by tumor suppressor genes halt progress through the cell cycle in response to DNA damage or growth-suppression signals from the extracellular environment. When tumor suppressor genes are mutated or inactivated, cells are unable to respond normally to cell-cycle checkpoints, or are unable to undergo programmed cell death if DNA damage is extensive. This leads to a further increase in mutations and to the inability of the cell to leave the cell cycle when it should become quiescent.

TABLE 16.2	Some Proto-oncogenes and Tumor Suppressor Genes		
Proto-oncogene	**Normal Function**	**Alteration in Cancer**	**Associated Cancers**
Ha-ras	Signal transduction molecule, binds GTP/GDP	Point mutations	Colorectal, bladder, many types
c-erbB	Transmembrane growth factor receptor	Gene amplification, point mutations	Glioblastomas, breast cancer, cervix
c-myc	Transcription factor, regulates cell cycle, differentiation, apoptosis	Translocation, amplification, point mutations	Lymphomas, leukemias, lung cancer, many types
c-kit	Tyrosine kinase, signal transduction	Mutation	Sarcomas
RARα	Hormone-dependent transcription factor, differentiation	Chromosomal translocations with PML gene, fusion product	Acute promyelocytic leukemia
E6	Human papillomavirus encoded oncogene, inactivates p53	HPV infection	Cervical cancer
Cyclins	Bind to CDKs, regulate cell cycle	Gene amplification, overexpression	Lung, esophagus, many types
CDK2, 4	Cyclin-dependent kinases, regulate cell-cycle phases	Overexpression, mutation	Bladder, breast, many types
Tumor Suppressor	**Normal Function**	**Alteration in Cancer**	**Associated Cancers**
p53	Cell-cycle checkpoints, apoptosis	Mutation, inactivation by viral oncogene products	Brain, lung, colorectal, breast, many types
RB1	Cell-cycle checkpoints, binds E2F	Mutation, deletion, inactivation by viral oncogene products	Retinoblastoma, osteosarcoma, many types
APC	Cell–cell interaction	Mutation	Colorectal cancers, brain, thyroid
Bcl2	Apoptosis regulation	Overexpression blocks apoptosis	Lymphomas, leukemias
BRCA2	DNA repair	Point mutations	Breast, ovarian, prostate cancers

When both alleles of a tumor suppressor gene are inactivated, and other changes in the cell keep it growing and dividing, cells may become tumorigenic.

The following are examples of proto-oncogenes and tumor suppressor genes that contribute to cancer when mutated. There are more than 200 oncogenes and tumor suppressor genes now known, and more will likely be discovered as cancer research continues.

The *ras* Proto-oncogenes

Some of the most frequently mutated genes in human tumors are those of the **ras gene family**. These genes are mutated in more than 40 percent of human tumors. The *ras* gene family encodes signal transduction molecules that are associated with the cell membrane and regulate cell growth and division. Ras proteins normally transmit signals from the cell membrane to the nucleus, stimulating the cell to divide in response to external growth factors. Ras proteins alternate between an inactive (switched off) and an active (switched on) state by binding either guanosine diphosphate (GDP) or guanosine triphosphate (GTP). When a cell encounters a growth factor (such as platelet-derived growth factor or epidermal growth factor), growth factor receptors on the cell membrane bind to the growth factor, resulting in autophosphorylation of the cytoplasmic portion of the growth factor receptor. This causes recruitment of proteins known as nucleotide exchange factors to the plasma membrane. These nucleotide exchange factors cause Ras to release GDP and bind GTP, thereby activating Ras. The active, GTP-bound form of Ras then sends its signals through cascades of protein phosphorylations in the cytoplasm (Figure 16–8). The end-point of these cascades is activation of nuclear transcription factors that stimulate expression of genes whose products drive the cell from quiescence into the cell cycle. Once Ras has sent its signals to the nucleus, it hydrolyzes GTP to GDP and becomes inactive. Mutations that convert the proto-oncogene *ras* to an oncogene prevent the Ras protein from hydrolyzing GTP to GDP and hence freeze the Ras protein into its "on" conformation, constantly stimulating the cell to divide.

The *p53* Tumor Suppressor Gene

The most frequently mutated gene in human cancers—occurring in more than 50 percent of all cancers—is the **p53 gene**. This gene encodes a nuclear protein that acts as a transcription factor that represses or stimulates transcription of more than 50 different genes.

Normally, the p53 protein is continuously synthesized but is rapidly degraded and therefore is present in cells at low levels. In addition, the p53 protein is normally bound to another protein called Mdm2, which prevents the phosphorylations and acetylations that convert the p53 protein from an inactive to an active form. Several types of events cause a rapid increase in the nuclear levels of activated p53 protein. These include chemical damage to DNA, double-stranded breaks in DNA induced by ionizing radiation, or the presence of DNA-repair intermediates generated by exposure of cells to ultraviolet light. Increases in the levels of activated p53 protein result

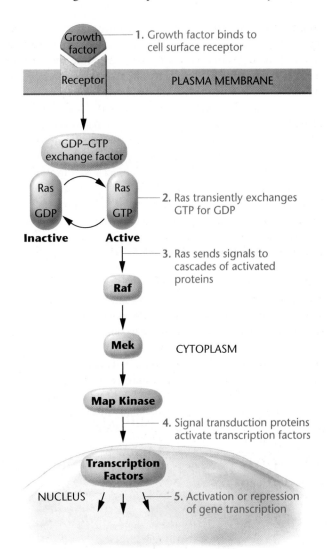

FIGURE 16–8 A signal transduction pathway mediated by Ras.

from increases in protein phosphorylation, acetylation, and p53 protein stability.

The p53 protein initiates two different responses to DNA damage: arrest of the cell cycle followed by DNA repair, or apoptosis and cell death if DNA cannot be repaired. Each of these responses is accomplished by p53 acting as a transcription factor that stimulates or represses the expression of genes involved in each of these responses.

In normal cells, p53 can arrest the cell cycle at several phases. To arrest the cell cycle at the G1/S checkpoint, activated p53 protein stimulates transcription of a gene encoding the p21 protein. The p21 protein inhibits the CDK4/cyclin D1 complex, hence preventing the cell from moving from the G1 phase into S phase. Activated p53 protein also regulates expression of genes that retard the progress of DNA replication, thus allowing time for DNA damage to be repaired during S phase. By regulating expression of other genes, activated p53 can block cells at the G2/M checkpoint, if DNA damage occurs during S phase.

Activated p53 can also instruct a damaged cell to commit suicide by apoptosis. It does so by activating the transcription of the *Bax* gene and repressing transcription of the *Bcl2* gene. In normal cells, the BAX protein is present in a heterodimer with the Bcl2 protein and the cell remains viable. When the levels of BAX protein increase in response to p53 stimulation of *Bax* gene transcription, BAX homodimers are formed and these homodimers activate the cellular changes that lead to cellular self-destruction (Figure 16–7). In cancer cells that lack functional p53, BAX protein levels do not increase in response to cellular damage and apoptosis may not occur.

Hence, cells lacking functional p53 are unable to arrest at cell-cycle checkpoints or to enter apoptosis in response to DNA damage. As a result, they move unchecked through the cell cycle, regardless of the condition of the cell's DNA. Cells lacking p53 have high mutation rates and accumulate the types of mutations that lead to cancer. Because of the importance of the *p53* gene to genomic integrity, it is often referred to as the "guardian of the genome."

The *RB1* Tumor Suppressor Gene

The loss or mutation of the **RB1** (**retinoblastoma 1**) tumor suppressor gene contributes to the development of many cancers, including those of the breast, bone, lung, and bladder. The *RB1* gene was originally identified as a result of studies on **retinoblastoma**, an inherited disorder in which tumors develop in the eyes of young children. Retinoblastoma occurs with a frequency of about 1 in 14,000 to 20,000 individuals. In the familial form of the disease, individuals inherit one mutated allele of the *RB1* gene and have an 85 percent chance of devel-

oping retinoblastomas as well as an increased chance of developing other cancers. All somatic cells of patients with hereditary retinoblastoma contain one mutated allele of the *RB1* gene. However, it is only when the second normal allele of the *RB1* gene is lost or mutated in certain retinal cells that retinoblastoma develops. In individuals that do not have this hereditary condition, retinoblastoma is extremely rare, as it requires at least two separate somatic mutations in a retinal cell in order to inactivate both copies of the *RB1* gene (Figure 16–9).

The **retinoblastoma protein (pRB)** is a tumor suppressor protein that controls the G1/S cell-cycle checkpoint. The pRB protein is found in the nuclei of all cell types at all stages of the cell cycle. However, its activity varies throughout the cell cycle, depending on its phosphorylation state. When cells are in the G0 phase of the cell cycle, the pRB protein is nonphosphorylated and binds to transcription factors such as E2F, inactivating them (Figure 16–10). When the cell is stimulated by growth factors, it enters G1 and approaches S phase. Throughout the G1 phase, the pRB protein becomes phosphorylated by the CDK4/cyclin D1 complex. Phosphorylated pRB is inactive and releases its bound regulatory proteins. When E2F and other regulators are released by pRB, they are free to induce the expression of over 30 genes whose products are required for the transition from G1 into S phase. After cells traverse S, G2, and M phases, pRB reverts to a nonphosphorylated state, binds to regulatory proteins such as E2F, and keeps them

(a) Familial retinoblastoma

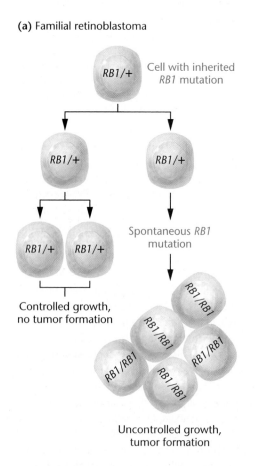

Cell with inherited *RB1* mutation

Spontaneous *RB1* mutation

Controlled growth, no tumor formation

Uncontrolled growth, tumor formation

(b) Sporadic retinoblastoma

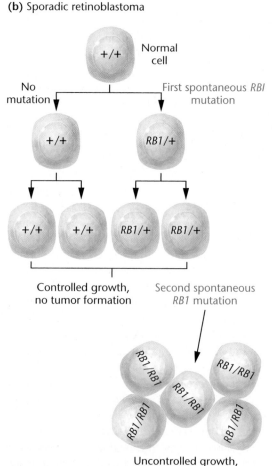

Normal cell

No mutation   First spontaneous *RB1* mutation

Controlled growth, no tumor formation   Second spontaneous *RB1* mutation

Uncontrolled growth, tumor formation

FIGURE 16–9 (a) In familial retinoblastoma, one mutation (designated as *RB1*) is inherited and present in all cells. A second mutation at the retinoblastoma locus in any retinal cell contributes to uncontrolled cell growth and tumor formation. (b) In sporadic retinoblastoma, two independent mutations in both alleles of the retinoblastoma gene within a single cell are acquired sequentially, also leading to tumor formation.

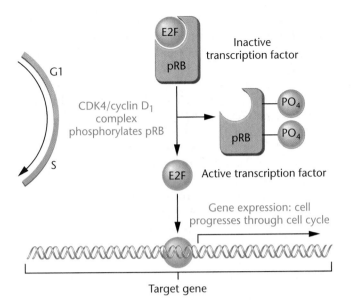

FIGURE 16–10 In the nucleus during G0 and early G1, pRB interacts with and inactivates transcription factor E2F. As the cell moves from G1 to S phase, a CDK4/cyclinD1 complex forms and adds phosphate groups to pRB. As pRB becomes phosphorylated, E2F is released and becomes transcriptionally active, allowing the cell to pass through S phase. Phosphorylation of pRB is transitory; as CDK/cyclin complexes are degraded and the cell moves through the cell cycle to early G1, pRB phosphorylation declines.

sequestered until required for the next cell cycle. In normal quiescent cells, the pRB protein is active and prevents passage into S phase. In many cancer cells, including retinoblastoma cells, both copies of the *RB1* gene are defective, inactive, or absent, and progression through the cell cycle is not regulated.

Now Solve This

Problem 23 on page 374 asks you to explain how in the inherited Li–Fraumeni syndrome, mutations in one allele of the *p53* gene can give rise to a wide variety of different cancers.

Hint: To answer this question, you might review the cellular functions regulated by the normal *p53* tumor suppressor gene. Consider how each of these functions could be affected if the *p53* gene product is either absent or defective. To understand how mutations in one allele of a tumor suppressor gene result in tumors that contain aberrations in both alleles, read about loss of heterozygosity in Section 16.6.

16.5 Cancer Is a Genetic Disorder Affecting Cell Adhesion

As discussed at the beginning of this chapter, uncontrolled growth alone is insufficient to create a malignant and life-threatening cancer. Cancer cells must also acquire the features of metastasis, which include the ability to disengage from the original tumor site, to enter the blood or lymphatic system, to

invade surrounding tissues, and to develop into secondary tumors. In order to leave the site of the primary tumor and invade other tissues, tumor cells must dissociate from other cells and digest components of the **extracellular matrix** and **basal lamina**, which normally contain and separate tissues. The extracellular matrix and basal lamina are composed of proteins and carbohydrates. They form the scaffold for tissue growth and normally inhibit the migration of cells.

The ability to invade the extracellular matrix is also a property of some normal cell types. For example, implantation of the embryo in the uterine wall during pregnancy requires cell migration across the extracellular matrix. In addition, white blood cells reach the site of infection by penetrating capillary walls. The mechanisms of invasion are probably similar in these normal cells and in cancer cells. The difference is that, in normal cells, the invasive ability is tightly regulated, whereas in tumor cells, this regulation has been lost.

Although less is known about the genes that control metastasis than about those controlling the cell cycle, it is likely that metastasis is controlled by a large number of genes, including those that encode **cell-adhesion molecules**, cytoskeleton regulators, and proteolytic enzymes. For example, epithelial tumors have a lower than normal level of the E-cadherin glycoprotein, which is responsible for cell–cell adhesion in normal tissues. Also, proteolytic enzymes such as **metalloproteinases** are present at higher than normal levels in highly malignant tumors and are not susceptible to the normal controls conferred by regulatory molecules such as **tissue inhibitors of metalloproteinases (TIMPs)**. It has been shown that the level of aggressiveness of a tumor correlates positively with the levels of proteolytic enzymes expressed by the tumor. Hence, inappropriately expressed cell adhesion and proteinase enzymes may assist malignant tumor cells by loosening the normal constraints on cell location and creating holes through which the tumor cells can pass into and out of the circulatory system.

Now Solve This

Problem 25 on page 374 describes the isolation of a gene that is mutated in metastatic tumors. The gene appears to be a member of the serine protease family. You are asked to conjecture how this gene might contribute to the development of highly invasive cancers.

Hint: As you learned in this section, metastatic cancer cells have the ability to escape their adhesions to other cells, travel through the circulatory system, and invade distant tissues. You might think about what types of metastatic processes are facilitated by having a protease molecule that lacks normal regulation. Also consider what types of mutations in this gene could lead to increased metastasis.

16.6 Predisposition to Some Cancers Can Be Inherited

Although the vast majority of human cancers are sporadic, a small fraction (1–2 percent) have an hereditary or familial component. At present, about 50 forms of hereditary cancer are known (Table 16.3).

TABLE 16.3	Inherited Predispositions to Cancer
Tumor Predisposition Syndromes	**Chromosome**
Early-onset familial breast cancer	17q
Familial adenomatous polyposis	5q
Familial melanoma	9p
Gorlin syndrome	9q
Hereditary nonpolyposis colon cancer	2p
Li-Fraumeni syndrome	17p
Multiple endocrine neoplasia, type 1	11q
Multiple endocrine neoplasia, type 2	22q
Neurofibromatosis, type 1	17q
Neurofibromatosis, type 2	22q
Retinoblastoma	13q
Von Hippel-Lindau syndrome	3p
Wilms tumor	11p

Most inherited cancer-susceptibility genes, though transmitted in a Mendelian dominant fashion, are not sufficient in themselves to trigger development of a cancer. At least one other somatic mutation in the other copy of the gene must occur in order to drive a cell toward tumorigenesis. In addition, mutations in other genes are usually necessary to fully express the cancer phenotype. As mentioned above, inherited mutations in the *RB1* gene predispose individuals to developing various cancers. Although the normal somatic cells of these patients are heterozygous for the *RB1* mutation, cells within their tumors contain mutations in both copies of the gene. The phenomenon whereby the second, wild-type, allele is mutated in a tumor is known as **loss of heterozygosity**. Although loss of heterozygosity is an essential first step in expression of these inherited cancers, further mutations in other proto-oncogenes and tumor suppressor genes are necessary for the passage of tumor cells to full malignancy.

The development of hereditary colon cancer illustrates how inherited mutations in one allele of a gene contribute only one step in the multistep pathway leading to malignancy.

About 1 percent of colon cancer cases result from a genetic predisposition to cancer known as **familial adenomatous polyposis (FAP)**. In FAP, individuals inherit one mutant copy of the *APC* (adenomatous polyposis) gene located on the long arm of chromosome 5. Mutations include deletions, frameshift, and point mutations. The normal function of the *APC* gene product is to act as a tumor suppressor controlling cell–cell contact and growth inhibition by interacting with the β-catenin protein. The presence of a heterozygous *APC* mutation causes the epithelial cells of the colon to partially escape cell-cycle control, and the cells divide to form small clusters of cells called **polyps** or adenomas. People who are heterozygous for this condition develop hundreds to thousands of colon and rectal polyps early in life. Although it is not necessary for the second allele of the *APC* gene to be mutated in polyps at this stage, in the majority of cases, the second *APC* allele becomes mutant in a later stage of cancer development. The relative order of mutations in the development of FAP is shown in Figure 16–11.

The second mutation in polyp cells that contain an *APC* gene mutation occurs in the *ras* proto-oncogene. The combined *APC* and *ras* gene mutations bring about the development of intermediate adenomas. Cells within these adenomas have defects in normal cell differentiation and will grow in culture, in the absence of contact with other cells and hence are described as **transformed**. The third step toward malignancy requires loss of function of both alleles of the *DCC* (*d*eleted in *c*olon *c*ancer) gene. The *DCC* gene product is thought to be involved with cell adhesion and differentiation. Mutations in both *DCC* alleles result in the formation of late-stage adenomas with a number of finger-like outgrowths (villi). When late adenomas progress to cancerous adenomas, they usually suffer loss of functional *p53* genes. The final steps toward malignancy involve mutations in an unknown number of genes associated with metastasis.

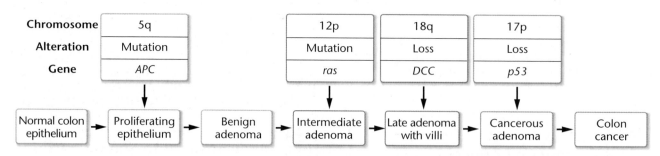

FIGURE 16–11 A model for the multistep development of colon cancer. The first step is the loss or inactivation of one allele of the *APC* gene on chromosome 5. In FAP cases, one mutant *APC* allele is inherited. Subsequent mutations involving genes on chromosomes 12, 17, and 18 in cells of benign adenomas can lead to a malignant transformation that results in colon cancer. Although the mutations on chromosomes 12, 17, and 18 usually occur at a later stage than those involving chromosome 5, the sum of changes is more important than the order in which they occur.

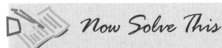

Problem 7 on page 372 asks you to explain why some tobacco smokers and some people with inherited mutations in cancer-related genes never develop cancer.

Hint: When considering the reasons why cancer is not an inevitable consequence of genetic or environmental conditions, you might think about the multistep nature of cancer, the random nature of the spontaneous mutations, and the types of genetically controlled functions that are abnormal in cancer cells. It might also be useful to consider how each individual's genetic background may affect mutation rates and differences in DNA repair functions.

16.7 Viruses Contribute to Cancer in Both Humans and Animals

Viruses that cause cancer in animals have played a significant role in the search for knowledge about the genetics of human cancer. Most cancer-causing animal viruses are RNA viruses known as **retroviruses**. Because they transform cells into cancer cells, they are known as acute transforming retroviruses. The first of these **acute transforming retroviruses** was discovered in 1910 by Francis Peyton Rous. Rous was studying sarcomas (solid tumors of muscle, bone, or fat) in chickens, and he observed that extracts from these tumors caused the formation of new sarcomas when they were injected into tumor-free chickens. Several decades later, the agent within the extract that caused the sarcomas was identified as a retrovirus and was named the **Rous sarcoma virus (RSV)**.

To understand how retroviruses cause cancer in animals, it is necessary to know how these viruses replicate in cells. When a retrovirus infects a cell, its RNA genome is copied into DNA by the **reverse transcriptase** enzyme, which is brought into the cell with the infecting virus. The DNA copy then enters the nucleus of the infected cell, where it integrates at random into the host cell's genome. The integrated DNA copy of the retroviral RNA is called a **provirus**. The proviral DNA contains powerful enhancer and promoter elements in its U5 and U3 sequences at the ends of the provirus (Figure 16–12). The U5 promoter uses the host cell's transcription proteins, directing transcription of the viral genes (*gag*, *pol*, and *env*). The products of these genes are the proteins and RNA genomes that make up the new retroviral particles. Because the provirus is integrated into the host genome, it is replicated along with the host's DNA during the cell's normal cell cycle. A retrovirus may not kill a cell, but it may continue to use the cell as a factory to replicate more viruses that will then infect surrounding cells.

A retrovirus may cause cancer in two different ways. First, the proviral DNA may integrate by chance near one of the cell's normal proto-oncogenes. The strong promoters and enhancers in the provirus then stimulate high levels or inappropriate timing of transcription of the proto-oncogene, leading to stimulation of host cell proliferation. Second, a retrovirus may pick up a copy of a host proto-oncogene and integrate it into its genome (Figure 16–12). The new viral oncogene may be mutated during the process of transfer into the virus, or it may be expressed at abnormal levels because it is now under control of viral promoters. Retroviruses that carry these viral oncogenes can infect and transform normal cells into tumor cells. In the case of RSV, the oncogene that was captured from the chicken genome was the *c-src* gene. Through the study of many acute transforming viruses of animals, scientists have identified dozens of proto-oncogenes.

So far, no acute transforming retroviruses have been identified in humans. However, RNA and DNA viruses contribute to the development of human cancers in a variety of ways. It is thought that, worldwide, about 15 percent of cancers are associated with viruses, making virus infection the second greatest risk factor for cancer, next to tobacco smoking.

The most significant contributors to virus-induced cancers are the **papillomaviruses (HPV 16 and 18)**, **human T-cell leukemia virus (HTLV-1)**, **hepatitis B virus**, and **Epstein-Barr virus** (Table 16.4). Like other risk factors for cancer, including hereditary predisposition to certain cancers, virus infection alone is not sufficient to trigger human cancers. Other factors, including DNA damage or the accumulation of mutations in one or more of a cell's oncogenes and tumor suppressor genes, are required to move a cell down the multistep pathway to cancer.

Because viruses are comprised solely of a nucleic acid genome surrounded by a proteinaceous coat, viruses must utilize the host cell's biosynthetic machinery in order to reproduce themselves. As most viruses need the host cell to be in an actively growing state in order to access the host's DNA synthetic enzymes, these viruses often contain genes encoding products that stimulate the cell cycle. If the virus does not kill the host cell, the potential exists for a loss of cell cycle control and the beginning of tumorigenesis.

Nonacute retrovirus

| R | U5 | *gag* | *pol* | *env* | U3 | R |

Genome of virus that can infect but not transform a cell

+

Cellular proto-oncogene in infected cell

| | *c-onc* | |

Transfer of proto-oncogene from host cell to viral genome

Acute transforming retrovirus

| R | U5 | *gag* | *pol* | *env* | *v-onc* | U3 | R |

FIGURE 16–12 The genome of a typical retrovirus is shown at the top of the diagram. The genome contains repeats at the termini (R), the U5 and U3 regions that contain promoter and enhancer elements, and the three major genes that encode viral structural proteins (*gag* and *env*) and the viral reverse transcriptase (*pol*). RNA transcripts of the entire viral genome comprise the new viral genomes. If the retrovirus acquires all or part of a host cell proto-oncogene (*c-onc*), this gene (now known as a *v-onc*) is expressed along with the viral genes, leading to overexpression or inappropriate expression of the *v-onc* gene. The *v-onc* gene may also acquire mutations that enhance its transforming ability.

TABLE 16.4	Human Viruses Associated with Cancer		
Virus	**Cancer**	**Oncogenes**	**Mechanism**
Human papillomavirus 16, 18	Cervical cancer	*E6, E7*	Inhibit p53 and pRB tumor suppressors
Hepatitis B virus	Liver cancer	*HBx*	Signal transduction, stimulates cell cycle
Epstein-Barr virus	Burkitt's lymphoma, naso-pharyngeal cancer	Unknown	Unknown
Human herpesvirus 8	AIDS-related Kaposi's sarcoma	Several possible	Unknown
Human T-cell leukemia virus	Adult T-cell leukemia	*pX*	Stimulates cell cycle

16.8 Environmental Agents Contribute to Human Cancers

Any substance or event that damages DNA has the potential to be carcinogenic. Unrepaired or inaccurately repaired DNA introduces mutations which, if they occur in proto-oncogenes or tumor suppressor genes, can lead to abnormal regulation of the cell cycle or disruption of controls over cell contact and invasion.

Our environment, both natural and man-made, contains abundant carcinogens. These include chemicals, radiation, some viruses, and chronic infections. Perhaps the most significant carcinogen in our environment is tobacco smoke. Epidemiologists estimate that about 30 percent of human cancer deaths are associated with cigarette smoking. Tobacco smoke contains a number of cancer-causing chemicals, some of which preferentially mutate proto-oncogenes such as *ras* and tumor suppressor genes such as *p53*.

Diet is often implicated in the development of cancer. Consumption of red meat and animal fat is associated with some cancers, such as colon, prostate, or breast cancer. The mechanisms by which these substances may contribute to carcinogenesis are not clear but may involve stimulation of cell division through hormones or creation of carcinogenic chemicals during cooking. Alcohol may cause inflammation of the liver and contribute to liver cancer.

Although most people perceive the man-made, industrial environment to be a highly significant contributor to cancer, it may account for only a small percentage of total cancers, and often in only specialized situations. Some of the most mutagenic agents, and hence potentially the most carcinogenic, are natural substances and natural processes. For example, **aflatoxin**, a component of a mold that grows on peanuts and corn, is one of the most carcinogenic chemicals known. Most chemical carcinogens, such as **nitrosamines**, are components of synthetic substances; however, many are naturally occurring. For example, natural pesticides and antibiotics found in plants may be carcinogenic, and the human body itself creates alkylating agents in the acidic environment of the gut. Nevertheless, these observations do not diminish the serious cancer risks to specific populations who are exposed to man-made carcinogens, such as synthetic pesticides or asbestos.

DNA lesions brought about by natural radiation (X-rays, ultraviolet light), natural dietary substances, and the external environment contribute the majority of environmentally caused mutations that lead to cancer. Normal metabolism creates oxidative end products that can damage DNA, proteins, and lipids. It is estimated that the human body suffers about 10,000 damaging DNA lesions per day due to the actions of oxygen free radicals. DNA repair enzymes deal successfully with most of this damage; however, some damage may lead to permanent mutations. The process of DNA replication itself is mutagenic. Hence, substances like growth factors or hormones that simulate cell division are ultimately mutagenic and perhaps carcinogenic. Chronic inflammation due to infection also stimulates tissue repair and cell division, resulting in DNA lesions accumulating during replication. These mutations may persist, particularly if cell-cycle checkpoints are compromised due to mutations or inactivation of tumor suppressor genes such as *p53* or *RB1*.

Both ultraviolet (UV) light and ionizing radiation (such as X-rays and gamma rays) induce DNA damage. UV damage in sunlight is well accepted as an inducer of skin cancers. Ionizing radiation has clearly demonstrated itself as a carcinogen in studies of populations exposed to neutron and gamma radiation from atomic blasts such as those in Hiroshima and Nagasaki. Another significant environmental component, radon gas, may be responsible for up to 50 percent of the ionizing radiation exposure for the U.S. population and could contribute to lung cancers in some populations.

Now Solve This

Problem 16 on page 374 asks you to provide your own estimate of what percentage of money spent on cancer research should be devoted to research and education on preventing cancer and what percentage should be devoted to research on cancer cures.

Hint: In answering this question, think about the relative rates of environmentally induced and spontaneous cancers. Second, consider the proportion of environmentally induced cancers that are due to lifestyle choices. (An interesting source of information on this topic can be found in Ames, B. N. et al. 1995. The causes and prevention of cancer. *Proc. Natl. Acad. Sci. USA* 92: 5258–5265.)

GENETICS, TECHNOLOGY, AND SOCIETY

Breast Cancer: The Double-Edged Sword of Genetic Testing

These are exhilarating times for genetics and biotechnology. Close on the heels of the completion of the Human Genome Project has come a rush of optimism about future applications of genetics. Scientists and the media predict that gene technologies will soon diagnose and cure diseases as diverse as diabetes, asthma, heart disease, and Parkinson disease.

The prospect of using genetics to prevent and cure a whole range of diseases is exciting. However, in our enthusiasm, we often forget that these new technologies have significant limitations and profound ethical concerns. The story of genetic testing for breast cancer illustrates how we must temper our high expectations with respect for uncertainty.

Breast cancer is the most common cancer among women and the second leading cause of all cancer deaths (after lung cancer). Each year, more than 190,000 new cases are diagnosed in the United States. Breast cancer is not limited to women; about 1400 men are also diagnosed with the disease each year. A woman's lifetime risk of developing breast cancer is about 12 percent, and the risk increases with age.

Approximately 5 to 10 percent of breast cancers are familial, defined by the appearance of several cases of breast or ovarian cancer among near blood relatives and the early onset of these diseases. In 1994, two genes were identified that show linkage to familial breast cancers. Germ-line mutations in these genes (*BRCA1* and *BRCA2*) are associated with the majority of familial breast cancers. The molecular functions of *BRCA1* and *BRCA2* are still uncertain, although they appear to be tumor suppressor genes whose products are involved in repairing damaged DNA. Women who bear mutations in *BRCA1* or *BRCA2* have a 36 to 85 percent lifetime risk of developing breast cancer and a 16 to 60 percent risk of developing ovarian cancer. Men with germ-line mutations in *BRCA2* have a 6 percent lifetime breast cancer risk—a hundredfold increase over the general male population.

BRCA1 and *BRCA2* genetic tests detect any of the over 2000 different mutations that are known to occur within the coding regions of these genes, but the tests have limitations. They do not detect mutations in regulatory regions outside the coding region—mutations that could cause aberrant expression of these genes. Also, little is known about how any particular mutation manifests itself in terms of cancer risk, and the effects of each mutation may be modified by environmental factors and by interactions with other susceptibility genes.

Many patients at risk for familial breast cancer opt to undergo genetic testing. These patients feel that test results will help them to prevent breast or ovarian cancers, will guide them in childbearing decisions, and allow them to inform family members at risk. But none of these benefits is clear-cut.

A woman whose *BRCA* test results are negative may be relieved and feel that she is not subject to familial breast cancer. However, her risk of developing breast cancer is still 12 percent (the population risk), and she should continue to monitor for the disease. Also, a negative *BRCA* genetic test does not eliminate the possibility that she bears an inherited mutation in another gene that increases breast cancer risk or that *BRCA1* or *BRCA2* gene mutations exist in regions of the genes that are inaccessible to current genetic tests.

A woman whose test results are positive faces difficult choices. Her treatment options are poor, consisting of close monitoring, prophylactic mastectomy or oophorectomy (removal of breasts and ovaries respectively), and taking drugs such as tamoxifen. Prophylactic surgery reduces her risks but does not eliminate them, as cancers can still occur in tissues that remain after surgery. Drugs such as tamoxifen reduce her risks but have serious side effects. Genetic tests affect not only the patient but also the patient's entire family. People often experience fear, anxiety, and guilt on learning that they are carriers of a genetic disease. Studies show that people who refuse genetic test results often suffer from even more anxiety than those who opt to learn the results. Confidentiality is also a major concern. Patients fear that their genetic test results may be leaked to insurance companies or employers, jeopardizing their prospects for jobs or affordable health and life insurance. One study shows that a quarter of eligible patients refuse *BRCA* gene testing because of concerns about cost, confidentiality, and potential discrimination.

Genetic testing is such a new development that the health system has lagged behind the science. Because genetic testing has both psychological and medical ambiguities, genetic counseling is imperative for patients and their families. However, there are insufficient numbers of genetic counselors with experience in genetic testing, and even in the most qualified hands, issues are complex and difficult. Physicians often have limited knowledge of human clinical genetics and feel inadequate to advise their patients. The federal government and the insurance industries have yet to develop comprehensive policies concerning genetic tests and genetic information. Given the unclear interpretation of *BRCA* genetic tests, the relatively ineffective treatment options, and the potential for psychological and societal side effects, it is not surprising that only about 60 percent of familial breast cancer patients and their families decide to take the genetic tests.

The unanswered questions about *BRCA1* and *BRCA2* genetic testing are many and important. What cancer risks are associated with which mutations? Should all people have access to *BRCA* tests or only those at high risk? How can we ensure that the high costs of genetic tests and counseling do not limit this new technology to only a portion of the population? As we develop genetic tests for more and more diseases over the next few decades, our struggle with these issues will continue to grow.

Reference

Surbone, A. 2001. Ethical implications of genetic testing for breast cancer susceptibility. *Crit. Rev. in Onc./Hem.* 40: 149–157.

Web Sites

Genetic Testing for *BRCA1* and *BRCA2*: It's Your Choice [online]. National Institutes of Health, 2002. **http://cis.nci.nih.gov/fact/3_62.htm**

CHAPTER SUMMARY

1. Cancer is a genetic disease, predominantly of somatic cells. About 1 percent of cancers are associated with germ-line mutations that increase the susceptibility to certain cancers.

2. Cancer cells show two basic properties: abnormal cell proliferation and a propensity to spread and invade other parts of the body (metastasis). Genes controlling these aspects of cellular function are either mutated or expressed inappropriately in cancer cells.

3. Cancers are clonal, meaning that all cells within a tumor originate from a single cell that contained a number of mutations.

4. The development of cancer is a multistep process, requiring mutations in several cancer-related genes.

5. Cancer cells show high rates of mutation, chromosomal abnormalities, and genomic instability. This leads to the accumulation of mutations in specific genes that control aspects of cell proliferation, apoptosis, differentiation, DNA repair, cell migration and cell-cell contact.

6. Cancer cells have defects in the regulation of cell-cycle progression, cell-cycle checkpoints, and signal transduction pathways.

7. Proto-oncogenes are normal genes that promote cell growth and division. When proto-oncogenes are mutated or mis-expressed in cancer cells, they are known as oncogenes.

8. Tumor suppressor genes normally regulate cell-cycle checkpoints and apoptosis. When tumor suppressor genes are mutated or inactivated, cells cannot correct DNA damage. This leads to accumulation of mutations that may cause cancer.

9. Inherited mutations in cancer-susceptibility genes are not sufficient to trigger cancer. A second somatic mutation, in the other copy of the gene, is necessary to trigger tumorigenesis. In addition, mutations in other cancer-related genes are necessary for the development of hereditary cancers.

10. RNA and DNA tumor viruses contribute to cancers by stimulating cells to proliferate, introducing new oncogenes, interfering with the cell's normal tumor suppressor gene products, or stimulating the expression of a cell's proto-oncogenes.

11. Environmental agents such as chemicals, radiation, viruses, and chronic infections contribute to the development of cancer. The most significant environmental factors that affect human cancers are tobacco smoke, diet, and natural radiation.

KEY TERMS

acute transforming retroviruses, 369

aflatoxin, 370

apoptosis, 363

basal lamina, 367

benign tumor, 359

Burkitt's lymphoma, 359

carcinogens, 360

caspases, 363

cell adhesion molecules, 367

cell proliferation, 359

chronic myelogenous leukemia (CML), 360

cyclin-dependent kinases (CDKs), 362

cyclins, 362

Epstein-Barr virus, 369

extracellular matrix, 367

familial adenomatous polyposis (FAP), 368

G1/S checkpoint, 362

G2/M checkpoint, 362

hepatitis B virus, 369

hereditary nonpolyposis colorectal cancer (HNPCC), 361

human T-cell leukemia virus (HTLV-1), 369

loss of heterozygosity, 368

M checkpoint, 362

malignant tumors, 359

metalloproteinases, 367

metastases, 359

metastasis, 359

mutator phenotype, 360

nitrosamines, 370

oncogene, 364

papilloma viruses (HPV 16, 18), 369

p53 gene, 365

Philadelphia chromosome, 360

polyp, 368

protein kinase, 360

proto-oncogenes, 364

provirus, 369

programmed cell death, 363

ras gene family, 365

RB1 (retinoblastoma 1), 366

retinoblastoma, 366

retinoblastoma protein (pRB), 366

retroviruses, 369

reverse transcriptase, 369

Rous sarcoma virus (RSV), 369

signal transduction, 362

tissue inhibitors of metalloproteinases (TIMPs), 367

transformed, 368

tumorigenesis, 360

tumor suppressor genes, 364

INSIGHTS AND SOLUTIONS

1. In disorders such as retinoblastoma, a mutation in one allele of the *RB1* gene can be inherited from the germ line, causing an autosomal dominant predisposition to the development of eye tumors. To develop tumors, a somatic mutation in the second copy of the *RB1* gene is necessary, indicating that the mutation itself acts as a recessive trait. Given that the first mutation can be inherited, in what ways can a second mutational event occur?

Solution: In considering how this second mutation arises, we must look at several types of mutational events, including changes in nucleotide sequence and events that involve whole chromosomes or chromosome parts. Retinoblastoma results when both copies of the *RB1* locus are lost or inactivated. With this in mind, you must first list the phenomena that can result in a mutational loss or the inactivation of a gene.

One way the second *RB1* mutation can occur is by a nucleotide alteration that converts the remaining normal *RB1* allele to a mutant form. This alteration can occur through a nucleotide substitution or by a frameshift mutation caused by the insertion or deletion of nucleotides during replication. A second mechanism involves the loss of the chromosome carrying the normal allele. This event would take place during mitosis, resulting in chromosome 13 monosomy and leaving the mutant copy of the gene as the only *RB1* allele. This mechanism does not necessarily involve loss of the entire chromosome; deletion of the long arm (*RB1* is on 13q) or an interstitial deletion involving the *RB1* locus and some surrounding material would have the same result. Alternatively, a chromosome aberration involving loss of the normal copy of the *RB1* gene might be followed by duplication of the chromosome carrying the mutant allele. Two copies of chromosome 13 would be restored to the cell, but the normal *RB1* allele would not be present. Finally, a recombination event followed by chromosome segregation could produce a homozygous combination of mutant *RB1* alleles.

2. Proto-oncogenes can be converted to oncogenes in a number of different ways. In some cases, the proto-oncogene itself becomes amplified up to hundreds of times in a cancer cell. An example is the *cyclin D1* gene, which is amplified in some cancers. In other cases, the proto-oncogene may be mutated in a limited number of specific ways leading to alterations in the gene product's structure. The *ras* gene is an example of a proto-oncogene that becomes oncogenic after suffering point mutations in specific regions of the gene. Explain why these two proto-oncogenes (*cyclin D1* and *ras*) undergo such different alterations in order to convert them into oncogenes.

Solution: The first step to solving this question is to understand the normal functions of these proto-oncogenes and to think about how either amplification or mutation would affect each of these functions.

The cyclin D1 protein regulates progression of the cell cycle from G1 into S phase, by binding to CDK4 and activating this kinase. The cyclin D1/CDK4 complex phosphorylates a number of proteins including pRB, which in turn activates other proteins in a cascade that results in transcription of genes whose products are necessary for DNA replication in S phase. The simplest way to increase the activity of cyclin D1 would be to increase the number of cyclin D1 molecules available for binding to the cell's endogenous CDK4 molecules. This can be accomplished by several mechanisms, including amplification of the *cyclin D1* gene. In contrast, a point mutation in the *cyclin D1* gene would most likely interfere with the ability of the cyclin D1 protein to bind to CDK4; hence, mutations within the gene would probably repress cell-cycle progression rather than stimulate it.

The *ras* gene product is a signal transduction protein that operates as an on/off switch in response to external stimulation by growth factors. It does so by binding either GTP (the "on" state) or GDP (the "off" state). Oncogenic mutations in the *ras* gene occur in specific regions that alter the ability of the Ras protein to exchange GDP for GTP. Oncogenic Ras proteins are locked in the "on" conformation, bound to GTP. In this way, they constantly stimulate the cell to divide. An amplification of the *ras* gene would simply provide more molecules of normal Ras protein, which would still be capable of on/off regulation. Hence, simple amplification of *ras* would be less likely to be oncogenic.

3. Explain why many oncogenic viruses contain genes whose products interact with tumor suppressor proteins.

Solution: In order to answer this question, it is useful to consider what viruses try to accomplish in cells and what the roles of tumor suppressors are in normal cells.

The goal of oncogenic viruses is not to cause cancer but to maximize the potential for viral replication. Viruses are relatively simple entities, consisting of only a nucleic acid genome—either RNA or DNA—and a relatively small number of structural and enzymatic proteins. They depend on their host cells for much of the biosynthetic machinery and structural components necessary to replicate their genomes and assemble new virus particles. For example, many viruses require the host cell's RNA polymerase II enzyme, transcription factors, and ribonucleotide precursors in order to transcribe their viral genes. They also require components of the host cell's DNA replication machinery to replicate viral genomes. Hence, the ideal host cell for a virus infection is one that is within the cell cycle, preferably in late G1 to early S phase.

Most cells in a higher eukaryote such as humans are quiescent (in G0 phase). In order to stimulate the infected cell to enter the cell cycle and become primed for DNA replication, many viruses contain genes that encode growth-stimulating proteins. As we learned in this chapter, tumor suppressor proteins are those involved in either restraining the cell cycle at checkpoints or in triggering the process of programmed cell death. Both of these functions are inhibitory to the goals of a typical virus; therefore, many viruses have evolved methods to inactivate tumor suppressors. One of the ways in which viral proteins can inactivate tumor suppressor proteins is to bind to them and inhibit their functions. The tumor suppressors p53 and pRB are common targets of viral regulatory proteins, such as the E6 and E7 proteins of HPV 16 and 18. By inactivating tumor suppressors, these viruses are able to maintain the cell within the cell cycle. For the host, however, this growth stimulation in the absence of functional cell-cycle checkpoints can lead to increased mutation accumulation and possible tumorigenesis.

PROBLEMS AND DISCUSSION QUESTIONS

1. As a genetic counselor, you are asked to assess the risk for a couple with a family history of retinoblastoma who are thinking about having children. Both the husband and wife are phenotypically normal, but the husband has a sister with familial retinoblastoma in both eyes. What is the probability that this couple will have a child with retinoblastoma? Are there any tests that you could recommend to help in this assessment?

2. What events occur in each phase of the cell cycle? Which phase is most variable in length?

3. Where are the major regulatory points in the cell cycle?

4. List the functions of kinases and cyclins, and describe how they interact to cause cells to move through the cell cycle.

5. (a) How does pRB function to keep cells at the G1 checkpoint? (b) How do cells get past the G1 checkpoint to move into S phase?

6. What is the difference between saying that cancer is inherited and saying that the predisposition to cancer is inherited?

7. Although tobacco smoking is responsible for a large number of human cancers, not all smokers develop cancer. Similarly, some people who inherit mutations in the tumor suppressor genes *p53* or *RB1* never develop cancer. Describe some reasons for these observations.

8. What is apoptosis, and under what circumstances do cells undergo this process?

9. Define tumor suppressor genes. Why is a mutation in a single copy of a tumor suppressor gene expected to behave as a recessive gene?

10. In the Rous sarcoma virus (RSV) genome, the host cell proto-oncogene is converted into an oncogene. How does this conversion occur?

11. Part of the Ras protein is associated with the plasma membrane, and part extends into the cytoplasm. How does the Ras protein transmit a signal from outside the cell into the cytoplasm? What happens in cases where the *ras* gene is mutated?

12. If a cell suffers damage to its DNA while in S phase, how can this damage be repaired before the cell enters mitosis?

13. Distinguish between oncogenes and proto-oncogenes. In what ways can proto-oncogenes be converted to oncogenes?

14. Of the two classes of genes associated with cancer, tumor suppressor genes and oncogenes, mutations in which group can be considered gain-of-function mutations? In which group are the loss-of-function mutations? Explain.

15. How do translocations such as the Philadelphia chromosome contribute to cancer?

16. Given that cancers can be environmentally induced and that some environmental factors are the result of lifestyle choices such as smoking, sun exposure, and diet, what percentage of the money spent on cancer research do you think should be devoted to research and education on preventing cancer rather than on finding cancer cures?

17. In CML, leukemic blood cells can be distinguished from other cells of the body by the presence of a functional BCR-ABL hybrid protein. Explain how this characteristic provides an opportunity to develop a therapeutic approach to a treatment for CML.

18. How do normal cells protect themselves from accumulating mutations in genes that could lead to cancer? How do cancer cells differ from normal cells in these processes?

19. Describe the difference between an acute transforming virus and a virus that does not cause tumors.

20. Explain how environmental agents such as chemicals and radiation cause cancer.

21. Radiotherapy (treatment with ionizing radiation) is one of the most effective current cancer treatments. It works by damaging DNA and other cellular components. In which ways could radiotherapy control or cure cancer, and why does radiotherapy often have significant side effects?

22. Assume that a young woman in a suspected breast cancer family takes the *BRCA1* and *BRCA2* genetic tests and receives negative results. That is, she does not test positive for the mutant alleles of *BRCA1* or *BRCA2*. Can she consider herself free of risk for breast cancer?

23. People with a genetic condition known as Li-Fraumeni syndrome inherit one mutant copy of the *p53* gene. These people have a high risk of developing a number of different cancers, such as breast cancer, leukemias, bone cancers, adrenocortical tumors, and brain tumors. Explain how mutations in one cancer-related gene can give rise to such a diverse range of tumors.

24. Explain the differences between a benign and malignant tumor.

25. As part of a cancer research project, you have discovered a gene that is mutated in many metastatic tumors. After determining the DNA sequence of this gene, you compare the sequence with those of other genes in the human genome sequence database. Your gene appears to code for an amino acid sequence that resembles sequences found in some serine proteases. Conjecture how your new gene might contribute to the development of highly invasive cancers.

26. A study by Bose and colleagues (1998. *Blood* 92: 3362–3367) and a previous study by Biernaux and others (1996. *Bone Marrow Transplant* 17: (Suppl. 3) S45–S47) showed that *BCR-ABL* fusion gene transcripts can be detected in 25 to 30 percent of healthy adults who do not develop chronic myelogenous leukemia (CML). Explain how these individuals can carry a fusion gene that is transcriptionally active and yet do not develop CML.

27. Those who inherit a mutant allele of the *RB1* gene are at risk for developing a bone cancer called osteosarcoma. You suspect that in these cases, osteosarcoma requires a mutation in the second *RB1* allele and have cultured some osteosarcoma cells and obtained a cDNA clone of a normal human *RB1* gene. A colleague sends you a research paper revealing that a strain of cancer-prone mice develop malignant tumors when injected with osteosarcoma cells, and you obtain these mice. Using these three resources, what experiments would you perform to determine (a) whether osteosarcoma cells carry two *RB1* mutations, (b) whether osteosarcoma cells produce any pRB protein, and (c) if the addition of a normal *RB1* gene will change the cancer-causing potential of osteosarcoma cells?

28. The compound benzo[*a*]pyrene is found in cigarette smoke. This compound chemically modifies guanine bases in DNA. Such abnormal bases are usually removed by an enzyme that hydrolyzes the base, leaving an apurinic site. If such a site is left unrepaired, an adenine is preferentially inserted across from the apurinic site. In a study of lung cancer patients (Harris, A. 1991. *Nature* 350: 377–378), tumor cells from 15 out of 25 patients had a G to T transversion in the *p53* gene, which has a known role in cancer formation. You are testifying as an expert witness in a court case in which the widow of a man who was a lifelong smoker and died of lung cancer is suing a company for manufacturing the

TABLE 16.5	Predisposing Mutations in *BRCA1*			
		Mutation		
Kindred	**Codon**	**Nucleotide Change**	**Coding Effect**	**Frequency in Control Chromosomes**
1901	24	−11 bp	Frameshift or splice	0/180
2082	1313	C → T	Gln → Stop	0/170
1910	1756	Extra C	Frameshift	0/162
2099	1775	T → G	Met → Arg	0/120
2035	NA*	?	Loss of transcript	NA*

Source: 1994. *Science* 266: 66–71.

*NA indicates not applicable, as the regulatory mutation is inferred, and the position has not been identified.

tobacco products that killed her husband. What would you tell the jury? (Science only please—no personal expositions on lawyers or the legal system!)

29. Table 16.5 summarizes some of the data that have been collected on *BRCA1* mutations in families with a high incidence of both early-onset breast and ovarian cancer. Table 16.6 shows neutral polymorphisms found in control families (with no increased frequency of breast and ovarian cancer). (a) Note the coding effect of the mutation found in kindred group 2082 in Table 16.5. This results from a single base-pair substitution. Draw the normal double-stranded DNA sequence for this codon (with the 5′ and 3′ ends labeled), and show the sequence of events that generated this mutation, assuming that it resulted from an uncorrected mismatch event during DNA replication. (b) Examine the types of mutations that are listed in Table 16.5 and determine if the

BRCA1 gene is likely to be a tumor suppressor gene or an oncogene. (c) Although the mutations in Table 16.5 are clearly deleterious and cause breast cancer in women at very young ages, each of the kindred groups had at least one woman who carried the mutation but lived until age 80 without developing cancer. Name at least two different mechanisms (or variables) that could underlie variation in the expression of a mutant phenotype and propose an explanation for the incomplete penetrance of this mutation. How do these mechanisms or variables relate to this explanation?

30. Examine Table 16.6. (a) What is meant by a neutral polymorphism? (b) What is the significance of this table in the context of examining a family or population for *BRCA1* mutations that predispose an individual to cancer? (c) Is the PM2 polymorphism likely to result in a neutral missense mutation or a silent mutation? (d) Answer part (c) for the PM3 polymorphism.

TABLE 16.6	Neutral Polymorphisms in *BRCA1*					
			Frequency in Control Chromosomes*			
Name	**Codon Location**	**Base in Codon†**	**A**	**C**	**G**	**T**
PM1	317	2	152	0	10	0
PM6	878	2	0	55	0	100
PM7	1190	2	109	0	53	0
PM2	1443	3	0	115	0	58
PM3	1619	1	116	0	52	0

*The number of chromosomes with a particular base at the indicated polymorphic site (A, C, G, or T) is shown.

17 Recombinant DNA Technology

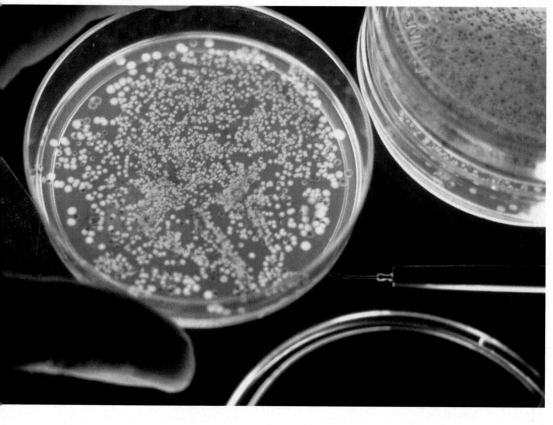

A Petri dish showing the growth of host cells, some of which have taken up recombinant plasmids.

CHAPTER CONCEPTS

- Recombinant DNA technology creates artificial combinations of DNA sequences.

- Recombinant DNA technology depends in part on the ability to cleave and rejoin DNA segments at specific base sequences.

- The most useful application of recombinant DNA technology is in cloning DNA.

- For cloning, DNA segments are inserted into vectors (such as plasmids), transferred into host cells (such as bacterial cells), where the recombinant molecules replicate as the host cells divide.

- Recombinant molecules can be cloned in prokaryotic or eukaryotic host cells.

- DNA segments can be quickly and efficiently amplified thousands of times using the polymerase chain reaction (PCR).

- Cloned DNA can be characterized in several ways, the most detailed being DNA sequencing.

The term **recombinant DNA** has two meanings in genetics. The more specific of the two refers to a DNA molecule formed in the laboratory by joining together DNA sequences from different sources. Such molecules are artificial laboratory creations and are not found in nature. The term *recombinant DNA* is also used more loosely to refer to the technology that is utilized to create and study these hybrid molecules constructed from DNA obtained from different biological sources. Such techniques to create, **clone** (replicate), and analyze recombinant DNA molecules were developed in the 1970s as a way for researchers to isolate and study specific DNA sequences. The power of recombinant DNA technology is astonishing, making it possible to identify and isolate a single gene or DNA segment of interest from the thousands or tens of thousands present in a genome. Subsequently, through cloning, huge quantities of identical copies of this specific DNA molecule can be produced. These identical copies can then be manipulated for numerous purposes, including research into the structure and organization of the DNA, or for the commercial production of its encoded protein. In this chapter, we review the basic methods of recombinant DNA technology. Following this introduction, in Chapter 18 we will consider comparative genomic analysis, which has been made possible by recombinant DNA techniques. Then, in Chapter 19, we will discuss some applications of this technology, sometimes called genetic engineering, to research, medicine, the legal system, agriculture, and industry.

How Do We Know?

In this chapter, we will focus on the essential techniques of recombinant DNA technology, including the construction and cloning of recombinant DNA molecules, and the ways in which clones of interest can be identified and analyzed. As you study this topic, you should try to answer several fundamental questions:

1. How do we know that restriction enzymes recognize palindromic sequences running in opposite directions on complementary strands of a DNA molecule?

2. How do we know that DNA fragments have been successfully incorporated into vectors such as plasmids?

3. How do we know that PCR requires two primers and a DNA polymerase?

4. How do we know that a cloned DNA fragment contains a specific gene?

17.1 An Overview of Recombinant DNA Technology

Recombinant DNA technology uses methods derived from nucleic acid biochemistry, coupled with genetic techniques originally developed for the study of bacteria and viruses. This technology is a powerful tool for isolating potentially unlimited quantities of a gene. A basic procedure for making and cloning recombinant DNA molecules uses the following steps:

1. The DNA to be cloned is purified from cells or tissues.

2. Proteins called **restriction enzymes** are used to generate specific DNA fragments. These enzymes recognize and cut DNA molecules at specific nucleotide sequences.

3. The fragments produced by restriction enzymes are joined to other DNA molecules that serve as **vectors**, or carrier molecules. A vector joined to a DNA fragment is a **recombinant DNA molecule**.

4. The recombinant DNA molecule is transferred into a host cell where it replicates, producing many identical copies, or clones, of the recombinant DNA.

5. As host cells replicate, the cloned DNA molecules within them are passed on to all their progeny, creating a population of host cells, each of which carries copies of the cloned DNA sequence.

6. The cloned DNA can be recovered from host cells and purified for use in research or commercial applications.

17.2 Recombinant DNA Molecules Are Constructed from Several Components

Recombinant DNA technology provides scientists with methods for isolating large quantities of specific genes and other DNA sequences from large and complex genomes. For example, the human genome contains more than 3 billion nucleotides and 20,000 to 25,000 genes. Restriction enzymes cut the genome into smaller fragments that can be manipulated, separated, copied, and studied individually. This allows researchers to investigate many aspects of gene organization and function, identify factors that regulate gene expression, and study the nature and functions of encoded proteins. We will begin our discussion of this technology by considering the steps used to construct recombinant DNA molecules.

Restriction Enzymes

Restriction enzymes are produced by bacteria, and restrict or prevent viral infection by degrading the DNA of invading viruses. More than 400 restriction enzymes have been identified, and about 100 of these are commonly used by researchers. A restriction enzyme binds to DNA and recognizes a specific nucleotide sequence called a **recognition sequence** and cuts both strands of the DNA within that sequence. The usefulness of restriction enzymes in cloning derives from their ability to accurately and reproducibly cut genomic DNA into fragments called **restriction fragments**. The size of restriction fragments is determined by how often a given restriction enzyme cuts the DNA. Enzymes with a four-base recognition sequence such as *Alu*I (AGCT) will cut, on average, every 256 base pairs ($4^n = 4^4 = 256$) if all four nucleotides are present in equal proportions, producing many small fragments. Enzymes like

Web Tutorial 17.1 Recombinant DNA Technology: Restriction Enzymes

Recognition site

5′ G-A-A-T-T-C 3′
3′ C-T-T-A-A-G 5′

↓

Treatment with *Eco*RI

↓

5′ G   A-A-T-T-C 3′
                  Complementary tails
3′ C-T-T-A-A   G 5′

FIGURE 17–1 The restriction enzyme *Eco*RI recognizes and binds to the palindromic nucleotide sequence GAATTC. Cleavage of the DNA at this site produces complementary single-stranded tails (often called sticky ends). These single-stranded tails can anneal with single-stranded tails from other DNA fragments to form recombinant DNA molecules (after ligation).

*Not*I have an eight-base recognition sequence (GCGGCCGC) and cut the DNA on average every 65,500 base pairs ($4^n = 4^8$), producing fewer, larger fragments. The actual fragment sizes produced by cutting DNA with a given restriction enzyme vary because the number and location of recognition sequences are not always distributed evenly in DNA.

One of the first restriction enzymes to be identified was isolated from an R strain of *Escherichia coli* and is designated *Eco*RI (pronounced echo-r-one or eeko-r-one). Its nucleotide recognition sequence and cleavage pattern are shown in Figure 17–1. Recognition sequences for some frequently used restriction enzymes are shown in Figure 17–2. The most common recognition sequences are four or six nucleotides long, but some enzymes have recognition sequences of eight or more nucleotides. Enzymes like *Eco*RI and *Hin*dIII make offset cuts in the DNA strands and produce fragments with single-stranded tails, while others, such as *Alu*I and *Bal*I cut both strands at the same nucleotide pair and produce blunt ends.

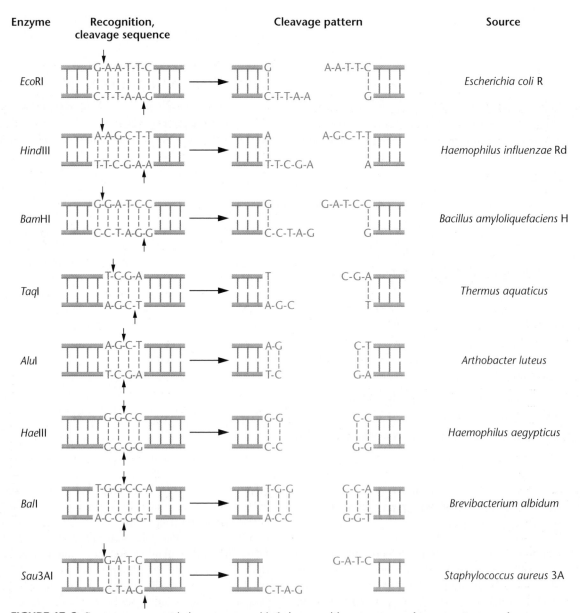

Enzyme	Recognition, cleavage sequence	Cleavage pattern		Source

*Eco*RI — G-A-A-T-T-C / C-T-T-A-A-G → G / C-T-T-A-A + A-A-T-T-C / G — *Escherichia coli* R

*Hin*dIII — A-A-G-C-T-T / T-T-C-G-A-A → A / T-T-C-G-A + A-G-C-T-T / A — *Haemophilus influenzae* Rd

*Bam*HI — G-G-A-T-C-C / C-C-T-A-G-G → G / C-C-T-A-G + G-A-T-C-C / G — *Bacillus amyloliquefaciens* H

*Taq*I — T-C-G-A / A-G-C-T → T / A-G-C + C-G-A / T — *Thermus aquaticus*

*Alu*I — A-G-C-T / T-C-G-A → A-G / T-C + C-T / G-A — *Arthobacter luteus*

*Hae*III — G-G-C-C / C-C-G-G → G-G / C-C + C-C / G-G — *Haemophilus aegypticus*

*Bal*I — T-G-G-C-C-A / A-C-C-G-G-T → T-G-G / A-C-C + C-C-A / G-G-T — *Brevibacterium albidum*

*Sau*3AI — G-A-T-C / C-T-A-G → / C-T-A-G + G-A-T-C / — *Staphylococcus aureus* 3A

FIGURE 17–2 Some common restriction enzymes, with their recognition sequences, cleavage patterns, and sources.

FIGURE 17–3 DNA from different sources is cleaved with *Eco*RI and mixed to allow annealing. The enzyme DNA ligase then chemically bonds these annealed fragments into an intact recombinant DNA molecule.

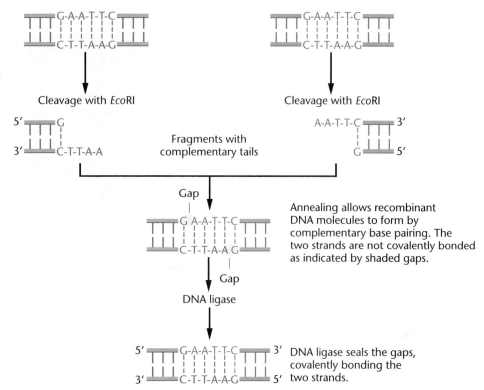

Cleavage with *Eco*RI

Cleavage with *Eco*RI

Fragments with complementary tails

Annealing allows recombinant DNA molecules to form by complementary base pairing. The two strands are not covalently bonded as indicated by shaded gaps.

Gap

DNA ligase

DNA ligase seals the gaps, covalently bonding the two strands.

DNA fragments produced by *Eco*RI digestion have over-hanging single-stranded tails ("sticky ends") that can base-pair with complementary single-stranded tails on other DNA fragments. When mixed together in the correct proportions, the single-stranded ends of DNA fragments from two sources can **anneal**, or stick together by base pairing of their single-stranded ends. By adding the enzyme **DNA ligase** to the solution, the fragments can be covalently linked to form recombinant DNA molecules (Figure 17–3).

Vectors

Fragments of DNA produced by restriction enzyme digestion cannot directly enter bacterial cells for cloning. However, when a DNA fragment is joined to a vector, it can gain entry to a host cell, where it can be replicated or cloned into many copies. Vectors are, in essence, carrier DNA molecules.

To serve as a vector, a DNA molecule must be able to independently replicate itself and the DNA segment it carries once it is inside a host cell. The vector should also contain restriction enzyme cleavage sites that allow insertion of DNA fragments to be cloned. To distinguish host cells that have taken up a vector from those that have not, vectors usually carry a selectable marker gene (antibiotic resistance genes, for example). Finally, the vector and its inserted DNA should be easy to recover from the host cell.

Genetically modified **plasmids** were the first vectors developed and are still widely used for cloning. Plasmids used as cloning vectors are derived from naturally occurring, extra-chromosomal, double-stranded DNA molecules that replicate autonomously within bacterial cells (Figure 17–4). The genetics of plasmids and their host bacterial cells are discussed in Chapter 8. In this section, we will emphasize the use of plasmids as vectors for cloning DNA. Plasmids have been exten-

sively modified by genetic engineering to serve as cloning vectors. Many genetically engineered plasmids are now available with a range of useful features. For example, although only a single plasmid may enter a host cell, once inside, some plasmids can increase their copy number so that several hundred copies are present. As vectors, these plasmids allow many copies of cloned DNA to be produced. For convenience, these vectors have also been engineered to contain a number of convenient restriction enzyme recognition sequences and marker genes.

One such plasmid is pUC18 (Figure 17–5), which has several useful features as a vector.

1. It is small (2686 bp), so it can carry relatively large DNA inserts.

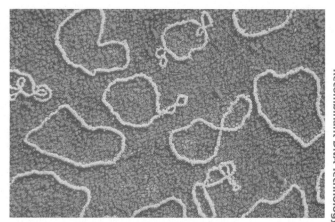

FIGURE 17–4 A color-enhanced electron micrograph of circular plasmid molecules isolated from *E. coli*. Genetically engineered plasmids are used as vectors for cloning DNA.

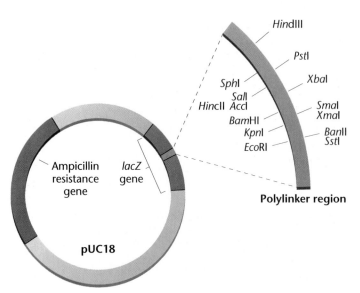

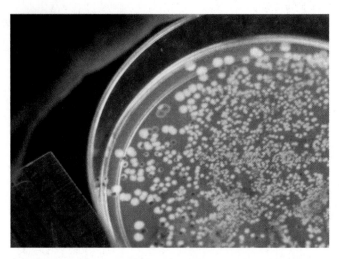

FIGURE 17–6 A Petri dish showing the growth of bacterial cells after uptake of recombinant plasmids. The medium on the plate contains a compound called Xgal. DNA cloned into the pUC18 vector disrupts the gene responsible for metabolism of Xgal and formation of blue colonies. As a result, it is easy to distinguish colonies carrying cloned DNA inserts. Cells in blue colonies contain vectors without cloned DNA inserts, whereas cells in white colonies contain vectors carrying DNA inserts.

FIGURE 17–5 A diagram of the plasmid pUC18 showing the polylinker region, located within a *lacZ* gene. DNA inserted into the polylinker region disrupts the *lacZ* gene, resulting in white colonies that allow direct identification of bacterial colonies carrying cloned DNA inserts.

2. In a host cell, it replicates up to 500 copies per cell, thus producing many copies of inserted DNA fragments.

3. A large number of restriction enzyme sites have been engineered into pUC18, conveniently clustered in one region called a **polylinker site**.

4. It allows recombinant plasmids to be easily identified. For example, pUC18 carries a fragment of the bacterial *lacZ* gene, and the polylinker is inserted into this fragment. Expression of *lacZ* causes bacterial host cells carrying

pUC18 to produce blue colonies when grown on medium containing a compound known as Xgal. If a DNA fragment is inserted anywhere in the polylinker site, the *lacZ* gene is disrupted and becomes inactive. Thus, a bacterial cell carrying pUC18 with an inserted DNA fragment forms white colonies on Xgal medium, whereas colonies carrying pUC18 plasmids without inserted DNA fragments form blue colonies (Figure 17–6).

The steps in generating a recombinant DNA molecule using pUC18 are illustrated in Figure 17–7.

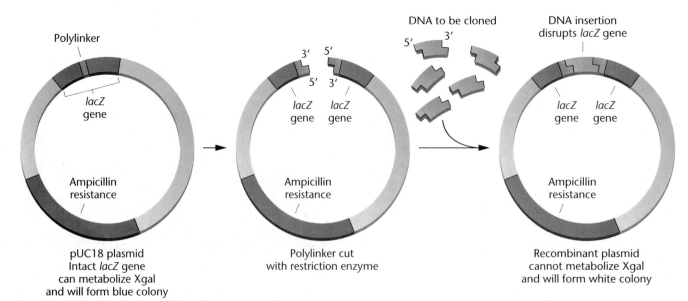

FIGURE 17–7 The plasmid vector pUC18 carries unique restriction enzyme cleavage sites in a polylinker region within the *lacZ* gene. Cleavage of the plasmid and DNA to be cloned with a restriction enzyme and insertion of a DNA fragment into the polylinker region disrupts the *lacZ* gene, making plasmids carrying inserts unable to metabolize Xgal, resulting in the formation of white colonies on nutrient plates containing Xgal.

Although many types of plasmid cloning vectors are available, other more specialized vectors have been developed from viral and bacterial genomes and from yeast plasmids. Some specialized vectors are **expression vectors**, engineered with control sequences that allow expression of inserted genes. Others such as **bacterial artificial chromosomes** (**BACs**, pronounced "backs") or **yeast artificial chromosomes** (**YACs**, pronounced as "yaks"), are designed to clone very long segments of DNA.

Like natural chromosomes, a YAC (Figure 17–8) has telomeres at each end, an origin of replication (which initiates DNA synthesis), and a centromere. These components are joined to selectable marker genes (*TRP1, SUP4,* and *URA3*) and a cluster of restriction sites for DNA inserts. Yeast chromosomes range from 230 kb to over 1900 kb, making it possible to clone DNA inserts from 100 to 1000 kb in YACs. The ability to clone large pieces of DNA into these vectors makes them an important tool in genome pro-

jects, including the Human Genome Project, which we will discuss in Chapter 18.

Although cloning in yeast vectors and host cells is currently the most advanced eukaryotic system used, other systems, including human artificial chromosome vectors using mammalian cells as hosts, are being developed.

17.3 Cloning in Host Cells

As discussed earlier, scientists use recombinant DNA technology to construct *and* replicate recombinant DNA molecules to make cloned copies of specific sequences. Replication takes place after transfer of recombinant molecules into host cells. This cloning method, known as cell-based cloning, was the first method developed and is widely used to clone DNA.

A variety of prokaryotic and eukaryotic cells can serve as hosts for recombinant vector replication. One of the most commonly used hosts is a laboratory strain of the bacterium *E. coli* known as K12. *E. coli* strains such as K12 are genetically well characterized, and can host a wide range of vectors. As an example, the following steps can be used to create recombinant DNA molecules and transfer them to an *E. coli* host cell (Figure 17–9) where they are cloned:

1. The DNA to be cloned is isolated and treated with a restriction enzyme to create fragments ending in a specific sequence.

2. The fragments are then linked to plasmid molecules that have been cut with the same restriction enzyme, creating a collection of recombinant vectors.

3. The recombinant vectors are transferred into *E. coli* host cells. Inside the host cell, a vector replicates to form many copies, or clones.

4. The bacteria are plated on nutrient medium, where they form colonies, and are screened to identify those that have taken up the recombinant plasmids.

Because the cells in each colony are derived from a single ancestral cell, all the cells in the colony and the plasmids they contain are genetically identical clones.

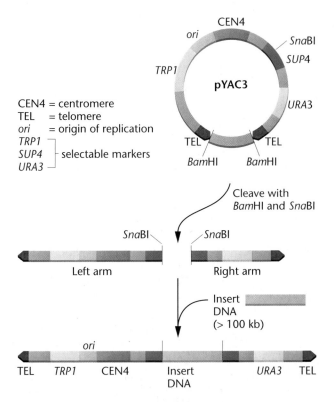

Now Solve This

Problem 12 on page 395 asks about preparing a library in a vector carrying two antibiotic resistance genes.

Hint: Inserting foreign DNA into the vector disrupts one of the resistance genes in the plasmid. Bacteria taking up recombinant plasmids will be able to grow on medium containing the antibiotic encoded by the non-disrupted gene.

FIGURE 17–8 The yeast artificial chromosome pYAC3 contains telomere sequences (TEL), a centromere (CEN4) derived from yeast chromosome 4, and an origin of replication (*ori*). These elements give the cloning vector the properties of a chromosome. *TRP1* and *URA3* are yeast genes that are selectable markers for the left and right arms of the chromosome. Within the *SUP4* gene is a restriction site for the enzyme *Sna*Bl. Two *Bam*H1 restriction sites flank a spacer segment. Cleavage with *Sna*Bl and *Bam*Hl breaks the artificial chromosome into two arms. The DNA to be cloned is treated with *Sna*Bl, producing a collection of fragments. The arms and fragments are ligated together, and the artificial chromosome is inserted into yeast host cells. Because yeast chromosomes are large, the artificial chromosome accepts inserts in the million base-pair range.

Although *E. coli* is the most commonly used prokaryotic host cell, the yeast *Saccharomyces cerevisiae* is extensively used as a host cell for the cloning and expression of eukaryotic genes for several reasons. (1) Although yeast is a eukaryotic organism, it can be grown and manipulated in much the same way as bacterial

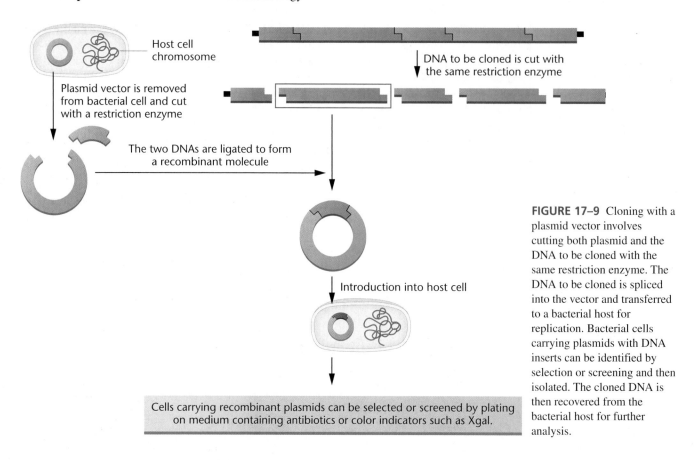

Host cell chromosome

Plasmid vector is removed from bacterial cell and cut with a restriction enzyme

DNA to be cloned is cut with the same restriction enzyme

The two DNAs are ligated to form a recombinant molecule

Introduction into host cell

Cells carrying recombinant plasmids can be selected or screened by plating on medium containing antibiotics or color indicators such as Xgal.

FIGURE 17–9 Cloning with a plasmid vector involves cutting both plasmid and the DNA to be cloned with the same restriction enzyme. The DNA to be cloned is spliced into the vector and transferred to a bacterial host for replication. Bacterial cells carrying plasmids with DNA inserts can be identified by selection or screening and then isolated. The cloned DNA is then recovered from the bacterial host for further analysis.

cells. (2) The genetics of yeast has been intensively studied, providing a large catalog of mutants and a highly developed genetic map. (3) To study the function of some eukaryotic proteins, it is necessary to use a host cell that can posttranslationally modify the protein to convert it into a functional form—bacterial host cells cannot carry out some of these modifications. (4) Yeast has been used for centuries in the baking and brewing industries and is considered to be a safe host for producing vaccines and therapeutic proteins. Table 17.1 lists some of the products of cloning in yeast.

17.4 The Polymerase Chain Reaction Copies DNA Without Host Cells

Recombinant DNA techniques were developed in the early 1970s and revolutionized research in genetics and molecular biology. These methods also gave birth to the booming biotechnology industry. However, cloning DNA using vectors

TABLE 17.1	Recombinant Proteins Synthesized in Yeast Host Cells

Hepatitis B virus surface protein
Malaria parasite protein
Insulin
Epidermal growth factor
Platelet-derived growth factor
α_1-Antitrypsin
Clotting factor XIIIA

and host cells is often labor-intensive and time-consuming. In 1986, another technique, called the **polymerase chain reaction (PCR)**, was developed. This advance again revolutionized recombinant DNA methodology and further accelerated the pace of biological research.

PCR is a rapid method of cloning DNA molecules that extends the power of recombinant DNA research and, in many cases, eliminates the need for host cells in DNA cloning. Although cell-based cloning is still widely used, PCR is the method of choice in many applications, including molecular biology, human genetics, evolution, development, conservation, and forensics.

PCR copies a specific DNA sequence through a series of *in vitro* reactions and can amplify target DNA sequences present in very small quantities in a population of other DNA molecules. As a prerequisite for PCR, some information about the nucleotide sequence of the target DNA is required. This sequence information is used to synthesize two oligonucleotide primers: one complementary to the 5′ end and one complementary to the 3′ end of the target DNA. The primers are added to a sample of DNA that has been converted into single strands. The primers bind to complementary nucleotides that flank the sequence to be cloned. A heat-stable DNA polymerase is added after hybridization, and it synthesizes a second strand of the target DNA (Figure 17–10). Repeating these steps makes more copies of the DNA.

In practice, the PCR reaction involves three steps. The amount of cloned DNA produced is theoretically limited only by the number of times these steps are repeated.

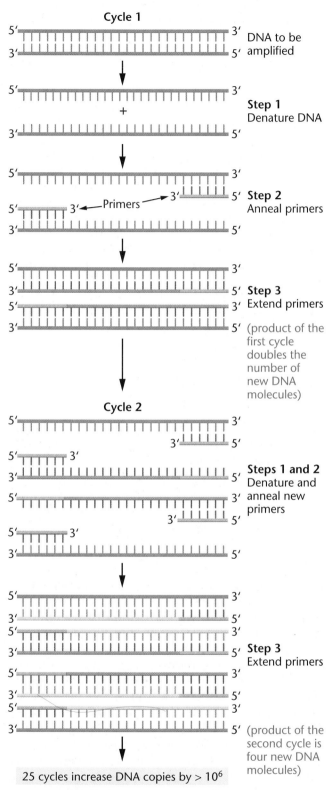

Cycle 1

5′ 3′ DNA to be amplified
3′ 5′

↓

5′ 3′ **Step 1**
+ Denature DNA
3′ 5′

↓

5′ 3′ **Step 2**
—Primers— 3′ 5′ Anneal primers
5′ 3′
3′ 5′

↓

5′ 3′ **Step 3**
3′ 5′ Extend primers
5′ 3′
3′ 5′ (product of the first cycle doubles the number of new DNA molecules)

↓

Cycle 2

5′ 3′
3′ 5′ **Steps 1 and 2**
5′ 3′ Denature and anneal new primers
5′ 3′
3′ 5′

↓

5′ 3′ **Step 3**
3′ 5′ Extend primers
5′ 3′
3′ 5′
5′ 3′
3′ 5′ (product of the second cycle is four new DNA molecules)

↓

25 cycles increase DNA copies by > 10⁶

FIGURE 17–10 In the polymerase chain reaction (PCR), the target DNA is denatured into single strands; each strand is then annealed to a short, complementary primer. DNA polymerase extends the primers in the 3′ direction, using the single-stranded DNA as a template. The result is two newly synthesized double-stranded DNA molecules with the primers incorporated into them. Repeated cycles of PCR can amplify the original DNA sequence by more than a millionfold.

1. The DNA to be cloned is *denatured* into single strands. This DNA can come from many sources including genomic DNA, mummified remains, fossils, or forensic samples such as dried blood, semen, single hairs, or dried samples from medical records. Heating to 90–95°C denatures the double-stranded DNA, which dissociates into single strands (in about 1 minute).

2. The temperature of the reaction is lowered to somewhere between 50°C and 70°C, and at this *annealing* temperature, the primers bind to the denatured DNA. As described above, these primers are synthetic oligonucleotides (15 to 30 nucleotides long) complementary to sequences flanking the target DNA. The primers serve as starting points for synthesizing new DNA strands complementary to the target DNA.

3. A heat-stable form of DNA polymerase (such as *Taq* polymerase*) is added to the reaction mixture. DNA synthesis is carried out at temperatures between 70°C and 75°C (the optimal temperature for *Taq* polymerase). The *Taq* polymerase *extends* the primers by adding nucleotides in the 5′-to-3′ direction, making a double-stranded copy of the target DNA.

Each set of three steps—**denaturation** of the double-stranded DNA, **primer annealing**, and **extension** by polymerase—is a cycle. PCR is a chain reaction because the number of new DNA strands is doubled in each cycle, and the new strands, along with the old strands, serve as templates in the next cycle. Each cycle, which takes 2-5 minutes, can be repeated, and in less than 3 hours, 25 to 30 cycles result in an over 1 millionfold increase in the amount of DNA (see Figure 17–10). The process is automated by machines called *thermocyclers* that can be programmed to carry out a predetermined number of cycles, yielding large amounts of a specific DNA sequence that can be used for many purposes, including cloning into plasmid vectors, sequencing, clinical diagnosis, and genetic screening.

PCR-based DNA amplification has several advantages over cell-based cloning. PCR is rapid and can be carried out in a few hours, rather than the days required for cell-based cloning. In addition, the design of PCR primers is done automatically with computer software, and the commercial synthesis of the oligonucleotides is also fast and economical. If desired, the products of PCR can be cloned into plasmid vectors for further use.

PCR is also very sensitive and amplifies target DNA from vanishingly small DNA samples, including the DNA in a single cell. This feature of PCR is invaluable in several areas, including genetic testing, forensics, and molecular paleontology. By carefully designing primers, DNA sequences that have been partially degraded, contaminated with other materials, or embedded in a matrix (such as amber) can be recovered and amplified, when conventional cloning would be difficult or impossible.

Limitations of PCR

Although PCR is a valuable technique, it does have limitations: some information about the nucleotide sequence of the target DNA must be known, and even minor contamination of

***Taq* polymerase is an enzyme from a bacterium, *Thermus aquaticus*, which lives in hot springs.

the sample with DNA from other sources can cause problems. For example, cells shed from the skin of a laboratory worker can contaminate samples gathered from a crime scene or taken from fossils, making it difficult to obtain accurate results. PCR reactions must always be performed in parallel with carefully designed and appropriate controls.

Other Applications of PCR

PCR DNA cloning is now one of the most widely used techniques in genetics and molecular biology. PCR and its variations have many other applications. It quickly identifies restriction-site variants as well as variations in tandemly repeated DNA sequences that can be used as genetic markers for gene-mapping studies and forensic identification. Gene-specific primers provide a way of screening for mutations in genetic disorders, allowing the location and nature of the mutation to be determined quickly. Primers can be designed to distinguish target sequences that differ by only a single nucleotide. This makes it possible to develop allele-specific probes for genetic testing. Random primers indiscriminately amplify DNA and are particularly advantageous when studying samples from single cells, fossils, or crime scenes where a single hair or even a saliva-moistened postage stamp is the source of the DNA. Using PCR, researchers also explore uncharacterized DNA regions adjacent to known regions and even sequence DNA. PCR has been used to enforce the worldwide ban on the sale of certain whale products and to settle arguments about the pedigree background of purebred dogs. In short, PCR is one of the most versatile techniques in modern genetics.

17.5 Recombinant Libraries Are Collections of Cloned Sequences

Only relatively small DNA segments result from cloning in conventional plasmid vectors, and these may represent only a single gene or a portion of a gene. As a result, a large collection of clones is needed to explore even a small fraction of an organism's genome. A set of DNA clones derived from a single individual or a single population is a cloned *library*. These libraries can represent an entire genome, a single chromosome, or a set of genes expressed in a single cell type.

Genomic Libraries

Ideally, a **genomic library** contains at least one copy of all the sequences in an organism's genome. Genomic libraries are constructed using host-cell cloning methods, since PCR-cloned DNA fragments are relatively small. In making a genomic library, DNA is extracted from cells or tissues, and cut with restriction enzymes, and the resulting fragments are inserted into vectors. Since some vectors (such as plasmids) carry only a few thousand base pairs of inserted DNA, selecting the vector so that it contains the whole genome in the smallest number of clones is an important consideration.

How big does a genomic library have to be to have a 95 or 99 percent chance of containing all the sequences in a genome? The number of clones required to carry all sequences in a genome depends on several factors, including

the average size of the cloned inserts, the size of the genome to be cloned, and the level of probability desired. The number of clones needed in a library can be calculated as

$$N = \frac{\ln(1 - P)}{\ln(1 - f)}$$

where N is the number of required clones, P is the probability of recovering a given sequence, and f is the fraction of the genome in each clone.

Suppose we want to prepare a library of the human genome large enough to have a 99 percent chance of containing all the sequences in the genome. Because the human genome is so large, the choice of vector is a primary consideration in making this library. If we construct the library using a plasmid vector with an average insert size of 5 kb, then over 2.4 million clones would be required for a 99 percent probability of recovering any given sequence from the genome. Because of its size, this library would be difficult to use efficiently. However, if the library was constructed with a YAC vector having an average insert size of 1 Mb (1 Mb = 1 million base pairs), then the library would only need to contain only about 14,000 YACs, making it relatively easy to use. Vectors with large cloning capacities such as YACs are an essential part of the Human Genome Project.

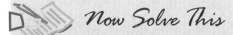

Now Solve This

Problem 13 on page 396 involves calculating how many clones it takes to make a *Drosophila* genomic library using a plasmid vector.

Hint: Remember, there are three parameters in this calculation: the size of the genome, the average size of the cloned inserts, and the probability of having a gene included in the library.

cDNA Libraries

To study specific events in development, cell death, cancer, and other biological processes, a library of the subset of the genome that is expressed in a given cell type at a given time can be a valuable tool. Genomic libraries contain all the genes in a genome, but these cannot be directly used to find genes that are transcriptionally active in a cell.

A **cDNA library** contains DNA copies made from the mRNA molecules present in a cell population at a given time and represents the genes expressed in a cell type at the time the library is made. It is called a cDNA library because the DNA is complementary to the nucleotide sequence of the mRNA.

Clones in a cDNA library are not the same as the clones in a genomic library. Eukaryotic mRNA is processed from pre-mRNA transcripts; intron sequences are removed during processing and a poly-A tail is added to the 3′ end of the molecule. In addition, an mRNA molecule does not include the sequences adjacent to the gene that regulate its activity.

A cDNA library is prepared by isolating mRNA from a population of cells. This is possible because almost all eukaryotic mRNA molecules contain a poly-A tail at their 3′ ends. mRNAs with poly-A tails are isolated and used as templates for the synthesis of **complementary DNA (cDNA)** molecules. The cDNA molecules are subsequently cloned into vectors to produce a cDNA library that is a snapshot of genes that were transcriptionally active at a given time.

To make a cDNA library, mRNA is mixed with oligo-dT primers, which anneal to the poly-A tails, forming a partially double-stranded product (Figure 17–11). The enzyme **reverse transcriptase** extends the primer and synthesizes a complementary DNA copy of the mRNA sequence. The product of this reaction is an mRNA–DNA double-stranded hybrid molecule. Action of the enzyme **RNAse H** introduces nicks in the RNA strand by

partially digesting the RNA. The remaining RNA fragments serve as primers for the enzyme DNA polymerase I. (This situation is similar to synthesis of the lagging strand of DNA in prokaryotes.) DNA polymerase I synthesizes a second DNA strand and removes the RNA primers, producing double-stranded cDNA.

The cDNA can be cloned into a vector by attaching linker sequences to the ends of the cDNA. Linker sequences are short double-stranded oligonucleotides containing a restriction enzyme recognition sequence (e.g., *Eco*RI). After attachment to the cDNAs, the linkers are cut with the enzyme, such as *Eco*RI, and ligated to vectors treated with the same enzyme. Many cDNA libraries are available from cells and tissues in specific stages of development, or different organs such as brain, muscle, kidney, and such. These libraries provide an instant catalog of all the genes active in a cell at a specific time.

A cDNA library can also be prepared using a variation of PCR called **reverse transcriptase PCR (RT-PCR)**. In this procedure, reverse transcriptase is used to generate single-stranded cDNA copies of mRNA molecules as described earlier. This reaction is followed by PCR to copy the single-stranded DNA into double-stranded molecules and then amplify these into many copies. The amplified cDNA is cloned into plasmid vectors to produce a cDNA library. RT-PCR is more sensitive than conventional cDNA preparation and is a powerful tool for identifying mRNAs that may be present in only one or two copies per cell.

17.6 Specific Clones Can Be Recovered from a Library

A genomic library often consists of several hundred thousand clones. To find a specific gene, we need to identify and isolate only the clone or clones containing that gene. We must also determine whether a given clone contains all or only part of the gene we are studying. Several methods allow us to sort through a library (called *screening* the library) to recover clones of interest. The choice of method often depends on the circumstances and available information about the gene being sought.

Probes Identify Specific Clones

Probes are used to screen a library to recover clones of a specific gene. A **probe** is any DNA or RNA sequence that has been labeled in some way and is complementary to some part of a cloned sequence present in the library.

When used in a hybridization reaction, the probe binds to any complementary DNA sequences present in one or more clones. Probes can be labeled with radioactive isotopes, or with compounds that undergo chemical or color reactions to indicate the location of a specific clone in a library.

Probes are derived from a variety of sources—even related genes isolated from other species can be used if enough of the DNA sequence has been conserved. For example, extrachromosomal copies of the ribosomal genes of the clawed frog *Xenopus laevis* can be isolated by centrifugation and cloned into plasmid vectors. Because ribosomal gene sequences are highly conserved, clones carrying the human ribosomal genes can be recovered from a human genomic library using cloned

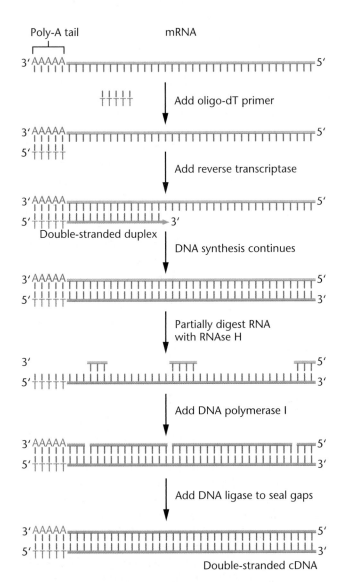

FIGURE 17–11 Producing cDNA from mRNA. Because many eukaryotic mRNAs have a polyadenylated tail (A) of variable length at their 3′ end, a short oligo-dT annealed to this tail serves as a primer for the enzyme reverse transcriptase. Reverse transcriptase uses the mRNA as a template to synthesize a complementary DNA strand (cDNA) and forms an mRNA/cDNA double-stranded duplex. The mRNA is digested with the enzyme RNAse H, producing gaps in the RNA strand. The 3′ ends of the remaining RNA serve as primers for DNA polymerase, which synthesizes a second DNA strand. The result is a double-stranded cDNA molecule that can be cloned into a suitable vector or used directly as a probe for library screening.

1. Colonies of the plasmid library are overlaid with a DNA-binding filter

Colonies transferred to filter

2. Colonies are transferred to filter, then lysed, and DNA is denatured

3. Filter is placed in a heat-sealed bag with a solution containing the labeled probe; the probe hybridizes with denatured DNA from colonies

4. Filter is rinsed to remove excess probe, then dried; X-ray film is placed over the filter for autoradiography

Film

Hybridization of the probe to one colony from the original plate is indicated by a spot on the X-ray film.

5. Using the original plate, cells are picked from the colony that hybridized to the probe.

6. Cells are transferred to a medium for growth and further analysis

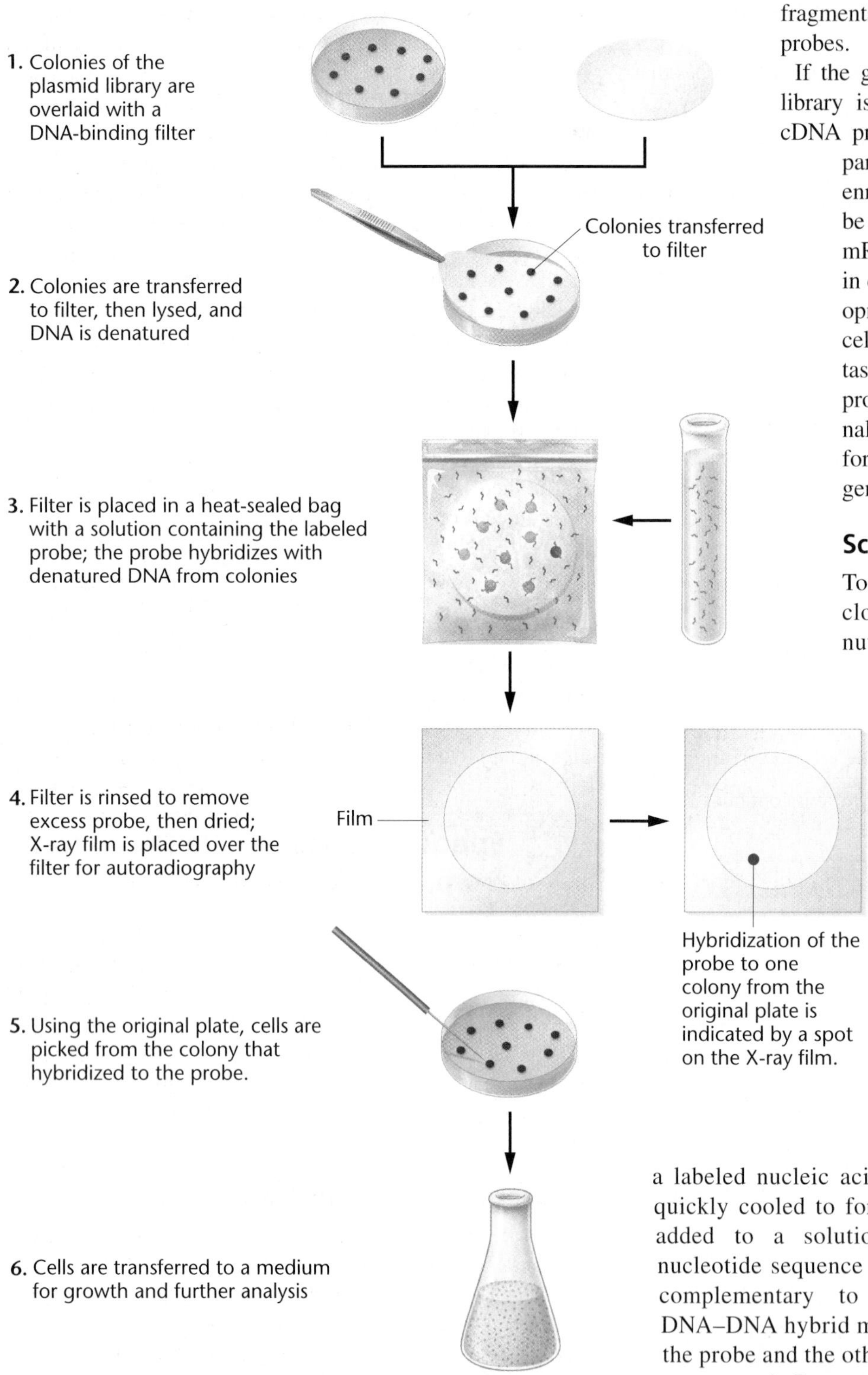

FIGURE 17–12 Screening a plasmid library to recover a cloned gene. The library, present in bacteria on Petri plates, is overlaid with a DNA binding filter, and colonies are transferred to the filter. Colonies on the filter are lysed, and the DNA is denatured to single strands. The filter is placed in a hybridization bag along with buffer and a labeled single-stranded DNA probe. During incubation, the probe forms a double-stranded hybrid with complementary sequences on the filter. The filter is removed from the bag and washed to remove excess probe. Hybrids are detected by placing a piece of X-ray film over the filter and exposing it for a short time. The film is developed, and hybridization events are visualized as spots on the film. Colonies containing the insert that hybridized to the probe are identified from the orientation of the spots. Cells are picked from this colony for growth and further analysis.

fragments of *Xenopus* ribosomal DNA as probes.

If the gene to be selected from a genomic library is expressed in certain cell types, a cDNA probe can be used. This technique is particularly helpful when purified or enriched mRNA for a gene product can be obtained. For example, β-globin mRNA is present in high concentrations in certain stages of red blood cell development. The mRNA purified from these cells can be copied by reverse transcriptase into cDNA molecules for use as a probe. In fact, a cDNA probe was originally used to recover the structural gene for human β-globin from a cloned genomic library.

Screening a Library

To screen a *genomic plasmid library*, clones from the library are grown on nutrient agar plates, where they form hundreds or thousands of colonies (Figure 17–12). A replica of the colonies on each plate is made by gently pressing a nylon or nitrocellulose filter onto the plate's surface; this transfers the pattern of bacterial colonies from the plate to the filter. The filter is processed to lyse the bacterial cells, denature the double-stranded DNA released from the cells into single strands, and bind these strands to the filter.

The DNA on the filter is screened by incubation with a labeled nucleic acid probe. The probe is heated and quickly cooled to form single-stranded molecules, and added to a solution containing the filter. If the nucleotide sequence of any of the DNA on the filter is complementary to the probe, a double-stranded DNA–DNA hybrid molecule will form (one strand from the probe and the other from the cloned DNA on the filter). For example, if a cDNA probe of β-globin is used, it will bind to cloned DNA encoding the β-globin gene. After incubation of the probe and the filter, unbound and/or excess probe molecules are washed away, and the filter is assayed to detect the hybrid molecules that remain. If a radioactive probe has been used, the filter is overlaid with a piece of X-ray film. Radioactive decay in the probe molecules bound to DNA on the filter will expose the film, producing dark spots on the film. These spots represent colonies on the plate containing the cloned gene of interest (Figure 17–12). Using the position of spots on the film as a guide, the corresponding colonies on the plate are identified and recovered. The cloned DNA they contain can be used in further experiments. With some nonradioactive probes, a chemical reaction emits photons of light (chemiluminescence) to expose the photographic film and reveal the location of colonies carrying the gene of interest.

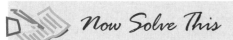

Now Solve This

Problem 18 on page 396 involves selecting a cloned gene from a cDNA library.

Hint: cDNA clones do not have all the sequences of a genomic library, but do have the coding sequences, and can be selected with the proper probe.

17.7 Cloned Sequences Can Be Analyzed in Several Ways

The recovery and identification of genes and other specific DNA sequences by cloning or by PCR is a powerful tool for analyzing genomic structure and function. In fact, much of the Human Genome Project is based on such techniques. In the following sections, we consider some of these methods, which are used to answer questions about the organization and function of cloned sequences.

Restriction Mapping

One of the first steps in characterizing a DNA clone is the construction of a **restriction map**. A restriction map establishes the number, order, and distance between restriction enzyme cleavage sites along a cloned segment of DNA. Restriction maps for different cloned DNAs are usually different enough to serve as an identity tag for that clone. Recall that restriction map units are expressed in **base pairs (bp)** or, for longer lengths, **kilobase (kb)** pairs. Restriction maps provide information about the length of a cloned insert and the location of restriction enzyme cleavage sites within the clone. These data can be used to re-clone fragments of a gene or compare its internal organization with that of other cloned sequences.

Fragments generated by cutting DNA with restriction enzymes can be separated by gel electrophoresis, a method that separates fragments by size, with the smallest pieces moving farthest. The fragments appear as a series of bands that can be visualized by staining the DNA with ethidium bromide and viewing under ultraviolet illumination.

Figure 17–13 shows the construction of a restriction map from a cloned DNA segment. For this map, let's begin with a cloned DNA segment 7.0 kb in length. Three samples of the cloned DNA are digested with restriction enzymes—one with *Hin*dIII, one with *Sal*I, and one with

both *Hin*dIII and *Sal*I. The fragments are separated by gel electrophoresis (discussed in Chapter 9). The resulting bands are photographed or scanned for analysis. The molecular weights of the fragments are measured by comparing their location on the gel to a set of molecular weight standards run in adjacent lanes. The restriction map is then constructed by analyzing the number and length of the fragments. When the DNA is cut with *Hin*dIII, two fragments (0.8 and 6.2 kb) are produced, confirming that the cloned insert is 7.0 kb in length and that there is only one cleavage site for this enzyme (located 0.8 kb from one end). When the DNA is cut with *Sal*I, two fragments (1.2 and 5.8 kb) are produced, indicating that this enzyme also has one cutting site, located 1.2 kb from one end of the cloned DNA segment.

These results show that each enzyme has one restriction site, but the relative position of the two restriction enzyme cleavage sites is unknown. From the information available, two different maps are possible. In Figure 17–13, model 1 shows the *Hin*dIII site located 0.8 kb from one end and the *Sal*I site 1.2 kb from the same end. The alternative map, model 2, locates the *Hin*dIII site 0.8 kb from one end and the *Sal*I site 1.2 kb from the other end.

The correct model is determined by analyzing the results from the sample digested with both *Hin*dIII and *Sal*I. Model 1 predicts that digestion with both enzymes will generate three fragments: 0.4, 0.8, and 5.8 kb in length; model 2 predicts that there will be three fragments of 0.8, 5.0, and 1.2 kb. The number and size of the fragments seen on the gel after digestion with both enzymes indicates that model 1 is correct (see Figure 17–13).

Restriction maps are an important way of characterizing cloned DNA, and they can be constructed in the absence of any other information about the DNA, including its coding capacity or function. In conjunction with other techniques, restriction mapping can define the boundaries of a gene, dissect the internal organization of a gene and its flanking regions, and locate mutational sites within genes.

Restriction digestion of clones plays an important role in mapping genes to specific human chromosomes and to defined regions of individual chromosomes. In addition, if a restriction site maps close to a mutant allele, this sequence can be used as a marker in genetic testing to identify carriers of recessively inherited disorders, or to prenatally diagnose a fetal genotype. This topic will be discussed in Chapter 19.

Nucleic Acid Blotting

Many of the techniques described in this chapter rely on hybridization between complementary nucleic acid (DNA or RNA) molecules. One of the most widely used methods for detecting such hybrids is called **Southern blotting** (after Edward Southern, who devised it). The Southern blot method can be used to identify which clones in a library contain a given DNA sequence (such as a ribosomal gene or a β-globin gene), and to characterize the size of the fragments, thus deriving a restriction map of the cloned DNA. Southern blots

Web Tutorial 17.5
Restriction Mapping

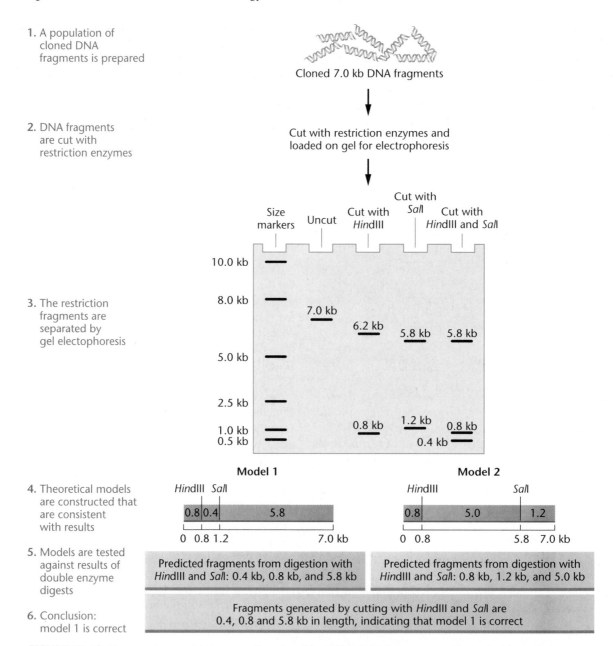

1. A population of cloned DNA fragments is prepared

Cloned 7.0 kb DNA fragments

2. DNA fragments are cut with restriction enzymes

Cut with restriction enzymes and loaded on gel for electrophoresis

3. The restriction fragments are separated by gel electophoresis

Size markers | Uncut | Cut with HindIII | Cut with SalI | Cut with HindIII and SalI

10.0 kb
8.0 kb
7.0 kb
6.2 kb
5.8 kb
5.8 kb
5.0 kb
2.5 kb
1.0 kb
0.5 kb
0.8 kb
1.2 kb
0.8 kb
0.4 kb

4. Theoretical models are constructed that are consistent with results

Model 1
HindIII SalI
0.8 | 0.4 | 5.8
0   0.8 1.2   7.0 kb

Model 2
HindIII   SalI
0.8 | 5.0 | 1.2
0   0.8   5.8 7.0 kb

5. Models are tested against results of double enzyme digests

Predicted fragments from digestion with HindIII and SalI: 0.4 kb, 0.8 kb, and 5.8 kb

Predicted fragments from digestion with HindIII and SalI: 0.8 kb, 1.2 kb, and 5.0 kb

6. Conclusion: model 1 is correct

Fragments generated by cutting with HindIII and SalI are 0.4, 0.8 and 5.8 kb in length, indicating that model 1 is correct

FIGURE 17–13 Constructing a restriction map. Samples of the 7.0 kb DNA fragments are digested with restriction enzymes: One sample is digested with HindIII, one with SalI, and one with both HindIII and SalI. The resulting fragments are separated by gel electrophoresis. The sizes of the separated fragments are measured by comparing them with molecular-weight standards in an adjacent lane. Cutting the DNA with HindIII generates two fragments: 0.8 kb and 6.2 kb. Cutting with SalI produces two fragments: 1.2 kb and 5.8 kb. Models are constructed to predict the fragment sizes generated by cutting with both HindIII and SalI. Model 1 predicts that 0.4, 0.8, and 5.8 kb fragments will result from cutting with both enzymes. Model 2 predicts that 0.8, 1.2, and 5.0 kb fragments will result. Comparing the predicted fragments with those observed on the gel indicates that model 1 is the correct restriction map.

can also be used to identify fragments carrying specific genes in genomic DNA digested with a restriction enzyme.

The Southern blotting technique has two components: separation of DNA fragments by gel electrophoresis and hybridization of the fragments using labeled probes. Gel electrophoresis can be used to characterize the number of fragments produced by restriction digestion and to estimate their molecular weights. Restriction enzyme digestion of large genomes, such as the human genome, with more than 3 bil-

lion nucleotides, will produce a large number of fragments that will run together on a gel to produce a continuous smear. The identification of specific fragments is accomplished in the next step. Hybridization characterizes the DNA sequences present in the fragments. The DNA to be characterized by Southern blot hybridization can come from several sources, including clones selected from a library or genomic DNA. Our discussion will use examples from both sources to show how Southern blots are used to characterize the number, size, organization, and sequence content of DNA fragments.

To make a Southern blot, DNA is cut into fragments with one or more restriction enzymes and the fragments are separated by gel electrophoresis (Figure 17–14). In preparation for hybridization, DNA in the gel is denatured with alkaline treatment to form single-stranded fragments. The gel is then overlaid with a DNA-binding membrane, usually nitrocellulose or a nylon derivative. The DNA fragments in the gel are transferred to the membrane by placing the membrane and the gel on a wick (often a sponge) sitting in a buffer solution. Layers of paper towels or blotting paper are placed on top of the filter and held in place with a weight. Capillary action draws the buffer solution up through the gel, transferring the DNA fragments to the membrane.

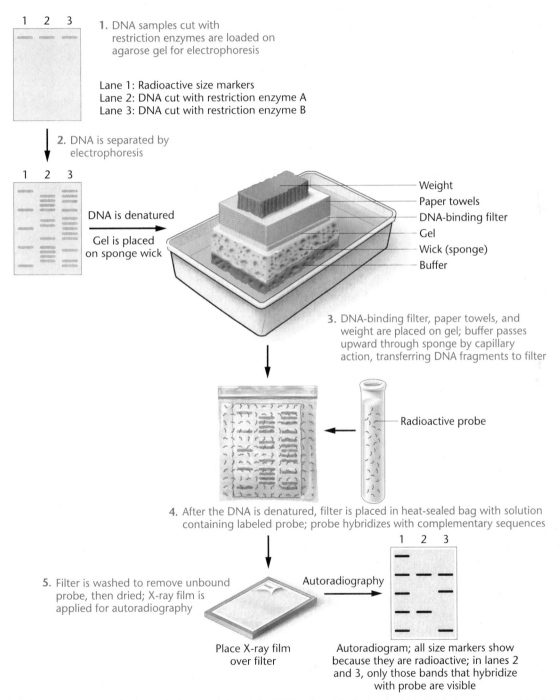

FIGURE 17–14 In the Southern blotting technique, samples of the DNA to be probed are cut with restriction enzymes and the fragments are separated by gel electrophoresis. The pattern of fragments is visualized and photographed under ultraviolet illumination by staining the gel with ethidium bromide. The gel is then placed on a sponge wick in contact with a buffer solution and covered with a DNA-binding filter. Layers of paper towels or blotting paper are placed on top of the filter and held in place with a weight. Capillary action draws the buffer through the gel, transferring the pattern of DNA fragments from the gel to the filter. The DNA fragments on the filter are then denatured into single strands and hybridized with a labeled DNA probe. The filter is washed to remove excess probe and overlaid with a piece of X-ray film for autoradiography. The hybridized fragments show as bands on the X-ray film.

(a)   **(b)**

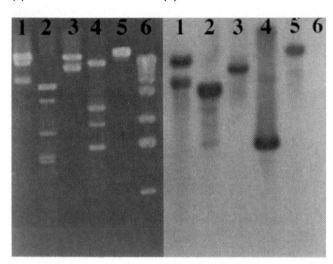

FIGURE 17–15 (a) Agarose gel stained with ethidium bromide to show DNA fragments. (b) Exposed X-ray film of a Southern blot prepared from the gel in part (a). Only those bands containing DNA sequences complementary to the probe show hybridization.

The filter is placed in a heat-sealed bag with a radioatively labeled, single-stranded DNA probe for hybridization. DNA fragments on the filter that are complementary to the probe's nucleotide sequence form double-stranded hybrids. Excess probe is washed away, and the hybridized fragments are visualized on a piece of film (Figure 17–14).

In Figure 17–15, researchers cut samples of DNA with several restriction enzymes. The pattern of fragments obtained for each restriction enzyme is shown in Figure 17–15(a). A Southern blot of this gel is shown in Figure 17–15(b). The probe hybridized only to bands containing complementary sequences, identifying fragments of interest.

In addition to characterizing cloned DNAs, Southern blots can be used to create restriction maps within and near a gene and to identify DNA fragments carrying all or parts of a single gene in a mixture of fragments. By comparing the pattern of bands in normal cells to those from patients with genetic disorders or cancer, Southern blots also detect rearrangements, deletions, and duplications in genes associated with human genetic disorders and cancers.

To determine whether a gene is transcriptionally active in a given cell or tissue type, a related blotting technique probes for the presence of mRNA complementary to a cloned gene. To do this, mRNA is extracted from a specific cell or tissue type and separated by gel electrophoresis. The resulting pattern of RNA bands is transferred to a membrane, as in Southern blotting. The membrane is then hybridized to a labeled single-stranded DNA probe derived from a cloned copy of the gene. If mRNA complementary to the DNA probe is present, it will be detected as a band on the film. Because the original procedure (DNA bound to a filter) is known as a Southern blot, this procedure (RNA bound to a filter) is called a **northern blot.** (Following this somewhat perverse logic, another procedure involving proteins bound to a filter is known as a **western blot.**)

Northern blots provide information about the expression of specific genes and are used to study patterns of gene expression in embryonic tissues, cancer, and genetic disorders. Northern blots also detect alternatively spliced mRNAs (multiple types of transcripts derived from a single gene) and are used to derive other information about transcribed mRNAs. If marker RNAs of known size are run as controls, northern blots can be used to measure the size of a gene's mRNA transcripts. In addition, the amount of transcribed RNA present in a cell type is related to the density of the RNA band on the film. Measuring band density gives the relative transcriptional activity. Thus, northern blots characterize and quantify the transcriptional activity of genes in different cells, tissues, and organisms.

17.8 DNA Sequencing: The Ultimate Characterization of a Clone

In a sense, a cloned DNA molecule, or any DNA, from a clone to a genome, is completely characterized only when its nucleotide sequence is known. The ability to sequence DNA has greatly enhanced our understanding of genome organization and our knowledge of genes, including structure, function, and mechanisms of regulation.

The most commonly used method of DNA sequencing was developed by James Sanger and his colleagues. In this procedure, a DNA molecule whose sequence is to be determined is converted to single strands that are used as a template for synthesizing a series of complementary strands. Each of these strands randomly terminates at a different,

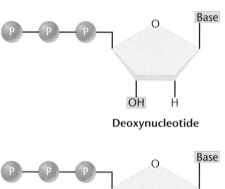

FIGURE 17–16 Deoxynucleotides (top) have an OH group at the 3′ position in the ribose molecule. Dideoxynucleotides (bottom) lack an OH group at this position, preventing formation of a phosphodiester bond with another nucleotide, terminating further elongation of the template strand.

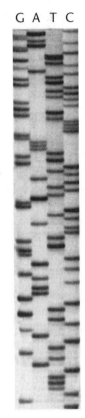

specific nucleotide. This produces a series of DNA fragments that are separated by electrophoresis and analyzed to reveal the sequence of the DNA. In the first step of this reaction, DNA is heated to denature it and form single strands. The single-stranded DNA is mixed with primers that anneal to the 3′ end of the DNA. Samples of the primer-bound single-stranded DNA are distributed into four tubes. In the next step, the four deoxyribonucleotide triphosphates (dATP, dCTP, dGTP, and dTTP) are added to each tube. In addition, each tube receives a small amount of one modified deoxyribonucleotide (Figure 17–16), called a **dideoxynucleotide** (e.g., ddATP, ddCTP, ddGTP, and ddTTP). Dideoxynucleotides have a 3′-H instead of a 3′-OH group. One of the deoxyribonucleotides or the primer is labeled with radioactivity for later analysis of the sequence. DNA polymerase is added to each tube, and the primer is elongated in the 5′ to 3′ direction, forming a complementary strand to the template.

As DNA synthesis takes place, the polymerase occasionally inserts a dideoxynucleotide instead of a deoxyribonucleotide into a growing DNA strand. Since the dideoxynucleotide has no 3′-OH group, it cannot form a 3′ bond with another nucleotide, and DNA synthesis terminates. For example, in the tube with added ddATP, the polymerase inserts ddATP instead of dATP, causing termination of chain elongation (Figure 17–17). As the reaction proceeds, the tube with ddATP will accumulate DNA molecules that terminate at all the different positions containing A. In the other tubes, reactions terminate at G, C, and T, respectively. The DNA fragments from each reaction tube (one for each dideoxynucleotide) are separated in adjacent lanes by gel electrophoresis. The result is a series of bands forming a ladderlike pattern that is visualized by developing film exposed to the gel (Figure 17–18). The nucleotide sequence is read directly from bottom to top, corresponding to the 5′ to 3′ sequence of the DNA strand complementary to the template.

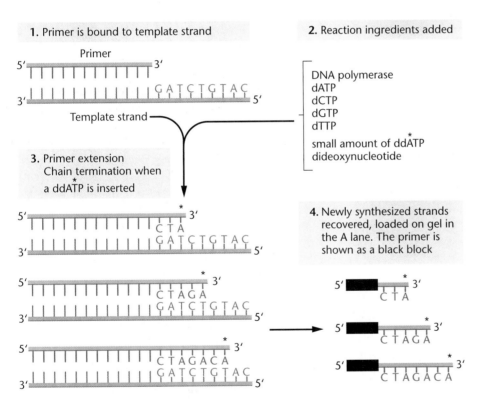

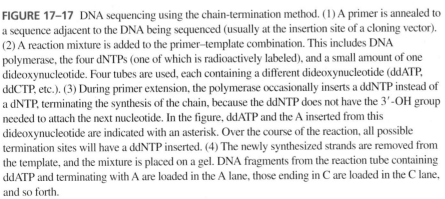

FIGURE 17–17 DNA sequencing using the chain-termination method. (1) A primer is annealed to a sequence adjacent to the DNA being sequenced (usually at the insertion site of a cloning vector). (2) A reaction mixture is added to the primer–template combination. This includes DNA polymerase, the four dNTPs (one of which is radioactively labeled), and a small amount of one dideoxynucleotide. Four tubes are used, each containing a different dideoxynucleotide (ddATP, ddCTP, etc.). (3) During primer extension, the polymerase occasionally inserts a ddNTP instead of a dNTP, terminating the synthesis of the chain, because the ddNTP does not have the 3′-OH group needed to attach the next nucleotide. In the figure, ddATP and the A inserted from this dideoxynucleotide are indicated with an asterisk. Over the course of the reaction, all possible termination sites will have a ddNTP inserted. (4) The newly synthesized strands are removed from the template, and the mixture is placed on a gel. DNA fragments from the reaction tube containing ddATP and terminating with A are loaded in the A lane, those ending in C are loaded in the C lane, and so forth.

FIGURE 17–18 DNA sequencing gel showing the separation of newly synthesized fragments in the four sequencing reactions (one per lane). To obtain the base sequence of the DNA fragment, the gel is read from the bottom, beginning with the lowest band in any lane, then the next lowest, then the next, and so on. For example, the sequence of the DNA on this gel begins with 5′-CG at the very bottom of the gel, proceeds upwards as 5′-CGCTTTCATGTCA, and so forth.

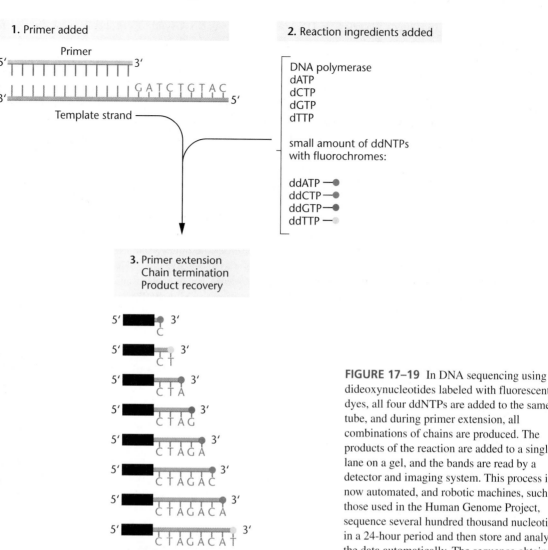

1. Primer added

Primer

2. Reaction ingredients added

DNA polymerase
dATP
dCTP
dGTP
dTTP

small amount of ddNTPs
with fluorochromes:

ddATP
ddCTP
ddGTP
ddTTP

Template strand

3. Primer extension
Chain termination
Product recovery

4. Electrophoresis, imaging, data analysis

FIGURE 17–19 In DNA sequencing using dideoxynucleotides labeled with fluorescent dyes, all four ddNTPs are added to the same tube, and during primer extension, all combinations of chains are produced. The products of the reaction are added to a single lane on a gel, and the bands are read by a detector and imaging system. This process is now automated, and robotic machines, such as those used in the Human Genome Project, sequence several hundred thousand nucleotides in a 24-hour period and then store and analyze the data automatically. The sequence obtained is by extension of the primer and is read from the newly synthesized strand, not the template strand. Thus, the sequence obtained begins with 5′-CTAGACATG.

DNA sequencing is now largely automated and uses machines that can sequence several hundred thousand nucleotides per day. In this procedure, each of the four dideoxynucleotide analogs is labeled with a different colored fluorescent dye (Figure 17–19), so that chains terminating in adenosine are labeled with one color, those ending in cytosine with another color, and so forth. All four labeled dideoxynucleotides are added to a single tube, and after primer extension by DNA polymerase, the reaction products are loaded into one lane on a gel. The gel is scanned with a laser, causing each band to fluoresce a different color. A detector in the sequencing machine reads the color of each band and determines whether it represents an A, T, C, or G. The data are represented as a series of colored peaks, each corresponding to one nucleotide in the sequence (Figure 17–20).

DNA Sequencing and Genome Projects

DNA sequencing is the heart of genome projects. Using a combination of recombinant DNA techniques and nucleotide sequencing, researchers have sequenced the genomes of more than 375 species of prokaryotes, with hundreds more projects underway. The Human Genome Project sponsored by the U.S. Department of Energy and the National Institutes of Health and a private project sponsored by the biotechnology company Celera used a combination of genomic and chromosome-specific libraries to finish sequencing the coding portion of the human genome in 2003. Genome projects for many other eukaryotic species have also been completed with the use of DNA sequencing technology, and many more are underway.

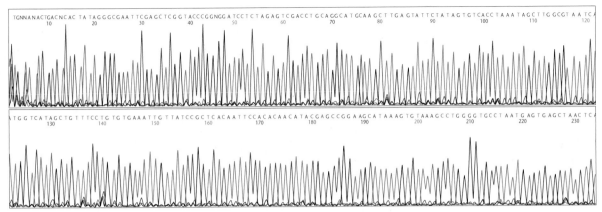

FIGURE 17–20 Automated DNA sequencing using fluorescent dyes, one for each base. Each peak represents the correct nucleotide in the sequence. The sequence extending from the primer (which is not shown here) starts at the upper left of the diagram and extends rightward. The bases labeled as N are ambiguous and cannot be identified with certainty. These ambiguous base readings are more likely to occur near the primer because the quality of sequence determination deteriorates the closer the sequence is to the primer. The separated bases are read in order along the axis from left to right. Thus, the sequence begins as 5′-TGNNANACTGACNCAC. Numbers below the bases indicate length of the sequence in base pairs.

GENETICS, TECHNOLOGY, AND SOCIETY

Beyond Dolly: The Cloning of Humans

The death of Dolly the sheep, in February 2003, marked a poignant end to the beginning of the human cloning debate.

Six years earlier, her birth took the world by surprise. Before Dolly, the idea that an animal could be cloned from the cells of an adult animal was science fiction—something from *Brave New World* or *The Boys from Brazil*. For decades, scientists believed that it would be impossible to clone mammals, as DNA from adult cells could not be reprogrammed to code for the development of a new, complete organism. But then Dolly appeared.

Dolly, who was cloned from a frozen udder cell of a long-dead sheep, was brought into being by a group of embryologists led by Ian Wilmut and Keith Campbell at the Roslin Institute in Scotland. Their goal was to clone transgenic farm animals that secrete pharmaceutical products, such as blood-clotting factors or insulin, into their milk. In this way, herds of identical animals might be used as bioreactors to synthesize large quantities of medically important proteins.

The cloning method that Wilmut and Campbell used to create Dolly—a procedure called *nuclear transfer*—was first suggested by embryologist Hans Spemann in 1938. The method is simply to replace the nucleus of an egg with the nucleus from an adult cell, thereby creating a hybrid zygote. In theory, the genetic information in the donor nucleus should direct all further embryonic development, and the new organism should be a genetic replica of the adult that donated the nucleus. Although the procedure sounds simple in theory, it proved to be extremely difficult in practice because adult nuclei express only a small subset of the genes required for embryonic development. Wilmut and Campbell overcame this limitation by reprogramming the adult nuclei before transferring them into recipient eggs. They did this by starving the donor cells so they became quiescent. In addition, they passed an electric current through the recipient egg. For unknown reasons, these procedures turned on the silent genes within the differentiated cell nucleus. To create Dolly, over 200 udder-cell nuclei were transferred into eggs. Of these nuclear transfers, only 29 developed into embryos; and 13 of them were implanted into surrogate mother ewes. One pregnancy resulted, which culminated in the birth of Dolly. Although Dolly was the first mammal to be cloned from adult cells, the method has since been used to clone mice, pigs, cattle, rabbits, goats, an ox, a mule, a horse, and a cat.

Dolly's birth not only shattered scientists' views of mammalian cloning but it triggered an avalanche of controversy. The idea that humans might be cloned was denounced as immoral, repugnant, and ethically wrong. Within days of the announcement of Dolly's birth, bills were introduced into the United States Congress to prohibit research into human cloning, and worldwide bans were called for. Frightening scenarios were proposed—rich and powerful people cloning themselves for reasons of vanity, people with serious illnesses cloning replicas to act as organ donors, and legions of human clones suffering loss of autonomy, individuality, and kinship ties.

Is it really possible to clone humans? And if so, *should* we create human clones? The answer to the first question is simple: The same technology used to create Dolly could be used to clone a human. Most of the technical procedures are already used for human *in vitro* fertilizations, and it seems likely that adult human nuclei could be reprogrammed similarly to the adult sheep nuclei that created Dolly. However, the cloning process is

(Cont. on the next page)

extremely inefficient. Nuclear transfers into hundreds of human eggs would be required to yield one full-term pregnancy. As yet, there have been no verified successful attempts to clone a human, despite a few high-profile claims by several doctors and a religious sect.

The answer to the second question is not as simple. To begin with, it is necessary to understand the differences between two kinds of cloning—reproductive and therapeutic. Reproductive cloning results in the creation of an animal, such as Dolly. In contrast, therapeutic cloning creates early-stage embryos for the purpose of harvesting stem cells—cells that have the potential to treat various diseases. As legislators and the public tend to equate these two forms of cloning, research into human therapeutic cloning has been affected by all-encompassing bans that target reproductive cloning.

Although most scientists believe that research into therapeutic cloning should continue, they conclude that we should refrain from human reproductive cloning for both scientific and ethical reasons. The most serious concerns are that many cloned animals appear to suffer developmental defects. About 12% of cloned mice and 38% of cloned goats show congenital abnormalities such as oversized internal organs, and respiratory, circulatory, and immunological defects. The ethical arguments against human reproductive cloning involve concerns about potential abuses of the technology and threats to the dignity of human procreation. Some people worry that human clones might be discriminated against or cannibalized for spare parts and even question whether they would have the same personhood as nonclones.

In contrast, proponents of human cloning suggest that the technology should be considered merely another fertility treatment, like *in vitro* fertilization—enabling infertile couples to have children that are genetically related to them or who do not carry a genetic defect of one parent.

As the debate over human cloning continues, Dolly will take her place among the important players in the history of modern science. Her remains have been preserved and are now on display in the National Museum of Scotland in Edinburgh. She is survived by several of her six lambs and the sadness of those who knew her and cared for her.

References

Jaenisch, R., and I. Wilmut, 2001. Don't clone humans! *Science* 291: 2552.

CHAPTER SUMMARY

1. The cornerstone of recombinant DNA technology is a class of enzymes called restriction enzymes, which cut DNA at specific recognition sites. The fragments thus produced are joined with DNA vectors to form recombinant DNA molecules.

2. Vectors replicate autonomously in host cells and facilitate the manipulation of the newly created recombinant DNA molecules. Vectors are constructed from many sources, including bacterial plasmids.

3. Recombinant DNA molecules are transferred into a host cell, and cloned copies are produced during host-cell replication. A variety of host cells may be used for replication, including bacteria and yeast. Cloned copies of foreign DNA sequences can be recovered, purified, and analyzed.

4. The polymerase chain reaction (PCR) is a method for amplifying a specific DNA sequence present in a population of DNA sequences, such as genomic DNA. PCR allows DNA to be amplified without host cells and is a rapid, sensitive method with wide-ranging applications.

5. Once cloned, DNA sequences are analyzed through a variety of methods, including restriction mapping and DNA sequencing. Other methods such as Southern blotting use hybridization to identify genes and flanking regulatory regions from restriction enzyme-digested genomic DNA.

KEY TERMS

anneal, 379
bacterial artificial
  chromosome (BAC), 381
base pair (bp), 387
cDNA library, 384
clone, 377
complementary DNA (cDNA), 385
denaturation, 383
dideoxynucleotide, 391
DNA ligase, 379
expression vectors, 381

extension, 383
genomic library, 384
kilobase (kb), 387
northern blot, 390
plasmid, 379
polylinker site, 380
polymerase chain reaction (PCR), 382
primer annealing, 383
probe, 385
recognition sequence, 377
recombinant DNA, 377

recombinant DNA molecule, 377
restriction enzyme, 377
restriction fragments, 377
restriction map, 387
reverse transcriptase, 385
reverse transcriptase PCR (RT-PCR), 385
RNAse H, 385
Southern blotting, 387
vector, 377
western blot, 390
yeast artificial chromosome (YAC), 381

INSIGHTS AND SOLUTIONS

The recognition site for the restriction enzyme *Sau*3AI is GATC (see Figure 17–3); the recognition site for the enzyme *Bam*HI is GGATCC, where the four internal bases are identical to the *Sau*3AI site. Therefore, the single-stranded ends produced by the two enzymes can be identical. Suppose you have a cloning vector that contains a *Bam*HI site and foreign DNA that you have cut with *Sau*3AI. (a) Can this DNA be ligated into the *Bam*HI site of the vector, and if so, why? (b) Can the DNA segment cloned into this site be cut from the vector with *Sau*3AI? With *Bam*HI? What potential problems do you see with the use of *Bam*HI?

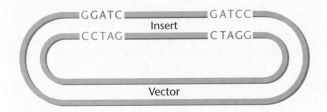

Solution:

(a) DNA cut with *Sau*3AI can be ligated into the vector's *Bam*HI site because the single-stranded ends generated by the two enzymes are identical.

(b) The DNA can be cut from the vector with *Sau*3AI because the recognition sequence for this enzyme (GATC) is maintained on each side of the insert. Recovering the cloned insert with *Bam*HI is more problematic. In the ligated vector, the conserved sequences are GGATC (left) and GATCC (right). The correct base will *follow* the conserved sequence (to produce GGATCC on the left) only about 25 percent of the time, and the correct base will *precede* the conserved sequence (and produce GGATCC on the right) about 25 percent of the time as well. Thus, *Bam*HI will be able to cut the insert from the vector (0.25 × 0.25 = 0.0625), or only about 6 percent of the time.

PROBLEMS AND DISCUSSION QUESTIONS

1. What roles do restriction enzymes, vectors, and host cells play in recombinant DNA studies?

2. Why is oligo-dT an effective primer for reverse transcriptase?

3. The human insulin gene contains a number of introns. In spite of the fact that bacterial cells do not excise introns from mRNA, explain how a gene like this can be cloned into a bacterial cell and produce insulin.

4. Restriction enzymes recognize palindromic sequences in intact DNA molecules and cleave the double-stranded helix at these sites. Inasmuch as the bases are internal in a DNA double helix, how is this recognition accomplished?

5. Although the potential benefits of cloning in higher plants are obvious, the development of this field has lagged behind cloning in bacteria, yeast, and mammalian cells. Can you think of a reason for this?

6. Using DNA sequencing on a cloned DNA segment, you recover the following nucleotide sequence. Does this segment contain a palindromic recognition site for a restriction enzyme? What is the double-stranded sequence of the palindrome? What enzyme would cut at this site? (Consult Figure 17–3 for a list of restriction enzyme recognition sites.)

CAGTATCCTAGGCAT

7. Restriction enzyme sites are palindromic; that is, they read the same in the 5′-to-3′ direction on each strand of DNA. What is the advantage of having restriction sites organized in this way?

8. List the advantages and disadvantages of using plasmids and YACs as cloning vectors.

9. Listed below are the cleavage sites for some restriction enzymes that recognize the sequences of four bases and six bases. Assuming random distribution and equal amounts of each nucleotide in the DNA to be cut, on average, how far apart are each of these restriction sites?

*Taq*I	TCGA	*Hind*III	AAGCTT
*Alu*I	AGCT	*Bal*I	TGGCCA
*Hae*III	GGCC	*Bam*HI	GGATCC

10. Some restriction enzymes have recognition sites that are specific and unambiguous, whereas others have sites that are ambiguous, such that any purine, pyrimidine, or nucleotide can occupy a given position in the cutting site. In the examples below, *Not*I has an unambiguous cutting site, *Hinf*I has an ambiguous cutting site (N = any nucleotide), and *Xho*II also has an ambiguous cutting site (Pu = any purine and Py = any purine). Assuming random distribution and equal amounts of each nucleotide in the DNA to be cut, on average, how far apart are each of these restriction cutting sites?

*Not*I	GCGGCCGC
*Hinf*I	GANTC
*Xho*II	PuGATCPy

11. What are the advantages of using a restriction enzyme with relatively few cutting sites? When would you use such enzymes?

12. An ampicillin-resistant, tetracycline-resistant plasmid, pBR322, is cleaved with *Pst*I, which cleaves within the ampicillin resistance gene. The cut plasmid is ligated with *Pst*I-digested *Drosophila* DNA to prepare a genomic library, and the mixture is used to transform *E. coli* K12. (a) Which antibiotic should be added to the medium to select cells that have incorporated a plasmid? (b) What growth pattern should be selected to obtain plasmids containing *Drosophila* inserts? (c) How can you explain the presence of colonies that are resistant to both antibiotics?

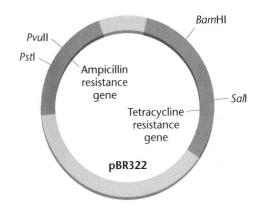

13. Plasmids isolated from the clones in Problem 12 are found to have an average length of 5 kb. Given that the *Drosophila* genome is 1.5×10^5 kb long, how many clones would be necessary to give a 99 percent probability that this library contains all genomic sequences?

14. In a control experiment, a plasmid containing a *Hin*dIII site within a kanamycin resistance gene is cut with *Hin*dIII, re-ligated, and used to transform *E. coli* K12 cells. Kanamycin-resistant colonies are selected, and plasmid DNA from these colonies is subjected to electrophoresis. Most of the colonies contain plasmids that produce single bands that migrate at the same rate as the original intact plasmid. A few colonies, however, produce two bands, one of original size and one that migrates much higher in the gel. Diagram the origin of this slow band as a product of ligation.

15. You have just created the world's first genomic library from the African okapi, a relative of the giraffe. No genes from this genome have been previously isolated or described. You wish to isolate the gene encoding the oxygen-transporting protein β-globin from the okapi library. This gene has been isolated from humans, and its nucleotide sequence and amino acid sequence are available in databases. Using the information available about the human β-globin gene, what two strategies can you use to isolate this gene from the okapi library?

16. When making cDNA, the single-stranded DNA produced by reverse transcriptase can be made double stranded by treatment with DNA polymerase I. However, no primer is required with the DNA polymerase. Why is this?

17. What should you consider in deciding which vector to use in constructing a genomic library of eukaryotic DNA?

18. You are given a cDNA library of human genes prepared in a bacterial plasmid vector. You are also given the cloned yeast gene that encodes EF-1a, a protein that is highly conserved in protein sequence among eukaryotes. Outline how you would use these resources to identify the human cDNA clone encoding EF-1a.

19. Once you have isolated the human cDNA clone for EF-1a in Problem 18, you sequence the clone and find that it is 1384 nucleotide pairs long. Using this cDNA clone as a probe, you isolate the DNA encoding EF-1a from a human genomic library. The genomic clone is sequenced and found to be 5282 nucleotides long. What accounts for the difference in length observed between the cDNA clone and the genomic clone?

20. You have recovered a cloned DNA segment and determine that the insert is 1300 bp in length. To characterize this cloned segment, you isolate the insert and decide to construct a restriction map. Using enzyme I and enzyme II, followed by gel electrophoresis, you determine the number and size of the fragments produced by enzymes I and II alone and in combination shown in the following table. Construct a restriction map from these data, showing the positions of the restriction sites relative to one another and the distance between them in units of base pairs.

Enzymes	Restriction Fragment Sizes (bp)
I	350, 950
II	200, 1100
I and II	150, 200, 950

21. To create a cDNA library, cDNA can be cloned into vectors. In analyzing cDNA clones, it is often difficult to find clones that are full length—that is, extending to the 5′-end of the mRNA. Why is this so?

22. Although the capture and trading of great apes has been banned in 112 countries since 1973, it is estimated that about 1000 chimpanzees are removed annually from Africa and smuggled into Europe, the United States, and Japan. This illegal trade is often disguised by simulating births in captivity. Until recently, genetic identity tests to uncover these illegal activities were not used because of the lack of highly polymorphic markers and the difficulties of obtaining chimpanzee blood samples. Recently, a study was reported in which DNA samples were extracted from freshly plucked chimpanzee hair roots and used as templates for PCR. The primers used in these studies flank highly polymorphic sites in human DNA that result from variable numbers of tandem nucleotide repeats. Several offspring and their putative parents were tested to determine whether the offspring are "legitimate" or the product of illegal trading. The data are shown in the following Southern blot.

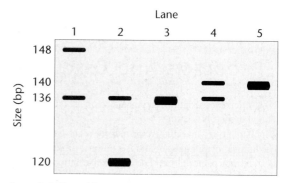

Lane 1: father chimpanzee
Lane 2: mother chimpanzee
Lanes 3–5: putative offspring A, B, C

Examine the data carefully, and choose the best conclusion.
(a) None of the offspring are legitimate.
(b) Offspring B and C are not the products of these parents and were probably purchased on the illegal market. The data are consistent with offspring A being legitimate.
(c) Offspring A and B are products of the parents shown, but C is not and was therefore probably purchased on the illegal market.
(d) There are not enough data to draw any conclusions. Additional polymorphic sites should be examined.
(e) No conclusion can be drawn because "human" primers were used.

23. The accompanying partial restriction map shows a recombinant plasmid, pBIO220, formed by cloning a piece of *Drosophila* DNA (striped box), including the gene *rosy* into the vector pBR322, which also contains the penicillin resistance gene, *pen*. The vector part of the plasmid contains only the two E sites shown, and no A or B sites. The gel shows several restriction digests of pBIO220.

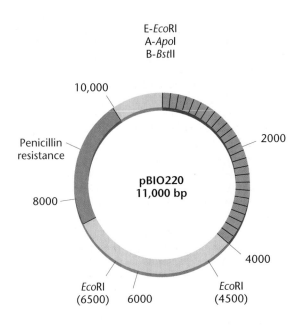

E-*Eco*RI
A-*Apo*I
B-*Bst*II

pBIO220
11,000 bp

10,000

2000

4000

6000

8000

*Eco*RI
(6500)

*Eco*RI
(4500)

Penicillin
resistance

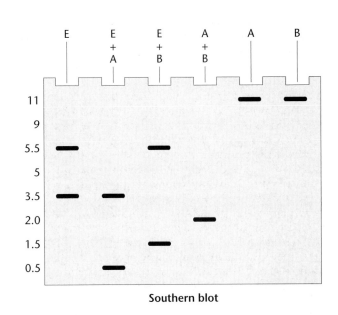

Southern blot

(a) Use the stained gel pattern to deduce where restriction sites
are located in the cloned fragment.

(b) A PCR-amplified copy of the entire 2000-bp *rosy* gene was
used to probe a Southern blot of the same gel. Use the Southern
blot results to deduce the locations of *rosy* in the cloned fragment. Redraw the map showing the location of the *rosy* gene.

24. One of the six restriction maps shown here is consistent with the
pattern of bands shown in the following gel after digestion with
several restriction endonucleases. The enzymes that were used
are shown.

(a) From your analysis of the pattern of bands on the gel, select
the correct map and explain your reasoning.

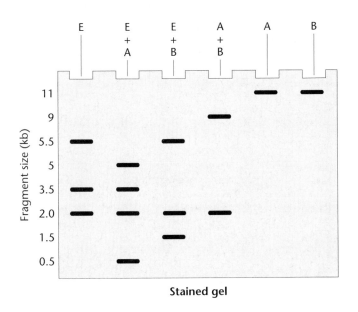

Stained gel

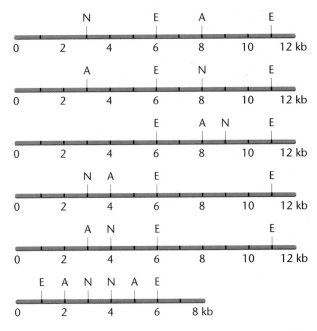

(b) In a Southern blot prepared from this gel, the highlighted bands (pink) hybridized with the gene *pep*. Where is the *pep* gene located?

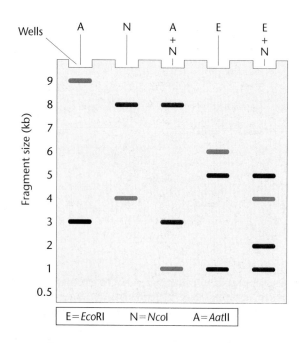

E = *Eco*RI   N = *Nco*I   A = *Aat*II

25. List the steps involved in screening a genomic library. What must be known before starting such a procedure? What are the potential problems with such a procedure, and how can they be overcome or minimized?

26. To estimate the number of cleavage sites in a particular piece of DNA with a known size, you can apply the formula $N/4^n$, where N is the number of base pairs in the target DNA and n is the number of bases in the recognition sequence of the restriction enzyme. If the recognition sequence for *Bam*HI is GGATCC and the phage λ DNA contains approximately 48,500 bp, how many restriction sites would you expect?

27. In a typical PCR reaction, what phenomena are occurring at temperature ranges (a) 90–95°C, (b) 50–70°C, and (c) 70–75°C?

28. A widely used method for calculating the annealing temperature for a primer used in PCR is 5 degrees below the $T_m(°C)$, which is computed by the equation $81.5 + 0.41(\%GC) - (675/N)$, where %GC is the percentage of G or C nucleotides in the oligonucleotide and N is the length of the oligonucleotide. Notice from the formula that both the GC content *and* the length of the oligonucleotide are variables. Assuming you have the oligonucleotide shown below as a primer, compute the annealing temperature for PCR. What is the relationship between $T_m(°C)$ and %GC? Why? (*Note:* In reality, this computation provides only a starting point for empirical determination of the most useful annealing temperature.)

5′-TTGAAAATATTTCCCATTGCC-3′

29. We usually think of enzymes as being most active at around 37°C, yet in PCR the DNA polymerase is subjected to multiple exposures of relatively high temperatures and seems to function appropriately at 70–75°C. What is special about the DNA polymerizing enzymes typically used in PCR?

30. How are dideoxynucleotides (ddNTPs) used in the chain termination method of DNA sequencing?

31. Assume you have conducted a standard DNA sequencing using the chain-termination method. You performed all the steps correctly and electrophoresed the resulting DNA fragments correctly, but when you looked at the sequencing gel, many of the bands were duplicated (in terms of length) in other lanes. What might have happened?

Genomics and Proteomics

Arabidopsis thaliana, one of the model organisms used to study genetics.

- Genomics relies on the methods of recombinant DNA technology and DNA sequencing to define and analyze the genomes of organisms.

- Genomics has several subfields, including structural genomics, functional genomics, and comparative genomics.

- The analysis of genomic sequence data is refining our ideas about the size and organization of prokaryotic and eukaryotic genomes.

- The human genome contains 20,000 to 25,000 protein-coding genes, many of them organized in clusters and separated by large spans of intergenic spacer DNA.

- Comparative genomics is useful in studying the evolutionary relationships and history among genes, gene families, and organisms.

- Proteomics is used to provide information about the protein content, structure, and function of cells and subcellular components.

The term **genome** was coined in 1920, at a time when geneticists began to focus on the information encoded in a haploid set of chromosomes (one definition of a genome). To systematically identify and map all the genes in an organism's genome, geneticists developed a two-part approach: (1) collect mutants and (2) generate genetic maps by linkage analysis using the mutant strains. Using this approach, geneticists gradually focused on a small number of organisms, including fruit flies, maize, mice, various bacteria, and yeast.

This tactic, developed just about 100 years ago, was the backbone of genetic analysis, and is still widely used. Mutational analysis does have some limitations, however. It requires the isolation of at least one mutation for each gene in order to identify all genes in a genome. Obtaining mutations can be time-consuming, and some of these have no clear phenotype, making it difficult or impossible to map the mutated gene.

In the 1980s, human geneticists began using recombinant DNA technology to map, isolate, clone, and sequence genes, including those for cystic fibrosis, neurofibromatosis, and dozens of other disorders. Although these methods identified one gene at a time, by the mid-1980s, more than 3500 genes and markers had been assigned to human chromosomes, putting geneticists on the threshold of producing high-resolution maps of all human chromosomes.

While recombinant DNA-based advances in genetic mapping were being developed, Fred Sanger and colleagues began the field of **genomics** (the study of genomes) by sequencing the 5400-nucleotide genome of the virus ϕX174. Other viral genomes were sequenced in short order, but the process was slow and labor-intensive, limiting its use to small genomes. Over the next decade, the development of faster, automated sequencing methods made it possible to consider sequencing the larger and more complex genomes of eukaryotes, including the 3.2 billion nucleotides that make up the haploid human genome, which is some 600,000 times larger than the ϕX174 genome.

With several hundred genomes from both prokaryotes and eukaryotes already sequenced and more than two thousand additional projects underway, genomics has grown into a major field with wide-ranging goals (Table 18.1). Genomics encompasses several subfields, including **structural genomics**, **functional genomics**, and **comparative genomics**. Structural genomics focuses on construction of genomic sequence data, gene discovery, and localization and the construction of gene maps. Functional genomics studies the biological function of genes, their regulation and their products. Comparative genomics compares gene or protein sequences from different genomes to elucidate functional and evolutionary relationships.

Gathering and analyzing genome information is a large-scale, labor-intensive endeavor requiring the coordinated effort of many laboratories. For this reason, geneticists often form collaborations, called genome projects. One of the largest and best-known of these is the **Human Genome Project (HGP)**, which we will discuss later in this chapter.

TABLE 18.1	Goals of Genomics

Compile the genomic sequences of organisms.

Establish the location of all genes in a genome.

Annotate the gene set in a genome.

Establish the function of all genes in a genome.

Generate gene expression profiles of cells under differing conditions.

Identify all the proteins that can be produced from a given genome.

Compare genes and proteins between organisms to establish evolutionary relationships.

Catalog sequence polymorphisms in a population of individuals from the same species.

An outgrowth of genomics is **proteomics**, the study of the proteins present in a cell at a given time under a given set of conditions. Proteomics includes (1) the identification of all proteins expressed in a cell under a given set of conditions, (2) the nature and extent of any posttranslational modification of proteins, (3) the nature and extent of protein–protein interactions in the cell, and (4) the subcellular location of proteins in various cellular compartments.

In this chapter, we describe the technology of genomics and review the significant findings from a variety of genome projects using prokaryotic and eukaryotic organisms. We will also discuss some applications of comparative genomics, including efforts to determine the minimum number of genes necessary for life, and the evolution of gene families. Finally, we will examine how proteomics provides insight into the structure and function of proteins and how this information helps us understand how proteins interact to produce the unique properties of different cell types.

How Do We Know

In this chapter, we will focus on the methods used in sequencing genomes, analysis of the sequence, and assignment of gene function. We also examine what is known about prokaryotic and eukaryotic genomes. As you study this topic, you should try to answer several fundamental questions:

1. How do we know that a sequenced genome is accurate?

2. How do we know when all the genes in a sequenced genome have been identified?

3. How do we know the function of proteins encoded in a genome?

4. How do we know the minimum gene set required for life?

5. How do we know which proteins are expressed in a cell at a given time?

18.1 Structural Genomics Uses DNA Sequencing to Study Genomes

Structural genomics uses DNA sequencing technology and software programs to generate, store, and analyze genomic sequence information. We will first discuss two approaches to genome sequencing and then briefly consider how this information is stored and analyzed.

Map-Based Sequencing

Geneticists use two different methods for sequencing genomes. The first to be developed was **map-based sequencing** [Figure 18–1(a)]. Map-based sequencing begins with the construction of a genomic library using vectors that can accommodate large fragments of an organism's genome. Next, the clones are assembled into genetic maps using recombinational analysis.

To increase map resolution, physical maps, based on direct analysis of DNA rather than recombinational frequency are constructed. Physical map distances are measured in base pairs (bp), usually expressed in thousands of base pairs (1 kb = 1000 base pairs) or millions of base pairs (1 Mb = 1,000,000 base pairs). One type of physical map is constructed to form a series of ordered, overlapping clones [Figure 18–1(a)]. Each clone is sequenced individually, and the sequence of the overlapping clones is assembled using software algorithms to produce the finished genome sequence. The map-based method was chosen for the publicly funded Human Genome Project sponsored by the NIH and the Department of Energy. Making large clones and generating restriction maps are time-limiting steps in this approach, and genome projects with this approach have long timelines for completion.

Shotgun Sequencing

The **shotgun sequencing method** [Figure 18–1(b)] was made possible by advances in sequencing technology and development of software to assemble sequences from many short overlapping clones into a continuous sequence. In this method, two or three genomic libraries are made by shearing the DNA into fragments. Libraries of short fragments (about 2

kb in length), medium-length fragments (about 15 to 20 kb), and long fragments (200 to 300 kb) are prepared. Clones are selected at random from each library and sequenced. Software is used to assemble long stretches of sequence from overlapping

(a) Map-based sequencing method

Chromosome 21

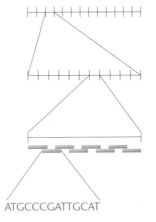

ATGCCCGATTGCAT

Genetic map of markers spaced about 1 Mb (1 million base pairs) apart. This map is derived from recombination studies

Physical map showing order, physical distance of markers on chromosomes. Spaced about 100,000 base pairs apart

Set of overlapping ordered clones, called contigs, each covering 0.5–1.0 kb

Each clone is sequenced and the overlapping sequences are assembled by computer into a final sequence for each chromosome

(b) Shotgun sequencing method

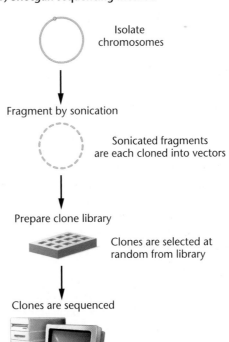

Isolate chromosomes

Fragment by sonication

Sonicated fragments are each cloned into vectors

Prepare clone library

Clones are selected at random from library

Clones are sequenced

Computer with compiler software

Assembler software used to assemble sequence

FIGURE 18–1 (a) In the map-based sequencing method, clones from a genomic library are organized into genetic and physical maps for each chromosome. After the clones are arranged into physical maps, they are broken into smaller, overlapping clones called contigs, which cover each chromosome. Each smaller clone is sequenced, and the genomic sequence is assembled by stringing together the nucleotide sequence of the clones. (b) In the shotgun sequencing method, a genomic library is constructed from sheared fragments of genomic DNA. Clones are selected from the library at random and sequenced. The sequence is assembled by looking for sequence overlaps between clones. This is done with a computer, using assembler software designed for genomic analysis.

short fragments from the libraries, using the sequences from the larger clones as a framework.

This method, developed by Craig Venter and his colleagues at the Institute for Genome Research (TIGR), was used to sequence the genome of the bacterium *Haemophilus influenzae* in 1995. This was the first complete genome from a free-living organism to be sequenced. After refining the method and using it to sequence the genomes of other prokaryotes, the shotgun method was used to sequence eukaryotic genomes, including fruit flies, dog, and humans. In a later section of this chapter, we will discuss the Human Genome Project in more detail and review some of the properties of the human genome discovered with these sequencing methods.

Genomic Analysis

Genome projects generate large amounts of DNA sequence information. These data are useful only when they have been analyzed and interpreted. Several types of analyses using software algorithms and databases are necessary to identify and characterize gene-coding regions of anonymous DNA sequences. Some of the goals of these analyses include:

1. Ensuring the sequence is complete and accurate.

2. Identifying protein-coding genes by finding open reading frames (ORFs).

3. Finding promoter sites, eukaryotic enhancers, silencers, transcription initiation sites, and translation initiation sites to confirm gene identity.

4. In eukaryotic genomes, finding splice sites, introns, and exons.

5. Converting the DNA sequence into the predicted amino acid sequence of the encoded protein.

The predicted protein sequences are analyzed to predict function. Protein function can often be inferred if sequences of homologous proteins with known function can be found in a database. The ultimate goal of sequence analysis is to derive a complete functional description of all genes in the genome of an organism.

18.2   Annotation Is Used to Find Genes in DNA Sequence Data

Once obtained, genomic sequence data must be stored and analyzed. This is a formidable task: The human genome sequence consists of more than 3 billion nucleotides, and genomes of some other eukaryotes are even larger. **Bioinformatics** is an emerging field that includes the use of computer hardware and software for the acquisition, storage, analysis, and visualization of genome sequence data. Public online databases for the storage and analysis of genome information are now essential tools for geneticists.

Before genome sequence data can be analyzed, it is necessary to ensure that the sequence is accurate. This is a preliminary step in analyzing genomic data.

Sequence Verification

To ensure that the nucleotide sequence of a genome is complete and error-free, the genome is sequenced more than once, a process known as multiple coverage. For example, using the shotgun method, researchers sequenced the 6.3 million nucleotides in the genome of the bacterium *Pseudomonas aeruginosa* seven times to ensure that the sequence was accurate. Even with this level of redundancy, the assembler software found 1604 regions requiring further clarification. These regions were reanalyzed and resequenced to improve accuracy. Finally, the shotgun sequence was compared with the sequence of two widely separated genomic regions obtained by conventional cloning, allowing the researchers to be confident in the accuracy of the genomic sequence. This level of care is not unusual; similar precautions are taken in all genome projects. Assembly of the genomic sequence from multiple sequencing runs is known as **compiling**.

Finding the Genes

After a genome has been sequenced and compiled, the sequence is analyzed using computer algorithms to search for genes and other functional sequences. This step, called **annotation**, identifies initiation sequences, protein-coding genes, their regulatory sequences, and termination sequences. Annotation also identifies nonprotein-coding genes (including ribosomal RNA, transfer RNA, and small nuclear RNAs) and finds the mobile genetic elements and repetitive-sequence families that may be present in the genome. We will focus our discussion on the identification of protein-coding genes.

Genes have several identifying features that can be detected by software analysis of the DNA sequence (remember the genetic code, however, is written in RNA nucleotides). For example, protein-coding genes contain one or more **open reading frames (ORFs)**, continuous sets of DNA nucleotide triplets that can be translated into the amino acid sequence of a protein. ORFs begin with an initiation sequence (usually ATG) and end with a termination sequence (TAA, TAG, or TGA).

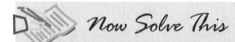

Now Solve This

Problem 7 on page 422 involves searching a genomic sequence for ORFs.

Hint: In setting limits for the size of ORFs that may represent genes, be aware of the average gene length in the organism's genome.

The organization of genes in eukaryotic genomes (including the human genome) makes direct searching for ORFs more difficult than in prokaryotic genomes. First, many eukaryotic genes have introns, noncoding regions that interrupt coding regions (exons). As a result, many, if not most, eukaryotic genes are not organized as continuous ORFs; instead, the

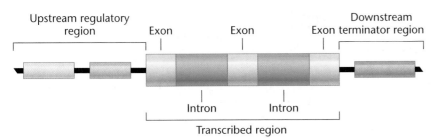

FIGURE 18–2 Most eukaryotic genes are organized into coding segments (exons) and noncoding segments (introns). In some cases, introns make up the majority of the gene sequence. In annotating a genome sequence, it is important to distinguish among introns, exons, and genes.

sequences consists of ORFs (exons) interspersed with introns (Figure 18–2). Second, genes in humans and other eukaryotes are often widely spaced, increasing the chances of finding false ORFs in the regions between gene clusters.

Figure 18–3(a) shows a portion of the human genome sequence. From a casual inspection, it is not clear whether this sequence contains any genes, and if so, how many. In analyzing this sequence, identifiable footprints of genes provide clues that reveal whether the sequence being analyzed contains a protein-coding gene. Control regions at the beginning of genes are marked by identifiable sequences. Splice sites between exons and introns have a predictable sequence (most introns begin with GT and end with AG), as does the end of the gene where a poly-A tail is added to the transcript. In addition, if a DNA sequence encodes a protein, after splicing the sequence contains one or more open reading frames.

Analysis of the sequence [Figure 18–3(b)] shows it contains a control region and three exons. The two unshaded regions between the exons represent introns, regions removed as the mRNA is processed. Using this sequence to search genomic databases shows this is the sequence of a single gene, the human β-globin gene [Figure 18–3 (c)].

18.3 Functional Genomics Classifies Genes and Identifies Their Functions

After a genomic sequence is annotated and ORFs have been identified, the next task is to assign functions to all the genes (in the form of ORFs) in the sequence. Some of the identified genes may already have functions assigned by the classic method of mutagenesis and linkage mapping, but many other genes have no identified function. One approach to assigning functions to these genes is the use of **homology searches**. This analysis has several components:

1. Use programs such as BLAST to search databases to find similar sequences already identified as genes in other organisms.

2. Compare the sequence of an ORF with that of a well-characterized gene from another organism.

3. Scan the ORF for functional motifs, regions of DNA that encode protein domains such as ion channels, DNA-binding regions, or secretion/export signals.

Let's examine how this strategy is used in the analysis of a prokaryotic genome.

(a)

```
gagccacacc  ctagggttgg  ccaatctact  cccaggagca  gggagggcag  gagccagggc
tgggcataaa  agtcagggca  gagccatcta  ttgcttacat  ttgcttctga  cacaactgtg
ttcactagca  acctcaaaca  gacaccatgg  tgcacctgac  tcctgaggag  aagtctgccg
ttactgccct  gtggggcaag  gtgaacgtgg  atgaagttgg  tggtgaggcc  ctgggcaggt
tggtatcaag  gttacaagac  aggtttaagg  agaccaatag  aaactgggca  tgtggagaca
gagaagactc  ttgggtttct  gataggcact  gactctctct  gcctattggt  ctatttttccc
acccttaggc  tgctggtggt  ctacccttgg  acccagaggt  tctttgagtc  ctttggggat
ctgtccactc  ctgatgctgt  tatgggcaac  cctaaggtga  aggctcatgg  caagaaagtg
ctcggtgcct  ttagtgatgg  cctggctcac  ctggacaacc  tcaagggcac  ctttgccaca
ctgagtgagc  tgcactgtga  caagctgcac  gtggatcctg  agaacttcag  ggtgagtcta
tgggaccctt  gatgttttct  ttccccttct  tttctatggt  taagttcatg  tcataggaag
gggagaagta  acagggtaca  gtttagaatg  ggaaacagac  gaatgattgc  atcagtgtgg
aagtctcagg  atcgttttag  tttctttttat  ttgctgttca  taacaattgt  tttctttttgt
ttaattcttg  ctttcttttt  ttttcttctc  cgcaattttt  actattatac  ttaatgcctt
aacattgtgt  ataacaaaag  gaaatatctc  tgagatacat  taagtaactt  aaaaaaaaac
tttacacagt  ctgcctagta  cattactatt  tggaatatat  gtgtgcttat  ttgcatattc
ataatctccc  tactttattt  tcttttattt  ttaattgata  cataatcatt  atacatattt
atgggttaaa  gtgtaatgtt  ttaatatgta  tacacatatt  gaccaaatca  gggtaatttt
gcatttgtaa  ttttaaaaaa  tgctttcttc  ttttaatata  cttttttgtt  tatcttattt
ctaatacttt  ccctaatctc  tttctttcag  ggcaataatg  atacaatgta  tcatgcctct
ttgcaccatt  ctaaagaata  acagtgataa  tttctgggtt  aaggcaatag  caatatttct
gcatataaat  atttctgcat  ataaattgta  actgatgtaa  gaggtttcat  attgctaata
gcagctacaa  tccagctacc  attctgcttt  tattttatgg  ttgggataag  gctggattat
tctgagtcca  agctaggccc  ttttgctaat  catgttcata  cctcttatct  tcctcccaca
gctcctgggc  aacgtgctgg  tctgtgtgct  ggcccatcac  tttggcaaag  aattcacccc
accagtgcag  gctgcctatc  agaaagtggt  ggctggtgtg  gctaatgccc  tggcccacaa
gtatcactaa  gctcgctttc  ttgctgtcca  atttctatta  aaggttcctt  tgttccctaa
gtccaactac  taaactgggg  gatattatga  agggccttga  gcatctggat  tctgcctaat
aaaaaacatt  tattttcatt  gcaatgatgt  atttaaatta  tttctgaata  ttttactaaa
```

(b)

```
gagccacacc  ctagggttgg  ccaatctact  cccaggagca  gggagggcag  gagccagggc
tgggcataaa  agtcagggca  gagccatcta  ttgcttacat  ttgcttctga  cacaactgtg
ttcactagca  acctcaaaca  gacaccatgg  tgcacctgac  tcctgaggag  aagtctgccg
ttactgccct  gtggggcaag  gtgaacgtgg  atgaagttgg  tggtgaggcc  ctgggcaggt
tggtatcaag  gttacaagac  aggtttaagg  agaccaatag  aaactgggca  tgtggagaca
gagaagactc  ttgggtttct  gataggcact  gactctctct  gcctattggt  ctatttttccc
acccttaggc  tgctggtggt  ctacccttgg  acccagaggt  tctttgagtc  ctttggggat
ctgtccactc  ctgatgctgt  tatgggcaac  cctaaggtga  aggctcatgg  caagaaagtg
ctcggtgcct  ttagtgatgg  cctggctcac  ctggacaacc  tcaagggcac  ctttgccaca
ctgagtgagc  tgcactgtga  caagctgcac  gtggatcctg  agaacttcag  ggtgagtcta
tgggaccctt  gatgttttct  ttccccttct  tttctatggt  taagttcatg  tcataggaag
gggagaagta  acagggtaca  gtttagaatg  ggaaacagac  gaatgattgc  atcagtgtgg
aagtctcagg  atcgttttag  tttctttttat  ttgctgttca  taacaattgt  tttctttttgt
ttaattcttg  ctttcttttt  ttttcttctc  cgcaattttt  actattatac  ttaatgcctt
aacattgtgt  ataacaaaag  gaaatatctc  tgagatacat  taagtaactt  aaaaaaaaac
tttacacagt  ctgcctagta  cattactatt  tggaatatat  gtgtgcttat  ttgcatattc
ataatctccc  tactttattt  tcttttattt  ttaattgata  cataatcatt  atacatattt
atgggttaaa  gtgtaatgtt  ttaatatgta  tacacatatt  gaccaaatca  gggtaatttt
gcatttgtaa  ttttaaaaaa  tgctttcttc  ttttaatata  cttttttgtt  tatcttattt
ctaatacttt  ccctaatctc  tttctttcag  ggcaataatg  atacaatgta  tcatgcctct
ttgcaccatt  ctaaagaata  acagtgataa  tttctgggtt  aaggcaatag  caatatttct
gcatataaat  atttctgcat  ataaattgta  actgatgtaa  gaggtttcat  attgctaata
gcagctacaa  tccagctacc  attctgcttt  tattttatgg  ttgggataag  gctggattat
tctgagtcca  agctaggccc  ttttgctaat  catgttcata  cctcttatct  tcctcccaca
gctcctgggc  aacgtgctgg  tctgtgtgct  ggcccatcac  tttggcaaag  aattcacccc
accagtgcag  gctgcctatc  agaaagtggt  ggctggtgtg  gctaatgccc  tggcccacaa
gtatcactaa  gctcgctttc  ttgctgtcca  atttctatta  aaggttcctt  tgttccctaa
gtccaactac  taaactgggg  gatattatga  agggccttga  gcatctggat  tctgcctaat
aaaaaacatt  tattttcatt  gcaatgatgt  atttaaatta  tttctgaata  ttttactaaa
```

(c)

ORF1   ORF2   ORF3

0.0   0.5   1.0   1.5   kb

FIGURE 18–3 Analysis of DNA sequence information. By convention, the sequence is presented in groups of ten nucleotides, although in reality, the sequence is continuous. (a) Human DNA sequence data. Location of genes, if any, in this sequence is not apparent without analysis. (b) Analyzed sequence showing the location of an upstream regulatory sequence (green) and three open reading frames (ORFs) (blue). (c) A diagram showing the size and location of open reading frames for the three exons of the human β-globin gene encoded by the sequence in (a).

Assigning Gene Functions

Table 18.2 shows the results of an ORF analysis in the genome of a bacterium, *Pseudomonas aeruginosa*. In this analysis, as in all functional genomic analyses, ORFs are classified on a scale according to confidence levels. Level 1 genes are those whose function is already known from conventional genetic analysis or are genes otherwise known to have a demonstrated function. Level 2 genes have a strong homology to genes with known functions in other organisms. Level 3 genes have a proposed function based on encoded motifs or limited homology to genes in other organisms. Level 4 genes have no known or proposed function. In a functional analysis of the *P. aeruginosa* genome (Table 18.3), about half the genes have no known or proposed function. This situation is not unique to *P. aeruginosa*; in almost all organisms whose genomes have been sequenced to date, about half of their genes have no known function. This demonstrates the extent of new information potentially available through future genomic analysis.

The *E. coli* Project

The genome of *E. coli* strain K12 was sequenced using the shotgun method in 1997 and was the second prokaryotic genome to be sequenced. *E. coli* has a genome of 4.6 Mb, and

---

TABLE 18.2   Analysis of the *Pseudomonas aeruginosa* Genome

General Features

Genome size (bp)	6,264,403
G + C content	66.6%
Coding regions	89.4%

Coding Sequences

Confidence level	ORFs	(%)	Definition
1	372	6.7	*P. aeruginosa* genes with demonstrated function
2	1059	19.0	Strong homology to genes with demonstrated function from other organisms
3	1590	28.5	Genes with proposed function based on motif searches or limited homology
4a	769	13.8	Homologs of reported genes of unknown function
4b	1780	32.0	No homology to any reported sequences
Total	5570	100	

Source: Stover *et al.* 2000. Complete genome sequence of *Pseudomonas aeruginosa* PA01, an opportunistic pathogen. *Nature* 406: 959–964. Table 1, p. 961.

TABLE 18.3   Functional Classes of Predicted *Genes* in *Pseudomonas*

Functional Class	ORFs Number	%
Adaptation, protection (e.g., cold shock proteins)	60	1.1
Amino acid biosynthesis and metabolism	150	2.7
Antibiotic resistance and suspectibility	19	0.3
Biosynthesis of cofactors, prosthetic groups, and carriers	119	2.1
Carbon compound catabolism	130	2.3
Cell division	26	0.5
Cell wall	83	1.5
Central intermediary metabolism	64	1.1
Chaperones and heat shock proteins	52	0.9
Chemotaxis	43	0.8
DNA replication, recombination, modification, and repair	81	1.5
Energy metabolism	166	3.0
Fatty acid and phospholipid metabolism	60	1.0
Membrane proteins	7	0.1
Motility and attachment	65	1.2
Nucleotide bisosynthesis and metabolism	60	1.1
Protein secretion/export apparatus	83	1.5
Putative enzymes	409	7.3
Related to phage, transposon, or plasmid	38	0.7
Secreted factors (toxins, enzymes, alginate)	58	1.0
Transcriptional regulators	403	7.2
Transcription, RNA processing, and degradation	45	0.8
Translation, posttranslational modification, and degradation	149	2.7
Transport of small molecules	555	10.0
Two-component regulatory systems	118	2.1
Hypothetical	1774	31.8
Unknown (conserved hypothetical)	757	13.6
Total	5570	100

Source: Stover *et al.* 2000. Complete genome sequence of *Pseudomonas aeruginosa* PA01, an opportunistic pathogen. *Nature* 406: 959–964.

FIGURE 18–4 (a) *E. coli*, one of the model organisms in genetics. (b) Status of functional gene assignments for the *E. coli* genome. The total is over 100% because of overlap between proteins with known functions and those whose 3D structure is known.

(a)

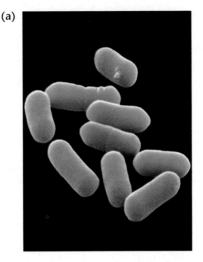

(b)

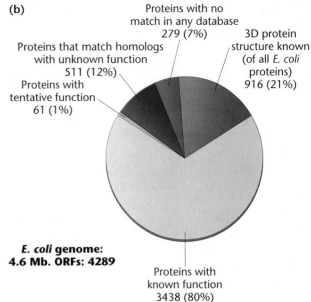

Proteins with no match in any database
279 (7%)

3D protein structure known (of all *E. coli* proteins)
916 (21%)

Proteins that match homologs with unknown function
511 (12%)

Proteins with tentative function
61 (1%)

Proteins with known function
3438 (80%)

E. coli genome:
4.6 Mb. ORFs: 4289

annotation shows that it contains 4289 ORFs organized as a single circular chromosome. ORFs occupy almost 88 percent of the genome, regulatory sequences about 11 percent, and repetitive sequences only about 0.7 percent. In spite of the fact that the genetics of *E. coli* had been intensively studied for almost 50 years using classic genetic methods, almost 38 percent of the annotated ORFs were for genes with no known functions.

The goal in the analysis of the *E. coli* genome, as in all genomic projects, is to move all of the genes up to Level 1 or 2 and to understand how these genes and their products interact in the biology of the organism. To reach the goal of assigning functions to all the genes in *E. coli*, researchers are creating (1) knockout mutations for every gene, (2) a clone for every ORF, (3) a gene expression database for patterns of expression under a variety of physiological conditions, and (4) a 3D structure of all proteins encoded in the genome. The current classifications of *E. coli* genes are shown in Figure 18–4. Genes with no known homology and no known functions now comprise 19 percent of the 4289 ORFs in the genome.

18.4 Prokaryotic Genomes Have Some Unexpected Features

Sequencing of more than four hundred prokaryotic and eukaryotic genomes has now been completed. The results indicate that there are significant differences in genome organization between prokaryotes and eukaryotes. Therefore, we will consider the organization of these genomes separately. Since most prokaryotes have small genomes amenable to shotgun cloning and sequencing, most of the completed projects have focused on prokaryotes and more than 900 additional projects to sequence prokaryotic genomes are under way. Many of the prokaryotic genomes already sequenced are from organisms that cause human diseases, such as cholera, tuberculosis, and leprosy.

Genome Size and Gene Number in Bacteria

Based on genome project results, we can make a number of generalizations about the size of bacterial genomes. Traditionally, the bacterial genome has been thought of as relatively small (less than 5 Mb) and contained within a single circular DNA molecule. *E. coli*, used as the protoypical bacterial **model organism** in genetics, has a genome with these characteristics. However, the flood of genomic information now available has challenged this viewpoint as typical for all bacteria (Table 18.4). Although most prokaryotic genomes are small, genome sizes vary across a surprisingly wide range. In fact, there is some overlap in size between larger bacterial genomes (30 Mb in *Bacillus megaterium*) and smaller eukaryotic genomes (12.1 Mb in yeast). Gene number in bacterial genomes is also variable. Gene numbers cover a ten-fold range, from less than 500 to more than 5,000 genes. In addition, although many bacteria have a single, circular chromosome, there is substantial variation in chromosome organization and number among bacterial species.

TABLE 18.4	Genome Size and Gene Number in Selected Prokaryotes	
	Genome Size (Mb)	**Number of Genes**
Archaea		
Archaeoglobus fulgidis	2.17	2437
Methanococcus jannaschii	1.66	1783
Thermoplasma acidophilium	1.56	1509
Eubacteria		
E. coli	4.64	4289
Bacillus subtilis	4.21	4779
H. influenzae	1.83	1738
Aquifex aeolicus	1.55	1749
Rickettsia prowazekii	1.11	834
Mycoplasma pneumoniae	0.82	680
Mycoplasma genitalium	0.58	483

TABLE 18.5	Bacterial Genome Organization	
Bacterial species	**Chromosome Number and Organization**	**Plasmid Number**
Agrobacterium tumefaciens	one linear + one circular	two circular
Bacillus subtilis	one circular	
Bacillus thuringiensis	one circular	six
Borrelia burgdorferi	one linear	>17 circular + linear
Buchnera sp.	one circular	two circular
Deinococcus radiodurans	two circular	two circular
Escherichia coli	one circular	
Rhodobacter sphaeroides	two circular	five circular
Sinorhizobium meliloti	one circular	two megaplasmids
Vibrio cholerae	two circular	

Chromosome Number

Although most prokaryotic genomes sequenced to date *are* single, circular DNA molecules, an increasing number of genomes composed of linear DNA molecules are being identified, including the genome of *Borrelia burgdorferi*, the organism that causes Lyme disease. Perhaps more important, genome data are blurring the distinction between plasmids and bacterial chromosomes and redefining our view of the bacterial genome as a single DNA molecule (Table 18.5). Plasmids are small, usually circular DNA molecules containing genes that are not essential to their bacterial host cell (discussed in Chapter 8).

Plasmids self-replicate and are distributed to daughter cells along with the host chromosome during cell division. Because plasmids carry nonessential genes and can be transferred from one cell to another, and because the same plasmid is often present in bacteria from different species, plasmid genes are usually not included as part of a bacterial genome. However, *B. burgdorferi* carries approximately 17 plasmids, which contain at least 430 genes. Some of these genes are essential to the bacterium, including genes for purine biosynthesis and membrane proteins, which in most species are carried on the bacterial chromosome.

Sequencing of the *Vibrio cholerae* (the organism responsible for cholera) genome revealed the presence of two circular chromosomes (Figure 18–5). Chromosome 1 (2.96 Mb) contains

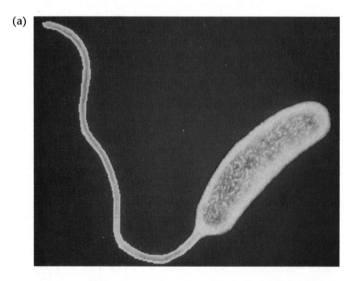

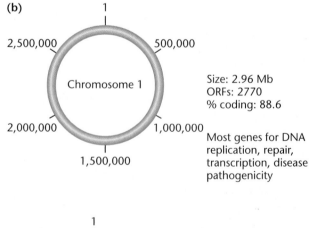

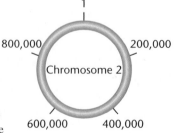

Size: 2.96 Mb
ORFs: 2770
% coding: 88.6

Most genes for DNA replication, repair, transcription, disease pathogenicity

Size: 1.07 Mb
ORFs: 1115
% coding: 86.3

Genes for metabolic pathways, DNA repair, as well as plasmid-like sequences

FIGURE 18–5 (a) *Vibrio cholerae*, the organism that causes cholera. (b) The *Vibrio* genome is contained in two chromosomes. The larger chromosome (chromosome 1) contains most of the genes for essential cellular functions and infectivity. Most of the genes on chromosome 2 (52% of 1115) are of unknown function. The bias in gene content and the presence of plasmid-like sequences on chromosome 2 suggest that this chromosome was a megaplasmid captured by an ancestral *Vibrio* species, or arose by transfer of chromosomal genes to a plasmid.

2770 ORFs, and chromosome 2 (1.07 Mb) contains 1115 ORFs, some of which encode essential genes, such as ribosomal proteins. Based on its nucleotide sequence, chromosome 2 is apparently derived from a plasmid captured by an ancestral species, or it could represent a plasmid with many inserted chromosomal genes. Two chromosomes are also present in other species of *Vibrio*.

Other bacteria also have genomes with two or more chromosomes, including *Agrobacterium tumefaciens*, *Rhodobacter sphaeroides*, and *Sinorhizobium meliloti* (Table 18.5). The finding that some bacterial species have multiple chromosomes raises questions about how replication and segregation of chromosomes during cell division are coordinated, and what undiscovered mechanisms of gene regulation may exist in bacteria. Answers to these and other questions raised by genomic discoveries will redefine some ideas about prokaryotic genomes, the nature of plasmids, and the differences between plasmids and chromosomes, and may provide clues about the evolution of multichromosome eukaryotic genomes.

Gene Distribution

We can make two generalizations about the organization of protein-coding genes in bacteria. First, gene density is very high, averaging about one gene per kilobase of DNA. For example, *E. coli* has a 4.6 Mb genome containing 4289 protein-coding genes. *Mycoplasma genitalium* has a small genome (0.58 Mb), with 483 genes, and its gene density is also close to one gene per kilobase pair. This close packing of genes in prokaryotic genomes means that a very high proportion of the DNA (approximately 85 to 90%) serves as coding DNA. Typically, only a small amount of a bacterial genome is noncoding DNA, often in the form of regulatory sequences or as transposable elements that can move from one place to another in the genome.

A second generalization we can make is that bacterial genomes contain operons (polygenic transcription units whose products are part of a common biochemical pathway). The organization of a small segment of the *E. coli* genome is shown in Figure 18–6. The arrows show genes organized into transcription units, one of which is organized as an operon. In *E. coli*, 27 percent of all genes are contained in operons (almost 600 operons). In other bacterial genomes, the organization of genes into transcriptional units is challenging our ideas about the nature of operons. In *Aquifex aeolicus* (Figure 18–7), one polygenic transcription unit contains six genes involved in several different cellular processes with no apparent common relationships: two genes for DNA recombination, one for lipid synthesis, one for nucleic acid synthesis, one for protein synthesis, and one that encodes a protein

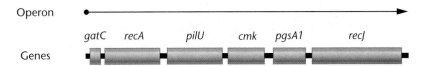

FIGURE 18–7 An operon from the *A. aeolicus* genome contains genes for protein synthesis (*gatC*), for DNA recombination (*recA* and *recJ*), for a motility protein (*pilU*), for nucleotide biosynthesis (*cmk*), and for lipid biosynthesis (*pgsA1*). This organization challenges the conventional idea that genes in an operon encode products that control a common biochemical pathway.

for cell motility. Other polygenic transcription units in this species also contain genes with an array of different functions. This finding, combined with similar results from other genome projects, raises questions about the consensus that operons encode products that control a single metabolic pathway in bacterial cells.

18.5 Eukaryotic Genomes Have Several Organizational Patterns

We now turn to a consideration of eukaryotic genomes. Although the basic features of the eukaryotic genome are similar in different species, genome size is highly variable (Table 18.6). Genome sizes range from about 10 Mb in fungi to over 100,000 Mb in some flowering plants (a ten-thousandfold range); the number of chromosomes per genome ranges from two into the hundreds (about a hundredfold range), but the number of genes varies much less dramatically than either genome size or chromosome number.

Unique Features of Eukaryotic Genomes

Eukaryotic genomes have several features not found in prokaryotes:

- **Gene density.** In prokaryotes, gene density is close to 1 gene/kb. In eukaryotic genomes, there is a wide range of gene density. In yeast, there is about 1 gene/2kb, in *Drosophila*, about 1 gene/13kb, and in humans, gene density varies greatly from chromosome to chromosome. Human chromosome 22 has about 1 gene/64kb, while on chromosome 13 there is 1 gene/155kb of DNA.

- **Introns.** Most eukaryotic genes contain introns (discussed in Chapter 12). There is wide variation in the number of introns across genomes and from gene to gene. The entire yeast genome has only 239 introns, while just a single gene in the human genome can contain more than 100 introns.

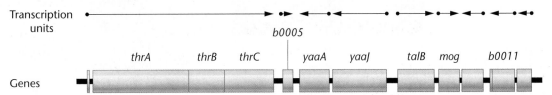

FIGURE 18–6 A portion of the *E. coli* chromosome showing a three-gene operon at the left, followed by 8 genes that are transcribed individually. A dot indicates the promoter for the operon and each gene. Arrows indicate the direction of transcription.

TABLE 18.6	Genome Size and Gene Number in Selected Eukaryotes	
Organism	Genome Size (Mb)	Number of Genes
S. cerevisiae (yeast)	12.1	≈6600
Plasmodium falciparum (malaria parasite)	30	≈5200
C. elegans (nematode)	97	≈19,000
Arabidopsis thalania (mustard plant)	125	≈25,000
D. melanogaster (fruit fly)	170	≈14,000
Oryza sativa (rice)	400	≈38,000
Canis familiaris (dog)	2400	≈19,000
Zea mays (maize)	2500	≈20,000
Homo sapiens (human)	3200	≈22,000
Horedeum vulgare (barley)	5300	≈20,000

- **Repetitive sequences.** The presence of introns and the existence of repetitive sequences are two major reasons for the wide range of genome sizes in eukaryotes. In some plants, such as maize, repetitive sequences are the dominant feature of the genome. The maize genome has about 2500 Mb of DNA, more than two-thirds of which is composed of repetitive DNA. In the human, about half the genome is repetitive DNA.

The Yeast Genome

The yeast (*Saccharomyces cerevisiae*) genome was the first eukaryotic genome sequenced. As outlined in Chapter 17, yeast can be grown and manipulated as easily as some prokaryotes, and its highly developed genetic map and large collection of mutants made it an excellent candidate for genome sequencing.

The yeast genome was sequenced chromosome-by-chromosome using a map-based approach. The genome contains 12.1 Mb of DNA, distributed over 16 chromosomes (Figure 18–8). In the years following publication of the yeast genome, the predicted number of ORFs has changed several times as analysis of the genome has proceeded. Originally thought to have about 6200 genes, the yeast gene count is currently just under 6600 genes, and this number will probably change again as more evidence is gathered. The number of characterized genes has risen to over 4300, and an additional 1400 or so genes have been assigned as Level 2 or 3 genes based on homology. This leaves about 800 genes that remain to be characterized, some of which may not turn out to be genes at all. A recent classification of genes in the yeast genome is shown in Figure 18–8.

The *Arabidopsis* Genome

To geneticists, the small flowering plant *Arabidopsis thaliana* is the fruit fly of the plant world [Figure 18–9(a)]. *Arabidopsis* is small (several hundred can be grown in the space occupied by this page), has a short generation time, and a relatively small genome distributed on five chromosomes. More importantly, the plant kingdom evolved independently from the animal kingdom, and analysis of flowering plants can provide insight into the ways in which evolutionary and coevolutionary adaptations have shaped their genomes. The *Arabidopsis* genome was sequenced using map-based sequencing. Its small genome, 125 Mb distributed in five chromosomes, contains an estimated 25,000 genes with a gene density of 1 gene per 5 kb [Figure 18–9(b)]. At least half the genes in the *Arabidopsis* genome are identical to or closely related to genes found in bacteria and humans, but it also contains genes encoding dozens of protein families that may be unique to plants.

Both genome and gene duplications have played a large role in the evolution of the *Arabidopsis* genome. Many of the 25,000 genes in its genome are duplicated, and there are fewer than 15,000 different genes. For example, chromosomes 2 and 4 of *Arabidopsis* contain many tandem gene duplications (239 duplications on chromosome 2, involving 539 genes) as well as larger duplications involving 4 blocks of DNA sequences, spanning 2.5 Mb. There is some interchromosomal gene duplication as well; genes on chromosome 4 are also present on chromosome 5.

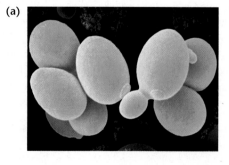

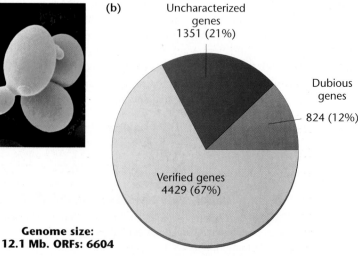

Genome size: 12.1 Mb. ORFs: 6604

Uncharacterized genes 1351 (21%)

Dubious genes 824 (12%)

Verified genes 4429 (67%)

FIGURE 18–8 (a) *Saccharomyces cerevisiae*, a model eukaryotic organism. (b) Functional analysis of protein coding genes as of 09-06-06.

(a)

(b)

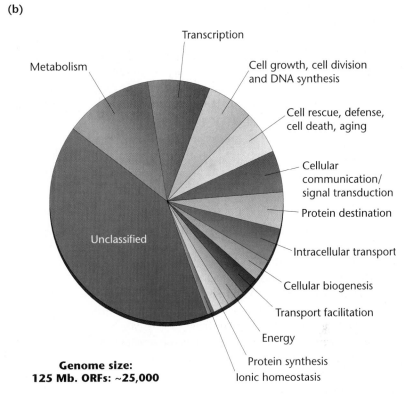

**Genome size:
125 Mb. ORFs: ~25,000**

FIGURE 18–9 *Arabidopsis thaliana* (a), a small plant used as a genetic model to study genome organization and development of flowering plants. (b) Assignment of *Arabidopsis* genes to functional categories based on homology searches.

Crop plants, such as maize, rice, and barley, have much larger genomes than *Arabidopsis*, but some have about the same number of genes (see Table 18.6). In contrast with *Arabidopsis*, genes in these large-genome plants are clustered in stretches of DNA separated by long stretches of intergenic spacer DNA. Collectively, these gene clusters occupy only about 12 to 24 percent of the genome. In maize, the intergenic DNA is composed mainly of transposons.

Sequencing of the rice genome was completed in 2005. Analysis shows that this cereal crop genome contains about 400 Mb of DNA, encoding approximately 37,500 genes on 12 chromosomes. About 15% of all rice genes are found in duplicated segments of the genome. Close to 90 percent of the genes in *Arabidposis* are found in rice, but only 70 percent of genes in rice are found in *Arabidopsis*, indicating that in addition to genes found in other flowering plants, cereal crops may have unique gene sets. Identification of these genes and their functions in rice and other cereal plants will be critical in helping improve crop yields to feed the Earth's growing population.

18.6 Mapping the Human Genome Was Coordinated by the Human Genome Project

The Human Genome Project (HGP) is a coordinated international effort to determine the sequence of the haploid human genome and identify all the genes it contains. The HGP extends the tradition of genetic mapping by mutational analy-

sis to identify and map all the genes in an organism's genome. This goal is shared by the HGP and all other genome projects, using recombinant DNA technology and DNA sequencing instead of mutational analysis.

The HGP has produced a plethora of information, much of which is still being analyzed and interpreted. What is clear so far is that genomic information from a wide range of organisms demonstrates that humans and all other species share a common set of genes essential for cellular function and reproduction, confirming that all living organisms arose from a common ancestor. These evolutionary relationships facilitate the analysis of the human genome and the development of model organisms to study inherited human diseases.

Origins of the Project

The publicly funded Human Genome Project began in 1990 under the direction of James Watson, the codiscoverer of the double-helix structure of DNA. Instead of finding and mapping markers and disease genes one at a time, the HGP set out to sequence all the DNA in the human genome, identify and map the thousands of genes to the 24 chromosomes (22 autosomes, plus X and Y), and establish the function of all genes. To deal with the impact that genetic information would have on society, the HGP set up the **ELSI program (Ethical, Legal, and Social Implications)** to ensure that genetic information would be safeguarded and not used in discriminatory ways.

Although not reflected in the name, the Human Genome Project also included projects to sequence the genomes of several model organisms used in experimental genetics. These include *E. coli*, *S. cerevisiae*, *C. elegans*, *D. melanogaster*, and *M. musculus*, the mouse.

In 1999, a privately-funded human genome project led by Craig Venter at Celera Corporation was announced. This project's goal was to use the shotgun method to more rapidly sequence the human genome. This announcement set off an intense competition between the two teams to be first with the human genome sequence.

Major Features

In June 2000, the leaders of the public and private genome projects met at the White House and jointly announced the completion of a draft sequence of the human genome. Figure 18–10 summarizes the genes and their assigned functions. In February 2001, they each published an analysis covering about 96 percent of the euchromatic region of the genome. The remaining work of completing the sequence by filling in the gaps was completed in 2003 by the HGP. The major features of the human genome are presented in Table 18.7. As you can see in this table, many interesting and unique observations have provided us with major insights into own genome.

Unfinished Tasks

More than 99 percent of the euchromatic portion of the human genome sequence has been finished. Gaps (unsequenced regions) in the euchromatic portion of the genome are associated with duplicated regions of the genome that are difficult to assemble. Gaps in the heterochromatic portion of the genome

include regions around centromeres, the ribosomal DNA genes, and the telomeres; these span more than 200 Mb of sequence. The heterochromatic regions of the human genome were excluded from the genome projects because they contain extensive sets of tandemly repeated DNA sequences with few if any genes. Gaps of both types are nonrandomly distributed in the genome, and because of differences in size and sequence content, several different strategies may be needed to fill in all the unsequenced regions.

The Chimpanzee Genome

Although not part of the HGP, the nucleotide sequence of the chimpanzee (*Pan troglodytes*) genome was completed in 2004. Overall, the chimp and human sequences differ by less than 2 percent. Comparisons between these genomes offer some interesting insights into what makes some primates humans and makes others chimpanzees.

The speciation events that separated humans and chimpanzee occurred less than 6.3 million years ago, and recent evidence from genomic analysis indicates that these species initially diverged, then exchanged genes again before separating completely. After the species separated, their genomes have been evolving separately. This can be seen by comparing differences between the sequence of chimpanzee chromosome 22 (chimps have 48 chromosomes and humans have 46, so the numbering is different) and its human **ortholog**, chromosome 21 (orthologous genes and chromosomes share a common ancestor). These chromosomes have accumulated nucleotide substitutions that total 1.44 percent of the sequence. The most surprising

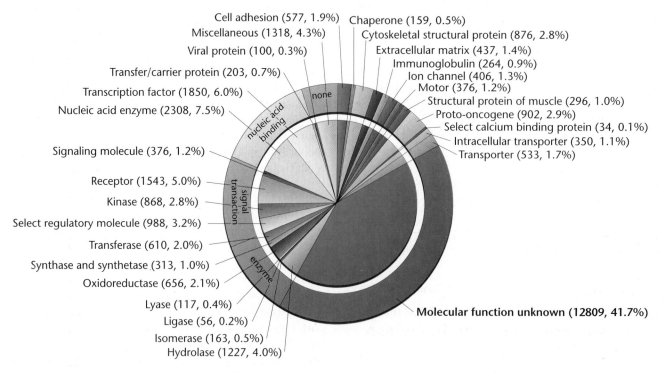

FIGURE 18–10 A preliminary list of assigned functions for genes in the human genome based on similarity to proteins of known function. Among the most common genes are those involved in nucleic acid metabolism (7.5% of all genes identified), transcription factors (6.0%), receptors (5%), protein kinases (2.8%), and cytoskeletal structural proteins (2.8%). A total of 12,809 predicted proteins (41%) have unknown functions, reflecting the work that is still needed to fully decipher our genome.

TABLE 18.7	Major Features of the Human Genome

- The human genome contains more than 3 billion nucleotides, but protein-coding sequences make up only about 5 percent of the genome.
- At least 50 percent of the genome is derived from transposable elements, such as LINE and *Alu sequences.*
- The human genome contains between 20,000 and 25,000 protein-coding genes, far fewer than the predicted number of 50,000–100,000 genes.
- More than 40 percent of the genes identified have no known molecular function.
- Genes are not uniformly distributed on the 24 human chromosomes. Gene-rich clusters are separated by gene-poor "deserts" that account for 20 percent of the genome. These deserts correlate with G bands seen in stained chromosomes. Chromosome 19 has the highest gene density, and chromosomes 13 and the Y chromosome have the lowest gene density.
- Human genes are larger and contain more and larger introns than genes in invertebrate genomes such as *Drosophila*. The largest known human gene encodes dystrophin, a muscle protein. This gene, associated in mutant form with muscular dystrophy, is 2.5 Mb in length (Chapter 14), larger than many bacterial chromosomes. Most of this gene is composed of introns.
- The number of introns in human genes ranges from 0 (histone genes) to 234 (in the gene for *titin*, a muscle protein).

difference is the discovery of 68,000 nucleotide insertions or deletions (called **indels**) in the chimp and human chromosomes, a frequency of 1 indel every 470 bases. Many of these are *Alu* insertions in human chromosome 21. Although the overall difference in the nucleotide sequence is small, there are significant differences in the encoded genes. Only 17 percent of the genes analyzed encode identical proteins in both chromosomes; the other 83 percent encode genes with one or more amino acid differences.

The differences between the chromosomes (Table 18.8) and their genes are only part of the story. Differences in the time and place of gene expression play a major role in differentiating the two primates. Using DNA microarrays (see Chapter 19), researchers compared expression patterns of 202 genes in human and chimp cells from brain and liver. They found more species-specific differences in expression of brain genes than liver genes. To further examine these differences, Svante Pääbo and colleagues compared expression of 10,000 genes in human and chimpanzee brains. They found that 10 percent of genes examined differ in expression in one or more regions of the brain. More importantly, these differences are associated with genes in regions of the human genome that have been duplicated following divergence of chimps and humans. This indicates that genome evolution, speciation, and gene expression are interconnected. Further work on these segmental duplications and the genes they contain may identify genes that help make us human.

TABLE 18.8	Comparisons Between Human Chromosome 21 and Chimpanzee Chromosome 22	
	Human 21	**Chimpanzee 22**
Size (bp)	33,127,944	32,799,845
G + C Content	40.94	41.01
CpG Islands	950	885
SINES (*Alu* elements)	15,137	15,048
Genes	284	272
Pseudogenes	98	89

18.7 Comparative Genomics Is a Versatile Tool

Comparative genomics compares the genome of one organism with that of another. It is a field with many research and practical applications, including gene discovery, and the development of model organisms to study human diseases. It also incorporates the study of gene and genome evolution and the relationship between organisms and their environment. Comparative genomics uses a wide range of techniques and resources, including the construction and use of nucleotide and protein databases containing nucleic acid and amino acid sequences, fluorescent *in situ* hybridization (FISH), as well as experimental methods. These resources are used to identify genetic differences and similarities between organisms, to determine how these differences contribute to differences in phenotype, life cycle, or other attributes, and to ascertain the evolutionary history of these genetic differences.

In the following sections, we will discuss applications of comparative genomics in model systems and evolution.

The Dog as a Model Organism

Analysis of the growing number of genome sequences confirms that all organisms are related and descended from a common ancestor. Similar gene sets are used in all organisms for basic cellular functions, such as DNA replication, transcription, and translation. These findings led to development and use of model organisms to study inherited human disorders and the interaction of genes and environment in complex diseases, such as cardiovascular disease and behavioral disorders. As mentioned earlier, these model organisms include yeast, fruit fly, and the mouse. A rough draft of the dog (*Canis familiaris*) genome was published in 2003 (Figure 18–11), and a more thorough analysis was finished in 2005, providing one of the most useful models with which to study our own genome.

The dog offers several advantages for studying heritable human diseases. Dogs share many genetic disorders with humans, including over 400 single-gene disorders, sex chromosome aneuploidies, multifactorial diseases including epilepsy, and genetic predispositions to cancer.

(a)

(b)

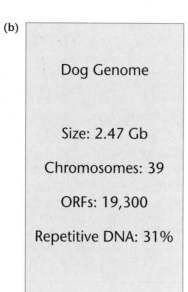

Dog Genome

Size: 2.47 Gb

Chromosomes: 39

ORFs: 19,300

Repetitive DNA: 31%

FIGURE 18–11 Dog genome. (a) Shadow, a poodle, whose DNA was used to sequence a rough draft of the dog genome. (b) Characteristics of the dog genome.

Comparative chromosome painting using fluorescently labeled probes from the dog genome hybridized to human chromosomes was used to visualize homologies between the genomes. About 90 conserved blocks of the dog genome can be mapped to human chromosomes by comparative FISH studies (Figure 18–12). These blocks contain DNA sequences with a high degree of similarity between the dog and the human genomes, reflecting the evolutionary relationship between dogs and our species.

Genomic analysis indicates that at least 60 percent of inherited diseases in dogs have similar or identical molecular causes, including point mutations and deletions, as in humans. In addition, at least 50 percent of the genetic diseases in dogs are breed-specific, so the mutant allele segregates in relatively homogeneous genetic backgrounds. Dog breeds resemble isolated human populations in having a small number of founders and a long period of relative genetic isolation. These properties facilitate the use of individual breeds as models of human genetic disorders.

In addition, differences in biology and behavior among dog breeds is well documented. Mapping DNA sequence differences (polymorphisms) among breeds may help identify candidate genes that contribute to both physiological and behavioral differences.

The Minimum Genome for Living Cells

What is the minimum number of genes necessary to support life? Although we do not yet know the minimum number of genes necessary for life in free-living organisms, we can use the small genomes of obligate parasites to speculate on the minimum number of genes *required to maintain life*. To do this, we can compare sequence information from the bacterial genomes of *Mycoplasma genitalium* and *M. pneumoniae*, both of which are human parasitic pathogens. These two closely related organisms are among the simplest self-replicating prokaryotes known

and can serve as model systems to understand the essential functions of a self-replicating cell. *M. genitalium* has a genome of 580 kb, and that of *M. pneumoniae* is 816 kb. These cause disease in a wide range of hosts, including insects, plants, and humans. (In humans, these bacteria cause genital and respiratory infections.)

The *M. genitalium* genome has 483 protein-coding genes, the smallest bacterial genome sequenced to date. The genome of a related species, *M. pneumoniae*, has those 483 genes and an additional 194, for a total of 677 protein-coding genes. In contrast, the 1.8 Mb genome of *Haemophilus influenzae* (the first bacterial genome sequenced) has 1783 genes.

The availability of genome sequences allows us to ask whether the 483 genes carried by *M. genitalium* (and shared with *M. pneumoniae*) are close to the minimum gene set needed for life. In other words, can we define life in terms of a number of specific genes? A combination of comparative and experimental methods can be used to answer this question. The comparative approach is based on the premise that genes shared by distantly related organisms are likely to be essential for life. By comparing the gene sets shared by different organisms, it should be possible to catalog those that are shared and develop a list of genes needed for life.

M. genitalium and *H. influenzae* last shared a common ancestor about 1.5 billion years ago. Genes shared between these species should represent those essential for life. By comparing the nucleotide sequences of the *M. genitalium* genes with the *H. influenzae* genes, researchers identified 240

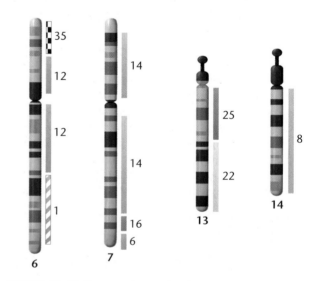

FIGURE 18–12 Interpretative results of chromosomal painting, a variation of fluorescent *in situ* hybridization (FISH), has been used to identify homologous segments on human and dog chromosomes. These homologies can be used to compare dog and human DNA sequences in the search for genes in both the dog and the human genome. Human chromosome 6 has sequences found on three dog chromosomes (1, 12, and 35). Human chromosome 7 is composed of sequences found on dog chromosomes 6, 14 and 16. The acrocentric human chromosomes 13 and 14 contain sequences found on dog chromosomes 22 and 25, and on dog chromosome 8, respectively.

orthologous genes between the two species. In addition to the 240 shared genes, 16 genes with different sequences but identical functions were identified. These represent essential functions that are performed by nonorthologous genes. Thus, comparative genomics estimates that 256 genes may represent the minimum gene set needed for life.

Craig Venter and his colleagues used an experimental approach to determine how many of the 480 *M. genitalium* genes are essential for life. They used transposons to selectively mutate genes in *M. genitalium*. Mutations in essential genes produced a lethal phenotype, but mutation of nonessential genes did not affect viability. They found that many of the 480 genes were nonessential and that the minimum gene set for *M. genitalium* is about 265 to 300 genes. This is close to the value of 256 derived from comparative genome analysis.

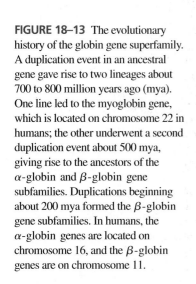

Now Solve This

Problem 20 on page 424 asks you to explain differential gene loss in members of the prokaryotic genus *Buchneria*.

Hint: Remember that symbiotic organisms coevolve with their hosts.

18.8 Comparative Genomics Reveals Evolution and Function in Multigene Families

Members of **multigene families** share similar, but not identical, DNA sequences through duplication and descent from a single ancestral gene. Their gene products frequently have similar functions, and the genes are often, but not always, found at a single chromosomal locus. A group of multigene families is called a superfamily. To see how analysis of multigene families provides insight into eukaryotic genome evolution and function, we first examine the superfamily of globin genes, which encode very similar but not identical polypeptide chains with closely related functions. Multiple proteins that arise from single-gene duplications are known as **paralogs**. The globin genes that encode the polypeptides in hemoglobin molecules exemplify a paralogous multigene superfamily that arose by duplication and dispersal to different chromosomal sites.

The Origin and Components of the Globin Gene Superfamily

The almost uninterrupted flow of sequence data from genome projects is providing evidence that multigene families are present in many, if not all, genomes. Molecular phylogenetics has traced the ancestry and relationships among members of gene families. One of the best-studied examples of divergent evolution is the **globin gene superfamily** (Figure 18–13). In this family, an ancestral gene encoding an oxygen transport protein was duplicated some 800 million years ago, producing two sister genes, one of which evolved into the modern-day myoglobin gene. (Myoglobin is an oxygen-carrying protein found in muscle.) The other gene underwent further duplication and divergence and some 500 million years ago formed prototypes of the **α-globin** and **β-globin** genes. These genes encode proteins found in hemoglobin, the oxygen-carrying molecule in red blood cells. Additional duplications within these genes occurred within the last 200 million years. Events subsequent to each duplication dispersed these gene subfamilies to different chromosomes, and in the human genome, each now resides on a separate chromosome.

Similar patterns of evolution are observed in other gene families, including the trypsin–chymotrypsin family of proteases, the homeotic selector genes of animals, and the rhodopsin family of visual pigments.

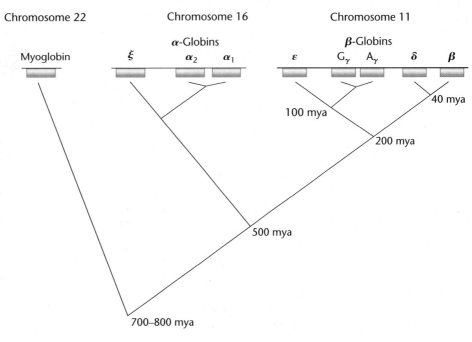

FIGURE 18–13 The evolutionary history of the globin gene superfamily. A duplication event in an ancestral gene gave rise to two lineages about 700 to 800 million years ago (mya). One line led to the myoglobin gene, which is located on chromosome 22 in humans; the other underwent a second duplication event about 500 mya, giving rise to the ancestors of the α-globin and β-globin gene subfamilies. Duplications beginning about 200 mya formed the β-globin gene subfamilies. In humans, the α-globin genes are located on chromosome 16, and the β-globin genes are on chromosome 11.

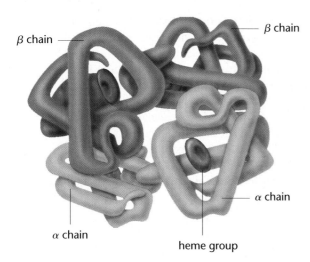

β chain

β chain

α chain

α chain

heme group

FIGURE 18–14 Each functional hemoglobin molecule consists of two alpha and two beta chains, each of which carries a heme group.

Adult hemoglobin is a tetramer, containing two α- and two β-polypeptides (Figure 18–14). Each polypeptide incorporates a heme group that reversibly binds oxygen. The α-globin gene cluster on chromosome 16 and the β-globin gene cluster on chromosome 11 share nucleotide sequence similarity, but the highest degree of sequence similarity is found within subfamilies (Figure 18–15).

The α-globin gene subfamily [Figure 18–16(a)] contains three genes: the ζ (zeta) gene, expressed only in early embryogenesis, and two copies of the α gene, expressed during the fetal ($α_1$) and adult stages ($α_2$). In addition, the cluster contains two pseudogenes (ζ and $α_1$). Pseudogenes are nonfunctional relatives of protein-coding genes. The pseudogenes in this family are designated by the prefix ψ (psi), followed by the symbol of the gene they most resemble. Thus, the designation $ψα_1$ indicates a pseudogene of the fetal $α_1$ gene.

The organization of the α-globin subfamily members and the location of their introns and exons reveal several features. First, as is common in eukaryotes, the DNA encoding the three functional α genes occupies only a small portion of the 30 kb region containing the subfamily. Most of the DNA in this region is intergenic spacer. Second, each functional gene in this subfamily contains two introns at precisely the same positions. Third, the nucleotide sequences within corresponding

exons are nearly identical in the ζ and α genes. Each of these genes encode polypeptide chains of 141 amino acids. However, their intron sequences are highly divergent, even though they are about the same size. Note that much of the nucleotide sequence of each gene is contained in these noncoding introns.

The human β-globin gene cluster contains five genes spaced over 60 kb of DNA [Figure 18–16(b)]. As with the α-globin gene subfamily, the order of genes on the chromosome parallels their order of expression during development. Three of the five genes are expressed before birth. The ε (epsilon) gene is expressed only during embryogenesis, while the two nearly identical γ genes ($G_γ$ and $A_γ$) are expressed only during fetal development. The polypeptide products of the two γ genes differ only by a single amino acid. The two remaining genes, δ and β, are expressed after birth and throughout life. A single pseudogene, $ψβ_1$, is present in this subfamily. All five functional genes in this cluster encode proteins with 146 amino acids and have two similar-sized introns at exactly the same positions. The second intron in the β-globin genes is significantly larger than its counterpart in the functional α-globin genes. These similarities reflect the evolutionary history of each subfamily and the events such as gene duplication, nucleotide substitution, and chromosome translocations that produced the present-day globin superfamily.

The Immunoglobulin Gene Family

The **immunoglobulin gene superfamily** is a large, diverse, ancient group of related genes found in all vertebrates, whose evolutionary lineage is not yet completely known. Many family members have functions unrelated to the immune system. Genes that encode proteins with immune system functions evolved through exon shuffling, gene duplication, and adaptation to new functions. The subfamilies encoding immune system genes appeared in the early stages of vertebrate evolution.

The immune system is an effective barrier against successful invasion by potentially harmful invading organisms or foreign substances. These invaders are recognized as nonself and are subsequently destroyed by the immune system. A highly specialized case of localized genomic alterations is a hallmark of maturation and function in the immune system. From a set of less than 300 genes, events—including recombination, deletions, and mismatching of rejoined DNA segments in imma-

α-globin V – L S P A D K T N V K A A W G K V G A H A G E Y G A E A L E R M F L S F P T T K T Y F P H F – D L S H
β-globin V H L T P E E K S A V T A L W G K V – – N V D E V G G E A L G R L L V V Y P W T Q R F F E S F G D L S T

α-globin – – – G S A Q V K G H G K K V A – D A L T N A V A H V D D – M P N A L S A L S D L H A H K L R V D P V N
β-globin A V M G N P K V K A H G K K V L – G A F S D G L A H L D N – L K G T F A T L S E L H C D K L H V D P E N

α-globin L L S H C L L V T L A A H L P A E F T P A V H A S L D K F L A S V S T V L T S K Y R – 141
β-globin L L G N V L V C V L A H H F G K E F T P P V Q A A Y Q K V V A G V A N A L A H K Y H – 146

FIGURE 18–15 The amino acid sequence of the alpha (α) and beta (β) globin genes using the single-letter abbreviations for the amino acids (see Figure 13–5). Shaded areas indicate identical amino acids. The two genes are descended from a common ancestor and diverged from each other about 500 million years ago.

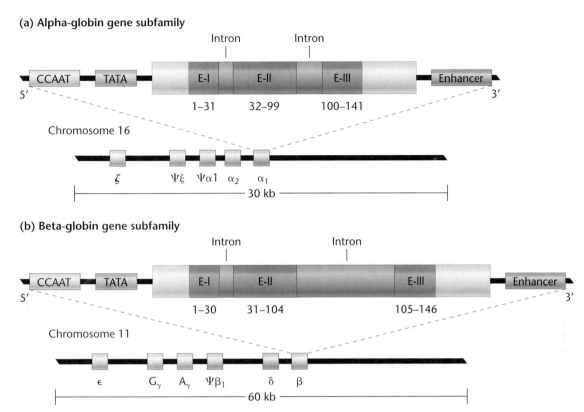

FIGURE 18–16 (a) Organization of the α-globin gene subfamily on chromosome 16, and (b) the β-globin gene subfamily on chromosome 11. Also shown is the internal organization of the α_1 gene and the β gene. Each gene contains three exons (E-I, E-II, E-III) and two introns. The numbers below the exons indicate the location of amino acids in the gene product encoded by each exon.

ture cells of the immune system—create a vast potential for immune response.

One part of the immune system produces antibodies against invading organisms or foreign substances. Anything that causes antibody production is termed an **antigen**. Many different molecules can act as antigens, including proteins, polysaccharides, and nucleic acids. **Antibodies** are proteins produced and secreted by the B cells of the immune system. Vertebrates can produce millions of different antibodies, each responding to a different antigen. In the next sections, we examine the molecular basis of antibody diversity.

Antibodies are produced by plasma B cells, a type of lymphocyte (white blood cell). There are five classes of antibodies or **immunoglobulins (Ig)**: IgM, IgD, IgG, IgE, and IgA

(Table 18.9). The IgG class represents about 80 percent of the antibodies found in the blood and is the most intensively characterized group of antibodies.

A typical antibody molecule (Figure 18–17) contains two copies of two different polypeptide chains held together by disulfide bonds. Each larger, or **heavy (H) chain**, molecule contains approximately 440 amino acids. The amino acid

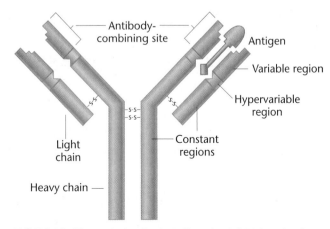

FIGURE 18–17 A typical antibody (IgG) molecule is Y-shaped and contains four polypeptide chains. The longer arms are H chains, and the shorter arms are L chains. The chains are joined by disulfide bonds. Each chain contains a variable region, a hypervariable region, and a constant region. The variable and hypervariable regions of a pair of L and H chains form a combining site that interacts with a specific antigen.

TABLE 18.9	Classes and Components of Immunoglobulins		
Ig Class	**Light Chain**	**Heavy Chain**	**Tetramers**
IgM	κ or λ	μ	$\kappa_2\mu_2, \lambda_2\mu_2$
IgD	κ or λ	δ	$\kappa_2\delta_2, \lambda_2\delta_2$
IgG	κ or λ	γ	$\kappa_2\gamma_2, \lambda_2\gamma_2$
IgE	κ or λ	ϵ	$\kappa_2\epsilon_2, \lambda_2\epsilon_2$
IgA	κ or λ	α	$\kappa_2\alpha_2, \lambda_2\alpha_2$

sequence at the N-terminus differs among heavy chains and is known as a variable region (V_H). The remaining C-terminal amino acids are the same in all H chains and make up the constant region (C_H). In humans, genes on the long arm of chromosome 14 encode the H chains.

Each **light (L) chain** contains 220 amino acids, half of which make up the variable region (V_L). The remaining amino acids at the C-terminus make up the constant region (C_L). Two types of L chains exist: κ-chains, encoded by genes on human chromosome 2, and λ-chains, encoded by genes on chromosome 22. Together, the variable regions of the heavy and light chains form the **antibody-combining site**. Each combining site has a unique structural conformation that allows it to bind to a specific antigen, like a key in a lock.

In response to an antigen, the immune system can generate cells that synthesize a unique antibody. Because billions of different antibodies can be made, it is impossible for separate genes to encode each of them; there is simply not enough DNA in the human genome to encode all of these gene products.

The key to understanding antibody diversity lies in understanding immunoglobulin gene structure. In humans, the κ-chain gene (Figure 18–18) has several components: the leader-variable (L-V) region, the joining (J) region, and a constant (C) region. There are 70 to 100 DNA segments in the

L-V region, each with a different nucleotide sequence and a promoter, but no enhancer. The J region contains six different segments, and the C region contains only one C segment. An enhancer, but no promoter, is located between the last J segment and the C segment. This ensures that segments cannot be transcribed individually.

During B-cell maturation, one of the 100 L-V regions ($L_2 - V_2$ in Figure 18–18) and its promoter are randomly joined via a recombination event to one of the six J regions (J_3 in the figure) and to the C region (and its enhancer). The recombination event forms a functional L-chain gene (the second line in Figure 18–18). All DNA between the selected L-V segment and the J_3 segment is excised and destroyed. This type of recombination differs from that seen in meiosis and involves different enzymes. The joining event between the V and J regions is imprecise and can occur over a span of about 6 bp. In addition, during this recombination event, the ends of the DNA are subjected to a **break-nibble-add mechanism**, in which a few bases are randomly removed and a few bases randomly added. The newly created gene contains three exons (L, V-J, and C) and two introns, one between L and V-J, another between J_3 and C. The unselected J segments (J_4-J_6 in this example), become part of the second intron. The L-chain gene is transcribed, processed, and translated to

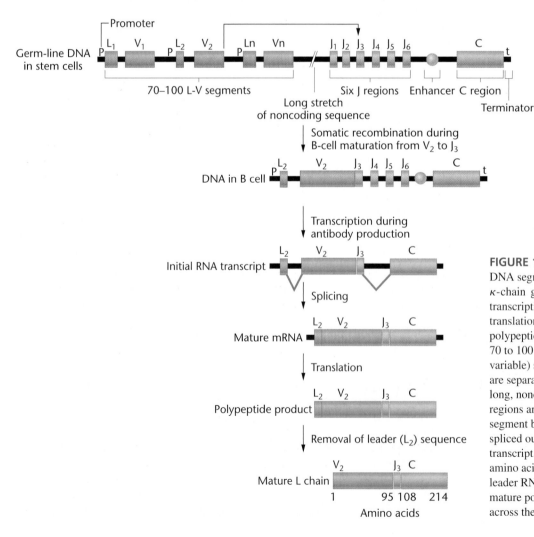

FIGURE 18–18 Formation of the DNA segments encoding a human κ-chain gene and the subsequent transcription, mRNA splicing, and translation leading to the final polypeptide chain. In germ-line DNA, 70 to 100 different L-V (leader-variable) segments are present. These are separated from the J regions by a long, noncoding sequence. The J regions are separated from a single C segment by an intron that must be spliced out of the initial mRNA transcript. Following translation, the amino acid sequence derived from the leader RNA is cleaved off as the mature polypeptide chain passes across the cell membrane.

form κ-L-chain proteins that become part of an antibody molecule. The rearranged gene is stable and is passed on to all progeny of the B cell in which this event occurs.

Antibody diversity in κ-L-chain production results from combining any of the 100 segments in the L-V regions with any of the 6 segments in the J regions, which generates about 600 different κ-chain genes. Since each joining reaction can occur over several base pairs, the number of possible genes increases to several thousand. Additional diversity comes from the break-nibble-add mechanism. Recombination events generate a similar number of different λ-chain genes.

The heavy-chain genes extend over a large span of DNA and include four types of regions: L-V, D, J, and C. During B-cell maturation, random recombination of H-chain components generates a large number of H-chain genes. If we assume there are 25 D segments, combining any of these with any of the 300 V segments and 6 J segments generates 45,000 H-chain genes. With imprecise joining and the break-nibble-add mechanism, the total number of different genes climbs markedly.

To summarize, antibody diversity is due to four features of the immunoglobulin gene system: (1) multiple numbers of variable segments, (2) multiple numbers of diversity and joining segments, (3) multiple splice locations with break-nibble-add joining, and (4) multiple combinations of L chains with H chains. As antibody-forming B cells mature, DNA recombination rearranges these genes so that each mature B lymphocyte encodes, synthesizes, and secretes only one specific type of antibody. Each mature B cell can make one type of light chain (κ or λ) and one type of heavy chain. When an antigen is present, it stimulates the B cells and populations of differentiated cells are produced, all of which synthesize one type of antibody that interacts with the antigen.

18.9 Proteomics Identifies and Analyzes the Proteins in a Cell

As genome projects provide information about the number and kinds of genes present in prokaryotic and eukaryotic genomes, the question of protein function is becoming a central issue in biology. As stated earlier, in most of the genomes sequenced to date, many newly discovered genes have no known function. Others have only presumed functions assigned by analogy with known genes made by sequence comparisons from databases, not experimental evidence from laboratory experiments. For example, in *E. coli* and *S. cerevisiae*, more than half of the genes encoded by their genomes have no known function, and in the human genome, about 40 percent of the genes are of unknown function.

Reconciling Gene Number and Protein Number

As more genomic sequence data become available for eukaryotes, it is apparent that the complexity of an organism is not necessarily related to the number of genes in its genome. In the pregenomic era, gene function was equated with identifying an encoded gene product. However, sequencing has revealed that the link between gene and gene product is often

complex. Genes can have multiple transcription start sites that produce several different types of transcripts. Alternative splicing and editing of pre-mRNA molecules can generate dozens of different proteins from a single gene. Altogether, it is estimated that 40 to 60 percent of human genes produce more than one protein by alternative splicing.

Once made, many gene products are modified by cleavage of end groups (such as signal sequences, propeptides, or initiator methionine residues), by the addition of chemical groups (e.g., methyl, acetyl, phosphoryl), or by linkage to sugars and lipids. Proteins are internally and externally cross-linked and in some cases, processed by removing internal amino acid sequences (called **inteins**). Over a hundred mechanisms of posttranslational modification are known. Analysis of protein function is complicated by the fact that many proteins work via protein–protein interactions or as part of a large molecular complex. Thus, the human genome, which has 20,000 to 25,000 protein-coding genes, has the capacity to produce several hundred thousand different gene products. Proteomics is used to reconcile the differences between the number of genes in a genome and the number of proteins observed in cells. The protein set in a cell is called the **proteome**. Proteomics provides information about a protein's function, structure, posttranslational modifications, protein–protein interactions, cellular localization, variants, and relationships (shared domains, evolutionary history) to other proteins—for every protein encoded in a genome.

Proteomics Technology

Proteomics uses a variety of techniques to separate and identify proteins isolated from cells. The most commonly used combination of techniques involves **two-dimensional gel electrophoresis (2DGE)** and **mass spectrometry (MS)**. In 2DGE (Figure 18–19), proteins extracted from cells grown under two different conditions are loaded onto polyacrylamide gels, and in the first dimension, proteins are separated according to their electrical charge. When completed, the gels are rotated 90°, and in the second dimension, proteins are separated according to their molecular weight. When the gels are stained, proteins are revealed as spots; typical gels show from 200 to 10,000 spots (Figure 18–20). The proteins isolated by 2DGE must then be identified. In the past, proteins were identified by their binding to an antibody, or by fragmentation and stepwise analysis to produce the amino acid sequence. These time-consuming and labor-intensive methods are not useful in analyzing the hundreds or thousands of proteins that can be isolated from a cell.

To speed up and automate the process, mass spectrometry permits the mass of large molecules to be accurately determined (Figure 18–21). One method, **matrix-assisted laser desorption and ionization (MALDI)** accomplishes this in several steps. First, a protein isolated from a 2D gel is digested with the enzyme trypsin, producing a characteristic set of peptide fragments. Each fragment is analyzed by MALDI, producing a peptide mass fingerprint. Second, a database search algorithm (e.g., BLAST) is used to search protein databases. In searching, the algorithim breaks each known protein into a series of predicted peptide fragments (a process known as virtual trypsin digests) and computes the mass of each

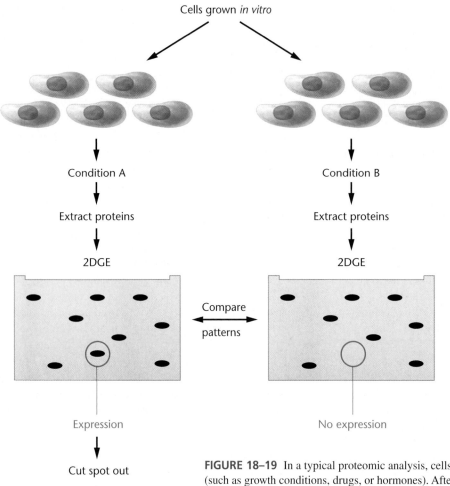

Cells grown *in vitro*

Condition A

Extract proteins

2DGE

Compare patterns

Condition B

Extract proteins

2DGE

Expression

No expression

Cut spot out

Mass spectrometry

FIGURE 18–19 In a typical proteomic analysis, cells are exposed to two different conditions (such as growth conditions, drugs, or hormones). After treatment, proteins are extracted and separated by 2DGE. The pattern of spots is then compared for evidence of differential gene expression. Spots of interest are cut out from the gel, digested into peptide fragments, and analyzed by mass spectrometry to identify the protein in the spot.

fragment. Masses of predicted fragments are compared to those produced by MALDI. The search algorithm produces a series of matches between protein fragments in the database and those in the protein being analyzed. A perfect match identifies a protein with no ambiguity. However, because of alternative splicing or posttranslational modification, there may not be a perfect match, and the identity of the protein may have to be confirmed by other mass spectrometry techniques. High-performance instruments can identify hundreds of proteins per day, and banks of spectrometers can process thousands of samples in a single day. Collectively, these techniques are referred to as **high-throughput technology**.

Now Solve This

In Problem 28 on page 424 a protein domain is used to query a database and recover proteins with a given sequence. You are asked to explain why the function of the query sequence is not related to the function of the other proteins.

Hint: Remember that protein domains may have related functions and that proteins can contain several different domains.

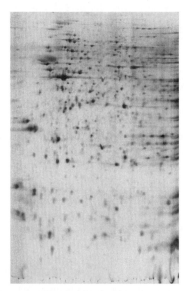

FIGURE 18–20 A two-dimensional protein gel, showing the separated proteins as spots. Several thousand proteins can be displayed on such gels.

Changes in the Bacterial Proteome

As outlined earlier, *M. genitalium*, with a genome of 480 genes, represents one of the simplest living organisms known. Valerie Wasinger and her colleagues used proteomics to provide a snapshot of which genes are expressed under two different growth conditions in *M. genitalium*: exponential growth and the stationary phase following rapid growth.

Using 2DGE, Wasinger's group identified 427 protein spots in exponentially growing cells. Of these, 201 were analyzed and identified by peptide digestion, mass spectrometry, and database searches. The analysis uncovered 158 known proteins (33 percent of the proteome) and 17 unknown proteins. The remaining spots included fragments derived from larger proteins, different forms of the same protein (isoforms), and posttranslationally modified products. The identified proteins included enzymes involved in energy metabolism, DNA replication, transcription, translation, and transport of materials across the cell membrane.

During the transition from exponential growth to the stationary phase, cell division in the culture slowed and then stopped. At this time, there was a 42 percent reduction in the number of proteins synthesized. In addition, some new proteins appeared, while other proteins underwent dramatic changes in abundance. These changes are apparently a conse-

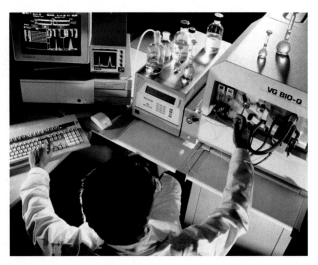

FIGURE 18–21 Proteins are digested and analyzed by mass spectrometry to identify the gene products present in a cell at a given time.

quence of nutrient depletion, increased acidity of the growth medium, and other environmental changes (Figure 18–22).

Wasinger's analysis helps establish the minimum number of expressed genes required for maintenance of the living state and the changes in gene expression that accompany the transition to the stationary phase.

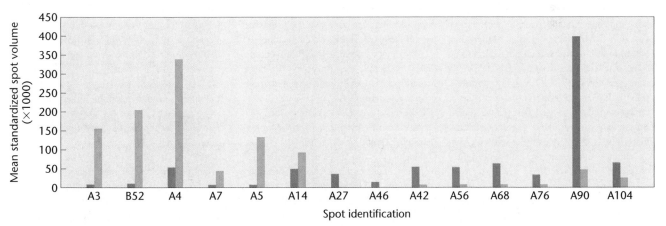

FIGURE 18–22 Changes in spot size for selected proteins during exponential growth (blue bars) and during stationary growth (orange bars). These differences represent changes in gene expression and the upregulation or downregulation of individual proteins as growth conditions change.

GENETICS, TECHNOLOGY, AND SOCIETY

Whose DNA Is It, Anyway?

Few things are more personal than our genetic material. Our genes establish the framework of our physical, mental and even emotional makeups. We pass on our genes to our offspring, thereby ensuring the continuation of our genetic traits into the future. There is something deep within our psyches that holds our genetic material at the center of our beings.

It may be jarring, then, to realize that DNA sequences derived from about 20% of our genes are owned by others and are a potential source of commercial profit.

Over the last 30 years, the U.S. Patent and Trademark Office has issued patents for more than 4,300 human gene sequences. These patents have been granted to universities, governments, and private corporations. DNA-related patents cover cloned genes, their regulatory sequences, small sequences such as ESTs (expressed sequence tags) and SNPs (single nucleotide polymorphisms), and various methods and diagnostic tools based on these sequences. Some human genes have multiple patents covering gene variants and fragments, as well as numerous potential applications, including diagnostic tests, or future uses of the sequences as drug-screening probes. Even more remarkably, there are over three million patent applications filed for genes, gene fragments and other genome-related materials. How many of these will end up as patents, and what the patents might cover, are still unknown.

This ever-increasing "land-grab" over the human genome has raised serious concerns about the nature of patents, the meaning of genetic information and the balance between commercial gain and public interest.

Proponents of gene patenting point out that patent protection is essential for the development of DNA-based diagnostics and pharmaceuticals. Without exclusive rights, no one would be willing to invest the years of research and millions of dollars required to develop and test new medicines. They also argue that patenting reduces secrecy, as patent holders are required to publicly disclose the details of their patent, thereby speeding research in related areas.

In contrast, many people are uncomfortable about the idea that human genes and gene sequences can be owned and used for commercial gain. They argue that the human genome is naturally occurring and unique, and cannot be treated like other inventions, such as mousetraps or works of art. Some worry that patent protection will stifle research and healthcare, as more and more genes, diagnostic tests and treatments based on DNA sequences are held by commercial interests and the fees for using these patented items are controlled by a few.

Supporters of gene patenting counter by explaining that patents over DNA sequences are not the same things as ownership over the genes from which the sequences are derived. Patents cover only isolated or cloned versions of sequences and not genes as they occur naturally in anyone's body—even though the sequences may match. In addition, patents give exclusive rights to someone to exploit the patented item only for a limited period of time—usually 20 years. During this time, patent holders can license the use of their patent to anyone, often for a fee, meaning that the exclusiveness of the patent is not written in stone.

Is there any evidence that DNA patents have restricted the progress of basic research or the development of new diagnostics and treatments?

Some recent data support the idea that patents have had neutral effects on basic research. A study conducted by the National Academy of Sciences in 2005 revealed that only about eight percent of university researchers conducted work on patented gene sequences. The remainder either did not know about any relevant patents or did not need to use patented reagents.

In contrast to basic research, the potential conflicts between DNA patents and medical diagnostics may be more problematic. Critics of gene patenting often cite the case of the breast cancer gene *BRCA1* and the patents held by Myriad Genetics Laboratories in Salt Lake City, Utah. In the 1990s, Myriad developed genetic tests for breast cancer, based on specific mutations in the *BRCA1* gene. Their initial patent claimed rights over the normal *BRCA1* sequence, various mutations, *BRCA1* diagnostic tests, the methods used to screen tumor samples,

any information derived from the *BRCA1* gene and all methods to diagnose and treat hereditary breast and ovarian cancer. Myriad has enforced many aspects of this patent, preventing other research centers from offering their own versions of the tests, at lower costs. For example, in Canada and Europe, Myriad required that all samples be sent to their laboratories in Utah for testing. Publicly-funded laboratories in Canada had been using their own *BRCA1* tests at lower cost, but stopped the practice after threats of legal action. Although one Canadian province continues to use its own *BRCA1* tests despite possible future law suits, other provinces now send their samples to Myriad Genetics for testing. Some scientists worry that Myriad will build up the world's largest bank of *BRCA1* samples and monopolize the information derived from these samples. In 1994, European countries challenged Myriad's patent claims and have won the right to bypass the *BRCA1* patents. Recently, the American College of Medical Genetics expressed concern that wide-ranging patent claims such as those held by Myriad Genetics may increase the costs of genetic testing, slow the development of quality control, and restrict the development of related gene tests.

Many questions raised by the race to patent the human genome remain unanswered: Should naturally derived materials such as DNA be patentable? Will patents over DNA sequences slow the development of competitive and cost-effective diagnostics and treatments? In the future, will the entire human genome be patented, increasing the costs of basic research? As more DNA sequences become patented, will we be able to reconcile the needs of private industry with those of the public good?

References

Stix, G. 2006. Owning the stuff of life. *Scientific American* 294: 76–83.

Web Site

Genetics and patenting. Human Genome Program, Ethical, Social and Legal Issues, U.S. Department of Energy, 2006. **http://www.ornl.gov/sci/techresources/ Human_Genome/elsi/patents.shtml**

CHAPTER SUMMARY

1. Genome sequences are analyzed to ensure that the sequence is accurate, to identify all encoded genes, and to classify known genes into functional categories. The sequences are then deposited into searchable databases.

2. Bacterial genomes have very high gene density, averaging one gene per kilobase pair of DNA. Typically, as much as 90 percent of the chromosome encodes genes. Many genes are organized into polycistronic transcription units defined as operons. In general, bacterial genes do not contain introns.

3. Eukaryotic genomes have a gene density that is much lower than that in bacteria. Genes typically are not organized into operons; rather, each is a separate transcription unit. Eukaryotic genes are often interrupted with introns.

4. Complex multicellular eukaryotes differ from the less complex yeast in a number of ways. Complex eukaryotes have more genes and much more DNA. This results in gene densities falling to 1 gene per 5 kb or even 1 gene per 10–20 kb or more. A higher proportion of multicellular eukaryotic genes have introns. The number of introns per gene increases, and the size of introns

increases as complexity increases from yeast to humans. Some plants, such as *Arabidopsis*, have a gene structure and organization that is indistinguishable from animals. Other plants, such as maize, have a different organization, with vast blocks of transposable elements separating islands of genes.

5. The human genome contains 3.2 billion nucleotides, but only about 5 percent of this DNA contains protein-coding genes. It is estimated that humans carry between 20,000 and 25,000 genes. Large differences exist in gene density on different chromosomes, with gene-rich regions alternating with gene-poor regions. The protein-coding genes of humans have larger and more numerous introns than those of other eukaryotes such as fruit flies.

6. Many eukaryotic genes have undergone duplication followed by sequence divergence, leading to multigene families. The globin gene clusters and the immunoglobulin gene families are prime examples of this phenomenon.

7. Proteomics is used to study the expression of genes in bacterial cells under different growth conditions and is providing insight into the gene sets that cells use during growth.

KEY TERMS

α-globin gene, 413

annotation, 402

antibody, 415

antibody-combining site, 416

antigen, 415

β-globin gene, 413

bioinformatics, 402

break-nibble-add mechanism, 416

comparative genomics, 400

compiling, 402

ELSI program, 409

functional genomics, 400

gene density, 407

genome, 400

genomics, 400

globin gene superfamily, 413

heavy (H) chain, 415

high-throughput technology, 418

homology searches, 403

Human Genome Project (HGP), 400

immunoglobulin (Ig), 415

immunoglobulin gene superfamily, 414

indels, 411

intein, 417

intron, 407

light (L) chain, 416

map-based sequencing, 401

matrix-assisted laser descripton and ionization (MALDI), 417

mass spectrometry, 417

model organisms, 405

multigene families, 413

open reading frames (ORFs), 402

ortholog, 410

paralog, 413

proteome, 417

proteomics, 400

repetitive sequence, 408

shotgun sequencing method, 401

structural genomics, 400

two dimensional gel electrophoresis, 417

INSIGHTS AND SOLUTIONS

1. How are gene duplications generated, and how do they contribute to genome evolution?

Solution: Gene duplications can be generated by unequal crossing over and DNA slippage. In addition, chromosomal aberrations such as inversions and translocations can lead to gene duplication. It is possible to duplicate genes through horizontal transfer (transfer between species) involving transposons, plasmids, and other transferred DNAs. Gene duplication contributes to evolution by providing mutational opportunities without compromising the health and survivorship of the organism containing the genome. Gene duplications allow for mutational experimentation and therefore evolution.

2. Recent sequencing of the heterochromatic regions (repeat-rich sequences concentrated in centromeric and telomeric areas) of the *Drosophila* genome indicates that within 20.7 Mb, there are 297 protein-coding genes (Bergman et al. 2002. *genomebiology3 (12)@genomebiology.com/2002/3/12/RESEARCH/0086).* Given that the euchromatic regions of the genome contain 13,379 protein-coding genes in 116.8 Mb, what general conclusion is apparent?

Solution: Gene density in euchromatic regions of the *Drosophila* genome is about one gene per 8730 base pairs, while gene density in heterochromatic regions is one gene per 70,000 bases (20.7 Mb/297). Clearly, a given region of heterochromatin is much less likely to contain a gene than the same-sized region in euchromatin.

PROBLEMS AND DISCUSSION QUESTIONS

1. Used in gene annotation, gene prediction programs allow researchers to identify likely coding regions in DNA sequences. Annotation is complicated when genes are complex, contain multiple initiation sites, or contain numerous exons. For example, Pavy and others (1999. *Bioinformatics* 15: 887–899) determined that even for the most well-studied organisms, such programs can predict correct exon boundaries only about 80 percent of the time. Given this percentage, what is the likelihood of determining the correct exon boundaries in a gene with five exons?

2. Recent genome-sequencing efforts have provided considerable insight into the molecular nature of living systems. However, it has become increasingly apparent that to fully comprehend the genome, reannotation, manual verification, and more advanced techniques will be needed. Haas and colleagues (2002. *genomebiology3(6)@genomebiology.com/2002/3/6/RESEARCH/0029*) recently reannotated the *Arabidopsis* genome and found 240 new genes, 92 of which are homologous to known proteins. In addition, they identified a new class of exons, called microexons, which vary in length from 3 to 25 base pairs. Based on this information, how would you qualify the statement that "to know the sequence of DNA is to know the blueprint of life"?

3. In a recent draft annotation and overview of the human genome sequence, Wright and others (2001. *genomebiology2(7)@genomebiology.com/2001/2/7/RESEARCH/0025*) presented a graph similar to the one shown below. The graph details the approximate number of embryo-specific genes for each chromosome. Review earlier information in the text on human chromosomal aneuploids, and correlate that information with the graph. Does this graph provide insight as to why some aneuploids occur and others do not?

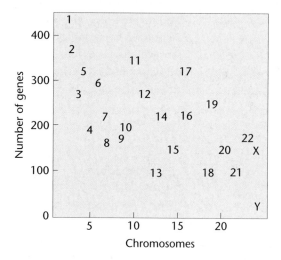

4. The process of annotating a sequenced genome is continual. In March 2000, the first annotated sequence of the *Drosophila* genome was released, which predicted 13,601 protein-coding genes within the euchromatic region of the genome. Shown in the next column are selected data from Release 2 and Release 3. (Modified from Misra et al. 2002. *genomebiology3(12)@genomebiology.com/2002/3/12/RESEARCH/0083*.) (a) Assuming a uniform distribution for Release 3, approximately how many base pairs of DNA are between protein-coding genes in *Drosophila*? (b) On average, approximately how many exons are there per gene for Release 3? (c) Approximately how many introns are there per gene? (d) What appears to be the most significant difference between Release 2 and Release 3?

Criteria	Release 2	Release 3
Total length of euchromatin	116.2 Mb	116.8 Mb
Total protein-coding genes	13,474	13,379
Protein-coding exons	50,667	54,934
Introns	48,381	48,257
Genes with alternative transcripts	689	2729

5. The data given in Problem 4 indicate that the more closely researchers examine genome sequences, the more complex the interpretations of those data will become. Misra and colleagues (2002) found that nested and overlapping genes are common in *Drosophila*. They determined that approximately 7.5 percent of all Release 3 genes were included within the introns of other genes and that the majority are transcribed from the opposite strand of the including gene. In addition, they found that about 15 percent of the annotated genes involve the overlap of mRNAs on opposite strands. What impact will this information have on genome annotation, and what clinical significance might it have?

6. In April 2003, scientists announced that the Human Genome Project had completed its mission and that the human genome was completely known. One of the leaders of the project was quoted as saying, "We have before us the instruction set that carries each of us from the one-cell egg through adulthood to the grave." However, from the beginning, the HGP excluded the heterochromatic regions at the tips of chromosomes and the regions surrounding the centromeres. The human genome is about 3000 Mb, and heterochromatic regions comprise about 15 percent of this total. If gene density in human heterochromatin is the same as it is in *Drosophila*, how many genes remain to be discovered in humans? Would you be comfortable stating that all human genes have been identified? See question 2 in the "Insights and Solutions" section.

7. One of the main problems in annotation is deciding how long an ORF must be before it is accepted as a gene. Shown on the next page are three different ORF scans of the *E. coli* genome region containing the *lacY* gene. Regions shaded in brown indicate ORFs. The scans have been set to accept ORFs of 50, 100, and 300 nucleotides as genes. How many putative genes are detected in each scan? The longest ORF covers 1254 bp; the next longest covers 234 bp; and the shortest covers 54 bp. How can you decide how many genes are actually in this region? In this type of ORF scan, is it more likely that the number of genes in the genome will be overestimated or underestimated? Why?

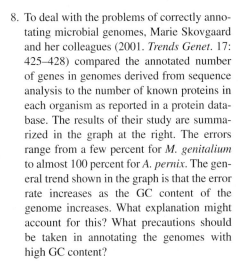

50
Sequenced strand

Complementary strand

100
Sequenced strand

Complementary strand

300
Sequenced strand

Complementary strand

9. What is functional genomics? How does it differ from comparative genomics?

10. What, if any, features do bacterial genomes share with eukaryotic genomes?

11. What is the definition of an operon? How does information from the *Aquifex aeolicus* genome alter our ideas about operons?

12. Plasmids can be transferred between species of bacteria, and most carry nonessential genes. For these and other reasons, plasmid genes have not been included as part of the genomes of bacterial species. The bacterium *B. burgdorferi* contains 17 plasmids carrying 430 genes, some of which are essential for life. Should plasmids carrying essential genes be considered as part of an organism's genome? What about other plasmids that do not carry such genes? In other words, how do we define an organism's genome in these cases?

13. Compare and contrast the chemical nature, size, and form assumed by the genetic material of bacteria and yeast.

14. Why might we predict that the organization of eukaryotic genetic material is more complex than that of viruses or bacteria?

15. Compare the organization of bacterial genes to that of eukaryotic genes. What are the major differences?

16. Based on the completion of the genome sequence of *Arabidopsis* in 2000, Simillion and others (2002. *Proc. Nat. Acad. Sci. [USA]* 99: 13627–13632) have estimated that the *Arabidopsis* genome has undergone three rounds of genome duplication or polyploidization events in the last 100 million years. Presently, of the approximately 25,000 genes in *Arabidopsis*, these researchers estimate that about 80 percent are in duplicated regions of the genome. Examination of Table 18.6 reveals that *Arabidopsis*, rice, maize, and barley have approximately the same number of genes, but that *Arabidopsis* has much less DNA. What type of genomic organization appears to account for the vast differences in DNA content with similar gene numbers in these species?

17. Annotation of the human genome sequence reveals that our genome contains 20,000 to 25,000 genes. Proteomic analysis indicates that human cells are capable of synthesizing more than 300,000 different proteins. How can this discrepancy be reconciled?

8. To deal with the problems of correctly annotating microbial genomes, Marie Skovgaard and her colleagues (2001. *Trends Genet.* 17: 425–428) compared the annotated number of genes in genomes derived from sequence analysis to the number of known proteins in each organism as reported in a protein database. The results of their study are summarized in the graph at the right. The errors range from a few percent for *M. genitalium* to almost 100 percent for *A. pernix*. The general trend shown in the graph is that the error rate increases as the GC content of the genome increases. What explanation might account for this? What precautions should be taken in annotating the genomes with high GC content?

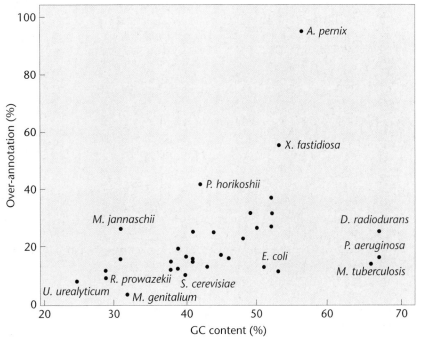

18. List some of the following general features of the human genome: size, how much codes for proteins, how much is composed of repetitive sequences, and how many genes it contains.

19. The discovery that *M. genitalium* has a genome of 0.58 Mb and only 470 protein-coding genes has sparked interest in determining the minimum number of genes needed for a living cell. In the search for organisms with smaller and smaller genomes, a new species of Archaea, *Nanoarchaeum equitans*, was discovered in a high-temperature vent on the ocean floor (Huber et al. 2002. *Nature* 417: 63–67). This prokaryote has one of the smallest cell sizes ever discovered, and its genome is only about 0.5 Mb. However, organisms such as *M. genitalium*, *N. equitans*, and other microbes with very small genomes are either parasites or symbionts. How does this affect the search for a minimum genome? Should the definition of the minimum genome size for a living cell be redefined?

20. In the search for the smallest bacterial genome and the minimum number of genes necessary for life, attention has turned to species of *Buchnera*, which live as intracellular symbionts in aphid intestinal cells. As symbionts, they need not maintain the genes necessary for infection and for evasion of the host's immune system as do parasites or pathogens and may have smaller genomes. The genome of one species of *Buchnera*, designated APS, has been sequenced. It has 564 genes in a circular chromosome of 640 kb (Shigenobou et al. 2000. *Nature* 407: 81–86). To determine whether *Buchnera* genome size is conserved across different groups of aphids, Gil and colleagues (2002. *Proc. Nat. Acad. Sci. [USA]* 99: 4454–4458) physically mapped the genome sizes of nine *Buchnera* genomes that were isolated from five aphid families. The genomes were sized by digestion with restriction enzymes, and separation of the resulting fragments was done by gel electrophoresis. The data for some *Buchnera* species are given in the following table. Although there are some discrepancies in sizes, the sum of the fragments corresponds to the size of the chromosome that appeared on the gel without restriction digestion. From your analysis of the data, is genome reduction in *Buchnera* still occurring? How do the genome sizes obtained for these species compare with the genome of *M. genitalium*? The APS species of *Buchnera* contains 564 coding genes in a 641 kb genome. How many genes should be present in species CCE? How does this compare to the number of genes in *M. genitalium*? Are there other ways to determine the minimum genome needed for life without searching for other bacterial species with small genomes?

21. What do V_L C_H IgG, J, and D represent in immunoglobulin structure?

22. If germ-line DNA contains 10 V, 30 D, 50 J, and 3 C segments, how many unique DNA sequences can be formed by recombination?

23. If there are 5 V, 10 D, and 20 J regions available to form a heavy-chain gene and 10 V and 100 J regions are available to form a L-chain gene, how many unique antibodies can be formed?

24. In addition to comparisons of nucleotide sequences for determining phylogenetic relationships among organisms, studies of gene order have become an informative source for genome studies. In addition to providing evidence of evolutionary relationships, gene-order data have been used to predict gene function and functional interactions of proteins. Tamames (2001) (*Genome Biology 2: RESEARCH0020.* E pub 2001 Jun 01) determined that in prokaryotes, loss of gene order occurs when phylogenetic distance increases, but contrary to expectation, significant conservation is maintained in distant groups. What factors might you expect to contribute to the conservation of gene order among distantly related species?

25. Genomic sequencing has opened doors to numerous studies that help us understand the evolutionary forces shaping the genetic makeup of organisms. Using databases containing the sequences of 25 genomes, Kreil and Ouzounis (2001) (*Nucl. Acids Res.* 29: 1608–1615) examined the relationship between GC content and global amino acid composition. They found that it is possible to identify thermophilic species on the basis of their amino acid compositions alone, which suggests that evolution in a hot environment selects for a certain whole-organism amino acid composition. In what way might evolution in extreme environments influence genome and amino acid composition? How might evolution in extreme environments influence the interpretation of genome sequence data?

26. What are pseudogenes, and how are they produced?

27. The β-globin gene family consists of 60 kb of DNA, yet only 5 percent of the DNA encodes gene products. Account for as much of the remaining 95 percent of the DNA as you can.

28. Annotation of the proteome attempts to relate each protein to function in time and space. Traditionally, protein annotation depended on an amino acid sequence comparison between a query protein and a protein with known function. If two proteins shared a considerable portion of their sequence, the query would be assumed to share the function of the annotated protein. Following is a representation of the "look-the-same" method of protein annotation involving a query sequence and three different human proteins (modified from Rigoutsos et al. 2002. *Nucl. Acids Res.* 30: 3901–3916). Note that the query sequence aligns to common domains within the three other proteins. What argument might you present to suggest that the function of the query is not related to the function of the other three proteins?

Region of amino acid sequence match to query

Size of DNA Fragments Produced by Restriction Enzymes, (kb)				
Buchnera	*Apa*I	*Rsr*II	*Apa*I + *Rsr*II	Total DNA Length kb
APS	286, 226, 73, 52, 3.4	277, 264, 99	240, 104, 99, 73, 52, 45, 24, 3.4	640±
THS	545	545	320, 234	544±
CCE	440	450	405, 46	448±
CCU	265, 135, 50, 25	475	200, 136, 64, 50, 30	476±

Applications and Ethics of Genetic Engineering

GeneChip® Probe Array

Hybridized Probe Cell

Single stranded, fluorescently labeled DNA target

Oligonucleotide probe

50 μm

Each field contains millions of copies of a specific oligonucleotide probe

1.28 cm

P53-4

DNA microarrays, such as this GeneChip, produced by Affymetrix, Inc., are used in clinical and research applications to detect mutations and analyze patterns of gene expression.

Over 50,000 different probes complementary to genetic information of interest

Image of Hybridized Probe Array

CHAPTER CONCEPTS

- Genetic engineering has revolutionized agriculture through the development of genetically modified crop plants that have enhanced nutritional value and are resistant to herbicides, environmental conditions, and pests.

- Genetically modified plants and animals can produce useful pharmaceutical products.

- Genetic technologies are used to diagnose genetic disorders and detect genetic predispositions to certain diseases.

- Gene therapy has been used to treat genetic disorders, so far with mixed results.

- Genetic technologies have enhanced the design of new drugs and will soon help to assess a person's reaction to drug treatments.

- DNA fingerprinting can identify specific individuals and is widely used in forensics, archaeology, and conservation biology.

- Genetic technologies and genetic modifications pose ethical problems that are being addressed by changes in public policy and laws.

Since the dawn of recombinant DNA technology in the 1970s, scientists have harnessed genetic engineering not only for biological research, but also for direct applications in medicine, agriculture, and biotechnology. The ability to manipulate DNA *in vitro* and to introduce genes into living cells has allowed scientists to generate new varieties of plants and animals with specific gene traits, to diagnose and treat genetic diseases, and to manufacture cheaper and more effective pharmaceutical products. Industry analysts estimate that genetic engineering will lead to U.S. commercial products worth over $45 billion by 2009.*

Along with the growth of applications for DNA technologies, serious concerns about our power to manipulate genes and apply gene technologies have emerged. Genetic engineering has the potential to significantly alter how humans deal with the natural world; hence, it raises ethical, social, and economic questions that few other technologies elicit.

Genetic engineering in the twenty-first century is a major topic that cannot be fully explored in the context of an introductory textbook. As a result, this chapter will present only a selection of applications that illustrate the power of genetic engineering. We will begin by explaining how genetic engineering has modified agriculturally important plants and animals. Next, we examine the impact of genetic technologies on the diagnosis and treatment of human diseases. In addition, we will briefly describe how genetic engineering has affected the production of pharmaceutical products, and how DNA technologies are now used for forensic applications. Finally, we explore some of the social, ethical, and legal implications of the new DNA technologies.

How Do We Know?

In this chapter we will focus on some of the many applications of DNA technologies. We will explore how genetic engineering and genomics have changed agriculture, medicine, and forensics. As you study this topic, you should try to answer several fundamental questions:

1. How do we determine whether genetically modified plants are safe for human consumption?

2. What experimental evidence confirms that we have introduced a useful gene into a transgenic organism and that it performs as we anticipate?

3. How do we verify that vaccines produced in plants are effective?

4. How do we know from DNA analysis that a human fetus has sickle cell anemia?

5. How do we know in DNA microarray analysis that specific genes are being expressed in a specific tissue?

6. How do we know whether a forensic DNA profile comes from only one individual?

---

* The terms **biotechnology**, and **genetic engineering** are sometimes used interchangeably, although their definitions are distinct. Genetic engineering relates to the alteration of the genetic makeup of living organisms, using recombinant DNA technologies. Organisms that result from genetic engineering are referred to as genetically modified (GM) or transgenic. Biotechnology is a more general term, encompassing the use of living organisms, or products derived from them, to carry out a wide range of processes, including the manufacture of food or industrial products.

19.1 Genetic Engineering Has Revolutionized Agriculture

For millennia, farmers have manipulated the genetic makeup of plants and animals to enhance food production. Until the advent of genetic engineering 30 years ago, these genetic manipulations were restricted to the selection and breeding of naturally occurring or mutagen-induced variants. Genetic engineering presents a powerful new tool for altering the genetic constitution of agriculturally important organisms. Scientists can now identify, isolate, and clone genes that confer desired traits, then specifically and efficiently introduce these into organisms. As a result, it is possible to quickly introduce insect resistance, herbicide resistance, or nutritional characteristics into farm plants and animals. The gene-based revolution in agriculture began in 1996 with the introduction of herbicide-resistant soybeans. Since then, dozens of transgenic agricultural crops and animals have entered agricultural use. Worldwide, over 200 million acres are planted with genetically engineered crops, particularly herbicide and pest-resistant soybeans, corn, cotton, and canola. These crops are planted by over 8 million farmers in 17 countries.

In this section, we examine some examples of genetically engineered organisms used in agriculture. The ethical and social concerns associated with these developments will be examined in Section 19.6.

Transgenic Crops and Herbicide Resistance

Damage from weed infestation destroys about 10 percent of crops worldwide. In an attempt to combat this problem, farmers often use herbicides to kill weeds prior to seeding a field crop. As the most efficient herbicides also kill crop plants, herbicide uses are limited. The creation of herbicide-resistant crops has opened the way to efficient weed control and increased yields of some major agricultural crops. At present, over 75 percent of soybeans and cotton in the United States are resistant to the herbicide glyphosate.

We will examine how resistance to one herbicide (glyphosate) was transferred to crops such as soybeans and maize. The herbicide **glyphosate** is effective at very low concentrations, is not toxic to humans, and is rapidly degraded by soil microorganisms. Glyphosate kills plants by inhibiting the action of a chloroplast enzyme called **EPSP synthase**. This enzyme is important in amino acid biosynthesis in both bacteria and plants. Without the ability to synthesize vital amino acids, plants wither and die.

The first step in producing a glyphosate-resistant crop plant was the isolation and cloning of an EPSP synthase gene from a glyphosate-resistant strain of *E. coli*. This EPSP gene was cloned into a vector between promoter sequences derived from a plant virus and transcription termination sequences derived from a plant gene. The recombinant vector was transformed into the bacterium *Agrobacterium tumifaciens* (Figure 19–1). The plasmid-carrying bacteria were then used to infect plant cells derived from plant leaves. The clumps of cells (calluses) that formed after infection with *A. tumifaciens* were tested for their ability to grow in the presence of glyphosate. Glyphosate-resistant calluses were grown into transgenic plants and sprayed with glyphosate at concentrations four times higher than that needed to kill wild-type plants. Transgenic plants that

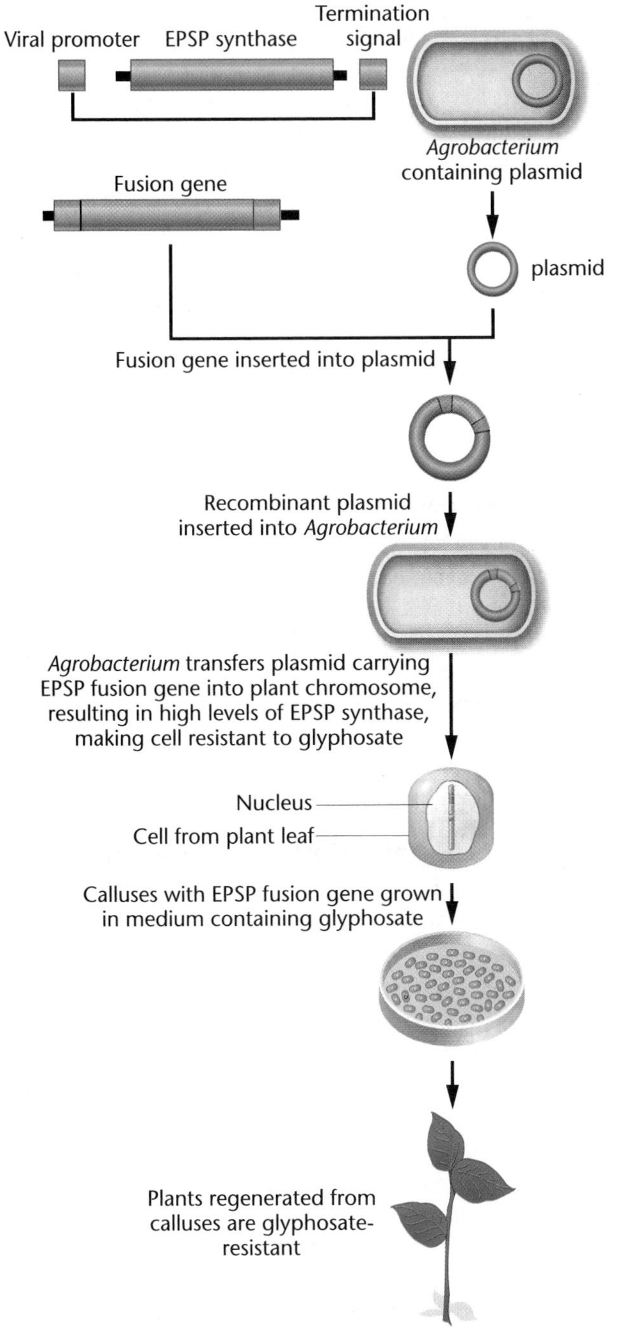

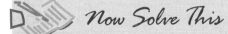

FIGURE 19–1 In gene transfer of glyphosate resistance, the EPSP synthase gene is fused to a promoter from the cauliflower mosaic virus. This fusion gene is then transferred to a plasmid vector, and the recombinant vector is transformed into an *Agrobacterium* host. *Agrobacterium* infection of cultured plant cells transfers the EPSP synthase fusion gene into a plant-cell chromosome. Cells that acquire the gene are able to synthesize large quantities of EPSP synthase, making them resistant to the herbicide glyphosate. Resistant cells are selected by growth in herbicide-containing medium. Plants regenerated from these cells are herbicide-resistant.

expressed EPSP synthase grew and developed, while the control plants withered and died.

Similar transgenic techniques have been used to make plants resistant to several other herbicides, as well to pathogens such as viruses and insect larvae. Plants with increased tolerance to drought and salty soils are now in development.

Now Solve This

Problem 5 on page 447 asks you to predict whether the gene conferring glyphosate resistance could escape from a transgenic crop into weed plants.

Hint: Consider the methods by which plants breed and the types of selective pressures on both wild plants and domestic crops.

Nutritional Enhancement of Crop Plants

In the last 50 to 100 years, genetic improvement of crop plants through the traditional methods of artificial selection and genetic crosses has resulted in dramatic increases in productivity and nutritional enhancement. For example, maize yields have increased fourfold over the last 60 years, and more than half of this increase is due to genetic improvement by artificial selection.

Gene transfer by recombinant DNA techniques offers a new way to enhance the nutritional value of plants. Many crop plants are deficient in some of the nutrients required in the human diet, and biotechnology is being used to produce crops that meet these dietary requirements. One example is the production of "**golden rice**" with enhanced levels of β-carotene, a precursor to vitamin A (Figure 19–2). Vitamin A deficiency is prevalent in many areas of Asia and Africa, and more than 500,000 children a year become permanently blind as a result of this deficiency. Rice is a major staple food in these regions but does not contain vitamin A.

To create golden rice, scientists transferred three genes encoding enzymes in the biosynthetic pathway leading to β-carotenoid synthesis into the rice genome, using methods of recombinant DNA technology. Two of these genes came from

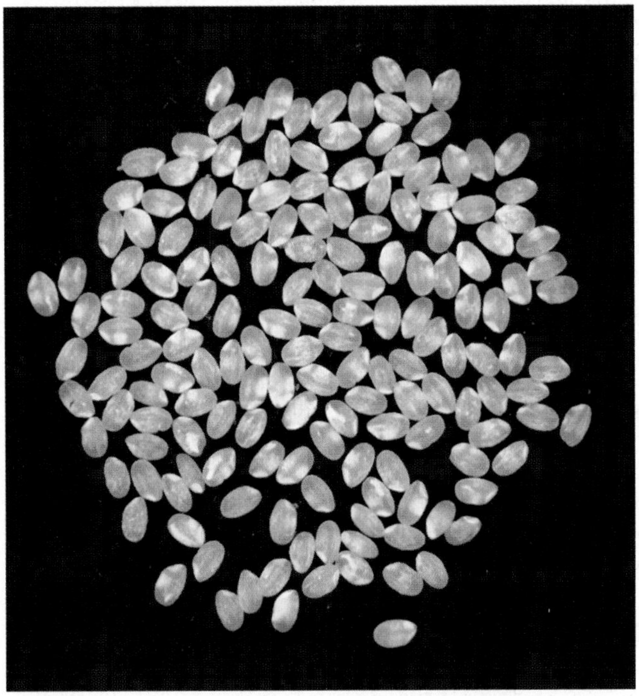

FIGURE 19–2 A photograph of golden rice, a strain genetically modified to produce β-carotene, a precursor to vitamin A. Many children in countries where rice is a dietary staple have diets deficient in vitamin A and develop blindness.

the daffodil and one from a bacterium. Although golden rice is currently available for planting, its levels of β-carotene are not sufficiently high to efficiently combat vitamin A deficiencies. To address this problem, new rice varieties with higher levels of β-carotene production are in development. Acceptance of genetically engineered varieties of rice and efficient distribution of golden rice remain challenges to its wider use.

Many other varieties of nutritionally enhanced food crops have been, and are being, developed. These include plants with augmented levels of key fatty acids, antioxidants, and other vitamins and minerals. These efforts are directed at addressing the dietary lack of nutrients affecting more than 40 percent of the world's population. Other future developments include decaffeinated teas and coffees, as well as crops enhanced for other characteristics such as taste, growth rates, yields, color, storage, and ripening characteristics.

Transgenic Animals with Enhanced Productivity

Although genetically engineered plants are major players in modern agriculture, transgenic animals are less widespread. Nonetheless, some high-profile examples of genetically engineered farm animals have created public interest and controversy in the field. The first genetically engineered livestock were created in 1985, using pro-nuclear injection methods. These methods involve the introduction of cloned DNA into a pronucleus of a fertilized egg. The injected DNA integrates at random into the host's genome, and the eggs are then introduced into surrogate mothers. The offspring that maintain the integrated DNA are selected and crossed in order to produce homozygous lines. This technique is also discussed in Chapter 17. Attempts to create farm animals containing transgenic growth hormone genes have not been successful, probably because growth is a complex, multigene trait. One notable exception is the transgenic Atlantic salmon, bearing copies of a Chinook salmon **growth hormone gene** driven by a constitutive promoter. These salmon mature quickly and appear to have no adverse health effects. These transgenic farmed salmon are currently under review by the Food and Drug Administration. Another successful transgenic farm animal is a pig that expresses the gene encoding the enzyme phytase. These pigs are able to break down dietary phosphorus, thereby reducing excretion of phosphorus, which is a major pollutant in pig farms. Other transgenic animals in development are those with resistance to infections and those lacking certain genes, such the prion protein *PrP* gene, that contribute to disease susceptibility.

Currently, the major uses for transgenic farm animals are as bioreactors to produce useful pharmaceutical products, as discussed in the next section.

19.2 Genetically Engineered Organisms Synthesize a Wide Range of Biological and Pharmaceutical Products

Perhaps the most successful and widespread use of recombinant DNA technology is for the production of **biopharmaceuticals**. Prior to the recombinant DNA era, biopharmaceutical

agents such as insulin, clotting factors, or growth hormones were purified from tissues such as pancreas, blood, or pituitary glands. Clearly, these sources were in limited supply, and the purification processes were expensive. In addition, products derived from these natural sources could be contaminated by disease agents such as viruses. Now that human genes encoding important therapeutic proteins can be cloned and expressed in a number of nonhuman host cell types, we have more abundant, safer, and less expensive sources of biopharmaceuticals.

In this section, we outline several examples of therapeutic products that are produced by expression of cloned genes in transgenic host cells and organisms. Table 19.1 lists a number of other important pharmaceutical products currently synthesized in transgenic plants and animals.

Insulin Production in Bacteria

Several therapeutic proteins have been produced by introducing human genes into bacteria. In most cases, the human gene is cloned into a plasmid and the recombinant vector is introduced into the bacterial host. After ensuring that the transferred gene is expressed, large quantities of the transformed bacteria are grown, and the human protein is recovered and purified from bacterial extracts.

The first human gene product manufactured by recombinant DNA technology was human insulin, which was licensed for therapeutic use in 1982. Previously, insulin was chemically extracted from the pancreas glands of cows and pigs obtained from slaughterhouses. **Insulin** is a protein hormone that regulates glucose metabolism. Individuals who cannot produce insulin have diabetes, a disease that, in its more severe form (type I), affects more than 2 million individuals in the United States. Although synthetic human insulin is now produced by another process, a look at the original genetic engineering method is instructive, as it shows both the promise and the difficulty of applying recombinant DNA technology.

Clusters of cells embedded in the pancreas synthesize a precursor polypeptide known as preproinsulin. As this polypeptide is secreted from the cell, amino acids are cleaved from the end and the middle of the chain. These cleavages produce the mature insulin molecule, which contains two polypeptide chains (the A and B chains), joined by disulfide bonds. The A subunit contains 21 amino acids, and the B subunit contains 30.

In the original bacterial process, synthetic genes that encode the A and B subunits were constructed by oligonucleotide synthesis (63 nucleotides for the A polypeptide and 90 nucleotides for the B polypeptide). Each synthetic oligonucleotide was inserted into a vector adjacent to the *lacZ* gene encoding the bacterial form of the enzyme β-galactosidase. When transferred to a bacterial host, the *lacZ* gene and the adjacent synthetic oligonucleotide were transcribed and translated as a unit. The product, a **fusion polypeptide**, contained the amino acid sequence for β-galactosidase attached to the amino acid sequence for one of the insulin subunits (Figure 19–3). The fusion proteins were purified from bacterial extracts and treated with cyanogen bromide, a chemical that cleaves the fusion protein from the β-galactosidase. When the fusion products were mixed, the two insulin subunits spontaneously united, forming an intact, active insulin molecule. The purified insulin was then packaged for use by diabetics.

TABLE 19.1	Some Genetically Engineered Pharmaceutical Products Now Available or Under Development	
Gene Product	**Condition Treated**	**Host Type**
Tissue plasminogen activator tPA	Heart attack, stroke	Cultured mammalian cells
Human growth hormone	Dwarfism	Cultured mammalian cells
Monoclonal antibodies against vascular endothelial growth factor (VEGF)	Cancers	Cultured mammalian cells
Human clotting factor VIII	Hemophilia A	Transgenic sheep, pigs
C1 inhibitor	Hereditary angiodema	Transgenic rabbits
Recombinant human antithrombin	Hereditary antithrombin deficiency	Transgenic goats
Hepatitis B surface protein vaccine	Hepatitis B infections	Cultured yeast cells
Immunoglobulin IgG1 to HSV-2 glycoprotein B	Herpesvirus infections	Transgenic soybeans
Recombinant monoclonal antibodies	Diagnosis and passive immunization against rabies	Transgenic tobacco, soybeans
Norwalk virus capsid protein	Norwalk virus infections	Potato (edible vaccine)
E.coli heat-labile enterotoxin	*E.coli* infections	Potato (edible vaccine)

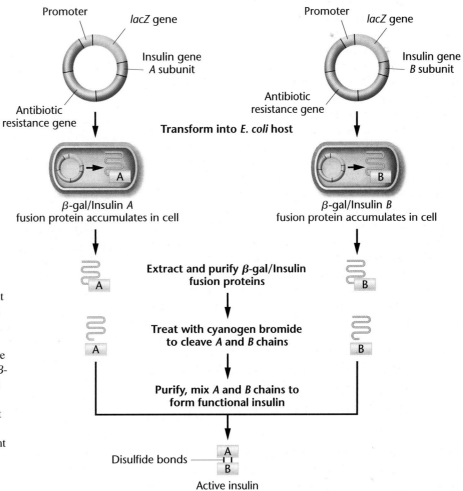

FIGURE 19–3 To synthesize recombinant human insulin, synthetic oligonucleotides encoding the insulin *A* and *B* chains were inserted at the tail end of a cloned *E. coli* *lacZ* gene. The recombinant plasmids were transformed into *E. coli* hosts, where the β-gal/insulin fusion protein was synthesized and accumulated in the host cells. Fusion proteins were then extracted from the host cells and purified. Insulin chains were released from β-galactosidase by treatment with cyanogen bromide. The insulin subunits were purified and mixed to produce a functional insulin molecule.

Transgenic Animal Hosts and Pharmaceutical Products

Although bacterial hosts have been used to produce therapeutic proteins, there are some disadvantages in using prokaryotic hosts to synthesize eukaryotic proteins. One problem is that bacterial cells cannot process and modify many eukaryotic proteins. As a result, they cannot add the sugars and phosphate groups to the proteins that are needed for full biological activity. In addition, eukaryotic proteins produced in prokaryotic cells often don't fold into the proper three-dimensional configuration and, as a result, are inactive. To overcome these difficulties and increase yields, many biopharmaceuticals are now produced in eukaryotic hosts. As seen in Table 19.1, eukaryotic hosts may include cultured eukaryotic cells (plant or animal) or transgenic farm animals. The therapeutic proteins produced by these hosts may then be purified from the host cells or from products such as milk.

An example of a biopharmaceutical product synthesized in transgenic animals is that of the human protein $\alpha 1$-antitrypsin. A deficiency of the enzyme $\alpha 1$-antitrypsin is associated with the heritable form of emphysema, a progressive and fatal respiratory disorder common among people of European ancestry. To produce $\alpha 1$-antitrypsin for use in treating this disease, the human gene was cloned into a vector at a site adjacent to a sheep promoter sequence that activates transcription in milk-producing cells. Genes placed next to this promoter are expressed only in mammary tissue. This fusion gene was microinjected into sheep zygotes fertilized *in vitro*. The fertilized zygotes were transferred to surrogate mothers. The resulting transgenic sheep developed normally and produced milk containing high concentrations of functional human $\alpha 1$-antitrypsin. This human protein is present in concentrations of up to 35 grams per liter of milk and can be easily extracted and purified. A small herd of lactating transgenic sheep can provide an abundant supply of this protein. Herds of other transgenic animals acting as biofactories are becoming part of the pharmaceutical industry. In fact, the famous sheep, Dolly, was cloned to facilitate the establishment of a flock of sheep that would consistently produce high levels of human proteins.

Human proteins produced in transgenic animals undergo clinical testing as a first step in the therapeutic use of recombinant human proteins. A recombinant human enzyme, α-glucosidase, produced in rabbit milk, is now in clinical tests for treating children with **Pompe disease**. This progressive and fatal metabolic disorder is caused by a lack of the enzyme α-glucosidase and is inherited as an autosomal recessive condition. In early-onset Pompe disease, infants have poor muscle tone and may never sit or stand. Most die before two years of age from respiratory or cardiac complications. In one of the first tests, the recombinant enzyme was given to affected children weekly, with no significant side effects. All of the children showed normal enzyme activity in the tissues analyzed, and all showed improvements in their symptoms. If large-scale trials are successful, recombinant α-glucosidase from transgenic animals will become the preferred method of treatment for this disorder.

Transgenic Plants and Edible Vaccines

One of the most promising applications of biotechnology may be the production of vaccines. Vaccines stimulate the immune system to produce antibodies against disease-causing organisms and thereby confer immunity against specific diseases. Traditionally, two types of vaccines have been used: inactivated vaccines, which are prepared from killed samples of the infectious virus or bacteria; and attenuated vaccines, which are live viruses or bacteria that can no longer reproduce but can cause a mild form of the disease. Genetic engineering is being used to produce a new type of vaccine called a **subunit vaccine**, which consists of one or more surface proteins from the virus or bacterium, but not the entire virus or bacterium. This surface protein acts as an antigen that stimulates the immune system to make antibodies that act against the organism from which it was derived.

One of the first subunit vaccines was made against the **hepatitis B virus**, which causes liver damage and cancer. The gene that encodes the hepatitis B surface protein was cloned into a yeast expression vector, and the cloned gene was expressed in yeast host cells. The protein was then extracted and purified from the host cells and packaged for use as a vaccine.

Vaccination programs in developing countries are faced with serious problems of manufacturing, transporting, and storing the vaccines. Most vaccines need refrigeration and must be injected under sterile conditions. In many rural areas, refrigeration and sterilization facilities are not available. To overcome these problems, scentists are attempting to develop vaccines

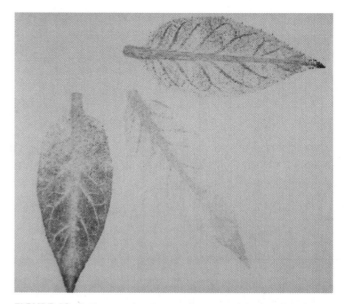

FIGURE 19–4 Transgenic tobacco plants carrying an antigenic subunit of the hepatitis B virus. Leaves from transgenic plants were treated with antibodies against hepatitis B antigen, and stained to show whether the plants produced the antigen. The central leaf in the photograph is a leaf from a normal tobacco plant and does not stain with the anti-hepatitis B antibodies. Tobacco plants are used in these experiments because they are a common experimental organism in plant molecular biology. For use as a vaccine, the antigenic subunit would be transferred to an edible crop plant.

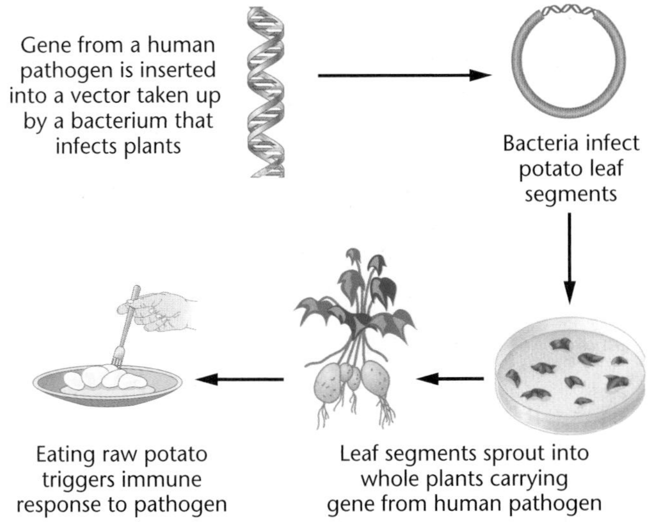

Gene from a human pathogen is inserted into a vector taken up by a bacterium that infects plants

Bacteria infect potato leaf segments

Leaf segments sprout into whole plants carrying gene from human pathogen

Eating raw potato triggers immune response to pathogen

FIGURE 19–5 To make an edible vaccine, a gene from a pathogen (a disease-causing agent, such as a virus or bacterium) is transferred into a vector which is then transferred into a bacterial host that infects plant cells. Infection of potato plant leaves transfers the vector and the pathogen's gene into the nuclei of potato leaf cells. The leaf segments are grown into mature potato plants that express the pathogenic gene. Eating the raw potato produced by these plants triggers an immune response to the protein encoded by the pathogen's gene, conferring immunity to infection by this pathogen.

that can be synthesized in edible food plants. These vaccines would be inexpensive to produce, would not require refrigeration, and would not have to be given under sterile conditions by trained medical personnel.

As a model system, the gene encoding an antigenic subunit of the hepatitis B virus has been introduced into the tobacco plant and expressed in its leaves (Figure 19–4). For use as a source of vaccine, the gene would have to be inserted into food plants such as grains or vegetables. Other edible vaccines are in clinical trials. A vaccine against a bacterium that causes diarrhea has been produced in genetically engineered potatoes (Figure 19–5) and used to successfully vaccinate human volunteers who ate small quantities (50–100 g) of the potatoes. In trials with another vaccine, transgenic spinach expressing rabies viral antigens was fed to volunteers, who showed significant increases in rabies-specific antibodies. Tests using vaccine-producing bananas are currently under way. Genetically engineered edible plants may soon be available to vaccinate infants, children, and adults against many infectious diseases. Edible vaccines are also discussed at length in the GTS essay on page 183.

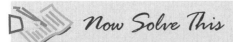

Now Solve This

Problem 4 on page 446 asks you to interpret data on the development of an edible vaccine.

Hint: Stimulation of antibody formation by the smallest possible portion of a protein is important to ensure vaccine specificity.

19.3 Genetic Engineering and Genomics Are Transforming Medical Diagnosis

Geneticists are now applying knowledge about the human genome and the genetic basis of many diseases to a wide range of medical applications. Gene-based technologies have already had a major impact on the diagnosis of disease and will soon revolutionize medical treatments and the development of specific and effective pharmaceuticals. As more information emerges from the Human Genome Project, researchers are identifying increasing numbers of genes involved in both single-gene diseases and complex genetic traits. Methods based on recombinant DNA technologies are speeding the identification of specific mutations that lead to hereditary disorders and somatic diseases such as cancers.

In this section, we provide a brief overview of some methods used to diagnose genetic diseases. In the next section, we will describe some exciting technologies that will transform medical treatments over the next few decades.

Genetic Tests Based on Restriction Enzyme Analysis

Using DNA-based tests, scientists can directly examine a patient's DNA for mutations associated with disease. Gene testing was one of the first successful applications of recombinant DNA technology, and currently more than 900 gene tests are in use. These tests usually detect DNA mutations that are associated with single-gene disorders that are inherited in a Mendelian fashion. Examples of such genetic tests are those that detect sickle-cell anemia, cystic fibrosis, Huntington disease, hemophilias, and muscular dystrophies. Other genetic tests have been developed for complex disorders such as breast and colon cancers. Gene tests are used to perform prenatal diagnosis of genetic diseases, to identify carriers, to predict the future development of disease in adults, and to confirm the diagnosis of a disease detected by other methods.

A classic method of genetic testing is **restriction enzyme analysis**. To illustrate this method, we examine the prenatal diagnosis of **sickle-cell anemia**. This disease is an autosomal recessive condition common in people with family origins in areas of West Africa, the Mediterranean basin, and parts of the Middle East and India. This disease is caused by a single amino acid substitution in the β-globin protein that is created by a single-nucleotide substitution in the β-globin gene. The single-nucleotide substitution also eliminates a cutting site in the β-globin gene for the restriction enzymes *Mst*II and *Cvn*I. As a result, the mutation alters the pattern of restriction fragments seen on Southern blots. These differences in restriction cutting sites are used to prenatally diagnose sickle-cell anemia and to establish the parental genotypes and the genotypes of other family members who may be heterozygous carriers of this condition. This method is also known as **restriction fragment length polymorphism (RFLP)** analysis.

FIGURE 19–6 For amniocentesis, the position of the fetus is first determined by ultrasound, and a needle is then inserted through the abdominal and uterine walls to recover fluid and fetal cells for genetic or biochemical analysis.

Placenta
Amniotic cells

Centrifuge →

Amniotic fluid
Amniotic cavity
Uterine wall

Fluid: Composition analysis

Cells: Karyotype, sex determination, biochemical, and recombinant DNA studies

Cell culture: Biochemical studies, chromosomal analysis

Analysis using recombinant DNA methods

For prenatal diagnosis, fetal cells are obtained by **amniocentesis** or chorionic villus sampling (Figure 19–6), and adult cells are obtained from blood samples. DNA is extracted from the samples and digested with *Mst*II. This enzyme cuts three times within a region of the normal β-globin gene, producing two small DNA fragments. In the mutant sickle-cell allele, the middle *Mst*II site is destroyed by the mutation, and one large restriction fragment is produced by *Mst*II digestion (Figure 19–7). The restriction-digested DNA fragments are separated by gel electrophoresis, transferred to a nylon membrane, and visualized by Southern blot hybridization, using a probe from this region.

Figure 19–7 shows the results of genetic testing for sickle-cell anemia in one family. The parents (I-1 and I-2) are both heterozygous carriers of the mutation. *Mst*II digestion of the parents' DNA produces a large band (the mutant allele) and two smaller bands (the normal allele) in each case. The parents' first child (II-1) is homozygous normal because she has only the two smaller bands. The second child (II-2) has sickle-cell anemia; he has only one large band and is homozygous for the mutant allele. The fetus (II-3) has a large band and two small bands and

is therefore heterozygous for sickle-cell anemia. He or she will be unaffected, but will be a carrier.

Only about 5 to 10 percent of all point mutations can be detected by restriction enzyme analysis because most mutations occur in regions of the genome that do not contain restriction enzyme cutting sites. However, if the gene of interest has been sequenced and the disease-associated mutations are known, geneticists can employ synthetic oligonucleotides to detect these mutations, as described next.

Now Solve This

Problem 20 on page 448 uses RFLP analysis to determine whether sisters in a family are carriers of hemophilia.

Hint: Differences in the number and location of restriction enzyme sites create RFLPs that can be used to determine genotypes.

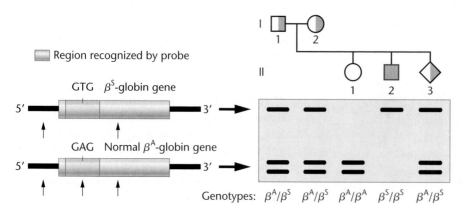

Region recognized by probe

GTG   β^S^-globin gene

5' ━━━━━━━━━━ 3'

GAG   Normal β^A^-globin gene

5' ━━━━━━━━━━ 3'

Genotypes: β^A^/β^S^   β^A^/β^S^   β^A^/β^A^   β^S^/β^S^   β^A^/β^S^

FIGURE 19–7 A Southern blot diagnosis of sickle-cell anemia, with arrows representing the location of restriction enzyme recognition sites. In the mutant β-globin allele (β^S^) a point mutation (GAG→GTG) has destroyed a restriction enzyme cutting site, resulting in a single large fragment on a Southern blot. In the pedigree, the family has one unaffected homozygous normal daughter (II-1), an affected son (II-2), and an unaffected carrier fetus (II-3). The genotype of each family member can be read directly from the blot and is shown below each lane.

Genetic Tests Using Allele-Specific Oligonucleotides

Another method of genetic testing involves the use of synthetic DNA probes known as **allele-specific oligonucleotides (ASOs)**. Scientists use these short single-stranded fragments of DNA to identify alleles that differ by as little as a single nucleotide. In contrast to restriction enzyme analysis, which is limited to cases for which a mutation changes a restriction site, ASOs detect single-nucleotide changes of all types, including those that do not affect restriction enzyme cutting sites. As a result, this method offers increased resolution and wider application. Under proper conditions, an ASO will hybridize only with its complementary DNA sequence and not with other sequences, even those that vary by as little as a single nucleotide.

Genetic tests using ASOs and PCR analysis are now available to screen for many disorders, including sickle-cell anemia. In this procedure, DNA is extracted and used to amplify a region of the β-globin gene by PCR. A small amount of the amplified DNA is spotted onto filters, and each filter is hybridized to an ASO synthesized to resemble the relevant sequence from a normal or mutant β-globin gene (Figure 19–8). The ASO is tagged with a molecule that is either radioactive or fluorescent, in order to allow visualization of the hybridized ASO on the filter. After visualization, the genotype can be read directly from the filters. This rapid, inexpensive, and highly accurate technique is used to diagnose a wide range of genetic disorders caused by point mutations.

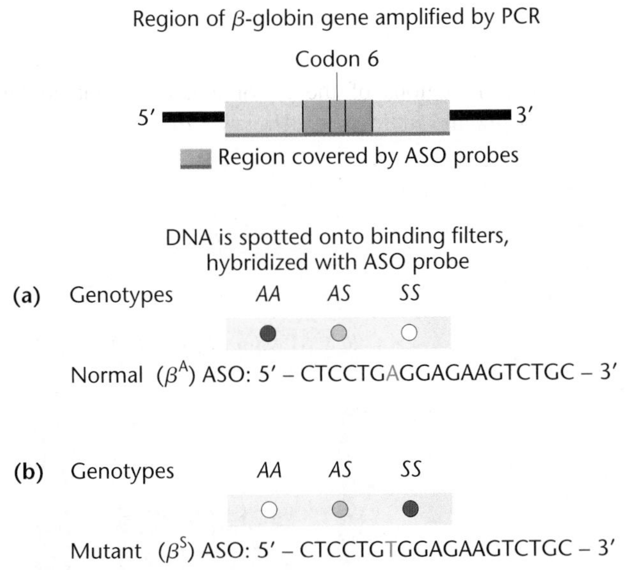

ASOs can also be used to screen for disorders that involve deletions instead of single-nucleotide mutations. An example is the use of ASOs to diagnose cystic fibrosis, which in many cases is caused by a small deletion called △508 in the cystic fibrosis transmembrane conductance regulator (*CFTR*) gene.

Now Solve This

Problem 19 on page 447 asks whether DNA sequences from normal and sickle-cell alleles of the β-globin gene will bind to a given ASO.

Hint: ASO analysis is done under conditions that allow only identical nucleotide sequences to hybridize to the ASO on the filter.

Genetic Tests Using DNA Microarrays and Genome Scans

Both restriction enzyme and ASO analyses are efficient methods to screen for gene mutations; however, they can only detect the presence of one or a few specific mutations whose identity and locations in the gene are known. There is also a need for genetic tests that detect complex mutation patterns or previously unknown mutations in genes that are associated with genetic diseases and cancers. For example, the gene that is responsible for cystic fibrosis (the *CFTR* gene) contains 27 exons and encompasses 250 kilobases of genomic DNA. There are more than 500 known mutations in the *CFTR* gene—point mutations, insertions, and deletions—that are associated with cystic fibrosis, and these are widely distributed through the gene. In addition, other *CFTR* mutations may be associated with cystic fibrosis but have not yet been characterized. Similarly, over 500 different mutations are known to occur within the tumor suppressor *p53* gene, and any of these mutations may be associated with, or predispose a patient to, a variety of cancers. In order to screen for mutations in these genes, comprehensive, **high-throughput methods** are required.

One emerging high-throughput screening technique is based on the use of DNA microarrays. **DNA microarrays** (also called DNA chips) are small, solid supports, such as glass or membrane, on which known fragments of DNA are deposited in a precise pattern. Each spot on a DNA microarray is called a **field**. The DNA fragments that are deposited in a DNA microarray field—called probes—are single-stranded and may be oligonucleotides synthesized *in vitro* or longer fragments of DNA created from cloning or PCR amplification. There are typically over a million identical molecules of DNA in each field (Figure 19–9). The numbers and types of DNA sequences on a microarray are dictated by the type of analysis that is required. For example, each field on a microarray might contain a DNA sequence derived from each member of a gene family, or sequence variants from one or several genes of interest, or a sequence derived from each gene in an organism's genome.

Figure

Region of β-globin gene amplified by PCR

Codon 6

5′ ————————— 3′

Region covered by ASO probes

DNA is spotted onto binding filters, hybridized with ASO probe

(a) Genotypes   AA   AS   SS

Normal (β^A) ASO: 5′ – CTCCTGAGGAGAAGTCTGC – 3′

(b) Genotypes   AA   AS   SS

Mutant (β^S) ASO: 5′ – CTCCTGTGGAGAAGTCTGC – 3′

FIGURE 19–8 To determine a genotype with allele-specific oligonucleotides (ASOs), the β-globin gene is amplified by PCR, using DNA extracted from white blood cells or cells obtained by amniocentesis. The amplified DNA is denatured and spotted onto strips of DNA-binding filters. Each strip is hybridized to a specific ASO and visualized on X-ray film after hybridization and exposure. If all three genotypes are hybridized to an ASO from the normal β-globin allele, the pattern in (a) will be observed: *AA*-homozygous individuals have normal hemoglobin that has two copies of the normal β-globin gene and will show heavy hybridization; *AS*-heterozygous individuals carry one normal β-globin allele and one mutant allele and will show weaker hybridization; *SS*-homozygous sickle-cell individuals carry no normal copy of the β-globin gene and will show no hybridization to the ASO probe for the normal β-globin allele. (b) The same genotypes hybridized to the probe for the sickle-cell β-globin allele will show the reverse pattern: no hybridization by the *AA* genotype, weak hybridization by the heterozygote (*AS*), and strong hybridization by the homozygous sickle-cell genotype (*SS*).

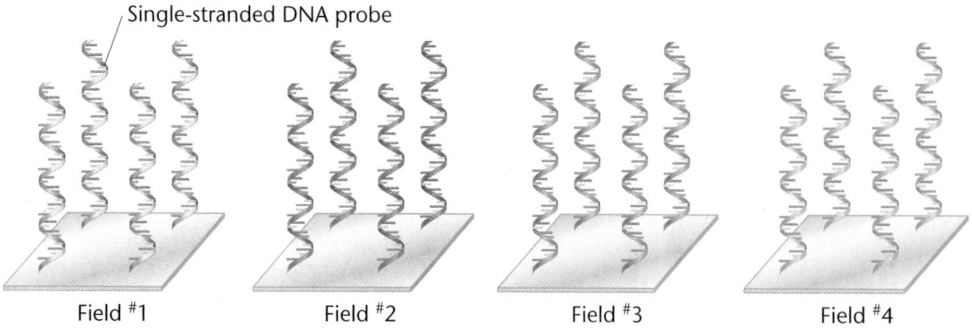

Single-stranded DNA probe

Field #1   Field #2   Field #3   Field #4

FIGURE 19–9 In making a DNA microarray, short, single-stranded DNA molecules of known sequence are attached to a glass substrate. Each cluster of identical molecules occupies an area known as a field on the microarray. Each field is about half the width of a human hair.

Scientists use DNA microarrays as platforms on which to hybridize a DNA (or RNA) sample to be analyzed.

What makes DNA microarrays so amazing is the immense amount of information that can be simultaneously generated from a single array. DNA microarrays the size of postage stamps (just over 1 cm square) can contain up to 500,000 different fields, each representing a different DNA sequence (Figure 19–10). Scientists are beginning to use DNA microarrays in a wide range of applications, including the detection of mutations in genomic DNA.

Geneticists often use a type of DNA microarray known as a **genotyping microarray** to detect mutations in specific genes. The probes on a genotyping microarray consist of short oligonucleotides about 20 nucleotides long, which are synthesized directly on the fields of the microarray. The probes on the microarray are designed to methodically scan through the gene of interest, one nucleotide at a time, in order to detect the presence of a mutation at each position in the gene. To do this, each position in the gene is tested by a set of five oligonucleotides (and hence five fields) that are identical in sequence except for one nucleotide. That position is occupied by either A, C, G, T, or a deletion (Figure 19–11).

To use a genotyping microarray, the DNA sample to be tested is cleaved into small fragments, and the DNA regions of interest are amplified by PCR, labeled with a fluorescent dye, and then hybridized to the microarray. DNA molecules with a sequence that exactly matches the sequence present in a microarray field will hybridize to that field. DNA molecules that differ by one or more nucleotides will hybridize less efficiently, or not at all, depending on the hybridization conditions. After washing off material that does not hybridize to the microarray, scientists analyze the microarray with a fluorescence scanner to determine which fields hybridized to the test DNA sample and which fields did not hybridize.

Under stringent hybridization conditions, the PCR-amplified DNA fragment will hybridize most efficiently to its exact complement, but less efficiently to the other possible sequence variants. Hence, the strongest hybridization signal to the five fields will be to the oligonucleotide probe that contains the correct nucleotide at the tested position. By reading the intensity of hybridization to each field, the sequence of the test DNA can be determined (Figure 19–12).

DNA microarrays have been designed to scan for mutations in many disease-related genes, including the *p53* gene, which is mutated in 60 percent of all cancers, and the *BRCA1* gene, which, when mutated, predisposes women to breast cancer.

In addition to testing for mutations in single genes, DNA microarrays can contain probes that detect markers known as **single-nucleotide polymorphisms (SNPs)**. SNPs occur randomly about every 100 to 300 nucleotides throughout the human genome, both inside and outside of genes. Scientists have discovered that certain SNP sequences at a specific locus are shared by certain segments of the population. In addition, certain SNPs cosegregate with genes associated with some disease conditions. By correlating the presence or absence of a particular SNP with a genetic disease, scientists are able to use the SNP as a genetic testing marker. The presence of SNPs on a DNA microarray allows scientists to simultaneously screen thousands of genes that might be involved in single-

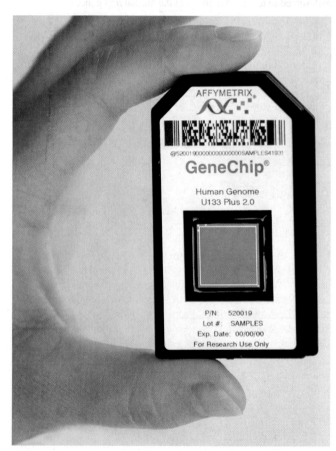

FIGURE 19–10 A commercially available DNA microarray, called a GeneChip, marketed by Affymetrix, Inc. This microarray contains over 50,000 different probes and allows scientists to simultaneously assess the expression levels of most of the genes in the human genome.

Region of wild-type
p53 gene to be tested: 5' – acttgtcatg gcgactgtcc acctttgtgc –

Set 1	Probe 1	3' – tgaacagtaa cgctgacagg – 5'
	Probe 2	3' – tgaacagtac cgctgacagg – 5'
	Probe 3	3' – tgaacagtat cgctgacagg – 5'
	Probe 4	3' – tgaacagtag cgctgacagg – 5'
	Probe 5	3' – tgaacagta– cgctgacagg – 5'
Set 2	Probe 1	3' – gaacagtac agctgacagg t – 5'
	Probe 2	3' – gaacagtac cgctgacagg t – 5'
	Probe 3	3' – gaacagtac tgctgacagg t – 5'
	Probe 4	3' – gaacagtac ggctgacagg t – 5'
	Probe 5	3' – gaacagtac –gctgacagg t – 5'
Set 3	Probe 1	3' – aacagtac cactgacagg tg – 5'
	Probe 2	3' – aacagtac ccctgacagg tg – 5'
	Probe 3	3' – aacagtac ctctgacagg tg – 5'
	Probe 4	3' – aacagtac cgctgacagg tg – 5'
	Probe 5	3' – aacagtac c–ctgacagg tg – 5'

FIGURE 19–11 Example of oligonucleotide probe design for use on a genotyping DNA microarray. Each probe would occupy a different field on the microarray. Each set of five probes is centered over the nucleotide position to be tested (highlighted in blue). Each of the probes in a set contains either an A, C, T, G or a deletion at the test position. One probe in each set is complementary to the wild-type DNA sequence; all other probes are complementary to potential mutations at the same position. Each nucleotide in the gene is tested with a set of five probes, and each probe set is offset from the previous set by one nucleotide. The pattern is repeated throughout the gene until every nucleotide has been tested. These particular probes are designed to test for mutations in the human *p53* gene.

gene diseases as well as those involved in disorders exhibiting multifactorial inheritance. This technique, known as **genome scanning**, makes it possible to analyze a person's DNA for dozens or hundreds of disease alleles, including those that might predispose the person to heart attacks, diabetes, Alzheimer disease, and other genetically defined disease subtypes. Although genome scanning is still being developed, it is likely to be available within a few years. Perhaps in decades to come, all newborns will be scanned to determine their lifetime risks for suffering from genetic disorders.

Genetic Analysis Using Gene Expression Microarrays

Another powerful application of DNA microarrays is to discover patterns of gene expression—for example, in normal cells or in cells from patients with cancer or genetic diseases. Microarrays used for this purpose are known as **gene expression microarrays.**

Gene expression microarrays differ somewhat from those used for genotyping. The probes on expression microarrays may be either cDNA fragments or synthetic oligonucleotides that represent the coding regions of genes to be profiled. An expression array may contain probes for only a few specific genes thought to be expressed differently in two cell types or may contain probes representing each gene in the genome.

In a typical expression microarray analysis, mRNA is isolated from two cell types—as an example, normal cells and cancer cells arising from the same cell type. The mRNA samples contain transcripts from each gene that is expressed in that cell type.

Some genes are expressed more efficiently than others; therefore, each type of mRNA is present at a different level. The level of each mRNA is characteristic of the cell type. In order to use the mRNA samples, scientists first convert the mRNA molecules into cDNA molecules, using reverse transcriptase. The cDNAs from the normal cells are tagged with a fluorescent dye (e.g., green), and the cDNAs from the cancer cells are tagged with a different fluorescent dye (e.g., red). The labeled cDNAs are mixed together and pumped over a DNA microarray bearing probes representing genes thought to be expressed in both normal and cancer cells. The cDNA molecules bind to complementary single-stranded probes on the microarray but not to other probes. After washing off the nonbinding cDNAs, scientists scan the microarray with a laser. The pattern of hybridization appears as a series of colored dots, with each dot corresponding to one field of the microarray.

The colors on the microarray fields provide a sensitive measure of the relative levels of each cDNA in the mixture. If an mRNA is present only in normal cells, the probe representing the gene encoding that mRNA will appear as a green dot,

(a)

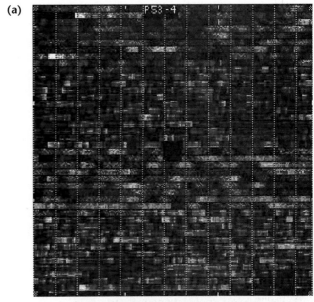

(b)

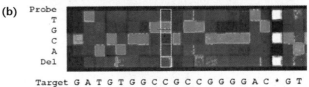

Target G A T G T G G C C G C C G G G G A C * G T

FIGURE 19–12 A *p53* GeneChip (Affymetrix, Inc.) after hybridization to *p53* PCR products. (a) This DNA microarray contains over 65,000 fields, each about 50 micrometers square. The microarray tests for mutations within the entire 1.2-kb coding sequence of the human *p53* gene. The dotted vertical lines are generated by control fields and allow for alignment of the DNA chip during scanner analysis. (b) An enlarged portion of the *p53* microarray. Each square is a field to which the test DNA has hybridized. Each row represents one of the four nucleotides or a deletion in the test position of the probe. Each column represents the test nucleotide. The column marked with an asterisk (*) is the alignment control. The sequence of the *p53* gene is shown under the photo.

because only "green" cDNAs have hybridized to it. Similarly, if an mRNA is present only in the cancer cells, the microarray probe for that gene will appear as a red dot. If both samples contain the same cDNA, in the same relative amounts, both cDNAs will hybridize to the same field, which will appear yellow (Figure 19–13). Intermediate colors indicate that the cDNAs are present at different levels in the two samples.

Expression microarray profiling has revealed that certain cancers have distinct patterns of gene expression and that these patterns correlate with factors such as the cancer's stage, clinical course, or response to treatment. In one such experiment, scientists examined gene expression in both normal white blood cells and in cells from a white blood cell cancer known as **diffuse large B-cell lymphoma (DLBCL)**. About 40 percent of patients with DLBCL respond well to chemotherapy and have long survival times. The other 60 percent respond poorly to therapy and have short survival. The investigators assayed the expression profiles of 18,000 genes and discovered that there were two types of DLBCL, with almost inverse patterns of gene expression (Figure 19–14). One type of DLBCL, called GC B-like, had an expression pattern dramatically different from that of a second type, called activated B-like. Patients with the activated B-like pattern of gene expression had much lower survival rates than patients with the GC B-like pattern. The researchers concluded that DLBCL is actually two different diseases with different outcomes. Once this new technology is translated into routine clinical use, it

may be possible to adjust therapies for each group of cancer patients and to identify new specific treatments based on gene expression profiles.

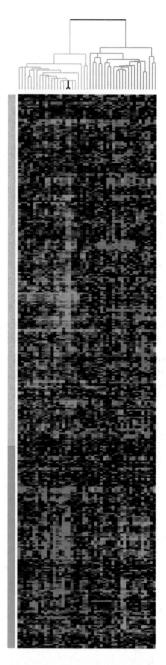

(a)

mRNA present only in normal cells

mRNA present only in cancer cells

mRNA present in equal amounts in normal and cancer cells

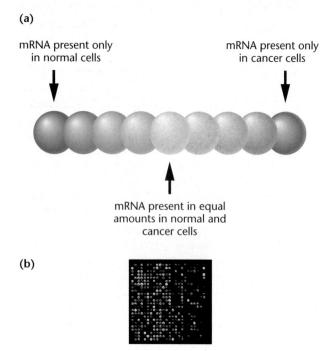

(b)

FIGURE 19–13 The color of dots on an expression DNA microarray represent levels of gene expression. (a) Green dots represent genes expressed only in one cell type (e.g., normal cells), and red dots represent genes expressed only in another cell type (e.g., cancer cells). Intermediate colors represent different levels of expression of the same gene in the two cell types. (b) A small portion of a DNA microarray, showing different levels of hybridization to each field.

FIGURE 19–14 Gene expression analysis generated from expression DNA microarrays that analyzed 18,000 genes expressed in normal and cancerous lymphocytes. Each row represents a summary of the gene expression from one particular gene; each column represents data from one cancer patients sample. The colors represent ratios of relative gene expression, compared to normal control cells. Red represents expression greater than the mean level in controls, green represents expression lower than the controls, and the intensity of the color represents the magnitude of difference from the mean. In this summary analysis, the cancer patients' samples are grouped by how closely their gene expression profiles resemble each other. The cluster of cancer patients' samples marked with orange at the top of the figure are GC B-like DLBCL cells. The blue cluster contains cancer patients' samples within the activated B-like DLBCL group.

Genetic Engineering and Genomics Promise New and Targeted Medical Therapies

In this chapter, we have described how recombinant DNA technologies are changing medical diagnosis and are allowing scientists to manufacture abundant and effective therapeutic proteins. In the near future, we will see even more transformative medical treatments, based on genomics and advanced DNA-based technologies. In this section, we will examine two exciting new methodologies that have the potential to cure genetic diseases and yield specific drug treatments.

Pharmacogenomics and Rational Drug Design

Every year, more than 2 million Americans experience serious side effects to medications, and more than 100,000 die from adverse drug reactions. In addition, most drugs are effective in only about 60 percent of the population. Until now, selection of effective medications for each individual has been a random, trial-and-error process. The new field of **pharmacogenomics** promises to lead to more specific, effective, and tailor-made drugs that are directed toward each person's individual genetic makeup.

Pharmacogenomics began in the 1950s, when scientists discovered that reactions to drugs had a hereditary component. We now know that many genes affect a person's reaction to drugs. These genes encode products such as cell-surface receptors that bind a drug and allow it to enter a cell, as well as enzymes that metabolize drugs. For example, liver enzymes encoded by the cytochrome P450 gene family affect the metabolism of many modern drugs, including those used to treat cardiovascular and neurological conditions. DNA sequence variations in these genes result in enzymes with different abilities to metabolize and utilize these drugs. For instance, gene variants that encode inactive forms of the cytochrome P450 enzymes are associated with a patient's inability to break down drugs in the body, leading to drug overdoses. A genetic test that recognizes some of these variants is currently being used to screen patients that are recruited into clinical trials for new drugs. Another example is the reaction of certain people to the thiopurine drugs used to treat childhood leukemias. Some individuals have sequence variations in the gene encoding the enzyme TPMT that breaks down thiopurines. In these individuals, thiopurine cancer drugs can build up to toxic levels. A genetic test is now available to detect these gene variants, allowing clinicians to tailor the drug dosage to the individual.

Several methods are being developed in order to expand the uses of pharmacogenomics. One promising method involves the detection of single-nucleotide polymorphisms (SNPs). Earlier in this chapter, we discussed SNPs in the context of detecting mutations using tests based on ASOs, as well as those based on genome scans. The hope is that researchers will be able to identify a shared SNP sequence in the DNA of people who also share a heritable reaction to

a drug. If the SNP segregates with a part of the genome containing the gene responsible for the drug reaction, it may be possible to devise gene tests based on the SNP, without even knowing the identity of the responsible gene. In the future, DNA microarrays may be used to screen a patient's genome for multiple drug reactions. Scientists predict that such gene tests—a type of genome scan—will be available by about 2010.

Knowledge from genetics and molecular biology is also contributing to the development of new drugs targeted to specific disease-associated molecules. Most drug development is currently based on trail-and-error testing of chemicals in lab animals, in the hope of finding one that affects the disease in question. In contrast, **rational drug design** involves the synthesis of specific chemical substances that affect specific gene products. An example of a rational drug design product is the new drug imatinib (**Gleevec**), used to treat chronic myelogenous leukemia (CML). Geneticists had discovered that CML cells contain the Philadelphia chromosome, which results from a reciprocal translocation between chromosomes 9 and 22. Gene cloning revealed that the t(9;22) translocation creates a fusion of the *C-ABL* proto-oncogene with the *BCR* gene. This *BCR-ABL* fusion gene encodes a powerful fusion protein that causes cells to escape cell-cycle control. This fusion protein, which acts as a tyrosine kinase, is not present in noncancer cells from CML patients.

To develop Gleevec, chemists used high-throughput screens of chemical libraries to find a molecule that bound to the BCR-ABL enzyme. After chemical modifications to make it bind more tightly, tests showed that Gleevec specifically inhibited BCR-ABL activity. Clinical trials revealed that Gleevec was effective against CML, with minimal side effects and a higher remission rate than that seen with conventional therapies. Gleevec is now used to treat CML and several other cancers. As scientists discover more genes and gene products that are associated with diseases, rational drug design promises to be a powerful technology within the next decade.

Gene Therapy

For more than two decades, gene products such as insulin have been used to treat genetic disorders. Although drug treatments are often effective in controlling symptoms, the ideal goal of medical treatment is to cure these diseases. In an effort to cure genetic diseases, scientists are actively investigating **gene therapy**—a therapeutic technique that aims to transfer normal genes into a patient's cells. In theory, the normal genes will be transcribed and translated into functional gene products which, in turn, will bring about a normal phenotype. Delivery of these normal structural genes and their regulatory sequences is accomplished by using a vector or gene transfer system.

In the first generation of gene therapy trials, scientists used genetically modified retroviruses as vectors. These vectors are based on a mouse virus called **Moloney murine leukemia virus (MLV)** (Figure 19–15). The vectors are created by removing a cluster of three genes from the virus, and inserting a cloned human gene. After packaging in a viral protein coat,

(a)

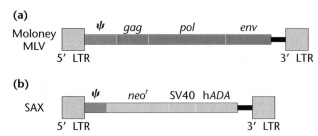

(b)

FIGURE 19–15 (a) The native Moloney MLV genome contains a sequence required for encapsulation (ψ), as well as genes that encode viral coat proteins (*gag*), an RNA-dependent DNA polymerase (*pol*), and surface glycoproteins (*env*). At each end, the genome is flanked by long terminal repeat (LTR) sequences that control transcription and integration into the host genome. (b) The SAX vector retains the LTR and ψ sequences and includes a bacterial neomycin resistance (*neo*ʳ) gene that can be used as a selection marker. As shown, the vector carries a cloned human adenosine deaminase (h*ADA*) gene, which is fused to an SV40 early region promoter–enhancer. The SAX construct is typical of retroviral vectors that are used in human gene therapy.

the recombinant vector is used to infect cells. Once inside a cell, the virus cannot replicate itself because of the missing viral genes. In the cell, the recombinant virus with the inserted human gene moves to the nucleus and integrates into a site on a chromosome, and becomes part of the genome. If the gene is expressed, it produces a normal gene product that may be able to correct the mutation carried by the affected individual.

Human gene therapy began in 1990 with the treatment of a young girl named Ashanti De Silva (Figure 19–16), who has a heritable disorder called **severe combined immunodeficiency (SCID)**. Individuals with SCID have no functional immune system and usually die from what would normally be minor infections. Ashanti has an autosomal form of SCID

FIGURE 19–16 Ashanti De Silva, the first person to be treated by gene therapy.

caused by a mutation in the gene encoding the enzyme **adenosine deaminase (ADA)**. Her gene therapy began when clinicians isolated some of her white blood cells, called T cells (Figure 19–17). These cells, which are part of the immune system, were mixed with a retroviral vector carrying an inserted copy of the normal *ADA* gene. The virus infected many of the T cells, and a normal copy of the *ADA* gene was inserted into the genome of some T cells. After mixing with the vector, the T cells were grown in the laboratory and analyzed to make sure that the transferred *ADA* gene was expressed. A billion or so genetically altered T cells were injected into her bloodstream. Some of these T cells migrated to Ashanti's bone marrow and began dividing, and producing ADA. She has ADA protein expression in 25 to 30 percent of her T cells, which is enough to allow her to lead a normal life.

In later trials, attempts were made to transfer the *ADA* gene into the bone marrow cells that form T cells, but these attempts were mostly unsuccessful. Although gene therapy was originally developed as a treatment for single-gene (monogenic) inherited diseases, this technique was rapidly adapted for the treatment of acquired diseases such as cancer, neurodegenerative diseases, cardiovascular disease, and infectious diseases, such as HIV. As a result, most gene therapy treatments and trials involve these disorders [Figure 19–18(a)].

In addition to retroviral vectors, other viruses and methods are being used to transfer genes into human cells [Figure 19–18(b)]. These methods include the use of viral vectors, chemically assisted transfer of genes across cell membranes, and fusion of cells with artificial vesicles containing cloned DNA sequences.

Over a 10-year period, from 1990 to 1999, more than 4000 people underwent gene transfer for a variety of genetic disorders. These trials often failed and thus led to a loss of confidence in gene therapy. Hopes for gene therapy plummeted even further in September 1999 when a teenager, Jesse Gelsinger, died during gene therapy. His death was triggered by a massive inflammatory response to the vector, a modified **adenovirus**. (Adenoviruses cause colds and respiratory infections.) The outlook for gene therapy brightened in 2000, when a group of French researchers reported the first large-scale success in gene therapy. Ten patients with a fatal X-linked form of SCID developed functional immune systems after being treated with a retroviral vector carrying a normal gene. However, three patients later developed leukemia-like disorders. Their cancer cells contained the retroviral vector, inserted near or into a gene called *LMO2*. This insertion activated the *LMO2* gene, causing uncontrolled white blood cell proliferation and development of leukemia. The Gelsinger and French X-SCID stories, as well as the future of gene therapy, are also discussed in the essay at the end of this chapter.

Most problems associated with gene therapy have been traced to the vectors used to transfer therapeutic genes into cells. These vectors, including MLV and adenovirus, have several serious drawbacks. First, integration of retroviral genomes (including the human therapeutic gene) into the host cell's genome occurs only if the host cells are replicating their DNA. In the body, only a small number of cells in any tissue

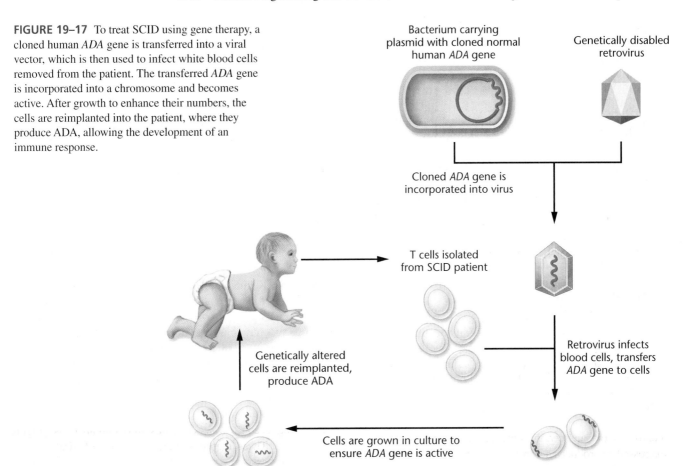

FIGURE 19–17 To treat SCID using gene therapy, a cloned human *ADA* gene is transferred into a viral vector, which is then used to infect white blood cells removed from the patient. The transferred *ADA* gene is incorporated into a chromosome and becomes active. After growth to enhance their numbers, the cells are reimplanted into the patient, where they produce ADA, allowing the development of an immune response.

are dividing and replicating their DNA. Second, most viral vectors are capable of causing an immune response in the patient, as happened in Jesse Gelsinger's case. Third, insertion of viral genomes into host chromosomes can activate or mutate an essential gene, as in the case of the three French patients. Fourth, retroviruses cannot carry DNA sequences much larger than 8 kb. Many human genes exceed this size. Finally, there is a possibility that a fully infectious virus could be created if the vector were to recombine with another viral genome already present in the host cell.

To overcome these problems, new viral vectors and strategies for transferring genes into cells are being developed. Researchers hope that the use of new gene delivery systems will circumvent the problems inherent

in earlier vectors, as well as allow regulation of both insertion sites and the levels of gene product produced from the therapeutic genes.

Now Solve This

Problem 11 on page 447 asks you to explain why diseases such as muscular dystrophy would be difficult to cure with gene therapy.

Hint: Consider the types of tissues that are affected in muscular dystrophy patients.

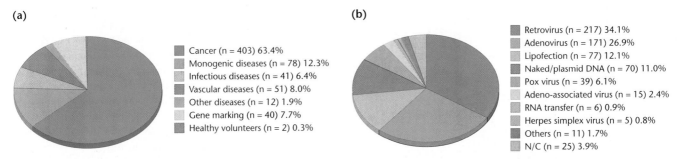

FIGURE 19–18 (a) Pie chart summarizing, by disease, more than 630 gene therapy trials under way worldwide. Most trials involve cancer treatment. (b) Gene therapy trials ranked by the vectors used. Retroviruses are the most widely used vectors, accounting for 34 percent of the total. Newer vectors, such as adeno-associated virus, account for only a small percentage of vectors. N/C = not classified

DNA Profiles Help Identify Individuals

As discussed previously (Figure 19–7), the presence or absence of restriction sites at specific locations within the human genome can be used as genetic markers. Another type of genetic marker, discovered in the mid-1980s, is based on variations in the length of repetitive DNA sequence clusters. The number of repeats within these clusters varies between individuals and between population groups. These polymorphisms in DNA serve as the basis for **DNA profiling**—also called **DNA fingerprinting** or DNA typing. DNA profiling is used in a wide variety of applications, such as paternity testing, forensics, identification of human or animal remains, archaeology, and conservation biology.

DNA Profiling Based on DNA Minisatellites (VNTRs)

The first DNA profiling method was developed in the United Kingdom by Sir Alec Jeffreys and was used in 1986 to convict the murderer of two English schoolgirls. This method involves the analysis of large repeated regions of DNA, called **minisatellites**, or **variable numbers of tandem repeats (VNTRs)**.

VNTRs are repeating clusters of about 10 to 100 nucleotides. For example, the VNTR

5'-GACTGCCTGCTAAGAT**GACTGCCTGCTAAGAT**-
GACTGCCTGCTAAGAT-3'

is composed of three tandem repeats of the 16 nucleotide sequence GACTGCCTGCTAAGAT. Clusters of such sequences are widely dispersed in the human genome. The number of repeats at each locus ranges from 2 to more than

100—each repeat length representing one allele of that locus. VNTRs are particularly useful for DNA profiling because there are about 30 different possible alleles (repeat lengths) at any VNTR within a population. This creates a large number of possible genotypes. For example, if one examined four different VNTRs within a population, and each VNTR had 20 possible alleles, there would be about 2 billion possible genotypes in this four-locus profile.

To create a VNTR profile, DNA is extracted from the tissue sample and is digested with a restriction enzyme that cleaves on either side of the VNTR repeat region. The digested DNA is separated by gel electrophoresis, transferred to a membrane, and hybridized with a radioactive probe that recognizes DNA sequences within the VNTR region. After exposing the membrane to X-ray film, the pattern of bands is measured, with larger VNTR repeat alleles migrating near the top of the gel and smaller VNTRs migrating closer to the bottom (Figure 19–19). The pattern of bands is always the same for a given individual, no matter what tissue is used as the source of the DNA. Because the pattern varies from individual to individual and there are a huge number of possible VNTRs and alleles, if enough VNTRs were analyzed, each person's DNA profile would be unique (except, of course, for identical twins). In practice, about five or six VNTR loci are analyzed to create a DNA profile.

A significant limitation of VNTR profiling is that it requires a relatively large sample of DNA (10,000 cells or about 50 μg of DNA)—more than is usually found at a typical crime scene—and the DNA must be relatively intact (nondegraded). As a result, VNTR profiling is used most frequently when large tissue samples are available—such as in paternity testing. Blood drawn from the child, mother, and alleged father provide an abundant source of fresh, intact cells for DNA extraction and analysis.

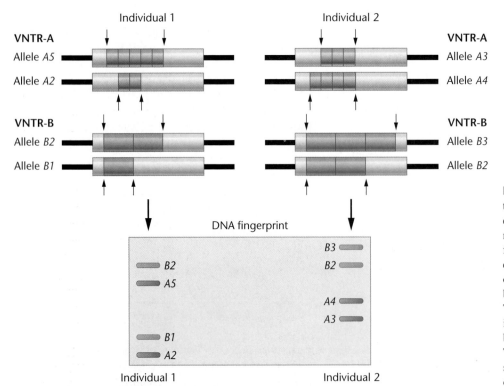

FIGURE 19–19 VNTR alleles at two loci (*A* and *B*) are shown for each individual. (Arrows mark restriction enzyme cutting sites flanking the VNTRs.) Restriction enzyme digestion produces a series of fragments that can be detected as bands on a Southern blot (bottom). The number of repeats at each locus is variable, so the overall pattern of bands is distinct for each individual. The DNA fingerprint (profile) shows that both individuals share one allele (*B2*).

DNA Profiling Based on DNA Microsatellites (STRs)

To overcome the problems of DNA sample size and conditions necessary for VNTR profiling, scientists have developed alternate techniques based on the use of PCR to amplify shorter polymorphic DNA markers. One of these methods examines sequences called **microsatellites** or **short tandem repeats (STRs)**. STRs are similar to VNTRs, but the repeated motif is shorter—between two and nine base pairs. In addition, there are only about 7 to 40 repeats in each STR locus. Although hundreds of STR loci are found in the human genome, only a subset is used for DNA profiling. The FBI and other law enforcement agencies have selected 13 STR loci to be used as a core set for forensic analysis. DNA profiles based on these loci are stored in a national DNA database, called the **Combined DNA Index System (CODIS)**. Profiles generated in criminal cases, as well as from missing persons and from samples collected from crime scenes are stored in the CODIS.

To create an STR profile, DNA is extracted from the sample, and the DNA is amplified using sets of primers that specifically hybridize to regions of DNA flanking each of the STR loci to be analyzed. Each primer set within the mix is labeled with a different fluorescent dye, making the PCR products amplified from each STR locus a different color. The sizes of the amplified fragments may be measured by standard gel electrophoresis (Figure 19–20); however, most STR profiles use faster, more sensitive measures of allele size. The most common method is to apply the amplified mixture to the top of a thin capillary tube filled with a gel-like substance and then pass an electric current through the tube. Larger DNA fragments will be retarded in the capillary, and smaller ones will pass through more quickly. This is known as **capillary gel electrophoresis**. At the bottom of the tube, a laser detects when each fluorescent DNA fragment exits the capillary tube and assigns a size to the fragment. All fragments are represented as peaks on a graph. The entire process of creating an STR profile is automated—often within self-contained kits that are used in research labs or in mobile crime units.

A major advantage of STR profiles is that they can be generated from trace samples (e.g., single hairs or saliva left on a cigarette butt) and even from samples that are old or degraded (e.g., a skull found in a field or an ancient Egyptian mummy). As a result, STRs have replaced VNTRs in most forensic laboratories. In addition, STR typing is less expensive, less labor-intensive, and much faster than VNTR analysis, so it is rapidly replacing VNTR typing in most paternity laboratories as well.

The results of STR analysis are interpreted using statistics, probability, and population genetics. The population frequency of each STR allele in the standard set has been measured in many groups of people across the United States and the world. Using this information, scientists can calculate the probability of having any combination of these 13 alleles. For example, if an allele of locus 1 is carried by 1 in 333 individuals, and an allele of locus 2 is carried by 1 in 83 individuals, the probability that someone will carry both alleles is equal to the product of their individual frequencies, or nearly 1 in 28,000 ($1/333 \times 1/83$). This overall probability may not be especially convincing, but if an allele of a third locus—carried by 1 in 100 individuals—and an allele of a fourth locus—carried by 1 in 25 individuals—are included in the calculations, the combined frequency becomes almost 1 in 70 million. That is, only about four individuals in the U.S. population will carry this combination of alleles by chance. When a profile containing all 13 CODIS STR loci is generated, the chance of anyone having the same profile is 1 in 10 billion. Since the planet's population is only about 6 billion, it is easy to see why STR analysis is so powerful in its ability to discriminate between samples from different individuals.

Forensic Applications of DNA Profiling

Over the last 20 years, DNA profiling has become a powerful tool to help convict the guilty, identify the remains of victims, and exonerate the innocent. Its power emerges from the extreme sensitivity of the technology and its ability to discriminate accurately between crime scene samples. It has rapidly replaced older, flawed techniques and is helping to transform the police and justice systems. Since 1992, over 170 convicted persons in the United States have been exonerated of their alleged crimes, based on DNA profile analysis of old crime samples. Many of these convicted individuals had served decades in prison for crimes they did not commit—some incarcerated on death row and at least one wrongly executed prior to having access to DNA profiling evidence. As national and state DNA databases grow, they are becoming important tools in solving criminal cases where no arrest has been made and no suspects have been identified—called **cold cases**. By matching DNA from crime scenes with DNA profiles in the database, more than 1600 cold cases were solved between 2000 and 2003.

Although forensic DNA profiling is a powerful technology, it is important to understand what it can, and cannot, tell us about guilt and innocence. For example, in a criminal case, if a suspect's DNA profile does not match that of the evidence sample, the suspect can be excluded *as the source of that sample*. Whether or not this finding indicates innocence would depend on all other evidence in the case. For instance, a suspect in a rape case may not have contributed the semen sample evidence but may have been involved in the crime by restraining the victim.

Interpretation may also be unclear when there is a match between an evidence sample and a suspect's DNA profile. When all 13 core STR loci are used to create DNA profiles,

Web Tutorial 19.2
DNA Fingerprinting

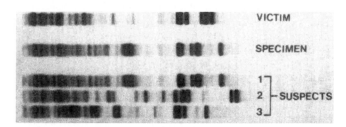

FIGURE 19–20 DNA profiles prepared using STRs reveal patterns from the victim, the evidence, and three suspects in a criminal case. In this profile, PCR-amplified DNA is visualized by gel electrophoresis.

there is less than one chance in 10 billion that two matching DNA profiles could have come from two different individuals. Although this sounds convincing, in some circumstances such matches might occur. For example, the two profiles could match, by chance. Although this is unlikely, it cannot be ruled out. There is a greater chance of a match between related individuals—siblings, parents, or other relatives. It is even possible that the two profiles may have arisen from identical twins who have identical genomes. It is possible to find a match between a crime scene sample and an innocent person's profile as a result of sample mixup, contamination, or even unintended or deliberate tampering. The very sensitivity of DNA profiling—sufficient to generate a profile from a single hair or saliva on the back of a postage stamp—makes it highly prone to contamination if sampling and analysis conditions are not stringently controlled.

19.6 Genetic Engineering and Genomics Create Ethical, Social and Legal Questions

Geneticists now use recombinant DNA technology to identify genes, diagnose and treat genetic disorders, produce commercial and pharmaceutical products, and solve crimes. Knowledge gained by sequencing the human genome will greatly advance our understanding of human genetics and will have an ever-increasing impact on biomedical research and health care. However, the applications that arise from our knowledge of genetic engineering raise ethical, social, and legal issues that must be identified, debated, and resolved. Resolutions often take the form of laws or public policy. We will present a brief overview of some current ethical debates surrounding the uses of gene technologies.

Concerns about Genetically Modified Organisms

Most genetically modified food products contain an introduced gene encoding a protein that confers a desired trait (for example, herbicide resistance or insect resistance). Much of the concern over genetically modified plants centers on issues of safety and environmental consequences. Are genetically modified plants containing the new protein safe to eat? In general, if the proteins are not found to be toxic or allergenic and do not have other negative physiological effects, they are not considered to be a significant hazard to health. One such case is that of plants containing EPSP synthase. EPSP synthase is quickly degraded by digestive fluids, is nontoxic to mice at doses thousands of times higher than any potential human exposure, and has no amino acid sequence similarity to known protein toxins or allergens. Standardized methods for evaluating proteins in genetically modified foods are being developed in the United States and Europe. In Europe, labeling of food containing genetically modified ingredients is mandatory, but in the United States such labeling is not required at the present time.

Another concern involves the environmental risks posed by releasing genetically modified organisms into the environment. Environmental risks include possible gene transfer by cross breeding with wild plants, toxicity, and invasiveness of the modified plant, resulting in loss of natural species (loss of biodiversity). Although laboratory and field studies suggest that cross-pollination and gene transfer can occur between some genetically engineered plants and wild relatives, there is little evidence that this has occurred in nature. If, for example, glyphosate resistance was transferred from cultivated plants such as canola into wild relatives, the herbicide-resistant weeds could make herbicide treatment ineffective. Biotechnology companies are now attempting to engineer transgenic plants into sterile forms that will be unable to transfer their genes into other plants. Built-in sterility would also ensure that farmers could not produce their own seed from genetically modified crops, guaranteeing that biotechnology companies would have exclusive distribution over the seed. This, in itself, is an ethical issue, particularly in underdeveloped countries with limited resources to purchase genetically modified seeds.

Genetic Testing and Ethical Dilemmas

We have considered examples of prenatal diagnosis, heterozygote screening of adults for single recessive disorders, and genome scanning. Although these technologies are valuable additions to medical diagnosis, they also present ethical problems that will not be easy to resolve. For example, what should people know before deciding to have a genome scan or a genetic test for a single disorder? How can we protect the information revealed by such tests? How can we define and prevent genetic discrimination? We know that heterozygotes for sickle-cell anemia are more resistant to malaria than people who are homozygous for the normal allele; this type of protection may be true of other genetic disorders as well. In such a case, how do we keep the beneficial aspects of some mutations as we strive to eliminate their destructive aspects?

Many of the potential risks and benefits of genetic testing are still unknown. We can test for many genetic diseases for which there are no effective treatments to cure or mitigate the clinical consequences. Should we test people for these disorders? With present technology, a negative result does not necessarily rule out future development of a disease; nor does a positive result always mean that an individual will get the disease. How can we effectively communicate the results of testing and the risks to those being tested? Lawmakers and other groups made up of scientists, health-care professionals, ethicists, and consumers are debating these issues and formulating policy options.

The Ethical Concerns Surrounding Gene Therapy

Gene therapy raises several ethical concerns, and many forms of therapy are still sources of intense debate. At present, all gene therapy trials are restricted to using somatic cells as targets for gene transfer. This form of gene therapy is called **somatic gene therapy**; only one individual is affected, and the therapy is done with the permission and informed consent of the patient or family.

Two other forms of gene therapy have not been approved, primarily because of the unresolved ethical issues surrounding them. The first is called **germ-line therapy**, whereby germ cells (the cells that give rise to the gametes—i.e., sperms and eggs) or mature gametes are used as targets for gene transfer. In this approach, the transferred gene is incorporated into all the cells of the body, including the germ cells. This means that individuals in future generations will also be affected, without their consent. Is this procedure ethical? Do we have the right to make this decision for future generations? Thus far, the concerns have outweighed the potential benefits, and such research is prohibited.

The second unapproved form of gene therapy—which raises an even greater ethical dilemma—is termed **enhancement gene therapy**, whereby people may be "enhanced" for some desired trait. This use of gene therapy is extremely controversial and is strongly opposed by many people. Should genetic technology be used to enhance human potential? For example, should it be permissible to use gene therapy to increase height, enhance athletic ability, or extend intellectual potential? Presently, the consensus is that enhancement therapy, like germ-line therapy, is an unacceptable use of gene therapy. However, there is an ongoing debate, and many issues are still unresolved. For example, the U.S. Food and Drug Administration now permits growth hormone produced by recombinant DNA technology to be used as a growth enhancer, in addition to its current medical use for the treatment of growth-associated genetic disorders. Critics charge that the use of a gene product for enhancement will lead to the use of transferred genes for the same purpose. The outcome of these debates may affect not only the fate of individuals but the direction of our society as well.

The Ethical, Legal, and Social Implications (ELSI) Program

When the Human Genome Project was first discussed, scientists and the general public raised concerns about how genome information would be used and how the interests of both individuals and society can be protected. To address these concerns, the **Ethical, Legal, and Social Implications (ELSI) Program** was established as an adjunct to the Human Genome Project. The ELSI Program considers a number of issues, including the impact of genetic information on individuals, the privacy and confidentiality of genetic information, implications for medical practice, genetic counseling, and reproductive decision making. Through research grants, workshops, and public forums, ELSI is formulating policy options to address these issues.

ELSI focuses on four areas: (1) privacy and fairness in the use and interpretation of genetic information, (2) ways of transferring genetic knowledge from the research laboratory to clinical practice, (3) ways to ensure that participants in genetic research know and understand the potential risks and benefits of their participation and give informed consent, and (4) public and professional education. It is hoped that, as the Human Genome Project moves from generating information about the genetic basis of disease to improving treatments, promoting prevention, and developing cures, these and other ethical issues will be identified and an international consensus will be developed on appropriate policies and laws.

GENETICS, TECHNOLOGY, AND SOCIETY

Gene Therapy—Two Steps Forward or Two Steps Back?

In September 1999, 18-year-old Jesse Gelsinger received his first dose of gene therapy. Large numbers of adenovirus vectors bearing the ornithine transcarbamylase (*OTC*) gene were injected into his hepatic artery. The virus vectors were expected to lodge in his liver, enter liver cells, and trigger the production of OTC protein. In turn, the OTC protein might correct his genetic defect and perhaps cure him of his liver disease. However, within hours, a massive immune reaction surged through Jesse's body. He developed a high fever, his lungs filled with fluid, multiple organs shut down, and he died four days later of acute respiratory failure. In the aftermath of the tragedy, several government and scientific inquiries were initiated. Investigators learned that clinical trial scientists had not reported other adverse reactions to gene therapy and that some of the scientists were affiliated with private companies that could benefit financially from the trials. They found that serious side effects in animal studies were not explained to patients during informed-consent discussions and some clinical trials were proceeding too quickly in the face of data that suggested caution. The U.S. Food and Drug Administration (FDA) scrutinized gene therapy trials across the country, halted a number of trials, and shut down several gene therapy programs. Other research groups voluntarily suspended their gene therapy studies.

Jesse's death dealt a severe blow to the struggling field of gene therapy—a blow from which it was still reeling when a second tragedy hit.

In April 2000, a French group announced that gene therapy for X-linked severe combined immunodeficiency (X-SCID) had succeeded. It was the first unequivocal success in the gene therapy field. In this study, the young patients' bone marrow cells were removed, treated with a retrovirus bearing the γc transmembrane protein gene, and transplanted back to the patients. Ten of 11 patients were cured of their immune deficiency and were able to lead normal lives. Published reports of the study were greeted with enthusiasm by the gene therapy community. But elation turned to despair in 2003, when it became clear that 2 of the 10 children who had been cured of X-SCID had developed leukemia as a direct result of

their therapy. The FDA immediately halted 27 similar gene therapy clinical trials and, once again, gene therapy underwent a profound reassessment. In 2005, a third child in the French X-SCID study developed leukemia, likely as a result of gene therapy.

Up until the apparent success of the French X-SCID clinical trials, gene therapy had suffered not only from the scandals and scrutiny that emerged from Jesse Gelsinger's death but also from the skepticism of scientists and the general public about the feasibility of this much-promoted therapeutic technique.

Since the first clinical trial for gene therapy in 1990, over 600 gene therapy clinical trials involving over 4000 patients have been initiated. These trials aim to cure cancers, inherited diseases such as hemophilia and cystic fibrosis, and infectious diseases such as AIDS. Despite high expectations and intense publicity, therapeutic benefits have been unclear at best, and more frequently, absent.

The most significant positive outcome of gene therapy came from the first clinical trial in 1990. Ashanti De Silva, who had received retroviral-transduced T cells for severe combined immunodeficiency (SCID), now leads a normal life, but the reasons for her success are not completely resolved. She had been given a new drug treatment that replaced her missing ADA enzyme prior to and after gene therapy. Hence, it is still not known how much of her cure is due to gene therapy and how much is due to drug treatment.

To date, no human gene therapy product has been approved for sale. Critics of gene therapy continue to criticize research groups for undue haste, conflicts of interest, and sloppy clinical trial management, and for promising much but delivering little. In the mid-1990s, a National Institutes of Health review committee concluded that significant problems remain in all basic aspects of gene therapy, including those aspects that led to Jesse Gelsinger's death and the X-SCID leukemias—adverse immune reactions to viral vectors and the side effects of retroviral vector integration into the host genome.

The question remains whether gene therapy can ever recover from these setbacks and fulfill its promise as a cure for genetic diseases. At present, hundreds of gene therapy clinical trials are under way in the United

States—most are for cancer and are in phase I trials (which examine safety and dosage, but not efficacy). Tighter restrictions on clinical trial protocols are designed to correct some of the procedural problems that emerged from the Gelsinger case. In addition, basic science is proceeding with development of safe, effective vectors, such as those that insert vector sequences into specific regions of the genome (reducing the possibility of cancer or gene mutation due to vector insertion) and those bearing receptors on their surfaces that allow them to infect specific cell types. However, the steps leading to successful gene therapy are many. There is a need to optimize tissue-specific expression of therapeutic genes, to efficiently transduce cells in culture, to predict and control immune reactions to vectors, and to develop better animal models in which to test gene therapies prior to clinical trials.

Many scientists feel that we should continue gene therapy research and clinical trials despite the setbacks. However, they have a more sober view of its progress. Clinical trials for any new therapy are potentially dangerous, and often animal studies will not accurately reflect the reaction of individual humans to a new drug or procedure. Inevitably, more adverse reactions to gene therapy will emerge in the clinical trials as methods become more effective. Those in the field now believe that the road ahead will be longer and more difficult than first imagined, but not impossible. Perhaps we should view gene therapy as we have antibiotics, organ transplants, and manned space travel. There will be setbacks and even tragedies, but step by small step, we will move toward a technology that could—someday—provide cures for many severe genetic diseases.

Reference

Thomas, C. E., Ehrhardt, A., and Kay, M. A. 2003. Progress and problems with the use of viral vectors for gene therapy. *Nat. Rev. Gen.* 4: 346–358.

Web Sites

Thompson, L. September/October 2000. *Human gene therapy: Harsh lessons, high hopes [on-line].* FDA Consumer Magazine.
http://www.fda.gov/fdac/features/2000/500_gene.html

CHAPTER SUMMARY

1. Scientists have applied genetic engineering technologies to improve crop plants by introducing genes that confer herbicide and insect resistance and enhance nutritional value. Gene transfer is also used to enhance the qualities of farm animals.

2. Genetically engineered organisms synthesize abundant quantities of useful biological and pharmaceutical products. Genes that encode these substances have been introduced into bacteria, plants, and animals.

3. Genetic tests directly examine a patient's DNA in order to detect mutations that are associated with genetic diseases. These genetic tests are used for prenatal diagnosis, to identify disease carriers, to confirm medical diagnoses, and to predict the course of genetic diseases.

4. Pharmacogenomics and rational drug design are emerging from our understanding of how genes affect reactions to drugs, and the genetic basis of diseases such as cancers. It will soon be possible to predict a patient's reaction to drugs prior to taking the medication and to tailor-make drugs to specifically target a patient's condition.

5. We may soon be able to directly correct gene defects by using gene therapies. Despite early setbacks, gene therapy clinical trials are underway to treat cancers and single-gene genetic diseases.

6. DNA fingerprinting is used for forensic applications as well as for paternity testing, identification of human and animal remains, and in a wide range of other fields, including archaeology, conservation biology, and public health.

7. Genetic engineering and genomics raise ethical, social, and legal concerns that must be addressed prior to the adoption of new technologies.

KEY TERMS

adenosine deaminase (ADA), 438
adenovirus, 438
allele-specific oligonucleotides (ASOs), 433
amniocentesis, 432
biopharmaceuticals, 428
biotechnology, 426
capillary gel electrophoresis, 441
cold cases, 441
Combined DNA Index System (CODIS), 441
diffuse large B-cell lymphoma (DLBCL), 436
DNA fingerprinting, 440
DNA microarrays, 433
DNA profiling, 440
enhancement gene therapy, 443
EPSP synthase, 426
Ethical, Legal, and Social
  Implications (ELSI) Program, 443

field, 433
fusion polypeptide, 428
gene expression microarrays, 435
gene therapy, 437
genetic engineering, 426
genome scanning, 435
genotyping microarray, 434
germ-line therapy, 443
Gleevec, 437
glyphosate, 426
golden rice, 427
growth hormone gene, 428
hepatitis B virus, 430
high-throughput methods, 433
insulin, 428
microsatellites, 441
minisatellites, 440

Moloney murine leukemia virus (MLV), 437
pharmacogenomics, 437
Pompe disease, 430
rational drug design, 437
restriction enzyme analysis, 431
restriction fragment length polymorphism
  (RFLP), 431
severe combined immunodeficiency
  (SCID), 438
short tandem repeats (STRs), 441
sickle-cell anemia, 431
single-nucleotide polymorphisms,
  (SNPs), 434
somatic gene therapy, 442
subunit vaccine, 430
variable numbers of tandem repeats
  (VNTRs), 440

INSIGHTS AND SOLUTIONS

1. Probes for DNA fingerprinting can be derived from a single locus or multiple loci. Two multiple-loci probes have been widely employed in both criminal and civil cases and derive from minisatellite loci on chromosome 1 (1cen-q24) and chromosome 7 (7q31.3). These probes, which are used because they produce a highly individual fingerprint, have determined paternity in thousands of cases over the last few years. The results of DNA fingerprinting of a mother (M), putative father (F), and child (C), using the aforementioned probes, are shown in the accompanying figure. The child has 6 maternal bands, 10 paternal bands, and 6 bands shared between the mother and the alleged father. Based on this fingerprint, can you conclude that this man is the father of the child?

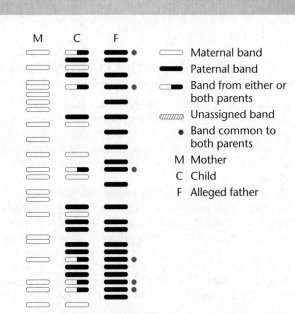

(Cont. on the next page)

Solution: All bands present in the child can be assigned as coming from either the mother or the father. In other words, all the bands in the child's DNA fingerprint that are not maternal are present in the father. Since the father and the child share 10 bands and the child has no unassigned bands, paternity can be assigned with confidence. In fact, the chance that this man is not the father is on the order of 10^{-13}.

2. The DNA fingerprints of a mother, a child, and the alleged father in a second case are shown in the figure below. The child has 8 maternal bands, 14 paternal bands, 6 bands that are common to both the mother and the alleged father, and 1 band that is not present in either the mother or the alleged father. What are the possible explanations for the presence of the last band? Based on your analysis of the band pattern, which explanation is most likely?

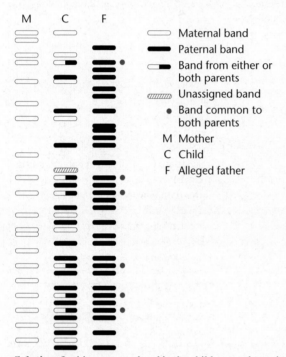

Solution: In this case, one band in the child cannot be assigned to either parent. Two possible explanations are that the child is mutant for one band or that the man tested is not the father. To estimate the probability of paternity, the mean number of resolved bands (*n*) is determined and the mean probability (*x*) that a band in individual *A* matches that in a second, unrelated individual *B* is calculated. In this case, because the child and the father share 14 bands in common, the probability that the man tested is not the father is very low (probably 10^{-7} or lower). As a result, the most likely explanation is that the child is mutant for a single band. In fact, in 1419 cases of genuine paternity resolved by the minisatellite probes on chromosomes 1 and 7, single bands in the children were recorded in 399 cases, accounting for 28 percent of all cases.

3. Infection by HIV-1 (human immunodeficiency virus) is responsible for the destruction of cells in the immune system and results in the symptoms of AIDS (acquired immunodeficiency syndrome). HIV infects and kills cells of the immune system that carry a cell-surface receptor known as CD4. An HIV surface protein known as gp120 binds to the CD4 receptor and allows the virus to enter into the cell. The gene encoding the CD4 protein has been cloned. How might this clone be used along with recombinant DNA techniques to combat HIV infection?

Solution: Several methods that use the *CD4* gene are being explored to combat HIV infection. First, because infection depends on an interaction between the viral gp120 protein and the CD4 protein, the cloned *CD4* gene has been modified to produce a soluble form of the protein (sCD4). The idea is that HIV can be prevented from infecting cells if the gp120 protein of the virus is bound up with extra molecules of the soluble form of the CD4 protein. Thus it is unable to bind to CD4 proteins on the surface of immune system cells. Studies in cell culture systems indicate that the presence of sCD4 effectively prevents HIV infection of tissue culture cells. However, studies in HIV-positive humans have been somewhat disappointing, mainly because the strains of HIV used in the laboratory are different from those found in infected individuals. Since HIV-infected cells carry the viral gp120 protein on their surface, the *CD4* gene has been fused with genes encoding bacterial toxins to kill them. The resulting fusion protein contains CD4 regions that bind to gp120 and toxin regions that should kill the infected cell. In tissue culture experiments, cells infected with HIV are killed by the fusion protein, whereas uninfected cells survive. Researchers hope that the targeted delivery of drugs and toxins can be used in therapeutic applications to treat HIV infection.

PROBLEMS AND DISCUSSION QUESTIONS

1. In attempting to vaccinate people against diseases by having them eat antigens (such as the cholera toxin), the antigen must reach the cells of the small intestine. What are some potential problems of this method? Why don't absorbed food molecules stimulate the immune system and make you allergic to the food you eat?

2. One of the main safety issues associated with genetically modified crops is the potential for allergenicity caused by introduction of an allergen or caused by changing the level of expression of a host allergen. Since common allergenic proteins often have identical stretches of a few (six or seven) amino acids in common, Kleter and Peijnenburg (2002. *BMC Struct. Biol.* 2: 8) developed a method for screening transgenic crops to evaluate potential allergenic properties. How do you think they accomplished this?

3. There are more than 1000 cloned farm animals in the United States. Within the near future, milk from cloned cows and their offspring (born naturally) may be available in supermarkets. These animals are clones but have not been transgenically modified, and they are no different than are identical twins. Should milk from such animals and their naturally born offspring be labeled as coming from cloned cows or their descendants? Why?

4. One of the major causes of sickness, death, and economic loss in the cattle industry is *Mannheimia haemolytica*, which causes bovine pasteurellosis (shipping fever). Noninvasive delivery of a vaccine using transgenic plants expressing immunogens would reduce labor costs and trauma to livestock. An early step toward developing an edible vaccine is to determine whether an injected version of an antigen (usually a derivative of the pathogen) is capable of stimulating the development of antibodies in a test

organism. The following table assesses the ability of a transgenic portion of a toxin (Lkt) of *M. haemolytica* to stimulate development of specific antibodies in rabbits. (Table modified from Lee et al. 2001. *Infect. and Immunity* 69: 5786–5793.) (a) What general conclusion can you draw from the data? (b) With regards to development of a usable edible vaccine, what work remains to be done?

Immunogen Injected	Antibody Production in Serum
Lkt50*—saline extract	+
Lkt50—column extract	+
Mock injection	−
Pre-injection	−

*Lkt50 is a smaller derivative of Lkt that lacks all hydrophobic regions. + indicates at least 50 percent neutralization of toxicity of Lkt; − indicates no neutralization activity.

5. Outline the steps involved in transferring glyphosate resistance to a crop plant. Do you envision that this trait can escape from the crop plant and make weeds glyphosate-resistant? Why or why not?

6. Now that enhancement therapy using one gene product has been approved, is this justification for using enhancement gene therapy?

7. Recombinant adenoviruses have been used in a number of pre-clinical studies to determine the efficacy of gene therapy for rheumatoid arthritis and osteoarthritis. Genes can be delivered by injection to the tissues that need them. Christopher Evans and colleagues (2001. *Arthritis Res.* 3: 142–146) estimated that approximately 20 percent of all human gene therapy trials have used adenoviruses for gene delivery. The death of a patient in 1999 after infusion of adenoviral vectors has caused concern. As you consider the use of viral vectors as therapy-delivery vehicles for human pathologies, what factors seem of paramount concern?

8. Define somatic gene therapy, germ-line therapy, and enhancement gene therapy. Which of these is currently in use?

9. *Transductional targeting* is a preferred route for the delivery of therapeutics for human diseases. It involves the development of tissue-specific interactions between the viral vector and a specific tissue. A genetic approach used by Ponnazhagan and colleagues (2002. *J. Virol.* 76: 12,900–07) involves engineering the capsid of an adeno-associated virus (type 2) vector to target specific human cell types. Nongenetic approaches are also possible. Speculate on problems associated with the genetic approach of capsid alteration and problems that might be associated with nongenetic approaches to transductional targeting.

10. The development of safe vectors for human gene therapy has been a goal since 1990. Of the variety of problems associated with viral-based vectors, many such viruses (i.e., SV40) have transformation properties thought to be mediated by binding and inactivating gene products such as p53, retinoblastoma protein (pRB), and others. Mark Cooper and colleagues (1997. *Proc. Nat. Acad. Sci. (USA)* 94: 6450–6455) developed SV40-based vectors that are deficient in binding p53, pRB, and other proteins. Why would you specifically want to avoid inactivating p53, pRB, and related proteins?

11. Gene therapy for human genetic disorders involves transferring a copy of the normal human gene into a vector and using the vector to transfer the cloned human gene into target tissues. Presumably, the gene enters the target tissue and becomes active, and the gene product relieves the symptoms. (a) Why are disorders such as muscular dystrophy difficult to treat by gene therapy? (b) What are the potential problems of using retroviruses as vectors? (c) Should gene therapy involve germ-line tissue instead of somatic tissue? What are some of the potential ethical problems associated with the former approach?

12. Sequencing the human genome and the development of microarray technology promises to improve our understanding of normal and abnormal cell behavior. In an article on the applications of microarray technology (2001. *Breast Can. Res.* 3: 158–175), Cooper stated that "data from such studies could revolutionize cancer diagnosis." (a) What is microarray technology? (b) In what way might this technology revolutionize our understanding of cancer?

13. What are the advantages of using STRs instead of VNTRs for DNA profiling?

14. In producing physical maps of markers and cloned sequences, what advantage does *Drosophila* offer that other organisms, including humans, do not?

15. Suppose you develop a screening method for cystic fibrosis that allows you to identify the predominant mutation △508 and the next six most prevalent mutations. What must you consider before using this method to screen a population for this disorder?

16. A couple with European ancestry seeks genetic counseling before having children, because of a history of cystic fibrosis (CF) in the husband's family. ASO testing for CF reveals that the husband is heterozygous for the △508 mutation and the wife is heterozygous for the *R117* mutation. You are the couple's genetic counselor. At a meeting with you, they express their conviction that they are not at risk for having an affected child because they each carry different mutations and cannot have a child who is homozygous for either mutation. What would you say to them?

17. Dominant mutations can be categorized according to whether they increase or decrease the overall activity of a gene or gene product. Although a loss-of-function mutation (a mutation that inactivates the gene product) is usually recessive, for some genes, one dose of the normal gene product, encoded by the normal allele, is not sufficient to produce a normal phenotype. In this case, a loss-of-function mutation in the gene will be dominant, and the gene is said to be *haploinsufficient*. A second category of dominant mutation is the gain-of-function mutation, which results in a new activity or increased activity or expression of a gene or gene product. The gene therapy technique currently used in clinical trials involves the "addition" to somatic cells of a normal copy of a gene. In other words, a normal copy of the gene is inserted into the genome of the mutant somatic cell, but the mutated copy of the gene is not removed or replaced. Will this strategy work for either of the two aforementioned types of dominant mutations?

18. Why are most recombinant human proteins produced in animal or plant hosts instead of bacterial host cells?

19. The DNA sequence surrounding the site of the sickle-cell mutation in the β-globin gene, for normal and mutant genes is shown as follows.

5′-GACTCCTGAGGAGAAGT-3′

3′-CTGAGGACTCCTCTTCA-5′

Normal DNA

5′-GACTCCTGTGGAGAAGT-3′

3′-CTGAGGACACCTCTTCA-5′

Sickle-cell DNA

Each type of DNA is denatured into single strands and applied to a filter. The paper containing the two spots is hybridized to an ASO of the sequence

$$5' - GACTCCTGAGGAGAAGT - 3'$$

Which spot, if either, will hybridize to this probe?

20. One form of hemophilia, an X-linked disorder of blood clotting, is caused by a mutation in clotting factor VIII. Many single-nucleotide mutations of this gene have been described, making the detection of mutant genes by Southern blots inefficient. There is, however, an RFLP for the enzyme *Hind*III contained in an intron of the factor VIII gene. The two restriction fragments generated from digestion of either wild-type or mutant DNA are shown below.

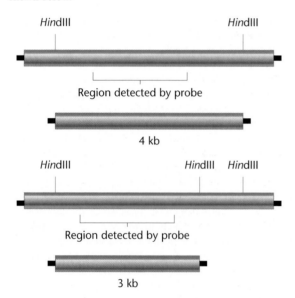

A female whose brother has hemophilia has a 50 percent risk of being a carrier of this disorder. To test her status, DNA is obtained from her white blood cells and those of family members, cut with *Hind*III, and the fragments are probed and visualized by Southern blotting. Using the results below, determine whether either of the females in generation II is a carrier for hemophilia.

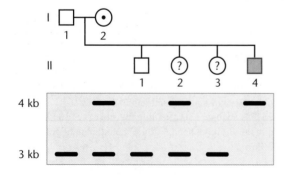

21. The human insulin gene contains introns. Since bacterial cells will not excise introns from mRNA, how can a gene like this be cloned into a bacterial cell that will produce insulin?

22. In mice transfected with the rabbit β-globin gene, the rabbit gene is active in several tissues, including the spleen, brain, and kidney. In addition, some mice suffer from thalassemia (a form of anemia) caused by an imbalance in the coordinate production of α and β-globins. Which problems associated with gene therapy are illustrated by these findings?

23. When genome scanning technologies become widespread, medical records will contain the results of such testing. Who should have access to this information? Should employers, potential employers, or insurance companies be allowed to have this information? Would you favor or oppose having the government establish and maintain a central database containing the results of genome scanning on members of the population?

24. What limits the use of differences in restriction enzyme sites as a way of detecting point mutations in human genes?

25. You are asked to assist with a prenatal genetic test for a couple, each of whom is found to be a carrier for a deletion in the β-globin gene that produces β-thalassemia when homozygous. The couple already has one child who is unaffected and is not a carrier. The woman is pregnant, and the couple wants to know the status of the fetus. You receive DNA samples obtained from the fetus by amniocentesis and from the rest of the family by extraction from white blood cells. Using a probe for the deletion, you obtain the blot shown below. Is the fetus affected? What is its genotype for the β-globin gene?

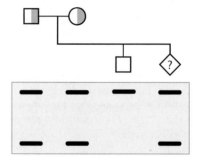

Developmental Genetics

This unusual four-winged Drosophila has developed an extra set of wings as a result of a homeotic mutation.

■ CHAPTER CONCEPTS

■ Gene action in development is based on differential transcription of selected genes.

■ Animals use a small number of signaling systems and regulatory networks to construct adult body forms from the zygote. These shared properties make it possible to use animal models to study human development.

■ Differentiation is controlled by cascades of gene action that follow the specification and determination of developmental fate.

■ Plants independently evolved some developmental mechanisms equivalent to those of animals.

■ In many organisms, cell–cell signaling programs the developmental fate of adjacent and distant cells.

In multicellular plants and animals, a fertilized egg undergoes a cycle of developmental events that ultimately give rise to an adult member of the species. Thousands, millions, or even billions of cells are organized into a cohesive and coordinated unit that we perceive as a living organism (Figure 20–1). The series of events through which organisms attain their final adult body plan is studied by developmental biologists. This area of study is perhaps the most intriguing in biology. The comprehension of developmental processes requires not only knowledge of many different biological disciplines—such as molecular, cellular, and organismal biology—but also an understanding of how biological processes change over developmental time.

Over the last two decades genetic analysis and molecular biology have shown that in spite of wide diversity in the size and shape of adult organisms, all multicellular organisms share many genes, genetic pathways, and molecular signaling mechanisms in the processes leading from the zygote to the adult. At the cellular level, development is marked by three important events: **specification**, when the first cues conferring a spatially discrete identity are established, **determination**, the time when a specific developmental fate for a cell becomes fixed, and **differentiation**, the process by which a cell achieves its final form and function. In addition, genomics has shown that many evolutionary and developmental relationships are shared among higher organisms. Genetic analysis has identified genes that regulate developmental processes, and we are beginning to understand how the action and interaction of these genes control basic developmental processes in eukaryotes.

In this chapter, the primary emphasis will be on how genetics has been used to study development. This area, called developmental genetics, has contributed tremendously to our understanding of developmental processes, because genetic information is required for the molecular and cellular functions mediating developmental events and contributes to the continually changing phenotype of the newly formed organism.

HOW DO WE KNOW ?

In this chapter, we will focus on both the large-scale and the inter- and intracellular events that take place during embryogenesis and the development of adult structures. As you study this topic, you should try to answer several fundamental questions:

1. How do we know how many genes control development in an organism like *Drosophila*?

2. How do we know that molecular gradients in the egg control development?

3. How do we know that a genetic program specifying a body part can be changed?

4. How do we know that chemical and physical signals between cells control developmental events?

20.1 Evolutionary Conservation of Developmental Mechanisms Can Be Studied Using Model Organisms

Genetic analysis of development in a wide range of organisms has demonstrated that a common set of developmental mechanisms and signaling systems are used in all animals. For example, eye formation in *Drosophila* and eye formation in humans are controlled by similar gene sets working through the same regulatory network. In fact, a "master regulator" gene for mouse eyes promotes the formation of *Drosophila* compound eyes when it is artificially expressed in *Drosophila* legs and antennae. As another example, most of the differences in shape between zebras and zebrafish are controlled by different patterns of expression in one gene set, called homeotic (abbreviated as *Hox*) genes, not by expression of many different genes. Genome sequencing projects have confirmed that genes from a wide range of organisms are conserved. This homology in genes and regulatory mechanisms means that many aspects of normal human embryonic development and associated genetic disorders can be studied in model organisms such as *Drosophila*, which unlike humans, can be genetically manipulated (see Chapter 1 for a discussion of model organisms in genetics).

Although many developmental mechanisms are similar among all animals, evolution has also generated new and different approaches to transform a zygote into an adult. These evolutionary changes can be attributed to several genetic mechanisms, including mutation, gene duplication and divergence, the assignment of new functions to old genes, and the recruitment of genes to new developmental pathways. For example, the turtle shell, a structure unique

(a)

(b)

FIGURE 20–1 (a) A *Drosophila* embryo and (b) the adult fly that develops from it.

among reptiles, is formed by adapting a signaling pathway to a new function. The shell forms by outgrowth of the dorsal body wall under control of the fibroblast growth factor 10 (*FGF-10*) gene, a component of a signal pathway that controls development of other outgrowths: limbs and lungs. The emphasis in this chapter, however, will be on the similarities among species.

Analysis of Developmental Mechanisms

In the space of this chapter, we cannot survey all aspects of development, nor can we explore the genetic analysis of all developmental mechanisms triggered by the fusion of sperm and egg. Instead, we will focus on several general processes in development:

- how the adult body plan of animals is laid down in the embryo

- the program of gene expression that turns undifferentiated cells into differentiated cells

- the role of cell–cell communication in development.

Interwoven with the discussion of these mechanisms will be several other topics: the role of master switch genes in controlling developmental pathways, cytoplasmic components in the oocyte that control early transcriptional patterns, and the signaling systems that alter patterns of gene expression.

We will use three model systems—*D. melanogaster*, *A. thaliana*, and *C. elegans*—to illustrate these developmental processes and the related topics. We will examine the role of differential gene expression in the progressive restriction of developmental options leading to the formation of the adult body plan in the model organisms *Drosophila* and *Arabidopsis*. We will then expand the discussion to include the selection of pathways that result in differentiated cells in plants and animals, and consider the role of cell–cell communication in development in *C. elegans*.

20.2 Genetic Analysis of Embryonic Development in *Drosophila* Reveals How the Body Axis of Animals Is Specified

How a certain cell turns specific genes on or off at specific stages of development is a central question in developmental biology. At present, there is no simple answer to this question. However, information derived from the study of model organisms gives us a starting point. In particular, the genetic and molecular analysis of embryonic development in *Drosophila* highlights the key role of molecular components in the oocyte cytoplasm in controlling gene expression.

Overview of *Drosophila* Development

The life cycle of *Drosophila* is about 10 days long and has several distinct phases: the embryo, three larval stages, the pupal stage, and the adult stage (Figure 20–2). Internally, the

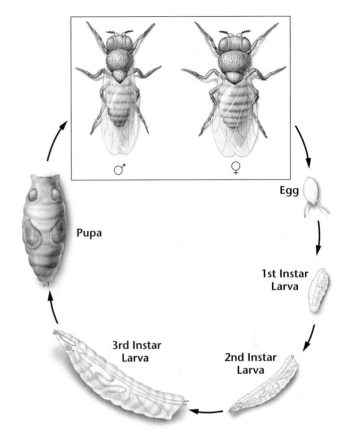

FIGURE 20–2 *Drosophila* life cycle.

egg cytoplasm is organized into a series of maternally-derived molecular gradients. These gradients, encoded by the maternal genome and placed in the egg during oogenesis, play a key role in establishing the developmental fates of nuclei located in specific regions of the embryo.

Immediately after fertilization, the zygote nucleus undergoes a series of nuclear divisions without cytokinesis [Figure 20–3(a) and (b)], forming a syncytial blastoderm (a syncytium is any cell with more than one nucleus). At about the tenth division, nuclei migrate to the periphery of the egg into cytoplasm containing localized gradients of maternally derived mRNA transcripts and proteins [Figure 20–3 (c)]. After several more divisions, the nuclei become enclosed in plasma membranes [Figure 20–3 (d)].

Germ cell formation at the posterior pole of the embryo demonstrates the regulatory role of these localized cytoplasmic components [Figure 20–3 (c) and (d)]. Nuclei transplanted from other regions of the embryo into the posterior cytoplasm (the pole plasm) will form germ cells. Hence, the cytoplasm of the posterior pole contains maternal components that direct nuclei to form germ cells.

The transcriptional programs triggered in the non-germ cell nuclei form the embryo's anterior–posterior (front to back) and dorsal–ventral (upper to lower) axes of symmetry, leading to the formation of a segmented embryo [Figure 20–3 (e)]. Under control of the *Hox* gene set (discussed in a later section), these segments give rise to the differentiated structures of the adult fly [Figure 20–3 (f)].

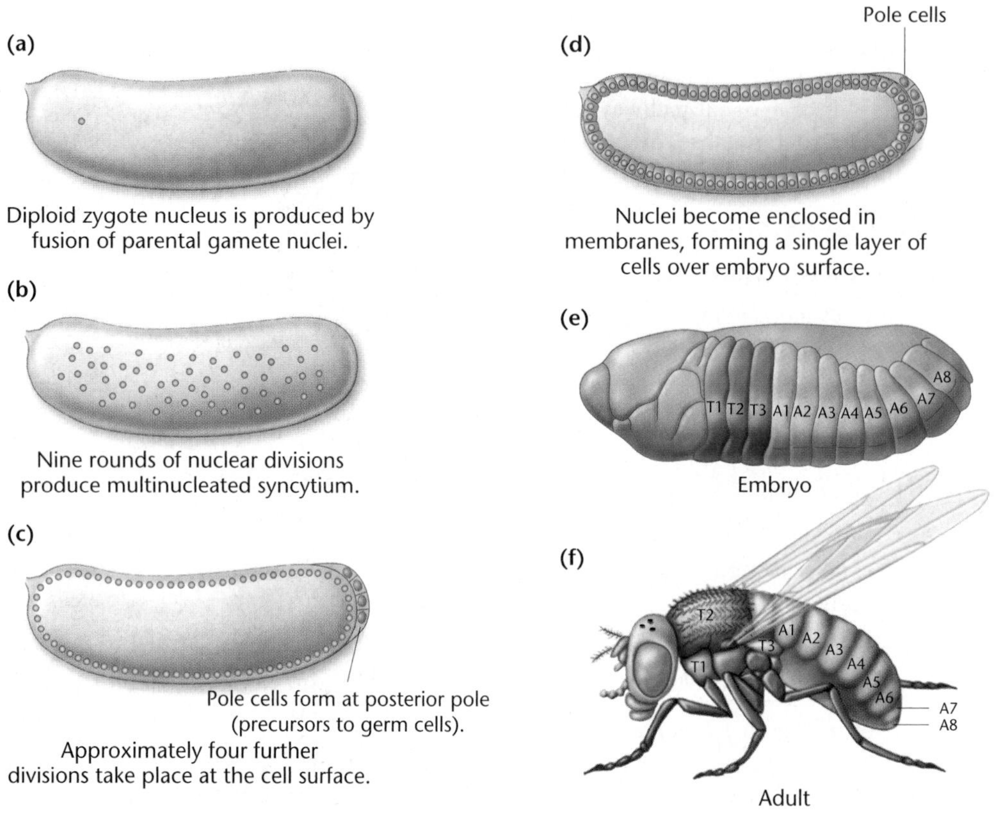

(a)

Diploid zygote nucleus is produced by
fusion of parental gamete nuclei.

(b)

Nine rounds of nuclear divisions
produce multinucleated syncytium.

(c)

Pole cells form at posterior pole
(precursors to germ cells).
Approximately four further
divisions take place at the cell surface.

(d)

Pole cells

Nuclei become enclosed in
membranes, forming a single layer of
cells over embryo surface.

(e)

T1 T2 T3 A1 A2 A3 A4 A5 A6 A7 A8

Embryo

(f)

T2
A1
T3 A2 A3
T1   A4
        A5
        A6
          A7
          A8

Adult

FIGURE 20–3 Early stages of embryonic development in *Drosophila*. (a) Fertilized egg with zygotic nucleus (2*n*), shortly after fertilization. (b) Nuclear divisions occur about every 10 minutes. Nine rounds of division produce a multinucleate cell, the syncytial blastoderm. (c) At the tenth division, the nuclei migrate to the periphery or cortex of the egg, and four additional rounds of nuclear division occur. A small cluster of cells, the pole cells, form at the posterior pole about 2.5 hours after fertilization. These cells will form the germ cells of the adult. (d) About 3 hours after fertilization, the nuclei become enclosed in membranes, forming a single layer of cells over the embryo surface, creating the cellular blastoderm. (e) The embryo at about 10 hours after fertilization. At this stage, the segmentation pattern of the body is clearly established. Behind the segments that will form the head, T1–T3 are thoracic segments, and A1–A9 are abdominal segments. (f) The segmented embryo and the adult structures that form from each segment.

Genetic Analysis of Embryogenesis

Two types of genes control embryonic development in *Drosophila*: maternal-effect genes and zygotic genes (Figure 20–4). Products of maternal-effect genes (mRNA and/or proteins) are deposited in the developing egg by the "mother" fly. Many of these products are distributed in a gradient or concentrated in specific regions of the egg cytoplasm. Female flies homozygous for recessive deleterious mutations in maternal-effect genes are sterile since none of their embryos receive wild-type gene products from their mother, and therefore develop abnormally. These maternal-effect genes encode transcription factors, receptors, and proteins that regulate translation. During embryonic development, these gene products activate or repress expression of the zygotic genome in a temporal and spatial sequence.

Zygotic genes are those transcribed from the newly formed nucleus after fertilization and expressed in the developing embryo. Flies with deleterious mutations in zygotic genes exhibit embryonic lethality. In a cross between two flies heterozygous for a recessive zygotic mutation, one-fourth of the embryos (the homozygotes) fail to develop normally and die.

In *Drosophila*, many zygotic genes are transcribed in specific regions of the embryo that depend on the distribution of maternal-effect proteins.

Our knowledge of the genes that regulate *Drosophila* development is based on the work of Christiane Nüsslein-Volhard, Eric Wieschaus, and Ed Lewis, who were awarded the 1995 Nobel Prize for Physiology or Medicine. Ed Lewis initially identified and studied one of these regulatory genes in the 1970s. In the late 1970s, Nüsslein-Volhard and Wieschaus devised a strategy to identify all the genes that control development in *Drosophila*. To do this, they examined thousands of offspring of mutagenized flies, looking for recessive embryonic lethal mutations with defects in external structures. The parents were thus identified as heterozygous carriers of these mutations, which they grouped into three classes: *gap, pair-rule*, and *segment polarity* genes. In 1980, Nüsslein-Volhard and Wieschaus proposed a model in which embryonic development is initiated by gradients of maternal-effect gene products. The positional information laid down by these molecular gradients is interpreted by two sets of zygotic genes: (1) **segmentation genes** (gap, pair-rule, and segment polarity genes), and (2)

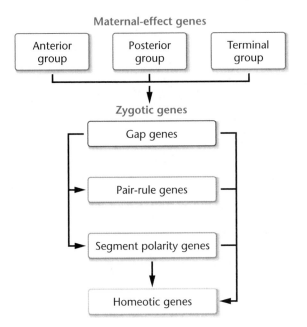

FIGURE 20–4 The hierarchy of genes involved in establishing the segmented body plan in *Drosophila*. Gene products from the maternal genes regulate the expression of the first three groups of zygotic genes (gap, pair-rule, and segment polarity, collectively called the segmentation genes), which in turn control expression of the homeotic genes.

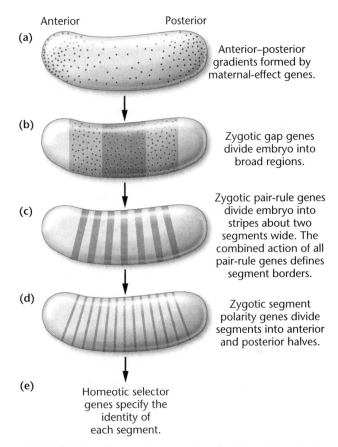

FIGURE 20–5 (a) Progressive restriction of cell fate during development in *Drosophila*. Gradients of maternal proteins are established along the anterior–posterior axis of the embryo. (b), (c), and (d) Three groups of segmentation genes progressively define the body segments. (e) Individual segments are given identity by the homeotic genes.

homeotic selector (*Hox*) genes. The segmentation genes divide the embryo into a series of stripes or segments and define the number, size, and polarity of each segment. The homeotic genes specify the fate of each segment and the adult structures formed from the segments (Figure 20–4).

The model is shown in Figure 20–5. Most maternal-effect gene products placed in the egg during oogenesis are activated immediately after fertilization and form the anterior–posterior axis of the embryo [Figure 20–5(a)]. Many maternal gene products encode transcription factors which activate transcription of the gap genes, whose expression divides the embryo into a series of regions corresponding to the head, thorax, and abdomen of the adult [Figure 20–5 (b)]. Gap proteins are transcription factors that activate pair-rule genes, whose products divide the embryo into smaller regions about two segments wide [Figure 20–5 (c)]. The combined action of all gap genes defines segment borders. The pair-rule genes in turn activate the segment polarity genes, which then divide each segment into anterior and posterior regions [Figure 20–5 (d)]. The collective action of the maternal genes that form the anterior–posterior axis and the segmentation genes define the field of action for the homeotic (*Hox*) genes [Figure 20–5 (e)].

Now Solve This

Problem 5 on page 468 involves screening for mutants that affect external structures of the embryo. You are asked to reconcile your results with those published in the literature.

Hint: In reconciling the results of your study with the conclusions of Wieschaus and Schüpbach, remember the differences between genes and alleles.

20.3 Zygotic Genes Program Segment Formation

Zygotic genes are activated or repressed in a positional gradient by the maternal-effect gene products. The expression of the three subsets of segmentation genes divides the embryo into a series of segments along the anterior–posterior axis. These segmentation genes are normally transcribed in the developing embryo, and their mutations have embryonic lethal phenotypes.

Over 20 segmentation genes have been identified (Table 20.1). They are classified on the basis of their mutant phenotypes: (1) mutations in gap genes delete a group of adjacent segments, (2) mutations in pair-rule genes affect every other segment and eliminate a specific part of each affected segment, and (3) mutations in segment polarity genes cause defects in homologous portions of each segment. We'll examine each group in greater detail.

Gap Genes

Transcription of **gap genes** is activated or inactivated by gene products previously expressed along the anterior–posterior axis and by other genes of the maternal gradient system. When

TABLE 20.1	Segmentation Genes in *Drosophila*	
Gap Genes	**Pair-Rule Genes**	**Segment Polarity Genes**
Krüppel	*hairy*	*engrailed*
knirps	*even-skipped*	*wingless*
hunchback	*runt*	*cubitis interruptus*[D]
giant	*fushi-tarazu*	*hedgehog*
tailless	*odd-paired*	*fused*
buckebein	*odd-skipped*	*armadillo*
caudal	*sloppy-paired*	*patched*
		gooseberry
		paired
		naked
		disheveled

mutated, these gap genes produce large gaps in the embryo's segmentation pattern. *hunchback* mutants lose head and thorax structures, *Krüppel* mutants lose thoracic and abdominal structures, and *knirps* mutants lose most abdominal structures. Transcription of gap genes divides the embryo into a series of regions (the head, thorax, and abdomen). Within these regions, different combinations of gene activity eventually specify both the type of segment that forms and the proper order of segments in the body of the larva, pupa, and adult. Expression of the gap genes in the embryo correlates roughly with their mutant phenotypes: *hunchback* at the anterior, *Krüppel* in the middle (Figure 20–6), and *knirps* at the posterior region of the embryo. Gap genes encode transcription factors that control the expression of pair-rule genes.

Pair-Rule Genes

Pair-rule genes divide the head–thorax–abdomen regions established by gap genes into sections about two segments wide. Mutations in pair-rule genes eliminate segment-size sections at every other segment. The pair-rule genes are expressed in narrow bands or stripes of nuclei that extend around the circumference of the embryo. The expression of this gene set first establishes the boundaries of segments and then establishes the developmental fate of the cells within each segment by controlling the segment

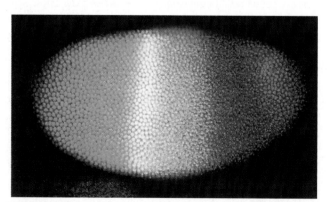

FIGURE 20–6 Expression of gap genes in a *Drosophila* embryo. The hunchback protein is shown in orange, and Krüppel is indicated in green. The yellow stripe is created when cells contain both hunchback and Krüppel proteins. Each dot in the embryo is a nucleus.

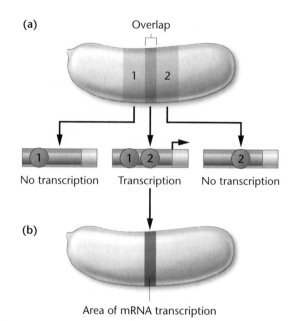

(a) Overlap

1   2

1 — No transcription   1 2 — Transcription   2 — No transcription

(b)

Area of mRNA transcription

FIGURE 20–7 New patterns of gene expression can be generated by overlapping regions containing two different gene products. (a) Transcription factors 1 and 2 are present in an overlapping region of expression. If both transcription factors must bind to the promoter of a target gene to trigger expression, the gene will be active only in cells containing both factors (most likely in the zone of overlap). (b) The expression of the target gene in the restricted region of the embryo.

polarity genes. At least eight pair-rule genes act to divide the embryo into a series of stripes. However, the boundaries of these stripes overlap, meaning that cells in the stripes express different combinations of pair-rule genes in an overlapping fashion (Figure 20–7). The transcription of the pair-rule genes is mediated by the action of gap gene products and maternal gene products, but the pattern is resolved into highly delineated stripes by the interaction among the gene products of the pair-rule genes themselves (Figure 20–8). All gap genes encode transcription factors.

Segment Polarity Genes

Expression of **segment polarity genes** is controlled by transcription factors encoded by pair-rule genes. Within each segment, segment polarity genes become active in a single band of cells that extends around the embryo's circumference (Figure 20–9). This divides the embryo into 14 segments, and the products of the segment polarity genes control the cellular identity within each segment.

Segmentation Genes in Mice and Humans

Segment formation in *Drosophila* depends on the action of three subsets of segmentation genes. Are these same genes found in humans and other mammals, and do they control aspects of embryonic development in these organisms? To answer these questions, let's examine one of the segmentation genes in *Drosophila*. *runt* is one of the principal pair-rule genes in *Drosophila* and, later in development, controls aspects of sex determination and formation of the fly's nervous system. The gene encodes a protein (called runt) that regulates transcription of its target genes. Runt contains a 128-amino-acid DNA-binding

(a)

(b)

FIGURE 20–8 Stripe pattern of pair-rule gene expression in *Drosophila* embryo. This embryo is stained to show patterns of expression of the genes *even-skipped* and *fushi-tarazu*; (a) low-power view, and (b) high-power view of the same embryo.

region called the runt domain that is highly conserved in mouse and human proteins. In fact, *in vitro* experiments show that the *Drosophila* and mouse runt proteins are functionally interchangeable. In the mouse, *runt* is expressed early in development and controls hematopoesis (formation of blood cells), osteogenesis (formation of bone), and formation of the genital system. In each case, whether in *Drosophila* or mouse, expression of *runt* specifies the fate of uncommitted cells in the embryo. As an illus-

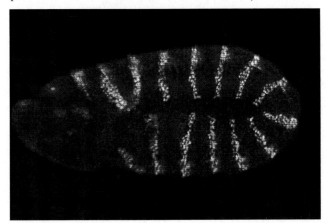

FIGURE 20–9 The 14 stripes of expression of the segment polarity gene *engrailed* in a *Drosophila* embryo.

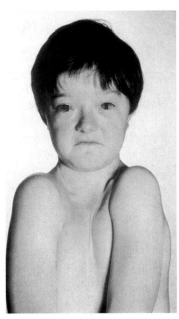

FIGURE 20–10 A boy affected with cleidocranial dystrophy (CCD). This disorder, inherited as an autosomal dominant trait, is caused by mutation in a human *runt* gene, *CBFA*. Affected heterozygotes have a number of skeletal defects, including a hole in the top of the skull where the infant fontanel fails to close, and collar bones that do not develop or form only small stumps. Because the collar bones do not form, CCD individuals can fold their shoulders across their chests. *Reprinted by permission from Macmillan Publishers Ltd.: Fig 1 on p244 from: British Dental Journal 195: 243–48 2003. Greenwood, M. and Meechan, J.G. "General medicine and surgery for dental practitioners." Copyright © Macmillan Magazines Limited.*

tration, let's look at the role of a runt-domain gene in controlling formation of the skeletal system in mice and humans. Mutation in *CBFA*, a human homolog of *runt*, causes cleidocranial dysplasia (CCD), an autosomal-dominantly-inherited trait. Those affected with CCD have a hole in the top of their skull because their fontanel does not close. Their collar bones (clavicles) do not develop, enabling them to fold their shoulders across their chest (Figure 20–10). Mice with one mutant copy of the *runt* homolog have a phenotype similar to that seen in humans; mice with two mutant copies of the gene have no bones at all. Their skeletons are made only of cartilage, much like sharks (Figure 20–11), emphasizing the role of the *runt* gene as a developmental switch controlling the formation of bone.

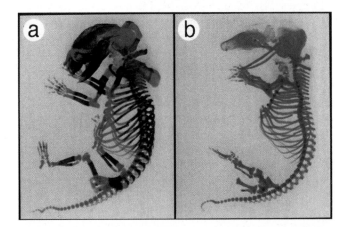

FIGURE 20–11 Bone formation in normal mice and mutants for the *runt* gene *Cbfa1*. (a) Normal mouse embryos at day 17.5 show cartilage (blue) and bone (brown). (b) The skeleton of a 17.5-day homozygous mutant embryo. Only cartilage has formed in the skeleton. There is complete absence of bone formation in the mutant mouse. Expression of a normal copy of the *Cbfa1* gene is essential for specifying the developmental fate of bone-forming osteoblasts.

20.4 Homeotic Selector Genes Specify Parts of the Adult Body

As segment boundaries are established by action of the segmentation genes, the homeotic (from the Greek word for "same") genes are activated as targets of the zygotic genes. Expression of homeotic selector genes determines which adult structures will be formed by each body segment. In *Drosophila*, this includes the antennae, mouth parts, legs, wings, thorax, and abdomen. Mutants of these genes are called **homeotic mutants** because the structure formed by one segment is transformed so that it is the same as that of an another segment. For example, the wild-type allele of *Antennapedia (Antp)* specifies formation of a leg on the second segment of the thorax. Dominant gain-of-function *Antp* mutations cause this gene to be expressed in the head as well, and mutant flies have a leg on their head in place of an antenna (Figure 20–12).

Hox Genes in *Drosophila*

The *Drosophila* genome contains two clusters of homeotic selector genes (*Hox* genes) on chromosome 3 (Table 20.2). One cluster, the *Antennapedia (ANT-C)* complex, contains

five genes that specify structures in the head and first two thoracic segments [Figure 20–13 (a)]. The second cluster, the *bithorax (BX-C)* complex, contains three genes that specify structures in the posterior portion of the second thoracic segment, the entire third thoracic segment, and abdominal segments [Figure 20–13 (b)].

The *Hox* genes have two properties in common. First, each *Hox* gene (listed in Table 20.2) encodes a transcription factor that includes a 180-bp domain known as a **homeobox**. (*Hox* is a contraction of homeobox.) The homeobox encodes a DNA-binding sequence of 60 amino acids known as a **homeodomain**. Second, expression of the genes is colinear. Genes at the 3′-end of a cluster are expressed at the anterior end of the embryo, those in the middle are expressed in the

(a)

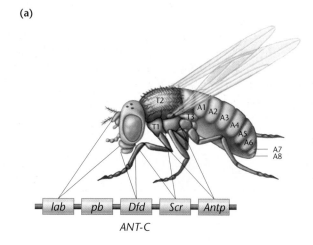

ANT-C

(b)

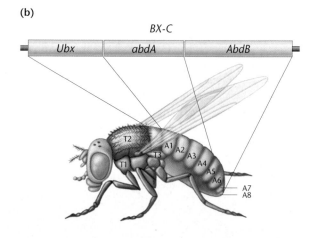

BX-C

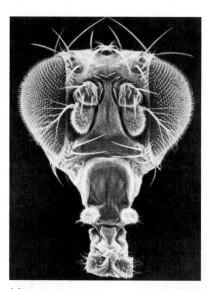

(a)

(b)

FIGURE 20–12
Antennapedia (Antp) mutation in *Drosophila*. (a) Head from wild-type *Drosophila*, showing the antenna and other head parts. (b) Head from an *Antp* mutant, showing the replacement of normal antenna structures with legs. This is caused by activation of the *Antp* gene in the head region.

FIGURE 20–13 Genes of the *Antennapedia* complex and the adult structures they specify. (a) In the *ANT-C* complex, the *labial (lab)* and *Deformed (Dfd)* genes control the formation of head segments. The *Sex comb reduced (Scr)* and *Antennapedia (Antp)* genes specify the identity of the first two thoracic segments. The remaining gene in the complex, *proboscipedia (pb)*, may not act during embryogenesis but may be required to maintain the differentiated state in adults. In mutants, the labial palps are transformed into legs. (b) In the *BX-C* complex, *Ultrabithorax (Ubx)* controls formation of structures in the posterior compartment of T2 and structures in T3. The two other genes, *abdominal A (abdA)* and *Abdominal B (AbdB)*, specify the segmental identities of the eight abdominal segments (A1–A8).

TABLE 20.2	*Hox* Genes of *Drosophila*
Antennapedia Complex	**Bithorax Complex**
labial	Ultrabithorax
Antennapedia	abdominal A
Sex combs reduced	Abdominal B
Deformed	
proboscipedia	

middle of the embryo, and genes at the 5′-end of a cluster are expressed at the embryo's posterior region (Figure 20–14). Although *Hox* genes were first identified in *Drosophila*, they are found in the genomes of most eukaryotes with segmented body plans, including zebrafish, *Xenopus* (African frog), chickens, mice, and humans (Figure 20–15).

To summarize, genes that control development in *Drosophila* act in a temporally and spatially ordered cascade, beginning with the genes that establish the anterior–posterior (and dorsal–ventral) axis of the egg and early embryo. Gradients of maternal mRNAs and proteins along the anterior–posterior axis activate the gap genes, which subdivide the embryo

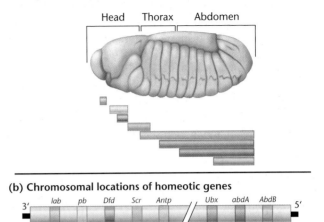

(a) Expression domains of homeotic genes

(b) Chromosomal locations of homeotic genes

FIGURE 20–14 The colinear relationship between the spatial pattern of expression and chromosomal locations of homeotic genes in *Drosophila*. (a) *Drosophila* embryo and the domains of homeotic gene expression in the embryonic epidermis and central nervous system. (b) Chromosomal location of homeotic selector genes. Note that the order of genes on the chromosome correlates with the sequential anterior borders of their expression domains.

FIGURE 20–15 Conservation of organization and patterns of expression in *Hox* genes. (Top) The structures formed in adult *Drosophila* are shown, with the colors corresponding to members of the *Hox* cluster that control their formation. (Middle) The reconstructed *Hox* cluster of the common ancestor to all bilateral organisms contains seven genes. (Bottom) The arrangement and expression pattern of the four clusters of *Hox* genes in an early human embyro. Note that some of the posterior genes are expressed in the limbs. The expression pattern is inferred from that observed in mice. As in *Drosophila*, genes at the (3′-end) of the cluster form anterior structures, and genes at the (5′-end) of the cluster form posterior structures. Because of duplications, genes with equal homology to an ancestral sequence are indicated by brackets.

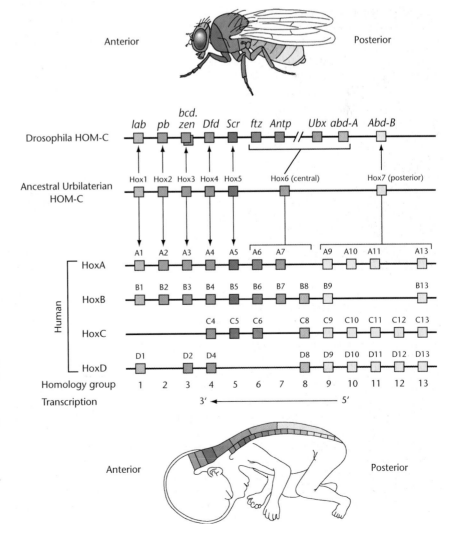

into broad bands. Gap genes in turn activate the pair-rule genes, which divide the embryo into segments. The final group of segmentation genes, the segment polarity genes, divides each segment into anterior and posterior regions arranged linearly along the anterior–posterior axis. The segments are then given identity by the *Hox* genes. Therefore, this progressive restriction of developmental potential of the *Drosophila* embryo's cells (all of which occurs during the first third of embryogenesis) involves a cascade of gene action, with regulatory proteins acting on transcription, translation, and signal transduction.

Hox Genes and Human Genetic Disorders

Although first described in *Drosophila*, *Hox* genes are found in the genomes of all animals where they play a fundamental role in shaping the body and its appendages. Because of the genome duplications associated with vertebrate evolution, most tetrapods, including humans, have four clusters of *Hox* genes (*HOXA*, *HOXB*, *HOXC*, and *HOXD*). The conservation of sequence, the order of genes in the *Hox* clusters, and their pattern of expression in vertebrates suggests that these gene sets function in bilaterally symmetrical animals, including flies and humans, to control the pattern of structures along the anterior–posterior axis (Figure 20–16), and in the development of appendages. Evidence for this comes from mutational studies in chicks and mice showing that *HOXD* genes near the 5′-end of the cluster play critical roles in limb development. This role for *HOXD* genes in humans was confirmed by the discovery that several inherited limb malformations are caused by specific mutations in *HOXD* genes. For example, mutations in *HOXD13* cause synpolydactlyly (SPD), a malformation characterized by extra fingers and toes, and abnormalities in bones of the hands and feet (Figure 20–17).

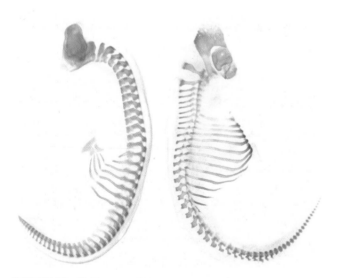

FIGURE 20–16 Patterns of *Hox* gene expression control the formation of structures along the anterior–posterior axis of bilaterally symmetrical animals in a species-specific manner. In the chick (left) and the mouse (right) expression of the same set of *Hox* genes is differentially programmed in time and space to produce different body forms.

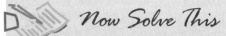

Now Solve This

Problem 15 on page 468 involves analysis of expression patterns for two genes in *Drosophila* embryos with different genetic backgrounds. You are asked to decide which gene regulates the other.

Hint: It is this genetic background that provides the clues to the timing of expression and regulatory patterns of the two genes in question.

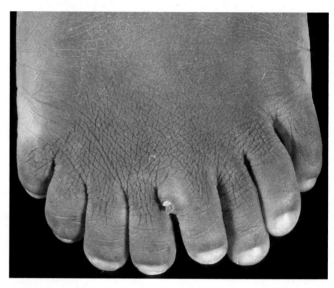

FIGURE 20–17 Mutations in posterior *Hox* genes (*HOXD13* in this case) in humans result in malformations of the limbs, shown here as extra toes. This condition is known as synpolydactlyly. Mutations in *HOXD13* are also associated with abnormalities of the bones in the hands and feet.

20.5 Cascades of Gene Action Control Differentiation

In addition to homeobox genes in the *Hox* gene clusters, there is a large and diverse family of other homeobox genes scattered throughout eukaryotic genomes. In *Drosophila*, one of these, *Distal-less (Dll)*, plays an important role in the development of appendages, including the antennae, mouthparts, legs, and wings. This gene, which maps to chromosome 2 and encodes a transcription factor, is one of the earliest known gene expressed in appendage formation. In the antennae, expression of the wild-type *Dll* allele has two roles: It directs cells to form an antenna instead of a leg, and it controls the proximal (closest to the body) to distal (farthest from the body) specification of antennal structures. Mutations in *Dll* produce a wide range of phenotypes, including transformation of the antennae into legs and the formation of shortened legs that are missing distal structures.

The antenna is the ear and olfactory center (nose) of the fly and has several components, including the arista and three antennal segments [Figure 20–18 (a)]. The arista transfers vibrations from sound waves through the first and second antennal segments to the antennal nerve and then to the brain. The third antennal segment is covered with olfactory receptors. Odors generate signals that are transferred to the first two antennal segments and then via a nerve to the brain.

Expression of *Dll* in the early pupal stage [Figure 20–18 (b)] is restricted to the second and third antennal segments and the arista, and is the first step in a cascade of gene expression associated with differentiation. At the larval-pupal transformation, five genes are activated by *Dll,* and each is expressed in a segment-specific pattern [Figure 20–18 (c)]. Four of these genes, *spalt (sal), spineless (ss), bric a brac (bab1),* and *aristaless (al),* encode transcription factors. Each of these genes in turn controls multiple target genes. The fifth gene, *dachshund (dac),* encodes a protein of unknown function that is confined to the nucleus. The genes activated by *Dll* expression initiate a cascade of segment-specific gene expression causing differentiation of the antenna and arista, but the details of how this is accomplished are still being studied.

In humans, there are six genes (*DLX1–DLX6*) in the *Distalless* gene family with a homeobox sequence closely related to the *Drosophila Dll* gene. The *Drosophila Dll* gene is expressed in the head and appendages of the developing fly. The human *DLX3* gene is expressed in the developing head,

and mutations in this gene are responsible for an inherited condition called trichodontoosseous syndrome (TDO). This autosomal dominant disorder causes deficiencies in the calcification of bones in the skull and enamel defects in teeth.

The discovery that TDO is caused by a mutation in a homeobox gene began with mutational analysis of *Drosophila* development and illustrates the valuable role that model organisms play in understanding human genetic disorders. Human *Distalless* genes were discovered by screening a human cDNA library with a cloned probe containing the homeobox sequence of the *Drosophila Dll* gene, and the gene for *DLX3* was mapped to chromosome 17 using fluorescent *in situ* hybridization (FISH) in 1995. TDO was originally described in 1966, but the nature of the gene and its location were unknown. In 1997, geneticists found that TDO was linked to markers on chromosome 17. Further work mapped TDO to the same locus as *DLX3*, and in 1998, analysis of mutations in affected individuals confirmed that TDO is caused by a 4bp deletion in *DLX3*.

20.6 Plants Have Evolved Developmental Systems That Parallel Those of Animals

Plants and animals diverged from a common unicellular ancestor about 1.6 billion years ago, after the origin of eukaryotes and probably before the rise of multicellular organisms. Genome sequencing and genetic analysis of mutants in plants and animals indicate that basic developmental mechanisms have evolved independently in plants and animals. We have already examined the genetic systems that control development and pattern formation in animals, using *Drosophila* as a model organism.

Flower development in *Arabidopsis thaliana* (Figure 20–19), a small plant in the mustard family, has been used to study pattern

(a)

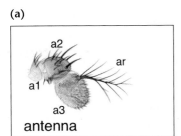

antenna

(b)

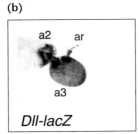

Dll-lacZ

(c)

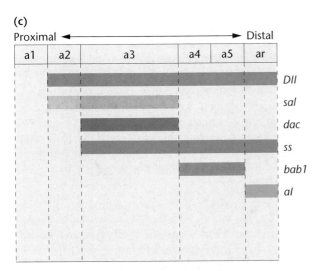

FIGURE 20–18 Action of the *Dll* gene in *Drosophila* produces (a) the wild-type antenna, divided into the arista (ar) and three antennal segments (a1, a2, a3). (b) *Dll* expression (shown in blue) in a late pupal antenna. Expression at this stage is limited to the arista, part of a2, and a3. (c) Activation of *Dll* target genes in antennal segments during the third larval instar just before the pupal stage. At this stage, *Dll* is expressed in all segments except a1. All of the target genes except *dac* encode transcription factors. Activation of these genes initiates a cascade of gene action in other gene sets.

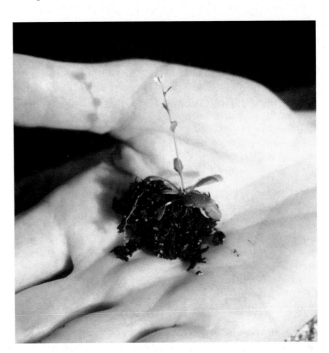

FIGURE 20–19 The flowering plant *Arabidopsis thalania,* used as a model organism in plant genetics.

(a)

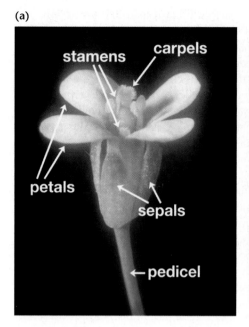

(b)

FIGURE 20–20 (a) Parts of the *Arabidopsis* flower. The floral organs are arranged concentrically. The sepals form the outermost ring, followed by petals and stamens, with carpels on the inside. (b) View of the flower from above.

formation in plants. A cluster of undifferentiated cells, called the *floral meristem*, gives rise to flowers (Figure 20–20). Each flower consists of four organs—sepals, petals, stamens, and carpels—that develop from concentric rings of cells within the meristem [Figure 20–21(a)]. Each organ develops from a different whorl.

Homeotic Genes in *Arabidopsis*

Three classes of floral homeotic genes control the development of these organs. Class A genes specify sepals, class A and class B genes specify petals, and class B and class C genes control stamen formation. Class C genes alone specify carpels [Figure 20–21 (b)]. The genes in each class are listed in Table 20.3. Class A genes are active in whorls 1 and 2 (sepals and petals), class B genes are expressed in whorls 2 and 3 (petals and stamens), and class C genes are expressed in whorls 3 and 4 (stamens and carpels). The organ formed depends on the expression pattern of the three gene classes. Expression of class A genes in whorl 1 causes sepals to form. Expression of class A *and* class B genes in whorl 2 leads to petal formation. Expression of class B and class C genes in

TABLE 20.3	Homeotic Selector Genes in *Arabidopsis**
Class A	*APETALA1 (AP1)*
	APETALA2 (AP2)
Class B	*APETALA3 (AP3)*
	PISTILLATA (P1)
Class C	*AGAMOUS (AG)*

*By convention, wild-type genes in *Arabidopsis* use capital letters.

whorl 3 leads to stamen formation. In whorl 4, expression of class C genes causes carpel formation.

As in *Drosophila*, mutations in homeotic genes cause organs to form in abnormal locations. For example, in *AP2* mutants (mutation of a class A gene), the order of organs is carpel, stamen, stamen, and carpel instead of the normal order, sepal, petal, stamen, and carpel [Figure 20–22 (a) and (b)]. In B loss-of-function mutants, petals become sepals, and stamens are transformed into carpels [Figure 20–22 (c)] and the order of organs becomes sepal, sepal, carpel, carpel.

Plants carrying a mutation for the class 3 gene AGAMOUS will have petals in whorl 3 (instead of stamens) and sepals in whorl 4 (instead of carpels), and the order of organs will be sepal, petal, petal, and sepal [Figure 20–22 (d)].

Evolutionary Divergence in Homeotic Genes

From genetic analysis of development in both *Drosophila* and *Arabidopsis*, it is clear that each organism uses a different set of master regulatory genes to establish the body axis and specify the identity of structures along the axis. In *Drosophila*, this task is accom-

(a)

1 — A genes
2 — B genes
3 —
4 — C genes

(b)

Sepal
Petal
Carpel
Stamen

FIGURE 20–21 Cell arrangement in the floral meristem. (a) The four concentric rings, or whorls, labeled 1–4, give rise to (b) arrangement of the sepals, petals, stamens, and carpels, respectively, in the mature flower.

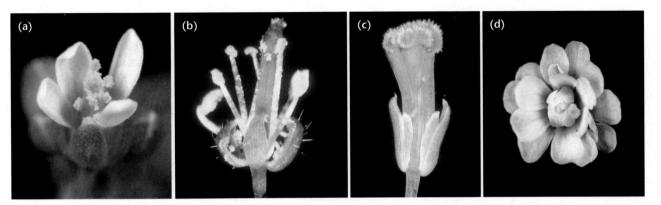

FIGURE 20–22 (a) Wild-type flowers of *Arabidopsis* have (from outside to inside) sepals, petals, stamens, and carpels. (b) Homeotic *APETALA2* mutant flower, with carpels, stamens, stamens, and carpels. (c) *PISTILLATA* mutants have sepals, sepals, carpels, and carpels. (d) *AGAMOUS* mutants have petals and sepals at places where stamens and carpels should form.

plished in part by the *Hox* genes, which encode a set of transcription factors sharing a homeobox domain. In *Arabidopsis*, however, the floral homeotic genes are members of a different family of transcription factors, called the **MADS-box proteins**. Each member of this family contains a common sequence of 58 amino acids with no similarity in amino acid sequence or protein structure with the *Hox* genes. Both gene sets encode transcription factors, both sets are master regulators of development expressed in a pattern of overlapping domains, and both specify identity of structures.

Reflecting their common evolutionary origins, the genomes of both organisms contain members of the homeobox and MADS-box genes, but these genes have been adapted for different uses in the plant and animal kingdoms, indicating that developmental mechanisms evolved independently in each group.

In both plants and animals, the action of transcription factors depends on changes in chromatin structure that make genes available for expression. Mechanisms of transcription initiation are conserved in plants and animals, as is reflected in homology of genes in *Drosophila* and *Arabidopsis* that maintain patterns of expression initiated by regulatory gene sets. Action of the floral homeotic genes is controlled by a gene called *CURLY LEAF*. This gene shares significant homology with members of the *Drosophila Polycomb* gene family, a group of genes that regulate homeobox genes during development in the fruit fly. Both of these genes encode proteins that alter chromatin conformation and shut off gene expression. Thus, although different genes are used to control development, both plants and animals use an evolutionarily conserved mechanism to regulate expression of these gene sets.

20.7 Cell–Cell Interactions in Development Are Modeled in *C. elegans*

During development in multicellular organisms, cell–cell interactions influence the transcriptional programs and developmental fate of surrounding cells. Cell–cell interaction is an important process in the embryonic development of most

eukaryotic organisms, including *Drosophila*, as well as vertebrates such as clawed frogs (*Xenopus*), mice, and humans.

Signaling Pathways in Development

In early development, animals use a number of signaling pathways to regulate development; after organogenesis begins, other signal pathways are added to those already in use. These newly activated pathways act both independently and in coordinated networks to elicit specific transcriptional responses. The signal networks establish anterior–posterior polarity and body axes, coordinate pattern formation, and direct the differentiation of tissues and organs. The signaling pathways used in early development and some of the developmental processes they control are listed in Table 20.4. After an

TABLE 20.4	Signaling Pathways Used in Early Embryonic Development

Wnt Pathway
Dorsalization of body
Female reproductive development
Dorsal–ventral differences

TGF-β Pathway
Mesoderm induction
Left–right asymmetry
Bone development

Hedgehog Pathway
Notochord induction
Somitogenesis
Gut/visceral mesoderm

Receptor Tyrosine Kinase Pathway
Mesoderm maintenance

Notch Signaling Pathway
Blood cell development
Neurogenesis
Retina development

Source: *Taken from* Gerhart, J. 1999. 1998 Warkany lecture: Signaling pathways in development. *Teratology* 60: 226–239.

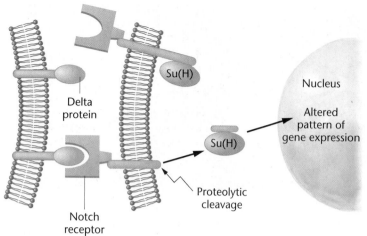

FIGURE 20–23 Components of the Notch signaling pathway in *Drosophila*. The cell carrying the Delta transmembrane protein is the sending cell; the cell carrying the transmembrane Notch protein receives the signal. Binding of Delta to Notch triggers a proteolytic-mediated activation of transcription. The fragment cleaved from the cytoplasmic side of the Notch protein combines with the Su(H) protein and moves to the nucleus where it activates a program of gene transcription.

introduction to the components and interactions of one of these systems—the **Notch signaling pathway**—we will briefly examine its role in the development of the vulva in the nematode, *Caenorhabditis elegans*.

The Notch Signaling Pathway

The genes in the Notch pathway are named after the *Drosophila* mutants that were used to identify components of this signaling pathway. Notch works through direct cell–cell contact to control the developmental fate of the interacting cells. The *Notch* gene (and the equivalent gene in other organisms) encodes a transmembrane signal receptor (Figure 20–23). The signal is another transmembrane protein encoded by the *Delta* gene (and its equivalents). Because both the signal and receptor are membrane-bound, the Notch signal system works only between adjacent cells. When the Delta protein binds to the Notch receptor, the cytoplasmic tail of the Notch protein cleaves off and binds to a cytoplasmic protein encoded by the *Su(H)* (suppressor of *Hairless*) gene. This protein complex moves into the nucleus and binds to transcriptional cofactors, activating transcription of a gene set that controls a specific developmental pathway (Figure 20–23).

Variations of this pathway control a number of different developmental processes in *Drosophila*, establishing the dorsal–ventral boundary in the wing disc and directing the fate of cells in the nervous system, muscle, and gut. One of the main roles of the Notch signal system is specifying the fate of equivalent cells in a population. In its simplest form, this interaction involves two neighboring cells that are developmentally equivalent. In humans, four members of the Notch family (*NOTCH1–NOTCH4*) have been identified. Mutations in these genes and other genes in the Notch pathway are responsible for a number of human developmental disorders, including Alagille syndrome (AGS) and spondylocostal dysostosis (SD). We will explore the role of the Notch signaling

system in development of the vulva in *C. elegans*, after a brief introduction to nematode embryogenesis.

Overview of *C. elegans* Development

The nematode *C. elegans* is widely used to study the genetic control of development. This organism has several advantages for such studies: (1) the genetics of the organism are well known, (2) the genome sequence is available, and (3) adults are formed from a small number of cells that follow a highly deterministic developmental program that is unchanged from individual to individual. Adult nematodes are about 1 mm long and mature from a fertilized egg in about two days (Figure 20–24). The life cycle consists of an embryonic stage (about 16 hours), four larval stages (L1 through L4), and the adult stage. Adults are of two sexes: XX self-fertilizing hermaphrodites that can make both eggs and sperm, and XO males. Self-fertilization of mutagen-treated hermaphrodites quickly results in homozygous stocks of mutant strains, and hundreds of mutants have been generated, catalogued, and mapped.

Adult hermaphrodites have 959 somatic cells (and about 2000 germ cells). The exact cell lineage from fertilized egg to adult has been mapped (Figure 20–25) and is invariant from individual to individual. Knowing the lineage of each cell, we can easily follow events resulting from mutations that alter cell fate or from

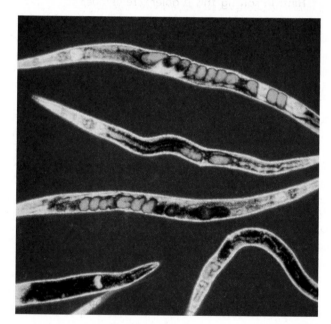

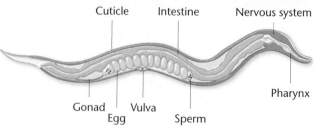

FIGURE 20–24 An adult *Caenorhabditis elegans* hermaphrodite. This nematode, about 1 mm in length, consists of 959 cells and has been used to study many aspects of the genetic control of development.

FIGURE 20–25 A truncated cell lineage chart for *C. elegans*, showing early divisions and the tissues and organs. Each vertical line represents a cell division, and horizontal lines connect the two cells produced. For example, the first cell division creates two new cells from the zygote, AB and P1. The cells in this chart refer to those present in the first-stage larva L1. During subsequent larval stages, further cell divisions will produce the 959 somatic cells of the adult hermaphrodite worm.

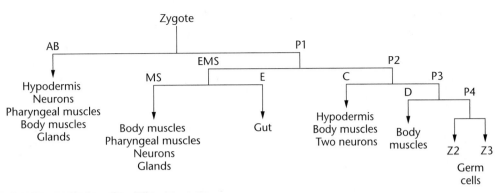

killing of cells with laser microbeams or ultraviolet irradiation. In *C. elegans* hermaphrodites, the fate of cells in the development of the reproductive system is determined by cell–cell interaction, giving us insight into how gene expression and cell–cell interaction work together to specify developmental outcomes.

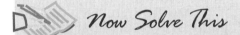

Now Solve This

Problem 26 on page 469 involves two genes that control sex determination in *C. elegans*. You are asked to analyze a model of sex determination.

Hint: In solving this problem, remember to consider the action of gene products and the effect of loss-of-function mutations on expression of other genes or the action of other proteins.

Genetic Analysis of Vulva Formation

C. elegans adult hermaphrodites lay eggs through the vulva, an opening located about midbody (Figure 20–24). The vulva is formed in stages during larval development via several rounds of cell–cell interactions.

In *C. elegans*, two developmentally equivalent neighboring cells, Z1.ppp and Z4.aaa, interact with each other so that one becomes the gonadal anchor cell and the other becomes a precursor to the uterus. The determination of which cell becomes which occurs during the second larval stage (L2) and is controlled by the Notch receptor gene, *lin-12*. In recessive *lin-12(0)* mutants (a loss-of-function mutant), both cells become anchor cells. The dominant mutation *lin-12(d)* (a gain-of-function mutation) causes both to become uterine precursors. Thus, it appears that the expression of the *lin-12* gene causes the selection of the uterine pathway, since in the absence of the LIN-12 (Notch) receptor, both cells become anchor cells.

As shown in Figure 20–26, however, the situation is more complex than it first appears. Initially, the two neighboring cells are developmentally equivalent. Each synthesizes low levels of the Notch signal protein (encoded by the *lag-2* gene), *and* the Notch receptor protein. This situation is unstable, and both cells cannot continue simultaneously sending and receiving developmental signals. By chance, the cell secreting more of the signal (LAG-2 or Delta protein) causes the neighboring cell to increase production of the receptor (LIN-12 protein). The cell producing more

receptor protein becomes the uterine precursor, and the other cell, producing more signal protein, becomes the anchor cell. The critical factor in this first round of cell–cell interaction is the balance between the LAG-2 (Delta) signal gene product and the LIN-12 (Notch) gene product.

A second round of cell–cell communication leads to formation of the vulva. This interaction involves the anchor cell (located in the gonad) and six precursor cells (located in the skin) adjacent to the gonad. The precursor cells are named P3.p to P8.p and collectively are called Pn.p cells. The fate of each Pn.p cell is

(a)

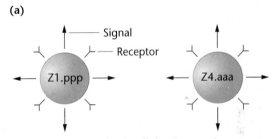

During L2, both cells begin secreting signal for uterine differentiation

(b)

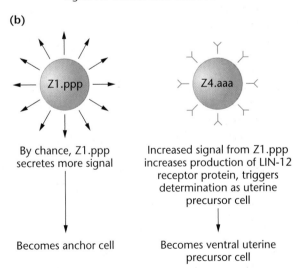

By chance, Z1.ppp secretes more signal

Becomes anchor cell

Increased signal from Z1.ppp increases production of LIN-12 receptor protein, triggers determination as uterine precursor cell

Becomes ventral uterine precursor cell

FIGURE 20–26 Cell–cell interaction in anchor cell determination. (a) During L2, two neighboring cells begin the secretion of chemical signals for the induction of uterine differentiation. (b) By chance, cell Z1.ppp produces more of these signals, causing cell Z4.aaa to increase production of the receptor for signals. The action of increased signals causes Z4.aaa to become the ventral uterine precursor cell and allows Z1.ppp to become the anchor cell.

specified by its position relative to the anchor cell. Figure 20–27 shows the developmental pathway described in the following paragraphs.

In larval stage 3, the *lin-3* gene is activated in the anchor cell. The gene product, LIN-3, is a signal protein related to vertebrate epidermal growth factor (EGF). All six precursor cells express a receptor encoded by *let-23*, a gene homologous to the vertebrate EGF receptor. The binding of LIN-3 to the LET-23 receptor triggers an intracellular cascade of events that determines whether the precursor cells will form the primary vulval precursor cell or secondary vulval cells. In other words, *let-23* establishes the primary and secondary fates of precursor cells. Recessive loss-of-function *let-23* mutations cause all Pn.p cells to act as if they have not received any signal, and as a result, no vulva forms.

Normally, the cell closest to the anchor cell (P6.p) receives the strongest signal initiated by LIN-3 binding to LET-23. This signal activates expression of the *Vulvaless (Vul)* gene (the gene is named for its mutant phenotype) in P6.p, and this cell becomes the primary vulval precursor cell—it then divides three times to produce vulva cells. The two neighboring cells (P5.p and P7.p) receive a lower amount of signal and initiate a secondary fate. These cells divide asymmetrically to form additional vulva cells.

To reinforce these developmental pathways, a third level of cell–cell interaction is used. In this pathway, the primary vulval cell (P6.p) activates *lin-12* in the two neighboring cells (P5.p and P7.p). This signal prevents P5.p and P7.p from adopting the division pattern of the primary cell. In other words, cells in which both *Vul* and *lin-12* are active cannot become primary vulva cells. The three remaining precursor cells (P3.p, P4.p, and P8.p) receive no signal from the anchor cell. In these cells the *Multivulva (Muv)* gene is expressed, *Muv* represses *Vul*, and the three cells develop as skin cells.

Thus, three levels of cell–cell interactions are used in the developmental pathway leading to vulva formation in *C. elegans*. First, two neighboring cells interact to establish the identity of the anchor cell. Second, the anchor cell interacts with three vulval precursor cells to establish the identity of the primary vulval precursor cell (usually P6.p) and two secondary cells (P5.p and P7.p). Third, the primary vulval cell interacts with the secondary cells to suppress their ability to adopt the pathway of the primary cell. Each interaction is accompanied by production of molecular signals and by reception and processing of these signals in neighboring cells.

Notch Signaling Systems in Humans

In humans, there are four Notch genes encoding receptors that play important roles in defining borders during segmentation and generating cells of the immune system and the bone marrow stem cells. Mutations in the Notch signaling pathway are responsible for several inherited disorders, including Alagille syndrome. This autosomal dominant disorder produces developmental abnormalities of the liver, the heart and circulatory system, and the skeleton.

This theme of cell–cell interactions acting in a spatial and temporal cascade to specify the developmental fates of individual cells is a developmental theme repeated over and over in organisms from prokaryotes to humans.

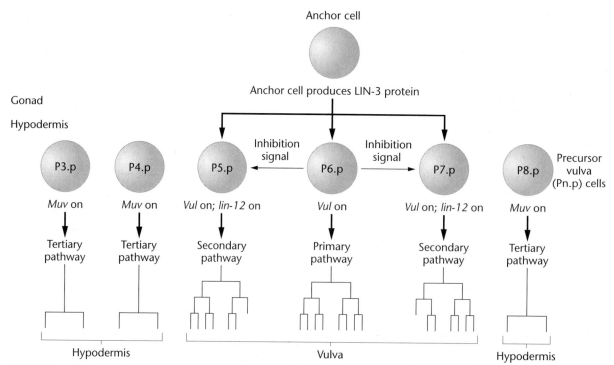

FIGURE 20–27 Cell lineage determination in *C. elegans* vulva formation. A signal from the anchor cell in the form of LIN-3 protein is received by three precursor vulval cells (Pn.p cells). The cells closest to the anchor cell become primary vulval precursor cells, and adjacent cells become secondary precursor cells. Primary cells produce a signal that activates the *lin-12* gene in secondary cells, preventing them from becoming primary cells. Flanking precursor cells, which receive no signal from the anchor cell, increase activity of the *Muv* gene and become skin (hypodermis) cells, instead of vulval cells.

GENETICS, TECHNOLOGY, AND SOCIETY

Stem Cell Wars

Stem cell research is at the center of a battle fought by scientists, politicians, advocacy groups, religious leaders, and ethicists. Proponents fight for the right to carry out stem cell research, claiming that it is revolutionary, if not miraculous—with the potential to cure diabetes, Parkinson disease, and spinal cord injuries, and also improve the quality of life for millions. Critics lobby for an end to stem cell research, warning that it will propel us down the slippery slope toward disregard for human life. Although stem cell research is the focus of presidential proclamations, highly publicized media campaigns, and legislative bans, few of us understand it sufficiently to evaluate its pros and cons. What is stem cell research, and why does it spark such intense controversy?

Stem cells are primitive cells that replicate indefinitely and have the unique capacity to differentiate into cells with specialized functions, such as those found in the heart, brain, liver, and muscle. Stem cells are the origin of all the cells that make up the approximately 200 distinct types of tissues in our bodies. In contrast to stem cells, mature, fully differentiated cells do not replicate or undergo transformations into different cell types. Some types of stem cells are defined as *totipotent*, meaning that they have the ability to differentiate into any mature cell type in the body. Other types of stem cells are pluripotent and are able to differentiate into several mature cell types. Still others are unipotent, and can form only one cell type.

In the last few years, several research teams have isolated and cultured human pluripotent stem cells. These cells remain undifferentiated and grow indefinitely in culture dishes. When treated with growth factors or hormones, these pluripotent stem cells differentiate into cells that have characteristics of neural, bone, kidney, liver, heart, or pancreatic cells.

The fact that pluripotent stem cells grow prolifically in culture and differentiate into more specialized cells has created great excitement. Some foresee a day when stem cells may be a cornucopia from which to harvest unlimited numbers of specialized cells to replace those in damaged and diseased tissues. Hence, stem cells could be used to treat Parkinson disease, type 1 diabetes, chronic heart disease, kidney and liver failure, Alzheimer disease, Duchenne muscular dystrophy, and spinal cord injuries. Some predict that stem cells will be genetically modified to eliminate transplant rejection or to deliver specific gene products, thereby correcting genetic defects or treating cancers.

The excitement about stem cell therapies has been fueled by reports of dramatically successful experiments in animals. For example, mice with spinal cord injuries regained their mobility and bowel and bladder control after they were injected with human stem cells. In 2005, Korean researchers claimed to have established patient-specific stem cell lines genetically matched to patients with spinal cord injuries, diabetes, and immune system disorders. The 2006 announcement that the publication describing this work was retracted and that the results were falsified has only intensified the furor about stem cell research.

Both proponents and critics of stem cell research agree that stem cell therapies could be revolutionary. Why, then, should stem cell research be so contentious? The answer to that question lies in the source of pluripotent stem cells. To date, all pluripotent stem cell lines have been derived from five-day embryonic blastocysts. Blastocysts at this stage consist of about 2110 cells, most of which will develop into placental and supporting tissues for the early embryo. The inner cell mass of the blastocyst consists of about 30 to 40 pluripotent stem cells that develop into all tissues of the embryo. *In vitro* fertilization clinics grow fertilized eggs to the five-day blastocyst stage prior to uterine transfer. Embryonic stem (ES) cell lines are created by dissecting out the inner cell mass of five-day blastocysts and growing the undifferentiated cells in culture dishes. All human ES cell lines have been derived from unused five-day blastocysts that were discarded by *in vitro* fertilization clinics.

The fact that early embryos are destroyed in the process of establishing human ES cell lines disturbs people who believe that preimplantation embryos are persons with rights; however, it does not disturb people who believe that these embryos are too primitive to have an inherent moral status. Both sides in the debate put forth lengthy arguments revolving around the fundamental question of what constitutes a human being.

Critics of ES cell research argue that we may be able to benefit from stem cell therapies without resorting to the use of ES cells. This argument is based on recent reports about the plasticity of adult stem cells. Adult stem cells are undifferentiated cells present in differentiated tissues such as blood and brain. They divide within the differentiated tissue and differentiate into mature cells that make up the tissue in which they are found. Adult stem cells have been found in bone marrow, blood, the retina, the brain, skeletal muscle, the liver, skin, and the pancreas. The best-known adult stem cells are hematopoietic stem cells (HSCs) which are found in bone marrow, peripheral blood, and umbilical cords. HSCs differentiate into mature blood cell types such as red blood cells, lymphocytes, and macrophages. HSCs have been used clinically for many years, as transplant material to reconstitute the immune systems of patients undergoing treatment for cancer and autoimmune diseases. Interestingly, recent studies suggest that adult stem cells may have the capacity to differentiate into other cell types.

Although these reports are intriguing, it is still too early to know whether adult stem cells will hold the same pluripotent promise as ES cells. Adult stem cells are rare, difficult to identify and isolate, and grow poorly, if at all, in culture. However, if these obstacles can be overcome, adult stem cells may provide an ethical alternative to ES cells and calm the raging debate. On the other hand, new philosophical dilemmas could be created. If adult stem cells are found to exhibit the same pluripotency as ES cells, they could have the same potential to create a human embryo—dragging critics and proponents of stem cell research back into the same moral quagmire. At the present time, it is impossible to predict whether either adult or embryonic stem cells will be as miraculous as predicted by scientists and the popular press. But if stem cell research

(Cont. on the next page)

progresses at its current rapid pace, we won't have long to wait.

References

Chong, S., and Normile, D. 2006. How young Korean researchers helped unearth a scandal. *Science* 311: 22–25.

Freed, C. R. 2002. Will embryonic stem cells be a useful source of dopamine neurons for transplant into patients with Parkinson's disease? *Proc. Nad. Acad. Sci. (USA)* 99: 1755–1757.

Robertson. J. A. 2001. Human embryonic stem cell research: ethical and legal issues. *Nat. Rev. Gen.* 2: 74–78.

Web Sites

National Institutes of Health. 2001. "Stem Cells: Scientific Progress and Future Research Directions."

http://stemcells.nih.gov/info/scireport/ (or for the .pdf version: **http://stemcells.nih.gov/info/scireport/PDFs/fullrptstem.pdf**)

National Institutes of Health. 2004. "Stem Cell Information." **http://stemcells.nih.gov/index.asp**

CHAPTER SUMMARY

1. The role of genetic information during development and differentiation is one of the major areas of study in biology. Geneticists are exploring this topic by isolating developmental mutations and identifying the genes involved in controlling developmental processes.

2. Determination is the regulatory event whereby cell fate becomes fixed during early development. Determination precedes the actual differentiation or specialization of distinctive cell types.

3. During embryogenesis, specific gene activity is controlled by the internal environment of the cell, including localized cytoplasmic components. In flies, the regulation of early events is mediated by the maternal cytoplasm, which then influences zygotic gene expression. As development proceeds, both the cell's internal environment and its external environment become further altered by the presence of early gene products and communication with other cells.

4. In *Drosophila*, both genetic and molecular studies have confirmed that the egg contains information that specifies the body plan of the larva and adult and that interactions of embryonic nuclei with the maternal cytoplasm initiate transcriptional cascades that progressively divide the embryo into segments and assign segmental identities to the developing body plan.

5. Extensive genetic analysis of embryonic development in *Drosophila* has led to the identification of maternal-effect genes that lay down the anterior–posterior axis of the embryo. In addition, these maternal-effect genes activate sets of zygotic segmentation genes, initiating a cascade of gene regulation that ends with the determination of segment identity by the homeotic selector genes. These same gene sets control aspects of embryonic development in all bilateral animals, including humans.

6. Flower formation in *Arabidopsis* is controlled by homeotic genes, but these gene sets are from a different gene family than the homeotic selector genes of *Drosophila* and other animals.

7. In *C. elegans*, the stereotyped lineage of all cells allows developmental biologists to study the cell–cell signaling required for organogenesis.

KEY TERMS

determination, 450

differentiation, 450

gap genes, 453

homeobox, 456

homeodomain, 456

homeotic selector (*Hox*) genes, 453

homeotic mutants, 456

MADS-box proteins, 461

Notch signaling pathway, 462

pair-rule genes, 454

segment polarity genes, 454

specification, 450

segmentation genes, 452

INSIGHTS AND SOLUTIONS

1. In the slime mold *Dictyostelium*, experimental evidence suggests that cyclic AMP (cAMP) plays a central role in the developmental program leading to spore formation. The genes encoding the cAMP cell-surface receptor have been cloned, and the amino acid sequence of the protein components is known. To form reproductive structures, free-living individual cells aggregate together and then differentiate into one of two cell types, prespore cells or prestalk cells. Aggregating cells secrete waves or oscillations of cAMP to foster the aggregation of cells and then continuously secrete cAMP to activate genes in the aggregated cells at later stages of development. It has been proposed that cAMP controls cell–cell interaction and gene expression. It is important to test this hypothesis by using several experimental techniques. What different approaches can you devise to test this hypothesis, and what specific experimental systems would you employ to test them?

Solution: Two of the most powerful forms of analysis in biology involve the use of biochemical analogs (or inhibitors) to block gene transcription or the action of gene products in a predictable way, and the use of mutations to alter genes and their products. These two approaches can be used to study the role of cAMP in the developmental program of *Dictyostelium*. First, compounds chemically related to cAMP, such as GTP and GDP, can be used to test whether they have any effect on the processes controlled by cAMP. In fact, both GTP and GDP lower the affinity of cell-surface receptors for cAMP, effectively blocking the action of cAMP. To inhibit the synthesis of the cAMP receptor, it is possible to construct a vector containing a DNA sequence that transcribes an antisense RNA (a molecule that has a base sequence complementary to the mRNA). Antisense RNA forms a double-stranded structure with the mRNA, preventing it from being transcribed. If normal cells are transformed with a vector that expresses antisense RNA, no cAMP receptors will be produced. It is possible to predict that such cells will fail to respond to a gradient of cAMP and, consequently, will not migrate to an aggregation center. In fact, that is what happens. Such cells remain dispersed and nonmigratory in the presence of cAMP. Similarly, it is possible to determine whether this response to cAMP is necessary to trigger changes in the transcriptional program by assaying for the expression of developmentally regulated genes in cells expressing this antisense RNA.

Mutational analysis can be used to dissect components of the cAMP receptor system. One approach is to use transformation with wild-type genes to restore mutant function. Similarly, because the genes for the receptor proteins have been cloned, it is possible to construct mutants with known alterations in the component proteins and transform them into cells to assess their effects.

2. In the sea urchin, early development may occur even in the presence of actinomycin D, which inhibits RNA synthesis. However, if actinomycin D is present early in development but removed a few hours later, all development stops. In fact, if actinomycin D is present only between the sixth and eleventh hours of development, events that normally occur at the fifteenth hour are arrested. What conclusions can be drawn concerning the role of gene transcription between hours 6 and 15?

Solution: Maternal mRNAs are present in the fertilized sea urchin egg. Thus, a considerable amount of development can take place without transcription of the embryo's genome. Because development past fifteen hours is inhibited by prior treatment with actinomycin D, it appears that transcripts from the embryo's genome are required to initiate or maintain these events. This transcription must take place between the sixth and fifteenth hours of development.

3. If it were possible to introduce one of the homeotic genes from *Drosophila* into an *Arabidopsis* embryo homozygous for a homeotic flowering gene, would you expect any of the *Drosophila* genes to negate (rescue) the *Arabidopsis* mutant phenotype? Why or why not?

Solution: The *Drosophila* homeotic genes belong to the *Hox* gene family, while *Arabidopsis* homeotic genes belong to the MADS-box protein family. Both gene families are present in *Drosophila* and *Arabidopsis*, but they have evolved different functions in the animal and the plant kingdoms. As a result, it is unlikely that a transferred *Drosophila* Hox gene would rescue the phenotype of a MADS-box mutant, but only an actual experiment would confirm this.

■ PROBLEMS AND DISCUSSION QUESTIONS

1. Carefully distinguish between the terms *differentiation* and *determination*. Which phenomenon occurs initially during development?

2. Nuclei from almost any source may be injected into *Xenopus* oocytes. Studies have shown that these nuclei remain active in transcription and translation. How can such an experimental system be useful in developmental genetic studies?

3. Distinguish between the syncytial blastoderm stage and the cellular blastoderm stage in *Drosophila* embryogenesis.

4. (a) What are maternal-effect genes? (b) When are gene products from these genes made, and where are they located? (c) What aspects of development do maternal-effect genes control? (d) What is the phenotype of maternal-effect mutations?

5. Suppose you initiate a screen for maternal-effect mutations in *Drosophila* affecting external structures of the embryo and your screen identifies more than 100 mutations that affect external structures. From their screening, Weischaus and Schüpbach concluded that there are about 40 maternal-effect genes. How do you reconcile these different results?

6. (a) What are zygotic genes, and when are their gene products made? (b) What is the phenotype associated with zygotic gene mutations? Does the maternal genotype contain zygotic genes?

7. List the main classes of zygotic genes. What is the function of each class of these genes?

8. Experiments have shown that any nuclei placed in the polar cytoplasm at the posterior pole of the *Drosophila* egg will differentiate into germ cells. If polar cytoplasm is transplanted into the anterior end of the egg just after fertilization, what will happen to nuclei that migrate into this cytoplasm at the anterior pole?

9. How can you determine whether a particular gene is being transcribed in different cell types?

10. You observe that a particular gene is being transcribed during development. How can you tell whether the expression of this gene is under transcriptional or translational control?

11. What are *Hox* genes? What properties do they have in common? Are all homeotic genes *Hox* genes?

12. The homeotic mutation *Antennapedia* causes mutant *Drosophila* to have legs in place of antennae and is a dominant gain-of-function mutation. What are the properties of such mutations? How does the *Antennapedia* gene change antennae into legs?

13. The *Drosophila* homeotic mutation *spineless aristapedia* (ss^a) results in the formation of a miniature tarsal structure (normally part of the leg) on the end of the antenna. What insight is provided by (ss^a) concerning the role of genes during determination?

14. Embryogenesis and oncogenesis (generation of cancer) share a number of features including cell proliferation, apoptosis, cell migration and invasion, formation of new blood vessels, and differential gene activity. Embryonic cells are relatively undifferentiated, and cancer cells appear to be undifferentiated or dedifferentiated. Homeotic gene expression directs early development, and mutant expression leads to loss of the differentiated state or an alternative cell identity. M. T. Lewis (2000. *Breast Can. Res.* 2: 158–69) suggested that breast cancer may be caused by the altered expression of homeotic genes. When he examined 11 such genes in cancers, eight were underexpressed while three were overexpressed compared with controls. Given what you know about homeotic genes, could they be involved in oncogenesis?

15. In *Drosophila*, both *fushi tarazu* (*ftz*) and *engrailed (eng)* genes encode homeobox transcription factors and are capable of eliciting the expression of other genes. Both genes work at about the same time during development and in the same region to specify cell fate in body segments. To discover if *ftz* regulates the expression of *engrailed*; if *engrailed* regulates *ftz*; or if both are regulated by another gene, you perform a mutant analysis. In *ftz*⁻ embryos (*ftz/ftz*) engrailed protein is absent; in *engrailed*⁻ embryos (*eng/eng*) *ftz* expression is normal. What does this tell you about the regulation of these two genes—does the *engrailed* gene regulate *ftz*, or does the *ftz* gene regulate *engrailed*?

16. Early development depends on the temporal and spatial interplay between maternally-supplied material and mRNA and the onset of zygotic gene expression. Maternally-encoded mRNAs must be produced, positioned, and degraded (Surdej and Jacobs-Lorena, 1998. *Mol. Cell Biol.* 18: 2892–2900). For example, transcription of the *bicoid* gene that determines anterior–posterior polarity in *Drosophila* is maternal. The mRNA is synthesized in the ovary by nurse cells and then transported to the oocyte, where it localizes to the anterior ends of oocytes. After egg deposition, *bicoid* mRNA is translated and unstable bicoid protein forms a decreasing concentration gradient from the anterior end of the embryo. At the start of gastrulation, *bicoid* mRNA has been degraded. Consider two models to explain the degradation of *bicoid* mRNA: (1) degradation may result from signals within the mRNA (intrinsic model), or (2) degradation may result from the mRNA's position within the egg (extrinsic model). Experimentally, how could one distinguish between these two models?

17. Formation of germ cells in *Drosophila* and many other embryos is dependent on their position in the embryo and their exposure to localized cytoplasmic determinants. Nuclei exposed to cytoplasm in the posterior end of *Drosophila* eggs (the pole plasm) form cells that develop into germ cells under the direction of maternally derived components. R. Amikura et al. (2001. *Proc. Nat. Acad. Sci. (USA)* 98: 9133–9138) consistently found mitochondria-type ribosomes outside mitochondria in the germ plasma of *Drosophila* embryos and postulated that they are intimately related to germ cell specification. If you were studying this phenomenon, what would you want to know about the activity of these ribosomes?

18. One of the most interesting aspects of early development is the remodeling of the cell cycle from rapid cell divisions, apparently lacking G1 and G2 phases, to slower cell cycles with measurable G1 and G2 phases and checkpoints. During this remodeling, maternal mRNAs that specify cyclins are deadenylated, and zygotic genes are activated to produce cyclins. Audic et al. (2001. *Mol. and Cell. Biol.* 21: 1662–1671) suggest that deadenylation requires transcription of zygotic genes. Present a diagram that captures the significant features of these findings.

19. In studying gene action during development, it is desirable to be able to position genes in a hierarchy or pathway of action to establish which genes are primary and in what order genes act. There are several ways of doing this. One is to make double mutants and study the outcome. The gene *fushi-tarazu* (*ftz*) is expressed in early embryos at the seven-stripe stage. All of the genes involved in forming the anterior–posterior pattern affect the expression of this gene, as do the *gap* genes. However, expression of segment-polarity genes is affected by *ftz*. What is the location of *ftz* in this hierarchy?

20. A number of genes that control expression of *Hox* genes in *Drosophila* have been identified. One of these homozygous mutants is *extra sex combs*, where some of the head and all of

the thorax and abdominal segments develop as the last abdominal segment. In other words, all affected segments develop as posterior segments. What does this phenotype tell you about which set of *Hox* genes is controlled by the *extra sex combs* gene?

21. The *apterous* gene in *Drosophila* encodes a protein required for wing patterning and growth. It is also known to function in nerve development, fertility, and viability. When human and mouse genes whose protein products closely resemble *apterous* were used to generate transgenic *Drosophila* (Rincon-Limas et al. 1999. *Proc. Nat. Acad. Sci. [USA]* 96: 2165–2170), the *apterous* mutant phenotype was *rescued*. In addition, the whole-body expression patterns in the transgenic *Drosophila* were similar to normal *apterous*. (a) What is meant by the term *rescued* in this context? (b) What do these results indicate about the molecular nature of development?

22. In *Arabidopsis*, flower development is controlled by sets of homeotic genes. How many classes of these genes are there, and what structures are formed by their individual and combined expression?

23. The floral homeotic genes of *Arabidopsis* belong to the MADS-box gene family, while in *Drosophila*, homeotic genes belong to the homeobox gene family. In both *Arabidopsis* and *Drosophila*, members of the *Polycomb* gene family control expression of these divergent homeotic genes. How do *Polycomb* genes control expression of two very different sets of homeotic genes?

24. Vulval development in *C. elegans* begins when two neighboring cells (Z1.ppp and Z4.aaa) interact with each other by cell–cell signaling involving two components: a membrane-bound signal molecule and a membrane-bound receptor. By chance, one cell produces more signal and causes its neighbor to produce more receptor. The signal-producing cell becomes the anchor cell, and the receptor-producing cell becomes the uterine precursor. This form of cell–cell interaction is called the Notch/Delta signaling system. Although it is a widely used signaling mechanism in metazoans, this pathway works only in adjacent cells. Why is this so, and what are the advantages and disadvantages of such a system?

25. The identification and characterization of genes that control sex determination has been another focus of investigators working with *C. elegans*. As with *Drosophila*, sex in this organism is determined by the ratio of X chromosomes to sets of autosomes. A diploid wild-type male has one X chromosome, and a diploid wild-type hermaphrodite has two X chromosomes. Many different mutations have been identified that affect sex determination. Loss-of-function mutations in a gene called *her-1* cause an XO nematode to develop into a hermaphrodite and have no effect on XX development. (That is, XX nematodes are normal hermaphrodites.) In contrast, loss-of-function mutations in a gene called *tra-1* cause an XX nematode to develop into a male. Deduce the roles of these genes in wild-type sex determination from this information.

26. Based on the information in Problem 25 and the analysis of the phenotypes of single- and double-mutant strains, a model for sex determination in *C. elegans* has been generated. This model proposes that the *her-1* gene controls sex determination by establishing the level of activity of the *tra-1* gene, which in turn, controls the expression of genes involved in generating the various sexually dimorphic tissues. Given this information, (a) does the *her-1* gene product have a negative or a positive effect on the activity of the *tra-1* gene? (b) What would be the phenotype of a *tra-1*, *her-1* double mutant?

21 Quantitative Genetics

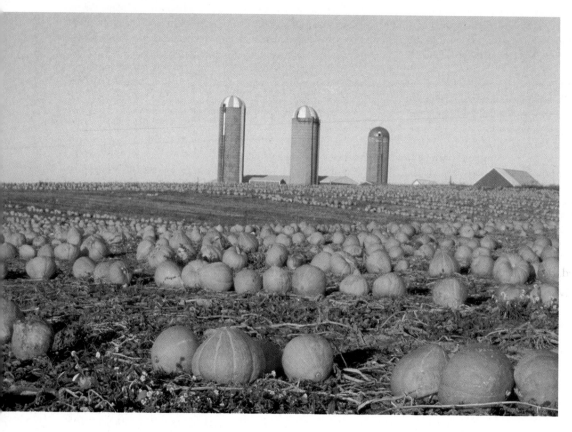

A field of pumpkins, where size is a quantitative trait under the influence of additive alleles.

CHAPTER CONCEPTS

- Quantitative inheritance results in a range of measurable phenotypes for polygenic traits.

- While most polygenic traits show continuous variation, not all do.

- Quantitative traits can be explained in Mendelian terms, whereby certain alleles provide an additive effect on the traits under study.

- The study of polygenic traits relies on statistical analysis.

- Heritability estimates the genetic contribution to phenotypic variability under specific environmental conditions.

- Twin studies allow an estimation of heritability in humans.

- Quantitative trait loci (QTLs) can be identified and mapped.

We have looked at examples of phenotypic variation that can be classified into distinct and separate categories: pea plants were tall or dwarf; squash fruit shape was spherical, disc shaped, or elongated; and fruit fly eye color was red or white. (See Chapter 4.) Traits such as these with a small number of discrete phenotypes show **discontinuous variation**. Typically in these traits, a genotype will produce a single identifiable phenotype, although phenomena such as variable penetrance and expressivity, pleiotropy, and epistasis can obscure the relationship between genotype and phenotype, even for simple discontinuous traits. Other traits, however, including many of medical or agricultural importance, are more complex. They show much more variation with a continuous range of phenotypes that cannot be easily classified into distinct categories. Examples of such traits showing **continuous variation** include human height or weight, milk or meat production in cattle, crop yield, and seed protein content. Continuous variation across a range of phenotypes is measured and described in quantitative terms, so this genetic phenomenon is known as **quantitative inheritance**. Because the varying phenotypes result from the input of genes at multiple loci, quantitative traits are sometimes also said to be **polygenic** (literally "many genes"). For traits showing continuous variation, the genotype generated at fertilization establishes the quantitative range within which a particular individual can fall. However, the final phenotype is often also influenced by environmental factors to which that individual is exposed. Human height, for example, is partly genetically determined, but it is also affected by environmental factors such as nutrition. Those phenotypes that result from gene action and environmental influences are sometimes termed **complex** or **multifactorial traits**.

In this chapter, we will examine examples of quantitative inheritance and some of the statistical techniques used to study complex traits. We will also consider how geneticists assess the relative importance of genetic versus environmental factors contributing to continuous phenotypic variation, and we will discuss approaches to identifying and mapping genes that influence quantitative traits.

How Do We Know?

In this chapter, we will focus on traits that exhibit quantitative phenotypic variation and that are under the genetic control of alleles whose influence is additive in nature. As you study this topic, you should try to answer several fundamental questions:

1. What observations made it apparent to early geneticists that quantitative traits must be under a different mode of inheritance than the more qualitative traits studied by Mendel?

2. What findings led geneticists to postulate the multiple-factor hypothesis that invoked the idea of additive alleles to explain inheritance patterns?

3. How do we know how many genes control a quantitative trait?

4. How do we assess environmental factors to determine if they impact the phenotype of a quantitatively inherited trait?

21.1 Not All Polygenic Traits Show Continuous Variation

When a trait demonstrates continuous variation in a population, individual measurements may form a gradation of phenotypes with no clear categories. Such a trait is most often the result of polygenic inheritance, and as we established above, polygenic traits are frequently multifactorial, with environmental factors contributing to the range of phenotypes observed.

In addition to continuous quantitative traits where phenotypic variation can fall at any point along a scale of measurement, there are two other classes of polygenic traits. **Meristic traits** are those in which the phenotypes are recorded by counting whole numbers. Examples of meristic traits include the number of seeds in a pod or the number of eggs laid by a chicken in a year. These are quantitative traits, but they do not have an infinite range of phenotypes. For example, a pod may contain 2, 4, or 6 seeds but not 5.75. **Threshold traits** are polygenic (and frequently environmental factors affect the phenotypes, so they are also multifactorial), but they are distinguished from continuous and meristic traits by having a small number of discrete phenotypic classes. Threshold traits are of interest to human geneticists as an increasing number of diseases are now suspected of showing this pattern of polygenic inheritance. One example is **Type II diabetes**, also known as adult-onset diabetes because it typically affects individuals who are middle aged or older. A population can be divided into just two phenotypic classes for this trait—individuals who have Type II diabetes and those who do not—so at first glance this may appear to more closely resemble a simple monogenic trait. However, no single adult-onset diabetes gene has been identified. Instead, the combination of alleles present at multiple contributing loci gives an individual a greater or lesser likelihood of developing the disease. These varying levels of liability form a continuous range. At one extreme are those at very low risk for Type II diabetes, while at the other end of the distribution are those whose genotypes make it highly likely they will develop the disease (Figure 21–1). As with many threshold traits, environmental factors also play a role in determining the final phenotype, with diet and lifestyle having significant impact on whether an individual with moderate to high genetic liability will actually develop Type II diabetes.

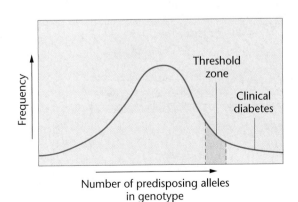

FIGURE 21–1 Illustration of a threshold trait—Type II diabetes.

Corolla

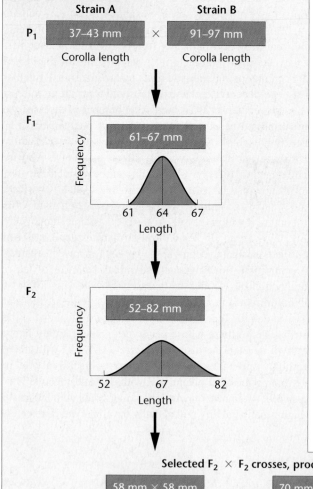

21.2 Quantitative Traits Can Be Explained in Mendelian Terms

The issue of whether continuous phenotypic variation could be explained in Mendelian terms caused considerable controversy in the early 1900s. Some scientists argued that, although Mendel's unit factors or genes explained patterns of discontinuous segregation with discrete phenotypic classes, they could not also account for the range of phenotypes seen in quantitative patterns of inheritance. However, geneticists William Bateson and George Udny Yule, adhering to a Mendelian explanation, proposed the **multiple-factor** or **multiple-gene hypothesis** in which many genes, each individually behaving in a Mendelian fashion, contribute to the phenotype in a *cumulative* or *quantitative* way.

The Multiple-Gene Hypothesis for Quantitative Inheritance

By 1920, the conclusions of several critical sets of experiments largely resolved the controversy and demonstrated that Mendelian factors could account for continuous variation. In one experiment, Edward M. East crossed two strains of the tobacco plant *Nicotiana longiflora*. The fused inner petals of the flower, or corollas, of strain A were decidedly shorter than the corollas of strain B. With only minor variation, each strain was true-breeding. Thus, the differences between them were clearly under genetic control.

When plants from the two strains were crossed, the F_1, F_2, and selected F_3 data demonstrated a distinct pattern (Figure 21–2). The F_1 generation displayed corollas that were intermediate in length compared with the P_1 varieties, and showed only minor variability among individuals. While corolla lengths of the P_1 plants were about 40 mm and 94 mm, the F_1

FIGURE 21–2 The F_1, F_2, and selected F_3 results of East's cross between two strains of *Nicotiana* with different corolla lengths. Plants of strain A vary from 37 to 43 mm, while plants of strain B vary from 91 to 97 mm. The photograph shows the flower and corolla of a tobacco plant.

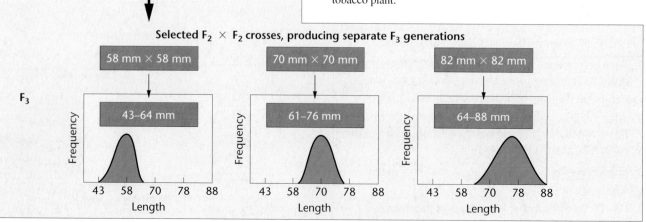

generation contained plants with corollas that were all about 64 mm. In the F_2 generation, lengths varied much more, ranging from 52 to 82 mm. The majority of individuals resembled their F_1 parents (64 mm), and as the deviation from this length increased, fewer and fewer plants were observed. When the data were plotted graphically (frequency versus length), a bell-shaped curve resulted.

East further experimented with this population by selecting F_2 plants of various corolla lengths and allowing them to produce separate F_3 generations. Several are shown in Figure 21–2. In each case, a bell-shaped distribution was observed, with most individuals similar in length to the selected F_2 parents but with considerable variation around this value.

East's experiments demonstrated that, although the variation in corolla length seemed continuous, experimental crosses resulted in the segregation of distinct phenotypic classes as observed in the three independent F_3 categories. This key finding was the basis for the multiple-factor hypothesis, which explains how traits can deviate considerably in their expression.

The multiple-gene hypothesis was strengthened by another key set of experimental results published by Hermann Nilsson-Ehle. He used grain color in wheat to test the concept that the cumulative effects of alleles at multiple loci produce the range of phenotypes seen in quantitative traits. In one set of experiments, wheat with red grain was crossed to wheat with white grain (Figure 21–3). The F_1 generation demonstrated an intermediate pink color, which at first sight suggested incomplete dominance of two alleles at a single locus. However, in the F_2, Nilsson-Ehle did not observe the 3:1 segregation typical of a monhybrid cross. Instead, approximately 15/16 of the plants showed some degree of red grain color, while 1/16 of the plants showed white grain color. Careful examination of the F_2 revealed that grain with color could be classified into four different shades of red. Because the F_2 ratio occurred in sixteenths, it appears that two genes, each with two alleles, control the phenotype and that they segregate independently from one another in a Mendelian fashion.

If each gene has one potential **additive allele** that contributes to the red grain color and one potential **nonadditive allele** that fails to produce any red pigment, we can see how the multiple-factor hypothesis could account for the various grain color phenotypes. In the P_1, both parents were homozygous; the red parent contains only additive alleles (*AABB* in Figure 21–3), while the white parent contains only nonadditive alleles (*aabb*). The F_1 is heterozygous (*AaBb*), contains two additive (*A* and *B*) and two nonadditive (*a* and *b*) alleles, and expresses the intermediate pink phenotype. In the F_2, each offspring has 4, 3, 2, 1, or 0 additive alleles. F_2 plants with no additive alleles are white (*aabb*) like one of the P_1 parents, while F_2 plants with 4 additive alleles are red (*AABB*) like the other P_1 parent. Plants with 3, 2, or 1 additive alleles constitute the other three categories of red color observed in the F_2. The greater the number of additive alleles in the genotype, the more intense the red color expressed in the phenotype, as each additive allele present contributes equally to the cumulative amount of pigment produced in the grain.

Nilsson-Ehle's results showed how continuous variation could still be explained in a Mendelian fashion, with additive alleles at multiple loci influencing the phenotype in a quantitative manner but each individual allele segregating according to Mendelian rules. As we saw in Nilsson-Ehle's initial cross, if two loci, each with two alleles, were involved, then five F_2 phenotypic categories in a 1:4:6:4:1 ratio would be expected. However, there is no reason why

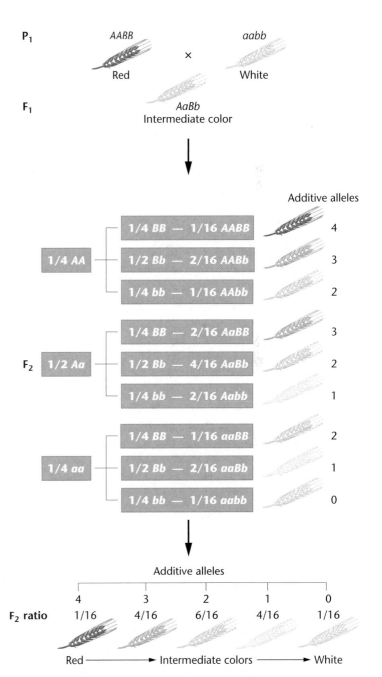

FIGURE 21–3 Illustration of how the multiple-gene hypothesis of polygenic inheritance accounts for the 1:4:6:4:1 phenotypic ratio of grain color when all alleles designated by an uppercase letter are additive and contribute an equal amount of pigment to the phenotype.

three, four, or more loci cannot function in a similar fashion in controlling various quantitative phenotypes. As more quantitative loci become involved, greater and greater numbers of classes appear in the F$_2$ in more complex ratios. The number of phenotypes and the expected F$_2$ ratios for crosses involving up to five gene pairs are illustrated in Figure 21–4.

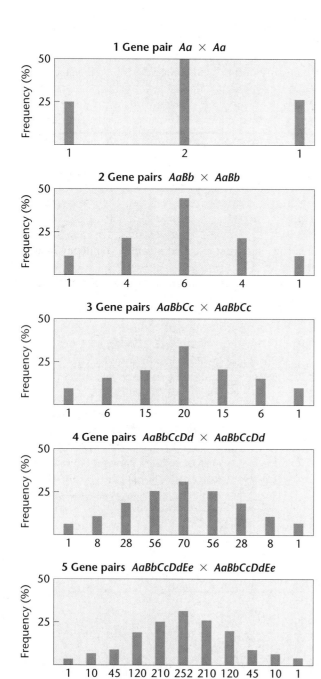

FIGURE 21–4 The results of crossing two heterozygotes when polygenic inheritance is in operation with one to five gene pairs. Each histogram bar indicates a distinct F$_2$ phenotypic class from one extreme (left end) to the other extreme (right end). Each phenotype results from a different number of additive alleles.

Additive Alleles: The Basis of Continuous Variation

The multiple-gene hypothesis embodies the following major points:

1. Phenotypic traits showing continuous variation can be quantified by measuring, weighing, counting, and so on.

2. Two or more gene loci, often scattered throughout the genome, account for the hereditary influence on the phenotype in an *additive way*. Because many genes may be involved, inheritance of this type is called *polygenic*.

3. Each gene locus may be occupied either by an *additive* allele, which contributes a constant amount to the phenotype, or by a *nonadditive* allele, which does not contribute quantitatively to the phenotype.

4. The contribution to the phenotype of each additive allele, though often small, is approximately equal.

5. Together, the additive alleles contributing to a single quantitative character produce substantial phenotypic variation.

Calculating the Number of Polygenes

Various formulas have been developed for estimating the number of polygenes contributing to a quantitative trait. For example, if the ratio of F$_2$ individuals resembling *either* of the two extreme P$_1$ phenotypes can be determined, the number of polygenes involved (*n*) may be calculated as follows:

$$1/4^n = \text{ratio of F}_2 \text{ individuals expressing}$$
$$\text{either extreme phenotype}$$

In the example of the red and white wheat grain color summarized in Figure 21–3, 1/16 of the progeny are either red *or* white like the P$_1$ phenotypes. This ratio can be substituted on the right side of the equation to solve for *n*:

$$\frac{1}{4^n} = \frac{1}{16}$$

$$\frac{1}{4^2} = \frac{1}{16}$$

$$n = 2$$

Table 21.1 lists the ratio and the number of F$_2$ phenotypic classes produced in crosses involving up to five gene pairs.

For low numbers of polygenes, it is sometimes easier to use the $(2n + 1)$ rule. If *n* equals the number of additive loci involved in the trait, then $2n + 1$ will determine the total number of possible phenotypes. For example, if $n = 2$, $2n + 1 = 5$ and each phenotype is the result of 4, 3, 2, 1, or 0 additive alleles. If $n = 3$, $2n + 1 = 7$ and each phenotype is the result of 6, 5, 4, 3, 2, 1, or 0 additive alleles.

It should be noted, however, that both these simple methods for estimating the number of polygenes involved in a quantitative trait assume that all the relevant alleles contribute equally and additively, and also that phenotypic expression in the F$_2$ is not affected significantly by environmental factors. As we will see later, for many quantitative traits, these assumptions may not be true.

TABLE 21.1	Determination of the Number of Gene Pairs (n) Involved in Polygenic Crosses	
n	Individuals Expressing an Extreme Phenotype	Distinctive Phenotypic Classes
1	1/4	3
2	1/16	5
3	1/64	7
4	1/256	9
5	1/1024	11

Now Solve This

Problem 3(a) on page 485 gives F_1 and F_2 ranges for a quantitative trait and asks you to calculate the number of polygenes involved.

Hint: Remember the ratio (not the number) of parental phenotypes reappearing in the F_2 is the key.

21.3 The Study of Polygenic Traits Relies on Statistical Analysis

One of the most important challenges facing genetic researchers working with quantitative traits is determining how much of the phenotypic variation observed in a population is due to genotypic differences among individuals and how much is due to environmental factors. Ways of partitioning variance are discussed in more detail in the next section of this chapter, but first we need to consider the basic statistical tools used with data derived from measuring quantitative phenotypic variation. It is not usually feasible to measure expression of a polygenic trait in every individual in a population, so a random subset of individuals is usually selected for measurement to provide a sample. It is important to remember that the accuracy of the information obtained depends on whether the sample is truly random and representative of the population from which it was drawn. Suppose, for example, that a student wants to determine the average height of the 100 students in his genetics class and for his sample he measures the two students sitting next to him, both of whom happen to be centers on the college basketball team. It is unlikely that this sample will provide a good estimate of the average height of the class for two reasons: first, it is too small; second, it is not a representative subset of the class (unless all 100 students are on the basketball team).

If the sample measured for expression of a quantitative trait is sufficiently large and also representative of the population from which it is drawn, we often find that the data form a

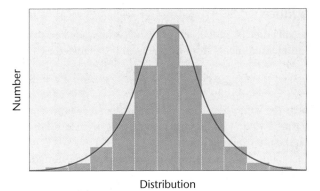
FIGURE 21–5 Normal frequency distribution characterized by a bell-shaped curve.

normal distribution with a characteristic bell-shaped curve when plotted as a frequency histogram (Figure 21–5). Several statistical methods are useful in the analysis of traits that exhibit a normal distribution, including the mean, variance, standard deviation, standard error of the mean, and covariance.

The Mean

The mean provides information about where the central point lies along a range of measurements for a quantitative trait. We can see from Figure 21–6 that the distributions of two sets of phenotypic measurements graphed cluster around a central value. This clustering is called a **central tendency**, and the central point is the **mean** ($\overline{X}$).

The mean is the arithmetic average of a set of measurements and is calculated as

$$\overline{X} = \frac{\Sigma X_i}{n}$$

where $\overline{X}$ is the mean, ΣX_i represents the sum of all individual values in the sample, and n is the number of individual values.

The mean provides a useful descriptive summary of the sample, but it tells us nothing about the range or spread of the data. As illustrated in Figure 21–6, a symmetrical distribution of values in the sample may, in one case, be clustered near the mean. Or a set of measurements may have the same mean but be distributed more widely around it. A second statistic, the variance, provides information about the spread of data around the mean.

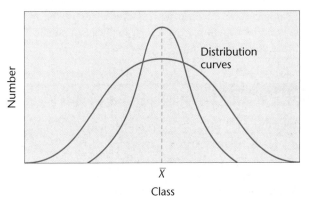
FIGURE 21–6 Two normal frequency distributions with the same mean but different amounts of variation.

Variance

The **variance** (s^2) for a sample is the average squared distance of all measurements from the mean. It is calculated as

$$s^2 = \frac{\Sigma(X_i - \overline{X})^2}{n - 1}$$

where the sum (Σ) of the squared differences between each measured value (X_i) and the mean ($\overline{X}$) is divided by one less than the total sample size $n - 1$.

As Figure 21–6 shows, two sets of sample measurements for a quantitative trait can have the same mean but different variances. Estimation of variance can be useful in determining the degree of genetic control of traits when the immediate environment also influences the phenotype.

Standard Deviation

Because the variance is a squared value, its unit of measurement is also squared (m², g², etc.). To express variation around the mean in the original units of measurement, we can use the square root of the variance, a term called the **standard deviation (s)**:

$$s = \sqrt{s^2}$$

Table 21.2 shows the percentage of individual values within a normal distribution that fall within different multiples of the standard deviation. One standard deviation to either side of the mean includes 68 percent of all values in the sample. More than 95 percent of all values are found within two standard deviations to either side of the mean. This means that the standard deviation s can also be interpreted in the form of a probability. For example, a sample measurement picked at random has a 68 percent probability of falling within the range of one standard deviation on either side of the mean.

Standard Error of the Mean

If multiple samples are taken from a population and measured for the same quantitative trait, we might find that their means vary. Theoretically, larger truly random samples will represent the population more accurately and their means will be closer to each other. To measure the accuracy of the sample mean we use the **standard error of the mean ($S_{\overline{X}}$)**, calculated as

$$S_{\overline{X}} = \frac{s}{\sqrt{n}}$$

where s is the standard deviation and $\sqrt{n}$ is the square root of the sample size. Because the standard error of the mean is computed by dividing s by $\sqrt{n}$, it is always a smaller value than the standard deviation.

TABLE 21.2	Sample Inclusion for Various s Values
Multiples of s	**Sample Included (%)**
$\overline{X} \pm 1s$	68.3
$\overline{X} \pm 1.96s$	95.0
$\overline{X} \pm 2s$	95.5
$\overline{X} \pm 3s$	99.7

Covariance

Often geneticists working with quantitative traits find that they have to consider two phenotypic characters simultaneously. For example, a poultry breeder might investigate the correlation between body weight and egg production in hens: Do heavier birds tend to lay more eggs? The **covariance** statistic measures how much variation is common to both quantitative traits. It is calculated by taking the deviations from the mean for each trait (just as we did for estimating variance) for each individual in the sample. This gives a pair of values that is multiplied together; the sum of all these individual products is then divided by one fewer than the number in the sample. Thus the covariance **cov$_{XY}$** of two sets of trait measurements, X and Y, is calculated as

$$\text{cov}_{XY} = \frac{\Sigma[(X_i - \overline{X})(Y_i - \overline{Y})]}{n - 1}$$

The covariance can be then be standardized as a further statistic, the **correlation coefficient (r)**. The calculation is

$$r = \text{cov}_{XY}/S_X S_Y$$

where S_X is the standard deviation of the first set of quantitative measurements X, and S_Y is the standard deviation of the second set of quantitative measurements Y. Values for the correlation coefficient r can range from -1 to $+1$. Positive values mean that an increase in measurement for one trait tends to be associated with an increase in measurement for the other, while negative r values mean that increases in one trait are associated with decreases in the other. Therefore, if heavier hens do tend to lay more eggs, a positive r value can be expected. A negative r value, on the other hand, suggests that greater egg production is more likely from less heavy birds. One important point to note about correlation coefficients is that r values close to $+1$ or -1 do not imply that a cause-and-effect relationship exists between two traits. Correlation simply tells us the extent to which variation in one quantitative trait is associated with variation in another, not what causes that variation.

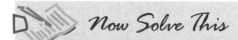 *Now Solve This*

Problem 15 on page 486 provides data for two quantitative traits in a flock of sheep. You are asked to determine if the traits are correlated.

Hint: First calculate the standard deviation for each trait, then the covariance.

Analysis of a Quantitative Character

One homozygous tomato variety produces fruit averaging 18 oz in weight, while fruit from a different homozygous variety averages 6 oz. The F_1 obtained by crossing these two varieties has fruit weights ranging from 10 to 14 oz. The F_2 population contains individuals that produce fruit ranging from 6 to 18 oz. The results characterizing both generations are shown in Table 21.3.

TABLE 21.3	Frequency Distribution of F_1 and F_2 Progeny												
							Weight (oz)						
		6	7	8	9	10	11	12	13	14	15	16	17	18
Number of Individuals	F_1					4	14	16	12	6				
	F_2	1	1	2	0	9	13	17	14	7	4	3	0	1

The mean value for the fruit weight in the F_1 generation can be calculated as

$$\overline{X} = \frac{\Sigma X_i}{n} = \frac{626}{52} = 12.04$$

Similarly, the mean value for fruit weight in the F_2 generation is calculated as

$$\overline{X} = \frac{\Sigma X_i}{n} = \frac{872}{72} = 12.11$$

Although these mean values are similar, the frequency distributions in Table 21.3 show more variation in the F_2 generation. This variation can be quantified as the sample variance s^2, calculated as the sum of the squared differences between each value and the mean, divided by one less than the total number of observations. So

$$s^2 = \frac{\Sigma(X_i - \overline{X})^2}{n - 1}$$

The variance is 1.29 for the F_1 generation and 4.27 for the F_2 generation. When converted to the standard deviation $\left(s = \sqrt{s^2}\right)$, the values become 1.13 and 2.06, respectively. Therefore, the distribution of tomato weight in the F_1 generation can be described as 12.04 ± 1.13, and in the F_2 generation it can be described as 12.11 ± 2.06.

Assuming that both tomato varieties are homozygous at the relevant loci and that the alleles controlling fruit weight act additively, we can estimate the number of polygenes involved in this trait. Since $1/72$ of the F_2 offspring have a phenotype that overlaps one of the parental strains (72 total F_2 offspring; one weighs 6 oz, one weighs 18 oz; see Table 21.3), the use of the formula $1/4^n = 1/72$ indicates that n is between 3 and 4, indicative of the number of genes that control fruit weight in these tomato strains.

21.4 Heritability Estimates the Genetic Contribution to Phenotypic Variability

The question most often asked by geneticists working with multifactorial traits is how much of the observed phenotypic variation in a population is due to genotypic differences among individuals and how much is due to environment. The term **heritability** is used to describe the proportion of total phenotypic variation in a population due to genetic factors.

For a multifactorial trait in a given population, a high-heritability estimate indicates that much of the variation can be attributed to genetic factors, with the environment having less impact on expression of the trait. With a low-heritability estimate, environmental factors are likely to have a greater impact on phenotypic variation within the population.

The concept of heritability is frequently misunderstood and misused. It should be emphasized that heritability does not indicate how much of a trait is genetically determined or the extent to which an individual's phenotype is due to genotype. In recent years, such misinterpretations of heritability for human quantitative traits have led to controversy, notably in relation to measurements of "intelligence quotient" or IQ. Variation in heritability estimates for IQ among different racial groups tested led to incorrect suggestions that unalterable genetic factors control differences in intelligence levels among human races. Such suggestions misrepresented the meaning of heritability and ignored the contribution of genotype-by-environment interaction (see page 478) to phenotypic variation in a population. Moreover, heritability is not fixed for a trait. For example, a heritability estimate for egg production in a flock of chickens kept in individual cages might be high, indicating that differences in egg output among individual birds is largely due to genetic differences inasmuch as they all share very similar environments. For a different flock kept outdoors, heritability for egg production might be much lower, since variation among different birds may also reflect differences in their individual environments. Such differences could include how much food each bird manages to find and whether it competes successfully for a good roosting spot at night. Thus a heritability estimate tells us the proportion of phenotypic variation that can be attributed to genetic variation *within a certain population in a particular environment*. If we measure heritability for the same trait among different populations in a range of environments, we frequently find that the calculated heritability values have large standard errors. This is an important point to remember when considering heritability estimates for traits in human populations. A mean heritability estimate of 0.65 for human height does not mean that your height is 65 percent due to your genes. Rather, it means that in the populations sampled, on average, *65 percent of the overall variation in height could be explained by genotypic differences among individuals.*

With this subtle, but important, distinction in mind, we will now consider how geneticists divide the phenotypic variation observed in a population into genetic and environmental components. As we saw in the previous section, this variation can be quantified as a sample variance: taking measurements of the trait in question from a representative sample of the population and determining the extent of the spread of those measurements around the sample mean. This gives us an estimate of the total **phenotypic variance** in the population (V_P). Heritability estimates are obtained by using different experimental and statistical techniques to partition V_P into **genotypic variance** (V_G) and **environmental variance** (V_E) components.

An important factor contributing to overall levels of phenotypic variation is the extent to which individual genotypes affect the phenotype differently among varying environments. For example, wheat variety A may yield an average of 20 bushels an acre on poor soil, while variety B yields an average of 17 bushels. On good soil, variety A yields 22 bushels, while variety B averages 25 bushels an acre. There are differences in yield between the two genotypically distinct varieties, so variation in wheat yield has a genetic component. Because both varieties yield more on good soil, yield is also affected by environment. However, we also see that the two varieties do not respond to better soil conditions equally: The genotype of wheat variety B achieves a greater increase in yield on good soil than does variety A. Thus, we have differences in the interaction of genotype with environment contributing to variation for yield in populations of wheat plants. This third component of phenotypic variation is **genotype-by-environment interaction variance** ($V_{G \times E}$) (Figure 21–7).

(a)

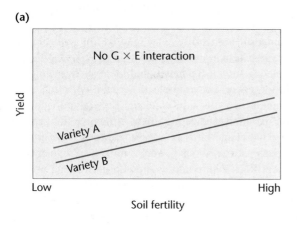

(b)

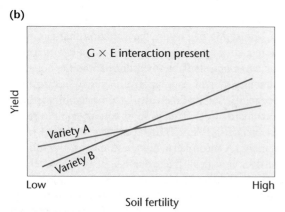

FIGURE 21–7 Differences in yield between two wheat varieties at different soil fertility levels. (a) No G × E interaction: The varieties show genetic differences in yield but respond equally to increasing soil fertility. (b) G × E interaction present: Variety A outyields B at low soil fertility, but B yields more than A at high-fertility levels.

We can now summarize all the components of total phenotypic variance V_P using the following equation:

$$V_P = V_G + V_E + V_{G \times E}$$

In other words, total phenotypic variance can be subdivided into genotypic variance, environmental variance, and genotype-by-environment interaction variance. When obtaining heritability estimates for a multifactorial trait, researchers often assume that the genotype-by-environment interaction variance is small enough that it can be ignored or combined with the environmental variance. However, it is worth remembering that this kind of approximation is another reason why heritability values are *estimates* for a given population, not a *fixed attribute* for a trait.

Animal and plant breeders use a range of experimental techniques to estimate heritabilities by partitioning measurements of phenotypic variance into genotypic and environmental components. One approach uses inbred strains containing genetically homogeneous individuals with highly homozygous genotypes. Experiments are then designed to test the effects of a range of environmental conditions on phenotypic variability. Variation *between* different inbred strains reared in a constant environment is due predominantly to genetic factors. Variation *among* members of the same inbred strain reared under different conditions is more likely to be due to environmental factors. Other approaches involve analyzing variance for a quantitative trait among offspring from different crosses, or comparing expression of a trait among offspring and parents reared in the same environment. More details of these experimental techniques can be found in some of the Selected Readings in Appendix A.

Broad-Sense Heritability

Broad-sense heritability (represented by the term H^2) measures the contribution of the genotypic variance to the total phenotypic variance. It is estimated as a proportion:

$$H^2 = \frac{V_G}{V_P}$$

Heritability values for a trait in a population range from 0.0 to 1.0. A value approaching 1.0 indicates that the environmental conditions have little impact on phenotypic variance, which is therefore largely due to genotypic differences among individuals in the population. Low values close to 0.0 indicate that environmental factors, not genotypic differences, are largely responsible for the observed phenotypic variation within the population studied. Few quantitative traits have very high or very low heritability estimates, suggesting that both genetics and environment play a part in the expression of most phenotypes in the range.

The genotypic variance component V_G used in broad-sense heritability estimates includes all types of genetic variation in the population. It does not distinguish between quantitative trait loci with alleles acting additively as opposed to those with epistatic or dominance effects. Broad-sense heritability estimates also assume that the genotype-by-environment variance component is negligible. Although broad-sense heritability estimates for a trait are of general genetic interest, these limitations mean that this kind of heritability is not very useful in breeding programs. Animal or plant breeders wishing to develop improved strains of livestock or higher yielding crop varieties need more precise heritability estimates for the traits they wish to manipulate in a population. Therefore, another type of estimate, narrow-sense heritability, has been devised that is of more practical use.

Narrow-Sense Heritability

Narrow-sense heritability (h^2) is the proportion of phenotypic variance due to additive genotypic variance alone. Genotypic variance can be divided into subcomponents representing the different modes of action of alleles at quantitative trait loci. As not all the genes involved in a quantitative trait affect the phenotype in the same way, this partitioning distinguishes between three different kinds of gene action contributing to genotypic variance. **Additive variance**, V_A, is the genotypic variance due to the additive action of alleles at quantitative trait loci. **Dominance variance**, V_D, is the deviation from the additive components that results when phenotypic expression in heterozygotes is not precisely intermediate between the two homozygotes. **Interactive variance**, V_I, is the deviation from the additive components that occurs when two or more loci behave epistatically. The amount of interactive variance is often negligible, and so this component may be excluded from calculations of total genotypic variance.

The partitioning of the total genotypic variance V_G is summarized in the equation

$$V_G = V_A + V_D + V_I$$

and a narrow-sense heritability estimate based only on that portion of the genotypic variance due to additive gene action becomes

$$h^2 = \frac{V_A}{V_P}$$

Omitting V_I and separating V_P into genotypic and environmental variance components, we obtain

$$h^2 = \frac{V_A}{V_E + V_A + V_D}$$

Heritability estimates are used in animal and plant breeding to indicate the potential response of a population to artificial selection for a quantitative trait. Narrow-sense heritability, h^2, provides a more accurate prediction of selection response than broad-sense heritability, H^2, and therefore h^2 is more widely used by breeders.

Now Solve This

Problem 12 on page 486 provides data for two quantitative traits in a herd of hogs and asks you to calculate broad- and narrow-sense heritabilities and then to choose the trait amenable to selection.

Hint: Consider which variance component's proportion is the best indicator of potential response to selection.

Artificial Selection

Artificial selection is the process of choosing specific individuals with preferred phenotypes from an initially heterogeneous population for future breeding purposes. Theoretically, if artificial selection based on the same trait preferences is repeated over multiple generations, a population can be developed containing a high frequency of individuals with the desired characteristics. If selection is for a simple trait controlled by just one or two genes subject to little environmental influence, generating the desired population of plants or animals is relatively fast and easy. However, many traits of economic importance in crops and livestock, such as grain yield in plants, weight gain or milk yield in cattle, and speed or stamina in horses, are polygenic and frequently multifactorial. Artificial selection for such traits is slower and more complex. Narrow-sense heritability estimates are valuable to the plant or animal breeder because, as we have just seen, they estimate the proportion of total phenotypic variance for the trait that is due to additive genetic variance. Quantitative trait alleles with additive action are those most easily manipulated by the breeder. Alleles at quantitative trait loci that generate dominance effects or interact epistatically (and therefore contribute to V_D or V_I) are less responsive to artificial selection. Thus narrow-sense heritability, h^2, can be used to predict the impact of selection. The higher the estimated value for h^2 in a population, the greater the change in phenotypic range for the trait that the breeder will see in the next generation after artificial selection.

Partitioning the genetic variance components to calculate h^2 and predict response to selection is a complex task requiring careful experimental design and analysis. The simplest approach is to select individuals with superior phenotypes for the desired quantitative trait from a heterogeneous population and to breed offspring from those individuals. The mean scores for the trait can then be compared to (1) the original population (M), (2) the selected individuals used as parents ($M1$), and (3) the offspring resulting from interbreeding the selected parents ($M2$). The relationship between these three means and h^2 is

$$h^2 = \frac{M2 - M}{M1 - M}$$

This equation can be further simplified by defining $M2 - M$ as the **selection response (R)** and $M1 - M$ as the **selection differential (S)**, so h^2 reflects the ratio of the response observed to the total response possible. Thus,

$$h^2 = \frac{R}{S}$$

A narrow-sense heritability value obtained in this way by selective breeding and measuring the response in the offspring is referred to as an estimate of **realized heritability**.

As an example of a realized heritability estimate, suppose that we measure the diameter of corn kernels in a population where the mean diameter M was 20 mm. From this population, we select a group with the smallest diameters, for which the mean $M1$ equals 10 mm. The selected plants are interbred, and the mean diameter $M2$ of the progeny kernels is 13 mm. We can calculate the realized heritability h^2 to estimate the potential for artificial selection on kernel size:

$$h^2 = \frac{M2 - M}{M1 - M}$$

$$h^2 = \frac{13 - 20}{10 - 20}$$

$$= \frac{-7}{-10}$$

$$= 0.70$$

This value for narrow-sense heritability indicates that the selection potential for kernel size is relatively high.

The longest running artificial selection experiment known is still being conducted at the State Agricultural Laboratory in Illinois. Since 1896, corn has been selected for both high and low oil content. After 76 generations, selection continues to result in increased oil content (Figure 21–8). With each cycle of successful selection, more of the corn plants accumulate a higher percentage of additive alleles involved in oil production. Consequently, the narrow-sense heritabili-

TABLE 21.4	Estimates of Heritability for Traits in Different Organisms
Trait	**Heritability (%) (h^2)**
Mice	
Tail length	60
Body weight	37
Litter size	15
Chickens	
Body weight	50
Egg production	20
Egg hatchability	15
Cattle	
Birthweight	51
Milk yield	44
Conception rate	3

ty h^2 of increased oil content in succeeding generations has declined (see parenthetical values at generations 9, 25, 52, and 76 in Figure 21–8) as artificial selection comes closer and closer to optimizing the genetic potential for oil production. Theoretically, the process will continue until all individuals in the population possess a uniform genotype that includes all the additive alleles responsible for high oil content. At that point, h^2 will be reduced to zero, and response to artificial selection will cease. The decrease in response to selection for low oil content shows that heritability for the trait is approaching this point.

Table 21.4 lists narrow-sense heritability estimates expressed as percentage values for a variety of quantitative traits in different organisms. As you can see, these h^2 values vary, but heritability tends to be low for quantitative traits that are essential to an organism's survival. Remember, this does not mean that there is no genetic contribution to the observed phenotypes for such traits. Instead, the low h^2 values show that natural selection has already largely optimized the genetic component of these traits during evolution. Egg production, litter size, and conception rate are examples where such physiological limitations on selection have already been established. Traits that are less critical to survival, such as body weight, tail length, and wing length, have higher heritabilities as more genotypic variation for such traits is still present in the population. Remember, too, any single heritability estimate provides information on only one population in a specific environment. Therefore, narrow-sense heritability estimates are more valuable as predictors of response to selection when they are based on data collected in many populations and environments and where a clear trend is established.

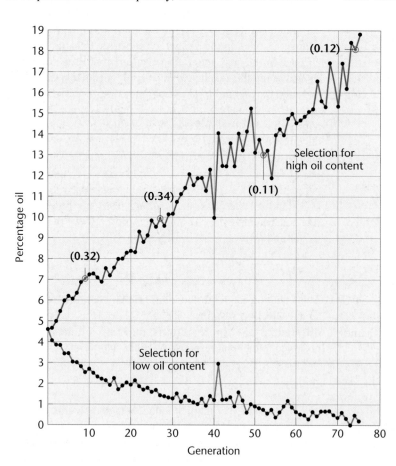

FIGURE 21–8 Response of corn selected for high and low oil content over 76 generations. The numbers in parentheses at generations 9, 25, 52, and 76 for the "high oil" line indicate the calculation of heritability at these points in the continuing experiment.

21.5 Twin Studies Allow an Estimation of Heritability in Humans

For obvious reasons, traditional heritability studies are not possible in humans. However, human twins can be useful subjects for examining how much variance for a multifactorial trait is genotypic as opposed to environmental. **Monozygotic (MZ)** or **identical twins**, derived from the division and splitting of a single egg following fertilization, are genotypically identical. **Dizygotic (DZ)** or **fraternal twins**, on the other hand, originate from two separate fertilization events and are as genetically similar as any other two siblings, with about 50 percent of their genes in common. For a given trait, therefore, phenotypic differences between pairs of identical twins will be equivalent to the environmental variance V_E (because the genotypic variance is zero). Phenotypic differences between dizygotic twins, however, represent both environmental variance V_E and approximately half the genotypic variance V_G. Comparison of phenotypic variances for the same trait in monozygotic and dizygotic sets of twins provides an estimate of broad-sense heritability for the trait.

Another approach used in twin studies has been to compare identical twin pairs reared together with pairs that were separated and raised in different settings. For any particular trait, average similarities or differences can be investigated. Twins are said to be **concordant** for a given trait if both express it or neither expresses it. If one expresses the trait and the other does not, the pair is said to be **discordant**. Comparison of the concordance values of MZ versus DZ twins reared together illustrates the potential value for heritability assessment. (See the following "Now Solve This" feature.)

Such data must be examined very carefully before any conclusions are drawn. If the concordance value approaches 90 to 100 percent in monozygotic twins, we might be inclined to interpret that value as indicating a large genetic contribution to the expression of the trait. In some cases—for example, blood types and eye color—we know that this is indeed true. In the case of contracting measles, however, a high concordance value merely indicates that the trait is almost always induced by a factor in the environment—in this case, a virus.

It is more meaningful to compare the *difference* between the concordance values of monozygotic and dizygotic twins. If these values are significantly higher for monozygotic twins than for dizygotic twins, we suspect that a strong genetic component is involved in determination of the trait. We reach this conclusion because monozygotic twins, with identical genotypes, would be expected to show a greater concordance than genetically related, but not genetically identical, dizygotic twins. In the case of measles, where concordance is high in both types of twins, the environment is assumed to contribute significantly. Such an analysis is useful because phenotypic characteristics that remain similar in different environments are likely to have a strong genetic component.

Interesting as they are, human twin studies contain some unavoidable sources of error. For example, identical twins are often treated more similarly by parents and teachers than are nonidentical twins, especially where the nonidentical pairs are of different sex. This may inflate the environmental variance for nonidentical twin pairs. Another possible error source is genotype-by-environment interaction, which can increase the total phenotypic variance for nonidentical twins compared to identical twins raised in the same environment. Heritability estimates for human traits based on twin studies should therefore be considered approximations and examined very carefully before any conclusions are drawn.

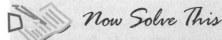

Now Solve This

Problem 24 on page 488 lists the percentages of monozygotic (MZ) and dizygotic (DZ) twins expressing the same phenotype for different traits and asks you to evaluate the relative importance of genetic and environmental factors.

Hint: Consider how MZ twins differ genetically from DZ twins. If each pair of twins was raised in the same environment, consider how that would affect expression of genetically determined versus environmentally determined phenotypes.

21.6 Quantitative Trait Loci Can Be Mapped

The kind of pedigree analysis we looked at earlier in the text (see Chapters 3 and 4) is of little use in identifying the multiple genes involved in quantitative traits. Environmental effects, interaction among segregating alleles, and the number of genes contributing to a polygenic phenotype make it difficult to isolate the effect of any one individual gene. However, because many quantitative traits are of economic or medical importance, it is useful to identify the genes involved and determine their location in the genome. Multiple genes contributing to a quantitative trait are known as **QTLs**. A single **quantitative trait locus** is designated as a **QTL**. Identifying and studying these loci allows the estimation of how many genes are involved in a quantitative trait and whether they all contribute equally to the trait or whether some genes influence the phenotype more strongly than others. Mapping reveals whether the QTLs associated with a trait are grouped on a single chromosome or scattered throughout the genome.

To find and map QTLs, researchers look for associations between particular DNA sequences within the genome and phenotypes falling within a certain range of the quantitative phenotype. One way to do this is to cross homozygous inbred lines that have different phenotypic extremes for the trait of interest to create an F_1 generation whose members will be heterozygous at most of the loci contributing to the trait. Additional crosses, either among F_1 individuals or between the F_1 and the inbred parent lines, result in F_2 generations with a high degree of segregation for different QTL genotypes and associated phenotypes. This segregating F_2 is known as the **QTL mapping population**. Researchers measure expression of the trait by individuals in the mapping population and identify different genotypes using DNA markers such as **restriction fragment length polymorphisms (RFLPs)** and microsatellites. Computer-based statistical analysis is then used to examine the correlation between marker genotypes and phenotypic variation for the trait. If a DNA marker is not linked to a QTL, then the phenotypic mean score for the trait will not vary among individuals with different genotypes at that marker locus. However, if a DNA marker is linked to a QTL, then genotypes differing at that marker locus will also differ in their expression of the trait. When this occurs, the marker locus and the QTL are said to *cosegregate*. Consistent cosegregation establishes the presence of a QTL at or near the DNA marker along the chromosome: The marker and QTL are linked. When numerous QTLs for a given trait have been located, a genetic map is created giving the positions of the genes involved on the different chromosomes.

DNA markers are now available for many organisms of agricultural importance, making possible systematic mapping of QTLs. For example, hundreds of RFLP markers have been located in the tomato. QTL analysis is performed by crossing plants with extreme, but opposite, phenotypes and following the crosses through several generations. Many loci have been identified that are responsible for quantitative traits such as fruit size, shape, soluble solid content, and acidity. These loci are distributed on all 12 chromosomes representing the haploid genome of this plant. Several chromosomes contain loci for multiple traits.

Mapping and characterizing QTLs in the tomato have been the focus of a highly successful research effort conducted by Steven Tanksley and his colleagues at Cornell University. Their research shows that approximately 30 QTLs contribute to tomato fruit size and shape. However, these QTLs are not equal in their effects: 10 genes scattered among seven chromosomes account for most of the observed phenotypic variation for these traits. Different alleles at just one of these loci, *fw2.2* on chromosome 2, can change fruit size by up to 30 percent. The *fw2.2* gene has been isolated, cloned, and transferred between plants, with interesting results. While the cultivated tomato can weigh up to 1000 grams, fruit from the related wild species thought to be the progenitor of the modern tomato weighs only a few grams. Two distinct alleles of *fw2.2*, identified as a result of RFLP mapping studies, appear to have played a major part in developing the domesticated tomato. One allele is present in all wild small-fruited varieties of tomatoes investigated. The other allele is present in all domesticated

large-fruited varieties. When the cloned *fw2.2* allele from small-fruited varieties is transferred to a plant that normally produces large tomatoes, the transformed plant produces fruits that are greatly reduced in weight (Figure 21–9). In the varieties studied by Tanksley's group, the reduction averaged 17 grams, a significant phenotypic change caused by the action of a single QTL. Further analysis of *fw2.2* revealed that the large-fruited allele at this locus codes for a protein acting as a negative repressor of cell division, resulting in increased mitosis in the cortical tissue that forms the fruit. The small-fruited *fw2.2* allele codes for the same protein but has mutations in the promoter region that reduce transcription levels, resulting in lower rates of cell division and much smaller fruit.

Marker-based mapping can identify chromosomal regions where QTLs are found, but identifying individual quantitative genes usually requires additional techniques. One approach is to develop DNA markers that are tightly linked to a gene of interest and then use recombinant DNA technology to identify the gene itself. Clearly, finding and characterizing all the genes involved in a quantitative trait is a long and complex task. At present, *fw2.2* is one of few quantitative genes that have been isolated, cloned, and studied in detail. However, mapping QTLs and defining the function of genes present in these regions in agriculturally important plants will enhance programs designed to improve crop yields. These studies also pave the way for similar research on animal genes.

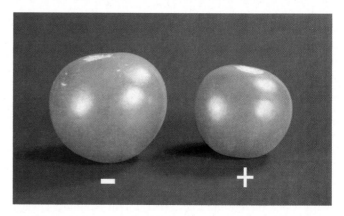

FIGURE 21–9 Phenotypic effect of the *fw2.2* transgene in the tomato. When the allele causing small fruit is transferred to a plant that normally produces large fruit, the fruit is reduced in size (+). A control fruit (−), which is normally larger, is shown for comparison.

CHAPTER SUMMARY

1. Quantitative inheritance results in a range of phenotypes for a trait due to the action of multiple genes combined with environmental factors.

2. Many polygenic quantitative traits show continuous variation that cannot be classified into discrete phenotypic classes. Some threshold traits are also polygenic, although they have a limited number of distinct phenotypic classes.

3. Statistical methods are used to analyze quantitative traits, including the mean, variance, standard deviation, standard error, and covariance.

4. Heritability estimates the relative contribution of genetic versus environmental factors to the range of phenotypic variation seen in a quantitative trait in a particular population.

5. Twin studies are used to estimate heritabilities for polygenic traits in humans.

6. Multiple genes contributing to a quantitative trait are known as quantitative trait loci or QTLs. Mapping techniques using DNA markers can be used to locate and characterize QTLs.

GENETICS, TECHNOLOGY, AND SOCIETY

The Green Revolution Revisited

Of the more than 6 billion people now living on Earth, about 750 million don't have enough to eat. And despite efforts to limit population growth, an additional 1 million people are expected to go hungry each year for the next several decades.

Will we be able to solve this problem? The past gives us some reasons to be optimistic. In the 1950s and 1960s, in the face of looming population increases, plant scientists around the world set about to increase the production of crop plants, including the three most important grains, wheat, rice, and maize. These efforts became known as the Green Revolution. The approach was three pronged: (1) to increase the use of fertilizers, pesticides, and irrigation water; (2) to bring more land under cultivation; and (3) to develop improved varieties of crop plants by intensive plant breeding. While highly successful initially, the rate of increase in grain yields has slowed. If food production is to keep pace with the projected increase in the world's population, plant breeders will have to depend more and more on the genetic improvement of crop plants to provide higher yields. But is this possible? Are we approaching the theoretical limits of yield in important crop plants? Recent work with rice suggests that the answer to this question is a resounding no.

Rice ranks third in worldwide production, just behind wheat and maize. About 2 billion people, fully one-third of Earth's population, depend on rice for their basic nourishment. The majority of the world's rice crop is grown and consumed in Asia, but it is also a dietary staple in Africa and Central America. The Green Revolution for rice began in 1960, with the establishment of the International Rice Research Institute (IRRI), headquartered at Los Baños, Philippines. The goal was to breed rice with improved disease resistance and higher yield. Breeders were almost too successful. The first high-yield varieties were so top-heavy with grain that they tended to fall over. (Plant breeders call this "lodging.") To reduce lodging, IRRI breeders crossed a high-yield line with a dwarf native variety to create semidwarf

lines, which were introduced to farmers in 1966. Owing in large part to the adoption of the semidwarf lines, the world production of rice doubled in the next 25 years.

Rice breeders cannot afford to rest on their laurels, however, as the yield of modern rice varieties has not improved much in recent years. Predictions suggest that a 70 percent increase in the annual rice harvest may be necessary to keep pace with anticipated population growth during the next 30 years. Breeders are now looking to wild rice varieties for further crop improvement. Leading the way are Susan McCouch, Steven Tanksley, and their coworkers at Cornell University. To test the hypothesis that wild-rice species carry genes that will improve the yield of cultivated varieties, they crossed cultivated rice (*Oryza sativa*) with a low-yield wild ancestor species (*Oryza rufipogon*) and then successively backcrossed the inter-specific hybrid to cultivated rice for three generations. In theory, this would create lines whose genomes were about 95 percent from *O. sativa* and 5 percent from *O. rufipogon*. When testing these backcrossed lines for grain yield, they found that several of them outproduced cultivated rice by as much as 30 percent. These results demonstrated strikingly that even though wild-rice relatives have low yields and appear to be inferior to cultivated rice, they still carry genes that will increase the yield of elite rice varieties. It will now be up to breeders to exploit the wild-rice relatives.

But introducing favorable genes from a wild relative into a cultivated variety by conventional breeding is a long and involved process, often requiring a decade or more of crossing, selection, backcrossing, and more selection. Future improvements in cultivated species must be quicker if crop yields are to keep up with population growth. Fortunately, modern gene-mapping techniques are now leading to the identification of QTLs that control complex traits such as yield and disease resistance. This makes possible a more direct approach to crop improvement, which has been termed the *advanced backcross QTL method*.

First, a cultivated variety is crossed with a wild relative, just as *O. sativa* was interbred with *O. rufipogon*. The hybrid is then backcrossed to the cultivated variety to generate lines that contain only a small fraction of the "wild" genome. The backcross lines with the best qualities (e.g., highest yield and most disease resistance) are selected, and the "wild QTLs" responsible for the superior performance are identified, using a detailed molecular linkage map. Once the beneficial QTLs are discovered by this method, they may be introduced into other cultivated varieties. In order for this strategy to succeed, it is essential that wild crop relatives be preserved as a storehouse of potentially useful genes. Efforts were begun in the 1970s to protect the existing crop relatives of many plants in their natural habitats and to preserve them in seed banks. As the work of McCouch and Tanksley and others has shown, it is not possible to predict which wild varieties may be needed decades or even centuries from now to contribute their beneficial alleles to cultivated varieties. To prevent the loss of superior genes, the widest possible spectrum of wild species must be preserved, even those that have no obvious favorable characteristics.

Almost 60 years ago, the great Russian plant geneticist N. I. Vavilov suggested that wild relatives of crop plants could be the source of genes to improve agriculture. In the coming century, Vavilov's vision may finally be realized, as genes from long-neglected wild crop relatives, identified by new molecular methods, spark a revitalized Green Revolution.

References

Mann, C. 1997. Reseeding the green revolution. *Science* 277: 1038–1043.

Ronald, P. C. 1997. Making rice disease-resistant. *Sci. Am.* (November) 277: 98–105.

Tanksley, S. D., and McCouch, S. R. 1997. Seed banks and molecular maps: Unlocking genetic potential from the wild. *Science* 277: 1063–1066.

Xiao, J., et al. 1996. Genes from wild rice improve yield. *Nature* 384: 223–224.

KEY TERMS

INSIGHTS AND SOLUTIONS

1. In a plant, height varies from 6 to 36 cm. When 6-cm and 36-cm plants were crossed, all plants were 21 cm. In the F_2 generation, a continuous range of heights was observed. Most were around 21 cm, and 3 of 200 were as short as the 6-cm P_1 parent.

(a) What mode of inheritance is illustrated, and how many gene pairs are involved?

Solution: Polygenic inheritance is illustrated where a continuous trait is involved and where alleles contribute additively to the phenotype. The 3/200 ratio of F_2 plants is the key to determining the number of gene pairs. This reduces to a ratio of 1/66.7, very close to 1/64. Using the formula $1/4^n = 1/64$ (where 1/64 is equal to the proportion of F_2 phenotypes as extreme as either P_1 parent), $n = 3$. Therefore, three gene pairs are involved.

(b) How much does each additive allele contribute to height?

Solution: The variation between the two extreme phenotypes is

$$36 - 6 = 30 \text{ cm}$$

Because there are six potential additive alleles (*AABBCC*), each contributes

$$30/6 = 5 \text{ cm}$$

to the base height of 6 cm, which results when no additive alleles (*aabbcc*) are part of the genotype.

(c) List all genotypes that give rise to plants that are 31 cm.

Solution: All genotypes that include five additive alleles will be 31 cm (5 alleles × 5 cm/allele + 6 cm base height = 31 cm). Therefore, *AABBCc*, *AABbCC*, and *AaBBCC* are the genotypes that will result in plants that are 31 cm.

(d) In a cross separate from the earlier mentioned F_1, a plant of unknown phenotype and genotype was testcrossed with the following results:

   1/4 11 cm

   2/4 16 cm

   1/4 21 cm

An astute genetics student realized that the unknown plant could be only one phenotype but could be any of three genotypes. What were they?

Solution: When testcrossed (with *aabbcc*), the unknown plant must be able to contribute either one, two, or three additive alleles in its gametes in order to yield the three phenotypes in the offspring. Since no 6-cm offspring are observed, the unknown plant never contributes all nonadditive alleles (*abc*). Only plants that are homozygous at one locus and heterozygous at the other two loci will meet these criteria. Therefore, the unknown parent can be any of three genotypes, all of which have a phenotype of 26 cm:

   AABbCc

   AaBbCC

   AaBBCc

For example, in the first genotype (*AABbCc*), *AABbCc* × *aabbcc* yields

   1/4 *AaBbCc* 21 cm

   1/4 *AaBbcc* 16 cm

   1/4 *AabbCc* 16 cm

   1/4 *Aabbcc* 11 cm

which is the ratio of phenotypes observed.

2. The results shown in the following table were recorded for ear length in corn:

Length of Ear in cm																	
	5	6	7	8	9	10	11	12	13	14	15	16	17	18	19	20	21
Parent A	4	21	24	8													
Parent B					3	11	12	15	26	15	10	7	2				
F_1				1	12	12	14	17	9	4							

For each of the parental strains and the F_1 generation, calculate the mean values for ear length.

Solution: The mean values can be calculated as follows:

$$\overline{X} = \frac{\Sigma X_i}{n} \qquad\qquad P_A : \overline{X} = \frac{\Sigma X_i}{n} = \frac{378}{57} = 6.63$$

$$P_B: \overline{X} = \frac{\Sigma X_i}{n} = \frac{1697}{101} = 16.80$$

$$F_1: \overline{X} = \frac{\Sigma X_i}{n} = \frac{836}{69} = 12.11$$

3. For the corn plant described in Problem 2, compare the mean of the F_1 with that of each parental strain. What does this tell you about the type of gene action involved?

Solution: The F_1 mean (12.11) is almost midway between the parental means of 6.63 and 16.80. This indicates that the effect of the genes in question may be additive.

4. The mean and variance of corolla length in two highly inbred strains of *Nicotiana* and their progeny are shown in the accompanying table.

One parent (P_1) has a short corolla, and the other parent (P_2) has a long corolla. Calculate the broad-sense heritability (H^2) of corolla length in this plant.

Strain	Mean (mm)	Variance (mm)
P_1 short	40.47	3.12
P_2 long	93.75	3.87
F_1 ($P_1 \times P_2$)	63.90	4.74
F_2 ($F_1 \times F_1$)	68.72	47.70

Solution: The formula for estimating heritability is $H^2 = V_G/V_P$, where V_G and V_P are the genetic and phenotypic components of variation, respectively. The main issue in this problem is obtaining some estimate of two components of phenotypic variation: genetic and environmental factors. V_P is the combination of genetic and environmental variance. Because the two parental strains are true breeding, they are assumed to be homozygous, and the variance of 3.12 and 3.87 is considered to be the result of environmental influences. The average of these two values is 3.50. The F_1 is also genetically homogeneous and gives us an additional estimate of the impact of environmental factors. By averaging this value along with that of the parents,

$$\frac{4.74 + 3.50}{2} = 4.12$$

we obtain a relatively good idea of environmental impact on the phenotype. The phenotypic variance in the F_2 is the sum of the genetic V_G and environmental (V_E) components. We have estimated the environmental input as 4.12, so 47.70 minus 4.12 gives us an estimate of V_G, which is 43.58. Heritability then becomes 43.58/47.70 or 0.91. This value, when viewed in percentage form, indicates that about 91 percent of the variation in corolla length is due to genetic influences.

PROBLEMS AND DISCUSSION QUESTIONS

1. What is the difference between continuous and discontinuous variation? Which of the two is most likely to be the result of polygenic inheritance?
2. Define the following: (a) polygene, (b) additive alleles, (c) correlation, (d) monozygotic and dizygotic twins, (e) heritability, and (f) QTL.
3. A homozygous plant with 20-cm diameter flowers is crossed with a homozygous plant of the same species that has 40-cm diameter flowers. The F_1 plants all have flowers 30 cm in diameter. In the F_2 generation of 512 plants, 2 plants have flowers 20 cm in diameter, 2 plants have flowers 40 cm in diameter, and the remaining 510 plants have flowers of a range of sizes in between.
   (a) Assuming that all alleles involved act additively, how many genes control flower size in this plant?
   (b) What frequency distribution of flower diameter would you expect to see in the progeny of a backcross between an F_1 plant and the large-flowered parent?
   (c) Calculate the mean, variance, and standard deviation for flower diameter in the backcross progeny from (b).
4. A dark-red strain and a white strain of wheat are crossed and produce an intermediate, medium-red F_1. When the F_1 plants are interbred, an F_2 generation is produced in a ratio of 1 dark-red: 4 medium-dark-red: 6 medium-red: 4 light-red: 1 white. Further crosses reveal that the dark-red and white F_2 plants are true breeding.
   (a) Based on the ratio of offspring in the F_2, how many genes are involved in the production of color?
   (b) How many additive alleles are needed to produce each possible phenotype?
   (c) Assign symbols to these alleles and list possible genotypes that give rise to the medium-red and the light-red phenotypes.

   (d) Predict the outcome of the F_1 and F_2 generations in a cross between a true-breeding medium-red plant and a white plant.
5. Height in humans depends on the additive action of genes. Assume that this trait is controlled by the four loci R, S, T, and U and that environmental effects are negligible. Instead of additive versus nonadditive alleles, assume that additive and partially additive alleles exist. Additive alleles contribute two units, and partially additive alleles contribute one unit to height.
   (a) Can two individuals of moderate height produce offspring that are much taller or shorter than either parent? If so, how?
   (b) If an individual with the minimum height specified by these genes marries an individual of intermediate or moderate height, will any of their children be taller than the tall parent? Why or why not?
6. An inbred strain of plants has a mean height of 24 cm. A second strain of the same species from a different geographical region also has a mean height of 24 cm. When plants from the two strains are crossed together, the F_1 plants are the same height as the parent plants. However, the F_2 generation shows a wide range of heights; the majority are like the P_1 and F_1 plants, but approximately 4 of 1000 are only 12 cm high and about 4 of 1000 are 36 cm high.
   (a) What mode of inheritance is occurring here?
   (b) How many gene pairs are involved?
   (c) How much does each gene contribute to plant height?
   (d) Indicate one possible set of genotypes for the original P_1 parents and the F_1 plants that could account for these results.
   (e) Indicate three possible genotypes that could account for F_2 plants that are 18 cm high and three that account for F_2 plants that are 33 cm high.

7. Erma and Harvey were a compatible barnyard pair but a curious sight. Harvey's tail was only 6 cm, while Erma's was 30 cm. Their F_1 piglet offspring all grew tails that were 18 cm. When inbred, an F_2 generation resulted in many piglets (Erma and Harvey's grandpigs), whose tails ranged in 4-cm intervals from 6 to 30 cm (6, 10, 14, 18, 22, 26, and 30). Most had 18-cm tails, while 1/64 had 6-cm tails and 1/64 had 30-cm tails.
   (a) Explain how tail length is inherited by describing the mode of inheritance, indicating how many gene pairs are at work and designating the genotypes of Harvey, Erma, and their 18-cm offspring.
   (b) If one of the 18-cm F_1 pigs were mated with one of the 6-cm F_2 pigs, what phenotypic ratio would be predicted if many offspring resulted? Diagram the cross.

8. In the following table, average differences of height, weight, and fingerprint ridge count between monozygotic twins (reared together and apart), dizygotic twins, and nontwin siblings are compared:

Trait	MZ Reared Together	MZ Reared Apart	DZ Reared Together	Sibs Reared Together
Height (cm)	1.7	1.8	4.4	4.5
Weight (kg)	1.9	4.5	4.5	4.7
Ridge count	0.7	0.6	2.4	2.7

Based on the data in this table, which of these quantitative traits have the highest heritability values?

9. What kind of heritability estimates (broad sense or narrow sense) are obtained from human twin studies?

10. List as many human traits as you can that are likely to be under the control of a polygenic mode of inheritance.

11. Corn plants from a test plot are measured, and the distribution of heights at 10-cm intervals is recorded in the following table:

Height (cm)	Plants (no.)
100	20
110	60
120	90
130	130
140	180
150	120
160	70
170	50
180	40

Calculate (a) the mean height, (b) the variance, (c) the standard deviation, and (d) the standard error of the mean. Plot a rough graph of plant height against frequency. Do the values represent a normal distribution? Based on your calculations, how would you assess the variation within this population?

12. The following variances were calculated for two traits in a herd of hogs.

Trait	V_P	V_G	V_A
Back fat	30.6	12.2	8.44
Body length	52.4	26.4	11.7

(a) Calculate broad-sense (H^2) and narrow-sense (h^2) heritabilities for each trait in this herd.
(b) Which of the two traits will respond best to selection by a breeder? Why?

13. The mean and variance of plant height of two highly inbred strains (P_1 and P_2) and their progeny (F_1 and F_2) are shown here:

Strain	Mean (cm)	Variance
P_1	34.2	4.2
P_2	55.3	3.8
F_1	44.2	5.6
F_2	46.3	10.3

Calculate the broad-sense heritability (H^2) of plant height in this species.

14. A hypothetical study investigated the vitamin A content and the cholesterol content of eggs from a large population of chickens. The variances (V) were calculated, as shown here:

Variance	Trait	
	Vitamin A	Cholesterol
V_P	123.5	862.0
V_E	96.2	484.6
V_A	12.0	192.1
V_D	15.3	185.3

(a) Calculate the narrow-sense heritability (h^2) for both traits.
(b) Which trait, if either, is likely to respond to selection?

15. The following table shows measurements for fiber lengths and fleece weight in a small flock of eight sheep.

	Sheep Fiber Length (cm)	Fleece Weight (kg)
1	9.7	7.9
2	5.6	4.5
3	10.7	8.3
4	6.8	5.4
5	11.0	9.1
6	4.5	4.9
7	7.4	6.0
8	5.9	5.1

(a) What are the mean, variance, and standard deviation for each trait in this flock?
(b) What is the covariance of the two traits?
(c) What is the correlation coefficient for fiber length and fleece weight?
(d) Do you think an increase in fiber length causes greater fleece weight? Why or why not?

16. In a herd of dairy cows, the narrow-sense heritability for milk protein content is 0.76, and for milk butterfat it is 0.82. The correlation coefficient between milk protein content and butterfat is 0.91. If the farmer selects for cows producing more butterfat in their milk, what will be the most likely effect on milk protein content in the next generation?

17. In an assessment of learning in *Drosophila*, flies may be trained to avoid certain olfactory cues. In one population, a mean of 8.5 trials was required. A subgroup of this parental population that was trained most quickly (mean = 6.0) was interbred, and their progeny was examined. These flies demonstrated a mean training value of 7.5. Calculate realized heritability for olfactory learning in *Drosophila*.

18. Suppose you want to develop a population of *Drosophila* that would rapidly learn to avoid certain substances the flies could detect by smell. Based on the heritability estimate you obtained in Problem 17, would it be worth doing this by artificial selection? Why or why not?

19. In a population of tomato plants, mean fruit weight is 60 g and h^2 is 0.3. Predict the mean weight of the progeny if tomato plants whose fruit averaged 80 g were selected from the original population and interbred.

20. In a population of 100 inbred genotypically identical rice plants, variance for grain yield is 4.67. What is the heritability for yield? Would you advise a rice breeder to improve yield in this strain of rice plants by selection?

21. A mutant strain of *Drosophila* was isolated and shown to be resistant to an experimental insecticide, whereas normal (wild-type) flies were sensitive to the chemical. Following a cross between resistant flies and sensitive flies, isolated populations were derived that had various combinations of chromosomes from the two strains. Each was tested for resistance, as shown in the figure. Analyze the data and draw any appropriate conclusion about which chromosome(s) contain a gene responsible for inheritance of resistance to the insecticide.

Chromosome

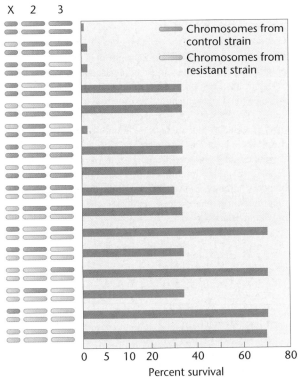

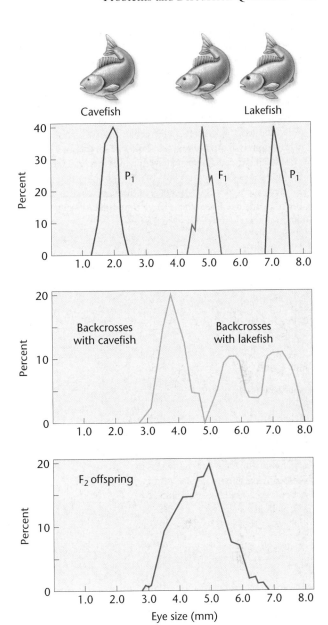

22. In 1988, Horst Wilkens investigated blind cavefish, comparing them with members of a sibling species with normal vision that are found in a lake. (We will call them cavefish and lakefish, respectively.) Wilkens found that cavefish eyes are about seven times smaller than lakefish eyes. F_1 hybrids have eyes of intermediate size. These data as well as the $F_1 \times F_1$ cross and those from backcrosses ($F_1 \times$ cavefish and $F_1 \times$ lakefish) are depicted as follows:

Examine Wilkens's results and respond to the following questions:

(a) Based strictly on the F_1 and F_2 results of Wilkens's initial crosses, what possible explanation concerning the inheritance of eye size seems most feasible?

(b) Based on the results of the F_1 backcross with cavefish, is your explanation supported? Explain.

(c) Based on the results of the F_1 backcross with lakefish, is your explanation supported? Explain.

(d) Wilkens examined about 1000 F_2 progeny and estimated that six to seven genes are involved in determining eye size. Is the sample size adequate to justify this conclusion? Propose an experimental protocol to test the hypothesis.

(e) A comparison of the embryonic eye in cavefish and lakefish revealed that both reach approximately 4 mm in diameter. However, lakefish eyes continue to grow, while cavefish eye size is greatly reduced. Speculate on the role of the genes involved in this problem. [*Reference:* Wilkens, H. (1988). *Ecol. Biol.* 23: 271–367.]

23. A 3-inch plant was crossed with a 15-inch plant and all F_1 plants were 9-inches. In the F_2, plants exhibited a "normal distribution" with heights of 3, 4, 5, 6, 7, 8, 9, 10, 11, 12, 13, 14, and 15 inches.
    (a) What ratio will constitute the "normal distribution" in the F_2?
    (b) What outcome will occur if the F_1 plants are testcrossed with plants that are homozygous for all nonadditive alleles?

24. The following table gives the percentage of twin pairs studied in which both twins expressed the same phenotype for a trait (concordance). Percentages listed are for different traits in monozygotic (MZ) and dizygotic (DZ) twins. Assuming that both twins in each pair were raised together in the same environment, what do you conclude about the relative importance of genetic versus environmental factors for each trait?

Trait	MZ%	DZ%
Blood types	100	66
Eye color	99	28
Mental retardation	97	37
Measles	95	87
Hair color	89	22
Handedness	79	77
Idiopathic epilepsy	72	15
Schizophrenia	69	10
Diabetes	65	18
Identical allergy	59	5
Cleft lip	42	5
Club foot	32	3
Mammary cancer	6	3

25. In a cross between a strain of large guinea pigs and a strain of small guinea pigs, the F_1 are phenotypically uniform with an average size about intermediate between that of the two parental strains. Among 1014 individuals, 3 are about the same size as the small parental strain and 5 are about the same size as the large parental strain. How many gene pairs are involved in the inheritance of size in these strains of guinea pigs?

26. Type A1B brachydactyly (short middle phalanges) is a genetically determined trait that maps to the short arm of chromosome 5 in humans. If you classify individuals as either having or not having brachydactyly, the trait appears to follow a single-locus, incompletely dominant pattern of inheritance. However, if one examines the fingers and toes of affected individuals, one sees a range of expression from extremely short to only slightly short. What might cause such variation in the expression of brachydactyly?

27. In a series of crosses between two true-breeding strains of peaches, the F_1 generation was uniform, producing 30-g peaches. The F_2 fruit mass ranges from 38 to 22 g at intervals of 2 g. (a) Using these data, determine the number of polygenic loci involved in the inheritance of peach mass. (b) Using gene symbols of your choice, give the genotypes of the parents and the F_1.

28. Students in a genetics laboratory began an experiment in an attempt to increase heat tolerance in two strains of *Drosophila melanogaster*. One strain was trapped from the wild six weeks before the experiment was to begin; the other was obtained from a *Drosophila* repository at a university laboratory. In which strain would you expect to see the most rapid and extensive response to heat-tolerance selection?

29. Consider a true-breeding plant, *AABBCC*, crossed with another true-breeding plant, *aabbcc*, whose resulting offspring are *AaBbCc*. If you cross the F_1s where independent assortment is operational, the expected fraction of offspring in each phenotypic class is given by the expression $N!/[M!(N - M!)]$, where N is the total number of alleles (six in this example) and M is the number of uppercase alleles. In a cross of *AaBbCc* $\times$ *AaBbCc*, what proportion of the offspring would be expected to contain two uppercase alleles?

30. Canine hip dysplasia is a quantitative trait that continues to affect most large breeds of dogs in spite of approximately 40 years of effort to reduce the impact of this condition. Breeders and veterinarians rely on radiographic and universal registries to enhance the development of breeding schemes to reduce its incidence. Recent data (Wood and Lakhani. 2003. *Vet. Rec.* 152: 69–72) indicate that there is a "month-of-birth" effect on hip dysplasia in Labrador retrievers and Gordon setters, whereby the frequency and extent of expression of this disorder varies depending on the time of year dogs are born. Speculate on how breeders attempt to "select" out this disorder and what the month-of-birth phenomenon indicates about the expression of polygenic traits?

Population Genetics

These lady-bird beetles, from the Chiricahua Mountains in Arizona, show considerable phenotypic variation.

CHAPTER CONCEPTS

- Allele frequencies in population gene pools can vary over time.

- The relationship between allele frequencies and genotype frequencies in an ideal population is described by the Hardy-Weinberg law.

- Natural selection is a major force driving allele frequency change and population evolution.

- Migration and drift can also cause changes in allele frequency.

- Nonrandom mating changes population genotype frequency but not allele frequency.

- Mutation creates new alleles in a population gene pool.

When Alfred Russel Wallace and Charles Darwin first identified natural selection as the mechanism of adaptive evolution in the mid-nineteenth century, there was no accurate model of the mechanisms responsible for variation and inheritance. Gregor Mendel published his work on the inheritance of traits in 1866, but it received little notice at the time. The rediscovery of Mendel's work in 1900 began a 30-year effort to reconcile Mendel's concept of genes and alleles with the theory of evolution. In a key insight, biologists realized that the frequency of phenotypic traits in a population is linked to the relative abundance of the alleles influencing those traits. This laid the foundation for the emergence of **population genetics**—the study of genetic variation in populations and how it changes over time. Population geneticists investigate patterns of genetic variation or **genetic structure** within and among groups of interbreeding individuals. As changes in genetic structure form the basis for evolution of a population, population genetics has become an important subdiscipline of evolutionary biology.

Early in the twentieth century a number of workers, including G. Udny Yule, William Castle, Godfrey Hardy, and Wilhelm Weinberg, formulated the basic principles of population genetics. Theoretical geneticists Sewall Wright, Ronald Fisher, and J. B. S. Haldane developed mathematical models to describe the genetic structure of populations. More recently, field researchers have been able to test these models using biochemical and molecular techniques that measure variation directly at the protein and DNA levels. These experiments examine allele frequencies and the forces that act on them, such as selection, mutation, migration, and genetic drift. In this chapter, we will consider general aspects of population genetics and discuss other areas of genetics that relate to evolution.

How Do We Know?

Population geneticists study changes in allele frequency in population gene pools over time, the nature and amount of genetic diversity in populations, and patterns of distribution of different genotypes. As you read this chapter think about these questions:

1. How do geneticists detect the presence of different alleles in a population?
2. How are different genotypes identified?
3. Are differences in morphological phenotype (the physical appearance of an individual) the only way that genetically distinct individuals in a population can be identified?

22.1 Allele Frequencies in Population Gene Pools Vary over Time

A **population** is a group of individuals sharing a common set of genes that live in the same geographic area and actually or potentially interbreed. All the alleles shared by these individuals constitute the **gene pool** for the population. Often when we

examine a single genetic locus in this population, we find that distribution of the alleles at this locus results in individuals with different genotypes. A key element of population genetics is the computation of frequencies of various alleles in the gene pool, frequencies of different genotypes in the population, and how these frequencies change from one generation to the next. Population geneticists use these calculations to address questions such as: How much genetic variation is present in a population? Are genotypes randomly distributed in time and space, or do discernible patterns exist? What processes affect the composition of a population's gene pool? Do these processes produce genetic divergence among populations?

Populations are dynamic; they expand and contract through changes in birth and death rates, migration, or contact with other populations. Often some individuals within a population will produce more offspring than others, contributing a disproportionate amount of their alleles to the next generation. Thus, the dynamic nature of a population can, over time, lead to changes in the population's gene pool.

22.2 The Hardy-Weinberg Law Describes the Relationship between Allele Frequencies and Genotype Frequencies in an Ideal Population

The theoretical relationship between the relative proportions of alleles in the gene pool and the frequencies of different genotypes in the population was elegantly described in the early 1900s in a mathematical model developed independently by the British mathematician Godfrey H. Hardy and the German physician Wilhelm Weinberg. This model, called the **Hardy-Weinberg law**, describes what happens to alleles and genotypes in an "ideal" population that is infinitely large and randomly mating, and that is not subject to any evolutionary forces such as mutation, migration, or selection. Under these conditions, the Hardy-Weinberg model makes two predictions:

1. The frequencies of the alleles in the gene pool do not change over time.

2. If two alleles at a locus, *A* and *a*, are considered, then as we will show later in this chapter, after one generation of random mating, the frequencies of genotypes *AA:Aa:aa* in the population can be calculated as

$$p^2 + 2pq + q^2 = 1$$

where p = frequency of allele *A* and q = frequency of allele *a*.

A population that meets these criteria, and in which the frequencies p and q of two alleles at a locus do result in the predicted genotypic frequencies, is said to be in Hardy-Weinberg equilibrium. As we will see later in this chapter, it is rare for a real population to conform totally to the Hardy-Weinberg model and for all allele and genotype frequencies to remain unchanged for generation after generation.

The Hardy-Weinberg model uses Mendelian principles of segregation and simple probability to explain the relationship between allele and genotype frequencies in a population. We can demonstrate how this works by considering the example of a single autosomal locus with two alleles, *A* and *a*, in a population where the frequency of *A* is 0.7 and the frequency of *a* is 0.3. Note that $0.7 + 0.3 = 1$, indicating that all the alleles for gene *A* present in the gene pool are accounted for. We assume that individuals mate randomly, following Hardy-Weinberg requirements, so for any one zygote, the probability that the female gamete will contain *A* is 0.7, and the probability that the male gamete will contain *A* is also 0.7. The probability that *both* gametes will contain *A* is $0.7 \times 0.7 = 0.49$. Thus we predict that genotype *AA* will occur 49 percent of the time. The probability that a zygote will be formed from a female gamete carrying *A* and a male gamete carrying *a* is $0.7 \times 0.3 = 0.21$, and the probability of a female gamete carrying *a* being fertilized by a male gamete carrying *A* is $0.3 \times 0.7 = 0.21$, so the frequency of genotype *Aa* is $0.21 + 0.21 = 0.42 = 42$ percent. Finally, the probability that a zygote will be formed from two gametes carrying *a* is $0.3 \times 0.3 = 0.09$, so the frequency of genotype *aa* is 9 percent. As a check on our calculations, note that $0.49 + 0.42 + 0.09 = 1.0$, confirming that we have accounted for all of the zygotes. These calculations are summarized in Figure 22–1.

We started with the frequency of a particular allele in a specific gene pool and calculated the probability that certain genotypes would be produced from this pool. When the zygotes develop into adults and reproduce, what will be the frequency distribution of alleles in the new gene pool? Under the Hardy-Weinberg law, we assume that all genotypes have equal rates of survival and reproduction. This means that in the next generation, all genotypes contribute equally to the new gene pool. The *AA* individuals constitute 49 percent of the population, and we can predict that the gametes they produce will constitute 49 percent of the gene pool. These gametes all carry allele *A*. Likewise, *Aa* individuals consti-

tute 42 percent of the population, so we predict that their gametes will constitute 42 percent of the new gene pool. Half (0.5) of these gametes will carry allele *A*. Thus, the frequency of allele *A* in the gene pool is $0.49 + (0.5)0.42 = 0.7$. The other half of the gametes produced by *Aa* individuals will carry allele *a*. The *aa* individuals constitute 9 percent of the population, so their gametes will constitute 9 percent of the new gene pool. These gametes all carry allele *a*. Thus, we can predict that the allele *a* in the new gene pool is $(0.5)0.42 + 0.09 = 0.3$. As a check on our calculation, note that $0.7 + 0.3 = 1.0$, accounting for all of the gametes in the gene pool of the new generation.

We have arrived back at the beginning, with a gene pool in which the frequency of allele *A* is 0.7 and the frequency of allele *a* is 0.3. For the general case of the Hardy-Weinberg law, we use variables instead of numerical values for the allele frequencies. Imagine a gene pool in which the frequency of allele *A* is *p* and the frequency of allele *a* is *q*, such that $p + q = 1$. If we randomly draw male and female gametes from the gene pool and pair them to make a zygote, the probability that both will carry allele *A* is $p \times p$. Thus, the frequency of genotype *AA* among the zygotes is p^2. The probability that the female gamete carries *A* and the male gamete carries *a* is $p \times q$ and that the female gamete carries *a* and the male gamete carries *A* is $q \times p$. Thus, the frequency of genotype *Aa* among the zygotes is $2pq$. Finally, the probability that both gametes carry *a* is $q \times q$, making the frequency of genotype *aa* among the zygotes q^2. Therefore, the distribution of genotypes among the zygotes is

$$p^2 + 2pq + q^2 = 1$$

This is summarized in Figure 22–2.

These calculations demonstrate the two main predictions of the Hardy-Weinberg law: Allele frequencies in our population do not change from one generation to the next, and genotype frequencies after one generation of random mating can be predicted from the allele frequencies. In other words, this population does not change or evolve with respect to the locus we have examined. Remember, however, the assumptions about the theoretical population described by the Hardy-Weinberg model:

Sperm

	fr(*A*) = 0.7	fr(*a*) = 0.3
Eggs fr(*A*) = 0.7	fr(*AA*) = 0.7 × 0.7 = 0.49	fr(*Aa*) = 0.7 × 0.3 = 0.21
fr(*a*) = 0.3	fr(*aA*) = 0.3 × 0.7 = 0.21	fr(*aa*) = 0.3 × 0.3 = 0.09

FIGURE 22–1 Calculating genotype frequencies from allele frequencies. Gametes represent samples drawn from the gene pool to form the genotypes of the next generation. In this population, the frequency of the *A* allele is 0.7, and the frequency of the *a* allele is 0.3. The frequencies of the genotypes in the next generation are calculated as 0.49 for *AA*, 0.42 for *Aa*, and 0.09 for *aa*. Under the Hardy-Weinberg law, the frequencies of *A* and *a* remain constant from generation to generation.

Sperm

	fr(*A*) = *p*	fr(*a*) = *q*
Eggs fr(*A*) = *p*	fr(*AA*) = p^2	fr(*Aa*) = *pq*
fr(*a*) = *q*	fr(*aA*) = *qp*	fr(*aa*) = q^2

FIGURE 22–2 The general case of allele and genotype frequencies under Hardy-Weinberg assumptions. The frequency of allele *A* is *p*, and the frequency of allele *a* is *q*. After mating, the three genotypes *AA*, *Aa*, and *aa* have the frequencies p^2, $2pq$, and q^2, respectively.

1. Individuals of all genotypes have equal rates of survival and equal reproductive success—that is, there is no selection.

2. No new alleles are created or converted from one allele into another by mutation.

3. Individuals do not migrate into or out of the population.

4. The population is infinitely large, which in practical terms means that the population is large enough that sampling errors and other random effects are negligible.

5. Individuals in the population mate randomly.

These assumptions are what make the Hardy-Weinberg law so useful in population genetics research. By specifying the conditions under which the population cannot evolve, the Hardy-Weinberg model identifies the real-world forces that cause allele frequencies to change. In other words, by holding certain conditions constant, Hardy-Weinberg isolates the forces of evolution and allows them to be quantified. Application of the Hardy-Weinberg model can also reveal "neutral genes" in a population gene pool—those not being operated on by the forces of evolution.

There are three additional important consequences of the Hardy-Weinberg law. First, it shows that dominant traits do not necessarily increase from one generation to the next. Second, it demonstrates that **genetic variability** can be maintained in a population since, once established in an ideal population, allele frequencies remain unchanged. Third, if we invoke Hardy-Weinberg assumptions, then knowing the frequency of just one genotype enables us to calculate the frequencies of all other genotypes at that locus. This is particularly useful in human genetics because we can calculate the frequency of heterozygous carriers for recessive genetic disorders even when all we know is the frequency of affected individuals. (See Section 22.4.)

Now Solve This

Problem 1 on page 512 asks you to calculate frequencies of a dominant and a recessive allele in a population where you know the phenotype frequencies.

Hint: Determine which allele frequency (*p* or *q*) you must estimate first when homozygous dominant and heterozygous genotypes have the same phenotype.

22.3 The Hardy-Weinberg Law Can Be Applied to Human Populations

To show how allele frequencies are measured in a real population and to demonstrate the kind of questions asked by population genetics researchers, we will consider a human population example drawn from research into genetic factors that influence an individual's susceptibility to infection by HIV-1, the virus responsible for the bulk of AIDS cases worldwide. Of particular interest to AIDS researchers is the small number of individuals who make high-risk choices (such as unprotected sex with HIV-positive partners), yet remain uninfected. In 1996, Rong Liu and colleagues discovered that two exposed, but uninfected, individuals were homozygous for a mutant allele of a gene called *CCR5*.

The *CCR5* gene, located on chromosome 3, encodes a protein called the C–C chemokine receptor-5, often abbreviated CCR5 (Figure 22–3). Chemokines are signaling molecules associated with the immune system. When white blood cells with CCR5 coreceptor proteins bind to chemokines, the cells respond by moving into inflamed tissues to fight infection.

The CCR5 protein is also a receptor for strains of HIV-1. To gain entry into cells, a protein called Env (short for envelope protein) on the surface of HIV-1 binds to the CD4 protein on the surface of the host cell. Binding to CD4 causes Env to change shape and form a second binding site. This second site binds to CCR5, which, in turn, initiates the fusion of the viral protein coat with the host-cell membrane. The merging of viral envelope with cell membrane transports the HIV-1 viral core into the host cell's cytoplasm.

The mutant allele of the *CCR5* gene contains a 32-bp deletion in one of its coding regions (Figure 22–4). The protein encoded by the mutant allele is shortened and made nonfunctional. The protein never makes it to the cell membrane, so HIV-1 cannot enter these cells. The gene's normal allele is called *CCR51* (we will designate this as *1*), and the allele with the 32-bp deletion is called *CCR5-Δ32* (we will designate this allele as *Δ32*). Both of the two uninfected individuals described by Liu had the genotype *Δ32/Δ32*. As a result, they had no CCR5 receptors on the surface of their cells and were resistant to infection by strains of HIV-1 that require CCR5 as a coreceptor.

At least three *Δ32/Δ32* individuals who are infected with HIV-1 have been found. Curiously, researchers have not discovered any adverse effects associated with this genotype. Heterozygotes with genotype *1/Δ32* are susceptible to HIV-1 infection, but evidence suggests that they progress more slowly

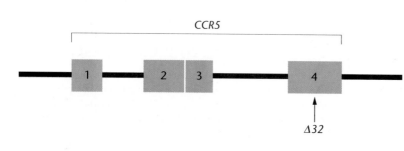

FIGURE 22–3 Organization of the *CCR5* gene in region 3p21.3. The gene contains 4 exons and 2 introns. The arrow shows the location of the 32-bp deletion in exon 4 that confers resistance to HIV-1 infection.

(a)
Allele 1

AGC TCT CAT TTT CCA TAC AGT CAG TAT CAA TTC TGG AAG AAT TTC CAG ACA TTA AAG ATA GTC

Ser Ser His Phe Pro Tyr Ser Gln Tyr Gln Phe Trp Lys Asn Phe Gln Thr Leu Lys Ile Val
    170   180

(b)
Allele Δ32

AGC TCT CAT TTT CCA TAC ATT AAA GAT AGT CAT CTT GGG   . . . . . Stop

Ser Ser His Phe Pro Tyr *Ile Lys Asp Ser His Leu Gly* (+24 aa)
    170

FIGURE 22–4 (a) Nucleotide sequence and amino acid sequence of a region in exon 4 of the normal (allele *1*) of the *CCR5* gene. (b) The deletion in the *Δ32* allele, associated with HIV-1 resistance. The first seven amino acids encoded out of frame as a results of this 32-bp deletion are shown in italics. The amino acid number for serine (170) is shown for orientation. The truncated protein of the *Δ32* allele is not inserted into the plasma membrane and cannot therefore act as a coreceptor for HIV-1.

TABLE 22.1	*CCR5* Genotypes and Phenotypes
Genotype	**Phenotype**
1/1	Susceptible to sexually transmitted strains of HIV-1
1/Δ32	Susceptible but may progress to AIDS slowly
Δ32/Δ32	Resistant to most sexually transmitted strains of HIV-1

to AIDS. Table 22.1 summarizes the genotypes possible at the *CCR5* locus and the phenotypes associated with each.

The discovery of the *CCR5-Δ32* allele and the fact that it provides some protection against AIDS generate two important questions: Which human populations harbor the *Δ32* allele, and how common is it? To address these questions, several teams of researchers surveyed a large number of people from a variety of populations. Genotypes were determined by direct analysis of DNA (Figure 22–5). We'll look at one sam-

ple population of 100 French individuals from a survey in Brittany in which 79 individuals have genotype *1/1*, 20 have genotype *1/Δ32*, and 1 has genotype *Δ32/Δ32*. This sample has 158 *1* alleles carried by the *1/1* individuals plus 20 *1* alleles carried by *1/Δ32* individuals, for a total of 178. The frequency of the *CCR51* allele in the sample population is thus $178/200 = 0.89 = 89$ percent. The *CCR5-Δ32* allele count shows 20 carried by *1/Δ32* individuals, plus 2 carried by the *Δ32/Δ32* individual, for a total of 22. The frequency of the *CCR5-Δ32* allele is thus $22/200 = 0.11 = 11$ percent. Notice that $p + q = 1$, confirming that we have accounted for the entire gene pool. Table 22.2 shows two methods for computing the frequencies of the *1* and *Δ32* alleles in the Brittany sample.

Figure 22–6 shows the frequency of the *CCR5-Δ32* allele in the 18 European populations surveyed. The studies show that populations in northern Europe around the Baltic Sea have the highest

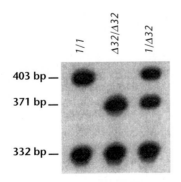

FIGURE 22–5 Allelic variation in the *CCR5* gene. Michel Samson and colleagues used PCR to amplify a part of the *CCR5* gene containing the site of the 32-bp deletion, cut the resulting DNA fragments with a restriction enzyme, and ran the fragments on an electrophoresis gel. Each lane reveals the genotype of a single individual. The *1* allele produces a 332-bp fragment and a 403-bp fragment; the *Δ32* allele produces a 332-bp fragment and a 371-bp fragment. Heterozygotes produce three bands.

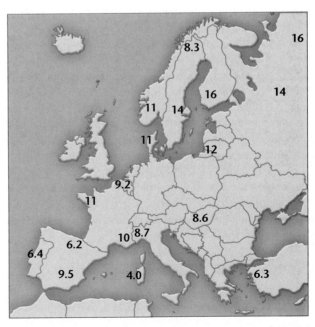

FIGURE 22–6 The frequency (percentage) of the *CCR5-Δ32* allele in 18 European populations.

| TABLE 22.2 | Methods of Determining Allele Frequencies from Data on Genotypes | | | |

A. Counting Alleles

Genotype	1/1	1/Δ32	Δ32/32	Total
Number of individuals	79	20	1	100
Number of 1 alleles	158	20	0	178
Number of Δ32 alleles	0	20	2	22
Total number of alleles	158	40	2	200

Frequency of *CCR51* in sample: 178/200 = 0.89 = 89%

Frequency of *CCR5-Δ32* in sample: 22/200 = 0.11 = 11%

B. From Genotype Frequencies

Genotype	1/1	1/Δ32	Δ32/32	Total
Number of individuals	79	20	1	100
Genotype frequency	79/100 = 0.79	20/100 = 0.20	1/100 = 0.01	1.00

Frequency of *CCR51* in sample: 0.79 + (0.5)0.20 = 0.89

Frequency of *CCR5-Δ32* in sample: (0.5)0.20 + 0.01 = 0.11

frequencies of the Δ32 allele. There is a sharp gradient in allele frequency from north to south, and populations in Sardinia and Greece have very low frequencies. In populations without European ancestry, the Δ32 allele is essentially absent. The highly patterned global distribution of the Δ32 allele presents an evolutionary puzzle that we'll return to later in the chapter.

Can we expect the *CCR5-Δ32* allele to increase in human populations in which it is currently rare? From what we now know about the Hardy-Weinberg law, we can say that if (1) individuals of all genotypes have equal rates of survival and reproduction, (2) there is no mutation, (3) no differential migration into or out of the population, (4) the population is extremely large, and (5) the individuals in the population choose their mates randomly with respect to the locus of interest, then the frequency of the Δ32 allele in human populations will not change. However, as we shall see in the last half of the chapter, when the assumptions of the Hardy-Weinberg law are not met—because of natural selection, mutation, migration, or genetic drift—the allele frequencies in a population may change from one generation to the next. Nonrandom mating does not, by itself, alter *allele* frequencies, but by altering *genotype* frequencies, it indirectly affects the course of evolution. The Hardy-Weinberg law tells geneticists where to look to find the causes of evolution in populations. We return to the *CCR5-Δ32* allele later in the chapter to see how a population genetics perspective has generated fruitful hypotheses for further research.

Testing for Hardy-Weinberg Equilibrium

One way we establish whether one (or more) of the Hardy-Weinberg assumptions does not hold in a given population is by determining whether the population's genotypes are in equilibrium. To do this, we first determine the frequencies of the genotypes, either directly from the phenotypes (if heterozygotes are recognizable) or by analyzing proteins or DNA sequences. We then calculate the allele frequencies from the genotype frequencies, as demonstrated earlier. Finally, we use the allele frequencies in the parental generation to predict the offspring's genotype frequencies. According to the Hardy-Weinberg law, the genotype frequencies are predicted to fit the $p^2 + 2pq + q^2 = 1$ relationship. If they do not, then one or more of the assumptions are invalid for the population in question.

We will use the *CCR5* genotypes of a population in Britain to demonstrate the Hardy-Weinberg law. The population includes 283 individuals, of which 223 have genotype *1/1*; 57 have genotype *1/Δ32*; and 3 have genotype *Δ32/Δ32*. These numbers represent genotype frequencies of $223/283 = 0.788$, $57/283 = 0.201$, and $3/283 = 0.011$, respectively. From the genotype frequencies, we compute the *CCR51* allele frequency as 0.89 and the frequency of the *CCR5-Δ32* allele as 0.11. From these allele frequencies, we can use the Hardy-Weinberg law to determine whether this population is in equilibrium. The allele frequencies predict the genotype frequencies as follows:

Expected frequency of genotype

$$1/1 = p^2 = (0.89)^2 = 0.792$$

Expected frequency of genotype

$$1/\Delta 32 = 2pq = 2(0.89)(0.11) = 0.196$$

Expected frequency of genotype

$$\Delta 32/\Delta 32 = q^2 = (0.11)^2 = 0.012$$

These expected frequencies are nearly identical to the observed frequencies. Our test of this population has failed to provide evidence that Hardy-Weinberg assumptions are being violated. The conclusion is confirmed by a χ^2 analysis. (See Chapter 3.) The χ^2 value in this case is tiny: 0.00023. To reject the null hypothesis at even the most generous, accepted level, $p = 0.05$, the χ^2 value would have to be 3.84. (In a test for Hardy-Weinberg equilibrium, the degrees of freedom are given by $k - 1 - m$, where k is the number of genotypes and m is the number of independent allele frequencies estimated from the data. Here, $k = 3$ and $m = 1$ since calculating only one allele frequency allows us to determine the other by subtraction. Thus, we have $3 - 1 - 1 = 1$ degree of freedom.

On the other hand, if the Hardy-Weinberg test demonstrates that the population is not in equilibrium, it will indicate that one or more assumptions are not being met. A group of researchers documented a real human population that is not in Hardy-Weinberg equilibrium. These researchers studied a small, isolated population of Native Americans in the American Southwest. They determined the genotype of each individual at two loci in the major histocompatibility complex (MHC). These genes, *HLA-A* and *HLA-B*, encode proteins that are involved in the immune system's discrimination between self and nonself. The immune systems of individuals heterozygous at MHC loci appear to recognize a greater diversity of foreign invaders and thus may be better able to fight disease. The results of this survey showed significantly more individuals who are heterozygous at both loci and significantly fewer homozygous individuals than would be expected under the Hardy-Weinberg law. Violation of either, or both, of two Hardy-Weinberg assumptions could explain the excess of heterozygotes among members of this population. First, fetuses, children, and adults who are heterozygous for *HLA-A* and *HLA-B* may have higher rates of survival than individuals who are homozygous (i.e., selection is occurring). Second, rather than choosing their mates randomly, individuals in this population may somehow prefer mates whose MHC genotypes differ from their own.

Now Solve This

Problem 6 on page 512 asks you to determine whether two populations are in Hardy-Weinberg equilibrium, based on their genotype frequencies.

Hint: Start by determining the allele frequencies based on these data.

22.4 The Hardy-Weinberg Law Can Be Used for Multiple Alleles and Estimating Heterozygote Frequencies

Having established the foundation of the Hardy-Weinberg law, we can now consider how it may be used in estimating allele frequencies within populations. We will examine calculations for situations involving multiple alleles and heterozygote frequencies.

Calculating Frequencies for Multiple Alleles in Hardy-Weinberg Populations

We commonly find several alleles of a single locus in a population. The ABO blood group in humans (discussed in Chapter 4) is such an example. The locus I (isoagglutinin) has three alleles I^A, I^B, and I^O, yielding six possible genotypic combinations $(I^A I^A, I^B I^B, I^O I^O, I^A I^B, I^A I^O, I^B I^O)$. Remember that in this case I^A and I^B are codominant alleles, and both of these are dominant to I^O. The result is that homozygous $I^A I^A$ and heterozygous $I^A I^O$ individuals are phenotypically identical, as are $I^B I^B$ and $I^B I^O$ individuals, so we can distinguish only four phenotypic combinations.

By adding another variable to the Hardy-Weinberg equation, we can calculate both the genotype and allele frequencies for the situation involving three alleles. Let p, q, and r represent the frequencies of alleles I^A, I^B, and I^O, respectively. Note that because there are three alleles

$$p + q + r = 1$$

Under Hardy-Weinberg assumptions, the frequencies of the genotypes are given by

$$(p + q + r)^2 = p^2 + q^2 + r^2 + 2pq + 2pr + 2qr = 1$$

If we know the frequencies of blood types for a population, we can then estimate the frequencies for the three alleles of the ABO system. For example, in one population sampled, the following blood-type frequencies are observed: A = 0.53, B = 0.13, O = 0.26. Because the I^O allele is recessive, the population's frequency of type O blood equals the proportion of the recessive genotype r^2. Thus,

$$r^2 = 0.26$$
$$r = \sqrt{0.26}$$
$$r = 0.51$$

Using r, we can calculate the allele frequencies for the I^A and I^B alleles. The I^A allele is present in two genotypes, $I^A I^A$ and $I^A I^O$ alleles. The frequency of the $I^A I^A$ genotype is represented by p^2 and the $I^A I^O$ genotype by $2pr$. Therefore, the combined frequency of type A blood and type O blood is given by

$$p^2 + 2pr + r^2 = 0.53 + 0.26$$

If we factor the left side of the equation and take the sum of the terms on the right, we get

$$(p + r)^2 = 0.79$$
$$p + r = \sqrt{0.79}$$
$$p = 0.89 - r$$
$$p = 0.89 - 0.51 = 0.38$$

Having calculated p and r, the frequencies of allele I^A and allele I^O, we can now calculate the frequency for the I^B allele:

$$p + q + r = 1$$
$$q = 1 - p - r$$
$$= 1 - 0.38 - 0.51$$
$$= 0.11$$

The phenotypic frequencies and genotypic frequencies for this population are summarized in Table 22.3.

Calculating Heterozygote Frequency

In another application, the Hardy-Weinberg law allows us to estimate the frequency of heterozygotes in a population. The frequency of a recessive trait can usually be determined by counting such individuals in a sample of the population. With this information and the Hardy-Weinberg law, we can then calculate the allele and genotype frequencies.

Cystic fibrosis, an autosomal recessive trait, has an incidence of about $1/2500 = 0.0004$ in people of northern European ancestry. Individuals with cystic fibrosis are easily distinguished from the population at large by such symptoms as extra-salty sweat, excess amounts of thick mucus in the lungs, and susceptibility to bacterial infections. Because this is a recessive trait, individuals with cystic fibrosis must be homozygous. Their frequency in a population is represented by q^2, provided that mating has been random in the previous generation. The frequency of the recessive allele is therefore

$$q = \sqrt{q^2} = \sqrt{0.0004} = 0.02$$

Since $p + q = 1$, then the frequency of p is

$$p = 1 - q = 1 - 0.02 = 0.98$$

In the Hardy-Weinberg equation, the frequency of heterozygotes is $2pq$. Thus,

$$2pq = 2(0.98)(0.02)$$
$$= 0.04 \text{ or 4 percent, or } 1/25$$

Thus, heterozygotes for cystic fibrosis are rather common in the population (about $1/25$, or 4%), even though the incidence of homozygous recessives is only $1/2500$, or 0.04 percent. Calculations such as these are estimates because the population may not meet all Hardy-Weinberg assumptions.

In general, the frequencies of all three genotypes can be estimated once the frequency of either allele is known and Hardy-Weinberg assumptions are invoked. The relationship between genotype and allele frequency is shown in Figure 22–7. It is important to note that heterozygotes increase rapidly in a population as the values of p and q move from 0 or 1 toward 0.5. This observation confirms our conclusion that when a recessive trait such as cystic fibrosis is rare, the majority of those carrying the allele are heterozygotes. In populations in which the frequencies of p and q are between 0.33 and 0.67, heterozygotes occur at higher frequency than either homozygote.

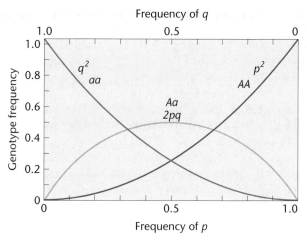

FIGURE 22–7 The relationship between genotype and allele frequencies derived from the Hardy-Weinberg equation.

TABLE 22.3	Calculating Genotype Frequencies for Multiple Alleles in a Hardy-Weinberg population Where the Frequency of Allele $I^A = 0.38$, Allele $I^B = 0.11$, and Allele $I^O = 0.51$		
Genotype	**Genotype Frequency**	**Phenotype**	**Phenotype Frequency**
$I^A I^A$	$p^2 = (0.38)^2 = 0.14$	A	0.53
$I^A I^O$	$2pr = 2(0.38)(0.51) = 0.39$		
$I^B I^B$	$q^2 = (0.11)^2 = 0.01$	B	0.12
$I^B I^O$	$2qr = 2(0.11)(0.51) = 0.11$		
$I^A I^B$	$2pq = 2(0.38)(0.11) = 0.084$	AB	0.08
$I^O I^O$	$r^2 = (0.51)^2 = 0.26$	O	0.26

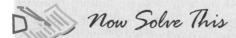

Problem 17 on page 512 asks you to calculate the frequency of heterozygous carriers of the recessive trait albinism in a human population.

Hint: First determine the frequency of the albinism allele in this population.

22.5 Natural Selection Is a Major Force Driving Allele Frequency Change

We have noted that the Hardy-Weinberg law establishes an ideal population that allows us to estimate allele and genotype frequencies at a given locus in populations in which the assumptions of large population size, lack of migration, random mating, absence of selection, absence of mutation, and equal viability hold. Obviously, it is almost impossible to find natural populations in which all these assumptions hold for all loci. In nature, populations are dynamic, and changes in size and gene pool are common. The Hardy-Weinberg law allows us to investigate populations that vary from the ideal. In this and the following sections, we discuss factors that prevent populations from reaching Hardy-Weinberg equilibrium or that drive populations toward a different equilibrium, as well as the relative contribution of these factors to evolutionary change.

Natural Selection

The first assumption of the Hardy-Weinberg law is that individuals of all genotypes have equal rates of survival and equal reproductive success. If this assumption does not hold, allele frequencies may change from one generation to the next. To see why, let's imagine a population of 100 individuals in which the frequency of allele A is 0.5 and that of allele a is 0.5. Assuming the previous generation mated randomly, we find that the genotype frequencies in the present generation are $(0.5)^2 = 0.25$ for AA, $2(0.5)(0.5) = 0.5$ for Aa, and $(0.5)^2 = 0.25$ for aa. Because our population contains 100 individuals, we have 25 AA individuals, 50 Aa individuals, and 25 aa individuals. Now suppose that individuals with different genotypes have different rates of survival: All 25 AA individuals survive to reproduce, 90 percent or 45 of the Aa individuals survive to reproduce, and 80 percent or 20 of the aa individuals survive to reproduce. When the survivors reproduce, each contributes two gametes to the new gene pool, giving us $2(25) + 2(45) + 2(20) = 180$ gametes. What are the frequencies of the two alleles in the surviving population? We have 50 A gametes from AA individuals, plus 45 A gametes from Aa individuals, so the frequency of allele A is $(50 + 45)/180 = 0.53$. We have 45 a gametes from Aa individuals, plus 40 a gametes from aa individuals, so the frequency of allele a is $(45 + 40)/180 = 0.47$.

These differ from the frequencies we started with. The frequency of allele A has increased, whereas the frequency of allele a has declined. A difference among individuals in survival or reproduction rate (or both) is called **natural selection**.

Natural selection is the principal force that shifts allele frequencies within large populations and is one of the most important factors in evolutionary change.

Fitness and Selection

Selection occurs whenever individuals with a particular genotype enjoy an advantage in survival or reproduction over other genotypes. However, selection intensity may vary from less than 1 percent to 100 percent for a lethal gene. In the previous example, selection was strong. Weak selection might involve just a fraction of a percent difference in the survival rates of different genotypes. Advantages in survival and reproduction ultimately translate into increased genetic contribution to future generations. An individual's genetic contribution to future generations is called its **fitness**. Thus, genotypes associated with high rates of reproductive success are said to have high fitness, whereas genotypes associated with low reproductive success are said to have low fitness.

Hardy-Weinberg analysis also allows us to examine fitness. By convention, population geneticists use the letter w to represent fitness. Thus, w_{AA} represents the relative fitness of genotype AA, w_{Aa} the relative fitness of genotype Aa, and w_{aa} the relative fitness of genotype aa. Assigning the values $w_{AA} = 1, w_{Aa} = 0.9$, and $w_{aa} = 0.8$ could mean, for example, that all AA individuals survive, 90 percent of the Aa individuals survive, and 80 percent of the aa individuals survive, as in the previous example.

Let's consider selection against deleterious alleles. Fitness values $w_{AA} = 1, w_{Aa} = 1$, and $w_{aa} = 0$ describe a situation in which a is a recessive lethal or sterility allele. As homozygous recessive individuals die without leaving offspring, the frequency of allele a will decline. The decline in the frequency of allele a is described by the equation

$$q_g = \frac{q_0}{1 + gq_0}$$

where q_g is the frequency of allele a in generation g, q_0 is the starting frequency of a (i.e., the frequency of a in generation zero), and g is the number of generations that have passed.

Figure 22–8 shows what happens to a lethal recessive allele with an initial frequency of 0.5. At first, because of the high percentage of aa genotypes, the frequency of allele a declines rapidly. The frequency of a is halved in only two generations. By the sixth generation, the frequency is halved again. By now, however, the majority of a alleles are carried by heterozygotes. Because a is recessive, these heterozygotes are not selected against. This means that as time continues to pass, the frequency of allele a declines ever more slowly. As long as heterozygotes continue to mate, it is difficult for selection to completely eliminate a recessive allele from a population.

Of course, a deleterious allele need not be recessive; many other scenarios are possible. A deleterious allele may be codominant, so that heterozygotes have intermediate fitness. Or an allele may be deleterious in the homozygous state but beneficial in the heterozygous state, like the allele for sickle-cell anemia. Figure 22–9 shows examples in which a deleterious allele is codominant, but the intensity of selection varies from strong to weak. In each case, the frequency of the

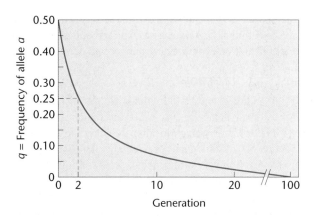

FIGURE 22–8 Change in the frequency of a lethal recessive allele, *a*. The frequency of *a* is halved in two generations, and halved again by the sixth generation. Subsequent reductions occur slowly because the majority of *a* alleles are carried by heterozygotes.

Generation	p	q	p²	2pq	q²
0	0.50	0.50	0.25	0.50	0.25
1	0.67	0.33	0.44	0.44	0.12
2	0.75	0.25	0.56	0.38	0.06
3	0.80	0.20	0.64	0.32	0.04
4	0.83	0.17	0.69	0.28	0.03
5	0.86	0.14	0.73	0.25	0.02
6	0.88	0.12	0.77	0.21	0.01
10	0.91	0.09	0.84	0.15	0.01
20	0.95	0.05	0.91	0.09	< 0.01
40	0.98	0.02	0.95	0.05	< 0.01
70	0.99	0.01	0.98	0.02	< 0.01
100	0.99	0.01	0.98	0.02	< 0.01

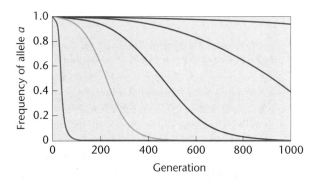

Selection Against Allele *a*				
Strong ◀				▶ Weak
───	───	───	───	───
w_{AA} 1.0	1.0	1.0	1.0	1.0
w_{Aa} 0.90	0.98	0.99	0.995	0.998
w_{aa} 0.80	0.96	0.98	0.99	0.996

FIGURE 22–9 The effect of selection on allele frequency. The rate at which a deleterious allele is removed from a population depends heavily on the strength of selection.

deleterious allele, *a*, starts at 0.99 and declines over time. However, the rate of decline depends heavily on the strength of selection. When only 90 percent of the heterozygotes and 80 percent of the *aa* homozygotes survive (red curve), the frequency of allele *a* drops from 0.99 to less than 0.01 in 85 generations. However, when 99.8 percent of the heterozygotes and 99.6 percent of the *aa* homozygotes survive (blue curve), it takes 1000 generations for the frequency of allele *a* to drop from 0.99 to 0.93. Two important conclusions can be drawn from this graph. First, given thousands of generations, even weak selection can cause substantial changes in allele frequencies; because evolution generally occurs over a large number of generations, selection is a powerful force of evolution. Second, for selection to produce rapid changes in allele frequencies, changes measurable within a human life span, the differences in fitness among genotypes must be large, or generation time must be short.

The manner in which selection affects allele frequencies allows us to make some inferences about the *CCR5-Δ32* allele that we discussed earlier. Because individuals with genotype *Δ32/Δ32* are resistant to most sexually transmitted strains of HIV-1, while individuals with genotypes *1/1* and *1/Δ32* are susceptible, we might expect AIDS to act as a selective force causing the frequency of the *Δ32* allele to increase over time. Indeed, it probably will, but the increase in frequency is likely to be slow in human terms.

We can do a rough calculation as follows: Imagine a population in which the current frequency of the *Δ32* allele is 0.10. Under Hardy-Weinberg assumptions, the genotype frequencies in this population are 0.81 for *1/1*, 0.18 for *1/Δ32*, and 0.01 for *Δ32/Δ32*. Imagine also that 1 percent of the *1/1* and *1/Δ32* individuals in this population will contract HIV and die of AIDS before reproducing.

Based on our assumptions, we can assign fitness levels to the genotypes as follows: $w_{1/1} = 0.99$; $w_{1/Δ32} = 0.99$; $w_{Δ32/Δ32} = 1.0$. Given the assigned fitness, we can predict that the frequency of the *CCR5-Δ32* allele in the next generation will be 0.100091. In fact, it will take about 100 generations (about 2000 years) for the frequency of the *Δ32* allele to reach just 0.11 (Figure 22–10). In other words, the frequency

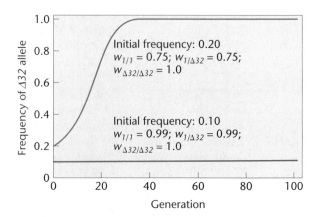

FIGURE 22–10 The rate at which the frequency of the *CCR5-Δ32* allele changes in hypothetical populations with different initial frequencies and different fitnesses.

of the Δ*32* allele will probably not change much over the next few generations in most populations that currently harbor it. A population genetic perspective sheds light on the *CCR5-Δ32* story in other ways as well. Two research groups have analyzed genetic variation at marker loci closely linked to the *CCR5* gene. Both groups concluded that most, if not all, present-day copies of the Δ*32* allele are descended from a single ancestral copy that appeared in northeastern Europe at most a few thousand years ago. In fact, one group estimates that the common ancestor of all Δ*32* alleles existed just 700 years ago. How could a new allele rise from a frequency of virtually zero to as high as 20 percent in roughly 30 generations?

It seems that there must have been strong selection in favor of the Δ*32* allele, most likely in the form of an infectious disease. The agent of selection cannot have been HIV-1 because HIV-1 moved from chimpanzees to humans too recently. Because selection occurred about 700 years ago, J. C. Stephens suggests that the agent of selection was bubonic plague. During the Black Death of 1346–1352, between a quarter and a third of all Europeans died from plague. Bubonic plague is caused by the bacterium *Yersinia pestis*. This bacterium manufactures a protein that kills some kinds of white blood cells. Stephens hypothesizes that the process through which the bacterial protein kills white cells involves the *CCR5* gene product. If true, some mechanism makes individuals homozygous for the Δ*32* allele more likely to survive plague epidemics.

Based on their study of the *CCR5* locus in Jewish and European populations, William Klitz and his colleagues have suggested that the Δ*32* allele may have arisen as early as the eighth century. They also suggest that smallpox may be the selective agent responsible for the rapid increase in the frequency of this allele. Both HIV and variola, the virus responsible for smallpox, use the CCR5 receptor for infecting cells, and with a fatality rate of 25 percent, waves of smallpox would be a powerful selective agent.

FIGURE 22–11 Pesticides used to control insects act as selective agents, changing allele frequencies of resistance genes in the populations exposed to the pesticide.

If past epidemics are responsible for the high frequency of the *CCR5-Δ32* allele in European populations, then the virtual absence of the allele in non-European populations is at first somewhat puzzling. Perhaps plague has been more common in Europe than elsewhere. Alternatively, perhaps other populations have different alleles of the *CCR5* gene that also confer protection against the plague. Teams of researchers looking for other alleles of the *CCR5* gene in various populations have found a total of 20 mutant alleles, including Δ*32*. Sixteen of these alleles encode proteins different in structure from that encoded by the *CCR51* allele. Some, like Δ*32*, are loss-of-function alleles. Some of the alleles appear confined to Asian populations, others to African populations. Their frequencies are as high as 3 to 4 percent. Together, these discoveries are consistent with the hypothesis that alteration or loss of the CCR5 protein protects against an as-yet-unidentified infectious disease or diseases.

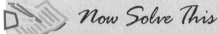

Now Solve This

Problem 8 on page 512 asks you to calculate the effect of varying fitness levels for different genotypes on allele frequencies.

Hint: First calculate the genotype frequencies in the next generation.

Selection in Natural Populations

Geneticists have done a great deal of research on the effect of natural selection on allele frequencies in both laboratory and natural populations. Among the most detailed studies of natural populations are those that involve insects exposed to pesticides. Christine Chevillon and colleagues, for example, studied the effect of the insecticide chlorpyrifos on allele frequencies in populations of house mosquitoes (Figure 22–11). Chlorpyrifos kills mosquitoes by interfering with the function of the enzyme acetylcholinesterase (ACE), which under normal circumstances breaks down the neurotransmitter acetylcholine. An allele of ACE called *Ace^R* encodes a slightly altered version of ACE that is immune to interference by chlorpyrifos.

Chevillon measured the frequency of the *Ace^R* allele in nine populations. In the first four locations, chlorpyrifos had been used to control mosquitoes for 22 years; in the last five locations, chlorpyrifos had never been used. Chevillon predicted that the frequency of *Ace^R* would be higher in the exposed populations. The researchers also predicted that aside from possible linkage effects, the frequencies of alleles for enzymes unrelated to the physiological effects of chlorpyrifos would show no such pattern. Among the control enzymes studied was aspartate amino transferase 1. The results appear in Figure 22–12. As the researchers predicted, on average, the frequency of the *Ace^R* allele was significantly higher in the exposed populations. Also as predicted, the frequencies of the most common alleles of the control enzyme gene showed no such trends. The explanation is that during the 22 years of exposure to the pesticide, mosquitoes had higher rates of

House mosquito, *Culex pipiens.*

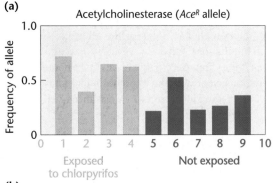

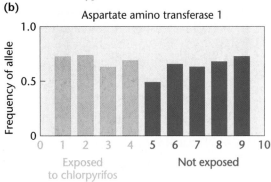

Geographic location

FIGURE 22–12 The effect of selection on allele frequencies in natural populations. (a) The frequency of the *Ace^R* allele, which confers resistance to the insecticide chlorpyrifos, is usually higher in house mosquito populations exposed to chlorpyrifos. (b) The frequency of an allele for an enzyme unrelated to chlorpyrifos metabolism (aspartate amino transferase 1) shows no such pattern.

survival if they carried the *Ace^R* allele. In other words, the *Ace^R* allele had been favored by natural selection.

In the case of mosquito populations exposed to chlorpyrifos, the rate of change in allele frequencies at the ACE locus will depend on the intensity of selection against mosquitoes that do not have the *Ace^R* allele. If selection is operating against an allele in the population with an initial frequency q, the new frequency q' after one generation can be calculated using the following equation:

$$q' = \frac{q(1 - sq)}{1 - sq^2}$$

where s is the selection coefficient against individuals homozygous for q. The selection coefficient is a measure of the amount by which the relative fitness is altered (increased or decreased). Thus, if the relative fitness of mosquitoes that do not have the

Ace^R allele and are not resistant to chlorpyrifos is 0.75 and the initial frequency of the normal version of the ACE allele is 0.95, then after one generation of exposure to the insecticide the new frequency q' of the normal allele will be:

$$q' = \frac{0.95(1 - 0.25 \times 0.95)}{1 - 0.25 \times 0.95^2} = 0.935$$

This shows that even with a relatively strong selection in its favor, a new allele occurring at a low frequency will require multiple generations to produce a significant change in the gene pool.

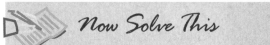

Now Solve This

Problem 22 on page 512 asks you to calculate the change in frequency of a resistance allele in an insect population feeding on a transgenic corn crop.

Hint: Start by answering these questions: Which allele is being selected against? What is the relative fitness of individuals homozygous for that allele? What is the selection coefficient?

Natural Selection and Quantitative Traits

As we saw in Chapter 21, many phenotypic traits are controlled not by alleles at a single locus but are quantitative: The phenotype is the result of the combined influence of the individual's genotype at many different loci and the environment. As with traits controlled by single genes, selection acting on these quantitative traits can be classified as (1) directional, (2) stabilizing, or (3) disruptive.

In **directional selection** (Figure 22–13) the genotype conferring one phenotypic extreme is selected, resulting in a change in the population mean over time. This form of selection is widely practiced in plant and animal breeding. If the trait is polygenic, the most extreme phenotypes that the genotype can express will appear in the population only after prolonged selection. An example of a long-running experiment in directional artificial selection is the study on high and low oil content in corn conducted at the State Agricultural Laboratory in Illinois, described in Chapter 21.

In nature, directional selection can occur when one of the phenotypic extremes becomes selected for or against, usually as a result of changes in the environment. A carefully documented example comes from research by Peter and Rosemary Grant and their colleagues, who used the medium ground finches (*Geospiza fortis*) of Daphne Major Island in the Galapagos Islands. The beak size of these birds varies enormously. In 1976, for example, some birds in the population had beaks less than 7 mm deep, while others had beaks more than 12 mm deep. Beak size is heritable, which means that large-beaked parents tend to have large-beaked offspring and small-beaked parents tend to have small-beaked offspring. In 1977, a severe drought on Daphne killed some 80 percent of the finches. Big-beaked birds survived at higher rates than small-beaked birds because when food became scarce, the big-beaked birds were able to eat a greater variety of seeds. When the drought

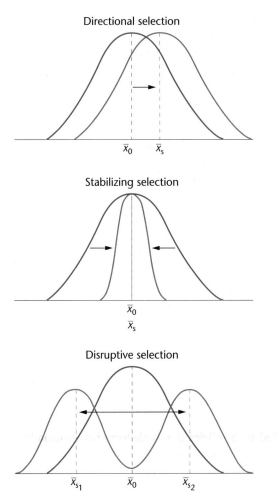

FIGURE 22–13   The impact of directional, stabilizing, and disruptive selection. In each case, the mean of an original population (green) and the mean of the population following selection (red) are shown.

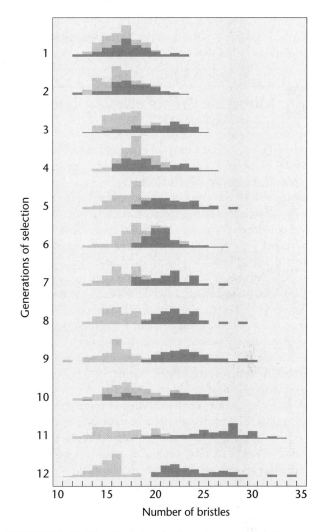

FIGURE 22–15   The effect of disruptive selection on bristle number in *Drosophila*. When individuals with the highest and lowest bristle number were selected, the population showed a nonoverlapping divergence in only 12 generations.

ended in 1978 and the survivors paired off and bred, the offspring inherited their parents' big beaks. Between 1976 and 1978, the beak depth of the average finch in the Daphne Major population increased by just over 0.5 mm, shifting the average beak size toward one phenotypic extreme.

   Stabilizing selection, in contrast, tends to favor intermediate types, with both extreme phenotypes being selected against. Over time, this will reduce the population variance

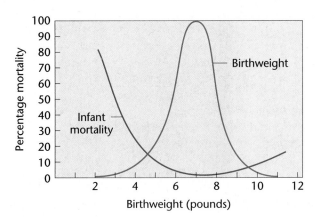

FIGURE 22–14   Relationship between birth weight and mortality in humans.

but without a significant shift in the mean. One of the clearest demonstrations of stabilizing selection is provided by the data of Mary Karn and Sheldon Penrose on human birth weight and survival for 13,730 children born over an 11-year period. Figure 22–14 shows the distribution of birth weight and the percentage of mortality at 4 weeks of age. Infant mortality increases on either side of the optimal birth weight of 7.5 pounds and quite dramatically so at the low end. At the genetic level, stabilizing selection acts to keep a population well adapted to its environment. In this situation, individuals closer to the average for a given trait will have higher fitness.

   Disruptive selection is selection against intermediates and for both phenotypic extremes. It can be viewed as the opposite of stabilizing selection because the intermediate types are selected against. This will result in a population with an increasingly bimodal distribution for the trait, as we can see in Figure 22–15. In one set of experiments, John Thoday applied disruptive artificial selection to a population of *Drosophila* on the basis of bristle number. In every generation, he allowed only the flies with high- or low-bristle numbers to breed. After several generations, most of the flies could be easily placed in a low- or high-bristle category. In natural populations, such a situation might exist for a population in a heterogeneous environment.

22.6 Mutation Creates New Alleles in a Gene Pool

Within a population, the gene pool is reshuffled each generation to produce new genotypes in the offspring. Because the number of possible genotypic combinations is so large, population members alive at any given time represent only a fraction of all possible genotypes. The enormous genetic reserve present in the gene pool allows Mendelian assortment and recombination to produce new genotypic combinations continuously. But assortment and recombination do not produce new alleles. **Mutation** alone acts to create new alleles. It is important to keep in mind that mutational events occur at random—that is, without regard for any possible benefit or disadvantage to the organism. In this section, we consider whether mutation is, by itself, a significant factor in causing allele frequencies to change.

To determine whether mutation is a significant force in changing allele frequencies, we measure the rate at which mutations are produced. As most mutations are recessive, it is difficult to observe mutation rates directly in diploid organisms. Indirect methods that use probability and statistics or large-scale screening programs are employed. For certain dominant mutations, however, a direct method of measurement can be used. To ensure accuracy, several conditions must be met:

1. The allele must produce a distinctive phenotype that can be distinguished from similar phenotypes produced by recessive alleles.

2. The trait must be fully expressed or completely penetrant so that mutant individuals can be identified.

3. An identical phenotype must never be produced by non-genetic agents such as drugs or chemicals.

Mutation rates can be stated as the number of new mutant alleles per given number of gametes. Suppose that for a given gene that undergoes mutation to a dominant allele, 2 out of 100,000 births exhibit a mutant phenotype. In these two cases, the parents are phenotypically normal. Because the zygotes that produced these births each carry two copies of the gene, we have actually surveyed 200,000 copies of the gene (or 200,000 gametes). If we assume that the affected births are each heterozygous, we have uncovered two mutant alleles out of 200,000. Thus, the mutation rate is 2/200,000 or 1/100,000, which in scientific notation is written as 1×10^{-5}. In humans, a dominant form of dwarfism known as **achondroplasia** fulfills the requirements for measuring mutation rates. Individuals with this skeletal disorder have an enlarged skull, short arms and legs, and can be diagnosed by X-ray examination at birth. In a survey of almost 250,000 births, the mutation rate (μ) for achondroplasia has been calculated as

$$\mu = 1.4 \times 10^{-5} \pm 0.5 \times 10^{-5}$$

Knowing the rate of mutation, we can estimate the extent to which mutation can cause allele frequencies to change from one generation to the next. We represent the normal allele as d and the allele for achondroplasia as D.

Imagine a population of 500,000 individuals in which everyone has genotype dd. The initial frequency of d is 1.0, and the initial frequency of D is 0. If each individual contributes two gametes to the gene pool, the gene pool will contain 1,000,000 gametes, all carrying allele d. While the gametes are in the gene pool, 1.4 of every 100,000 d alleles mutate into a D allele. The frequency of allele d is now $(1,000,000 - 14)/1,000,000 = 0.999986$, and the frequency of D is $14/1,000,000 = 0.000014$. From these numbers, it will clearly be a long time before mutation, by itself, causes any appreciable change in the allele frequencies in this population.

More generally, if we have two alleles, A with frequency p and a with frequency q, and if μ represents the rate of mutations converting A into a, and we assume that the rate of mutation of a to A is zero, then the frequencies of the alleles in the next generation are given by

$$p_{g+1} = p_g - \mu p_g \quad \text{and} \quad q_{g+1} = q_g + \mu p_g$$

where p_{g+1} and q_{g+1} represent the allele frequencies in the next generation, and p_g and q_g represent the allele frequencies in the present generation.

Figure 22–16 shows the replacement rate (change over time) in allele A for a population in which the initial frequency of A is 1.0 and the rate of mutation (μ) converting A into a is 1.0×10^{-5}. At this mutation rate, it will take about 70,000 generations to reduce the frequency of A to 0.5. Even if the rate of mutation increases through exposure to higher levels of radioactivity or chemical mutagens, the impact of mutation on allele frequencies will be extremely weak. The ultimate source of the genetic variability, mutation provides the raw material for evolution, but *by itself* plays a relatively insignificant role in changing allele frequencies. Instead, the fate of alleles created by mutation is more likely to be determined by natural selection (discussed previously) and genetic drift (discussed later).

An evolutionary perspective on mutation can lead to medical discoveries. The autosomal recessive disease cystic fibrosis, which we discussed previously in relation to calculating heterozygote frequencies, is caused by a loss-of-function

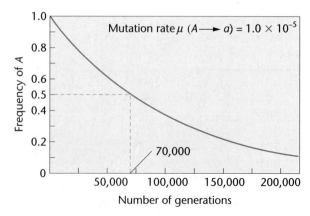

FIGURE 22–16 Replacement rate of an allele by mutation alone, assuming an average mutation rate of 1.0×10^{-5}.

mutation in the gene for a cell-surface protein called the cystic fibrosis transmembrane conductance regulator (CFTR).

The frequency of the mutant alleles causing cystic fibrosis is about 2 percent in European populations. Until recently, most individuals with two mutant alleles died before reproducing, meaning that selection against homozygous recessive individuals was rather strong. This creates a puzzle: In the face of selection against them, what has maintained the mutant alleles at an overall frequency of 2 percent?

The mutation-selection balance hypothesis posits that mutation is constantly creating new alleles to replace the ones being eliminated by selection. However, for this scenario to work, the rate of mutations creating new alleles would have to be rather high, on the order of 5×10^{-4} to counteract the effect of selection. Many evolutionary geneticists instead prefer an alternative explanation, the heterozygote superiority hypothesis. According to the heterozygote superiority hypothesis, selection against homozygous mutant individuals is counterbalanced by selection in favor of heterozygotes. The most popular agent of selection in favor of heterozygotes is resistance to an as-yet-unidentified disease.

Recent work suggests that cystic fibrosis heterozygotes may have enhanced resistance to typhoid fever. Typhoid fever is caused by the bacterium *Salmonella typhi*, which infiltrates cells of the intestinal lining. In laboratory studies, mouse intestinal cells that were heterozygous for *CFTR-*Δ*508*, the analog of the most common cystic fibrosis mutation in humans, acquired 86 percent fewer bacteria than did cells homozygous for the wild-type allele. Whether humans heterozygous for *CFTR-*Δ*508* also enjoy resistance to typhoid fever remains to be established. If they do, then cystic fibrosis will join sickle-cell anemia as an example of heterozygote superiority, where the fitness of heterozygotes is superior to that of either homozygous genotype.

22.7 Migration and Gene Flow Can Alter Allele Frequencies

Occasionally, a species divides into populations that are separated geographically. Various evolutionary forces, including selection, can establish different allele frequencies in such populations. **Migration** occurs when individuals move between the populations. Imagine a species in which a single locus has two alleles, *A* and *a*. There are two populations of this species, one on a mainland and one on an island. The frequency of *A* on the mainland is represented by p_m, and the frequency of *A* on the island is p_i. Under the influence of migration from the mainland to the island, the frequency of *A* in the next generation on the island (p_i') is given by

$$p_i' = (1 - m)p_i + mp_m$$

where *m* represents migrants from the mainland to the island.

Under these conditions, the frequency of *A* in the next generation on the island (p_i') will be affected by migration. For example, assume that $p_i = 0.4$ and $p_m = 0.6$ and that 10 percent of the parents of the next generation are migrants

from the mainland, so that $m = 0.1$. In the next generation, the frequency of allele *A* on the island will be

$$p_i' = [(1 - 0.1) \times 0.4] + (0.1 \times 0.6)$$
$$= 0.36 + 0.06$$
$$= 0.42$$

In this case, migration from the mainland has changed the frequency of *A* on the island from 0.40 to 0.42 in a single generation.

If either *m* is large or p_m is very different from p_i, then a rather large change in the frequency of *A* can occur in a single generation. If migration is the only force acting to change the allele frequency on the island, then an equilibrium will be attained only when $p_i = p_m$. These calculations reveal that the change in allele frequency attributable to migration is proportional to the differences in allele frequency between the donor and recipient populations and to the rate of migration. As *m* can have a wide range of values, the effect of migration can substantially alter allele frequencies in populations, as shown for the *B* allele of the ABO blood group in Figure 22–17. Although migration can be difficult to quantify, it can often be estimated.

Migration can also be regarded as the flow of genes between populations that were once, but are no longer, geographically isolated. Esteban Parra and colleagues measured allele frequencies for several different DNA sequence polymorphisms in African-American and European-American populations and in African and European populations representative of the ancestral populations from which the two American populations are descended. One locus they studied, a restriction site polymorphism called *FY-NULL*, has two alleles, *FY-NULL*1 and *FY-NULL*2. Figure 22–18 shows the frequency of *FY-NULL*1 in each population. The frequency of this allele is 0 in the three African populations and 1.0 in the three European populations, but lies between these extremes in all African-American and European-American populations. The simplest explanation for these data is that genes have mixed between American populations with predominantly African ancestry and American populations with predominantly European ancestry. Based on *FY-NULL* and several other loci, researchers estimate that African-American populations derive between 11.6 and 22.5 percent of their ancestry from Europeans and that European-American populations derive between 0.5 and 1.2 percent of their ancestry from Africans.

Now Solve This

Problem 24 on page 513 asks you to calculate the change in frequency of a mutant allele in a population of bighorn sheep that has been augmented by artificially introducing individuals from another population.

Hint: If the introduced sheep came from a population that lacked the mutation, would you expect the allele frequency in the augmented population to increase or decrease?

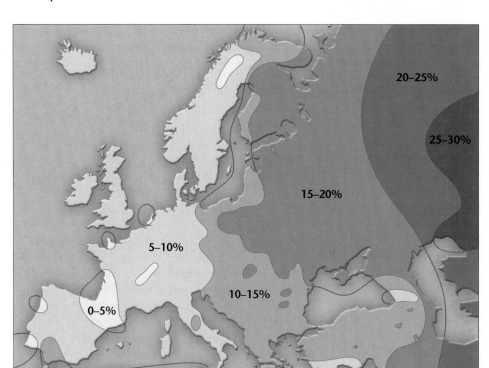

FIGURE 22–17 Migration as a force in evolution. The *B* allele of the ABO locus is present in a gradient from east to west. This allele shows the highest frequency in Central Asia, and the lowest is in northeastern Spain. The gradient parallels the waves of Mongol migration into Europe following the fall of the Roman Empire and is a genetic relic of human history.

Frequency of the *FY-NULL*1* Allele (a restriction-site polymorphism) in Various Human Populations

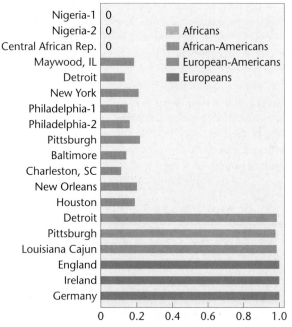

FIGURE 22–18 Frequency histogram of a restriction site polymorphism allele in several populations. These and other data demonstrate that African-American and European-American populations are of mixed ancestry.

22.8 Genetic Drift Causes Random Changes in Allele Frequency in Small Populations

In small populations, significant random fluctuations in allele frequencies are possible by chance deviation. The degree of fluctuation increases as the population size decreases, a situation known as **genetic drift**. In addition to small population size, drift can arise through the **founder effect**, which occurs when a population originates from a small number of individuals. Although the population may later increase to a large size, the genes carried by all members are derived from those of the founders (assuming no mutation, no selection, or migration and the presence of random mating). Drift can also arise via **genetic bottleneck**. Bottlenecks develop when a large population undergoes a drastic but temporary reduction in numbers. Even though the population recovers, its genetic diversity has been greatly reduced.

Genetic drift occurs when the number of reproducing individuals in a population is too small to ensure that all the alleles in the gene pool will be passed on to the next generation in their existing frequencies. Drift is defined as a change in allele frequency resulting from random gamete sampling. To visualize this, imagine a sexually reproducing diploid population consisting of 25 males and 25 females. Even if all individuals in the population reproduce as 25 separate mating pairs (an unlikely scenario in real life) to generate 25 offspring, the number of gametes forming the gene pool of the next generation will be just 50. It is unlikely that such a small number of gametes will accurately reflect the genetic structure of the parent population. Consequently, individual alleles may be over- or underrepresented in the gamete pool, resulting in random changes in frequency from one generation to the next. If we consider a single locus with two alleles, *A* and *a*, genetic drift may result in one of the alleles eventually disappearing, while the other becomes fixed—in other words, it becomes the only version of that gene present in the gene pool of the population. A simple correlation describes the likelihood of the fixation or the loss of an allele. The probability that an allele will be fixed through drift is the same as its initial frequency. So, if $p(A) = 0.8$, the probability of *A* becoming fixed is 0.8, or 80 percent; the probability of *A* being lost through drift is $(1 - 0.8) = 0.2$, or 20 percent.

To study genetic drift in laboratory populations of *Drosophila melanogaster*, Warwick Kerr and Sewall Wright set up over 100 lines, with four males and four females as the parents for each line. Within each line, the frequency of the sex-linked bristle mutant *forked* (*f*) and its wild-type allele (f^+) was 0.5. In each generation, four males and four females were chosen at random to parent the next generation. After 16 generations, the complete loss of one allele and the fixation of the other had occurred in 70 lines—29 in which only the *forked* allele was present and 41 in which the wild-type allele had become fixed. The remaining lines were still segregating the two alleles or had gone extinct. If fixation had occurred randomly, then an equal number of lines should have become fixed for each allele. In fact, the experimental results do not differ statistically from the expected ratio of 35:35, demonstrating that alleles can spread through a population and eliminate other alleles by chance alone.

Founder Effects in Human Populations

Allele frequencies in certain human populations demonstrate the role of genetic drift as an evolutionary force in natural populations. Native Americans living in the southwestern United States have a high frequency of oculocutaneous albinism (OCA). In the Navajo, who live primarily in northeast Arizona, albinism occurs with a frequency of 1 in 1500–2000, compared with whites (1 in 36,000) and African-Americans (1 in 10,000). There are four different forms of OCA (OCA1–4), all with varying degrees of melanin deficiency in the skin, eyes, and hair. The OCA phenotype in the Navajo overlaps those of OCA2 and OCA4. OCA2 is caused by mutations in the *P* gene, which encodes a plasma membrane protein; OCA4 is caused by mutations in a gene called *MAPT*. To investigate the genetic basis of albinism in the Navajo, Murray Brilliant and his colleagues screened for mutations in the *P* and *MAPT* genes. In their study, no mutations in the *MAPT* gene were found. All Navajo with albinism were homozygous for a 122.5-kb deletion in the *P* gene, spanning exons 10–20 (Figure 22–19). This deletion allele was not present in 34 individuals belonging to other Native American populations.

After molecular analysis of the breakpoints, the investigators developed a set of PCR primers to identify homozygous affected individuals and heterozygous carriers (Figure 22–20). Using these primers, they surveyed 134 normally pigmented Navajo and 42 members of the Apache, a tribe closely related to the Navajo. Based on this sample, the heterozygote frequency in the Navajo is estimated to be 4.5 percent. No carriers were found in the Apache population that was studied.

The 122.5-kb deletion allele causing OCA2 was found only in the Navajo population and not in members of other Native American tribes in the southwestern United States. This suggests that the mutant allele is specific to the Navajo and may have arisen in a single individual who was one of a small number of founders of the Navajo population. Such founder mutations originate on a single chromosome and have a characteristic set of closely linked flanking markers. Over time, mutation and recombination modify and/or replace this set of markers. By knowing the rate of recombination and the rate of mutation in this region of the genome, the origin of the founder mutation can be dated. Using these parameters and

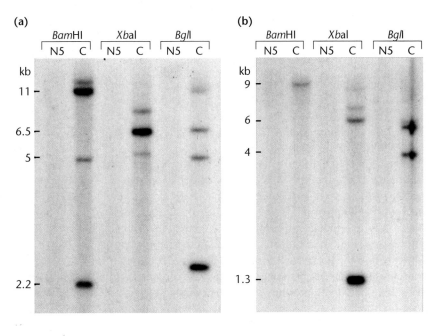

FIGURE 22–19 Genomic DNA digests from a Navajo affected with albinism (N5) and a normally pigmented individual (C). (a) Hybridization with a probe covering exons 11–15 of the *P* gene; there are no hybridizing fragments detected in N5. (b) Hybridization with a probe covering exons 15–20 of the *P* gene; there are no hybridizing fragments detected in N5. This confirms the presence of a deletion in affected individuals.
Courtesy of Murray Brilliant, "A 122.5 kilobase deletion of P gene underlies the high prevalence of oculocutaneous albinism type 2 in the Navajo population." From: American Journal Human Genetics 72: 62–72, fig. 1 p. 65. Published by University of Chicago Press.

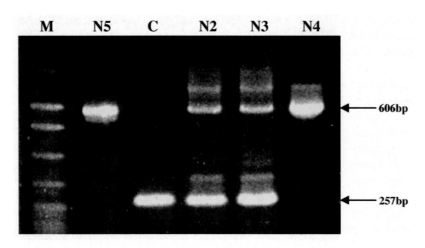

FIGURE 22–20 PCR screens of Navajo affected with albinism (N4 and N5) and the parents of N4 (N2 and N3). Affected individuals (N4 and N5), heterozygous carriers (N2 and N3), and a homozygous normal individual (C) each give a distinctive band pattern, allowing detection of heterozygous carriers in the population. Molecular size markers (M) are in the first lane.
Courtesy of Murray Brilliant, "A 122.5 kilobase deletion of P gene underlies the high prevalence of oculocutaneous albinism type 2 in the Navajo population." From: American Journal Human Genetics 72: 62–72, fig. 3 p. 67. Published by University of Chicago Press.

studying flanking alleles in deletion and nondeletion chromosomes, Brilliant and his colleagues estimated the age of the mutation to be between 400 and 11,000 years. To narrow this range, they relied on tribal history. Navajo oral tradition indicates that the Navajo and Apache became separate populations between 600 and 1000 years ago. Because the deletion is not found in the Apaches, it probably arose in the Navajo population after the tribes split. On this basis, the deletion is estimated to be 400 to 1000 years old and probably arose as a founder mutation.

Allele Loss during a Bottleneck

Although many populations have undergone well-documented bottlenecks with presumptive loss of genetic diversity, the amount of genetic variability present in the pre-bottleneck population—and therefore the amount of genetic variation lost in the bottleneck—often cannot be measured. In Illinois, the greater prairie chicken population was estimated in the millions in the 1860s. As a result of human activity, the population declined to 25,000 birds in 1933 and to fewer than 50 in 1993. Fitness, as measured by egg hatching, also declined over that time period, from a 93 percent hatch rate in 1935 to an estimated 56 percent in 1990. To estimate allele variation in pre- and post-bottleneck populations of the prairie chicken, Juan Bouzat and his colleagues used PCR to amplify alleles at six microsatellite loci. Samples were obtained from pre-bottleneck museum specimens and from present-day Illinois populations. In addition, allele frequencies were measured in prairie chicken populations from Kansas, Minnesota, and Nebraska, which have no known history of bottlenecks. A PCR analysis is shown in Figure 22–21, and the results of the allele survey are shown in Table 22.4.

For each locus, the alleles in the Illinois population are a subset of those in other populations and in the pre-bottleneck Illinois population. Because all alleles in the present-day Illinois population are shared with the other populations, they were probably present in the Illinois population before the bottleneck. Six alleles at four loci found in museum specimens are missing from the Illinois population, but most of them are common in the other populations. The

lost alleles represent a significant loss of genetic diversity caused by the bottleneck. In addition to allele loss, the Illinois population has significantly lower levels of heterozygosity. The outcomes are consequences of genetic drift, in this case caused by a population bottleneck. The small population size during the bottleneck and the resulting inbreeding are probably responsible for the decline in fitness currently observed in the Illinois population.

22.9 Nonrandom Mating Changes Genotype Frequency but Not Allele Frequency

We have explored how violations of the first four assumptions of the Hardy-Weinberg law, in the form of selection, mutation, migration, and genetic drift, can cause allele frequencies to change. The fifth assumption is that the members of a population mate at random; in other words, any one genotype has an equal probability of mating with any other genotype in the population. Nonrandom mating can change the frequencies of genotypes in a population. Subsequent selection for or against certain genotypes has the potential to affect the overall frequencies of the alleles they contain, but it is important to note that nonrandom mating *does not itself directly change allele frequencies.*

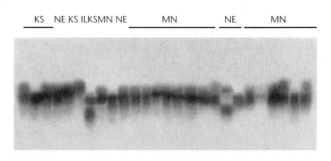

FIGURE 22–21 Alleles of the microsatellite locus *ADL42* detected in DNA samples of prairie chickens from Kansas (KS), Nebraska (NE), Illinois (IL), and Minnesota (MN).

TABLE 22.4	Microsatellite Alleles in Prairie Chicken Populations					
	Microsatellite Loci					
Population	*ADL42*	*ADL23*	*ADL44*	*ADL146*	*ADL162*	*ADL230*
Illinois	ABC	ABCD	ADFH	ABC	BE	CEFGHI
Kansas	ABCD	BCDEF	ACDEFGH	ABCDE	ABCDE	ABCDEFGHI
Minnesota	ABCD	ABCD	ABCDEFGH	ABCE	BCDE	BCDEFGHI
Nebraska	ABCD	ABCDE	ABCDEFGH	BCEF	BCDE	BCDEFGHIJK
Museum specimens	AB	BCDE	ADF	ABCE	BCD*E*FG	BCDEFG I *L*

Note: Highlighted letters indicate alleles found in the museum specimens which have been lost as a result of the prairie chickens' demographic contraction. Italic letters indicate alleles unique to the Illinois population prior to its demographic contraction.

Nonrandom mating can take one of several forms. In **positive assortive mating** similar genotypes are more likely to mate than dissimilar ones. This often occurs in humans: A number of studies have indicated that many people are more attracted to individuals who physically resemble them (and are therefore more likely to be genetically similar as well). **Negative assortive mating** occurs when dissimilar genotypes are more likely to mate; some plant species have inbuilt pollen/stigma recognition systems that prevent fertilization between individuals with the same alleles at key loci. However, the form of nonrandom mating most commonly found to affect genotype frequencies in population genetics is **inbreeding**.

Inbreeding

Inbreeding occurs when mating individuals are more closely related than any two individuals drawn from the population at random; loosely defined, inbreeding is mating among relatives. For a given allele, inbreeding increases the proportion of homozygotes in the population. A completely inbred population will theoretically consist only of homozygous genotypes. To demonstrate this, let us consider the most extreme form of inbreeding, **self-fertilization**, which though rare in animals is widespread in plant species. Figure 22–22 shows the results of four generations of self-fertilization, starting with a single individual heterozygous for one pair of alleles. By the fourth generation, only about 6 percent of the individuals are still heterozygous, and 94 percent of the population is homozygous. Note, however, that the frequencies of alleles *A* and *a* remain unchanged at 50 percent.

Not all inbreeding in populations occurs through self-fertilization, and different degrees of inbreeding exist. To describe the intensity of inbreeding in a population, geneticist Sewall Wright devised the **coefficient of inbreeding** (*F*). *F* quantifies the probability that the two alleles of a given gene in an individual are identical *because they are descended from the same single copy of the allele in an ancestor*. If *F* = 1, all individuals in the population are homozygous, and both alleles in every individual are derived from the same ancestral copy.

If *F* = 0, no individual has two alleles derived from a common ancestral copy.

One simple method of estimating *F* for a population is based on the inverse relationship between inbreeding and the frequency of heterozygotes: As the level of inbreeding increases, the proportion of heterozygotes declines. Therefore, *F* can be calculated as

$$F = \frac{H_e - H_o}{H_e}$$

where H_e is the expected heterozygosity based on the Hardy-Weinberg law and H_o is the observed heterozygosity in a population. Note that in a completely random mating population the expected and observed levels of heterozygosity will be equal and *F* = 0.

A different method can be used to estimate *F* for an individual. Figure 22–23 shows a pedigree of a first-cousin marriage. The fourth-generation female (shaded pink) is

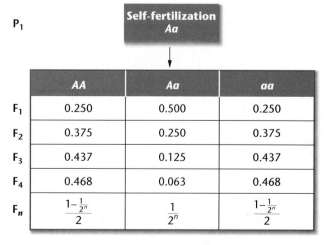

	AA	Aa	aa
F_1	0.250	0.500	0.250
F_2	0.375	0.250	0.375
F_3	0.437	0.125	0.437
F_4	0.468	0.063	0.468
F_n	$\dfrac{1 - \frac{1}{2^n}}{2}$	$\dfrac{1}{2^n}$	$\dfrac{1 - \frac{1}{2^n}}{2}$

P_1 — Self-fertilization *Aa*

FIGURE 22–22 Reduction in heterozygote frequency brought about by self-fertilization. After *n* generations, the frequencies of the genotypes can be calculated according to the formulas in the bottom row.

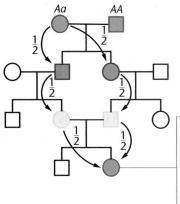

The chance that this female will inherit two copies of her great-grandmother's *a* allele is

$$F = \frac{1}{2} \times \frac{1}{2} \times \frac{1}{2} \times \frac{1}{2} \times \frac{1}{2} \times \frac{1}{2} = \frac{1}{64}$$

Because the female's two alleles could be identical by descent from any of four different alleles,

$$F = 4 \times \frac{1}{64} = \frac{1}{16}$$

FIGURE 22–23 Calculating the coefficient of inbreeding *(F)* for the offspring of a first-cousin marriage.

the daughter of first cousins (yellow). Suppose her great-grandmother (green) was a carrier of a recessive lethal allele, *a*. What is the probability that the fourth-generation female will inherit two copies of her great-grandmother's lethal allele? For this to happen, (1) the great-grandmother had to pass a copy of the allele to her son, (2) her son had to pass it to his daughter, and (3) his daughter had to pass it to her daughter (the pink female). Also, (4) the great-grandmother had to pass a copy of the allele to her daughter, (5) her daughter had to pass it to her son, and (6) her son had to pass it to his daughter (the pink female). Each of the six necessary events has an individual probability of 1/2, and they *all* have to happen, so the probability that the pink female will inherit two copies of her great-grandmother's lethal allele is $(1/2)^6 = 1/64$. To calculate an overall value of *F* for the pink female as a child of a first-cousin marriage, remember that she could also inherit two copies of any of the other three alleles present in her great-grandparents. Because any of four possibilities would give the pink female two alleles identical by descent from an ancestral copy,

$$F = 4 \times (1/64) = 1/16$$

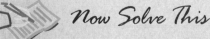

Now Solve This

Problem 21 on page 512 asks you to calculate the inbreeding coefficient of a population based on DNA marker data.

Hint: First, work out the frequencies of the DNA marker alleles. If this population is in Hardy-Weinberg equilibrium, how many heterozygotes would you expect to see?

Genetic Effects of Inbreeding

Inbreeding results in the production of individuals homozygous for recessive alleles that were previously concealed in heterozygotes. Because many recessive alleles are deleterious when homozygous, one consequence of inbreeding is an increased chance that an individual will be homozygous for a recessive deleterious allele. Inbred populations often have a lowered mean fitness. **Inbreeding depression** is a measure of the loss of fitness caused by inbreeding. In domesticated plants and animals, inbreeding and selection have been used for thousands of years, and these organisms already have a high degree of homozygosity at many loci. Further inbreeding will usually produce only a small loss of fitness. However, inbreeding among individuals from large, randomly mating populations can produce high levels of inbreeding depression. This effect can be seen by examining the mortality rates in offspring of inbred animals in zoo populations (Table 22.5). The potential impact of inbreeding on populations of threatened and endangered species is discussed further in Chapter 24.

In humans, inbreeding increases the risk of spontaneous abortions, neonatal deaths, congenital deformities, and recessive genetic disorders. Although less common than in the past, inbreeding occurs in many regions of the world where social customs favor marriage between first cousins. Alan Bittles and James Neel analyzed data from numerous studies on different cultures. They found that the rate of child mortality (i.e., death in the first several years of life) varies dramatically from culture to culture. No matter what the baseline mortality rate for children of unrelated parents, however, children of first cousins virtually always have a higher death rate—typically by about 4.5 percentage points.

TABLE 22.5	Mortality in Offspring of Inbred Zoo Animals			
Species	**n**		**Noninbred**	**Inbred**	**Inbreeding Coefficient**
Zebra	32	Lived:	20	3	0.250
		Died:	7	2	
Eld's deer	24	Lived:	13	0	0.250
		Died:	4	7	
Giraffe	19	Lived:	11	2	0.250
		Died:	3	3	
Oryx	42	Lived:	35	0	0.250
		Died:	2	5	
Dorcas gazelle	92	Lived:	36	17	0.269
		Died:	14	25	

It is important to note that inbreeding is not always harmful. Indeed, inbreeding has long been recognized as a useful tool for breeders of domesticated plants and animals. When an inbreeding program is initiated, homozygosity increases, and some breeding stocks become fixed for favorable alleles and others for unfavorable alleles. By selecting the more viable and vigorous plants or animals, the proportion of individuals carrying desirable traits can be increased.

If members of two inbred lines are mated, hybrid offspring are often more vigorous in desirable traits than either of the parental lines. This phenomenon is called **hybrid vigor**. When such an approach was used in breeding programs established for maize, crop yields increased tremendously. However, hybrid vigor is highest in the F_1 generation and typically declines in subsequent generations due to additional allelic segregation and recombination. Consequently, the F_1 hybrids must be regenerated each time by crossing the original inbred parental lines.

Hybrid vigor has been explained in two ways. The first theory, the **dominance hypothesis**, incorporates the obvious reversal of inbreeding depression, which inevitably must occur in outcrossing. Consider a cross between two strains of maize with the following genotypes:

Strain A		Strain B		F_1
aaBBCCddee	×	*AAbbccDDEE*	$\longrightarrow$	*AaBbCcDdEe*

The F_1 hybrids are heterozygotes at all loci shown. Any deleterious recessive alleles present in the homozygous form in the parents are masked by the more favorable dominant alleles in the hybrids. Such masking is thought to cause hybrid vigor.

The second theory, **overdominance**, holds that in many cases the heterozygote is superior to either homozygote. This may relate to the fact that in the heterozygote two forms of a gene product may be present, providing a form of biochemical diversity. Thus, the cumulative effect of heterozygosity at many loci accounts for the hybrid vigor. Most likely, hybrid vigor results from a combination of phenomena explained by both hypotheses.

We have seen that nonrandom mating can drive genotype frequencies in a population away from their expected values under the Hardy-Weinberg law. This can indirectly affect the course of evolution. As in stocks purposely inbred by animal and plant breeders, inbreeding in a natural population may increase the frequency of homozygotes for a deleterious recessive allele. With domestic stocks, this increases the efficiency with which selection removes the deleterious allele from the population. Consequently, once deleterious genes are removed, inbreeding no longer causes problems.

CHAPTER SUMMARY

1. Populations evolve as a result of changes in allele frequency at a number of loci over a period of time. Population genetics studies the factors driving change in allele frequencies and the amount and distribution of genetic variation in populations.

2. The Hardy-Weinberg law provides a simple mathematical model describing the relationship between allele frequency and genotype frequency in a population. This allows prediction of allele or genotype frequencies at a given locus in a population under a set of simple assumptions. If there is no selection, mutation, or migration, if the population is large, and if individuals mate at random, then allele frequencies will not change from one generation to the next.

3. The Hardy-Weinberg formula can be used to investigate whether or not a population is in evolutionary equilibrium at a given locus and to estimate the frequency of heterozygotes in a population from the frequency of homozygous recessives.

4. By specifying the conditions under which allele frequencies will not change, the Hardy-Weinberg law identifies the forces that can drive evolution in a population. Selection, mutation, migration, and genetic drift can cause change in allele frequencies.

5. Nonrandom mating alters genotype frequency but not allele frequency in a population. Inbreeding, or mating between relatives, is the form of nonrandom mating with the most significant impact: It increases the frequency of homozygotes in a population and decreases the frequency of heterozygotes.

6. Natural selection is the most powerful force affecting allele frequency. The rate of change under natural selection depends on initial allele frequencies, selection intensity, and the relative fitness of different genotypes.

7. Mutation and migration introduce new alleles into a population, but usually have only small effects on allele frequencies. To a large extent, the frequency of new alleles in a population depends on the fitness they confer and the action of selection.

8. Genetic drift produces random change in allele frequencies as a result of gamete sampling. It can have a major impact in small populations.

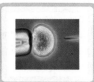

GENETICS, TECHNOLOGY, AND SOCIETY

Tracking Our Genetic Footprints out of Africa

Where did we come from? Are we one human family with minor differences, or are we separated into races with profound and ancient roots? For millennia, our efforts to answer these questions invoked legends, mythologies, and the creation stories of our many religions. Over the last century, paleoanthropologists have applied a variety of increasingly sophisticated scientific tools to explore our origins and human kinships. The evolving story of our beginnings, based on modern genetics, is as fascinating and controversial as any creation myth.

Based on the physical traits and distribution of hominid fossils, most paleoanthropologists agree that a large-brained, tool-using hominid named *Homo erectus* appeared in East Africa about 2 million years ago. This species used simple stone tools, hunted but did not fish, did not build houses or fireplaces, and lacked ritual burial practices. About 1.7 million years ago, *H. erectus* spread into Eurasia and South Asia. Most scientists also agree that *H. erectus* likely developed into several hominid types including Neanderthals (in Europe) and Peking man or Java man (in Asia). These hominids were anatomically robust, with large heavy skeletons and skulls. Neanderthals and other *H. erectus* groups disappeared 50,000 to 30,000 years ago—around the same time that anatomically modern humans (*H. sapiens*) appeared all over the world.

It is at this point in our history—when ancient hominids gave way to lighter-skeletoned, anatomically modern humans—that controversy arises.

At present, there are two dominant hypotheses to explain our origins: the multiregional and the out-of-Africa hypotheses. The multiregional hypothesis is based primarily on archaeological and fossil evidence. It proposes that *H. sapiens* developed gradually and simultaneously all over the world from existing *H. erectus* groups, including Neanderthals. Interbreeding between these groups eventually made *H. sapiens* a genetically homogeneous species. Natural selection over 1.5 million years then created the regional variants (races) that we see today. In the multiregional view, our genetic makeup should include contributions from Neanderthals

and other *H. erectus* groups. In contrast, the out-of-Africa hypothesis, based primarily on genetic analyses of modern human populations, contends that *H. sapiens* evolved from *H. erectus* in sub-Saharan Africa about 200,000 to 400,000 years ago. A small band of *H. sapiens* (fewer than 10,000) then left Africa, expanded, and migrated into Europe and Asia around 100,000 years ago. By about 60,000 years ago, populations of *H. sapiens* reached Australia and later migrated into North America. In the out-of-Africa model, *H. sapiens* replaced all the preexisting *H. erectus* types, without interbreeding. In this way, *H. sapiens* became the only species in the genus by about 30,000 years ago.

Although still contentious, most genetic evidence supports the out-of-Africa hypothesis. Humans all over the globe are remarkably similar genetically. DNA sequences from any two people chosen at random are 99.9 percent identical. There is more genetic identity between two persons chosen at random from a human population than there is between two chimpanzees chosen at random from a chimpanzee population. Interestingly, about 90 percent of the genetic differences that do exist occur between individuals rather than between populations. This unusually high degree of genetic relatedness in all humans around the world supports the idea that our species arose recently from a small founding group of humans.

Studies of mitochondrial DNA sequences from current human populations reveal that the highest levels of genetic variation occur within African populations. Africans show twice the mitochondrial DNA-sequence diversity of non-Africans. This implies that the earliest branches of *H. sapiens* diverged in Africa and had a longer time to accumulate mitochondrial DNA mutations, which are thought to accumulate at a constant rate over time.

DNA sequences from mitochondrial, Y-chromosome, and chromosome-21 markers support the idea that our roots are in East Africa and that the migration out of Africa occurred through Ethiopia, along the coast of the Arabian peninsula, and outward to Eurasia and Southeast Asia. Recent data based on DNA-sequence diversity in nuclear microsatellites further support the notion that

humans migrated out of Africa and dispersed throughout the world from a small founding population. Sub-Saharan African populations show the highest levels of microsatellite heterozygosity, followed by those in the Middle East, Europe, East Asia, Oceania, and the Americas—in that order. Native American populations show about 15 percent less microsatellite heterozygosity than that seen in Africans.

By comparing DNA-sequence differences between populations around the world and by extrapolating back to a time when all sequences would have been the same, paleoanthropologists propose that modern *H. sapiens* developed from a small group in Africa between 200,000 and 400,000 years ago. The time of the out-of-Africa migration is calculated to be 50,000 to 100,000 years ago.

The recent sequencing of Neanderthal mitochondrial DNA shows that it is so different from ours that Neanderthals were likely a separate species and that Neanderthals and *H. sapiens* diverged about 600,000 years ago. Hence, it appears unlikely that Neanderthals or other *H. erectus* groups such as Peking man contributed significantly to the *H. sapiens* gene pool.

To trace the migration routes of *H. sapiens* and to answer other questions about the origins of our species, a privately funded program, called the Genographic Project, has been organized. Under the direction of the anthropologist Spencer Wells, the Genographic Project seeks to obtain DNA samples from 100,000 individuals. Interested individuals can purchase kits containing cheek swabs for collecting cheek cells for DNA extraction. Analysis will focus on Y chromosome and mitochondrial markers that have been used to provide the outlines of how Earth was colonized by our species. The results can be downloaded from the project website four to six weeks after the swabs are submitted.

So, if all people on Earth are so similar genetically, how did we come to have such a range of physical differences, which some describe as racial differences? Many geneticists believe that the genetic changes responsible for these characteristics such as skin color and facial features could accumulate over short periods of time, especially if these characteristics are adaptive to particular climatic and geographic conditions.

As with any explanation of human origins, the out-of-Africa hypothesis is actively debated and may undergo mutation—or even extinction—over time. As methods to sequence DNA from ancient fossils improve, it may be possible to fill the gaps in our genetic pathway leading out of Africa and help us to resolve those age-old questions about our origins.

Reference

Cavalli-Sforza, L. L., and Feldman M. W., 2003. The application of molecular genetic approaches to the study of human evolution. *Nat. Gen. (Suppl.)* 33: 266–275.

Wells, S. 2003. *The Journey of Man: A Genetic Odyssey.* Princeton, N.J.: Princeton University Press.

Web sites

Johanson, D. 2001. *Origins of modern humans: Multiregional or out of Africa?* [online] **http://www. actionbioscience. org/evolution/ johanson.html**

KEY TERMS

INSIGHTS AND SOLUTIONS

1. Tay-Sachs disease is caused by loss-of-function mutations in a gene on chromosome 15 that encodes a lysosomal enzyme. Tay-Sachs is inherited as an autosomal recessive condition. Among Ashkenazi Jews of central European ancestry, about 1 in 3600 children is born with the disease. What fraction of the individuals in this population are carriers?

Solution: If we let p represent the frequency of the wild-type enzyme allele and q the total frequency of recessive loss-of-function alleles, and if we assume that the population is in Hardy-Weinberg equilibrium, then the frequencies of the genotypes are given by p^2 for homozygous normal, $2pq$ for carriers, and q^2 for individuals with Tay-Sachs. The frequency of Tay-Sachs alleles is thus

$$q = \sqrt{q^2} = \sqrt{\frac{1}{3600}} = 0.017$$

Since $p + q = 1$, we have

$$p = 1 - q = 1 - 0.017 = 0.983$$

Therefore, we can estimate that the frequency of carriers is

$$2pq = 2(0.983)(0.017) = 0.033 \text{ or } 1 \text{ in } 30$$

2. *Eugenics* is the term employed for the selective breeding of humans to bring about improvements in populations. As a eugenic measure, it has been suggested that individuals suffering from serious genetic disorders should be prevented (sometimes by force of law) from reproducing (by sterilization, if necessary) in order to reduce the frequency of the disorders in future generations. Suppose that such a recessive trait were present in the population at a frequency of 1 in 40,000 and that affected individuals did not reproduce. In

10 generations, or about 250 years, what would be the frequency of the condition? Are the eugenic measures effective in this case?

Solution: Let q represent the frequency of the recessive allele responsible for the disorder. Because the disorder is recessive, we can estimate that

$$q = \frac{1}{40,000} = 0.005$$

If all affected individuals are prevented from reproducing, then in an evolutionary sense, the disorder is lethal: Affected individuals have zero fitness. This means that we can predict the frequency of the recessive allele 10 generations in the future by using the following equation:

$$q_g = q_0/(1 + gq_0)$$

Here, $q_0 = 0.005$, and $g = 10$, so we have

$$q_{10} = \frac{(0.005)}{[1 + (10 \times 0.005)]}$$
$$= 0.0048$$

If $q_{10} = 0.0048$, then the frequency of homozygous recessive individuals will be roughly

$$(q_{10})^2 = (0.0048)^2 = 0.000023 = \frac{1}{43,500}$$

The frequency of the genetic disorder has been reduced from 1 in 40,000 to 1 in 43,500 in 10 generations, indicating that this eugenic measure has limited effectiveness.

PROBLEMS AND DISCUSSION QUESTIONS

1. The ability to taste the compound PTC is controlled by a dominant allele *T*, while individuals homozygous for the recessive allele *t* are unable to taste PTC. In a genetics class of 125 students, 88 can taste PTC and 37 cannot. Calculate the frequency of the *T* and *t* alleles in this population and the frequency of the genotypes.

2. Calculate the frequencies of the *AA*, *Aa*, and *aa* genotypes after one generation if the initial population consists of 0.2 *AA*, 0.6 *Aa*, and 0.2 *aa* genotypes and meets the requirements of the Hardy-Weinberg relationship. What genotype frequencies will occur after a second generation?

3. Consider rare disorders in a population caused by an autosomal recessive mutation. From the frequencies of the disorder in the population given, calculate the percentage of heterozygous carriers:
   (a) 0.0064
   (b) 0.000081
   (c) 0.09
   (d) 0.01
   (e) 0.10

4. What must be assumed in order to validate the answers in Problem 3?

5. In a population where only the total number of individuals with the dominant phenotype is known, how can you calculate the percentage of carriers and homozygous recessives?

6. Determine whether the following two sets of data represent populations that are in Hardy-Weinberg equilibrium (use χ^2 analysis if necessary):
   (a) *CCR5* genotypes: *1/1*, 60 percent; *1/Δ32*, 35.1 percent; *Δ32/Δ32*, 4.9 percent
   (b) Sickle-cell hemoglobin: *AA*, 75.6 percent; *AS*, 24.2 percent; *SS*, 0.2 percent

7. If 4 percent of a population in equilibrium expresses a recessive trait, what is the probability that the offspring of two individuals who do not express the trait will express it?

8. Consider a population in which the frequency of allele *A* is $p = 0.7$ and the frequency of allele *a* is $q = 0.3$, and where the alleles are codominant. What will be the allele frequencies after one generation if the following occurs?
   (a) $w_{AA} = 1$, $w_{Aa} = 0.9$, and $w_{aa} = 0.8$
   (b) $w_{AA} = 1$, $w_{Aa} = 0.95$, $w_{aa} = 0.9$
   (c) $w_{AA} = 1$, $w_{Aa} = 0.99$, $w_{aa} = 0.98$
   (d) $w_{AA} = 0.8$, $w_{Aa} = 1$, $w_{aa} = 0.8$

9. If the initial allele frequencies are $p = 0.5$ and $q = 0.5$ and allele *a* is a lethal recessive, what will be the frequencies after 1, 5, 10, 25, 100, and 1000 generations?

10. Under what circumstances might a lethal dominant allele persist in a population?

11. Determine the frequency of allele *A* in an island population after one generation of migration from the mainland under the following conditions:
    (a) $p_i = 0.6$; $p_m = 0.1$; $m = 0.2$
    (b) $p_i = 0.2$; $p_m = 0.7$; $m = 0.3$
    (c) $p_i = 0.1$; $p_m = 0.2$; $m = 0.1$

12. Assume that a recessive autosomal disorder occurs in 1 of 10,000 individuals (0.0001) in the general population and that in this population about 2 percent (0.02) of the individuals are carriers for the disorder. Estimate the probability of this disorder occurring in the offspring of a marriage between first cousins. Compare this probability to the population at large.

13. What is the basis of inbreeding depression?

14. Describe how inbreeding can be used in the domestication of plants and animals. Discuss the theories underlying these techniques.

15. Evaluate the following statement: Inbreeding increases the frequency of recessive alleles in a population.

16. In a breeding program to improve crop plants, which of the following mating systems should be employed to produce a homozygous line in the shortest possible time?
    (a) self-fertilization
    (b) brother–sister matings
    (c) first-cousin matings
    (d) random matings
    Illustrate your choice with pedigree diagrams.

17. If the albino phenotype occurs in 1/10,000 individuals in a population at equilibrium and albinism is caused by an autosomal recessive allele *a*, calculate the frequency of
    (a) the recessive mutant allele
    (b) the normal dominant allele
    (c) heterozygotes in the population
    (d) matings between heterozygotes

18. One of the first Mendelian traits identified in humans was a dominant condition known as *brachydactyly*. This gene causes an abnormal shortening of the fingers or toes (or both). At the time, it was thought by some that the dominant trait would spread until 75 percent of the population would be affected (because the phenotypic ratio of dominant to recessive is 3:1). Show that the reasoning is incorrect.

19. Achondroplasia is a dominant trait that causes a characteristic form of dwarfism. In a survey of 50,000 births, five infants with achondroplasia were identified. Three of the affected infants had affected parents, while two had normal parents. Calculate the mutation rate for achondroplasia and express the rate as the number of mutant genes per given number of gametes.

20. A prospective groom, who is normal, has a sister with cystic fibrosis (CF), an autosomal recessive disease. Their parents are normal. The brother plans to marry a woman who has no history of CF in her family. What is the probability that they will produce a CF child? They are both Caucasian, and the overall frequency of CF in the Caucasian population is 1/2500—that is, 1 affected child per 2500. (Assume the population meets the Hardy-Weinberg assumptions.)

21. A botanist studying water lilies in an isolated pond observed three leaf shapes in the population: round, arrowhead, and scalloped. Marker analysis of DNA from 125 individuals showed the round-leafed plants to be homozygous for allele *r1*, while the plants with arrowhead leaves were homozygous for a different allele at the same locus, *r2*. Plants with scalloped leaves showed DNA profiles with both the *r1* and *r2* markers. Frequency of the *r1* marker was estimated at 0.81. If the botanist counted 20 plants with scalloped leaves in the pond, what is the inbreeding coefficient *F* for this population?

22. A farmer plants transgenic Bt corn that is genetically modified to produce its own insecticide. Of the corn borer larvae feeding on these Bt corn plants, only 10 percent survive unless they have at least one copy of the dominant resistance allele *B* that confers resistance to the Bt insecticide. When the farmer first plants Bt corn, the frequency of the *B* resistance allele in the corn borer

population is 0.02. What will be the frequency of the resistance allele after one generation of corn borers fed on Bt corn?

23. In an isolated population of 50 desert bighorn sheep, a mutant recessive allele c has been found to cause curled coats in both males and females. The normal dominant allele C produces straight coats. A biologist studying these sheep counts four with curled coats and takes blood samples for DNA marker analysis, which reveals that 17 of the straight-coated sheep are carriers of the c allele. What is the inbreeding coefficient F for this population?

24. To increase genetic diversity in the bighorn sheep population described in Problem 23, 10 sheep are introduced from a population in a different region where the c mutation is absent. Assuming that random mating occurs between the original and the introduced sheep, and that the c allele is selectively neutral, what will be the frequency of c in the next generation?

25. A form of dwarfism known as Ellis-van Creveld syndrome was first discovered in the late 1930s, when Richard Ellis and Simon van Creveld shared a train compartment on the way to a pediatrics meeting. In the course of conversation, they discovered that they each had a patient with this syndrome. They published a description of the syndrome in 1940. Affected individuals have a short-limbed form of dwarfism and often have defects of the lips and teeth, and polydactyly (extra fingers). The largest pedigree for the condition was reported in an Old Order Amish population in eastern Pennsylvania by Victor McKusick and his colleagues (1964). In that community, about 5 per 1000 births are affected, and in the population of 8000, the observed frequency is 2 per 1000. All affected individuals have unaffected parents, and all affected cases can trace their ancestry to Samuel King and his wife, who arrived in the area in 1774. It is known that neither King nor his wife was affected with the disorder. There are no cases of the disorder in other Amish communities, such as those in Ohio or Indiana.

(a) From the information provided, derive the most likely mode of inheritance of this disorder. Using the Hardy-Weinberg law, calculate the frequency of the mutant allele in the population and the frequency of heterozygotes, assuming Hardy-Weinberg conditions.

(b) What is the most likely explanation for the high frequency of the disorder in the Pennsylvania Amish community and its absence in other Amish communities?

26. The following graph shows the variation in the frequency of a particular allele (A) that occurred over time in two relatively small, independent populations exposed to very similar environmental conditions. A student has analyzed the graph and concluded that the best explanation for these data is that the selection for allele A is occurring in Population 1. Is the student's conclusion correct? Explain.

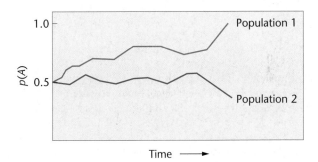

23 Evolutionary Genetics

Light and dark forms of the peppered moth Biston betularia on light tree bark. These moths are often cited as examples of rapid evolutionary change resulting from selective pressures during the industrial revolution in Great Britain.

■ CHAPTER CONCEPTS

- Speciation can occur by transformation of a single species over time, or by splitting gene pools so that one species evolves into two or more new species.

- Genetic variation is present in populations and species.

- The genetic structure of populations changes over time. Some of this variation is the product of natural selection.

- Several factors, including reduced gene flow, selection, or genetic drift can lead to speciation.

- Evolutionary history can often be reconstructed by studying genetic differences among populations or species.

In Chapter 22, we described how populations evolve by changes in allele frequencies. We also outlined forces that cause allele frequencies to change. Mutation, migration, selection, and drift—individually and collectively—bring about evolutionary divergence and species formation. This process involves not only genetic divergence but also the presence of environmental or ecological diversity. If a population is spread over a geographic range that contains a number of subenvironments or niches, populations can adapt to the conditions in these niches and become genetically differentiated. Differentiated populations are dynamic: They may continue to exist, become extinct, merge with the parental population, or diverge from the parental population until they become reproductively isolated and form a new species. This process can modify a species over time, transforming it into another species, or it can result in one species splitting into two or more species.

A **species** can be defined as a group of interbreeding or potentially interbreeding organisms that is reproductively isolated in nature from all other such groups. In sexually reproducing organisms, **speciation** divides a single gene pool into two or more separate gene pools. Changes in morphology, physiology, and adaptation to an ecological niche may also occur but are not necessary components of the speciation event. Speciation can take place gradually or within a few generations, but it is difficult to define the exact moment when a new species forms.

Because population and species divergence is accompanied by genetic differentiation, we can use patterns of genetic differences to reconstruct evolutionary history. After exploring the genetic structure of populations, their divergence across space and time, and the process of speciation, we'll discuss how genetic data can be used to answer questions that have an evolutionary context.

? How Do We Know ?

In this chapter, we will focus on how forces such as selection and drift act on genetic variation to bring about evolutionary change in populations and the formation of new species. As you study this topic, you should try to answer several fundamental questions:

1. How do we know how much genetic variation is in a population?

2. How do we know when populations have diverged to the point that they form two different species?

3. How do we know whether the genetic structure of a population is static or dynamic?

4. How do we know how much time is required to form species?

5. How do we know the age of the last common ancestor shared by two species?

23.1 Defining A Species Is a Challenge for Evolutionary Biology

Before we consider how genetic divergence can lead to speciation, let's examine how biologists in various disciplines define a species. Charles Darwin and other naturalists of the late nineteenth century relied on morphology to define a species. Closely related species are often separated by distinct differences in morphological characters. This approach is useful with living species as well as those from the fossil record, but it also poses several problems. Even within recognized species there are often large gaps in characters and, conversely, there are many examples of morphologically very similar species that nonetheless can not interbreed and must be regarded as separate species. In addition, where there is conflict in classification, it is often not clear which morphological characteristics should be used to classify the species.

Our growing understanding of population genetics and evolution in the mid-twentieth century led to the biological species concept. A biological species is defined as *a group of interbreeding or potentially interbreeding populations reproductively isolated in nature from all other such groups*. As such, these species have distinct gene pools, so that speciation occurs when one gene pool divides into two or more separate gene pools. Changes in morphology, physiology, and adaptation to an ecological niche may also occur, but are not considered necessary components of speciation, which can take place gradually or within a few generations.

Zoologists in particular favor the biological species concept since it is easily applied to animals, most of whom have distinct sexes and exchange genes through sexual reproduction. However, this definition also has problems. Many species, such as most prokaryotes, do not reproduce sexually and so individuals within the species do not interbreed. Yet clearly, these individuals are not all different species! Also, about 20 percent of plant species and 10 percent of bird and butterfly species can interbreed yet are recognized as distinct species. The presence in genomes of genes resulting from horizontal gene transfer also suggests that hybridization across species is not uncommon, and argues against defining species solely by reproductive isolation. In addition, it is impossible to use the biological species concept for extinct species since we are not able to determine the presence of reproductive isolation in fossils. Even for living species it may be impossible to test for reproductive isolation in a cluster of closely related putative species.

With the establishment of databases that contain a wealth of molecular and genomic sequence information, our ability to identify distinct species has been enhanced. This has resulted in a third way of defining species, based on phylogeny. According to this definition, a species consists of those organisms that are monophyletic (belong to the same taxonomic group) and share one or more uniquely derived characteristics. This approach has its problems as well, because we cannot be certain that we have determined the correct relationships among all possible relatives in a phylogeny to identify these monophyletic groups (called clades).

Finally, some biologists use an ecological species concept, in which species are defined as groups of individuals that live in the same type of environment and have shared ecological requirements. The main problem here is that there are examples in which populations that seem to be the same species (i.e., they readily interbreed with one another) have very distinct ecological requirements.

The scientific community is engaged in an ongoing dialogue about how to define a species. This is one of the most important and challenging questions in evolutionary biology, but for now, has no clear answer. The biological species concept is most often used by population geneticists studying evolution, and will be used in the following discussions about the process of species formation. Its emphasis on reproductive isolation not only helps define species, but also provides a focus for thinking about the process of speciation.

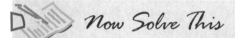

Now Solve This

In Problem 4 on page 535 you are asked how phenotypically similar species can be identified as separate species, rather than as the same species with some phenotypic variation.

Hint: The species concept that you rely upon will influence the answer to this question.

23.2 Speciation Can Occur by Transformation or by Splitting Gene Pools

Figure 23–1 shows an evolutionary tree, or **phylogeny**, that describes the history of several hypothetical lizard species. The passage of time is plotted horizontally as changes in the phenotype of the lizards occur and the lineages diverge or separate.

The history of these lizards begins with species 1. For some time, that species experiences evolutionary **stasis**; that is, it does not change. Species 1 then undergoes a process of steady transformation, called **phyletic evolution**, or **anagenesis**, and becomes species 2. During this process, only one species is present at all times, and it becomes difficult to identify the precise time at which species 1 becomes 2. After species 2 forms, it undergoes **cladogenesis**, whereby it gives rise to two

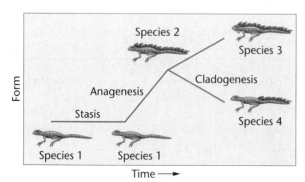

FIGURE 23–1 In phyletic evolution, or anagenesis, one species is transformed over time into another species. At all times, only one species exists. In cladogenesis, one species splits into two or more species.

distinct and independent daughter species. Further transformation of these daughter species produces the species we see today, species 3 and species 4. These extant species are in many cases experimentally verifiable. That is, they may be tested to determine their reproductive compatibility (one species) or incompatibility (two species).

In his 1859 book, *On the Origin of Species*, Charles Darwin amassed evidence that all species derive from a single common ancestor by transformation and speciation:

> All living things have much in common, in their chemical composition, their germinal vesicles, their cellular structure, and their laws of growth and reproduction.... Therefore I should infer ... that probably all the organic beings which have ever lived on this earth have descended from some one primordial form.

Everything biologists have since learned supports Darwin's conclusion that only one tree of life exists. To understand evolution, we must understand the mechanisms that transform one species into another and that split one species into two or more.

Chapter 22 covered the mechanisms responsible for the transformation of species. Chief among them is natural selection, discovered independently by Alfred Russel Wallace and Darwin. The Wallace–Darwin concept of natural selection can be summarized as follows:

1. Individuals of a species exhibit variations in phenotype—for example, differences in size, agility, coloration, defenses against enemies, ability to obtain food, courtship behaviors, flowering times, and so on.

2. Many of these variations, even small and seemingly insignificant ones, are heritable and passed on to offspring.

3. Organisms tend to reproduce in an exponential fashion. More offspring are produced than can survive. This causes members of a species to engage in a struggle for survival, competing with other members of the community for scarce resources. Offspring also must avoid predators, and in sexually reproducing species, adults must compete for mates.

4. In the struggle for survival, individuals with particular phenotypes will be more successful than individuals with others, allowing the former to survive and reproduce at higher rates.

As a consequence of natural selection, species change. The phenotypes that confer improved ability to survive and reproduce become more common, and the phenotypes that confer poor prospects for survival and reproduction may eventually disappear. Under certain conditions, populations that at one time could interbreed may lose that capability, thus segregating their adaptations into particular niches. If this adaptive selection is imposed on the hybrids produced by parents from different populations, selection favors the formation of two species where there was once only one.

Although Wallace and Darwin proposed that natural selection explains how evolution occurs, they could not explain either the origin of the variations that provided the raw material for evolution or how such variations are passed from parents to offspring. In the twentieth century, as biologists applied the principles of Mendelian genetics to populations, the source of variation (mutation) and the mechanism of inheritance (segregation of alleles) were both explained. We now view evolution as due to changes in allele frequencies in populations over time. This union of population genetics with the theory of natural selection generated a new view of the evolutionary process, called *neo-Darwinism*.

In this chapter, we consider the mechanisms responsible for speciation. As with natural selection, our understanding of speciation is built on key insights about genetic variation.

23.3 Genetic Variation Is Present in Most Populations and Species

At first glance, it seems that members of a well-adapted population should be highly homozygous because the most favorable allele at each locus has become fixed. Certainly, an examination of most populations of plants and animals reveals many phenotypic similarities among individuals. However, a large body of evidence indicates that most populations contain a high degree of heterozygosity. This built-in genetic diversity is concealed, so to speak, because it is not necessarily apparent in the phenotype; hence, detecting it is not a simple task. Nevertheless, with the use of the techniques discussed next, such investigation has been successful.

Artificial Selection Reveals Variation

One way to determine whether genetic variation affects a phenotypic character is to impose artificial selection on the character. A phenotype with no genetic variation will not respond to selection; if there is genetic variation, the phenotype will change over a few generations. A dramatic example of this is the domestic dog. The broad array of sizes, shapes, colors, and behaviors seen in different breeds of dogs all arose from selection on the genetic variation present in wild wolves, from which all domestic dogs are descended. Genetic and archeological evidence indicates that the domestication of dogs took place at least 14,000 years ago, and possibly much earlier. On a shorter time scale, laboratory selection experiments on the fruit fly *Drosophila melanogaster* have caused significant changes over a few generations in almost every phenotype imaginable, including size, shape, developmental rate, fecundity, and behavior.

Variations in Amino Acid Sequence

Gel electrophoresis separates protein molecules on the basis of differences in size and electrical charge. If a nucleotide change in a structural gene causes the substitution of a charged amino acid (such as glutamic acid) for an uncharged amino acid (such as glycine), the protein's net electrical charge will change. This difference in charge can be detected as a change in the rate at which proteins migrate through an electrical field. In the mid-1960s, John Hubby and Richard Lewontin used gel electrophoresis to measure protein variation in natural populations of *Drosophila*, and Harry Harris used the same techniques to measure human variation. In subsequent years, researchers have used the technique to study genetic variation in a wide range of organisms (Table 23.1), although subsequently developed DNA techniques have in many cases replaced the use of protein electrophoresis.

The electrophoretically distinct forms of an enzyme produced by different alleles are called *allozymes*. As shown in the table, a surprisingly large percentage of loci from diverse species produce distinct allozymes. Of the populations listed, approximately 30 loci per species were examined, and about 30 percent of the loci were detectably polymorphic, with an average of 10 percent allozyme heterozygosity per diploid genome. These estimates clearly support the notion that organismal genotypes harbor vast amounts of DNA-based variability.

These values apply only to genetic variation detectable by altered protein migration in an electric field. Electrophoresis probably detects only about 30 percent of the actual variation due to amino acid substitutions, because many substitutions do not change the net electric charge on the molecule. Richard Lewontin estimated that about two-thirds of all loci in a population are polymorphic. In any individual within the population, about one-third of the loci exhibit genetic variation in the form of heterozygosity. The significance of genetic variation detected by electrophoresis is controversial. Some argue that allozymes are functionally equivalent and therefore do not play any role in adaptive evolution. We address this argument later in this section.

TABLE 23.1	Allozyme Heterozygosity at the Protein Level			
Species	**Populations Studied**	**Loci Examined**	**Polymorphic Loci* per Population (%)**	**Heterozygotes per Locus (%)**
Homo sapiens (humans)	1	71	28	6.7
Mus musculus (mouse)	4	41	29	9.1
Drosophila pseudoobscura (fruit fly)	10	24	43	12.8
Limulus polyphemus (horseshoe crab)	4	25	25	6.1

* A polymorphic locus is one for which a population harbors more than one allele.
Source: From R.C. Lewontin, New York. Columbia University Press. *The Genetic Basic of Evolutionary Change*, copyright 1974, p. 117.

Variations in Nucleotide Sequence

The most direct way to estimate genetic variation is to compare the nucleotide sequences of genes and intergenic regions carried by individuals in a population. With the development of techniques for PCR, cloning and sequencing DNA, and with the proliferation of genome projects, nucleotide sequence variations have been catalogued for an increasing number of genes and genomes. In an early study, Alec Jeffreys used restriction enzymes to detect polymorphisms in 60 unrelated individuals. His aim was to estimate the total number of DNA sequence variants in humans. His results showed that within the genes of the β-globin cluster, 1 in 100 base pairs showed polymorphic variation. If that region is representative of the genome, this study indicates that at least 3×10^7 nucleotide variants per genome are possible.

In another study, Martin Kreitman examined the *alcohol dehydrogenase* locus (*Adh*) in *Drosophila melanogaster* (Figure 23–2). This locus encodes two allozymic variants: the *Adh-f* and the *Adh-s* alleles. These differ by only a single amino acid (thr versus lys at codon 192). To determine whether the amount of genetic variation detectable at the protein level (one amino acid difference) corresponds to the variation at the nucleotide level, Kreitman cloned and sequenced *Adh* loci from five natural populations of *Drosophila*.

The 11 cloned loci contained 43 nucleotide variations from the consensus *Adh* sequence of 2721 base pairs. These variations are distributed throughout the gene: 14 in the exon coding regions, 18 in the introns, and 11 in the untranslated and flanking regions. Of the 14 variations in coding regions, only one leads to an amino acid replacement—the one in codon 192, producing the two observed electrophoretic variants. The other 13 coding region nucleotide substitutions do not lead to amino acid replacements. Are the differences in the number of allozyme and nucleotide variants the result of natural selection? If so, of what significance is this fact? Let's examine these questions.

Among the most intensively studied loci in humans is the locus encoding the cystic fibrosis transmembrane conductance regulator (CFTR). Recessive loss-of-function mutations in the *CFTR* locus cause **cystic fibrosis**, a disease with symptoms including salty skin and the production of an excess of thick mucus in the lungs, which leads to susceptibility to bacterial infections. Geneticists have examined the *CFTR* locus and have found more than 1500 different mutations that can cause the disease. Among these mutations are missense mutations, amino acid deletions, nonsense mutations, frameshifts, and splice defects.

Figure 23–3 shows a map of the 27 exons in the *CFTR* locus, with most exons identified by function. The histogram above the map shows the locations of some of the disease-causing mutations and the number of copies of each mutation that have been found. One mutation, a 3-bp deletion in exon 10 called Δ508, accounts for about 70 percent of all mutant cystic fibrosis alleles. In populations of European ancestry, between 1 in 44 and 1 in 20 individuals are heterozygous carriers of mutant alleles. Note that Figure 23–3 includes only mutations that alter the function of the CFTR protein. There are undoubtedly many more *CFTR* alleles with silent mutations that do not change the structure of the protein and that do not affect its function.

Studies of other organisms, including the rat, the mouse, and the mustard plant *Arabadopsis thaliana*, have produced similar estimates of nucleotide diversity in various genes. These studies indicate that there is an enormous reservoir of genetic variability within most populations and that, at the level of DNA, most, and perhaps all, genes exhibit diversity from individual to individual.

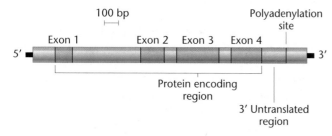

FIGURE 23–2 Organization of the *Adh* locus of *Drosophila melanogaster*.

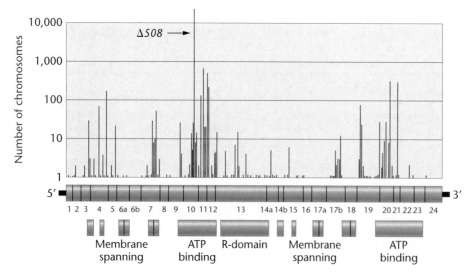

FIGURE 23–3 The locations of some disease-causing mutations in the cystic fibrosis gene. The histogram shows the frequency of each mutation geneticists have found. (The vertical axis is on a logarithmic scale.) The genetic map below the histogram shows the locations and relative sizes of the 27 exons of the *CFTR* locus. The boxes at the bottom indicate the functions of different domains of the CFTR protein.

Problem 11 on page 535 asks you to determine what types of nucleotide substitutions will *not* be detected by electrophoretic studies of proteins.

Hint: Enzymes are detected in gels by their activity in converting their natural substrate into a detectable product. Allozymes that migrate at different speeds in the gel matrix do so because of their variable charge, caused by differences in amino acid sequence.

23.4 How Can We Explain High Levels of Genetic Variation?

As mentioned earlier, the finding that populations harbor considerable genetic variation at the amino acid and nucleotide levels came as a surprise to many evolutionary biologists. The early consensus was that selection would favor a single optimal (wild-type) allele at each locus and that, as a result, populations would have high levels of homozygosity. This expectation was obviously wrong, and considerable research and argument ensued concerning the forces that maintain genetic variation.

The **neutral theory** of molecular evolution, proposed by Motoo Kimura, argues that mutations leading to amino acid substitutions are usually detrimental, with a very small fraction that are favorable. Some mutations are neutral or functionally equivalent to the allele that is replaced. Those polymorphisms that are favorable or detrimental are preserved or removed from the population, respectively, by natural selection. However, the frequency of the neutral alleles in a population will be determined by mutation rates and random genetic drift. Some neutral mutations will drift to fixation in the population; other neutral mutations will be lost. At any given time, the population may contain several neutral alleles at any particular locus. The diversity of alleles at most polymorphic loci does not, however, reflect the action of natural selection, but instead is a function of population size (larger populations have more variation) and the fractions of mutations that are neutral.

The alternative explanation for the high levels of variation is **natural selection**. There are several examples in which enzyme or protein polymorphisms are maintained by adaptation to certain environmental conditions. The well-known advantage of sickle-cell anemia heterozygotes when infected by malarial parasites is such an example, and another was discussed in Chapter 22, when we considered evidence that polymorphism at the *CFTR* locus may reflect the superior fitness of heterozygotes in areas where typhoid fever is common.

Fitness differences of a fraction of a percent would be sufficient to maintain a polymorphism, but this would be difficult to measure. Current data are therefore insufficient to determine what fraction of molecular genetic variation is neutral and what fraction is subject to selection. The neutral theory nonetheless serves a crucial function: By pointing out that some genetic variation is expected simply as a result of muta-tion and drift, the neutral theory provides a working hypothesis for studies of molecular evolution. In other words, biologists must find positive evidence that selection is acting on allele frequencies at a particular locus before they can reject the simpler assumption that only mutation and drift are at work.

23.5 The Genetic Structure of a Species Can Vary From Population to Population

As population geneticists discovered that most populations harbor considerable genetic diversity, they also found that the genetic structure of populations varies across space and time. To illustrate this idea, let's look at studies on *Drosophila pseudoobscura* conducted by Theodosius Dobzhansky and his colleagues. This species is found over a wide range of environmental habitats, including the western and southwestern United States. Although the flies throughout this range are morphologically similar, Dobzhansky's team discovered that populations of this species from different geographical locations vary in the arrangement of genes on chromosome 3. They found several different inversions in this chromosome that can be detected by loop formations in larval polytene chromosomes. (See Chapter 6 for a description of polytene chromosomes.) Each inversion sequence is named after the locale in which it was first discovered (e.g., AR = Arrowhead, British Columbia, and CH = Chiricahua Mountains, Arizona). The inversion sequences were compared with one standard sequence, arbitrarily designated ST.

Figure 23–4 compares the frequencies of three arrangements at different elevations in the Sierra Nevada Mountains in California. The ST arrangement is most common at low elevations; at 8000 feet, AR is the most common and ST the least common. In the populations studied, the frequency of the CH arrangement gradually increases with elevation, a phenomenon that is probably the result of natural selection and that parallels the gradual environmental changes occurring at higher elevations, such as decreasing air temperature.

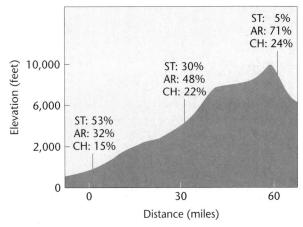

FIGURE 23–4 Inversions in chromosome 3 of *D. pseudoobscura* at different elevations in the Sierra Nevada range near Yosemite National Park.

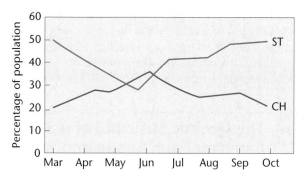

FIGURE 23–5 Changes in the ST and CH arrangements in *D. pseudoobscura* throughout the year.

Dobzhansky's team also found that if populations are collected at a single site throughout the year, inversion frequencies change as well. That is, cyclic variations in chromosome arrangements occur as the seasons change, as shown in Figure 23–5. This variation was consistently observed over a period of several years. In spring, the frequency of ST always declines and that of CH increases.

To test the hypothesis that this cyclic change is a response to natural selection, Dobzhansky and his group devised a laboratory experiment. They constructed large population cages from which samples of *D. pseudoobscura* could be removed periodically and studied. They began with a population that was 88 percent CH and 12 percent ST. The flies were maintained at 25°C and sampled over a 1-year period. As shown in Figure 23–6, the frequency of ST increased gradually from 12 to 70 percent. At that point, an equilibrium between ST and CH was reached. When the same experiment was performed at 16°C, no change in inversion frequency occurred. The researchers concluded that the equilibrium reached at 25°C was in response to the elevated temperature, the only variable in the experiment.

The results of the study indicate that a balance in the frequency of the two inversions and the gene arrangements they contain is superior to either inversion by itself. This is why the increase did not continue, but showed a leveling off. The equilibrium reached in the experiment presumably represents the highest mean fitness in the population under controlled laboratory conditions. This interpretation of the experiment sug-

gests that natural selection is the driving force maintaining the diversity in chromosome 3 inversions.

In a more extensive study, Dobzhansky and his colleagues sampled *D. pseudoobscura* populations over a wide geographic range. They found 22 different chromosome arrangements in populations from 12 locations. In Figure 23–7, the frequencies of five of these inversions are shown according to geographic location. Most populations differ only in the relative frequencies of inversions. Collectively, Dobzhansky's data show that the genetic structure of *D. pseudoobscura* populations is different from place to place and from one time to another. At least some of this variation in population genetic structure is the result of natural selection.

Another example of how the genetic structure of a species varies among populations is provided by the work of Dennis Powers and Patricia Schulte on the mummichog (*Fundulus heteroclitus*), a small fish (5 to 10 cm long) that lives in inlets, bays, and estuaries along the Atlantic coast of North America from Florida to Newfoundland. These workers measured allele frequencies at the locus encoding the enzyme lactate dehydrogenase-B (LDH-B), which is made in the liver, heart, and red skeletal muscle. LDH-B converts lactate to pyruvate and is thus pivotal in both the manufacture of glucose and aerobic metabolism. Two allozymes of

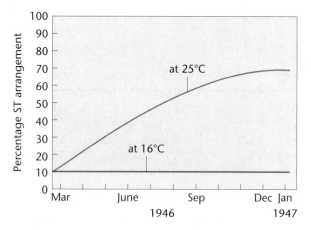

FIGURE 23–6 Increase in the ST arrangement of *D. pseudoobscura* in population cages under laboratory conditions.

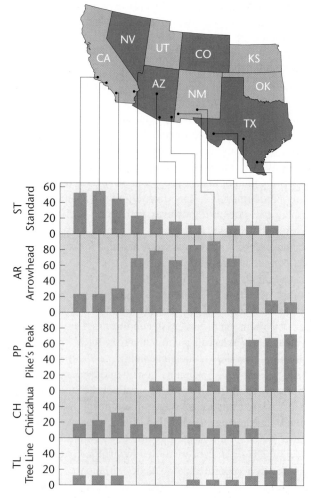

FIGURE 23–7 Relative frequencies (percentages) of five chromosomal inversions in *D. pseudoobscura* in different geographic regions.

LDH-B are seen on gels; they differ at two amino acid positions. The alleles encoding the allozymes are $Ldh\text{-}B^a$ and $Ldh\text{-}B^b$.

Frequencies of the $Ldh\text{-}B$ alleles vary dramatically among mummichog populations [Figure 23–8(a)]. In northern populations, where the mean water temperature is about 6°C, $Ldh\text{-}B^b$ predominates. In southern populations, where the mean water temperature is about 21°C, $Ldh\text{-}B^a$ predominates. Between the geographic extremes, allele frequencies are intermediate.

To determine whether the geographic variation in $Ldh\text{-}B$ allele frequencies is due to natural selection, Powers and Schulte studied the biochemical properties of the LDH-B allozymes. They found that the enzyme encoded by $Ldh\text{-}B^b$ has higher catalytic efficiency at low temperatures, whereas the gene product of $Ldh\text{-}B^a$ is more efficient at high temperatures [Figure 23–8(b)]. A mixture of the two forms has intermediate efficiency at all temperatures.

In addition to the functional differences between the LDH-B allozymes, the $Ldh\text{-}B$ alleles have nucleotide sequence differences in their regulatory regions. One such difference results in the $Ldh\text{-}B^b$ allele's rate of transcription being more than twice that of the $Ldh\text{-}B^a$ allele. As a result, northern fish have higher concentrations of the LDH-B enzyme in their cells.

Differences in the catalytic efficiency and transcription rate of the two $Ldh\text{-}B$ alleles are consistent with the hypothesis that mummichog populations are adapted to the temperatures at which they live. The fish are ectotherms: Their body temperature is determined by their environment. In general, low body temperatures slow an ectotherm's metabolic rate. The higher transcription rate of the $Ldh\text{-}B^b$ allele and the superior low-temperature catalytic efficiency of its gene product appear to help northern fish compensate for the tendency of their cold environment to reduce their metabolic rate. The superior high-temperature catalytic efficiency and lower transcription rate associated with the $Ldh\text{-}B^a$ allele appear to allow southern fish to economize on the resources devoted to glucose production and aerobic metabolism. On the basis of this and other evidence, Powers and Schulte suggest that differences among mummichog populations in $Ldh\text{-}B$ allele frequencies are the result of natural selection.

23.6 Reduced Gene Flow, Selection, and Genetic Drift Can Lead to Speciation

We have learned that most populations harbor considerable genetic variation and that different populations within a species may have different alleles or allele frequencies at a variety of loci. The genetic divergence of these populations can be caused by natural selection, genetic drift, or both. In Chapter 22, we saw that the migration of individuals between populations, together with the gene flow that accompanies that migration, tends to homogenize allele frequencies among populations. In other words, migration counteracts the tendency of populations to diverge.

When gene flow between populations is reduced or absent, the populations may diverge to the point that members of one population are no longer able to interbreed successfully with members of the other. When populations reach the point where they are reproductively isolated from one another, they have become different species, according to the biological species concept.

The biological barriers that prevent or reduce interbreeding between populations are called **reproductive isolating mechanisms**, classified in Table 23.2. These mechanisms may be ecological, seasonal, mechanical, or physiological.

Prezygotic isolating mechanisms prevent individuals from mating in the first place. Individuals from different populations may not find each other at the right time, may not recognize each other as suitable mates, or may try to mate but find that they are unable to do so.

Postzygotic isolating mechanisms create reproductive isolation even when the members of two populations are willing and able to mate with each other. For example, genetic divergence may have reached the stage where the viability or fertility of hybrids is reduced. Hybrid zygotes may be formed, but all or most may be inviable. Alternatively, the hybrids may be viable, but be sterile or suffer from reduced fertility. Yet again, the hybrids themselves may be fertile, but their progeny may have lowered viability or fertility. In all these situations, hybrids will not reproduce, and are genetic dead-ends. These postzygotic mechanisms act at or beyond the level of the zygote and are generated by genetic divergence.

(a)

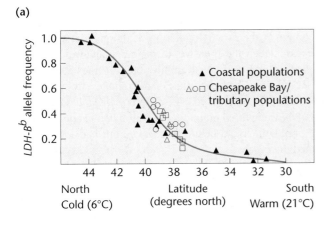

(b)

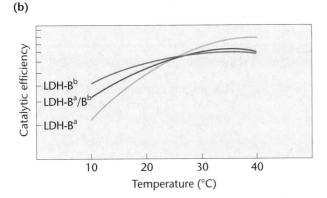

FIGURE 23–8 Variation in genetic structure among mummichog populations. (a) Frequencies of the $LDH\text{-}B^b$ allele in populations along the Atlantic coast of North America. Solid symbols represent coastal populations, and open symbols represent populations from Chesapeake Bay and its tributaries. (b) Catalytic efficiency of LDH-B allozymes as a function of temperature.

TABLE 23.2	Reproductive Isolating Mechanisms

Prezygotic Mechanisms (prevent fertilization and zygote formation)

1. **Geographic or ecological:** The populations live in the same regions but occupy different habitats.

2. **Seasonal or temporal:** The populations live in the same regions but are sexually mature at different times.

3. **Behavioral (only in animals):** The populations are isolated by different and incompatible behavior before mating.

4. **Mechanical:** Cross-fertilization is prevented or restricted by differences in reproductive structures (genitalia in animals, flowers in plants).

5. **Physiological:** Gametes fail to survive in alien reproductive tracts, or fertilization does not occur even if two gametes meet.

Postzygotic Mechanisms (fertilization takes place and hybrid zygotes are formed, but these are nonviable or give rise to weak or sterile hybrids)

1. **Hybrid nonviability or weakness**

2. **Developmental hybrid sterility:** Hybrids are sterile because gonads develop abnormally or meiosis breaks down before completion.

3. **Segregational hybrid sterility:** Hybrids are sterile because of abnormal segregation into gametes of whole chromosomes, chromosome segments, or combinations of genes.

4. **F_2 breakdown:** F_1 hybrids are normal, vigorous, and fertile, but the F_2 contains many weak or sterile individuals.

Source: From G. Ledyard Stebbins, *Processes of organic evolution,* 3rd ed., copyright 1977, p. 143. Reprinted by permission of Prentice Hall, Upper Saddle River, NJ.

In some models of speciation, postzygotic isolation evolves first and is followed by prezygotic isolation. Postzygotic isolating mechanisms waste gametes and zygotes and lower the reproductive fitness of hybrid survivors. Selection will therefore favor the spread of alleles that reduce the formation of hybrids, leading to the development of prezygotic isolating mechanisms, which in turn prevent interbreeding and the formation of hybrid zygotes and offspring. In animal evolution, one of the most effective prezygotic mechanism is behavioral isolation, involving courtship behavior.

Observing Speciation

In this section, we consider two examples of speciation, one from a laboratory study and the other from a field study. Diane Dodd and her colleagues studied the evolution of digestive physiology in *Drosophila pseudoobscura*. They collected flies from a wild population and established several separate laboratory populations. Some of the laboratory populations were raised on starch-based medium and others on maltose-based medium. Both food sources were stressful for the flies. It was only after several months of evolution by natural selection that the populations adapted to their artificial diets and began to thrive.

Dodd wanted to know whether the starch-adapted populations and the maltose-adapted populations, which diverged under strong selection and in the absence of gene flow, had become different species. Roughly a year after the populations were established, a series of mating trials were performed. For each trial, 48 flies were placed in a bottle: 12 males and 12 females from a starch-adapted population and 12 males and 12 females from a maltose-adapted population. She then noted which flies mated.

Dodd predicted that if the populations adapted to different media had speciated, then the flies would prefer to mate with members of their own population. If the populations had not speciated, then the flies would mate at random. The results appear in Table 23.3. Roughly 600 of the 900 matings observed were between males and females from the same population. In other words, the differently adapted fly populations showed partial premating isolation. Dodd concluded that the populations had begun to speciate but had not yet completed the process.

The Isthmus of Panama, which created a land bridge connecting North and South America and simultaneously separated the Caribbean Sea from the Pacific Ocean, formed roughly 3 million years ago. Nancy Knowlton and colleagues took advantage of a natural experiment that the formation of the Isthmus of

TABLE 23.3	Incipient Speciation in Laboratory Populations of *Drosophila pseudoobscura* (Number of Matings Involving Each Kind of Male–Female Pair)	
	Female	
	Starch-adapted	Maltose-adapted
Male		
Starch-adapted	290	153
Maltose-adapted	149	312

Source: Compiled from Dodd, D.M.B. 1989. Reproductive isolation as a consequence of adaptive divergence in *Drosophila pseudoobscura. Evolution* 43: 1308–1311.

Panama had performed on several species of snapping shrimps (Figure 23–9). After identifying seven Caribbean species of snapping shrimp, they matched each one with a Pacific species to form a pair. Members of each pair were more similar to each other in structure and appearance than either was to any other species in its own ocean. Analysis of allozyme allele frequencies and mitochondrial DNA sequences confirmed that the members of each pair were one another's closest genetic relatives.

The Knowlton team's interpretation of these data is that, prior to the formation of the isthmus, the ancestors of each pair were a single species. When the isthmus closed, the seven ancestral species were each divided into two separate populations, one in the Caribbean and the other in the Pacific.

Meeting in a dish in Knowlton's lab for the first time in 3 million years, would Caribbean and Pacific members of a species pair recognize each other as suitable mates? Knowlton placed males and females of a species pair together and noted their behavior toward each other. She then calculated the relative inclination of Caribbean–Pacific couples to mate versus that of Caribbean–Caribbean or Pacific–Pacific couples. For three of the seven species pairs, transoceanic couples refused to mate altogether. For the other four species pairs, transoceanic couples were 33, 45, 67, and 86 percent as likely to mate with each other as were same-ocean pairs. Of the same-ocean couples that mated, 60 percent produced viable clutches of eggs. Of the transoceanic couples that mated, only 1 percent produced viable clutches. We can conclude from these results that 3 million years of separation has resulted in complete or nearly complete speciation, involving strong pre- and postzygotic isolating mechanisms for all seven species pairs.

Minimum Genetic Divergence for Speciation

How much genetic separation is required between two populations before they become different species? We shall consider two examples, one from an insect and the other from a plant, which demonstrate that in some cases the answer is "not very much."

Researchers estimate that *Drosophila heteroneura* and *D. silvestris*, found only on the island of Hawaii, diverged from a common ancestral species only about 300,000 years ago. The

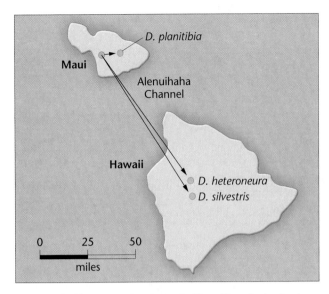

FIGURE 23–10 Proposed pathway for Hawaii's colonization by members of the *D. planitibia* species. The purple circle represents a population ancestral to the three present-day species.

two species are thought to be descended from *D. planitibia* colonists from the older island of Maui (Figure 23–10). The two species are clearly separated from each other by different and incompatible courtship and mating behaviors (a prezygotic isolating mechanism), by morphology, and by body and wing pigmentation (Figure 23–11).

(a)

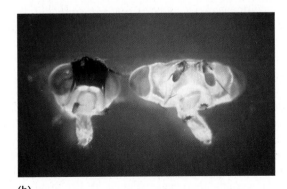

(b)

FIGURE 23–11 (a) Differences in pigmentation patterns in *D. sylvestris* (left) and *D. heteroneura* (right). (b) Head morphology in *D. sylvestris* (left) and *D. heteroneura* (right).

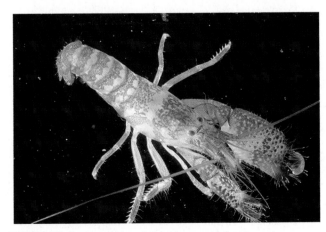

FIGURE 23–9 A snapping shrimp (genus *Alpheus*).

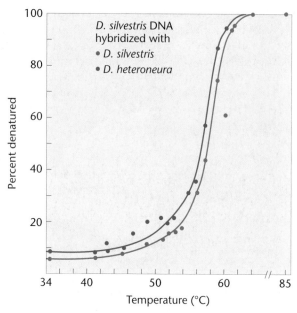

FIGURE 23–12 Nucleotide sequence diversity in the *Drosophila planitibia* species complex. The shift to the left by the heterologous hybrids indicates the degree of nucleotide sequence divergence.

In spite of their morphological and behavioral divergence, demonstrating significant differences between these species in chromosomal inversion patterns or protein polymorphisms is difficult. DNA hybridization studies carried out on the two species indicate that the sequence diversity between them is only about 0.55 percent (Figure 23–12). Thus, nucleotide sequence diversity precedes the development of protein or chromosomal polymorphisms.

Genetic evidence suggests that the differences between *D. heteroneura* and *D. silvestris* are controlled by a relatively small number of genes. For example, as few as 15 to 19 major loci may be responsible for the morphological differences between the species, demonstrating that the process of speciation can involve only a small number of genes.

Studies using two closely related species of the monkey flower, a plant that grows in the Rocky Mountains and areas west, confirm that species can be separated by only a few genetic differences. One species, *Mimulus cardinalis*, is fertilized by hummingbirds and does not interbreed with *Mimulus lewisii*, which is fertilized by bumblebees. H. D. Bradshaw and his colleagues studied genetic differences related to reproduction in the two species—namely, flower shape, size, and color and nectar production. For each trait, a difference in a single gene provided at least 25 percent of the variation observed among laboratory-created hybrids. *M. cardinalis* makes 80 times more nectar than does *M. lewisii*, and a single gene is responsible for at least half the difference. A single gene also controls a large part of the differences in flower color between the two species (Figure 23–13). In this case, as in the Hawaiian *Drosophila*, species differences can be traced to a relatively small number of genes.

The Rate of Speciation

How much time is required for speciation? In many cases, speciation takes place slowly over a long period. In other cases, however, speciation can be surprisingly rapid.

The Rift Valley lakes of East Africa support hundreds of species of cichlid fish. Lake Victoria (Figure 23–14), for example, has more than 400 species. Cichlids are highly specialized for different niches (Figure 23–15). Some eat algae floating on the water's surface, whereas others are bottom feeders, insect feeders, mollusk eaters, and predators on other fish species. Lake Tanganyika has a similar array of species. Genetic analyses indicate that the species in a given lake are all more closely related to each other than to species from other lakes. The implication is that most or all the species in, say, Lake Tanganyika are descended from a single common ancestor and that they diverged within their home lake.

Lake Victoria is between 250,000 and 750,000 years old, and there is evidence that the lake may have dried out nearly completely less than 14,000 years ago. Is it possible that the 400 species in the lake today evolved from a common ancestral species in less than 14,000 years?

In a study of cichlid origins in Lake Tanganyika, Norihiro Okada and colleagues examined the insertion of a novel family of repetitive DNA sequences called short interspersed elements (SINES) (discussed in Chapter 12) into the genomes of

(a)

(b)

FIGURE 23–13 Flowers of two closely related species of monkey flowers, (a) *Mimulus cardinalis* and (b) *Mimulus lewisii*.

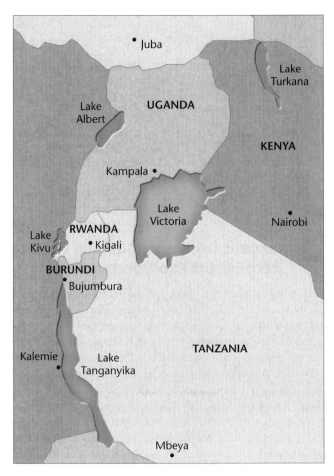

FIGURE 23–14 Lake Victoria in the Rift Valley of East Africa is home to more than 400 species of cichlids.

all the sites tested, indicating that the species in each tribe are descended from a single ancestral species (Figure 23–16). If further research using SINES and other molecular markers produces similar results with the Lake Victoria species, it means that the 400 cichlid species in this lake evolved in less than 14,000 years. If confirmed, this finding would represent the fastest evolutionary divergence of species ever documented in vertebrates.

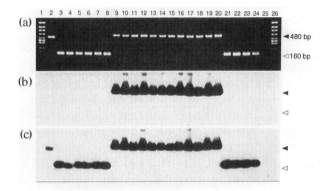

FIGURE 23–16 Using repetitive DNA elements to trace species relationships in cichlids from Lake Tanganyika. (a) Photograph of an agarose gel showing PCR fragments generated from primers that flank the AFC family of SINES. Large fragments containing this family are present in all samples of genomic DNA from members of the Lamprologini tribe of cichlids (lanes 9–20). DNA from the species in lane 2 has a similar but shorter fragment, which may represent another repetitive sequence. DNA from other species (lanes 3–8 and 21–24) produce short, non-SINE-containing fragments. (b) A Southern blot of the gel from (a) probed with DNA from the AFC family of SINES. AFC is present in DNA from all species in the Lamprologini tribe (lanes 9–20), but not in DNA from other species (lanes 2–8 and 21–24). (c) A second Southern blot of the gel from (a) probed with the genomic sequence at which the AFC SINE inserts. All species examined (lanes 2–24) contain the insertion sequence. The larger fragments in Lamprologini DNA (lanes 9–20) correspond to those in the previous blot, showing that the SINE is inserted at the same site in all tribal species. In sum, these data show that the AFC SINE is present only in members of this tribe, is present in all member species, and is inserted at the same site in all cases. These results are interpreted as showing a common origin for the species of the tribe in question.

cichlid species in Lake Tanganyika. SINES are a type of retroposon, and the random integration of a SINE at a locus is most likely an irreversible event. If a SINE is present at the same locus in the genome of all species examined, this is strong evidence that all of those species descend from a common ancestor. Using a SINE called AFC, Okada's team screened 33 species of cichlids belonging to four groups (called species tribes). In each tribe, the SINE was present at

FIGURE 23–15 Cichlids occupy a diverse array of niches, and each species is specialized for a distinct food source.

Even faster speciation is possible through the mechanism of polyploidy. The formation of animal species by polyploidy is rare, but polyploidy is an important factor in plant evolution. It is estimated that one-half of all flowering plants have evolved by this mechanism. One form of polyploidy is allopolyploidy (see Chapter 6), produced by doubling the chromosome number in a hybrid formed by crossing two species.

If two species of related plants have the genetic constitution SS and TT (where S and T represent the haploid set of chromosomes in each species), the hybrid would have the chromosome constitution ST. Normally, such a plant would be sterile, because few or no homologous chromosomes would be present at meiosis. However, if the hybrid undergoes a spontaneous doubling of its chromosome number, a tetraploid SSTT plant would be produced. This chromosomal aberration might occur during mitosis in somatic tissue, giving rise to a partially tetraploid plant that produces some tetraploid flowers. Alternatively, aberrant meiotic events may produce ST gametes that, when fertilized, would yield SSTT zygotes. The SSTT plants would be fertile because they would possess homologous chromosomes which could be segregated into ST gametes. This new, true-breeding tetraploid would have a combination of characteristics derived from the parental species and would be reproductively isolated from them because hybrids would be triploids and consequently sterile.

The tobacco plant *Nicotiana tabacum* (2n = 48) is the result of a doubling of the chromosome number in the hybrid between *N. otophora* (2n = 24) and *N. silvestris* (2n = 24) (Figure 23–17). The origin of *N. tabacum* is an example of virtually instantaneous speciation.

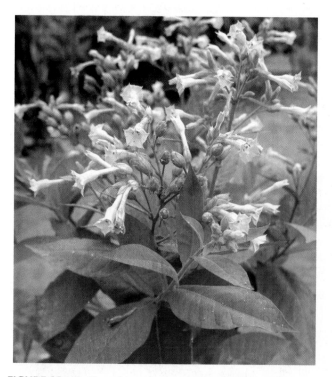

FIGURE 23–17 The cultivated tobacco plant *Nicotiana tabacum* is the result of hybridization between two other species.

Now Solve This

In Problem 26 on page 536 you are asked to interpret whether neutral sequence polymorphisms found among the cichlid species in African lakes help to explain their evolution.

Hint: If a sequence is neutral, then it should not be subject to any form of selection through time and should only be subject to random mutation and genetic drift.

23.7 Genetic Differences Can Be Used to Reconstruct Evolutionary History

Early in this chapter, we noted that the course of evolutionary history involves the transformation of one species into another and the splitting of species into two or more new species. Our examples have demonstrated that speciation is associated with changes in the genetic structure of populations and with genetic divergence. Therefore, we should be able to use genetic differences among species to reconstruct their evolutionary histories.

In an important early example of phylogeny reconstruction, W. M. Fitch and E. Margoliash assembled data on the amino acid sequence for cytochrome c in a variety of organisms. **Cytochrome c** is a respiratory molecule found in the mitochondria of eukaryotes, and its amino acid sequence has evolved very slowly. For example, the amino acid sequence in humans and chimpanzees is identical, and humans and rhesus monkeys show only one amino acid difference. We should expect little evolution in a protein whose function is essential for fitness, but such close similarity is remarkable, considering that the fossil record indicates that the lines leading to humans and monkeys diverged from a common ancestral species approximately 20 million years ago.

Column (a) of Table 23.4 shows the number of amino acid differences between cytochrome c in humans and a variety of other species. The table is broadly consistent with our intuitions about how closely related we are to these other species. For example, we are more closely related to other mammals than we are to insects, and we are more closely related to insects than we are to fungi. Likewise, our cytochrome c differs in 10 amino acids from that of dogs, in 24 amino acids from that of moths, and in 38 amino acids from that of yeast.

However, more than one nucleotide change may be required to change a given amino acid. When the nucleotide changes necessary for all amino acid differences observed in a protein are totaled, the **minimal mutational distance** between the genes of any two species is established. Column (b) in Table 23.4 shows such an analysis of the gene encoding cytochrome c. As expected, these values are larger than the corresponding number of amino acids separating humans from the other nine organisms listed.

Fitch used data on the minimal mutational distances between the cytochrome c genes of 19 organisms to reconstruct their

TABLE 23.4	Amino Acid Differences and the Minimal Mutational Distances between Cytochrome c in Humans and Other Organisms	
	(a)	**(b)**
	Amino Acid	**Minimal Mutational**
Organism	**Differences**	**Distance**
Human	0	0
Chimpanzee	0	0
Rhesus monkey	1	1
Rabbit	9	12
Pig	10	13
Dog	10	13
Horse	12	17
Penguin	11	18
Moth	24	36
Yeast	38	56

Source: From W.M. Fitch and E. Margoliash, Construction of phylogenetic trees, *Science* 155: 279–284, January 20, 1967. Copyright 1967 by the American Association for the Advancement of Science.

evolutionary history. The result is an estimate of the evolutionary tree, or phylogeny, which unites the species (Figure 23–18). The black dots on the tips of the branches represent existing species, which are connected to the inferred common ancestors represented by red dots. The ancestral species evolved and diverged to produce the modern species. The common ancestors are connected to still earlier common ancestors, culminating in a single common ancestor for all the species on the tree, represented by the red dot on the extreme left.

Constructing Evolutionary Trees

Many methods use data on genetic differences to estimate phylogenies. It is beyond the scope of this chapter to review all of them. Instead, we shall present just one method, called the *u*nweighted *p*air *g*roup *m*ethod using *a*rithmetic averages, or UPGMA. This method is not the most powerful one for estimating phylogenies from genetic data, but, in spite of its lengthy name, it is intuitively straightforward. Furthermore, UPGMA works reasonably well under many circumstances.

The starting point for UPGMA is a table of genetic distances (a measure of genetic divergence) among a group of species [Figure 23–19(a)]. To demonstrate the technique, we will use data from DNA hybridization studies by Charles Sibley and Jon Alquist, who computed genetic distances between humans and four species of apes: the common chimpanzee, the gorilla, the siamang, and the common gibbon.

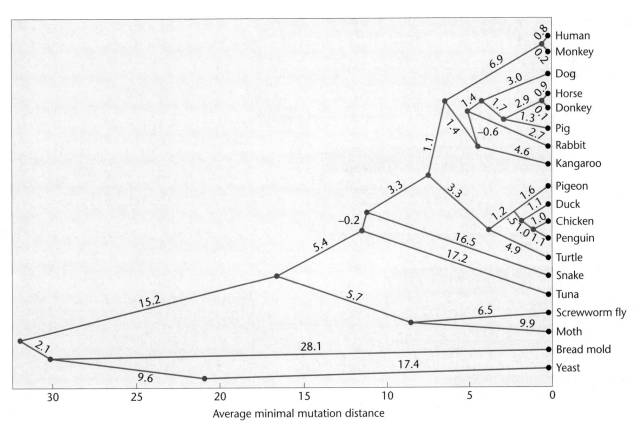

FIGURE 23–18 Phylogeny constructed by comparing homologies in cytochrome c amino acid sequences. *Reprinted with permission from Fitch, W.M., and Margoliash E. 1967. Construction of phylogenetic trees. Science 279:279–84. Figure 2. Copyright 1967 AAAS.*

The method uses the following steps:

1. Search for the smallest genetic distance between any pair of species. In Figure 23–19(a), this distance is 1.628, the distance between human and chimpanzee. Once identified, the corresponding species pair is placed on neighboring branches of an evolutionary tree [Figure 23–19(b)]. The length of each branch is half the genetic distance between the species (1.628/2), so the branches connecting human and chimpanzee to their common ancestor are each about 0.81.

2. Recalculate the genetic distances between this species pair and all other species [Figure 23–19(c)]. The genetic distance between the human–chimpanzee (Hu-Ch) cluster* and the other species is the average of the distances between each member of the cluster and the other species. For example, the genetic distance between the human–chimp cluster and the gorilla is the average of the human–gorilla distance and the chimp–gorilla distance, or

$$(2.267 + 2.21)/2 = 2.2385$$

3. Repeat steps 1 and 2 until all the species have been added to the tree.

---

*We say "cluster" even though what we really have is a pair; the reason is that we can (and will) form larger groups and it is convenient to call them by the same, more inclusive name—ergo *cluster*.

The smallest distance in the recalculated table is 1.95, the distance between siamang and gibbon [Figure 23–19(c)]. To build the next section of the tree, siamang and gibbon are placed on neighboring branches, with lengths equal to half of 1.95, or 0.98 [Figure 23–19(d)].

Recalculating the table, we now find that there are two clusters plus the gorilla [Figure 23–19(e)]. The genetic distance between the human–chimp cluster and the siamang–gibbon cluster is the average of four distances: human–siamang, human–gibbon, chimp–siamang, and chimp–gibbon.

Now the smallest genetic distance is 2.239, the distance between the human–chimp cluster and the gorilla. We therefore add a gorilla branch to the tree and connect it to the common ancestor of the human–chimp cluster [Figure 23–19(f)]. The branches are drawn so that the distance between the tips of any two branches in the human–chimp–gorilla cluster is 2.239.

Recalculating the table for the final time [Figure 23–19 (g)], we find that the genetic distance between the human–chimp–gorilla cluster and the siamang–gibbon cluster is the average of six genetic distances: human–siamang, human–gibbon, chimp–siamang, chimp–gibbon, gorilla–siamang, and gorilla–gibbon. This distance is 4.778, which allows us to complete our evolutionary tree [Figure 23–19(h)]. The tree indicates that humans and chimpanzees are one another's closest relatives. That is not to say that

(a)

	Human	Chimp	Gorilla	Siamang	Gibbon
Human	–				
Chimp	1.628	–			
Gorilla	2.267	2.21	–		
Siamang	4.7	5.133	4.543	–	
Gibbon	4.779	4.76	4.753	1.95	–

(c)

	Hu-Ch	Gorilla	Siamang	Gibbon
Hu-Ch	–			
Gorilla	2.2385	–		
Siamang	4.9165	4.543	–	
Gibbon	4.7695	4.753	1.95	–

(e)

	Hu-Ch	Gorilla	Si-Gi
Hu-Ch	–		
Gorilla	2.239	–	
Si-Gi	4.843	4.648	–

(g)

	Hu-Ch-Go	Si-Gi
Hu-Ch-Go	–	
Si-Gi	4.778	–

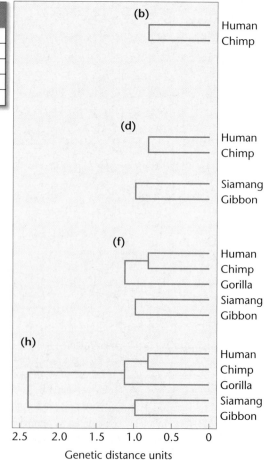

FIGURE 23–19 Phylogeny reconstruction by UPGMA.

humans evolved from chimpanzees; rather, humans and chimpanzees share a more recent common ancestor than either shares with other species on the tree.

The chief shortcoming of UPGMA is that it provides no means of determining how well its trees fit the data, compared with other possible trees. For example, how much better does the tree we just created fit our data than a tree that shows chimpanzees and gorillas as closest relatives?

Evolutionary geneticists have developed several techniques for searching the set of all possible trees connecting a group of species and calculating the relative performance of each. One set of techniques, called **parsimony** methods, compares trees by using the minimum number of evolutionary changes each requires and then selects the simplest possible tree. Often, however, there is a set of equally parsimonious trees, so these methods often must be combined with other criteria to develop the most accurate phylogeny. Another set of analytical tools, called **maximum-likelihood** methods, starts with a model of the evolutionary process, calculates how likely it is that evolution will produce each possible tree under the model, and selects the most likely tree. It is important to keep in mind that a phylogeny produced from data on genetic distances is not necessarily the way species actually evolved, but instead is a reasonable estimate of their evolu-

tionary histories, based on the methods used. Phylogenies are considered more reliable if multiple methods produce the same tree.

Molecular Clocks

In many cases, we would like to estimate not only which members of a set of species are most closely related, but also when their common ancestors lived. Sometimes we can do so, thanks to **molecular clocks**—amino acid sequences or nucleotide sequences in which evolutionary changes accumulate at a constant rate over time.

Research by Walter M. Fitch and colleagues on the influenza A virus shows how molecular clocks are used. Fitch and his associates sequenced part of the hemagglutinin gene from flu viruses that had been isolated at different times over a 20-year period. They calculated the number of nucleotide differences among the various viruses and constructed an evolutionary tree [Figure 23–20(b)]. Most strains have gone extinct, leaving no descendants among the more recently isolated viruses. Fitch's group then plotted the number of nucleotide substitutions between the first virus and each subsequent virus against the year in which the virus was isolated [Figure 23–20(a)]. The points all fall very close to a straight line, indicating that nucleotide substitutions in this gene have accumulated at a steady rate. The hemagglutinin

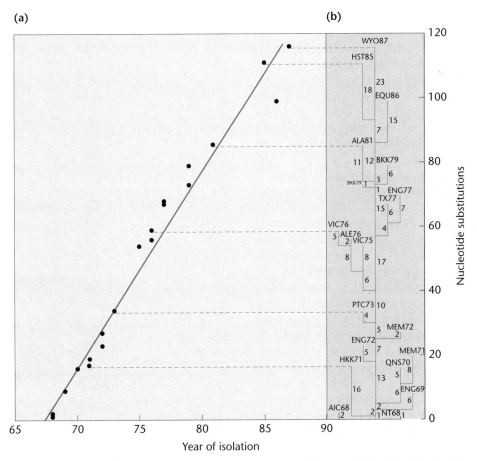

FIGURE 23–20 Molecular clock in the influenza A hemagglutinin gene. (a) Number of nucleotide differences between the first isolate and each subsequent isolate as a function of year of isolation. (b) Estimate of the phylogeny of the isolates.

gene thus serves as a molecular clock. Molecular clocks are used to compare the sequences of new flu viruses as they appear each year and to estimate the time that has passed since each diverged from a common ancestor.

Molecular clocks must be carefully calibrated and used with caution. For example, Fitch's data indicate that strains of influenza A that jump from birds to humans have evolved much more rapidly than strains that have remained in birds. Hence, a molecular clock calibrated from human strains of the virus would be highly misleading if applied to bird strains.

23.8 Reconstructing Evolutionary History Allows Us to Answer Many Questions

Evolutionary analysis addresses a wide range of questions, some of which relate to evolution directly, whereas others deal with contemporary issues and even criminal activity.

Transmission of HIV

In late 1986, a Florida dentist tested positive for HIV. Several months later, he was diagnosed with AIDS. He continued to practice general dentistry for two more years, until one of his patients, a young woman with no known risk factors, discovered that she, too, was infected with HIV. When the dentist publicly urged his other patients to have themselves tested, several more were found to be HIV positive. Did this dentist transmit HIV to his patients, or did the patients become infected by some other means?

At first glance, this situation appears to be a long way from a discussion of evolution; however, the movement of a virus from one individual to another is similar to the founding of a new island population by a small number of migrants. Gene flow between the ancestral population and the new population is nonexistent, and the populations are free to diverge, as a result of genetic drift, adaptation to different environments, or both.

If the dentist passed his HIV infection to his patients, then the viral strains isolated from these patients should be more closely related to each other and to the dentist's strain than to strains from any other individuals living in the same area. Following this line of reasoning, Chin-Yih Ou and colleagues sequenced portions of the gene for the HIV envelope protein from viruses collected from the dentist, 10 of his patients, and several other HIV-infected individuals from the area who served as local controls.

An evolutionary tree for the HIV strains produced from these sequence data is shown in Figure 23–21. The viruses isolated from the participants in the study are at the tips of the branches; branch points join these to the inferred common ancestors. The HIV samples from patients A, B, C, E, G, and I all share a more recent common ancestor with each other and with the dentist's strains than with any of the HIV

samples from any of the other local individuals. The evolutionary relationship among these HIV strains indicates that the dentist did, indeed, transmit his infection to those patients. In contrast, the viruses taken from patients D, F, H, and J are all more closely related to strains from local controls (LC) than they are to the dentist's strains. These patients got their infection from someone other than the dentist. Patient J, in fact, seems to have acquired his infection from two different sources. The Centers for Disease Control, which tracks HIV cases, has not been able to determine how the dentist-to-patient transmission took place. There have been no other documented cases of provider-to-patient transmission in a health-care setting.

Neanderthals and Modern Humans

Paleontological evidence indicates that the Neanderthals, *Homo neanderthalensis*, lived in Europe and western Asia some 300,000 to 30,000 years ago. For at least 30,000 years, Neanderthals coexisted with anatomically modern humans (*H. sapiens*) in several areas. Questions about Neanderthals and modern humans remain unanswered: (1) Can Nean-

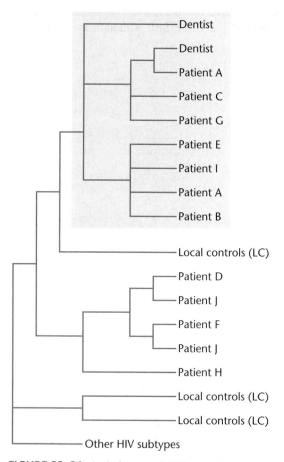

FIGURE 23–21 A phylogeny of HIV strains taken from a dentist, his patients, and several local controls (LC). The group of viral strains in the shaded portion of the phylogeny are all derived from a common ancestral strain.

derthals be regarded as direct ancestors of modern humans? (2) Did Neanderthals and *H. sapiens* interbreed, so that descendants of the Neanderthals are alive today? Or did the Neanderthals die off and become extinct?

To resolve these issues, several groups have focused their attention on the recovery and analysis of DNA extracted from Neanderthal bones. In 1997, Svante Pääbo, Matthias Krings, and their colleagues extracted fragments of mitochondrial DNA from a Neanderthal skeleton found in Feldhofer cave near Düsseldorf, Germany. After confirming that they had indeed isolated Neanderthal gene fragments, the researchers placed the Neanderthal sequences on a phylogenetic tree together with sequences from more than 2000 modern humans [Figure 23–22(a)]. On the basis of Pääbo and Krings' analysis, Neanderthals appear to be a distant relative of modern humans. Using a molecular clock calibrated with chimpanzees and humans, the researchers calculated that the last common ancestor of Neanderthals and modern humans lived roughly 600,000 years ago, four times as long ago as the last common ancestor of all modern humans. However, considerable caution is required when drawing conclusions based on a single Neanderthal specimen.

More recently, Igor Ovchinnikov and his colleagues analyzed mitochondrial DNA recovered from Neanderthal remains discovered in Mezmaiskaya cave in the Caucasus Mountains east of the Black Sea. Although the two Neanderthal sequences (i.e., Pääbo and Krings', on the one hand, and Ovchinnikov's, on the other) are from individuals from different geographic regions more than 1000 miles apart, they vary by only about 3.5 percent, indicating that they derive from a single gene pool. Furthermore, the amount of variation between the two Neanderthal sequences is comparable to that seen among modern humans. Phylogenetic analysis places the two Neanderthals in a group that is distinct from modern humans [Figure 23–22(b)]; the conclusion to be drawn from the two studies is that although Neanderthals and humans have a common ancestor, the Neanderthals were a separate hominid line and did not contribute mitochondrial genes to *H. sapiens*. Taken together, the studies suggest that when the Neanderthals disappeared, their lineage went extinct with them. However, it is still possible that some Neanderthal males mated with human females, but not

the other way around. To resolve this question, an effort is underway to recover and sequence fragments of Neanderthal nuclear genes.

Origin of Mitochondria

Mitochondria are cellular organelles that contain their own genome in the form of a circular DNA molecule. In humans, the mitochondrial genome encodes 13 proteins required for oxidative phosphorylation and ATP production, 22 tRNA molecules, and 2 rDNA genes. The organization and function of mitochondrial genomes bear many similarities to those of bacterial genomes. Therefore, mitochondria are thought to derive from bacteria; that is, they are descended from free-living prokaryotic organisms that became intracellular symbionts in nucleated host cells. Once incorporated into a nucleated cell, the genome of the mitochondrion became smaller over time, and the reduced genome we now see in mitochondria is thought to be the product of gene loss and gene transfer to the nucleus of the host cell. However, many questions about the origin and evolution of mitochondria remain. Among them is one relating to the bacterial origins of mitochondria: Which group of bacteria are they descended from?

Researchers investigating this question have used information derived from nucleotide sequencing of the mitochondrial genes for small-subunit ribosomal RNAs (SSU rRNA). These sequences are under strong functional constraint and so are highly conserved. Phylogenetic analysis of the sequences identified a group of bacteria known as the α-proteobacteria (purple bacteria) as the closest living relatives of mitochondria. More recent work has divided this group into two subclasses, one of which is called the rickettsial subdivision.

Based on the SSU phylogeny, the rickettsia have been identified as the bacteria most closely related to mitochondria. Interestingly, rickettsia, like mitochondria, live only inside eukaryotic cells.

Unlike mitochondria, however, rickettsia are disease-causing parasites, responsible for typhus, one of the most serious diseases in the history of our species. Genomic sequencing of one rickettsial species, *Rickettsia prowazekii*, provides additional insight into the relationship between these bacteria and

FIGURE 23–22 (a) A phylogeny estimated from mitochondrial DNA sequences of one Neanderthal specimen from Western Germany (Feldhofer) and over 2000 modern humans. (b) Analysis of Neanderthal DNA from a specimen recovered in a cave east of the Black Sea (Mezmaiskaya).

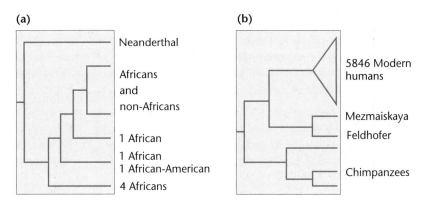

(a)

Neanderthal

Africans
and
non-Africans

1 African

1 African
1 African-American

4 Africans

(b)

5846 Modern
humans

Mezmaiskaya
Feldhofer

Chimpanzees

mitochondria. Using the genome sequence information for genes involved in ATP synthesis derived from *R. prowazekii*, other bacteria, and mitochondria from several sources, Siv Andersson and colleagues constructed a phylogenetic tree (Figure 23–23). Their tree indicates a close evolutionary rela-tionship between *R. prowazekii* and mitochondria. Evidence from SSU analysis results in a similar tree. Thus, two lines of research, from SSUs and proteins involved in ATP synthesis, provide strong evidence that the mitochondria we carry in our cells derive from an ancient ancestor of the rickettsia.

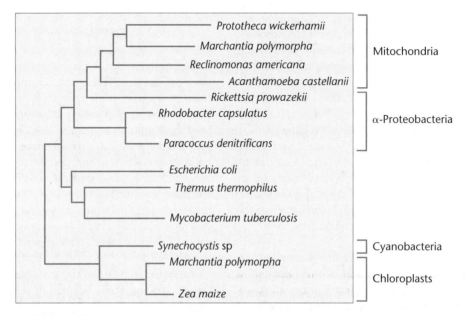

FIGURE 23–23 The evolutionary relationships of mitochondria, α-proteobacteria, cyanobac-teria, and chloroplasts. Note that, although *R. prowazekii* is an α-proteobacterium, it is more closely related to mitochondria than to other α-proteobacteria.

CHAPTER SUMMARY

1. Today's species are the products of an evolutionary history that included the transformation, splitting, and divergence of lineages. Alfred Russel Wallace and Charles Darwin for-mulated the theory of natural selection, which provides a major mechanism for the transformation of lineages. The genetic basis of evolution and the role of natural selection in changing allele frequencies were discovered in the twentieth century.

2. When population geneticists began studying the genetic struc-ture of populations, they found that most populations harbor considerable genetic diversity, which becomes apparent in elec-trophoretic studies of proteins and at the level of DNA se-quences. Whether the genetic diversity of populations is maintained primarily by mutation plus genetic drift or by natural selection is a matter of some debate.

3. The geographic ranges of most species encompass a degree of environmental diversity. As a result of both adaptation to differ-ent environments and genetic drift, different populations within a species may have different alleles or allele frequencies at many loci.

4. Gene flow among populations tends to homogenize their genetic composition. When gene flow is reduced, genetic drift and adap-tation to different environments can cause populations to diverge. Eventually, populations may become so different that the individuals in one population either will not or cannot mate with the individuals in the other. At this point, the divergent pop-ulations have become different species.

5. Because speciation is associated with genetic divergence, we can use the genetic differences among species to infer their evolu-tionary history. By comparing amino acid or nucleotide se-quences, we can determine the genetic distances among species. We can then use the genetic distances to reconstruct evolutionary trees. The simplest methods for reconstructing phylogenies are based on the assumption that the least divergent species in terms of DNA and protein sequences are one another's closest relatives.

6. The reconstruction of evolutionary trees is a key technique in answering a diversity of interesting questions. Examples include tracing the route of transmission of a disease in an epidemic, deciphering recent events in human evolution, and determining the evolutionary relationships among all organisms.

Genetics, Technology, and Society

What Can We Learn from the Mistakes of the Eugenics Movement?

The eugenics movement had its origins in the ideas of the English scientist Francis Galton, who became convinced from his study of the appearance of geniuses within families (including his own) that intelligence is inherited. Galton concluded in his 1869 book *Hereditary Genius* that it would be "quite practicable to produce a highly gifted race of men by judicious marriages during several consecutive generations." The term *eugenics*, coined by Galton in 1883, refers to the improvement of the human species by such selective mating. Once Mendel's principles were rediscovered in 1900, the eugenics movement flourished.

The eugenicists believed that a wide range of human attributes were inherited as single-gene Mendelian traits, including many aspects of behavior, intelligence, and moral character. Their overriding concern was that the presumed genetically "feebleminded" and immoral in the population were reproducing faster than the genetically superior, and that this differential birthrate would result in the progressive deterioration of the intellectual capacity and moral fiber of the human race. Several remedies were proposed. Positive eugenics called for the encouragement of especially "fit" parents to have more children. More central to the goals of the eugenicists, however, was the negative eugenics approach aimed at discouraging the reproduction of the genetically inferior or, better yet, eliminating it altogether.

In the United States, the eugenics movement enjoyed wide popular support for a time and had a significant impact on public policy. Partially at the urging of prominent eugenicists, 30 states passed laws compelling the sterilization of criminals, epileptics, and inmates in mental institutions; most states enacted laws invalidating marriages between the "feebleminded" and others considered eugenically unfit. The crowning legislative achievement of the eugenics movement, however, was the passage of the Immigration Restriction Act of 1924, which severely limited the entry of immigrants from Eastern and Southern Europe due to their perceived mental inferiority.

Throughout the first two decades of the century, most geneticists passively accepted the views of eugenicists, but by the 1930s critics recognized that the goals of the eugenics movement were determined more by racism, class prejudice, and anti-immigrant sentiment than by sound genetics. Increasingly, prominent geneticists began to speak out against the eugenics movement, among them William Castle, Thomas Hunt Morgan, and Hermann Muller. When the horrific extremes to which the Nazis took eugenics became known, a strong reaction developed that all but ended the eugenics movement.

Paradoxically, the eugenics movement arose at the same time that basic Mendelian principles were being developed, principles that eventually undermined the theoretical foundation of eugenics. Today, any student who has completed an introductory course in genetics should be able to identify several fundamental mistakes the eugenicists made.

1. They assumed that complex human traits such as intelligence and personality were strictly inherited, completely disregarding any environmental contribution to the phenotype. Their reasoning was that because certain traits ran in families, they must be genetically determined.

2. They assumed that these complex traits were determined by single-gene two-allele systems, with dominant and recessive alleles. This belief persisted despite research showing that multiple genes contribute to many phenotypes.

3. They assumed that a single ideal genotype existed in humans. Presumably, such a genotype would be highly homozygous in order to be sustained. This precept runs counter to current evidence suggesting that a high level of homozygosity is often deleterious, supporting the superiority of the heterozygote.

4. They assumed that the frequency of recessively inherited defects in the population could be significantly lowered by preventing homozygotes from reproducing. In fact, for recessive traits that are relatively rare, most of the recessive alleles in the population are carried by asymptomatic heterozygotes who are spared from such selection. Negative eugenic practices, no matter how harsh, are relatively ineffective at eliminating such traits.

5. They thought that those deemed genetically unfit in the population might out-reproduce those thought to be genetically fit. This is the exact reverse of the Darwinian concept of fitness, which equates reproductive success with fitness. (Galton, being Darwin's first cousin, should have understood this principle!)

More than seven decades have passed since the eugenics movement was in full bloom. We now have a much more sophisticated understanding of genetics, as well as a greater awareness of its potential misuses. But the application of current genetic technologies makes possible a "new eugenics" of a scope and power that Francis Galton could not have imagined. In particular, prenatal genetic screening and *in vitro* fertilization enable the selection of children according to their genotype, a power that will dramatically increase as more and more genes are associated with inherited diseases, normal physiology, and perhaps even behaviors.

As we move into this new genetic age, we must not forget the mistakes of the early eugenicists. We must remember that the phenotype is a complex interaction between the genotype and the environment, and we must not lapse into an approach that treats a person as only a collection of genes. We must keep in mind that many genes may contribute to a particular phenotype, whether a disease or a behavior, and that the alleles of these genes may interact in unpredictable ways. We must not fall prey to the assumption that there is an ideal genotype. The success of all populations in nature is believed to be

(Cont. on the next page)

enhanced by genetic diversity. And most of all, we must not use genetic information to advance ideological goals. We may find that there is a fine line between the legitimate uses of genetic technologies, such as having healthy children, and other eugenic practices. It will be up to us to decide exactly where the line falls.

References

Allen, G. E., Jacoby, R., and Glauberman, N. (eds.). 1995. Eugenics and American social history, 1880–1950. *Genome* 31: 885–889.

Hartl, D. L. 1988. A primer of population genetics, 2nd ed. Sunderland, MA: Sinauer.

Kevles, D. J. 1985. In the name of eugenics: Genetics and the uses of human heredity. Berkeley: University of California Press.

KEY TERMS

anagenesis, 516

cladogenesis, 516

cystic fibrosis, 518

cytochrome c, 526

gel electrophoresis, 517

maximum-likelihood, 529

minimal mutational distance, 526

molecular clocks, 529

natural selection, 519

neutral theory, 519

parsimony, 529

phyletic evolution, 516

phylogeny, 516

postzygotic isolating mechanisms, 521

prezygotic isolating mechanisms, 521

reproductive isolating mechanisms, 521

speciation, 515

species, 515

stasis, 516

INSIGHTS AND SOLUTIONS

1. Sequence analysis of DNA can be accomplished by a number of techniques. Protein sequencing, on the other hand, is made more complex by the fact that 20 different subunits, as opposed to just the four nucleotides of DNA, need to be unambiguously identified and enumerated. Because of their unique properties, the N-terminal and C-terminal amino acids in a protein are easy to identify, but the array in between offers a difficult challenge because many proteins contain hundreds of amino acids. How is it that protein sequencing is accomplished?

Solution: The strategy for protein sequencing is the same as that for DNA sequencing: divide and conquer. To accomplish this, specific enzymes are used that reproducibly cleave proteins between certain amino acids. The use of different enzymes produces overlapping fragments, each of which is isolated, and each amino acid sequence is determined by chemical means. Sequences from overlapping fragments are then assembled to give a sequence for the entire protein. Alternatively, researchers can first sequence the gene that encodes the protein and then use the genetic code to infer the protein's amino acid sequence.

2. A single plant twice the size of others in the same population suddenly appears. Normally, plants of that species reproduce by self-fertilization and by cross-fertilization. Is this new giant plant simply a variant, or could it be a new species? How would you determine which it is?

Solution: One of the most widespread mechanisms of speciation in higher plants is polyploidy, the multiplication of entire sets of chromosomes. The result of polyploidy is usually a larger plant with larger flowers and seeds. There are two ways of testing the new variant to determine whether it is a new species. First, the giant plant should be crossed with a normal-sized plant to see whether the giant plant produces viable, fertile offspring. If it does not, then the two different types of plants would appear to be reproductively isolated. Second, the giant plant should be cytogenetically screened to examine its chromosome complement. If it has twice the number of its normal-sized neighbors, it is a tetraploid that may have arisen spontaneously. If the chromosome number differs by a factor of two and the new plant is reproductively isolated from its normal-sized neighbors, it is a new species.

PROBLEMS AND DISCUSSION QUESTIONS

1. What is the neo-Darwinian definition of evolution?
2. Define speciation. How does it differ from evolution?
3. Wallace and Darwin identified natural selection as a force acting on phenotypic variation. What two processes related to phenotypic variation were they unable to explain?
4. Two closely related species may have very few phenotypic differences. How can we justify classifying them as two different species instead of one species with a small range of phenotypic variation?
5. Natural selection can lead to speciation. Is this the only way species can form?
6. The maintenance of allozymic diversity in natural populations remains a controversial topic. Nevo (2001. *Proc. Natl. Acad. Sci. [USA]* 98: 6233–6240) examined polymorphisms among 111 above-ground and 132 subterranean mammalian species and found that subterranean mammals are less polymorphic than those living above ground. Provide a possible explanation.
7. Discuss the rationale behind the statement that inversions in chromosome 3 of *D. pseudoobscura* represent genetic variation.
8. Dobzhansky's studies on populations of *D. pseudoobscura* showed changes in the frequency of the ST and CH chromosome arrangements throughout the year. Why hasn't one of these arrangements been eliminated over a long period of time?
9. Describe how populations with substantial genetic differences can form. What is the role of natural selection?
10. Price et al. (1999. *J. Bacteriol.* 181: 2358–2362) conducted a genetic study of the toxin transport protein (PA) of *Bacillus anthracis*, the bacterium that causes anthrax in humans. Within the 2294-nucleotide gene in 26 strains they identified five point mutations—two missense and three synonyms—among different isolates. Necropsy samples from an anthrax outbreak in 1979 revealed a novel missense mutation and five unique nucleotide changes among ten victims. The authors concluded that these data indicate little or no horizontal transfer between different *B. anthracis* strains. (a) Which types of nucleotide changes (missense or synonyms) cause amino acid changes? (b) What is meant by horizontal transfer? (c) On what basis did the authors conclude that evidence of horizontal transfer is absent from their data?
11. What types of nucleotide substitutions will not be detected by electrophoretic studies of a gene's protein product?
12. A recent study examining the mutation rates of 5669 mammalian genes (17,208 sequences) indicates that, contrary to popular belief, mutation rates among lineages with vastly different generation lengths and physiological attributes are remarkably constant (Kumar, S., and Subramanian S. 2002. *Proc. Natl. Acad. Sci. [USA]* 99: 803–808). The average rate is estimated at 12.2×10^{-9} per bp per year. What is the significance of this finding in terms of mammalian evolution?
13. Discuss the arguments supporting the neutral theory of molecular evolution. What counterarguments are proposed by the selectionists? Of what value is the debate concerning the neutral theory?
14. What genetic changes take place during speciation?
15. The mummichog shows *Ldh-B* allele frequency differences from cold northern waters to warm southern waters. (a) Is there a correlation between allele type and allele frequency and water temperature? (b) Why are differences in catalytic efficiency *and* differences in transcription rates important in this adaptive strategy?
16. List the barriers that prevent interbreeding and give an example of each.
17. What are the two groups of reproductive isolating mechanisms? Which of these is regarded as more efficient, and why?
18. In the tobacco plant *N. tabacum*, formed as a hybrid between *N. otophora* and *N. silvestris*, would you expect to find higher levels of heterozygosity in the polyploid species than in the diploid parental species? Is this an advantage or disadvantage?
19. Why are many species of flowering plants polyploid, but only a few animal species polyploid?
20. Some critics have warned that the use of gene therapy to correct genetic disorders will affect the course of human evolution. Evaluate this criticism in light of what you know about population genetics and evolution, distinguishing between somatic gene therapy and germline gene therapy.
21. Shown below are two homologous lengths of the α and β chains of human hemoglobin. Consult the genetic code dictionary (Figure 12–7) and determine how many amino acid substitutions may have occurred as a result of a single nucleotide substitution. For any that cannot occur as the result of a single change, determine the minimal mutational distance.

α:	Ala	Val	Ala	His	Val	Asp	Asp	Met	Pro
β:	Gly	Leu	Ala	His	Leu	Asp	Asn	Leu	Lys

22. Determine the minimal mutational distances between these amino acid sequences of cytochrome c from various organisms. Compare the distance between humans and each organism.

Human:	Lys	Glu	Glu	Arg	Ala	Asp
Horse:	Lys	Thr	Glu	Arg	Glu	Asp
Pig:	Lys	Gly	Glu	Arg	Glu	Asp
Dog:	Thr	Gly	Glu	Arg	Glu	Asp
Chicken:	Lys	Ser	Glu	Arg	Val	Asp
Bullfrog:	Lys	Gly	Glu	Arg	Glu	Asp
Fungus:	Ala	Lys	Asp	Arg	Asn	Asp

23. The data on the next page represent short DNA sequences from five species: human, chimpanzee, gorilla, orangutan, and baboon. The sequences are each 50 bp long. They represent a short piece of a gene for testis-specific protein Y, located on the Y chromosome. The complete human sequence is given; the sequences for the other four species are shown only where they differ from the human sequence. Calculate the genetic difference between each pair of species. (For example, chimps and gorillas differ in 2 out of 50 bases, or 4 percent; thus, the genetic difference between chimps and gorillas is 0.04.) Then use UPGMA to reconstruct the phylogeny for these five species. Is your phylogeny consistent with that shown in Figure 23–19?

```
Human:        A G A G G T T T T T C A G T G A A T G A A G C T A T T T T T A A G G G A G T G T G A T T G C T G C C
Chimpanzee:                                           C
Gorilla:                          T            C                                                        C
Orangutan:        T            G T     C       C                              C                         C
Baboon:       C   C   G    G T            C               G          G C           C         C   G
```

24. The genetic difference between two *Drosophila* species, *D. heteroneura* and *D. sylvestris*, as measured by nucleotide diversity, is about 1.8 percent. The difference between chimpanzees (*P. troglodytes*) and humans (*H. sapiens*) is about the same, yet the latter species are classified in different genera. In your opinion, is this valid? Explain why.

25. The use of nucleotide sequence data to measure genetic variability is complicated by the fact that the genes of higher eukaryotes are complex in organization and contain 5' and 3' flanking regions as well as introns. Researchers have compared the nucleotide sequence of two cloned alleles of the γ-*globin* gene from a single individual and found a variation of 1 percent. Those differences include 13 substitutions of one nucleotide for another and 3 short DNA segments that have been inserted in one allele or deleted in the other. None of the changes takes place in the gene's exons (coding regions). Why do you think this is so, and should it change our concept of genetic variation?

26. In a recent study of cichlid fish inhabiting Lake Victoria in Africa, Nagl et al. (1998. *Proc. Natl. Acad. Sci. [USA]* 95: 14,238–14,243) examined suspected neutral sequence polymorphisms in noncoding genomic loci in 12 lake species and their putative riverine ancestors. At all loci, the same polymorphism was found in nearly all of the tested species from Lake Victoria, both lacustrine and riverine. Different polymorphisms at these loci were found in cichlids at other African lakes. (a) Why would you suspect neutral sequences to be located in noncoding genomic regions? (b) What conclusions can be drawn from these polymorphism data in terms of cichlid ancestry in these lakes?

27. Given that there are approximately 400 cichlid species in Lake Victoria and that it dried up almost completely about 14,000 years ago, what evidence indicates that extremely rapid evolutionary adaptation rather than extensive immigration occurred?

28. Comparisons of Neanderthal mitochondrial DNA with that of modern humans indicate that they are not related to modern humans and did not contribute to our mitochondrial heritage. However, because Neanderthals and modern humans are separated by at least 25,000 years, this does not rule out some forms of inter-breeding causing the modern European gene pool to be derived from both Neanderthals and early humans (called Cro-Magnons). To resolve this question, Caramelli et al. (2003. *Proc. Natl. Acad. Sci. [USA]* 100: 6593–6597) analyzed mitochondrial DNA sequences from 25,000-year-old Cro-Magnon remains and compared them to four Neanderthal specimens and a large data set derived from modern humans. The results are shown in the graph.

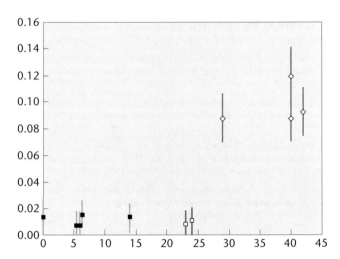

The *x*-axis represents the age of the specimens in thousands of years; the *y*-axis represents the average genetic distance. Modern humans are indicated by filled squares; Cro-Magnons, open squares; and Neanderthals, diamonds. (a) What can you conclude about the relationship between Cro-Magnons and modern Europeans? What about the relationship between Cro-Magnons and Neanderthals? (b) From these data, does it seem likely that Neanderthals made any contributions to the Cro-Magnon gene pool or the modern European gene pool?

Conservation Genetics

Cryogenically preserved seeds of rare crop varieties at the U.S. Department of Agriculture National Center for Genetics Resources Preservation.

■ CHAPTER CONCEPTS

- ■ Species require genetic diversity for long-term survival and adaptation.

- ■ Small, isolated populations are particularly vulnerable to genetic effects.

- ■ Endangered populations that recover in size may not redevelop genetic diversity.

- ■ Conservation and breeding efforts focus on retaining genetic diversity for long-term species survival.

As the twenty-first century progresses, the diversity of life on Earth is under increasing pressure from the direct and indirect effects of explosive human population growth. Approximately 10 million *Homo sapiens* lived on the planet 10,000 years ago. This number grew to 100 million 2000 years ago and to 2.5 billion by 1950. Within the span of a single lifetime, the world's human population more than doubled to 5.5 billion in 1993 and is projected to reach as high as 19 billion by 2100 (Figure 24–1).

The effect of accelerating human population growth on other species has been dramatic. Data from the 2006 World Conservation Union (IUCN) Red List of Threatened Species show that, globally, the long-term survival of 23 percent of all mammals, 12 percent of birds, 51 percent of reptiles, 31 percent of amphibians, and 40 percent of fish species is threatened. Plants are not faring much better: The 2006 IUCN Red List also includes over 8300 vascular plant species worldwide. Not just wild species are at risk. Genetic diversity in domesticated plants and animals is also being lost as many traditional crop varieties and livestock breeds disappear. The Food and Agriculture Organization (FAO) estimates that since 1900, 75 percent of the genetic diversity in agricultural crops has been lost. Out of approximately 5000 different breeds of domesticated farm animals worldwide, more than one-third are at risk.

Why should we be concerned about losing **biodiversity**—that is, the biological variation represented by these different plants and animals? As the fossil record shows, unrelated to human influences, many different plants and animals that once inhabited the planet have become extinct over millions of years, and others have taken their places. Biologists are concerned, however, at the accelerated rate of species extinctions we are witnessing today, all of which can be ascribed to direct or indirect human impacts. Deliberate hunting or harvesting of plants and animals by

FIGURE 24–2 The brown tree snake (*Boiga irregularis*).

humans and habitat destruction through human development activities have greatly reduced the populations of many species. An additional problem in many parts of the world is the deliberate or accidental introduction by humans of invasive nonnative plants or animals that prey on or compete with native species, further jeopardizing their survival. For example, the brown tree snake, *Boiga irregularis* (Figure 24–2), was accidentally transported by ship in the late 1940s to the island of Guam. An aggressive predator, the brown tree snake has so far caused the extinction of 12 bird and 4 lizard species that were previously native to Guam. This introduced species continues to threaten native biodiversity on the island.

One of the greatest threats to biodiversity, however, may be climate change due to global warming. Scientists are increasingly concerned that this will lead to the extinction of many species, especially those adapted to live in colder environments. For example, in 2006 the IUCN reclassified the 22,000 remaining polar bears (Figure 24–3) as vulnerable due to fears that their habitat will be lost from global warming.

As a result of species extinction, biologists fear that **ecosystems**—the complex webs of interdependent, but diverse, plants and animals found together in the same environment—may collapse if key sustaining species are lost. This portends possible consequences for our own long-term survival. Other scientists have pointed out that we lose unknown economic potential of unexploited plants and animals if we allow them to become extinct. Rare species sometimes turn out to be immensely valuable. For example, an obscure bacterium, *Thermus aquaticus*, was first discovered in a thermal pool in Yellowstone National Park and is known to exist in only a few hot springs throughout the world. DNA polymerase isolated from this microorganism functions at very high temperatures and is essential to the polymerase chain reaction (PCR). This research technique is an essential one in the multibillion dollar global biotechnology industry. Quite apart from practical benefits, scientists and nonscientists alike have made the case that human life will be diminished both

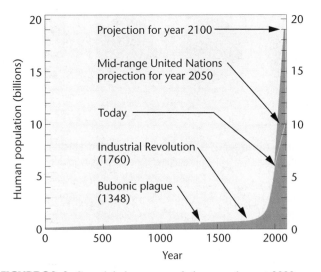

FIGURE 24–1 Growth in human population over the past 2000 years and projected through 2100.

FIGURE 24–3 The polar bear (*Ursus maritimus*).

spiritually and aesthetically if we do not strive to maintain the fascinating, often beautiful variety of living organisms that share the planet with us.

Conservation biologists work to understand and maintain biodiversity, studying the factors that lead to species decline and the ways species can be preserved. The new field of **conservation genetics** has emerged in the last 20 years as scientists have begun to recognize that genetics will be an important tool in maintaining and restoring population viability. The applications of genetics to conservation biology are multifaceted; in this chapter, we will explore just a few of them. Underlying the increasingly important role of genetics in conservation biology is the recognition that biodiversity depends on genetic diversity and that maintaining biodiversity in the long term is unlikely if genetic diversity is lost.

24.1 Genetic Diversity Is at the Heart of Conservation Genetics

While biodiversity encompasses the variation represented by all existing species of plants and animals on this planet at any given time, **genetic diversity** is not as easy to define and study. Genetic diversity can be considered on two levels: **interspecific diversity** and **intraspecific diversity.**

Diversity between species, or *interspecific diversity*, is reflected in the number of different plant and animal species present in an ecosystem. Some ecosystems have a very high level of species diversity, such as a tropical rainforest in which hundreds of different plant and animal species may be found within a few square meters [Figure 24–4(a)]. Other ecosystems, especially where plants and animals must adapt to a harsh environment, may have much lower levels of inter-specific diversity [Figure 24–4(b)]. Lists of the different plant and animal species found in a particular environment are used to compile inventories of species. The inventories identify diversity hot-spots: geographic areas with especially high levels of interspecific diversity where conservation efforts can be focused. Conservation biologists working at the ecosystem level are interested in preventing species from being lost and in restoring species that were once part of the system but are no longer present. The reestablishment of the gray wolf (*Canis lupus*) in Yellowstone National Park and the release of captive-bred California condors (*Gymnogyps californianus*) into the mountain ranges they previously occupied in the southwestern United States are examples of current attempts by conservation biologists to restore missing species to their ecosystems.

Intraspecific diversity—the diversity within a species—is reflected in the level of genetic variation occurring between individuals within a single population of a given species (*intrapopulation diversity*) or between different populations

How Do We Know?

In this chapter we will focus on the genetic approaches used during conservation efforts. Conservation geneticists assess genetic diversity and work to maintain population numbers for long-term species survival. As you study this topic, you should try to answer several fundamental questions:

1. How do we know the extent of genetic diversity in a species?

2. How do we know that diminished genetic diversity is detrimental to species survival?

3. How do we determine the most effective approach to countering decreased population size?

4. How do we attempt to "conserve" existing genetic diversity?

(a) **(b)**

FIGURE 24–4 (a) A tropical rainforest is an ecosystem with high interspecific diversity. (b) A coastal marsh in North Carolina exemplifies an ecosystem with low interspecific diversity.

of the same species (*interpopulation diversity*). Genetic variation within populations can be measured as the frequency of individuals in the population that are heterozygous at a given locus or as the number of different alleles at a locus that are present in the population gene pool. When DNA-profiling techniques are used, the percentage of polymorphic loci—those represented by varying bands on the DNA profile in different individuals—can be calculated to indicate the extent of genetic diversity in a population. In outbreeding species, most intraspecific genetic diversity is found at the intrapopulation level.

Significant interpopulation diversity can occur if populations are separated geographically and there is no migration or exchange of gametes between them. Similarly, predominantly inbreeding species (such as self-fertilizing plants) tend to have greater levels of interpopulation than intrapopulation diversity. A limited number of genotypes dominates an individual population, but there is greater differentiation among populations. Understanding the breeding system and the distribution of genetic variation in an endangered species is important for its conservation, not only to ensure continued production of offspring but also to determine the best strategy for maintaining intraspecific diversity. For example, would it be more effective to preserve a few large populations or many small, distinct ones? This information can then be used to guide conservation or restoration efforts.

Loss of Genetic Diversity

Loss of genetic diversity in nondomesticated species is usually associated with a reduction in population size. This may be due to excessive hunting or harvesting. For example, biologists blame commercial overfishing for the collapse in the early 1990s of the deep-sea cod populations off the Newfoundland coast. This was once one of the most productive fisheries in the world, but despite being protected from fishing since 1992 these cod populations have not fully recovered. Habitat loss is also a major cause of population decline. As the global human population increases, more land is developed for housing and transport systems or is put into agricultural production, reducing or eliminating areas that were once home to wild plants and animals. The shrinking available habitat reduces populations of wild species and often also isolates them from each other as individual populations become trapped in pockets of undeveloped land surrounded by areas taken over for agriculture, urban development, or other human uses. This process is known as **population fragmentation**. When populations are no longer in contact with each other, gene flow through migration or gamete exchange between them ceases, and an important mechanism for maintaining genetic variation is lost.

In domesticated species, loss of genetic diversity is not usually the result of habitat loss or collapsing population numbers; there is little risk that cows or corn as species will become extinct any time soon. Reduction in diversity within domesticated species can instead be traced to changes in agricultural practice and consumer demand. Modern farming techniques have greatly increased production levels, but they have also led to greater genetic uniformity. As farmers switch to new crop varieties or improved livestock strains on a large scale, they abandon cultivation of many older local types, which may then disappear if efforts are not made to preserve them.

For example, in 1900, more than 100 different kinds of potatoes were available in the United States. Today, three quarters of commercial potato production in this country depends on just nine varieties. A single type, Russet Burbank, makes up 45 percent of the total acreage planted, with many farmers also growing related potato varieties such as Ranger Russet and Russet Norkotah. Modern varieties such as these are better adapted for modern agricultural production, but older types often contain useful genes that can still play a vital role in survival functions, such as resistance to disease, cold, or drought. Indeed, there is concern that loss of genetic diversity in commercially grown U.S. potatoes makes them vulnerable to emerging new strains of the deadly fungal disease late blight (*Phytophthera infestans*) that caused widespread famine when it destroyed potato crops in Ireland in the 1840s. Agricultural researchers are now searching for genes conferring resistance to late blight in older potato varieties that have been conserved.

Identifying Genetic Diversity

For many years, population geneticists based estimates of intraspecific diversity on phenotypic differences between individuals, such as different colors of seeds or flowers or variation in markings (Figure 24–5). As techniques were developed to analyze genes and gene products at the molecular level, identifying genetic diversity became more precise. For example, a simple molecular method that has been widely used is allozyme analysis. **Allozymes** are multiple versions of a single enzyme occurring within a single species. Allozymes catalyze the same reaction, but differences in the polypeptide chains forming the bulk of the molecule occur because of slight variations in the allelic DNA sequences at the locus coding for the subunits of the enzyme. This results in forms of the protein that differ in size or net charge and which can be separated by electrophoresis. The presence of allozyme variation in a population can be used as an indicator of the lower

FIGURE 24–5 Phenotypic variation in seed color and markings in the common bean (*Phaseolus vulgaris*) reveals high levels of intraspecific diversity.

limit of genetic variation because the different versions of the molecule indicate that different alleles are present at a locus.

DNA analysis is a more modern and direct molecular approach used to detect and quantify genetic differences between individuals. Nuclear, mitochondrial, and chloroplast DNA can all be analyzed to determine levels of genetic diversity. We have already described in Chapter 19 applications of several of these techniques that use restriction enzymes and PCR during analysis. Conservation biologists have used similar techniques to examine levels and distribution of genetic variation both within and between populations and to subsequently guide conservation efforts.

One technique that has been widely used to analyze genetic diversity in plant and animal populations focuses on **STRs (short tandem repeats)**, also known as **microsatellites**. STRs consist of short sequences (two to nine bases) repeated a variable number of times and are found throughout the genome. The number of times a repeat is present at a given STR locus often varies between individuals. Although the STR regions themselves do not contain expressed genes, conservation biologists can use microsatellite variation among individuals within populations as an indirect estimate of the amount of overall genetic variation present.

For example, researchers from the University of Arizona recently used microsatellite analysis to compare genetic variation in the three remaining populations of the Mauna Loa silversword, an endangered plant found only in Hawaii. Biologists working to conserve this plant needed to know whether they should try to increase genetic diversity by redistributing seed among the populations. However, STR analysis not only revealed unexpected amounts of diversity still present within populations, but also showed that the three populations were genetically quite different from each other. The seed redistribution plan was abandoned in favor of preserving the unique identity of each population.

In another recent study utilizing DNA markers, **restriction fragment length polymorphisms (RFLPs)** were analyzed in the only remaining population of a critically endangered plant species (*Limonium cavanillesii*) growing on the Mediterranean coast of Spain. Researchers found very low levels of diversity but did detect genetically distinct individual plants within the population from which seeds could be collected.

In still another recent RFLP study, DNA marker analysis of the southwestern willow flycatcher, an endangered migratory bird species that nests in the southwestern United States, found significant levels of genetic diversity within populations at different breeding sites but little interpopulation diversity. The scientists studying the flycatcher therefore concluded that migration of individual birds between breeding populations is important in maintaining diversity in the species and that conservation efforts should be directed toward preserving nesting sites that allow such migration to continue.

Another application of DNA markers in conservation involves **DNA fingerprinting**. The use of such a DNA-profiling technique in forensic science was discussed in Chapter 19. Similar techniques can be used to uncover illegal trade in endangered plant and animal species that are protected by law. For example, since 1986, an international moratorium on commercial whaling has been in place, with only limited hunting of a few species permitted. In 1994, scientists based in New Zealand and Hawaii used PCR to amplify mitochondrial DNA (mtDNA) sequences extracted from meat on sale in markets in Japan, where whale meat is prized as a delicacy. The DNA fingerprints revealed several meat samples from humpback and fin whales, which are protected species. Two meat samples were not from whales at all but were dolphin meat! This information assisted Japanese customs officials in tracking down sources of illegal whale meat imports.

Unlawful trade in animal products also occurs in the United States. In a recent study by scientists at the University of Florida, amplification of mtDNA sequences from turtle meat sold in Louisiana and Florida revealed that one quarter of the samples were actually alligator, not turtle. Most of the rest were from small freshwater turtles, indicating that populations of freshwater turtle species were declining through overharvesting and thus were in need of protection. Further refinements to DNA extraction and profiling techniques allowed accurate identification of illegally harvested sea turtle eggs cooked and served in restaurants, and led to prosecution of the poachers.

New methods for analyzing DNA marker data also enable estimation of past changes in populations, based on the rate at which mtDNA is known to accumulate spontaneous point mutations. In a controversial recent study, scientists at Stanford University led by Dr. Stephen Palumbi measured sequence diversity at key points in mtDNA from humpback whales and compared their findings with the amount of diversity expected for a given population size. The Stanford researchers found much more mtDNA sequence diversity than expected, and concluded that before being hunted almost to extinction in the nineteenth and twentieth centuries the global humpback whale population must have numbered over one million, ten times higher than historical estimates based on records from whaling ships. As current international agreements could permit hunting of whales to resume when populations have recovered to 54 percent of what is thought to be their original carrying capacity, Palumbi's findings are clearly important for the future management of this species.

24.2 Population Size Has a Major Impact on Species Survival

Some species have never been numerous, especially those that are adapted to survive in unusual habitats. Biologists refer to such species as *naturally rare*. *Newly rare* species, on the other hand, are those whose numbers are in decline because of pressures such as habitat loss. Populations of such species may not only be small in number but also fragmented and isolated from other populations. Both decreased population size and increased isolation have important genetic consequences, leading to an increased risk of loss of diversity.

How small must a population be before it is considered endangered? It varies somewhat with species, but small populations can quickly become vulnerable to genetic phenomena that increase the risk of extinction. In general, a population of fewer than 100 individuals is considered extremely sensitive to these problems, which include genetic drift, inbreeding, and reduction in gene flow. The effects of such problems on species survival are substantial. Studies of bighorn sheep, for example, have shown that populations of fewer than 50 are highly likely to become extinct within 50 years. Projections based on computer models show that for all species, populations of fewer than 10,000 are likely to be limited in adaptive genetic variation and at least 100,000 individuals must be present if a population is to show long-term sustainability.

Determining the number of individuals a population must contain in order to have long-term sustainability is complicated by the fact that not all members of a population are equally likely to produce offspring: some will be infertile, too young, or too old. The **effective population size** (N_e) is defined as the number of individuals in a population having an equal probability of contributing gametes to the next generation. N_e is almost always smaller than the **absolute population size** (N). The effective population size can be calculated in different ways, depending on the factors that are preventing all individuals in a population from contributing equally to the next generation. In a sexually reproducing population that contains different numbers of males and females, for example, the effective population size is calculated as

$$N_e = \frac{4(N_m N_f)}{N_m + N_f}$$

where N_m is the number of males and N_f the number of females in the population. Hence a population of 100 males and 100 females would have an effective size of $4(100 \times 100)/(100 + 100) = 200$. In contrast, if there were 180 males and only 20 females, the effective population size would be $4(180 \times 20)/(180 + 20) = 72$.

Effective population size is also influenced by fluctuations in absolute population size from one generation to the next. Here, the effective population size is the harmonic mean of the numbers in each generation, so

$$N_e = 1 \left/ \frac{1}{t}\left(\frac{1}{N_1} + \frac{1}{N_2} \cdots + \frac{1}{N_t}\right)\right.$$

where t is the total number of generations being considered. For example, if a population went through a temporary reduction in size in generation 2, so that $N_1 = 100$, $N_2 = 10$, and $N_3 = 100$, then

$$N_e = 1 \left/ \frac{1}{3}\left(\frac{1}{100} + \frac{1}{10} + \frac{1}{100}\right)\right. = \frac{1}{0.04} = 25$$

In this case, although the actual mean number of individuals in the population over three generations was 70, the effective population size during that time was only 25. A severe temporary reduction in size such as this is known as a **population bottleneck**. Bottlenecks occur when a population or species is reduced to a few reproducing individuals whose offspring then increase in numbers over subsequent generations to reestablish the population. Although the number of individuals may be restored to healthier levels, genetic diversity in the newly expanded population is often severely reduced because gametes from the handful of surviving individuals functioning as parents only represent a subset of the original gene pool.

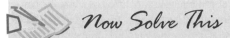

Now Solve This

Question 13 on page 552 asks you to calculate effective population size N_e for an endangered gorilla population.

Hint: Consider what unusual feature of the social structure of the gorilla population will affect N_e.

Captive-breeding programs, in which a few surviving individuals from an endangered species are removed from the wild and their offspring raised in a protected environment to rebuild the population, inevitably create population bottlenecks. Bottlenecks also occur naturally when a small number of individuals from one population migrate to establish a new population elsewhere. When a new population derived from a small subset of individuals has significantly less genetic diversity than the original population, it exhibits the **founder effect**. Reduced levels of genetic diversity due to a founder effect can persist for many generations, as shown by studies of two species of Antarctic fur seal (*Arctocephalus gazella* and *Arctocephalus tropicalis*). Seal hunters in the eighteenth and nineteenth centuries had severely reduced populations of these species, eliminating them from parts of their natural range in the Southern Ocean.* Although the number of Antarctic fur seals has now rebounded and the two species have recolonized much of their original habitat, mtDNA fingerprinting reveals that a founder effect can still be detected in *A. gazella*, with reduced genetic variation in current populations descended from a handful of surviving individuals.

The cheetah (*Acinonyx jubatus*), shown in Figure 24–6, is another well-studied example of a species with reduced genetic variation resulting from at least one severe population bottleneck that occurred in its recent history. Allozyme studies of South African cheetah populations have shown levels of genetic variation that are less than 10 percent of those found in other mammals. The abnormal spermatozoa and poor reproductive

---

*In 2000, this fifth world ocean was delimited from the southern portions of the Atlantic, Indian, and Pacific oceans.

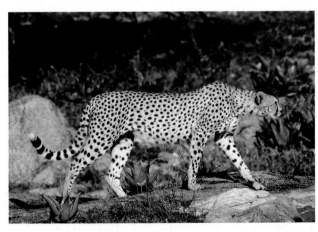

FIGURE 24–6 The cheetah (*Acinonyx jubatus*), a species with reduced genetic variation following population bottlenecks.

rates commonly observed in cheetahs are thought to be linked to the lack of genetic diversity, although when and how the population bottleneck occurred in this species is still unclear.

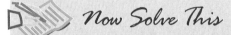

Now Solve This

Question 22 on page 553 ask you to compare allozyme data for cheetah populations with similar data from other cat species.

Hint: Consider what genetic factors potentially reduce the levels of genetic variation reflected in the data. Which species will be most affected by these factors?

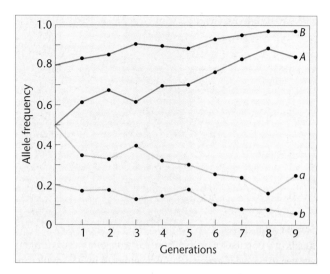

FIGURE 24–7 Change in frequencies over 10 generations for two sets of alleles, A/a and B/b, in a theoretical population subject to genetic drift.

24.3 Genetic Effects Are More Pronounced in Small, Isolated Populations

Small isolated populations, such as those found in threatened and endangered species, or those affected by population fragmentation, are especially vulnerable to genetic drift, inbreeding, and reduction in gene flow. These phenomena act on the gene pool in different ways but ultimately have similar effects in that they all can further reduce genetic diversity and long-term species viability.

Genetic Drift

If the number of breeding individuals in a population is reduced in size, fewer gametes will form the next generation. The alleles carried by these gametes may not be a representative sample of all those present in the population; purely by chance, some alleles may be underrepresented or not present at all, which will cause changes in allele frequency over time, resulting in **genetic drift**. A serious result of genetic drift in populations with a small effective population size is the loss of genetic variation. Genetic drift is a random process, so either deleterious or advantageous alleles can become fixed within a small population. In other words, one allele becomes the only version of that gene present in the gene pool of the population. This means a useful allele can be lost even if it has the potential to increase fitness or long-term adaptability. If we consider a single locus with two alleles, A and a, genetic drift may result in one of the alleles eventually disappearing, while the other becomes fixed. A simple correlation describes the likelihood of the fixation or the loss of an allele. The probability that an allele will be fixed through drift is the same as its initial frequency. So, if the initial frequency of A is 0.8, the probability of A becoming fixed is 0.8, or 80 percent, and the probability of A being lost through drift is 0.2 or 20 percent. Figure 24–7 is a graph of the effect of genetic drift on a theoretical population.

Inbreeding

In small populations, the chance of **inbreeding** (matings between closely related individuals) is greater. As described in Chapter 22, inbreeding increases the proportion of

homozygotes in a population, thus increasing the possibility that an individual may be homozygous for a deleterious allele. The **inbreeding coefficient** (F) measures the extent of inbreeding occurring in a population; F measures the probability that two alleles of a given gene in an individual are derived from a common ancestral allele. The inbreeding coefficient is inversely related to the frequency of heterozygotes in the population and can be calculated as

$$F = \frac{2pq - H}{2pq}$$

where $2pq$ is the expected frequency of heterozygotes based on the Hardy–Weinberg law and H is the actual frequency of heterozygotes in the population.

In a declining population that has become small enough for drift to occur, heterozygosity (H) will decrease with each generation. The smaller the effective population size, the more rapid the decrease in H and the resulting increase in F, as can be seen from the equation

$$H_t/H_0 = \left(1 - \frac{1}{2N_e}\right)^t$$

where H_0 is the initial frequency of heterozygotes, H_t is the frequency of heterozygotes after t generations, and N_e is the effective population size. Figure 24–8 compares rates of increase in F for different effective population sizes.

Now Solve This

Question 1 on page 552 asks you to calculate effective population sizes, a heterozygote frequency, and an inbreeding coefficient for a jackal population that crashed and then recovered.

Hint: Calculate N_e for each generation first, and remember that heterozygote frequency and inbreeding coefficient are inversely related.

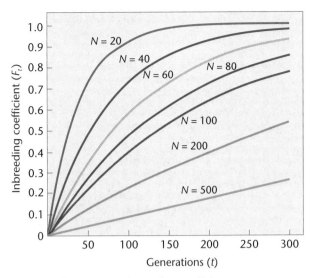

FIGURE 24–8 Increase in inbreeding coefficient.

What effect does inbreeding have on the long-term survival of a population? In some cases, inbreeding may not immediately reduce the amount of genetic variation present in the overall gene pool of a species in which numbers of individuals remain high. Self-pollinating plants, for example, often show high levels of homozygosity and relatively little genetic variation within a single population. However, they tend to have considerable variation between different populations, each of which has adapted to slightly different local environmental conditions. On the other hand, in most outbreeding species, including all mammals, inbreeding is associated with reduced fitness and lower survival rates among offspring. This **inbreeding depression** can result from increased homozyosity for deleterious alleles. The number of deleterious alleles present in the gene pool of a population is called the **genetic load** (or genetic burden).

In some species, inbreeding accompanied by selection against less fit individuals homozygous for deleterious alleles has resulted in the elimination of these alleles from the gene pool, a process known as *purging the genetic load*. Species that have successfully purged their genetic load do not show continued reduction in fitness, even after many generations of inbreeding. This is true of numerous domesticated species, especially self-pollinating plants such as wheat. However, computer simulation experiments have shown that it may take 50 generations or more to complete the purging process, during which time inbreeding depression will still occur.

Alternatively, inbreeding depression can result from heterozygous individuals having a higher level of fitness than either of the corresponding homozygotes. In this case, the long-term survival of the population requires that inbreeding be avoided and that the levels of all alleles in the gene pool be maintained. Both of these requirements can be difficult in a species that has already suffered a significant reduction in population size.

The effects of inbreeding depression and loss of genetic variation in a small isolated population have been documented in the case of the Isle Royale gray wolves (Figure 24–9). Around 1950, a pair of gray wolves apparently crossed an ice

bridge from the Canadian mainland to Isle Royale in Lake Superior. The island had no other wolves and had an abundance of moose, which became the wolves' main food source. By 1980, the Isle Royale wolf population had increased to over 50 individuals. Over the next decade, however, wolf numbers declined to fewer than a dozen with no new litters being born, despite plentiful food and no apparent sign of disease. The genetic variation of the remaining wolves was examined by mtDNA analysis and nuclear DNA fingerprinting. It was found that the Isle Royale wolves had levels of homozygosity that were twice as high as those of wolves in an adjacent mainland population. Furthermore, the wolves all possessed the same mtDNA genotype, consistent with descent from the same female. Hence, the degree of relatedness between individual Isle Royale wolves was equivalent to that of full siblings. This finding suggests that the wolves' reproductive failure was due to inbreeding depression, a phenomenon that has also been seen in captive-wolf populations.

Reduction in Gene Flow

Gene flow, the gradual exchange of alleles between two populations, is brought about by the dispersal of gametes or the migration of individuals. It is an important mechanism for introducing new alleles into a gene pool and increasing genetic variation. Migration is the main route for gene flow in animals. In plants, gene flow occurs not through movement of individuals but as a result of cross-pollination between different populations and through seed dispersal. Isolation and fragmentation of populations in rare and declining species significantly reduce gene flow and the potential for maintaining genetic diversity. As we have already discussed, habitat loss is a major threat to species survival. It is not unusual for a threatened or endangered species to be restricted to small separate pockets of the remaining habitat. This isolates and fragments the surviving populations so that movement of individuals can no longer occur between them, thus preventing gene flow.

The term **metapopulation** is used to describe a population consisting of spatially separated subpopulations with limited gene flow, especially if local extinctions and replacements of some of the subpopulations occur over time. One well-studied metapopulation is that of the endangered red-cockaded wood-

FIGURE 24–9 Isle Royale gray wolf (*Canis lupus*).

pecker (*Picoides borealis*) (Figure 24–10), which was once common in pinewoods throughout the southeastern United States. Habitat loss because of logging has reduced the remaining populations of this bird to small scattered sites isolated from each other, with virtually no migration between them. Studies using allozymes and DNA profiling show that the smallest surviving woodpecker populations (where $N = <100$) have suffered the greatest loss of genetic diversity and are at most risk of inbreeding depression, compared with larger populations where $N = >100$. Management of the red-cockaded woodpecker now includes efforts to increase the genetic diversity of the smallest populations by introducing birds from different larger populations to artificially recreate the gene flow by migration that would have occurred in the original unfragmented distribution of the species.

Another species in which the effects of gene-flow reduction have been studied is the North American brown bear, *Ursus arctos*. A team of Canadian and U.S. researchers measured heterozygosity in different brown bear populations using amplified microsatellite markers to create DNA profiles for individual bears. The researchers found that levels of heterozygosity in the brown bear population living in and around Yellowstone Park were only two-thirds as high as those in brown bear populations from Canada and mainland Alaska. They concluded that this was due to the isolation and reduced migration of the Yellowstone bears. The habitat of the Canadian and Alaskan bear populations was much less fragmented, allowing migration of individuals and consequent gene flow between populations. The researchers also examined an island population of brown bears on the Kodiak archipelago off the Alaskan coast. Here, heterozygosity was even lower—less than one-half that of the mainland populations—providing further evidence for the effect of small population size, restricted migration, and restricted gene flow on the reduction of genetic diversity.

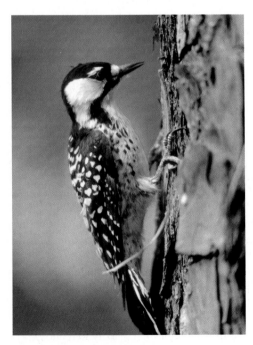

FIGURE 24–10 The red-cockaded woodpecker (*Picoides borealis*).

24.4 Genetic Erosion Diminishes Genetic Diversity

The loss of previously existing genetic diversity from a population or species is referred to as **genetic erosion**. Why does it matter if a population loses genetic diversity, especially if the numbers of individuals remain high? Genetic erosion has two important effects on a population. First, it can result in the loss of potentially useful alleles from the gene pool, thus reducing the ability of the population to adapt to changing environmental conditions and increasing its risk of extinction. Several decades before the dawn of modern genetics, Charles Darwin recognized the importance of diversity to long-term species survival and evolutionary success. In *On the Origin of Species* (1859), Darwin wrote:

> *The more diversified the descendants of any one species … by so much will they be better enabled to seize on many widely diversified places in the polity of nature, and so enabled to increase in numbers.*

While the importance of allelic diversity for population survival under changing environmental conditions can be presumed for an endangered species, it has actually been demonstrated in several weed species whose gene pools contain alleles conferring resistance to chemical herbicide. If the weeds are not sprayed, the plants with these resistance alleles enjoy no particular selective advantage, and the resistance allele often persists at a relatively low frequency. However, if herbicide is applied, the individuals with the resistance allele are much more likely to survive and generate a new resistant population, while populations whose gene pools lack the resistance allele will be eliminated.

Second, genetic erosion resulting in allele loss will reduce levels of heterozygosity. At the population level, reduced heterozygosity will be seen as an increase in the number of individuals homozygous at a given locus. At the individual level, a decrease in the number of heterozygous loci within the genotype of a particular plant or animal will occur. As we have seen, loss of heterozygosity is a common consequence of reduced population size. Alleles may be lost through genetic drift or because individuals carrying them die without reproducing. Smaller populations also increase the likelihood of inbreeding, which inevitably increases homozygosity.

Obviously, once an allele is lost from a gene pool, the potential for heterozygosity is greatly reduced, or it is completely eliminated if there are only two alleles at the locus in question and one is now fixed. The level of homozygosity that can be tolerated varies with species. Studies of populations showing higher-than-normal levels of homozygosity have documented a range of deleterious effects, including reduced sperm viability and reproductive abnormalities in African lions, increased offspring mortality in elephant seals, and reduced nesting success in woodpeckers.

As yet, no evidence has emerged from field studies conclusively linking the extinction of a wild population to genetic erosion. However, laboratory studies using *Drosophila melanogaster* to model evolutionary events show that loss of genetic variation does reduce the ability of a population to

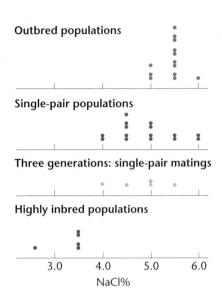

FIGURE 24–11 Effects of bottlenecks in various populations on evolutionary potential in *Drosophila*, as shown by distributions of tolerance of NaCl concentrations at extinction.

adapt to changing environmental conditions. Fruit flies, with their small size and rapid generation time of only 10 to 14 days, are a useful model organism for studying evolutionary events, especially because large populations can be readily developed and maintained through multiple generations.

In one set of experiments carried out by scientists at Macquarie University in Sydney, Australia, *Drosophila* populations were reduced to a single pair for up to three generations to simulate a population bottleneck and were then allowed to increase in number. The capacity of the bottlenecked populations to tolerate increasing levels of sodium chloride was compared with that of normal outbred populations. Researchers found that the bottlenecked populations became extinct at lower salt concentrations (Figure 24–11).

Other experiments comparing inbred and outbred *Drosophila* populations exposed to environmental stresses, such as high temperatures and ethanol presence, have shown that populations with even a low level of inbreeding have a much greater probability of extinction at lower levels of environmental stress than outbred populations. Experimental results such as these indicate that genetic erosion does indeed reduce the long-term viability of a population by reducing its capacity to adapt to changing environmental conditions.

24.5 Conservation of Genetic Diversity Is Essential to Species Survival

Scientists working to maintain biological diversity face several dilemmas. Should they focus on preserving individual populations, or should they take a broader approach by trying to conserve not just one species but all the interdependent plants and animals in an ecosystem? How can genetic diversity be

maintained in a species whose numbers are declining? Can genetic diversity lost from a population be restored? Early conservation efforts often focused only on the population size of an endangered species. Biologists now recognize that a complex interplay of different factors must be considered during conservation efforts, including the need to examine the habitat and role of a species within an ecosystem, as well as the importance of genetic variation for long-term survival.

Ex Situ Conservation: Captive Breeding

Ex situ (Latin for off-site) **conservation** involves removing plants or animals from their original habitat to an artificially maintained location such as a zoo or botanic garden. This living collection can then form the basis of a captive-breeding program.

These programs, where a few surviving individuals from an endangered species are removed from the wild and their offspring are raised in a protected environment to rebuild the population, have been instrumental in bringing a number of species back from the brink of extinction. However, such programs can have undesirable genetic consequences that potentially jeopardize the long-term survival of the species even after population numbers have been restored. A captive-breeding program is rarely initiated until very few individuals are left in the wild, when the original genetic diversity of the species is already depleted. The breeding program is frequently based on a small number of captured individuals, so genetic diversity is further reduced by the founder effect. Since captive-breeding facilities often accommodate only a small number of individuals in the breeding group, inbreeding is difficult to avoid, especially for animal species that form harems (breeding groups dominated by a single male), so that N_e is considerably smaller than N. Finally, unintended selection for genotypes more suited to captive-breeding conditions over time can reduce the overall capacity of the recovered population to adapt and survive in the wild.

How can this type of program be managed to minimize these genetic effects? As we have already seen, loss of genetic diversity measured as heterozygosity over t generations can be expressed as

$$H_t/H_0 = \left(1 - \frac{1}{2N_e}\right)^t$$

This equation indicates that the loss of genetic diversity H_t/H_0 will be greater with a smaller effective population size N_e and a larger number of generations t. We can expand this equation for a captive-breeding population as follows:

$$H_t/H_0 = [1 - (1/2N_{fo})]\{1 - 1/[2N(N_e/N)]\}^{t-1}$$

where N_{fo} is the effective size of the founding population, N is the mean absolute population size, and N_e is the mean effective population size of the group over t generations. Examination of this equation shows that maximum genetic diversity in a captive-breeding group will be maintained by (1) using the largest possible number of founding individuals to maximize N_{fo}; (2) maximizing N_e/N so that as many individuals as possible produce offspring each generation; and (3) minimizing the number of generations in captivity to reduce t. The importance of maximizing N_{fo} to increase allelic diversity in the founding population can also be seen in Figure 24–12. Suc-

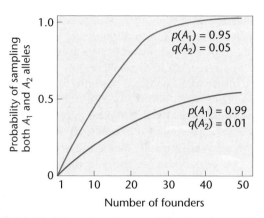

FIGURE 24–12 Effect of captive-population founder number on the probability of maintaining both A_1 and A_2 alleles at a locus.

cessful programs of this sort for endangered species are managed with these three goals in mind. In addition, keeping good pedigree records to avoid matings between relatives and exchanging individuals between different breeding programs when possible will reduce inbreeding.

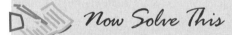

Now Solve This

Question 14 on page 552 asks you to estimate the potential for genetic erosion in a captive population of rare red pandas.

Hint: If N_e/N is maintained at 0.42 and actual population size at 50, you can calculate N_e. Then use the equation given on the previous page.

Captive Breeding: The Black-Footed Ferret

The black-footed ferret, *Mustela nigripes* (Figure 24–13), is an excellent example of a species that has been rescued through captive breeding. This ferret was once widespread throughout the plains of the western United States, but by the 1970s it was considered to be extinct. Decades of trapping and poisoning of the ferret and its main prey, the prairie dog, had decimated their populations. However, in 1981 a small surviving colony of black-footed ferrets was discovered on a ranch near Meeteetse, Wyoming. Conservation biologists first tried to conserve this ferret population *in situ*, but in 1985 the colony was infected with canine distemper, which nearly wiped them out. Of the 18 ferrets that were saved and transferred to a captive facility, only 8 were considered to be sufficiently unrelated to become the founders. The captive-breeding program has produced more than 3000 ferrets, and the species is now being reintroduced to parts of its original range. The goal is to establish populations of 1500 ferrets in sustainable wild colonies by 2010. By early 2006, more than 400 captive-bred ferrets had been released, and the first ferret kits born in the wild were reported from a reintroduction site in Colorado.

Although the recovery of the black-footed ferret appears to be a success, the small founder group of eight individuals caused a severe bottleneck for this species. Additional loss of genetic diversity in the captive-breeding program occurred because of drift, limited reproduction from several of the founder females, and a high rate of breeding by one founder male. The risk of inbreeding and further genetic erosion in this type of program is still a problem.

Genetic management strategies, such as using DNA markers to identify the most genetically varied individuals, maintaining careful pedigree records to avoid mating closely related animals, and developing techniques for artificial insemination and sperm cryopreservation, have helped to conserve the genetic diversity remaining in the species. Conservation geneticists working on the recovery project estimate that all existing black-footed ferrets share about 12 percent of their genome. This is roughly the equivalent of being full cousins. The effects of this degree of genetic similarity in the expanding ferret population are unclear. Occasional abnormalities such as webbed feet and kinked or short tails have been observed in the captive population, but without a noninbred population available for comparison, it is unclear whether this is evidence that inbreeding is increasing the homozygosity for deleterious alleles.

Scientists disagree as to whether the long-term future of the black-footed ferret is jeopardized by the severe bottleneck and subsequent loss of genetic diversity the species has experienced. Some geneticists suggest that because the one surviving Meeteetse population was so isolated, it may have already become sufficiently inbred to purge any deleterious alleles. Other researchers point out that studies of black-footed ferret DNA extracted from museum specimens show that the ferrets alive today have lost significant genetic diversity compared with earlier pre-bottleneck populations, and they suggest that loss of fitness because of inbreeding will inevitably be seen over time.

Ex Situ Conservation and Gene Banks

Another form of *ex situ* conservation is provided by establishing **gene banks**. In contrast to housing entire animals or plants, these collections instead provide long-term storage and preservation for reproductive components, such as sperm, ova, and frozen embryos in the case of animals, and seeds, pollen,

FIGURE 24–13 The black-footed ferret (*Mustela nigripes*).

and cultured tissue in the case of plants. Many more individual genotypes can be preserved for longer periods in a gene bank than in a living collection. Cryopreserved gametes or seeds can be used to reconstitute lost or endangered animals or plants after many years in storage (see the chapter opening photograph on p. 537.) Because they are expensive to construct and maintain, most gene banks are used to conserve these components of domesticated species having economic value.

Gene banks have been established in many countries to help preserve genetic material of agricultural importance, such as traditional crop varieties that are no longer grown or old livestock breeds that are becoming rare. One of the most important *ex situ* collections in the United States is the National Center for Genetic Resources Preservation, a Department of Agriculture (USDA) facility in Fort Collins, Colorado, which maintains more than 350,000 different crop varieties and related wild species. Some of the accessions are stored as seeds and others as cryogenically preserved tissue from which whole plants can be regenerated. Animal genetic resources, including frozen semen and embryos from endangered livestock breeds, are also preserved at this facility.

Ex situ conservation using gene banks, through often vital, has several disadvantages. A major problem with gene banks is that even large collections cannot contain all the genetic variation that is present in a species. Conservation geneticists attempt to address this problem by identifying a **core collection** for a species. The core collection is a subset of individual genotypes that represent as much as possible of the genetic variation within a species; preserving the core collection takes priority over randomly collecting and preserving large numbers of genotypes. Another disadvantage of *ex situ* conservation is that the artificial conditions under which a species is preserved in a living collection or gene bank often create their own selection pressures. When seeds of a rare plant species are maintained in cold storage, for example, selection may occur for those genotypes better adapted to withstand the lower temperatures, and genotypes may be lost that would actually have greater fitness in the plant's natural environment. Yet another problem posed by *ex situ* conservation is that while the greatest biological diversity in both domesticated and nondomesticated species is frequently found in underdeveloped countries, most *ex situ* collections are situated in developed countries that have the resources to establish and maintain them. This leads to conflict over who owns and has access to the potentially valuable genetic resources maintained in such collections.

In Situ Conservation

In situ (Latin for on-site) **conservation** attempts to preserve the population size and biological diversity of a species while it is maintained in its original habitat. The use of species inventories to identify diversity hot-spots is an important tool for determining the best places to establish parks and reserves where plants and animals can be protected from hunting or collecting and where their habitat can be preserved.

For domesticated species, there is increasing interest in "on-farm" preservation, whereby farmers are encouraged with additional resources and financial incentives to maintain traditional crop varieties and livestock breeds. Nondomesticated species with economic potential have also been targeted for *in situ* preservation. In 1998, the USDA established its first *in situ* conservation sites for a wild plant to protect populations of the native rock grape (*Vitis rupestris*) in several eastern states. Winegrowers prize the rock grape not for its fruit, but for its roots; grape vines grafted onto wild rock grape rootstock are resistant to phylloxera, a serious pest of wine grapes.

The advantage of *in situ* conservation for the rock grape, as with other species, is that larger populations with greater genetic diversity can be maintained. Species conserved *in situ* will also continue to live and reproduce in the environments to which they are adapted, reducing the likelihood that novel selection pressures will produce undesirable changes in allele frequency. However, as the global human population continues to rise, setting aside suitable areas for *in situ* conservation becomes a greater challenge. For example, in the case of the North American brown bear, even in large preserves like Yellowstone National Park, a species' migration and gene flow may be eliminated with the consequent loss of genetic diversity. This problem is even more acute in smaller, more fragmented areas of protected habitat.

Population Augmentation

What genetic considerations should accompany efforts to restore populations or species that are in decline? As we have already seen, populations that go through bottlenecks continue to suffer from low levels of genetic diversity, even after their numbers have recovered. We have also seen that inbreeding and drift contribute to genetic erosion in small populations, and fragmentation interrupts migration and gene flow, further reducing diversity. Captive-breeding programs designed to restore a critically endangered species from a few surviving individuals risk genetic erosion in the renewed population from founder effect and inbreeding, as in the case of the black-footed ferret.

An alternative strategy used by conservation biologists is **population augmentation**—boosting the numbers of a declining population by transplanting and releasing individuals of the same species captured or collected from more numerous populations elsewhere. Attempts to reestablish gene flow in severely fragmented populations of the endangered red-cockaded woodpecker by augmenting the smallest populations were described earlier. This method has also been employed with the Florida panther, *Felis coryi* (Figure 24–14), which was reduced to an isolated population of less than 50 animals confined to the area around the Big Cypress Swamp and Everglades National Park in south Florida. As we will discuss in the Genetics, Technology, and Society essay at the end of this chapter, DNA-profiling patterns show high levels of inbreeding in the Florida panther, with reduced fitness because of severe reproductive abnormalities and increased susceptibility to parasite infec-

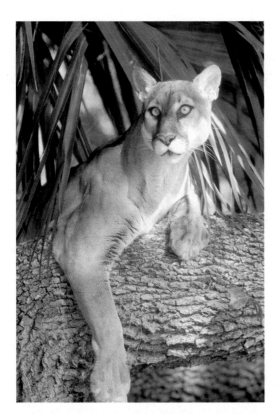

FIGURE 24–14 The Florida panther (*Felis coryi*).

tions. In this effort, seven unrelated animals from a captive population of South and North American panthers were released into the Everglades in the 1960s and allowed to interbreed with the Florida population, thus producing more genetically diverse family groups. The case of the Florida

panther shows how augmentation can increase population numbers, as well as genetic diversity, if the transplanted individuals are unrelated to those in the population to which they are introduced. However, the Florida panther also demonstrates a potential problem with population augmentation, namely, genetic swamping. This occurs when the gene pool of the original population is overwhelmed by different genotypes from the transplanted individuals and loses its identity. Further augmentation of the Florida panther population remains controversial, as some biologists argue that the unique features allowing the Florida panther to be classified as a separate subspecies are lost when breeding takes place with other panthers.

Another difficulty with population augmentation can be caused by **outbreeding depression**, where reduced fitness occurs in the progeny from matings between genetically diverse individuals. Outbreeding depression in the F_1 generation occurs when the heterozygous offspring are less well adapted to local environmental conditions than either parent. This phenomenon has been documented in some annual plants in which seeds of the same species—but from a different location—were used to revegetate a damaged area and cross-pollinated with the remaining local plants. Outbreeding depression in the F_2 and later generations is due to the disruption of coadapted gene complexes—groups of alleles that have evolved to work together to produce the best level of fitness in an individual. This type of outbreeding depression has been documented in F_2 hybrid offspring from matings between fish from different salmon populations in Alaska. These studies suggest that restoring the most beneficial type and amount of genetic diversity in a population is more complicated than previously thought, reinforcing the argument that the best long-term strategy for species survival is to prevent the loss of diversity in the first place.

CHAPTER SUMMARY

1. Biodiversity is lost as increasing numbers of plants and animal species are threatened with extinction. Conservation genetics applies principles of population genetics to the preservation and restoration of threatened species. A major concern of conservation geneticists is the maintenance of genetic diversity.

2. Genetic diversity includes interspecific diversity, which is reflected by the number of different species present in an ecosystem, and intraspecific diversity, which is reflected by genetic variation within a population or between different populations of the same species. Genetic diversity can be measured by examining different phenotypes in the population or, at the molecular level, by using allozyme analysis or DNA-profiling techniques.

3. Major declines in a species' population numbers reduce genetic diversity and contribute to the risk of extinction as a result of genetic drift, inbreeding, or loss of gene flow. Populations that suffer severe reductions in effective size and then recover have passed through a population bottleneck and often show reduced genetic diversity.

4. Loss of genetic diversity reduces the capacity of a population to adapt to changing environmental conditions because useful alleles may disappear from the gene pool. Reduced genetic diversity also results in greater levels of homozygosity in a population, often leading to the manifestation of deleterious alleles and inbreeding depression.

5. Conservation of genetic diversity depends on *ex situ* methods, such as living collections, captive-breeding programs, and gene banks, as well as *in situ* approaches, such as the establishment of parks and preserves.

6. Population augmentation, in which individuals are transplanted into a declining population from a more numerous population of the same species located elsewhere, can be used to increase numbers and genetic diversity. However, the risk of outbreeding depression and genetic swamping accompanies this process.

Gene Pools and Endangered Species: The Plight of the Florida Panther

In the last 400 years, more than 700 of Earth's animal and plant species have become extinct. In the United States alone, at least 30 species have suffered extinction in the last decade. In addition, hundreds of genetically distinct plant and animal species are now endangered. This dramatic loss of biological diversity is a direct consequence of human activity. As humans increase in number and spread over Earth's surface, we harness more and more of Earth's resources for our use. As a consequence of these activities, we have cleared the forests, polluted the water, and permanently altered Earth's natural balance.

Although the destruction continues, we are beginning to understand the implications of our actions and to make efforts to save some of the more endangered life forms. But despite our best efforts to save threatened species, we often intercede too late—after population numbers have suffered severe declines and after important ecosystems have been permanently compromised. As a result, much of the genetic diversity that existed in these species is lost and the populations must be rebuilt from reduced gene pools. This genetic uniformity can reduce fitness, as well as expose genetic diseases. Reduced fitness may further deplete the numbers of the threatened plant or animal, and the population may spiral downward to extinction.

The story of the Florida panther provides a dramatic example of an animal brought to the verge of extinction and the challenges that must be overcome to restore it to a healthy place in the ecosystem. The Florida panther is one of 30 subspecies of cougar and is one of the most endangered mammals in the world. The panthers once roamed the southeastern corner of the United States, from South Carolina and Arkansas to the southern tip of Florida. As people settled in the panthers' habitats, the panthers were considered to be a threat to livestock and humans. So the panthers were killed by hunting, poisoning, highway collisions,

and loss of the habitat that supports their prey—primarily wild deer and hogs. By 1967, Florida panthers were listed as endangered, and only about 30 Florida panthers were left. Population estimates predicted that the Florida panther would be extinct by the year 2055.

Because of about 20 generations of geographical isolation and inbreeding, Florida panthers had the lowest levels of genetic heterozygosity of any subspecies of cougar. The loss of genetic diversity manifested itself in the appearance of some severe genetic defects. For example, almost 80 percent of panther males born after 1989 in the Big Cypress region showed a rare, heritable (autosomal dominant or X-linked recessive) condition known as cryptorchidism—failure of one or both testicles to descend. This defect is associated with low testosterone levels and reduced sperm count. Life-threatening congenital heart defects also appeared in the Florida panther population, possibly because of an autosomal dominant gene defect. In addition, some immune deficiencies emerged, making the small panther population more susceptible to diseases and further contributing to the population's decline. Other less serious genetic features appeared, such as a kink in the tail and a whorl of fur on the back.

Over the last two decades, a faint glimmer of hope has appeared for the Florida panther. Federal and state agencies, as well as private individuals, have implemented a Florida Panther Recovery Program. The program's goal is to exceed 500 breeding animals by the year 2010. If successful, the panther species would be granted a 95 percent probability of survival, while retaining up to 90 percent of its genetic diversity. The plan includes initiating a captive-breeding program, enforcing strict protection, increasing and improving the panther's habitat, and educating the public and private landowners. Wildlife underpasses have been constructed along highways in panther territory, and these have significantly reduced pan-

ther highway fatalities (which account for about half of panther deaths). By 2002, Florida panther numbers had increased to about 100 animals.

To retard the detrimental effects of inbreeding, eight wild female Texas panthers (a related subspecies from western Texas) were released into Florida panther territory in 1995. Three of the females died prior to breeding; however, the remaining Texas females gave birth to litters of healthy kittens. It is estimated that 40 to 70 of the 100 panthers in South Florida are now hybrids of Texas panthers and Florida panthers. None of the hybrids appears to have the kinked tail or other traits characteristic of the inbred Florida panthers, and hybrid kittens are three times more likely to survive to adulthood. The hybrid cats are more diverse genetically than purebred Florida panthers, with up to 20 to 30 percent of their genetic material being contributed by the Texas panthers. This is of concern to some biologists, who worry that the Texas cats may genetically swamp the distinct Florida panther population.

Paradoxically, the success of the restoration program has become a problem. Now that the population has reached about 100 animals, the Florida panther has almost exceeded the capacity of its existing habitat. Biologists estimate that the population must reach at least 250 individuals in order to be self-sustaining. To reach this number, the panthers will need to expand their territory, again putting them in direct competition with human expansion in Florida. Biologists are now evaluating potential new territories for the growing population of Florida panthers— including parts of Louisiana, Arkansas, and South Carolina.

Despite the recent success of the restoration program, the survival of the Florida panther is far from certain. The panther's comeback will require years of monitoring and frequent intervention. In addition, people must be willing to share their land with wild creatures that are dangerous and do not directly further human interests. However, public support for the return of the Florida panther has been strong, so there may be hope for this unique, impressive animal.

References

Maehr, D. S., and Lacy, R. C. 2002. Avoiding the lurking pitfalls in Florida panther recovery. *Wildlife Soc. Bull.* 30(3): 971–978.

Mansfield, K. G., and Land, E. D. 2002. Cryptorchidism in Florida panthers: Prevalence, features and influence of genetic restoration. *J. of Wildlife Diseases* 38(4): 693–698.

Pimm, S. L., Dollar, L., and Bass., O. L. 2006. The genetic rescue of the Florida panther. *Anim. Conserv.* 9:115–122.

Web Site

Florida Panther Net [online]
http://www.panther.state.fl.us/

■ KEY TERMS

INSIGHTS AND SOLUTIONS

1. Is a rare species found as several fragmented subpopulations more vulnerable to extinction than an equally rare species found as one larger population? What factors should be considered when managing fragmented populations of a rare species?

Solution: A rare species in which the remaining individuals are divided among smaller isolated subpopulations can appear to be less vulnerable. If one subpopulation becomes extinct through local causes, such as disease or habitat loss, then the remaining subpopulations may still survive. However, genetic drift will cause smaller populations to experience more rapidly increasing homozygosity over time compared with larger populations. Even with random mating, the change in heterozygosity from one generation to the next because of drift can be calculated as

$$H_1 = H_0(1 - 1/2N)$$

where H_0 is the frequency of heterozygotes in the present generation, H_1 is the frequency of heterozygotes in the next generation, and N is the number of individuals in the population.

Thus, in a small population of 50 individuals with an initial heterozygote frequency of 0.5, heterozygosity will decline to $0.5(1 - 1/100) = 0.495$, a loss of 0.5 percent in just one generation. In a larger population of 500 individuals and the same initial heterozygote frequency, after one generation, heterozygosity will be $0.5(1 - 1/1000) = 0.4995$, a loss of only 0.05 percent.

Therefore, smaller populations are likely to show the effects of homozygosity for deleterious alleles sooner than larger populations, even with random mating. If populations are fragmented so that movement of individuals or gametes between them is prevented, management options could include transplanting individuals from one subpopulation to another to enable gene flow to occur. Establishment of "wildlife corridors" of undisturbed habitat that connect fragmented populations could be considered. In captive populations, exchange of breeding adults (or their gametes through shipment of preserved semen or pollen) can be undertaken. Management for increased population numbers, however, is vital to prevent further genetic erosion through drift.

PROBLEMS AND DISCUSSION QUESTIONS

1. A wildlife biologist studied four generations of a population of rare Ethiopian jackals. When the study began, there were 47 jackals in the population, and analysis of microsatellite loci from these animals showed a heterozygote frequency of 0.55. In the second generation, an outbreak of distemper occurred in the population, and only 17 animals survived to adulthood. These jackals produced 20 surviving offspring, which in turn gave rise to 35 progeny in the fourth generation. (a) What was the effective population size for the four generations of this study? (b) Based on this effective population size, what is the heterozygote frequency of the jackal population in generation 4? (c) What is the inbreeding coefficient in generation 4, assuming an inbreeding coefficient of $F = 0$ at the beginning of the study, no change in microsatellite allele frequencies in the gene pool, and random mating in all generations?

2. Chondrodystrophy, a lethal form of dwarfism, has recently been reported in captive populations of the California condor and has killed embryos in 5 out of 169 fertile eggs. Chondrodystrophy in condors appears to be caused by an autosomal recessive allele with an estimated frequency of 0.09 in the gene pool of this species. (a) How do you think California condor populations should be managed in the future to minimize the effect of this lethal allele? (b) What are the advantages and disadvantages of attempting to eliminate it from the gene pool?

3. A geneticist is studying three loci, each with one dominant and one recessive allele, in a small population of rare plants. She estimates the frequencies of the alleles at each of these loci as $A = 0.75$, $a = 0.25$; $B = 0.80$, $b = 0.20$; $C = 0.95$, $c = 0.05$. What is the probability that all the recessive alleles will be lost from the population through genetic drift?

4. How are genetic drift and inbreeding similar in their effects on a population? How are they different?

5. You are the manager of a game park in Africa with a native herd of just 16 black rhinos, an endangered species worldwide. Describe how you would manage this herd to establish a viable population of black rhinos in the park. What genetic factors would you take into consideration in your management plan?

6. Compare the causes and effects of inbreeding depression and outbreeding depression.

7. Cloning, using the techniques similar to those pioneered by the Scottish scientists who produced Dolly the sheep, has been proposed as a way to increase the numbers of some highly endangered mammalian species. Discuss the advantages and disadvantages of using such an approach to aid long-term species survival.

8. In a population of wild poppies found in a remote region of the mountains of eastern Mexico, almost all of the members have pale yellow flowers, but breeding experiments show that pale yellow is recessive to deep orange. Using the tools of the conservation geneticists described in this chapter, how could you experimentally determine whether the prevalence of the recessive phenotype among the eastern Mexican poppy population is due to natural selection or simply to the effects of genetic drift and/or inbreeding?

9. Contrast *ex situ* conservation techniques with *in situ* conservation techniques.

10. Describe how the captive-breeding *ex situ* conservation approach is applied to a severely endangered species.

11. Explain why a low amount of genetic diversity in a species is a detriment to the survival of that species.

12. Contrast allozyme analysis with RFLP analysis as measures of genetic diversity.

13. A population of endangered lowland gorillas is studied by conservation biologists in the wild. The biologists count 15 gorillas but observe that the population consists of two harems, each dominated by a different single male. One harem contains eight females and the other has five. What is the effective population size, N_e, of the gorilla population?

14. Twenty endangered red pandas are taken into captivity to found a breeding group. (a) What is the probability that at least one of the captured pandas has the genotype B_1/B_2 if $p(B_1) = 0.99$? (b) Careful management keeps the N_e/N ratio for the red panda breeding population at 0.42. If the overall size of the captive population is maintained at 50 individuals, what proportion of the heterozygosity present in the founding population will still be present after five generations in captivity?

15. Use your analysis of the red panda in Problem 14 to make suggestions for managing the captive-breeding population of red pandas to maintain as much genetic diversity as possible.

16. Przewalski's horse (*Equus przewalskii*), thought by some biologists to represent the ancestral species from which modern horses were domesticated, is classified as a separate species from the domestic horse, although members of the two species can mate and produce fertile offspring. Przewalski's horse was hunted to extinction in its native habitat in the Asian steppes by the 1920s. The few hundred Przewalski's horses alive today are descended from 12 surviving individuals that had been taken into captivity. The founder breeding group also included a domestic mare. Currently, there is interest in reintroducing Przewalski's horse to its original range. (a) What genetic factors associated with the animals available for reintroduction do you think should be considered before a decision is reached? (b) What advice would you give to the conservation biologists managing the reintroduction project?

17. DNA profiles based on different kinds of molecular markers are increasingly used to measure levels of genetic diversity in populations of threatened and endangered species. Do you think these marker-based estimates of genetic diversity are reliable indicators of a population's potential for survival and adaptation in a natural environment? Why, or why not?

18. Seed banks provide managed protection for the conservation of species of economic and noneconomic importance. Schoen and colleagues (*1998. Proc. Natl. Acad. Sci. [USA]* 95: 394–399) quantified considerable fitness decay in seeds maintained in long-term storage. When germination falls below 65–85 percent, regeneration (planting and seed collection) of a finite sample of the stored seed type is recommended. What genetic consequences might you expect to accompany the conservation practice of long-term seed storage and regeneration?

19. According to Holmes (*2001. Proc. Natl. Acad. Sci. [USA]* 98: 5072–5077), "One of the first questions a resource manager asks about threatened and endangered species is: How bad is it?" More formally, the question seeks a population viability

analysis that includes estimates of extinction risk. What factors would you consider significant in providing an estimate of extinction risk for a species?

20. Microsatellite loci are short (2-to-5 base pair) tandem repeats, which are abundantly and somewhat randomly distributed in all eukaryotic chromosomes. They show high mutability such that new alleles are produced more frequently than in traditional protein-coding genes. It is likely that most microsatellites are selectively neutral and highly heterozygous in natural populations. Knowing that cheetahs underwent a population bottleneck approximately 12,000 years ago, North American pumas 10,000 years ago, and Gir Forest lions 1000 years ago, in which of these species would you expect to see the highest degree of microsatellite polymorphism? the lowest?

21. Considering the behavior and evolution of microsatellites described in Problem 20, provide a graph that relates microsatellite polymorphism (variance) with the number of years since a species' bottleneck. Select variances from 1.0 to 9.0, where increasing variance represents increasing polymorphism, and set the range of years since the bottleneck from 10,000 to 50,000.

22. Allozymes are electrophoretically distinct forms of a particular protein, while microsatellites and minisatellites are repetitive DNA sequences that have been found in all eukaryotes studied to date. Such repetitive sequences are rarely associated with coding sequences of DNA. The data shown here represent the percent heterozygosity of allozymes and microsatellite and minisatellite DNAs in nuclear genomes of four species of felines (modified from Driscoll et al., *Genome Res.* 2002. 12: 414–423). (a) Which species appears to contain the greatest genetic variability? Why might this species be so variable? (b) Which species appears to have the least genetic variability? (c) Why are allozymes less variable than minisatellite or microsatellite DNAs?

	% Heterozygosity			
Marker	Cheetah	Lion	Puma	Domesticated Cat
Allozyme	1.4	0.0	1.8	8.2
Minisatellite	43.3	2.9	10.3	44.9
Microsatellite	46.7	7.9	14.7	68.1

23. If we assume that one of the species in Problem 22 had undergone a recent population crisis in which the number of effective breeders reached critically low levels, which species do you think it would be?

24. For many conservation efforts, scientists lack sufficient data to make definitive conservation decisions. Yet the allocation of habitats for conservation cannot be delayed until such data are available. In these cases, conservation efforts are often directed toward three classes of species: flagships (high-profile species), umbrellas (species requiring large areas for habitat), and biodiversity indicators (species representing diverse, especially productive habitats) (Andelman and Fagan, 2000. *Proc. Natl. Acad. Sci. [USA]* 97: 5954–5959). In terms of protecting threatened species, what advantages and disadvantages might accompany investing scarce talent and resources in each of these classes?

Selected Readings

Chapter 1 Introduction to Genetics

Barnum, S. R. 2005. *Biotechnology*, 2nd ed. Belmont, CA: Brooks-Cole.

Bilen, J., and Bonini, N. M. 2005. *Drosophila* as a model for human neurodegenerative disease. *Annu. Rev. Genet.* 39: 153–171.

Dale, P. J., Clarke, B., and Fontes, E. M. G. 2002. Potential for the environmental impact of transgenic crops. *Nature Biotech.* 20: 567–574.

Fortini, M., and Bonini, N. M. 2000. Modeling human neurodegenerative diseases in *Drosophila. Trends Genet.* 16: 161–167.

Lonberg, N. 2005. Human antibodies from transgenic animals. *Nature Biotech.* 23: 1117–1125.

Lurquin, P. 2002. *High tech harvest.* Boulder, CO: Westview Press.

Pearson, H. 2006. What is a gene? *Nature* 441: 399–401.

Potter, C. J., Turenchalk, G. S., and Xu, T. 2000. *Drosophila* in cancer research: An expanding role. *Trends Genet.* 16: 33–39.

Pray, C. E., Huang, J., Hu, R., and Rozelle, S. 2002. Five years of Bt cotton in China—The benefits continue. *The Plant Journal* 31: 423–430.

Primrose, S. B., and Twyman, R. M. 2004. *Genomics: Applications in human biology.* Oxford: Blackwell Publishing.

Weinberg, R. A. 1985. The molecules of life. *Sci. Am.* (Oct.) 253: 48–57.

Wisniewski, J-P., Frange, N., Massonneau, A., and Dumas, C. 2002. Between myth and reality: Genetically modified maize, an example of a sizeable scientific controversy. *Biochimie* 84: 1095–1103.

Chapter 2 Mitosis and Meiosis

Alberts, B., et al. 2002. *Molecular biology of the cell*, 4th ed. New York: Garland.

Brachet, J., and Mirsky, A. E. 1961. *The cell: Meiosis and mitosis*, Vol. 3. Orlando, FL: Academic Press.

DuPraw, E. J. 1970. *DNA and chromosomes.* New York: Holt, Rinehart & Winston.

Glover, D. M., Gonzalez, C., and Raff, J. W. 1993. The centrosome. *Sci. Am.* (June) 268: 62–68.

Golomb, H. M., and Bahr, G. F. 1971. Scanning electron microscopic observations of surface structures of isolated human chromosomes. *Science* 171: 1024–1026.

Hartwell, L. H., and Karstan, M. B. 1994. Cell cycle control and cancer. *Science* 266: 1821–1828.

Hartwell, L. H., and Weinert, T. A. 1989. Checkpoint controls that ensure the order of cell cycle events. *Science* 246: 629–634.

Mazia, D. 1961. How cells divide. *Sci. Am.* (Jan.) 205: 101–120.

———. 1974. The cell cycle. *Sci. Am.* (Jan.) 235: 54–64.

McIntosh, J. R., and McDonald, K. L. 1989. The mitotic spindle. *Sci. Am.* (Oct.) 261: 48–56.

Westergaard, M., and von Wettstein, D. 1972. The synaptinemal complex. *Annu. Rev. Genet.* 6: 71–110.

Chapter 3 Mendelian Genetics

Bennett, R. L., et. al. 1995. Recommendations for standardized human pedigree nomenclature. *Am. J. Hum. Genet.* 56: 745–752.

Carlson, E. A. 1987. *The gene: A critical history*, 2nd ed. Philadelphia: Saunders.

Dunn, L. C. 1965. *A short history of genetics.* New York: McGraw-Hill.

Henig, R. M. 2001. *The monk in the garden: The lost and found genius of Gregor Mendel, the father of genetics.* New York: Houghton-Mifflin.

Miller, J. A. 1984. Mendel's peas: A matter of genius or of guile? *Sci. News* 125: 108–109.

Olby, R. C. 1985. *Origins of Mendelism*, 2nd ed. London: Constable.

Orel, V. 1996. *Gregor Mendel: The first geneticist.* Oxford: Oxford University Press.

Peters, J., ed. 1959. *Classic papers in genetics.* Englewood Cliffs, NJ: Prentice-Hall.

Sokal, R. R., and Rohlf, F. J. 1987. *Introduction to biostatistics*, 2nd ed. New York: W. H. Freeman.

Stern, C., and Sherwood, E. 1966. *The origins of genetics: A Mendel source book.* San Francisco: W. H. Freeman.

Stubbe, H. 1972. *History of genetics: From prehistoric times to the rediscovery of Mendel's laws.* Cambridge, MA: MIT Press.

Sturtevant, A. H. 1965. *A history of genetics.* New York: Harper & Row.

Tschermak-Seysenegg, E. 1951. The rediscovery of Mendel's work. *J. Hered.* 42: 163–172.

Welling, F. 1991. Historical study: Johann Gregor Mendel 1822–1884. *Am. J. Med. Genet.* 40: 1–25.

Chapter 4 Modification of Mendelian Ratios

Bartolomei, M. S., and Tilghman, S. M. 1997. Genomic imprinting in mammals. *Annu. Rev. Genet.* 31: 493–525.

Brink, R. A., ed. 1967. *Heritage from Mendel.* Madison: University of Wisconsin Press.

Bultman, S. J., Michaud, E. J., and Woychik, R. P. 1992. Molecular characterization of the mouse *agouti* locus. *Cell* 71: 1195–1204.

Carlson, E. A. 1987. *The gene: A critical history*, 2nd ed. Philadelphia: Saunders.

Cattanach, B. M., and Jones, J. 1994. Genetic imprinting in the mouse: Implications for gene regulation. *J. Inherit. Metab. Dis.* 17: 403–420.

Choman, A. 1998. The myoclonic epilepsy and ragged-red fiber mutation provides new insights into human mitochondrial function and genetics. *Am. J. Hum. Genet.* 62: 745–751.

Dunn, L. C. 1966. *A short history of genetics.* New York: McGraw-Hill.

Feil, R., and Kelsey, G. 1997. Genomic imprinting: A chromatin connection. *Am. J. Hum. Genet.* 61: 1213–1219.

Foster, M. 1965. Mammalian pigment genetics. *Adv. Genet.* 13: 311–339.

Freeman, G., and Lundelius, J. W. 1982. The developmental genetics of dextrality and sinistrality in the gastropod *Lymnaea Peregra. Wilhelm Roux Arch.* 191: 69–83.

Grant, V. 1975. *Genetics of flowering plants.* New York: Columbia University Press.

Harper, P. S., et al. 1992. Anticipation in myotonic dystrophy: New light on an old problem. *Am. J. Hum. Genet.* 51: 10–16.

Howeler, C. J., et al. 1989. Anticipation in myotonic dystrophy: Fact or fiction? *Brain* 112: 779–797.

Morgan, T. H. 1910. Sex-limited inheritance in *Drosophila. Science* 32: 120–122.

Peters, J. A., ed. 1959. *Classic papers in genetics.* Englewood Cliffs, NJ: Prentice-Hall.

Phillips, P. C. 1998. The language of gene interaction. *Genetics* 149: 1167–1171.

Race, R. R., and Sanger, R. 1975. *Blood groups in man.* 6th ed. Oxford: Blackwell.

Sapienza, C. 1990. Parental imprinting of genes. *Sci. Am.* (Oct.) 363: 52–60.

Siracusa, L. D. 1994. The *agouti* gene: Turned on to yellow. *Cell* 10: 423–428.

Sturtevant, A. H. 1923. Inheritance of the direction of coiling in *Limnaea. Science* 58: 269–270.

Wallace, D. C. 1997. Mitochondrial DNA in aging and disease. *Sci. Am.* (Aug.) 277: 40–59.

———. 1999. Mitochondrial diseases in man and mouse. *Science* 283: 1482–1488.

Wallace, D. C., et al. 1988. Familial mitochondrial encephalomyopathy (MERRF): Genetic, pathophysiological and biochemical characterization of a mitochondrial DNA disease. *Cell* 55: 601–610.

Watkins, M. W. 1966. Blood group substances. *Science* 152: 172–181.

Yoshida, A. 1982. Biochemical genetics of the human blood group ABO system. *Am. J. Hum. Genet.* 34: 1–14.

Chapter 5 Sex Determination and Sex Chromosomes

Amory, J. K., et al. 2000. Klinefelter's syndrome. *Lancet* 356: 333–335.

Avner, P., and Heard, E. 2001. X-chromosome inactivation: Counting, choice, and initiation. *Nature Reviews* 2: 59–67.

Burgoyne, P. S. 1998. The mammalian Y chromosome: A new perspective. *Bioessays* 20: 363–366.

Carrel, L., and Willard, H. F. 1998. Counting on Xist. *Nature Genetics* 19: 211–212.

Court-Brown, W. M. 1968. Males with an XYY sex chromosome complement. *J. Med. Genet.* 5: 341–359.

Davidson, R., Nitowski, H., and Childs, B. 1963. Demonstration of two populations of cells in human females heterozygous for glucose-6-phosphate dehydrogenase variants. *Proc. Natl. Acad. Sci. USA* 50: 481–485.

Erickson, J. D. 1976. The secondary sex ratio of the United States, 1969–71: Association with race, parental ages, birth order, paternal education and legitimacy. *Ann. Hum. Genet.* (London) 40: 205–212.

Hodgkin, J. 1990. Sex determination compared in *Drosophila* and *Caenorhabditis. Nature* 344: 721–728.

Hook, E. B. 1973. Behavioral implications of the humans XYY genotype. *Science* 179: 139–150.

Irish, E. E. 1996. Regulation of sex determination in maize. *BioEssays* 18: 363–369.

Jacobs, P. A., et al. 1974. A cytogenetic survey of 11,680 newborn infants. *Ann. Hum. Genet.* 37: 359–376.

Jegalian, K., and Lahn, B. T. 2001. Why the Y is so weird. *Sci. Am.* (Feb.) 284: 56–61.

Koopman, P., et al. 1991. Male development of chromosomally female mice transgenic for *Sry. Nature* 351: 117–121.

Lahn, B. T., and Page, D. C. 1997. Functional coherence of the human Y chromosome. *Science* 278: 675–680.

Lucchesi, J. 1983. The relationship between gene dosage, gene expression, and sex in *Drosophila. Dev. Genet.* 3: 275–282.

Lyon, M. F. 1972. X-chromosome inactivation and developmental patterns in mammals. *Biol. Rev.* 47: 1–35.

———. 1988. X-chromosome inactivation and the location and expression of X-linked genes. *Am. J. Hum. Genet.* 42: 8–16.

———. 1998. X-chromosome inactivation spreads itself: Effects in autosomes. *Am. J. Hum. Genet.* 63: 17–19.

McMillen, M. M. 1979. Differential mortality by sex in fetal and neonatal deaths. *Science* 204: 89–91.

Penny, G. D., et al. 1996. Requirement for Xist in X chromosome inactivation. *Nature* 379: 131–137.

Pieau, C. 1996. Temperature variation and sex determination in reptiles. *BioEssays* 18: 19–26.

Westergaard, M. 1958. The mechanism of sex determination in dioecious flowering plants. *Adv. Genet.* 9: 217–281.

Whiting, P. W. 1939. Multiple alleles in sex determination in *Habrobracon. J. Morphology* 66: 323–355.

Witkin, H. A., et al. 1996. Criminality in XYY and XXY men. *Science* 193: 547–555.

Chapter 6 Chromosome Mutations: Variation in Number and Arrangement

Antonarakis, S. E. 1998. Ten years of genomics, chromosome 21, and Down syndrome. *Genomics* 51: 1–16.

Ashley-Koch, A. E., et al. 1997. Examination of factors associated with instability of the *FMR1* CGG repeat. *Am. J. Hum. Genet.* 63: 776–785.

Beasley, J. O. 1942. Meiotic chromosome behavior in species, species hybrids, haploids, and induced polyploids of *Gossypium. Genetics* 27: 25–54.

Blakeslee, A. F. 1934. New jimson weeds from old chromosomes. *J. Hered.* 25: 80–108.

Boue, A. 1985. Cytogenetics of pregnancy wastage. *Adv. Hum. Genet.* 14: 1–58.

Carr, D. H. 1971. Genetic basis of abortion. *Ann. Rev. Genet.* 5: 65–80.

Croce, C. M. 1996. The *FHIT* gene at 3p14.2 is abnormal in lung cancer. *Cell* 85: 17–26.

DeArce, M. A., and Kearns, A. 1984. The fragile X syndrome: The patients and their chromosomes. *J. Med. Genet.* 21: 84–91.

Feldman, M., and Sears, E. R. 1981. The wild gene resources of wheat. *Sci. Am.* (Jan.) 244: 102–112.

Galitski, T., et al. 1999. Ploidy regulation of gene expression. *Science* 285: 251–254.

Gersh, M., et al. 1995. Evidence for a distinct region causing a cat-like cry in patients with 5p deletions. *Am. J. Hum. Genet.* 56: 1404–1410.

Hassold, T. J., et al. 1980. Effect of maternal age on autosomal trisomies. *Ann. Hum. Genet.* (London) 44: 29–36.

Hassold, T. J., and Hunt, P. 2001. To err (meiotically) is human: The genesis of human aneuploidy. *Nat. Rev. Gen.* 2: 280–291.

Hassold, T., and Jacobs, P. A. 1984. Trisomy in man. *Annu. Rev. Genet.* 18: 69–98.

Hecht, F. 1988. Enigmatic fragile sites on human chromosomes. *Trends Genet.* 4: 121–122.

Hulse, J. H., and Spurgeon, D. 1974. *Triticale. Sci. Am.* (Aug.) 231: 72–81.

Jacobs, P. A., et al. 1974. A cytogenetic survey of 11,680 newborn infants. *Ann. Hum. Genet.* 37: 359–376.

Kaiser, P. 1984. Pericentric inversions: Problems and significance for clinical genetics. *Hum. Genet.* 68: 1–47.

Lewis, E. B. 1950. The phenomenon of position effect. *Adv. Genet.* 3: 73–115.

Lewis, W. H., ed. 1980. *Polyploidy: Biological relevance.* New York: Plenum Press.

Lynch, M., and Conery, J. S. 2000. The evolutionary fate and consequences of duplicate genes. *Science* 290: 1151–1154.

Madan, K. 1995. Paracentric inversions: A review. *Hum. Genet.* 96: 503–515.

Ohno, S. 1970. *Evolution by gene duplication.* New York: Springer-Verlag.

Oostra, B. A., and Verkerk, A. J. 1992. The fragile X syndrome: Isolation of the *FMR-1* gene and characterization of the fragile X mutation. *Chromosoma* 101: 381–387.

Patterson, D. 1987. The causes of Down syndrome. *Sci. Am.* (Aug.) 257: 52–61.

Shepard, J., et al. 1983. Genetic transfer in plants through interspecific protoplast fusion. *Science* 21: 683–688.

Shepard, J. F. 1982. The regeneration of potato plants from protoplasts. *Sci. Am.* (May) 246: 154–166.

Strickberger, M. W. 2000. *Evolution,* 3rd ed. Boston: Jones and Bartlett.

Sutherland, G. 1985. The enigma of the fragile X chromosome. *Trends Genet.* 1: 108–111.

Taylor, A. I. 1968. Autosomal trisomy syndromes: A detailed study of 27 cases of Edwards syndrome and 27 cases of Patau syndrome. *J. Med. Genet.* 5: 227–252.

Tjio, J. H., and Levan, A. 1956. The chromosome number of man. *Hereditas* 42: 1–6.

Wilkins, L. E., Brown, J. X., and Wolf, B. 1980. Psychomotor development in 65 home-reared children with cri-du-chat syndrome. *J. Pediatr.* 97: 401–405.

Yunis, J. J., ed. 1977. *New chromosomal syndromes.* Orlando, FL: Academic Press.

Chapter 7 Linkage and Mapping in Eukaryotes

Allen, G. E. 1978. *Thomas Hunt Morgan: The man and his science.* Princeton, NJ: Princeton University Press.

Chaganti, R., Schonberg, S., and German, J. 1974. A manyfold increase in sister chromatid exchange in Bloom syndrome lymphocytes. *Proc. Natl. Acad. Sci.* 71: 4508–4512.

Creighton, H. S., and McClintock, B. 1931. A correlation of cytological and genetical crossing over in Zea mays. *Proc. Natl. Acad. Sci.* 17: 492–497.

Douglas, L., and Novitski, E. 1977. What chance did Mendel's experiments give him of noticing linkage? *Heredity* 38: 253–257.

Ellis, N. A. et al. 1995. The Bloom's syndrome gene product is homologous to RecQ helicases. *Cell* 83: 655–666.

Ephrussi, B., and Weiss, M. C. 1969. Hybrid somatic cells. Sci. Am. (Apr.) 220: 26–35.

Latt, S. A. 1981. Sister chromatid exchange formation. *Annu. Rev. Genet.* 15: 11–56.

Lindsley, D. L., and Grell, E. H. 1972. *Genetic variations of Drosophila melanogaster.* Washington, DC: Carnegie Institute of Washington.

Morgan, T. H. 1911. An attempt to analyze the constitution of the chromosomes on the basis of sex-linked inheritance in *Drosophila. J. Exp. Zool.* 11: 365–414.

Morton, N. E. 1955. Sequential test for the detection of linkage. *Am. J. Hum.* Genet. 7: 277–318.

————. 1995. LODs—Past and present. *Genetics* 140: 7–12.

Neuffer, M. G., Jones, L., and Zober, M. 1968. The mutants of maize. Madison, WI: Crop Sci. Soc. of America.

Perkins, D. 1962. Crossing over and interference in a multiply marked chromosome arm of *Neurospora. Genetics* 47: 1253–1274.

Ruddle, F. H., and Kucherlapati, R. S. 1974. Hybrid cells and human genes. *Sci. Am.* (July) 231: 36–49.

Stahl, F. W. 1979. *Genetic recombination.* New York: W. H. Freeman.

Stern, C. 1936. Somatic crossing over and segregation in *Drosophila melanogaster. Genetics* 21: 625–631.

Sturtevant, A. H. 1913. The linear arrangement of six sex-linked factors in *Drosophila,* as shown by their mode of association. *J. Exp. Zool.* 14: 43–59.

————. 1965. *A history of genetics.* New York: Harper & Row.

Voeller, B. R., ed. 1968. *The chromosome theory of inheritance: Classical papers in development and heredity.* New York: Appleton-Century-Croft.

Wolff, S., ed. 1982. *Sister chromatid exchange.* New York: Wiley-Interscience.

Chapter 8 Genetic Analysis and Mapping in Bacteria and Bacteriophages

Adelberg, E. A. 1960. *Papers on bacterial genetics.* Boston: Little, Brown.

Birge, E. A. 1988. *Bacterial and bacteriophage genetics—An introduction.* New York: Springer-Verlag.

Brock, T. 1990. *The emergence of bacterial genetics.* Cold Spring Harbor, NY: Cold Spring Harbor Press.

Broda, P. 1979. *Plasmids.* New York: W. H. Freeman.

Bukhari, A. I., Shapiro, J. A., and Adhya, S. L., eds. 1977. *DNA insertion elements, plasmids, and episomes.* Cold Spring Harbor, NY: Cold Spring Harbor Press.

Cairns, J., Stent, G. S., and Watson, J. D., eds. 1966. *Phage and the origins of molecular biology.* Cold Spring Harbor, NY: Cold Spring Harbor Press.

Campbell, A. M. 1976. How viruses insert their DNA into the DNA of the host cell. *Sci. Am.* (Dec.) 235: 102–113.

Fox, M. S. 1966. On the mechanism of integration of transforming deoxyribonucleate. *J. Gen. Physiol.* 49: 183–196.

Hayes, W. 1968. *The genetics of bacteria and their viruses.* 2nd ed. New York: Wiley.

Hershey, A. D., and Rotman, R. 1949. Genetic recombination between host range and plaque-type mutants of bacteriophage in single cells. *Genetics* 34: 44–71.

Hotchkiss, R. D., and Marmur, J. 1954. Double marker transformations as evidence of linked factors in deoxyribonucleate transforming agents. *Proc. Natl. Acad. Sci. (USA)* 40: 55–60.

Jacob, F., and Wollman, E. L. 1961. Viruses and genes. *Sci. Am.* (June) 204: 92–106.

Kohiyama, M., et al. 2003. Bacterial sex: Playing voyeurs 50 years later. *Science* 301: 802–803.

Kruse, H., and Sorum, H. 1994. Transfer of multiple drug resistance plasmids between bacteria of diverse origins in natural microenvironments. *Appl. Environ. Microbiol.* 60: 4015–4021.

Lederberg, J. 1986. Forty years of genetic recombination in bacteria: A fortieth anniversary reminiscence. *Nature* 324: 627–628.

Luria, S. E., and Delbruck, M. 1943. Mutations of bacteria from virus sensitivity to virus resistance. *Genetics* 28: 491–511.

Lwoff, A. 1953. Lysogeny. *Bacteriol. Rev.* 17: 269–337.

Miller, J. H. 1992. *A short course in bacterial genetics.* Cold Spring Harbor, NY: Cold Spring Harbor Press.

Miller, R. V. 1998. Bacterial gene swapping in nature. *Sci. Am.* (Jan.) 278: 66–71.

Morse, M. L., Lederberg, E. M., and Lederberg, J. 1956. Transduction in *Escherichia coli* K12. *Genetics* 41: 141–156.

Novick, R. P. 1980. Plasmids. *Sci. Am.* (Dec.) 243: 102–127.

Smith-Keary, P. F. 1989. *Molecular genetics of* Escherichia coli. New York: Guilford Press.

Stahl, F. W. 1987. Genetic recombination. *Sci. Am.* (Nov.) 256: 91–101.

Stent, G. S. 1963. *Molecular biology of bacterial viruses.* New York: W. H. Freeman.

———. 1966. *Papers on bacterial viruses.* 2nd ed. Boston: Little, Brown.

Wollman, E. L., Jacob, F., and Hayes, W. 1956. Conjugation and genetic recombination in *Escherichia coli* K12. *Cold Spring Harb. Symp. Quant. Biol.* 21: 141–162.

Zinder, N. D. 1958. Transduction in bacteria. *Sci. Am.* (Nov.) 199: 38–46.

Chapter 9 DNA Structure and Analysis

Adleman, L. M. 1998. Computing with DNA. *Sci. Am.* (Aug.) 279: 54–61.

Alloway, J. L. 1933. Further observations on the use of pneumococcus extracts in effecting transformation of type *in vitro. J. Exp. Med.* 57: 265–278.

Avery, O. T., MacLeod, C. M., and McCarty, M. 1944. Studies on the chemical nature of the substance inducing transformation of pneumococcal types: Induction of transformation by a desoxyribonucleic acid fraction isolated from pneumococcus type III. *J. Exp. Med.* 79: 137–158. (Reprinted in Taylor, J. H. 1965. *Selected papers in molecular genetics.* Orlando, FL: Academic Press.)

Britten, R. J., and Kohne, D. E. 1968. Repeated sequences in DNA. *Science* 161: 529–540.

Chargaff, E. 1950. Chemical specificity of nucleic acids and mechanism for their enzymatic degradation. *Experientia* 6: 201–209.

Dawson, M. H. 1930. The transformation of pneumococcal types: I. The interconvertibility of type-specific *S. pneumococci. J. Exp. Med.* 51: 123–147.

Dickerson, R. E., et al. 1982. The anatomy of A-, B-, and Z-DNA. *Science* 216: 475–485.

Dubos, R. J. 1976. *The professor, the Institute and DNA: Oswald T. Avery, his life and scientific achievements.* New York: Rockefeller University Press.

Felsenfeld, G. 1985. DNA. *Sci. Am.* (Oct.) 253: 58–78.

Franklin, R. E., and Gosling, R. G. 1953. Molecular configuration in sodium thymonucleate. *Nature* 171: 740–741.

Griffith, F. 1928. The significance of pneumococcal types. *J. Hyg.* 27: 113–159.

Guthrie, G. D., and Sinsheimer, R. L. 1960. Infection of protoplasts of *Escherichia coli* by subviral particles. *J. Mol. Biol.* 2: 297–305.

Hershey, A. D., and Chase, M. 1952. Independent functions of viral protein and nucleic acid in growth of bacteriophage. *J. Gen. Phys.* 36: 39–56. (Reprinted in Taylor, J. H. 1965. *Selected papers in molecular genetics.* Orlando, FL: Academic Press.)

Levene, P. A., and Simms, H. S. 1926. Nucleic acid structure as determined by electrometric titration data. *J. Biol. Chem.* 70: 327–341.

McCarty, M.. 1985. *The transforming principle: Discovering that genes are made of DNA.* New York: W. W. Norton.

Olby, R. 1974. *The path to the double helix.* Seattle: University of Washington Press.

Pauling, L., and Corey, R. B. 1953. A proposed structure for the nucleic acids. *Proc. Natl. Acad. Sci. USA* 39: 84–97.

Rich, A., Nordheim, A., and Wang, A. H.-J. 1984. The chemistry and biology of left-handed Z-DNA. *Annu. Rev. Biochem.* 53: 791–846.

Spizizen, J. 1957. Infection of protoplasts by disrupted T2 viruses. *Proc. Natl. Acad. Sci. USA* 43: 694–701.

Stent, G. S., ed. 1981. *The double helix: Text, commentary, review, and original papers.* New York: W. W. Norton.

Varmus, H. 1988. Retroviruses. *Science* 240: 1427–1435.

Watson, J. D. 1968. *The double helix.* New York: Atheneum.

Watson, J. D., and Crick, F. C. 1953a. Molecular structure of nucleic acids: A structure for deoxyribose nucleic acids. *Nature* 171: 737–738.

———. 1953b. Genetic implications of the structure of deoxyribose nucleic acid. *Nature* 171: 964.

Wilkins, M. H. F., Stokes, A. R., and Wilson, H. R. 1953. Molecular structure of desoxypentose nucleic acids. *Nature* 171: 738–740.

Chapter 10 DNA Replication and Synthesis

Blackburn, E. H. 1991. Structure and function of telomeres. *Nature* 350: 569–572.

Davidson, J. N. 1976. *The biochemistry of nucleic acids*, 8th ed. Orlando, FL: Academic Press.

DeLucia, P., and Cairns, J. 1969. Isolation of an *E. coli* strain with a mutation affecting DNA polymerase. *Nature* 224: 1164–1166.

Gilbert, D. M. 2001. Making sense of eukaryotic DNA replication origins. *Science* 294: 96–100.

Greider, C.W. 1996. Telomeres, telomerase, and cancer. *Sci. Am.* (Feb.) 274: 92–97.

———. 1998. Telomerase activity, cell proliferation, and cancer. *Proc. Natl. Acad. Sci. (USA)* 95: 90–92.

Holliday, R. 1964. A mechanism for gene conversion in fungi. *Genet. Res.* 5: 282–304.

Holmes, F. L. 2001. *Meselson, Stahl, and replication of DNA: A history of the "most beautiful experiment in biology."* New Haven, CT: Yale University Press.

Kim, J., Kaminker, P., and Campisi, J. 2002. Telomeres, aging and cancer: In search of a happy ending. *Oncogene* 21: 503–511.

Kornberg, A. 1960. Biological synthesis of DNA. *Science* 131:1503–1508.

Kornberg, A. 1974. *DNA synthesis.* New York: W. H. Freeman.

Kornberg, A., and Baker, T. 1992. *DNA replication,* 2nd ed. New York: W. H. Freeman.

Meselson, M., and Stahl, F. W. 1958. The replication of DNA in *Escherichia coli. Proc. Natl. Acad. Sci. (USA)* 44: 671–682.

Mitchell, M. B. 1955. Aberrant recombination of pyridoxine mutants of *Neurospora. Proc. Natl. Acad. Sci. (USA)* 41: 215–220.

Okazaki, T., et al. 1979. Structure and metabolism of the RNA primer in the discontinuous replication of prokaryotic DNA. *Cold Spring Harbor Symp. Quant. Biol.* 43: 203–222.

Radman, M., and Wagner, R. 1988. The high fidelity of DNA duplication. *Sci. Am.* (Aug.) 259: 40–46.

———. 1987. Genetic recombination. *Sci. Am.* (Feb.) 256: 90–101.

Taylor, J. H., Woods, P. S., and Hughes, W. C. 1957. The organization and duplication of chromosomes revealed by autoradiographic studies using tritium-labeled thymidine. *Proc. Natl. Acad. Sci. (USA)* 48: 122–128.

Wang, J. C. 1982. DNA topoisomerases. *Sci. Am.* (July) 247: 94–108.

———. 1987. Recent studies of DNA topoisomerases. *Biochim. Biophys. Acta* 909: 1–9.

Whitehouse, H. L. K. 1982. *Genetic recombination: Understanding the mechanisms.* New York: Wiley.

Chapter 11 Chromosome Structure and DNA Sequence Organization

Angelier, N., et al. 1984. Scanning electron microscopy of amphibian lampbrush chromosomes. *Chromosoma* 89: 243–253.

Beerman, W., and Clever, U. 1964. Chromosome puffs. *Sci. Am.* (Apr.) 210: 50–58.

Carbon, J. 1984. Yeast centromeres: Structure and function. *Cell* 37: 352–353.

Chen, T. R., and Ruddle, F. H. 1971. Karyotype analysis utilizing differential stained constitutive heterochromatin of human and murine chromosomes. *Chromosoma* 34: 51–72.

DuPraw, E. J. 1970. *DNA and chromosomes.* New York: Holt, Rinehart & Winston.

Gall, J. G. 1981. Chromosome structure and the *C*-value paradox. *J. Cell Biol.* 91: 3s–14s.

Hewish, D. R., and Burgoyne, L. 1973. Chromatin substructure. The digestion of chromatin DNA at regularly spaced sites by a nuclear deoxyribonuclease. *Biochem. Biophys. Res. Comm.* 52: 504–510.

Korenberg, J. R., and Rykowski, M. C. 1988. Human genome organization: *Alu,* LINES, and the molecular organization of metaphase chromosome bands. *Cell* 53: 391–400.

Kornberg, R. D. 1975. Chromatin structure: A repeating unit of histones and DNA. *Science* 184: 868–871.

Kornberg, R. D., and Klug, A. 1981. The nucleosome. *Sci. Am.* (Feb.) 244: 52–64.

Lorch, Y., Zhang, N, and Kornberg, R. D. 1999. Histone octamer transfer by a chromatin-remodeling complex. *Cell* 96: 389–392.

Luger, K., et al. 1997. Crystal structure of the nucleosome core particle at 2.8 resolution. *Nature* 389: 251–256.

Moyzis, R. K. 1991. The human telomere. *Sci. Am.* (Aug.) 265: 48–55.

Olins, A. L., and Olins, D. E. 1974. Spheroid chromatin units (*n* bodies). *Science* 183: 330–332.

———. 1978. Nucleosomes: The structural quantum in chromosomes. *Am. Sci.* 66: 704–711.

Singer, M. F. 1982. SINES and LINES: Highly repeated short and long interspersed sequences in mammalian genomes. *Cell* 28: 433–434.

Wolfe, A. 1998. *Chromatin: Structure and Function,* 3rd ed. San Diego: Academic Press.

Chapter 12 The Genetic Code and Transcription

Barrell, B. G., Air, G., and Hutchinson, C. 1976. Overlapping genes in bacteriophage ϕX174 *Nature* 264: 34–40.

Barrell, B. G., Banker, A. T., and Drouin, J. 1979. A different genetic code in human mitochondria. *Nature* 282: 189–194.

Bass, B. L., ed. 2000. *RNA editing.* Oxford: Oxford University Press.

Brenner, S., Jacob, F., and Meselson, M. 1961. An unstable intermediate carrying information from genes to ribosomes for protein synthesis. *Nature* 190: 575–580.

Brenner, S., Stretton, A. O. W., and Kaplan, D. 1965. Genetic code: The nonsense triplets for chain termination and their suppression. *Nature* 206: 994–998.

Cattaneo, R. 1991. Different types of messenger RNA editing. *Annu. Rev. Genet.* 25: 71–88.

Cech, T. R. 1986. RNA as an enzyme. *Sci. Am.* (Nov.) 255(5): 64–75.

———. 1987. The chemistry of self-splicing RNA and RNA enzymes. *Science* 236: 1532–1539.

Chambon, P. 1981. Split genes. *Sci. Am.* (May) 244: 60–71.

Cramer, P., et al. 2000. Architecture of RNA polymerase II and implications for the transcription mechanism. *Science* 288: 640–649.

Crick, F. H. C. 1962. The genetic code. *Sci. Am.* (Oct.) 207: 66–77.

———. 1966a. The genetic code: III. *Sci. Am.* (Oct.) 215: 55–63.

———. 1966b. Codon–anticodon pairing: The wobble hypothesis. *J. Mol. Biol.* 19: 548–555.

Crick, F. H. C., Barnett, L., Brenner, S., and Watts-Tobin, R. J. 1961. General nature of the genetic code for proteins. *Nature* 192: 1227–1232.

Darnell, J. E. 1983. The processing of RNA. *Sci. Am.* (Oct.) 249: 90–100.

———. 1985. RNA. *Sci. Am.* (Oct.) 253: 68–87.

Dickerson, R. E. 1983. The DNA helix and how it is read. *Sci. Am.* (Dec.) 249: 94–111.

Dugaiczk, A., et al. 1978. The natural ovalbumin gene contains seven intervening sequences. *Nature* 274: 328–333.

Fiers, W., et al. 1976. Complete nucleotide sequence of bacteriophage MS2 RNA: Primary and secondary structure of the replicase gene. *Nature* 260: 500–507.

Gamow, G. 1954. Possible relation between DNA and protein structures. *Nature* 173: 318.

Hall, B. D., and Spiegelman, S. 1961. Sequence complementarity of T2-DNA and T2-specific RNA. *Proc. Natl. Acad. Sci. (USA)* 47: 137–146.

Hamkalo, B. 1985. Visualizing transcription in chromosomes. *Trends Genet.* 1: 255–260.

Khorana, H. G. 1967. Polynucleotide synthesis and the genetic code. *Harvey Lectures* 62: 79–105.

Miller, O. L., Hamkalo, B., and Thomas, C. 1970. Visualization of bacterial genes in action. *Science* 169: 392–395.

Nirenberg, M. W. 1963. The genetic code: II. *Sci. Am.* (Mar.) 190: 80–94.

O'Malley, B., et al. 1979. A comparison of the sequence organization of the chicken ovalbumin and ovomucoid genes. In *Eucaryotic gene regulation*, R. Axel et al., eds., pp. 281–299. Orlando, FL: Academic Press.

Reed, R., and Maniatis, T. 1985. Intron sequences involved in lariat formation during pre-mRNA splicing. *Cell* 41: 95–105.

Sharp, P. A. 1994. Nobel lecture: Split genes and RNA splicing. *Cell* 77: 805–815.

Steitz, J. A. 1988. Snurps. *Sci. Am.* (June) 258(6): 56–63.

Volkin, E., and Astrachan, L. 1956. Phosphorus incorporation in *E. coli* ribonucleic acids after infection with bacteriophage T2. *Virology* 2: 149–161.

Watson, J. D. 1963. Involvement of RNA in the synthesis of proteins. *Science* 140: 17–26.

Woychik, N. A., and Jampsey, M. 2002. The RNA polymerase II machinery: Structure illuminates function. *Cell* 108: 453–464.

Chapter 13 Translation and Proteins

Anfinsen, C. B. 1973. Principles that govern the folding of protein chains. *Science* 181: 223–230.

Bartholome, K. 1979. Genetics and biochemistry of phenylketonuria—Present state. *Hum. Genet.* 51: 241–245.

Beadle, G. W., and Tatum, E. L. 1941. Genetic control of biochemical reactions in *Neurospora. Proc. Natl. Acad. Sci. USA* 27: 499–506.

Beet, E. A. 1949. The genetics of the sickle-cell trait in a Bantu tribe. *Ann. Eugenics* 14: 279–284.

Brenner, S. 1955. Tryptophan biosynthesis in *Salmonella typhimurium. Proc. Natl. Acad. Sci. USA* 41: 862–863.

Doolittle, R. F. 1985. Proteins. *Sci. Am.* (Oct.) 253: 88–99.

Frank, J. 1998. How the ribosome works. *Amer. Scient.* 86: 428–439.

Garrod, A. E. 1902. The incidence of alkaptonuria: A study in chemical individuality. *Lancet* 2: 1616–1620.

———. 1909. *Inborn errors of metabolism.* London: Oxford University Press. (Reprinted 1963, Oxford University Press, London.)

Garrod, S. C. 1989. Family influences on A. E. Garrod's thinking. *J. Inher. Metab. Dis.* 12: 2–8.

Ingram, V. M. 1957. Gene mutations in human hemoglobin: The chemical difference between normal and sickle-cell hemoglobin. *Nature* 180: 326–328.

Koshland, D. E. 1973. Protein shape and control. *Sci. Am.* (Oct.) 229: 52–64.

Lake, J. A. 1981. The ribosome. *Sci. Am.* (Aug.) 245: 84–97.

Maniatis, T., et al. 1980. The molecular genetics of human hemoglobins. *Annu. Rev. Genet.* 14: 145–178.

Neel, J. V. 1949. The inheritance of sickle-cell anemia. *Science* 110: 64–66.

Nirenberg, M. W., and Leder, P. 1964. RNA codewords and protein synthesis. *Science* 145: 1399–1407.

Nomura, M. 1984. The control of ribosome synthesis. *Sci. Am.* (Jan.) 250: 102–114.

Pauling, L., Itano, H. A., Singer, S. J., and Wells, I. C. 1949. Sickle-cell anemia: A molecular disease. *Science* 110: 543–548.

Ramakrishnan, V. 2002. Ribosome structure and the mechanism of translation. *Cell* 108: 557–572.

Rich, A., and Houkim, S. 1978. The three-dimensional structure of transfer RNA. *Sci. Am.* (Jan.) 238: 52–62.

Rich, A., Warner, J. R., and Goodman, H. M. 1963. The structure and function of polyribosomes. *Cold Spring Harbor Symp. Quant. Biol.* 28: 269–285.

Richards, F. M. 1991. The protein-folding problem. *Sci. Am.* (Jan.) 264: 54–63.

Rould, M. A., et al. 1989. Structure of *E. coli* glutaminyl-tRNA synthetase complexed with tRNA^{gln} and ATP at 2.8 resolution. *Science* 246: 1135–1142.

Scott-Moncrieff, R. 1936. A biochemical survey of some Mendelian factors for flower colour. *J. Genet.* 32: 117–170.

Scriver, C. R., and Clow, C. L. 1980. Phenylketonuria and other phenylalanine hydroxylation mutants in man. *Annu. Rev. Genet.* 14: 179–202.

Srb, A. M., and Horowitz, N. H. 1944. The ornithine cycle in *Neurospora* and its genetic control. *J. Biol. Chem.* 154: 129–139.

Warner, J., and Rich, A. 1964. The number of soluble RNA molecules on reticulocyte polyribosomes. *Proc. Natl. Acad. Sci. USA* 51: 1134–1141.

Wimberly, B. T., et al. 2000. Structure of the 30S ribosomal subunit. *Nature* 407: 327–333.

Yusupov, M. M., et al. 2001. Crystal structure of the ribosome at 5.5A resolution. *Science* 292: 883–896.

Chapter 14 Gene Mutation, DNA Repair, and Transposition

Ames, B. N., McCann, J., and Yamasaki, E. 1975. Method for detecting carcinogens and mutagens with the *Salmonella*/mammalian microsome mutagenicity test. *Mut. Res.* 31: 347–364.

Bates, G., and Leharch, H. 1994. Trinucleotide repeat expansions and human genetic disease. *BioEssays* 16: 277–284.

Cairns, J., Overbaugh, J., and Miller, S. 1988. The origin of mutants. *Nature* 335: 142–145.

Cleaver, J. E. 1990. Do we know the cause of xeroderma pigmentosum? *Carcinogenesis* 11: 875–882.

Comfort, N. C. 2001. *The tangled field: Barbara McClintock's search for the patterns of genetic control.* Cambridge, MA: Harvard University Press.

Friedberg, E. C., Walker, G. C., and Siede, W. 1995. *DNA repair and mutagenesis.* Washington, DC: ASM Press.

Jiricny, J. 1998. Eukaryotic mismatch repair: An update. *Mutation Research* 409: 107–121.

Landers, E. S. et al., 2001. Initial sequencing and analysis of the human genome. *Nature* 409: 860–921.

Miki, Y. 1998. Retrotransposal integration of mobile genetic elements in human disease. *J. Human Genet.* 43: 77–84.

O'Hare, K. 1985. The mechanism and control of *P* element transposition in *Drosophila. Trends Genet.* 1: 250–254.

Radman, M., and Wagner, R. 1988. The high fidelity of DNA duplication. *Sci. Am.* (Aug.) 259: 40–46.

Chapter 15 Regulation of Gene Expression

Becker, P. B., and Hörz, W. 2002. ATP-dependent nucleosome remodeling. *Annu. Rev. Biochem.* 71: 247–273.

Beckwith, J. R., and Zipser, D., eds. 1970. *The lactose operon.* Cold Spring Harbor, NY: Cold Spring Harbor Laboratory Press.

Bertrand, K., et al. 1975. New features of the regulation of the tryptophan operon. *Science* 189: 22–26.

Black, D. L. 2000. Protein diversity from alternative splicing: a challenge for bioinformatics and postgenomic biology. *Cell* 103: 367–370.

Black, D. L. 2003. Mechanisms of alternative pre-mRNA splicing. *Annu. Rev. Biochem.* 72: 291–336.

Butler, J. E. F., and Kadonaga, J. T. 2002. The RNA polymerase II core promoter: A key component in the regulation of gene expression. *Genes Dev.* 16: 2583–2592.

Dillon, N., and Sabbattini, P. 2000. Functional gene expression domains: Defining the functional unit of eukaryotic gene regulation. *Bioessays* 22: 657–665.

Gilbert, W., and Müller-Hill, B. 1966. Isolation of the *lac* repressor. *Proc. Natl. Acad. Sci. USA* 56: 1891–1898.

———. 1967. The *lac* operator is DNA. *Proc. Natl. Acad. Sci. USA* 58: 2415–2421.

Jackson, D. A. 2003. The anatomy of transcription sites. *Curr. Opin. Cell Biol.* 15: 311–317.

Jacob, F., and Monod, J. 1961. Genetic regulatory mechanisms in the synthesis of proteins. *J. Mol. Biol.* 3: 318–356.

Lee, T. I., and Young, R. A. 2000. Transcription of eukaryotic protein-coding genes. *Ann. Rev. Genet.* 34: 77–137.

Meister, G., and Tuschl, T. 2004. Mechanisms of gene silencing by double-standard RNA. *Nature* 431: 343–349.

Mello, C. C., and Conte, D., Jr. 2004. Revealing the world of RNA interference. *Nature* 431: 338–342.

Turner, B. M. 2000. Histone acetylation and an epigenetic code. *Bioessays* 22: 836–845.

Yanofsky, C. 1981. Attenuation in the control of expression of bacterial operons. *Nature* 289: 751–758.

Yanofsky, C., and Kolter, R. 1982. Attenuation in amino acid biosynthetic operons. *Annu. Rev. Genet.* 16: 113–134.

Chapter 16 Cell-Cycle Regulation and Cancer

Ames, B. N., Gold, L. S., and Willett, W. C. 1995. The causes and prevention of cancer. *Proc. Natl. Acad. Sci. USA* 92: 5258–5265.

Bernards, R., and Weinberg, R. A. 2002. A progression puzzle. *Nature* 418: 823.

Brown, M. A. 1997. Tumor suppressor genes and human cancer. *Adv. Genet.* 36: 45–135.

Cavenee, W. K., and White, R. L. 1995. The genetic basis of cancer. *Sci. Am.* (Mar.) 272: 72–79.

Compagni, A., and Christofori, G. 2000. Recent advances in research on multistage tumorigenesis. *Brit. J. Cancer* 83: 1–5.

Cornelis, J. F., et al. 1998. Metastasis. *Am. Scient.* 86: 130–141.

Fearon, E. R. 1997. Human cancer syndromes: Clues to the origin and nature of cancer. *Science* 278: 1043–1050.

Futreal, P. A., et al. 2004. A census of human cancer genes. *Nature Reviews Cancer* 4: 177–183.

Hartwell, L. H., and Kastan, M. B. 1994. Cell cycle control and cancer. *Science* 266: 1821–1827.

Lengauer, C., Kinzler, K. W., and Vogelstein, B. 1997. Genetic instability in colorectal cancer. *Nature* 386: 623–627.

Nurse, P. 1997. Checkpoint pathways come of age. *Cell* 91: 865–867.

Raff, M. 1998. Cell suicide for beginners. *Nature* 396: 119–122.

Sherr, C. J. 1996. Cancer cell cycles. *Science* 274: 1672–1677.

Vogelstein, B., and Kinzler, K.W. 1993. The multistep nature of cancer. *Trends in Genetics* 9: 138–141.

Weinberg, R. A. 1995. The retinoblastoma protein and cell cycle control. *Cell* 81: 323–330.

———. 1996. How cancer arises. *Sci. Am.* (Sept.) 275: 62–70.

Chapter 17 Recombinant DNA Technology

Burger, S. L., and Kimmel, A. R. 1987. *Guide to molecular cloning techniques. Methods in enzymology.* Vol. 152. San Diego: Academic Press.

Flotte, T. R. 2005. Adeno-associated virus-based gene therapy for inherited disorders. *Pediatr. Res.* 58: 1143–1147.

McKusick, V. A. 1988. The new genetics and clinical medicine. *Hosp. Pract.* 23: 177–191.

Metzker, M. L. 2005. Emerging technologies in DNA sequencing. *Genome Res.* 15: 1767–1776.

Mullis, K. B. 1990. The unusual origin of the polymerase chain reaction. *Sci. Am.* (Apr.) 262: 56–65.

Pingoud, A., Fuxreiter, M., Pingoud, V., and Wende, W. 2005. Type II restriction endonucleases: structure and mechanism. *Cell Mol. Life Sci.* 62: 685–707.

Sambrook, J., and Russell, D. 2001. *Molecular cloning: A laboratory manual*, 3rd ed. Cold Spring Harbor, NJ: Cold Spring Harbor Press.

Southern, E. 1975. Detection of specific sequences among DNA fragments separated by gel electrophoresis. *J. Mol. Biol.* 98: 503–507.

Chapter 18 Genomics and Proteomics

Andersen, J., et al. 2002. Directed proteomic analysis of the human nucleolus. *Curr. Biol.* 12: 1–11.

Butler, J. M. 2006. Genetics and genomics of core short tandem repeat loci used in human identity testing. *J. Forensic Sci.* 51: 253–265.

Chandonia, J. M. 2006. The impact of structural genomics: expectations and outcomes. *Science* 311: 347–351.

Domon, B., and Aebersold, R. 2006. Mass spectrometry and protein analysis. *Science* 312: 212–217.

Feuk, L., Carson, A. R., and Scherer, S. W. 2006. Structural variation in the human genome. *Nat. Rev. Genet.* 7: 85–97.

Fleming, K., et al. 2006. The proteome: structure, function and evolution. *Philos. Trans. R. Soc. Lond. B Biol. Sci.* 361: 441–451.

Harrison, P. M., et al. 2002. A question of size: the eukaryotic proteome and the problem in defining it. *Nucl. Acids Res.* 30: 1083–1090.

International Chimpanzee Chromosome 22 Consortium. 2004. DNA sequence and comparative analysis of chimpanzee chromosome 22. *Nature* 429: 383–388.

International Human Genome Sequencing Consortium. 2001. Initial sequencing and analysis of the human genome. *Nature* 409: 860–921.

Joyce, A. R., and Palsson, B. O. 2006. The model organism as a system integrating "omics" data sets. *Nat. Rev. Mol. Cell Biol.* 7: 901–909.

Lonberg, N. 2005. Human antibodies from transgenic animals. *Nat. Biotechnol.* 23: 1117-1125.

Nei, M., and Rooney, A. P. 2005. Concerted and birth-and-death evolution of multigene families. *Annu. Rev. Genet.* 39: 197–218.

Ochman, H., and Davilos, L. M. 2006. The nature and dynamics of bacterial genomes. *Science* 311: 1730–1733.

Rabinowicz, P. D., and Bennetzen, J. L. 2006. The maize genome as a model for efficient sequence analysis of large plant genomes. *Curr. Opin. Plant Biol.* 9: 49–56.

Switnoski, M., Szczerbal, I., and Nowacka, J. 2004. The dog genome map and its use in mammalian comparative genomics. *J. Appl. Physiol.* 45: 195–214.

Venter, J. C., et al. 2001. The sequence of the human genome. *Science* 291: 1304–1351.

Chapter 19 Applications and Ethics of Genetic Engineering

Atherton, K. T. 2002. Safety assessment of genetically modified crops. *Toxicol.* 81/182: 421–426.

Cavanna-Calvo, M., et al. 2000. Gene therapy of severe combined immunodeficiency (SCID)-XI disease. *Science* 288: 669–672.

Dale, P. J., Clarke, B., and Fontes, E. M. G. 2002. Potential for the environmental impact of transgenic crops. *Nat. Biotechnol.* 20: 567–574.

Engler, O. B., et al. 2001. Peptide vaccines against hepatitis B virus: From animal model to human studies. *Mol. Immunol.* 38: 457–465.

Friend, S. H., and Stoughton, R. B. 2002. The magic of microarrays. *Sci. Am.* (Feb) 286: 44–50.

Hunter, C. V., Tiley, L. S., and Sang, H. M. 2005. Developments in transgenic technology: applications for medicine. *Trends Mol. Med.* 11: 293–300.

Leung, Y. F., Lam, D. S. C., and Pang, C. P. 2001. When drug development meets genomics. *Pharmacogenomics Journal* 1: 237–238.

O'Connor, T. P., and Crystal, R. G. 2006. Genetic medicines: treatment strategies for hereditary disorders. *Nature Reviews Genetics* 7: 261–276.

Pray, C. E., Huang, J., and Rozelle, S. 2002. Five years of Bt cotton in China: The benefits continue. *The Plant J.* 31: 423–430.

Schillberg, S., Fischer, T., and Emans, N. 2003. Molecular farming of recombinant antibodies in plants. *Cell Mol. Life Sci.* 60: 433–445.

Shastry, B. S. 2006. Pharmacogenetics and the concept of individualized medicine. *Pharmacogenetics Journal* 6: 16–21.

Varsha, M. S. 2006. DNA fingerprinting in the criminal justice system: An overview. *DNA Cell Biol.* 25: 181–188.

Chapter 20 Developmental Genetics

Goodman, F. 2002. Limb malformations and the human *HOX* genes. *Am. J. Med. Genet.* 112: 256–265.

Gridley, T. 2003. Notch signaling and inherited human diseases. *Hum. Mol. Genet.* 12: R9–R13.

Krizek, B. A., and Fletcher, J. C. 2006. Molecular mechanisms of flower development: An armchair guide. *Nat. Rev. Genet.* 6: 688–698.

Kumar, J. P., and Moses, K. 2001. EGF receptor and Notch signaling act upstream of eyeless/PAX6 to control eye specification. *Cell* 104: 687–697.

Lall, S., and Patel, N. H. 2001. Conservation and divergence in molecular mechanisms of axis formation. *Annu. Rev. Genet.* 35: 407–437.

Moens, C. B., and Selleri, L. 2006. Hox cofactors in vertebrate development. *Dev. Biol.* 291: 193–206.

Nüsslein-Volhard, C., and Weischaus, E. 1980. Mutations affecting segment number and polarity in *Drosophila*. *Nature* 287: 795–801.

Reyes, J. C. 2006. Chromatin modifiers that control plant development. *Curr. Opin. Plant Biol.* 9: 21–27.

Rudel, D., and Sommer, R. J. 2003. The evolution of developmental mechanisms. *Dev. Biol.* 264: 15–37.

Stathopoulos, A., and Levine, M. 2005. Genomic regulatory networks and animal development. *Dev. Cell* 9: 449–462.

Sundaram. M. V. 2005. The love-hate relationship between Ras and Notch. *Genes Dev.* 19: 1825–1839.

Tickle, C. 2006. Making digit patterns in the vertebrate limb. *Nat. Rev. Mol. Cell Biol.* 7: 45–53.

Verakasa, A., Del Campo, M., and McGinnis, W. 2000. Developmental patterning genes and their conserved functions: from model organisms to humans. *Mol. Genet. Metabol.* 69: 85–100.

Wang, M., and Sternberg, P. W. 2001. Pattern formation during *C. elegans* vulval induction. *Curr. Top. Dev. Biol.* 51: 189–220.

Chapter 21 Quantitative Genetics

Browman, K. W. 2001. Review of statistical methods of QTL mapping in experimental crosses. *Lab Animal* 30: 44–52.

Crow, J. F. 1993. Francis Galton: Count and measure, measure and count. *Genetics* 135: 1.

Dudley, J. W. 1977. 76 generations of selection for oil and protein percentage in maize. In *Proc. Intern. Conf. on Quant. Genet.,* E. Pollack, O. Kempthorne, and T. Bailey, eds. pp. 459–473. Ames: Iowa State University Press.

Falconer, D. S., and Mackay, F. C. 1996. *Introduction to quantitative genetics,* 4th ed. Essex, England: Longman.

Farber, S. 1980. *Identical twins reared apart.* New York: Basic Books.

Feldman, M. W., and Lewontin, R. C. 1975. The heritability hangup. *Science* 190: 1163–1166.

Frary, A., et al. 2000. *fw2.2*: A quantitative trait locus key to the evolution of tomato fruit size. *Science* 289: 85–88.

Haley, C. 1991. Use of DNA fingerprints for the detection of major genes for quantitative traits in domestic species. *Anim. Genet.* 22: 259–277.

———. 1996. Livestock QTLs: Bringing home the bacon. *Trends Genet.* 11: 488–490.

Lander, E., and Botstein, D. 1989. Mapping Mendelian factors underlying quantitative traits using RFLP linkage maps. *Genetics* 121: 185–199.

Lander, E., and Schork, N. 1994. Genetic dissection of complex traits. *Science* 265: 2037–2048.

Lynch, M., and Walsh, B. 1998. *Genetics and analysis of quantitative traits.* Sunderland, MA: Sinauer Associates.

Macy, T. F. C. 2001. Quantitative trait loci in *Drosophila*. *Nature Reviews Genetics* 2: 11–19.

Newman, H. H., Freeman, F. N., and Holzinger, K. T. 1937. *Twins: A study of heredity and environment.* Chicago: University of Chicago Press.

Paterson, A., Deverna, J., Lanini, B., and Tanksley, S. 1990. Fine mapping of quantitative traits loci using selected overlapping recombinant chromosomes in an interspecific cross of tomato. *Genetics* 124: 735–742.

Tanksley, S. D. 2004. The genetic, developmental and molecular bases of fruit size and shape variation in tomato. *Plant Cell* 16: S181–S189.

Zar, J. H. 1999. *Biostatistical analysis,* 4th ed. Upper Saddle River, NJ: Prentice Hall.

Chapter 22 Population Genetics

Ansari-Lari, M. A., et al. 1997. The extent of genetic variation in the *CCR5* gene. *Nature Genetics* 16: 221–222.

Ballou, J., and Ralls, K. 1982. Inbreeding and juvenile mortality in small populations of ungulates: A detailed analysis. *Biol. Conserv.* 24: 239–272.

Bauduer, F., Feingold, J., and Lacombe, D. 2005. The Basques: Review of population genetics and Mendelian disorders. *Hum. Biol.* 77: 619–637.

Bittles, A. H., and Neel, J. V. 1994. The costs of human inbreeding and their implications for variations at the DNA level. *Nature Genet.* 8: 117–121.

Carrington, M., Kissner, T., et al. 1997. Novel alleles of the chemokine-receptor gene *CCR5*. *Am. J. Hum. Genet.* 61: 1261–1267.

Chevillon, C., et al. 1995. Population structure and dynamics of selected genes in the mosquito *Culex pipiens*. *Evolution* 49: 997–1007.

Dudley, J. W. 1977. 76 generations of selection for oil and protein percentage in maize. In *Proc. Intern. Conf. on Quant. Genet.*, E. Pollack et al., eds., pp. 459–473. Ames: Iowa State University Press.

Fisher, R. A. 1930. *The genetical theory of natural selection*. Oxford: Clarendon Press. (Reprinted Dover Press, 1958)

Freeman, S., and Herron, J. C. 2001. *Evolutionary analysis*. Upper Saddle River, NJ: Prentice Hall.

Freire-Maia, N. 1990. Five landmarks in inbreeding studies. *Am. J. Med. Genet.* 35: 118–120.

Grant, P. R. 1986. *Ecology and evolution of Darwin's finches*. Princeton, NJ: Princeton University Press.

Karn, M. N., and Penrose, L. S. 1951. Birth weight and gestation time in relation to maternal age, parity and infant survival. *Ann. Eugen.* 16: 147–164.

Kerr, W. E., and Wright, S. 1954. Experimental studies of the distribution of gene frequencies in very small populations of *Drosophila melanogaster*. *Evolution* 8: 172–177.

Kwiatkowski, D. P. 2005. How malaria has affected the human genome and what human genetics can teach us about malaria. *Am. J. Hum. Genet.* 77: 171–192.

Laikre, L., Ryman, N., and Thompson, E.A. 1993. Hereditary blindness in a captive wolf (*Canis lupus*) population: Frequency reduction of a deleterious allele in relation to gene conservation. *Conservation Biol.* 7: 592–601.

Leibert, F., et al. 1998. The *DCCR5* mutation conferring protection against HIV-1 in Caucasian populations has a single and recent origin in northeastern Europe. *Hum. Molec. Genet.* 7: 399–406.

Liu, R., et al. 1996. Homozygous defect in HIV-1 coreceptor accounts for resistance in some multiply-exposed individuals to HIV-1 infection. *Cell* 86: 367–377.

Markow, T., et al. 1993. HLA polymorphism in the Havasupai: Evidence for balancing selection. *Am. J. Hum. Genet.* 53: 943–952.

Parra, E. J., et al. 1998. Estimating African American admixture proportions by use of population-specific alleles. *Am. J. Hum. Genet.* 63: 1839–1851.

Pier, G. B., et al. 1998. *Salmonella typhi* uses CFTR to enter intestinal epithelial cells. *Nature* 393: 79–82.

Woodworth, C. M., Leng, E. R., and Jugenheimer, R. W. 1952. Fifty generations of selection for protein and oil in corn. *Agron. J.* 44: 60–66.

Chapter 23 Evolutionary Genetics

Adcock, G. J., Dennis, E. S., Easteal, S., Huttley, G. A., Jermiin, L. S., Peacock, W. J., and Thorne, A. 2001. Mitochondrial DNA sequences in ancient Australians: Implications for modern human origins. *Proc. Nat. Acad. Sci.* 98: 537–542.

Anderson, W., Dobzhansky, T., Pavlovsky, O., Powell, J., and Yardley, D. 1975. Genetics of natural populations: XLII. Three decades of genetic change in *Drosophila pseudoobscura*. *Evolution* 29: 24–36.

Ayala, F. J., 1984. Molecular polymorphism: How much is there, and why is there so much? *Dev. Genet.* 4: 379–391.

Bradshaw, H. D., Jr., Otto, K. G., and Frewen, B. E. 1998. Quantitative trait loci affecting differences in floral morphology between two species of monkey flower (*Mimulus*). *Genetics* 149: 367–382.

Dobzhansky, T. 1947. Adaptive changes induced by natural selection in wild populations of *Drosophila*. *Evolution* 1: 1–16.

Freeman, S., and Herron, J. C. 2001. *Evolutionary analysis*. Upper Saddle River, NJ: Prentice Hall.

Hardison, R. 1999. The evolution of hemoglobin. *Amer. Scient.* 87: 126–137.

Ke, Y., et al. 2001. African origin of modern humans in East Asia: A tale of 12,000 Y chromosomes. *Science* 292: 1151–1153.

Kimura, M. 1989. The neutral theory of molecular evolution and the world view of the neutralists. *Genome* 31: 24–31.

King, M. C., and Wilson, A. C. 1975. Evolution at two levels: Molecular similarities and biological differences between humans and chimpanzees. *Science* 188: 107–116.

Knowlton, N., Weigt, L. A., Solorzano, L. A., Mills, D. K., and Bermingham, E. 1993. Divergence in proteins, mitochondrial DNA, and reproductive compatibility across the Isthmus of Panama. *Science* 260: 1629–1632.

Kreitman, M. 1983. Nucleotide polymorphism at the alcohol dehydrogenase locus of *Drosophila melanogaster*. *Nature* 304: 412–417.

Lamb, R. S., and Irish, V. F. 2003. Functional divergence within the *APETALA3/PISTILLATA* floral homeotic gene lineages. *Proc. Natl. Acad. Sci.* 100: 6558–6563.

Lewontin, R. C., and Hubby, J. L. 1966. A molecular approach to the study of genic heterozygosity in natural populations: II. Amount of variation and degree of heterozygosity in natural populations of *Drosophila pseudoobscura*. *Genetics* 54: 595–609.

Ou, C.-Y., et al. 1992. Molecular epidemiology of HIV transmission in a dental practice. *Science* 256: 1165–1171.

Relethford, J. H. 2001. Ancient DNA and the origin of modern humans. *Proc. Nat. Acad. Sci.* 98: 390–391.

Stiassny, M. L. J., and Meyer, A. 1999. Cichlids of the Rift Lakes. *Sci. Am.* (Feb.) 280: 64–69.

Storz, J. F. 2005. Using genome scans of DNA polymorphism to infer adaptive population divergence. *Mol. Ecol.* 14: 671–688.

Takahashi, K., 1998. A novel family of short interspersed repetitive elements (SINES) from cichlids: The pattern of insertion of SINES at orthologous loci support the proposed monophyly of four major groups of cichlid fishes in Lake Tanganyika. *Mol. Biol. Evol.* 15: 391–407.

Chapter 24 Conservation Genetics

Baker, C. S., and Palumbi, S. R. 1994. Which whales are hunted? A molecular genetic approach to monitoring whaling. *Science* 265: 1538–1539.

Daniels, S. J., and Walters, J. R. 2000. Inbreeding depression and its effects on natal dispersal in red-cockaded woodpeckers. *Condor* 102: 482–491.

Dobson, A., and Lyles, A. 2000. Black-footed ferret recovery. *Science* 288: 985.

Frankham, R. 1995. Conservation genetics. *Ann. Rev. Genet.* 29: 305–327.

Friar, E. A., Boose, D. L., LaDoux, T., Roalson, E. H., and Robichaux., R. H. 2001. Population structure in the endangered Mauna Loa silversword, *Argyroxiphium kauense*, and its bearing on reintroduction. *Molecular Ecology* 10: 1657–1663.

Gharrett, A. J., and Smoker, W. W. 1991. Two generations of hybrids between even-year and odd-year pink salmon (*Oncorhynchus gorbuscha*): A test for outbreeding depression? *Canadian J. Fish. Aquat. Sci.* 48: 426–438.

Hedrick, P. W. 2001. Conservation genetics: Where are we now? *Trends Ecol. Evol.* 16: 629–636.

Lacy, R. C. 1997. Importance of genetic variation to the viability of mammalian populations. *J. Mammalogy* 78: 320–335.

Moore, M. K., Bemiss, J. A., Rice, S. M., Quattro, J. M., and Woodley, C. M. 2003. Use of restriction fragment length polymorphisms to identify sea turtle eggs and cooked meats to species. *Conserv. Genet.* 4: 95–103.

Paetkau, D., et al. 1998. Variation in genetic diversity across the range of North American brown bears. *Conserv. Biol.* 12: 418–429.

Ralls, K., et al. 2000. Genetic management of chondrodystrophy in California condors. *Animal Conserv.* 3: 145–153.

Roman J., and Bowen, B. W. 2000. The mock turtle syndrome: Genetic identification of turtle meat purchased in the southeastern United States of America. *Animal Conserv.* 3: 61–65.

Roman, J. and Palumbi, S. R. 2003. Whales before whaling in the North Atlantic. *Science* 301: 508–510.

Wayne, R. K., et al. 1991. Conservation genetics of the endangered Isle Royale gray wolf. *Conserv. Biol.* 5: 41–51.

Wilson, E. O., ed. 1988. *Biodiversity*. Washington, DC: National Academy of Sciences.

Wynen, L. P., et al. 2000. Postsealing genetic variation and population structure of two species of fur seal. *Mol. Ecol.* 9: 299–314.

Solutions to Problems and Discussion Questions

Chapter 1 Introduction to Genetics

2. Based on the parallels between Mendel's model of heredity and the behavior of chromosomes, the chromosome theory of inheritance emerged. It states that inherited traits are controlled by genes residing on chromosomes that are transmitted by gametes.

4. A gene variant is called an allele. There can be many such variants in a population, but for a diploid organism, only two such alleles can exist in any given individual.

6. *Genes*, linear sequences of nucleotides, usually exert their influence by producing proteins through the process of transcription and translation. Genes are the functional units of heredity. They often associate with proteins to form *chromosomes*.

8. The central dogma of molecular genetics refers to the relationships among DNA, RNA, and protein.

10. Restriction enzymes (endonucleases) cut double-stranded DNA at particular base sequences. When a vector is cleaved with the same enzyme, complementary ends are created, such that ends, regardless of their origin, can be combined and ligated to form intact double-stranded structures.

12. Supporters of organismic patenting argue that it is needed to encourage innovation and allow the costs of discovery to be recovered. Others argue that natural substances should not be privately owned and that once owned by a small number of companies, free enterprise will be stifled. Individuals and companies needing vital, but patented products, may have limited access.

14. Model organisms are not only useful, but necessary, for understanding genes that influence human diseases. Given that genetic/molecular systems are highly conserved across broad phylogenetic lines, what is learned in one organism is usually applied to all organisms. Most model organisms have peculiarities, such as ease of growth, genetic understanding, or abundant offspring, that make them especially informative in genetic studies.

Chapter 2 Mitosis and Meiosis

2. Chromosomes that are homologous share many properties, including overall length; position of the centromere (metacentric, submetacentric, acrocentric, telocentric); banding patterns; type and location of genes; and autoradiographic pattern.
Diploidy is a term often used in conjunction with the symbol 2*n*. It means that both members of a homologous pair of chromosomes are present. *Haploidy* specifically refers to the fact that each haploid cell contains *one chromosome of each homologous pair of chromosomes.*

6. Metacentric (a) centromere in the middle, submetacentric (b) centromere off-center, acrocentric (c) centromere toward one end, telocentric (d), centromere at one end. Anaphase configurations are influenced by centromere placement as indicated below.

8. Major divisions of the cell cycle include interphase and mitosis. Interphase is composed of four phases: G1, G0, S, and G2. During the S phase, chromosomal DNA doubles. Karyokinesis involves nuclear division, while cytokinesis involves division of the cytoplasm.

10. Compared with mitosis, which maintains a chromosomal constancy, meiosis provides for a reduction in chromosome number and an opportunity for exchange of genetic material between homologous chromosomes.

12. Sister chromatids are genetically identical, except where mutations may have occurred during DNA replication. Nonsister chromatids are genetically similar when on homologous chromosomes or genetically dissimilar when on nonhomologous chromosomes. If crossing over occurs, then chromatids attached to the same centromere will no longer be identical.

14. (a) 8 tetrads.
(b) 8 dyads.
(c) 8 monads migrating to *each* pole.

16. First, through independent assortment of chromosomes at anaphase I of meiosis, daughter cells (secondary spermatocytes and secondary oocytes) may contain different sets of maternally and paternally derived chromosomes. Second, crossing over, which happens at a much higher frequency in meiotic cells as compared to mitotic cells, allows maternally and paternally derived chromosomes to exchange segments, thereby increasing the likelihood that daughter cells (that is, secondary spermatocytes and secondary oocytes) are genetically unique.

18. There would be 16 combinations with the addition of another chromosome pair.

20. $(1/2)^{10}$

22. In angiosperms, meiosis results in the formation of microspores (male) and megaspores (female) that give rise to the haploid male and female gametophyte stage. Micro- and megagametophytes produce the pollen and the ovules, respectively. Following fertilization, the sporophyte is formed.

24. The folded-fiber model is based on each chromatid consisting of a single fiber wound like a skein of yarn. Each fiber consists of DNA and protein. A coiling process occurs during the transition of interphase chromatin to more condensed chromosomes during prophase of mitosis or meiosis. Such condensation leads to a 5000-fold contraction in the length of the DNA within each chromatid. The transition is at the end of interphase and the beginning of prophase when the chromosomes are in the condensation process. This eventually leads to the typically shortened and "fattened" metaphase chromosome.

26. Duplicated chromosomes A^m, A^p, B^m, B^p, C^m, and C^p will align at metaphase, with the centromeres dividing and sister chromatids going to opposite poles at anaphase.

28. As long as you have accounted for eight possible combinations in the previous problem, there would be no new ones added in this problem.

30. See the products of nondisjunction of chromosome C at the end of meiosis I, as follows.

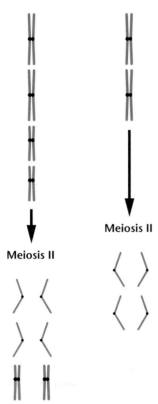

At the end of meiosis II, assuming that, as the problem states, the C chromosomes separate as dyads instead of monads during meiosis II, you would have monads for the A and B chromosomes, and dyads (from the cell on the left) for both C chromosomes as one possibility. However, another possibility exists, as shown below for the products of meiosis II.

Chapter 3 Mendelian Genetics

2. (a) The fact that they produce an albino child requires that each parent provides an *a* gene to the albino child; thus the parents must both be heterozygous (*Aa*). **(b)** The female must be *aa*. Since all the children are normal, one would consider the male to be *AA* instead of *Aa*. However, the male could be *Aa*. Under that circumstance, the likelihood of having six children, all normal, is 1/64.

6. Symbolism:

w = wrinkled seeds g = green cotyledons
W = round seeds G = yellow cotyledons

P$_1$: *WWGG*   ×   *wwgg*

Gametes produced: One member of each gene pair is "segregated" to each gamete.

F$_1$: *WwGg*

F$_1$ × F$_1$: *WwGg* × *WwGg*

9/16 *W_G_* round seeds, yellow cotyledons
3/16 *W_gg* round seeds, green cotyledons
3/16 *wwG_* wrinkled seeds, yellow cotyledons
1/16 *wwgg* wrinkled seeds, green cotyledons

Seed Shape	Cotyledon Color		Phenotypes
	3/4 yellow ⟶		9/16 round, yellow
3/4 round			
	1/4 green ⟶		3/16 round, green
	3/4 yellow ⟶		3/16 wrinkled, yellow
1/4 wrinkled			
	1/4 green ⟶		1/16 wrinkled, green.

8. In Problem 7, (c) fits this description.

10. 1. Factors (genes) occur in pairs.
   2. Some genes have dominant and recessive alleles.
   3. Alleles segregate from each other during gamete formation. When homologous chromosomes separate from each other at anaphase I, alleles will go to opposite poles of the meiotic apparatus.
   4. One gene pair separates independently from other gene pairs. Different gene pairs on the same homologous pair of chromosomes (if far apart) or on nonhomologous chromosomes will separate independently from each other during meiosis.

12. Homozygosity refers to a condition where both genes of a pair are the same (i.e., *AA* or *GG* or *hh*), whereas heterozygosity refers to the condition where members of a gene pair are different (i.e., *Aa* or *Gg* or *Bb*).

14. (a) 4: *AB, Ab, aB, ab*
   (b) 2: *AB, aB*
   (c) 8: *ABC, ABc, AbC, Abc, aBC, aBc, abC, abc*
   (d) 2: *ABc, aBc*
   (e) 4: *ABc, Abc, aBc, abc*
   (f) $2^5 = 32$

16. The offspring will occur in a typical 1:1:1:1 as

1/4 *WwGg* (round, yellow)

1/4 *Wwgg* (round, green)

1/4 *wwGg* (wrinkled, yellow)

1/4 *wwgg* (wrinkled, green)

18.

Expected ratio	Observed (o)	Expected (e)
9/16	315	312.75
3/16	108	104.25
3/16	101	104.25
1/16	32	34.75

$$\chi^2 = 0.47$$

χ^2 value is associated with a probability greater than 0.90 for 3 degrees of freedom. The observed and expected values do not deviate significantly. To deal with parts **(b)** and **(c)**, it is easier to see the observed values for the monohybrid ratios if the phenotypes are listed:

smooth, yellow	315
smooth, green	108
wrinkled, yellow	101
wrinkled, green	32

For the smooth: wrinkled *monohybrid component*, the smooth types total 423 (315 + 108), while the wrinkled types total 133 (101 + 32).

Expected ratio	Observed (o)	Expected (e)
3/4	423	417
1/4	133	139

The χ^2 value is 0.35 for 1 degree of freedom, which gives a p value greater than 0.50 and less than 0.90. We fail to reject the null hypothesis and are confident that the observed values do not differ significantly from the expected values.

(c) For the yellow:green portion of the problem, see that there are 416 yellow plants (315 + 101) and 140 (108 + 32) green plants.

Expected ratio	Observed (o)	Expected (e)
3/4	416	417
1/4	140	139

The χ^2 value is 0.01 for 1 degree of freedom, and the p value is greater than 0.90. We fail to reject the null hypothesis and are confident that the observed values do not differ significantly from the expected values.

20. Use of the $p = 0.10$ as the "critical" value for rejecting or failing to reject the null hypothesis instead of $p = 0.05$ would allow more null hypotheses to be rejected. As the critical p value is increased, it takes a smaller χ^2 value to cause rejection of the null hypothesis. It would take less difference between the expected and observed values to reject the null hypothesis. Therefore the stringency of failing to reject the null hypothesis is increased.

22. The probability of getting *aabbcc* from the *AaBbCC* × *AABbCc* mating is zero.

24. Genes that are recessive can skip generations and exist in a carrier state in parents. For example, notice that II-4 and II-5 produce a female child (III-4) with the affected phenotype. On these criteria alone, the gene must be viewed as being recessive.

I-1 (*Aa*), I-2 (*aa*), I-3 (*Aa*), I-4 (*Aa*)

II-1 (*aa*), II-2 (*Aa*), II-3 (*aa*), II-4 (*Aa*), II-5 (*Aa*),

II-6 (*aa*), II-7 (*AA* or *Aa*), II-8 (*AA* or *Aa*)

III-1 (*AA* or *Aa*), III-2 (*AA* or *Aa*), III-3 (*AA* or *Aa*),

III-4 (*aa*), III-5 (probably *AA*), III-6 (*aa*)

IV-1 through IV-7 (all *Aa*).

26. The gene is inherited as an autosomal recessive.

I-1 (*aa*), I-2 (*Aa* or *AA*), I-3 (*Aa*), I-4 (*Aa*)

II-1 (*Aa*), II-2 (*Aa*), II-3 (*Aa*), II-4 (*Aa*), II-5 (*aa*),

II-6 (*AA* or *Aa*), II-7 (*AA* or *Aa*)

III-1 (*AA* or *Aa*), III-2 (*aa*), III-3 (*AA* or *Aa*)

28. (a) First consider that each parent is homozygous (true-breeding in the question) and since in the F_1 only round, axial, violet, and full phenotypes were expressed, they must each be dominant. Because all genes are on nonhomologous chromosomes, independent assortment will occur.

(b) Round, axial, violet and full would be the most frequent phenotypes: 3/4 × 3/4 × 3/4 × 3/4

(c) Wrinkled, terminal, white, and constricted would be the least frequent phenotypes: 1/4 × 1/4 × 1/4 × 1/4

(d) (3/4 × 3/4 × 3/4 × 3/4) + (1/4 × 1/4 × 1/4 × 1/4) = 82/256

(e) There would be 16 different phenotypes in the testcross offspring, just as there are 16 different phenotypes in the F_2 generation.

30. (a) Data indicate that *straight* is dominant to *curled* and *short* is dominant to *long*.

Possible symbols would be (using standard *Drosophila* symbolism):

straight wings = w^+ curled wings = w
short bristles = b^+ long bristles = b

(b) Cross 1: w^+/w; b^+/b × w^+/w; b^+/b

Cross 2: w^+/w; b/b × w^+/w; b/b
Cross 3: w/w; b/b × w^+/w; b^+/b
Cross 4: w^+/w^+; b^+/b × w^+/w^+; b^+/b
            (one parent could be w^+/w)
Cross 5: w/w; b^+/b × w^+/w; b^+/b.

Chapter 4 Modification of Mendelian Ratios

2. *Incomplete dominance* can be viewed more as a quantitative phenomenon where the heterozygote is intermediate (approximately) between the limits set by the homozygotes.
Codominance can be viewed in a more qualitative manner where both of the alleles in the heterozygote are expressed.

4. $Pp × Pp$

1/4 *PP* (lethal)

2/4 *Pp* (platinum)

1/4 *pp* (silver)

Therefore, the ratio of surviving foxes is 2/3 platinum, 1/3 silver. The *P* allele behaves as a recessive in terms of lethality (seen only in the homozygote) but as a dominant in terms of coat color (seen in the homozygote).

6. Flower color: *RR* = red; *Rr* = pink; *rr* = white
Flower shape: *P* = personate; *p* = peloric
(a) *RRpp* × *rrPP* ⟶ *RrPp*
(b) *RRPP* × *rrpp* ⟶ *RrPp*

(c) *RrPp* × *RRpp* ⟶
$$\begin{cases} RRPp \\ RRpp \\ RrPp \\ Rrpp \end{cases}$$

(d) *RrPp* × *rrpp* ⟶
$$\begin{cases} rrPp \\ rrpp \\ RrPp \\ Rrpp \end{cases}$$

In the cross of the F_1 of (a) to the F_1 of (b), both of which are double heterozygotes, one would expect the following:

$$RrPp × RrPp$$

/ 3/4 personate   ⟶ 3/16 red, personate
1/4 red
\ 1/4 peloric   ⟶ 1/16 red, peloric
/ 3/4 personate   ⟶ 6/16 pink, personate
2/4 pink
\ 1/4 peloric   ⟶ 2/16 pink, peloric
/ 3/4 personate   ⟶ 3/16 white, personate
1/4 white
\ 1/4 peloric   ⟶ 1/16 white, peloric

8. This is a case of gene interaction (novel phenotypes) where the yellow and black types (double mutants) interact to give the cream phenotype, and epistasis where the *cc* genotype produces albino.
(a) *AaBbCc* ⟶ gray (*C* allows pigment).
(b) *A_B_Cc* ⟶ gray (*C* allows pigment).
(c) Use the forked-line method for this portion.

/ 1/2 *Cc* ⟶ 9/32 gray
3/4 *B_*
/   \ 1/2 *cc* ⟶ 9/32 albino
3/4 *A_*
\   / 1/2 *Cc* ⟶ 3/32 yellow
1/4 *bb*
\ 1/2 *cc* ⟶ 3/32 albino

$$/ \; 1/2 \; Cc \longrightarrow 3/32 \text{ black}$$
$$3/4 \; B\_$$
$$/ \qquad \setminus 1/2 \; cc \longrightarrow 3/32 \text{ albino}$$
$$1/4 \; aa$$
$$\setminus \qquad / \; 1/2 \; Cc \longrightarrow 1/32 \text{ cream}$$
$$1/4 \; bb$$
$$\setminus 1/2 \; cc \longrightarrow 1/32 \text{ albino}$$

Combining the phenotypes gives (always count the proportions to see that they add up to 1.0):

16/32 albino;
9/32 gray;
3/32 yellow;
3/32 black;
1/32 cream

10. (a) 1/4
   (b) 1/2
   (c) 1/4
   (d) zero

12. RR = red, Rr = red in females,
   Rr = mahogany in males,
   rr = mahogany.
   P_1: female: RR (red) × male: rr (mahogany)
   F_1: Rr = females red; males mahogany

   1/2 females (red)
   1/2 males (mahogany)

   F_2: 1/4 RR; 2/4 Rr; 1/4 rr

Because half of the offspring are males and half are females, one could, for clarity, rewrite the F_2 as:

	1/2 females	1/2 males
1/4 RR	1/8 red	1/8 red
2/4 Rr	2/8 red	2/8 mahogany
1/4 rr	1/8 mahogany	1/8 mahogany

14. Symbolism: Normal wing margins = X^+; scalloped = X^{sd}
   (a) P_1: $X^{sd}X^{sd} \times X^+/Y$ ⌐
     F_1: 1/2 X^+X^{sd} (female, normal)
        1/2 X^{sd}/Y (male, scalloped)
     F_2: 1/4 X^+X^{sd} (female, normal)
        1/4 $X^{sd}X^{sd}$ (female, scalloped)
        1/4 X^+/Y (male, normal)
        1/4 X^{sd}/Y (male, scalloped)

   (b) P_1: $X^+/X^+ \times X^{sd}/Y$ ⌐
     F_1: 1/2 X^+X^{sd} (female, normal)
        1/2 X^+/Y (male, normal)
     F_2: 1/4 X^+X^+ (female, normal)
        1/4 X^+X^{sd} (female, normal)
        1/4 X^+/Y (male, normal)
        1/4 X^{sd}/Y (male, scalloped)

If the *scalloped* gene were not X-linked, then all of the F_1 offspring would be wild (phenotypically) and a 3:1 ratio of normal to scalloped would occur in the F_2.

16. (a) P_1: $X^vX^v; +/+ \times X^+/Y; b^r/b^r$

     F_1: 1/2 $X^+X^v; +/b^r$ (female, normal)
        1/2 $X^vY; +/b^r$ (male, vermilion)

F_2:

Eye color (X)	Eye color (autosomal)	
1/4 females, normal	3/4 normal	(3/16)
	1/4 brown	(1/16)
1/4 females, vermilion	3/4 normal	(3/16)
	1/4 brown	(1/16)
1/4 males, normal	3/4 normal	(3/16)
	1/4 brown	(1/16)
1/4 males, vermilion	3/4 normal	(3/16)
	1/4 brown	(1/16)

   3/16 females, normal
   1/16 females, brown
   3/16 females, vermilion
   1/16 females, white
   3/16 males, normal
   1/16 males, brown
   3/16 males, vermilion
   1/16 males, white

(b) P_1: $X^+X^+; b^r/b^r \times X^v/Y; +/+$ ⌐
     F_1: 1/2 $X^+X^v; +/b^r$ (female, normal)
        1/2 $X^+/Y; +/b^r$ (male, normal)
     F_2: 6/16 females, normal
        2/16 females, brown
        3/16 males, normal
        1/16 males, brown
        3/16 males, vermilion
        1/16 males, white

(c) P_1: $X^vX^v; b^r/b^r \times X^+/Y; +/+$ ⌐
     F_1: 1/2 $X^+X^v; +/b^r$ (female, normal)
        1/2 $X^v/Y; +/b^r$ (male, vermilion)
     F_2: 3/16 females, normal
        1/16 females, brown
        3/16 females, vermilion
        1/16 females, white
        3/16 males, normal
        1/16 males, brown
        3/16 males, vermilion
        1/16 males, white

18. (a) One might hypothesize that two gene pairs are involved in the inheritance of one trait while one gene pair is involved in the other.
   (b) Croaking is due to one (dominant/recessive) gene pair, while eye color is due to two gene pairs. Because there is a 9:4:3 ratio regarding eye color, some gene interaction (epistasis) is indicated.
   (c) Symbolism: Croaking: $R\_$ = rib-it; rr = knee-deep
     Eye color: Since the most frequent phenotype is blue eye, let $A\_B\_$ represent the genotypes. For the purple class, "a 3/16 group" use the A_bb genotypes. The "4/16" class (green) would be the $aaB\_$ and the $aabb$ groups.
   (d) $AABBrr \times AAbbRR$ which would produce an F_1 of $AABbRr$ which would be blue-eyed and rib-it. The F_2 will follow a pattern of a 9:3:3:1 ratio because of homozygosity for the A locus and heterozygosity for both the B and R loci.

   9/16 $AAB\_R\_$ = blue-eyed, rib-it
   3/16 AAB_rr = blue-eyed, knee-deep
   3/16 $AAbbR\_$ = purple-eyed, rib-it
   1/16 $AAbbrr$ = purple-eyed, knee-deep

20. $A\_B\_$ = (solid white) 9/16
   $aaB\_$ = (solid white) 3/16
   A_bb = (black-and-white spotted) 3/16
   $aabb$ = (solid black) 1/16

The selection of "*bb*" as giving the spotted phenotype is arbitrary. One could obtain *AAbb* true-breeding black and white-spotted cattle.

22. (a) One can envision two pathways leading to the production of green pigment:

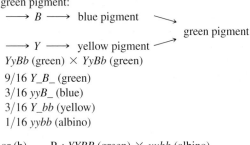

$YyBb$ (green) $\times$ $YyBb$ (green)

9/16 $Y\_B\_$ (green)
3/16 $yyB\_$ (blue)
3/16 Y_bb (yellow)
1/16 $yybb$ (albino)

or (b) P_1: *YYBB* (green) $\times$ *yybb* (albino)

YYbb $\times$ *yyBB*
$\downarrow$
YyBb (green)

Crossing these F_1's gives the observed ratios in the F_2.

24. (a,b) In looking at the pedigrees, one can see that the condition cannot be dominant because it appears in the offspring (II-3 and II-4) and not the parents in the first two cases. The condition is therefore *recessive*. In the second cross, note that the father is not shaded, yet the daughter (II-4) is. If the condition is recessive, then it must also be *autosomal*.
(c) II − 1 = *AA or Aa*
    II − 6 = *AA or Aa*
    II − 9 = *Aa*

26. Symbolism:

$A\_B\_$ = black
A_bb = golden
$aabb$ = golden
$aaB\_$ = brown

The combination of *bb* is epistatic to the *A* locus.
(a) $AAB\_ \times aaBB$ (other configurations possible but each must give all offspring with *A* and *B* dominant alleles)
(b) $AaB\_ \times aaBB$ (other configurations are possible)
(c) $AABb \times aaBb$
(d) $AABB \times aabb$
(e) $AaBb \times Aabb$
(f) $AaBb \times aabb$
(g) $aaBb \times aaBb$
(h) $AaBb \times AaBb$
Those genotypes which will breed true will be as follows:
black = *AABB*
golden = all genotypes which are *bb*
brown = *aaBB*

28. (a) This is a case of incomplete dominance in which, as shown in the third cross, the heterozygote (palomino) produces a typical 1:2:1 ratio. Therefore one can set the following symbols:
$C^{ch}C^{ch}$ = chestnut
C^cC^c = cremello
$C^{ch}C^c$ = palomino
(b) The F_1 resulting from matings between cremello and chestnut horses would be expected to be all palomino. The F_2 would be expected to fall in a 1:2:1 ratio as in the third cross in part (a).

30. Cross #1 = (c)
    Cross #2 = (d)
    Cross #3 = (b)
    Cross #4 = (e)
    Cross #5 = (a)

Given that each parental/offspring grouping can only be used once, there are no other combinations.

32. With extranuclear inheritance, the phenotype is determined by the nuclear (maternal effect) or cytoplasmic (organelle or infectious) condition of the parent that contributes the bulk of the cytoplasm to the offspring. Standard Mendelian ratios (3:1) are usually not present. In general, the results of reciprocal crosses differ. In X-linked inheritance, the pattern is often from grandfather through carrier mother to son. Patterns of extrachromosomal inheritance are often not influenced by the sex of the offspring.

34. The female parent must have been *Dd*, while the other must have been *dd*.

36. (a) The presence of bcd^-/bcd^- males can be explained by the maternal effect: mothers were bcd^+/bcd^-.
(b) The cross female bcd^+/bcd^- × male bcd^-/bcd^- will produce an F_1 with normal embryogenesis because of the maternal effect. In the F_2, any cross having bcd^+/bcd^- mothers will have phenotypically normal embryos. Any cross involving homozygous bcd^-/bcd^- mothers will have problems with embryogenesis.

Chapter 5 *Sex Determination and Sex Chromosomes*

4. Sexual differentiation is the response of cells, tissues, and organs to signals provided by the genetic mechanisms of sex determination. In other words, genes are present that signal developmental pathways whereby the sexes are generated. Sexual differentiation is the complex set of responses to those genetic signals.

6. In *Drosophila* it is the balance between the number of X chromosomes and the number of haploid sets of autosomes that determines sex. In humans there is a small region on the Y chromosome that determines maleness.

8. In *primary* nondisjunction, half of the gametes contain two X chromosomes while the complementary gametes contain no X chromosomes. Fertilization, by a Y-bearing sperm cell, of those female gametes with two X chromosomes would produce the XXY Klinefelter syndrome. Fertilization of the "no-X" female gamete with a normal X-bearing sperm will produce the Turner syndrome.

10. No. Since the Y chromosome cannot be detected in these crosses, there is no way to distinguish the two modes of sex determination.

12. One would see daughters with the white eye phenotype and sons with the miniature wing phenotype.

14. Because synapsis of chromosomes in meiotic tissue is often accompanied by crossing over, it would be detrimental to sex-determining mechanisms to have sex-determining loci on the Y chromosome transferred, through crossing over, to the X chromosome.

16. There is a simple formula for determining the number of Barr bodies in a given cell: N-1, where N is the number of X chromosomes.

Klinefelter syndrome (XXY)	= 1
Turner syndrome (XO)	= 0
47, XYY	= 0
47, XXX	= 2
48, XXXX	= 3

18. Unless other markers, cytological or molecular, are available, one cannot test the Lyon hypothesis with homozygous X-linked genes. The test requires identification of allelic alternatives.

20. Males normally have only one X chromosome; therefore such mosaicism cannot occur. Females normally have two X chromosomes. There are cases of male calico cats that are XXY.

22. In mammals, the scheme of sex determination is dependent on the presence of a piece of the Y chromosome. In *Bonellia*

viridis, the female proboscis contains a substance that triggers a morphological, physiological, and behavioral developmental pattern producing males. To elucidate the mechanism, one could attempt to isolate and characterize the active substance by testing different chemical fractions of the proboscis. Mutant analysis usually provides critical approaches into developmental processes. Depending on characteristics of the organism, one could attempt to isolate mutants that lead to changes in male or female development. Third, by using microtissue transplantations, one could attempt to determine which anatomical "centers" of the embryo respond to the chemical cues of the female.

24. One could account for the significant departures from a 1:1 ratio of males to females by suggesting that at anaphase I of meiosis, the Y chromosome more often goes to the pole that produces the more viable sperm cells. One could also speculate that the Y-bearing sperm has a higher likelihood of surviving in the female reproductive tract, or that the egg surface is more receptive to Y-bearing sperm. At this time the mechanism is unclear.

26.

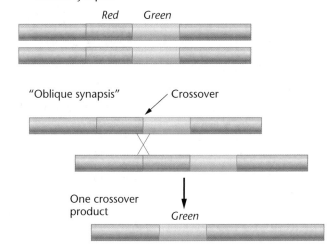

"Normal synapsis"

Red   Green

"Oblique synapsis"   Crossover

One crossover product

Green

28. (a) Something is missing from the male-determining system of sex determination either at the level of the genes, gene products, receptors, and so on.

(b) The *SOX9* gene or its product is probably involved in male development. Perhaps it is activated by *SRY*.

(c) There is probably some evolutionary relationship between the *SOX9* gene and *SRY*. There is considerable evidence that many other genes and pseudogenes are also homologous to *SRY*.

(d) Normal female sexual development does not require the *SOX9* gene or gene product(s).

30. In snapping turtles, sex determination is strongly influenced by temperature, such that males are favored in the 26–34° C range. Lizards, on the other hand, appear to have their sex determined by factors other than temperature in the 20–40° C range.

32. The white patches of CC are due to an autosomal gene *S* for white spotting that prevents pigment formation in the cell lineages in which it is expressed. Homozygous *SS* cats have more white than heterozygous *Ss* cats, and there is no absolute pattern of patches due to the *S* allele. So the distribution of white patches would be expected to be different from Rainbow. In addition, since X chromosome inactivation is random, CC would have a different patch pattern from her genetic mother on the random X inactivation basis alone.

Chapter 6 Chromosome Mutations: Variation in Number and Arrangement

4. The fact that there is a significant maternal age effect associated with Down syndrome indicates that nondisjunction in older females contributes disproportionately to the number of Down syndrome individuals. In addition, certain genetic and cytogenetic marker data indicate the influence of female nondisjunction.

6. Because an allotetraploid has a possibility of producing bivalents at meiosis I, it would be considered the most fertile of the three. Having an even number of chromosomes to match up at the metaphase I plate, autotetraploids would be considered to be more fertile than autotriploids.

8. It is likely that an interspecific hybridization occurred followed by chromosome doubling. These events probably produced a fertile amphidiploid (allotetraploid).

10. While there is the appearance that crossing over is suppressed in inversion "heterozygotes," the phenomenon extends from the fact that the crossover chromatids end up being abnormal in genetic content. As such, they fail to produce viable (or competitive) gametes or lead to zygotic or embryonic death.

12. In a work entitled *Evolution by Gene Duplication*, Ohno suggests that gene duplication has been essential in the origin of new genes. If gene products serve essential functions, mutation and therefore evolution would not be possible unless these gene products could be compensated for by products of duplicated, normal genes. The duplicated genes, or the original genes themselves, would be able to undergo mutational "experimentation" without necessarily threatening the survival of the organism.

14. A Turner syndrome female has the sex-chromosome composition of XO. If the father had hemophilia, it is likely that the Turner syndrome individual inherited the X chromosome from the father and no sex chromosome from the mother. If nondisjunction occurred in the mother, either during meiosis I or meiosis II, an egg with no X chromosome could be the result.

16. Given the basic chromosome set of nine unique chromosomes (a haploid complement) other forms with the "*n* multiples" are forms of polyploidy. Organisms with 27 chromosomes ($3n$) are more likely to be sterile because there are trivalents at meiosis I that cause a relatively high number of unbalanced gametes to be formed.

18. The cross would be as follows: *WWWW* × *wwww*
F$_1$: *WWww*
F$_2$: 35*W* and 1*w*

20. Since two *Gl*$_1$ alleles and two *ws*$_3$ alleles are present in the triploid, they must have come from the pollen parent. By the wording of the problem, it is implied that the pollen parent contributed an unreduced (2n) gamete; however, another explanation, dispermic fertilization is possible. In this case two *Gl*$_1$ *ws*$_3$ gametes could have fertilized the ovule.

22. In a paracentric inversion, one recombinant chromatid is dicentric (two centromeres), while the other is acentric (lacks a centromere). In a pericentric inversion, recombinant chromatids have duplications and deletions; however, no acentric or dicentric chromatids are produced.

24. In plants, gametes with aberrant unbalanced genetic complements usually fail to develop normally, leading to aborted pollen or ovules. Therefore, genetic imbalance is revealed prior to fertilization and inviable seeds result. In animals, aberrant or unbalanced gametes tend to function; however, significant embryonic (and subsequent) developmental abnormalities are likely to occur.

26. (a, b, c) It is likely that the translocation is the cause of the miscarriages. Segregation of the chromosomal elements will produce approximately half unbalanced gametes. The chance of a normal child is approximately one in two, however; half of the normal children will be translocation carriers. Should she abandon her attempts to have a child of her own? The answer to this question is more one of personal choice than science. It is the task of the scientific community to provide accurate information within the limits of technology. Generally speaking, this information is provided to individuals so that they can make informed decisions. In this case, the woman has been given information that probably fits her circumstance. It is up to her to make such a personal decision.

28. (a) The father must have contributed the abnormal X-linked gene.
    (b) Since the son is XXY and heterozygous for anhidrotic dysplasia, he must have received both the defective gene and the Y chromosome from his father. Thus nondisjunction must have occurred during meiosis I.
    (c) This son's mosaic phenotype is caused by X-chromosome inactivation, a form of dosage compensation in mammals.

30. Below is a description of breakage/reunion events which illustrate such a translocation in relatively small, similarly sized, chromosomes 19 (metacentric) and 20 (metacentric/submetacentric). The case described here is shown occurring before S phase duplication. The same phenomenon is shown in the text as occurring after S phase. Since the likelihood of such a translocation is fairly small in a general population, inbreeding probably played a significant role in allowing the translocation to "meet itself."

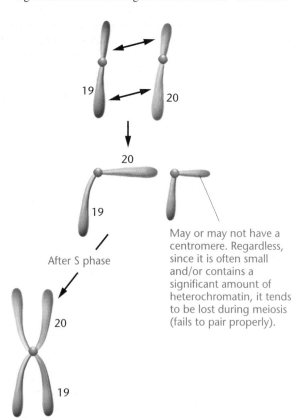

After S phase

May or may not have a centromere. Regardless, since it is often small and/or contains a significant amount of heterochromatin, it tends to be lost during meiosis (fails to pair properly).

Chapter 7 Linkage and Chromosome Mapping in Eukaryotes

2. First, in order for chromosomes to engage in crossing over, they must be in proximity. It is likely that the side-by-side pairing that occurs during synapsis is the earliest time during the cell cycle that chromosomes achieve that necessary proximity. Second, chiasmata are visible during prophase I of meiosis and it is likely that these structures are intimately associated with the genetic event of crossing over.

4. Because crossing over occurs at the four-strand stage of the cell cycle (that is, after S phase), notice that each single crossover involves only two of the four chromatids.

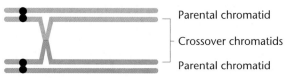

Parental chromatid

Crossover chromatids

Parental chromatid

6. Positive interference occurs when a crossover in one region of a chromosome interferes with crossovers in nearby regions. Such interference ranges from zero (no interference) to 1.0 (complete interference). Interference is often explained by a physical rigidity of chromatids such that they are unlikely to make sufficiently sharp bends to allow crossovers to be close together.

8.
$$dp \text{ -- } cl \text{ --------------- } ap$$
$$3 \qquad\qquad 39$$

10. The question is whether the arrangement in the parents is *coupled* ($RY/ry \times ry/ry$) or *not coupled* ($Ry/rY \times ry/ry$). Notice that the most frequent phenotypes in the offspring, the parentals, are colored, green (88) and colorless, yellow (92). This indicates that the heterozygous parent in the testcross is coupled ($RY/ry \times ry/ry$). There would be 10 map units between the loci.

12. The cross would be $PZ/pz \times pz/pz$. Adding the crossover percentages together (6.9 + 7.1) gives 14%, which would be the map distance between the two genes.

14.

	female A	female B	Frequency
NCO	3, 4	7, 8	first
SCO	1, 2	3, 4	second
SCO	7, 8	5, 6	third
DCO	5, 6	1, 2	fourth

16. (a) $y\,w\,+/+\,+\,ct \quad \times \quad y\,w\,+/Y$
    (b)
    $$y \text{ ---------- } w \text{ ---------------------- } ct$$
    $$0.0 \qquad 1.5 \qquad\qquad 20.0$$
    (c) There were $0.185 \times 0.015 \times 1000 = 2.775$ double crossovers expected.
    (d) Because the cross to the F_1 males included the normal (wild-type) gene for *cut wings*, it would not be possible to unequivocally determine the genotypes from the F_2 phenotypes for all classes.

18. The results would change because of no crossing over in males.

20. (a,b)
$$a - b = \frac{32 + 38 + 0 + 0}{1000} \times 100$$
$$= 7 \text{ map units}$$
$$b - c = \frac{11 + 9 + 0 + 0}{1000} \times 100$$
$$= 2 \text{ map units}$$

(c) The progeny phenotypes that are missing are $++c$ and $a\,b+$, which, of 1000 offspring, 1.4 would be expected. Perhaps by chance or some other unknown selective factor, they were not observed.

22. These observations as well as the results of other experiments indicate that the synaptonemal complex is required for crossing over.

24. Since the genetic map is more accurate when relatively small distances are covered and when large numbers of offspring are

scored, this map would probably not be too accurate with such a small sample size.

26. $A\_B\_$ 33/64

A_bb 15/64

$aaB\_$ 15/64

$aabb$ 1/64

28. By having microscopically visible markers on the chromosomes, Creighton and McClintock were able to show that homologous chromosomal material physically exchanged segments during crossing over.

30. $a\ b\ c$ = 168

$+ + +$ = 168

$a + +$ = 20

$+ b\ c$ = 20

$+ + c$ = 10

$a\ b +$ = 10

$+ b +$ = 2

$a + c$ = 2

The map distances would be computed as follows:

$$a\text{–}b = \frac{20 + 20 + 2 + 2}{400} \times 100$$

$$= 11 \text{ map units}$$

$$b\text{–}c = \frac{10 + 10 + 2 + 2}{400} \times 100$$

$$= 6 \text{ map units.}$$

32. B^+ = wild-type eye shape

B = Bar eye shape

m^+ = wild-type wings

m = miniature wings

e^+ = wild-type body color

e = ebony body color

Mapping the distance between B and m would be as follows:

$$\frac{(57 + 64)}{(226 + 218 + 57 + 64)} \times 100 = \frac{121}{565} \times 100 = 21.4 \text{ map units}$$

We would conclude that the *ebony* locus is either far away from B and m (50 map units or more) or it is on a different chromosome. In fact, *ebony* is on a different chromosome.

Chapter 8 Genetic Analysis and Mapping in Bacteria and Bacteriophages

2. (a) The requirement for physical contact between bacterial cells during conjugation was established by placing a filter in a U-tube such that the medium could be exchanged but the bacteria could not come in contact. Under this condition, conjugation does not occur.

(b) By treating cells with streptomycin, an antibiotic, it was shown that recombination would not occur if one of the two bacterial strains was inactivated. However, if the other was similarly treated, recombination would occur. Thus, directionality was suggested, with one strain being a donor strain and the other being the recipient.

(c) An F^+ bacterium contains a circular, double-stranded, structurally independent DNA molecule which can direct recombination.

4. Bacteria that are F^+ possess the F factor, while those that are F^- lack the F factor. In Hfr cells, the F factor is integrated into the bacterial chromosome, and in F' bacteria, the F factor is free of the bacterial chromosome yet possesses a piece of the bacterial chromosome.

6. In an Hfr $\times$ F^- cross, the F factor is directing the transfer of the donor chromosome. It takes at least 90 minutes to transfer the entire chromosome. Because the F factor is the last element to be transferred and the conjugation tube is fragile, the likelihood for complete transfer is low.

8. Transformation requires *competence* on the part of the recipient bacterium, meaning that only under certain conditions are bacterial cells capable of being transformed. Transforming DNA must be *double-stranded* to begin with, yet is converted to a single-stranded structure upon insertion into the host cell. The most efficient length of the transforming DNA is about 1/200 of the size of the host chromosome. Transformation is an energy-requiring process, and the number of sites on the bacterial cell surface is limited.

10. Notice that the incorporation of loci a^+ or b^+ occurs much more frequently than the incorporation of b^+ and c^+ together (210 to 1) and the incorporation of all three genes $a^+b^+c^+$ occurs relatively infrequently. If a and b loci are close together and both are far from locus c, then fewer crossovers would be required to incorporate the two linked loci compared to all three loci. If all three loci were close together, then the frequency of incorporation of all three would be similar to the frequency of incorporation of any two contiguous loci, which is not the case.

14. A single plaque is a clearing of bacteria resulting from the lytic action of millions of bacteriophage.

16. A lytic cycle occurs as bacteriophage enter a bacterial host and form progeny phage after a relatively short period of time. *Lysogeny* is a complex process whereby certain temperate phage can enter a bacterial cell and instead of following a lytic developmental path, integrate their DNA into the bacterial chromosome.

18. In their experiment a filter was placed between the two auxotrophic strains, which would not allow contact. F-mediated conjugation requires contact, and without that contact, such conjugation cannot occur. The treatment with DNase showed that the filterable agent was not naked DNA.

20. Cotransduction of genes in generalized transduction allows linkage relationships to be determined because the closer two genes are to each other, the higher the likelihood that they will be physically "linked" together in a single DNA strand during transdution.

22. Viral recombination occurs when there is a sufficiently high number of infecting viruses, so that there is a high likelihood that more than one type of phage will infect a given bacterium. Under this condition, phage chromosomes can recombine by crossing over.

24. (a) The original concentration is greater than 10^5.

(b) The original concentration is about 1.4×10^7.

(c) Remembering that 0.1 ml is typically used in the plaque assay, the initial concentration of phage is less than 10^7. Coupling this information with the calculations in part (b) above, it would appear that the initial concentration of phage is around 1×10^7 and the failure to obtain plaques in this portion of the experiment is expected and due to sampling error.

26. Because the frequency of double transformants is quite high (compare the trp^+tyr^+ transformants in A and B experiments), one may conclude that the genes are quite closely linked together.

28. (a) Rifampicin eliminates the donor strain which is rif^s.

(b) $\underline{\quad b\ \ a \qquad\qquad\qquad\qquad c \qquad F\quad}$

(c) An additional antibiotic selection should be added and recombinants must be replated on a rifampicin medium to determine which ones are sensitive.

30. Since g cotransforms with f, it is likely to be in the $c\ b\ f$ "linkage group" and would be expected to cotransform with each. One would not expect transformation with a, d or e.

32. (a) No, all functional groups do not impact similarly on conjugative transfer of R27. Regions 1, 2, and 4 appear to be least influenced by mutation because transfer is at 100%.

(b) Regions 3, 5, 6, 8, 9, 10, 12, 13, and 14 appear to have the most impact on conjugation because when mutant, conjugation is abolished.

(c) Regions 7 and 11, when mutant, only partially abolish conjugation; therefore they probably have less impact on conjugation than those listed in part (b).

(d) The data in this problem provide some insight into the complexity of the genetic processes involved in bacterial conjugation. The regions that have the most impact on conjugation fall into three different functional groups. In addition, notice that regions 1, 2, 4, 7, and 11, those that appear to have little if any impact on conjugation, are functionally related as indicated by their shading.

Chapter 9 DNA Structure and Analysis

2. While some experiments examined the possible nature of the hereditary material, abundant knowledge of the structural and enzymatic properties of proteins generated a bias that worked to favor proteins as the hereditary substance. In addition, proteins were composed of as many as 20 different subunits (amino acids), thereby providing ample structural and functional variation for the multiple tasks that must be accomplished by the genetic material. The tetranucleotide hypothesis (structure) provided insufficient variability to account for the diverse roles of the genetic material.

4. Transformation is dependent on a macromolecule (DNA) that can be extracted and purified from bacteria. During such purification, however, other macromolecular species may contaminate the DNA. Specific degradative enzymes—proteases, RNase, and DNase—were used to selectively eliminate components of the extract, and, if transformation was concomitantly eliminated, then the eliminated fraction is the transforming principle. DNase eliminates DNA and transformation. Therefore it must be the transforming principle.

6. Actually, phosphorus is found in approximately equal amounts in DNA and RNA. Therefore, labeling with ^{32}P would "tag" both RNA and DNA. However, the T2 phage, in its mature state, contains very little if any RNA. Therefore DNA would be interpreted as being the genetic material in T2 phage.

8. The early evidence would be considered indirect, in that at no time was there an experiment, like transformation in bacteria, in which genetic information in one organism was transferred to another using DNA. Rather, by comparing DNA content in various cell types (sperm and somatic cells) and observing that the *action* and *absorption* spectra of ultraviolet light were correlated, DNA was considered to be the genetic material. This suggestion was supported by the fact that DNA was shown to be the genetic material in bacteria and some phage. Direct evidence for DNA being the genetic material comes from a variety of observations, including gene transfer that has been facilitated by recombinant DNA techniques.

10. Linkages among the three components require the removal of water (H_2O).

12. Guanine: 2-amino-6-oxypurine
Cytosine: 2-oxy-4-aminopyrimidine
Thymine: 2,4-dioxy-5-methylpyrimidine
Uracil: 2,4-dioxypyrimidine

16. Because in double-stranded DNA, A = T and G = C (within limits of experimental error), the data presented would have indicated a lack of pairing of these bases in favor of a single-stranded structure or some other nonhydrogen-bonded structure. Alternatively, from the data it would appear that A = C and T = G, which would negate the chance for typical hydrogen bonding since opposite charge relationships do not exist. Therefore, it is quite unlikely that a tight helical structure would form at all.

18. Three main differences between RNA and DNA are the following:
1. Uracil in RNA replaces thymine in DNA.
2. Ribose in RNA replaces deoxyribose in DNA.
3. RNA often occurs as both single- and partially double-stranded forms, whereas DNA most often occurs in a double-stranded form.

20. The nitrogenous bases of nucleic acids (nucleosides, nucleotides, and single- and double-stranded polynucleotides) absorb UV light maximally at wavelengths 254 to 260 nm. Using this phenomenon, one can often determine the presence and concentration of nucleic acids in a mixture. Since proteins absorb UV light maximally at 280 nm, this is a relatively simple way of dealing with mixtures of biologically important molecules.

UV absorption is greater in single-stranded molecules (hyperchromic shift) as compared to double-stranded structures. Therefore, one can easily determine, by applying denaturing conditions, whether a nucleic acid is in the single- or double-stranded form. In addition, A-T rich DNA denatures more readily than G-C rich DNA. Therefore, one can estimate base content by denaturation kinetics.

22. A *hyperchromic effect* is the increased absorption of UV light as double-stranded DNA (or RNA for that matter) is converted to single-stranded DNA. If one monitors the UV absorption with a spectrophotometer during the melting process, the hyperchromic shift can be observed. The T_m is the point on the profile (temperature) at which half (50%) of the sample is denatured.

24. For curve A, there is evidence for a rapidly renaturing species (repetitive) and a slowly renaturing species (unique). The fraction which reassociates faster than the *E. coli* DNA is highly repetitive and the last fraction (with the highest $C_ot_{1/2}$ value) contains primarily unique sequences. Fraction B contains mostly unique, relatively complex DNA.

26. In one sentence of their paper in *Nature*, Watson and Crick state,
It has not escaped our notice that the specific pairing we have postulated immediately suggests a possible copying mechanism for the genetic material.

28. MS-2 = 200 base pairs

E. coli = 2×10^6 base pairs

30. Left side (a) = right, right side (b) = left.

32. Under this condition, the hydrolyzed 5-methyl cytosine becomes thymine.

34. (a) Heat application would yield a hyprochromic shift if the DNA is double-stranded. One could also get a rough estimation of the GC content from the kinetics of denaturation and the degree of sequence complexity from comparative renaturation studies.

(b) Determination of base content by hydrolysis and chromatography could be used for comparative purposes and could also provide evidence as to the strandedness of the DNA.

(c) Antibodies for Z-DNA could be used to determine the degree of left-handed structures, if present.

(d) Sequencing the DNA from both viruses would indicate sequence homology. In addition, through various electronic searches readily available on the Internet (Web site: **blast @ ncbi.nlm.nih.gov**, for example) one could determine whether similar sequences exist in other viruses or in other organisms.

Chapter 10 DNA Replication and Synthesis

2. By labeling the pool of nitrogenous bases of the DNA of *E. coli* with the heavy isotope ^{15}N, it would be possible to "follow" the "old" DNA.

4. (a) Under a conservative scheme, all of the newly labeled DNA will go to one sister chromatid, while the other sister chromatid will remain unlabeled. **(b)** Under a dispersive scheme, all of the newly labeled DNA will be interspersed with unlabeled DNA. Both sister chromatids would appear as evenly labeled structures.

6. The *in vitro* replication requires a DNA template, a primer to give a double-stranded portion, a divalent cation (Mg^{++}), and all four of the deoxyribonucleoside triphosphates: dATP, dCTP, dTTP, and dGTP. The lower case "d" refers to the deoxyribose sugar.

8. Several analytical approaches showed that the products of DNA polymerase I were probably copies of the template DNA. *Base composition* was used initially to compare both templates and products. Within experimental error, those data strongly suggested that the DNA replicated faithfully.

10. *Biologically active* DNA implies that the DNA is capable of supporting typical metabolic activities of the cell or organism and is capable of faithful reproduction.

12. None can *initiate* DNA synthesis on a template, but all can *elongate* an existing DNA strand, assuming there is a template strand as shown in the figure below. Polymerization of nucleotides occurs in the 5′ to 3′ direction where each 5′ phosphate is added to the 3′ end of the growing polynucleotide.

All three enzymes are large complex proteins with a molecular weight in excess of 100,000 daltons, and each has 3′ to 5′ exonuclease activity.

DNA polymerase I:
   polymerization
   3′-5′ exonuclease activity
   5′-3′ exonuclease activity
   present in large amounts
   relatively stable
   removal of RNA primer

DNA polymerase II:
   polymerization
   3′-5′ exonuclease activity
   possibly involved in repair function

DNA polymerase III:
   polymerization
   3′-5′ exonuclease activity
   essential for replication
   complex molecule

14.

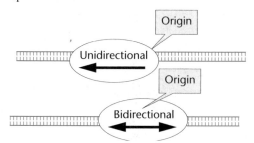

16. *Okazaki fragments* are relatively short (1000 to 2000 bases in prokaryotes) DNA fragments that are synthesized in a discontinuous fashion on the lagging strand during DNA replication. DNA *ligase* is required to form phosphodiester linkages in gaps that are generated when DNA polymerase I removes RNA primer and meets newly synthesized DNA ahead of it. *Primer RNA* is formed by RNA primase to serve as an initiation point for the production of DNA strands on a DNA template.

18. Eukaryotic DNA is replicated in a manner that is very similar to that of *E. coli*. Synthesis is bidirectional, continuous on one strand and discontinuous on the other, and the requirements of synthesis (four deoxyribonucleoside triphosphates, divalent cation, template, and primer) are the same. Okazaki fragments of eukaryotes are about one-tenth the size of those in bacteria. Because there is a much greater amount of DNA to be replicated and DNA replication is slower, there are multiple initiation sites for replication in eukaryotes (and increased DNA polymerase per cell) in contrast to the single replication origin in prokaryotes. Replication occurs at different sites during different intervals of the S phase.

20. (a) about 4,000,000 bp.
   (b) 1.3 mm

22. Gene conversion is now considered a result of heteroduplex formation that is accompanied by mismatched bases. When these mismatches are corrected, the "conversion" occurs.

24. Telomerase activity is present in germ-line tissue to maintain telomere length from one generation to the next.

26. If replication is conservative, the first autoradiographs would have label distributed only on one side (chromatid) of the metaphase chromosome as shown below.

28. If the DNA contained parallel strands in the double helix and the polymerase would be able to accommodate such parallel strands, there would be continuous synthesis and no Okazaki fragments. The telomere problem would only be at one end. Several other possibilities exist. If the DNA strands were replicated as complete single strands, the synthesis could begin at the opposite free ends. In addition, if the DNA existed only as a single strand, the same results would occur.

30. Conservative replication can be eliminated.

32. First, given that label is distributed on both ends of the structure, replication must be bidirectional. Second, since both upper and lower portions contain label, no restrictions to synthesis are apparent. The distribution of low-density grains in the center would indicate that replication begins in the middle and proceeds to the outward areas in both directions rather evenly.

Chapter 11 Chromosome Structure and DNA Sequence Organization

2. By having a circular chromosome, no free ends present the problem of linear chromosomes, namely, complete replication of terminal sequences.

4. Mitochondrial DNA varies in size from 16 kb in humans to over 350 kb in some plants and codes for ribosomal, transfer and messenger RNAs. The protein-synthesizing apparatus and many components of cellular respiration are jointly formed from nuclear and mitochondrial genes. Chloroplast DNA codes for ribosimal RNAs, as well as tRNAs and mRNAs for ribosomal proteins and photosynthesis. Nuclear genes also contribute to the pool of functional proteins in chloroplasts.

6. Like ribosomal RNA, mitochondria and chloroplasts contain proteins more similar to bacterial forms than those found in eukaryotes. These findings support the endosymbiont hypothesis for the origin of such organelles from bacteria.

8. Since eukaryotic chromosomes are "multirepliconic" in that there are multiple replication forks along their lengths, one would expect to see multiple clusters of radioactivity.

10. Long interspersed elements (LINEs) are repetitive transposable DNA sequences in humans. The most prominent family, designated **L1**, is about 6.4 kb each and is represented about 100,000 times. LINEs are often referred to as retrotransposons because their mechanism of transposition resembles that used by retroviruses.

12. Because of the diverse cell types of multicellular eukaryotes, a variety of gene products is required, which may be related to the increase in DNA content per cell. In addition, the advantage of diploidy automatically increases DNA content per cell. However, when the question is seen in another way, it is likely that a much higher *percentage* of the genome of a prokaryote is actually involved in phenotype production than in a eukaryote.

Eukaryotes have evolved the capacity to obtain and maintain what appears to be large amounts of "extra" (perhaps "junk") DNA. Prokaryotes, on the other hand, with their relatively short life cycle, are extremely efficient in their accumulation and use of their genome. In addition, the eukaryotic genome is divided into separate entities (chromosomes) to perhaps facilitate the partitioning process in mitosis and meiosis.

14. Nucleosomes are octomeric structures of two molecules of each histone (H2A, H2B, H3, and H4) except H1. Between the nucleosomes and complexed with linker DNA is histone H1. A 146-base pair sequence of DNA wraps around the nucleosome.

16. *Heterochromatin* is chromosomal material that stains deeply and remains condensed when other parts of chromosomes, euchromatin, are otherwise pale and decondensed. Heterochromatic regions replicate late in S phase and are relatively inactive in a genetic sense because there are few genes present, or if they are present, they are repressed. Telomeres and the areas adjacent to centromeres are composed of heterochromatin.

18. Volume of DNA:
$3.14 \times 10 \text{ Å} \times 10 \text{ Å} \times (50 \times 10^4 \text{ Å}) = 1.57 \times 10^8 \text{ Å}^3$
Volume of capsid:
$4/3 \ (3.14 \times 400 \text{ Å} \times 400 \text{ Å} \times 400 \text{ Å}) = 2.67 \times 10^8 \text{ Å}^3$
Because the capsid head has a greater volume than the volume of DNA, the DNA will fit into the capsid.

20. Volume of the nucleus $= 5.23 \times 10^{11} \text{ nm}^3$
Volume of the chromosome $= 1.9 \times 10^{11} \text{ nm}^3$
Therefore, the percentage of the volume of the nucleus occupied by the chromatin is

$$\frac{1.9 \times 10^{11} \text{ nm}^3}{5.23 \times 10^{11} \text{ nm}^3} \times 100 = \text{about } 36.3\%$$

22. Chromosomes are not randomly distributed within nuclei. Homologous chromosomes tend to distribute themselves opposite each other and in an antiparallel manner, meaning that their positions are in reverse order on opposite sides of the nucleus. Assuming that such patterns are maintained throughout the entire cell cycle, it is possible that chromosomal positions may influence gene function and/or chromosomal behavior during mitosis and/or meiosis. If gene function is influenced not only by gene position in a chromosome but also by gene position in a nucleus, then an alternative explanation for position effect exists.

24. Nucleosomes follow a *dispersive* pattern with each daughter chromatid containing a mixture of old and original nucleosomes. One could test the distribution of nucleosomes by conducting an autoradiographic experiment similar to Tayler-Woods-Hughes, but instead of labeling the DNA with ^{3}H-thymidine, one would label some or all the histones H2A, H2B, H3, and H4 in nucleosomes.

26. Bacteriophage lambda is capable of forming a closed, double-stranded circular molecule because of a 12-base pair, single-stranded, complementary "overhanging" sequence at the 5′ end of each single strand.

28. The general frequency and pattern of various trinucleotide repeat motifs are similar in all taxonomic groups. Within-gene trinucleotide repeats are the most frequent repeat motif in all taxonomic groups followed by hexanucleotide repeats. One explanation might be that various microsatellite types (mono, di, tri, etc.) are generated at different rates in different genomic regions (within and between genes). A second possibility is that selection acts differentially depending on the type and location of a repeat. The correlation between the high frequency of tri- and hexanucleotide repeats within genes and a triplet code specifying particular amino acids within genes may not be coincidental.

30. If microsatellites in general are flanked by a conserved sequence, those conserved sequences may be involved in the generation and/or maintenance of the microsatellite. Alternatively, the microsatellite may generate the nonmicrosatellite region. Any hypothesis presented is in need of additional investigation before definitive statements can be made.

Chapter 12 The Genetic Code and Transcription

2. (a) The reason that $(+++)$ or $(---)$ restored the reading frame is because the code is triplet. By having the $(+++)$ or $(---)$, the translation system is "out of phase" until the third "+" or " − " is encountered. If the code contained six nucleotides (a sextuplet code), then the translation system is "out of phase" until the sixth " + " or " − " is encountered. In this case, the "out of phase" region would probably be more extensive and likely cause more amino acid alterations; however, the reading frame would eventually be established. (b) Given a sextuplet code, restoration of the reading frames would only occur with the addition or loss of 6 nucleotides. Lay out a sequence such as CATDOGPIGOWLCATDOGPIGOWLCAT. . . and test parts "a" and "b" in the problem.

4. There are two possibilities for establishing the reading frames: ACA, if one starts at the first base, and CAC, if one starts at the second base. These would code for two different amino acids (ACA = threonine; CAC = histidine) and would produce repeating polypeptides that would alternate *thr-his-thr-his.* . . or *his-thr-his-thr.* . . Given the sequence CUACUACUACUA, notice the different reading frames producing three different sequences each containing the same amino acid.

Codons:	CUA	CUA	CUA	CUA. . .
Amino Acids:	leu	leu	leu	leu. . .
	UAC	UAC	UAC	UAC. . .
	tyr	tyr	tyr	tyr. . .
	ACU	ACU	ACU	ACU. . .
	thr	thr	thr	thr. . .

If a tetranucleotide is used, such as ACGUACGUACGU...

Codons:	ACG	UAC	GUA	CGU	ACG
Amino Acids:	thr	tyr	val	arg	thr
	CGU	ACG	UAC	GUA	CGU
	arg	thr	tyr	val	arg
	GUA	CGU	ACG	UAC	GUA
	val	arg	thr	tyr	val
	UAC	GUA	CGU	ACG	UAC
	tyr	val	arg	thr	tyr

Notice that the sequences are the same except that the starting amino acid changes.

6. From the repeating polymer ACACA... one can say that threonine is either CAC or ACA. From the polymer CAACAA... with ACACA..., ACA is the only codon in common. Therefore, threonine would have the codon ACA.

8. The basis of the technique is that if a trinucleotide contains bases (a codon) that are complementary to the anticodon of a charged tRNA, a relatively large complex is formed containing the ribosome, the tRNA, and the trinucleotide. This complex is trapped in the filter, whereas the components by themselves are not trapped. If the amino acid on a charged, trapped tRNA is radioactive, then the filter becomes radioactive.

10. Apply the most conservative pathway of change.

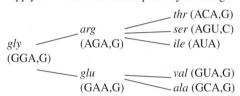

12. Because Poly U is complementary to Poly A, double-stranded structures will be formed. In order for an RNA to serve as a messenger RNA, it must be single-stranded, thereby exposing the bases for interaction with ribosomal subunits and tRNAs.

14. (a)

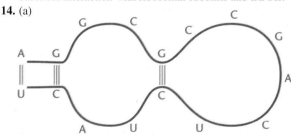

(b) TCCGCGGCTGAGATGA (use complementary bases, substituting T for U)

(c) GCU

(d) Assuming that the AGG... is the 5′ end of the mRNA, then the sequence would be *arg-arg-arg-leu-tyr*

16. (a) Starting from the 5′ end and locating the AUG triplets, one finds two initiation sites leading to the following two sequences:
met-his-thr-tyr-glu-thr-leu-gly
met-arg-pro-leu-asp (or glu)

(b) In the shorter of the two reading sequences (the one using the internal AUG triplet), a UGA triplet was introduced at the second codon. While not in the reading frames of the longer polypeptide (using the first AUG codon) the UGA triplet eliminates the product starting at the second initiation codon.

18. The central dogma of molecular genetics, and to some extent all of biology, states that DNA produces, through transcription, RNA, which is "decoded" (during translation) to produce proteins.

20. RNA polymerase from *E. coli* is a complex, large (almost 500,000 daltons) molecule composed of subunits ($\alpha,\beta,\beta',\sigma$) in the proportion $\alpha 2,\beta,\beta',\sigma$ for the holoenzyme. The β subunit provides catalytic function, while the sigma (σ) subunit is involved in recognition of specific promoters. The core enzyme is the protein without the sigma.

22. While some folding (from complementary base pairing) may occur with mRNA molecules, they generally exist as single-stranded structures that are quite labile. Eukaryotic mRNAs are generally processed such that the 5′ end is "capped" and the 3′ end has a considerable string of adenine bases. It is thought that these features protect the mRNAs from degradation. Such stability of eukaryotic mRNAs probably evolved with the differentiation of nuclear and cytoplasmic functions. Because prokaryotic cells exist in a less stable environment (nutritionally and physically, for example) than many cells of multicellular organisms, rapid genetic response to environmental change is likely to be adaptive. To accomplish such rapid responses, a labile gene product (mRNA) is advantageous. A pancreatic cell, which is developmentally stable and exists in a relatively stable environment, could produce more insulin on stable mRNAs for a given transcriptional rate.

24. Proline: C_3, and one of the C_2A triplets
Histidine: one of the C_2A triplets
Threonine: one C_2A triplet, and one A_2C triplet
Glutamine: one of the A_2C triplets
Asparagine: one of the A_2C triplets
Lysine: A_3

26. (a, b) Use the code table to determine the number of triplets that code each amino acid. Then construct a graph and plot such as this one below:

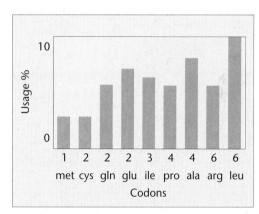

(c) There appears to be a weak correlation between the relative frequency of amino acid usage and the number of triplets for each.

(d) To continue to investigate this issue one might examine additional amino acids in a similar manner. In addition, different phylogenetic groups use code synonyms differently. It may be possible to find situations in which the relationships are more extreme. One might also examine more proteins to determine whether such a weak correlation is stronger with different proteins.

28. (a, b, c) Both nucleic acids include the expected bases: A, T, G, C for DNA and A, U, G, C for RNA. The DNA sample is compatible with a double-stranded structure because the purine/pyrimidine ratio for DNA is 1.0 and is therefore consistent with a Watson-Crick model for DNA. The purine/pyrimidine ratio is not 1.0 for RNA, so the RNA must not be double stranded.

(d) If all the DNA is transcribed as stated in the problem, the (A + U)/(C + G) ratio in RNA should equal the (A + T)/(C + G) ratio in DNA and it does (1.2). To show that it doesn't matter if one or two strands are copied, draw out a double-stranded DNA with a 1.2 ratio of (A + T)/(C + G).

Make RNA copies from one or both strands to see the resulting $(A + U)/(C + G)$ RNA ratios. Notice that if both strands are copied, the purine/pyrimidine ratio $(A + G)/(U + C)$ should be 1.0 and it is not. Therefore, it is likely that only one of the two strands is copied and the strand that is copied is richer in Ts and/or Cs compared with As and/or Gs. This would produce more As and Gs in the numerator of the equation thereby giving the ratio of 1.3.

30. The advantage would be that if sequence homologies can be identified for a variety of HIV isolates, then perhaps a single or a few vaccines could be developed for the multitude of subtypes that infect various parts of the world. In other words, the wider the match of a vaccine to circulating infectives, the more likely the efficacy. On the other hand, the more finely aligned a vaccine is to the target, the more likely it is that new or previously undiscovered variants will escape vaccination attempts.

32. (a) Alternative splicing occurs when pre-mRNAs are spliced in more than one way to yield various combinations of exons in the final mRNA product. Upon translation of a group of alternatively spliced mRNAs, a series of related proteins, called isoforms, are produced.

(b) It is likely that alternative splicing evolved to provide a variety of functionally related proteins in a particular tissue from one original source.

Some tissues might be more prone to develop alternative splicing if they depend on a number of related protein functions. In addition, if genes found in certain tissues have more exons in their active genes, alternative splicing would be expected. While some information is available concerning the mechanisms of alternative splicing, at this time little is known about the underlying forces that drove the evolution of tissue-specific alternative splicing.

Chapter 13 Translation and Proteins

2. Transfer RNAs are "adaptor" molecules in that they provide a way for amino acids to interact with sequences of bases in nucleic acids. Amino acids are specifically and individually attached to the $3'$ end of tRNAs that possess a three-base sequence (the anticodon) to base-pair with three bases of mRNA. Messenger RNA, on the other hand, contains a copy of the triplet codes that are stored in DNA. The sequences of bases in mRNA interact, three at a time, with the anticodons of tRNAs.

Enzymes involved in transcription include the following: RNA polymerase (*E. coli*), and RNA polymerase I, II, III (eukaryotes). Those involved in translation include the following: aminoacyl tRNA synthetases, peptidyl transferase, and GTP-dependent release factors.

4. The sequence of base triplets in mRNA constitutes the sequence of codons. A three-base portion of the tRNA constitutes the anticodon.

6. An amino acid in the presence of ATP, Mg^{++}, and a specific aminoacyl synthetase produces an amino acid-AMP enzyme complex $(+ PP_i.)$ This complex interacts with a specific tRNA to produce the aminoacyl tRNA.

8. While too much phenylalanine and its derivatives cause PKU in phenylketonurics, too little will restrict protein synthesis.

10. Tyrosine is a precursor to melanin, skin pigment. Individuals with PKU fail to convert phenylalanine to tyrosine and even though tyrosine is obtained from the diet, at the population level, individuals with PKU have a tendency for less skin pigmentation.

12.

14. The fact that enzymes are a subclass of the general term *protein*, a *one-gene:one-protein* statement might seem to be more appropriate. However, some proteins are made up of subunits, each different type of subunit (polypeptide chain) being under the control of a different gene. Under this circumstance, the *one-gene:one-polypeptide* might be more reasonable. It turns out that many functions of cells and organisms are controlled by stretches of DNA, which either produce no protein product (operator and promoter regions, for example) or have more than one function as in the case of overlapping genes and differential mRNA splicing. A simple statement regarding the relationship of a stretch of DNA to its physical product is difficult to justify.

16. Each chain represents a primary structure of amino acids connected by covalent peptide bonds. Secondary structures are determined by hydrogen bonding between components of the peptide bonds. Alpha helices and pleated sheets result. Tertiary structures are formed from interactions of the amino acid side chains, while the quaternary level results from the associations of chains.

18. In the late 1940s Pauling demonstrated a difference in the electrophoretic mobility of HbA and HbS (sickle-cell hemoglobin) and concluded that the difference had a chemical basis. Ingram determined that the chemical change occurs in the primary structure of the globin portion of the molecule using the fingerprinting technique. He found a change in the 6th amino acid in the β chain.

20. Dividing 20 by 0.34 gives the number of nucleotides (about 59) occupied by a ribosome. Dividing 59 by three gives the approximate number of triplet codes, approximately 20.

22. Since all higher levels of protein structure are dependent on the sequence of amino acids (primary structure), it is the primary structure that is most influential in determining protein structure and function.

24. Enzymes function to regulate catabolic and anabolic activities of cells. They influence (lower) the *energy of activation,* thus allowing chemical reactions to occur under conditions that are compatible with living systems.

26. When an expectant mother returns to consumption of phenylalanine in her diet, she subjects her baby to higher than normal levels of phenylalanine throughout its development. Since increased phenylalanine is toxic, many (approximately 90%) newborns are severely and irreversibly retarded at birth.

28. Even though three gene pairs are involved, notice that because of the pattern of mutations, each cross may be treated as monohybrid (a) or dihybrid (b,c).

(a) F_1: $AABbCC$ = speckled
$\quad F_2$: 3 AAB_CC = speckled
$\qquad$ 1 $AAbbCC$ = yellow
(b) F_1: $AABbCc$ = speckled
$\quad F_2$: 9 $AAB\_C\_$ = speckled
$\qquad$ 3 AAB_cc = green
$\qquad$ 3 $AAbbC\_$ = yellow $\Big\}$4
$\qquad$ 1 $AAbbcc$ = yellow
(c) F_1: $AaBBCc$ = speckled
$\quad F_2$: 9 $A\_BBC\_$ = speckled
$\qquad$ 3 A_BBcc = green
$\qquad$ 3 $aaBBC\_$ = colorless $\Big\}$4
$\qquad$ 1 $aaBBcc$ = colorless

30. The normal glutamic acid is a negatively charged amino acid, whereas valine carries no net charge and lysine is positively charged. Given these significant charge changes, one would predict some, if not considerable, influence on protein structure and function. Such changes could stem from internal changes in folding or interactions with other molecules in the RBC, especially other hemoglobin molecules.

32.

```
        c        b        a
pink - - ▶ rose - - ▶ orange - - ▶ purple
```

The above hypothesis could be tested by conducting a backcross as given below:

$$AaBbCc \times aabbcc$$

The cross should give a

4(pink):2(rose):1(orange):1(purple) ratio

34. The antibacterial action of evernimicin is probably at the A site on the ribosome.

Chapter 14 Gene Mutation, DNA Repair, and Transposition

2. When conducting genetic screens, one assumes that all the cells of an organism are genetically identical. Therefore, the organism responds to the screen and enables detection. If a somatic cell of a multicellular organism is mutated, it is highly unlikely that the organism will be sufficiently altered to respond to a screen.

4. Each gene and its product function in an environment that has evolved, or co-evolved. A coordinated output of each gene product is required for life. Deviations from the norm, caused by mutation, are likely to be disruptive because of the complex and interactive environment in which each gene product must function. However, on occasion a beneficial variation occurs.

6. A *conditional* mutation is one that produces a wild-type phenotype under one environmental condition and a mutant phenotype under a different condition.

8. All three of the agents are mutagenic because they cause base substitutions. Deaminating agents oxidatively deaminate bases such that cytosine is converted to uracil and adenine is converted to hypoxanthine. Uracil pairs with adenine, and hypoxanthine pairs with cytosine. Alkylating agents donate an alkyl group to the amino or keto groups of nucleotides, thus altering base-pairing affinities. 6-ethyl guanine acts like adenine, thus pairing with thymine. Base analogs such as 5-bromouracil and 2-amino purine are incorporated as thymine and adenine, respectively, yet they pair with guanine and cytosine, respectively.

10. X-rays are of higher energy and shorter wavelength than UV light. They have greater penetrating ability and can create more disruption of DNA.

12. *Photoreactivation* can lead to repair of UV-induced damage. An enzyme, photoreactivation enzyme, will absorb a photon of light to cleave thymine dimers. *Excision repair* involves the products of several genes, DNA polymerase I, and DNA ligase to clip out the UV-induced dimer, fill in, and join the phosphodiester backbone in the resulting gap. The excision repair process can be activated by damage that distorts the DNA helix. *Recombinational repair* is a system that responds to DNA that has escaped other repair mechanisms at the time of replication. If a gap is created on one of the newly synthesized strands, a "rescue operation or SOS response" allows the gap to be filled. Many different gene products are involved in this repair process: *recA, lexA*. In SOS repair, the proofreading by DNA polymerase III is suppressed, and this therefore is called an "error-prone system."

14. Each involves amplification of trinucleotide repeats. As the degree of amplification increases, so does the degree of expression of the abnormal phenotype.

16. *Xeroderma pigmentosum* is a form of human skin cancer caused by perhaps several rare autosomal genes that interfere with the repair of damaged DNA. Studies with heterokaryons provided evidence for complementation, indicating that as many as seven different genes may be involved.

The photoreactivation repair enzyme appears to be involved.

18. Each organism mentioned in the problem possesses a variety of transposable elements. Bacteria possess insertion sequences (about 800 to 1500 base pairs in length) as well as transposons that are larger. Both are mobile in bacterial, viral, and plasmid DNAs, and both have repeated base sequences at their ends. In maize, Barbara McClintock described the genetic behavior of mobile elements (*Ds* and *Ac*). *Ds* can move if *Ac* is present, thus *transposable controlling elements* exist.

An *Ac* element is 4563 base pairs long and similar in structure to some bacterial transposons. Transposons often code for transposase enzymes, which are essential for transposition. *Copia* elements in *Drosophila* may be present in numerous copies in the genome and contain direct and inverted terminal repeats. *P* elements, also in *Drosophila,* are responsible for a phenomenon called hybrid dysgenesis.

Humans possess a variety of transposable elements including the *Alu* family of short interspersed elements (SINES), which are between 200 and 300 base pairs long and may exist in 300,000 copies per genome. Long interspersed elements (LINES) also occur in the human genome and seem to be capable of movement. Such elements share common structural features, are often mobile, and may influence gene activity.

20. It is likely that the reverse transcriptase, in making DNA, provides a DNA segment that is capable of integrating into the yeast chromosome as other types of DNA are known to do.

22. It is possible that through the reduction of certain environmental agents that cause mutations, mutation rates might be reduced. On the other hand, certain industrial and medical activities actually concentrate mutagens (radioactive agents and hazardous chemicals). Unless human populations are protected from such agents, mutation rates might increase. If one asks about the accumulation of mutations (not rates) in human populations as a result of improved living conditions and medical care, then it is likely that as the environment becomes less harsh (through improvements), more mutations will be tolerated as selection pressure decreases. However, as individuals live longer and have children at a later age, some studies indicate that older males accumulate more gametic mutations.

24. The *dystrophin* gene is very large and composed of 97 exons and 2.6 Mb in length. Given the size of this gene and the number of exons/introns, many opportunities exist for mutational upset.

26. There are several ways in which an unexpected mutant gene may enter a pedigree. If a gene is incompletely penetrant, it may be present in a population and only express itself under certain conditions. It is unlikely that the gene for hemophilia behaved in this manner. If a gene's expression is suppressed by another mutation in an individual, it is possible that offspring may inherit a given gene and not inherit its suppressor. Such offspring would have hemophilia. Since all genetic variations must arise at some point, it is possible that the mutation in the Queen Victoria's family was new, arising in the father.

Lastly, it is possible that the mother was heterozygous and by chance, no other individuals in her family were unlucky enough to receive the mutant gene.

28. Your study should include examination of the following short-term aspects: immediate assessment of radiation amounts distributed in a matrix of the bomb sites as well as a control area not receiving bomb-induced radiation, radiation exposure as measured by radiation sickness and evidence of radiation poisoning from tissue samples, abortion rates, birthing rates, and chromosomal studies.

Long-term assessment should include sex-ratio distortion (males being more influenced by X-linked recessive lethals than females), chromosomal studies, birth and abortion rates, cancer frequency and type, and genetic disorders. In each case data should be compared to the control site to see if changes are bomb-related. In addition, to attempt to determine cause-effect, it is often helpful to show a dose response. Thus, by comparing the location of individuals at the time of exposure to the matrix of radiation amounts, one may be able to determine whether those most exposed to radiation suffer the most physiologically and genetically. If a positive correlation is observed, then statistically significant conclusions may be possible.

30. (a) For those organisms that generate energy by aerobic respiration, a process occurs that involves the reduction of molecular oxygen. Partially reduced species are produced as intermediates and by-products of such molecular action: O_2^-, H_2O_2, and OH^-. These species are potent electrophilic oxidants that escape mitochondria and attack numerous cellular components. Collectively, these are called reactive oxygen species (ROS).

(b)

oxoGuanine

Cytosine   Guanine

When casually examining the structures in the above diagrams, it is not immediately obvious that oxoG:A pairs should occur. However, hydrogen bonding can occur to any other base, including self pairs. Homopurine (A:A, G:G) and heteropurine (A:G) pairs represent anomalous base-pairing possibilities even with nonaltered bases. While G:C is undoubtedly the most stable, several mispairs are actually stronger than the A:T pair. Base pairing is complicated by the fact that the purines possess two H-bonding faces: the Watson-Crick face, involving ring positions 1 and 6 for Adenine and, 1, 2, and 6 for Guanine; and the Hoogsteen face involving ring positions 6 and 7. The typical pairing mode is indicated as *wc*, where pairing occurs on the Watson-Crick face in the normal orientation, even for the mispair A:G. Alteration of pairing and favoring of the Hoogsteen face can occur with the alteration generated by oxoGuanine. Indeed, triple helix configurations commonly involve the Hoogsteen face.

(c) If not repaired (see below), the first round of replication involves the pairing of oxoG to Adenine (see problem 30(b)), while in the next round of replication, Adenine pairs with its normal Thymine. Therefore, if one starts with a G:C pair, one ends up with an A:T pair.

(d) It turns out that G:G>T:A transversions are quite commonly found in human cancers and are especially prevalent in the tumor suppressor gene *p53*. Thus, the cellular defense system has been extensively studied. One component is a triphosphatase that cleanses the nucleotide precursor pool by removing the two outermost phosphates from oxo-dGTP. Another involves a DNA glycosylase that initiates repair of misreplicated oxoG:A by hydrolyzing the glycosidic bond linking the adenine base to the sugar. Another is a DNA gycosylase/lyase system that recognizes oxoG opposite cytosine. Of the three systems, the DNA glycosylases are probably the most effective.

32. Individuals with xeroderma pigmentosum (XP) are much more likely to contract skin cancer in youth than non-XP individuals. By age 20, approximately 80 percent of the XP population has skin cancer compared with approximately 4 percent in the non-XP group. XP individuals lack one or more genes involved in DNA repair.

Chapter 15 Regulation of Gene Expression

2. Under *negative* control, the regulatory molecule interferes with transcription, while in *positive* control, the regulatory molecule stimulates transcription. Negative control requires a molecule to be removed from the DNA for transcription to occur. Positive control requires a molecule to be added to the DNA for transcription to occur.

4. (a) Due to the deletion of a base early in the *lacZ* gene, there will be a shift of all the reading frames downstream from the deletion. It is likely that either premature chain termination of translation will occur (from the introduction of a nonsense triplet in a reading frame) or the normal chain termination will be ignored. Regardless, a mutant condition for the *Z* gene will be likely. If such a cell is placed on a lactose medium, it will be incapable of growth because β-galactosidase is not available. (b) If the deletion occurs early in the *A* gene, one might expect impaired function of the *A* gene product, but it will not influence the use of lactose as a carbon source.

6. $I^+O^+Z^+$ = Because of the function of the active repressor from the I^+ gene, and no lactose to influence its function, there will be **No Enzyme Made**.

$I^+O^cZ^+$ = There will be a **Functional Enzyme Made** because the constitutive operator is in *cis* with a *Z* gene. The lactose in the medium will have no influence because of the constitutive operator. The repressor cannot bind to the mutant operator.

$I^-O^+Z^-$ = There will be a **Nonfunctional Enzyme Made** because with I^- the system is constitutive, but the *Z* gene is mutant. The absence of lactose in the medium will have no influence because of the nonfunctional repressor. The mutant repressor cannot bind to the operator.

$I^-O^+Z^-$ = There will be a **Nonfunctional Enzyme Made** because with I^- the system is constitutive, but the *Z* gene is mutant. The lactose in the medium will have no influence because of the nonfunctional repressor. The mutant repressor cannot bind to the operator.

$I^-O^+Z^+/F'I^+$ = There will be **No Enzyme Made** because in the absence of lactose, the repressor product of the I^+ gene will bind to the operator and inhibit transcription.

$I^+O^cZ^+/F'O^+$ = Because there is a constitutive operator in *cis* with a normal Z gene, there will be **Functional Enzyme Made**. The lactose in the medium will have no influence because of the mutant operator.

$I^+O^+Z^-/F'I^+O^+Z^+$ = Because there is lactose in the medium, the repressor protein will not bind to the operator and transcription will occur. The presence of a normal Z gene allows a **Functional and Nonfunctional Enzyme to be Made**. The repressor protein is diffusible, working in *trans*.

$I^-O^+Z^-/F'I^+O^+Z^+$ = Because there is no lactose in the medium, the repressor protein (from I^+) will repress the operators and there will be **No Enzyme Made**.

$I^sO^+Z^+/F'O^+$ = With the product of I^s there is binding of the repressor to the operator and therefore **No Enzyme Made**. The lack of lactose in the medium is of no consequence because the mutant repressor is insensitive to lactose.

$I^+O^cZ^+/F'O^+Z^+$ = The arrangement of the constitutive operator (O^c) with the Z gene will cause a **Functional Enzyme to be Made**.

8. Catabolite repression is a mechanism whereby glucose, a catabolite of lactose, inhibits the synthesis of β-galactosidase. When glucose and lactose are both present, glucose is preferentially used as the energy source. When glucose is exhausted, β-galactosidase synthesis occurs and lactose is metabolized. Thus, catabolite repression balances the use of glucose and lactose through regulation of β-galactosidase synthesis.

10. (a) With no lactose and no glucose, the operon is off because the *lac* repressor is bound to the operator, and although CAP is bound to its binding site, it will not override the action of the repressor.
   (b) With lactose added to the medium, the *lac* repressor is inactivated and the operon is transcribing the structural genes. With no glucose, the CAP is bound to its binding site, thus enhancing transcription.
   (c) With no lactose present in the medium, the *lac* repressor is bound to the operator region, and since glucose inhibits adenyl cyclase, the CAP protein will not interact with its binding site. The operon is therefore "off."
   (d) With lactose present, the *lac* repressor is inactivated; however, since glucose is also present, CAP will not interact with its binding site. Under this condition transcription is severely diminished and the operon can be considered to be "off."

12. Attenuation functions to reduce the synthesis of tryptophan when it is in full supply. It does so by reducing transcription of the *tryptophan* operon. The same phenomenon is observed when *tryptophan* activates the repressor to shut off transcription of the tryptophan operon.

14. First, notice that in the first row of data, the presence of *tm* in the medium causes the production of active enzyme from the wildtype arrangement of genes. From this one would conclude that the system is *inducible*. To determine which gene is the structural gene, look for the *IE* function and see that it is related to C. Therefore, C codes for the **structural gene**. Because when B is mutant, no enzyme is produced, B must be the **promoter**.

Notice that when genes A and D are mutant, constitutive synthesis occurs; therefore, one must be the operator and the other gene codes for the repressor protein. To distinguish these functions, one must remember that the repressor operates as a diffusible substance and can be on the host chromosome or the F factor (functioning in *trans*). However, the operator can only operate in

cis. In addition, in *cis*, the constitutive operator is dominant to its wild-type allele, while the mutant repressor is recessive to its wild-type allele.

Notice that the mutant A gene is dominant to its wild-type allele, whereas the mutant D^- allele is recessive (behaving as wild type in the first row). Therefore, the A locus is the **operator**, and the D locus is the **repressor** gene.

16. Because the operon by itself (when mutant as in strain #3) gives constitutive synthesis of the structural genes, the *cis*-acting system is supported. The *cis*-acting element is most likely part of the operon.

18. If one could develop an assay for the other gene products under SOS control, with a *lex*A$^-$ strain, the other gene products should be present at induced levels.

22. Transcription factors are proteins that are *necessary* for the initiation of transcription. However, they are not *sufficient* for the initiation of transcription. To be activated, RNA polymerase II requires a number of transcription factors. Transcription factors contain at least two functional domains: one binds to the DNA sequences of promoters and/or enhancers, while the other interacts with RNA polymerase or other transcription factors. Some transcription factors bind to other transcription factors without themselves binding to DNA.

24. The work of Cleveland and colleagues allowed selective changes to be made in the *met-arg-glu-lys* sequence. Only the engineered mRNA sequences that caused an amino acid substitution negated the autoregulation, indicating that it is the sequence of the amino acids, not the mRNA that is critical in the process of autoregulation. The model that depicts binding of factors to the nascent polypeptide chain is supported. One might stabilize the proposed MREI-protein complex with "crosslinkers," treat with RNAse to digest mRNA and to break up polysomes, then isolate individual ribosomes. One may use some specific antibody or other method to determine whether tubulin subunits contaminate the ribosome population.

26.

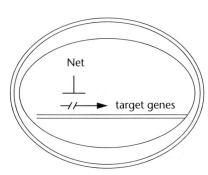

Neutral Conditions

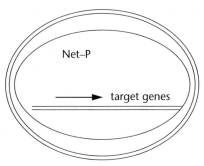

Phosphorylated Net

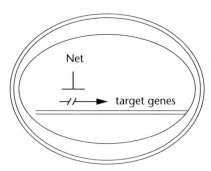

Neutral Conditions

28. Methylation of CpGs causes a reduction in luciferase expression, which is somewhat proportional to the amount of methylation and patch size. Methylation within the transcription unit more drastically reduces luciferase expression compared with methylation outside the transcription unit. A high degree of methylation outside the transcription unit (593 CpGs) has as great of an impact on depressing transcription as the same degree of methylation within the transcription unit.

30. If the exon 45 deletion causes a reading frameshift leading to reduced dystrophin production and the removal of exon 46 reestablishes the reading frame, then one would expect enhanced production of dystrophin.

Chapter 16 Cell-Cycle Regulation and Cancer

2. The G1 stage begins after mitosis and is involved in the synthesis of many cytoplasmic elements. In the S phase DNA synthesis occurs. G2 is a period of growth and preparation for mitosis. Most cell cycle time variation is caused by changes in the duration of G1. G0 is the nondividing state.

4. Kinases regulate other proteins by adding phosphate groups. Cyclins bind to the kinases, switching them on and off. CDK4 binds to cyclin D, moving cells from G1 to S. At the G2/mitosis border a CDK1 (cyclin-dependent kinase) combines with another cyclin (cyclin B). Phosphorylation occurs, bringing about a series of changes in the nuclear membrane via caldesmon, cytoskeleton, and histone H1.

6. To say that a particular trait is inherited conveys the assumption that when a particular genetic circumstance is present, it will be revealed in the phenotype. When one discusses an inherited predisposition, one usually refers to situations where a particular phenotype is expressed in families in some consistent pattern. However, the phenotype may not always be expressed or may manifest itself in different ways.

8. Apoptosis or programmed cell death is a genetically controlled process that leads to death of a cell. It is a natural process involved in morphogenesis and a protective mechanism against cancer formation. During apoptosis, nuclear DNA becomes fragmented, cellular structures are disrupted, and the cells dissolve.

10. There are a number of ways in which proto-oncogenes are converted to oncogenes: point mutations in which a mutant gene acts as a positive "switch" in the cell cycle, translocations where a hybrid gene might be formed, and overexpression where a gene might acquire a new promoter and/or enhancer. In the case of RSV, an oncogene (*c-src*) was captured from the chicken genome.

12. Various kinases can be activated by breaks in DNA. One kinase, called ATM and/or a kinase called Chk2, phosphorylates BRCA1 and p53. The activated p53 arrests replication during the S phase to facilitate DNA repair. The activated BRCA1 protein, in conjunction with BRCA2, mRAD51, and other nuclear proteins is involved in repairing the DNA.

14. Proto-oncogenes are those that normally function to promote or maintain cell division. In the mutant state (oncogenes), they induce or maintain uncontrolled cell division; that is, there is a gain-of-function. Generally this gain-of-function takes the form of increased or abnormally continuous gene output. On the other hand, loss-of-function is generally attributed to tumor suppressor genes that function to halt passage through the cell cycle. When such genes are mutant, they have lost their capacity to halt the cell cycle.

16. Unfortunately, it is common to spend enormous amounts of money dealing with diseases after they occur rather than concentrating on disease prevention. Too often pressure from special interest groups or lack of political stimulus retards advances in education and prevention. Obviously, it is less expensive, both in terms of human suffering and money, to seek preventive measures for as many diseases as possible. However, having gained some understanding of the mechanisms of disease, in this case cancer, it must also be stated that no matter what preventive measures are taken it will be impossible to completely eliminate disease from the human population. It is extremely important, however, that we increase efforts to educate and protect the human population from as many hazardous environmental agents as possible.

18. Normal cells are often capable of withstanding mutational assault because they have checkpoints and DNA repair mechanisms in place. When such mechanisms fail, cancer may be a result. Through mutation, such protective mechanisms are compromised in cancer cells, and as a result they show higher than normal rates of mutation, chromosomal abnormalities, and genomic instability.

20. Certain environmental agents such as chemicals and X-rays cause mutations. Since genes control the cell cycle, mutations in cell-cycle control genes, or those that impact on cell-cycle control, can lead to cancer.

22. No, she will still have the general population risk of about 10 percent. In addition, it is possible that genetic tests won't detect all breast cancer mutations.

24. A benign tumor is a multicellular cell mass that is usually localized to a given anatomical site. Malignant tumors are those generated by cells that have migrated to one or more secondary sites.

26. As with many forms of cancer, a single gene alteration is not the only requirement. The authors (Bose *et al.*) state: "but only infrequently do the cells acquire the additional changes necessary to produce leukemia in humans." Some studies indicate that variations (often deletions) in the region of the breakpoints may influence expression of CML.

28. Any agent that causes damage to DNA is a potential carcinogen since cell-cycle control is achieved by gene (DNA) products, proteins. Since cigarette smoke is known to contain an agent that changes DNA, in this case 28 by *transversions*, numerous modified gene products (including cell-cycle controlling proteins) are likely to be produced. The fact that many cancer patients have such transversions in *p53* strongly suggests that cancer is caused by agents in cigarette smoke.

30. (a, b) Even though there are changes in the *BRCA1* gene, they don't always have physiological consequences. Such neutral polymorphisms make screening difficult in that one can't always be certain that a mutation will cause problems for the patient.

(c) The polymorphism in *PM2* is probably a silent mutation because the third base of the codon is involved.

(d) The polymorphism in *PM3* is probably a neutral missense mutation because the first base is involved.

Chapter 17 Recombinant DNA Technology

2. *Reverse transcriptase* is often used to promote the formation of cDNA (complementary DNA) from a mRNA molecule. Eukaryotic mRNAs typically have a 3′ polyA tail, as indicated in the diagram below. The poly-dT segment provides a double-stranded section that serves to prime the production of the complementary strand.

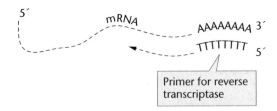

4. It is believed that the protein interacts with the major groove of the DNA helix. This information comes from the structure of the proteins that have been sufficiently well studied to suggest that the DNA major groove and "fingers" or extensions of the protein form the basis of interaction.

6. This segment contains the palindromic sequence of GGATCC, which is recognized by the restriction enzyme *Bam*HI. The double-stranded sequence is the following

CCTAGG
GGATCC

8. Because of their small size, plasmids are relatively easy to separate from the host bacterial chromosome, and they have relatively few restriction sites. They can be engineered fairly easily (i.e. polylinkers and reporter genes added). YACs (yeast artificial chromosomes) contain telomeres, an origin of replication, and a centromere and are extensively used to clone DNA in yeast. With selectable markers (*TRP1* and *URA3*) and a cluster of restriction sites, DNA inserts ranging from 100 kb to 1000 kb can be cloned and inserted into yeast. Since yeast, being eukaryotes, undergo many of the typical RNA and protein processing steps of other, more complex eukaryotes, the advantages are numerous when working with eukaryotic genes.

10. This problem can be solved by the following expressions:

*Not*I 4^8
*Hinf*I $4 \times 4 \times 1 \times 4 \times 4$
*Xho*II $2 \times 4 \times 4 \times 4 \times 4 \times 2$

12. (a) Bacteria that have been transformed with the recombinant plasmid will be resistant to tetracycline, and therefore tetracycline should be added to the medium.

(b) Colonies that grow on a tetracycline medium should be tested for growth on an ampicillin medium either by replica plating or some similar controlled transfer method. Those bacteria that do not grow on the ampicillin medium probably contain the *Drosophila* DNA insert.

(c) Resistance to both antibiotics by a transformed bacterium could be explained in several ways. First, if cleavage with the *Pst*I was incomplete, then no change in biological properties of the uncut plasmids would be expected. Also, it is possible that the cut ends of the plasmid were ligated together in the original form with no insert.

14. For the antibiotic resistance to be present, the ligation will reform the plasmid into its original form. However, two of the plasmids can join to form a dimer by each rejoining to form a single complex.

16. Because of complementary base pairing, the 3′ end of the DNA strand often loops back onto itself, thereby providing a primer for DNA polymerase I.

18. A filter is used to bind the DNA from the colonies containing recombinant plasmids. A labeled probe is constructed from the protein sequence of EF-1a. Since it is highly conserved, it should show considerable complementation to the human EF-1a cDNA. It is used to detect, through hybridization, the DNA of interest. Cells with the desired clone are then picked from the original plate, and the plasmid is isolated from the cells.

20.

$$\underline{\quad 200 \underset{150}{\overset{\text{II} \quad \text{I}}{\big|\big|}} \quad 950 \quad\quad}$$

22. Option (b) fits the expectation because the thick band in the offspring probably represents the bands at approximately the same position in both parents. The likelihood of such a match is expected to be low in the general population.

24. (a) Pattern #5 is the likely choice. Notice that digest A + N breaks up the 6-kb E fragment. (b) The only place of consistent overlap to the probe is the 1-kb fragment between A and N.

26. There would be approximately 4096 base pairs between sites. Given that lambda DNA contains approximately 48,500 base pairs, there would be about 11.8 sites (48,500/4096).

28. $T_m(°C)$ = about 63°C. Subtracting 5°C gives us a good starting point of about 58°C for PCR with this primer. As the %GC and length increase, the $T_m(°C)$ increases. GC pairs contain three hydrogen bonds rather than two as between AT pairs.

30. ddNTPs are analogs of the "normal" deoxyribonucleotide triphosphates (dNTPs), but they lack a 3′-hydroxyl group. As DNA synthesis occurs, the DNA polymerase occasionally inserts a ddNTP into a growing DNA strand. Since there is no 3′-hydroxyl group, chain elongation cannot take place, and resulting fragments are formed, which can be separated by electrophoresis.

Chapter 18 Genomics and Proteomics

2. Knowing the sequence of DNA in an organism is only the beginning. Annotating the DNA is a significant challenge. Even at that, knowing how gene products interact in time and space (proteomics) will take additional rounds of technological advances as yet unconsidered.

4. (a) Assuming an average gene size of 5,000 base pairs, there would be about 6.7×10^7 base pairs comprising genes. Subtracting this value from 116.8 Mb gives 49.8 Mb between genes. Dividing 49.8 Mb by 13,379 genes gives about 3700 bases between genes.

(b) 54,934/13,379 = 4.11 exons

(c) 48,257/13,379 = 3.61 introns

(d) There is a marked increase in the number of genes involved in alternative transcripts.

6. Perhaps as many as 650 genes are located in human heterochromatin; therefore, a considerable number of genes are yet to be identified. In addition, because these genes are located in heterochromatin, they may possess unique properties.

8. Notice that the percentage of GC pairs compared to AT pairs is quite low in punctuation triplets. Therefore, when scanning DNA sequences for ORFs with high AT content, many short sequences are obtained which are clearly not likely to be involved in protein production. However, when DNA is GC rich, the likelihood of long ORFs similar to protein-coding size is increased. Therefore, the likelihood of falsely considering a sequence "protein-coding" increases with increasing GC content as indicated in the figure.

10. Genomes of both are composed of double-stranded DNA (larger in eukaryotes) associated with proteins (more transient in prokaryotes). Both contain open reading frames, but those of prokaryotes are more densely packed. Both have some genes in clusters, but these are much more pronounced in prokaryotes (operons). There are a few repetitive sequences in prokaryotes, but these are much more common in eukaryotes. Both contain informational sequences, but those of eukaryotes are often interrupted (introns).

12. In all likelihood, an organism's genome will probably come to encompass all genetic elements that can be shown to be stable cellular inhabitants.

14. The diverse cell types of multicellular eukaryotes require a variety of gene products that may be related to the increase in DNA content per cell. In addition, the advantage of diploidy automatically increases DNA content per cell. However, seeing the question in another way, it is likely that a much higher *percentage* of the genome of a prokaryote is actually involved in phenotype production than in a eukaryote. Eukaryotes have evolved the capacity to obtain and maintain what appears to be large amounts of "extra," perhaps "junk," DNA.

In contrast, prokaryotes with their relatively short life cycle, are extremely efficient in their accumulation and use of their genome. Given the larger amount of DNA per cell in eukaryotes and the requirement that the DNA be partitioned in an orderly fashion to daughter cells during cell division, certain mechanisms and structures (mitosis, nucleosomes, centromeres, etc.) have evolved for *packaging* the DNA.

16. The relatively small compact genome of *Arabidopsis* resembles those of *Drosophila* and *C. elegans*. Both coding and noncoding DNA loss can account for the compact genome of *Arabidopsis*. Compared with rice, maize, and barley, there are fewer "gene-empty" repetitive DNA regions in intergenic areas and many copies of transposable elements (intergenic also) have been lost.

18. The human genome is composed of over 3 billion nucleotides in which about 5 percent code for genes. At least 50 percent of the genome is derived from transposable elements. Genes are unevenly distributed over chromosomes with clusters of gene-rich regions separated by gene-poor ones (deserts). Human genes tend to be larger and contain more and larger introns than invertebrates such as *Drosophila*. Hundreds of genes have been transferred from bacteria into vertebrates. Duplicated regions are common, which may facilitate chromosomal rearrangement. The human genome appears to contain approximately 20,000 to 25,000 protein-coding genes.

20. Assuming that the APS strain is the ancestral strain, the remaining strains appear to have smaller genome sizes, indicating genome reduction. The smallest *Buchnera* genome is approximately 448 kb compared to a genome size of *M. genitalium* of about 600 kb with about 480 protein-coding genes. The APS genome codes for about 564 genes in its 641 kb genome. A gene is coded every 1136 bp (641,000/564) for the APS strain and every 1250 bp (600,000/480) for *M. genitalium*. Given these data, the CCE species should code for approximately: (448,000/1193) = 375.5 genes. (*Note:* 1193 was obtained as the average gene spacing of the two bacterial species mentioned above.) Using these calculations, the CCE strain would contain fewer genes than *M. genitalium*.

There are other possible approaches to determine minimum genome size to sustain life. Among them would include computational studies whereby one might estimate the number of essential chemical reactions that are needed for life. Another would be to take an organism with a small number of genes and then systematically mutate genes to see if elimination of genes caused reduced survival. By eliminating individual and groups of genes by mutation, the minimum number might be obtainable.

22. 10 V $\times$ 30 D $\times$ 50 J $\times$ 3 C = 45,000.

24. Selection for gene order might occur as a result of any one or combination of the following: functional and/or structural interaction of proteins coded in a gene cluster, similar localization of gene transcripts in a cell, and operon structure or common regulation. In addition, horizontal gene transfer resulting from transformation and/or transduction disseminates gene combinations in forms different from those generated by typical vertical transfer (parent to offspring).

26. Pseudogenes are nonfunctional versions of genes that resemble gene sequences but contain significant nucleotide changes that prevent their expression. They are formed by gene duplication and subsequent mutation.

28. Since the *in vivo* function of such a protein is determined by secondary and tertiary structures as well as local surface chemistries in active or functional sites, the nonidentical sequences may have considerable influence on function. A number of other factors suggesting different functions include associations with other molecules (cytoplasmic, membrane, or extracellular), chemical nature and position of binding domains, post-translational modification, and signal sequences.

Chapter 19 Applications and Ethics of Genetic Engineering

2. Kleter and Peijnenburg used the BLAST tool from the **http://www.ncbi.nlm.nih.gov/BLAST** website to conduct a series of alignment comparisons of transgenic sequences with sequences of known allergenic proteins. Of 33 transgenic proteins screened for identities of at least six contiguous amino acids found in allergenic proteins, 22 gave positive results.

4. (a) Both the saline and column extracts of Lkt50 appear to be capable of inducing at least 50 percent neutralization of toxicity when injected into rabbits. (b) In order for a successful edible vaccine to be developed, numerous hurdles must be overcome. First, the immunogen must be stably incorporated into the host plant hereditary material, and the host must express only that immunogen. During feeding, the immunogen must be transported across the intestinal wall unaltered, or altered in such a way as to stimulate the desired immune response. There must be guarantees that potentially harmful by-products of transgenesis have not been produced. In other words, broad ecological and environmental issues must be addressed to prevent a transgenic plant from becoming an unintended vector for harm to the environment or any organisms feeding on the plant (directly or indirectly).

6. Enhancement therapy using gene products benefits from the application of modern biotechnology without being burdened by alteration of the genome. Enhancement gene therapy, however, opens the door to a variety of ethical issues. What limits can/should be imposed on individuals or institutions seeking to improve human qualities? What qualities should be open for enhancement? Gene therapy is not without medical and ethical risks.

8. Somatic gene therapy involves attempts to alter the genetic material in non-germ-line cells. Clinical trials are currently underway. Germ-line therapy, though certainly more efficient (while perhaps more difficult technically), alters the germ line and is transmitted to offspring. There are considerable ethical problems associated with germ plasm therapy. It recalls previous attempts of the eugenics movements of past decades which

involved the use of selective breeding to purify the human stock. Some present-day biologists have said publicly that germ-line gene therapy will *not* be conducted.

10. p53 and pRB are tumor suppressor proteins and are required by the cell to effectively monitor the cell cycle. Reduction in their activity would diminish normal cell-cycle controls and most likely lead to cancer.

12. (a) A microarray is a solid support containing an orderly arrangement of DNA samples. A typical array contains thousands of DNA spots that may be small oligonucleotides, cDNAs, or short genomic sequences. Labeled sequences hybridize to the immobilized DNAs by standard base pairing. Such technology allows a method for monitoring RNA expression levels of thousands of genes in virtually any cell population.

   (b) Using microarray technology, researchers can observe the overall behavior of the genome in cancer and normal cells and by comparison, determine which genes are active or inactive under various circumstances. It is possible to identify the set of genes whose expression or lack thereof defines the properties of each tumor type. This application can therefore lead to precise diagnosis and refine possible therapies. In addition, microarray profiling can be used to determine the efficacy of particular therapies.

14. *Drosophila* is a unique experimental organism in that there is a vast knowledge of its genetics, it is easily cultured and genetically manipulated, and it contains unique chromosomes, polytene chromosomes, which allow visual landmarks. Coupled with probe-labeling (sequence tagged sites), the visible landmarks (chromomeres) and ease of manipulation, one can actually see where important genes are located in chromosomes. *Drosophila* also contains P elements that allow sequence markers to be inserted into the genome. Microdissection of chromosomes is also useful in developing specific clones for sequencing. In addition, techniques have been developed (*in situ* hybridization) to allow scientists to actually determine the distributions of gene activities in all tissues of the organism.

16. With both parents heterozygous, each child born will have a 25 percent chance of developing CF.

18. In general, bacteria do not process eukaryotic proteins in the same manner as eukaryotes. Transgenic eukaryotes are more likely to correctly process eukaryotic proteins, thus increasing the likelihood of their normal biological activity.

20. The answer provided here is based on the condition that individual I-2 is a carrier and the son, II-4, has the disorder. The 3 kb fragment occurs in the normal I-1 father and the normal son II-1. The affected son, II-4, has the 4 kb fragment. One daughter, II-2, is a carrier while the other daughter, II-3, is not a carrier.

22. The first problem is the localization of the introduced DNA into the target tissue and target location in the genome. Inappropriate targeting may have serious consequences. In addition, it is often difficult to control the output of introduced DNA. Genetic regulation is complicated and subject to a number of factors, including upstream and downstream signals, as well as various post-transcriptional processing schemes. Artificial control of these factors will prove difficult.

24. Using restriction enzyme analysis to detect point mutations in humans is a tedious, trial-and-error, process. Given the size of the human genome in terms of base sequences and the relatively low number of unique restriction enzymes, the likelihood of matching a specific point mutation, separate from other normal sequence variations, to a desired gene is low.

Chapter 20 Developmental Genetics

2. Because the egg represents an isolated, "closed" system that can be mechanically, environmentally, and to some extent biochemically manipulated, various conditions may be developed that allow one to study facets of gene regulation.

4. (a-d) Genes that control early development are often dependent on the deposition of their products (mRNA, transcription factors, various structural proteins, etc.) in the egg by the mother. Such maternal-effect genes control early events such as defining anterior-posterior polarity. Such products are placed in eggs during oogenesis and are activated immediately after fertilization. The phenotypes of maternal-effect mutations vary from determination of general body plan to eye pigmentation and direction of shell coiling.

6. (a,b) Zygotic genes are activated or repressed depending on their response to maternal-effect gene products. Three subsets of zygotic genes divide the embryo into segments. These segmentation genes are normally transcribed in the developing embryo, and their mutations have embryonic lethal phenotypes. The maternal genotype contains zygotic genes, and these are passed to the embryo as with any other gene.

8. Because the polar cytoplasm contains information to form germ cells, one would expect such a transplantation procedure to generate germ cells in the anterior region. Work done by Illmensee and Mahowald in 1974 verified this expectation.

10. First, one may determine whether levels of hnRNA are consistent among various cell types of interest. If the hnRNA pools for a given gene are consistent in various cell types, then transcriptional control can be eliminated as a possibility. Support for translational control can be achieved directly by determining, in different cell types, the presence of a variety of mRNA species with common sequences. This can be accomplished only in cases where sufficient knowledge exists for specific mRNA trapping or labeling. Clues as to translational control via alternative splicing can sometimes be achieved by examining the amino acid sequence of proteins. Similarities in certain structural/functional motifs may indicate alternative RNA processing.

12. The "gain-of-function" *Antp* mutation causes the wild-type *Antennapedia* gene to be expressed in the head, and mutant flies have legs on the head in place of antenna. Such mutations are often dominant.

14. Homeotic genes encode DNA binding domains that influence gene expression, and any factor that influences gene expression may, under some circumstances, influence cell-cycle control.
   However attractive this model is, there have been no homeotic transformations noted in mammary glands, so the typical expression of mutant homeotic genes in insects is not revealed in mammary tissue according to Lewis (2000). A substantial number of experiments will be needed to establish a functional link between homeotic gene mutation and cancer induction. Mutagenesis and transgenesis experiments are likely to be most productive in establishing a cause-effect relationship.

16. Two coupled approaches might be used. First, one could make transgenic flies that contain a series of deletions spanning all segments of the *bicoid* mRNA: the coding region, 5' and 3' untranslated regions. Comparison of stabilities of individual, deleted mRNAs with controls would indicate whether a particular segment of the mRNA contains a degradation signal sequence. If a degradation-sensitive region or signal sequence is located by deletion, that same intact region, when ligated to a noninvolved, nondegraded mRNA (like a ribosomal protein or tubulin mRNA) should foster degradation in a manner similar to the *bicoid* mRNA.

18. Pre-cell-cycle remodeling:

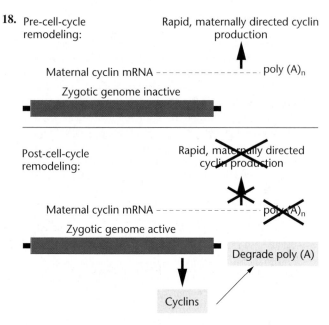

20. It is likely that this gene normally controls the expression of *BX-C* genes in all body segments. The wild-type product of *esc* stored in the egg may be required to interpret the information correctly stored in the egg cortex.

22. Three classes of flower *homeotic* genes are known, which are activated in an overlapping pattern to specify various floral organs. Class *A* genes give rise to sepals. Expression of *A* and *B* class genes specifies petals, *B* and *C* genes control stamen formation, and expression of *C* genes gives rise to carpels.

24. Because signal-receptor interactions depend on membrane-bound structures, the pathway can only work with adjacent cells. The advantage of such a system is that only cells in a certain location will be influenced—those in contact. A disadvantage will occur if large groups of cells are to be induced into a particular developmental pathway or if cells not in contact need to be induced.

26. (a,b) If the *her-1*$^+$ product acts as a negative regulator, then when the gene is mutant, suppression over *tra-1*$^-$ is lost and hermaphroditism will be the result. The double mutant should be male because even though there is no suppression from *her-1*$^-$, there is no *tra-1*$^+$ product to support hermaphrodite development.

Chapter 21 Quantitative Genetics

2. (a) *Polygenes* are those genes involved in determining continuously varying or multiple-factor traits.

(b) *Additive alleles* are those alleles that account for the hereditary influence on the phenotype in an additive way.

(c) *Correlation* is a statistic that varies from -1 to $+1$ and describes the extent to which variation in one trait is associated with variation in another. It does not imply that a cause-and-effect relationship exists between two traits.

(d) *Monozygotic twins* are derived from a single fertilized egg and are thus genetically identical to each other. *Dizygotic twins* arise from two eggs fertilized by two sperm cells. They have the same genetic relationship as siblings.

(e) *Heritability* is a measure of the degree to which the phenotypic variation of a given trait is due to genetic factors.

(f) QTL stands for Quantitative Trait Loci, which are multiple genes that contribute to a quantitative trait.

4. If you add the numbers given for the ratio, you obtain the value of 16, which is indicative of a dihybrid cross. The distribution is that of a dihybrid cross with additive effects.

(a) Because a dihybrid result has been identified, two loci are involved in the production of color. There are two alleles at each locus for a total of four alleles.

(b, c) Because the description of red, medium-red, etc., gives us no indication of a *quantity* of color in any form of units, we would not be able to actually quantify a unit amount for each change in color. We can say that each gene (additive allele) provides an equal unit amount to the phenotype and the colors differ from each other in multiples of that unit amount. The number of additive alleles needed to produce each phenotype is given below:

$$1/16 = \text{dark red} \qquad = AABB$$
$$4/16 = \text{medium-dark red} = 2AABb$$
$$\qquad\qquad\qquad\qquad\qquad\qquad 2AaBB$$
$$6/16 = \text{medium red} \qquad = AAbb$$
$$\qquad\qquad\qquad\qquad\qquad\qquad 4AaBb$$
$$\qquad\qquad\qquad\qquad\qquad\qquad aaBB$$
$$4/16 = \text{light red} \qquad = 2aaBb$$
$$\qquad\qquad\qquad\qquad\qquad\qquad 2Aabb$$
$$1/16 = \text{white} \qquad = aabb$$

(d) $F_1 = $ all light red

$F_2 = 1/4$ medium red

$\qquad 2/4$ light red

$\qquad 1/4$ white

6. (a, b) There are four gene pairs involved.

(c) 3 cm

(d) A typical F_1 cross that produces a "typical" F_2 distribution would be where all gene pairs are heterozygous ($AaBbCcDd$), independently assorting, and additive. Many possible sets of parents that would give an F_1 of this type.

An example for the parents: $\qquad AABBccdd \times aabbCCDD$

(e) Since the *aabbccdd* genotype gives a height of 12 cm and each upper-case allele adds 3 cm to the height, there are many possibilities for an 18 cm plant:

AAbbccdd,

AaBbccdd,

aaBbCcdd, etc.

Any plant with seven upper-case letters will be 33 cm tall:

AABBCCDd,

AABBCcDD,

AABbCCDD, for examples.

8. Ridge count and height have the highest heritability values.

10. Height, general body structure, skin color, and perhaps most common behavioral traits including intelligence

12. (a) *For backfat*:

Broad-sense heritability $= H^2 = 12.2/30.6 = 0.398$

Narrow-sense heritability $= h^2 = 8.44/30.6 = 0.276$

For body length:

Broad-sense heritability $= H^2 = 26.4/52.4 = 0.504$

Narrow-sense heritability $= h^2 = 11.7/52.4 = 0.223$

(b) Of the two traits, selection for back fat would produce more response.

14. (a) For Vitamin A

$\qquad h_A^2 = V_A/V_P = V_A/(V_E + V_A + V_D) = 0.097$

For Cholesterol

$\qquad h_A^2 = 0.223$

(b) Cholesterol content should be influenced to a greater extent by selection.

16. Given that both narrow-sense heritability values are relatively high, it is likely that a farmer would be able to alter both milk protein content and butterfat by selection. The value of 0.91 for the correlation coefficient between protein content and butterfat suggests that if one selects for butterfat, protein content will increase. However, correlation coefficients describe the extent to which variation in one quantitative trait is associated with variation in another and does not reveal the underlying causes of such variation. Assuming that these dairy cows had been selected for high butterfat in the past and increased protein content followed that selection (for butterfat), it is likely that selection for butterfat would continue to correlate with increased protein content. However, there may well be a point where physiological circumstances change and selection for high butterfat may be at the expense of protein content.

18. Given the realized heritability value of 0.4, it is unlikely that selection experiments would cause a rapid and/or significant response to selection. A minor response might result from intense selection.

20. Since the rice plants are genetically identical, V_G is zero and $H^2 = V_G/V_P$ = zero. Broad-sense heritability is a measure to which the phenotypic variance is due to genetic factors. In this case, with genetically identical plants, H^2 is zero, and the variance observed in grain yield is due to the environment. Selection would not be effective in this strain of rice.

22. (a) The most direct explanation would involve two gene pairs, with each additive gene contributing about 1.2 mm to the phenotype.
    (b) The fit to this backcross supports the original hypothesis.
    (c) These data do not support the simple hypothesis provided in part (a).
    (d,e) With these data, one can see no distinct phenotypic classes suggesting that the environment may play a role in eye development or that there are more genes involved.

24. For traits including blood type, eye color, and mental retardation, there is a fairly significant difference between MZ and DZ groups and therefore a high genetic component. However, for measles, the difference is not as significant, indicating a greater role of the environment. Hair color has a significant genetic component, as do idiopathic epilepsy, schizophrenia, diabetes, allergies, cleft lip, and club foot. The genetic component to mammary cancer is present but minimal according to these data.

26. As with many traits that are caused by numerous loci acting additively, some genes have more influence on expression than others. In addition, environmental factors may play a role in the expression of some polygenic traits. In the case of brachydactyly, numerous modifier genes in the genome can influence brachydactyly expression. Examination of OMIM (*Online Mendelian Inheritance of Man*) through **http://www.ncbi.nlm.nih.gov/** will illustrate this point.

28. One would expect a greater response to selection in the wild population.

30. Breeders attempt to "select" out this disorder by first maintaining complete and detailed breeding records of afflicted strains. Second, they avoid breeding dogs whose close relatives are afflicted. The molecular-developmental mechanism that causes the "month of birth" effect in canine hip dysplasia is unknown. However, with many, perhaps all quantitative traits, it is clear that there is a significant environmental influence on both the penetrance and/or expression of the phenotype. With many genes acting in various ways to influence a phenotype, there are opportunities for varied molecular and developmental intraorganismic microenvironments. Stated another way, the longer and more complex the molecular distance from the genome to the phenotype, the greater the likelihood for environmental factors to be involved in expression.

Chapter 22 Population Genetics

2. $p = 0.5$

   $q = 1 - p = 0.5$

   Frequency of AA = 0.25 or 25%

   Frequency of Aa = 0.5 or 50%

   Frequency of aa = 0.25 or 25%

4. In order for the Hardy-Weinberg equations to apply, the population must be in equilibrium.

6. (a) The equilibrium values will be as follows:
   Frequency of $l/l = p^2$ = 0.6014 or 60.14%
   Frequency of $l/\Delta 32 = 2pq$ = 0.3482 or 34.82%
   Frequency of $\Delta 32/\Delta 32 = q^2$ = 0.0504 or 5.04%
   Comparing these equilibrium values with the observed values strongly suggests that the observed values are drawn from a population in equilibrium.
   (b) The equilibrium values will be as follows:

   Frequency of $AA = p^2$ = 0.7691 or 76.91%
   Frequency of $AS = 2pq$ = 0.2157 or 21.57%
   Frequency of $SS = q^2$ = 0.0151 or 1.51%
   Comparing these equilibrium values with the observed values suggests that the observed values may be drawn from a population that is not in equilibrium. Notice that there are more heterozygotes than predicted and fewer SS types. To test for a Hardy-Weinberg equilibrium, apply the chi-square test as follows.

   $$\chi^2 = \frac{\Sigma(o - e)^2}{e}$$
   $$= 1.47$$

   In calculating degrees of freedom in a test of gene frequencies, the "free variables" are reduced by an additional degree of freedom because one estimated a parameter (p or q) used in determining the expected values. Therefore, there is one degree of freedom even though there are three classes. Entering the χ^2 table with 1 degree of freedom gives a value of 3.84 at the 0.05 probability level. Since the χ^2 value calculated here is smaller, the null hypothesis (the observed values fluctuate from the equilibrium values by chance and chance alone) should not be rejected. Thus the frequencies of AA, AS, SS sampled a population that is in equilibrium.

8. The following formula calculates the frequency of an allele in the next generation for any selection scenario, given the frequencies of a and A in this generation and the fitness of all three genotypes.

   $$q_{g+1} = [w_{Aa}p_gq_g + w_{aa}q_g^2]/[w_{AA}p_g^2 + w_{Aa}2p_gq_g + w_{aa}q_g^2]$$

   where q_{g+1} is the frequency of the a allele in the next generation, q_g is the frequency of the a allele in this generation, p_g is the frequency of the A allele in this generation, and each "w" represents the fitness of their respective genotypes.
   (a)
   $$q_{g+1} = [.9(.7)(.3) + .8(.3)^2]/[1(.7)^2 + .9(2)(.7)(.3) + .8(.3)^2]$$
   $\qquad q_{g+1} = 0.278 \qquad p_{g+1} = 0.722$
   (b) $q_{g+1} = 0.289 \quad p_{g+1} = 0.711$
   (c) $q_{g+1} = 0.298 \quad p_{g+1} = 0.702$
   (d) $q_{g+1} = 0.319 \quad p_{g+1} = 0.681$

10. Since a dominant lethal gene is highly selected against, it is unlikely that it will exist at too high a frequency, if at all. However, if the gene shows incomplete penetrance or late age of onset (after reproductive age), it may remain in a population.

12. Given the frequency of the disorder in the population as 1 in 10,000 individuals (0.0001), then $q^2 = 0.0001$, and $q = 0.01$.

The frequency of heterozygosity is *2pq* or approximately 0.02 as also stated in the problem. The probability for one of the grandparents to be heterozygous would therefore be 0.02 + 0.02 or 0.04 or 1/25. If one of the grandparents is a carrier, then the probability of the offspring from a first-cousin mating being homozygous for the recessive gene is 1/16. Multiplying the two probabilities together gives $1/16 \times 1/25 = 1/400$.

Following the same analysis for the second-cousin mating gives $1/64 \times 1/25 = 1/1600$. Notice that the population at large has a frequency of homozygotes of 1/10,000; therefore, one can easily see how inbreeding increases the likelihood of homozygosity.

14. Because heterozygosity tends to mask expression of recessive genes that may be desirable in a domesticated animal or plant, inbreeding schemes are often used to render strains homozygous so that such recessive genes can be expressed. In addition, assume that a particularly desirable trait occurs in a domesticated plant or animal. The best way to increase the frequency of individuals with that trait is by self-fertilization (not often possible) or by matings to blood relatives (inbreeding). In theory, one increases the likelihood of a gene "meeting itself" by various inbreeding schemes. There are disadvantages to increasing the degree of homozygosity by inbreeding. *Inbreeding depression* is a reduction in fitness often associated with an increase in homozygosity.

16. The quickest way to generate a homozygous line of an organism is to *self-fertilize* that organism. Because this is not always possible, brother-sister matings are often used.

18. A population is in equilibrium when the distribution of genotypes occurs at or around the $p^2 + 2pq + q^2$ expression. Equilibrium does not mean 25% *AA*, 50% *Aa*, and 25% *aa*. This confusion often stems from the 1:2:1(or 3:1) ratio seen in Mendelian crosses.

20. The overall probability of the couple producing a CF child is $98/2500 \times 2/3 \times 1/4$.

22. The frequency of the *b* allele after one generation of corn borers fed on Bt corn would be computed as follows:

$q' = q(1 - sq)/(1 - sq^2)$

$q' = 0.98[1 - (0.9)(0.98)]/1 - [(0.9)(0.98)(0.98)]$

$q' = 0.852$

Therefore, $p' = 0.148$ (*B* allele)

24. The equation for determining the impact of immigration on the gene pool of an existing population is estimated by the following equation: $p_i' = (1 - m)p_i + mp_m$

$p_i' = (1 - 0.2)(0.75) + (0.2)(1.0)$

$p_i' = 0.8$

26. Given small populations and very similar environmental conditions, it is more likely that "sampling error" or genetic drift is operating. Since the same gene is behaving differently under similar environmental conditions, selection is an unlikely explanation.

Chapter 23 Evolutionary Genetics

2. Speciation is the process that leads to the formation of species. Evolution is the change in a population over time. Speciation is one of many results of evolution.

4. Organisms may appear to be similar but be reproductively isolated for a sufficient period to justify their species identity. If significant genetic differences occur, they can be considered separate species.

6. The subterranean niche is less broad and less dynamic than above ground. The underground microhabitat consists of a narrower range of climatic changes, thus genetic polymorphism is not selected for. This conclusion has been supported by additional studies indicating that genetic diversity is positively correlated with niche width.

8. Results from laboratory studies indicated that there was a selective advantage in having the two inversions present rather than either one. Thus, natural selection favored the maintenance of both inversions over the loss of either.

10. (a) Missense mutations cause amino acid changes.
  (b) Horizontal transfer refers to the process of passing genetic information from one organism to another without producing offspring. In bacteria, plasmid transfer is an example of horizontal transfer.
  (c) The fact that none of the isolates shared identical nucleotide changes indicates that there is little genetic exchange among different strains. Each alteration is unique, most likely originating in an ancestral strain and maintained in descendents of that strain only.

12. The approximate similarity of mutation rates among genes and lineages should provide more credible estimates of divergence times of species and allow for broader interpretations of sequence comparisons. It also provides for increased understanding of the mutational processes that govern evolution among mammalian genomes. For instance, if the rate of mutation is fairly constant among lineages or cells that have a more rapid turnover, it indicates that replication-related errors do not make a significant contribution to mutation rates.

14. In general, speciation involves the gradual accumulation of genetic changes to a point where reproductive isolation occurs. Depending on environmental or geographic conditions, genetic changes may occur slowly or rapidly. They can involve point or chromosomal changes.

16. Reproductive isolating mechanisms are grouped into the categories Prezygotic and Postzygotic and include those listed below. See the text for specific examples and illustrations.
  • Geographic or ecological
  • Seasonal or temporal
  • Behavoral
  • Mechanical
  • Physiological
  • Hybrid inviability or weakness
  • Developmental hybrid sterility
  • Segregational hybrid sterility
  • F_2 breakdown

18. Polyploid plants that result from hybridization of two species would be expected to be more heterozygous than the diploid parental species because two distinct genomes are combined. Generally, genetic variation is an advantage unless a significant degree of that variation is outside acceptable physiological tolerance.

20. Somatic gene therapy, like any therapy, allows some individuals to live more normal lives than those not receiving therapy. As such, the ability of such individuals to contribute to the gene pool increases the likelihood that less fit genes will enter and be maintained in the gene pool. This is a normal consequence of therapy, genetic or not, and in the face of disease control and prevention, societies have generally accepted this consequence. Germ-line therapy could, if successful, lead to limited, isolated, and infrequent removal of a gene from a gene lineage. However, given the present state of the science, its impact on the course of human evolution will be diluted and negated by a host of other factors that afflict humankind.

22. Approach this problem by writing the possible codons for all the amino acids (except Arg and Asp, which show no change) in the human cytochrome c chain. Then determine the minimum number of nucleotide substitutions required for each changed amino acid in the various organisms. Once listed, count up the numbers for each organism: horse, 3; pig, 2; dog, 3; chicken, 3; bullfrog, 2; fungus, 6.

24. The classification of organisms into different species is based on evidence (morphological, genetic, ecological, etc.) that they are reproductively isolated. That is, there must be evidence that gene flow does not occur among the groups being called different species. Classifications above the species level (genus, family, etc.) are not based on such empirical data. Indeed, classification above the species level is somewhat arbitrary and based on traditions that extend far beyond DNA sequence information. In addition, recall that DNA sequence divergence is not always directly proportional to morphological, behavioral, or ecological divergence. While the genus classifications provided in this problem seem to be invalid, other factors, well beyond simple DNA sequence comparison, must be considered in classification practices. As more information is gained on the meaning of DNA sequence differences (ΔT_m) in comparison to morphological factors, many phylogenetic relationships will be reconsidered, and it is possible that adjustments will be needed in some classification schemes.

26. (a) Since noncoding genomic regions are probably silent genetically, it is likely that they contribute little, if anything, to the phenotype. Selection acts on the phenotype; therefore, such noncoding regions are probably selectively neutral. (b) These polymorphism data indicate that all the Lake Victoria area (lake and contributing rivers) cichlids are related by recent ancestry, whereas those from neighboring lakes are more distantly related. In addition, since Lake Victoria dried out about 14,000 years ago, it is likely that it was repopulated by a relatively small sample of cichlids.

28. The pattern of genetic distances through time indicates that from the present to about 25,000 years ago, modern humans and Cro-Magnons show an approximately constant number of differences. Conversely, there is an abrupt increase in genetic distance seen in comparing modern humans and Cro-Magnons with Neanderthals. The results indicate a clear discontinuity between modern humans, Cro-Magnons, and Neanderthals with respect to genetic variation in the mitochondrial DNAs sampled. Assuming that the sampling and analytical techniques used to generate the data are valid, it appears that Neanderthals made little, if any, genetic contributions to the Cro-Magnon or modern European gene pool. It could be argued that the absence of Neanderthal mtDNA lineages in living humans is a consequence of random drift or lineage extinction since the disappearance of Neanderthals. However, the examination of mtDNA in ancient Cro-Magnon mtDNA shows no evidence of a historical relationship and suggests that Neanderthals were not genetically related to the ancestors of modern humans.

Chapter 24   Conservation Genetics

2. First, if detailed records are kept of the breeding partners of the captive birds, then knowledge of heterozygotes should be available. Breeding programs could be established to restrict matings between those carrying the lethal gene. Such "kinship management" is often used in captive populations. If kinship records are not available, it is often possible to establish kinship using genetic markers such as DNA microsatellite polymorphisms. Using such markers, one can often identify mating partners and link them to their offspring.

By coupling knowledge of mating partners with the likelihood of producing a lethal genetic combination, selective matings can often be used to minimize the influence of a deleterious gene. In addition, such markers can be used to establish matings that optimize genetic mixing, thus reducing inbreeding depression.

4. Both genetic drift and inbreeding tend to drive populations toward homozygosity. Genetic drift is more common when the effective breeding size of the population is low. When this condition prevails, inbreeding is also much more likely. They are different in that inbreeding can occur when certain population structures or behaviors favor matings between relatives, regardless of the effective size of the population. Inbreeding tends to increase the frequency of both homozygous classes at the expense of the heterozygotes. Genetic drift can lead to fixation of one allele or the other, thus producing a single homozygous class.

6. Inbreeding depression, over time, reduces the level of heterozygosity, usually a selectively advantageous quality of a species. When homozygosity increases (through loss of heterozygosity), deleterious alleles are likely to become more of a load on a population. Outbreeding depression occurs when there is a reduction in fitness of progeny from genetically diverse individuals. It is usually attributed to offspring being less well-adapted to the local environmental conditions of the parents. Even though forced outbreeding may be necessary to save a threatened species, where population numbers are low, it significantly and permanently changes the genetic makeup of the species.

8. Often, molecular assays of overall heterozygosity can indicate the degree of inbreeding and/or genetic drift. An allele whose frequency is dictated by inbreeding will not be uniquely influenced. That is, other alleles would be characterized by decreased heterozygosity as well. So, if the genome in general has a relatively high degree of heterozygosity, the gene is probably influenced by selection rather than inbreeding and/or genetic drift.

10. Generally, threatened species are captured and bred in an artificial environment until sufficient population numbers are achieved to ensure species survival. Next, genetic management strategies are applied to breed individuals in such a way as to increase genetic heterozygosity as much as possible. If plants are involved, seed banks are often used to facilitate long-term survival.

12. Allozymes are variants of a given allele often detected by electrophoresis. Such variation may or may not impact on the fitness of an individual. The greater the allozyme variation, the more genetically heterogeneous the individual. It is generally agreed that such genetic diversity is essential for long-term survival. All other factors being equal, allozyme variation is more likely to reflect physiological variation than RFLP variation because RFLP regions are not necessarily found in protein-coding regions of the genome. RFLP allows one to detect very small amounts of genetic diversity in a population and is unlikely to encounter an organism that is not in some way variable compared to other organisms (within and among species).

14. (a) 0.396 (b) 0.886

Therefore there is a loss of approximately 11.4 percent heterozygosity after five generations.

16. (a, b) First, it will be necessary to determine whether the native habitat in the Asian steppes of the 1920s is suitable to any introduction. If the original range is supportive of reintroduction, care must be taken to introduce horses with maximum genetic diversity possible. To do so one might monitor RFLP patterns. Since the founder breeding group included a domestic mare, it may be desirable to select those for reintroduction which are least like the domestic mare genetically. It might be desirable to release

reasonably sized breeding groups in separate locations within the range to enhance eventual genetic diversity.

18. From a physiological standpoint, cryogenic preservation in liquid nitrogen can allow 100 years or more of seed storage for some species. However, such elaborate storage can only be offered to a small fraction of the world's seeds. Thus, seeds of most species undergo storage loss, which decreases genetic diversity. Seeds of tropical plants are somewhat intolerant to cold storage and must be regenerated frequently. Such a practice is prone to a loss of genetic diversity arising from genetic drift. Only a finite number of seeds can be used in each regeneration procedure, and the restriction of sample size (often fewer than 100 plants) reduces genetic diversity. To somewhat counteract this problem, plants are grown under optimum conditions to reduce selection. Another problem associated with preserved seeds is the accumulation of deleterious mutations both as a result of seed storage and regeneration. Some studies indicate increased frequencies of chromosomal and mtDNA lesions, chlorophyll deficiency mutations, and decreased DNA polymerase activity associated with long-term seed storage.

20. The longest bottleneck-to-present interval occurred with cheetahs, and one would expect cheetahs to show the highest degree of microsatellite polymorphism. The shortest bottleneck-to-present interval occurred with the Gir Forest lions so it would be expected to have the least polymorphism. Data from Driscoll et al. (2002 Genome Research 12: 414–423) include the following estimates of microsatellite polymorphism in the three feline groups mentioned above: cheetahs (84.1%), pumas (42.9%), and Gir Forest lions (19.3%).

22. (a) The species with the greatest genetic variability, as estimated by these markers, is the domesticated cat. Domesticated cats share in an immense and variable gene pool. Their staggering numbers and outbreeding behaviors allow them to maintain a high degree of genetic variability.

(b) The lion has the least genetic variability.

(c) Since allozymes code for proteins and proteins often provide a significant function in an organism, selection is stronger and mutations are less tolerated. Selection would be expected to be more harsh on DNA segments that are related to function. In addition, by their very nature, microsatellites and minisatellites are more mutable.

24. While flagship species (often large mammals) may make it possible to gather considerable public support and funding, they may reduce support for indicator species that may have a greater impact on a community of species. Primary producers (plants) are a necessary component of a diverse and supportive habitat. If one focuses on a flagship species within an area, it is possible that other areas will suffer more dramatically because foundational species are lost. Using umbrella species to protect a large geographic area in hopes of protecting other species in that area is a reasonable approach. However, the size of an area is not necessarily a primary factor in determining species success. Diversity and productivity of a habitat are major contributors to species success. Since land is at a premium, it may be wiser in the long run to select umbrella species in diverse and productive habitats rather than on the basis of land size. By selecting sets of species that show considerable biodiversity, one increases the likelihood of protecting a sufficiently rich habitat to support many species. Such habitats are often of considerable economic value, thereby making their availability limited.

Glossary

abortive transduction An event in which transducing DNA fails to be incorporated into the recipient chromosome. See also *transduction*.

acentric chromosome Chromosome or chromosome fragment with no centromere.

acquired immunodeficiency syndrome (AIDS) An infectious disease caused by a retrovirus named the human immunodeficiency virus (HIV). The disease is characterized by a gradual depletion of T lymphocytes, recurring fever, weight loss, multiple opportunistic infections, and rare forms of pneumonia and cancer associated with collapse of the immune system.

acridine dyes A class of organic compounds that bind to DNA and intercalate into the double-stranded structure, producing local disruptions of base pairing. These disruptions result in nucleotide additions or deletions in the next round of replication.

acrocentric chromosome Chromosome with the centromere located very close to one end. Human chromosomes 13, 14, 15, 21, and 22 are acrocentric.

active site That portion of a protein, usually an enzyme, whose structural integrity is required for function (e.g., the substrate binding site of an enzyme).

adaptation A heritable component of the phenotype that confers an advantage in survival and reproductive success. The process by which organisms adapt to the current environmental conditions.

additive genes See *polygenic inheritance*.

additive variance Genetic variance that is attributed to the substitution of one allele for another at a given locus. This variance can be used to predict the rate of response to phenotypic selection in quantitative traits.

A-DNA An alternative form of the right-handed double-helical structure of DNA in which the helix is more tightly coiled, with 11 base pairs per full turn of the helix instead of 10. In the A form, the bases in the helix are displaced laterally and tilted in relation to the longitudinal axis. It is not yet clear whether this form has biological significance.

albinism A condition caused by the lack of melanin production in the iris, hair, and skin. It is most often inherited as an autosomal recessive trait in humans and can be caused by mutations at a number of loci.

aleurone layer In seeds, the outer layer of the endosperm.

alkaptonuria An autosomal recessive condition in humans caused by the lack of the enzyme homogentisic acid oxidase. Urine of homozygous individuals turns dark upon standing because of oxidation of excreted homogentisic acid. The cartilage of homozygous adults blackens from deposition of a pigment derived from homogentisic acid. Affected individuals often develop arthritic conditions.

allele One of the possible mutational states of a gene, distinguished from other alleles by phenotypic effects.

allele frequency Measurement of the proportion of individuals in a population carrying a particular allele.

allele-specific oligonucleotide (ASO) Synthetic nucleotides, usually 15 to 20 bp in length, that under carefully controlled conditions will hybridize only to a perfectly matching complementary sequence. Under these conditions, ASOs with a one-nucleotide mismatch will not hybridize.

allelic exclusion In a plasma cell heterozygous for an immunoglobulin gene, the selective action of only one allele.

allelism test See *complementation test*.

allolactose A lactose derivative that acts as the inducer for the *lac* operon.

allopatric speciation Process of speciation associated with geographic isolation.

allopolyploid Polyploid condition formed by the union of two or more distinct chromosome sets with a subsequent doubling of chromosome number.

allosteric effect Conformational change in the active site of a protein brought about by interaction with an effector molecule.

allotetraploid Diploid for two genomes derived from different species.

allozyme An allelic form of a protein that can be distinguished from other forms by electrophoresis.

alpha fetoprotein (AFP) A 70-kDa glycoprotein synthesized during embryonic development by the yolk sac. High levels of this protein in the amniotic fluid are associated with neural tube defects such as spina bifida; lower-than-normal levels may be associated with Down syndrome.

alternative splicing Generation of different protein molecules from the same pre-mRNA by changing the number and order of exons in the mRNA product.

***Alu* sequence** An interspersed DNA sequence of approximately 300 bp found in the genome of primates that is cleaved by the restriction enzyme *Alu*I. These sequences are composed of a head-to-tail dimer. The first monomer is approximately 140 bp and the second, approximately 170 bp. In humans, they are dispersed throughout the genome and are present in 300,000 to 600,000 copies, constituting some 3 to 6 percent of the genome. See *short interspersed elements*.

amber codon The codon UAG, which does not code for an amino acid but for chain termination.

Ames test An assay developed by Bruce Ames to detect mutagenic and carcinogenic compounds, using reversion to histidine independence in the bacterium *Salmonella typhimurium*.

amino acid Any of the subunit building blocks that are covalently linked to form proteins.

aminoacyl tRNA Covalently linked combination of an amino acid and a tRNA molecule.

amniocentesis A procedure used to test for fetal defects in which fluid and fetal cells are withdrawn from the amniotic layer surrounding the fetus.

amphidiploid See *allotetraploid*.

anabolism The metabolic synthesis of complex molecules from less complex precursors.

anagenesis The transformation of one species into another species over evolutionary time.

analog A chemical compound structurally similar to another, but differing by a single functional group (e.g., 5-bromodeoxyuridine is an analog of thymidine).

anaphase Stage of cell division in which chromosomes begin moving to opposite poles of the cell.

anaphase I The stage in the first meiotic division during which homologous chromosome separate from one another.

aneuploidy A condition in which the chromosome number is not an exact multiple of the haploid set.

angstrom (Å) Unit of length equal to 10^{-10} meters.

annotation Analysis of genomic nucleotide sequence data to identify the protein-coding genes, the nonprotein-coding genes, their regulatory sequences, and their function(s).

antibody Protein (immunoglobulin) produced in response to an antigenic stimulus with the capacity to bind specifically to an antigen.

anticipation A phenomenon first observed in myotonic dystrophy, where the severity of the symptoms increases from generation to generation and the age of onset decreases from generation to generation. This phenomenon is caused by the expansion of trinucleotide repeats within or near a gene.

anticodon The nucleotide triplet in a tRNA molecule that is complementary to and binds with the codon triplet in an mRNA molecule.

antigen A molecule, often a cell-surface protein, that is capable of eliciting the formation of antibodies.

antiparallel Describing molecules in parallel alignment but running in opposite directions. Most commonly used to describe the opposite orientations of the two strands of a DNA molecule.

apoptosis A genetically controlled program of cell death, activated as part of normal development or as a result of cell damage.

ascospore A meiotic spore produced in certain fungi.

ascus In fungi, the sac enclosing the four or eight ascospores.

asexual reproduction Production of offspring in the absence of any sexual process.

assortative mating Nonrandom mating between males and females of a species. Selection of mates with the same genotype is positive; selection of mates with opposite genotypes is negative.

ATP Adenosine triphosphate. A nucleotide that is the main energy source in cells.

attached-X chromosome Two conjoined X chromosomes that share a single centromere.

attenuator A nucleotide sequence between the promoter and the structural gene of some operons that regulates the transit of RNA polymerase, reducing transcription of the related structural gene.

autogamy A process of self-fertilization resulting in homozygosis.

autoimmune disease The production of antibodies that results from an immune response to one's own molecules, cells, or tissues. Such a response results from the inability of the immune system to distinguish self from nonself. Diseases such as arthritis, scleroderma, systemic lupus erythematosus, and juvenile-onset diabetes are examples of autoimmune diseases.

autonomously replicating sequences (ARS) Origins of replication, about 100 nucleotides in length, found in yeast chromosomes. ARS elements are also present in organelle DNA.

autopolyploidy Polyploid condition resulting from the duplication of one diploid set of chromosomes.

autoradiography Production of a photographic image by radioactive decay; used to localize radioactively labeled compounds within cells and tissues.

autosomes Chromosomes other than the sex chromosomes. In humans, there are 22 pairs of autosomes.

autotetraploid An autopolyploid condition composed of four similar genomes. In this situation, genes with two alleles (*A* and *a*) can have five genotypic classes: *AAAA* (quadraplex), *AAAa* (triplex), *AAaa* (duplex), *Aaaa* (simplex), and *aaaa* (nulliplex).

auxotroph A mutant microorganism or cell line that requires a substance for growth that can be synthesized by wild-type strains.

backcross A cross involving an F_1 heterozygote and one of the P_1 parents (or an organism with a genotype identical to one of the parents).

bacteriophage A virus that infects bacteria (also, *phage*).

bacteriophage λ A member of the lambdoid family of viruses that attach to, infect, and replicate within bacterial cells, destroying the host cell in the process. Genetically modified lambda phages are used as vectors in recombinant DNA research.

bacteriophage μ A group of phages whose genetic material behaves as an insertion sequence that can inactivate host genes and rearrange host chromosomes.

balanced lethals Recessive, nonallelic lethal genes, each carried on different homologous chromosomes. When organisms carrying balanced lethal genes are interbred, only organisms with genotypes identical to the parents (heterozygotes) survive.

balanced polymorphism Genetic polymorphism maintained in a population by natural selection.

Barr body Densely staining nuclear mass seen in the somatic nuclei of mammalian females. Discovered by Murray Barr, this body represents an inactivated X chromosome.

base analog See *analog*.

base substitution A single-base change in a DNA molecule that produces a mutation. There are two types of substitutions: *transitions*, in which a purine is substituted for a purine or a pyrimidine for a pyrimidine; and *transversions*, in which a purine is substituted for a pyrimidine, or vice versa.

β-galactosidase A bacterial enzyme encoded by the *lacZ* gene that converts lactose into galactose and glucose.

bidirectional replication A mechanism of DNA replication in which two replication forks move in opposite directions from a common origin of replication.

biodiversity The genetic diversity present in populations and species of plants and animals.

bioinformatics The use of computer hardware and software in the storage, management, and visualization of biological information such as genome sequence data.

biometry The application of statistics and statistical methods to biological problems.

biotechnology Commercial and/or industrial processes that utilize biological organisms or products.

bivalents Synapsed homologous chromosomes in the first prophase of meiosis.

Bombay phenotype A rare variant of the ABO system in which affected individuals do not have A or B antigens and thus appear as blood type O, even though their genotype may carry unexpressed alleles for the A and/or B antigens.

bottleneck Fluctuation in allele frequency that occurs when a population undergoes a temporary reduction in size.

BrdU (5-bromodeoxyuridine) A mutagenically active analog of thymidine in which the methyl group at the 5 position in thymine is replaced by bromine; also abbreviated BUdR.

BSE (bovine spongiform encephalopathy) A fatal, degenerative brain disease of cattle caused by prion infection. This infection is transmissible to humans and other animals. Also known as mad cow disease.

buoyant density A property of particles (and molecules) that depends upon their actual density, as determined by partial specific volume and degree of hydration. It provides the basis for density gradient separation of molecules or particles.

CAAT box A highly conserved DNA sequence found in the untranslated promoter region of eukaryotic genes. This sequence is recognized by transcription factors.

CAP Catabolite activator protein; a protein that binds cAMP and regulates the activation of inducible operons.

carcinogen A physical or chemical agent that causes cancer.

carrier An individual heterozygous for a recessive trait.

catabolism A metabolic reaction in which complex molecules are broken down into simpler forms, often accompanied by the release of energy.

catabolite activator protein See *CAP*.

catabolite repression The selective inactivation of an operon by a metabolic product of the enzymes encoded by the operon.

cdc mutation A class of cell division cycle mutations in yeasts that affect the timing of and progression through the cell cycle.

cDNA DNA synthesized from an RNA template by the enzyme reverse transcriptase.

cDNA library A collection of cloned cDNA sequences.

cell cycle Sum of the phases of growth of an individual cell type; divided into G1 (gap 1), S (DNA synthesis), G2 (gap 2), and M (mitosis).

cell-free extract A preparation of the soluble fraction of cells, made by lysing cells and removing the particulate matter, such as nuclei, membranes, and organelles. Often used to carry out the synthesis of proteins by the addition of specific, exogenous mRNA molecules.

CEN In yeasts, fragments of chromosomal DNA, about 120 bp in length, that when inserted into plasmids confer the ability to segregate during mitosis. These segments contain at least three types of sequence elements associated with centromere function.

centimeter (cm) A unit of length equal to 10^{-2} meter.

centimorgan (cM) A unit of distance between genes on chromosomes. One centimorgan represents a value of 1 percent crossing over between two genes.

central dogma The concept that information flow progresses from DNA to RNA to proteins. Although exceptions are known, this idea is central to an understanding of gene function.

centric fusion See *Robertsonian translocation*.

centriole A cytoplasmic organelle composed of nine groups of microtubules, generally arranged in triplets. Centrioles function in the generation of cilia and flagella and serve as foci for the spindles in cell division.

centromere Specialized region of a chromosome to which sister chromatids remain attached after replication and the site to which spindle fibers attach during cell division. Location of the centromere determines the shape of the chromosome during the anaphase portion of cell division. Also known as the primary constriction.

centrosome Region of the cytoplasm containing a pair of centrioles.

chaperone A protein that regulates the folding of a polypeptide into a functional three-dimensional shape.

character An observable phenotypic attribute of an organism.

charon phages A group of genetically modified lambda phages designed to be used as vectors (carriers) for cloning foreign DNA. Named after the ferryman in Greek mythology who carried the souls of the dead across the River Styx.

chemotaxis Negative or positive response to a chemical gradient.

chiasma (pl., chiasmata) The crossed strands of nonsister chromatids seen in diplonema of the first meiotic division. Regarded as the cytological evidence for exchange of chromosomal material, or crossing over.

chi-square (χ^2) analysis Statistical test to determine if an observed set of data fits a theoretical expectation.

chloroplast A cytoplasmic self-replicating organelle containing chlorophyll. The site of photosynthesis.

chorionic villus sampling (CVS) A technique of prenatal diagnosis that intravaginally retrieves chorionic fetal cells and uses them to detect cytogenetic and biochemical defects in the embryo.

chromatid One of the longitudinal subunits of a replicated chromosome; it is joined to its sister chromatid at the centromere.

chromatin The complex of DNA, RNA, histones, and nonhistone proteins that make up uncoiled chromosomes characteristic of the eukaryotic interphase nucleus.

chromatography Technique for the separation of a mixture of solubilized molecules by their differential migration over a substrate.

chromocenter An aggregation of centromeres and heterochromatic elements of polytene chromosomes.

chromomere A coiled, beadlike region of a chromosome most easily visualized during cell division. The aligned chromomeres of polytene chromosomes are responsible for their distinctive banding pattern.

chromosomal aberration Any change resulting in the duplication, deletion, or rearrangement of chromosomal material.

chromosomal mutation See *chromosomal aberration*.

chromosomal polymorphism Alternative structures or arrangements of a chromosome that are carried by members of a population.

chromosome In prokaryotes, one or more DNA molecules containing the genome; in eukaryotes, a DNA molecule complexed with RNA and proteins to form a threadlike structure containing genetic information arranged in a linear sequence and visible during mitosis and meiosis.

chromosome banding Technique for the differential staining of mitotic or meiotic chromosomes to produce a characteristic banding pattern, or selective staining of certain chromosomal regions such as centromeres, the nucleolus organizer regions, and GC- or AT-rich regions—not to be confused with the banding pattern present in polytene chromosomes, which is produced by the alignment of chromomeres.

chromosome map A diagram showing the location of genes on chromosomes.

chromosome puff A localized uncoiling and swelling in a polytene chromosome, usually regarded as a sign of active transcription.

chromosome theory of inheritance The idea put forward independently by Walter Sutton and Theodore Boveri that chromosomes are the carriers of genes and the basis for the Mendelian mechanisms of segregation and independent assortment.

chromosome walking A method for analyzing long stretches of DNA, in which the end of a cloned segment of DNA is subcloned and used as a probe to identify other clones that overlap the first clone.

cis configuration The arrangement of two genes or two mutant sites within a gene on the same homolog, such as

$$\frac{a^1 \quad a^2}{+ \quad +}$$

Contrasts with a *trans* arrangement, where the mutant alleles are located on opposite homologs.

cis–trans test A genetic test to determine whether two mutations are located within the same cistron.

cistron That portion of a DNA molecule coding for a single polypeptide chain; defined by a genetic test as a region within which two mutations cannot complement each other.

cladogenesis The transformation of one species into two distinct and independent daughter species over evolutionary time.

cline A gradient of genotype or phenotype distributed over a geographic range.

clonal selection Theory of the immune system that proposes that antibody diversity precedes exposure to the antigen and that the antigen functions to select the cells containing its specific antibody to undergo proliferation.

clone Identical molecules, cells, or organisms derived from a single ancestor by asexual or parasexual methods, such as a DNA segment that has been enzymatically inserted into a plasmid or chromosome of a phage or a bacterium and replicated to form many copies.

cloned DNA library A collection of cloned DNA molecules representing all or part of an individual's genome.

code See *genetic code*.

codominance Condition in which the phenotypic effects of a gene's alleles are fully and simultaneously expressed in the heterozygote.

codon A triplet of nucleotides that specifies or encodes the information for a single amino acid. Sixty-one codons specify the amino acids used in proteins, and three codons called stop codons signal termination of growth of the polypeptide chain.

coefficient of coincidence A ratio of the observed number of double crossovers divided by the expected number of such crossovers.

coefficient of inbreeding The probability that two alleles present in a zygote are descended from a common ancestor.

coefficient of selection (*s*) A measurement of the reproductive disadvantage of a given genotype in a population. For example, if for genotype *aa* only 99 of 100 individuals reproduce, then the selection coefficient is 0.01.

colchicine An alkaloid compound that inhibits spindle formation during cell division. It is used in the preparation of karyotypes to collect a large population of cells inhibited at the metaphase stage of mitosis.

colinearity The linear relationship between the nucleotide sequence in a gene (or the RNA transcribed from it) and the order of amino acids in the polypeptide chain specified by the gene.

comparative genomics Analysis of genomic information from different species. It is used to study the origin and evolution of organisms and genomes, and the evolution of gene function.

competence In bacteria, the transient state or condition during which the cell can bind and internalize exogenous DNA molecules, making transformation possible.

complementarity Chemical affinity between nitrogenous bases as a result of hydrogen bonding; responsible for the base pairing between the strands of the DNA double helix.

complementation test A genetic test to determine whether two mutations occur within the same gene. If two mutations are introduced into a cell simultaneously and produce a wild-type phenotype (i.e., they complement each other), they are often nonallelic. If a mutant phenotype is produced, the mutations are noncomplementing and are often allelic.

complete linkage A condition in which two genes are located so close to each other that no recombination occurs between them.

complexity The total number of nucleotides or nucleotide pairs in a population of nucleic acid molecules as determined by reassociation kinetics.

complex locus A gene within which a set of functionally related pseudoalleles can be identified by recombinational analysis (e.g., the *bithorax* locus in *Drosophila*).

complex trait A trait whose phenotype is determined by the interaction of multiple genes and environmental factors.

concatemer A chain or linear series of subunits linked together. The process of forming a concatemer is called concatenation (e.g., multiple units of a phage genome produced during replication).

concordance Pairs or groups of individuals identical in their phenotype. In twin studies, a condition in which both twins exhibit or fail to exhibit a trait under investigation.

conditional mutation A mutation that expresses a wild-type phenotype under certain (permissive) conditions and a mutant phenotype under other (restrictive) conditions.

conjugation Temporary fusion of two single-celled organisms for the sexual transfer of genetic material.

consanguineous Related by a common ancestor within the previous few generations.

consensus sequence A basically common, though not necessarily identical, sequence of nucleotides in DNA or amino acids in proteins.

conservation genetics The branch of genetics concerned with the preservation and maintenance of wild species of plants and animals in their natural environments.

continuous variation Phenotype variation exhibited by quantitative traits distributed from one phenotypic extreme to another in an overlapping or continuous fashion.

cosmid A vector designed to allow cloning of large segments of foreign DNA. Cosmids are composed of the *cos* sites of phage λ inserted into a plasmid. In cloning, the recombinant DNA molecules are packaged into phage protein coats, and after infection of bacterial cells, the recombinant molecule replicates and can be maintained as a plasmid.

coupling conformation See *cis configuration*.

covalent bond A nonionic chemical bond formed by the sharing of electrons.

Creutzfeldt-Jakob disease (CJD) A progressive degenerative and fatal disease of the brain and nervous system caused by mutations in the prion protein gene on chromosome 20 that produce aberrant forms of the encoded protein. CDJ is inherited as an autosomal dominant trait.

cri-du-chat syndrome A clinical syndrome in humans produced by a deletion of a portion of the short arm of chromosome 5. Afflicted infants have a distinctive cry that sounds like that of a cat.

crossing over The exchange of chromosomal material (parts of chromosomal arms) between homologous chromosomes by breakage and reunion. The exchange of material between nonsister chromatids during meiosis is the basis of genetic recombination.

cross-reacting material (CRM) Nonfunctional form of an enzyme, produced by a mutant gene, that is recognized by antibodies made against the normal enzyme.

C-terminal amino acid The terminal amino acid in a peptide chain that carries a free carboxyl group.

C terminus The end of a polypeptide that carries a free carboxyl group of the last amino acid. By convention, the structural formula of polypeptides is written with the C terminus at the right.

C **value** The haploid amount of DNA present in a genome.

C **value paradox** The apparent paradox that there is no relationship between the size of the genome and the evolutionary complexity of species. For example, the *C* value (haploid genome size) of amphibians varies by a factor of 100.

cyclic adenosine monophosphate (cAMP) An important regulatory molecule in both prokaryotic and eukaryotic organisms.

cyclins A class of proteins found in eukaryotic cells that are synthesized and degraded in synchrony with the cell cycle and regulate passage through stages of the cycle.

cytogenetics A branch of biology in which the techniques of both cytology and genetics are used to study heredity.

cytokinesis The division or separation of the cytoplasm during mitosis or meiosis.

cytological map A diagram showing the location of genes at particular chromosomal sites.

cytoplasmic inheritance Non-Mendelian form of inheritance involving genetic information transmitted by self-replicating cytoplasmic organelles such as mitochondria and chloroplasts.

cytoskeleton An internal array of microtubules, microfilaments, and intermediate filaments that confers shape and the ability to move on a eukaryotic cell.

dalton (Da) A unit of mass equal to that of the hydrogen atom, which is 1.67×10^{-24} grams. A unit used in designating molecular weights.

Darwinian fitness See *fitness*.

deficiency A chromosomal mutation involving the loss or deletion of chromosomal material.

degenerate code The genetic code, where a given amino acid may be represented by more than one codon. For example, some amino acids (leucine) have six codons, whereas others (isoleucine) have three.

deletion See *deficiency*.

deme A local interbreeding population.

denatured DNA DNA molecules that have been separated into single strands.

de novo Newly arising; synthesized from less complex precursors rather than having been produced by modification of an existing molecule.

density gradient centrifugation A method of separating macromolecular mixtures by the use of centrifugal force and solutions of varying density. In buoyant density gradient centrifugation using cesium chloride, the cesium solution establishes a gradient under the influence of the centrifugal field and a mixture of macromolecules such as DNA sediment in the gradient until the density of the cesium chloride solution equals their own, separating them by differences in density.

deoxyribonuclease A class of enzymes that breaks down DNA into oligonucleotide fragments by introducing single-stranded breaks into the double helix. *See DNase.*

deoxyribonucleic acid (DNA) A macromolecule usually consisting of antiparallel polynucleotide chains held together by hydrogen bonds, in which the sugar residues are deoxyribose. DNA is the primary carrier of genetic information.

deoxyribose The five-carbon sugar associated with the deoxyribonucleotides found in DNA.

dermatoglyphics The study of the surface ridges of the skin, especially of the hands and feet.

determination A regulatory event that establishes a specific pattern of future gene activity and developmental fate for a given cell.

diakinesis The final stage of meiotic prophase I in which the chromosomes become tightly coiled and compacted and move toward the periphery of the nucleus.

dicentric chromosome A chromosome having two centromeres.

dideoxynucleotide A nucleotide containing a dexoyribose sugar lacking a 3′ hydroxyl group. Stops further chain elongation when incorporated into a growing polynucleotide; used in the Sanger method of DNA sequencing.

differentiation The process of complex changes by which cells and tissues attain their adult structure and functional capacity.

dihybrid cross A genetic cross involving two characters in which the parents possess different forms of each character (e.g., yellow, round × green, wrinkled peas).

diploid A condition in which chromosomes exist in pairs; having two of each chromosome.

diplonema A stage of meiotic prophase immediately after pachynema. In the diplotene stage, one pair of sister chromatids begins separating from the other, and chiasmata become visible. These overlaps move laterally toward the ends of the chromatids (terminalization).

directional selection A selective force that changes the frequency of an allele in a given direction, either toward fixation or toward elimination.

discontinuous replication of DNA The synthesis of DNA in discontinuous fragments on the lagging strand of the replication fork. The fragments, known as Okazaki fragments, are joined by DNA ligase to form a continuous strand.

discontinuous variation Phenotypic data that fall into two or more distinct, nonoverlapping classes.

discordance In twin studies, a situation where one twin expresses a trait but the other does not.

disjunction The separation of chromosomes at the anaphase stage of cell division.

disruptive selection Simultaneous selection for phenotypic extremes in a population, usually resulting in the production of two phenotypically discontinuous strains.

dizygotic twins Twins produced from separate fertilization events; two ova fertilized independently. Also known as fraternal twins.

DNA See *deoxyribonucleic acid*.

DNA fingerprinting A molecular method for identifying an individual member of a population or species. The pattern of DNA fragments is obtained by restriction enzyme digestion, followed by Southern blot hybridization using minisatellite probes. See also *STR sequences*.

DNA footprinting See *footprinting*.

DNA gyrase One of the DNA topoisomerases that acts during DNA replication to reduce molecular tension caused by supercoiling. DNA gyrase produces, then seals double-stranded breaks.

DNA ligase An enzyme that forms a covalent bond between the 5′ end of one polynucleotide chain and the 3′ end of another polynucleotide chain. It is also called polynucleotide-joining enzyme.

DNA microarray An ordered arrangement of DNA molecules of known sequence on a substrate, such as glass or silicon. DNA or RNA target molecules labeled with fluorescent dyes are added to arrays, allowing analysis of thousands of target molecules in a single experiment. Also called DNA chips®.

DNA polymerase An enzyme that catalyzes the synthesis of DNA from deoxyribonucleotides and a template DNA molecule.

DNase Deoxyribonuclease; an enzyme that degrades or breaks down DNA into fragments or constitutive nucleotides.

dominance The expression of a trait in the heterozygous condition.

dominant suppression A form of epistasis in which a dominant allele at one locus suppresses the effect of a dominant allele at another locus, resulting in a 13:3 phenotypic ratio.

dosage compensation A genetic mechanism that regulates the levels of gene products at certain loci on the X chromosome in mammals such that males and females have equal amounts of a gene product. In mammals, this is accomplished by random inactivation of one X chromosome.

double crossover Two separate events of chromosome breakage and exchange occurring within the same tetrad.

double helix The model for DNA structure proposed by James Watson and Francis Crick, involving two antiparallel hydrogen-bonded polynucleotide chains wound into a right-handed helical configuration, with 10 base pairs per full turn of the double helix. Often called B-DNA.

Duchenne muscular dystrophy An X-linked recessive genetic disorder caused by a mutation in the gene for dystrophin, a protein found in muscle cells. Affected males show a progressive weakness and wasting of muscle tissue. Death ensues by about age 20 caused by respiratory infection or cardiac failure.

duplication A chromosomal aberration in which a segment of the chromosome is repeated.

dyad The products of tetrad separation or disjunction at the first meiotic prophase. It consists of two sister chromatids joined at the centromere.

dystrophin A protein that attaches to the inside of the muscle cell plasma membrane and stabilizes the membrane during muscle contraction. Mutations in the gene encoding dystrophin cause Duchenne and Becker muscular dystrophy. See *Duchenne muscular dystrophy*.

effective population size The number of individuals in a population with an equal probability of contributing gametes to the next generation.

effector molecule Small, biologically active molecule that regulates the activity of a protein by binding to a specific receptor site on the protein.

electrophoresis A technique that separates a mixture of molecules by their differential migration through a stationary medium (such as a gel) in an electrical field.

endocytosis The uptake by a cell of fluids, macromolecules, or particles by pinocytosis, phagocytosis, or receptor-mediated endocytosis.

endomitosis Chromosomal replication that is not accompanied by either nuclear or cytoplasmic division.

endonuclease An enzyme that hydrolyzes internal phosphodiester bonds in a polynucleotide chain or nucleic acid molecule.

endoplasmic reticulum A membranous organelle system in the cytoplasm of eukaryotic cells. The outer surface of the membranes may be ribosome-studded (rough ER) or smooth ER.

endopolyploidy The increase in chromosome sets that results from endomitotic replication within somatic nuclei.

endosymbiont theory The proposal that self-replicating cellular organelles such as mitochondria and chloroplasts were originally free-living prokaryotic organisms that entered into a symbiotic relationship with eukaryotic nucleated cells.

enhancer Originally identified as a 72-bp sequence in the genome of a virus, SV40, that increases the transcriptional activity of nearby structural genes. Similar sequences that enhance transcription have been identified in the genomes of eukaryotic cells. Enhancers can act over a distance of thousands of base pairs and can be located upstream, downstream, 5', 3', or internal to the gene they affect, and thus are different from promoters.

environment The complex of geographic, climatic, and biotic factors within which an organism lives.

enzyme A protein or complex of proteins that catalyzes a specific biochemical reaction.

epigenesis The idea that an organism develops by the appearance and growth of new structures. Opposed to preformationism, which holds that development is the growth of structures already present in the egg.

episome A circular genetic element in bacterial cells that can replicate independently of the bacterial chromosome or integrate and replicate as part of the chromosome.

epistasis Nonreciprocal interaction between genes such that one gene interferes with or prevents the expression of another gene. In *Drosophila* for example, the recessive gene *eyeless*, when homozygous, prevents the expression of eye color genes present in the genome.

epitope That portion of a macromolecule or cell that acts to elicit an antibody response; an antigenic determinant. A complex molecule or cell can contain several such sites.

equational division A division of each chromosome into longitudinal halves that are distributed into two daughter nuclei. Chromosome division in mitosis is an example of equational division.

equatorial plate See *metaphase plate*.

euchromatin Chromatin or chromosomal regions that are lightly staining and are relatively uncoiled during the interphase portion of the cell cycle. Euchromatic regions contain most of the structural genes.

eugenics The improvement of the human species by selective breeding. Positive eugenics refers to the promotion of breeding of people with favorable genes, and negative eugenics refers to the discouragement of breeding among those with undesirable traits.

eukaryotes Organisms having true nuclei and membranous organelles and whose cells demonstrate mitosis and meiosis.

euphenics Medical or genetic intervention to reduce the impact of defective genotypes.

euploid Polyploid with a chromosome number that is an exact multiple of a basic chromosome set.

evolution The origin of plants and animals from preexisting types; descent with modifications.

excision repair Removal of damaged DNA segments followed by repair. Excision can include the removal of individual bases (base repair) or a stretch of damaged nucleotides (nucleotide repair). The gap created by excision is filled by DNA polymerase, and the ends are ligated to form an intact molecule.

exon (extron) The DNA segments of a gene that are transcribed and translated into proteins.

exonuclease An enzyme that breaks down nucleic acid molecules by breaking the phosphodiester bonds at the 3'- or 5'-terminal nucleotides.

expressed sequence tags (ESTs) All or part of the nucleotide sequence of cDNA clones; used as markers in construction of genetic maps.

expression vector Plasmids or phages carrying promoter regions designed to cause expression of inserted DNA sequences.

expressivity The degree or range in which a phenotype for a given trait is expressed.

extranuclear inheritance Transmission of traits by genetic information contained in cytoplasmic organelles such as mitochondria and chloroplasts.

F^- cell A bacterial cell that does not contain a fertility factor and that acts as a recipient in bacterial conjugation.

F^+ cell A bacterial cell containing a fertility factor and that acts as a donor in bacterial conjugation.

F factor An episome in bacterial cells that confers the ability to act as a donor in conjugation (also, *fertility factor*).

F′ factor A fertility factor that contains a portion of the bacterial chromosome.

F_1 generation First filial generation; the progeny resulting from the first cross in a series.

F_2 generation Second filial generation; the progeny resulting from a cross of the F_1 generation.

F pilus See *pilus*.

familial trait A trait transmitted through and expressed by members of a family.

fate map A diagram of an embryo showing the location of cells whose development fate is known.

fertility factor See *F factor*.

filial generations See *F_1 generation*; *F_2 generation*.

fingerprint The unique pattern of ridges and whorls on the tip of a human finger. Also, the pattern obtained by enzymatically cleaving a protein or nucleic acid and subjecting the digest to two-dimensional chromatography or electrophoresis. See also *DNA fingerprinting*.

FISH See *fluorescence* in situ *hybridization*.

fitness A measure of the relative survival and reproductive success of a given individual or genotype.

fixation In population genetics, a condition in which all members of a population are homozygous for a given allele.

fluctuation test A statistical test developed by Salvadore Luria and Max Delbrück to determine whether bacterial mutations arise spontaneously or are produced in response to selective agents.

fluorescence in situ hybridization (FISH) A method of *in situ* hybridization that utilizes probes labeled with a fluorescent tag, causing the site of hybridization to fluoresce when viewed in ultraviolet light under a microscope.

flush–crash cycle A period of rapid population growth followed by a drastic reduction in population size.

f-met See *formylmethionine*.

folded-fiber model A model of eukaryotic chromosome organization in which each sister chromatid consists of a single fiber, composed of double-stranded DNA and protein, which is wound like a tightly coiled skein of yarn.

footprinting A technique for identifying a DNA sequence that binds to a particular protein, based on the idea that the phosphodiester bonds in the region covered by the protein are protected from digestion by deoxyribonucleases (also, *DNA footprinting*).

formylmethionine (f-met) A molecule derived from the amino acid methionine by attachment of a formyl group to its terminal amino group. This is the first amino acid inserted in all bacterial polypeptides. Also known as *N*-formyl methionine.

founder effect A form of genetic drift. The establishment of a population by a small number of individuals whose genotypes carry only a fraction of the different kinds of alleles in the parental population.

fragile site A heritable gap or nonstaining region of a chromosome that can be induced to generate chromosome breaks.

fragile X syndrome A genetic disorder caused by the expansion of a CGG trinucleotide repeat and a fragile site at Xq27.3 within the *FMR-1* gene. Fragile X syndrome is the most common form of mental retardation. Affected males have distinctive facial features and are mentally retarded. Carrier females lack physical symptoms but as a group have higher rates of mental retardation than normal individuals.

frameshift mutation A mutational event leading to the insertion of one or more base pairs in a gene, shifting the codon reading frame in all codons that follow the mutational site.

fraternal twins See *dizygotic twins*.

functional genomics The study of the functions and interrelationships among all the genes and nongene components in a genome.

G1 checkpoint A point in the G1 phase of the cell cycle when a cell becomes committed to initiate DNA synthesis and continue the cycle or withdraw into the G0 resting stage.

G0 A point in the G1 phase where cells withdraw from the cell cycle and enter a nondividing but metabolically active state.

gamete A specialized reproductive cell with a haploid number of chromosomes.

gap genes Genes expressed in contiguous domains along the anterior–posterior axis of the *Drosophila* embryo that regulate the process of segmentation in each domain.

gene The fundamental physical unit of heredity whose existence can be confirmed by allelic variants and which occupies a specific chromosomal locus. A DNA sequence coding for a single polypeptide.

gene amplification The process by which gene sequences are selected and differentially replicated either extrachromosomally or intrachromosomally.

gene conversion The process of nonreciprocal recombination by which one allele in a heterozygote is converted into the corresponding allele.

gene duplication An event in replication leading to the production of a tandem repeat of a gene sequence.

gene flow The gradual exchange of genes between two populations; brought about by the dispersal of gametes or the migration of individuals.

gene frequency The percentage of alleles of a given type in a population.

gene interaction Production of novel phenotypes by the interaction of alleles of different genes.

gene mutation See *point mutation*.

gene pool The total of all alleles possessed by reproductive members of a population.

gene therapy The transfer of normal genes into organisms carrying a defective allele or alleles of a given gene. Somatic gene therapy uses somatic cells as targets, germ cell therapy transfers genes into germ cells, altering the genetic status of offspring.

generalized transduction The transduction of any gene in the bacterial genome by a phage.

genetically modified organism (GMO) A plant or animal that carries a gene from another species transferred to its genome using recombinant DNA technology, and where the gene is expressed to produce a gene product.

genetic anticipation The phenomenon of a progressively earlier age of onset and increasing severity of symptoms in successive generations for a genetic disorder. See also *anticipation.*

genetic background All genes carried in the genome other than the one being studied.

genetic code The nucleotide triplets that code for the 20 amino acids or for chain initiation or termination.

genetic counseling Analysis of risk for genetic defects in a family and the presentation of options available to avoid or ameliorate possible risks.

genetic drift Random variation in allele frequency from generation to generation, most often observed in small populations.

genetic engineering The technique of altering the genetic constitution of cells or individuals by the selective removal, insertion, or modification of individual genes or gene sets.

genetic equilibrium Maintenance of allele frequencies at the same value in successive generations. A condition in which allele frequencies are neither increasing nor decreasing.

genetic erosion The loss of genetic diversity from a population or a species.

genetic fine structure Intragenic recombinational analysis that provides mapping information at the level of individual nucleotides.

genetic load Average number of recessive lethal genes carried in the heterozygous condition by an individual in a population.

genetic polymorphism The stable coexistence of two or more discontinuous genotypes in a population. When the frequencies of two alleles are carried to an equilibrium, the condition is called *balanced polymorphism.*

genetics The branch of biology that deals with heredity and the expression of inherited traits.

genome The set of genes carried by an individual.

genomic imprinting A condition where the expression of a gene depends on whether the gene has been inherited from a male or a female parent.

genomics The study of genomes, including nucleotide sequence, gene content, organization, and gene number.

genotype The specific allelic or genetic constitution of an organism; often, the allelic composition of one or a limited number of genes under investigation.

germ line An embryonic cell lineage that forms the reproductive cells (eggs and sperm).

germ plasm Hereditary material transmitted from generation to generation.

Goldberg-Hogness box A short nucleotide sequence 20 to 30 bp upstream from the initiation site of eukaryotic genes to which RNA polymerase II binds. The consensus sequence is TATAAAA. Also known as TATA box.

graft-versus-host disease (GVHD) In transplants, reaction by immunologically competent cells of the donor against the antigens present on the cells of the host. It is often a fatal condition in human bone marrow transplants.

green revolution A program that resulted in a two- to threefold increase in crop yields generated by the development of new varieties of cereal plants with shorter stems and increased disease resistance.

gynandromorph An individual composed of cells with both male and female genotypes.

gyrase One of a class of enzymes known as topoisomerases. Gyrase converts closed circular DNA to a negatively supercoiled form prior to replication, transcription, or recombination. See also *DNA gyrase.*

haploid A cell or organism having a single set of unpaired chromosomes; the gametic chromosome number.

haplotype The set of alleles from closely linked loci carried by an individual and usually inherited as a unit.

Hardy-Weinberg law The principle that both gene and genotype frequencies will remain in equilibrium in an infinitely large population in the absence of mutation, migration, selection, and nonrandom mating.

heat shock A transient response following exposure of cells or organisms to elevated temperatures. The response involves activation of a small number of loci, inactivation of some previously active loci, and selective translation of heat-shock mRNA. It appears to be a nearly universal phenomenon observed in organisms ranging from bacteria to humans.

helicase An enzyme that participates in DNA replication by unwinding the double helix near the replication fork.

helix–turn–helix motif The structure of a region of DNA-binding proteins in which a turn of four amino acids holds two α helices at right angles to each other.

hemizygous Conditions where a gene is present in a single dose in an otherwise diploid cell. Usually applied to genes on the X chromosome in heterogametic males.

hemoglobin (Hb) An iron-containing, oxygen-carrying protein occurring chiefly in the red blood cells of vertebrates.

hemophilia An X-linked trait in humans associated with defective blood-clotting mechanisms.

heredity Transmission of traits from one generation to another.

heritability A measure of the degree to which observed phenotypic differences for a trait are genetic.

heterochromatin The heavily staining, late-replicating regions of chromosomes that are condensed in interphase. Thought to be devoid of structural genes.

heteroduplex A double-stranded nucleic acid molecule in which each polynucleotide chain has a different origin. These structures may be produced as intermediates in a recombinational event or by the *in vitro* reannealing of single-stranded, complementary molecules.

heterogametic sex The sex that produces gametes containing unlike sex chromosomes.

heterogeneous nuclear RNA (hnRNA) The collection of RNA transcripts in the nucleus, representing precursors and processing intermediates to rRNA, mRNA, and tRNA. Also represents RNA transcripts that will not be transported to the cytoplasm, such as snRNA.

heterokaryon A somatic cell containing nuclei from two different sources.

heterozygote An individual with different alleles at one or more loci. Such individuals will produce unlike gametes and therefore will not breed true.

Hfr A strain of bacteria exhibiting a high frequency of recombination. These strains have a chromosomally integrated F factor that is able to mobilize and transfer part of the chromosome to a recipient F^- cell.

histocompatibility antigens See *HLA.*

histones Proteins complexed with DNA in the nucleus. They are rich in the basic amino acids arginine and lysine, and function in coiling DNA to form nucleosomes.

HLA Cell-surface proteins, produced by histocompatibility loci, involved in the acceptance or rejection of tissue and organ grafts and transplants.

hnRNA See *heterogeneous nuclear RNA*.

Holliday structure An intermediate in bidirectional DNA recombination seen in the transmission electron microscope as an X-shaped structure showing four single-stranded DNA regions.

homeobox A sequence of about 180 nucleotides that encodes a sequence of 60 amino acids called a *homeodomain*, which is part of a DNA-binding protein that acts as a transcription factor.

homeotic mutation A mutation that causes a tissue normally determined to form a specific organ or body part to alter its differentiation and form another structure.

homogametic sex The sex that produces gametes that do not differ with respect to sex chromosome content; in mammals, the female is homogametic.

homologous chromosomes Chromosomes that synapse or pair during meiosis. Chromosomes that are identical with respect to their genetic loci and centromere placement.

homozygote An individual with identical alleles at one or more loci. These individuals will produce identical gametes and will therefore breed true.

homunculus The miniature individual imagined by preformationists to be contained within the sperm or egg.

H substance The carbohydrate group present on the surface of red blood cells. When unmodified, it results in blood type O; when modified by the addition of monosaccharides, it results in types A, B, and AB.

human immunodeficiency virus (HIV) A human retrovirus associated with the onset and progression of AIDS. See *acquired immunodeficiency syndrome*.

hybrid An individual produced by crossing two parents of different genotypes.

hybrid vigor The superiority of a heterozygote over either homozygote for a given trait.

hydrogen bond An electrostatic attraction between a hydrogen atom bonded to a strongly electronegative atom such as oxygen or nitrogen and another atom that is electronegative or contains an unshared electron pair.

hypervariable regions The regions of antibody molecules that attach to antigens. These regions have a high degree of diversity in amino acid content.

identical twins See *monozygotic twins*.

Ig See *immunoglobulin*.

imaginal disc Discrete groups of cells set aside during embryogenesis in holometabolous insects, which are determined to form the external body parts of the adult.

immunoglobulin (Ig) The class of serum proteins having the properties of antibodies.

inborn error of metabolism A genetically controlled biochemical disorder; usually an enzyme defect that produces a clinical syndrome.

inbreeding Mating between closely related organisms.

inbreeding depression A decrease in viability, vigor, or growth in progeny after several generations of inbreeding.

incomplete dominance Expression of heterozygous phenotype that is distinct from and often intermediate to that of either parent.

incomplete linkage Occasional separation of two genes on the same chromosome by a recombinational event.

indel An insertion or deletion of nucleotides in DNA or amino acids in a protein.

independent assortment The independent behavior of each pair of homologous chromosomes during their segregation in meiosis I. The random distribution of maternal and paternal homologs into gametes.

inducer An effector molecule that activates transcription.

inducible enzyme system An enzyme system under the control of a regulatory molecule, or inducer, which acts to block a repressor and allow transcription.

initiation codon The triplet of nucleotides (AUG) in an mRNA molecule that codes for the insertion of the amino acid methionine as the first amino acid in a polypeptide chain.

insertion sequence See *IS element*.

in situ hybridization A technique for the cytological localization of DNA sequences complementary to a given nucleic acid or polynucleotide.

intein An internal stretch of amino acids removed from polypeptides during post-translational processing.

intercalating agent A compound that inserts between bases in a DNA molecule, disrupting the alignments and pairing of bases in the complementary strands (e.g., acridine dyes).

interference A measure of the degree to which one crossover affects the incidence of another crossover in an adjacent region of the same chromatid. Negative interference increases the chances of another crossover; positive interference reduces the probability of a second crossover event.

interphase That portion of the cell cycle between divisions.

intervening sequence See *intron*.

intron A portion of DNA between coding regions in a gene that is transcribed but does not appear in the mRNA product.

inversion A chromosomal aberration in which the order of a chromosomal segment has been reversed.

inversion loop The chromosomal configuration resulting from the synapsis of homologous chromosomes, one of which carries an inversion.

in vitro Literally, *in glass*; outside the living organism; occurring in an artificial environment such as a test tube.

in vivo Literally, *in the living*; occurring within the living body of an organism.

IS element A mobile DNA segment that is transposable to any of a number of sites in the genome.

isoagglutinogen An antigenic factor or substance present on the surface of cells that is capable of inducing the formation of an antibody.

isochromosome An aberrant chromosome with two identical arms and homologous loci.

isolating mechanism Any barrier to the exchange of genes between different populations of a group of organisms. In general, isolation can be classified as spatial, environmental, or reproductive.

isotopes Forms of chemical elements that have the same number of protons and electrons but differ in the number of neutrons contained in the atomic nucleus.

isozyme Any of two or more distinct forms of an enzyme that have identical or nearly identical chemical properties but differ in some property such as net electrical charge, pH optima, number and type of subunits, or substrate concentration.

κ particles   DNA-containing cytoplasmic particles found in certain strains of *Paramecium aurelia*. When these self-reproducing particles are transferred into the growth medium, they release a toxin, paramecin, that kills other sensitive strains. A nuclear gene, *K*, is responsible for maintaining kappa particles in the cytoplasm.

karyokinesis   The process of nuclear division.

karyotype   The chromosome complement of a cell or an individual. Often used to refer to the arrangement of metaphase chromosomes in a sequence according to length and position of the centromere.

kilobase (kb)   A unit of length consisting of 1000 nucleotides.

kinetochore   A fibrous structure with a size of about 400 nm, located within the centromere. It appears to be the site of microtubule attachment during division.

Klinefelter syndrome   A genetic disorder in human males caused by the presence of an extra X chromosome. Klinefelter males are XXY instead of XY. This syndrome is associated with enlarged breasts, small testes, sterility, and occasionally, mild mental retardation.

knockout mice   In producing knockout mice, a cloned normal gene is inactivated by the insertion of a marker, such as an antibiotic resistance gene. The altered gene is transferred to embryonic stem cells, where the altered gene will replace the normal gene (in some cells). These cells are injected into a blastomere embryo, producing a mouse that is bred to yield mice homozygous for the mutated gene.

***lac* repressor protein**   A protein that binds to the operator in the *lac* operon and blocks transcription.

lagging strand   In DNA replication, the strand synthesized in a discontinuous fashion, 5′ to 3′ away from the replication fork. Each short piece of DNA synthesized in this fashion is called an Okazaki fragment.

lampbrush chromosomes   Meiotic chromosomes characterized by extended lateral loops, which reach maximum extension during diplonema. Although most intensively studied in amphibians, these structures occur in meiotic cells of organisms ranging from insects to humans.

lariat structure   A structure formed by an intron via a 2′ to 5′ bond during processing and removal of that intron from an mRNA molecule.

leader sequence   That portion of an mRNA molecule from the 5′ end to the beginning codon; may contain regulatory or ribosome binding sites.

leading strand   During DNA replication, the strand synthesized continuously 5′ to 3′ from the origin of replication toward the replication fork.

leptonema   The initial stage of meiotic prophase I, during which the chromosomes become visible and are often arranged in a bouquet configuration, with one or both ends of the chromosomes gathered at one spot on the inner nuclear membrane.

lethal gene   A gene whose expression results in death.

leucine zipper   A structural motif in a DNA-binding protein that is characterized by a stretch of leucine residues spaced at every seventh amino acid residue, with adjacent regions of positively charged amino acids. Leucine zippers on two polypeptides may interact to form a dimer that binds to DNA.

linkage   Condition in which two or more nonallelic genes tend to be inherited together. Linked genes have their loci along the same chromosome; they do not assort independently but can be separated by crossing over.

linkage group   A group of genes that have their loci on the same chromosome.

linking number   The number of times that two strands of a closed, circular DNA duplex cross over each other.

locus (pl., loci)   The site or place on a chromosome where a particular gene is located.

lod score   A statistical method used to determine whether two loci are linked or unlinked. A lod (log of the odds) score of 4 indicates that linkage is 10,000 times more likely than nonlinkage. By convention, lod scores of 3–4 are signs of linkage.

long interspersed elements (LINES)   Repetitive sequences found in the genomes of higher organisms, such as the 6-kb L1 sequences found in primate genomes.

long terminal repeat (LTR)   Sequence of several hundred base pairs found at the ends of retroviral DNAs.

Lyon hypothesis   The random inactivation of the maternal or paternal X chromosome in somatic cells of mammalian females early in development. All daughter cells will have the same X chromosome inactivated, producing a mosaic pattern of expression of genes on the X chromosome.

lysis   The disintegration of a cell brought about by the rupture of its membrane.

lysogenic bacterium   A bacterial cell carrying the DNA of a temperate bacteriophage integrated into its chromosome.

lysogeny   The process by which the DNA of an infecting phage becomes repressed and integrated into the chromosome of the bacterial cell it infects.

lytic phase   The condition in which a temperate bacteriophage loses its integrated status in the host chromosome (becomes induced), replicates, and lyses the bacterial cell.

major histocompatibility (MHC) loci   In humans, the HLA complex; and in mice, the H2 complex.

mapping functions   Map distance estimates from recombination when the recombination frequency in a region exceeds 15 to 20 percent and double crossovers are undetectable.

map unit   A measure of the genetic distance between two genes, corresponding to a recombination frequency of 1 percent. See also *centimorgan*.

maternal effect   Phenotypic effects on the offspring produced by the maternal genome. Factors transmitted through the egg cytoplasm that produces a phenotypic effect in the progeny.

maternal influence   See *maternal effect*.

maternal inheritance   The transmission of traits via cytoplasmic genetic factors such as mitochondria or chloroplasts.

mean   The arithmetic average.

median   The value in a group of numbers below and above which there are equal numbers of data points or measurements.

meiosis   The process in gametogenesis or sporogenesis during which one replication of the chromosomes is followed by two nuclear divisions to produce four haploid cells.

melting profile (T_m)   The temperature at which a population of double-stranded nucleic acid molecules is half-dissociated into single strands. This is taken to be the melting temperature for that species of nucleic acid.

merozygote   A partially diploid bacterial cell containing, in addition to its own chromosome, a chromosome fragment introduced into the cell by transformation, transduction, or conjugation.

messenger RNA (mRNA)   An RNA molecule transcribed from DNA and translated into the amino acid sequence of a polypeptide.

metabolism The sum of chemical changes in living organisms by which energy is generated and used.

metacentric chromosome A chromosome with a centrally located centromere, producing chromosome arms of equal lengths.

metafemale In *Drosophila*, a poorly developed female of low viability in which the ratio of X chromosomes to sets of autosomes exceeds 1.0. Previously called a superfemale.

metamale In *Drosophila*, a poorly developed male of low viability in which the ratio of X chromosomes to sets of autosomes is less than 0.5. Previously called a supermale.

metaphase The stage of cell division in which condensed chromosomes lie in a central plane between the two poles of the cell and during which the chromosomes become attached to the spindle fibers.

metaphase plate The arrangement of mitotic or meiotic chromosomes at the equator of the cell during metaphase.

methylation Enzymatic transfer of methyl groups from S-adenosylmethionine to biological molecules including phospholipids, proteins, RNA, and DNA. Methylation of DNA is associated with reduction in gene expression and with epigenetic phenomena such as imprinting.

MHC See *major histocompatibility loci*.

micrometer (μm) A unit of length equal to 1×10^{-6} meter. Previously called a micron.

micron See *micrometer*.

migration coefficient An expression of the proportion of migrant genes entering the population per generation.

millimeter (mm) A unit of length equal to 1×10^{-3} meter.

minimal medium A medium containing only those nutrients that will support the growth and reproduction of wild-type strains of an organism.

minisatellite Short tandem repeats of 10 to 100 nucleotides widely dispersed in the genome of eukaryotes. The number of repeats at each locus is variable; these loci are known as variable number tandem repeats (VNTRs). Each variation represents a VNTR allele, and many loci have dozens of alleles. VNTRs are used in DNA fingerprinting. See also *STR sequences*.

mismatch repair A process of excision repair, during which an unpaired base or bases is excised, followed by the synthesis of a new segment, using the complementary strand as a template.

missense mutation A mutation that alters a codon to that of another amino acid, causing an altered translation product to be made.

mitochondrial DNA (mtDNA) Double-stranded, self-replicating circular DNA found in mitochondria. mtDNA encodes mitochondrial ribosomal RNAs, transfer RNAs, and proteins used in oxidative respiratory functions of the organelle.

mitochondrion Found in the cells of eukaryotes, a cytoplasmic, self-reproducing organelle that is the site of ATP synthesis.

mitogen A substance that stimulates mitosis in nondividing cells (e.g., phytohemagglutinin).

mitosis A form of cell division resulting in the production of two cells, each with the same chromosome and genetic complement as the parent cell.

mode In a set of data, the value occurring in the greatest frequency.

model organism An organism such as *Escherichia coli* or yeast, with relatively short life cycles whose genetics can be studied in a cost-effective and convenient way.

molecular clock The hypothesis that for a given gene, mutations accumulate at a constant rate in all organisms in a given lineage.

By comparing the number of mutations accumulated in a given gene, researchers can construct evolutionary histories among such organisms.

monohybrid cross A genetic cross between two individuals involving only one character (e.g., $AA \times aa$).

monosomic An aneuploid condition in which one member of a chromosome pair is missing; having a chromosome number of $2n - 1$.

monozygotic twins Twins produced from a single fertilization event; the first division of the zygote produces two cells, each of which develops into an embryo. Also known as identical twins.

mRNA See *messenger RNA*.

mtDNA See *mitochondrial DNA*.

multigene family A gene set descended from a common ancestor by duplication and subsequent divergence from a common ancestor. The globin genes are an example of a multigene family.

multiple alleles Three or more alleles of the same gene.

multiple-factor inheritance See *polygenic inheritance*.

multiple infection Simultaneous infection of a bacterial cell by more than one bacteriophage, often of different genotypes.

mu (μ) phage A phage group in which the genetic material behaves like an insertion sequence; capable of insertion, excision, transposition, inactivation of host genes, and induction of chromosomal rearrangements. See also *bacteriophage μ*.

mutagen Any agent that causes an increase in the rate of mutation.

mutant A cell or organism carrying an altered or mutant gene.

mutation The process that produces an alteration in DNA or chromosome structure; the source of most alleles.

mutation rate The frequency with which mutations take place at a given locus or in a population.

muton The smallest unit of mutation in a gene, corresponding to a single-base change.

nanometer (nm) A unit of length equal to 1×10^{-9} meter.

natural selection Differential reproduction of some members of a species resulting from variable fitness conferred by genotypic differences.

neutral mutation A mutation with no immediate adaptive significance or phenotypic effect.

nonautonomous transposon A transposable element that lacks a functional transposase gene.

noncrossover gamete A gamete that contains no chromosomes that have undergone genetic recombination.

nondisjunction An error during cell division in which homologous chromosomes (in meiosis) or the sister chromatids (in mitosis) fail to separate and migrate to opposite poles; responsible for defects such as monosomy and trisomy.

nonsense codon The nucleotide triplet in an mRNA molecule that signals the termination of translation. Three such termination codons are known: UGA, UAG, and UAA.

nonsense mutation A mutation that changes an amino acid codon into a termination codon: UAG, UAA, or UGA. Leads to premature termination during translation of mRNA.

NOR See *nucleolar organizer region*.

normal distribution A probability function that approximates the distribution of random variables. The normal curve, also known as a Gaussian or bell-shaped curve, is the graphic display of the normal distribution.

northern blot A technique in which RNA molecules are separated by electrophoresis and transferred by capillary action to a nylon or nitrocellulose membrane. Specific RNA molecules can be identified by hybridization to a labeled nucleic acid probe.

N-terminal amino acid The terminal amino acid in a peptide chain that carries a free amino group.

N terminus The free amino group of the first amino acid in a polypeptide. By convention, the structural formula of polypeptides is written with the N terminus at the left.

nuclease An enzyme that breaks bonds in nucleic acid molecules.

nucleoid The DNA-containing region within the cytoplasm in prokaryotic cells.

nucleolar organizer region (NOR) A chromosomal region containing the genes for rRNA; most often found in physical association with the nucleolus.

nucleolus A nuclear organelle that is the site of ribosome biosynthesis; usually associated with or formed in association with the NOR.

nucleoside A purine or pyrimidine base covalently linked to a ribose or deoxyribose sugar molecule.

nucleosome A complex of four histone molecules, each present in duplicate, wrapped by two turns of a DNA molecule. One of the basic units of eukaryotic chromosome structure.

nucleotide A nucleoside covalently linked to a phosphate group. Nucleotides are the basic building blocks of nucleic acids. The nucleotides commonly found in DNA are deoxyadenylic acid, deoxycytidylic acid, deoxyguanylic acid, and deoxythymidylic acid. The nucleotides in RNA are adenylic acid, cytidylic acid, guanylic acid, and uridylic acid.

nucleotide pair The pair of nucleotides (A and T or G and C) in opposite strands of a DNA molecule that are hydrogen-bonded to each other.

nucleus The membrane-bound cytoplasmic organelle of eukaryotic cells that contains the chromosomes and nucleolus.

null allele A mutant allele that produces no functional gene product; usually inherited as a recessive trait.

null hypothesis Used in statistical tests, it states that there is no difference between the observed and expected data sets. Statistical methods such as chi-square analysis are used to test the probability of this hypothesis.

nullisomic Describes an individual with a chromosomal aberration in which both members of a chromosome pair are missing.

Okazaki fragments The small, discontinuous strands of DNA produced on the lagging strand during DNA synthesis.

oligonucleotide A linear sequence of about 10 to 20 nucleotides connected by $5'-3'$ phosphodiester bonds.

oncogene A gene whose activity promotes uncontrolled proliferation in eukaryotic cells.

open reading frame (ORF) A nucleotide sequence organized as triplets that encodes the amino acids of a protein. Located between an initiation codon and a termination codon within a gene.

operator region A region of a DNA molecule that interacts with a specific repressor protein to control the expression of an adjacent gene or gene set.

operon A genetic unit consisting of one or more structural genes encoding polypeptides, and an adjacent operator gene that controls the transcriptional activity of the structural gene or genes.

origin of replication (ori) Sites along the length of a chromosome where DNA replication begins.

outbreeding depression Reduction in fitness in the offspring produced by mating genetically diverse parents. It is thought to result from a lowered adaptation to local environmental conditions.

overdominance The phenomenon where heterozygotes have a phenotype that is more extreme than either homozygous genotype.

overlapping code A genetic code first proposed by George Gamow in which any given nucleotide is shared by three adjacent codons.

pachynema The stage in meiotic prophase I when the synapsed homologous chromosomes split longitudinally (except at the centromere), producing a group of four chromatids called a tetrad.

pair-rule genes Genes expressed as stripes around the blastoderm embryo during development of the *Drosophila* embryo.

palindrome A word, number, verse, or sentence that reads the same backward or forward (e.g., *tis Ivan on a visit*). In nucleic acids, a sequence in which the base pairs read the same on complementary strands in the $5'$ to $3'$ direction. For example:

$$5'\text{-GAATTC-}3'$$
$$3'\text{-CTTAAG-}5'$$

These often occur as sites for restriction endonuclease recognition and cutting.

pangenesis A discarded theory of development that postulated the existence of pangenes—small particles from all parts of the body that concentrated in the gametes, passing traits from generation to generation, blending the traits of the parents in the offspring.

paracentric inversion A chromosomal inversion that does not include the centromere.

parental gamete See *noncrossover gamete.*

parthenogenesis Development of an egg without fertilization.

partial diploids See *merozygote.*

partial dominance See *incomplete dominance.*

patroclinous inheritance A form of genetic transmission in which the offspring have the phenotype of the father.

pedigree In human genetics, a diagram showing the ancestral relationships and transmission of genetic traits over several generations in a family.

P element Transposable DNA element found in *Drosophila* responsible for hybrid dysgenesis.

penetrance The frequency, expressed as a percentage, with which individuals of a given genotype manifest at least some degree of a specific mutant phenotype associated with a trait.

peptide bond The covalent bond between the amino group of one amino acid and the carboxyl group of another amino acid.

pericentric inversion A chromosomal inversion that involves both arms of the chromosome and thus involves the centromere.

phage See *bacteriophage.*

phenotype The observable properties of an organism that are genetically controlled.

phenylketonuria (PKU) A hereditary condition in humans associated with the inability to metabolize the amino acid phenylalanine. The most common form is caused by the lack of the enzyme phenylalanine hydroxylase.

Philadelphia chromosome The product of a reciprocal translocation that contains the short arm of chromosome 9 carrying the *C-ABL* oncogene and the long arm of chromosome 22 carrying the *BCR* gene.

phosphodiester bond In nucleic acids, the covalent bond between a phosphate group and adjacent nucleotides, extending from the 5′ carbon of one pentose (ribose or deoxyribose) to the 3′ carbon of the pentose in the neighboring nucleotide. Phosphodiester bonds form the backbone of nucleic acid molecules.

photoreactivation enzyme (PRE) An exonuclease that catalyzes the light-activated excision of ultraviolet-induced thymine dimers from DNA.

photoreactivation repair Light-induced repair of damage caused by exposure to ultraviolet light. Associated with an intracellular enzyme system.

phyletic evolution The gradual transformation of one species into another over time; vertical evolution.

phylogenetic tree A diagram summarizing the evolutionary relationship among groups of organisms. Often constructed from amino acid or nucleotide sequence data.

pilus A filament-like projection from the surface of a bacterial cell. Often associated with cells possessing F factors.

plaque A clear area on an otherwise opaque bacterial lawn, caused by the growth and reproduction of phages.

plasmid An extrachromosomal, circular DNA molecule (often carrying genetic information) that replicates independently of the host chromosome.

pleiotropy Condition in which a single mutation simultaneously affects several characters.

ploidy Term referring to the basic chromosome set or to multiples of that set.

point mutation A mutation that can be mapped to a single locus. At the molecular level, a mutation that results in the substitution of one nucleotide for another.

polar body A cell produced in females at either the first or second meiotic division, which contains almost no cytoplasm as a result of an unequal cytokinesis.

polycistronic mRNA A messenger RNA molecule that encodes the amino acid sequence of two or more polypeptide chains in adjacent structural genes.

polygenic inheritance The transmission of a phenotypic trait whose expression depends on the additive effect of a number of genes.

polylinker A segment of DNA that has been engineered to contain multiple sites for restriction enzyme digestion. Polylinkers are usually found in engineered vectors such as plasmids.

polymerase chain reaction (PCR) A method for amplifying DNA segments that uses cycles of denaturation, annealing to primers, and DNA polymerase-directed DNA synthesis.

polymerases Enzymes that catalyze the formation of DNA and RNA from deoxynucleotides and ribonucleotides, respectively.

polymorphism The existence of two or more discontinuous, segregating phenotypes in a population.

polynucleotide A linear sequence of more than 20 nucleotides, joined by 5′–3′ phosphodiester bonds. See also *oligonucleotide*.

polypeptide A molecule made up of amino acids joined by covalent peptide bonds. This term is used to denote the amino acid chain before it assumes its functional three-dimensional configuration.

polyploid A cell or individual having more than two sets of chromosomes.

polyribosome See *polysome*.

polysome A structure composed of two or more ribosomes associated with mRNA, engaged in translation. Formerly called polyribosome.

polytene chromosome A chromosome that has undergone several rounds of DNA replication without separation of the replicated chromosomes, thus forming a giant, thick chromosome with aligned chromomeres producing a characteristic banding pattern.

population A local group of individuals belonging to the same species, which are actually or potentially interbreeding.

position effect Change in expression of a gene associated with a change in the gene's location within the genome.

postzygotic isolation mechanism A factor that prevents or reduces inbreeding by acting after fertilization to produce nonviable, sterile hybrids or hybrids of lowered fitness.

preadaptive mutation A mutational event that later gains adaptive significance.

preformationism The discredited idea that an organism develops by growth of structures already present in the egg or sperm.

prezygotic isolation mechanism A factor that reduces inbreeding by preventing courtship, mating, or fertilization.

Pribnow box A 6-bp sequence upstream from the beginning of transcription in prokaryotic genes, to which the σ subunit of RNA polymerase binds. The consensus sequence for this box is TATAAT.

primary protein structure The sequence of amino acids in a polypeptide chain.

primary sex ratio Ratio of males to females at fertilization.

primer In nucleic acids, a short length of RNA or single-stranded DNA required for the functioning of polymerases.

prion An infectious pathogenic agent devoid of nucleic acid and composed of a protein, PrP, with a molecular weight of 27,000 to 30,000 Da. Prions are known to cause scrapie, a degenerative neurological disease in sheep; bovine spongiform encephalopathy (BSE or mad cow disease) in cattle; and similar diseases in humans, including kuru and Creutzfeldt-Jakob disease.

probability Ratio of the frequency of a given event to the frequency of all possible events.

proband An individual in whom a genetically determined trait of interest is first detected. Formerly known as a propositus.

probe A macromolecule such as DNA or RNA that has been labeled and can be detected by an assay such as autoradiography or fluorescence microscopy. Probes are used to identify target molecules, genes, or gene products.

product law In statistics, the law that holds that the probability of two independent events occurring simultaneously is the product of their independent probabilities.

progeny The offspring produced from a mating.

prokaryotes Organisms lacking nuclear membranes, meiosis, and mitosis. Bacteria and blue-green algae are examples of prokaryotic organisms.

promoter Region having a regulatory function and to which RNA polymerase binds prior to the initiation of transcription.

proofreading A molecular mechanism for correcting errors in replication, transcription, or translation. Also known as editing.

prophage A phage genome integrated into a bacterial chromosome. Bacterial cells carrying prophages are said to be lysogenic.

propositus (female, **proposita**) See *proband*.

protein A molecule composed of one or more polypeptides, each composed of amino acids covalently linked together.

proteomics The study of the expressed proteins present in a cell at a given time.

protooncogene A gene that normally functions to initiate or maintain cell division. Protooncogenes can be converted to oncogenes by alterations in structure or expression.

protoplast A bacterial or plant cell with the cell wall removed. Sometimes called a spheroplast.

prototroph A strain (usually microorganisms) that is capable of growth on a defined, minimal medium. Wild-type strains are usually regarded as prototrophs.

pseudoalleles Genes that behave as alleles to one another by complementation, but can be separated from one another by recombination.

pseudoautosomal inheritance Inheritance of alleles located within the regions of the Y chromosome that are homologous to the X chromosome. Because these alleles are located on both the X and Y chromosome, their pattern of inheritance is indistinguishable from that of autosomal inheritance.

pseudodominance The expression of a recessive allele on one homolog caused by the deletion of the dominant allele on the other homolog.

pseudogene A nonfunctional gene with sequence homology to a known structural gene present elsewhere in the genome. They differ from their functional relatives by insertions or deletions and by the presence of flanking direct repeat sequences of 10 to 20 nucleotides.

puff See *chromosome puff.*

punctuated equilibrium A pattern in the fossil record of long periods of species stability, punctuated with brief periods of species divergence.

quantitative inheritance See *polygenic inheritance.*

quantitative trait loci (QTL) Two or more genes that act on a single polygenic trait.

quantum speciation Formation of a new species within a single or a few generations by a combination of selection and drift.

quaternary protein structure Types and modes of interaction between two or more polypeptide chains within a protein molecule.

race A genotypically or geographically distinct subgroup within a species.

rad A unit of absorbed dose of radiation with an energy equal to 100 ergs per gram of irradiated tissue.

radioactive isotope One form of an element, differing in atomic weight and possessing an unstable nucleus that emits ionizing radiation during decay.

random amplified polymorphic DNA (RAPD) A PCR method that uses random primers about 10 nucleotides in length to amplify unknown DNA sequences.

random mating Mating between individuals without regard to genotype.

reading frame Linear sequence of codons in a nucleic acid.

reannealing Formation of double-stranded DNA molecules from dissociated single strands.

recessive A term describing an allele that is not expressed in the heterozygous condition.

reciprocal cross A paired cross in which the genotype of the female in the first cross is present as the genotype of the male in the second cross, and vice versa.

reciprocal translocation A chromosomal aberration in which non-homologous chromosomes exchange parts.

recombinant DNA A DNA molecule formed by joining two heterologous molecules. Usually applied to DNA molecules produced by *in vitro* ligation of DNA from two different organisms.

recombinant gamete A gamete containing a new combination of genes produced by crossing over during meiosis.

recombination The process that leads to the formation of new gene combinations on chromosomes.

recon A term coined by Seymour Benzer to denote the smallest genetic units between which recombination can occur.

reductional division The chromosome division that halves the diploid chromosome number. The first division of meiosis is a reductional division.

redundant genes Gene sequences present in more than one copy per haploid genome (e.g., ribosomal genes).

regulatory site A DNA sequence that is involved in the control of expression of other genes, usually involving an interaction with another molecule.

rem Radiation equivalent in man; the dosage of radiation that will cause the same biological effect as one roentgen of X-rays.

renaturation The process by which a denatured protein or nucleic acid returns to its normal three-dimensional structure.

repetitive DNA sequences DNA sequences present in many copies in the haploid genome.

replicating form (RF) Double-stranded nucleic acid molecule-present as an intermediate during the reproduction of certain viruses.

replication The process of DNA synthesis.

replication fork The Y-shaped region of a chromosome associated with the site of replication.

replicon A chromosomal region or free genetic element containing the DNA sequences necessary for the initiation of DNA replication.

replisome The term used to describe the complex of proteins, including DNA polymerase, that assembles at the bacterial replication fork to synthesize DNA.

repressible enzyme system An enzyme or group of enzymes whose synthesis is regulated by the intracellular concentration of certain metabolites.

repressor A protein that binds to a regulatory sequence adjacent to a gene and blocks transcription of the gene.

reproductive isolation Absence of interbreeding between populations, subspecies, or species. Reproductive isolation can be brought about by extrinsic factors, such as behavior, and intrinsic barriers, such as hybrid inviability.

resistance transfer factor (RTF) A component of R plasmids that confers the ability for cell–cell transfer of the R plasmid by conjugation.

resolution In an optical system, the shortest distance between two points or lines at which they can be perceived to be two points or lines.

restriction endonuclease Nuclease that recognizes specific nucleotide sequences in a DNA molecule and cleaves or nicks the DNA at those sites. Derived from a variety of microorganisms, those enzymes that cleave both strands of the DNA are used in the construction of recombinant DNA molecules.

restriction fragment length polymorphism (RFLP) Variation in the length of DNA fragments generated by a restriction endonuclease. These variations are caused by mutations that create or abolish cutting sites for restriction enzymes. RFLPs are inherited in a codominant fashion and can be used as genetic markers.

restrictive transduction See *specialized transduction*.

retrovirus Viruses with RNA as genetic material that utilize the enzyme reverse transcriptase during their life cycle.

reverse transcriptase A polymerase that uses RNA as a template to transcribe a single-stranded DNA molecule as a product.

reversion A mutation that restores the wild-type phenotype.

R factor (R plasmid) Bacterial plasmids that carry antibiotic resistance genes. Most R plasmids have two components: an r-determinant, which carries the antibiotic resistance genes, and the resistance transfer factor (RTF).

RFLP See *restriction fragment length polymorphism*.

Rh factor An antigenic system first described in the rhesus monkey. Recessive *r/r* individuals produce no Rh antigens and are Rh negative, while *R/R* and *R/r* individuals have Rh antigens on the surface of their red blood cells and are classified as Rh positive.

ribonucleic acid (RNA) A nucleic acid characterized by the sugar ribose and the pyrimidine uracil, usually a single-stranded polynucleotide. Several forms are recognized, including ribosomal RNA, messenger RNA, transfer RNA, and heterogeneous nuclear RNA.

ribose The five-carbon sugar associated with the ribonucleotides found in RNA.

ribosomal RNA (rRNA) The RNA molecules that are the structural components of the ribosomal subunits. In prokaryotes, these are the $16S$, $23S$, and $5S$ molecules; in eukaryotes, they are the $18S$, $28S$, and $5S$ molecules.

ribosome A ribonucleoprotein organelle consisting of two subunits, each containing RNA and protein. Ribosomes are the site of translation of mRNA codons into the amino acid sequence of a polypeptide chain.

RNA See *ribonucleic acid*.

RNA editing Alteration of the nucleotide sequence of an mRNA molecule after transcription and before translation. There are two main types of editing: substitution editing, which changes individual nucleotides, and insertion/deletion editing, in which individual nucleotides are added or deleted.

RNA polymerase An enzyme that catalyzes the formation of an RNA polynucleotide strand using the base sequence of a DNA molecule as a template.

RNase A class of enzymes that hydrolyzes RNA.

Robertsonian translocation A form of chromosomal aberration that involves the fusion of the long arms of acrocentric chromosomes at the centromere.

roentgen (R) A unit of measure of the amount of radiation corresponding to the generation of 2.083×10^9 ion pairs in one cubic centimeter of air at $0°C$ at an atmospheric pressure of 760 mm of mercury.

rolling circle model A model of DNA replication in which the growing point or replication fork rolls around a circular template strand; in each pass around the circle, the newly synthesized strand displaces the strand from the previous replication, producing a series of contiguous copies of the template strand.

R point The point during the G1 stage of the cell cycle when either a commitment is made to DNA synthesis and another cell cycle or the cell withdraws from the cycle and becomes quiescent. Also known as the restriction point.

rRNA See *ribosomal RNA*.

RTF See *resistance transfer factor*.

S_1 nuclease A deoxyribonuclease that cuts and degrades single-stranded molecules of DNA.

satellite DNA DNA that forms a minor band when genomic DNA is centrifuged in a cesium salt gradient. This DNA usually consists of short sequences repeated many times in the genome.

SCE See *sister chromatid exchange*.

secondary protein structure The α-helical or β-pleated-sheet form of a protein molecule brought about by the formation of hydrogen bonds between amino acids.

secondary sex ratio The ratio of males to females at birth.

secretor An individual having soluble forms of the blood group antigens A and/or B present in saliva and other body fluids. This condition is caused by a dominant, autosomal gene unlinked to the *ABO* locus (*I* locus).

sedimentation coefficient See *Svedberg coefficient unit*.

segment polarity genes Genes that regulate the spatial pattern of differentiation within each segment of the developing *Drosophila* embryo.

segregation The separation of homologous chromosomes into different gametes during meiosis.

selection The force that brings about changes in the frequency of alleles and genotypes in populations through differential reproduction.

selection coefficient (*s*) A quantitative measure of the relative fitness of one genotype compared with another. See also *coefficient of selection*.

selfing In plant genetics, the fertilization of ovules of a plant by pollen produced by the same plant; reproduction by self-fertilization.

semiconservative replication A model of DNA replication in which a double-stranded molecule replicates in such a way that the daughter molecules are composed of one parental (old) and one newly synthesized strand.

semisterility A condition in which a percentage of all zygotes is inviable.

sex chromatin body See *Barr body*.

sex chromosome A chromosome, such as the X or Y in humans, which is involved in sex determination.

sexduction Transmission of chromosomal genes from a donor bacterium to a recipient cell by the F factor.

sex-influenced inheritance Phenotypic expression that is conditioned by the sex of the individual. A heterozygote may express one phenotype in one sex and the alternate phenotype in the other sex.

sex-limited inheritance A trait that is expressed in only one sex even though the trait may not be X-linked.

sex ratio See *primary and secondary sex ratio*.

sexual reproduction Reproduction through the fusion of gametes, which are the haploid products of meiosis.

Shine-Dalgarno sequence The nucleotides AGGAGG present in the leader sequence of prokaryotic genes that serve as a ribosome binding site. The $16S$ RNA of the small ribosomal subunit contains a complementary sequence to which the mRNA binds.

short interspersed elements (SINES) Repetitive sequences found in the genomes of higher organisms, such as the 300-bp *Alu* sequence.

shotgun experiment The cloning of random fragments of genomic DNA into a vehicle such as a plasmid or phage, usually to produce a library from which clones of specific interest can be selected.

sibling species Species that are morphologically almost identical but that are reproductively isolated from one another.

sickle-cell anemia A genetic disease in humans caused by an autosomal recessive gene, fatal in the homozygous condition if untreated. Caused by an alteration in the amino acid sequence of the β chain of globin.

sickle-cell trait The phenotype exhibited by individuals heterozygous for the sickle-cell gene.

sigma (σ) factor A polypeptide subunit of the RNA polymerase that recognizes the binding site for the initiation of transcription.

single-stranded binding proteins (SSBs) In DNA replication, proteins that bind to and stabilize single-stranded regions of DNA that result from the action of unwinding proteins.

sister chromatid exchange (SCE) A crossing-over event that can occur in meiotic and mitotic cells; involves the reciprocal exchange of chromosomal material between sister chromatids joined by a common centromere. Such exchanges can be detected cytologically after BrdU incorporation into the replicating chromosomes.

site-directed mutagenesis A process that uses a synthetic oligonucleotide containing a mutant base or sequence as a primer for inducing a mutation at a specific site in a cloned gene.

small nuclear RNA (snRNA) Species of RNA molecules ranging in size from 90 to 400 nucleotides. The abundant snRNAs are present in 1×10^4 to 1×10^6 copies per cell, associated with proteins, and form RNP particles known as snRNPs or *snurps*. Six uridine-rich snRNAs known as U1–U6 are located in the nucleoplasm, and their complete nucleotide sequence is known. snRNAs have been implicated in the processing of pre-mRNA and may have a range of cleavage and ligation functions.

snurps See *small nuclear RNA.*

solenoid structure A level of eukaryotic chromosome structure generated by the supercoiling of nucleosomes.

somatic cell genetics The use of cultured somatic cells to investigate genetic phenomena by parasexual techniques involving the fusion of cells from different organisms.

somatic cells All cells other than the germ cells or gametes in an organism.

somatic mutation A nonheritable mutational event occurring in a somatic cell.

somatic pairing The pairing of homologous chromosomes in somatic cells.

SOS response The induction of enzymes to repair damaged DNA in *Escherichia coli.* The response involves activation of an enzyme that cleaves a repressor, activating a series of genes involved in DNA repair.

Southern blot A technique developed by Edward Southern in which DNA fragments produced by restriction enzyme digestion are separated by electrophoresis and transferred by capillary action to a nylon or nitrocellulose membrane. Specific DNA fragments can be identified by hybridization to a labeled nucleic acid probe.

spacer DNA DNA sequences found between genes, usually repetitive DNA segments.

specialized transduction Genetic transfer of only specific host genes by transducing phages.

speciation The process by which new species of plants and animals arise.

species A group of actually or potentially interbreeding individuals that is reproductively isolated from other such groups.

spheroplast See *protoplast.*

spindle fibers Cytoplasmic fibrils formed during cell division that are involved with the separation of chromatids at anaphase and their movement toward opposite poles in the cell.

spliceosome The nuclear macromolecule complex within which splicing reactions occur to remove introns from pre-mRNAs.

spontaneous mutation A mutation that is not induced by a mutagenic agent.

spore A unicellular body or cell encased in a protective coat that is produced by some bacteria, plants, and invertebrates; it is capable of survival in unfavorable environmental conditions; and it can give rise to a new individual upon germination. In plants, spores are the haploid products of meiosis.

SRY The sex-determining region of the *Y* chromosome found near the pseudoautosomal boundary of the Y chromosome. Accumulated evidence indicates that this gene is the testis-determining factor (TDF).

stabilizing selection Preferential reproduction of those individuals having genotypes close to the mean for the population. A selective elimination of genotypes at both extremes.

standard deviation A quantitative measure of the amount of variation in a sample of measurements from a population.

standard error A quantitative measure of the amount of variation in a sample of measurements from a population.

sterile The condition of being unable to reproduce. Also, free from contaminating microorganisms.

strain A group with common ancestry that has physiological or morphological characteristics of interest for genetic study or domestication.

STR sequences Short tandem repeats of two to nine base pairs found within minisatellite sequences. These sequences are used to prepare DNA profiles in forensics, paternity identification, and other applications.

structural gene A gene that encodes the amino acid sequence of a polypeptide chain.

structural genomics The identification of genes and their physical mapping done by sequencing an organism's genome.

sublethal gene A mutation causing lowered viability, with less than 50 percent of the individuals that carry the gene dying before maturity.

submetacentric chromosome A chromosome with the centromere placed so that one arm of the chromosome is slightly longer than the other.

subspecies A morphologically or geographically distinct interbreeding population of a species.

sum law The law holding that the probability of one or the other of two mutually exclusive events occurring is the sum of their individual probabilities.

supercoiled DNA A form of DNA structure in which the helix is coiled upon itself. Such structures can exist in stable forms only when the ends of the DNA are not free, as in a covalently closed circular DNA molecule.

superfemale See *metafemale.*

supermale See *metamale.*

suppressor mutation A mutation that acts to completely or partially restore the function lost by a previous mutation at another site.

Svedberg coefficient unit (S) A unit of measure for the rate at which particles (molecules) sediment in a centrifugal field. This unit is a function of several physicochemical properties, including size and shape. A sedimentation value of 1×10^{-13} sec is defined as one Svedberg coefficient unit.

symbiont An organism coexisting in a mutually beneficial relationship with another organism.

sympatric speciation Process of speciation involving populations that inhabit, at least in part, the same geographic range.

synapsis The pairing of homologous chromosomes at meiosis.

synaptonemal complex (SC) An organelle consisting of a tripartite nucleoprotein ribbon that forms between the paired homologous chromosomes in the pachytene stage of the first meiotic division.

syndrome A group of signs or symptoms that occur together and characterize a disease or abnormality.

synkaryon The nucleus of a zygote that results from the fusion of two gametic nuclei. Also used in somatic cell genetics to describe the product of nuclear fusion.

syntenic test In somatic cell genetics, a method for determining whether two genes are on the same chromosome.

TATA box See *Goldberg-Hogness box.*

tautomeric shift A reversible isomerization in a molecule, brought about by a shift in the localization of a hydrogen atom. In nucleic acids, tautomeric shifts in the bases of nucleotides can cause changes in other bases at replication and are a source of mutations.

TDF (testis-determining factor) The product of the *SRY* gene on the Y chromosome that controls the developmental switch point for the development of the indifferent gonad into a testis.

telocentric chromosome A chromosome in which the centromere is located at the end of the chromosome.

telomerase The enzyme that adds short, tandemly repeated DNA sequences to the ends of eukaryotic chromosomes.

telomere The terminal chromomere of a chromosome.

telophase The stage of cell division in which the daughter chromosomes reach the opposite poles of the cell and re-form nuclei. Telophase ends with the completion of cytokinesis.

telophase I In the first meiotic division, when duplicated chromosomes reach the poles of the dividing cell.

temperate phage A bacteriophage that can become a prophage and confer lysogeny upon the host bacterial cell.

temperature-sensitive mutation A conditional mutation that produces a mutant phenotype at one temperature range and a wild-type phenotype at another temperature range.

template The single-stranded DNA or RNA molecule that specifies the nucleotide sequence of a strand synthesized by a polymerase molecule.

terminalization The movement of chiasmata toward the ends of chromosomes during the diplotene stage of the first meiotic division.

tertiary protein structure The three-dimensional structure of a polypeptide chain brought about by folding upon itself.

testcross A cross between an individual whose genotype at one or more loci may be unknown and an individual who is homozygous recessive for the genes in question.

tetrad The four chromatids that make up paired homologs in the prophase of the first meiotic division. The four haploid cells produced by a single meiotic division.

tetrad analysis Method for the analysis of gene linkage and recombination, using the four haploid cells produced in a single meiotic division.

tetranucleotide hypothesis An early theory of DNA structure proposing that the molecule was composed of repeating units, each consisting of the four nucleotides containing adenine, thymine, cytosine, and guanine.

theta (θ) structure An intermediate in the bidirectional replication of circular DNA molecules. At about midway through the cycle of replication, the intermediate resembles the Greek letter theta.

thymine dimer A pair of adjacent thymine bases in a single polynucleotide strand that have chemical bonds formed between carbon atoms 5 and 6. This lesion, usually caused by exposure to ultraviolet light, inhibits DNA replication unless repaired by the appropriate enzymes.

T_m See *melting profile.*

topoisomerase A class of enzymes that convert DNA from one topological form to another. During DNA replication, these enzymes facilitate the unwinding of the double-helical structure of DNA.

totipotent The ability of a cell or embryo part to give rise to all adult structures. This capacity is usually progressively restricted during development.

trait Any detectable phenotypic variation of a particular inherited character.

trans **configuration** The arrangement of two mutant sites on opposite homologs, such as

$$\frac{a^1 \quad +}{+ \quad a^2}$$

Contrasts with a *cis* arrangement, where the sites are located on the same homolog.

transcription Transfer of genetic information from DNA by the synthesis of an RNA molecule copied from a DNA template.

transcriptome The set of mRNA molecules present in a cell at any given time.

transdetermination Change in the developmental fate of a cell or group of cells.

transduction The transfer of virally mediated genes from one bacterium to another or the transfer of eukaryotic genes mediated by retrovirus.

transfer RNA (tRNA) A small ribonucleic acid molecule that contains a three-base segment (anticodon) that recognizes a codon in mRNA, a binding site for a specific amino acid, and recognition sites for interaction with the ribosomes and the enzyme that links it to its specific amino acid.

transformation Heritable change in a cell or an organism brought about by exogenous DNA.

transgenic organism An organism whose genome has been modified by the introduction of external DNA sequences into the germline.

transition A mutational event in which one purine is replaced by another or one pyrimidine is replaced by another.

translation The derivation of the amino acid sequence of a polypeptide from the base sequence of an mRNA molecule in association with a ribosome.

translocation A chromosomal mutation associated with the transfer of a chromosomal segment from one chromosome to another. Also used to denote the movement of mRNA through the ribosome during translation.

transmission genetics The field of genetics concerned with the mechanisms by which genes are transferred from parent to offspring.

transposable element A DNA segment that translocates to other sites in the genome, essentially independent of sequence homology. Usually such elements are flanked by short inverted repeats of 20 to 40 base pairs at each end. Insertion into a structural gene can produce a mutant phenotype. Insertion and excision of transposable elements depend on two enzymes, transposase and resolvase. Such elements have been identified in both prokaryotes and eukaryotes.

transversion A mutational event in which a purine is replaced by a pyrimidine or a pyrimidine is replaced by a purine.

trinucleotide repeat A tandemly repeated cluster of three nucleotides (such as CTG) in or near a gene, which undergoes an expansion in copy number, resulting in a disease phenotype.

triploidy The condition in which a cell or organism possesses three haploid sets of chromosomes.

trisomy The condition in which a cell or organism possesses two copies of each chromosome, except for one, which is present in three copies. The general form for trisomy is therefore $2n + 1$.

tRNA See *transfer RNA*.

tumor suppressor gene A gene that encodes a gene product that normally functions to suppress cell division. Mutations in tumor suppressor genes result in the activation of cell division and tumor formation.

Turner syndrome A genetic condition in human females caused by a 45,X genotype. Such individuals are phenotypically female but are sterile because of undeveloped ovaries.

unequal crossing over A crossover between two improperly aligned homologs, producing one homolog with three copies of a region and the other with one copy of that region.

unique DNA DNA sequences that are present only once per genome.

universal code The assumption that the genetic code is used by all life forms. In general, this is true; some exceptions are found in mitochondria, ciliates, and mycoplasmas.

unwinding proteins Nuclear proteins that act during DNA replication to destabilize and unwind the DNA helix ahead of the replicating fork.

variable number tandem repeats (VNTRs) Short, repeated DNA sequences (2 to 20 nucleotides) present as tandem repeats between two restriction enzyme sites. Variations in the number of repeats create DNA fragments of differing lengths following restriction enzyme digestion.

variable region Portion of an immunoglobulin molecule that exhibits many amino acid sequence differences between antibodies of differing specificities.

variance A statistical measure of the variation of values from a central value, calculated as the square of the standard deviation.

variegation Patches of differing phenotypes, such as color, in a tissue.

vector In recombinant DNA, an agent such as a phage or plasmid into which a foreign DNA segment will be inserted.

viability The measure of the number of individuals in a given phenotypic class that survive, relative to another class (usually wild type).

virulent phage A bacteriophage that infects and lyses the host bacterial cell.

VNTRs See *variable number tandem repeats*.

western blot A technique in which proteins are separated by gel electrophoresis and transferred by capillary action to a nylon membrane or nitrocellulose sheet. A specific protein can be identified through hybridization to a labeled antibody.

wild type The most commonly observed phenotype or genotype, designated as the norm or standard.

wobble hypothesis An idea proposed by Francis Crick, stating that the third base in an anticodon can align in several ways to allow it to recognize more than one base in the codons of mRNA.

writhing number The number of times that the axis of a DNA duplex crosses itself by supercoiling.

W, Z chromosomes Sex chromosomes in species where the female is the heterogametic sex (WZ).

X chromosome Sex chromosome present in species where females are the homogametic sex (XX).

X inactivation In mammalian females, the random cessation of transcriptional activity of one X chromosome. This event, which occurs early in development, is a mechanism of dosage compensation. Molecular basis of inactivation is unknown but involves a region called the X-inactivation center (XIC) on the proximal end of the p arm. Some loci on the tip of the short arm of the X can escape inactivation. See also *Barr body; Lyon hypothesis*.

XIST A locus in the X-chromosome inactivation center that may control inactivation of the X chromosome in mammalian females.

X-linkage The pattern of inheritance resulting from genes located on the X chromosome.

X-ray crystallography A technique to determine the three-dimensional structure of molecules through diffraction patterns produced by X-ray scattering by crystals of the molecule under study.

YAC A cloning vector in the form of a yeast artificial chromosome, constructed using chromosomal elements including telomeres (from a ciliate), centromeres, origin of replication, and marker genes from yeast. YACs are used to clone long stretches of eukaryotic DNA.

Y chromosome Sex chromosome in species where the male is heterogametic (XY).

Y-linkage Mode of inheritance shown by genes located on the Y chromosome.

Z-DNA An alternative structure of DNA in which the two antiparallel polynucleotide chains form a left-handed double helix. Z-DNA has been shown to be present in chromosomes and may have a role in regulation of gene expression.

zein Principal storage protein of maize endosperm, consisting of two major proteins, with molecular weights of 19,000 and 21,000 Da.

zinc finger A DNA-binding domain of a protein that has a characteristic pattern of cysteine and histidine residues that complex with zinc ions, throwing intermediate amino acid residues into a series of loops or fingers.

zygote The diploid cell produced by the fusion of haploid gametic nuclei.

zygonema A stage of meiotic prophase I in which the homologous chromosomes synapse and pair along their entire length, forming bivalents. The synaptonemal complex forms at this stage.

Credits

Chapter 10 DNA Replication and Synthesis

CO10 Dr. Gopal Murti/Science Photo Library/Photo Researchers, Inc. **F10.5 (top)** Walter H. Hodge/Peter Arnold, Inc. **F10.5(a–b)** Figure from "Molecular Genetics," Pt. 1 pp. 74–75, J.H. Taylor (ed). Copyright ©1963 and renewed 1991, reproduced with permission from Elsevier Science Ltd. **F10.6** Reprinted from CELL, Vol. 25, 1981, pp 659, Sundin and Varshavsky, (1 figure), with permission from Elsevier Science. Courtesy of A. Varshavsky. **F10.14** Reproduced by permission from H.J. Kreigstein and D.S. Hogness, Proceedings of the National Academy of Sciences 71:136 (1974), p. 137, Fig. 2. **F10.15** Dr. Harold Weintraub, Howard Hughes Medical Institute, Fred Hutchinson Cancer Center/"Essential Molecular Biology" 2e, Freifelder & Malachinski, Jones & Bartlett, Fig. 7–24, pp. 141. **F10.18** David Dressler, Oxford University, England

Chapter 11 Chromosome Structure and DNA Sequence Organization

CO11 Science VU/BMRL/Visuals Unlimited **F11.1(a)** Dr. M. Wurtz/Biozentrum, University of Basel/SPL/Dr. M. Wurtz/Biozentrum, University of Basel/Science Photo Library/Photo Researchers, Inc. **F11.2** Science Source/Photo Researchers, Inc. **F11.3** Dr. Gopal Murti/Science Photo Library/Photo Researchers, Inc. **F11.4** Don W. Fawcett/Kahri/Dawid/Science Source/Photo Researchers, Inc. **F11.6** Dr. Richard D. Kolodnar/Dana-Farber Cancer Institute **F11.7** Image courtesy of Brian Harmon and John Sedat, University of California, San Francisco **F11.8** Science Source/Photo Researchers, Inc. **F11.9(b)** Omikron/Photo Researchers, Inc.

Chapter 12 The Genetic Code and Transcription

CO12 Prof. Oscar L. Miller/Science Photo Library/Photo Researchers, Inc. **F12.10** Bert W. O'Malley, M.D., Baylor College of Medicine **F12.14** O.L. Miller, Jr. Barbara A. Hamkalo, C.A. Thomas, Jr. Science 169:392–395, 1970 by the American Association for the Advancement of Science. F:2.

Chapter 13 Translation and Proteins

CO13 Reprinted from the front cover of Science, Vol. 292, May 4, 2001 "Crystal structure of a Thermus thermophilus 70S ribosome containing three bound transfer RNAs (top) and exploded views showing its different molecular components (middle and bottom). Image provided by Dr. Albion Baucom (baucom@biology.ucsc.edu). Copyright American Association for the Advancement of Science. **F13.9(a)** "The Structure and Function of Polyribosomes." Alexander Rich, Jonathan R. Warner and Howard M. Goodman, 1963. Reproduced by permission of the Cold Spring Harbor Laboratory Press. Cold Spring Harbor Symp. Quant. Biol. 28 (1963) fig. 4C **(top)**, p. 273 ©1964 **F13.9(b)** E.V. Kiseleva/E.V. Kiseleva **F13.13(a)** Dennis Kunkel/Phototake NYC **F13.13(b)** Francis Leroy/Biocosmos/Science Photo Library/Photo Researchers, Inc. **F13.18** Kenneth Eward/BioGrafx/Science Source/Photo Researchers, Inc.

Chapter 14 Gene Mutation, DNA Repair, and Transposition

CO14 Francis Leroy/Biocosmos/Science Photo Library/Photo Researchers, Inc. **F14.15(a–b)** W. Clark Lambert, M.D., Ph. D./University of Medicine & Dentistry of New Jersey **F14.21** Sue Ford/Science Photo Library/Photo Researchers, Inc. **F14.22** Mary Evans Picture Library/Photo Researchers, Inc.

Chapter 15 Regulation of Gene Expression

CO15 Science Lewis, et al/Johnson Research Foundation

Chapter 16 Cell-Cycle Regulation and Cancer

CO16 SPL/Photo Researchers, Inc. **F16.1(a)** Courtesy of Hesed M. Padilla-Nash, Antonio Fargiano, and Thomas Ried. Affiliation is Section of Cancer Genomics, Genetics Branch, Center for Research, National Cancer Institute, National Institutes of Health, Bethesda, MD 20892 **F16.1(b)** Courtesy of Hesed M. Padilla-Nash and Thomas Ried. Affiliation is Section of Cancer Genomics, Genetics Branch, Center for Research, National Cancer Institute, National Institutes of Health, Bethesda, MD 20892 **F16.2(a–b)** Courtesy of Professor Manfred Schwab, DKFZ, Heidelberg, Germany. **F16.7(a)** Dr. Gopal Murti/Photo Researchers, Inc.

Chapter 17 Recombinant DNA Technology

CO17 Michael Gabridge/Visuals Unlimited **F17.4** K.G. Murti/Visuals Unlimited **F17.6** Michael Gabridge/Visuals Unlimited **F17.15(a–b)** Laurie Ann Achenbach, Southern Illinois University at Carbondale **F17.18** Dr. Suzanne McCutcheon

Chapter 18 Genomics and Proteomics

CO18 Dr. Jeremy Burgess/Science Photo Library/Photo Researchers, Inc. **F18.4(a)** David McCarthy/SPL/Photo Researchers, Inc. **F18.5(a)** Moredun Animal Health Ltd./Photo Researchers, Inc. **F18.8(a)** SciMAT/Photo Researchers, Inc. **F18.9(a)** Holt Studios International/Photo Researchers, Inc. **F18.11(a)** The Institute for Genomic Research **F18.20** Dr. Carl Merril/Dr. Carl Merril/Laboratory of Biochemical Genetics of the National Institute of Mental Health, NIH **F18.21** Geoff Tompkinson/Science Photo Library/Photo Researchers, Inc.

Chapter 19 Applications and Ethics of Genetic Engineering

CO19 Affymetrix, Inc. **F19.2** Reprinted with permission from "New Genes Boost Rice Nutrients" by I. Potrykus and P. Beyer, Science, August 13, 1999, Vol. 285, pp 994. **F19.4** Charles J. Arntzen, Florence Ely Nelson Distinguished Professor and Founding Director, Arizona Biomedical Institute at Arizona State University **F19.10** GeneChip® Human Genome U133 Plus 2.0 Array. Courtesy of Affymetrix, Inc. **F19.12(a)** Fig. 1b and c from Clinical Chemistry 46:10 2000, Fredrik P. Wikman et al. Evaluation of the performance of a p53 sequencing microarray chip using 140 previously sequenced bladder tumor samples. **F19.12(b)** From Fig. 1d from Clinical Chemistry 46:10 2000, Fredrik P. Wikman et al. Evaluation of the performance of a p53 sequencing microarray chip using 140 previously sequenced bladder tumor samples. **F19.13(b)** Cancer Genetics Branch/National Human Genome Research Institute/NIH **F19.14** Reprinted with permission from Ash Alizadeh, Nature Magazine 2000:403, pgs. 503–511, figure 4 left panel. Copyright 2000 Macmillian Magazines Limited. **F19.16** Van De Silva **F19.20** Courtesy of Orchid BioSciences, Inc.

Chapter 20 Developmental Genetics

CO20 Edward B. Lewis, California Institute of Technology **F20.1(a)** F.R. Turner/Visuals Unlimited **F20.1(b)** R. Calentine/Visuals Unlimited **F20.6** Jim Langeland, Stephen Paddock, and Sean Carroll, University of Wisconsin at Madison **F20.8(a)** Peter A. Lawrence and P. Johnston, "Development", 105, 761–767 (1989). **F20.8(b)** Peter A. Lawrence "The Making of a Fly", Blackwells Scientific, 1992. **F20.9** Jim Langeland, Stephen Paddock, and Sean Carroll, University of Wisconsin at Madison **F20.10** Reprinted by permission from Macmillan Publishers Ltd.: Fig 1 on p244 from:

British Dental Journal 195:243–48 2003. Greenwood, M. and Meechan, J.G. "General medicine and surgery for dental practitioners." Copyright © Macmillan Magazines Limited. **F20.11(a–b)** Reproduced by permission from Fig. 2 on p. 766 in Cell 89:765–71, May 30, 1997. Copyright © by Elsevier Science Ltd. Image courtesy of Mike J. Owen. **F20.12(a–b)** Reproduced by permission from T. Kaufmann, et al., Advanced Genetics 27:309–362, 1990. Image courtesy of F. Rudolf Turner, Indiana University. **F20.16** Reproduced with permission from [Figure 1a on page 332 from "Kmita, M. and Duboule, D. 2003. Organizing Axes in Time and Space; 25 Years of Colinear Tinkering. Science 2003 301:331–333".]. Copyright American Association for the Advancement of Science. **F20.17** P. Barber/Custom Medical Stock Photo, Inc. **F20.19** Elliot M. Meyerowitz, California Institute of Technolgy, Division of Biology **F20.20(a–b)** Max-Planck-Institut fur Entwicklungsbiologie **F20.22(a–d)** Dr. Jose Luis Riechmann, Division of Biology, California Institute of Technolgy. From Science 2002:295, pp. 1482–85. **F20.24(top)** James King-Holmes/Photo Researchers, Inc.

Chapter 21 Quantitative Genetics
CO21 Ed Reschke/Peter Arnold, Inc. **F21.2** Norm Thomas/Photo Researchers, Inc. **F21.9** Courtesy of Dr. Steven D. Tanksley, Cornell University

Chapter 22 Population Genetics
CO22 Edward S. Ross, California Academy of Sciences **F22.5** Michel Samson, Frederick Libert, et al., "Resistance to HIV-1 infection in Caucasian individuals bearing mutant alleles of the CCR-5 chemokine receptor gene." Reprinted with permission from Nature [vol. 382, 22 August 1996, p. 725, Fig. 3]. Copyright 1996 Macmillan Magazines Limited **F22.11** Larsh K. Bristol/Visuals Unlimited **F22.12** Hans Pfletschinger/Getty Images Inc - Science Faction **F22.19(a–b)** Courtesy of Murray Brilliant, "A 122.5 kilobase deletion of P gene underlies the high prevalence of oculocutaneous albinism type 2 in the Navajo population": From: American Journal Human Genetics 72:62–72, fig. 1 p. 65. Published by University of Chicago Press **F22.20** Courtesy of Murray Brilliant, "A 122.5 kilobase deletion of P gene underlies the high prevalence of oculocutaneous albinism type 2 in the Navajo population": From: American Journal Human Genetics 72:62–72, fig. 3 p. 67. Published by University of Chicago Press **F22.21** From Bouzat, Jual L., et al. 1998. Genetic evaluation of a demographic bottleneck in the greater prairie chicken. Conservation Biology 12:863–43, pg 839, fig. 2, Part A.

Chapter 23 Evolutionary Genetics
CO23 Breck P. Kent/Animals Animals/Earth Scenes **F23.9** Carl C. Hansen/Nancy Knowlton/Smithsonian Institution Photo Services **F23.11(a–b)** Kenneth Kaneshiro/University of Hawaii/CCRT **F23.13(a–b)** Courtesy of Toby Bradshaw and Doug W. Schemske, University of Washington. Photo by Jordan Rehm. **F23.15(a–b)** Dr. Paul V. Loiselle **F23.16** Fig. 5 in Molecular Biology Evolution 15 (4) pages 391–407, Kazuhiko Takahasi/Norihiro Okada. The Society for Molecular Biology and Evolution. **F23.17** C.B. & D.W. Frith/Bruce Coleman Inc.

Chapter 24 Conservation Genetics
CO24 Scott Bauer/USDA/ARS/Agricultural Research Service **F24.2** John Mitchell/Photo Researchers, Inc. **F24.3** Daniel J Cox/Getty Images Inc. - Stone Allstock **F24.4(a)** David Austen/David R. Austen

Photography Group **F24.4(b)** Sarah Ward/Department of Soil and Crop Services/Colorado State University **F24.5** Sarah Ward/Department of Soil and Crop Services/Colorado State University **F24.6** Johnny Johnson/DRK Photo **F24.9** Art Wolfe/Getty Images Inc. - Stone Allstock **F24.10** Tim Thompson/Corbis/Bettmann **F24.13** Jim Brandenburg/Minden Pictures **F24.14** Lynn Stone/Animals Animals/Earth Scenes

FIGURES AND TEXT

Chapter 7 Linkage and Chromosome Mapping in Eukaryotes
p157 From Stig Blixt "Why Didn't Gregor Mendel Find Linkage" *Nature* (1975) 256–206

Chapter 15 Regulation of Gene Expression
F15.25 From Black, D. L. 2000. Protein diversity from alternative splicing: A challenge for bioinformatics and post-genome biology. Cell 103:367–370, Figure 1, p.368 with permission from Elsevier. **F15.27** From Hannon, G. J. 2002. RNA Interface. *Nature* 481:244–251, Figure 2, p.246. **F15.28** From Matzk, M. et al., 2004. Genetic analysis of RNA mediated transriptional gene silencing. Biochim. Biophys. Acta 1677:127–141, Figure 1, p.1678 with permission from Elsevier.

Chapter 16 Cell-Cycle Regulation and Cancer
F16.4 From Aaltonen *et al*., Clues to the pathogenesis of familial colorectal cancer, *Science* (1993) 260:812–819. Reprinted with permission from AAAS and author.

Chapter 18 Genomics and Proteomics
F18.4 Public Domain **F18.8** From Nucleic Acids Research (2006) Vol.34, Database issue D442–D445. Reprinted with permission. **F18.9** Arabidopsis Genome Initiative (2000) Analysis of the genome sequence of the flowering plant *Arabidopsis Thaliana*. *Nature 408*:796–815, Fig 2a, p.799. **F18.12** Reprinted from Genomics (61) 2, Breen, M. et al., Reciprocal chromosome painting reveals detailed regions of conserved synteny between the karyotypes of the domestic dog (conis familiaris) and human, pp145–155 (1999) with permission from Elsevier. **F18.22** From Wasinger, V. et al., (2000) The Proteome of *Mycoplasma Genitalium*. Eur J. Biochem. *267*:1571–82, Fig 5., p1575. Reprinted with permission from the author.

Chapter 19 Applications and Ethics of Genetic Engineering
F19.5 Reprinted with permission from Biodesign Institute, Arizona State University, Box 5001, Tempe, AZ 85287-5001.

Chapter 22 Population Genetics
F22.6 From Leibert, F. et al., 1998. The deltaccr5 mutation conferring protection against HIV-1 in Caucasian populations has a single and recent origin in Notheastern Europe. *Human Molecular Genetics (7)*:399–406. Reprinted with permission from Oxford University Press and author.

Chapter 23 Evolutionary Genetics
F23.3 Reprinted from Trends in Genetics (8):392–98 Tsui L., The spctrum of cystic fibrosis mutations (1992) with permission from Elsevier. **F23.8(a)** Reprinted with permission from the *Annual Review of Genetics*, Volume 25 copyright (1991) by Annual Reviews www.annualreview.org **F23.18** From Fitch, W.M. et al., Positive Darwinian evolution in human influenza A viruses, Proceedings of the National

Academy of Sciences (1991) 88:4270–73, Figure 1. Reprinted with permission from AAAS and author. **F23.19(b)** From Sibley, C.G. And Ahlquist, J. E. (1987) DNA hybridization evidence of hominoid phylogeny: results from an expanded data set. J. Mol. Evol. 26:99–121, Table 1, p101 with permission from Springer-Verlag GmbH. **F23.20** From Fitch, W.M. et al., 1991. Positive Darwinian evolution in human influenza A viruses, Proceedings of the National Academy of Sciences (1991) 88:4270–73, Figure 1. Reprinted with permission from AAAS and author. **F23.21** Reprinted from Current Biology (7) 3. Hillis, Phylogenic Analysis, ppR129–R131, Figure 1 (1998) with permission from Elsevier. **F23.22** Reprinted from Cell (90), Krings, M. *et al.*, pp.19–30, Figure 7a, p.26. (1997) with permission from Elsevier.

Index

Note: Bold page number indicates a figure; italic page number indicates a table.

A

A antigens, 64–65, **65**
AB antigens, 18
Aberrations, chromosomal, 115, 310
  heterozygous for the, 123
ABO blood groups, 64–**65**, **68**, **69**, 311
Abortive transduction, 180
Absolute population size, 542
Absorption of ultraviolet (UV) light, 205, **205**
Absorption spectrum, 196, **196**
Ac-Ds system in maize, 318–19, **318**, **319**
Acentric chromatid, 127
Acetylated histones, 341
Acetylation, 245, 341
Acetylcholinesterase (ACE), 499, **500**
Achondroplasia, *54*, 502
Acinonyx jubatus, 542–43, **542**
Acridine dyes, 309–10
Acridine orange, 310
Acrocentric chromosome, 20, **21**
Actin, 6, 295
Actin-derived microfilaments, 19
Action spectrum, 196, **196**
Activation energy, 296, **296**
Activator (Ac) mutation. *See Ac-Ds* system in maize
Activators, 347, **347**
Active site, 296
Acute transforming retroviruses, 369
ADA immune deficiency, **10**
Adaptation hypothesis, 165
Additive alleles, **470**, 473, **473**, 474
Additive variance, 479
Adelberg, Edward, 172
Adenine (A), 5, 197, **197**, **308**
Adenine methylase, 313
Adenomas, 368
Adenosine deaminase (ADA), 438
Adenosine triphosphate (ATP), 19, 198, **199**
Adenovirus, 438
Adenyl cyclase, 336, **336**
Adjacent segregation, **128**
A-DNA, 203, **204**
Adrenoleukodystrophy (ALD), **10**
Adsorption complex, 194–95
Adult stem cells, 465
Advanced backcross QTL method, 483
Aflatoxin, 370
Agarose gels, 207
Aging, 231
Agouti coat phenotypes, 66, 70
Agriculture
  genetic engineering in, 9, 426–28
    nutritional enhancement of crop plants, 427–28, **427**
    transgenic crops and herbicide resistance, 426–27, **427**
  Green Revolution in, 483
Agrobacterium-mediated transformation, 183
Agrobacterium tumefaciens, *406*, 407, 426–27, **427**
AIDS, 272, 492–93
  HIV transmission and, 530
Alagille syndrome, 462, 464

Alanine, chemical structure and designation of, **292**
Albinism, 53, *54*, 67, **68**, **286**
Alcohol dehydrogenase locus (Adh), 518
Alexandra, Tsarina, 85, 322
Aliquot, 177
Alkaptonuria, *54*, 286, **286**, 287
Alkylating agents, 309–10, **309**
Allele frequencies. *See also* Population genetics
  calculating, 490–92, **491**, *494*
  genetic drift and, 504–6
    founder effect and, 504, 505–6, **505**
  mutation and, 502–3, **502**
  natural selection as driving force in changing, 497–502, **498**, **499**
  nonrandom mating and, 506–10
Alleles, 4, 22, 41, 49, 62
  additive, **470**, 473, **473**, 474
  in altering phenotypes, 62–63
  lethal, 65–67, **66**
  multiple, 64–65
  mutant, 62
  nonadditive, 473
  null, 62
  symbols for, 41
  wild-type, 62, 63, 138
Allele-specific oligonucleotides (ASO), 433, **433**, **435**
Allium cepa, 22
Allolactose, 332*n*
Allopolyploidy, *115*, 120, 121–22
Allosteric, 332
Allotetraploid, 121
Alloway, Lionel J., 192
Allozymes, 517, *517*
  identifying genetic diversity with, 540–41
α-Globin genes, 7, 413–14, **413**, **415**
α-Glucosidase, 430
α Helix, 293, **293**
Alphoid family, 248
Alquist, Jon, 527
Alternate segregation, **128**
Alternation of generations, **33**
Alternative splicing, 109, 270
  pathways for mRNA, 348
  protein production and, 348
Alu element, 248–49, 320
*Alu*I, **378**
Alu insertions, 411
Alzheimer disease (AD), **10**
  Down syndrome and, 118
American College of Medical Genetics, 420
Ames test, 312–13
  in assessing mutagenicity of compounds, 312–13, **313**
Amino acids, 6
  degenerate coding for, 260
  methylation and phosphorylation of, 245
  N-terminus, 295
  structure of, **292**
Amino acid sequence, variations in, 517, *517*
Aminoacyladenylic acid, 281
Aminoacyl (A) sites, 283
Aminoacyl tRNA synthetases, **240**, 281
Amino group, 293
2-Amino purine (2-AP), 309
Amniocentesis, 119, 432, **432**
Amphidiploidy, *115*, 121, **121**

Ampicillin, resistance to, 174
Amplicon, 103
Ampliconic region, 103
Amyloidosis, **10**
Amyotrophic lateral sclerosis (ALS), **10**
Anabolism, 296
Anagenesis, 516, **516**
Anaphase, **3**, 23, **23**, 25–26, **25**
Anaphase I, 29, **30**
Anaphase II, 29, **31**
Anderson, Anna, 85
Andersson, Siv, 532
Anemia, sickle-cell, 6–7, **6**, **10**, 11, *54*, 209, 289–91, **290**, 303
  allele-specific oligonucleotide testing for, 433, **433**
  prenatal diagnosis of, 431–32, **432**
Aneuploidy, 115, *115*
  in plant kingdom, 116–17
  viability in human, 119–20
Angelman syndrome (AS), 80
Anhidrotic ectodermal dysplasia, 106
Animals
  cloned, 9, **9**, 393
  transgenic, 9, **10**, 196
Annealing, 383
Annotation, 402–3
Antennapedia (Antp-C) complex, 456, **456**
Antennapedia (Antp) gene, 456, **456**
Antibodies, 415. *See also* Immunoglobulin gene superfamily
  diversity of, 416–17
Antibody-combining site, 416
Anticancer drugs, 231
Anticodon, 257, 278
Anticodon loop, 280
Antigens, 415
  A, 64–65, **65**
  AB, 18
  B, 64–65, **65**
  histocompatibility, 18
  MN, 18
Antiparallel chains, 200–201
Antiparallel strands in DNA synthesis, 221, *222*, 223
Antirrhinum majus, 22
Antisense oligonucleotides, 272
Antisense RNA, 205, 272
Antiterminator hairpin, 338, **339**
Antithrombin, recombinant human, *429*
α1-Antitrypsin, 430
AP1 transcription factor, 346
APC, *364*, 368
AP endonuclease, 315, **315**
Apoptosis, 363–66, **363**
Apurinic sites, 307
Apyrimidinic (AP) site, 315
Aquifex aeolicus, **405**, 407, **407**
Arabidopsis, 22, 238, **399**, 518
  flower development in, 459–60, **460**, **461**
  genomes of, 408–9, *408*, **409**
  homeotic selector genes in, *460*
  as model organism, 12, **13**, 320
Archaea, genome size and gene number in, *405*
Archaeoglobus fulgidis, *405*
Arctocephalus gazella, 542
Arctocephalus tropicalis, 542